Gene Expression

VOLUME 2 EUCARYOTIC CHROMOSOMES

Gene Expression

SECOND EDITION

BENJAMIN LEWIN
Editor, *Cell*

A WILEY-INTERSCIENCE PUBLICATION

JOHN WILEY & SONS
New York • Chichester • Brisbane • Toronto • Singapore

Library of Congress Cataloging in Publication Data:
Lewin, Benjamin M.
　Gene expression.

　"A Wiley-Interscience publication."
　Includes bibliographies and indexes.
　CONTENTS:　　　　　　　　　　　v. 2.　Eucaryotic
chromosomes.
　1. Gene expression.　2. Molecular genetics.
I. Title.
QH450.L48　　　　1980　　　574.87'322　　80-10849
ISBN 0-471-01977-1 (v. 2)
ISBN 0-471-01976-3 pkb. (v. 2)

Printed in the United States of America

10 9 8 7 6 5

Notwithstanding Nicholas

Preface

Any relationship between this book and the previous edition is almost coincidental. During the past three years of writing, at times it has seemed that research might advance more rapidly than the results could be incorporated into the text. This has indeed been a period of such extraordinary progress that only a completely new book could do justice to the subject. This volume therefore shares with its predecessor only the same general aim: to provide a contemporary account of our knowledge of the organization and expression of eucaryotic genes. This book is, of course, the second part of the series *Gene Expression*, in which I hope eventually to provide a full and unifying account of the structure and function of the genetic apparatus of both procaryotes and eucaryotes, plasmids and viruses. The first and third volumes, *Procaryotic Genomes* and *Plasmids and Phages*, consider the fundamental processes of nucleic acid and protein synthesis and their control in bacteria, both uninfected and infected.

Eucaryotic Chromosomes is intended to delineate present knowledge of the structure of the chromosome and its constituent genetic units and to define the operations by which genes are expressed in the form of protein. The principal questions that are addressed remain the same as in the first edition of six years ago: what is the structure of the chromosome at the level of interaction between DNA and protein; how is chromosome behavior to be explained at the molecular level; why is there so much DNA and what is the nature of the genetic unit; how is expression of the gene controlled? But the answers are different. The discovery of the nucleosome as the basic subunit of chromatin should lead eventually to the definition of the ultrastructure of the chromosome itself in terms of macromolecular organization. The striking discovery that genes are not continuous sequences of DNA but have interrupted, mosaic structures leads to a new view of the nature of the genetic unit and the pathways for its expression. Together, these and other substantial advances that have been overshadowed only by this dramatic extension of horizons, have created a sense of momentum reminiscent of the now classical molecular biology of the sixties.

The rapidity of progress has left a pressing need for a comprehensive treatment to bring together these new views. In this book I hope to provide a critical assessment of the state of the art created by these advances. To define the limits of present knowledge I have in this book drawn on original data as extensively as in previous volumes. Too often molecular biology is regarded as a didactic series of conclusions—in simple terms the story of how DNA makes RNA makes protein. But it seems impossible to show what weight is to be attached to conclusions without scrutiny of experimental data; and so I have tried to show sample data that are representative and may serve to indicate the general relationship between datum and conclusion. I believe it is the relative success of this attempt to provide a critical tour d'horizon that will mark whether this volume achieves what seems to have been the usefulness of previous volumes in providing a vantage point from which to assess future advances.

I have felt it more important than ever to maintain my usual policy of relying upon data of papers published in principal research journals rather than obscure symposia or other unreviewed series. In recent years this blight on science has grown to the point where particular information may appear in overlapping form in several places; except in very rare cases where these irregular volumes provide the only pertinent data, I have confined my attention to primary research communications. Even so, the attempt to provide access just to the mainstream of research has generated a very substantial list of references. Here it should be noted that to make these more useful, papers are cited in two ways: experiments revealing principles, breaking new ground, are quoted in some detail in the text and figures; supporting data are cited as additional references in the tables and footnotes. Indeed, this is a general policy, with the book written in a sense on two levels. The text and figures attempt to define supportable conclusions; but some of the more eclectic caveats, details of further systems, discussion of peripheral issues, and so on, are to be found in the tables and footnotes. In this way I hope that the book will be equally useful to the reader who wishes to obtain a critical overview and to the reader who needs to pursue a particular topic in more detail.

It is naturally a pleasure to thank colleagues too many to name individually for stimulating discussions and for providing access to unpublished data; in particular Dr. Howard Green commented upon several chapters. Finally it is a peasure to acknowledge the support of my family through the elephantine gestation period, notably Ann for her belief that the book might indeed one day be completed.

BENJAMIN LEWIN

Cambridge, Massachusetts
May 1980

Contents

Gene Expression

PART 1

Cell Structure and Genetics

CHAPTER 1

Introduction:
The Genetic Apparatus

This book is concerned with two principal questions. What is the structure of the genetic apparatus of the eucaryotic cell? And what determines the state of genetic expression of a cell? Both questions have roots in the classical cytogenetic and developmental studies of the past century, but only very recently has it become possible to pose them in molecular terms. These are broad issues and to begin upon answers—all that is possible today—it is necessary to delve into the wider questions of cell structure and function as well as the direct problems of the nature of the genetic apparatus itself. Before approaching the organization and expression of the genetic material, we must therefore consider its dependence upon the circumstances in which it finds itself, that is to say, its relationship with the other components of the cell. Our general aim is to account for the heredity of the eucaryotic cell in terms of the properties of its constituent macromolecules.

The genetic apparatus of the eucaryotic cell includes not only the genetic material but also an array of components necessary to maintain its structure and to undertake its reproduction and expression. The genetic material is segregated from the cell cytoplasm by the nuclear membrane, the feature that distinguishes eucaryotes from procaryotes. As the repository of genetic information, the nucleus is the site of nucleic acid synthesis, whereas protein synthesis takes place in the cytoplasm.

The structure of the genetic apparatus is determined by the state of the cell in which it resides, and depends in particular upon the extent of gene expression and on whether the cell is dividing. Cell division requires a doubling of all components, including the genetic material, and culminates in a profound, though transient, reorganization of structure, during which the distinction between nucleus and cytoplasm is lost. This allows the duplicates of the genetic material to be segregated into daughter cells by virtue of their interaction with the division apparatus. This interaction is necessary for both cell reproduction and the formation of gametes and is therefore a central feature in the perpetuation of the genetic information of all eucaryotic cells. It is only recently that the components of the division apparatus have been defined and their relationship with other cellular structures recognized. We can therefore now begin an account of cell division in terms of these components instead of in the descriptive terms of cytology.

3

The genetic information contained in the nucleus is the same in all the cells of a given eucaryote. But because of differences in gene expression, each cell displays a characteristic phenotype, which is established by the proteins comprising its cytoplasm. In any eucaryotic cell, the condition of the nucleus must therefore be viewed in the perspective of the surrounding cytoplasm; and the interactions between nucleus and cytoplasm thus represent the state of gene expression. Indeed, this has long been recognized in the form of the traditional question of developmental biology: what controls development—the nucleus or the cytoplasm? We can now supersede this question with an account of the reciprocal interplay between nucleus and cytoplasm in terms of the passage of macromolecules.

In this first section of the book, we shall be concerned with the molecular architecture of the cell, in particular with the structural features involved in its reproduction, and with the nature of the interactions between nucleus and cytoplasm. We shall attempt to define the parameters that govern cell division and to consider the mechanisms by which the division is accomplished. This requires an explanation of the behavior of the genetic apparatus during division in terms of the changes in its structure and its association with other cellular components. As a prerequisite for studies of nuclear function, it is necessary to establish the extent to which the content of genetic information can be characterized by mutation. To consider directly the interactions between nucleus and cytoplasm, it is necessary to turn to situations in which the nucleus finds itself in an unusual cytoplasm, one to which normally it would not be exposed. Together these approaches contribute to a view of the genetic apparatus as a structure whose behavior is determined by the proteins associated with the genetic material. This opens the way to consider in further sections the molecular structure of the genetic apparatus, the nature of its store of genetic information, and the interactions that control its state of expression.

The Eucaryotic Chromosome

It has been known since the turn of the century that the chromosomes observed in dividing cells carry the genetic information. Somatic cells of each species possess a characteristic number of chromosomes which can be divided into two identical sets. The contents of the set define the haploid complement of the organism, which consists of one copy of each *homologue*. The only exception is provided by the sex chromosomes, for which one sex may be XX and the other XY. The diploid number of chromosomes $(2n)$ is maintained in all somatic cells of an organism by the mitotic cycle in which the chromosomes first are replicated and the cell then divides to generate two diploid daughter cells.

Gametes are produced by a meiotic division in which the number of chromosomes is halved so that each sperm or egg receives only the haploid complement (*n*). Union of the gametes reestablishes the diploid complement, so that each zygote obtains one copy of each chromosome from each parent. The diploid complement thus comprises one haploid set of paternal origin and one of maternal origin. Meiosis is responsible for the segregation of paternal and maternal alleles into different germ cells and for the independent inheritance of genes located on different chromosomes predicted by Mendel's laws.

Genes located on the same chromosome can be separated only by a recombination event in which there is physical exchange of material between homologues. The frequency of recombination is taken to represent the distance apart of linked genes. However, the size of the eucaryotic chromosome means that genetic exchange occurs often enough between genes located some distance apart to render their inheritance effectively independent. But by measuring their linkage to intervening markers, it is possible to show that both reside on one chromosome. This allows each gene to be placed on a linkage group. In cases in which it has been possible to map enough genes, the genetic map corresponds well with the visible features of the chromosomes: the number and the relative sizes of the linkage groups are the same as those of the chromosomes of the haploid set.

The eucaryotic chromosome therefore comprises a discrete genetic entity, a unique set of genes organized into a characteristic structure and in addition to the genetic material itself, each chromosome possesses the features necessary to fulfill its role as a unit of inheritance in cell division.

Cytological studies of chromosomes have been confined largely to observations of their behavior during cell division. Although *interphase*, to use the cytological description of the period between mitoses, is a misleading term in the sense that most of the synthetic activities of the cell take place during this period, an interphase cell appears quiescent compared with the spectacular changes in cell morphology seen at mitosis. An interphase nucleus is bounded by the nuclear membrane and the twisted filaments of interphase *chromatin* are distributed throughout the nucleus. Individual chromosomes cannot be distinguished, although *chromocenters* may be seen where the chromatin appears more condensed. Nucleoli often are visible and tend to be associated with the chromocenter(s).

The start of mitosis is defined when individual chromosomes can first be distinguished as they appear as extended threads. At this point, each chromosome appears to consist of two coiled threads, the *chromatids*. Prior to the start of visible division, the chromosomes have been duplicated and the chromatids represent the two daughter copies of each chromosome. Since each homologue consists of two chromatids, the cell has four copies of the haploid set.

As mitosis continues, the nuclear membrane dissolves and the former cell compartments are replaced by the spindle. The chromosomes become aligned on the equator in a highly compact form; then the sister chromatids of each homologue separate and move to opposite poles of the cell to form two diploid sets. Nuclear membranes form around these, and the cytoplasm is divided into two separate cells.

Chromosomes appear in their most distinctive forms when they are highly condensed and it is only at this point in mitosis that the features of the individual members of the chromosome set can be distinguished. The criteria used to describe the chromosomes are size and position of the centromere, the point at which the chromosome is attached to the spindle and by which it is pulled to the poles. Chromosomes in which the centromere is located at one end take a rodlike structure and are said to be *telocentric. Acrocentric* chromosomes appear similar but possess a small arm on one side of the centromere. (There has been some controversy about whether chromosomes may be truly telocentric or must always have a short arm, even if this is too small to detect cytologically.) *Metacentric* chromosomes have arms of about equal sizes because their centromeres are located at a central position; these take up a "V" shape at metaphase. When the two arms are of unequal length, the product is an "L"-shaped *submetacentric* chromosome. This classification of chromosomes, although useful in distinguishing the members of a set, probably has no functional significance.

The termini of chromosomes—the *telomeres*—appear to be differentiated from internal regions. When chromosomes are broken by X irradiation, the resulting segments may be free to fuse together again; but they do not appear to do so with the telomeres of unbroken chromosomes (McClintock, 1940; Muller and Herskowitz, 1954). It is therefore possible that the telomere may have a structure that precludes its attachment to other chromosomes or parts of chromosomes.

Little is known about the underlying structure of the chromosome during mitosis and meiosis. During the early stages when chromosome structure is less compact, it can be seen to take the form of a coiled fiber, the *chromonema*. This may show alternating wide and narrow regions which give the chromosome the appearance of beads on a string. The beadlike structures are known as *chromomeres* and the position of each chromomere is relatively constant for each chromosome. Probably they reflect the local manner of coiling of the chromosome. During the most condensed stages, a characteristic series of "bands" can be generated in each chromosome by treatment with several types of agent. Although presumably this reflects the organization of the chromosome, the nature of the interactions responsible for band formation is not clear. However, this does represent a way in which the chromosomes of a set may be distinguished unequivocally; and banding techniques have superseded earlier distinctions based on centromere position and chromosome size.

Some chromosome regions remain condensed during interphase. These identify the *heterochromatin*, contrasted with the much greater length of euchromatin, which uncoils and swells during the same period. Heterochromatin often appears to be located at the centromere and may be in close contact with the nucleolus. Heterochromatic regions on different chromosomes may aggregate to form a chromocenter. Heterochromatin appears to represent regions that are more highly condensed than euchromatin; and in chromosomes which possess both types of region the chromonemata can be seen to run continuously between them. But little is known about the basis of the more condensed state of heterochromatin.

In defining the structure of the chromosome, it is therefore necessary not only to account for the changes in the state of the genetic material during the cell cycle, but also to recognize that there may be regional differences in organization. It is necessary further to account for the acquisition of the centromeric structure by which segregation is accomplished, and for the other structural features apparent during mitosis.

The Genetic Material

It has been known virtually since their discovery that chromosomes consist of both nucleic acid and protein. Identification of the nucleic acid as DNA followed from the development of the Feulgen reaction, which was later developed into a quantitative method for estimating the nuclear content of DNA by spectrophotometry (Feulgen and Rossenbeck, 1924; Caspersson, 1936). Direct demonstrations that the genetic information resides in DNA have been possible only with procaryotic and viral systems and are discussed in Volume 3 of Gene Expression. But the equation of genetic information with DNA led to what might be described as the central tenet of modern cell biology: every somatic cell of a given eucaryote has the same content of DNA, and this is double the content of the gametes (Boivin, Vendrely and Vendrely, 1948; Mirsky and Ris, 1949, 1951; Swift, 1950). There are, of course, several well-characterized exceptions to this rule, involving the loss of DNA or variations in amount in certain plant and insect systems, as well as the selective amplification of certain genes in insects and amphibia; but these are exceptions and, certainly in mammalian systems, available data support the concept that (within experimental error) the content of genetic material of the somatic cell is invariant (see Bedi and Goldstein, 1976). This means that it is to changes in the expression of this fixed genetic complement that we must look for an explanation of the properties of each cell phenotype.

Determination of the DNA content of the genome requires the choice of a suitable cell, one in which the ploidy of the nucleus is known and there is little interference from cytoplasmic (organelle) DNA. Spermatozoa may be used to determine haploid DNA values directly; nucleated erythrocytes are suitable as examples of diploid cells not engaged in division, and they share with spermatozoa the property that the cell number may accurately be counted. Dividing cell populations are not appropriate, because some cells will have more than the diploid content of DNA, and the compensations that must be made for this inevitably introduce inaccuracies. A singularly inappropriate tissue is mammalian liver, since some cells are polyploid, and measurements of either relative or absolute DNA contents of such cells cannot be regarded as significant.

The method originally used to determine absolute DNA contents relied upon chemical measurements of phosphorus content. Although measurements in different laboratories, or even of different tissues in one laboratory, gave a range of results generally extending over about a 10% variation for the genome size, many of these data have not been superseded by more accurate direct measurements. Other methods are based upon photometric assays. The diphenylamine reaction was for some time the most accurate of these; and again, data obtained with this technique still stand as the determinations of genome content in several instances. More recent methods, which may be more accurate, include reactions with diaminobenzoic acid or ethidium bromide.

Comparisons of DNA content between different species have obvious evolutionary implications and may prove to be important in considering the relationship between DNA organization and function. The Feulgen technique has been developed into a reliable photometric method for making such comparisons. This has allowed the relative DNA contents of mammals, reptiles, birds, and plants to be investigated in some detail (Atkin et al., 1965; Bennett and Smith, 1976). It is now therefore possible to determine the ratio of DNA contents of two organisms with some accuracy, probably ±5%. But to use such data for determining absolute values, it is necessary to have a standard whose DNA content has been determined by chemical means. Unfortunately, in many cases an extensive series of determinations of DNA by Feulgen content has been based upon a standard whose accuracy is not certain. These genome values must therefore be regarded as only approximate, say within ±20%.

Data on the DNA contents of the haploid genomes of a representative array of eucaryotes, including in particular those that are discussed in this book, are summarized in Appendix 1. The limitations of these data, and references to other determinations of DNA content, are discussed in the legend. We shall come in Part 3 to the organization of eucaryotic DNA sequences; and so shall describe the analysis of DNA by renaturation kinetics in Chapter 18, where the relationship between sequence components and DNA content is discussed.

On the basis of the haploid genome values given in Appendix 1, it is clear that the range of DNA contents is extensive, varying over some five orders of magnitude. Compared with the bacterial genome of about 4×10^6 base pairs, the smallest eucaryotes, represented by unicellular organisms such as the yeast S. cerevisiae, are only slightly larger, about 2×10^7 base pairs. The aggregating eucaryotes, such as the slime mold Dictyostelium, are little larger at about 3×10^7 base pairs. The smallest fully multicellular eucaryote whose DNA has been characterized appears to be the nematode worm, C. elegans, with a genome of 8×10^7 base pairs. The fruit fly D. melanogaster is only slightly larger, with a size of 1.4×10^8 base pairs, some 30 times the genome size of E. coli.

Then there is an increase in the minimum genome size found in any genus more or less in correspondence with evolutionary complexity up to the mammals (see Figure 17.3). The range of genome sizes in a species most generally covers about an order of magnitude; for example, in insects from about 10^8 to about 5×10^9 base pairs. However, much larger variations are seen in some species; the most extensive appear to be those of the amphibia, from about 6×10^8 to about 8×10^{11} base pairs (probably the largest eucaryotic genome). The mammals show an unusually compact range, all with haploid genomes rather close to 3×10^9 base pairs, about 1000 times that of E. coli.

The wide variation in the content of DNA in the eucaryotic genome raises two general questions. How is this DNA organized within the chromosome and is this structure the same in all eucaryotes? What is the function of the DNA: does it all code for protein, which would imply a variation of 10^5-fold in the number of genes in the eucaryotes, or does only some of it represent proteins, with the remainder having some other role? In this first part of the book we shall consider the general organization of the genetic apparatus in terms of its ability to perpetuate itself through cell division. This is to treat the genetic apparatus in terms of the chromosome. In the second part we shall be concerned with the molecular structure of the chromosome, with the interactions between DNA and protein that may be common to all eucaryotes. In the last two parts we shall be able to consider the organization of DNA sequences and their expression.

One of the most remarkable features of the eucaryotic genetic apparatus is the enormous amount of DNA that may be contained in the nucleus of each cell. The diploid nucleus of a human cell, for example, contains some 5-6 picograms, which organized in a linear duplex would stretch for some 174 cm. Within the nucleus, it must therefore be organized in an extraordinarily compact package.

The DNA of a eucaryotic cell is not, of course, a single duplex molecule, for it is organized into many individual chromosomes. Although eucaryotic chromosomes vary greatly in size, the quantity of DNA in even a small chromosome may be considerable. The smallest of the 23 human chromosomes contains 0.046 picograms (4.6×10^7 base pairs), equivalent to a linear duplex length of 1.4 cm and some ten times the amount of DNA in the E. coli genome. The largest human chromosome contains about 0.235 picograms (2.35×10^8 base pairs), equivalent to a linear duplex of 7.3 cm (DuPraw, 1970). But the lengths of these chromosomes observed at metaphase are of the order of $2-10 \times 10^{-4}$ cm. In their most compact condition, the chromosomes are therefore some 10,000 times shorter than the distance their DNA would occupy if extended. Although a greater amount of DNA is involved in the eucaryotes, compact packaging is a general property that is displayed also by bacteria and viruses. For example, the DNA of E. coli is equivalent to an extended duplex of 0.14 cm and is contained in a cell of diameter of about 10^{-6} cm.

The enormous compression of DNA in the chromosome raises many topological questions. First of all, how is DNA physically arranged within a chromosome and how much precision is there in its organization? Is there only a general structure for most chromosome regions or do particular sequences of DNA interact with specific proteins? How does this structure change during the cell cycle? A second set of questions concerns the activities of DNA as a template for nucleic acid synthesis. Assuming that its replication and transcription occur in a manner similar to that established in bacteria, how does the double helix unwind within the restrictions of the nuclear volume so that it may be replicated; and how do the daughter duplexes separate to form the replica chromosomes that are distributed to the daughter cells at mitosis?

Of course, not all the genetic information of a eucaryotic cell is contained in the DNA of the chromosomes in the nucleus: mitochondria and chloroplasts also contain DNA. Examples of non-Mendelian inheritance are generated when cells are heterozygous for genes carried by organelle DNA. In this case, reproduction does not take place by the precise segregation of mitosis and meiosis but by a more approximate distribution of the organelles at division. When the egg provides the sole supply of mitochondria for a developing embryo, mitochondrial genes display maternal inheritance (see Chapter 21).

The amount of DNA in cytoplasmic organelles is small compared with that in the nucleus. Mitochondrial DNAs of animal cells are usually found as circular molecules of duplex DNA of about 2×10^4 base pairs. This might be expected to code for only a few functions, many of which are likely to be part of the protein synthetic apparatus of the organelle. The mitochondrial genomes of other eucaryotes, for example yeast, may be larger; and chloroplast DNA may carry a somewhat greater amount of genetic information. But the overall contribution of these cytoplasmic DNAs to the genetic information of the cell is relatively small, and appears to be concerned with coding for specialized functions of the organelle (see Chapter 21).

CHAPTER 2

The Cell Skeleton: Microtubule Assemblies

In spite of the enormous range of variations in the phenotype of eucaryotic cells—so wide, indeed, that there cannot be said to exist any "typical" eucaryotic cell—all cells face common problems in achieving division. Mitosis and meiosis can be seen as the unitary mechanisms by which diploid cell reproduction and haploid gamete formation are accomplished. Naturally there are differences in the detailed mechanisms of division in each cell type—some differences are particularly well characterized between plants and animals—but the same general course of events is followed. Prior to division itself, the chromosomes have been replicated. Segregation of new sets occurs by the directed movements of highly condensed chromosomes on the spindle which replaces the nuclear and cytoplasmic compartments. Then nuclei reform and the progeny cells separate.

During the long history of cytology, division has been studied as a unique process, one unconnected with the rest of the cell cycle in the sense that the components of the division apparatus appeared peculiar to division and unrelated to components of the interphase cell. But recently it has become apparent that the components from which the division apparatus is constructed are present also in cells during interphase. The spindle fibers responsible for chromosome movement are assembled from microtubules, which now can be seen to be present in the form of a network across the interphase cell. The filaments responsible for separating the progeny cells at cytokinesis constitute a contractile system including actin; and actin and other proteins found in the contractile apparatus of muscle cells now can be visualized in the form of a cytoplasmic network in many nonmuscle cells. The relationship between the division apparatus and these frameworks in interphase cells is clearly critical in attempting to define the events responsible for the perpetuation of the eucaryotic cell.

The proteins that constitute the cytoplasmic and mitotic fibers and filaments appear to be present in eucaryotic cells of diverse evolutionary origins. Remarkable similarities across evolutionary barriers are to be found also in the histone proteins which interact with DNA to form the basic structure of the eucaryotic genetic apparatus. This suggests the view that the genetic apparatus and the components responsible for cell division in eucaryotes date from a very early stage of evolution and have been subject to great selective pressure against variation. In this chapter and the next we shall be concerned with the state of the components of the fibers and filaments in interphase cells, where they form networks that might be regarded as comprising a cell "skeleton." Then in Chapter 4 we shall turn to the role of these components in accomplishing the division of one cell into two. In the second part of the book we shall consider the constitution of the genetic material itself in terms of a structure that may be common to all eucaryotes.

Occurrence of Microtubules

The microtubule comprises the basic entity from which several structures concerned with cell shape or motility are constructed. Early experiments associating microtubules with the maintenance of cell shape showed that protozoan axopodia are unstable under pressure or upon exposure to low temperature; the disaggregation of the microtubules constituting the axopodia is accompanied by the loss of cell shape (Tilney and Porter, 1965, 1967; Tilney, Hiramoto and Marsland, 1966). In plant cells, microtubules appear to play a structural role, and in animal cells they form a network of cytoplasmic fibers associated with the acquisition of cell shape. Cilia and flagella are microtubule systems that have been extensively studied in the protozoa; they are especially amenable to analysis of the structure of the microtubules and their associated components. Microtubules of some sperm tails also are concerned with motility. Mammalian brain contains a large number of microtubules organized as neurotubes, the dendrites and neurites that run parallel to the long axis of the neuronal cell and which are concerned with sensory functions. Brain is a particularly rich source of microtubules, where the constituent protein may amount to 1–2% of the total protein and to as much as 5–10% of the protein that is soluble after homogenization in sucrose. And the fibers of the mitotic and meiotic spindles in all cells are assembled from bundles of microtubules; they have been best characterized in the eggs of sea urchins and some other marine organisms. The spindle fibers, of course, are concerned with chromosome movement. They emanate from centrioles, structures themselves constructed from microtubules, which resemble the basal bodies from which cilia and flagella project.

This is by no means a complete list of structures constructed from microtubules, of which there are many other examples. The microtubule is a structure ubiquitous to the eucaryotic cell, then, appearing in the form of many specialized organelles as well as in spindle fibers at each division. First we shall be concerned with the structure and assembly of the microtubule itself, drawing upon the different systems that have been investigated to answer different problems. The occurrence of microtubules in so many systems of such evolutionary diversity poses the obvious question of their relationship. Do all microtubules have the same basic structure? Are they all constructed from the same protein components? How is the interaction of protein subunits controlled to generate the particular structure of each system? We shall see that while all microtubules appear to be assembled from the same types of subunit into similar types of structure, they differ in their stability and in the extraneous components with which they are associated. With these points established, we may then turn to examine the role of microtubules in the cytoplasmic networks of interphase cells. Later we shall consider their function in the mitotic apparatus (see Chapter 4).

Microtubule Ultrastructure

Microtubules are straight or (sometimes) curved hollow cylinders, some 24 ± 2 nm (240 Å) in outside diameter with an interior core of diameter 15 nm (for review see Olmsted and Borisy, 1973b; Snyder and McIntosh, 1976). The early results of André and Thiery (1963) and Pease (1963) first suggested that microtubules might be composed of longitudinal filaments, although the number was measured accurately only some time later. In *singlet* microtubules from virtually all sources there have proved to be 13 protofilaments running longitudinally. This includes plant meristem cells, sea urchin sperm and protozoan flagella, sea urchin mitotic apparatus, mammalian brain, and frog axon (Ledbetter and Porter, 1964; Phillips, 1966; Gall, 1966; Ringo, 1967a; Tilney et al., 1973b). Viewed in cross section the microtubule is a hollow cylinder with a wall consisting of 13 globular subunits each some 4–5 nm in diameter. Figure 2.1 shows an electron micrograph of a

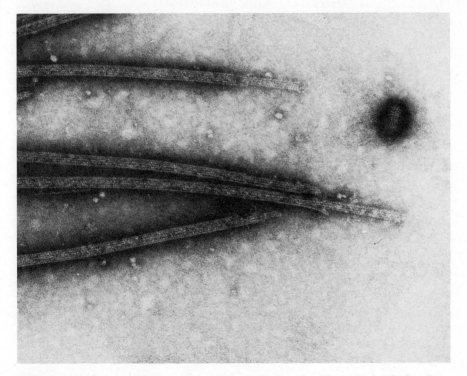

Figure 2.1. Singlet microtubules. These are A tubules dissociated from the tip of a flagellum. 120,000. Photograph kindly provided by Dr. A. V. Grimstone.

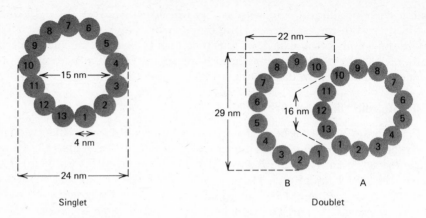

Figure 2.2. Dimensions of singlet and doublet microtubules in cross section.

singlet microtubule and Figure 2.2 illustrates its general structure. From the dimensions of the subunits each might be expected to be of the order of 50,000 daltons of protein.

In addition to the typical singlet fibers, microtubules are found as doublet and triplet rings. One of the best characterized structures containing microtubules is the flagellar axoneme, the axial filament of the flagellum, which comprises an outer circle of 9 microtubules surrounding 2 inner microtubules, often therefore known as the *9 + 2* structure. The same structure is found in echinoderm sperm flagella and protozoan flagella, the two systems on which the most work has been done. The flagella contain two types of microtubule. While the two inner microtubules represent singlet rings when viewed in cross section, the nine outer microtubules take the form of double rings and are known therefore as the *outer* doublets. The two rings of the doublet have different structures. One, the *A fiber*, displays the same circular cross section of 13 protofilaments as the typical singlet microtubule structure seen in the two inner tubules. The other, the *B fiber*, is more elliptical and appears to consist of 10 or 11 separate protofilaments fused with three of the A fiber filaments that thus form a common wall between the two rings as shown in Figure 2.2 (Ringo, 1967; Tilney et al., 1973b; Warner and Satir, 1973.)*

The flagellum projects from a *basal body*, a structure which appears closely

---

*The microtubules of the flagellar axoneme are connected by arms and the structure is completed by coarse fibers that surround the outer doublets. Arms of *dynein*, which possess an ATPase activity, are attached to the A fibers. The B fibers possess sites where the distal ends of the dynein arms may be attached to generate bridges between adjacent doublets. These bridges may form transiently to enable the doublets to slide past each other, an action dependent on the hydrolysis of ATP and responsible for flagellar movement. Radial links join the A fibers to a central sheath that surrounds the inner singlets. By resisting the sliding action, these may translate the movement of the microtubules into the characteristic flagellar bending. This is known as the *sliding filament* model for flagellar action (Gibbons, 1965; Warner, 1967; Hopkins, 1970; Summers and Gibbons, 1971, 1973).

related to the centrioles found at the poles of animal cells during mitosis (see Chapter 4). Basal bodies consist of 9 triplet microtubules, with an overall diameter of some 200 nm and a length of about 400 nm. (Electron micrographs of centrioles are presented as Figures 4.6 and 4.7). A transition from the triplet structure of the basal body to the doublet structure of the flagellar axoneme is made at the junction between them. The basal body probably is anchored to the cell membrane at its other end and has been characterized the most extensively in the alga Chlamydomonas reinhardii (Gibbons and Grimstone, 1960; Ringo, 1967b; Wolfe, 1970).

The existence of different classes of microtubules is suggested by variations in stability against disruption by certain agents (Behnke and Forer, 1967b; Tilney and Gibbons, 1968; Stephens, 1970; Olmsted et al., 1971). For example, most cytoplasmic microtubules are labile and dissociate rapidly upon exposure to pressure, cold, or the drug colchicine; whereas ciliary and flagellar microtubules are stable upon such treatment. Labile microtubules appear to be in equilibrium with a pool of free subunits, whereas stable microtubules, once assembled, do not dissociate into subunits (see below). There are differences also in the stability of the component fibers of the flagellar outer doublets, because upon warming the subunits of the B fiber become soluble somewhat more readily than those of the A fiber.

Differences in the structures associated with microtubules mean that microtubules from different sources may be attached to varying sets of extraneous proteins. For example, we have seen already that microtubules of the flagellar outer doublet are attached to arms; by contrast, the neurotubes of brain cells of many species have quite different distinguishing features, such as the presence of filamentous elements on the surface (Lane and Treherne, 1970; Yamada et al., 1970, 1971; Burton and Fernandez, 1973). It is therefore impossible to say whether differences in stability reside in the microtubules themselves or are conferred by the proteins associated with them. One approach to determining the basis of differences in stability lies in the characterization of the protein subunits of the microtubule; and this, and in particular the effect of colchicine, is discussed later.

Two features of the microtubule are basic in constructing a model for its structure. The number of protofilaments is 13; and the protofilament appears to be a stable entity, as can be seen during the disaggregation of microtubules into their component threads. The microtubule is now known to be constructed from a protein heterodimer which consists of two subunits of similar size (denoted α and β and discussed below). If the microtubule is to be formed by a series of quasi repetitive interactions, that is, if α and β subunits always are to bear the same relationship to one another, this means that the protofilament must comprise a series of repeating heterodimers. The alternative in which each protofilament consists of one type of protein subunit, so that the heterodimer lies in a horizontal plane, is excluded by the odd number of protofilaments, which would make the horizontal alternation of α and β subunits impossible.

The electron microscopy of Grimstone and Klug (1966) and Barnicott

(1966) showed that the protofilaments take the form of a series of globular beads about 4 nm in diameter. In addition to the 4 nm axial repeat, there is a repeat of 8 nm which is rather variable in relative intensity. The most likely interpretation is that the 4 nm component represents a protein subunit, while the 8 nm repeat corresponds to the heterodimer. Although this does not explain the variable relationship between the 4 nm and 8 nm axial repeats, it is consistent with the size of the monomeric protein subunits, whose 55,000 daltons would correspond reasonably well with globules of 4 nm diameter.

From the optical analysis of electron micrographs, an approach that allows reconstruction of the surface lattice from the superimposition of near-side and farside images, Amos and Klug (1974) constructed the model illustrated in Figure 2.3. This visualizes the protofilaments as consisting of 8 nm repeating units which are shaped rather like dumbbells and are joined end to end, slightly tilted with regard to the axis of the filament. The 4 nm repeat is produced by the components of the dumbbell. The dimers are staggered in adjacent protofilaments so that the lateral bonds in the tubule all connect α with β subunits. This generates a surface lattice with helical symmetry. The same approach led Erickson (1974, 1975) to construct a similar model. X-ray diffraction patterns do not provide sufficient detail to construct a precise model, but identify a 4 nm unit as the principal repeating element and thus are consistent with the optical analysis (Cohen et al., 1971, 1975; Mandelkow, Thomas and Cohen, 1977). An extension of this approach suggests that subunits of different types alternate in the protofilament, but identifies laterally adjacent subunits as of the same type (Crepau, McEwen and Edelstein, 1978).

The number of 13 protofilaments means that a microtubule does not display symmetry when viewed in cross section.* There must therefore be asymmetry in the attachment of structures such as bridges to the tubule, although presumably any given extraneous protein always attaches to the microtubule subunit in the same way.

In earlier work, apparent variations from the number of 13 protofilaments seem to have been caused by artifacts of the staining techniques, but more recently two exceptions have been identified. Crayfish sperm have microtubules comprising 15 protofilaments, while crayfish and lobster axons have microtubules of 12 protofilaments (contrasted with the usual 13 present in glial cytoplasmic microtubules) (Nagano and Suzuki, 1975; Burton et al.,

*Fujiwara and Tilney (1975) have speculated upon the evolution of this asymmetric structure by suggesting that the construction of a tubule from any prime number of protofilaments should result in some instability from the attempt to achieve symmetry (a presumed feature of biological assembly systems) in a structure unable to do so. With a smaller prime number of protofilaments (5, 7, or 11) perhaps the strain on a rigid bridge would be too great and might prevent the structure from assembling, whereas with a larger prime number (17, 19, or 23) the approach to symmetry would be closer, allowing too great a stability. Microtubules in many systems are labile aggregates which polymerize and depolymerize, so their stability may be a critical feature in selection. However, an important feature in stability is the nature of extraneous proteins associated with the microtubules of any particular system (see text below).

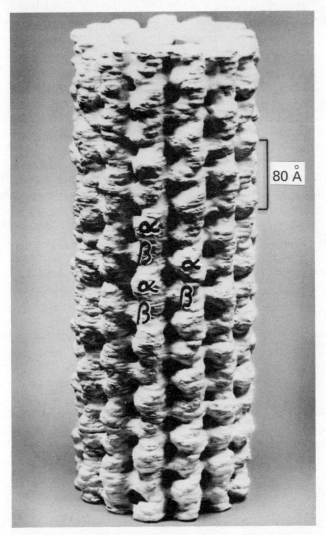

Figure 2.3. Model for microtubule construction from α and β subunits. From Amos and Klug (1974).

1975). The basis for this variation has not yet been established; and it will be particularly interesting to determine whether all three types of crayfish microtubule are constructed from the same subunits. At present these are the only known variations in microtubule construction and it remains to be seen how they are related to the particular uses of these microtubules. Given the possibility for variation, however, it is perhaps puzzling that the number of protofilaments is almost universally 13, because microtubules are found in many different situations in which they are associated with different extraneous structures and have characteristic abilities. Would it not

be advantageous to vary the structure to fit each situation? The strong evolutionary pressure to conserve the microtubule structure therefore remains
to be explained.

Components of Tubulin

The first approach to isolating the protein subunits of the microtubule
took advantage of the drug colchicine, which inhibits mitosis by preventing
the formation of spindle fibers. Use of colchicine labeled with ^{3}H allowed
a drug-binding protein sedimenting at 6S to be isolated from several different types of cell, including mammalian tissue culture lines, mammalian
brain, and sea urchin sperm tails and mitotic apparatus (Borisy and Taylor,
1967a,b; Shelanski and Taylor, 1967; Bibring and Baxandall, 1971; Wilson
and Meza, 1973). The same protein is found in both singlets and outer doublets of flagella, although it cannot be isolated by colchicine binding (Shelanski and Taylor, 1968; Renaud, Rowe and Gibbons, 1968). Subsequent
methods for purification have made use of the ability of tubulin to polymerize and depolymerize in vitro under appropriate conditions (see below).
The protein isolated from all sources has a molecular weight in the range
110,000-120,000 and is dissociated by guanidium HCl into two subunits
of the same weight (Weisenberg, Borisy, and Taylor, 1968; Kirkpatrick et
al., 1970; Lee, Frigon, and Timasheff, 1973). This suggests that all microtubules may be composed of the same type of protein, a dimer consisting
of two subunits of the same or very similar size. The present best estimate
for its mass is 55,000 daltons for the monomer, the value that we shall use
here for further calculations. We have seen that the size of the monomer fits
with that expected of the 4 nm diameter observed for the morphological unit
by electron microscopy.

    In gel systems that effect separation on the basis of charge, the subunits
form two bands. These appear to be present in microtubules isolated from
all sources, including mammalian and avian brain, sea urchin eggs and
sperm tails, and a variety of protozoan cilia and flagella (Bryan and Wilson,
1971; Feit, Slusarek, and Shelanski, 1971; for review see Bryan, 1974). The
monomeric subunits now are known as α tubulin and β tubulin and they
appear always to be present in 1:1 stoichiometry (Bibring and Baxandall,
1974). At one time it was thought that the A and B fibers of outer doublet
microtubules might contain different types of subunit (α in one and β in the
other), but now it is clear that the A and B fibers both contain both types
of subunit (Olmsted et al., 1971; Meza, Huang, and Bryan, 1972; Witman,
Carlson and Rosenbaum, 1972). Basal bodies, representing triplet microtubules, also consist of α and β tubulin (Gould, 1975).

    Are the two tubulin subunits organized solely as an $\alpha\beta$ heterodimer or
does the 6S dimer preparation represent a mixture of $\alpha\alpha$ and $\beta\beta$ homodimers? Generally it has been assumed that there is only one type of unit,

the $\alpha\beta$ heterodimer, from which microtubules are constructed; the existence of homodimers would mean that microtubules must consist of some arrangement of the two different dimers, which is difficult to visualize in terms of the odd number of protofilaments (see above).

One approach to defining the association of subunits is to treat dimeric tubulin with a protein cross-linking agent. Using diimidate reagents, Luduena, Shooter, and Wilson (1977) observed that initially there is a high proportion of α-β cross links, but an increasing proportion of homo cross links appears with time. Although not unequivocal, probably this means that in a new preparation only intradimer (α-β) cross links are generated, but that as tubulin generates aggregates with time (see below), interdimer crosslinking occurs. The addition of colchicine or vinblastine to prevent tubulin polymerization results in an appreciable increase in the proportion of α-β cross links. This supports the idea that the dimeric interaction is between α and β subunits.

The existence of α and β tubulin subunits in all microtubule systems shows that there is evolutionary constraint on the structure of the microtubule. Antigenic cross reactions between tubulins of different sources show the conservation of certain features, although differences in the immunological reaction cannot be interpreted clearly (Fulton, Kane, and Stephens, 1971; Morgan, Holladay, and Spooner, 1979). The conservation of function is displayed strikingly in the ability of tubulin from one source to polymerize onto microtubules of another species (see below). But there are differences as well as similarities between the various microtubule systems. This prompts the question of what variation there may be in the tubulins comprising the different microtubule systems of a given organism: are flagella and spindle fibers, for example, constructed from the same α and β tubulins or are these related, but distinct, products of different genes?

Some evidence has been adduced for polymorphism in the tubulin components of different sea urchin tissues. Bibring et al. (1976) observed that the α tubulin of sea urchin mitotic apparatus can be separated by gel electrophoresis into two bands, $\alpha1$ and $\alpha2$, whose combined mass equals that of β tubulin. Tubulin of the A fiber of ciliary doublets also has two α gel bands, but their mobilities do not appear to be identical with those of the mitotic apparatus. Sperm tails appear to have only the $\alpha2$ type of α tubulin. Both α and β tubulins of C. reinhardii and starfish (A. amurensis) flagella have been fractionated into several bands by gel electrophoresis or isoelectric focusing (Piperno and Luck, 1976; Kobayashi and Mohri, 1977). While this raises the possibility that there may be multiple types of tubulin, in lieu of sequence data it is possible that the differences in mobility are due to modifications such as phosphorylation or methylation rather than variations in primary sequence. Some preliminary data suggest that there may be differences in the cyanogen bromide peptides of the mitotic apparatus and sperm tail tubules, which would be expected if there were differences in sequence.

Conservation of the amino acid sequences of tubulin has been demonstrated for the N-terminal region of the α and β proteins of chick brain and sea urchin sperm tail outer doublet. In the first 25 residues of the two α tubulins, Luduena and Woodward (1973) found no differences; in the first 25 residues of the two β tubulins there is only a single amino acid substitution (isoleucine for methionine). About 700 million years are thought to have passed since the divergence of the chordates (including birds) and echinoderms (including sea urchins). If the immediate N-terminal region is typical of the rest of the protein (it represents only about 5% of the length), then both α and β tubulin must evolve extremely slowly, much more slowly, in fact, than other proteins, with the exception of the histones that also represent an example of extreme conservation of sequence (see Tables 11.2 and 11.3). Cyanogen bromide maps of these tubulins suggest that this conclusion may be valid for the entire molecule. Since the chick microtubules are presumably derived from neurotubes and the sea urchin from outer doublets of the sperm tail, there is a difference in microtubule type as well as in species. The sequences also demonstrated a relationship between the α and β tubulins, since in about half of the first 25 positions the same amino acid had been conserved. If α and β tubulins have diverged from a common ancestral protein, however, their separation must have occurred very early in eucaryote evolution.

Polymerization of Tubulin

The ability to assemble tubulin into microtubules in vitro has been developed only comparatively recently. Weisenberg (1972) found that rat brain tubulin can polymerize in the presence of:

ATP or GTP

Mg^{2+} ions

a chelator to remove Ca^{2+} ions

Olmsted and Borisy (1973a) reported that the polymerization of porcine brain tubulin is highly temperature dependent; maximum polymerization is achieved in 15–20 minutes at $37°C$, but there is no polymerization at all below $15°C$. This is consistent with the behavior shown by labile microtubules in vivo; for example, spindle fibers disappear upon exposure to cold. And polymerization of tubulin is inhibited both in vitro and in vivo by colchicine. We should note that the experiments on tubulin polymerization discussed in this and the next section have been carried out with mammalian brain tubulin; while tubulin from several mammalian brains appears to display the same properties, this cannot be taken as evidence that tubulin from other sources will necessarily follow the same rules.

Two moles of guanine nucleotide—either GDP or GTP—can be bound to each tubulin dimer; its presence stabilizes the dimeric structure and the ability to bind colchicine, which otherwise is lost with time (Stevens, Renaud and Gibbons, 1967; Shelanski and Taylor, 1968; Weisenberg, Borisy and

Taylor, 1968). In these and other experiments the proportion of guanine nucleotide in the isolated protein has varied in the range of 0.5–0.8 moles/ dimer. It is likely that this value is less than 1 mole/dimer due to losses in preparation or difficulties in extraction (Weisenberg, Deery, and Dickinson, 1976). When GDP or GTP is added, another molecule of guanine nucleotide can be bound to each dimer (again the measured level is about 0.5–0.8 moles/ dimer) and this binding is freely reversible (Berry and Shelanski, 1972; Jacobs, Smith, and Taylor, 1974). The most likely explanation is that in each dimer one monomeric subunit has a tightly bound GDP/GTP, while the other has a binding site at which GDP/GTP can be reversibly bound. The two binding sites are described as the N site (for nonexchangeable binding) and the E site (for exchangeable binding).

What role does guanine nucleotide play in the polymerization of tubulin? Two obvious possibilities are that hydrolysis of GTP is energetically coupled to polymerization or that nucleotide binding is needed for some allosteric change in the protein that is necessary for polymerization. Conflicting reports have been made both about the need for hydrolysis of GTP and on whether binding at the N or E site is involved.

The ability of ATP—which does not bind to tubulin—to support polymerization suggests that the presence of GTP rather than GDP may be necessary and that transphosphorylation may occur. Tubulin preparations appear to contain a nucleoside diphosphokinase activity; and their ability to undertake transphosphorylation correspondingly declines with the degree of purification of the tubulin. Either ATP or GTP can be used to donate a phosphate group to endogenous GDP.

After considerable confusion about the roles of the N and E sites, it has become clear that the N site contains GTP which is not hydrolyzed upon the polymerization of tubulin. The GTP present at the E site may be hydrolyzed, however, and then constitutes the substrate for the transphosphorylation system. The nucleotide bound at the E site ceases to be exchangeable in the polymerized tubulin (Weisenberg, Deery, and Dickinson, 1976; Penningroth and Kirschner, 1977). Contrasted with the idea that GTP hydrolysis occurs during polymerization, GMP–PCP (a nonhydrolyzable analogue) can support polymerization if all competing GDP is removed. (The presence of GDP prevents the GMP–PCP analogue from binding to tubulin and this reaction may have been responsible for earlier observations that the presence of GMP–PCP inhibits polymerization). The GTP bound at the N site is stable within the cell (Spiegelman, Penningroth, and Kirschner, 1977). Although it is not hydrolyzed during polymerization, its presence as a cofactor may be necessary. The need for the presence at the E site of a guanine nucleotide with three phosphate groups is shown by the acceptability of GTP and GMP–PCP but unacceptability of GDP. The presence of GMP–PCP implies that although hydrolysis may occur when GTP is present, it is not essential for polymerization; the role of the guanine nucleotide therefore remains to be clarified. Of course, these results do not exclude the possibility that hydrolysis of the GTP at the E site is necessary for microtubule assembly in vivo, but that the necessity is circumvented by conditions prevailing in vitro.

In measurements of the effect of calcium ions on assembly in vitro, half maximal inhibition of polymerization was caused at a concentration of 10^{-3} M (Olmsted and Borisy, 1975). Since physiological levels are thought to be of the order of 10^{-6} M, this would argue against the idea that calcium ions might be involved directly in controlling polymerization in vivo. Both Mg^{2+} ions and GTP appeared to be needed in stoichiometric amounts (10^{-3} M in these experiments). Lee and Timasheff (1977) reported that one additional Mg^{2+} ion becomes bound for every tubulin dimer incorporated into a growing microtubule. The level of Mg^{2+} ions necessary for polymerization has been controversial. Olmsted and Borisy (1973a, 1975) found maximum activity at about 10^{-3} M, which is close to the apparent physiological level, and observed half maximal inhibition of assembly by levels tenfold greater. But Lee and Timasheff (1977) and Herzog and Weber (1977) found maximum polymerization at about 10^{-2} M, although in different conditions (see below).

The caveat that conditions in vitro may not exactly reflect those in vivo, especially concerning the need for nucleotides, is borne out by the development of a method for polymerizing tubulin in the absence of added nucleotides. In sucrose or glycerol solutions, tubulin undergoes reversible polymerization without added GTP or ATP (Shelanski, Gaskin, and Cantor, 1973). As usual, assembly requires high temperature (37°C) and is inhibited by colchicine. Abnormal assembly forms appear under some conditions— for example, with increases in the concentration of tubulin—suggesting that misassembly may occur. This implies that the environment may be important in driving polymerization along the pathway that leads to formation of the microtubule rather than to other structures. It means further that the products of polymerization should be scrutinized by electron microscopy and not followed only by methods that detect overall aggregate formation (such as viscometry and turbidity).

Initiation of Microtubule Assembly

An important consequence of the ability to polymerize and depolymerize brain tubulin is that purification can be effected by using successive cycles of assembly and disassembly. This separates the active tubulin dimers from any inactive molecules present in the population (since these are unable to enter the polymerized form) as well as from any proteins that are associated, but do not copolymerize, with tubulin. In the course of purifying tubulin in this way, it turned out that preparations of isolated 6S dimers cannot polymerize; they require the presence of nucleating centers, which appear to comprise larger aggregates of tubulin also containing certain of the proteins associated with microtubules.

The first indication that initiation and elongation of assembly are separate reactions was provided by the observation of Borisy and Olmsted (1972) that centrifugation at high speed renders tubulin preparations unable to polymerize. The centrifugation removes disc shaped structures from the preparation, and this suggests that these may act as nucleating centers. Shelanski, Gaskin, and Cantor (1973) showed that both pellet and supernatant consist largely of tubulin, which suggests that the discs may represent tubulin organized in the form of some initiating structure. This implies that free tubulin dimers do not initiate the assembly of larger aggregates, although they provide the subunits from which the nucleating centers are constructed, as well as representing the form in which tubulin is added to the growing microtubule.

This concept is supported by observations that tubulin polymerization in vitro passes through two stages (Gaskin, Cantor, and Shelanski, 1974). Using turbidity to follow the formation of aggregates allows initiation to be distinguished from elongation because the technique is sensitive to formation of tubular structures but not to their length. The first stage is represented by a lag period, the second by the appearance of aggregates. The lag depends upon tubulin concentration, aggregates forming above a critical level which increases as the temperature decreases. The rate limiting, concentration-dependent first step can be equated with the formation of nucleating centers, which then are elongated relatively rapidly by the addition of dimers.

Upon disaggregation of brain microtubules by low temperature or addition of calcium ions, Kirschner et al. (1974) identified two types of product: 6S dimers; and larger aggregates, sedimenting around 36S, but somewhat heterogeneous in size. When examined in the electron microscope, three types of structure are visible: small particles (6S dimers of 6–7 nm diameter); *double rings* of about 43 nm and 29 nm diameter for the outer and inner circles (comprising about 23 of the small particles); and spiral structures whose average contour length is close to the total circumference of the double ring. Presumably the double ring and spiral structures are formed by coiling of the microtubule protofilaments during depolymerization.

Both the 6S and 36S preparations are constituted almost entirely (>90%) of α and β tubulin. Treatment with NaCl converts the 36S fraction into 6S dimers, which suggests that the 36S particles consist of tubulin dimers. This reaction is reversible. Treatment with high pH (10.6) irreversibly converts the 36S particles into 6S dimers. Weingarten et al. (1974) found that the 36S fraction is rich in the double rings and suggested that these may correspond to the nucleation discs. Only tubulin preparations containing the double rings can polymerize; the 6S dimers can be assembled into tubules in the presence but not in the absence of the rings. The 36S double rings can be converted into 20S single rings that are inactive in supporting tubule formation, for example by treatment at high pH (Doenges, Biedert, and Pawaletz, 1976). Other aggregate forms also have been identified and a detailed picture of their relationships is emerging (Marcum and Borisy, 1978; Vallee and Borisy, 1978). The 6S and 36S fractions appear to correspond to two preparations characterized in previous work as X tubulin (able to bind to colchicine) and Y tubulin (unable to bind the drug). Models for the initiation of microtubule assembly in vitro, in particular for the relationship between the double rings and the mature tubular structure, have been discussed by Kirschner et al. (1974, 1975). Alternatives have been considered by Bryan (1976) and Johnson and Borisy (1977). The principal conclusion to be drawn from these studies is that initiation requires the formation of a nucleating center composed largely of tubulin and that this may be the rate limiting step in microtubule assembly in vitro. However, this does not prove that the rings themselves may be equated with the nucleating centers; it remains possible that they contain components necessary for nucleation, which might, for example, be provided by dissociation of the rings and formation of some other nucleating structure. This raises the obvious question of what form the nucleating center may take in vivo. Very probably this depends upon the microtubule system; and in Chapter 4 we shall discuss the events responsible for initiating the formation of spindle fibers in the mitotic cell.

Initiation of tubule assembly appears to depend on the presence of proteins associated with tubulin; and these may be contained in the 36S double rings. High salt concentrations convert the 36S particles to 6S dimers; and when the salt is removed by dialysis, the reaction is reversed. Free 6S dimers added to the dissociated mixture are incorporated into 36S particles when the salt is removed. That the effect of high salt may be to remove a protein from the 36S particles that is necessary for polymerization is suggested by the results of experiments in which Weingarten et al. (1975) passed tubulin preparations through phosphocellulose columns. The tubulin is not adsorbed, but about 5% of the total protein (representing much of the nontubulin protein species) is retained. Tubulin recovered from the column is unable to polymerize, but addition of the adsorbed material restores its ability. The activity has been obtained as a crude protein preparation described as the *tau factor*. The principal activity has been purified as a set of four closely related proteins of 55,000–62,000 daltons, highly elongated in shape, whose relationship remains to be defined (Cleveland et al., 1977a,b).* Their action is stoichiometric rather than catalytic and they comprise about 2% of the mass of tubulin, a level that does not change over three cycles of assembly and disassembly.

When mammalian brain tubulin is prepared by cycles of polymerization and depolymerization, it contains high molecular weight proteins: these were originally described as the *MAP* (for microtubule associated proteins).

---

*The set of four proteins formally is defined as *tau-1*. The previous crude tau preparations also contain other proteins active in promoting microtubule assembly; these have not been investigated and are designated *tau-2*.

They represent up to about 20% of the total protein. The most prominent are MAP1 and MAP2, of roughly 350,000 and 300,000 daltons (Sloboda et al., 1975). The term MAP since has been used to describe both this preparation (the HMW MAP) and the phosphocellulose-stripped proteins (the tau MAP).

The HMW MAP preparation from mammalian brain appears to be involved in the initiation of polymerization in vitro. Tubulin purified away from the MAP components is unable to polymerize; and the formation of tubules increases with the amount of MAP added up to a molar ratio of about 1:10 (that is with 0.3 MAP/tubulin dimer on a weight basis) (Murphy and Borisy, 1975; Sloboda, Dentler and Rosenbaum, 1976). There appears to be no relationship between the HMW and tau preparations.*

The presence of microtubule associated proteins probably is common to all microtubule systems. Obviously some of the MAP fraction may represent the extraneous structures associated with the microtubules of particular types; but it is reasonable to suppose that proteins persistently copurifying with tubulin might include components that are part of the structure of the microtubule itself. The embarrassment of having two different fractions each apparently involved in the initiation of microtubule formation has yet to be resolved: their properties have been investigated in some detail in vitro, but whether either or both have the same functions in vivo is of course unknown. Usually the two types of preparation are obtained in different protocols; however, in one set of experiments, Herzog and Weber (1978b) have obtained both HMW and tau and shown that either is able to support initiation. From several series of experiments to characterize the initiating activities, it is clear that they share the feature that increasing amounts of MAP reduce the critical concentration of tubulin that is required for assembly. A possible difference in the initiating activity is indicated by the observation that HMW proteins promote the formation of 30S double rings while tau proteins stimulate production of 20S single rings (Vallee and Borisy, 1978).

An important question is whether the proteins necessary for initiation are involved only at this stage or participate also in elongation. One way to test this is to assay the ability of highly purified tubulin to polymerize when it is added to a preparation of nucleating centers. The source of nucleating centers has been sonicated fragments of microtubules reassembled from mammalian brain tubulin or isolated flagellar axonemes. Tubulin stripped of its associated initiating activity by passage through phosphocellulose or DEAE-cellulose was unable to elongate the nucleating structures, in contrast with the ability to do so of less purified tubulin (Bryan,

---

*One of the proteins present in crude tau preparations has been purified and described as TAP (tubulin assembly protein). It is 68,000 daltons in size and stimulates the initiation of microtubule assembly in vitro (Lockwood, 1978). However, a very similar protein is a component of brain 10 nm neurofilaments and has been shown to be present in microtubule preparations in the form of contaminating fragments of neurofilaments (Runge, Detrich, and Williams, 1979). The role of this protein in microtubule assembly is dubious. It remains to be seen whether the association is significant or fortuitous.

1976; Witman et al., 1976). But the addition of tau factor restored the ability of the stripped tubulin dimers to assemble onto the nucleating centers. The final yield of assembled microtubules as well as the rate of assembly depended on the amount of tau added, which suggests that the tau factor may be a structural component of the microtubule.

Tubulin purified by high-speed centrifugation to remove the ring structures, although unable to initiate polymerization, can elongate nucleating centers in vitro. This implies that it contains sufficient tau factor activity to support elongation, although this is not adequate to allow the formation of ring aggregates. Thus complete removal of the tau factor activity from preparations of tubulin dimers may require treatment with phosphocellulose or similar agents. The difference between the initiation and elongation activities of tubulin before its passage through phosphocellulose implies that a greater concentration of tau is needed for the formation of the nucleating centers than is necessary to support elongation. (However, this conclusion does rest upon the assumption that it is the same protein which is responsible for conferring upon phosphocellulose stripped tubulin the abilities to initiate and elongate. Until these activities are proven to reside in the same protein by a complete purification of the tau factor, it remains possible that different proteins are responsible.)

In systems in which microtubule assembly in vitro depends on the provision of HMW initiating factors, elongation can be supported by purified tubulin in the absence of initiating factors. However, the rate and extent of elongation is stimulated by the HMW preparations. Murphy, Johnson, and Borisy (1977) suggested that this occurs because the HMW proteins associated with the elongated microtubules have a stabilizing effect; the reduction in disassembly is seen as an increase in the overall progress of elongation.

Both tau and HMW proteins are associated with microtubules in vivo. Antisera prepared against either type of fraction stain the same structures in vivo that react with antisera against tubulin (Connolly et al., 1977, 1978). Microtubules assembled in vitro react periodically, every 32 nm, with an antiserum directed against MAP2 (Kim, Binder, and Rosenbaum, 1979). Both of the putative initiating activities are therefore associated with the entire length of microtubules, not just with nucleating centers.

Protocols have been developed which make it possible for brain tubulin dimers to aggregate into tubular structures in the absence of any additional protein or nucleating centers. These involve the establishment of conditions that are unlikely to resemble the physiological state.* One approach involves the use of high concentrations of glycerol (which has also been used to stabilize the isolated mitotic apparatus; see Chapter 4). Another

---

*Protocols involving high Mg^{2+} $(10^{-2}$ M) and the presence of 1–4 M glycerol were developed by Frigon and Timasheff (1975a,b) and Lee and Timasheff (1975, 1977); dextran has been introduced by Erickson and Voter (1976) and Herzog and Weber (1978a); 10% DMSO was used by Himes et al. (1976).

relies upon the provision of polycations, which perhaps may substitute for the MAP by causing the effective concentration of tubulin (a strongly anionic protein) to increase in their vicinity (Erickson and Voter, 1976). This raises the possibility that one activity necessary for initiation may be to obtain a great enough concentration of tubulin to cause local aggregation. Self assembly of brain tubulin also has been observed in conditions more akin to the physiological. The tubulin must be fresh (implying that the self assembly capacity is lost with aging), its concentration must be high (effectively about 10^{-5} M), and a high concentration of Mg^{2+} ions is necessary (Herzog and Weber, 1977). What form the nucleation reaction takes under these conditions is not yet known. The implication of these studies is that tubulin dimers possess an intrinsic ability to self assemble into tubular structures. To what extent this is modulated by other proteins in vivo remains to be established.

Many early attempts with the conventional procedures to polymerize tubulin from sources other than brain were unsuccessful. Subsequently, however, tubulin from a variety of sources was first polymerized by protocols involving high glycerol concentrations, whose function seems to be to attenuate the inhibitory effects of factors present in the crude preparations which include, but do not exclusively comprise, RNA. When the tubulin is increasingly purified, the dependence on glycerol is reduced. That it may be the presence of these inhibitors that is responsible for the failure of non-brain tubulin to polymerize in vitro is suggested by a rough assay showing that their amount correlates inversely with the ability of tubulin from each cell type to polymerize. Another factor is that accessory proteins are present in smaller amounts in these preparations than in brain tubulin; also it should be noted that different accessory proteins are present in different cells, in particular there being a dearth in nonbrain cells of the high molecular weight fraction. More recent experiments have supported the assembly of tubules by more usual protocols from a variety of sources, extending even to sea urchin sperm tail outer doublet.* It is interesting that the tubules assembled from the sea urchin tubulin show the same instability to temperature and colchicine that is displayed by tubules reassembled from brain. This contrasts with the much greater stability of the sperm tail microtubules in vivo and implies that it may be due to the characteristic accessory proteins that are present rather than to modifications in the tubulin itself.

To summarize, it seems likely that the role of the microtubule-associated proteins may be concerned with generating an environment in which tubulin dimers are able to interact in the manner required for microtubule as-

---

*A cultured glial cell line has been used by Wiche and Cole (1976a) and Wiche, Honig, and Cole (1979); other established mammalian cell lines, especially Ehrlich ascites and HeLa, by Doenges et al. (1977, 1979), Nagle, Doenges, and Bryan (1977), Weatherbee, Luftig, and Weihing (1978), and Bulinsky and Borisy (1979); sea urchin sperm tail outer doublet (solubilized by sonication) by Farrell and Wilson (1978), Farrell, Morse, and Wilson (1979), and Binder and Rosenbaum (1978).

sembly. This might involve inducing a conformational change to expose active site(s) in the dimer, stabilizing the structure of some particular array of dimers, or bringing dimers into juxtaposition in the correct manner. Treatments that allow tubulin to polymerize in the absence of MAP presumably succeed by creating an environment that allows tubulin dimers to interact directly to form and then to extend nucleating centers. The basic information for generating the structure of the microtubule must therefore reside with the tubulin dimer. It remains to be established how the polymerization of tubulin depends on its interaction with other proteins and whether the same or different reactions are involved in the initial formation of nucleating centers and their subsequent elongation.

Microtubule Elongation

The elongation of microtubules in vivo or in vitro appears to take place largely, but not exclusively, from one end. Following recovery from amputation, protozoan flagella appear to grow principally at the tips rather than at the base (Rosenbaum and Child, 1967; Rosenbaum, Molder, and Ringo, 1969). This unidirectional growth would imply that tubulin subunits must be transferred up the hollow shaft of the microtubule to be added at the tip. But more recent studies show that the two central microtubules terminate in a distal cap, which remains present when flagellar resorption is induced by treatment with high salt. When regeneration occurs, only the outer doublets are extended by distal addition of tubulin; the two central microtubules appear to be extended at their proximal ends (which are free in the region of transition to the basal body, where the outer doublet microtubules lack free ends and join the basal body). Removal of the cap allows the distal ends of the central microtubules to be elongated (Dentler and Rosenbaum, 1977). It is not clear how elongation of central microtubules proximally is coordinate with extension of outer microtubules distally. However, the general implication of these results is that the unidirectional extension of microtubules that had been thought to be an instrinsic property of the structure must be regarded with more caution; it may be determined by the availability of the ends. Other systems are not amenable to such analysis; in some cases microtubules appear to grow from well-localized organizing centers in one direction, but this does not prove the nature of the underlying mechanism of extension. Cytoplasmic microtubules are discussed below and mitotic microtubules in Chapter 4.

When examined in vitro, chick brain neurotubes appear to grow largely in one direction only. Dentler et al. (1974) obtained highly labeled fragments of brain microtubules which they incubated with unlabeled brain tubulin subunits. When tubulin polymerizes onto the fragments (which provide nucleating centers), the labeled segment should lie at one end of the tubule if assembly is unidirectional but in the center if it is bidirectional.

Labeled fragments always were located at or very close to one end of the extended tubule. This suggests that growth occurs principally in only one direction; any growth in the opposite direction must be very much slower.

A striking demonstration of the evolutionary conservation of tubulin structure is provided by the polymerization in vitro of chick or porcine brain tubulin onto basal bodies from Chlamydomonas or onto flagellar axonemes from Chlamydomonas or sea urchins (Allen and Borisy, 1974; Snell et al., 1974; Binder, Dentler, and Rosenbaum, 1975). The neurotubes assembled in vitro can be distinguished from the microtubules provided to form the nucleating centers, for example by their greater sensitivity to dissociation upon reduction of temperature or addition of calcium ions. The distal end of the nucleating microtubule can be distinguished from the proximal end by the frayed appearance where the bundle of microtubules dissociates into its component tubules. An example is shown in Figure 2.4. Growth in vitro is unidirectional; about 25 μm is added in 10 minutes at 37°C in a concentration of 1.5–2.0 mg/ml tubulin with 0.5–1.0 mg/ml of axoneme protein. At low concentrations of tubulin, growth occurs largely or entirely on the

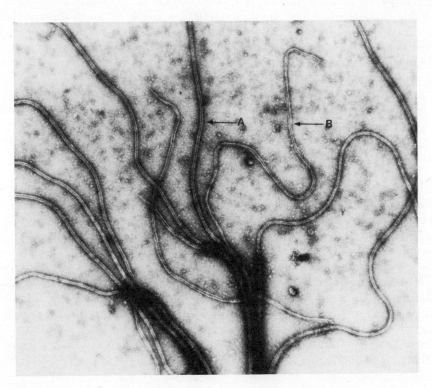

Figure 2.4. Neurotubes reassembled in vitro onto A and B tubules of flagellar microtubules. The arrows indicate neurotubes extended from A and B tubules, respectively. ×24,000. Data of Binder, Dentler, and Rosenbaum (1975).

distal ends; an increase in concentration allows the proximal end also to be extended, although much more slowly than the distal. Whatever sites may differ between mammalian and protozoan or echinoderm tubulin, then, the assembly site for interactions between subunits must have been conserved. However, although both the A and B fibers of doublet microtubules could be extended, only individual tubules were produced (in the latter case of reduced diameter), suggesting that tubulin of neurotubes does not have the ability to form doublet structures.

Tubulin of one species is able also to interact with the microtubules of the mitotic apparatus isolated from another species. The stabilization of the mitotic apparatus by added tubulin, and the heterospecific ability of tubulin to extend the spindle fibers, is discussed in Chapter 4.

Interactions of Microtubules with Mitotic Inhibitors

Three alkaloid drugs that interact with tubulin first came to attention as preparations used for medical purposes. Colchicine, podophyllotoxin, and vinblastine all since have been characterized as inhibitors of mitosis that act by preventing the assembly of spindle fibers. (It is not clear whether the medical effects of these drugs depend upon their ability to bind tubulin.) We shall discuss the inhibition of mitosis in Chapter 4 and shall come to the means of establishing cell synchrony by this method in Chapter 7. Of course, these alkaloids constitute only one class of mitotic poisons: the spindle inhibitors. Other compounds act upon other stages of mitosis, and we shall discuss these in Chapters 3 and 4. Here we are concerned with the nature of the interactions between tubulin and the alkaloid spindle inhibitors, whose chemical structures (which are unrelated) are shown in Figure 2.5.

By far the best characterized spindle inhibitor is colchicine, which is derived from seeds of plants of the genus Colchicum, and especially from the meadow saffron, Colchicum autumnale. Colchicum preparations have been used to treat ailments such as gout and rheumatism since ancient times. The isolation of crystalline colchicine was accomplished late in the nineteenth century and its chemical structure has been known since 1940 (for review see Eigsti and Dustin, 1955).

Colchicine has been known as an inhibitor of mitosis in both plant and animal cells for more than 40 years (Dustin, Havas, and Lits, 1937). It has been apparent since then that colchicine blocks cells in mitosis by causing the spindle fibers to disaggregate. More recently its effect has been equated with the ability to bind to tubulin to prevent the assembly of microtubules. Many compounds related to colchicine can be prepared from Colchicum and in general these are less effective as inhibitors of mitosis, probably due to a reduced affinity for tubulin. Colcemid, whose effects upon the cell are the same as those of colchicine, differs only in the substitution of a CH_3 group for a $COCH_3$ moiety.

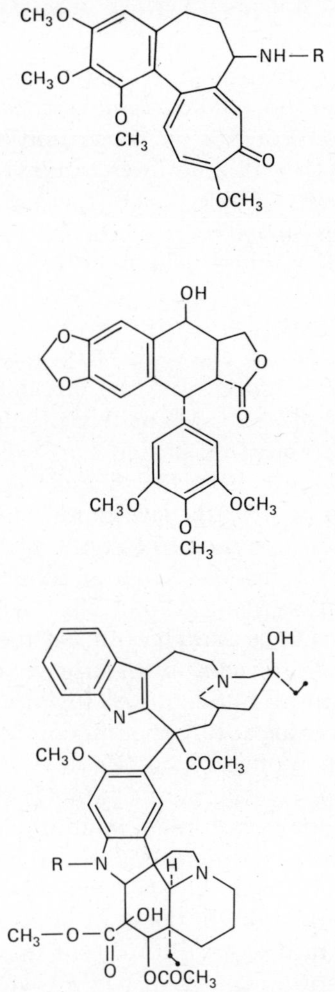

Colchicine: R = COCH₃
Colcemid: R = CH₃

Podophyllotoxin

Vinblastine: R = CH₃
Vincristine: R = CHO

Figure 2.5. Chemical structures of alkaloid inhibitors of mitosis.

Podophyllotoxin is derived from the root of a poisonous North American herb, the may apple, Podophyllum peltatum. Other species of Podophyllum also may synthesize the drug and again there are analogues available which vary in their efficiency of action. In the form of podophyllin, a crude preparation of resin that contains other components in addition to the toxin (some of them also active), extracts from Podophyllum have been used as a purgative or emetic, with a medical history of several hundred years (Bentley, 1861; Lloyd, 1910). The action of podophyllotoxin in inhibiting mitosis is identical with that of colchicine (Cornman and Cornman, 1951; Kelley and Hartwell, 1954). We shall see that in fact it binds to the same site on tubulin.

The periwinkle, Vinca, is the source of vinblastine, which inhibits mitosis by acting upon tubulin at a site different from that involved in the action of colchicine or podophyllotoxin. Vincristine is related to vinblastine by the substitution of a CHO for a CH_3 group and behaves in an identical manner.

While our interest here is concerned with the interaction of these drugs with tubulin, we should note that they also may exercise other, independent effects upon cell metabolism. For example, colchicine and podophyllotoxin inhibit nucleoside transport in some cells, whereas the vinca alkaloids appear to display this as well as other effects upon RNA synthesis (see Plagemann, 1970; Wilson, 1975).

All three alkaloids are derived from plants. All three inhibit mitosis in a wide range of eucaryotes, of which the best characterized include mammalian, echinoderm, and plant cells. One question raised by this universal action is how cell division is accomplished in the plants that produce the drugs. Early work showed that several varieties of Colchicum are resistant to inhibition of mitosis by colchicine (Levan, 1940; Levan and Steineggar, 1947; Cornman, 1942). It is difficult to quantitate the level of resistance from these data, but it seems clear that mitosis can proceed in bulbs of C. autumnale placed in 0.1 M colchicine, under conditions when a solution of 10^{-5} M colchicine is able to inhibit mitosis in the plant Allium. The level of colchicine in C. autumnale has been estimated to be as high as 10^{-2} M; these data therefore imply that the plant should be resistant to the drug that it produces. Similarly, it appears that Vinca rosea, although sensitive to colchicine at 2.5×10^{-4} M, is refractory to vinblastine at 10^{-2} M, although 10^{-4} M vinblastine is adequate to inhibit mitosis in other plants (Kramers and Stebbings, 1977).

There are two ways that resistance in each plant might be accomplished. The plant may produce an antagonist to the antimitotic activity of the drug that it produces; or the assembly of its spindle fibers may be immune from interference by the drug because the tubulin does not bind it. No experiments appear to have been performed with cultured cells that may answer these points; nor have there been any attempts to determine whether tubulin from these plants can bind the drugs that they synthesize. The basis of resistance is therefore unknown. Obviously any specific variation in microtubule structure that conferred resistance to an alkaloid inhibitor would be of great interest. In this context, we may note briefly that attempts to select cell mutants resistant to colchicine so far largely have generated mutants in uptake rather than in microtubule function (see Chapter 6).

What is the basis for the inhibitory effect of colchicine on mitosis? The spindle fibers disappear when colchicine is added to mitotic cells, which become blocked in metaphase, a condition described as the C-metaphase. This suggests that colchicine has caused the microtubules that comprise the spindle fibers to disaggregate. Exposure to low temperature has the same effect (see above). Sensitivity to treatment with colchicine or low temperature defines the labile class of microtubules, which includes those of the mitotic apparatus of mammals, plants and echinoderms. Labile micro-tubules appear to be in equilibrium with a pool of free tubulin subunits. The effects of colchicine and the other alkaloids are mediated by their inter-actions with the members of this pool; they do not interact directly with as-sembled microtubules. We have seen that tubulin was first isolated by virtue of its ability to bind colchicine; and the binding reaction has been charac-terized in the most detail for tubulins of mammalian neurotubes and sea urchin and mammalian mitotic apparatus.

The stable class of microtubules is represented by protozoan flagella, which are not sensitive to disruption by colchicine or low temperature. The stable microtubules are not transient structures in equilibrium with a subunit pool, but instead represent organelles of the cell. A long standing question has been the basis of this difference in stability, which might in principle reside either in the structure of the tubulin subunit or in the other proteins associated with the microtubules. Because of the indifference of stable micro-tubules to colchicine, the tubulin of the protozoan tubules has been purified by means other than drug binding. However, tubulin purified from the stable outer doublet microtubules of the sea urchin sperm tail is able to bind to colchicine with an affinity close to that displayed by tubulin from labile microtubules (Wilson and Meza, 1973). And the microtubules reassembled in vitro from this tubulin show the same sensitivity to colchicine as labile microtubules (see above). This suggests that the difference between labile and stable microtubules are not determined by differences in their tubulins. Similar experiments with tubulin from protozoan flagella would be useful to decide whether their insensitivity to colchicine has a similar basis or re-sults from the lack of a colchicine binding site (as has sometimes been as-sumed previously). Colchicine sensitivity in microtubules has been reviewed by Margulis (1973).

Binding of Colchicine to Tubulin

If the binding of colchicine to tubulin is regarded as a reversible interaction, it can be described by the law of mass action as:

$$\text{tubulin} + \text{colchicine} \underset{k_{-1}}{\overset{k_1}{\rightleftharpoons}} \text{tubulin-colchicine}$$

so that the equilibrium constant for association can be described as:

$$K_a = \frac{k_1}{k_{-1}} = \frac{[\text{tubulin-colchicine}]}{[\text{tubulin}][\text{colchicine}]}$$

The constant K_a describes the affinity of tubulin for colchicine. It may be measured either by determining the equilibrium concentrations of the com-ponents of the reaction when tubulin is incubated with colchicine or by determining the forward and reverse kinetic rate constants, k_1 and k_{-1}.

The binding of colchicine to mammalian tubulin was originally followed by incubating ^{3}H-labeled colchicine with tubulin and separating the bound and unbound colchicine, for example on Sephadex columns, in order to determine their concentrations. Early experiments gave similar results with this method and with the binding of colchicine to cultured cells, with values for K_a of 1.1×10^6 M^{-1} and 4.0×10^6 M^{-1}, respectively (Taylor, 1965; Borisy and Taylor, 1967a). This implies that the action of colchicine upon living cells is mediated in the same manner as its binding to tubulin in vitro.

Similar values for the affinity for tubulin have been obtained for tubulin from several sources. Analysis of tubulin from sea urchin eggs by the same means gives a somewhat lower association constant, measured as $2.9 - 6.5 \times 10^5$ M^{-1} (Wilson and Meza, 1973; Pfeffer, Asnes, and Wilson, 1976). Another technique that can be used to follow the binding of colchicine takes advantage of the observation that colchicine fluoresces when bound to tubulin but not when in the free state (Bhattacharyya and Wolff, 1974, 1975, 1976). This has the advantage that it is not necessary to separate bound and unbound colchicine. The association constant for mammalian brain tubulin measured by this means is 3.2×10^6 M^{-1}; and the stoichiometry of binding suggests that up to 1 colchicine molecule binds per dimer (measured values are 0.83–0.90). The binding of colchicine is highly dependent upon temperature; it is usually measured at 37°C, and no binding can be detected at 0°C.

A difficulty in determining the affinity of tubulin for colchicine by equilibrium measurements is that the colchicine-binding activity of tubulin decays with time. The loss of colchicine-binding activity can be inhibited by guanine nucleotides (see above) or by vinblastine (see below). The effect of the decay is to cause the amount of tubulin that is available for reaction to be overestimated, which leads to an underestimate for the value of K_a. An alternative approach is to use kinetic analysis, in which the forward (k_1) and reverse (k_{-1}) rate constants are measured. For mammalian brain tubulin, these have been estimated as 0.37×10^6 M^{-1} hr^{-1} and 0.009 hr^{-1}, respectively (Sherline, Leung, and Kipnis, 1975). This gives a value for K_a of 4.1×10^7 M^{-1}, about an order of magnitude greater than the association constant measured by equilibrium analysis. Free tubulin, and tubulin bound to colchicine, decay into the inactive form with half lives of 5.2 and 12.5 hours, respectively.

The dissociation of bound tubulin to release free (active) tubulin and colchicine occurs with a half-life ($t_{1/2}$) of 77 hours ($t_{1/2} = \ln 2/k_{-1}$). Consistent with the idea that this reaction is slow, labeled colchicine bound to tubulin exchanges only very slowly with free colchicine; Wilson (1975) stated a $t_{1/2}$ for this reaction of about 30 hours. Inhibition of mitosis by colchicine is reversible; no $t_{1/2}$ has been calculated for the reaction in vivo, but usually the inhibition can be reversed within some hours (see Chapter 4).

If only one colchicine molecule is bound per tubulin dimer, only one of the subunits should carry a colchicine-binding site. One approach to investigating the binding site is to use an affinity label, that is, to use instead of colchicine a modified compound that carries a moiety whose binding to the tubulin subunit can be followed. Schmitt and Atlas (1976) synthesized a bromocolchicine which appears to mimic the action of colchicine and which binds at the same tubulin site, as judged by its ability to inhibit colchicine binding. Bromocolchicine binds to both the α and β tubulin subunits, but the only irreversible binding appears to be to the α subunit.

The antimitotic activities of podophyllotoxin are virtually identical with those of colchicine (Cornman and Cornman, 1951; Kelley and Hartwell, 1954). Podophyllotoxin competes with colchicine for binding to tubulin of mammalian or avian brain or of sea urchin sperm tail (Bryan, 1972; Wilson, 1970; Wilson and Meza, 1973). From the inhibition reaction, the K_a for podophyllotoxin has been measured in the range of 0.8–1.4×10^6 M^{-1}, which is about twofold greater than the values measured for colchicine by equilibrium analysis in the same experiments (Pfeffer, Asnes, and Wilson, 1976). The competitive inhibition suggests that podophyllotoxin binds to the same site on the tubulin dimer that is recognized by colchicine. But its mechanism of binding is not identical with that of colchicine, for it is rapidly reversible, implying a greater value for the rate constant k_{-1} (Wilson, 1975).

Formation of Paracrystals with Vinblastine

Vinblastine has the same effect as colchicine and podophyllotoxin in causing the reversible disappearance of spindle fibers when cells are treated with low concentrations of the drug. All three alkaloids inhibit mitosis in mammalian cells at concentrations that depend upon the species, but which generally are of the order of 10^{-7} M. Bensch and Malawista (1969) observed that greater concentrations of vinblastine generate paracrystalline aggregates—large arrays consisting of regularly arranged subunits. The minimum concentration of vinblastine required for this effect was not established, but concentrations of more than 10^{-5} M appeared to be effective in L cell fibroblasts. The paracrystals formed in either L cells or sea urchin eggs contain α and β tubulins in the usual 1:1 ratio (Bryan, 1972). Presumably the crystalline structure represents tubulin dimers that have aggregated in some manner different from their usual interaction to form microtubules. A counterpart to the formation of paracrystals in vivo appears to be the precipitation of tubulin from solution in vitro by high concentrations of vinblastine (Olmsted et al., 1970). This effect appears to be less specific than the in vivo effect since other proteins may be precipitated as well as tubulin (Wilson et al., 1970).

Measurements of the affinity of tubulin for vinblastine have produced varying estimates for the association constant(s).* There appear to be two

---

*In some experiments two binding sites are found with different affinities for vinblastine, in others both binding sites appear to have the same affinity as seen in a linear Scatchard plot (Bhattacharyya and Wolff, 1976; Wilson, Creswell, and Chin, 1975; Wilson, Morse, and Bryan, 1978).

binding sites per dimer; although these have not yet been localized, a reasonable working model is to suppose that one is located on each subunit. Their occupancy by vinblastine has not yet been related to the effects of the drug. What is the relationship between the vinblastine-binding sites and the other drug binding sites on tubulin? Vinblastine binding is independent of colchicine or podophyllotoxin binding and stabilizes the loss of colchicine-binding activity that occurs in free tubulin. Thus each dimer possesses two vinblastine-binding sites and one colchicine/podophyllotoxin-binding site. Tubulin precipitated by vinblastine suffers a conformational change as shown both by an immediate reduction in its affinity for vinblastine itself and by a change in binding of GTP (Berry and Shelanski, 1972; Wilson, Creswell, and Chin, 1975). The binding of vinblastine differs from that of colchicine in lacking temperature dependence; the same association constant prevails at 37°C and at 0°C.

Poisoning Action of Alkaloids

The binding sites for both colchicine and vinblastine appear to be unavailable in assembled microtubules. This suggests that the alkaloids may disrupt mitosis by acting upon members of the pool of free tubulin dimers, thus interfering with the equilibrium between microtubules and their subunits. This is consistent, of course, with the idea that the effect of colchicine depends upon the stability of the microtubule.

Two models may be proposed for the action of the alkaloids in preventing polymerization. The simplest is to suppose that colchicine and vinblastine act by removing free tubulin dimers from the pool available for polymerization. With labile microtubules, this would push the equilibrium between the assembled microtubules and the pool of dimers toward disassembly. This model predicts that half-maximum inhibition of polymerization should be achieved when half of the tubulin dimers are bound to alkaloid. An alternative model is to suppose that the alkaloids may have a "poisoning" effect and need bind to only a small proportion of the tubulin dimers to achieve inhibition. This predicts that correspondingly lower concentrations of alkaloid should be effective in inhibiting polymerization.

The concentration of alkaloid needed to bind a given proportion of the tubulin dimers available in vitro can be calculated from the mass action equation. Olmsted and Borisy (1973a) found that a concentration of 1.0×10^{-6} M colchicine was sufficient to block the polymerization (to a level of >80%) of tubulin present in the form of a crude preparation (about 25% of a suspension of protein at 12 mg/ml, corresponding to 27×10^{-6} M tubulin). Yet at this alkaloid level, only 3.5% of the tubulin dimers should be bound, taking the value for K_a (10^6 M^{-1}) and assuming that all of the tubulin dimers are active. (Taking a more sensitive part of the inhibition curve, 0.5×10^{-6} M colchicine blocked polymerization 50%, but should have bound to only 2%

of the tubulin dimers.) This led Olmsted and Borisy to infer the existence of a poisoning effect in vitro.

How does this small proportion of tubulin-colchicine complexes block polymerization? One model is to suppose that colchicine binds directly to the dimers at the growing ends of the microtubules, perhaps because their affinity for the drug is increased, to block addition of further dimers. An alternative is that tubulin-colchicine complexes are formed in the free tubulin pool; and it is then their binding to the microtubule ends that halts growth. These possibilities were distinguished by Margolis and Wilson (1977) in experiments in which nucleation centers or free tubulin were preincubated with colchicine before mixing to initiate the polymerization reaction. Treatment of the nucleation centers has no effect on elongation. But the addition of colchicine to the dimer preparation inhibits polymerization. Inhibition reaches 60% when little more than 2% of the tubulin dimers are bound to the drug; it is complete with only 10% of the tubulin dimers inactivated.

Do the other alkaloids also act by this substoichiometric poisoning? Both podophyllotoxin and vinblastine appear also to block tubulin polymerization at concentrations at which only a small proportion of the dimers should be bound to the drug (Wilson, Anderson, and Chin, 1976). In both cases, inhibition is obtained when the drug level is only 1% of the tubulin level. All three alkaloids therefore seem to distort the structure of the tubulin dimer in such a way that its addition to the microtubule prevents the addition of further tubulin dimers.

Is this action responsible for inhibiting mitosis? The drug concentration needed to inhibit spindle fiber assembly in vivo is more difficult to determine, because the effective concentration of alkaloid in the cell need not necessarily be the same as that supplied in the medium. However, in those experiments where alkaloid effectiveness has been compared in vivo and in vitro, the same concentrations are needed for half maximum inhibition (Taylor, 1965; Wilson, 1975; Bhattacharyya and Wolff, 1976). This suggests that the limiting step in drug action may be the same in vivo and in vitro, that is, it lies with binding to tubulin. The conclusion that poisoning occurs in vivo is borne out also by the observation of Taylor (1965) that only 3–5% of the number of available colchicine-binding sites were filled when mitosis in cultured cells was completely inhibited.

A paradox is inherent in the poisoning effect. Alkaloids inhibit the assembly only of labile microtubules, and by so doing they cause assembled microtubules to disassemble. But if the disassembly takes place at the same site to which the tubulin-drug complexes have bound, their release should allow the addition of further dimers, most of which are not complexed with the drug. In other words, substoichiometric poisoning cannot occur unless the addition of the tubulin-drug complexes irreversibly inhibits the addition of dimers to the microtubule. This can happen only if disassembly does not occur at this site. The disruption of labile microtubules by the alkaloids then must imply that disassembly is occurring at the other end of the microtubule.

This leaves two possibilities for the mechanism by which the microtubule-tubulin equilibrium is established. Assembly may occur at one end of the microtubule and disassembly at the other. By blocking assembly, the drugs then allow disassembly only to continue. Or, while assembly occurs only at one end, disassembly may occur at both; in this case, drug binding to the assembly end must also inhibit disassembly there. Margolis and Wilson (1978) tested these models. One approach is to follow the incorporation of a label into microtubules. Using ^{3}H GTP, they found a linear addition of label until a plateau was reached. This implies that addition of dimers and loss of dimers occur at different sites, since a steady state would be established very rapidly if these events coincided in place. Another experiment was to label microtubules uniformly in the presence of ^{3}H GTP and then compare the rate of loss of the label in the absence and presence of podophyllotoxin. The rate of loss remained the same, which implies that the drug inhibits only assembly, since if it also affected disassembly there should be a reduction in the rate of loss. The possibility that disassembly occurs along the entire length of the microtubule was excluded by comparing the rate of loss of ^{3}H GTP from sheared and unsheared microtubules; the rate was proportional to the number of ends rather than to the total amount of material. This suggests that net assembly and disassembly of the microtubule occur at opposite ends; the same mechanism appears to hold when microtubules reconstructed from outer doublet tubulin are placed in equilibrium with the free dimer (Farrell, Kassis, and Wilson, 1979). This suggests that this feature may be inherent in the construction of microtubules; and this may be important in considering models for the formation and disruption of the mitotic spindle (see Chapter 4). These results suggest that labile microtubules may be "treadmills" in which tubulin is constantly migrating from one end to the other, so that net assembly or disassembly can be stimulated by blocking the action at the appropriate end.

Cytoplasmic Network of Microtubules

One approach to investigating the function of microtubules is the use of antisera prepared against tubulin. Among the first experiments of this nature were those of Fulton, Kane, and Stephens (1971) and Dales (1972), which showed that an antiserum prepared against tubulin of one source— flagellar axonemes of sea urchin sperm or vinblastine-induced paracrystals of murine origin, respectively—reacts with microtubules of cells of many different origins. This implies that a similar antigenic site must be present on tubulins of different species; whether it is identical in each case is not apparent from these results.

The reacting structures can be visualized in the cell by the use of indirect immunofluorescence. In this technique, a rabbit antiserum against tubulin is reacted with the target cells; then an antibody against the rabbit immunoglobulin—a goat antibody tagged with fluorescein—is added. By using antiserum prepared against microtubules extracted from the outer doublets of sea urchin sperm, Weber, Pollack, and Bibring (1975) and Weber, Bibring, and Osborn (1975) were able to visualize a network of cytoplasmic fibers crossing the entire cell. Similar cytoplasmic networks were observed with mouse 3T3 cells, monkey BSC-1 cells, and secondary fibroblasts of human, hamster, rat, and chick origin. An example of a 3T3 network is shown in Figure 2.6. The fibers disappear upon treatment with colchicine (at a concentration of 3×10^{-5} M) or with lowering of the temperature, that is, they show the same responses as the spindle fibers of mitotic cells. This demonstrates that microtubules are present not only in mitotic cells but also in the form of a cytoplasmic network in interphase cells, showing the same lability in both cases.

The antitubulin antibody reacts also with the vinblastine-induced paracrystals of HeLa or 3T3 cells, demonstrating that the same antigenic site is available in free tubulin, cytoplasmic microtubules, and the paracrystals. The antibody also reacts with mitotic HeLa cells. Essentially the same results have been obtained with antiserum prepared against purified mammalian brain tubulin, which confirms that it is indeed the tubulin protein

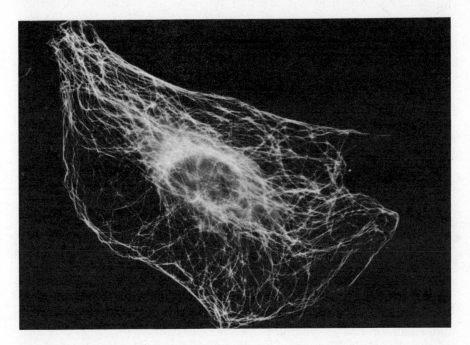

Figure 2.6. Indirect immunofluorescence of cytoplasmic microtubules in a 3T3 cell. Photograph kindly provided by Professor Klaus Weber.

and not some other microtubule protein against which the reaction occurs (Brinkley, Fuller, and Highfield, 1975; Wiche and Cole, 1976b; Frankel, 1976). One of the most striking features of these results is the evolutionary conservation of the antigenic site on tubulin across so many species; and, of course, this is consistent with other demonstrations of the selective pressure against change in this protein (see above).

The relationship between the cytoplasmic microtubules of the interphase cells and the spindle fibers of mitotic cells is an obvious question. Fuller, Brinkley, and Boughter (1975) observed that the cytoplasmic fibers disappear early in mitosis. Their disaggregation begins in early prophase (judged by the condensation of chromatin) and is complete by metaphase, when all of the fluorescence is localized on the mitotic spindle. The cytoplasmic microtubules reappear at the end of telophase or beginning of interphase. This course of events suggests that there may be a cyclic reutilization of tubulin, which is disassembled from cytoplasmic microtubules, becomes repolymerized into spindle fibers, which then in turn decay to regenerate the pool of tubulin from which the cytoplasmic microtubules grow (see Chapter 4). The total amount of tubulin in fibroblasts, measured by a radio-immunoassay, is some 3–4% of total protein; this is a little more than twice the amount of tubulin that would be required to generate 120–150 microtubules per cell of average length 50 μm (this last value assumes that immunofluorescence detects individual microtubules) (Hiller and Weber, 1978). It is not known whether the size of the free tubulin pool changes during mitosis.

What is responsible for organizing the cytoplasmic microtubules of interphase cells? Osborn and Weber (1976a,b) and Frankel (1976) observed that the tubules appear to extend radially toward the plasma membrane from a focus located in the perinuclear region. Tubulin-containing cylindrical structures about 3 μm long, from which the tubules may radiate, are located in this area. Although incubation at low temperature or treatment with agents such as colchicine abolishes the cytoplasmic array of tubules, the organizing structure remains intact. The responsibility of these structures for the extension of microtubules is suggested by the apparent regrowth from them that occurs following recovery from treatment with colcemid. An obvious question concerns their relationship with the centriole, but this remains to be resolved.

The nature of the organizing centers appears to be a characteristic of the cell phenotype. More detailed studies of both mouse and human fibroblasts have suggested that there may be several centers, all located in the perinuclear region (Spiegelman et al., 1979a,b). Although all appear equally active in promoting recovery from treatment with colchicine, at least two types of center can be distinguished by the response to griseofulvin, another mitotic inhibitor. First microtubules regenerate from one center; then after a lag of about an hour they grow from the other centers. Possibly the first is a primary center and the others are subsidiaries that are more sensitive to griseofulvin. The same use of multiple organizing centers is seen in undifferentiated mouse neuroblastoma cells; but upon serum deprivation, the cell shape changes from its rounded structure to an asymmetrical form, in which there is only a single organizing center. This appears to be generated by the aggregation of the multiple centers that were previously present. This suggests both that the organizing center may be more complex in structure than the centriole per se and also that the form of the microtubule network is adjusted to the particular needs of the cell by controlling the centers from which it emanates.

What is the function of the cytoplasmic tubular network? The microtubule systems of many specialized cells are concerned with cell motility or shape and it seems likely that there is a similar role for the networks of these interphase cells. That microtubules may be important for cell structure was indicated as long ago as 1964 by the observations of Robbins and Gonatas that morphological changes occur when interphase cells are treated with colchicine. At one time it seemed that only cells spread out on a substrate possessed longitudinally organized microtubules, since a much reduced array was apparent in transformed cells (which differ from the "normal" cells in possessing a more rounded shape) (Brinkley, Fuller, and Highfield, 1975; Miller, Fuseler, and Brinkley, 1977). However, the same cytoplasmic network is present in such cells, although its visualization is made more difficult by the rounded shape (Osborn and Weber, 1977b; De Mey et al., 1978; Turner et al., 1978). In cells that are rounded up, the microtubules seem to spread from the perinuclear region to the plasma membrane. During the spreading out of cells able to attach to the substratum, microtubules appear to be aligned almost perpendicular to the membrane. Then, as the cells flatten out, microtubules run along the cell parallel to the plasma membrane, as in the example of Figure 2.6 (Osborn and Weber, 1976b). Their organizing center may be aligned with the direction in which the cell moves (Albrecht-Buehler, 1977). The role of microtubules and of microfilaments in cell motility is discussed further in Chapter 3.

Supporting the idea that microtubules are involved in functions of cell shape and motility, there are observations of a connection between microtubules and proteins located in the cell membrane. The fluid membrane model views proteins located on the surface of the cell as being free to diffuse in a lipid matrix. The mobility of proteins on the surfaces of several types of cell, in particular including the lymphocyte which has perhaps been the most extensively studied, is influenced by lectins such as concanavalin A (for review see Nicolson, 1974). In low concentrations (<5 μg/ml), Con A stimulates the aggregation of several types of receptor molecule into the polar region, where a "cap" is formed that then may be taken into the interior of the cell by phagocytosis. At high concentrations (generally >50 μg/ml), however, Con A inhibits cap formation. These effects depend upon the type of cell and also on the temperature.

Microtubules have been implicated in the rearrangement of surface receptors by observations that these processes may be sensitive to concentrations of colchicine similar to those used to block mitosis. Formally, this shows that a colchicine-binding protein is involved, and presumably this represents tubulin constituting labile cytoplasmic microtubules. The inhibition of cap formation that is caused by high concentrations of Con A is overcome by the addition of colchicine, which allows receptor mobility to be restored to lymphocytes (Edelman, Yahara, and Wang, 1973; Yahara and Edelman, 1975). The capping that is caused by addition of low concentrations of Con A is prevented if colchicine is added to leucocytes (Ukena and Berlin, 1972; Ukena et al., 1974; Oliver, Ukena, and Berlin, 1974). These experiments were performed at 37°C; in both cases a reduction in temperature to 4°C has effects similar to the addition of colchicine. This is consistent with the idea that these rearrangements of surface receptors involve labile microtubules.

The connection of microtubules with cap formation has been shown directly in rabbit ovarian granulosa cells. Following treatment with Con A, microtubules can be visualized in an array perpendicular to the cap, and a fluorescent derivative of colchicine binds exclusively in the region of the cap (Albertini and Clark, 1975). The exact nature of the connection between the microtubules and the receptors located in the membrane is not clear, but it may be that the tubules "anchor" the receptors in the membrane and/or are required for their directed movements. Thus colchicine might be needed to release receptors from associated microtubules in order to allow cap formation in the presence of high Con A concentrations; and might block the formation or internalization of the cap in low Con A concentrations because microtubules are involved in the phagocytic entry of the receptors into the cell. In addition to the microtubules, however, contractile filaments may be involved in these processes, for the drug cytochalasin B which blocks their action also influences capping (Ukena et al., 1974; De Petris, 1975; see Chapter 3).

CHAPTER 3

The Cell Skeleton: The Contractile Network

The idea that the proteins of muscle might be present also in nonmuscle cells originated not long after the identification of actin and myosin as the contractile proteins, with the suggestion of Loewy (1952) that a protein with some properties of actomyosin was present in the plasmodium of a myxomycete. This implied that contractile proteins might be concerned with cell motility. Some time later, the presence of an actomyosinlike protein—then called thrombosthenin—was demonstrated in human blood platelets, a nonmuscle vertebrate cell type (Bettex-Galland and Luscher, 1959; Bettex-Galland et al., 1962). With the subsequent observation that actin and myosin extracted from slime molds closely resemble the proteins of vertebrate muscle, it became clear that contractile proteins are not confined to muscle cells alone (Hatano and Oosawa, 1966; Hatano and Tazawa, 1968; Adelman and Taylor, 1969a,b).

The presence of actin and myosin in cells other than those of muscle has now been demonstrated in a wide variety of cells of different phenotypes and from different organisms. This ubiquity raises two principal questions. What is the function of the contractile proteins in nonmuscle cells? And what is the structure into which these proteins are organized, and how is it related to the structure of muscle itself? We should also ask, of course, whether the contractile proteins of nonmuscle cells are identical with those of muscle tissue in the same organism or whether they are related proteins expressed from different genes. Related to this question is the issue of the extent of evolutionary divergence of contractile proteins of muscle and nonmuscle cells in different organisms.

In this chapter we shall be concerned principally with the status of contractile proteins in mammalian nonmuscle cells. Recently it has become clear that it is not only actin and myosin that are represented in nonmuscle cells, but that other of the muscle proteins also may be ubiquitous. These include α actinin as well as the control proteins tropomyosin and troponin and certain other, minor components of muscle. The presence together of all of these proteins implies that the organization and function of the contractile system in nonmuscle cells may be similar in its general nature to that of muscle. However, although it is clear that actin and myosin exist in the form of filaments in nonmuscle cells, and that the other components are associated with them, very little is known about the interactions between the components of what may be regarded as a contractile network. Although this network seems certain to be involved in cell motility, therefore, and in certain aspects of cell structure, the relationship between its organization and function has not yet been defined. Here we shall discuss the contractile network of cells in interphase; and in Chapter 4 we shall consider the functions of the contractile proteins in the completion of cell division.

Protein Components of Muscle

The contraction of muscle represents a system for converting chemical energy (released by hydrolysis of ATP) into mechanical energy (accomplished by fiber contraction). The basic unit of the contractile apparatus is the *sarcomere*, which is repeated many times in series and in parallel in both skeletal and cardiac muscle, generating the characteristic striated appearance. Figure 3.1 illustrates the structure of the sarcomere, which consists of partially overlapping thin and thick filaments. The thin filaments consist of polymerized actin, associated with the control proteins tropomyosin and troponin, and the thick filaments comprise polymerized myosin (for review see Huxley, 1969; Taylor, 1972; Mannherz and Goody, 1976).

The figure shows the sarcomere in the resting, or relaxed, state. Contraction is accomplished by the sliding past each other of the overlapping filaments; this is known as the sliding filament model for muscle. In the contracted state, the sarcomere has a length of about 1.8 μm, contrasted with the resting value of 2.6 μm. When a muscle contracts, all the adjacent sarcomeres experience the same contraction synchronously.

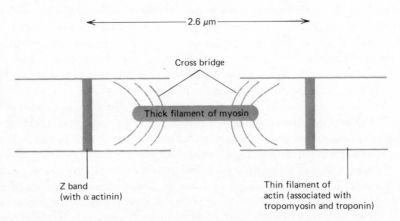

Figure 3.1. Diagrammatic representation of structure of muscle.

The thick and thin filaments are linked by *cross bridges* that are part of the myosin component. It is the interaction of these bridges with actin that generates the force for contraction. The release of energy for contraction via the hydrolysis of ATP depends upon the ATPase activity of myosin, which appears to reside in the part of the molecule that includes the bridges, but the details of this process are not yet fully elucidated. Actin and myosin may interact in vitro, or may be isolated from muscle, in the form of the *actomyosin* complex, which possesses this ATPase activity.

The thin filaments of actin are anchored in the Z bands, where α-actinin is localized. The polarity of the filaments is constant with respect to the Z band, that is, each sarcomere has a bipolar structure. Thus each thick filament consists of two halves in which the myosin is arranged in opposite orientations.

Regulation of contraction is mediated by reversible inhibition of the actin-myosin binding reaction, which responds to the level of calcium ions. In resting muscle the Ca^{2+} concentration is typically $\sim 10^{-7}$ M; under this condition the control systems prevent actin-myosin interaction. When the Ca^{2+} concentration is increased to $\sim 10^{-5}$ M, the inhibition of actin-myosin interaction is released and contraction is triggered. There appear to be two types of control of contraction. In molluscan muscle (and that of certain other invertebrates), regulation resides in the myosin molecule, which responds directly to the level of calcium ions. In vertebrates, the thin actin filaments contain a complex of control proteins, troponin-tropomyosin, which undertake the reaction with calcium ions. (Troponin is absent from molluscan muscle; tropomyosin is present but does not appear to play a regulating role.)

Both actin and myosin filaments remain the same length when muscle contracts, and in sliding past each other the two filaments bring adjacent Z bands closer together. It is this contraction that generates the pulling force of muscle. Two aspects of this process are especially relevant to the function of contractile proteins in nonmuscle cells. An obvious point is that for a force to be generated by the movement of the myosin filaments relative to the actin filaments, one set of filaments must be anchored to some cellular structure. In nonmuscle cells, we may therefore look for some linkage between filaments of contractile proteins and the cell plasma membrane. The sliding between actin and myosin in muscle appears to depend on a cyclical mechanism in which the cross bridges of myosin pull an actin filament a distance of some 5 nm, then releasing it to reattach nearer to the Z band so that the action can be repeated. The movement appears to be developed as the result of a change in the angle of attachment of the head of the myosin molecule at the actin filament; the remaining length of the myosin does not appear to be involved. This means that it may be possible to vary the arrangement of myosin but retain the same mechanism for contraction, so long as the relationship between the myosin head and the actin

Table 3.1 Major protein

Protein	Molecular weight	Constituent subunits in rabbit skeletal muscle
Myosin	470,000	2 heavy of 200,000 4 light of 20,000
Actin	41,785	single polypeptide chain
α-Actinin	190,000	2 identical subunits of 95,000 daltons
Tropomyosin	70,000	α subunit of 36,000 β subunit of 34,000 subunits very similar in sequence and present in ratio 4α:1β
Troponin	76,000	TnC of 18,000 daltons TnI of 21,000 daltons TnT of 37,000 daltons

Other proteins present in muscle include C protein (140,000 daltons; binds to mero-myosin), β-actinin (45,000 daltons; reduces length of actin filaments), filamin (250,000 daltons; present in smooth chick muscle but absent from skeletal), paramyosin

filament is maintained (see Huxley, 1973). Indeed, the striated appearance generated by the alignment of sarcomeres is seen only in skeletal and cardiac muscle of vertebrates; whereas smooth muscle lacks sarcomeres, yet appears to contract by the same basic mechanism. This raises the possibility that cytoplasmic actin filaments may be able to interact with myosin in non-muscle cells also in a similar manner, in spite of the absence from these cells of the thick filaments characteristic of myosin in the sarcomere.

The properties of the major proteins of the contractile system of muscle are summarized in Table 3.1. The proteins of rabbit skeletal muscle have been characterized in the most detail, but they appear to take rather similar forms in all vertebrate muscles.

In monomeric form, myosin is a large protein, consisting of two heavy chains (of about 200,000 daltons each) and a varying number (up to four) of light chains. The heavy chains appear constant in molecular weight but vary in amino acid sequence in different muscle systems. Variation in the structure as well as number of the light chains is extensive, and values have been reported from 15,000 to 28,000 daltons for the (nonidentical) chains of

components of muscle

Structure in other muscle systems	Properties
may be 2, 3 or 4 light chains of 15,000–28,000 daltons	rod shaped molecule of 135 nm with globular head containing N terminal regions of light chains; has ATPase activity
weight generally of 42,000–46,000 daltons	G actin is monomeric form which binds Ca^{2+} ions to generate helical filaments of polymeric F actin form; ATP hydrolyzed to ADP in polymerization
	rod shaped molecule of 30 × 2 nm; forms complex with F actin that is inhibited by tropomyosin; function not clear
all subunit(s) of 35,000 daltons	α helical rod of length 40 nm; binds to actin; binds to tropomyosin as part of control system
structure is similar, but molecular weights of subunits may vary	TnC binds Ca^{2+} ions; TnI inhibits actin activity TnT binds tropomyosin; activation by Ca^{2+} ions causes troponin-tropomyosin complex to release inhibition of contraction

(100,000 daltons; constituent of molluscan muscle). For references on protein constituents of muscle see Taylor (1972) and Mannherz and Goody (1976).

different systems. The myosin monomer takes the form of a rod some 135 nm in length, comprising the C terminal parts of both heavy chains, ending in a globular structure which includes the N terminal regions of the heavy chains and apparently all of the light chain sequences. The globular region provides the cross bridges. Digestion of myosin with trypsin divides the molecule into two parts: heavy meromyosin (HMM) is about 350,000 daltons, is derived from the globular region, and retains the ability to interact with actin; light meromyosin (LMM) is about 150,000 daltons and is a rodlike structure. Digestion with papain yields the heavy meromyosin subfragment 1 (S1), which is about 115,000 daltons and has the ability to bind to actin.

Actin appears to be a well conserved protein. Each monomer is able to bind a single Ca^{2+} ion and one ATP or ADP moiety. Actin commonly is prepared by extraction from an acetone powder of cells, in which most other proteins (but not actin) are denatured. Then the actin can be purified by extraction with ATP and a reducing agent. Alternatively, extraction may be made with KCl. In the monomeric form actin is described as *G-actin*; it polymerizes in vitro into filaments, described as *F-actin*, upon addition of

KCl or $MgCl_2$. Polymerization is accompanied by hydrolysis of the ATP residue bound to each actin chain.

The tropomyosin-troponin complex contains several polypeptide chains. Tropomyosin appears to consist of two polypeptides of about the same molecular weight, probably of closely related sequences, which form a rod-shaped structure predominantly comprising an α helix. This lies along the length of the actin filament. Troponin is a complex of three polypeptides. Although their properties are not yet completely resolved, TnC binds calcium ions, TnT binds to tropomyosin, and TnI inhibits the actomyosin ATPase activity. The tropomyosin-troponin complex inhibits contraction, but is prevented from so doing by the binding of calcium ions. The troponin complex is situated at intervals of about 40 nm along the thin actin filament; this corresponds to binding to one point on the tropomyosin molecule.

Occurrence of Cytoplasmic Contractile Proteins

All of the contractile proteins described in Table 3.1 now have been found in nonmuscle cells. In order to distinguish them from the components of muscle cells, they are often described as cytoplasmic contractile proteins.

Proteins with the physicochemical attributes of actin have been prepared from a wide variety of cells, including chick and bovine brain, chick fibroblasts, 3T3 cells and HeLa cells, erythrocytes and human platelets. Estimates for the amount of actin vary, but are in the range of 2–8% of total protein for cultured cell lines, fibroblasts and erythrocytes. It seems likely that actin is a universal component of the eucaryotic cell.[*]

Proof that this protein is actin requires two types of experiment: visualization of actin in filamentous form by specific cytochemical reactions; or determination of the amino acid sequence or tryptic map. Cytological studies demonstrate that actin is the component of the cellular microfilaments (see below). Biochemical studies show that there are multiple forms of actin, which appear to be closely related but nonidentical, and which may occur to characteristic extents in each cell type.

All forms of actin have a molecular weight in the same range, close to 43,000 daltons. The use of two dimensional electrophoresis or isoelectric focusing has allowed three forms of avian or mammalian actin to be distinguished; these are described as α, β, and γ, where α is the most acidic (Garrels and Gibson, 1976; Whalen et al., 1976; Rubenstein and Spudich, 1977). Skeletal and cardiac muscle contain largely or solely actin of α mobility, which is sometimes therefore described as sarcomeric actin. Prior to fusion, myoblasts contain all three forms of actin, but the α form becomes predominant in the differentiated tissue. The predominant form in smooth

---

[*]References: Adelstein, Pollard and Kuehl (1971), Fine and Bray (1971), Puszkin and Berl (1972), Yang and Perdue (1972), Bray and Thomas (1975, 1976), Gruenstein et al. (1975), Kane (1975), Tilney and Detmers (1975); for review see Pollard and Weihing (1974).

muscle (exemplified by chick gizzard) has the γ mobility. A wide variety of nonmuscle cells, including nerve cells, fibroblasts, kidney, brain, platelets, contain only the β and γ actins (with the exception of one experiment in which α actin was detected in fibroblasts).

What is the relationship between the multiple forms of actin? Are they derived from differing modifications of a single polypeptide chain or is each different in amino acid sequence? The tryptic peptide maps of the rat actins show that 30 of the 40 peptides are identical in all three species; α actin contains 4 peptides not present in β or γ, which contain 6 peptides not present in α. The patterns of the β and γ actins are almost identical (Garrels and Gibson, 1976). A similar difference between the sarcomeric and cytoplasmic actins is seen in the chick, where there are 7 differences in the tryptic peptides of actins extracted from muscle or from brain. Similarly the actins of 3T3 and HeLa cells have peptide maps very similar to, but not identical with, that of chick skeletal muscle actin, in contrast with the identity of chick and mammalian muscle actin maps (Gruenstein and Rich, 1975; Gruenstein, Rich and Weihing, 1975).

Supporting the idea that sarcomeric and cytoplasmic actins are the products of different genes, the cyanogen bromide peptide maps of human platelet and cardiac muscle actins reveal a difference in at least one amino acid, a substitution between threonine and valine (Elzinga, Moran and Adelstein, 1976). An increased number of actin proteins is distinguished by amino acid sequence analysis. In a study of bovine actins, Vandekerckhove and Weber (1978a,b,c) showed that sarcomeric (α) actins fall into distinct skeletal and heart muscle types that are not distinguished by isoelectric focusing: these differ in a reversal of the order of two (charged) amino acids present at positions 2 and 3 and also by neutral exchanges elsewhere. Two forms of smooth muscle actin are found in the aorta: these also are closely related in N-terminal sequence, although distinct from the sarcomeric types. One of these is the same as the major actin of chick gizzard, which behaves indistinguishably from γ actin on isoelectric focusing, but whose sequence is distinct. Finally the two cytoplasmic actins, β and γ, differ slightly from each other in amino acid sequence, and show 24–25 amino acid replacements compared with sarcomeric actin. Three different types of actin can therefore be recognized as constituents of sarcomeric muscle, smooth muscle, and cell cytoplasm: each consists of at least two different gene products. Complete sequencing and identification of minor components may lead to an increase in this estimate. The actin genes may therefore constitute a family of related, but not identical, sequences. This has been seen directly in the identification of multiple actin genes in Dictyostelium; and in the isolation of mRNA preparations from chick muscle or brain, and from rat myoblasts before or after fusion, which direct the synthesis in vitro of proteins with the respective mobilities characteristic of the actins of these tissues. This again suggests that different primary protein products are involved and that their synthesis is controlled at the level of transcription (McKeown and Firtel, 1978; Storti, Coen, and Rich, 1976; Storti and Rich, 1976; Hunter and Garrels, 1977).

Actin functions as a contractile protein only in the presence of myosin, of course, and so it is reasonable to expect cells that possess actin also to possess myosin. Proteins with the physicochemical properties of myosin have been isolated from a variety of cell types, including human blood platelets (where contraction is involved in the process of clot retraction), rat kidney fibroblasts, 3T3 and L cells. Myosin represents about 2.5% of total fibroblast protein and about 0.5% of brain protein.*

Because of the greater variability in the structure of myosin chains of muscle tissues, it is difficult to determine the relationship of cytoplasmic myosin to muscle myosin. However, the cytoplasmic myosin displays the usual characteristics of two heavy chains associated with light chains of about 10% the size. In addition to the variations in the light chains, some differences are evident in the tryptic and chymotryptic fingerprints of the heavy chains of different chick tissues, raising the possibility that these too may vary in sequence (Burridge and Bray, 1975).

Tropomyosin has been found in human blood platelets, chick brain, calf brain, pancreas and platelets, and mouse fibroblasts. In each case it displays a molecular weight of 30,000 daltons, compared with the 35,000 daltons displayed by all vertebrate muscle tissues (Cohen and Cohen, 1972; Fine et al., 1973; Fine and Blitz, 1975). All the muscle tropomyosins display similar chymotryptic maps; and all the cytoplasmic tropomyosins are similar by this criterion. This raises the possibility that there may be two types of tropomyosin, one in muscle tissue and one in the cytoplasms of other cell types. However, differences in electrophoretic mobility have been detected between the tropomyosins of different chick muscles (Izant and Lazarides, 1977).

The proteins of the contractile apparatus appear, therefore, to be represented in closely similar forms in both muscle and other cells (for review see Pollard and Weihing, 1974). The extent of variation between the two systems in each case remains to be determined, but it seems reasonable to suppose that each protein plays a similar, if not identical, role in both muscle and nonmuscle cells. This would imply that the control of the myosin-actin interaction in the cytoplasmic system again lies with the tropomyosin-troponin complex, and presumably may be responsive to the level of calcium ions.

At present, not all of the contractile proteins have been purified biochemically from nonmuscle cells, but the presence of α actinin and filamin has been demonstrated immunologically, as has that of actin, myosin, and tropomyosin (see below). The abilities of antibodies prepared against muscle proteins to react with the corresponding cytoplasmic contractile pro-

---

*References: Adelstein, Pollard and Kuehl (1971), Adelstein et al. (1972), Ostlund, Pastan and Adelstein (1974), Chi, Fellini and Holtzer (1975), Bray and Thomas (1975), Burridge and Bray (1975); for reviews see Pollard and Weihing (1974), Korn (1978).

teins provides further evidence of homology, since this means that at least the same antigenic site must be present in each case.

Presence of Actin in Cytoplasmic Microfilaments

Three types of filamentous structures have been identified in eucaryotic cells. Microtubules have a diameter of about 24 nm, intermediate filaments are about 10 nm in width, and microfilaments are some 6 nm across (see Goldman and Follett, 1969). Arrays of cytoplasmic fibers were first recognized in fixed cells by Heidenhain (1889); and Lewis and Lewis (1924) observed these *stress fibers* in living cells. Current studies of their structure and function began with the investigations of Buckley and Porter (1967), which showed that the stress fibers of cultured rat embryo cells comprise bundles of microfilaments. The fibers can be observed by phase contrast, polarized light, or Nomarski optics, in cells such as rat embryo or 3T3 lines which become well spread upon their growth substrates. Although not immediately obvious in BHK21 cells that are not well spread, treatment with colchicine causes the cells to spread out more extensively, and then the fibers can be seen (Goldman and Knipe, 1972).

A spread out cell in which the stress fibers are evident at the points of contact with the substrate is shown in Figure 3.2. Electron microscopy of thin sections of fixed cells shows that stress fibers consist of many very closely packed microfilaments, arranged in parallel, as can be seen from the example in Figure 3.3. The microfilament bundles lie close below the plasma membrane, but microfilament threads may permeate deeper into the cytoplasm in the form of a somewhat looser framework. As implied by their name, the stress fibers are spread out in the direction of physical stress. Their location in the region where the cell contacts its substrate immediately suggests that they are involved in the establishment of cell shape and motility.

Microfilaments display the same dimensions as the filaments generated by the polymerization of actin; and their identity with actin was first demonstrated by a cytological technique. This depends upon the mild proteolysis of myosin to yield the light meromyosin (LMM) representing part of the rodlike region and the heavy meromyosin fragment (HMM) which includes the gobular end (see above). Upon further digestion, the S1 subfragment is cleaved from the HMM fragment. Both HMM and S1 have the property of reacting in a specific manner with actin filaments, to generate what is known as an *arrowhead complex* in which "arrowheads" appear at a characteristic periodicity along the filament. These can be visualized by negative staining.

The first demonstration of the state of actin in nonmuscle cells resulted from the adaptation of this technique to allow its use on intact cells. Ishikawa, Bischoff, and Holtzer (1969) found that when cells are extracted with glycerol, their microfilaments become able to bind HMM; the procedure used is to place cells in a high concentration of glycerol, which is then re-

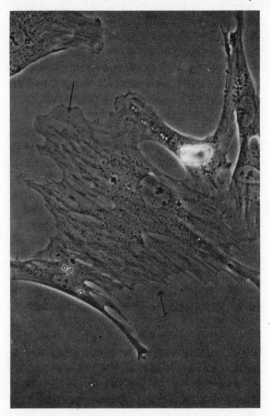

Figure 3.2. Stress fibers visualized in living rat embryo cells by phase contrast. Fibers are evident in the more spread out regions indicated by the arrows. ×350. Data of Goldman, Schloss, and Starger (1976).

duced. Probably this creates pores in the membranes that allow the entry of HMM. By this means, they were able to observe arrowhead complexes in fibroblasts as well as in muscle. The subsequent application of this technique has shown that the cytoplasms of many types of cells contain microfilaments constituted of actin.* An example of an array of actin filaments visualized by arrowhead formation is shown in Figure 3.4.

A limitation of this technique is that visualization of the arrowheads requires use of the electron microscope, making it difficult to view the array of fibers in a whole cell. However, Sanger (1975b) developed a technique in which HMM is coupled to fluorescein, which allows the light microscope to be used. This approach made it possible to see an array of parallel fibers which correspond to the stress fibers seen by phase contrast in living cells. The equivalence of the two arrays was shown also by Goldman (1975), who

*The cells examined include human blood platelets, mouse neuroblastoma cells, newt eggs, *Amoeba proteus* (Behnke et al., 1971; Pollard and Korn, 1971; Perry, John and Thomas, 1971; Chang and Goldman, 1973; Perdue, 1973).

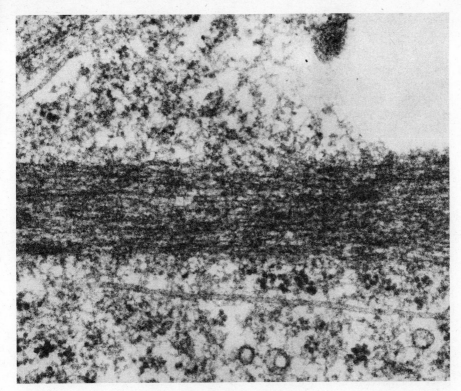

Figure 3.3. Electron micrograph of stress fiber showing the close packed bundle of microfilaments. x64,000. Data of Goldman, Schloss, and Starger (1976).

Figure 3.4. Negatively stained actin filaments. (a) A filament formed from purified Acanthamoeba actin. (b) An Amoeba proteus thin filament decorated with rabbit muscle heavy meromyosin to form the characteristic arrowheads. The bars equal 0.1 μm. Data of Pollard (1975).

modified the HMM binding technique to allow its use with cells spread out on the substrate (the earlier glycerol extraction technique caused cells to lift off the surface). One caveat in interpreting the results of the original experiments with HMM is that HMM can cause monomeric actin to polymerize in vitro; it is therefore necessary to show that the fibers existed before addition of the heavy meromyosin. But the equivalence of the actin filaments with stress fibers, and the use of indirect immunofluorescence (see below) both imply that HMM reacts with preexisting fibers and does not induce them.

Microfilaments therefore comprise polymerized actin and may be a universal feature of eucaryotic cells. In addition to their role in establishing the shape and motility of interphase cells, the actin filaments appear to be responsible for forming a contractile ring which accomplishes the final separation of the two daughter cells following mitosis. Reaction with HMM identifies actin filaments in the cleavage furrow between the potential daughter cells (Forer and Behnke, 1972a,b; Schroder, 1973; Hinckley and Telser, 1974). The properties of the contractile ring are discussed in Chapter 4.

A network of fibers of contractile proteins in the cytoplasm of nonmuscle cells has been visualized by indirect immunofluorescence, in which the cells are reacted successively with a rabbit antibody directed against the contractile protein and a fluorescently tagged goat antibody directed against rabbit immunoglobulin. The application of this technique to microtubules has been discussed in Chapter 2. Early attempts to prepare antibodies specific for individual contractile proteins encountered problems caused by difficulties in purifying the proteins sufficiently. Later attempts were more successful because of the introduction of the technique of preparing antibody against protein purified by SDS gel electrophoresis; in spite of the denaturation of the protein, the antibody preparation possesses appropriate specificity.

In the first application of this technique, Lazarides and Weber (1974) visualized a network of filaments containing actin in mouse 3T3 and hamster BHK cells. Two types of pattern are observed and these are shown in Figures 3.5 and 3.6. The fibers may be arranged in parallel arrays along the long axis of the cell, or the filaments may converge on focal points. This is the same network observed by the use of HMM and corresponds with the distribution seen in living cells by phase contrast microscopy (Goldman et al., 1975). The stress fibers therefore can be seen to take the form of bundles of actin filaments, each bundle generally displaying a diameter of 0.5–0.7 μm (and thus containing of the order of 100 filaments).

Actin filaments share with cytoplasmic microtubules the property of reversible assembly and disassembly in response to changing conditions. The organization of both sets of fibers changes during cell division; cytoplasmic microtubules are replaced by spindle fibers at mitosis; and actin filaments are reorganized as threads on the spindle and then are replaced by the contractile ring at cytokinesis. The reversibility between assembled microfilaments and free subunits raises the question of what controls the equilibrium

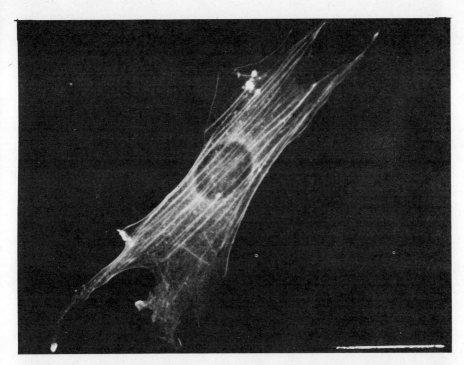

Figure 3.5. Actin filaments organized parallel to the long axis of a spread out BHK 21 cell, visualized by indirect immunofluorescence. The bar is 50 μm. Data of Lazarides and Weber (1974).

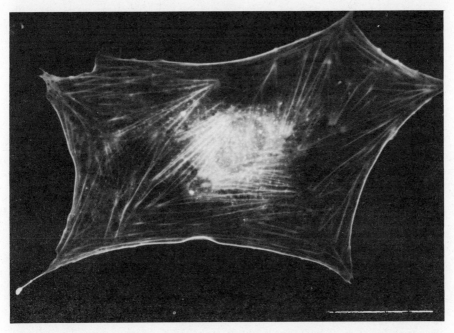

Figure 3.6. Actin filaments converging on focal points in a 3T3 cell, visualized by indirect immunofluorescence. The bar is 50 μm. Data of Lazarides and Weber (1974).

and in what state any free subunits are stored. An important question is
therefore what proportion of the cellular actin is in the polymerized state
at any time (for references see first footnote). As extracted from the cell,
purified (free) actin often is able to polymerize spontaneously in vitro under
conditions that are thought to be physiological (the exact conditions may
vary with the source of the actin), which suggests that the cell possesses
other proteins that inhibit polymerization or stimulate depolymerization.

An interesting reaction in which actin is involved is the formation of a
gel when extracts prepared with sucrose or glycerol from several types of
cell are warmed.* A common feature is that actin is always the major protein component; other proteins also are present, among which there is often
a high molecular weight actin binding protein. If a sufficient content of
myosin is present, the gel may contract. The structure of the gel is a three-
dimensional network of cross-linked actin filaments. The ease of its formation and its precise properties—for example, the response to Ca^{2+} ions—
vary with the source of the extract, presumably depending on both the
properties of the actin itself and the other proteins present. An interesting
question is to what extent such structures may form in vivo; the marine egg
cortex is a probable site for such gels. Indeed, this system may now offer one
approach to the wider question of how the organization of actin is controlled
in the cell by the other proteins that associate with it.

A system for visualizing the spatial and temporal control of the appearance of actin filaments that has been developed by Tilney et al. (1973a) is the
acrosomal reacton of echinoderm sperm. When sperm comes into contact with egg jelly, a process of < 100 nm in diameter and about 90 μm in
length is generated in less than 30 seconds. The process consists of microfilaments of actin; the reaction represents polymerization of preexisting
subunits and is not due to de novo synthesis. Where is the nonfilamentous
precursor form of actin stored? Tilney (1976a,b) observed that unpolymerized actin is sequestered during spermatogenesis as a "cup," associated
with specialized regions of the nuclear envelope. The stored nonfilamentous
actin can be isolated from the sperm after membranes have been removed
with Triton and then is associated with two large proteins of 250,000 and
230,000 daltons. This is reminiscent of the presence of actin in erythrocytes in a nonfilamentous form associated with spectrin (of similar large
size) in the form of a meshwork of the two proteins (Tilney and Detmers,
1975). So possibly the role of the proteins is to prevent polymerization of
actin. The polymerization reaction seems to be initiated by a rise in pH,
following which the actin is no longer associated with its companion proteins (Tilney et al., 1978).

Formation of the acrosomal process is initiated in a dense region, termed

*Gelation has been followed in extracts of sea urchin eggs (Kane, 1975, 1976), Acanthamoeba
(Pollard, 1976), D. discoideum (Condeelis and Taylor, 1977), Xenopus eggs (Clark and Mer-
riam, 1978; Merriam and Clark, 1978), rabbit macrophages (Stossel and Hartwig, 1976), HeLa
(Weihing, 1977).

the actomere, that is located within the cup of nonfilamentous actin (Tilney, 1978; Tilney and Kallenbach, 1979). In one species (Thyone) the actomere consists of a small bundle of filaments; in another (Pisaster) it is amorphous. In either case, the acrosomal process consists of filaments polarized in one direction only, as visualized by the formation of arrowheads with myosin S1; the arrowheads point only toward the actomere (whose constituent filaments in Thyone are themselves organized in the same direction, although their visualization is more difficult, presumably because of the presence of other proteins as well as actin). This analysis suggests that actin filaments may share with microtubules the need for a nucleation center to initiate the assembly reaction. Another instance in which this has been observed is the regeneration of actin filaments from the microvillous border of Salamander intestine cells after microvilli have been stripped by hydrostatic pressure (Tilney and Cardell, 1970). The filaments grow from dense sites on the membrane, but with the opposite polarity from that seen in Echinoderm sperm, that is with arrowheads pointing away from the nucleating center (see below). Whether the actual addition of actin monomers is unidirectional or bidirectional is not known, although when short actin filaments decorated with HMM are incubated with actin in vitro, growth occurs preferentially in the direction opposed to the arrowhead. In a manner similar to the results obtained with microtubules, this is the sole direction of growth at low concentrations of actin monomer; but slower growth in the other direction occurs at higher concentrations (Woodrum, Rich, and Pollard, 1975).

Network of Contractile Proteins

Are the other contractile proteins part of the network of actin filaments. By preparing an antibody against chick smooth muscle myosin, Weber and Groeschel-Stewart (1974) were able to visualize a pattern of fibers similar to that revealed by the actin antibody. A difference is that when visualized by myosin antibody, the fibers appear to be striated, that is, there are interruptions contrasted with the continuous fibers seen with actin antibody. This does not unambiguously reveal the arrangement of myosin on the fiber; myosin is a large molecule and which part of it possesses the antigenic site is not known. Nor is it yet clear in what way it is associated with the microfilaments, since this apparently is not in the form of the thick fibers seen in muscle. But this difference does not preclude the development of a contractile force by an actin-myosin interaction similar to that of the sarcomere (see above). The same cytoplasmic array of actin-myosin fibers is seen in human, mouse, rat and chick fibroblasts, which argues that it may be a common feature of mammalian, avian, and presumably also other cells.

   Myosin appears to be located at the surface of the fibroblast. It is able to migrate to form patches, a property possessed by proteins of the plasma membrane; aggregation of trypsin-dissociated cells may be prevented by antiserum against myosin; and the myosin heavy chain can be labeled by lactoperoxidase-catalyzed iodination of L cells, a technique specific for surface proteins (Willingham, Ostlund and Pastan, 1974; Gwynn et al., 1974; Olden, Willingham and Pastan, 1976). Since the myosin molecule has a length of some 135 nm, it could easily traverse the 10 nm thickness of the plasma membrane; these results therefore show simply that at least some part(s) of the myosin molecule must protrude onto the cell surface. Another technique that localizes myosin is the use of ferritin-labeled antibodies; an antibody directed against human smooth muscle myosin labels WI38 (human) fibroblasts in a region some 50–80 nm from the cytoplasmic surface of the plasma membrane (Painter, Sheetz, and Singer, 1975). These results therefore raise the possibility that the microfilament network is connected to the surface via the myosin component.

Antibodies against tropomyosin and α actinin also reveal the same network of fibers seen with the actin and myosin antisera. Antibody against tropomyosin purified from chick skeletal muscle reacts against tropomyosin of chick muscle, rabbit muscle and mouse fibroblast, implying that at least the antigenic site displays evolutionary stability. With human skin fibroblasts, the network of fibers visualized by indirect immunofluorescence is very similar to the actin network (Lazarides, 1975a,b). A difference is that the fluorescence is periodic: the average fluorescing length is 1.2 μm, these segments being separated by an axial distance of about 0.4 μm.

The component of the muscle Z line, α actinin, also can be located in the microfilaments. Indirect immunofluorescence of rat embryo cells treated with antibody against skeletal muscle α actinin reveals the usual network of fibers (Lazarides and Burridge, 1975). These take the form of a periodic fluorescence with 0.4 μm fluorescent lengths separated by nonfluorescing lengths of 1.2 μm (although these distances are variable, unlike muscle where the Z lines are a constant 2.6 μm apart). The α actinin periodicity is the reverse of the tropomyosin periodicity, raising the possibility that the microfilaments consist of a continuous array of actin fibers, associated alternately with α actinin and tropomyosin.

Similar experiments with gerbil fibroma cells have suggested also that α actinin may alternate with myosin, a form of organization that would be analogous to the sarcomere drawn in Figure 3.1 (Gordon, 1978). The same alternating arrangement has been found in the sheath of fibers that underlies the upper surface of spread human fibroblasts (Zigmond, Otto and Bryan, 1979). It is possible that tropomyosin may be located at the same sites as myosin. Given the diverse nature of the systems in which the microfilament arrays have been characterized, it is hard to compare directly the results obtained with different cells. But it is possible that there is a consistent underlying arrangement analogous to that of the sarcomere, although not involving the formation of thick myosin filaments.

In addition to the network of fibers, there are scattered fluorescent patches, representing sites where groups of filaments terminate or converge to focal points of α actinin. It is possible that α actinin may participate in attachment of actin filaments to the membrane (see below).

Another contractile protein visualized in nonmuscle cells is filamin, a 20,000 dalton constituent of chick smooth muscle that is absent from skeletal muscle. Indirect immunofluorescence with an antibody against filamin reveals a network similar to the myosin network in chick embryo and mouse 3T3 cells (Wang, Ash, and Singer, 1975).

When cells are treated with trypsin, they become detached from the substrate and round up to resemble mitotic cells. Actin filaments disappear with the detachment; and treatment by immunofluorescence then reveals only a diffuse uniform fluorescence. Upon replating, the cells attach to the substrate and reacquire the characteristic spread out appearance. During the first hours of spreading, about 40% of the cells develop a regular polygonal network of fibers, and this may represent an intermediate in the assembly of the bundles of filaments (Lazarides, 1976). By 4 hours after spreading, the first fibers become visible by fluorescence. At this time, actin antibody reveals foci about 3.7 μm apart, connected by bundles of filaments; α actinin antibody reveals only foci, and tropomyosin fluorescence remains diffuse. By 8 hours the actin and α actinin patterns cover the whole cell and tropomyosin forms a pattern which appears to be complementary to α actinin; it shows an array of fibers but lacks fluorescence at the foci connecting them. Figure 3.7 displays an example of the polygonal network stained with antibody against actin. In addition to the polygonal network, fibers attached to vertices of the polygon extend all the way to the edge of the spread-

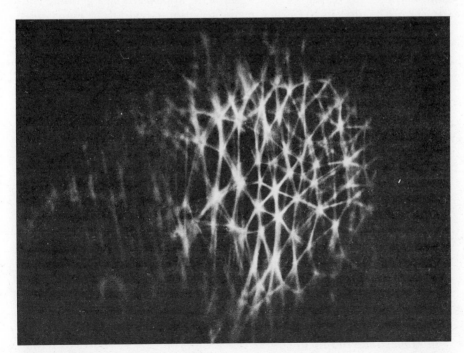

Figure 3.7. Indirect immunofluorescence of a rat embryo cell visualized by actin antibody 8.5 hours after spreading. Data of Lazarides (1976).

ing membrane. These fibers may represent the start of the establishment of the usual bundles of filaments organized parallel to the cell axis. They appear to contain actin, α actinin and tropomyosin (showing its usual periodicity here, though not in the fibers of the polygonal network). The immunofluorescence patterns suggest that the polygonal network is constructed from fibers of actin and tropomyosin connected by foci at which α actinin is located. The network also can be seen in living cells by phase contrast microscopy, supporting the idea that it is not an artefact of immunofluorescence. Rearrangement of α actinin could be an important stage in converting the polygonal network to arrays of stress fibers.

The visualization of actin filaments concerns the cytoplasm. Is actin also a constituent of the nucleus? Its presence in preparations of nuclear proteins, for example as a major component together with other structural proteins in the chromosomal nonhistone proteins, is difficult to interpret (see Chapter 12). Does this mean that actin filaments are important in nuclear or chromosomal structure; or is their presence trivial, for example a result of contamination? One approach to this question has been to assay the actin content of nuclei isolated by micromanipulation instead of by cell fractionation. This has been done directly with Xenopus oocytes and indirectly with Amoeba, in which nuclei have been transferred from labeled to unlabeled cells after which the content of labeled actin has been determined (Clark and Merriam, 1977; Goldstein, Rubin and Ko, 1977; Rubin, Goldstein and Ko, 1978). In both cases actin appears to be a genuine constituent of the nucleus. Its structure and function remain to be established.

Stress Fibers and Cell Motility

Cells established in culture assume a wide range of morphologies that depend upon the origin of the cell and the conditions of growth. We shall turn to the question of growth control in Chapter 7; here we shall be concerned with the relationship between cell movement and the stress fibers. Much recent work has been performed upon the stress fibers of cultured mammalian cells of fibroblastic origin; and it is in such systems that the relationship between filament systems and growth control has been studied. We shall therefore be concerned largely with these systems; the involvement of actomyosin contractile systems in the motility of the lower eucaryotes has been reviewed by Pollard and Weihing (1974).

In order to grow in culture, fibroblasts must be provided with a solid substrate to which to attach. The cells then become "anchored" to the substrate and grow to form a *monolayer* (which may be up to 3-4 cells in thickness). This contrasts with cells able to grow well in suspension, that is, without being attached to a solid surface. The need for a solid substrate is described as *anchorage dependence*.

Fibroblasts are motile in culture (for reviews of early work see Abercrombie, 1961; Ambrose, 1961). The leading edge of a fibroblast moving on a smooth surface is flattened, with a thickness typically of only some 110–160 nm. This thin region is characterized by the presence of *ruffles* (Abercrombie et al., 1970a,b,c; 1971). These are transient linear projections into the medium which lie roughly parallel to the front edge of the cell. Most ruffles are sustained for less than 1 minute, and they move backward relative to the substratum. The ruffles, and some similar structures seen elsewhere on the cell, are sometimes described as *lamellipodia*. Ruffles appear to reflect the rapid assembly of new cell surface material at the leading edge, a process that may be connected to cell movement.

Attachment to the substratum is achieved in the form of many points of close contact, sometimes described as *plaques*. These are about 1–2 μm^2 in area and there appear to be 10–100 such regions beneath each cell. Movement of the cell obviously demands that these contacts should be made and broken rapidly. New contacts are made at the leading edge, while previous contacts are broken at the tail, in a process in which the tail moves rapidly and may be readsorbed into the main body of the cell. Microfilaments can be seen to run close to the plaques, and this suggests that they may provide the force for breaking the contacts with the substratum. Microfilament bundles also are prominent along the axis of the tail (for review see Goldman, Schloss, and Starger, 1976). Albrecht-Buehler (1977) has suggested that they may comprise "rails" along which the body of the cell is able to move. Contacts between fibroblasts appear to be made by the same mechanism that is involved in establishing attachment to the substratum (Heaysman and Pegrum, 1973).

Actin filaments in muscle are anchored to the Z bands, a feature necessary for the generation of contractile force. In nonmuscle cells, actin filaments often can be seen to be associated with the membrane at sites involved in motility. What is the nature of the attachment between the microfilaments and the membrane? Intense immunofluorescence is seen in these regions with anti α actinin, which suggests that they could be a counterpart to the Z band (Lazarides and Burridge, 1975). Differences in myosin and filamin antibody staining also are seen at regions of the cell periphery: microspikes, ruffles, areas of cell-cell contact all lack myosin and possess filamin (Heggeness, Wang, and Singer, 1977). A protein present at the termini of microfilaments in plaques of cultured chicken cells has been identified by immunofluorescence using antibody directed against a 130,000 dalton protein present in chick gizzard (Geiger, 1979). How this organization is related to contractile motion remains to be established.

The attachment of actin filaments to membranes has been examined in some detail in chick intestinal epithelial cells and in Mytilus and Limulus sperm (Mooseker and Tilney, 1975; Tilney and Mooseker, 1976). The epithelial cells possess a brush border consisting of microvilli which contain bundles of actin filaments; and the sperm possess acrosomal filament bundles. In both cases the actin filaments are attached to membranes at one end, in a dense structure. The use of the myosin subfragment S1 allowed arrowheads to be generated on the filaments; and in each case all the filaments displayed the same polarity, with the arrowheads pointing away from the membrane. Thus the attachment of actin filaments to the membrane differs from their relationship with the Z bands since the bipolarity of muscle is replaced by a unidirectional organization in the microvilli and acrosomal systems. Whether this form of organization is common to other nonmuscle actin filaments remains to be established.

In the microvillus, the actin filaments run from their attachment to the membrane at the tip into the cell body where they end in the terminal web, a structure consisting of microfilaments which is itself attached to the membrane. The filaments in the microvillus are attached also to the membrane periodically along their length. The use of indirect immunofluorescence shows that of the contractile proteins only actin may be found in the microvillus: myosin, tropomyosin, filamin, and α actinin all are located in the terminal web (Bretscher and Weber, 1978; Geiger, Tokuyasu, and Singer, 1979; Mooseker, Pollard, and Fujiwara, 1978). Although this means that myosin is in an appropriate location to interact with the actin filaments to generate a contractile force, it also excludes what had been a popular model that α actinin might be responsible for linking the actin filaments to the membrane both at the tip and periodically. Presumably some other protein exercises this function; and its characterization will be interesting.

The extent to which a cell becomes spread out on the substratum depends on its type; for example, 3T3 cells characteristically become more extensively spread out than BHK 21 cells (see above). The occurrence of stress fibers is correlated with spreading out. This relationship is reflected in a striking manner by the changes that occur when anchorage-dependent cells are transformed by SV40 or polyoma virus. Transformation releases some of the restrictions upon the growth of cultured fibroblasts. The transformed cells are no longer anchorage dependent, so that the flat spread out structure gives way to a more rounded form of organization; cell proliferation is not inhibited by contact with other cells, so that cells may pile up on each other instead of stopping at a monolayer; and the dependence on serum may be much reduced. Correlated with this change in structure is a decline in the number of stress fibers (McNutt, Culp, and Black, 1973). Indirect immunofluorescence with actin and myosin antibodies shows that when 3T3 or rat embryo cells are transformed by SV40, the parallel sheaths of microfilaments are replaced by a diffuse fluorescence. This disruption applies to both actin and myosin (Pollack, Osborn, and Weber, 1975). Since the total amount of actin per cell remains unchanged, the disruption must reflect a redistribution of the contractile protein. Revertants of the transformed lines can be isolated which regain some or all of the properties of anchorage dependence, contact inhibition, and serum dependence. Only those revertants regaining anchorage dependence display the lines of actin and myosin fibers. A similar loss of the actin cables is seen when chick embryo fibroblasts are transformed by the Rous sarcoma (RNA) virus (Edelman and Yahara, 1976; Wang and Goldberg, 1976). The correlation between ability to adhere to the medium and the presence of microfilaments suggests their involvement in cell motility (Willingham et al., 1977).

Microfilaments and Microtubules in the Cell Skeleton

The idea that there may be a direct connection between the actin filaments within the cell and proteins on the cell surface has been mooted on several occasions. Changes in the transformed cell phenotype involve events that may be controlled by feedbacks from the surface. For example, in dense cultures chick embryo fibroblasts do not display ruffles, but these appear very shortly after transformation, preceding the change in shape from flat to rounded (Ambros, Chen, and Buchanan, 1975). The absence of actin cables in anchorage-dependent cell lines is correlated with the production of plasminogen activator, which is released by cultured cells into the medium to activate the protease, plasmin (Pollack and Rifkin, 1975). This suggests that a protease acting from outside the cell can influence the internal pattern of actin filaments.

Changes in cell surface proteins occur upon transformation; and one of the most prominent is the disappearance from the surface of the large glycoprotein, fibronectin (for review see Hynes, 1974). Addition of fibronectin to transformed cells restores the morphology of the parental line, in which fibronectin is organized in an extracellular network and actin cables can be recognized within the spread out structure (Mautner and Hynes, 1977; Ali et al., 1977; Yamada, 1978). A suggestion that surface fibronectin may be directly connected to intracellular actin has been based on observations of coincidence between the networks, especially at attachment plaques (Hynes and Destree, 1978; Singer, 1979).

Some cell lines respond to cyclic AMP by changes in motility and shape that are the reverse of those seen upon viral transformation. Two examples are CHO cells (derived from Chinese hamster ovary) and XC cells (derived from a sarcoma induced by the Rous sarcoma virus); both normally grow as compact cells in random orientation. Treatment with dibutyryl cyclic AMP (a derivative able to enter the cell more readily) causes a morphological change to the elongated fibroblastic type of organization, in which cells are rather well spread out and grow in an orientated monolayer (Hsie and Puck, 1971; Hsie, Jones, and Puck, 1971; Johnson, Friedman, and Pastan, 1971). Colchicine at concentrations of $\sim 10^{-6}$ M prevents the acquisition of the elongated state, suggesting that a reorganization of microtubules is involved. Cytochalasin B at the same concentration has the same effect as colchicine, implying that microfilaments also are necessary (Puck, Waldren, and Hsie, 1972; see below).

The effects of cyclic AMP have since proved to be somewhat more complex than represented by the "reverse transformation," varying with the type of cell. However, the implication of the ability of both colchicine and cytochalasin B to counteract the reverse transformation is that maintenance of the elongated structure requires both the cytoplasmic network of microtubules and the bundles of microfilaments. This conclusion is supported by direct observations of the effects of dibutyryl cyclic AMP upon these fibers. Treatment with dibutyryl cyclic AMP causes cellular processes to be extended, and bundles of microfilaments are associated with the plasma membrane of the processes. Microtubules accumulate in arrays parallel to the microfilaments, but somewhat further away from the surface (Porter et al., 1974; Willingham and Pastan, 1975).

Both microfilament and microtubule systems appear to be involved in the mobility of proteins located in the plasma membrane. The effect of colchicine upon the mobility of Con A binding sites on the cell surface suggests that surface receptors may be linked to the intracellular network of microtubules (see Chapter 2). Microfilaments are implicated by the similar effects displayed by cytochalasin B and also by observations that actomyosin-containing fibers are organized in an array related to the distribution of Con A binding sites (Ukena et al., 1974; De Petris, 1975; Ash and Singer, 1976). The accumulation of actin, myosin, and α actinin under sites of capping is striking (Geiger and Singer, 1979; Bourguignon, Tokuyasu, and Singer, 1978). Consistent with the idea that the mobility of surface receptors depends upon intracellular networks, viral transformation is accompanied by changes in mobility, taking the form of a change in the distribution of receptors and a decrease in the ability of Con A to restrict movement (Rosenblith et al., 1973; Nicolson, 1973; Edelman and Yahara, 1976; Ash, Vogt, and Singer, 1976). This is consistent with the idea that the disruption of these networks upon transformation releases the surface receptors from their association with the microtubules and microfilaments. However, in at least some cases the redistribution of Con A occurs predominantly by direct endocytosis followed by intracellular accumulation at the pole, rather than by surface movement. The endocytosis is prevented by cytochalasin B, suggesting that microfilaments are involved in this process (Storrie and Edelson, 1977). It should be noted also that there may be more than one mechanism for capping: Braun et al (1978a,b) have divided molecules subject to capping into two classes, depending on whether the reaction follows the path described, or represents an alternative, less active process that does not seem to rely on actomyosin.

Disruption of Microfilaments by Cytochalasin B

Derived from the fungal mold Helminthosporium dematioideum, cytochalasin B was shown by Carter (1967) to abolish the motility of L cells and to prevent cleavage following mitosis. Its chemical structure is shown in Figure 3.8. Both effects are seen at low drug concentrations, about 0.5 μg/ml ($\sim 10^{-6}$ M). The inhibition of motility is specific for contractile systems;

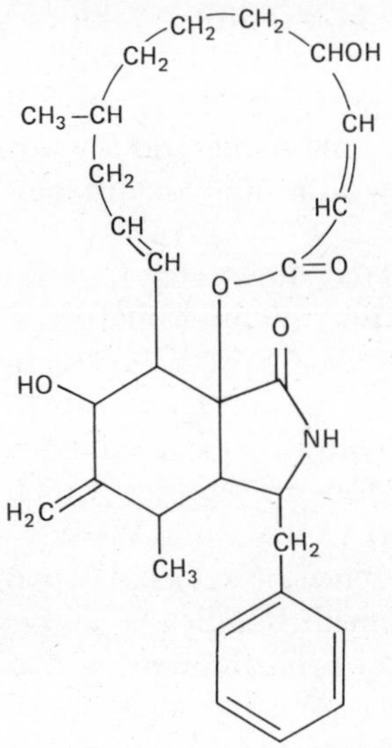

Figure 3.8. Chemical structure of cytochalasin B.

the motility of ciliates, which resides in microtubule rather than microfilament systems, is not inhibited. The inhibition of division results from disruption of the cleavage furrow; nuclear division is not inhibited, with the result that consecutive nuclear divisions may take place without cleavage, to produce multinucleate cells. A more drastic effect is seen at higher drug concentrations, generally >10 μg/ml, when a high proportion of the cells extrude their nuclei (extrusion begins to occur at about 1 μg/ml). The use of this effect in preparing enucleated cells is discussed in Chapter 8.

In addition to these effects on cell motility and morphology, cytochalasin B inhibits the transport of hexose sugars into mammalian and avian cells. The doses required for this effect are low and it has been suggested that transport inhibition can be displayed before effects on motility become apparent. At all events, the effect on transport appears to be independent of that on motility (Estensen and Plagemann, 1972; Mizel and Wilson, 1972; Kletzien and Perdue, 1973). Further evidence for the independence of the effects is provided by the failure of the related compounds dihydrocytochalasin B and cytochalasin D to inhibit transport while continuing to inhibit motility (Atlas and Lin, 1978; Lin et al., 1978).

The multiple effects of cytochalasin B are paralleled by the existence of cellular binding sites with different affinities for the drug. Scatchard plot analysis identifies a minimum of two classes of binding sites in HeLa cells or bovine platelets: high affinity with a K_a of about 10^7 M^{-1}; and low affinity with a K_a of about 10^5 M^{-1} (Lin, Santi, and Spudich, 1974). The inhibition of hexose transport occurs at drug levels consistent with binding to the high affinity sites, which appear to comprise membrane proteins involved in sugar transport (in red blood cells) (Lin and Spudich, 1974). Because it is difficult to quantitate the effects of cytochalasin B on motility, cytokinesis and nuclear extrusion, it is not possible to equate binding to the low affinity sites with particular effects. This leaves as an open question the issue of whether the effects on motility and cleavage on the one hand, and nuclear extrusion on the other, are mediated by binding to the same or different types of site.

Treatment of a variety of cells with cytochalasin B has shown that its effect upon motility and cleavage resides in the disruption of microfilaments: intermediate filaments and microtubules are unaffected. Chick glial cells in culture display bundles of microfilaments similar to those of rat embryo fibroblasts; upon treatment with cytochalasin B, the bundles are replaced by a mass of densely packed matrix with a fine filamentous appearance (which may or may not actually contain filaments) (Spooner, Yamada, and Wessells, 1971). The cessation of cell movement in the presence of cytochalasin B suggests that motility may depend on the microfilament bundles.

The disruption of microfilaments by cytochalasin B has been visualized directly by the application of immunofluorescence. With both lympho-blastoid cells and fibroblasts, addition of cytochalasin B sees the replacement of the filamentous pattern with an amorphous array (Norberg, Lidman, and Fagreus, 1975). The actin disaggregated from the microfilaments may become localized in star-shaped patches, where it is probably in a nonfilamentous condition (Weber et al., 1976). The reaction is reversible.

Sometimes an intermediate stage in the disruption can be seen in which the actin filaments disaggregate into short rods. Similar rods, of about 5 μm in length, are the predominant product when cells are induced to contract by treatment with cytochalasin A. This raises the possibility that their formation in both cases may reflect the existence of a periodicity between labile sites in the microfilaments (Rathke et al., 1977).

For some time there have been questions about whether the effect of cytochalasin B on microfilaments is direct or indirect. Weber et al. (1976) observed that the rearrangement of actin filaments does not occur in glycerinated cells. This suggests that the disruption might not be due to a direct specific action of the drug on one of the contractile proteins. One suggested possibility for an indirect mode of action is that cytochalasin B might disrupt the attachment of microfilaments to the membrane. On the other hand, two observations argue for a more direct effect. Cytochalasin B is able to disrupt the gels that actin may form in vitro; and a cellular complex isolated for its ability to bind the drug is able to induce polymerization of monomeric actin in an action that is inhibited by cytochalasin B (Hartwig and Stossel, 1976; Weihing, 1976; Lin and Lin, 1979).

Evidence that microfilaments are involved in contacts between cells as well as in movement over the substrate is provided by studies of the effect of cytochalasin B on cell interactions. Stoker (1975) took metabolic cooperation to provide a criterion for cell contact, using a system in which a polyoma-transformed BHK cell line is cultured together with 3T3 cells. The BHK line lacks the enzyme hypoxanthine phosphoribosyl transferase (HGPRT) and therefore cannot incorporate ^{3}H-labeled hypoxanthine into DNA. The 3T3 cells incorporate the purine, and BHK cells in contact with them also are able to do so, as revealed by autoradiography. But the addition of cytochalasin B at concentrations >1 μg/ml abolishes this cooperation. The implication is that effective contact requires functional microfilaments.

Intermediate Filaments

Intermediate filaments are found in many cell types, but little is known about their origin or function. Indeed, all filaments with a diameter of ~10 nm, distinct from the 6 nm microfilaments and the 24 nm microtubules, have been described as "intermediate filaments"; and until recently there has been no evidence on whether the similarity in the dimensions of the threads seen in different cells reflects a relationship between them or is coincidental. Prominent examples of this class of filaments are the neurofilaments of neurones, the glial filaments of glial cells, the tonofilaments of epithelial cells, the analogous filaments of mesenchymal cells, and the 10 nm filaments of smooth muscle.

The major component of the filaments of glial cells is the glial acidic fibrillary protein, a species of about 51,000 daltons (Dahl and Bignami, 1973, 1976; Schachner et al., 1977). Neurofilaments are a prominent component of nerve axons, where they may be involved in axoplasmic transport. Early experiments to identify the protein component led to the isolation instead of the glial fibrillary protein, present as a contaminant (Yen et al., 1976; Blose, Shelanski, and Chacko, 1977). This resulted in suggestions that the glial and neurofilaments might be identical. But the neurofilaments of rat peripheral nerve and spinal cord as well as other sources now have been shown to have three principal protein components of about 200,000, 150,000 and 69,000 daltons (Schlaepfer and Freeman, 1978; Liem et al., 1978). Thus neurofilaments and glial filaments are quite different systems.

The tonofilaments of epithelial cells were first shown to consist of fibers of keratin in human epidermal cells (Sun and Green, 1978). The keratins comprise a family of related proteins, in a size range that varies with the source but is roughly from 43,000 to 65,000 daltons. A similar constitution underlies the filaments of rat kangaroo Pt K2 cells (an epithelial line derived from kidney), where the filaments have been visualized by indirect immunofluorescence with sera from normal rabbits (Osborn, Franke, and Weber, 1977). This suggests that the rabbit contains auto antibodies directed against a component that is present also in the Pt K2 cell. These intermediate filaments form both straight fibers and also a network with branches and interconnections. Its organization is not influenced by addition of colcemid or cytochalasin B.

Another filamentous system is present in cells of mesenchymal origin. The intermediate filaments of BHK 21 cells are organized as a perinuclear ring, and may form into a "juxtanuclear cap" as a transient stage during spreading on substratum. The filaments also accumulate in a cap upon treatment with colchicine; the same system is involved in each case as shown by the characterization of the principal proteins of about 55,000 and 54,000 daltons (Starger et al., 1978). The use of antibodies shows that a similar network is present in 3T3 cells and human endothelial cells (Franke et al., 1978, 1979b). The name "vimentin" has been suggested for the protein, in these cases with a molecular weight estimated as 57,000 daltons. HeLa cells contain both the keratin tonofilaments and the vimentin-containing network, although the latter can be recognized only after colchicine has been used to induce the perinuclear whorl (Franke et al., 1979a). This suggests that the production of these intermediate filament types is not mutually exclusive. The nature of the effect that colchicine has on the vimentin-containing filaments has not yet been established. The same network has been visualized in several cell types with an autoimmune antibody (Gordon, Bushnell, and Burridge, 1978).

The subunit of the intermediate filaments of chick smooth muscle has a size of 50,000 daltons and is resolved into two closely related spots by two-dimensional electrophoresis (Izant and Lazarides, 1977; Lazarides and Balzer, 1978). It has been identified also in guinea pig smooth muscle, where it is a major cell constituent, corresponding to about 15–25% of the intracellular structural protein; its content relative to actin is 40–80% (Small and Sobieszek, 1977). The intermediate filaments remain in position when cells are extracted with detergent to generate ghosts; since the filaments appear therefore to be part of the cell skeleton, the name "skeletin" was suggested for the subunit. An antibody against the chick smooth muscle subunit reacts also with skeletal and cardiac muscle, where the filaments appear to connect the Z bands of adjacent myofibrils and possibly may connect the contractile system with the membrane (Lazarides and Hubbard, 1976). Because the intermediate filaments often are associated with desmosomes (which are involved in attachments between adjacent cells) the name "desmin" has been suggested for the subunit. The network of filaments can be seen in embryo cardiac cells as well as mature muscle (Lazarides, 1978). These intermediate filaments are unrelated to those of brain (Bennett et al., 1978).

To summarize, intermediate filaments constitute a third intracellular network, analogous to the microfilaments and microtubules in the sense that the network in any cell appears to be constructed from a small number of proteins. But there are at least five types of intermediate filament, whose structures and presumably functions are different.

Detergent Resistant Cytoskeleton

A large proportion of the membranes and proteins of the cell can be removed by treatment with nonionic detergents to leave a *cytoskeleton* which appears to retain many of the characteristic structural features of the cell. This implies that it may be the components of this cytoskeleton that are responsible for maintaining the shape and organization of the cell.

The procedure of exposure to hypertonic medium followed by extraction with Triton X-100 was introduced by Lenk at al. (1977). This treatment removes the plasma membrane, endoplasmic reticulum, mitochondria, and the nuclear envelope. In spite of the removal of virtually all of the membrane and of about 75% of the cytoplasmic protein, a thin section through the remaining skeleton shows a recognizable morphology. Figure 3.9 shows that the densely packed cytoplasm previously evident is replaced by a much emptier structure in which filaments and polyribosomes can be seen. The nucleus is still clearly outlined.

This protocol is performed at 0°C, so that microtubules are absent from

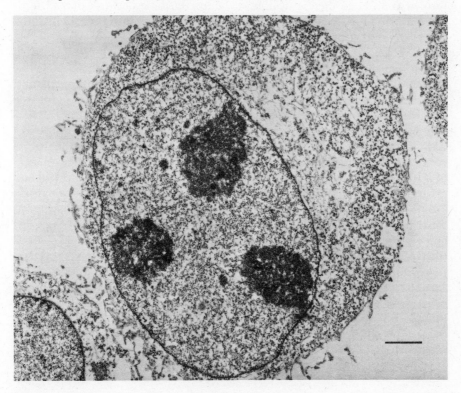

Figure 3.9. Cytoskeleton formed by 2 min of exposure to hypertonic buffer followed by addition of Triton. The bar indicates 1 μm. Data of Lenk et al. (1977).

the cytoskeleton. However, intermediate filaments can be seen in the vicinity of the nucleus and microfilaments can be seen near the cell boundary. The retention of a large proportion of the cellular polyribosomes suggests that they may be attached to the cytoskeleton. They can be dissociated from it if the cells are treated with cytochalasin B before the skeleton is prepared; under the conditions of treatment used, the cytochalasin does not seem to disrupt the structure of the skeleton, leaving open the question of the nature of its effect upon the polysome attachment.

The present presence of actin in the detergent-resistant cytoskeleton of 3T3 or chick embryo cells has been examined by indirect immunofluorescence. Osborn and Weber (1977a) found that most of the microfilament bundles remain part of the cytoskeleton. When the proteins remaining in the cytoskeleton are examined by gel electrophoresis, the most prominent are some of the histones, a protein with the size of actin, and a protein of 58,000 daltons that may be the subunit of the intermediate filaments. Tubulin is absent and indirect immunofluorescence with antitubulin shows that only fragments of microtubules appear to remain. These experiments were performed at 37°C, so the disruption of the microtubule network may be a direct effect and not due to the use of conditions in which it is labile. Centrioles, however, are retained.

The general conclusion suggested by these studies on the cytoskeleton is that the cellular structure may be maintained by protein networks; these structural features lack membranes. Thus the presence of microfilaments and intermediate filaments in the cytoskeleton suggests that an important role of these networks may be the establishment of cell structure.

Related extraction procedures have been used to isolate nuclear proteins. Exposure of rat liver nuclei to a series of steps of which the critical stages are treatment with nucleases and extraction with Triton leads to the isolation of a fraction consisting largely of protein, with about 18% nucleic acid and 1% lipid. Three major proteins in the size range of 60,000–70,000 daltons are prominent components. The properties of this fraction have been variously interpreted. On the one hand Berezney and Coffey (1977) and Comings and Okada (1976) have suggested that it constitutes a flexible nuclear matrix, fibrilar in structure, which extends within the nuclear volume as well as representing a surface feature. On the other hand, Dwyer and Blobel (1976) and Grace, Blum, and Blobel (1978) have shown that the proteins reside at nuclear pores within the lamina. Their distribution becomes diffuse during mitosis. Possible functions in which these proteins may be involved include the attachment of chromatin to nuclear structures and the nucleocytoplasmic transport of RNA.

CHAPTER 4

Chromosome Segregation: Mitosis

The evolutionary importance of a mechanism to ensure that a cell distributes its replicated chromosomes evenly to the two daughter cells of a division needs no emphasis. Mitosis is a continuous process, of course, but cell division can usefully be considered in three stages. First there is a drastic reorganization of the structure of the cell, in which the usual compartments of nucleus and cytoplasm disappear, to be replaced by a spindle which consists of fibers running between the two poles of the cell. The establishment of the mitotic apparatus appears to take place by rearrangement of cell components rather than by synthesis de novo of new constituents. Then chromosomes are attached to the spindle fibers by their centromeres, become aligned on the equatorial plane, and move toward the poles to become segregated into two diploid sets. Finally the two chromosome sets, each now becoming encased in a nuclear membrane, must find themselves in separate cytoplasms. This last step is accomplished by the formation of a contractile ring, which tightens to pinch off the two daughter cells, which may then move apart.

Mitosis represents the culmination of events that have taken place during the preceding growth cycle of the cell. During interphase the cell doubles in size and both reproduces its chromosomes and synthesizes the components needed to form the mitotic apparatus. In contrast to the emphasis placed on the mechanics of mitosis by cytological observations, it is in fact a short interruption—usually occupying less than 5% of the division cycle—in the normal functioning of the cell. Protein and RNA synthesis are much reduced during the process of division; and DNA has previously been synthesized.

The Mitotic Apparatus

Early observations of mitosis in the last century suggested the concept that each chromosome of the diploid set splits longitudinally into two sister chromatids, which are then segregated to the daughter cells by the subsequent steps of division. The term *prophase* was introduced originally to describe the early stage of mitosis that precedes the visible division of chromosomes into sister chromatids; and *metaphase* describes the succeeding stage, after chromosome division has become evident. Of course, it is now clear that the chromosomes are reproduced well before the beginning of mitosis; and, indeed, they may now be visualized as double structures from the start of prophase, when two elongated threads appear to be coiled around each other. During prophase the chromosomes contract; and in metaphase the condensed chromosomes become aligned along the equatorial plane of the cell.

The end of prophase and the beginning of metaphase—sometimes described as *prometaphase*—is marked by dissolution of the nuclear membrane and the appearance of faint lines running between the two poles of the cell. These *spindle fibers* appear to terminate in regions with a starlike structure of many radiating lines (the *asters*). The chromosomes are attached to the spindle fibers by their kinetochores, dense structures located at the centromeres. *Anaphase* is defined by the period when the chromosomes move along the spindle fibers toward the two poles of the cell; one member of each sister chromatid pair moves toward each pole. *Telophase* describes the final stage when nuclear membranes form around the groups of chromosomes at each pole; this is a reversal of prophase, during which the chromosomes lose their individual structures to become part of the network of interphase chromatin.

Spindle fibers were seen in stained sections of cells in early studies of mitosis; and the controversy about their nature was settled only when Inoué (1953) observed the birefringence that they cause when living metaphase plates are viewed in polarized light. The pattern of birefringence corresponds to the organization of many fine filaments parallel to the long axis of the cell (see Inoué, 1964).

The first suggestion that microtubules might be important in mitosis came from Ledbetter and Porter (1963); more recent work has been reviewed by Luykx (1970). Electron microscopy suggests that the spindle fibers are composed of microtubules loosely organized in sheets and bundles. The number and distribution of microtubules usually corresponds well with the intensity and pattern of birefringence seen in the polarizing microscope (Inoué and Sato, 1967; Goldman and Rebhun, 1969; LaFountain, 1974). This suggests that microtubules are the sole source of the birefringence. However, in some cases a discrepancy has been noted between birefringence and microtubule concentration; or treatments that reduce birefringence may appear to leave microtubules unaffected (Behnke and Forer, 1967b; Forer, Kalnins and Zimmerman, 1976). This has led to suggestions that other types of fiber, for example, actin filaments, might contribute to the birefringence. However, biophysical estimates of the parameters of fiber structure required to generate the observed birefringence correspond well with those of the microtubule fibers (Sato, Ellis and Inoué, 1975). And, of course, the birefringence is abolished by treatments that disrupt microtubules, such as cold temperature or exposure to colchicine (see below). Although none of these results excludes the possibility that fibrous structures other than microtubules contribute to the birefringence in minor amount, it is now generally accepted that microtubules represent by far the predominant component.

The mitotic apparatus can be extracted en masse from dividing sea urchin eggs; spindle fibers, asters, centrioles, all are visible in the preparation and chromosomes are retained at least through the earlier stages of isolation (Mazia and Dan, 1952). The original methods of isolation used alcohol-fixed cells, with agents such as digitonin to disperse the cytoplasm. These were succeeded by protocols including compounds that protect S—S bonds in the apparatus from reduction to —SH groups. This approach allowed the mitotic apparatus to be obtained by removing the membranes of sea urchin eggs and shaking the cells in sucrose with dithioglycol (Mazia and Dan, 1952; Mazia et al., 1960; for review see Mazia, 1961). A later technique that also allowed the mitotic apparatus to be extracted from living cells without prior fixation was the use of hexanediol (Kane, 1962; Rebhun and Sharpless, 1964).

Some difficulty was at first experienced in establishing which protein(s) constitute the mitotic apparatus. Early experiments with mitotic apparatus isolated by conventional means identified proteins with various properties (Sakai, 1966; Kiefer et al., 1966; Kane, 1967; Stephens, 1967). The difficulty in concluding that a given preparation represents the principal protein can be overcome by showing that its selective extraction causes the disappearance of microtubules; and, of course, it should be able to bind the mitotic inhibitor, colchicine. By applying organic mercurial, which depolymerizes outer doublet microtubules of sea urchin sperm flagella, Bibring and Baxandall (1971) selectively extracted a protein from sea urchin mitotic apparatus that has the same properties as the tubulin of flagellar microtubules. An antiserum against the intact outer doublet microtubule reacted with this protein. It is now clear that all microtubules are assembled from α and β tubulins (see Chapter 2). Indirect immunofluorescence with antitubulin preparations demonstrates that microtubules may be found in cultured cells in the form of a cytoplasmic network during interphase and are organized as spindle fibers during mitosis (see below). The implication that spindle fibers are generated by reorganization of microtubules and not by de novo synthesis of new tubulin subunits is consistent with earlier studies of colchicine-binding activity. When HeLa cells growing on ^{14}C leucine were incubated with ^{3}H colchicine, no association occurred between the radioactive labels when the ^{14}C leucine was given at metaphase. But the binding of ^{3}H colchicine to ^{14}C protein labeled during interphase was proportional to the total protein synthesis of the cell (Robbins and Shelanski, 1969). This implies that tubulin is synthesized continuously during interphase (presumably then entering cytoplasmic microtubules via a pool of free tubulin), becoming reorganized into spindle fibers at mitosis.

Development of the Centromere

Centromeres are not observed in interphase nuclei, so they must develop as morphological entities during the early stages of chromosome preparation for mitosis and meiosis. Possession of a centromere is essential if a chromosome is to behave properly on the spindle; chromosome fragments that lack a centromere behave abnormally at division and may be lost from the spindle or left in the equatorial plane. In this sense, the only essential part of the chromosome at mitosis is the centromere; and Mazia (1961) has suggested the concept that genes may be organized into a small number of chromosomes principally in order to make possible their even segregation at mitosis. This views the chromosome as an arrangement for attaching many genes to each centromere.

The centromeric region is one of the most conspicuous features of linear differentiation along the mitotic chromosome. It is especially clear at metaphase when it can be seen as a pronounced narrowing at the junction of the sister chromatids. This is known as the *primary constriction*. Whole mount preparations of chromosomes show that the number of chromatin fibers in the centromeric region is much reduced; and it is this reduction that presumably generates the primary constriction. An example is shown in Figure 4.1 and chromosome structure is discussed further in Chapter 10.

In some preparations, especially those fixed with OsO_4, the centromeric region of each chromatid appears to contain a small intensely staining granule, sometimes termed the spindle spherule. Spindle fibers appear to attach to this granule rather than to the whole region included in the primary constriction. The term *kinetochore* is often used to describe the granule. Loosely speaking, the terms centromere and kinetochore have been used to describe the same general region, but generally the term centromere now is understood in a wider sense to include the entire area of primary constriction seen in the light microscope, while the kinetochore describes only the specialized structure seen in the electron microscope by which the chromosome is attached to the spindle.

The kinetochore first appears as a patch of condensed filaments packaged in irregular bundles at a small constriction on each side of the chromatid pair at early prophase. According to the studies reviewed by Brinkley and Stubblefield (1970), each patch is a spherical structure of about 0.4 μm in diameter, apparently consisting of filaments of diameter 5–8 nm. Even during early prophase the sister centromeres are probably structurally separate from each other; and they appear to develop synchronously during the subsequent stages of division. As prophase continues, there is a transition in the structure at the centromere from these dense spherical masses to give less dense bands which extend for 0.1–0.5 μm, one along each side of the chromatid pair.

Studies of the ultrastructure of the centromeric region at this stage, such as those of Jokelainen (1967) with rat cells, reveal a dense inner layer, some

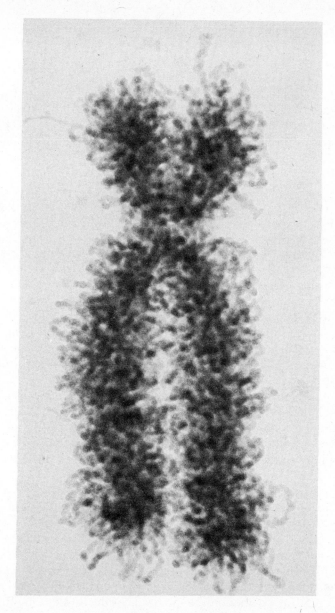

Figure 4.1. Whole mount electron micrograph of human chromosome 2. ×44,460. Kindly provided by Professor E. J. DuPraw (1970); from *DNA and Chromosomes* (Holt, Rinehart, and Winston).

0.15–0.30 μm in depth, of fibers that are continuous with the main body of the chromosome. Next there is a middle layer of about the same depth but of low electron density. This in turn is surrounded by an outer layer, which may appear amorphous or may seem constructed of fine filaments. This represents the kinetochore and is usually about 0.3–0.45 μm in depth. Seen at the surface of the chromosome, the kinetochore appears to take the form of bands of length 0.2–0.25 μm or of a flat circular structure of this diameter, located within the primary constriction. During mitosis in plant cells, the kinetochore appears to be closer to a spherical structure of diameter 0.2–0.25 μm, like the form it takes early in mammalian cell mitosis (for review see DuPraw, 1970). The underlying structure of the kinetochore is controversial and has been variously interpreted as composed of discs or of coiled coils. Nothing is known about its composition and about how it is attached to the chromatin fibers.

The orientation of the kinetochores appears to be responsible for the characteristically different behavior of chromosomes at mitosis and at meiosis. In mitosis, sister kinetochores separate on the spindle so that the sister chromatids of each pair move to opposite poles. This may result from the orientation of the kinetochores on opposite sides of the sister chromatids, which causes one kinetochore to face each pole from the beginning of mitosis. Figure 4.2 shows a section through a mitotic chromosome pair in which the attachment of spindle fibers on opposite sides can be seen. Chromosome segregation at mitosis is an accurate process; Luykx (1970) has calculated that the frequency of mitotic nondisjunction, when homologues fail to separate, is less than 0.03% in the mouse. (Although this value is a maximum estimate for this aberration, it does not include any errors that may result from failures of other types, such as spindle malformation.) Similarly low frequencies are found for meiotic nondisjunction; in D. melanogaster females, the overall frequency is ~0.18% (see Table 5.4).

In the first division of meiosis (by contrast with mitotic division), the two sister kinetochores of each homologue pair of the bivalent do not separate, but move together to one pole. The two sister kinetochores of the other homologue pair move together to the other pole. Then at the second division the sister kinetochores of each homologue pair separate as they do at mitosis. This behavior again appears to be the consequence of kinetochore orientation, for at the first meiotic division the two kinetochores of each chromatid pair lie together on the same side of the bivalent. By the second meiotic division, the sister kinetochores have separated so that they lie on opposite sides. The molecular basis for this movement is not known. Apart from the difference in orientation, there seems no reason to suppose that the structure of the kinetochores of meiotic chromosomes is different from that of the mitotic kinetochores.

Connection of microtubules to the kinetochores usually commences at about the time of dissolution of the nuclear membrane. In micromanipula-

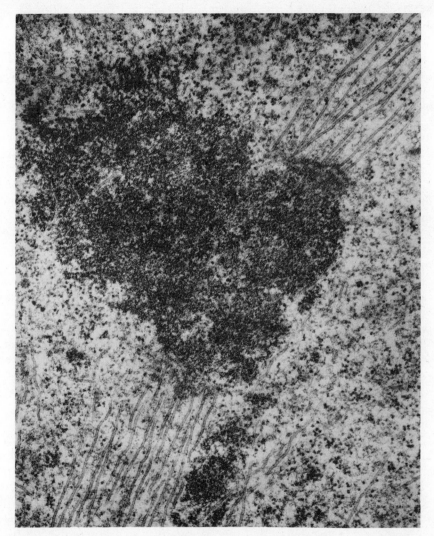

Figure 4.2. Small subtelocentric chromosome of rat kangaroo fibroblast at metaphase. Microtubules are attached to the sister kinetochores. ×51,000. Data of Brinkley and Stubblefield (1970).

tion experiments with spermatocytes of the grasshopper Melanoplus, Nicklas and Staehly (1967) and Nicklas (1967) found that the application of pressure with a needle tip can stretch the meiotic bivalent without detaching it from the spindle. This shows that the chromosome is firmly attached to the spindle; and this firm attachment continues until the end of anaphase. Experiments in which a bivalent was detached and reversed in orientation by a 180° turn showed that reattachment occurs and then is followed by movement toward the pole that the kinetochore faces (see also Nicklas et al., 1979).

Chromosomes appear to move individually during mitosis, so that localized forces must act on individual spindle fibers and kinetochores (Nicklas, 1972; Nicklas and Koch, 1972). During the history of cytology, many types of long range or short distance forces have been postulated to act on chromosomes during mitosis. Mazia (1961) has dispelled this mystical view of chromosome movement and has pointed out that pulling on the kinetochore provides an adequate explanation. The force acting at the kinetochore appears to be applied from the direction in which the chromosome is to move. This pulling force varies in magnitude during division; in general, it seems to be large as metaphase starts, declines gradually during metaphase, and then increases abruptly to mark the onset of anaphase.

Pulling seems to take place from both poles. The chromosomes follow an irregular path as they line up on the equatorial plane during metaphase; the equatorial alignment generally is thought to be the result of equal and opposite forces applied from the two poles on the opposing kinetochores of each chromosome pair. The cessation of movement at the equators would be explained if the force applied to each kinetochore at this stage were proportional to its distance from the pole to which it is connected (Ostergren, 1950; Luyx, 1970). In contrast with the stationary alignment of chromosomes, cellular particles may move toward the poles at metaphase. These may include chromosome fragments, mitochondria, persistent nucleoli, which presumably have become attached to the spindle fibers. This suggests that the pulling force acts not only through fibers attached to the chromosomes, but also on other fibers (for review see DuPraw, 1968).

The separation of sister chromatids and their subsequent movement apart toward the opposing poles defines the period of anaphase. During anaphase, only the centromeres show active movement, the rest of the chromosome appearing to be dragged passively by virtue of its attachment to the centromere. Figure 4.3 shows the anaphase movement of chromosomes of PtKl cultured cells. The active role of the centromere is used to characterize chromosomes by the shapes that they acquire during mitosis—metacentric, submetacentric, acrocentric and telocentric (see Chapter 1). The attribution of this function to the centromere alone is supported also by the unusual configurations acquired by abnormal chromosomes, such as fusion products with two centromeres; again these correspond to pulling applied at the centromere.

Distribution of Spindle Fibers

Following the formation of the spindle and the alignment of chromosomes on the metaphase plate, two events are responsible for accomplishing the

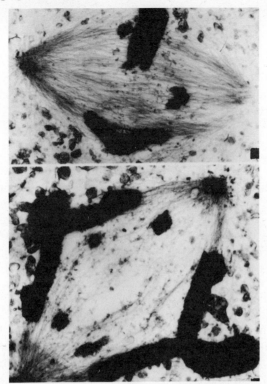

Figure 4.3. Electron micrograph of PtKl cells in mid anaphase (upper) and late anaphase (lower). Data of McIntosh et al. (1975).

segregation of the two chromosome sets. Sister chromatids separate and move toward the opposite poles. And the spindle elongates, so that the poles themselves move further apart. This is a necessary preparation for the formation of two separate nuclei at telophase and the subsequent separation of cytoplasm into two cells at cytokinesis.

Chromosome movement and spindle elongation generally have been regarded as functions of the spindle fibers. These can be divided into two general classes by the criterion of kinetochore attachment. *Continuous fibers* run from pole to pole. *Kinetochore fibers* (also known as chromosomal fibers) run from one pole to a kinetochore. Fibers also may attach to other parts of the chromosomes and these are known as *neocentric fibers.* In addition to the spindle fibers, there may be *astral fibers* that radiate from the asters but do not run through the spindle.

Spindle fibers must change in length during anaphase. As the distance between the poles increases with spindle elongation, continuous fibers must increase in length. As chromosomes move closer to the poles, kinetochore fibers must decrease in length. How these types of microtubules are distinguished to make possible these contrasting activities is not known.

The distance between the poles of a mitotic cell is usually of the order of 10–30 μm and chromosome movement occupies between 2 and 60 minutes depending on the cell type (for review see DuPraw, 1968; Forer, 1969). The rate of chromosome movement is generally in the range of 0.2–5.0 μm per minute. Thus in a cell where the chromosomes must move 10 μm from the equatorial plane toward the poles, movement would occupy some 5–10 minutes at a rate of 1–2 μm per minute. This would correspond to a mitosis of about 30 minutes.

The most work to define the dimensions and composition of the mitotic apparatus has been performed with cultured mammalian cells and sea urchin eggs. In cultured rat kangaroo fibroblasts, for example, the pole to pole distance is some 10 μm at metaphase and, as the spindle elongates, this increases to 20 μm by telophase. The dimensions of the spindle of cultured Chinese hamster cells are roughly half of these lengths (Brinkley and Cartwright, 1971). This means that the chromosomes are attached to the metaphase plate by microtubules that are about 5 μm long. During anaphase the daughter chromosomes first move to within about 3 μm of the poles and then move at the same rate as the poles themselves during the final stage. The mitotic apparatus of the sea urchin egg is somewhat larger, with a spindle length of some 24 μm at metaphase (Cohen and Rebhun, 1970).

The distribution of microtubules at mitosis can be determined by taking serial sections through a cell. At metaphase the greatest number of spindle fibers generally is found on either side of the equatorial plane (with a dip in the plane itself), decreasing toward the poles in each half spindle. In the spindle of WI38 (human diploid) fibroblasts, there are about 800 continuous microtubules and 800 kinetochore microtubules per half spindle (McIntosh and Landis, 1971). This corresponds to about 15 microtubules per kinetochore. In PtK1 (rat kangaroo) fibroblasts, a section taken close to the poles identifies about 1000 microtubules, there are some 2000 in the middle of the half spindle, and only about 1000 (the continuous fibers) cross the metaphase plate (Snyder and McIntosh, 1975). In another study of these cells, the total number of microtubules in the vicinity of the metaphase plate was 1500, with about 25 microtubules attached to each kinetochore. In Chinese hamster cells, the total number is about 500 (Brinkley and Cartwright, 1971). In the egg of the sea urchin Arbacia punctulata, the number of spindle fibers is about 2100 per half spindle (Cohen and Rebhun, 1970). The spindle fibers are most easily counted at metaphase, for by late anaphase, when the chromosomes have completed their movement toward the poles, the number of microtubules in the half spindle is much reduced and they begin to become disorientated. The continuous fibers running through the interzone between the developing daughter cells are maintained in parallel array through telophase.

Cross bridges between microtubules have been seen in a variety of systems (McIntosh, 1964). The presence of links between mitotic microtubules was reported by Hepler, McIntosh, and Cleland (1970) and Brinkley and

Cartwright (1971). The cross bridges seem to be about 5 nm in diameter and 25–60 nm in length. Even when they are not evident, the spindle microtubules tend to keep a fairly constant spacing, typically with a center to center distance of about 50 nm. Cross bridges may be a feature of cytoplasmic as well as mitotic microtubules; Bhisey and Freed (1971) reported that when interphase HeLa cells are cooled to 4°C to depolymerize the microtubules, electron opaque projections corresponding to cross bridges make a transient appearance during the cooling and subsequent warming. Of course, a problem inherent in the visualization of the cross bridges is that it is difficult to exclude the possibility of an artefact; however, if indeed they are a part of a matrix of microtubules, their structure and function in either interphase or mitotic cells remain unknown. One model for microtubule action at mitosis suggests that cross bridges may allow adjacent tubules to slide past each other in a process analogous to the functioning of muscle cross bridges (McIntosh, Hepler, and Van Wie, 1969; see below).

Chromosome movement toward the poles and the elongation of the spindle usually occur together and generally have been considered as part of a single type of process, although involving shortening of kinetochore microtubules and extension of continuous microtubules. An interesting system in which the two events may be temporally distinguished has been reported by Inoué and Ritter (1978). In the protozoan Barbulanympha, anaphase occurs in two distinct stages. First the chromosomes are pulled to the centrioles via fibers attached to their kinetochores (which remain embedded in a persistent nuclear membrane); and then the central spindle elongates to separate the two groups of chromosomes, whose distance from the centrioles remains constant. This demonstrates a difference in the properties of the two types of fiber and raises the possibility that different motive forces are responsible for the two processes.

Many interesting variations in mitosis indeed are found in the protozoa (for review see Kubai, 1975). It is not possible here to discuss all of these, but one exception to the usual course of events is provided by division in Syndinium, a dinoflagellate. Ris and Kubai (1974) reported that the chromosomes remain attached through the entire cell cycle to microtubules that run from their kinetochores to the centrioles. These kinetochore fibers do not change in length even during mitosis, so that the chromosomes behave somewhat as though permanently connected to the centrioles by rather rigid rods. During interphase each kinetochore is attached to both centrioles; after the kinetochores have duplicated to generate sisters, each is connected to a separate centriole. The nuclear membrane persists during division and the kinetochores appear as dense disc-shaped structures inserted in it. Early in mitosis microtubules form between the bases of the two centrioles; these fibers then elongate and it appears to be this elongation alone that is responsible for the movement apart of the two chromosome sets.

A possible speculation about the relationship of Syndinium division to the more usual mitosis is that it represents an earlier stage in the evolution of division, one in which chromosomes are attached by kinetochores to the spindle, but where no separate mechanism for their movement yet has evolved. Preceding this stage would be the direct attachment of chromosomes to the membrane; that is, prior to the evolution of the kinetochore, some less-well-differentiated site on the chromosome would be attached to the membrane and the daughter sites would segregate at division. This is seen in procaryotes and in some dinoflagellates (Kubai and Ris, 1969). Some interesting variations of mitosis are found in diatoms, in which the spindle may start as an extranuclear structure that later penetrates the nucleus to effect the separation of chromosomes by an indirect means that does not appear to involve attachment of microtubules to kinetochores; the chromosomes interact with a "collar" of electron dense material that may in turn interact with a central spindle formed by the overlap of microtubules emanating from each pole (Tippit and Pickett-Heaps, 1977; McDonald et al., 1977; Pickett-Heaps and Tippit, 1978). An intermediate stage in the evolution of the kinetochore may be represented by the protozoan Trychonympha, in which the chromosomes appear to segregate on the nuclear membrane early in mitosis, but then kinetochores become attached to microtubules later in the division (Kubai, 1973).

A central question in analyzing these variations of mitosis is whether there are differences in the fundamental mechanisms of chromosome segregation or whether a common mechanism takes different forms. Taking the usual, latter view, in summary the evolution of the spindle may have occurred first by its establishment in cells retaining the nuclear compartment during division, then progressing to dissolution of the nucleus and reorganization of the cell. The spindle may first have been an intranuclear structure, later becoming extranuclear (see Pickett-Heaps, 1975). From the perspective of the chromosome, the evolution of the spindle may be viewed as passing from an initial stage of chromosome attachment to membranes, through the development of a kinetochore and its capacity to attach to microtubules, finally leading to the establishment of a system for chromosome movement separate from that responsible for the elongation of the spindle.

Dynamic Equilibrium of Microtubules and Tubulin

It is clear that the spindle fibers are essential in maintaining the integrity of the mitotic apparatus and that this is a prerequisite for chromosome movement. But the extent of their function in accomplishing the movement itself has become an issue. Two important questions may be asked about their role. How do the continuous and kinetochore fibers function in the changes in spindle structure that occur during anaphase? And is it the attachment of the microtubules to the kinetochores that is responsible for chromosome movement; or do the microtubules guide the chromosomes on the spindle but not themselves apply the motive force?

Two types of activity might be envisaged to account for the changes in spindle fiber distribution during mitosis. The fibers might have a contractile action in which their content remains unaltered while there is a change in protein conformation. Or their structural organization might remain unaltered, but the length could be changed by addition or removal of subunits at the ends. The second concept is accommodated by the proposal of Inoué and Sato (1967) that the mitotic apparatus is in dynamic equilibrium between assembled microtubules and a pool of free tubulin.

Spindle fibers are assembled and disassembled in each mitotic cycle. During mitosis they do not appear to contract and elongate by the folding and unfolding of polypeptide chains, for fiber birefringence remains unchanged during anaphase movement (Inoué, 1960). This suggests that mitotic movement may be accompanied by assembly and disassembly of subunits. The concept that mitosis reflects the state of an equilibrium between microtubules and tubulin subunits is supported by the response of mitotic cells to treatments that disrupt microtubules. The equilibrium is temperature dependent, with dissociation favored at low temperatures; this is consistent with the behavior of labile microtubules in vitro (see Chapter 2). Pressure also causes a reversible decrease in spindle birefringence, an effect expected from the features of the dynamic equilibrium previously characterized (Salmon, 1975a,b,c).

The classical approach to interfering with the equilibrium of the spindle, of course, is provided by treatment with colchicine. This causes mitosis to halt at metaphase, equally effectively with animal, plant, and marine cells (Dustin, Havas, and Lits, 1937; for review see Eigsti and Dustin, 1955). Depending upon the concentration of colchicine and the time of treatment, the metaphase-arrested cells may disintegrate, reconstitute a nucleus containing both of the diploid chromosome sets, or recover to proceed through a cell division. The use of colchicine in inducing the formation of polyploid cells was realized at an early stage (Gavaudan and Pomriaskinski-Kobozieff, 1937; Eigsti, 1938; Levan, 1938; Nebel and Ruttle, 1938); its relevance to studies of the mechanism of chromosome duplication is discussed in Chapter 10.

Metaphase arrest results from the inhibition of spindle fiber formation with the consequent loss of the metaphase alignment of the chromosomes. The chromosomes in the metaphase-arrested state may become disorganized or may take up characteristic patterns, such as the starlike arrangement in which all the centromeres lie at one focus. The effect of colchicine is reversible; when cells that have been treated with the drug are resuspended in medium lacking it, spindle fibers reform (Brinkley, Stubblefield and Hsu, 1967; Rizzoni and Palitti, 1973). This implies that colchicine causes the microtubules to disaggregate into free subunits that can repolymerize when the drug is removed. The action of colchicine appears to result from a poisoning effect, in which the tubulin dimers that have bound colchicine then prevent further dimers from adding to the growing ends of microtubules; this leaves the microtubules able only to disassemble at their other ends (Taylor, 1965; Olmsted and Borisy, 1973a; Wilson, Creswell and Chin, 1975; Margolis and Wilson, 1977, 1978). We have discussed the action of colchicine, and of the other alkaloid inhibitors of mitosis, podophyllotoxin and vinblastine, in Chapter 2.

The existence of a dynamic equilibrium between microtubules and tubulin is supported by the success of attempts to isolate the mitotic apparatus by the use of buffer medium containing tubulin. Early attempts to isolate the mitotic apparatus were based upon the concept of stabilizing it by lysing cells into a medium providing only a poor solvent for protein (see above). But although the spindle isolated by such means may maintain its birefringence for some hours, it is not capable otherwise of reproducing its behavior in the cell. For example, it does not respond to changes in temperature and pressure with the expected changes in birefringence (see Forer and Zimmerman, 1976a,b). More recently, however, the mitotic apparatus has been isolated in a functional state by lysing cells into medium containing tubulin (Cande et al., 1974; Inoué, Borisy, and Kiehart, 1974; Rebhun et al., 1974; Snyder and McIntosh, 1975).

At a sufficient concentration of tubulin, generally of the order of 3×10^{-5} M, the mitotic apparatus of cultured mammalian cells or marine oocytes remains stable for at least an hour and exhibits changes in birefringence upon cooling and warming. In the PtK1 rat kangaroo cells, for example, this concentration of tubulin maintains the usual number of about 2000 spindle microtubules; this increases to about 5000 tubules at 7×10^{-5} M tubulin but decreases to some 500 tubules at 10^{-5} M tubulin. Most of the change is in nonkinetochore microtubules, which therefore seem to be more sensitive to tubulin concentration, that is, are less inherently stable. At lower concentrations of tubulin, the isolated mitotic apparatus is not stable but disintegrates. Treatments that prevent the polymerization of tubulin—addition of 5 mM $CaCl_2$ or of colchicine, or removal of GTP—prevent the added tubulin from stabilizing the spindle, which then disintegrates as though no tubulin had been added.

Two features of these experiments are especially interesting. The first is that the tubulin need not be from the same source as the mitotic apparatus; for example, chick or mammalian brain tubulin seems to be equally effective in stabilizing the mitotic apparatus of marine organisms. Tubulin originally derived presumably from neurotubes, and of a different species, therefore seems interchangeable with the subunits of various mitotic microtubules. The second is that in some of these experiments the mitotic spindle was able to elongate but not to contract. This suggests that tubulin can be added to microtubules in the in vitro systems but cannot be removed. This is consistent with a model in which the mitotic apparatus provides nucleating centers for microtubule extension but does not maintain all its usual regulatory mechanisms. One further point is that in some control experiments anaphase cells were lysed into buffer containing glycerol and nucleotides; chromosome movement then was able to continue for a short time, in spite of the absence of added tubulin, until the apparatus began to decay. If it is true that the mitotic apparatus isolated in this way does not retain a pool of free tubulin, this implies that anaphase movement does not require the addition of subunits to the microtubules.

One of the most important questions about mitosis is how the assembly and disassembly of microtubules is controlled. A feature of the equilibrium between labile microtubules and tubulin that may be significant in this context is the occurrence of assembly and disassembly at opposite ends of the tubule (see Chapter 2). If labile microtubules are "treadmills" in which tubulin is constantly migrating from one end to the other, localized actions on the appropriate end may result in net assembly or disassembly. A protein that may be involved in the process is calmodulin, the calcium dependent regulator protein, which is able to stimulate the disaggregation of microtubules in vitro in response to increases in the calcium level (Marcum et al., 1978). This protein is a low molecular weight species with a variety of properties that may be pertinent to the function of filamentous systems in the cell skeleton. It is an abundant protein that regulates several enzymatic activities, including cyclic nucleotide phosphodiesterase and actomyosin ATPase activity; its sequence is related to troponin C, a calcium-dependent regulator of muscle, although the two proteins are quite distinct.[*] Immuno-fluorescence studies show that calmodulin is present in the interphase cell in a pattern identical to the stress fibers; and in mitotic cells it is located specifically in the area of the spindle between the chromosomes and the poles (Welsh et al., 1978, 1979). This raises the possibility that it may be involved in regulating either microtubules or actin filaments located in this region.

Microtubules and Actin Filaments in Chromosome Movement

Changes in the lengths of the spindle microtubules must accompany both the elongation of the spindle and the movement of chromosomes. One view is that microtubules in fact are responsible for both processes. According to the model of Inoué and Sato (1967), they do not generate but do transmit a motive force that acts upon the chromosome kinetochores. The force is generated with the removal of subunits from the kinetochore fibers that link the chromosomes to the poles. Of course, other models also have been proposed for microtubule function. The sliding filament model supposes that, rather than moving coordinately to the poles, adjacent microtubules may slide past each other in an action dependent on cross bridges (McIntosh, Hepler, and Van Wie, 1969). An alternative view is to suppose that the pulling force has an origin independent of the microtubules. The possibility that this might lie in contractile filaments

[*]References: Teo and Wang (1973), Stevens et al. (1976), Dedman et al. (1977a,b; 1978).

is raised by recent demonstrations that actin and myosin are present in the mitotic spindle.

Formation of the spindle demands a reorganization of previously existing structures rather than a de novo synthesis of new components. During interphase the microtubules constitute a cytoplasmic network, whereas during mitosis they are organized in the form of spindle fibers. In both conditions they are labile and therefore may lie in an equilibrium with a pool of tubulin. What is responsible for effecting the cyclic changes in microtubule organization with each cell cycle? At the beginning of mitosis, spindle fibers can be seen to grow from polar regions close to the centrioles. Does this mean that the change is mediated by a shift in the equilibrium, so that at the beginning and end of each mitosis the microtubules disaggregate into subunits that then repolymerize from new nucleating centers? Or may some microtubules remain intact, but change in orientation within the cell? Similar questions may be asked about the relationship between the actin filaments of the interphase cell and those of the dividing cell.

Indirect immunofluorescence with antibody against tubulin shows that the cytoplasmic network of microtubules disappears during prophase. By metaphase the fluorescence has become localized to the spindle (in PtKl cells). The development of the spindle during prometaphase starts with a bright spot of fluorescence surrounded by the chromosomes; this then gives way to a picture in which the poles are highly fluorescent and both kinetochore and continuous fibers run along the spindle axis. By telophase the fluorescence has decreased appreciably (Fuller, Brinkley, and Boughter, 1975; Weber, Bibring, and Osborn, 1975). Figure 4.4 shows mitosis in HeLa cells visualized with antibody directed against tubulin. These results are consistent with the idea that the cytoplasmic microtubules disintegrate during prophase so that spindle fibers may be regenerated from the subunits that are released.

From these results and from previous analyses of the spindle fibers it is clear that microtubules connect the poles together and connect the chromosomes to the poles. However, actin filaments also connect the chromosomes to the poles. There are two difficulties in visualizing the contractile elements during mitosis. One is that the rounding up of cells during division makes it impossible to view the structure in one plane. The second is that only some of the contractile elements may be organized as part of the spindle, the rest providing an amorphous background that makes visualization difficult.

Cells of the PtK2 rat kangaroo line may not round up completely during mitosis, but instead remain partially flattened on the dish. This enabled Sanger (1975c) to apply a fluorescent HMM preparation that reacts with actin to cells in which the positions of the chromosomes can be seen and the stages of division identified. At the beginning of prophase the long cytoplasmic actin fibers are replaced by a diffuse fluorescence over the cytoplasm. At metaphase fluorescent fibers can be seen to join the chromosomes

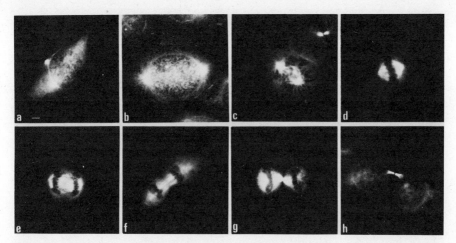

Figure 4.4. HeLa cells stained with tubulin antibody during mitosis. (a, b) During prophase the nuclear membrane is present and the fluorescence is localized in the region of the centrioles. (c) At prometaphase the nuclear membrane disappears and the fluorescence radiates from aster-like structures; chromosomes are seen as compact black bodies. (d) At metaphase the chromosomes can be seen on the equatorial plane of the fluorescent spindle. (e) At anaphase the chromosomes move to the poles. (f, g) During telophase two daughter nuclei are reconstituted. (h) Daughter cells are moving apart, but still connected by microtubules. Data of Weber, Bibring, and Osborn (1975).

to the polar regions. Some 5–9 fibers can be seen to run to each pole; they appear to be some 200 nm in diameter and some 7 μm in length. The lengths of these fibers shorten as the chromosomes move to the poles at anaphase. Similar results have been obtained with rat embryo and with PtK1 cells (Schloss, Milsted, and Goldman, 1977). Fibers lying between the separating chromosomes as well as between the chromosomes and the poles have been identified in HeLa and PtK2 cells by Herrman and Pollard (1978, 1979). The reason for the discrepancy is unknown.

Actin filaments also can be visualized at mitosis by indirect immuno-fluorescence. The results depend upon the method by which the cells are prepared. Cande, Lazarides, and McIntosh (1977) found that when PtK1 cells are fixed and then visualized with an antiactin preparation, interphase cells display the usual filaments but metaphase cells show the spindle as a bright object against a bright background. No details can be seen in these mitotic cells. But if the cells are lysed into a buffer containing tubulin before fixation, the general bright fluorescence is lost and it is possible to see a fluorescent halo at the poles with fibers running along the spindle. This suggests that the mitotic cell contains actin in an amorphous state, which is responsible for the general fluorescence of unlysed cells. Lysis allows this actin to be lost, thus revealing the smaller amount present in the form of fibers.

The actin fibers develop with the spindle. At prophase, the cytoplasmic

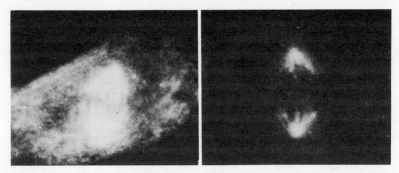

Figure 4.5. PtK1 cells visualized by indirect immunofluorescence with antibody against actin during metaphase (left) and anaphase (right).   Data of Cande, Lazarides, and McIntosh (1977).

actin filaments disappear and the level of amorphous staining in the cyto-plasm increases. The polar region around the centrioles begins to stain intensely for actin at the time of nuclear breakdown; this increases during prometaphase and extends into fibers by the time of metaphase. The num-ber of actin-containing fibers is small. The maximum seen in these con-ditions was 6 per half spindle, but since the fibers are located in different focal planes the total number is probably greater. It is possible that there is one such fiber per kinetochore. Examples of cells stained with the anti-actin preparation in metaphase and anaphase are shown in Figure 4.5.

The use of fluorescent techniques does not reveal how many actin fila-ments there may be in each of these fibers. But the low number of fibers and their apparently small diameter, and the presence of large amounts of amorphous actin in the mitotic cell, suggest that only part of the total actin of the cell becomes associated with the spindle in fibrous form. The amorphous actin presumably is generated by disaggregation of the inter-phase microfilaments. The growth of actin fibers seen on the spindle by the fluorescent techniques is consistent with the idea that all the micro-filaments are disaggregated, with some of the freed subunits repolymerizing into fibers on the spindle.

If cells are treated with colcemid before lysis, to prevent the development of the spindle, no fluorescent fibers are seen in the spindle region but the poles still fluoresce. This suggests that the development of the actin fibers depends upon the construction of the spindle from microtubules. Consis-tent with this view is the observation that the actin fibers in the spindle do not run in a straight line from pole to kinetochore, but follow the curved path of the microtubules.

These results do not formally exclude the possibility that cytoplasmic actin has been trapped in the spindle. But as the cells proceed into anaphase the lengths of the fibers shorten, consistent with the idea that they do con-nect the kinetochores to the poles. Unlike microtubules, none of the fibers crosses the interzone, again consistent with the view that they play a spe-cific role in chromosome movement but are not involved in spindle elongation. Late in anaphase, amorphous staining is seen in the interzone and by telophase only the cleavage furrow fluoresces with anti actin antibody. This suggests that all the actin of the cell participates in the formation of the contractile ring responsible for cytokinesis (see below).

Are the actin fibers of the spindle associated with other proteins of the contractile network? Fujiwara and Pollard (1976, 1978) have investigated the presence of myosin, using a directly fluorescent antibody that reacts with the rod part of the molecule. When flat interphase HeLa cells round up in prophase, the cytoplasm stains uniformly, while the nucleus is unstained. When the nuclear membrane breaks down, the chromosomes can be seen as dark objects against a brightly stained background. In metaphase and anaphase there is an intense staining between the chromosomes and the poles, which seems to represent a concentration of myosin, although individual fibers cannot be resolved. By late mitosis the cleavage furrow becomes stained. This shows that myosin is present in the part of the spindle where the actin fibers are located; although this does not prove the existence of a contractile interaction, certainly it renders plausible the idea that actomyosin contraction may occur during mitosis.

The idea that contractile elements might be responsible for chromosome movement is not new. Such proposals have been based on observations that under certain circumstances chromosome movement can be divorced from the activity of spindle fibers (see Forer, 1966; Forer and Behnke, 1972a; Rickards, 1975). Earlier observations also have been made of the presence of microfilaments in the spindle (Bajer and Molé-Bajer, 1969; McIntosh and Landis, 1971; Muller, 1972). But much more force is lent to the idea that actomyosin contraction is involved in chromosome movement by the demonstrations that actin fibers can be visualized in the locations to be expected if they connect kinetochores to the poles. Their possible presence elsewhere also must be explained.

If the actin filaments are to provide some or all of the force for chromosome movement, the fibers must be connected to each kinetochore and must be anchored at the poles. Nothing is known about either the kinetochore attachment or the constitution of any polar structure in which they may terminate. The nature of any interaction between these fibers and myosin or any other contractile proteins also remains to be characterized. The role of actomyosin-mediated contraction in mitotic chromosome movement remains an open question.

The attachment of kinetochores to microtubules that terminate close to or at the centrioles in the polar regions is, of course, well established. This means that even if the actin fibers rather than the microtubules are responsible for pulling the chromosomes, still it is necessary for the microtubules, which constitute a comparatively rigid structure, to shorten as the chromosomes move. Thus it is possible that actin filaments provide the motive force and that this is controlled by the ability of the microtubules to be shortened by disassembly of subunits. This would allow the elongation of the spindle and the movement of chromosomes at anaphase to be responsive to the same system. Whether or not the microtubules provide any of the force that acts on the chromosomes, the absence of any pole to pole actin fibers means that the continuous microtubules alone connect the opposite poles. This means that the microtubules alone must be responsible for elongation of the spindle. These continuous microtubules must increase in length during anaphase, unlike the kinetochore microtubules which must become shorter. The nature of the distinction between these two sets of microtubules is not clear; but the difference in behavior demanded of them implies that more than a simple change in equilibrium between assembled and free subunits must be involved (see below).

Structure of the Centriole

The characteristics of the metaphase plate vary with the type of cell. But all mitotic cells contain two spindle poles, the regions toward which the anaphase chromosomes migrate. Each pole of the spindle in animal cells typically is identified by the presence of a pair of centrioles, occasionally only a single centriole. Centrioles appear to be absent from higher plant cells but may be present in mitotic cells of some lower plants. The centriole together with its surrounding (pericentriolar) material, which sometimes appears as a halo of dense granules, is known as the *centrosome*. The *centrosphere* describes the region around the centrosome, including microtubules and membrane material. The *aster* includes the centrosphere and the outlying radial arrangement of microtubules, membranes, rows of mitochondria, and other components. The astral rays often are a predominant feature in light microscope studies of mitotic cells (though not in mammals).

The centriole is a small hollow cylinder about 0.5–0.7 μm in length and some 0.25 μm in diameter. The wall of the cylinder consists of nine triplet microtubules which give a pinwheel appearance when viewed in cross section (Stubblefield and Brinkley, 1967). Figure 4.6 shows that each of the nine triplets consists of three fused microtubules; only the inner microtubule of each triplet is strictly circular (Ross, 1968). The three microtubules of each triplet lie almost in one plane, so that each triplet forms a blade that runs the full length of the centriole.

The same, or a very similar, triplet structure is found in the basal bodies (also sometimes described as kinetosomes) of protozoan cilia and flagella (see Chapter 2). A functional relationship between the basal body and the centriole is indicated by the ability of basal bodies from Chlamydomonas or Tetrahymena to sponsor aster formation when injected into Xenopus eggs (see below). The structure of the microtubules of the centriole is not yet entirely resolved, principally because of the difficulty in purifying centrioles. However, it is clear that flagella extend from basal bodies by a

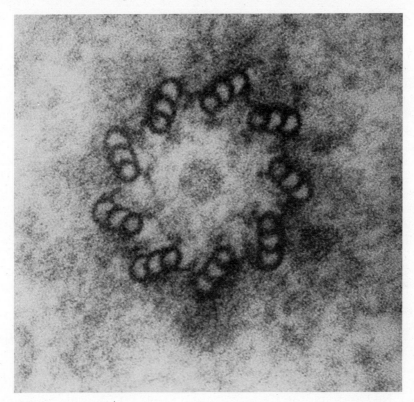

Figure 4.6. Cross section through the centriole of a human lymphocyte. ×240,000. Data of Ross (1968).

transition from the triplet structure into the doublets and singlets of the 9 + 2 structure (see Chapter 2). The structural relationship between basal bodies and flagella suggests that both are constituted from tubulin; and this is borne out by the observation that isolated basal bodies contain α and β tubulin (Gould, 1975). It seems likely that centrioles have the same composition.

It has been thought for some time that basal bodies contain nucleic acid; and, by inference, this has been taken to be a component also of the centriole (see DuPraw, 1970). This idea originated in the report of Randall and Disbrey (1965) that the basal bodies of Tetrahymena contain DNA. Subsequent experiments confirmed that isolated pellicles of Paramecium (a preparation containing basal bodies and some associated structures) also contain DNA (Smith-Sonneborn and Plaut, 1967, 1969).

Attempts to characterize the DNA associated with pellicles of Tetrahymena suggested that it does not correspond to the rather restricted set of sequences that would be expected if it were a specific component (Flavell and Jones, 1971). This makes it likely that the DNA is a contaminant, apparently derived from both the mitochondrial and nuclear compartments.

The amount of DNA apparently associated with the basal body is so small that it is very difficult to provide a definite answer on its origin by experiments on its structure. Another approach is to determine the susceptibility of basal body structure or function to treatments that destroy DNA. While the failure of such treatments to have an effect is not conclusive (the DNA might be inaccessible), a loss of structure or function would provide good evidence for a role for the nucleic acid. However, DNAase has no visible effect upon Paramecium basal bodies, while RNAase appears sometimes to have an effect and pronase destroys them (Dippell, 1976). The presence of RNA in Tetrahymena pellicles has been suggested by cytochemical analysis of basal bodies isolated by a method yielding purer preparations than were available previously. Acridine orange generates the yellow-green fluorescence characteristic of RNA (the dye would display a different color reaction with DNA); and this is destroyed by treatment with RNAase but not with DNAase (Hartman, Puma, and Gurney, 1974). Most of the RNA appears to be of ribosomal origin, which again opens the question of whether it is a genuine component of the basal body or an artefact of isolation.

The problems inherent in excluding the possibility of contamination mean that the only satisfactory proof that a nucleic acid is part of the basal body or centriole must be the characterization of the role it plays in the structure or a demonstration that its presence is necessary for proper function. No structural role has been ascribed to any nucleic acid; and nor has it been possible to obtain any specific preparation whose sequence might be investigated. But the ability of basal bodies to generate asters upon injection into Xenopus eggs provides a system in which function can be assayed. Heidemann, Sander, and Kirschner (1977) have demonstrated that RNAase abolishes the activity of Chlamydomonas or Tetrahymena basal bodies, whereas DNAase has no effect (see below). This suggests that RNA is a genuine component of the basal body. Labeling with $^{32}PO_4$ suggests that each basal body contains about 3×10^8 daltons of RNAase sensitive material. (This is about twice the amount of DNA previously estimated to be present.)

The number of centrioles doubles once during the cell cycle, from the two gained at the preceding mitosis to give the four that will be divided into two sets at the next mitosis. Daughter centrioles first become visible as short cylinders orientated perpendicularly to the parental centrioles and close to their proximal ends. These *procentrioles* are similar to, but only about half the length of, the mature centrioles. A similar form of reproduction is displayed by basal bodies of Chlamydomonas. Basal body pairs can be isolated in which the bodies are perpendicular to each other, each still attached to the flagellar axoneme. Reproduction takes place by the extension of a probasal body, which first becomes evident as a structure attached by two fibers to the proximal end of the basal body whose partner it will be after cell division (Gould, 1975).

Centriole reproduction commences at a characteristic point of the cell cycle, with HeLa cells at about the time that DNA synthesis begins, with L cells somewhat earlier, during G1 phase (Robbins, Jentzsch, and Micali, 1968; Rattner and Phillips, 1973). Inhibition of protein synthesis with cycloheximide prevents the appearance of the procentrioles, which must therefore depend upon the synthesis of new protein; but cycloheximide does not prevent the subsequent extension of the procentrioles into the mature centrioles (Phillips and Rattner, 1976). This implies that sufficient protein has been synthesized by the time of appearance of the procentriole; its extension into the mature centriole is therefore subject to some control other than provision of subunits.

Figure 4.7 shows the reproduction of one centriole during the L cell cycle. During interphase, each centriole becomes associated with a procentriole. At prophase, the mitotic cell therefore possesses two centriole-procentriole pairs. One pair becomes established at each pole, although this may not necessarily precede nuclear breakdown, but may take place at any point in the midprophase-prometaphase period (Roos, 1973; Rattner and Berns, 1976). Migration of the centrioles to the poles is (reversibly) prevented by colcemid (Brinkley, Stubblefield, and Hsu, 1967). It seems that only the parental centriole of each pair participates in the mitosis; the daughter procentriole plays no part. When the nucleus reforms about each chromosome set at telophase, a centriole pair is present in the surrounding cyto-

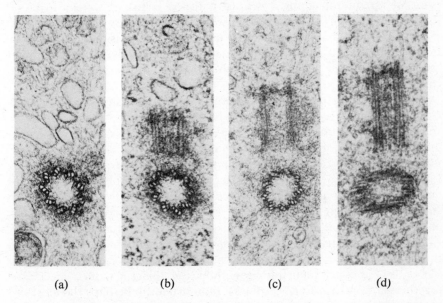

(a)   (b)   (c)   (d)

Figure 4.7. Centriole reproduction during the L cell cycle. (a) Early G_1; (b) S phase; (c) prophase; (d) metaphase. ×60,590. Data of Rattner and Phillips (1973).

plasm. During mitosis, or early in the succeeding division cycle, the procentriole is extended into a mature assembly. What is responsible for the association of centriole with procentriole, and how this is released so that parental and daughter centrioles may separate, is quite unknown.

The nature of the events involved in the reproduction of centrioles has been a vexed issue for some time. The idea that centrioles are reproduced by self replication, rather than by assembly de novo, originated with these observations that a daughter centriole appears to be assembled alongside the parental centriole (see Brinkley and Stubblefield, 1970). But while this does indeed suggest a connection between construction of the procentriole and the parental centriole, no changes can be visualized in the parental structure to support the idea that the centriole actually duplicates and then divides. This has led to suggestions that some structural information necessary for assembly might be reproduced by a templatelike process which relies upon the structure of the pre existing parental centriole. This might take the form, for example, of duplication of part of the parental structure to generate a nucleating center that is transported to the site where the daughter centriole is to be assembled. A more plausible proposal is that some part of the parental centriole is extended to act as a nucleating center. But in contrast with these ideas, there is evidence in several cell types that centrioles may be assembled de novo in the absence of pre existing centrioles (Turner, 1968; DuPraw, 1970; Miki-Noumura, 1977). This suggests that assembly may be analogous to the processes displayed by other macromolecular structures, such as the ribosome, in which the components are capable of assembling along a defined pathway. While it seems, therefore, that the more fanciful concepts of self-reproduction can be dismissed, it remains possible that in some cells a nucleating center is provided by some function dependent on either the structure or some activity of the parental centriole, although in other cells this may be the product of de novo construction.

Although the predominant component of the centriole is the triplet microtubule, many questions remain to be answered about both the construction of the microtubule itself and the nature of the other components that maintain the relationship between the nine triplets. The assembly of singlet microtubules now can be initiated and elongated in vitro (see Chapter 2), but nothing is known about the interactions responsible for assembly of doublet and triplet microtubules. Since there appears to be a pool of tubulin subunits in the cell (in equilibrium with the labile cytoplasmic network or spindle fibers), it does not seem likely that availability of tubulin controls centriole synthesis. Questions of obvious importance, in addition to that of what controls the initiation event, are what limits the length to which the procentriole and centriole grow, how the adjacent triplets are linked together, and what role RNA plays in the structure. Perhaps the most important question, however, is that of the relationship between the centriole and the spindle fibers.

When a centriole pair is present at a pole, it often lies in a plane approximately perpendicular to the spindle axis, although the axis of the centrioles does not appear to be related per se to the axis of the spindle. The exact position of the centrioles varies in cells of different species. But in all cells with centrioles, the polarity of the spindle is determined either by them or by material associated with them. In cells that lack centrioles, the spindle fibers must presumably interact with some other component. However, the ubiquity of the spindle as a mechanism for chromosome segregation, the universality of microtubules on the spindle, and the generally common features seen in all mitoses, suggest that all eucaryotic cells use the same (or very similar) devices for establishing the poles of the spindle; apparent variations in morphology probably result from differences in cellular organization and need not be taken to reflect any fundamentally different mechanisms of division.

Microtubule Organizing Centers

The basic feature of all spindles is the presence of microtubules spreading out from a focus at each polar region. Some microtubules run from pole to pole, some are connected to the kinetochores, and others may have free ends. A critical and controversial question is the nature of the structures from which the microtubules elongate. That there must be such *microtubule organizing centers* (MTOC) is implied by cytological observations of the growth of spindle fibers and also is consistent with the characteristics of microtubule assembly in vitro (see Chapter 2). These centers may be involved in terminating as well as initiating the growth of microtubules. Continuous fibers presumably are initiated at one pole and terminated at the other. Whether kinetochore fibers are initiated at the poles, or at the kinetochores, or both, has been a subject of debate for some time. How a growing microtubule finds its proper site of termination, and how termination at the MTOC may be related to initiation, remain to be established. And, of course, it is not necessary to emphasize that the relationship between the MTOC and the microtubule is dynamic: initially microtubules grow from the MTOC, but this process does not cease with establishment of the spindle, since continuous fibers are elongated and kinetochore fibers are shortened during the later stages of mitosis.

There appear to be two types of microtubule organizing center in the cell, located at the kinetochore and centriole. This concept allows the difference between the behavior demanded of the continuous and kinetochore fibers to be explained by the activity of the kinetochore in supporting microtubule growth. Luykx (1970) has suggested that throughout metaphase subunits continue to be added to the microtubules at the sister kinetochores, thus relieving the chromosomes from any pulling force exerted by the spindle fibers. In their reviews of chromosome movements at mitosis, Bajer and Molé-Bajer (1970, 1972) have suggested that the kinetochore fibers stop growing at anaphase, with the result that the chromosomes are able to move toward the poles (either because the lack of growth creates a motive force or because it allows contractile fibers to exercise their action without restraint by the microtubules). Thus during anaphase the continuous fibers may continue to be extended at the poles, causing the elongation of the spindle, while kinetochore fibers cease growth and then (presumably) shorten in length. Although it is possible that the continuous and kineto-chore fibers are distinguished at the poles, so that one may lengthen while the other shortens, it seems more plausible to suppose that their characteristic behaviors may depend on their origins. Perhaps the simplest model along these lines is to suppose that kinetochore fibers are initiated at the kinetochore and terminate at the poles, whereas continuous fibers are initiated (and subsequently elongated) at one pole, terminating at the other.

Early cytological observations that microtubules appear to grow from the kinetochores have been supported by more recent demonstrations that addition of tubulin can take place at kinetochores in vitro. When cultured mammalian cells are lysed into medium containing tubulin, the spindle may be maintained (see above). When the cells have been treated first with colcemid to cause metaphase arrest, the growth of microtubules by addition of tubulin can be followed in the in vitro system. Microtubules are extended from both the poles and the kinetochores, although the capacity of the kinetochores (judged by the length of the assembled tubules) is somewhat lower than that of the poles (McGill and Brinkley, 1975; Snyder and McIntosh, 1975). An increase in the capacity of the polar regions to initiate microtubule formation is seen as cells progress from prophase to metaphase. When HeLa cell mitotic chromosomes are isolated by gentle centrifugation, and incubated with chick brain tubulin, microtubules are extended from the kinetochores (Telzer, Moses, and Rosenbaum, 1975). An example of this growth is seen in Figure 4.8. From these results it seems clear that kineto-chores can act as microtubule-organizing centers, but it appears that their capacity is much lower than that of the polar organizing centers. Whether this capacity is responsible for the growth of all or only some of the kinet-ochore microtubules remains to be established.

The source of the microtubules that appear to extend from the poles has been highly controversial. The tubules clearly originate from the general region of the centriole, but there has been intensive discussion on whether they are promoted by the centriole itself or by some component of the peri-centriolar material. Spindle fibers sometimes appear to be attached to the base of the centriole, running perpendicular to its long axis. (This contrasts with the relationship between the basal body and the flagellar axoneme, in which the flagellar microtubules appear to be direct extensions of those in the basal body, and therefore are extended parallel to the long axis.) Sometimes, however, spindle fibers do not seem to be attached to the centriole but apparently terminate in the amorphous material of the pericentriolar

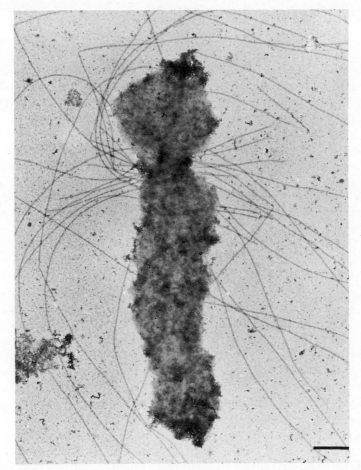

Figure 4.8. Metaphase HeLa chromosome isolated after incubation in tubulin. Microtubules extend from the kinetochores. The bar is 1 μm. Data of Telzer, Moses, and Rosenbaum (1975).

region. It is difficult to know whether these differences are real or reflect artefacts of fixation techniques. Mitosis and meiosis also have been observed in cells apparently lacking centrioles. The best examples are higher plants; but the lack of centrioles, or the termination of spindle fibers at foci well distant from them, as well as experiments suggesting that centrioles can be damaged without impeding mitosis, have been reported in other species also.*

There seem to be two critical questions about the nature of the polar microtubule-organizing centers. In cells possessing centrioles is it the cen-

*References: Pickett-Heaps (1969, 1971), Lovlie and Braten (1970), Braten and Nordby (1973), Szollosi, Calarco, and Donahue (1972), Berns et al. (1977), Brenner et al. (1977).

triole itself that is responsible for sponsoring microtubule growth? And in cells lacking centrioles, what is the form of the MTOC and how is it related to the MTOC of cells possessing centrioles? It is clear, of course, that mere possession of a centriole is not sufficient, since spindle fibers start to grow only at the beginning of mitosis and the initiating capacity then increases until metaphase (see above). An activation event must therefore take place at this time. It seems unlikely that all microtubules could attach directly to the centriole (since there are of the order of 1000 at the pole); this discrepancy may mean that the function of the centriole as an MTOC is more complex than simply providing a structure for microtubule attachment. One final point in this context is that cytoplasmic microtubules also appear to originate from an organizing center; the nature of this center, and its relationship with the mitotic MTOC, is not clear (see Chapter 2).

The best evidence that the centriole functions as a microtubule-organizing center is provided not by the ambiguous direct studies of mitotic cells, but by the ability of isolated basal bodies to sponsor aster formation when injected into Xenopus eggs. The ability of an injected microtubule nucleating center to promote division in amphibian eggs was first implied by nineteenth century developmental studies, which suggested that division depends upon the use of the centriole brought in by the sperm. Subsequent studies of marine species, including the sea urchin, led to the development of methods to induce activation, which showed that these eggs must possess endogenous nucleating centers. But frog egg development always fails at the formation of the first cleavage spindle, unless a "cleavage-initiating agent" is present. Usually this requirement is provided by pricking with an activating needle that has been dipped in blood.

The need for activation in some species and for a cleavage-initiating agent in others makes it possible to assay cellular components for their abilities to provide nucleating centers. One approach is to attempt to identify the MTOC in vitro. Activated surf clam eggs contain a component able to provide a nucleating center for tubulin polymerization in vitro; this is absent from eggs that have not been activated. At present the active component is available only in the form of a crude homogenate or low-speed supernatant (Weisenberg and Rosenfeld, 1975). Another approach is to use the Xenopus egg as an assay system. Material extracted from brain or from sperm of either Xenopus or L. pictus allows Xenopus egg development to proceed to gastrulation (Maller et al., 1976). When fractions were prepared from the sperm, only those carrying centrioles were active. Cytoplasmic preparations from mouse, chick or hamster brain support aster formation in Xenopus eggs, and cleavage furrows then develop (Heidemann and Kirschner, 1976). This demonstrates very strikingly the lack of species specificity in the MTOC. Basal bodies purified from Chlamydomonas or Tetrahymena are effective in inducing aster formation. An example is shown in Figure 4.9. Although it is difficult to exclude the possibility of contamination with other (active) components, the structural similarity between basal body and centriole at once suggests that it is some feature common to both

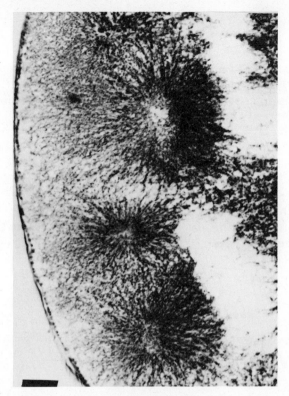

Figure 4.9. Asters induced by injecting basal bodies into a Xenopus egg. The bar is 30 μm. Data of Heidemann, Sander, and Kirschner (1977).

that is responsible for aster formation. And it may indeed be that the two structures are interchangeable.

When the isolated basal bodies are treated with RNAase, they lose their ability to promote aster formation (Heidemann, Sander, and Kirschner, 1977). However, treatment with RNAase does not prevent the basal body from acting as a nucleating center for the extension of microtubules in vitro. This suggests that RNA is a component of the basal body, and by inference of the centriole; but while it is necessary for aster formation, it is not needed for direct extension of microtubules.* This leaves open the question of whether its role is structural, or (perhaps unlikely) one of acting as a template; one possibility is that the role of the RNA is concerned with establishing the centrospheric cloud, from which the tubules might emanate.

The ability of components of the centrosome to initiate microtubule formation has been observed in vitro. Gould and Borisy (1977) reported

---

*This conclusion leaves open the question of whether RNA may be expected to be a component of the other type of MTOC, i.e. the kinetochore. The presence of ribonucleoprotein at newt kinetochores has been reported on the basis of staining experiments (Rieder, 1978). However, no functional test has yet been developed to determine the effect of removing the RNA with ribonuclease, so whether its location is fortuitous or significant remains to be seen.

that centrosomes of CHO cells arrested in metaphase with colcemid and then lysed in Triton X-100 and EGTA can act as nucleating centers when tubulin is added. Large numbers of microtubules elongate from the centrosomes; 100–250 microtubules per centrosome can be found upon incubation in 2×10^{-5} M tubulin. The omission of EGTA from the lysis medium causes the centrioles to become separated from the pericentriolar material. In these conditions, a comparatively small number of microtubules is elongated from the centrioles, preferentially from the end distal from the daughter procentriole. Patches of isolated pericentriolar material also are able to act as centers for elongating microtubules; the origin of these centers, in particular their relationship with the centriole, remains to be established. Similar observations have been reported by Telzer and Rosenbaum (1979), who found that most microtubules extending from isolated centrosomes extend from the pericentriolar material; only a few are elongated from the centrioles themselves, as direct extensions of the microtubules comprising the triplet blades. The ability of the pericentriolar material to nucleate microtubule assembly appears to be acquired only with mitosis; fractions isolated from interphase cells are inactive. Thus the development of the mitotic spindle may depend upon the acquisition of ability to sponsor microtubule assembly from these sites. The spindle pole bodies of S. cerevisiae, which lack centrioles, also are able to act as nucleating centers in vitro (Hyams and Borisy, 1978).

Cytokinesis

Cell division is completed by the separation of the two daughter cells, a process that is accomplished by the development of a *cleavage furrow* between the two nuclei as they reform at telophase. Early discussions on the nature of cytokinesis centered on two views. One was that the furrow is *pushed* inward by a growth or expansion of the cortical protoplasm (see Chambers, 1938; Swann, 1952; Mitchison, 1952). The other was that the furrow is *pulled* inwards by a contraction of the cortical protoplasm (see Marsland, 1938, 1950; Lewis, 1938, 1951). This model was put into its modern form by Marsland and Landau (1954), who proposed that development of the furrow is dictated by the structural state—that is, the contractile capacity—of the gelated cortical protoplasm in the furrow regions. This structure, represented by the presence of contractile material in the cleavage furrow, has since been described as the *cortical ring*.

The mitotic apparatus has been known for a long time to control the position of the cleavage furrow in dividing eggs; that is, the position of the spindle-aster complex dictates the position of the furrow (see Wilson, 1925). Early studies showed that this interaction is completed at a specific time; before this point in division, centrifugation or microsurgery to alter the location of the spindle and asters in turn changes the position of the furrow; but after this point, such treatment has no effect (see, for example, Harvey, 1935; Hiramoto, 1956, 1965, 1968; Dan and Kojima, 1963). Thus once cleavage has been initiated it proceeds independently of the mitotic apparatus (for review see Rapaport, 1971). It is possible to induce eggs to cleave before the mitotic apparatus appears, by the combination of centrifugation and hydrostatic pressure (Marsland, Zimmerman, and Auclair, 1960). This implies that the necessary components are present before mitosis and need only to be activated, not synthesized.

The first information about the nature of the contractile ring was provided by the observation of a system of filaments embedded in the dense cortex of furrow walls (see Mercer and Wolpert, 1958; Tilney and Marsland, 1969; Selman and Perry, 1970). In the HeLa cell an equatorial arrangement of fine filaments could be seen beneath the plasma membrane of the cleavage furrow; these are not present at metaphase (Schroeder, 1970). This appears to correspond to the contractile ring; it is about 10 μm around and 0.15 μm thick beneath the membrane. Development of the filament system is prevented by cytochalasin B, which suggests that it may represent microfilaments. In amphibian eggs, however, the contractile ring is not disrupted by cytochalasin B; the reason for this difference is not clear. The microfilaments of the contractile ring have the dimensions usually seen in actin filaments and bind HMM (Schroeder, 1973).

When actin filaments are visualized in dividing cells by reaction with HMM or specific antibodies, the usual array of cytoplasmic microfilaments is replaced by a diffuse fluorescence amongst which actin filaments can be seen (see above). With fluorescent HMM, the actin appears to become localized in the developing cleavage furrow by telophase (Sanger, 1975a). After the completion of division, there is a decrease in staining of the furrow and an increase at the poles (where the pseudopods form that enable the cells to move apart). However, when antiactin is used to visualize the location of actin, a diffuse fluorescence is seen over the entire dividing cell during cytokinesis (Herman and Pollard, 1979). Although actin filaments undoubtedly are present in the contractile ring, this suggests that their formation may depend on a reorganization of actin previously present in this region rather than a recruitment of all the cellular actin from its previous locations. Generally these results support the earlier analysis of Forer and Behnke (1972a,b), in which actin filaments in the meiotic spindle of the crane fly were visualized with HMM. At stages up to anaphase, the filaments take the form of orientated bundles lying beneath the plasma membrane; but then they become arranged in an equatorial array that develops into the cleavage furrow. This suggests that it is actin filaments present in the cell before division that are reorganized to form the cleavage furrow. The conclusion that a contractile system is involved in cleavage is supported further by the demonstration that myosin also is found in the cleavage furrow (Fujiwara and Pollard, 1976, 1978).

CHAPTER 5

Chromosome Segregation: Meiosis

Meiosis comprises a special form of nuclear division in which the progeny cells contain only a haploid complement of chromosomes instead of the diploid set present in all somatic cells. This reduction of the chromosome complement ensures that each germ cell gains only one allele representing each genetic locus. There is, of course, an equal probability that any given germ cell will gain the allele of maternal or paternal origin. Reassortment of genes carried on different chromosomes is accomplished by the independent segregation of nonhomologous chromosomes. But genes carried on the same chromosome are linked: the probability that they will enter the same gamete is taken to reflect the distance between them. Recombination between linked genes occurs during meiosis by a physical exchange between maternal and paternal homologues.

Each meiotic event may in principle produce four haploid gametes (or, more usually, gamete progenitors). Of course, whether all four haploid cells give rise to mature gametes depends upon the species and the sex. Here we shall be concerned with the events of meiosis itself, and in particular with the generation of new genetic combinations, and it will not be possible to discuss the subsequent events of spermatogonial or oocyte development.

Prior to meiosis, the last diploid forerunner of the gametes replicates its chromosomes, so that the cell enters division with the same two diploid sets with which it would enter a mitosis. The four copies of each chromosome are distributed to different progeny cells by two successive divisions, known as *meiosis I* and *meiosis II*. The first division segregates two diploid sets of chromosomes; the second division sees the production of two haploid sets from each of these diploid sets.

The first division of meiosis is characterized by a very lengthy prophase, which in some organisms may last even for several weeks or months. At the beginning of prophase, the chromosomes are seen as very fine, extended threads. During this stage, *leptotene*, they often have the appearance of beads on a string in which the chromomeres are clearly displayed. Their prior replication is not evident from their appearance, since only a single thread is apparent for each parental homologue. This led to some confusion in earlier cytological studies of meiosis (see Darlington, 1932).

Zygotene describes the stage when homologous threads approach each other and begin to pair side by side. This *synapsis* increases until the chromosomes are paired along their entire length; the paired structures are called *bivalents*. Synapsis is completed by the stage of *pachytene*, when the chromosomes appear as somewhat thicker threads.

The double nature of the two components of each homologous pair is clearly displayed at the subsequent stage of *diplotene*, when the homologues separate along much of their length but remain attached at certain restricted points. At this time, each of the homologues can be seen under the microscope to consist of two threads (*chromatids*). Each bivalent must therefore contain four chromatids and this is therefore sometimes described as the four strand stage of meiosis. The two chromatids of each homologue represent duplicate copies of the chromosome and are known as *sister chromatids*.

The points where the homologues fail to separate from each other in the diplotene cell represent the *chiasmata* at which genetic exchange appears to take place. The appearance of the chiasmata suggests that they represent sites where two chromatids are tightly associated. The last stage of the extended prophase is *diakinesis*, when the tetrad structures coil up to form shorter and thicker chromosomes. Chiasmata which at diplotene are located at sites between the centromere and the telomere appear to move progressively toward the ends of the bivalent during diakinesis in the process described as terminalization. Figure 5.1 shows two mouse homologues held together at the end of the subsequent metaphase by terminal chiasmata. Chromatin fibers cross between the chromatids at these locations (Burkholder et al., 1972).

During diakinesis any nucleoli disappear and the nuclear membrane dissolves so that prophase I is succeeded by metaphase I. Each bivalent becomes aligned at the equatorial plane between the poles. The centromeres of sister chromatids lie adjacent to each other, rather than opposed as at mitosis (see Chapter 4). Each bivalent thus consists of four chromatids organized in two chromatid pairs, the two centromeres of one pair facing one pole, and the two centromeres of the other pair facing the opposing pole. During anaphase I the two chromatid pairs of each bivalent therefore move toward the opposite poles of the spindle. At telophase I, nuclear membranes form briefly around each of the diploid chromosome groups at the two poles.

A short interphase when the chromosomes appear temporarily to lose some of their individual features is succeeded by prophase II, which leads to the second division. At metaphase II the two nuclear membranes dissolve and two new spindles form. At this second division, the behavior of the chromosomes is similar to their movements in mitosis: they line up on the equatorial plane of the metaphase plate with the centromeres of the sister chromatids of each pair aligned toward opposing poles. During anaphase II, the homologues separate and move to the poles; during telophase II a membrane surrounds each of the four haploid nuclei produced

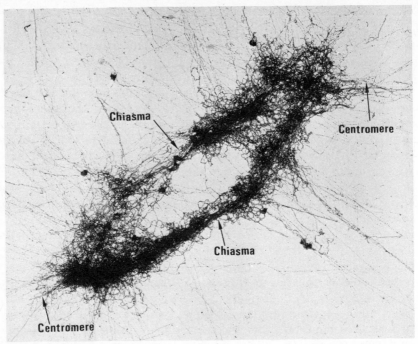

Figure 5.1. Whole mount preparation of mouse oocyte metaphase I bivalent with complete terminalization of chiasmata. The positions of the centromeres may be inferred from the attachment of microtubules. The positions of the chiasmata are marked by the passage of chromatin fibers between the non sister chromatids. ×7,650. Data of Burkholder et al. (1972).

by the two successive divisions. Then any or all of these nuclei may give rise to haploid gametes.

Formation of the Synaptonemal Complex

The critical events in recombination between genes carried on homologous chromosomes take place during the extended prophase of the first meiotic division. Recombination produces reciprocal copies of the chromosome, each of which is in part derived from one of the parental homologues and in part from the other. One of the most remarkable features of the process is that it involves breakage and crosswise reunion of DNA duplexes at precisely corresponding nucleotide sequences.* In procaryotic and viral sys-

*Inaccuracy in recombination would change the nucleotide base pair sequence by generating a deletion in one recombinant and a duplication in the other. The objection to this rests on the assumption that recombination may take place at any site, so that alterations resulting from imprecision would have a deleterious effect on the function of the loci in which they occur. This difficulty would be overcome if it were supposed that recombination might take place at particular sites whose exact sequences are irrelevant to genome function (for example,

tems, recombination involves exchange between DNA genomes per se; nothing is known about the means by which the corresponding sequences find each other. In eucaryotic cells, the breakage and reunion event between DNA molecules is preceded by the synapsis of homologous regions; and this pairing is the function of the chromosome structure as such rather than of its DNA content.

Chromosome pairing takes place in three stages. First, the homologous chromosomes (each of which has previously replicated) approach each other. We know virtually nothing about the attractive forces that must ensure that only homologues become associated. But the specificity of synapsis implies that each homologue can be distinguished from the other members of the set; and this in itself excludes models relying on generalized "attractive forces." Presumably each homologue pair has some component that enables each member to recognize only the other member when the bivalents are formed at zygotene.

Next the two homologue pairs become intimately associated with each other in the form of a *synaptonemal complex*, which is first generated at isolated points, usually including the telomeres, and then extends until the entire length of each chromosome is tightly synapsed. Formation of this complex brings the homologous regions of each chromosome into apposition.

But the distance between the synapsed homologues remains appreciable in molecular terms. Presumably there must therefore be a final stage before genetic exchange can occur, one in which the corresponding nucleotide sequences are brought into close apposition. Nothing is known about the mechanisms that may be involved.

The synaptonemal complex is visible only in the electron microscope; light microscopy reveals that the homologous chromosomes have become closely associated, but lacks sufficient resolution to define their structure. Since the exact stage of meiosis is not apparent from electron micrographs of chromosomes in prophase I, cells must be followed by both light and electron microscopy to correlate the formation and dissolution of the synaptonemal complex with progress through division. The complex can be found only in pachytene cells and disappears during diplotene at the early stages of disjunction (and see footnote in next section).

The complex was first observed by Fawcett (1956) and Moses (1956, 1958) as two dense parallel lines, the *lateral elements*, separated from each other by a less dense central region containing a somewhat finer dense filament, the *central element*. The triplet of parallel dense strands lies in a single plane

---

these might be members of a class of the repetitive sequence elements described in Chapter 18). However, it is no easier to imagine mechanisms for recombination that allow small errors than those that are precise. Evidence that recombination depends on the formation of *hybrid DNA* by base pairing between single strands of the recombining duplex molecules is discussed for both procaryotes and eucaryotes in Volume 1 of Gene Expression; and the processes of exchange as characterized in bacteriophages are discussed in Volume 3.

that curves and twists along its axis. The partner homologues of each bivalent lie one on each side of the complex, each associated with one lateral element (Moses, 1968).

The dimensions of the synaptonemal complex have by now been determined in an enormous variety of organisms (Westergaard and Von Wettstein, 1972; Gillies, 1975; see Table 5.1). In general, the lateral element has a diameter in the range of 30–50 nm. The central region separating the two lateral elements commonly has a diameter in the range 60–80 nm, although it tends to have a more tenuous appearance and therefore is not visualized as clearly as the lateral elements. In many cases both the lateral and central elements have an amorphous appearance; in other instances, either or both may have a banded appearance. An example of the synaptonemal complex is shown in Figure 5.2. Each homologue consists of a network of chromatin fibers associated with the lateral elements; their constituent sister chromatids are not visible in the complex and thus each lateral element and its associated chromatin appears to comprise a single structure. The synaptonemal complex joins homologous chromosomes over their entire length, including the centromeric regions, and extends without interruption from euchromatin into heterochromatin (Moses and Coleman, 1964; Moses, 1968).

Although the triplet of parallel strands is the form typically taken by the complex in most types of cell, other types of structure also may be seen at synapsis. In some insects, the complex appears to consist of multiple elements, whose structure and function have been discussed by Moens (1969b) and Rasmussen (1975). During the later stages of prophase in many mammals, the XY bivalent of male cells forms a structure sometimes termed the sex vesicle. Under the light microscope this appears as a dense body comparable to the nucleolus, consisting of two condensed homologous elements in general not paired with each other (Solari 1970a, 1974; Moses, 1977b; Tres, 1977). In some species, particular regions of the XY bivalent may pair, apparently taking the form of a synaptonemal complex at one terminus, while the remaining regions of both chromosomes remain without partners.

In general, synapsis takes place only between homologues and appears highly specific. But synaptonemal complexes can be observed in hybrids between two plants that are not well enough related to support genetic crossover (Comings and Okada, 1972). Synaptonemal complexes also have been found in haploid barley, where up to 60% of the length of the chromosome set may pair by forming intra- or interchromosomal complexes. Again this nonhomologous pairing is not followed by chiasma formation. This suggests that a generalized attraction between chromosomes may be able to support pairing in the absence of more specific interactions that might preempt it. (Although this does not apply, of course, to the unpaired regions of sex bivalents in mammals.) Although necessary, synaptonemal complex formation cannot be sufficient for crossing over.

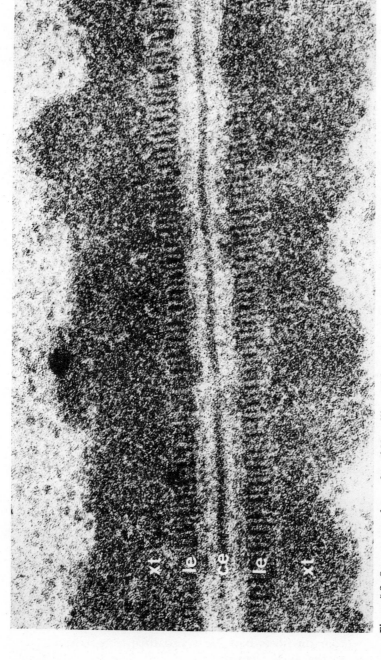

Figure 5.2. Synaptonemal complex of Neotellia. This section through a pachytene bivalent reveals a lateral element (le) of about 50 nm in diameter, which consists of alternating thick and thin bands embedded in the surface of the chromatin (xt). The electron dense central element (ce) has a diameter of about 18 nm. The distance between the two lateral elements is about 120 nm. Photograph ×61,200. Data of Westergaard and Von Wettstein (1970).

The traditional view of chromosome pairing derived from light microscope studies is that only two homologues are paired at any point. In triploid cells of the lilly, the switch of pairing partners seen in light microscopy is accompanied at the level of the electron microscope by a switch in the association of lateral elements (Moens, 1969a). But in other instances the axial elements of all three chromosomes appear to come into close contact. Meiotic cells of triploid chickens may display synaptonemal complexes that have three lateral elements separated by two central elements (Comings and Okada, 1971a). Tetraploid cells of yeast may display quadrivalent structures in which a pair of complexes run side by side and/or edge to edge (transitions between the two arrangements representing twists of one complex relative to the other) (Byers and Goetsch, 1975). This shows that some variation in the usual structure of the complex can occur, presumably due to the formation in the inner sister chromatid pairs of a lateral element on each side of the chromatin instead of only on one side.

The structure of the lateral and central elements is not well defined. In some cell types they appear to consist of filamentous structures, but in others their appearance is somewhat amorphous. However, the ubiquity of the synaptonemal complex in the form of chromatin fibers separated by the two lateral elements and single central element suggests that there may be components common to many species. It is clear that protein is the major constituent of both the lateral elements and the central element; what other components may be present is unclear. Treatment with DNAase disrupts the chromatin fibers but leaves the lateral elements and central element intact (Comings and Okada, 1972; Solari, 1972; Solari and Moses, 1973). It is possible that longitudinal filaments at the lateral element are digested, but these may be part of the chromatin. In contrast to earlier reports that the lateral and central elements are resistant to RNAase, Westergaard and Von Wettstein (1972) reported susceptibility; it is possible therefore that RNA is a component of the fibers that constitute these elements. Whatever the molecular composition of the elements, however, it seems clear that they represent structures independent of the chromosome threads themselves; rather must they provide a framework with which the chromatin fibers associate.

The components of the synaptonemal complex first become visible during leptotene when each unpaired chromosome—that is, each sister chromatid pair—takes the form of a network of chromatin fibers associated with a single axial element. This component gives rise to the lateral element, perhaps by the association of further material with the initially ill defined structure (Moens, 1968; Rasmussen, 1976). Before pairing the axial element may be located in the center of the chromatin network, but in the synaptonemal complex it is found on one side. Only a small proportion of the chromatin is intimately associated with the lateral element. In some whole mount preparations, the chromatin fibers have been visualized as a series of lateral loops attached to the axial elements in bunches (Comings and Okada, 1971b,c).

Both ends of the axial element of each unpaired leptotene chromosome may be attached to the nuclear envelope. Synapsis starts when some unknown mechanism brings the ends of the homologues to within about 300 nm of each other, perhaps involving a movement of the telomeres on the nuclear envelope. At all events, synapsis appears to start at the telomeres on the nuclear membrane. The axial elements of homologous chromosomes begin to pair with each other during zygotene; and pairing continues as the synaptonemal complex spreads from the telomeric starting points (Solari, 1970b; Comings and Okada, 1972; Westergaard and Von Wettstein, 1972).

The origin of the central element is not clear. In some cases it appears to be formed by filaments that extend from the two lateral elements to fuse in the central region (Moens, 1968; Comings and Okada, 1972; Solari and Moses, 1973). An alternative model is to suppose that it consists of pieces synthesized elsewhere and assembled between the lateral elements to provide a framework with which they can associate to form the synaptonemal complex (Westergaard and Von Wettstein, 1972; Rasmussen, 1976). That there are some differences in the protein components of the lateral and central elements is suggested by observations that they may respond differently to some staining techniques. And in whole mount preparations, the central element seems to be lost more readily than the lateral elements (Solari, 1972; Counce and Meyer, 1973).

Reconstruction of the Set of Synaptonemal Complexes

Whole mount electron microscopy allows the entire set of synaptonemal complexes to be visualized (compared with the individual regions seen in thin sections) (Counce and Meyer, 1973). An example is shown in Figure 5.3.* At leptotene and zygotene, the axial elements are attached to the nuclear envelope in what appear to be dense thickenings. The initiation of synapsis is not entirely synchronous, but the synaptonemal complexes are almost completely formed by late zygotene. Central elements have appeared; and at pachytene the bivalents are spread around the perimeter of the nucleus. The complexes often appear to be twisted, but there is no correlation between these twists and the individual chromosome, the stage of meiosis, or the frequency of recombination events (Moses, 1977a).

In whole mount studies, the kinetochore region appears marked by a short densely stained differentiation of both lateral elements. This is consistent with previous studies with sectioning techniques in which the positions of the kinetochores can be identified at pachytene (Moens, 1973; Gillies,

---

*A more recent technique that allows visualization with the light microscope is based on silver staining of meiotic cells; this also may be useful for karyotyping (Fletcher, 1979; Pathak and Hsu, 1979).

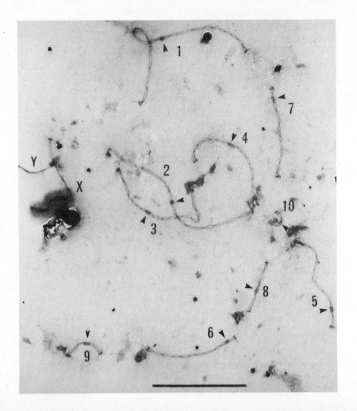

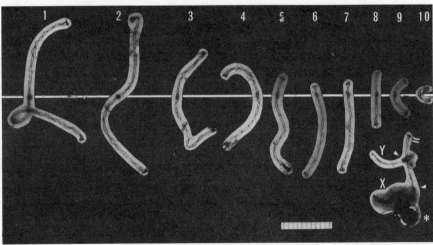

Figure 5.3. Synaptonemal complexes of the complete Chinese hamster complement visualized by whole mount electron microscopy. The upper set shows the complement spread out at late pachytene. The bivalents are numbered and kinetochores are indicated by arrows. The lower set shows the bivalents arranged as a karyotype. The bar is 10 μm. Data of Moses et al. (1977).

Rasmussen and Von Wettstein, 1973). Usually centromeres can be seen only during late mitosis and meiosis, making it difficult to assess their structure and function (if any) during the preceding interphase and first part of division. In premeiotic interphase cells of the onion, however, centromeres can be seen by electron microscopy; while apparently clustered at one pole, they are not associated in homologous pairs or attached to the nuclear membrane (Church and Moens, 1976). This suggests that, at least in this instance, premeiotic associations of centromeres are not involved in establishing the alignment of homologues.

The technique of serial reconstruction allows the entire karyotype to be visualized in the form of synaptonemal complexes. By taking serial sections of a meiotic cell, the three dimensional array of complexes can be determined and each complex can be equated with the appropriate member of the chromosome set (by comparing relative lengths with those of the mitotic set). This allows the effects of particular genetic changes to be followed; for example, what happens to the synaptonemal complex at an inversion can be determined. Such reconstructions now have been performed with several types of meiotic cell and Table 5.1 summarizes these data.

All bivalents of the karyotype are found in the form of the synaptonemal complex, which is continuous along the length of each paired structure (see also Wettstein and Sotelo, 1967; Gillies, 1975). This excludes only any unpaired sex chromosomes. One end of each synaptonemal complex ends at the nuclear membrane; in some cell types the other end also terminates on the membrane, in others it may be free, at least in some cases ending in a differentiated structure akin to a heterochromatic knob (Carpenter, 1975b; Church, 1976; Goldstein and Moens, 1976).

The length of the complex may change during meiosis; for example, in Sordaria it increases from 45 μm at leptotene to 60 μm at pachytene. The values shown in the Table represent measurements taken as soon as its formation is complete. Comparison between the total length of the complexes and the DNA content of the haploid genome makes it clear that in each case there is a large length of DNA per unit length of complex; the most common packing ratios seem to lie in the range of 500–1000. Thus although the synaptonemal complex usually is considered to bring homologues into apposition in order to achieve the close contact presumably necessary for crossing over, its formation clearly does not bring homologous sequences of DNA into an extended apposition. This is evident also from the width of the network of chromatin that lies adjacent to each lateral element; only a small proportion of the chromatin fiber can be in close contact with the lateral element. And, of course, there is a gap of some 70 nm that the central region occupies between each chromosome pair. The structure of the complex therefore leaves to be resolved the question of how homologous nucleotide sequences find each other; and if breakage and reunion indeed occurs at this stage, it must be necessary for some threads of chromatin to cross from one lateral element to the other.

Table 5.1. Characteristics of reconstructed

| | | Width of complex (nm) | | |
Species and cell type	Bivalent number	Total width	Lateral element	Central element
Saccharomyces cerevisiae (spore)	15–17	160	25	20
Neurospora crassa (ascus)	7	200	40	20
Sordaria macrospora (ascus)	7	230	70	20
Chlamydomonas reinhardii (zygote)	18–20	180	35	—
Drosophila melanogaster (oocyte)	4	130	20	32
Ascaris lumbricoides (oocyte) (spermatocyte)	12	119	27	25
Bombyx mori (oocyte)	28	130	25	20
Locusta migratoria (spermatocyte)	11	200	50	100
Zea mays (anther)	10	180	40	40

The bivalent number includes only paired chromosomes, that is, it omits any un-paired sex chromosomes; however, the DNA content given is that of the whole genome, so that when there is a part of the genome not involved in bivalent forma-tion, the calculated packing ratios may be too high (for example, in Ascaris, where several small sex chromosomes are not paired). The length of the complex is that measured by serial reconstruction experiments in which each bivalent can be iden-tified, except for Chlamydomonas and Saccharomyces, where the chromosome number is not certain; the range of lengths identifies the smallest and largest bivalents and the total gives the sum of the set. In each case the synaptonemal complex con-sists of a central element separated by a space on each side from a lateral element;

The only link evident between the two sides of the synaptonemal complex at the structural level is provided by electron dense bodies that appear to be spherical or cylindrical structures lying adjacent to the central element and spanning its width. These *nodes* were first observed in Neurospora and other fungi (Gillies, 1972, 1979; Zickler, 1973, 1977; Byers and Goetsch, 1975; Storms and Hastings, 1977). An example of a node in the synaptonemal complex of Sordaria macrospora is shown in Figure 5.4. Similar structures have been observed in the synaptonemal complex of D. melanogaster oocytes, where they take the form of spheres of about 100 nm in diameter lying along-side the synaptonemal complex (Carpenter, 1975a). These have been de-

synaptonemal complex karyotypes

Length (μm)		DNA content		
Range	Total of set	Length of DNA (mm)	Packing ratio	Ref.
0.2–4.7	28	3	100	A
4–11	50	16	320	B
4–10	45	—	—	C
1–8	75–81	—	—	D
—	110	55	500	E
5–20	138	140	1000	F
4–14	101		1400	
5–11	212	170	850	G
13–55	312	—	—	H
21–46	325	1200	3700	I

the only exception is Chlamydomonas, where it is not clear whether a central element exists. Chromatin is associated with the entire length of each lateral element; the packing ratio gives the length of DNA per unit length of synaptonemal complex.

References: A—Byers and Goetsch (1975); B—Gillies (1972); C—Zickler (1977); D—Storms and Hastings (1977); E—King (1970), Carpenter (1975b); F—Goldstein and Moens (1976); G—Rasmussen (1976); H—Moens (1969c, 1973), Counce and Meyer (1973); I—Gillies (1973), Gillies et al. (1973).

scribed as *recombination nodules*. A diagrammatic representation of the synaptonemal complex at such a site is shown in Figure 5.5.

There is no direct evidence to suggest that these nodes or nodules are involved in genetic exchange, but the idea that they correspond to sites of crossing over is consistent with observations that in Drosophila, Neurospora and yeast they occur with the same frequency and distribution as the chiasmata. There are up to 5 nodules in the Drosophila pachytene oocyte, compared with an average of 5–6 genetic exchanges; they lie in euchromatic and not in heterochromatic regions; and are located proximal and distal when there are 2 per chromosome, in the middle when there is only one.

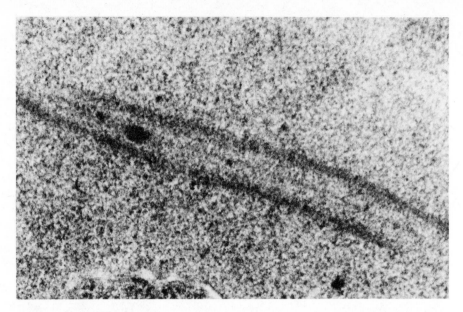

Figure 5.4. Longitudinal section through a synaptonemal complex of Sordaria macrospora at pachytene with a nodule evident as a thickening on the central element of the complex.   Data of Zickler (1977).

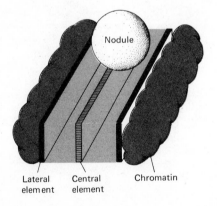

Lateral   Central   Chromatin
element   element

Figure 5.5. Illustration of the synaptonemal complex of the meiotic oocyte of Drosophila melanogaster.

In Neurospora, the average number of 19 nodes compares with a total genetic map length of 700–1000 units, predicting 14–20 genetic exchanges. Again the distribution of nodes is consistent with the expected pattern of crossovers. The average distance between nodes is 0.6 μm in yeast, compared with the average spacing between exchanges of 0.4 μm that is to be expected from the typical 70 exchanges per nucleus. In Drosophila, they are transient structures present only in mid pachytene. In Sordaria they are formed during zygotene and persist until diplotene; the total number at diplotene may correspond to the number of chiasmata. It seems doubtful whether structural studies will be able to resolve the question of whether they are involved in recombination, but an approach that may prove fruitful will be to see whether they are influenced by mutations that affect meiosis in Drosophila (see below).

No relationship can be seen between the synaptonemal complex and crossing over since chiasma formation is not observed at pachytene in either the light or electron microscope. When the pairing of homologues is terminated and replaced by repulsion at early diplotene, the synaptonemal complex is shed from the bivalent. Synaptonemal complexes generally appear to disintegrate gradually during diplotene (Westergaard and Von Wettstein, 1972). First the central element is lost and then any fibers apparently crossing from the lateral element to the central element. In many organisms the separated axial elements disappear at once, although in some instances they may return to their previous position in the center of the chromatin network before dissolution. At early diplotene, chiasmata seem to consist of short stretches of the synaptonemal complex which have been retained. However, we do not know whether the chiasma holds the bivalents together, and thus protects the synaptonemal complex from decay at these sites, or whether it is the retention of the synaptonemal complex that protects the chiasma at this time. Whatever the order of events, however, it is clear that the homologous chromosomes remain intimately associated at those sites that presumably identify the locations where DNA has suffered breakage and reunion.*

The genome is replicated before meiosis commences, but a small amount of DNA synthesis may take place during the early stages of division. In meiotic cells of the lilly, DNA equivalent to about 0.3% of the genome content appears to be synthesized during zygotene (Hotta and Stern, 1971, 1976). This synthesis appears to represent semi conservative replication to produce segments of DNA that remain comparatively short in length until pachytene. Inhibition of this synthesis prevents cells from proceeding through meiosis (Stern and Hotta, 1969). One possible speculation is that this synthesis might be concerned with pairing of DNA sequences for genetic exchange; this would be consistent with models that suppose the sites of crossover are determined *before* formation of the synaptonemal complex, that is, by events involving a limited pairing between rather short homol-

---

*The idea that the synaptonemal complex is the site of genetic recombination originated with the probable temporal coincidence of its occurrence with crossing over in meiotic prophase. A correlation at a different time has been reported in D. melanogaster. Heat treatment may induce recombination in normal genomes if applied during premeiotic S phase. This is also the period of sensitivity to high temperature for a temperature sensitive mutant deficient in recombination (Grell, 1978). Day and Grell (1976) have reported that a synaptonemal complex can be found in the oocytes at this time. This suggests that events influencing recombination may occur before meiosis; and the formation of a synaptonemal complex may be initiated in connection with DNA synthesis rather than during meiosis.

ogous sequences (see Stern, Westergaard and Von Wettstein, 1975). There also occurs what seems to be a repairlike synthesis of DNA during pachytene, which again might be involved in recombination. Meiotic DNA synthesis has been reported also in the mouse (Hotta, Chandley and Stern, 1977). Certainly it is true that current models for recombination are consistent with the synthesis of a somewhat limited amount of DNA at the site of exchange; and, indeed, there is evidence for such an association in some phage recombination events (see Volume 3 of Gene Expression). But it is necessary to remember that the amount of DNA synthesized per exchange may be small (of the order of 10^6 daltons in the phage systems), so that it would not be unreasonable to suppose that often it might be undetectable. And, of course, the role of any such synthesis in eucaryotic systems remains to be established.

Crossing Over and Chiasmata

The construction of linkage maps implies that each chromosome can be regarded as a linear array of genes, a concept that was developed in the early years of this century. The idea that many genes must be carried on each chromosome, and that there might be exchanges of alleles between homologues, originated with Sutton and with De Vries in 1903. With the discovery of linkage by Morgan in 1911, it became apparent that the proportion of recombinants is characteristic for any pair of genetic loci, and that it is independent of the particular alleles present. This led Sturtevant (1913) to propose that the extent of recombination between two genes can be taken as a measure of their distance apart, a concept that was accompanied by the realization that recombination units are additive.

Map distance is usually expressed in terms of the per cent recombination. The maximum frequency with which two loci may separate is 50%—this corresponds to the independent segregation of genes carried on different chromosomes and is also displayed by loci far apart on one chromosome. Linkage is reflected by the reduction of this frequency. For loci lying within a span of up to 50 units, the frequency of recombination between two loci *a* and *c* is close to the sum of the frequencies measured for *a-b* and *b-c*, where *b* is an intervening locus. The deviation from additivity is due to the occurrence of *double crossovers*. When recombination occurs in both of the intervals *a-b* and *b-c*, the parental combination of the outside markers *a* and *c* is restored in spite of the occurrence of recombination. The result is to reduce the measured *a-c* frequency below the sum of the frequencies *a-b* and *b-c*; this effect becomes more pronounced as the span of the outside markers increases towards the limit of 50 units.

Over short distances the occurrence of double crossovers is countered by *positive interference*. This describes the reduction observed in the frequency of double crossovers between outside markers *a-c* compared with the expected value (the product of the values for *a-b* and *b-c*). This interference implies that the presence of one genetic exchange inhibits the occurrence of another in the same region (Muller, 1916). In D. melanogaster, for example, interference is displayed over a distance of about 40% recombination within a chromosome arm, but does not extend across the centromere (Stevens, 1936). The reduction in double crossovers over short intervals extends the distance over which recombination frequencies are directly additive; and this has the practical consequence of aiding the construction of genetic maps. In all events, however, linkage between loci can be demonstrated directly only for those close enough to suffer less than 50% recombination. Loci lying farther apart on the same chromosome can be related by demonstrating that both are linked to some intervening marker(s). In this way, the genetic map of a chromosome can be extended into a linkage group of any length. The distance between loci is described in *map units*, where the map unit corresponds to 1% recombination, but is subject to compensation to allow for the limits on directly measured frequencies.* This unit is often described as the centi Morgan (that is, 1% recombination equals 1 cM).

The concept that a physical exchange of chromosome parts accompanies a genetic exchange was confirmed by following the effects of crossing over between two homologues that could be distinguished cytologically (by the presence of translocations) as well as by their different alleles: physical rearrangement was correlated with genetic recombination (Creighton and McClintock, 1931; Stern, 1931). The proposal that the basis of this rearrangement lies in the crossing over seen at the chiasmata in diplotene cells had been made previously by Janssens in 1909 (when he introduced the term *chiasma* to describe the cross like structures seen to connect homologues at this stage. The term *crossing over* was introduced by Morgan and Catell, 1912).

In good diplotene preparations of amphibian spermatocytes, Janssens observed that only two of the four chromatids appear to cross at a chiasma, the other two remaining uninvolved. This interpretation of the structure of the chiasma demands that crossing over takes place at some point after the chromosomes have duplicated to generate the two pairs of sister chromatids. The prediction that crossing over occurs at the four strand stage can be distinguished from the alternative that it occurs during the two strand stage preceding duplication by the genetic consequences of such crossovers. Crossing over at the four strand stage predicts that two of the four gametes

---

*The compensation that must be made to convert measured frequencies into map units is very much increased in bacteriophage systems, where there is *negative interference* over the short distances that are involved. Negative interference describes the increase in probability that more than one exchange will occur in the same vicinity. This results in extreme non additivity of directly measured recombination frequencies over very short intervals, of the order of size, say, of the gene. Recombination in phages is discussed in Volume 3 of Gene Expression; and the molecular interactions between recombining DNA molecules are discussed also in Volume 1.

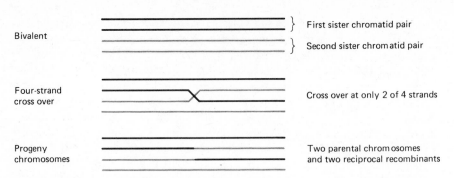

Figure 5.6. Four strand crossing over. The bivalent consists of two pairs of sister chromatids. Each pair comprises the duplicate copies of one of the two homologues. Each of the four "strands" therefore is a duplex DNA molecule. Crossing over between two nonsister chromatids corresponds to a chiasma and generates two reciprocal recombinants and two parental genomes.

of one meiosis should be reciprocal recombinants while two remain parental; at the two strand stage, a crossover would generate two reciprocal recombinants, whose subsequent duplication would yield four recombinants, two of each type (for review see Whitehouse, 1973). The structure generated by four strand crossing over is illustrated in Figure 5.6. To have visible consequences, the recombination event must involve non sister chromatids; exchanges that occur between the duplicate copies comprising each of the chromatid pairs are described as *sister chromatid exchanges* (and presumably would have simpler topological demands.)

Evidence that chiasmata are associated with physical exchange of chromosome material has been gained by autoradiography. This relies upon a protocol in which chromosomes are labeled with ^{3}H thymidine during the last mitotic S phase before germ line cells prepare for meiosis. This generates cells in which each chromosome contains a duplex of DNA one strand of which is labeled. Then the subsequent replication that occurs prior to meiosis generates sister chromatids one member of which is unlabeled and one of which is labeled (in one strand of its DNA). Introduced by Taylor (1965) with the grasshopper Romalea microptera, this technique since has been applied to several other species, mostly insects (Church and Wimber, 1969; Craig-Cameron and Jones, 1970; Peacock, 1970; Jones, 1971). Crossing over during meiosis should result in an exchange of label between nonsister chromatids whenever a chiasma involves one labeled and one unlabeled chromatid (which should represent 50% of the crossovers). Sister chromatid exchanges should be revealed by a switch in label from one sister chromatid to the other. The labeling patterns caused by combinations of chiasmata and sister chromatid exchanges also can be predicted. In spite of technical difficulties, the results are consistent with the idea that a chiasma involves exchange of material between nonsister chromatids. In contrast to the localized occurrence of chiasmata (see below), sister chromatid exchanges occur apparently at random sites; their frequency varies from very low in Schistocerca gregaria to 70% of the chiasma number in Stethophyma grossum. However, there has been discussion over the extent to which they may be induced by radiation. Another problem with autoradiographic techniques is the limited resolution afforded. More precision can be obtained with the technique of incorporating BUdR followed by staining with fluorescence and Giemsa; its application to studies of sister chromatid exchange in mitosis of cultured cells is discussed in Chapter 10. The labeling is more difficult to perform in vivo, but its application to meiosis of Locusta now has confirmed the results of earlier analyses by autoradiography (Tease and Jones, 1978).

The genetic predictions of four strand crossing over can be tested directly only when it is possible to isolate all the products of one meiosis. Such tetrad analysis is possible with some fungi and enabled Lindegren (1932, 1933) to confirm this model. Previously, crossing over had been shown to take place at the four strand stage in Drosophila by examining the genotypes of flies that had gained two copies of the X chromosome from the mother. These experiments utilized XXY or XXX progeny or involved attached X chromosomes, but in each case the consequence was that the progeny fly gained both reciprocal recombinants of the sex chromosome; formally this can be described as a half tetrad analysis (Bridges, 1916; Anderson, 1925; Bridges and Anderson, 1925). The involvement of only two of the four chromatids in a crossover immediately raises the question of the relationship between the pairs of strands involved in two adjacent crossovers. Different genetic consequences are expected depending upon whether the same or different chromatids are involved; and it is clear that which pair of chromatids is involved at one site has no influence on which pair participates at the next chiasma (Emerson and Beadle, 1933).

If the chiasmata represent the sites of genetic exchange, they should occur with a frequency and distribution that corresponds with that of recombination. However, while it is possible to show that qualitatively the occurrence of chiasmata is consistent with their expected role in genetic exchange, it has not yet been possible to establish this relationship on the basis of a quantitative correlation.

The average number of genetic exchanges per meiosis can be calculated from the total length of the genetic map, taking 50 units to represent one exchange. In principle, this can be compared with the number of chiasmata. However, the only eucaryote for which there is a detailed genetic map is D. melanogaster, which happens to be a difficult organism in which to visualize chiasmata. Genetic maps of some length have been constructed for yeast, mouse and maize, but in each case the number of loci is comparatively small, leaving open the likelihood that they correspond to only part of the genome. Tables 5.2 and 5.3 summarize available data, from which it seems that the number of chiasmata is reasonably close to the number of

Table 5.2 Genetic map sizes in eucaryotic genomes

Organism	Number of loci mapped	Haploid chromosome number	Range of map lengths	Total map lengths	DNA per 50 map units
N. crassa	400	7		900	3.6×10^5 bp
S. cerevisiae	175	17	20–300	3000	2.5×10^5 bp
S. pombe	118	3	300–600	1300	5.8×10^5 bp
D. melanogaster		4	0–111	280	2.5×10^7 bp
M. musculus	287	20		1500	9.0×10^7 bp
Z. mays		10	55–155	1120	1.0×10^8 bp

The haploid chromosome number agrees with the number of linkage groups for the higher eucaryotes; for yeast, only the number of linkage groups is known. The range of map lengths indicates the smallest and largest linkage groups (in cM); the total map length sums the known linkage groups. In all cases the discovery of further loci is likely to increase the genetic map length. Note that the two yeast species differ greatly in the number of linkage groups but show only a 2 fold discrepancy in map length. The amount of DNA per 50 map units gives an indication of the distance between genetic recombination events; by this measure the frequency of recombination is 100 times greater in yeast than in Drosophila. Whether there is any significance in the further decline in recombination frequency per unit DNA with increase in genome size is not known.

References: Neurospora—Perkins and Barry (1977); yeast—Mortimer and Hawthorne (1973, 1975); Kohli et al. (1977); fly—Mather (1936), Lefevre (1976); mouse—Carter (1955), Henderson and Edwards (1968), Lyon (1976); maize—Darlington (1934), Rhodes (1955), Whitehouse (1973), Neuffer and Coe (1974).

Table 5.3 Frequencies of genetic exchanges in eucaryotic genomes

Organism	Total no. of chiasmata	Range of chiasmata/ chromosome	Total number of exchanges per meiosis	DNA per exchange	Haploid DNA content (base pairs)
S. cerevisiae	not known	not known	16 + 60 = 76	2.0×10^5 bp	1.5×10^7
S. pombe	not known	not known	2 + 26 = 28	5.4×10^5 bp	1.5×10^7
D. melanogaster	6	0–2	3 + 6 = 9	1.5×10^7 bp	1.4×10^8
M. musculus	25		19 + 25 = 44	6.1×10^7 bp	2.7×10^9
Z. mays	27	2–4	9 + 27 = 36	9.7×10^7 bp	3.5×10^9
H. sapiens	50	1–4	22 + 50 = 72	4.6×10^7 bp	3.3×10^9

The total number of exchanges per meiosis is calculated as the sum of that due to segregation of chromosomes $(n - 1)$ and that due to chiasma formation (except for yeast, where the number of chiasmata is substituted by the number of genetic exchanges expected on the basis of 50 map units per exchange). Data are from the same sources cited in Table 5.2 except for those on man (Hulten, 1974).

expected genetic exchanges. But the paucity of data do not allow any proper correlation to be drawn.

The overall frequency of genetic reassortment is a characteristic of each species and includes the recombination resulting from chromosome segregation as well as that caused by chiasmata. In yeast, recombination is comparatively intense, two orders of magnitude greater than in the other organisms, where there is one exchange for every 10^7–10^8 base pairs of DNA. Put another way, if a gene is of the order of size of 5000 base pairs of DNA, in these species there is roughly one recombination event in a stretch of DNA equivalent to a length of more than some 10,000 genes. However, since the number and average size of genes is not known for any of these organisms, it is not possible to express the rate of recombination relative to genetic content. (The problem of eucaryotic gene numbers is the subject of Chapter 17 and gene size is discussed in Chapter 26.)

Each homologue of a given chromosome set forms a characteristic number of chiasmata that is related to its relative size: smaller homologues of the complement display low average numbers and larger homologues possess more chiasmata. In the human chromosome set, for example, smaller homologues may usually have only one chiasma, while the larger members of the set may have up to four chiasmata (Hulten, 1974). In this case, the occurrence on every bivalent of at least one chiasma suggests that the formation of at least one crossover per chromosome may be obligatory at spermatogenesis. In other species, this does not happen; the locust Schistocerca gregaria is an example that has been well studied and in which the small bivalents may have no chiasmata while the largest may have up to four (Henderson, 1963; Fox, 1973).

The general relationship between chromosome length and chiasma number is what would be expected of a situation in which recombination frequency is at least approximately constant across the chromosomes. This predicts that the relative sizes of the genetic linkage groups should correspond with the relative physical sizes of the chromosomes; and data from the limited number of cases where such a comparison can be made are consistent with this conclusion. [This is not to say that the genetic map exactly represents the physical locations of genes; indeed, distortions in this representation are clear in, for example, the map of D. melanogaster (see below). Apart from these distortions, however, there is no evidence to suggest that the rate of recombination differs appreciably according to the chromosome.]

The number of chiasmata displayed by a given bivalent is not fixed, but falls within a characteristic range; it is therefore the *average* number that can be correlated with the relative size of the chromosome in the set. The total number of chisamata per meiosis also is subject to some variation; in human cells, most meioses display a total number within the range 40–55, whereas in Schistocerca gregaria the range seems to be about 16–20. Fox (1973) observed that in S. gregaria spermatocytes displaying larger numbers of chiasmata, the autosomes tend to be longer at diplotene; since the unpaired X univalents also are longer, this suggests that the lengths of the chromosomes in the meiotic cell may determine the number of chiasmata, rather than vice versa.

The idea that the number of chiasmata is subject to genetic control (although it may also be influenced by environmental effects) is supported by experiments in which it has been possible to select for an increase or decrease in the total number. Starting with a line of Schistocerca gregaria that displays an average number of 20.4 chiasmata per meiosis, Shaw (1972, 1974) was able to select a low line (average 17.7) and a high line (average 21.7) over only 5 generations. Of course, over such a short period this may represent the isolation of extremes previously existing as variants in the population, rather than evolution in response to the selection. But the results suggest that the variation in chiasma number is subject to genetic control. We shall discuss meiotic mutants in the next section, but should note here that a large number of mutants has been found with drastic effects leading to the abolition or very great reduction of recombination. The implication is that many genes are involved, at one stage or another, in the formation of recombinants.

How the number of chiasmata is controlled is unknown. Early analyses demonstrated that their distribution along the chromosomes does not correspond to the Poisson distribution expected for random location. This excludes simple models supposing that it is sufficient to produce a rate limiting amount of some necessary component. Haldane (1931) and Mather (1933) showed that the deviation from random distribution is what would be expected if the occurrence of one chiasma reduces the likelihood that another will occur in the vicinity.

Attempts to define the distribution of chiasmata originated with Mather (1936, 1938, 1940), who proposed that the location of the first chiasma to form on any chromosome is relatively fixed while that of the second (if any) is determined by an interference exhibited by the first.* This allows the distribution to be defined in terms of the two variables d and i. The parameter d describes the distance of the first chiasma on each chromosome from some fixed point, that is, the centromere or telomere; i describes the distance between chiasmata on chromosomes long enough to have more than one. The interference distance is expected to be constant for all members of the complement.

The concept that an interference effect restricts the closeness with which chiasmata can form is consistent with data gained on meiosis in several species. Working with S. gregaria spermatocytes, Henderson (1963) reported that if the first chiasma is located at the telomere (that is, $d = 0$), then i is fairly constant at 7.3 μm for chromosomes long enough to have two chiasmata.

---

*Another idea proposed by Mather, that there may be competition for chiasmata among the bivalents, has been refuted by more recent studies. For example, in a study of grasshoppers that are trisomic—possess an additional chromosome—Hewitt (1964, 1967) and Hewitt and John (1965) found that there is no change in the chiasma frequency of the other chromosomes; this implies the absence of interchromosomal competition for chiasmata.

(The chromosomes vary in length at diplotene from 3–20 μm, with a total length for the set of 94 μm; the interference distance is therefore 7.8% of the total length. To accommodate all chiasmata within this model, it is necessary also to suppose that in 30–40% of the cases chiasmata are initiated simultaneously at both telomeres.) The results of Fox (1973) are consistent with a model in which the first chiasma forms close to the telomere and subsequent chiasmata (if any) form sequentially along the chromosome at intervals determined by the interference effect. In human spermatocytes, Hulten (1974) has assigned a value for i of 3.5% of the total autosomal length. In telocentric chromosomes—such as those of S. gregaria—interference clearly may extend along the entire chromosome. In metacentric and submetacentric chromosomes, it may be seen to be an effect restricted within chromosome arms; this may correspond with the failure of genetical positive interference to cross the centromere.

The reduction in the observed number of chiasmata lying close together (compared with that expected from a random distribution) makes it clear that there is a physical interference between the formation of adjacent chiasmata in all those species so far examined (see also Southern, 1967; Shaw and Knowles, 1976). Qualitatively, this should inhibit the occurrence of double crossovers; whether this effect indeed corresponds quantitatively with the positive interference measured in genetic crosses cannot be determined, because it is not possible to equate the physical interference distance with a genetic map distance. Whether the first chiasma always forms at a fixed location is more contentious, but at least in some cases there seems to be evidence for a preferential location. Again, because of the paucity of genetic data, it is not possible to say whether this is reflected in the expected distortion of the genetic map.

The construction of genetic maps generally is taken to assume that exchange events are randomly located. But although (where data are available) there is a general concordance between the relationship among linkage groups and that among chromosomes, it is clear that variations exist in individual areas. A well characterized case is the reduction in crossing over seen in the vicinity of the centromeres of D. melanogaster (for review see Shalet and Lefevre, 1976). This is made particularly clear when crossing over is examined in translocation strains, in which a chromosome segment has been transferred from its usual position near a centromere to a more distal location. Recombination frequencies within the segment are increased by the translocation. The consequence of the centromeric inhibition of crossing over is to reduce the length of the genetic map corresponding to this region. This distortion in the relationship between map distance and physical distance can be seen by comparing the genetic and salivary chromosome maps (see Becker, 1976). This point is discussed further in Chapter 16. In no other eucaryote can such a comparison be made. Thus, while Lyon (1976) has shown that the known genetic loci of the mouse do not follow the Poisson distribution expected if genes are evenly distributed in the genome and chiasmata form randomly, it is not clear whether the discrepancy lies with the distribution of genes or chiasmata (or both).

Genetics of Meiosis

The function of meiosis is to produce haploid gametes that are genetic recombinants between the two sets of parental genetic information. Meiosis has in common with mitosis the general reorganization of cellular structure discussed in Chapter 4 in which the gelatinous spindle replaces the divisions of nucleus and cytoplasm. Peculiar to meiosis, there are two critical features: the accomplishment of recombination between homologues; and the reduction of the chromosome number. Following the analysis of mitosis, it should become possible to describe the mechanics of meiosis, the orientated movements of the chromosomes on the two successive spindles, in terms of the forces applied by spindle fibers to the centromeres. Indeed, such analysis has been the basis for inferences about mitosis and meiosis for several decades, although only now is it possible to extend this to the molecular level. But this does not define the properties of the meiotic centromeres—and in particular the differences from mitosis—in terms more detailed than microscopic description. Such description of recombination is restricted to the visible properties of the synaptonemal complex and of the chiasmata, which is some distance removed from the interactions between DNA molecules (or perhaps between chromatin threads) that constitute the genetic exchange. Genetic analysis of crossing over again reveals general features, such as the time of occurrence, but does not approach the level of macromolecular interactions.

An approach that offers a better prospect of analyzing the process of genetic exchange is the isolation of mutants that are defective in meiosis. Of course, this analysis is limited by the criteria that can be applied to select the mutants. Defects generally can be classified as abnormalities in genetic recombination or in chromosome segregation. Mutations that reduce or abolish recombination, or distort the distribution of exchange events, may lie either in the control of recombination or in the exchange process itself. Interference with crossing over may cause defects in chromosome segregation; and so such mutants may be useful in analyzing the influence of chiasma formation upon subsequent events. Mutations that directly cause failures in chromosome segregation may be expected to lie either with the function of the centromere or with the operation of the spindle itself. Such mutations may be difficult or impossible to obtain, because they may also act at mitosis; a mutation that renders a proportion of gametes defective need not be inviable, but if it has the same effect on mitosis the development of the organism may be made impossible. A further limitation is that any mutation completely abolishing the production of active gametes can be analyzed only if it is recessive, since otherwise the lack of progeny makes it impossible to maintain. By analogy with bacterial genetics, this difficulty might be overcome if it were possible to obtain conditional lethal mutants, which could be propagated under permissive conditions but would display the defect in nonpermissive circumstances. With higher eucaryotes, these problems of inviability of mutations also affecting mitosis or dominantly abolishing gamete formation cannot at present be obviated. But with species that can be propagated by mitotic division as well as by the sexual life cycle, reproduction in the asexual mode in effect provides the permissive set of conditions. With the lower eucaryotes, mutants may therefore be isolated by virtue of their reduced fertility and may be characterized for defects in mitosis as well as meiosis.

Meiotic mutants have been isolated in a variety of lower eucaryotes, including Saccharomyces, Neurospora, Aspergillus, Ustilago, Ascobolus, Podospora, and Schizophyllun, and in higher plants (for review see Baker et al., 1976b; Sears, 1976). Of these, the yeast S. cerevisiae is perhaps the best characterized. From this work is seems that recombination is regulated at two levels: there is a general control of the overall frequency of crossing over (which may be reduced by mutation without affecting the distribution of crossovers); and there are more localized controls which influence the occurrence of crossing over only in particular regions. Mutations also are known which affect the subsequent stages of chromosome segregation; most of these appear to identify functions acting at the first meiotic division. In addition to these mutants in meiosis itself, because of the existence of the asexual life cycle it is possible to seek mutants unable to initiate meiosis. However, while it seems likely that these will help elucidate the pathway that leads to gamete formation, it is possible that whatever controls they reveal will be applicable to this class of organisms rather than of general application to meiosis in higher eucaryotes.

A detailed investigation of meiotic mutants has begun in Drosophila, an organism especially suitable for this purpose by virtue of the ease with which the necessary genetic manipulations can be performed. Meiotic mutants generally can be recognized by the distortions that they cause in the genetic constitutions of the progeny. Then the defect(s) in the mutant can be examined by genetic and cytological means, although there remains the problem of defining the nature of the mutations at the molecular level. Thus while the general properties of mutations that reduce recombination show that chiasma formation is necessary for proper segregation, the nature of the relationship is not revealed. And while the properties of one mutant suggest that sister centromeres are held together between the two meiotic divisions, this gives no information about the structures involved. This emphasizes the degree to which available experimental techniques limit the level at which meiosis can be investigated.

Meiotic Mutants of Drosophila

D. melanogaster has four chromosomes. The X chromosome is telocentric and has a large amount of heterochromatin adjacent to the centromere; this is known as the basal heterochromatin and it appears to be involved in pairing with the much smaller Y chromosome. Chromosomes II and III are large metacentric autosomes, with centromeric heterochromatin that aggregates with that of the X chromosome into the chromocenter seen in salivary glands (see Chapter 16). Chromosome IV is a very much smaller autosome.

Meiosis in females follows the usual sequence of events. Crossovers occur in all chromosome pairs except IV, which shows no recombination. The genetic map does not faithfully reflect the physical map visualized in the polytene chromosomes of the salivary glands, but shows a reduction in crossing over in the regions of centromeric heterochromatin (see above). Thus the basal heterochromatin of the X chromosome physically represents a much larger amount of material than is reflected in the minimal distance it occupies on the genetic map.

Meiosis in Drosophila males is unusual: there is no genetic recombination. In the males of most Drosophila species (including D. melanogaster), the stages of meiosis at which crossing over is thought to occur are absent; leptotene, zygotene and pachytene cannot be visualized. Autosomes pair to form bivalents, but these do not display chiasmata at metaphase (see Cooper, 1950).

Mutants that affect meiosis in the male can be identified, therefore, only by their effects upon chromosome segregation. Mutants in the female might in principle be obtained by selecting for alterations in either recombination or chromosome segregation. Apart from some mutants isolated by chance some years ago, most of the meiotic mutants of Drosophila were isolated in two series of experiments conducted by Sandler et al. (1968) and Baker and Carpenter (1972). In the first a natural population was examined for the presence of mutations on chromosomes II and III; and in the second, X chromosomes subjected to treatment with the mutagen EMS were screened for meiotic mutants. Other, individual mutants since have been isolated, but these remain the only systematic searches for meiotic mutants (for review see Baker and Hall, 1976). In both cases, the criterion used to select mutants was an increase in the level of abnormal chromosome segregation. (An attempt in the first set of experiments to isolate mutants by their ability to depress recombination on the X chromosome was unsuccessful.) From 180 natural II/III chromosomes, 13 mutants were isolated; and from 209 mutagenized X chromosomes, 31 mutants were obtained. It is therefore clear that there must be many genes affecting meiosis (no doubt including many that would not be detected by this approach). The significance of the occurrence of mutants at such a high frequency in a natural population is not clear.

Interference with chromosome segregation can be classified into three categories. *Nondisjunction* describes the failure of chromosomes to separate (disjoin) at one of the meiotic divisions. A failure at the first division is described as *reductional nondisjunction* and at the second may be known as *equational nondisjunction*. These may be distinguished by their different genetic consequences. The result of nondisjunction is that one of the gametes contains two copies of one homologue while another has none. Thus nondisjunction generates equal numbers of diplo- and nullo-gametes. *Chromosome loss* occurs when one (or more) chromosomes fail for some reason to enter the gametes. This is therefore recognized by an excess of nullo-gametes (that is, without a corresponding number of diplo-gametes). *Meiotic drive* was defined by Sandler, Hiraizumi, and Sandler (1957) as any force that may cause the frequencies with which the two types of gametes are produced by a heterozygote to deviate from the 1:1 ratio. This is therefore recognized by an abnormal ratio of the two genotypes. Meiotic drive may, of course, be caused by chromosome loss.

The spontaneous frequency of errors in chromosome segregation at meiosis is low. Table 5.4 summarizes data on nondisjunction in D. melanogaster (any chromosome loss would be included in these figures). In the female, for which complete data are available, the total frequency of nondisjunction appears to be <0.2%. Meiotic mutants therefore have been isolated in the first instance as lines that display nondisjunction for any particular chromosome at >1% (that is, a tenfold increase). These putative mutants then may be rescreened to identify the nature of the defect.

With two exceptions, all the known meiotic mutants of D. melanogaster interfere with meiosis I in either the male or female, but not in both. This is generally taken to imply that the control of meiosis I is different in the two sexes, which is consistent with the occurrence of recombination only in females. But, of course, the stages following exchange display similarities. Thus a more accurate interpretation may be that mutants have been isolated only in those features peculiar to each sex, because features that are common

Table 5.4 Frequency of spontaneous nondisjunction at meiosis in D. melanogaster

	Percent nondisjunction	
Chromosome	Females	Males
X(Y)	0.08	0.06
II	0.02	0.03
III	0.01	<0.03
IV	0.07	not determined

Data for female nondisjunction are from Hall (1972) and for male nondisjunction from Frost (1961). Somewhat similar values for some chromosomes have been reported by Sandler et al. (1968). The values for the two sexes cannot be directly compared because different experimental protocols were used. However, Baker and Carpenter (1972) measured female XX nondisjunction as 0.07% compared with male XY nondisjunction as 0.23%; and chromosome IV nondisjunction was 0.16% in females and 0.38% in males. This would suggest that nondisjunction may be more frequent at spermatogenesis than at oogenesis.

to both may be common also to mitosis, in which mutations may be lethal. This may also explain the absence of mutations in meiosis II in Drosophila and in other species, since the second division more closely resembles mitotic division.

The properties of mutations affecting meiosis I in Drosophila males are summarized in Table 5.5. The two mutants *SD* (segregation distorter) and *RD* (recovery disrupter) have been known for some time as examples of meiotic drive (for review see Zimmering, Sandler, and Nicoletti, 1970; Zimmering, 1976; Hartl and Hiraizumi, 1976). Both distort the genotypic constitutions of progeny by causing the absence of sperm of particular genotypes. Studies of the temperature sensitivities of the effects suggest that each is due to an action that occurs in early meiosis.

Copies of chromosome II that are *SD* carry two alleles, both of which are necessary for distortion to occur. The mutant allele *Sd*, located on the left arm close to the centromere, appears to specify a product that acts upon the mutant allele *Rsp* (previously known as Rsp^{sens}), located in the centromeric heterochromatin of the right arm. Meiotic drive occurs at the *Rsp* locus, but Sd^+ acts as a recessive suppressor of drive. Thus Rsp/Rsp^+ heterozygotes produce an excess of sperm carrying *Rsp* unless the males are homozygous for Sd^+. In *Rsp/Rsp* homozygotes, only about 50% of the sperm are functional if there is heterozygosity Sd/Sd^+ at the other locus; there are virtually no functional sperm in the double homozygote *Sd Rsp/Sd Rsp*. Another locus that may be concerned in distortion is *E(SD)*, located in or close to the centromeric heterochromatin of the left arm, at which mutations may greatly reduce the ability to distort.

Electron microscopic studies of spermatogenesis show that in males heterozygous for *SD* (i.e., *Sd Rsp/Sd⁺ Rsp⁺*), upto about half of the spermatid nuclei appear to have incompletely condensed chromatin, a feature that may be associated with defects in spermatid maturation (Tokuyasu, Peacock, and Hardy, 1976). These results therefore suggest that the *SD* system may identify loci concerned with spermatogenesis rather than with meiosis per se. It is possible that the same conclusion may be true of *RD*.

The mutants *mei-S8*, *mei-081* and *mei-269* etc. all cause nondisjunction at the first meiotic division. In these experiments, nondisjunction of the XY pair and of chromosome IV was examined. The mutants *mei-S8* and *mei-269* proved to have abnormal segregation of one but not the other. This has been taken to suggest that at male meiosis the behavior of each chromosome is under separate control; but to confirm this conclusion rigorously, it would be necessary to show that these mutations do not affect nondisjunction of chromosomes II and III; and it should be possible to isolate mutants that display nondisjunction of these autosomes. Because the X-linked mutations *mei-269*, etc., have properties similar to those of deletions in the basal heterochromatin, it seems likely that they specify products involved in the necessary pairing of this region. The mutant *mei-081* displays non-

Table 5.5 Mutations affecting male meiosis in D. melanogaster

Mutant locus	Origin, location, and dominance	Observed effects	Cause of action
SD	natural population; chromosome II; dominant	SD/+ ♂ transmit SD mutant chromosome to >50% of progeny (up to 95%); SD/SD♂ are sterile	SD inactivates sperm carrying SD^+; unknown action directed against SD^+ chromosome in early meiosis
RD	natural population; chromosome X; hemizygous	production of excess ♀ progeny	causes fragmentation of Y-bearing sperm; sensitive period is in early meiosis
mei-S8	natural population; chromosome IV; recessive	nondisjunction of IV but not of XY; II/III not examined; 60% of sperm are nullo or diplo IV with excess of nullo type	nondisjunction and possibly also loss occurs at meiosis I; not yet known whether affects only IV or also other autosomes
mei-081	natural population; chromosome III; recessive	causes independent nondisjunction of IV (7.8%) and Y (5%); II/III not examined	acts at meiosis I
mei-269 (and 19 others)	EMS-induced; chromosome X; hemizygous	nondisjunction of XY occurs at 1–10% with excess of nullo-XY; do not affect IV; II/III not examined	nondisjunction at meiosis I is similar to that of X chromosomes deleted in the basal heterochromatin; loci may be involved in establishing pairing in this region
pal	EMS-induced; chromosome II; recessive	nonindependent loss of paternal chromosomes at or subsequent to meiosis and in somatic cells of progeny of pal ♂	pal^+ may specify a component of the chromosome necessary for division, such as a centromeric component
Mr	natural population; chromosome II; semidominant	causes recombination to occur in ♂; rate is same on II and III with loci 50 map units apart showing about 1% recombination	nonreciprocal exchanges occur in clusters and reflect physical distance rather than ♀ genetic map; may be due to premeiotic events rather than induction of meiotic crossing over

References: SD—Sandler, Hiraizumi, and Sandler (1957), Hartl, Hiraizumi, and Crow (1968), Hartl (1974, 1975), Hartl and Hiraizumi (1976), Ganetzky (1977); *RD*—Erickson (1965); *mei-S8* and *mei-081*—Sandler et al (1968); *mei-269*—Baker and Carpenter (1972); *pal*—Baker (1975); *Mr*—Hiraizumi et al. (1973), Slatko and Hiraizumi (1975), Slatko (1978a,b).

disjunction of both the sex chromosomes and chromosome IV, but, again, to see whether it influences chromosomes II and III requires further data.

The mutation *pal* causes loss of paternal chromosomes not only at meiosis, but also at the first mitoses of the embryo. All paternal chromosomes may be lost, but at differing rates. The pattern of loss seems to be that the paternal chromosomes are unstable from spermatogenesis until the second or third embryonic division; but if they have not been lost by then, they become stable, and subsequent mitoses are normal. Loss of different paternal homologues is correlated; that is, the frequency with which two (or more) paternal chromosomes are lost at one division is somewhat greater than would be expected as the product of their individual frequencies of loss. Such behavior is common also in the meiotic mutants that cause nondisjunction in the female (see below). The loss only of paternal chromosomes at both meiosis and embryonic mitosis seems to exclude models in which the pal^+ locus specifies some component that is transferred extrachromosomally to the egg and is needed for proper division. This suggests that the pal^+ gene codes for some component of the chromosome, one that is necessary for meiosis and for the first couple of embryonic mitoses, but which becomes unnecessary (or is replaced by a component coded by the maternal genome).

Two other mutants also cause chromosome loss at early mitotic division. The mutation ca^{nd} causes nondisjunction and chromosome loss at oogenesis (see below); it also causes loss of maternal chromosomes during the first embryonic mitosis (for review see Hall, Gelbart, and Kankel, 1976). This again suggests that some component of the chromosome necessary for the first mitotic division must be made at the preceding meiosis. The mutation *mit* acts maternally and causes loss of both maternal and paternal chromosomes at the third or fourth mitotic divisions. It is not clear why a maternal function should be expressed only at this stage. The importance of these mutations is that *pal* and ca^{nd} provide instances in Drosophila in which a meiotic mutant also substantially affects mitosis; and the specific times of action of these and of the *mit* mutation demonstrate that there may be differences between mitoses occurring during different early developmental periods (although presumably later mitoses become independent of these effects).

The absence of recombination in males of D. melanogaster poses an immediate question: is it possible to isolate mutants in which genetic exchange occurs at spermatogenesis?* It may be significant that Baker and Carpenter (1972) found no such mutants among 209 mutagenized X chromosomes: this implies that the probability that a single mutation

---

*In the context of inducing recombination in males, Reddi, Reddy and Rao (1965) reported in a very brief note that an extract from ovaries induced crossing over when injected into males. Boiling inactivated its ability. But this provocative finding has not been succeeded by any detailed studies to provide confirmation.

can accomplish such a switch is rather low. But a mutation that causes recombination in males was isolated in the form of the second chromosome T007 in a natural population by Hiraizumi et al. (1973). Other such chromosomes since have been found naturally at appreciable frequencies, although their effects have not been characterized in the same detail (Matthews and Hiraizumi, 1978).

The major locus of T007 responsible for male recombination has been identified as *Mr*, but other subsidiary elements also appear to be involved. The characteristics of the resulting recombination appear to be more like those of mitotic recombination (see below) than the usual recombination in females. This is consistent also with the effects of T007 on female meiotic recombination. It seems likely that the process it causes does not take place at meiosis; it may be premeiotic, possibly involving the induction of chromosome breaks. A particularly interesting feature of chromosome T007 is its apparent ability to pass on the activity responsible for conferring male recombination. This may be acquired by third chromosomes passaged through a female in company with the T007 second chromosome, although T007 does not lose its ability. The nature of this effect, reminiscent of transposition seen in bacterial systems, remains to be established; one practical consequence is that it may explain the difficulties that have been encountered in attempting to map *Mr*.

Mutants that are defective in meiosis in the female are described in Table 5.6. All the known female mutants appear to act on the entire chro-

Table 5.6 Mutations affecting female meiosis in D. melanogaster

Mutant locus	Origin, location and dominance	Observed effects	Cause of action
c(3)G	c(3)G^{17} is spontaneous; c(3)G^{68} is EMS-induced; point mutation; chromosome III; recessive	nonuniform loss of recombination to 0.1% of control; 20–40% nonindependent nondisjunction at meiosis I; no synaptonemal complex; heterozygotes show increase in recombination	locus controls some stage of meiosis prior to synapsis and acts at precondition for crossing over
mei -218, -251, -S282, -352	218, 251, 352, EMS induced, chromosome X; S282, natural population, III; recessive	nonuniform loss of recombination to 10–60% of control; 50% nonindependent nondisjunction at meiosis I; interchromosomal effect absent or reduced by 218 & S282	primary effect is at precondition for exchange (C is increased where measurable); nondisjunction occurs primarily in nonexchange tetrads and may be secondary effect
mei-41	EMS-induced; chromosome X; recessive	as *mei-218* etc., but also has increased sensitivity to MMS-mutation in ♀ and ♂ where chromosome breakage occurs in somatic cells	as *mei-218* etc.; and MMS-sensitivity due to defect in post replication repair

Table 5.6 Mutations affecting female meiosis in D. melanogaster

Mutant locus	Origin, location and dominance	Observed effects	Cause of Action
mei-S51	natural population; loci on II & III; recessive	nonuniform 50% reduction in recombination; similar nonindependent nondisjunction at meiosis I	affects precondition for exchange, but must be mutant at two loci, so difficult to study
mei-9	EMS-induced; chromosome X; recessive	uniform 10 fold reduction in recombination (to 16% of control in mei-9b); 50% nonindependent nondisjunction at meiosis I; interchromosomal effect absent; defective in excision repair of pyrimidine dimers; chromosome breakage and mitotic recombination occur in somatic cells	defective in exchange process itself (C is unaltered); locus may be involved in repair as well as recombination; nondisjunction is a secondary effect
mei -38, -99, -160	EMS-induced; chromorome X; recessive	nonuniform reduction in recombination; increases nonindependent nondisjunction at both meiosis I and II	loci must act at several stages (including preexchange); or improper function at early stage leaves chromosomes or cell defective throughout meiosis
ca^{nd}	X-irradiation; chromosome III; recessive	loss and nondisjunction of all chromosomes at meiosis I is nonindependent; somatic loss also occurs	centromeres or spindle defective at meiosis I
nod	EMS-induced; chromosome X; recessive	causes independent nondisjunction of nonexchange chromosomes at meiosis I	affects disjunction due to distributive pairing so that chromosomes move independently to poles
abo	spontaneous; chromosome II; recessive point mutation	maternal effect causes egg mortality as function of amount of X heterochromatin; meiotic effect causes nonuniform decrease in recombination and nonindependent increase in nondisjunction at meiosis I	acts at precondition for exchange (although C is not altered)

References: c(3)G—Gowen (1933); Smith and King (1968), Hall (1972), Baker and Hall (1976); mei-218, etc.—Baker and Carpenter (1972), Carpenter and Sandler (1974), Baker et al. (1976a), Baker and Hall (1976), Parry (1973); mei-41—Baker and Carpenter (1972) Boyd et al. (1976); mei-S51—Sandler et al. (1968), Robbins (1971); mei-9—Baker and Carpenter (1972), Carpenter and Sandler (1974), Baker et al. (1976a), Boyd, Golino and Setlow (1976); mei-38, etc.—Baker and Carpenter (1972); ca^{nd}—Sturtevant (1929), Lewis and Gencarella (1952), Davis (1969); nod—Carpenter (1973); abo—Carpenter and Sandler (1974).

mosome complement, unlike the (putatively) chromosome-specific effects of the male mutants (for review see Baker and Hall, 1976). They can be divided into two general classes: those which reduce recombination; and those which interfere with chromosome segregation. All the mutations that affect recombination also cause increased nondisjunction of all chromosomes. This generally is thought to imply that exchange is necessary for proper segregation of chromosomes X, II and III (but not IV, which does not experience crossing over). Certainly it is true that the nondisjunction of the major chromosomes caused by these mutations appears to be a secondary effect, a consequence of their primary effect on crossing over; for the nondisjunction occurs primarily or exclusively in those cases in which exchange has failed. And the characteristics of this nondisjunction remain the same in mutants that affect genetic exchange in different ways (see below). However, this does not prove conclusively that all mutations reducing recombination will have this effect, especially since in most instances the meiotic mutants have been selected by virtue of their effects upon nondisjunction rather than for the ability to reduce recombination. While perhaps not likely, it remains possible that other selection procedures might yield mutants defective in recombination but not in chromosome segregation.

Relationship of Recombination and Nondisjunction

Mutations that inhibit genetic exchange might in principle identify loci acting either prior to the exchange or might lie in the process of exchange itself. Events necessary for making exchange possible may be concerned with functions such as ensuring close pairing of homologues. An argument originally due to Bridges (1915) proposes that these two types of mutation may be distinguished if recombination can be regarded as a two stage process. In the first stage, possible sites of exchange are established in an appropriate distribution; at the second stage, there is a fixed probability that an exchange actually will occur at any given site. This model can be assessed by its predictions concerning the coefficient of coincidence (C), which is a measure of the likelihood that exchanges occur in both of two given regions; when applied to two adjacent regions, it becomes an inverse measure of positive interference. Sandler et al. (1968) demonstrated that the two stage model predicts that C should to some extent be independent of map distance and confirmed that this applies to recombination in Drosophila. Mutations that affect preconditions for exchange should alter C, whereas mutations in the exchange process should leave it invariant. With the exception of *mei-9*, all the mutations reducing recombination cause an increase in C, where the coefficient can be measured. This suggests that they influence the establishment of sites of exchange, but do not concern the act of exchange.

However, because these mutations reduce recombination, often it is difficult to obtain sufficient data to measure C. Carpenter and Sandler (1974) therefore suggested that the effect of the mutations upon the distribution of exchanges may itself be a more reliable measure of the stage of crossing over that is impeded. Again with the sole exception of *mei-9*, all the mutations that reduce recombination do so in a nonuniform manner, in spite of considerable variations in the extent to which recombination is reduced. In normal meiosis, the genetic map deviates from the physical map in showing a compression in the region of the centromeres; this means that recombination events occur less rarely in these regions (see above). The nonuniform reduction in recombination makes the genetic map more closely resemble the physical map, by reducing recombination to a relatively greater extent in regions distal from the centromeres. The correlation between nonuniformity and the increase in C (where measurable) supports the idea that these mutations affect preconditions for exchange. This is to take the view that the distribution and frequency of crossovers is controlled by events occurring before exchange itself takes place. The exceptional mutation, *mei-9*, may lie in the process of exchange itself. Since many mutants appear to be deficient in preconditions for exchange, while only *mei-9* is deficient in crossing over per se, it may be true that more genes are involved in establishing the conditions necessary for recombination than in coding for the recombination functions. Or mutants of the latter type may be invariable because they display pleiotropic effects; in this context it is interesting that *mei-9* appears to cause defects in repair as well as in recombination (in obvious analogy with the overlap of these systems in bacteria; see Volume 1 of Gene Expression).

Another criterion by which the effects of meiotic mutations can be assessed is the *interchromosomal effect*. This describes the increase in crossing over that a heterozygous inversion in one of the major chromosome pairs causes in the other pairs. The increase in crossing over occurs largely in the region of the centromeric heterochromatin and at the distal ends of chromosome arms, with intervening regions very much less affected (for reviews see Lucchesi, 1976; Roberts, 1976). Similar effects may be caused by translocations and by compound chromosomes (which represent an intact pair of chromosomes attached to one centromere). The interchromosomal effect is abolished by the mutations *mei-S282*, *mei-218* and *mei-9*, which suggests that the wild type alleles at these loci are involved in creating it (Parry, 1973; Carpenter and Sandler, 1974). This implies that the effect depends not only on the system involved in establishing preconditions for pairing, but also upon the process of exchange; and, correspondingly, it may become a useful tool for investigating the basis upon which the distribution of crossovers is set.

The nondisjunction caused by mutations that reduce or eliminate exchange has the same characteristics in every case: it applies to all the chromosomes; and the nondisjunction of different homologues is not independent. When nondisjunction of one chromosome pair has occurred, there is a much increased probability that there will also have been a nondisjunction of another pair(s). This is true both of mutations apparently influencing preconditions for exchange and of the *mei-9* mutation in exchange (Carpenter and Sandler, 1974). The consistency of this effect supports the idea that the nondisjunction is a secondary defect caused by the failure of exchange (see above). However, a complication in this view is that the increased frequency of nondisjunction applies to chromosome IV as well as to the other pairs, although the fourth pair does not undergo exchange and therefore cannot rely upon it for disjunction (see Sandler and Szauter, 1978).

Two models have been proposed to explain the disjunction of chromosome IV and also of any other chromosomes that may not have undergone exchange. The *distributive pairing* model proposed by Grell (1962a,b; 1964a,b; 1976) postulates that all nonexchange chromosomes enter a pool in which chromosomes may pair with any other member. Usually the distributive pairing system will be responsible for accomplishing disjunction of the chromosome IV pair. Studies with flies of unusual genotypes, such as those carrying an additional chromosome or bearing translocations between chromosomes, suggest that distributive pairing takes place between nonhomologues as well as between homologues; the principal criterion for establishing pairing appears to be similarity in size, although in the absence of a chromosome of similar size with which to pair, a given nonexchange chromosome may pair with any other nonexchange chromosome. The *chromocenter* model proposed by Novitski (1964, 1975) postulates that interactions between the centromeres of chromosomes not suffering exchange are responsible for their subsequent disjunction. A critical difference between the two models is that distributive pairing relies upon a postexchange process whereas chromocentral interactions occur preexchange.

When genetic exchange does not occur, one consequence therefore may be that the non exchange chromosomes must be segregated by means of the system that usually disjoins the chromosome IV pair. The nondisjunction caused by mutations reducing recombination generally has been explained on the basis of the distributive pairing model. Given the similarities in size between the major chromosomes of D. melanogaster, failure in exchange may open the way to size-dependent nonhomologous pairing by the distributive system. Genetic data confirm that in non exchange tetrads, the X chromosome may disjoin from the major autosomes instead of from its homologue (see Baker and Hall, 1976). The result of this is nondisjunction of homologues. To accommodate the nondisjunction of chromosome IV, it is necessary then to postulate that the presence in the distributive pool of the major chromosomes in some way interferes with the pairing of the fourth pair. (Although since chromosome IV does not seem to disjoin from the X chromosome, the interference does not take the form of stable pairing.) In a sense, the distributive pairing system may be "overloaded." The occurrence of nonhomologous pairing explains the lack of independence that is seen in the nondisjunction of different homologues. Support for the idea that a separate system is responsible for the disjunction of non exchange chromosomes is provided by the isolation by Carpenter (1973) of the *nod* mutation, which causes nondisjunction specifically of chromosomes that have not undergone genetic exchange. Nondisjunction of different homologues is independent, as would be expected if there is a failure in the distributive system such that chromosomes move independently to the poles. (Obviously these ideas raise the question of whether spontaneous nondisjunction occurs only in non exchange tetrads; but spontaneously nondisjoined X chromosomes often have suffered exchange. See Merriam and Frost, 1964).

Although exchange may be necessary for nondisjunction, it is not sufficient. However, in contrast with the large number of mutations affecting the occurrence of crossovers, Table 5.6 is able to include only one mutation that affects disjunction as (presumably) its primary defect. This is claret-nondisjunctional (ca^{nd}), which like almost all of the recombinational mutants affects the first meiotic division. There are two difficulties in turning to the obvious explanation of such a mutant: that it is defective in centromere or spindle function. Since the defect does not apply to meiosis II, it seems unlikely to lie in the spindle, and if in the centromere must affect some function necessary only for the first meiotic division. However, this does not explain the nonindependent nondisjunction of homologs, since a simple defect in the centromere would lead to independent nondisjunction of affected chromosomes. An answer to this problem may lie in the mechanism of segregation, if there is some relationship between the way the different centromeres segregate (see Baker and Hall, 1976).

Of the mutants specific for defects in oogenesis, only *mei-38*, *mei-99*, and *mei-160* affect the second as well as the first division. The level of the defect is not yet clear. Although it is possible that these mutations identify functions directly involved in both meiotic divisions, another possibility is that they act only at meiosis I, but failure then may create defective chromosomes that behave abnormally in the subsequent division.

Two mutants have been obtained that cause defects in both oogenesis and spermatogenesis (see Table 5.7). The mutant *mei-S332* displays its defect at meiosis II, when there is nondisjunction or loss of chromosome pairs. However, it turns out that the mutation does not affect meiosis II directly, but identifies a function acting between the meiotic divisions. This appears to be concerned with holding together the sister centromeres, since Davis (1971) found that in *mei-S332* the centromeres separate prematurely, before meiosis II commences. This causes chromosomes to segregate abnormally on the subsequent spindle. Whether the mutation identifies a component of the centromere itself, or some other product that acts upon it, remains to be seen.

The mutant *ord* appears to have its primary defect in females at the level of exchange, since recombination is greatly reduced (although not in the

Table 5.7   Mutations affecting both male and female meiosis in D. melanogaster

Mutant locus	Origin, location and dominance	Observed effects	Cause of action
mei-S332	natural population; chromosome II; semi dominant	causes independent nondisjunction of all chromosome pairs at meiosis II; excess of nullo types suggests chromosome loss occurs also; homozygotes show 30–40% of gametes aberrant for any given chromosome, heterozygotes show aberrant frequencies of 1–2%	sister centromeres separate precociously; locus may specify product required to hold centromeres together between meiosis I and meiosis II; gene product might be component of centromere or involved in action occurring at time of separation
ord	EMS induced; chromosome II; recessive	recombination reduced to 8% of control in *ord/ord* and to 84% of control in *ord/ +* ; reduction is nonuniform but does not restore linearity with physical map; independent nondisjunction of all chromosomes occurs at both meiosis I and II in both ♂ and ♀ that are *ord/ord* (no effect seen in *ord/ +*); nondisjunction for any given chromosome is ~40% in ♀ (30% at I, 10% at II) and is ~ 30% in ♂ (10% at I, 20% at II)	locus must be necessary for exchange in ♀ and some nondisjunction at meiosis I may be caused by lack of exchange; but nondisjunction at meiosis II in ♀ and at both divisions in ♂ must be due to some other effect; level of action may be at pairing, necessary for subsequent proper orientation

References: *mei-S322*—Sandler et al. (1968), Davis (1971); *ord*—Mason (1976)

same manner as either the nonuniform or the uniform reduction seen in the other mutants). Nondisjunction occurs at both meiotic divisions; at least some of the nondisjunction at the female meiosis I may be a consequence of the failure in recombination, but this does not explain the defect at meiosis II or the defects at both meiotic divisions of spermatogenesis. One possibility suggested by Mason (1976) is that the mutation may affect pairing in males as well as females in such a way that the subsequent orientation of the chromosomes is incorrect at both divisions.

To summarize the properties of the meiotic mutants of D. melanogaster, virtually all affect only the first meiotic division in either spermatogenesis or oogenesis (see Figure 5.7). In the male, this is seen in increased nondisjunction, which in the case at least of the sex chromosomes appears to be specific for the chromosome (X) carrying the mutations *mei-269*, etc. In the female, almost all mutations appear to act at the level of preconditions necessary for exchange, distorting the usual (non linear) pattern of recombination by causing a relatively greater reduction of exchange in regions distal to the centromeres. These mutations, and the *mei-9* mutation inhibiting

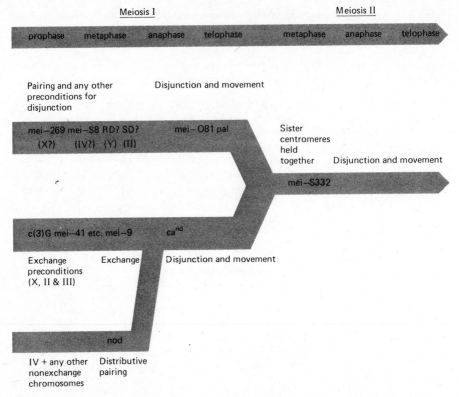

Figure 5.7. Pathway for genetic control of meiosis showing the probable times at which the known mutant loci act (and upon which particular chromosomes, where appropriate).

exchange itself, cause nondisjunction to occur at the first division. This appears to be the result of the lack of exchange, which allows nonexchange chromosomes to enter the distributive pool and thus to pair on the basis of size instead of homology. The only mutations acting directly on nondisjunction at present appear to be ca^{nd} (which also affects the first embryonic mitoses) and *nod* (which is specific for the distributive system).

Only two mutations affect meiosis in both male and female. While both result in nondisjunction at the second division, in each case this results from some earlier action. In *mei-S332* this concerns precocious separation of sister centromeres between the divisions (presumably a function specific for meiosis); and in *ord* there may be some earlier cause since a reduction in recombination (in the female) and first division nondisjunction (in both sexes) also occur.

Almost all meiotic mutants are specific for the reductional division. The failure to isolate mutants equally defective in both meiosis and mitosis may be because they would be inviable. This may explain the absence of mutations in the second meiotic division, since it may be that these also would affect mitosis. Of the three mutations that have strong effects on mitosis, ca^{nd} and *pal* were isolated as meiotic mutations and *mit* for its effect specific to mitosis. All act in this way only upon the very first embryonic divisions. This suggests that components of the chromosomes that are necessary for these divisions may be synthesized at the preceding meiosis and are subsequently replaced.

However, many of the meiotic mutants do display effects on mitosis of lesser magnitude. One explanation is that the mutants are leaky and that mitosis is (for unknown reasons) less sensitive to disruption than meiosis. Another possibility is that the loci may specify products whose major role is at the first meiotic division, but which are involved in a minor capacity in mitosis. The defects are observed as increased frequencies of chromosome loss or nondisjunction and are associated also with increased mitotic recombination, implying that this results from causes distinct from the origins of meiotic recombination (Baker, Carpenter, and Ripoll, 1978). Loci *mei-41*, *mei-9* and *c(3)G* increase mitotic chromosome instability, probably by leaving breaks in the chromosomes unrepaired. This is consistent with the idea that these loci specify functions also involved in repair. The type and distribution of chromosome breaks is characteristic for *mei-41*, *mei-9*, and also three mutagen-sensitive mutants, *mus-102*, *-105*, and *-109* (Gatti, 1979). Many other *mei* and *mus* mutants do not have this effect. Thus although chromosome instability is seen in *mei-352*, it appears unrelated to repair ability. The mutants apparently specific for disjunction, *mei-S332*, ca^{nd} and *ord* all increase mitotic chromosome instability; the relationship of this effect to their specific meiotic defects is not clear. As a general conclusion, however, it seems that some functions involved in meiosis may also be concerned with maintaining chromosome integrity in somatic cells. The mutants *nod* and *pal* do not affect mitotic stability, a result which for *pal* demonstrates strikingly the specificity of its effect upon early mitoses.

Mitotic Recombination

The lengthy prophase during which pairing and genetic exchange takes place is peculiar to the first meiotic division. But recombination can occur also during mitotic division. This is a rare event and sometimes is described as *somatic crossing over*. Mitotic recombination was first recognized by Stern (1936) in Drosophila, where it may result in mosaic spots of altered genotype. It has since been well characterized in Drosophila and in fungi, although presumably it occurs also in other species (for review see Becker, 1976).

The consequence of mitotic crossing over is to generate two daughter cells that are reciprocal recombinants. When this occurs early in the development of a somatic tissue, the descendants should constitute clones displaying the reciprocal genotypes. This predicts the occurrence of "twin spots," one representing each of the reciprocal types. However, in Drosophila single spots often are seen. This might be due to the death of one progeny type, the occurrence of double crossovers, or non reciprocal recombination.

A difficulty in analyzing the recombinant chromosomes generated by mitotic crossing over is that they are present only in somatic cells of one individual, and thus cannot be passed to the next generation. However, mitotic crossing over also may occur in spermatogonial or oogonial cells, in which case the resulting recombinants are visualized in the form of the progeny.

The occurrence of meiotic recombination at the four strand stage, rather than at the preceding stage of two strands, was shown by relying upon the following segregation of all four centromeres (see above). The lack of any such reduction after mitotic recombination makes it impossible to tell whether it occurs at the two strand (pre replication) or four strand (post replication) stage. *Somatic pairing* describes the juxtaposition of homologues that is sometimes seen at mitotic prophase, but since it is not known whether this is the time at which somatic crossing over occurs, it is not clear whether this is involved in mitotic recombination.

The frequency of somatic crossing over has not been estimated in terms of events per mitosis, but can be quantitated by the number of spots per 100 flies. By this measure, the frequency of spontaneous mitotic recombination can be increased by upto about an order of magnitude by irradiation with X rays or ultraviolet or by treatment with mutagens such as nitrogen mustard (Friesen, 1973; Puro, 1964; Haendle, 1971a,b; Auerbach, 1946; Prudhommeau, 1972; Becker, 1975).

What is the relationship between mitotic recombination and meiotic recombination? One difference is indicated by the occurrence of mitotic recombination at the same frequency in both males and females. Another is revealed by the relative frequencies measured for mitotic recombination between pairs of loci: mitotic recombination shows a relatively much greater occurrence of crossing over in the region of the centromere. In fact, a comparison of mitotic recombination maps with the physical maps of the salivary gland chromosomes shows a much better correspondence than is seen with the usual meiotic recombination maps (Garcia-Bellido, 1972; Becker, 1974, 1976). This suggests that mitotic recombination is not subject to the control of crossover distribution that applies to meiosis, but may occur at random.

Does mitotic recombination take place by exchange between precisely homologous sites (as in meiosis) or is it less specific? The induction of mitotic recombination by treatments that cause chromosome breaks is consistent with the idea that the process may depend on breakage and reunion events in which the rejoined ends are not necessarily exactly matching (see Becker, 1976). In this case, the two recombinant progeny cells would not have reciprocal genotypes. One might carry a duplication and the other a deletion (presumably causing lethality). This is one possible model for explaining the predominance of single spots; but, of course, the presence of twin spots does demonstrate that in at least some instances both products of the mitotic recombination are viable.

The occurrence of mitotic recombination may be greatly increased by a large number of mutations known as *Minutes*. These are recessive lethals with certain common dominant developmental effects. The *Minutes* have been thought to cause an increase in the spontaneous frequency of mitotic recombination, one that may occur only in the chromosome arm in which the mutation is located (Stern, 1936; Kaplan, 1953). However, in contrast with this conclusion, Ferrus (1975) has reported that *Minutes* do not cause any increase in the spontaneous frequency, but do cause an increase in the mitotic recombination frequency induced by X rays; this appears to be a general effect applying to the entire complement. This would imply that the effect of the *Minute* mutations might be to increase sensitivity to X-rays, perhaps interfering with the repair process in such a way that recombination becomes more likely.

In general, it does not seem that mitotic recombination is related to the process of genetic exchange at meiosis. Its response to treatments that induce chromosome breaks raises the possibility that it may be a consequence of the operation of repair systems.

CHAPTER 6

Somatic Cell Mutants

Genetic analysis of the bacterial genome has provided a penetrating approach for identifying the loci that are concerned with the expression of a given gene and for defining their roles as structural or regulator functions. Isolation of the gene products makes it possible to correlate changes in the protein with mutations in the gene and to follow the actions of regulator proteins in vitro. First from studies with the lactose operon, and then also from studies on other operons, the now classical model proposed by Jacob and Monod in 1961 has been confirmed in some detail. We shall come later to the question of the extent to which models of this type may account for the control of eucaryotic gene expression (see Chapter 28). Here it is pertinent to rely on the concept that the same general classes of genetic functions may be involved. Structural genes code for proteins with enzymatic or structural functions in the cell. Regulator genes code for proteins concerned with determining whether the structural genes are expressed. The expression of genes may be controlled by adjacent elements in the DNA whose role is to be recognized by proteins and which should therefore exert cis dominant control of the contiguous functions. In this chapter we shall be concerned with the extent to which an approach similar to that successfully applied to bacteria can be utilized with eucaryotic cells in culture to investigate the organization of the genome.

The genetic systems of most higher eucaryotes, especially the mammals, make it difficult or impossible to obtain a series of genotypes identical except for allelic differences at defined loci. Isolation of mutants is limited by the low number of individuals that can be screened and by difficulties in devising selective techniques. The nature of the mutants that can be obtained is restricted by the effects that mutation may have upon the viability of the organism. Many known mutants can be described only by their phenotype; the function of the mutated gene remains unknown. Even where mutants have been obtained, the resolution of genetic mapping is limited by the low number of progeny per mating and the (comparatively) long gestation period. In view of these constraints, perhaps it is not surprising that detailed genetic maps are available for few eucaryotes. Excluding the fungi, maps with sufficient markers to make possible the location of any new mutation are available for D. melanogaster and M. musculus. And only in Drosophila is it possible to correlate the genetic map with the cytological map.

Some of these difficulties can be overcome by utilizing somatic cell lines established in culture. This makes it possible to treat the system in terms analogous to those used with bacteria: a parental cell line can be cloned— that is, a large number of cells can be grown from a single ancestor—and this ensures that any mutants that are isolated can be compared directly with the parental line. Since large numbers of cells can be handled in this way, it is possible to isolate variants that occur at the low frequencies characteristic of mutation. Selective techniques can be applied; and it is possible to work with mutations in essential cell functions by establishing conditional lethal systems in which the mutation is effective under nonpermissive growth conditions, but is ineffectual in permissive conditions that allow perpetuation of the cell line. In principle, it should be possible to isolate mutations in any function for which a selective technique can be devised, although obviously this is limited to those functions that are expressed in cultured cells. What proportion this may be of the total genetic information utilized by a species through all stages of development is not clear.

Since there is no natural system for genetic recombination between somatic cells, in order to map the loci of mutation it is necessary to rely upon the formation of somatic cell hybrids, in which two parental cells are fused into a single daughter cell. This system allows complementation analysis to be performed; and by correlating the disappearance of a genetic marker with the loss of a chromosome from the hybrid, it is possible to assign mutant loci to chromosomes, although often it is not possible to make any assignment more detailed than this.

Comparatively sophisticated methods for isolating mutants of cultured cells must therefore be linked with rather primitive methods for mapping the sites of mutation. This limits the present usefulness of somatic cell lines for elucidating the organization of the genome. We shall come in Chapters 8 and 9 to the properties and uses of somatic cell hybrids. Here we shall be concerned with the use of somatic cell lines to obtain mutations in the mammalian genome. The main direction of this effort is to distinguish structural gene mutations from regulatory mutations; from this it should be possible to define the complexity of the systems governing the expression of structural genes. At present, the division of the eucaryotic genome into its constituent genetic functions is largely undefined. It is not clear what proportion of the genome is concerned with coding for proteins; nor is there any basis for assessing the proportion of such loci serving as structural or regulator genes. Nothing is known of the cis dominant recognition elements that are presumed to lie adjacent to the genes. The nature of the eucaryotic gene is the subject of Chapter 17, but here we shall try to assess the extent to which somatic cell genetics can contribute to its elucidation.

Somatic Cell Lines

One of the restrictions imposed on the use of somatic cells in culture is that only a limited number of cell lines has been established. Many are of fibroblastic origin. Primary cultures of fibroblasts have a uniform diploid karyotype, but unfortunately have a finite life span in culture (Hayflick, 1965). This makes it necessary to rely upon the "established" cell lines, some of whose origins are summarized in Table 6.1. The fibroblast lines grow in either monolayer or suspension culture and appear capable of indefinite propagation (for review see Thompson and Baker, 1973).

The principal difficulty posed for genetic studies with established cell lines is their instability of karyotype. No line has remained a true diploid. Some, including HeLa and L cells, are extensively heteroploid, whereas others, including CHO, V79, and DonC, have been described as pseudo diploid. But in all cases changes in the chromosome constitution appear to occur at least from time to time, even though stability may prevail for periods of many generations.

It is not clear to what extent the content of genetic information varies in cultured cell lines. Kraemer et al. (1973) have suggested that the total DNA content of HeLa cells remains more constant than would be expected from the variation in chromosome number. This has been taken to imply that

Table 6.1 Established somatic cell lines

Name	Organism and tissue of origin	Chromosome complement of species ($2n$)	Current mode and range of chromosome number in cell line
HeLa	human cervical carcinoma (fibroblast)	46	heteroploid ~71 (38–106)
L	mouse connective tissue (fibroblast)	40	heteroploid ~70 (40–115)
3T3	mouse embryo (fibroblast)	40	heteroploid ~76 (70–80)
S49	mouse lymphoma	40	pseudo diploid (39–41)
Friend	mouse erythroleukemia	40	pseudo diploid ~39 (35–43)
CHO	Chinese hamster ovary (fibroblast)	22	pseudo diploid ~21 (20–22)
V79	Chinese hamster lung (fibroblast)	22	pseudo tetraploid
Don DonC	Chinese hamster lung (fibroblast)	22	pseudo diploid ~23
BHK 21	Syrian hamster kidney (fibroblast)	44	pseudo diploid ~44 (42–45)

Table 6.1 (Continued)

Name	Organism and tissue of origin	Chromosome complement of species (2n)	Current mode and range of chromosome number in cell line
BSC 1	African green monkey kidney (epithelioid; SV40 infected)	60	pseudo tetraploid ~115
RAG	mouse renal adenocarcinoma (epithelioid)	40	heteroploid ~74
—	Rana pipiens embryo	26	haploid
S2	D. melanogaster embryo (epithelial)	4	diploid ♀ complement with some haplo-IV

The characteristic mode and range of chromosome numbers found in these lines is based on counts of the number of chromosomes seen at metaphase. With some mouse lines, this may be an under estimate because the mouse complement is entirely telocentric whereas metacentric chromosomes are formed by fusion in the cell lines; a better estimate would therefore be given by counting chromosome arms. In all cases the estimates reflect particular sub lines; others will be similar, but not necessarily identical.

References: HeLa—Gey et al. (1952), Syverton and McClaren (1957), Kraemer et al. (1971); L—Sanford et al. (1948), Littlefield (1963), Kraemer et al. (1971); 3T3—Todaro and Green (1963), Peterson and Weiss (1972); S-49—Horibata and Harris (1970); Friend—Friend et al. (1966); CHO—Puck et al. (1958), Tjio and Puck (1958); V79—Ford and Yerganian (1958); Don—Hsu and Zenzes (1964), Stubblefield (1966); BHK 21—McPherson and Stoker (1962), Stocker and McPherson (1964), Kraemer et al. (1971); BSC 1—Hopps et al. (1963); RAG—Klebe et al. (1970); R. pipiens—Freed and Mezger-Freed (1970), Mezger-Freed (1977); S2—Schneider (1972).

variations in the chromosome number may reflect rearrangements of genetic material rather than changes in content. But a critical test of this hypothesis requires measurements of absolute DNA contents of cell lines related by clear changes in the chromosome complement. Although it is certainly possible that heteroploid lines retain a constant overall content of DNA, in any case this does not imply that the same sequences are present; nor would the retention of a DNA content close to the diploid level necessarily imply that the pseudo diploid lines remain constant in genetic content.

Even in cell lines that retain a constant modal chromosome number, rearrangement of material appears to be common. Banding studies of the chromosomes of one line of CHO cells showed that only 8 of the 22 Chinese hamster chromosomes remained unaltered. Rearrangements of the parental complement could be recognized in other chromosomes (Deaven and Petersen, 1973). Further, but lesser, variations are seen in comparisons of sublines (Worton, Ho, and Duff, 1977). But the precision of this technique is somewhat limited (see Chapter 15). The genetic status of these lines therefore remains uncertain, notwithstanding proposals that the recognition of rearrangements implies some constancy in overall genetic content. It is not clear to what extent deletions and duplications may have occurred during the extensive reorganizations of chromosomal material. Even in pseudo diploid lines, then, the content and overall organization of genetic material cannot be assumed to be typical of the diploid organism.

The lack of certainty about the genetic content of established cell lines has several important consequences. It makes it difficult to establish a true mutation rate, since the number of copies in the cell of any given gene is not certain. When mutations occur representing the loss of some function, it is necessary to establish whether a gene is present in mutant form or is absent from the cell line because of the loss of genetic material. Dosage effects may be expected, which may make difficulties in studying regulatory functions.

Mutation and Epigenetic Variation

Perhaps the principal problem in investigating somatic cell mutants lies in the question of how a mutation is to be distinguished from an epigenetic variation. A mutation, of course, constitutes a change in the nucleotide sequence of DNA; this may or may not result in a change in the phenotype. Phenotypic changes that do not result from alterations in the genotype are attributed to epigenetic variation. By this definition, epigenetic variation would include the different phenotypic states that are produced by changes in gene expression during development. Indeed, this is the obvious paradigm for the cause of epigenetic variation, since it has become a truism to say that all somatic cells of an organism contain the same genetic content, although each has a different and characteristic phenotype.

The limited extent to which it has been possible to perpetuate cells in culture argues that only some types of cell may be able to grow in this way; and this at least raises the possibility that adaptations in gene expression may be necessary for survival in culture. The stability of the initial phenotype therefore cannot be taken for granted. Whenever a change occurs in the phenotype it is necessary to ask: is this the consequence of a mutation that has altered the genetic constitution of the cell, or does it result from an epigenetic event that alters gene expression but not the content of genetic information?

What criteria should be satisfied by putative mutations? A general expectation is that mutations (both forward and reverse) should occur at rates comparable to those seen with bacterial genes. This rests on the assumption that procaryotes and eucaryotes suffer mutational events at similar frequencies (including both hits by external agents and errors in replication) and have repair systems of broadly similar efficiencies (and see footnote to Chapter 17). If eucaryotic loci are taken to comprise targets comparable to bacterial genes, spontaneous mutations should arise at a rate of $\sim 10^{-6}$ per cell generation at any given locus. The mutant phenotype should breed true: it should be stably transmitted under nonselective conditions. This can be expressed as a requirement that the phenotype should not revert to that of the parent at a rate greater than is attributable to spontaneous mutation: this again means that it is expected to be $< 10^{-6}$ per cell generation. Nonselective conditions are used to test this, since mutations usually are isolated under conditions that make the mutant phenotype essential for survival. However, while the occurrence and reversion of a given phenotypic alteration at these rates is consistent with the anticipated properties of mutational events, this congruence alone provides no proof of mutation. So little is known about epigenetic variation that it is not possible to exclude the alternative explanation that a stable epigenetic event occurs and is reversed at frequencies the same as those expected of mutation.

Any demonstration that treatment with agents known to be mutagenic in bacteria increases the rate of a phenotypic alteration provides more suggestive evidence that a mutation is involved. In this case, increasing doses of the mutagen should increase the rate of occurrence of the mutation. Specificity in mutagenesis is difficult to demonstrate with somatic cells, but in principle a mutation caused by one sort of mutagen (base substitution or frameshift) should be reverted by mutagens of the same class.

Conclusive proof that an alteration in phenotype results from mutation can in principle be provided in two ways. For any mutant, it should be possible to map the mutation to a specific genetic locus that behaves in a Mendelian manner. Of course, this cannot be tested fully with somatic cell systems, but the demonstration that the alteration is associated with a given chromosome is acceptable. For any mutant in a gene coding for protein, a critical test of mutation is to demonstrate an alteration in the structure of the protein product. This is the first, crucial step in characterizing any gene system; for once a structural gene has been identified, further mutations can be examined by genetic or biochemical means to determine whether they reside in the same gene or represent another function. A major purpose of somatic cell genetics thus is to identify the structural genes among mutant functions.

Isolation of Auxotrophic Mutants

The range of mutations that can be isolated in somatic cells depends largely upon the selective systems that can be devised. Mutant lines have so far been isolated for auxotrophy, deficiencies in enzymes of nucleotide metabolism, resistance to drugs and other cytotoxic agents, and temperature sensitivity for growth, which includes defects in protein synthesis and in cell division. Of course, the total number and scope of the mutants appears very restricted compared with the panoply of variations available in bacteria. Perhaps the most striking difference between the eucaryotic and procaryotic systems is that in only a very small number of instances so far has it been possible to identify the molecular basis of the alteration in a somatic cell line.

Auxotrophic mutants have been isolated by screening cells for inability to grow on medium lacking some metabolite, contrasted with ability to grow on complete medium. The principle of this approach is illustrated in Figure 6.1. A population of cells is plated on a medium in which auxotrophs cannot grow and an agent then is applied to kill all dividing cells.* Puck and Kao (1967) introduced the use of BUdR followed by exposure to light. The BUdR is incorporated into DNA as an analog of thymidine and its disintegration by visible light then introduces lethal breaks into the genetic material. Only cells unable to grow on the deficient medium survive this treatment. These auxotrophs may be recovered and grown on complete medium, in which their deficiency is overcome by provision of whatever metabolite(s) they are unable to synthesize.

By using ethylmethane sulfate (EMS) and nitrosoguanidine (MNNG) to induce mutations before applying the selection protocol, Kao and Puck (1968, 1969, 1972) used the BUdR killing method to select several classes of auxotrophs from CHO (Chinese hamster ovary) cells. These include gly^- cells unable to survive unless glycine is provided, ade^- variants for whose growth adenine is necessary, ino^- cells requiring inositol, and variants with double requirements for adenine and thymine (AT^-) or with triple demands for glycine, adenine and thymine (GAT^-). The original CHO line from which these auxotrophs all have been isolated happened to be auxotrophic for proline, so all these sublines carry the marker pro^- in addition to their other deficiencies.

In some of the auxotrophic series the different isolates have been divided into complementation groups by use of the hybrid cell techniques described in Chapter 9. For example, the gly^- variants fall into four complementation groups ($glyA$-D), while the ade^- series falls into classes $adeA$-I (Kao, Chasin and Puck, 1969; Patterson, 1975). The step in which each type of variant is deficient can be identified by determining the ability of the cell line to grow on metabolic intermediates representing different stages in a meta-

---

*The technique of killing cells unable to grow under given selective conditions was originally developed with bacteria, of course, and is discussed in Volume 3 of Gene Expression. It was first applied to HeLa cells to select auxotrophs by DeMars and Hooper (1960), who made use of the greater toxicity of aminopterin toward growing compared with nongrowing cells. But the selective power of this system was limited and technical difficulties prevented full quantitation of the results. Chu et al. (1969) used this approach with more success.

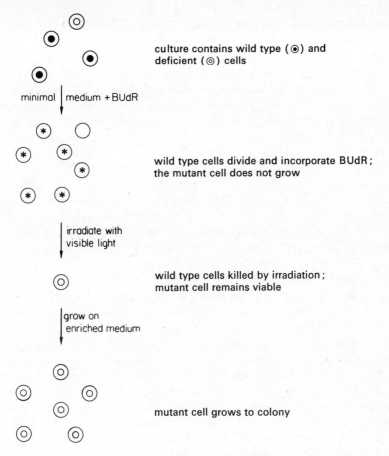

culture contains wild type (⊙) and
deficient (◎) cells

minimal | medium + BUdR

wild type cells divide and incorporate BUdR;
the mutant cell does not grow

irradiate with
visible light

wild type cells killed by irradiation;
mutant cell remains viable

grow on
enriched medium

mutant cell grows to colony

Figure 6.1. Protocol for selection of nutritional mutants. Cells lacking the ability to synthesize some essential growth factor cannot grow on minimal medium and therefore are immune
from the effects of incorporation of BUdR and irradiation with visible light. Method of Kao
and Puck (1968).

bolic pathway, by identifying the intermediates that may accumulate when
cells are grown on a precursor whose subsequent metabolism is blocked,
and ultimately by demonstrating a deficiency in some enzyme activity. Thus
the *glyA* line is deficient in the enzyme serine hydroxymethylase, which is
needed for the metabolic utilization of glycine (see Table 6.6); the defects of
the other *gly⁻* classes are not known. The steps at which several of the *ade⁻*
lines are defective have been identified by following the accumulation of
intermediates; these variants prove to lie in eight of the stages of the pathway for de novo purine biosynthesis (Patterson, Kao, and Puck, 1974; Patterson, 1975, 1976a,b).

Selection of Drug-Resistant Mutants

The mutants that in principle are the most straightforward to isolate are those that are resistant to drugs or other cytotoxic agents. The resistance of a cell line to any cytotoxic agent can be described in the form of a *drug response curve*. Cells are plated in the presence of varying concentrations of the agent and the proportion of survivors is determined for each drug concentration. Usually the data are plotted on a log scale, as in the example of Figure 6.2. The general form of the curve is a steep fall off of survivors from a value of 1 (all cells survive) to about 10^{-4} survival, as the drug concentration increases. A plateau is often reached at a surviving proportion of $< 10^{-4}$ and this represents drug-resistant variants. These colonies then may be picked and tested for drug resistance in fresh cultures. Not all will be resistant, due to metabolic variations, but any mutants should be present in this fraction. Formally this may be described as selection of resistant variants in the presence of a drug concentration sufficient to leave only a proportion of $< 10^{-4}$ survivors.

The degree of resistance attained by the surviving variants may be characterized by a dose-response curve obtained in the presence of greater concentrations of the drug. Essentially the resistant variant should form a response curve parallel to that of the parental line, but displaced from it to higher drug concentrations. The extent of the displacement is a measure of the degree of resistance of the variant line compared with the parent.

The usual parameter used to describe the resistance of any cell to a given drug is the D_{10}, the concentration of drug needed to reduce survival to 10% (that is, a proportion of 10^{-1}). From the example shown in Figure 6.2 it is evident that a survival level of 1% falls in a more suitable part of the parental curve for comparison. However, in many instances the resistant lines

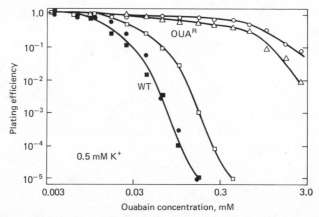

Figure 6.2. Dose response curve to plating in ouabain for wild type cells (● and ■) and three mutant lines (□, Δ, and ○). The mutants □ and ○ correspond to the classes summarized in Table 6.3. Data of Baker et al. (1974).

are sufficiently resistant to make it impractical to complete a dose response curve to a survival level of 1%, because the concentration of drug (theoretically) needed would be too high to test. An example is given in the figure. Indeed, sometimes a response curve cannot be taken even to the 10% survival level. The usual quantitation—only an approximate measure—of the extent of resistance in a variant line is given by the ratio of the resistant D_{10} value to that of the parental line. In general it is reasonable to expect this to be at least one order of magnitude if the resistant line is to be accepted as a genuine variant.

Drug-resistant lines may be obtained in two ways. In single-step protocols, acquisition of resistance is determined following a single exposure to a given concentration of the drug. Depending upon the nature of the variation, a series of resistant lines may have varying levels of resistance, that is they display various D_{10} values, although all are established in a range determined by the drug concentration used for selection. We should note, therefore, that a single-step protocol does not ensure an all or nothing response. When different concentrations of a drug are compared in selection protocols, the resistant survivors may characteristically display appropriate ranges of resistance. But "complete" resistance usually is achieved by a drug level in excess of that needed to reach the 10^{-4} survival level.

Sometimes it is impossible to acquire a high level of resistance in a single step protocol (all the parental cells die when the attempt is made). In such instances, a multistep protocol may be successful. In this approach the parental line is exposed to a given concentration of drug and then the resistant variants are exposed to a higher concentration. This may be repeated. Assuming that the basis of the resistance lies in mutation, this implies that several mutations are necessary to achieve high resistance. This means that in a single-step protocol the chance occurrence of a multiple event is very rare, too rare to detect, but that by isolating cells after each event it becomes possible to obtain cells that have gained all the necessary mutations. From the perspective of genetic characterization, usually only single-step protocols are acceptable, since the occurrence of multiple events means that it becomes too complicated to identify the genetic basis of the alterations.

Any compound that is cytotoxic can be used to select resistant variants. Table 6.2 summarizes the selective agents that have been used and the basis of their lethal effects. In several cases resistance has been located in a specific function in the variant. With nucleotide analogs, which are lethal when incorporated into DNA, most commonly the enzyme activity responsible for their conversion into the DNA precursor may be lost. With guanine analogs, the enzyme hypoxanthine guanine phosphoribosyl transferase (HGPRT) is needed to generate the nucleotide (see Figure 6.3); with adenine analogs the corresponding enzyme is adenine phosphoribosyl transferase (APRT); and with bromodeoxyuridine (BUdR), thymidine kinase is necessary for this conversion. Of course, other types of variant that display resistance also may be obtained, for example in uptake. A comparable type of

Table 6.2 Selection of variant somatic cells in culture

Selective agent	Lethal effect of agent	Basis of change in variant phenotype	Ref
8-Azaguanine 6-Thioguanine	incorporated into DNA in place of guanine	deficiency in level of HGPRT activity	A
Bromodeoxyuridine	incorporated into DNA in place of thymidine	deficiency in level of thymidine kinase activity	B
8-Azaadenine 2,6-Diaminopurine	incorporated into DNA in place of adenine	deficiency in level of APRT activity	C
Toyocamycin Tubercidin	not known (they are analogs of adenosine)	deficiency in level of adenosine kinase activity	D
Hydroxyurea	inhibits DNA synthesis	change in kinetic properties or level of ribonucleotide reductase	E
Aminopterin Methotrexate	inhibit folate metabolism	increase in level of dihydro folate reductase (DHFR) activity	F
α-Amanitin	inhibits rRNA transcription	RNA polymerase II becomes resistant to inhibition	G
Ouabain	inhibits plasma membrane Na^+-K^+ activated ATPase	not known	H
PHA	(all) lectins cause cell agglutination	deficient in level of N acetyl D glucosaminyl transferase activity	I
Colchicine	inhibits mitosis by binding to tubulin	change in cellular permeability to colchicine and other agents	J
Diphtheria toxin	inhibits translation by adding ADP to elongation factor 2	change in cell permeability; or failure of EF2 to become ADP-ribosylated	K
Trichoderm	inhibits translation	60S ribosome subunits bind less inhibitor	L
Emetine	inhibits translation	40S ribosome subunits altered to resist inhibition of chain elongation	M
L-Azetidine-2-Carboxylic acid	incorporated into protein in place of proline	less inhibition by AZCA of glu–pro conversion in biosynthetic pathway	N
Thialysine	incorporated into protein in place of lysine	deficiency in level of asn-tRNA synthetase	O

Table 6.2 *(Continued)*

Selective agent	Lethal effect of agent	Basis of change in variant phenotype	Ref
5-Fluorotryptophan	incorporated into protein in place of tryptophan	reduction in uptake by tryptophan transport s, stem	P
^{3}H amino acids	incorporated into protein	reduction in affinity of $\alpha\alpha$-tRNA synthetases for amino acids	Q
PALA (N phosphonoacetyl-L-aspartate)	inhibits synthesis of pyrimidines	increase in levels of carbamyl-P-synthetase, aspartate transcarbamylase and dihydroorotase	R
Cycloleucine	inhibits synthesis of S-adenosyl methionine	increase in level of methionine adenosyl transferase	S
2-Deoxygalactose	inhibits Leloir metabolic pathway	deficiency in level of galactokinase activity	T

References: A—Littlefield (1963), Gillin et al. (1972), Sato et al. (1972), Sharp et al. (1973); B—Kit et al. (1963), Roufa et al. (1973); C—Jones and Sargent (1974), Chasin (1974); D—Gupta and Siminovitch (1978f); E—Lewis and Wright (1978, 1979); F—Littlefield (1969), Nakamura and Littlefield (1972), Chang and Littlefield (1976); G—Chan et al. (1972), Somers et al. (1975), Ingles et al. (1976); H—Baker et al. (1974), Mankovitz et al. (1974); I—Stanley et al. (1975a,b); J—Ling and Thompson (1974), Bach-Hansen et al. (1976); K—Moehring and Moehring (1977), Gupta and Siminovitch (1978e); L—Gupta and Siminovitch (1978c); M—Gupta and Siminovitch (1976, 1977, 1978b); N—Wasmuth and Caskey (1976a); O—Wasmuth and Caskey (1976b); P—Taub and Englesberg (1976, 1978); Q—Thompson et al. (1975, 1977); R—Kempe et al. (1976); S—Caboche and Mulsant (1978); T—Whitfield et al. (1978).

agent is the amino acid analog, which is lethal by virtue of its incorporation into protein. Resistant variants may have changes in the protein synthetic apparatus (usually in the synthetase needed to charge tRNA with the analog), or may be unable to take up the analog from the medium. Drugs that act on specific steps of transcription or translation are potentially useful in obtaining variants in these processes; some that have been successfully exploited are α-amanitin (which acts on RNA polymerase II), emetine and trichoderm (which act on the ribosome), and diphtheria toxin (which acts on elongation factor 2). Other selective agents that have been introduced recently show that this approach can be applied to processes as various as cell division (resistance to colchicine), agglutination (resistance to lectins), membrane functions (resistance to ouabain), and other metabolic interactions (including resistance to inhibitors of folate metabolism or

Table 6-3 Characteristics of drug resistant cell lines selected in single step

Cell line	Drug resistance selected	D_{10} of parent	D_{10} of resistant variant	Relative resistance (variant D_{10}/ parent D_{10})	Drug concentration used for selection	Ref.
L human diploid fibro-blast	8 azaguanine	0.1 μg/ml (7×10^{-7} M)	9.0 μg/ml (6×10^{-5} M)	90	3.0 μg/ml (2×10^{-5} M)	A
	"	2×10^{-6} M	$>10^{-3}$ M	>5000	8×10^{-6} M	
CHL	BUdR	1.0×10^{-5} M	$>10^{-3}$ M	>100	1.0×10^{-5} M	B
CHO	methotrexate	1.2×10^{-8} M	2.1×10^{-7} M 6.6×10^{-7} M	17 55	1.0×10^{-7} M (survival of $<10^{-5}$)	C
CHO	α-amanitin	0.2 μg/ml	0.8 μg/ml >6.0 μg/ml	4 30	0.5-1.0 μg/ml (survival of 10^{-4})	D
CHO	ouabain	3.0×10^{-5} M	1.8×10^{-4} M 3.0×10^{-3} M	6 100	1-3 $\times 10^{-5}$ M (survival of 10^{-4})	E
CHO	PHA	3.5 μg/ml	>1000 μg/ml	>250	10-25 μg/ml (survival of $<10^{-4}$)	F
CHO	colchicine	1.0×10^{-7} M	1.0×10^{-6} M 4.0×10^{-6} M	11 40	2.5×10^{-7} M (survival of 10^{-4})	G
CHO	emetine	3.2×10^{-8} M	4.1×10^{-7} M 2.0×10^{-6} M	13 60	1-2 $\times 10^{-7}$ M (survival of $<10^{-6}$)	H
	(second step)	4.1×10^{-7} M	4.5×10^{-5} M	110	5.0×10^{-6} (survival of $<10^{-6}$)	
CHL	AZCA	4.0×10^{-4} M	$>8.0 \times 10^{-4}$ M	>2	6.0×10^{-4} M	I
CHO	hydroxyurea	10 μg/ml	50 μg/ml 250 μg/ml	5 25	25 μg/ml (survival of $<10^{-3}$)	J
S49	dexamethasone	4.0×10^{-8} M	$>5.0 \times 10^{-7}$ M	>10	1.0×10^{-7} M	K
CHO & human diploid fibroblasts	diphtheria toxin	2×10^{-4} LF/ml	10 LF/ml	$>50,000$	0.01–20 LF/ml	L

References: A—Littlefield (1963), Albertini and DeMars (1973); B—Roufa et al. (1973); C—Flintoff et al. (1976); D—Ingles et al. (1976); E—Baker et al. (1974); F—Stanley et al. (1975b); G—Ling and Thompson (1974); H—Gupta and Siminovitch (1976, 1978a); I—Wasmuth and Caskey (1976a); J—Wright and Lewis (1974); K—Sibley and Tomkins (1974a); L—Gupta and Siminovitch (1978d,e).

pyrimidine synthesis). As we have remarked before, provided that a sufficiently specific selective technique can be devised, any cellular enzyme can be the subject of a protocol to obtain variants.

The levels of drug resistance that have been acquired in those resistant lines whose properties have been quantitated are summarized in Table 6.3. In most cases there is a substantial increase in resistance, usually at least ten fold and often 100 fold, according to the ratio of the variant and parental D_{10} values. The level of resistance may be different in each independently isolated variant. Where independently isolated variants tend to fall into discrete classes with different levels of resistance, an example of each class is given in the Table.

Isolation of Temperature-Sensitive Mutants

Conditional lethal systems are necessary to allow isolation of mutants whose defects render the cells nonviable. Nonsense mutations have been used very successfully with phage/bacterial systems and offer the great advantage that any function can be mutated (see Volume 3 of Gene Expression). The nonsense mutation is expressed when the mutant gene finds itself in a cell in which the nonsense codons retain their usual function of causing the termination of polypeptide synthesis. But the mutation is suppressed when the gene resides in a cell that carries a mutation causing the production of a transfer RNA able to respond to the nonsense codon by inserting an amino acid into the polypeptide chain (see Volume 1 of Gene Expression). The mutation is expressed under the first set of *nonpermissive* conditions; but the mutant genomes can be perpetuated under the second set of *permissive* conditions (although the phenotype may not be completely wild type). Two conditions are necessary for such analysis. There must be available suppressor genes coding for mutant tRNAs that recognize nonsense codons. And it must be possible to place a given nonsense mutation in genetic backgrounds that are isogenic apart from the alleles present at the suppressor locus. Neither is yet generally fulfilled with virus/eucaryotic cell systems.

Another type of conditional lethal system is provided by temperature-sensitive mutations. Usually the mutant phenotype is expressed at high temperature, but the cells remain wild type at low temperature. The reverse situation is provided by the somewhat rarer cold sensitive mutants, in which high temperature represents the permissive conditions and low temperature the nonpermissive conditions. Since such responses are not to be expected of changes in the sequence of a recognition element in DNA, the isolation of temperature-sensitive mutants is taken as evidence that the mutation resides in a sequence coding for protein.

Temperature-sensitive mutants display the mutant and wild-type phenotypes under different sets of conditions, whereas with nonsense mutants different sets of cells must be used; this is no problem with viral nonsense mutants that can be grown on either wild-type or suppressor cells, or with mutations in nonessential cellular functions, but means that nonsense mutations do not provide a conditional lethal system for analyzing mutations that are lethal to the cell. The disadvantage of temperature-sensitive mutations is that it is not always possible to devise a selection procedure for isolating a given type of mutation. One approach is to isolate mutants at high temperature and then to screen these at low temperature for reversion to the wild phenotype. But this is difficult with eucaryotic cells in which it is often possible to obtain only small numbers of mutants. Another complication is that the temperature span within which somatic cells can be grown in culture is rather narrow, generally in the range of 34 to 39°C, which reduces the power of the selective technique because mutants isolated at high temperature often are fairly leaky and possess some proportion of the wild-type function.

Protocols to isolate mutants in essential cellular functions generally have taken the form of isolating variants that are unable to grow at high temperature, but which can be perpetuated at low temperature. Usually the cells are treated with a mutagen and, after an intervening period, are suspended at high temperature in the presence of agents that kill cells able to grow. Thompson et al. (1970, 1971) introduced the technique of adding highly tritiated thymidine to kill cells during DNA synthesis, followed by addition of ara-C to eliminate any cells that are resistant to the ^{3}H-TdR killing. Meiss and Basilico (1972) used treatment with FUdR to kill wild-type cells able to grow at high temperature; Patterson, Waldren, and Walker (1976) used the BUdR visible light technique. The difficulty with all of these protocols is that the mutations are isolated simply for their inability to grow at high temperature; any defect with this consequence may be included. To identify the mutant functions, it then becomes necessary to examine the temperature sensitive clones individually.

Table 6.4 summarizes data on the defects that have been identified. They fall into three general categories. Mutants inhibited in protein synthesis are unable to divide at high temperature and several defects of this sort, involving the production of either ribosomes or charged tRNA, have been found. Mutants unable to replicate DNA because they cannot enter S phase are likely to be the most common cell cycle variants. This happens simply because the lethal treatment at high temperature is directed against cells in S phase, so that mutants unable to enter it will be recovered more frequently than those blocked within or subsequent to it (which may tend to synthesize sufficient DNA to suffer the lethal effects). Mutants in S phase entry are potentially important in identifying genes involved in control of the cell cycle, although they may also include genes coding for components of the replication apparatus. The third class of mutants represents defects in cell division itself, either in mitosis or cytokinesis, and again may be expected to identify functions concerned with the control or mechanism of division.

Table 6.4 Characteristics of temperature-sensitive cell mutants

Parental cell line	Reported nature of defect	Ref.
BHK-21	deficient in production of 28S rRNA, possibly in processing	A
LY5178Y murine leukemia	require alanine; ala-tRNA synthetase more heat labile in vitro	B
CHO	protein synthesis inhibited; $\alpha\alpha$-tRNA synthetases more heat labile in vitro (several different mutants)	C
CHO	α amanitin resistance of RNA polymerase is temperature sensitive	D
CHO	HGPRT activity	E
BHK-21	cells blocked in G1 and unable to enter S phrase	F
CH Wg-la	DNA synthesis ceases; cells may be unable to enter S phase	G
L	DNA replication inhibited; DNA synthesis occurs but small intermediate chains accumulate; cells accumulate in G1 or S	H
L	DNA synthesis defective; cells accumulate in S or G2	I
CHO	colchicine resistance at high temperature; unable to enter S phase at low temperature; relationship of effects doubtful	J
HM-1	cells accumulate in metaphase or anaphase; unable to complete mitosis or cytokinesis (several different mutants)	K
CHO	cytokinesis blocked	L

References: A—Toniolo et al. (1973); B—Sato (1975); C—Thompson et al. (1973, 1978); D—Ingles (1978); E—Fenwick and Caskey (1975); F—Burstin et al. (1975); G—Roscoe et al. (1973); H—Sheinin (1976); I—Setterfield et al. (1978); J—Ling (1977); K—Wang and Yin (1976); Wissinger and Wang (1978); L—Smith and Wigglesworth (1972); Thompson and Lindl (1976).

Measurement of Mutation Rates

The occurrence of mutations can be quantitated by two parameters: the frequency and the rate. The frequency of occurrence is simply a measure of the proportion of cells in a given population that displays the variant phenotype. In principle, when a cell line is cloned and then examined later for some variant type, the frequency of variation should increase with generations of growth as new variants accumulate (although eventually an equilibrium should be attained when the generation of new variants is balanced by reversion). The frequency of variation is therefore an unsatisfactory measure, since it depends upon the conditions of growth. To calculate the rate of occurrence from the frequency, it is necessary to know how many cell divisions there have been since the line was propagated from a clone known not to contain any variants. It is necessary to determine how many independent mutational events have been responsible for generating an observed number of mutants.

The most effective ways to perform such an analysis are based upon the *fluctuation test* that Luria and Delbruck (1943) introduced to determine the source of bacterial mutations. There had been discussion about whether selective techniques lead to the isolation of mutants previously present in the cell population, or are responsible for inducing the mutations. (This is discussed in Volume 3 of Gene Expression.) The Luria-Delbruck fluctuation test consists of determining the number of mutational events in each of a large number of independent cultures. The protocol is to establish replicas in each of which a small initial number of cells (usually 100–200) is grown to yield a large population (generally $\sim 10^6$ cells); the number of mutants is then determined for each replica by applying an appropriate selective technique.

If mutations are induced only after exposure to selective conditions, the same results should be obtained in all cultures, since events occurring prior to the selection are irrelevant; it is only the number of cells plated under the selective condition that determines the number of mutants. But if mutations occur at random with respect to time, the number of mutants in each culture should depend upon the time at which mutation happened to occur. Mutations should occur early in some cultures, so that many mutant progeny are generated by the subsequent divisions that occur before the mutants are selected; mutants may arise at late times in other cultures, so that a smaller number of progeny results from each event. In some cultures there may be no mutational events. The statistical significance of the observed fluctuation in the number of mutants found in the replicate cultures can be determined by comparison with a control that represents samples taken simply from one culture. The range of the number of mutants per culture should be very much greater in the experimental replicas than in the control replicas, as in the example described in Table 6.5. This can be expressed by a much increased ratio of the variance to the mean (the control value should be close to 1).

The rate of mutation is expressed as the number of events occurring per cell generation. In an unsynchronized culture of growing cells, the total number of generations that has occurred is given by the expression:

$$\frac{N_f - N_i}{\ln 2}$$

where N_f and N_i are the final and initial numbers of cells. When the final number is very large compared with the initial number, which is usually the case, the total number of cell generations is given by:

$$\frac{N_f}{\ln 2}$$

If m is the total number of mutations occurring when a small inoculum containing no mutants is grown to a large population of cells, then the mutation rate is:

$$\frac{m \cdot \ln 2}{N_f} \quad \text{cell-generation}^{-1}$$

Table 6.5 Luria-Delbruck fluctuation test for spontaneous occurrence of oubain resistance in L cells

Conditions of Experiment
Number of replicate cultures = 28
Initial number of cells per replica = 100
Final number of cells per replica = 3×10^7

Experimental Results
Number of replicas with N *oua^r* colonies in:

N	Experimental replicas	Control replicas
0	6	0
1	2	0
2	9	0
3–4	5	0
5–8	2	10
9–16	2	19
17–32	1	1
33–64	1	0
>64	0	0

Summary of Occurrence of Mutations

Range of number of mutants/replica	0–34	6–18
Median number of mutants/replica	2	11
Mean number of mutants/replica	4.6	10.5
Variance	92	9.1
Variance/mean	20	0.9

Calculation of Mutation Rate
Proportion of cultures with no mutants, $P_0 = 6/28 = 0.21$
Average number of mutants per culture, $m = \ln P_0 = 1.5$
Total number of cell generations per culture = $3 \times 10^7/\ln 2 = 4.3 \times 10^7$
Rate of mutation = $1.5/4.3 \times 10^7 = 3.5 \times 10^{-8}$ cell-generation^{-1}

Data of Baker et al. (1974).

The difficulty in determining the number of independent mutational events is that what is observed in the series of experimental replicas of the fluctuation test is the number of mutants that has accumulated in each culture. This is much greater than the number of mutations, of course, because each spontaneously occurring mutant will have divided a number of times that depends upon its time of origin. It is therefore necessary to find a method of estimating the number of mutations from the observed number of mutants.

One approach utilizing the fluctuation test is to rely upon the number of cultures in which no mutants were found. The occurrence of mutations in the replica cultures should follow the Poisson distribution. In this case, the equation

$$P_0 = e^{-m}$$

relates the probability that no mutational event will occur (P_0) to the average number of mutational events per culture (m). Since P_0 is given directly by the proportion of experimental replicas lacking mutant colonies, m can be estimated as $-\ln P_0$. This method has the advantage that it is not biased by possible differences in the growth rates of mutant and wild type cells. It has the disadvantage that it makes a rather inefficient use of the data gained in the fluctuation test and is accurate only when conditions are such that the proportion of cultures lacking mutants is neither too large nor too small.

Other methods for calculating the number of mutations make use of the mean or median number of mutants observed in the experimental replicas. A method developed by Luria and Delbruck relies upon the observation that once cultures have progressed to a size large enough to make mutation a likely event, the number of mutants tends to the same value irrespective of time of occurrence. However, mutation rates calculated by this method are generally too great, because of the bias caused by the occurrence of some mutations at very early times, before the assumption becomes valid. Other statistical methods for estimating the mutation rate from the fluctuation test were developed by Lea and Colson (1949). However, none excludes the problems caused by differential growth of mutant and wild-type cells. Usually the fluctuation tests therefore are performed under conditions allowing the number of mutations per culture to be estimated from the number of null mutant plates.

Another approach is possible in cases where selection can be applied directly to plates containing the large populations grown from small inocula. Each mutant clone observed on the plate then can be taken to represent the descendents of a single spontaneous mutant, so that the number of mutant clones is equal to the number of mutational events (although a difficulty may be caused when mutant cells migrate from one colony to start another). This method follows an analysis developed for bacteria by Newcombe (1949), but has the disadvantage that in many eucaryotic cell systems it cannot be applied, because it is necessary to respread the cultures for the selective protocol.

A final point is that the use of the fluctuation test establishes only the random nature with reference to time of the event causing the variant phenotype. Thus it excludes the possibility that the selective agent has induced the variant. This does not in itself distinguish, however, between events of genetic and epigenetic origin.

Mutation Rates in Somatic Cells

Available data on rates of mutation in cultured somatic cells are summarized in Table 6.6. When Luria-Delbruck fluctuation tests have been performed, the rate suggested by the null method usually lies in the range of 10^{-6}–10^{-8} per cell per generation. Rates of reversion from the mutant phenotype have been obtained in very few cases; but the frequencies with which the wild type cells reappear in the mutant populations are low enough to demonstrate that the mutants have the stability expected of genetic changes. In almost no instance has the nature of the revertants been identified. The isolation of revertants, and the demonstration that they regain the original function that had been lost, is an important criterion for distinguishing point mutations from deletions. Point mutations may revert at the mutant site or may be suppressed by intragenic or extragenic secondary mutations. Deletions, however, cannot regain the lost function (although sometimes it may be possible to acquire the wild phenotype by substituting some other function for the deleted gene).

Spontaneous mutation rates should be increased by exposure to agents known to induce nucleotide changes in DNA. The Table mentions the mutagens shown to increase a given rate (or frequency) by at least an order of magnitude. Reversion also should be increased by mutagenesis but the data here are very scanty. Mutagens should induce characteristic types of defect, whose reversion should be stimulated by the same, but not by different, classes of mutagen. Point mutagens such as EMS or MNNG should be distinct in their effects from frameshift mutagens such as the acridines. But again the data are too few to allow a proper assessment of whether these mutagens behave in eucaryotic cells in the manner expected from their effects upon the bacterial genome.

Mutagenesis usually is performed by treating cells with an appropriate agent at a concentration and for a duration that increases the spontaneous rate of appearance of mutants by up to 100-fold (for review see Thompson and Baker, 1973). The mutagenic effect is influenced by both the dose of mutagen and duration of exposure. After mutagenesis, time is allowed for cells to recover from sublethal effects and to fix and express any mutations. Generally this means that the cell number is allowed to increase some 10-fold (about three generations of growth). Then the cells are transferred to nonpermissive conditions to select mutants. Since many cells may be killed

Table 6.6　Rates of occurrence and

Cell line	Mutant phenotype	Protein defect	Dominance
CHO	*glyA*	lack serine hydroxymethylase activity	recessive
CHO	*pro⁻*	cannot convert glutamic acid to semi-aldehyde	recessive
CHL (V79)	*gln⁻*	not known	recessive
CHL (V79)	*aza^r*	lack HGPRT activity	recessive
human lymphoblasts	*thg^r*	lack HGPRT activity	not known
CHL	*BUdR^r*	lack thymidine kinase activity	not known
Hepa-1 (mouse hepatoma)	*BaP^r* (benzo pyrene)	lack aryl hydrocarbon hydroxylase activity	not known
CHO	*oua^r*	in membrane transport	codominant
CHO	*eme^r*	in 40S ribosome subunit	recessive
CHO	*dap^r*	not known	not known
A9	*5F-trp^r*	not known	not known
A9	*phe^r*	not known	not known
human diploid fibroblast	*dip^r*	protein synthesis	not known

The spontaneous mutation rates are events per cell per generation, calculated from the data of Luria-Delbruck fluctuation tests by the null method. Where rates are not known, frequencies are given. Where no mutations have been observed in a given number of cells or cell-generations, the appropriate upper limit is given for the frequency or rate. The mutation rates given here have been calculated from the fluctuation data given in the references cited and may differ from the rates calculated by the authors.

by the mutagenesis protocol, the number of mutants is expressed as a proportion of the survivors.

The conditions of selection are important. One problem in selecting for drug resistance, for example, is that the drug in the medium may become depleted because it is taken up by sensitive cells; to ensure adequate selective pressure it may therefore be necessary to provide changes of medium.

reversion of somatic cell mutants

Spontaneous rate of occurrence	Mutagens increasing rate ($>10x$)	Spontaneous rate of reversion	Nature of revertants
not known	not characterized	frequency 3×10^{-8}	regain enzyme activity
not known	not characterized	frequency 1.2×10^{-6}	not known
not known	not characterized	1.4×10^{-7}	not known
1.5×10^{-8}	EMS, NG, X ray, BUdR/light	frequency 7.0×10^{-8}	not known
frequency 1.6×10^{-6}	EMS, NG	frequency $<1.5 \times 10^{-7}$	not known
not known	EMS, NG	frequency 4.0×10^{-7}	not known
2.3×10^{-7}	not characterized	not characterized	
3.5×10^{-8}	EMS	not characterized	
4.9×10^{-8}	EMS	not characterized	
$<1.0 \times 10^{-7}$	not characterized	$<1.0 \times 10^{-7}$	not known
8.7×10^{-7}	EMS, NG	not characterized	
4.8×10^{-5}	EMS, NG	not characterized	
6.5×10^{-7}	EMS, NG, ICR	not characterized	

References: *glyA*—Kao and Puck (1968), Chasin et al. (1974); *pro*—Kao and Puck (1967, 1968, 1969); *gln*—Chu and Malling (1968); *aza*—Chu et al. (1969, 1972), Chu (1971); *thg*—Sato et al. (1972); *BUdR*—Roufa et al. (1973); *BaP*—Hankinson (1979); *oua*—Baker et al. (1974); *eme*—Gupta and Siminovitch (1976, 1978a,b); *dap*—Chasin (1974); *trp*—Taub and Englesberg (1976); *phe*—Englesberg et al. (1976); *dip*—Gupta and Siminovitch (1978d).

Another difficulty is that metabolic cooperation may occur between cells; when the cell density is great enough, azaguanine-resistant mutants lacking HGPRT activity cannot be isolated, because contact between wild-type and mutant cells suppresses the deficiency in phenotype, probably by transferring nucleotides.* This means that there must be a compromise between

*References: Subak-Sharpe et al. (1969), Cox et al. (1970), Clements and Subak-Sharpe (1975b), Van Zeeland et al. (1972), Bols and Ringertz (1979).

using enough cells to detect rare mutants but few enough to avoid densities great enough to support metabolic cooperation.

Detailed dose-response curves have been obtained for some mutagens, where the expected relationship is found in which increasing doses of mutagen cause linear increases in the occurrence of mutants, at least over parts of the mutagen concentration range (Hsie et al., 1975; Friedrich and Coffino, 1977). The rates at which azaguanine resistance is induced by X rays lie in the range of 10^{-7}–10^{-6} per rad per locus for cultured mammalian cells (Bridges and Huckle, 1970; Bridges et al; 1970; Chu, 1971, 1974; Albertini and DeMars, 1973). This contrasts with the rate of mutation induced in E. coli, of about 10^{-9} per rad per locus. The discrepancy of two orders of magnitude raises the question of whether the same process is involved in procaryotic and eucaryotic cells.

A final point is that in comparing spontaneous and induced mutation rates it is necessary, of course, to avoid selective treatments that interfere with the mutation rate. This difficulty may be encountered by the BUdR light irradiation technique, for Chu, Sun, and Chang (1972) reported that this is mutagenic (increasing the spontaneous rate of azg^r variants by about ten fold under their conditions). This does not detract from the usefulness of the technique in isolating auxotrophic mutants, of course, but does mean that spontaneous mutation rates measured after this procedure may be in error.

An important caveat about the data in Table 6.6 is that in most cases they represent the rate (or frequency) of mutation to a given phenotype. Although in some cases the principal cause of this phenotype has been shown to lie with a defect in a given enzyme, usually it is not possible to say that all of the isolated mutants have this defect. For example, resistance to 8-azaguanine may be acquired by loss of the enzyme HGPRT; but among the resistant cells there are some that have retained the enzyme, and thus presumably have acquired resistance in some other way, perhaps involving uptake (see below). This means that the rate of mutations that cause loss of HGPRT may be lower than the rate of mutation to azaguanine resistance. As a general rule, therefore, the rate of mutation to a given phenotype cannot be equated with the mutation rate for a single locus, unless it is known that the selective technique is precise enough to allow survival only of a single type of mutant. It might seem that reversion of point mutations provides a more specific assay—restoration of the activity of a known gene product—which allows calculation of a rate applicable to a single locus. But distinguishing revertants at the mutant site from those in which there has been a second, compensating mutation (intragenic or extragenic) requires more detailed genetic mapping than usually is possible with somatic cells. Within these restrictions, the number of loci for which mutation and reversion rates can be calculated is very small. Perhaps only in the case of the HGPRT locus has sufficient characterization been possible to allow calculation of locus-specific rates.

The lack of sufficient experimental data to determine a mutation rate per locus raises a general question. Can any estimate be made of the rate of mutation to be expected in somatic cells? The general approach to this problem has been to assume that the value per haploid locus should be of the same order as that seen in bacteria, that is, $< 10^{-6}$ events per cell per generation. Following this line of argument, any value substantially greater than this, say $> 10^{-4}$ cell-generation^{-1}, is taken to reflect the occurrence of epigenetic rather than genetic events. Yet the rate of mutation depends on the intrinsic operations of the cell (as is shown by the identification of *mutator* genes that can alter the rate of mutation in procaryotes; see Volume 1 of Gene Expression). Relevant functions may include the accuracy of replication, the activity and specificity of repair systems, and perhaps other events influencing DNA metabolism. In view of the unknown relationship between the corresponding systems of eucaryotic and procaryotic cells, is it possible to say definitely that the mutation rate per locus should be similar in cultured somatic cells and bacteria? Indirect evidence that this may be so is provided by a somewhat similar dependence of the increase in mutation rates upon some known mutagens. Insufficient evidence is available to justify any firm assertion about the rate of mutation in the germ lines of higher eucaryotes (that is, as seen in populations). The concept that the rate at which genetic changes should occur in somatic cells is of the order of 10^{-6} per locus per generation thus remains a supposition—albeit a reasonable one—based on the analogy with bacteria.

Mutation and Gene Dosage in Cells of Different Ploidies

The rate at which detectable mutations occur in somatic cells should be influenced by the dominance of the mutation and the number of copies of the parental gene. For a dominant mutation, only one allele need be altered; the rate of occurrence therefore should increase in proportion to the number of alleles that provide targets for mutation. The occurrence of dominant mutations may be expected to be rare, however, since most mutational events introduce a defect in the function affected. The lack of a function is recessive; and in order for the cell to display a mutant phenotype it then becomes necessary for all the wild type copies of the gene to be mutated. In a diploid cell, if the rate of mutation per locus is 10^{-6} per generation, the rate of appearance of a recessive mutant phenotype should be 10^{-12} per generation. For a tetraploid cell the rate becomes 10^{-24} per generation. Even the rate for diploid cells is so low that it is reasonable to expect a complete failure of attempts to isolate recessive somatic cell mutants. In general, it is possible to examine at most only some 10^8–10^9 cell generations in an experiment and most reports concern numbers of $< 10^7$ cell generations. The implication is that only X-linked recessive mutations should be detectable in cultured diploid cells (for it appears that only a single X chromosome remains active in cultured female cells). Yet as the data of Table 6.6 show, putative mutations obtained in several cell lines have proved to be recessive when tested in hybrid cells. Of these, only HGPRT is X-linked.

Three types of explanation can in principle be proposed. The rate of mutation may be much greater than expected on the basis of the comparison with bacteria, so that the probability of double events becomes great enough to observe. The chromosome constitutions of the cultured cells may not be diploid for the locus mutated, so that only one wild type allele is present in the parental culture. This model would imply, in fact, that recessive mutations can be identified only in those instances where all copies of the gene but one have been lost, to generate hemizygosity. Or these "mutations" may not be genetic events but may be epigenetic in origin (for review see Siminovitch, 1976).

One approach to investigating the basis of variation in somatic cells is to compare the rates of mutation in cell lines known to differ in the number of copies of a given gene. Two types of protocol have been used. One is to use cell lines that differ in the number of copies of the entire chromosome set, that is, in their ploidy. Another is to compare the rates with which mutant genotypes $(-/-)$ can be obtained from cells that are homozygous $(+/+)$ or heterozygous $(+/-)$ for a given marker.

Two early series of experiments to investigate the effects of ploidy showed little reduction in the frequency of occurrence of mutant phenotypes when the chromosome number was increased. Mezger-Freed (1972, 1974, 1975) took advantage of the development of a haploid cell line derived from embryos of the frog Rana pipiens to show that resistance to BUdR occurs at a rate of 3×10^{-8} per cell-generation in the haploid state and at 7×10^{-7} per cell-generation in pseudo diploid cells. In neither case was the rate increased by treatment with mutagens. This was taken to suggest the occurrence of epigenetic events. Another attempt to compare rates of mutation in cells of different ploidy was made by Harris (1971, 1973) with diploid, tetraploid, and octaploid series of the CHL V79 cell line. Spontaneous mutants selected for resistance to azaguanine occurred at rates of 2.2×10^{-5}, 4.7×10^{-5}, and 1.3×10^{-5} per cell-generation with increasing ploidy. These are not significantly different and do not fit the expectations for either dominant or recessive mutation. Mutagenesis with nitrosoguanidine was equally effective with both diploid and tetraploid cells, that is, the dose response was the same.

Several problems confront the interpretation of these results. In both cases, it is not certain that the same (and recessive) mutations were selected in the different ploidy cell series: drug resistance can occur in more than one way. The possibility that genes have been lost rather than mutated, by chromosome loss or translocation, was not excluded. Some technical difficulties in assessing the rates involved have been pointed out by DeMars (1974): the absolute numbers of resistant cells were rather small, so that the spontaneous mutation rates may have been subject to the bias of small samples; and the cell concentrations may have been within the range that would permit metabolic cooperation.

In short, there are three principal factors that must be satisfied in order to determine the effects of ploidy on the rate of observed mutation. It must be possible to obtain sufficient numbers of mutant cells to make possible an accurate determination of the rate (and this is likely to be difficult in cases of increased ploidy). It must be possible to establish that the same types of mutation are involved in the mutant cells of different ploidies (which means that the nature of the defect must be characterized, so that only numbers of like mutants are compared). And karyotype analysis must establish the relationship between the chromosome complements of the mutant cells of each ploidy.

A further series of experiments has suggested that more than one type of event may be involved in generating mutants with a recessive phenotype from parental cells carrying two copies of the wild type gene. Chasin (1973) used EMS to induce mutants resistant to thioguanine in pseudo diploid CHO cells and in a tetraploid line induced by colcemid. The frequencies of induced mutation were 2.5×10^{-4} for the diploid cells and 9.3×10^{-6} for the tetraploid cells. If the diploid cells carry a single active X chromosome and the tetraploid cells carry two active X chromosomes, the predicted frequency of mutation for the tetraploid line is $(2.5 \times 10^{-4})^2 = 6.25 \times 10^{-8}$; this is 150 times less than the observed frequency. The mutations occurring in cells of both ploidies appear to be comparable, because in both series of mutants, assays of HGPRT activity showed less than 1% of wild-type activity. In both series, fusions to form hybrid cells confirmed that the mutations were recessive. A useful comparison can be made between these frequencies and those obtained in a series of experiments to measure the frequencies of reversion from gly^- to gly^+ induced by EMS. These mutations should be dominant. They occurred at frequencies of 3.4×10^{-6} for diploid cells and 7.1×10^{-6} for tetraploid cells, a relationship close to what might be expected from a doubling of the number of target copies of the gly^- gene.

Why is the occurrence of the recessive phenotype in the tetraploid cells so much greater than expected? One possible explanation is that drug resistance in the tetraploid cells is acquired by a combination of two mechanisms. If the tetraploid cells have two active copies of the X chromosome, resistance might be accomplished by loss of one X chromosome and mutation of the wild-type HGPRT gene on the other. The frequency with which the recessive phenotype occurs then should be the product of the frequency of chromosome loss from the tetraploid line and the frequency of mutation seen in the diploid line.

The rate of chromosome loss can be measured in hybrid tetraploid cells generated by fusion of two diploids, and it is reasonable to assume that a similar rate of loss may apply to the tetraploid cells created by colcemid treatment. Chasin (1972) previously showed that tetraploid hybrids that are heterozygous for thg^r/thg^s acquire resistance to thioguanine at a rate of $5 \times 10^{-5} - 3 \times 10^{-4}$ per cell-generation. If this rate—which is somewhat greater than the apparent rates of mutation to thioguanine resistance—is due to chromosome loss, then it appears that the thg^r X chromosome is lost at an average rate of 10^{-4} per generation. This predicts that the tetraploid cells lose one of the two X chromosomes at a rate of about 2×10^{-4} per cell-generation.

Mutagenesis of the colcemid induced tetraploids was carried out after 150 generations of tetraploid growth, so cells carrying only one X chromosome should be present at a frequency of 3×10^{-2}. Then the frequency of resistant cells should be $3 \times 10^{-2} \times 2.5 \times 10^{-4} = 7.5 \times 10^{-6}$, close to the frequency obtained. This calculation shows that the relationship between the frequencies of resistant cells in the diploid and tetraploid series is consistent with the involvement of chromosome loss in the tetraploids, although of course it does not prove that this is the mechanism.

This model predicts that the thioguanine resistant mutants of the tetraploid series should have only one active X chromosome. In principle this could best be tested by determining karyotypes, but it is not always possible to distinguish the chromosomes of the set with sufficient precision. Most tetraploid cells prove to have lost more than one chromosome (up to 5), which makes it more difficult to distinguish the chromosome missing in common. A less direct alternative is to test for dosage effects in the expression of genes known to be carried on the X chromosome. The gene for G6PDH (glucose-6-phosphate dehydrogenase) is an example. Chasin and Urlaub (1975) used this approach by constructing tetraploid hybrids that were HGPRT$^+$ G6PDH$^+$/HGPRT$^+$ G6PDH$^-$. If chromosome loss is a prerequisite for acquisition of thioguanine resistance, half the resistant mutants should be G6PDH$^+$ and half should be G6PDH$^-$, depending on which X chromosome happened to be lost and which mutated. Results were close to this expectation. Another prediction is that the frequency of segregants should increase with time after establishment of the tetraploids, due to continued chromosome loss. This happens, although at a rate somewhat greater than expected for X chromosome loss alone.

Direct evidence that chromosome loss can be responsible for the disappearance of HGPRT activity was gained in an experiment in which an HGPRT$^+$ Chinese hamster line was fused with an HGPRT$^-$ cell line. Because of translocations, the X chromosomes of the two parents could be distinguished. When the hybrids were selected for resistance to thioguanine, the resistant phenotype was associated with loss of part or the whole X chromosome contributed by the HGPRT$^+$ parent (Farrell and Worton, 1977).

Other results that may be explained by a combination of chromosome loss and mutation have been obtained in analyzing the resistance of mouse lymphoma cells to the cytolytic effects of cyclic AMP and glucocorticoid steroids. (The lethal response to these agents is a characteristic of lymphoid cells.) Coffino, Bourne, and Tomkins (1975) found that mutants of the S49 line resistant to dibutyryl cyclic AMP arise at rates of 2.1×10^{-7} cell generation^{-1} in the original pseudo diploid line and at 2.7×10^{-7} cell generation^{-1} in a tetraploid derivative. Diploid and tetraploid parental cells contain similar levels of cyclic AMP binding activity per unit of cell protein; resistant clones lack or have very little binding activity. Similarly, mutants of the S49 line resistant to 10^{-7} M dexamethasone arise at the same rate of 3.5×10^{-6} cell generation^{-1} in populations of pseudo-diploid and tetraploid sensitive cells (Sibley and Tomkins, 1974a,b). The frequency of occurrence is increased by both 9-aminoacridine and MNNG. Some 80% of the mutants are defective in the ability to bind the steroid; the remainder have defects in stages subsequent to uptake.

Similar results are obtained with the mouse thymoma line W7. Bourgeois and Newby (1977) took advantage of its sensitivity to dexamethasone to use low steroid concentrations (5×10^{-9} M) to isolate partially resistant variants. These arise at frequencies in the range $6.5 \times 10^{-8} - 2 \times 10^{-6}$, with a mean of 3×10^{-7}. Cells resistant to 10^{-5} M dexamethasone can be isolated from them at frequencies of $4 \times 10^{-6} - 1.5 \times 10^{-5}$; in one case the rate was 2×10^{-6} cell generation^{-1}. This contrasts with the failure to isolate directly any fully resistant mutants from the parental line (rate $<1.5 \times 10^{-9}$ cell generation^{-1}). The parental W7 cell line has 30,000 glucocorticoid receptors per cell; whereas the partially resistant lines and the S49 lymphoma have 15,000 sites per cell. (Somewhat lower values have been obtained by Gehring and Tomkins, 1974). This suggests that the W7 line may have two alleles for the glucocorticoid receptor; partially resistant mutants may have only one functional allele and the fully resistant mutants have lost both. The similarity between the partially resistant W7 mutants and the S49 parental line suggests that one of the two original alleles may have been lost or spontaneously mutated before this line was established. This explains the effectiveness of single step mutation protocols in isolating resistant variants. Tetraploids then should have only two functional alleles and could generate resistance by losing one and mutating one.

The influence upon segregation of the number of chromosomes in the cell was assessed in experiments in which Harris and Whitmore (1977) fused pha^r or aza^r (recessive) mutants with diploid or tetraploid cells that were sensitive to phytohemagglutinin or azaguanine. The first type of fusion generates sensitive quasi tetraploid cells, from which the drug resistant variants are recovered at rates of 5×10^{-5} and 1×10^{-5} cell-generation^{-1}, respectively. The second class of fusion generates sensitive quasi-hexaploid cells, from which resistant variants are recovered at rates of about 2×10^{-3} cell-generation^{-1} for each marker. This corresponds to increases of 40 fold and 200 fold compared with the quasi-tetraploid series. If the appearance of resistant cells is caused by segregation of the chromosomes carrying the wild type (sensitive) alleles, this means that the probability of loss must be much greater in hexaploid compared with tetraploid cells.

The rate at which chromosomes or parts of chromosomes might be lost in diploid or pseudo diploid cells is not known. But these results raise the question of whether such events may occur frequently enough to convert some of the supposedly diploid cell lines into hemizygotes for some loci. This would explain the isolation of recessive mutants at rates of the order expected from mutations of single loci. There are several indications that this may be the case. Segregation analysis has been applied to the markers *emt^r* and *thg^r* by fusing resistant CHO cells with various sensitive Chinese hamster cell lines (Gupta, Chan, and Siminovitch, 1978a). This gives pseudo tetraploid cells that are sensitive to the drug, but which generate resistant variants, presumably by loss of the chromosome carrying the resistant allele. The segregation of the *thg^r* marker provides a control, since it is carried by the (single) X chromosome and the hybrids should always possess only two active alleles. This gives a rate for chromosome loss of 4×10^{-4} to 15×10^{-4} for crosses between *thg^r* CHO and various sensitive cell lines. A similar rate of 2.3×10^{-4} is found for segregation of *emt^r* when the sensitive parent is also a CHO line. But the rate declines to 0.16×10^{-4} in crosses with other sensitive parents. This could be explained if the CHO line happens to be hemizygous for the *emt* locus (as it must be for *thg*); but other cell lines have retained two copies of the (wild type) *emt^s* allele, thereby reducing the segregation rate. This is supported by the observation that mutation to emetine resistance occurs in CHO cells at a rate that is 50–100 times greater than that seen with other Chinese hamster cell lines, compared with variations of only 2–5 fold for mutation to *thg^r* or *oua^r* (Campbell and Worton, 1979). Similarly, variations over three orders of magnitude have been seen in the frequencies at which resistance to toyocamycin occurs in several Chinese hamster cell lines (Gupta and Siminovitch, 1978f). The discrepancies are not great enough to represent the difference between independent mutations of one versus two copies of the gene, but would be consistent with the presence of one allele in some cells and two in others if segregation as well as mutation is invoked in the latter case.

Another criterion has been applied to examine cells for ploidy of the *ama* locus, at which codominant resistance to α-amanitin may be acquired. In *ama^r* CHO cells, all of the RNA polymerase II becomes resistant; but in two other Chinese hamster cell lines only half of the RNA polymerase II appears to resist inhibition (Gupta, Chan, and Sminovitch, 1978b). This suggests that the CHO line may be hemizygous, while the others have retained two alleles of the *ama* locus.

It is not yet possible to assess the extent to which hemizygosity occurs in CHO or other cell lines. These instances of hemizygous loci contrast with cases in which two alleles have been shown to be present, such as the case of the genes for APRT at which resistance to azaadenine is acquired (see next section). But these results raise a prima facie case that changes in the number of gene copies in cultured cell lines may be important or even essential in isolating recessive mutants. Such changes might well most commonly occur during chromosomal rearrangements rather than by wholesale loss.

Mutation Rates in Homozygous and Heterozygous Cells

Attempts to compare mutation rates in homozygous and heterozygous diploid cells support the idea that mutants may be isolated more readily when only one wild type gene is present. But because of the difficulties in quantitating the low frequencies involved with the homozygous cells, usually it is not possible to make a significant comparison of the two rates. When Clive et al. (1972) selected for resistance to BUdR in a mouse lymphoma line, they obtained only a single clone from a sample of 4.4×10^8 cells. If this clone represents the conversion of a $(+/+)$ parental cell into a $(-/-)$ mutant, the rate for the double mutation takes the value of $<1.6 \times 10^{-9}$. From the $(-/-)$ mutant line, it is possible to isolate $(+/-)$ revertants; and then from these in turn, further $(-/-)$ mutants can be isolated. This allows the spontaneous rate of mutation to be measured for the conversions $(-/-) \longrightarrow (+/-)$ and $(+/-) \longrightarrow (-/-)$. These data are summarized in Table 6.7. That the events observed in these experiments correspond to mutations of thymidine kinase is suggested by measurements of relative enzyme activity of 0, 44, and 100 in the $(-/-)$, $(+/-)$ and $(+/+)$ cells respectively.

The rate at which the $(-/-)$ mutants arise from the heterozygote is 6×10^{-8} cell generation^{-1}, which predicts that the mutation rate for the homozygote should be $(6 \times 10^{-8})^2 = 3.6 \times 10^{-15}$ cell generation^{-1}. Obviously this is very much lower than might be expected to permit the isolation of even the single clone observed in the selection protocol applied to the homozygote.* These results therefore do not support the hypothesis that recessive mutants are generated from homozygous diploids by two mutational events, each occurring independently at the same rate seen in the heterozygote.

The rate of reversion from the $(-/-)$ mutant to the $(+/-)$ heterozygote is tenfold lower than the forward rate of mutation measured in the heterozygote. A difference close to two orders of magnitude is seen in the other data summarized in the Table. This is not surprising, since the specificity

*The authors originally calculated that the number of cell generations in the homozygous selection had been 2×10^{10}, which would correspond to a mutation rate of 5×10^{-11} cell generation^{-1}. This would require the growth of more cells than is generally possible; the correct value is given in Table 6.7. But although the incorrect value is much smaller than the rate of mutation from the heterozygote, it remains several orders of magnitude greater than expected from the mutation rate seen in heterozygotes. Thus even this value does not allow the conclusion that there have been independent mutational events.

Table 6.7 Comparison of mutation rates in homozygous and heterozygous cells

Cell line	Phenotype	Mutation rate in homozygote	Mutation rate in heterozygote	Rate of reversion	Origin of mutations
CHL (V79)	BUdR^r (TK$^-$)	frequency $<2 \times 10^{-8}$	frequency 2×10^{-4}	frequency 10^{-5}-10^{-7} with average 3.5×10^{-6}	spontaneous
Mouse lymphoma	"	$<1.6 \times 10^{-9}$	6×10^{-8}*	6×10^{-9}*	spontaneous
CHO	DAP^r (APRT$^-$)	frequency 2×10^{-7}	frequency 2.0×10^{-4}	frequency 1.3×10^{-6}	EMS-induced
Mouse thymoma W7	dex^r	frequency $<1.5 \times 10^{-9}$ for full resistance; partial resistance varies but has mean of 3×10^{-7}	frequency 2.5×10^{-5} from partial to full resistance	not known	spontaneous

Values marked with an asterisk (*) were quoted by the authors but have not been confirmed by analysis of original data. References: Roufa et al. (1973), Clive et al. (1972), Chasin (1974), Bourgeois and Newby (1977).

of reversion may be greater than that of mutation. Any mutation that inactivates the gene product is sufficient to satisfy the selection for absence of an enzyme activity. Since many mutations may have this effect, this is likely to be a more frequent event than a recovery of activity, which must be due to exact reversion or to suppression of the previous mutation; only a limited number of sites may be available in this capacity. By the same criteria, it is impossible to use reversion rates for calculations about the expected occurrence of recessive forward mutations in either homozygotes or heterozygotes; although the rate of reversion is, of course, a more precise measure of the occurrence of mutational events.

Similar experiments with this system have been performed by Roufa, Sadow, and Caskey (1973). Again, no accurate estimate could be made of the mutation rate in homozygotes, except to say that the mutant with which these experiments started had occurred at a frequency of $<2 \times 10^{-8}$. However, $(+/-)$ revertant heterozygotes arose from the $(-/-)$ mutant with a frequency of 2×10^{-4}, certainly much greater than the frequency with which they occur in homozygous $(+/+)$ diploids. Without determining mutation rates, it is not possible to say whether these data support the idea that two events occur in the homozygote, compared with one in the heterozygote, but certainly they are consistent with it. A caveat, however, is that the frequencies were not increased by treatment with EMS, MNNG, or UV.

The $(+/-)$ revertants isolated from the $(-/-)$ mutant line had regained sensitivity to BUdR and displayed thymidine kinase activities ranging from 2 to 17% of wild type. These revertants were used to select a large number of $(-/-)$ clones during the measurement of the forward heterozygous mutation rate. These $(-/-)$ clones reverted to the $(+/-)$ condition at a variety of rates which is what would be expected if they contain different mutations inactivating the thymidine kinase enzyme.

Another system that has been used to compare mutation rates in homozygotes and heterozygotes is the acquisition of resistance to 2,6-diaminopurine, an adenine analog. Cells become dap^r by losing the activity of the enzyme APRT. Chasin (1974) found that spontaneous mutation rates were too low to measure; but the frequencies with which mutations were induced by EMS could be compared. The large difference between the homozygous frequency of 2×10^{-7} and the heterozygous frequency of 2×10^{-4} is consistent with the idea that two mutations must take place to generate resistance in the homozygotes, whereas one event will suffice for the heterozygotes. The reversion frequency, measured by selecting for the presence of APRT by plating the mutants in medium containing adenine and azaserine (which blocks endogenous purine synthesis), was two orders of magnitude below that of the forward rate per locus.

A similar approach allowed Jones and Sargent (1974) to suggest that CHO cells contain two genes for APRT which may be independently mutated. Using 8-azaadenine to select for lack of the enzyme activity, they found that resistance is acquired in two steps. The first step yields cells with an intermediate resistance level and these possess about 35% of the wild type APRT activity. Fully resistant mutants may be selected from these and have <6% of wild type enzyme activity. The second step occurs at a spontaneous rate of 3×10^{-7} cell generation^{-1}; the rate is increased tenfold by treatment with EMS. Revertants selected from the fully resistant cells show the phenotype of intermediate resistance. This suggests that mutation of one allele leads to partial resistance, mutation of the second allows complete resistance, and reversion of one of the mutant alleles regenerates cells of intermediate resistance.

Identification of the Structural Gene for HGPRT

One of the principal aims of somatic cell genetics is to identify mutations in structural genes. The HGPRT system is one of the very few in which this has been accomplished. But the properties of the mutants and revertants for this enzyme illustrate many of the difficulties that are encountered in establishing whether a given mutation lies in the structural gene.

The absence of an enzyme activity may result in principle from any one of three types of genetic alteration: the structural gene may be mutated; there may be a deletion of this locus (either within the chromosome or by loss of the chromosome); or there may be a mutation in some regulator function necessary for expression of the structural gene. Much weight has been attached to the isolation of revertants as a criterion for distinguishing point mutations (in either the structural gene or a regulator gene) from deletions. Point mutations should take the form of missense, nonsense, or frameshift alterations and should revert or be suppressed appropriately. Although deletions should not revert, the absence of reversion cannot be taken as evidence for deletion, first because in some instances the rates may be too low to detect, and second because of the unusual conditions in which some apparently nonrevertable lines have been induced to revert.

Structural gene mutants should have a protein product of altered amino acid sequence. Missense mutants should have a protein of normal size, but are deficient in enzyme activity because of their amino acid substitution. Nonsense mutants (and often frameshift mutants also) should synthesize a truncated protein, due to premature termination of protein synthesis. A protein product should therefore be present in all missense mutants and in those nonsense (and frameshift) mutants located toward the end of the gene. The short polypeptide produced by nonsense mutants near the start of the gene may be undetectable or may be degraded very rapidly.

Some mutants retain enzyme activity at a level lower than that of the wild type. If the deficiency is caused by a reduction in the level of enzyme due to a control mutation, the protein present in the mutant cells should have the same amino acid sequence as wild type. If the mutation lies in the structural gene, the enzyme may display altered kinetics or other physico-chemical properties in vitro. However, such alterations do not completely exclude the possibility that there is a deficiency in some function that modifies the primary protein product, or perhaps that processes it from a precursor.

Mutants lacking the enzyme activity completely can be tested by immunological assay for the presence of cross-reacting material (CRM) that represents inactive enzyme. The presence of protein related to the enzyme but lacking its catalytic activity suggests that the mutation lies in the structural gene. Purification of the mutant protein should allow its structure to be compared with the wild type protein; and ultimately this should make possible the determination of the site of mutation (and of reversion) within the protein. (The absence of CRM does not prove that the gene product is missing, since it might be present but have lost the antigenic site by mutation.)

A gene that is essential for HGPRT activity has been located by studies of a human metabolic disease. Lesch and Nyhan (1964) first characterized a disorder that is responsible for neurological abnormalities and is associated with a high rate of excretion of uric acid. The disease shows sex-linked inheritance and so the locus responsible must be located on the X chromosome. Blood cells taken from patients with Lesch-Nyhan disease lack HGPRT activity (Seegmiller et al., 1967; Choi and Bloom, 1970).

Early results identified cross-reacting material from five patients with Lesch-Nyhan disease (Rubin et al., 1971). However, in more recent experiments with a series of 16 patients, only one possessed CRM, the remainder displaying <3% of the wild-type level (Ghangas and Milman, 1975; Upchurch et al., 1975). The positive reaction obtained in previous experiments may therefore have been due to insufficient specificity in the antibody preparation, which could have reacted with some other protein component.

Under the usual conditions of assay, no HGPRT activity can be detected in extracts from the single patient possessing CRM. But by changing the assay conditions, McDonald and Kelley (1971) were able to identify an activity with a K_m altered from that of the wild type. This suggests that in this case the mutation may reside in the structural gene. The nature of the mutations in the remaining cases, which appear to be the more common, cannot be defined from these results.

Other evidence consistent with the idea that the X chromosome carries the structural gene for HGPRT is provided by mapping studies with hybrid cells; for loss of the human X chromosome can be correlated with loss of the ability to produce human HGPRT (see Chapter 9). (This does not in itself constitute a formal proof, because it could be that the X chromosome carries a gene whose product is necessary for expression of an HGPRT gene located elsewhere.) The simplest, and therefore the most likely, explanation of these results, however, is that the X chromosome carries the structural gene for HGPRT and that the loss of this enzyme is caused by mutations at this locus, located either in the structural gene or in some control element. Since the same sex linkage appears to be maintained in other mammals, this makes the HGPRT locus especially suitable for studies with somatic cells, for the presence of only a single active X chromosome (see above) avoids the difficulties involved in characterizing recessive mutations in diploid cells.

Isolation of Mutants Deficient in HGPRT Activity

Analogs of nucleotides are powerful agents for selecting cell lines with deficiencies in certain enzymes of nucleotide metabolism. Their lethal effects may be mediated by incorporation into DNA or by feedback inhibition of nucleotide biosynthesis. Purine and pyrimidine triphosphates both may be synthesized either by a de novo pathway or by "salvage" of the nucleotides. Growth in medium containing an analog selects against the salvage enzyme that makes possible the lethal utilization of the drug. Growth in medium

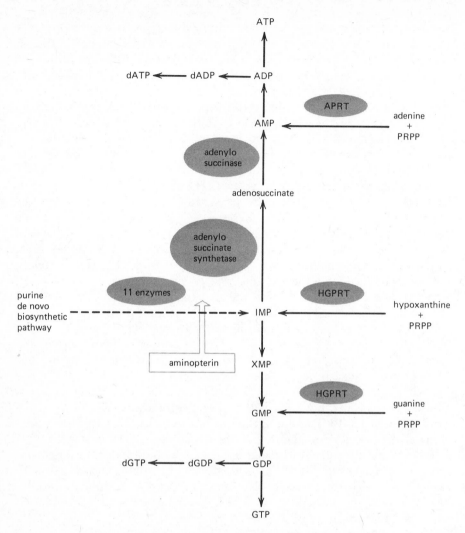

Figure 6.3. Pathways for interconversions of purine nucleotides.

containing the corresponding nucleoside and inhibitors that block its de novo synthesis provides counter selection for the enzyme.

One of the first analogs to be used was 8-azaguanine. As the metabolic pathway of Figure 6.3 shows, the first step in the utilization of guanine or hypoxanthine is phosphorylation by the enzyme HGPRT (hypoxanthine guanine phosphoribosyl transferase). In the absence of HGPRT, 8-azaguanine cannot enter the nucleotide pool and so its lethal effect is blocked. The lack of HGPRT is not harmful because the synthesis of purine triphosphates continues via the de novo biosynthetic pathway. Among variants resistant to 8-azaguanine (or to the similar analog 6-thioguanine), there-fore, should be those lacking HGPRT activity. From variants with this de-ficiency, cells that have regained the enzyme activity can be selected by plat-ing on HAT medium, so called for its active components of hypoxanthine, aminopterin, and thymidine. Because aminopterin blocks de novo nucleo-tide biosynthesis (by interfering with folate metabolism), utilization of the exogenous hypoxanthine via HGPRT then becomes essential for cell sur-vival. Figure 6.4 summarizes the protocols for selecting for and against the salvage enzyme.

Similar procedures may be applied to other enzymes of nucleotide me-

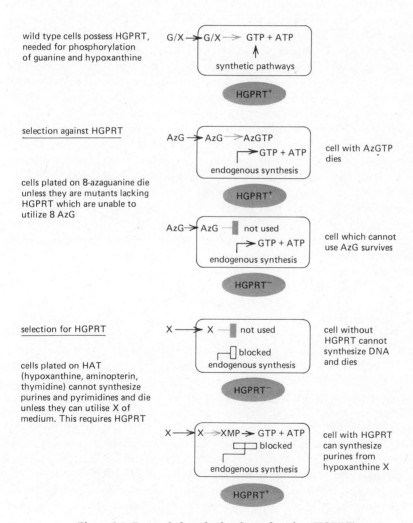

Figure 6.4. Protocols for selecting for and against HGPRT.

tabolism. Selection against APRT is achieved by growth in the presence of adenine analogs such as 2,6-diaminopurine or 8-azaadenine. Selection for APRT is accomplished by blocking de novo synthesis of adenine with inhibitors such as alanosine, azaserine, or aminopterin, so that the cells are compelled to utilize exogenous adenine via APRT (Kusano, Long, and Green, 1971; Chasin, 1974; Jones and Sargent, 1974). Selection against thymidine kinase is effected by growth in the presence of BUdR (which the enzyme converts into the lethal precursor); selection for the enzyme results from growth on HAT medium (which blocks the de novo pathway) (Kit et al., 1963; Littlefield, 1964b; Weiss and Green, 1967).

In the first application of this technique to HGPRT, Szybalski, Szybalska, and Ragni (1962) obtained variants of human cell lines resistant to azaguanine and isolated revertants from them able to grow on HAT. However, the frequencies with which these variants occurred were rather high and were not increased by mutagens. Littlefield (1963) isolated azaguanine-resistant L cells and showed that they possessed no detectable HGPRT activity and could not incorporate labeled hypoxanthine into DNA. In addition to this highly deficient line (which did not revert), mutants with lower resistance and with partial HGPRT activity were isolated, but these reverted at rates much higher than expected of mutants. In addition to 8-azaguanine and 6-thioguanine, 6-mercaptopurine, 8-azaguanosine, 8-aza-hypoxanthine, and other analogs, have been used to select variants lacking HGPRT from a wide variety of cell types.

In established cell lines that lack HGPRT activity, the frequency of reversion often is rather low. In some cases, no reversion has been observed, leading to suggestions that the structural gene might have been deleted. However, it appears that the frequency of reversion can be increased by exposing cells to the protocols used to obtain somatic cell hybrids. This was first noticed in experiments using a system in which rat or mouse cells lacking HGPRT were fused with human cells carrying the enzyme; and the hybrids were grown on HAT medium to select for the presence of the human enzyme. In one set of experiments, the hybrids contained both rat and human HGPRT (Croce et al., 1973; Bakay et al., 1975). In another series, murine cells grew on the HAT medium and appeared to be revertants possessing the mouse HGPRT enzyme (Watson et al., 1972). Reappearance of the rodent enzyme in the hybrid cells could be explained by supposing that a deficiency in some control protein was overcome by the provision of the partner genome following fusion. But this does not explain the reappearance of the enzyme in cells that apparently have not undergone a fusion. In further experiments, mouse A9 cells (a derivative of the L cell line) and RAG cells (a renal adenocarcinoma) reverted to regain HGPRT activity at frequencies of 1.8×10^{-8} and 9×10^{-8}, respectively, when subjected to cell fusion protocols; but they did not revert when plated directly upon HAT medium (Shin et al., 1973).

These events appear to involve the absence or presence of HGPRT protein because the A9 and RAG lines both lack CRM as well as enzyme activity; and the revertants of both cell lines possess HGPRT activity that reacts with the antiserum (Shin, 1974). While the nature of these events is not clear, two types of explanation have been proposed. One is that revertants are obtained more readily under the conditions of these protocols (which involve bringing the cells to a high concentration and giving a thermal shock). Since this treatment is not mutagenic, it may be that in some way it serves to make possible the growth of revertants that otherwise fail to be isolated. Another idea is that HGPRT activity is regained not by reversion of the gene that was originally mutated (or deleted) but by some event (perhaps epigenetic in nature) that causes the expression of a second HGPRT gene, one that otherwise fails to synthesize the protein. One source of a second gene might be the inactive X chromosome of female cells. The alternative is to postulate the existence of a second, inactive HGPRT gene in the haploid genome, one that is expressed only after the insults of the cell fusion protocol. It is difficult to devise experiments to test these hypotheses, which remain speculative.

Mutations in the Structural Gene for HGPRT

The types of mutants that are isolated in drug resistance depend on the selection protocols that are applied. Selection at low levels of 8-azaguanine (<0.3 μg/ml) leads to the isolation of lines with partial (30–50%) HGPRT activity. Selection of lines lacking HGPRT requires the use of higher levels of the analog (Morrow, 1970). Selection in the presence of 8-azaguanine at 3 μg/ml leads to the isolation of roughly equal numbers of mutants completely lacking HGPRT and mutants possessing at least some enzyme activity (Gillin et al., 1972). Selection for resistance to 3 μg/ml 8-azaguanine + 6 μg/ml 6-thioguanine leads to the isolation only of mutants lacking HGPRT activity (Sharp, Capecchi, and Capecchi, 1973). Thioguanine may be a more effective selective agent than azaguanine because it has a greater affinity for the HGPRT enzyme (Van Diggelen, Donahue, and Shin, 1979). These results emphasize the difficulties in establishing selection protocols that lead to the isolation of a single type of mutant, a prerequisite for obtaining enough independent mutants to make estimates of the rate of mutation.

Mutations that cause complete loss of HGPRT activity can be distinguished from those that leave the cell with some enzyme activity by plating on HAT medium. Cells that have lost HGPRT activity should fail to survive, but those that have retained at least some enzyme activity may be able to do so. Using this assay, Gillin et al. (1972) confirmed that azaguanine-resistant mutants of the V79 CHL cell line that cannot grow on HAT medium lack HGPRT activity when assayed for the enzyme in vitro. Most of these mutants lack CRM, but two display the immunological cross reaction and thus seem likely to be structural gene mutants (Beaudet, Roufa, and Caskey, 1973). Revertants able to plate on HAT medium arise at frequencies varying from 10^{-4} to 10^{-7}, which is consistent with heterogeneity in the sites of the mutation. The enzyme activities of the revertants vary widely, from 40 to 120% of wild type; all possess CRM. Similar results were obtained by Wahl, Hughes, and Capecchi (1975), who found that 40% of the mutants resistant to both azaguanine and thioguanine (and with <0.1% of wild type HGPRT activity) display CRM. There is wide variation in the extent of the immunological reaction and also in the kinetics of thermal inactivation of the CRM. This would be consistent with different mutations in the structural gene, perhaps missense in type since the protein was unaltered in size.

The azaguanine-resistant mutants characterized by Gillin et al. (1972) that are able to grow on HAT medium possess levels of HGPRT activity that vary from as little as 2% of wild type to more than the parental level. This suggests that even very low HGPRT levels make growth possible on HAT (which implies that this should be a sensitive test for reversion from total lack of HGPRT). Do these mutants possess an HGPRT protein in some way altered so as to cause resistance to the azaguanine? The properties of three of these mutants have been investigated by Fenwick et al. (1977). Two have HGPRT activities that are, respectively, the same and 1.5 times that of wild type in the in vitro assay; in both the enzyme is unaltered in size but has a K_m value for the substrate PRPP that is increased by about an order of magnitude, implying a decrease in the affinity of the enzyme for this substrate. How this causes azaguanine resistance remains to be seen. The other mutant has 25% of wild-type HGPRT activity in vitro; but its kinetics are little different from wild type; however, the electrophoretic mobility of the enzyme is reduced by an amount corresponding to about 1000 daltons.* The functional basis of the azaguanine-resistance in the size variant also remains to be determined. However, the demonstration that in each of these three cases there has been an alteration associated with HGPRT kinetics or size suggests that these mutations may lie in the structural gene (for review see Caskey and Kruth, 1979).

Another mutant to which this conclusion may apply lacks HGPRT activity but has CRM corresponding to a protein about 1000 daltons smaller than the wild type; revertants possess an enzyme of wild-type size. The

---

*There has been some difficulty in establishing the structure of the HGPRT enzyme. It appears to be a homomultimer, with small subunits reported as 25,000 daltons in the Chinese hamster, 27,000 daltons in the mouse, and 24,000 daltons in human. The apparent size of the active enzyme has varied, however; more recent estimates are for a trimer in Chinese hamster and mouse but a tetramer in man. Irrespective of the outcome of this discussion, the main point for genetic studies is that there appears to be only a single size of subunit, probably comprising a single type of polypeptide (Olsen and Milman, 1974; Hughes et al., 1975; Holden and Kelley, 1978).

conclusion that this size variant is a structural gene mutant is strengthened by the observation that in hybrid cells the mutant and wild-type enzymes are found in equal proportion. A thioguanine-resistant mutant of CHO cells isolated by Chasin and Urlaub (1976) has both an altered electrophoretic mobility and an increased (100-fold) K_m for PRPP. This also appears to be expressed codominantly since in hybrid cells the enzyme kinetics are those expected of a hybrid between the wild type and mutant enzymes; this suggests that mixing of the subunits has occurred.

An obvious question is whether the drug-resistant mutants that have not been characterized in such detail also lie in the structural gene. This can be answered by complementation analysis. Chasin and Urlaub screened 1000 independently isolated thg^r mutants for ability to complement mutants with the altered enzyme; none was able to do so. This suggests that the commonly isolated mutants all lie in either the structural gene or an adjacent dominant control locus. When mutants isolated for resistance to azaguanine was screened for ability to complement, 1 of 8 was able to do so. This also appears to be a structural gene mutation since it displays about 10% of HGPRT activity, has an increased (20-fold) K_m for PRPP, and displays an altered electrophoretic mobility. This raises the possibility that the complementation is interallelic, that is, with the two types of mutant subunit forming a hybrid enzyme with properties more like the wild type.

One approach to isolating missense mutants in a structural gene is to select for temperature sensitivity in the protein product. Fenwick and Caskey (1975) have therefore isolated mutants that are resistant to thioguanine at 39°C but which are able to grow on HAT medium at 33°C. The mutants have an HGPRT activity that is reduced in vitro at the higher temperature; and the mutant cells show a decline in incorporation of ^{14}C-hypoxanthine into DNA when the temperature is raised (whereas wild-type cells show an increase). The HGPRT activity of mutant and wild-type cells is equally well precipitated by antiserum, which is consistent with mutation of the structural gene. Other temperature-sensitive azg^r mutants isolated by Harris and Whitmore (1974) retain an unaltered HGPRT activity and may represent mutants in purine transport.

The best proof of structural gene mutation is direct evidence for a change in amino acid sequence. This is becoming possible because of recent developments of methods for increased purification of HGPRT. Milman et al. (1976) identified the spot corresponding to HGPRT on a two dimensional gel of HeLa cell extracts (using isoelectric focusing in one dimension and SDS gel electrophoresis in the other). Some 0.02% of the total soluble protein is found in a spot of 26,000 daltons with a pI of 6. In 24 mutants lacking HGPRT activity, the spot disappears. In a mutant with <0.01% of wild-type enzyme activity but with a wild-type level of CRM, the spot is observed in an altered position. Five independent revertants of this mutant (H23) all give a pattern in which there are two spots, one in the wild type position and one in the H23 position. This suggests that the revertants are synthesizing two forms of HGPRT, raising again the possibility that there might be more than one gene for the enzyme.

By using a micromethod for tryptic mapping, Milman, Krauss, and Olsen (1977) compared the fingerprints of the enzyme from wild type and H23 cells. With a lysine label (which should be present in one copy in each peptide, since trypsin splits bonds on the C-terminal side of either lysine or arginine), a new peptide appears in the mutant fingerprint. Although the wild-type peptide from which this is derived has not been identified, this change suggests that H23 arose by a mutation in the structural gene. The most likely form for this mutation is a missense change, although it is possible that it might be a nonsense mutation close to the end of the gene.

In another comparison of the tryptic fingerprints of a mutant and wild-type enzyme, Capecchi et al. (1977) have identified a nonsense mutation. Four peptides are present when Chinese hamster HGPRT is labeled with methionine (when a single labeled amino acid is used, only those peptides containing it are seen). One of these is altered in the fingerprint of CRM material of a mutant lacking HGPRT activity; it is the C-terminal peptide and probably is shorter in the mutant. A striking demonstration that this alteration has been caused by an ochre mutation was provided by injecting tRNA into the mutant cells. (The protocol for injection involves loading red blood cells with the tRNA and then fusing them with the CHL cell; this method is discussed in Chapter 8.) The microinjection of tRNA extracted from wild type or amber suppressor strains of E. coli or yeast has no effect; but tRNA from ochre suppressor strains of E. coli or yeast restores HGPRT activity to the mutant cell. This mutant therefore provides one of the very few instances in which a structural gene mutation can be identified and its basis established.

Characterization of Structural and Regulator Gene Mutants

The defects that are found in somatic cell mutants selected for resistance to several types of agent are summarized in Table 6.2. In most of these instances, although the defect has been localized it has not yet been shown whether this identifies the primary consequence of mutation or represents some secondary effect. In some cases, however, there is evidence that the mutation lies in the structural gene for the defective product, although none of these has been investigated in as much detail as the HGPRT system.

One of earliest auxotrophic mutants was the *glyA* derivative of the CHO line, which has a very low level of serine hydroxymethylase activity, the enzyme responsible for converting serine to glycine. (There are two SHM activities, one located in the mitochondrial fraction and one in the cytosol.

The former is missing in the mutant. Although located in the mitochondrion, the enzyme appears to be coded by a nuclear gene.) Chasin et al. (1974) demonstrated that a revertant from $glyA^-$ to $glyA^+$ regained SHM activity to about a third of the wild-type level. The enzyme of the revertant is more heat labile than the wild type, which is consistent with the idea that the $glyA$ mutation lies in the structural gene and has been overcome by either intragenic or extragenic suppression. The same technique of showing that there are physicochemical changes in the enzyme of a revertant isolated from a mutant altogether lacking the activity has been applied to glucose-6-phosphate dehydrogenase (Rosenstraus and Chasin, 1975, 1977).

A series of *ade* auxotrophs deficient in purine biosynthesis has been classified into complementation groups; and the step in the pathway that is defective has been identified in several examples (see above). In $adeA^-$ and $adeC^-$ cells the enzymes PRPP amidotransferase and GAR synthetase, respectively, are absent (Oates and Patterson, 1977). The first three enzyme activities of the pyrimidine biosynthetic pathway all are absent from $urdA^-$ cells that require uridine for growth; these probably reside in the same 200,000 dalton polypeptide chain, which is reduced in size in the mutant (Patterson and Carnwright, 1977; Davidson and Patterson, 1979). The same set of enzyme activities is affected in mutants selected for resistance to PALA (see Table 6.2). A similar situation may be responsible for the absence of both orotate phosphoribosyltransferase and orotidylate decarboxylase from mutants resistant to 6-azauridine or 5-fluorouracil (Levinson, Ullman, and Martin, 1979). Since each of the various auxotrophic mutants is deficient in a particular metabolic step, sometimes identified with a particular enzyme as described, these are appropriate for investigating further as with $glyA$ to determine the nature of the mutations associated with each particular deficiency.

Mutants of the S49 mouse lymphoma line resistant to cyclic AMP respond to neither endogenous nor exogenous cyclic AMP, which suggests that they may be defective in the ability to bind the cyclic nucleotide. Cyclic AMP binding activity resides in the cAMP-dependent protein kinase, which appears to be defective or absent in the mutants (Bourne, Coffino, and Tomkins, 1975). The mutants fall into three phenotypic classes according to their response to cyclic AMP. The nomenclature used to describe these has been somewhat erratic, changing apparently in response to views of the nature of the mutations. The *kinA* mutants require a 10-fold increase in cyclic AMP to activate the enzyme; they are also known as K_a type mutants. The *kinB* mutants have a reduced level of kinase activity; the maximum is upto about half of wild type, and these are described also as V_{max} mutants (although the implications of the name have not been substantiated, for example by kinetic analysis). The *kinC* mutants have no detectable activity and are completely resistant to the cyclic nucleotide; they are known also as kin^- or negative mutants.

Mutants of all three types are induced by EMS, MNNG, and ICR 191 (an acridine); EMS and MNNG are more effective in inducing the *kinA* class, whereas ICR 191 induces more of the *kinB* and *kinC* classes (Friedrich and Coffino, 1977). Although providing no proof of the basis of the mutations, this is consistent with the idea that *kinA* mutants result from point mutations altering the response of the enzyme to cyclic AMP, whereas *kinB* and *kinC* mutants may be produced by other types of change that include frameshifts.

The kinase consists of two types of subunit: the R (regulatory) subunit binds cyclic AMP; and the C (catalytic) subunit phosphorylates appropriate protein substrates when the R subunit dissociates from it upon binding cyclic AMP. In the *kinA* mutants the K_a value of the enzyme for nucleotide is increased 10-fold. When the kinases of sensitive and resistant cells are dissociated into subunits which are cross-reconstituted, the response of the heterologous enzyme is determined by the R subunit (Hochman et al., 1975).

Two-dimensional electrophoresis allows the proteins of wild-type and mutant cells to be compared; and Steinberg et al. (1977) used this technique to identify the R subunit of wild-type and *kinA* cells (and more extreme examples of this phenotype classified as *kinD*). The subunit occurs in two electrophoretic forms, differing in charge; one contains $^{32}PO_4$ and so probably carries a single phosphate group. They represent 0.05% of the total cell protein and are identified as the object of mutation by their change in position in extracts of mutant cells. In most mutants the spots are shifted to a more acidic position, by <1 charge unit; in two cases there is a basic shift of about 1.6 charge units. The phosphorylated and nonphosphorylated forms are equally affected, suggesting that the mutation does not lie in the phosphorylating activity. However, it might still represent some other form of modification, so it is not yet certain that the mutation lies in the structural gene for the R polypeptide. Both the mutant and wild-type subunits are seen in all mutant extracts, which suggests that the cells are diploid for sensitive and resistant alleles. There is always a greater amount of the mutant subunits, which may mean that the mutation becomes partly dominant, explaining its identification in diploids.

The rarest class of mutants is the *kin⁻* (*kinC*), which altogether lack kinase activity (level <0.03 of wild type), although they do retain 15–20% of the wild-type ability to bind cyclic AMP. Steinberg et al. (1978) observed that in hybrid cells the mutant gene is dominant over the wild type, which excludes the obvious explanation that the mutation inactivates the C subunit. Mixing experiments showed that there is no inhibitor of kinase activity produced specifically by the mutant cells; and the addition of extracts from *kin⁻* and *kin⁺* cells has the same effect on catalytic subunit activity, which excludes the possibility that the R subunit has a mutation preventing its dissociation from the C subunit on activation by cyclic AMP. The reverse model, in which such a mutation lies in C, is excluded by the demonstration that free R subunits are released from mutant cell extracts by cyclic AMP, although they appear to be present in amounts reduced from the wild type level. It is therefore likely that the *kin⁻* mutation is dominant because it resides in a function that acts in trans upon R and/or C gene expression, either to control synthesis of the subunit or possibly to make a critical post translational modification. The *kin⁻* mutation behaves as the only present example of what formally appears to be a dominant regulatory element in higher eucaryotic cells.

Selective agents that are highly specific for cell functions are provided by inhibitors of protein synthesis. The effect of the drugs on wild type and mutant protein synthetic systems can be characterized in vitro; and the ratio of doses necessary to inhibit protein synthesis to 50% in wild type and mutant generally is similar to the ratio of the D_{10} values determined by effect on cell viability. The drugs emetine and trichoderm act on the 40S and 60S ribosomal subunits, respectively; since these may be isolated in quantity, there is the prospect of identifying the ribosomal proteins involved in conferring resistance. Mutants with successively greater resistance to emetine may be obtained by means of two step isolation protocols; and in this case more than one locus may be involved (see Table 6.3). In just one of several *eme^r* CHO cell lines, the electrophoretic mobility of a single ribosomal protein, S20, is altered slightly; whether this is due to a change in modification or to mutation in the structural gene is not known, but the S20 alteration cosegregates with emetine resistance in cell hybrids (Boersma et al., 1979a,b). In other *eme^r* mutants no change in S20 (or any other ribosomal protein) can be detected; the locus of the mutations therefore may be, but is not necessarily, the same.

Diphtheria toxin has the highly specific function of ADP-ribosylating elongation factor 2. Resistant mutants fall into two classes; low level resistance is conferred by permeability mutants, while a level of resistance so high it cannot be quantitated is conferred by loss of susceptibility to the reaction. In different experiments, mutants of the latter class have taken two forms (Moehring and Moehring, 1977, 1979; Moehring et al., 1979; Gupta and Siminovitch, 1978a). The first shows complete failure of the reaction; but in the second only half of the EF2 molecules fails to be ADP-ribosylated, while the other half is inactivated. One possible explanation is that the cell line used for the first experiments is hemizygous, having lost one of its alleles; the second cell line has retained both alleles, only one of which has been mutated to confer resistance to ADP-ribosylation. Due to dosage effects, this variation in gene numbers could explain some anomalous behavior of the mutant alleles in dominance tests.

In several lines of cells isolated for resistance to α-amanitin, the activity of RNA polymerase II is less inhibited by the drug in vitro than is the wild type enzyme (Chan, Whitmore, and Siminovitch, 1972; Somers, Pearson, and Ingles, 1975). The reduction in inhibition in vitro correlates with the extent of resistance shown by each CHO mutant cell line in vivo (Ingles et al., 1976). In hybrid cells the resistance to α-amanitin is codominantly expressed, for the plating efficiency of the hybrids lies between the resistant and sensitive values. In a crude system, the RNA polymerase shows about 50% inhibition (Lobban and Siminovitch, 1975). This suggests that a mutation has occurred in a structural gene, rendering the RNA polymerase able to function in the presence of the drug.

Resistance to α-amanitin also has been obtained in an organism, D. melanogaster. Greenleaf et al. (1979) obtained mutant flies by recovering the adults that hatched from larvae grown on medium containing α-amanitin. Male flies were mutagenized with EMS; and from 3×10^6 mutagenized sperm, 16 putative mutant adults were obtained. Only 4 of these carried the trait when their progeny were retested; only one of the them has an RNA polymerase II enzyme with altered sensitivity to the drug. This strain has an X linked mutation that confers about 250-fold increased resistance to α-amanitin on the enzyme.

A system in which mutations may change the amount of enzyme is represented by resistance to aminopterin and methotrexate, which block folate metabolism. Littlefield (1969) found that after a stepwise selection of BHK cells for aminopterin resistance, mutant lines possessed increased levels of folate reductase activity. Nakamura and Littlefield (1972) observed that a BHK mutant line resistant to a concentration of methotrexate of 10^4 times the wild-type lethal level has a 140-fold increase in the specific activity of the enzyme dihydrofolate reductase (DHFR). The absence of any detectable physicochemical changes in the enzyme suggests that the effect may be due to synthesis of increased amounts.

In two quite different cell lines resistant to methotrexate, the amount of mRNA coding for DHFR is much increased. In a BHK mutant, translation of polysomal mRNA is vitro produces about 70 fold more DHFR activity than is synthesized from wild-type message (Chang and Littlefield, 1976). In a resistant mouse sarcoma line, the ability of mRNA to synthesize DHFR in vitro was proportional to the amount of DHFR synthesized in vivo, assayed by an immunological reaction specific for the protein (Alt et al., 1976; Kellems et al., 1976). This suggests that an increase in synthesis of DHFR may be a common mechanism for acquiring resistance to methotrexate.

The event responsible for the increased production of DHFR in the mouse sarcoma cells is an amplification of the number of copies of the structural gene for the enzyme. Alt et al. (1978) used a cDNA probe complementary to the DHFR mRNA to determine the amount of corresponding DNA in the cellular genome. The kinetics of DNA hybridization correspond to a 200-fold increase in the representation of complementary sequences in the mutant; and a reduction to a value 10-fold greater than wild-type occurs in a partial revertant. The revertant was obtained by continued growth in the absence of methotrexate, which leads to a decline in the level of DHFR activity. In contrast with the instability of the increase in these cells, a stable mutant of a mouse lymphoma line retains a 40-fold increase in both DHFR enzyme and gene level.

The methotrexate mutants were selected by multistep protocols, so that a series of events may have led to the gene amplification. Possible mechanisms include gene duplication and (subsequently) unequal crossing over. If each individual event doubles the number of gene copies, about 8 successive doublings would be required to generate 200 copies, or a somewhat greater number of events if each increases the number of copies by less than 100%. This implies a fairly frequent rate of occurrence by comparison with more conventional rates of point mutation. It will be interesting to see whether all multiple copies are arranged in a tandem array and whether they are integrated at the usual chromosomal site. An indication that this may be so is the appearance in one amplified line of a large chromosomal region that appears homogeneous in structure compared with the bands generated elsewhere by Giemsa. It is this region which reacts with DHFR cDNA probe upon in situ hybridization (Nunberg et al., 1978). The region corresponds to about 3.5% of the CHO karyotype, roughly 2×10^8 base pairs; if it represents a 200-fold amplification, the repeating sequence must be about 10^6 base pairs, which is very much longer than the DHFR structural gene. What other sequences may have been amplified together with DHFR, and what effect this may have on the cell, is an interesting question. While this amplification is unprecedented in cell genetics, and nothing yet is known about its molecular mechanism, it is difficult to believe that this will prove to be unique to the system. One speculation on why it has been observed here is that usually DHFR production may occur at the maximum possible level, excluding other mechanisms for increasing its synthesis. At all events, gene amplification appears to be the sole source of the increase in DHFR production, since the number of gene copies is proportional to the amount of enzyme protein.

The enzyme activity per se as well as its level may be subject to change. Flintoff, Davidson, and Siminovitch (1976) obtained two types of mutant in response to selection of CHO cells for resistance to methotrexate. The mtx^{RI} resistant cells possessed an enzyme with 10-fold increased resistance to inhibition by the drug in vitro. The mtx^{RII} cells were deficient in uptake of methotrexate. When the mtx^{RI} cells were subjected to a second round of selection for methotrexate, in the presence of 2×10^{-6} M drug compared with the 10^{-7} M used in the first step, the mtx^{RIII} mutant was obtained; this has a 10 fold increase in the level of reductase activity, without any apparent alteration in the properties of the enzyme. The alteration could be due to an increase in the activity of the enzyme or to increased synthesis. The cumulative effect of the two mutations, however, is to increase resistance to methotrexate beyond that achieved by the first step mutation (see Table 6.3).

The usual result of drug selection is the isolation of survivors that have acquired resistance; this may occur by an enzyme deficiency (either in a necessary metabolic conversion or in utilization of the drug for macromolecular synthesis), by an increase in enzyme activity, by a detoxification mechanism, or by an inability to take up the lethal agent. Another survival mechanism is the acquisition of dependence on the drug. An early example of drug dependence in bacteria is provided by the response to streptomycin, which distorts translation by acting on the ribosome. The distortion is counteracted by ribosomal mutations, which may make the bacterium resistant (able to grow with or without the drug) or dependent (able to grow normally only when the drug is provided, because the distortion has become necessary to release the ribosomes from what otherwise would be too stringent translation). This is discussed in Volume 1 of Gene Expression. In this case, the resistant and dependent effects are closely related and may even stem from the same mutations, depending on the genetic background. In eucaryotic cells, an example of dependence is provided by a response to BUdR. Here the acquisition of dependence occurs by a mechanism quite different from the loss of thymidine kinase which confers resistance.

A line dependent upon BUdR was derived from Syrian hamster melanoma cells that were grown first in 10^{-5} M BUdR and then were transferred to 10^{-4} M BUdR, in which they have since been maintained. Davidson and Bick (1973) characterized this line (B4) as dependent in view of the 50% reduction in growth rate that occurs upon removal of BUdR. In the presence of BUdR, about half of the thymidine residues in DNA are replaced by bromodeoxyuridine. By growing the B4 cells in medium containing hypoxanthine and aminopterin as well as 10^{-5} M BUdR, Bick and Davidson (1974) isolated the HAB subline in which all (>99.8%) of the thymidine residues in DNA are substituted by BUdR. Over the first 50 generations of growth, this line showed a slight increase in the G content of its DNA, but it has since displayed stable growth up to 275 generations (Bick and Davidson, 1976). This seems surprising in view of the known mutagenic effects of BUdR, which are thought to result from increased mispairing with G, leading to transitions from A-T to G-C base pairs. Presumably the mechanism for dependence involves some suppression of such effects. When BUdR is removed, the base composition of the HAB line returns to normal and the cells grow more slowly. Dependence is therefore only partial. This allows the characteristics of growth to be compared in the presence and absence of BUdR.

The original melanoma cells are highly malignant, showing such typical properties in culture such as lack of contact inhibition. This characteristic is maintained when the B4 line is grown in the presence of BUdR; but in the absence of BUdR the dependent cells become contact inhibited. Several other, although not all, of the properties that distinguish "transformed" and "nontransformed" cells respond in the same way (Davidson and Horn, 1974; Horn and Davidson, 1975). The dependent cells therefore require BUdR to maintain their usual pattern of growth. BUdR has been shown to influence the expression of differentiated functions in several cell types and it remains to be seen how the mechanism of dependence may be related to such effects on gene expression.

CHAPTER 7

The Cell Division Cycle

Cytological observation of the cell divides its life into two parts: interphase and mitosis. The cell starts its life cycle with the diploid set of chromosomes gained at telophase. For the duration of interphase, the chromosomes exist in the form of a network of interphase chromatin with few discernible features. No cytological changes are observed until prophase of the next mitosis, when two diploid sets of chromosomes are present. Replication of the chromosomes takes place during a restricted part of interphase, a feature revealed by the observation of Howard and Pelc (1953) that DNA synthesis is confined to a short period during the interphase of cells of Vicia faba. By using autoradiography to follow the incorporation of ^{32}P into DNA, they were able to show that the label enters DNA for a period representing only about 6 hours of the total 30 hours of the cell cycle. This period is separated in time from mitosis.

Thus interphase can be divided into three parts according to the status of the chromosomes. The daughter cells generated by mitosis enter the G_1 *period*, which is defined by the time during which they retain a diploid set of chromosomes. Cells leave G_1 when they begin to replicate their chromosomes. The period during which DNA synthesis takes place is termed *S phase*. At the completion of S phase, a cell has replicated the entire complement of genetic material and therefore possesses two diploid sets of genetic information. Cells remain in this state for the G_2 *period* until mitosis commences. S phase was so called as the synthetic period when DNA is replicated, G_1 and G_2 standing for the two gaps in the cell cycle when no DNA synthesis occurs.

Both the DNA and protein components of the chromosome are synthesized during S phase, so that by its completion the amount of chromatin in the cell has doubled (although without visible change in structure). The daughter chromosomes remain intimately associated with each other until they segregate at the subsequent mitosis. RNA and protein synthesis increase continuously throughout interphase, in contrast with the restriction of DNA synthesis to a limited part of the cell cycle.

Somatic cells that have ceased to divide usually are found in the G_1 state, with a diploid complement of chromosomes. This is true of most differentiated cells and of cells in culture that display contact inhibition (Todaro, Lazar, and Green, 1965). Cells that have ceased to divide sometimes are

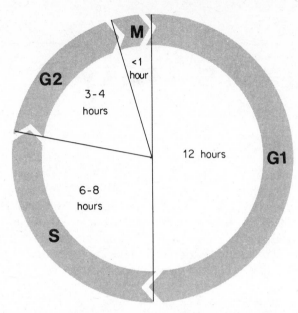

Figure 7.1. Cell cycle of L cells growing in culture at 24 hours per doubling. Data of Stanners and Till (1960).

described as being in the G_0 state. However, it has been controversial whether G_1 and G_0 represent genuinely different states of the cell or whether G_0 is a special (and reversible) example of the G_1 state.*

The amount of time spent in each phase of the cycle is characteristic of the particular cell type. For mammalian cells growing in culture, a typical cell cycle is that observed by Stanners and Till (1960) with mouse L cells. Figure 7.1 shows that the generation time is about 24 hours, some half of which is occupied by G_1, with S phase lasting for 6–8 hours and the G_2 period for 3–4 hours. Variations in the cell cycle have been reviewed by Zetterberg (1970) and Mitchison (1971).

The most variable period in the cell cycle is G_1, which ranges from zero in some cells to an indefinite length in others. To a large extent, the total duration of the cell cycle is therefore dictated by the length of G_1. This con-

---

*For example, after cultured cells have entered a quiescent state upon depletion of the medium, upon continued incubation they take an increasing length of time to recommence division when later stimulated with new medium (Augenlicht and Baserga, 1974). One view is that this should be considered to reflect increased commitment to the G_0 state. Another might be that it results from progress towards cell death (which also is the ultimate fate of non dividing differentiated cells). In this context, one might ask whether for cultured cells G_0 may not perhaps be equated with cell degeneration. In an organism in which the individual cell remains viable until other systems fail, the concept that G_0 may be distinct from G_1 may find more application.

clusion is supported both from surveys of the lengths of the phases of the cell cycle in different mammalian cells and from experiments in which the length of the cycle of a cell has been varied in culture.

The duration of S phase depends upon the cell phenotype and does not appear to be related to its content of DNA (see Table 20.1). Most studies on the cell cycle have been performed with cultured cells. With adult somatic cells of mammals, S phase most commonly occupies about 6–8 hours (Defendi and Manson, 1963). But it may sometimes take much longer. Restriction of the minimum length of S phase to around 6 hours cannot be due simply to the large amount of DNA to be replicated, for embryonic cells may display very much shorter S periods than the somatic cells of the adult. One example is S phase in early embryogenesis of X. laevis, which may be completed in less than 15 minutes, although it occupies some 5 hours in adult somatic cells (Graham and Morgan, 1966). The same tendency for S phase to be comparatively rapid in embryos, becoming longer later in development, is shown also in mammals, but is somewhat less pronounced, with the generation time of embryonic cells just a few hours short of that of adult cells. One factor permitting the short cell cycles of early embryogenesis may be that the cells (or nuclei) can divide without the need for appreciable growth. But the basis for the difference lies largely in changes in the organization of the units of replication rather than with any variation in the speed of DNA synthesis (see Chapter 20).

The length of the G_2 period may vary in cells with different generation times, but generally G_2 is regarded as comprising the time that the cell requires to prepare for mitosis after it has completed duplication of the chromosomes. This may involve physical changes in the condensation of the chromosomes as well as the possible accumulation of a mitotic inducer (see Chapter 8). The period of G_2 rarely occupies more than 6 hours and, in general, G_2 is the most constant period when the cycles of different cells are compared. But at least one instance is known in which G_2 is omitted in a cultured cell, so that in this case the functions usually performed in this period must take place during S phase (Liskay, 1977).

Assembly of organelles—such as ribosomes and mitochondria—probably takes place continuously during interphase (see Chapter 21). All components of the cell must double in the period between mitoses; and it is generally assumed that both new and old components become mixed in the cytoplasm and are partitioned about equally between the two daughter cells at the ensuing division.

In this chapter we shall be concerned with the growth of the cell during interphase and with the control of its transition from G_1 to S phase, that is, with the decision on whether it is to enter a new division cycle. Within S phase, the replication of DNA is an ordered process whose control is clearly important for progression through the cell cycle and this is discussed in Chapter 20.

Methods for Synchronizing Cell Division

It is difficult to follow events in the cell cycle with much precision of timing. Although a single cell can be followed microscopically, this technique naturally is limited in the parameters that can be measured. Most of the experiments discussed below which concern events occurring at particular stages of the cell cycle have therefore utilized synchronized populations of cells growing in culture. One drawback inherent in the use of synchronized populations is that the cells generally remain synchronized for only a short period of time, usually about two generations. By the third generation there is often an appreciable loss of synchrony, so that cyclical events can be followed through only two cell divisions.

The methods that are used to achieve synchrony fall into two general classes: selection synchrony and induction synchrony (for review see Mitchison, 1971). When cells are selected at some particular stage of the cycle, they may be subcultured as a synchronous population. Selection methods offer the important advantage that the cells are obtained in their "normal" state. One method used for selection is to isolate the new daughter cells as they are preferentially released from a solid surface at mitosis. But although this yields well synchronized cultures, it suffers from the disadvantage that the cultures are limited to small numbers of cells (Terasima and Tolmach, 1963; Tobey et al., 1967b). Also it cannot be used with suspension cultures. Much greater quantities of cells can be obtained by centrifuging a randomly grown culture so that cells in different stages of the cycle are separated according to their size. Various methods can be used to separate the cells, including velocity sedimentation, gradient or zonal centrifugation, and centrifugal elutriation (see Ayad et al., 1969; Schindler et al., 1970; Shall, 1973; Meistrich, Meyn, and Balogie, 1977). However, this does not yield very good synchrony because of the inherent limitation that the sizes of individual cells may vary somewhat at any given stage of the cell cycle.

Induction methods rely on treatment of an asynchronous culture to delay cells more advanced in the cycle either in division or in replication, so that the entire population accumulates at the same stage. A common method used to induce synchrony is to treat a culture with colchicine or colcemid, causing the cells to accumulate in a metaphase-like state (see Chapter 4). When the cells are transferred to medium lacking colchicine, the inhibition is rapidly reversed and all the cells enter G_1 together. Because colchicine may damage cells during long incubations (which would be necessary to cause all cells to accumulate at metaphase), a brief period of inhibition is often used to achieve a partial synchrony; then this is succeeded by the application of some other method to achieve full synchrony. The general disadvantage of using colchicine is the possibility that the drug may induce abnormal behavior in the cells.

Inhibition of DNA synthesis is commonly used to cause cells to accumulate at the beginning of S phase. High concentrations of thymidine halt DNA synthesis in mammalian cells, because of feedback effects from the increased dTTP pool that inhibit the formation of other nucleotides. This treatment usually is used in the form of a double thymidine block (Petersen and Anderson, 1964; Tobey et al., 1967a). Cells are treated with excess thymidine for a period equivalent to the sum of $G_2 + M + G_1$. Any cells that are in S phase when the thymidine is added halt DNA synthesis immediately. Cells in G_2, M, or G_1 can continue through the subsequent stages of the cycle until they conclude G_1 and enter S phase; they are halted at this point. This first treatment therefore produces a population of cells some of which are halted within various parts of S phase, others of which are blocked at the entry to S phase. The block is then lifted for a period longer in duration than S phase. This allows all the cells to complete S phase, no matter at which point they were inhibited. A second block of thymidine then is administered for long enough to allow the cells—almost all of which now are in G_2—to continue through the cell cycle until they conclude the next G_1. The entire population therefore becomes halted at the entry to S phase. But there may be appreciable escape from the thymidine block since cells can be detected in the inhibited population with DNA contents greater than the G_1 (diploid) value and there is some synthesis of DNA (Tobey and Crissman, 1972; Meyn et al., 1973).

Other inhibitors of DNA synthesis in common use include amethopterin (an antagonist of folic acid) and FUdR (5-fluorodeoxyuridine, a nucleotide analog). Both these compounds act on the thymidine utilization pathway and amethopterin also interferes with purine biosynthesis; their effects can be reversed by the addition of thymidine. Hydroxyurea is another effective inhibitor of eucaryotic DNA synthesis, probably by its action on ribonucleotide reductase (see Table 6.2). All these agents often are used to improve the synchrony achieved by a short incubation with colchicine. They may also be used in conjunction with isoleucine deprivation, which causes cells to arrest within G_1 (compared with the partial entry into S phase allowed by thymidine block). Tobey and Crissman (1972) developed a two-part method for synchrony in which cells are halted early in G_1 by the provision of only limiting amounts of isoleucine and then are released from the block in the presence of hydroxyurea, which prevents entry into S phase. Some synchrony persists into a second cycle, which follows the first with the usual periodicity; that is, the inhibition does not allow accumulation of intermediates sufficient to support two S phases in rapid succession (Hamlin and Pardee, 1976). This suggests that preparations for an S phase depend upon completion of the previous cycle. However, a disadvantage of these inhibitor treatments is that, although they prevent S phase, they may allow the cells to continue with other synthetic activities. Mitchison (1971) has pointed out that this may halt the DNA synthetic cycle while

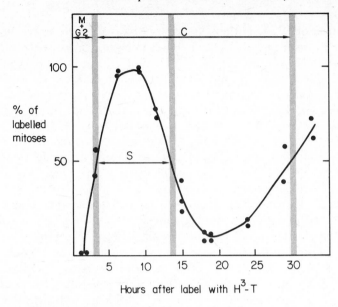

Hours after label with H³-T

Figure 7.2. Measurement of the cell cycle. HeLa cells were labeled with ³H-thymidine for 30 min and the proportion of labeled metaphase plates among the mitotic cells was determined after various periods of incubation in unlabeled medium. The first labeled mitotic cells to appear must have been completing S phase when they were labeled. The proportion of labeled cells increases as cells that were labeled during S phase pass through mitosis. The subsequent decrease represents cells that were in G_1 during the labeling period and proceed through S and G_2 to mitosis without being labeled. The increase after 20 hours represents the second division of cells originally labeled during the earlier S phase. The shaded vertical lines indicate the points at which measurements are made, when 50% of the cells become or cease to become labeled. The interval to the first ascending 50% line measures $M + G_2 = 3.5$ hr; the period of 10.5 hr between the 50% ascending and 50% descending lines measures S phase; and the delay between subsequent 50% ascending lines corresponds to the cycle $C = G_2 + M + G_1 + S = 26.5$ hr. Thus $G_1 = 26.5 - 10.5 - 3.5 = 12.5$ hr. Data of Baserga and Wiebel (1969).

the growth cycle continues. This means that the relationship of DNA synthesis to the cell cycle may be changed; results obtained by induction synchrony therefore must be interpreted with some caution.

The phases of the cell cycle usually are defined by taking advantage of the ability of cells to incorporate a label into DNA only while in S phase. This can be measured for the population en masse or may be determined for individual cells by autoradiography. Because individual chromosomes cannot be recognized during interphase, replication within the complement must be followed by labeling the cells with radioactive precursors— usually ³H-thymidine—during S phase, and then examining the labeled chromosomes at the subsequent metaphase. A short pulse dose of 10–20 minutes incubation with tritiated thymidine usually is given, after which metaphases are analyzed for the presence of the label after increasing intervals.

The first labeled metaphase plates to appear must represent division of cells that were at the very end of S phase when the label was added. The interval between addition of the label and its appearance in metaphases is therefore a measure of G_2. The last labeled metaphase plates to be detected must represent cells that were at the very beginning of S phase when the label was added. The time taken for their appearance must provide a measure of $S + G_2$. As Figure 7.2 shows, these periods usually are measured by taking the times when 50% of the cells first exhibit a label and when 50% of the cells then fail to exhibit a label (see Quastler and Sherman, 1959; Baserga and Wiebel, 1969; Mitchison, 1971). The difference between these two times corresponds to the length of S phase; and the interval between the appearance of two successive peaks of labeled metaphases measures the overall length of the cell cycle. The appearance of label in individual chromosomes in the metaphase cells reveals the order of their replication.

Protein Synthesis and Cell Growth During Interphase

Cellular synthesis of nucleic acids and proteins is largely confined to interphase. By following the growth of individual mouse fibroblasts with cytochemical assays, Killander and Zetterberg (1965a) measured the extent of DNA, RNA and protein synthesis as a function of time elapsed since the last mitosis. The amount of protein is measured by the cellular dry mass (which can be assayed by microinterferometry) and the amount of RNA and DNA is revealed by optical extinction at 265 nm and 546 nm after staining with feulgen.

The results of Figure 7.3 show that the cycle of these cells comprises a G_1 of 8 hours, an S phase of 6 hours, and a G_2 of 5 hours. (The amount of DNA fails to double exactly because the later age groups are contaminated by a few G_1 cells). Synthesis of DNA is discontinuous; RNA and protein synthesis occur continually. The technique does not allow measurements sufficiently precise to define the exact manner of growth; RNA content may increase in either a linear or exponential manner. By following the increase in protein through the use of labeled amino acids, Zetterberg and Killander (1965) confirmed that the cellular content of protein doubles during the growth cycle.

These results show the growth of the cell as a whole. Zetterberg (1966a,b) extended these studies to follow the synthesis and accumulation of protein in the cytoplasm and nucleus of the growing cell. The nuclear dry mass was determined by microinterferometry; subtraction from the total cell mass gives the mass of the cytoplasm. Figure 7.4 shows that the cytoplasmic mass increases predominantly during the first half of interphase. The predominant increase in nuclear mass takes place during the second part of interphase, its start coinciding roughly with S phase. Both nuclear and cytoplasmic dry mass double during the cycle, so that the relative propor-

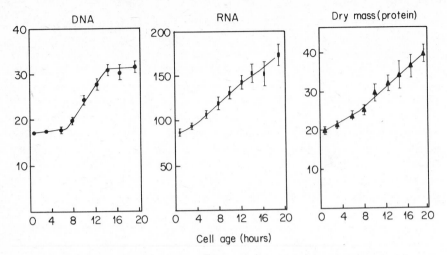

Figure 7.3. Accumulation of DNA, RNA, and protein during cycle of mouse fibroblasts. G_1 lasts for 8 hours; the S phase period is defined by the increase in DNA content that takes place from 8–14 hr after mitosis; and during G_2 the cells possess double the G_1 content of DNA. RNA and protein (given by dry mass) increase continuously during all three phases of the cycle. Data of Killander and Zetterberg (1965a).

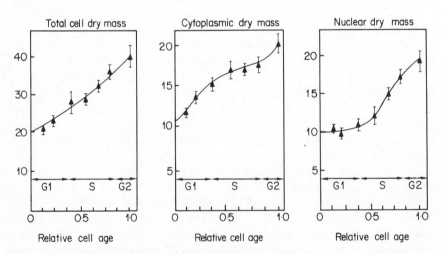

Figure 7.4. Accumulation of protein during cycle of mouse fibroblasts. Total dry mass—that is protein content—increases continuously. Much of the increase in cytoplasmic mass takes place during G_1 when the nuclear mass remains constant. Nuclear mass increases during S phase and G_2, when the cytoplasmic mass increases more slowly. Both nuclear and cytoplasmic mass double during the cycle. Data of Zetterberg (1966a).

tions of the cell are the same at the end of G_2 as at the beginning of G_1, with about half the mass in the cytoplasm and half in the nucleus.

The level of protein synthetic activity relative to cytoplasmic mass is constant throughout the cell cycle, which implies that the total level of protein synthesis increases in proportion to the total content of the cytoplasm. During G_1, most of the newly synthesized protein accumulates in the cytoplasm, generating the pattern of cytoplasmic growth during the first part of the cycle. During S phase, cytoplasmic protein synthesis continues at the same relative rate, but the mass accumulates in the nucleus instead of in the cytoplasm. This implies that much of the net protein synthesized during the second part of the cell cycle must be transported to the nucleus. The growth cycle of the cell has been reviewed by Zetterberg (1970).

The transfer of proteins from cytoplasm to nucleus takes place throughout interphase and is not confined to the later part of the cycle when the nucleus grows in size. Zetterberg (1966c) gave cultures of L cells a pulse dose of ^{3}H leucine and then followed the fate of the label by autoradiography. As time passes, more of the label is seen in the nucleus. Since this transfer takes place at a constant rate throughout interphase, but no protein accumulates in the nucleus during G_1, there must be a simultaneous movement of proteins from nucleus to cytoplasm in exchange during the early part of interphase.

This suggests a mobile view of the cell compartments, in which proteins move continually in both directions between nucleus and cytoplasm. During G_1, only the mass of the cytoplasm increases, for the numbers of proteins entering and leaving the nucleus appear to be in equilibrium. During S and G_2, much of the new protein content of the cell accumulates within the nucleus, by transport from its sites of synthesis in the cytoplasm. Analogous results have been obtained in other types of cell, including the lower eucaryotes (for review see Mitchison, 1971).

The accumulation of proteins that commences in the nucleus at S phase depends upon DNA synthesis. Auer et al. (1973) observed that when FUdR is used to inhibit DNA synthesis in L cells, total cellular mass continues to increase; synthesis of proteins is therefore independent of DNA synthesis. But the mass accumulates in the cytoplasm. The transport of proteins from cytoplasm to nucleus that causes the increase in nuclear mass during the latter part of the cell cycle therefore depends on DNA synthesis. (Only some of this increase can be due to the duplication of the chromatin, so much must represent proteins involved in other nuclear functions.)

The ability of proteins to enter nuclei rapidly is well documented, especially for situations that involve the reactivation of a previously inert nucleus. Transplantation of an adult somatic nucleus of Xenopus laevis into an enucleated egg is followed by the rapid entry of cytoplasmic proteins as it swells, and a similar situation is seen also when the inert nucleus of a chick erythrocyte is transferred by cell fusion into the cytoplasm of a HeLa cell (see Chapter 8). In the latter case, the reconstitution of an active chick nucleus is accomplished by human proteins, so that at least some nucleocytoplasmic interactions are not species specific. These experiments represent a situation in which the nucleus finds itself in an environment to which usually it would not be exposed. But the several instances in which the accumulation of proteins in the nucleus is correlated with activation and DNA synthesis implies that this movement may reflect a physiological process occurring during the normal cell cycle.

Although cells seem to vary somewhat in the age at which they enter S phase (relative to the last mitosis), their mass may be more constant at this time. This observation suggests that DNA synthesis may be initiated when the cell mass passes some critical level, perhaps in a manner analogous to the control of replication in bacteria (see Volume 1 of Gene Expression). By analyzing several different populations of mouse L cells during growth, Killander and Zetterberg (1965a,b) found that the mass of the cell doubles each generation, but that the variation of initial mass—that is immediately succeeding each mitosis—is quite wide between the different populations. But there is much less variation in mass at the beginning of S phase, which is consistent with the idea that replication is linked to the accumulation of a critical cell mass. In agreement with this idea, cells of populations which enter G_1 from mitosis with a low mass spend longer before initiating S phase than do populations which start the cycle with greater mass. Such a mechanism might evolve to ensure that the cell population retains the same mass through successive generations to counteract any uneven distribution of mass between daughter cells at mitosis. But on the other hand, data on the time at which S phase is initiated relative to the last mitosis do not fit this model (see next section).

Control of Entry into S Phase

The critical event in the cell cycle is the decision whether to enter S phase. Once a cell has begun to synthesize DNA, it is almost inevitably committed to complete replication and then to continue through G_2 and mitosis. (Exceptions are provided by partially or completely polyploid cells of tissues such as Dipteran salivary gland or mammalian liver.) A cell that does not enter S phase may remain in the G_1 (or G_0) state indefinitely, that is, until its death. The central question to be asked about the cell cycle is therefore what determines the commitment of a cell in G_1 to enter S phase. This question encompasses not only the issue of what comprises the differences in control between dividing and nondividing cells, but also that of what determines the length of the cell cycle of the members of a dividing population.

Initiation of DNA synthesis in the nucleus is controlled by signals from the cytoplasm. When a nucleus of an adult somatic cell that has ceased replication is placed in a new cytoplasmic environment, either by transplantation or by cell fusion, it may be stimulated to synthesize DNA and then to divide (see Chapter 8). These signals seem to be of low specificity, since the cytoplasm of one species may activate the nucleus of another.

In a large number of mammalian cell types, it has been shown that RNA and protein synthesis must be allowed to take place in G_1 phase as a prerequisite for entry to S phase. Such synthesis must take place in each G_1 phase to allow the subsequent S phase to commence. The effect on DNA replication of inhibiting RNA or protein synthesis appears to vary in different cell types and even with different culture conditions. In general, it seems that inhibition of protein synthesis at any time during G_1 will inhibit or delay the occurrence of S phase.* Only occasionally has there been any observation of a critical point in G_1 beyond which the cell appears committed to S phase and is not susceptible to inhibition of protein synthesis. In spite of results that certainly are less than clear cut, it seems likely that during G_1 there occurs some continual process that is necessary for entry into S phase. But, of course, such experiments do not reveal whether this process is concerned with the synthesis of enzymes necessary for DNA replication or represents a control circuit, although it does seem that the events that are involved are those concerned with the initiation of replication and not with subsequent steps in DNA synthesis.

Two general types of view have been taken of the cell cycle. The traditional model supposes that all phases of the cycle are of fixed length. Some event occurs at a determined time in G_1 and it is this that is responsible for committing the cell to enter S phase. If this time does not coincide with the beginning of S phase, the G_1 period can be divided into subphases of pre- and postcommitment. The event responsible for commitment might be discontinuous—such as the synthesis at this time of some control molecule—or might be the result of the continuous accumulation (or dilution) of cell components during G_1. This view is exemplified by the treatment of the cell cycle given by Prescott (1976).

An alternative model of more recent origin is to suppose that G_1 is not fixed in duration, but that during this phase, or during part of it, there is at all times a constant probability that a cell becomes committed to entry into S phase. While the length of S, G_2, and M (and possibly part of G_1) is the same for all cells, the length of the remaining part of the cycle cannot be predicted for any individual cell, although the distribution of lengths in the population can be calculated. This model was proposed in a somewhat restricted form by Burns and Tannock (1970) and, independently, in a generally applicable form by Smith and Martin (1973).

The two models make quite different predictions about the nature of

*References: Taylor (1965), Mueller and Kajiwara (1966b), Terasima and Yasukawa (1966), Fujiwara (1967), Weiss (1969), Highfield and Dewey (1972), Jakob (1972); for reviews see Baserga (1968), Mitchison (1971).

the variation in the length of G_1 among the members of a given population. If G_1 is of fixed duration, but with a scatter in its length in the population due to the inevitable variations between individuals, the values for the population should fall on a normal distribution. The breadth of the distribution depends upon the extent of individual variation; in a sense this might be regarded as the degree of looseness permissible in the control mechanism. The age of the cells at the time of division then would follow the kinetics plotted in the upper part of Figure 7.5. On the other hand, if entry to S phase occurs with fixed probability with respect to time, the kinetics should be exponential as depicted in the lower part of the figure. Since only part of the cell cycle is variable in length, there is a lag phase before any division occurs, which represents the constant part of the cell cycle; and then the number of cells that has not yet divided declines with

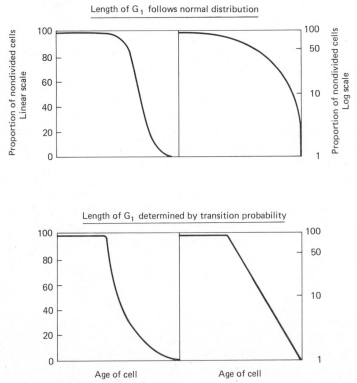

Figure 7.5. Theoretical kinetics of cell division. Each curve shows the proportion of cells remaining in interphase, that is, which has not yet divided, as a function of cell age. The upper panels assume that the length of G_1 is fixed and that in individual members of the population there is variation that falls on a normal distribution. The lower panels assume that there is no fixed duration of G_1, which is determined by the constant probability with time that there will be a transition into S phase. This gives an exponential decline with time of the number of nondivided cells. The predicted curves are plotted on the left on a linear scale and on the right on a logarithmic scale. After Smith and Martin (1973).

exponential kinetics. Experimental data to compare with these curves can be obtained by timing individual cell cycles through time lapse cinematography, by counting cell numbers at intervals after seeding a population (that is, when division cycles should occur synchronously), or by plotting a cumulative total of the fraction of labeled mitoses after labeling a cell population.

Both types of model share the assumption that $S + G_2 + M$ comprises a constant length of the cycle and that variation in the rate of cell division is caused by controlling the length of G_1 phase. The conventional model supposes that when the length of the cycle is changed, the duration of G_1 is altered from one fixed value to another fixed value. The probability model supposes that it is the likelihood of commitment to division that is altered. In neither case is it necessary to postulate that the entire length of G_1 is subject to variation in length. Part of G_1 may be variable, whereas part reamins constant. After a mitosis there may be a fixed period of G_1 before the variable period commences; and after commitment to division, there may be a fixed delay before the start of S phase. In these terms, the cell cycle can be divided into two phases. Phase B is constant in length and includes S, G_2, M and perhaps part of G_1. Phase A represents part or all of G_1 and identifies the part of the cycle whose duration is subject to control, either shifting between fixed values or decaying into state B with a probability of transition that is established by the conditions of growth.

The transition probability model allows the cell cycle to be determined in terms of two parameters, T_B and P. T_B defines the length of the constant phase, B. It is given by the lag period before any division starts and is best calculated from curves of the form of Figure 7.5d. Thus T_B represents the minimum intermitotic time. P is the probability of transition from phase B to phase A and is given by the slope of the exponential decay curve. The lower the value of P, the longer will be the cell cycle, (or, more precisely, since the cycle cannot be defined in these terms for an individual cell, the longer will be the time required for the population to double).

If N is the initial number of cells in a population and N_d is the number of cells that has divided by time T elapsed since the last division, the proportion of cells remaining in interphase, α, is given by the expression:

$$\alpha = \frac{N - N_d}{N}$$

The loss of cells from the interphase state is determined by P, so that:

$$\text{probability of transition} = \frac{\text{number of cells dividing in unit time}}{\text{number of cells not yet divided}}$$

Or in mathematical phraseology:

$$P = -\frac{1}{\alpha} \cdot \frac{d\alpha}{dt}$$

Rewriting the expression as:

$$\frac{d\alpha}{dt} = -\alpha P$$

integration with respect to time gives:

$$\frac{\alpha_t}{\alpha_0} = e^{-Pt}$$

and since at time zero, $\alpha = 1$,

$$\alpha_t = e^{-Pt}$$

or:

$$P = \frac{-\ln \alpha_t}{t}$$

This expression describes the exponential decay of cells from the non-committed into the committed state and explains why the plot of α against t represented in Figure 7.5d is linear when α is plotted on a logarithmic scale. From the exponential decline, P may be calculated by using the equation in either differentiated or integrated form. In the first case the decline is measured over a suitable interval and calculated as $(\alpha_{t1} - \alpha_{t2})/\alpha_{t1} \cdot (t2 - t1)$; in the second it is given by $-\ln \alpha_t/t$, but in this case it is necessary to remember that t is measured from the beginning of the A phase, that is, it corresponds to $T - T_B$.

Data for several cultured cell systems are plotted in Figure 7.6. All curves conform to the expectation of Figure 7.5d except for the occurrence of an initial downward inflection before the exponential decline is reached. This can be explained as variation in the length of T_B, presumably due to individual scatter, and seems generally to be in the range of 10–20%.

Formally these data can be said to fit cycles in which there is a constant period of length T_B (with limited scatter) and a variable period of length T_A, determined by the transition probability constant P. This does not locate the periods A and B with respect to the more usual division of the cycle into the compartments G_1, S, G_2, and M. Given the relative constancy of $S - G_2 - M$ and the relative variability of G_1, it is reasonable to equate these with B and A, respectively. But this leaves open an important question: what part of G_1 is occupied by A? The relative durations of B and of $S - G_2 - M$ might be established by determining these directly for the same population. The only instance in which suitable data are available appears to be curve c of Figure 7.6, which replots the data shown in Figure 7.3. In this case the duration of $S + G_2 + M$ can be calculated as 12 hours, which compares with the value for T_B of 15 hours. This would be consistent with the idea that the B phase may include part of the beginning or end of G_1, although, of course, more data would be required to establish such a hypothesis. At present there is no evidence on the nature of the $B - A$ transi-

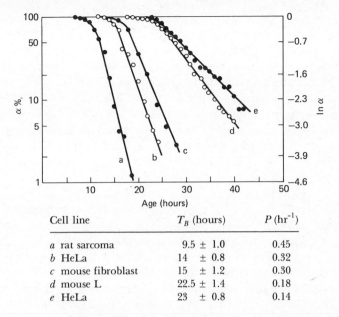

Cell line	T_B (hours)	P (hr^{-1})
a rat sarcoma	9.5 ± 1.0	0.45
b HeLa	14 ± 0.8	0.32
c mouse fibroblast	15 ± 1.2	0.30
d mouse L	22.5 ± 1.4	0.18
e HeLa	23 ± 0.8	0.14

Figure 7.6. Dependence on cell age of the proportion of cells remaining in interphase. The proportion of cells not yet divided, α, is plotted against the age of the cells as observed by time lapse cinematography. α is shown as a percent on the left and the corresponding natural logarithms are given on the right. Data calculated by Smith and Martin (1973).

tion, and in particular on the point of whether it precedes or coincides with the start of S phase.

Attempts to produce synchronous cell populations by either induction or selection rarely are successful beyond two generations; and even the second division usually is appreciably less well synchronized than the first. Generally this has been attributed to a combination of the existence of some scatter in the length of the cell cycle and the technical inability to achieve complete synchronization by any of the available methods. An alternative explanation follows from the transition probability model. Data such as those of Figure 7.2 are obtained by labeling a population with ^{3}H-thymidine and then identifying the labeled cells as they pass through mitosis after a period equivalent to G_2 and part of S. The first peak of labeled cells therefore represents events occurring during the constant B phase and may be interpreted in the same way irrespective of the view taken of the nature of the variable A phase. But the occurrence of a second peak is treated in a different way. According to the transition probability model, the cells dividing at each point in the first peak will have a distribution of division times set by the sum of the constant T_B and the variable T_A. The sum of these distributions emanating from all points of the first peak should give a second peak which is broader in time than the first and which fails to reach the same level. The width of the second peak is a function of T_B and P; its height is a function of P. The loss of synchrony in this second peak therefore reflects the variation intrinsic in the A phase. The population doubling time also is a complex function of T_B and P, and the evaluation of such observed data in terms of the transition probability model has been discussed by Smith and Martin (1973).

Although the data shown in Figure 7.6, and other data also, fit the transition probability model, obviously other models might be devised in which there is a lag followed by an exponential decay of cells from interphase into division. But the transition probability model makes another prediction that is less easily accommodated by other models. This concerns the behavior of sibling cells. After a division, neither of the daughter cells can divide until after the minimum intermitotic period, T_B. Then one of the daughter cells will divide first. Since the transition probability is constant with time, the probability that the other daughter cell will divide in the next hour is P. This means that in a population, the decay from interphase into division of the second sibling cell should follow the first according to the same equation that governs the periodicity of phase A, that is, $d\beta/dt = -\beta P$, where β is defined as the difference in the intermitotic times of the members of a sibling pair. A cumulative plot to show β as a function of time should therefore have the same slope as the α curve; but the β curve lacks the lag period because it is measured from a time that starts after T_B has been completed, that is, the division of the first sibling (Minor and Smith, 1974; Shields and Smith, 1977). An example is shown in Figure 7.7, which compares the β curve obtained by following sibling pairs with the α curve obtained by following the population as a whole. Both curves have the the same slope corresponding to $P = 0.46$ hr^{-1}; but the β curve has no lag, compared with the lag of the α curve equivalent to $T_B = 14$ hr. The absence of a lag makes the β curve especially suitable for the determination of P, since there is no problem in determining the point from which the exponential decay commences. Formally, the exponential form of the β curve demonstrates that sibling cells have the same constant component of the cell cycle (if any) and also share an exponential component.

What is the effect upon the transition probability of altering the conditions under which a cell culture is grown? One treatment whose effects upon cell growth are well characterized derives from the dependence of cultured cells upon the provision of serum in the medium. Thus when the serum concentration is reduced over a range in which it is limiting, there is a decrease in the rate of growth of the population. Shields and Smith (1977) showed that this reduction is accomplished by a decrease in P. For the mouse 3T6 line, P changes from about 0.96 hr^{-1} for cells growing in 5% serum to about 0.29 hr^{-1} in 0.5% serum, whereas T_B increases very slightly from 11.25 to 12 hr. For the mouse Balb-c 3T3 line, P declines from about 0.29 hr^{-1} in 5% serum to about 0.07 hr^{-1} in 1% serum. For two lines that grow at very different rates, therefore, changes in the growth rate are accomplished by changes in the probability of transition.

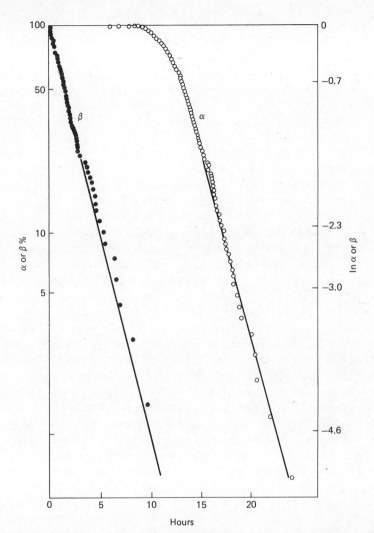

Figure 7.7. α and β curves for a population of SV3T3 cells. The α curve shows the proportion of cells with an intermitotic time equal to or greater than the plotted cell age. The β curve shows the fraction of sibling pairs with a difference in intermitotic times equal to or greater than the plotted time. Data of Shields and Smith (1977).

Quiescent and Proliferating Cultures

What happens when a cell population enters a quiescent state? Do some or all cells enter a G_0 condition distinct from the usual compartments of the cell cycle? The data of Shields and Smith show that as the serum concentration is reduced in a growing culture, the transition probability for all cells declines with the reduction in growth rate. In other words, only a single kinetic component is seen. In these experiments the probability curves extended to division of 95–97% of the cells, so that formally they support the conclusion that any cells behaving differently must represent less than 3–5% of the culture. In cultures that have achieved confluence, the rate of growth is very much less. With confluent Balb-c 3T3 and 3T6 cultures maintained in 0.5% serum, an α curve was followed for 11 days, a time long enough to see division of 40% of the cells. This gave a value for T_B of the order of 12 hours and a value for P of about 0.004 hr^{-1}. The demonstration that exponential decay of a single kinetic class is seen in confluent as well as growing cultures suggests that quiescence may be accomplished by reducing P to very low values rather than by removal of some or all cells into a separate non cycling G_0 compartment.

These results show that the transition probability depends upon the conditions of cell growth: indeed, it is by virtue of the ability to influence the probability that external factors are able to exert their efforts upon the length of the cell cycle. A brief report that the transition probability declines continuously with increases in cell density in a mouse mammary tumor line (S115), and that it is altered also by provision of testosterone, to which these cells respond, supports this conclusion (Robinson et al., 1976). How do these changes in probability occur: do they take the form of discontinuous responses to the environmental changes, effected either immediately or at some later time; or is there a gradual change in transition probability until a new stable value is established? Brooks (1975) reported in a brief note that when serum is added to stimulate quiescent cells into growth, the probability changes abruptly if adenosine is present, but alters only gradually in the absence of exogenous purines. This would seem to suggest that the probability may be adjusted instantly in response to environmental change, but that the mechanism by which this is accomplished in some way depends upon purine metabolism, and may be limited therefore by the absence of a sufficient supply of purines. Further evidence on the nature of the system responsible for determining the probability of transition is provided by studies of the effect of cycloheximide on the response of quiescent cells to serum. Even low doses of cycloheximide reduce the rate at which cells recommence division and this effect appears to be mediated by a reduction in the transition probability; the reduction is related to the dose of the inhibitor (Brooks, 1977). This is not the only effect of cycloheximide, for the drug also increases the lag time before any division commences. The reduction in transition probability and the extension of the lag period both seem to be proportional to the extent to which given doses of cycloheximide reduce the level of protein synthesis. The transition probability and lag period respond differently to changes in the serum concentration, however, since the change induced in the probability contrasts with the lack of any effect upon the duration of the lag. This difference seems to exclude any simple explanation of the effects, although it does not prevent the general conclusion that stimulation of quiescent cells requires protein synthesis and that this is necessary both during a lag before S phase can commence and to establish the transition probability thereafter.

This raises the question of the biological basis of the transition probability. It is difficult to see what cellular mechanism might be responsible for a random process. One type of model is to suppose that transition from the *B* state to the *A* state depends upon a protein(s) whose level fluctuates because of feedback loops that control its synthesis or degradation. If transition occurs when the fluctuations take the level over or under some critical amount, the frequency with which this occurs might be determined by controlling the extent of fluctuation. Obviously there must be a random element in the fluctuation, perhaps due to background "noise." The level of fluctuation must also be connected to the conditions of cell growth; one way it might achieve this responsiveness would be dependence on the rate of protein synthesis (but although the utility of this is evident for cultured somatic cells, it is not clear that it would be effective in differentiating systems). The consequence of this network should be that the time at which any individual cell makes the transition is not predictable, but that the rate of transition can be predicted for the population as a whole. Finally, it is scarcely necessary to observe that all the results supporting the transition probability model have been obtained with somatic cells in culture. Formally this does not exclude the possibility that the model describes the ability of cells to grow under these conditions when they exist as individuals, but does not pertain to the control of the cell cycle in the organism from which they are derived. Also it is necessary to note that the support for the model resides in the exponential decay with which cells leave interphase for division. The techniques used to follow this decay have been applied only to a single cell cycle; obviously it would be useful to see a demonstration that the values for T_B and P remain the same over a period of some generations in a dividing culture.

That the transition probability is valid for at least some biological situations is implied by its successful application to the growth of unicellular organisms, such as Euglena (see Minor and Smith, 1974). That it may not be valid at least for very early embryogenesis is suggested by the precise timing of synchronous divisions that may occur following fertilization of the egg; this implies the occurrence of events at predetermined times. On the other hand, the critical question about the control of the cell cycle during subsequent development is not the cycle length per se, but the number of descendents that a given cell will generate in a certain period. In this regard, the transition probability model encounters no difficulty.

A different view of the cell cycle has been suggested by observations that there may be a specific point in G_1 at which quiescent cells are halted. Pardee (1974) demonstrated that application of any one of a variety of nutritional blocks—such as deprivation for serum, isoleucine, glutamine, phosphate—causes BHK21 cells to halt at the same point in G_1, as judged by the time that it takes before the first S phase occurs upon subsequent removal of the block. After removal of one block, imposition of another is ineffective, implying that both must act at the same time. Yen and Pardee (1977, 1978) have taken this analysis further by introducing a microfluorimetric method for following cell cycles in non synchronized populations. Essentially the method involves measurement of the DNA content of cells to determine whether they are in G_1, S, or G_2. When an exponentially growing culture of 3T3 cells was shifted into medium containing only low serum, cells became halted in G_1 at a point about 3.3 hours through this phase, which was of some 5.5 hours in duration. After this point in G_1, cells are immune to serum deprivation and can continue through S phase, G_2, and mitosis into the next G_1. Deprivation of isoleucine appears to impose two blocks, one at 5.4 hours which corresponds to the very end of G_1 or beginning of S phase, and one at about 75% of the way through S phase. (This contrasts with the single block imposed in BHK21 cells.) When serum-arrested cells are returned to high levels of serum, proliferation resumes, but the first S phase does not occur until 11.5 hours after the shift of medium. This implies that the quiescent cells had left the normal cell cycle to enter a G_0 state in which a time longer than the normal duration of G_1 is required for re-entry into the division cycle. Isoleucine must be provided if these cells are to traverse the isoleucine-dependent step at the end of G_1. Restriction points also may be found in temperature-sensitive cell cycle mutants, by determining the point in the cell cycle at which mutants are able to continue in spite of an increase in temperature (Ashihara, Chang, and Baserga, 1978). The nature of these blocks to the cycle has been reviewed by Pardee et al. (1978).

The general conclusions suggested by these experiments are that commitment to enter S phase may occur at some point within G_1 prior to the initiation of replication and that quiescent cells may enter a nonproliferating pool distinct from that of cells that are in the cycling condition. Are these conclusions consistent with the transition probability model? Because most experiments on the nature of cell cycle blocks and on transition probability have been performed in somewhat different ways, it is not possible to compare them directly. Generally, however, the simplest form of the transition probability model requires that P is the only parameter subject to control. Thus blocks to the cycle should function by reducing P to a very low value and should not be associated with any specific point in the conventional G_1–G_2 cycle. But if the model is complicated by the admission of a second control, then it becomes possible to argue that the blocks might prevent transition between the two (A and B) phases, or, indeed, might act at some constant point in A that lies within G_1.

There have been few studies in which the existence of blocks and the kinetics of entry into division have been followed in the same system. Talavera and Basilico (1978) have shown that restoration of permissive conditions following serum deprivation, isoleucine starvation, or high temperature incubation of *ts* mutants of BHK cells, sees a lag followed by an exponential decay into division. The transition probability for the decay is influenced by the treatments to which the cells have been subjected and appears to apply to the entire cell population. The blocks do not all have the same effect, as shown by double deprivation treatments.

Another system is presented by Balb/c-3T3 cells rendered quiescent by density inhibition; two sets of serum factors are required for the resumption of division. A platelet growth factor(s) renders cells apparently "competent" to resume reproduction; and a plasma factor(s) causes competent cells to commence DNA synthesis after a lag of some hours (Pledger et al., 1977, 1978; Vogel et al., 1978). At least two points can therefore be recognized at which the cell cycle may be blocked. A similar relationship, but different absolute timing, is seen in the cycle of growing Swiss mouse 3T3 cells (Yen and Riddle, 1979). The acquisition of competence appears to be independent of the nutritional state of the cells, but the progression toward S phase permitted by plasma factor requires the presence of amino acids (Stiles et al., 1979). Judged by the lag between treatment and DNA synthesis, in Balb/c cells the competence control point lies about 12 hours prior to S phase and the progression control point about 6 hours before the start of DNA synthesis.

But the rate at which the cells enter S phase is exponential, with the value for P determined by the concentration of plasma factor(s). The recovery of these cells from quiescence therefore seems to involve removal from a G_0 compartment into a G_1 (or A) compartment, from which exit to S phase depends on transition probability. Two features of both the BHK and Balb/c systems may be set in apposition: the existence of a lag before resumption of replication implies the existence of a constant component due to the block, since cells cannot immediately start division with probability P; yet when division commences, all cells appear to form a single compartment as predicted by the transition probability model.

The nature of the blocks to recovery from deprivation or quiescence seems to depend on the cell system. Their existence suggests that under these circumstances control of the cell cycle can be exercised in more than one way, that is, by changing the ability to proceed past fixed points instead of or as well as changing the transition probability. The detrimental effects of the treatments needed to induce the blocks leave open the question of whether what has been revealed is a natural stage of the cell cycle or an aberration induced by the treatment. It is also hard to say whether the quiescent state induced in vitro reflects the situation of nondividing cells in vivo or is an unusual condition of the cell. The existence of blocks to recovery does run counter to the proposal that entry into quiescence is accomplished solely by a reduction in the transition probability, with no entry into a separate cell compartment.

Another situation which is difficult to explain in terms of the transition probability model occurs when cells in different stages of the cell cycle are fused together. Fusion of cells in S phase with cells in G_1 causes the G_1 nuclei to commence DNA synthesis sooner than would have been expected in the original cytoplasm (see Chapter 9). Fusions between cells at different stages of G_1 produce results consistent with the idea that inducer(s) of DNA synthesis accumulate continuously during G_1, so that replication commences when they reach a critical level (Rao, Sunkara, and Wilson, 1977). According to the probability model, however, all cells in the variable A phase possess an equivalent probability of making the transition to B phase, irrespective of the time they have been in this condition. There is therefore no reason to expect fusion to delay or advance either nucleus. Another result difficult to correlate with the probability model is the maintenance of a constant cell size through cycles of division. The growth curves of Figures 7.3 and 7.4 suggest that S phase is correlated with cell growth. This seems explicable in terms of the need to avoid wide fluctuations in cell size. But it is not consistent with the view that a transition may occur at any point during the variable A phase, unless cell growth is limited to the constant B phase, which does not appear to be the case.

Synthesis of Proteins During S Phase

A question that has been asked for many years is whether protein synthesis is subject to cell cycle controls. Are proteins necessary for DNA replication synthesized during S phase or are they present throughout the cycle? Are other proteins synthesized specifically during G_1 or G_2? And are proteins that are synthesized continuously through interphase subject to gene dosage controls—what happens when the number of copies of a gene is doubled during S phase?

Several technical problems make it difficult to follow the synthesis of replication enzymes. A general problem in examining fluctuations in enzyme levels during the cell cycle is that synchronization of cells with inhibitors—often necessary to obtain enough material to assay enzyme activities—may divorce the replication and cell growth cycles. Thus although cells may be compelled to accumulate in metaphase or at the beginning of S phase, cellular activities other than DNA synthesis may not be inhibited in the same way as replication itself.

Another caveat is that changes in enzyme activities may not reflect changes in de novo synthesis, but may be caused by activation or inhibition of preexisting enzyme molecules. In eucaryotic cells it has rarely been possible to assay the amounts of enzyme protein; often fluctuations in the enzyme level can be followed only by assay of its activity in vitro.

These difficulties are reflected in the information available on DNA polymerase activities. Three eucaryotic enzymes have been distinguished; but none can be obtained readily in pure condition and their relative functions are not yet clear (see Chapter 20). Early experiments to measure the DNA polymerase activities of different cell compartments in cultured animal cells or sea urchin embryos led to suggestions that there might be a movement of enzyme molecules from cytoplasm to nucleus during S phase (Littlefield, McGovern and Margeson, 1963; Friedman, 1970; Fansler and Loeb, 1972). More recent results have shown, however, that one of the enzymes leaks readily from the nucleus into cytoplasmic fractions; probably variations in its leakage are responsible for the observed differences in localization in the cell cycle. Attempts to follow enzyme activities during the cycle of synchronized HeLa cells have suggested that only one enzyme activity increases in S phase (Spadari and Weissbach, 1974b; Chiu and Baril, 1975). But difficulties in distinguishing the enzyme activities on the basis of their responses to altered assay conditions leave this only a tentative conclusion. It remains to be seen which DNA polymerase activity is the replicase.

There have been reports that several of the enzymes concerned with the production of precursors to DNA are subject to cyclic synthesis. These include ribonucleotide reductase (which converts CDP to dCDP), dCMP deaminase, deoxycytidine kinase, and thymidine kinase. In most of these cases, the enzyme activity has been measured during a single cell cycle following a synchronization procedure and has been shown to increase during S phase; the increase usually is prevented by actinomycin and may be enhanced by the inhibitor of replication, hydroxyurea. But in many cases the enzyme activity was not followed even until a decline in G_2; and in no case was it followed into a second cell cycle.* The results therefore leave open the possibility that the increase seen during this cell cycle is a consequence of the synchronization procedure or perhaps mimics events that occur when quiescent cells are stimulated into growth, that is, represents a single occurrence and not a cyclic phenomenon. More recent experiments with cultured human cells have allowed thymidine kinase activity to be followed through two cell cycles; and here it seems clear that an increase in the level during S phase is succeeded by a decrease during G_2 and then by a further increase at the next S phase (Bello, 1974). The usual susceptibility to actinomycin is seen and this is consistent with the idea that in this case there may be cyclic enzyme synthesis.

Although there is little evidence to suggest a general conclusion that the enzymes of DNA synthesis and metabolism are synthesized cyclically during S phase, there is some evidence that their production may be altered in nondividing cells. When dividing avian erythroblasts mature to nondividing polychromatic erythrocytes, DNA polymerase and thymidine kinase activities decline, disappearing entirely by the time mature erythrocytes are produced (Williams, 1972a,b). The reverse series of changes is seen when human lymphocytes that have ceased division are stimulated to

---

*References: Brent, Butler and Crathorn (1965), Stubblefield and Murphree (1967), Murphree, Stubblefield and Moore (1969), Gelband, Kim and Perez (1969), Brent (1971).

recommence growth by the mitogen PHA; DNA polymerase, thymidine kinase and TMP kinase activities all increase before replication of DNA starts (Loeb et al., 1970). Analogous changes in the activities of some DNA polymerases may occur when quiescent cultured cells enter the growth phase (Chang et al., 1973). If it is true that cells that have ceased division suffer declines in the levels of such enzymes, whereas dividing cells do not show comparable changes during their cycles, these inactivities may be considered markers of a G_0 state.

It would be interesting to know the point at which the enzymes of DNA synthesis and metabolism disappear from cells in which replication has been suspended indefinitely. Does this occur immediately following the last S phase or is commitment to a G_0 state a more gradual affair? In any case, reversal of such changes is naturally essential if quiescent cells later are stimulated to divide. But the disappearance and reappearance of enzyme in these conditions makes no implication that any similar events occur during the normal cell cycle. And although the activities of some DNA synthetic or metabolic enzymes may be correlated with replication, it is unlikely that any fluctuations in synthetic activities or in the levels of nucleotide precursors provides a control of replication; there is no evidence to support the concept that these activities may be limiting on replication. It seems reasonable to suppose that any enzyme activities correlated with replication are controlled by systems related to those which control replication.

An important case in which protein synthesis is completely correlated with DNA synthesis is that of the histones. This means that both the nucleic acid and protein components of the chromosomes are synthesized at the same time. Histone synthesis depends upon DNA synthesis, since the inhibition of replication brings the production of histones to a halt. The synthesis of DNA and of the histones is discussed in Chapters 20 and 28.

CHAPTER 8

Hybrid Cells: Nucleus and Cytoplasm

The construction of a somatic cell hybrid represents a dramatic perturbation of the organization of the cell in which a nucleus is placed in a new environment. Several techniques have been developed for the reconstruction of cells from their major compartments, and, depending upon the method used and the cell types to which it is applied, the hybrids may possess a single nucleus in a new cytoplasm, the nuclei of two parents in the cytoplasm of one, or both nuclei in mixed parental cytoplasm.

The earliest situations in which a nucleus exchanged one environment for another were provided by micromanipulation, in which the nucleus of one cell was transplanted into the enucleated cytoplasm of another. Such nuclear transplantations have been performed with unicellular eucaryotes and with amphibia, in which the nuclei of somatic cells can be transferred by pipette into the rather large eggs. Mostly these experiments were concerned with studies of the developmental potential of the nuclei of differentiated cells, although this approach since has been extended to analyze the behavior of macromolecules injected directly into egg cytoplasm.

More recently it has become possible to use the drug cytochalasin B to cause nuclei to become extruded from the cytoplasms of cultured cells. By applying centrifugation, this allows preparations of nuclei and enucleated cytoplasms to be obtained. Their recombination into viable cells allows hybrids to be formed in which the nucleus is derived from one cell type and the cytoplasm from another.

An earlier technique with similar consequences is the fusion of two somatic cells mediated by the Sendai virus, to generate a heterokaryon that contains both nuclei in a common cytoplasm. Among the descendents of this initial heterokaryon may be somatic cell hybrids, which possess a single nucleus containing the chromosome complements of both parents in a mixed cytoplasm.

While micromanipulation experiments have used intraspecific transfers, the techniques of cytochalasin B treatment or cell fusion allow a nucleus of one species to be combined with the cytoplasm of another, allowing interspecific as well as intraspecific nucleocytoplasmic interactions to be examined. In all these situations, the influence of its new environment upon the nucleus provides an important approach for examining the relationship between nucleus and cytoplasm. The effects of the new cytoplasm upon the expression of nuclear genes provide information about the specificity and general nature of control mechanisms; and the movements of proteins between cytoplasm and nucleus can be followed. In some interspecific somatic cell hybrids, the chromosomes of one parent are lost over a period of some generations. By correlating chromosome loss with the disappearance of some protein product, it is possible to map genes to chromosomes.

In this chapter we shall be concerned with the formation of hybrid cells and with their use for following nucleocytoplasmic interactions, in particular in the situation in which an inert nucleus is reactivated by formation of the hybrid. In the next chapter we shall consider the use of hybrid cells for mapping structural genes and analyzing the control of their expression, for changing chromosome structure by suddenly altering the phase of the cell cycle; and finally we shall come to the possibilities for transferring genes into somatic cells by fusion techniques or by treatment with isolated chromosomes or DNA.

Reciprocity of Nucleocytoplasmic Interactions

A question commonly asked in the early days of developmental biology was whether nucleus or cytoplasm controls development. Such a question now seems to lack subtlety, of course, for we understand development to constitute a series of cyclic interactions between nucleus and cytoplasm. It is the cytoplasm of the egg which contains the information necessary to elicit from the nucleus the response appropriate for the start of embryonic development; and the products of the genes expressed at this time in turn modify the cytoplasm so that new patterns of gene expression are induced in the nucleus during subsequent development.

Each nucleus in a multicellular eucaryote is restricted so that it expresses only certain genes. The nature of this restriction is one of the critical questions of developmental biology. The nucleus of the fertilized egg is said to be *totipotential:* although itself expressing only some genes, when its descendents subsequently generate the differentiated functions of each cell phenotype, they express all the genes utilized by the organism. As embryonic development proceeds, however, the range of gene functions that may be expressed in the descendents of any nucleus is progressively reduced as it enters pathways of increasing cell specialization.

Whereas the pattern of gene expression in an embryonic cell is transient in the sense that its descendents may express sets of genes different from those used in the embryonic state, a fully differentiated cell (and its descendents, if any) may maintain a fixed state of gene expression. Transplantation of nuclei from differentiated cells to embryonic cytoplasm has been used to determine whether nuclei that usually do not express certain genes nonetheless retain the potential to do so. These experiments show that the changes that occur in nuclear gene expression during differentiation are reversible.

Information in the cytoplasm must be molecular in the sense that it consists of regulator molecules that may enter the nucleus to influence its pattern of gene activation. The nature of this information can be studied by following the molecules which enter a nucleus after transplantation. The specificity of these signals can be deduced from the properties of hybrid cells produced by fusing two different somatic cells, in which genes of one parent may respond to cytoplasmic signals from the other.

Positional information also must be present in the cytoplasm of the egg and early embryo. Development requires the differentiation of nuclei with characteristic patterns of gene expression. This could not happen if all descendents of the first nucleus continue to respond to the cytoplasm in the same way. One model to explain the development of nuclei expressing different genes is to suppose that regulator molecules are not dispersed uniformly through the cytoplasm, but are concentrated in such a way that different nuclei are exposed to different regulator molecules (or to different quantities of them) and therefore start upon different developmental pathways. We know very little about how positional information is established and maintained.

Nucleocytoplasmic Interactions in Acetabularia

The reciprocal interplay of nucleus and cytoplasm is well illustrated by early experiments performed with the unicellular organism Acetabularia. The zygote of Acetabularia germinates to form a stalk which grows to a total length of some 3-5 cm in about three months. During the next month, a *cap* of characteristic morphology develops at the end of the stalk, with a diameter of about 1 cm compared with the stalk diameter of 0.3-0.4 cm. At the lower end of the stalk, a *rhizoid* forms to contain the single nucleus. This nucleus increases considerably in size during development, from some 4×10^{-9} mm^3 in the zygote to about 1×10^{-3} mm^3 when the cap is fully developed.

Information about the state of the cap must be transmitted to the nucleus, for in cells with mature caps the primary giant nucleus begins to disintegrate, extruding a small diploid secondary nucleus which divides mitotically many times (for review see Hammerling, 1963). The secondary nuclei produced by these divisions then are transported by cytoplasmic streaming to the cap, where additional mitoses are followed by meiosis. The final number of secondary nuclei is of the order of 7000-15,000. Formation of secondary nuclei depends upon signals from the cap, for nuclear division can be completely inhibited by removing the full grown cap before the nucleus begins its reproductive cycle. This process can be extended indefinitely by removing the cap each time it regenerates.

The development of the cap in turn depends upon the nucleus. Experiments with Acetabularia provided a direct demonstration of the role of the nucleus in heredity and of the need for a messenger molecule to transport its information to the cytoplasm. When the nucleus is removed from a cell, the anucleate cytoplasm may survive for several months. Its morphogenetic capacity is proportional to the length of the stalk; in fact, there is a concentration gradient of capacity from cap to rhizoid (for review see Hammerling, 1953). This implies that messengers are produced by the nucleus and transported to the cap where they accumulate in stable form.

Each of the several varieties of Acetabularia forms a cap of characteristic appearance. After the nucleus of one plant has been removed, that of another type may be inserted in its place. When nuclei are exchanged between A. mediterranea and A. crenulata, each hybrid plant develops a cap of the type characteristic of its new nucleus. Intermediate caps, showing characteristics of both the old and new nuclei, form in some instances, presumably because of mixing between the messengers synthesized by the new nucleus and those remaining from the previous nucleus. But when these caps are removed, the cap that regenerates displays the appearance of that coded by the new nucleus. When nuclei are added to instead of replacing the original nucleus of a plant, the type of cap that is formed depends on the ratio of the different classes of nuclei. The experiments thus demonstrate that each nucleus carries its own specific hereditary information, which can be expressed in the cytoplasm of another variety of the plant. The responses elicited from the nucleus by the state of development of the cytoplasm are species specific and determine the phenotype.

Transplantation of Xenopus Nuclei

Whether irreversible changes take place in the nuclei of somatic cells during differentiation is a question that has occupied developmental biologists since the last century. In most organisms, no change is apparent in the genetic material; all somatic cells appear to possess the same diploid chromosome content (see Chapter 1). Nucleic acid hybridization studies suggest that all differentiated cells contain the same sequences of DNA; there seems to be no addition of sequences representing particular genes in the chromosomes of the specialized cells in which they are expressed, nor any loss of functions that are not utilized (see Chapter 24). Of course, the precision of the hybridization technique is too low to prove that all genes are invariant in representation, but these results make it unlikely that any substantial changes in genetic content occur during differentiation.* Selective expres-

---

*We should note two exceptions to the invariance of genetic information. The genes for ribosomal RNA are amplified in certain amphibia during oogenesis to yield a satellite of ribosomal DNA; and they may be magnified in Drosophila to compensate for mutations reducing their number. The structure of genes for rRNA in both the normal and amplified condition is discussed in Chapter 27. In some tissues of certain organisms, cells may become polyploid as the

sion of genes alone must therefore be responsible for the development of differentiated phenotypes (for review see Davidson, 1976).

The general rule that genetic content is invariant does not preclude the possibility that other changes in the genetic material may occur, involving not the loss of genetic information but the irreversible inactivation in each cell line of the genes that are not to be expressed. Possible mechanisms would be the modification of nucleotide sequences, for example, by methylation, or the irreversible binding of proteins to DNA.

To decide whether a differentiated nucleus is totipotential or restricted in its developmental capacity requires placing it in a situation in which it has the opportunity to express functions that have been turned off in its differentiated state. A critical test of the ability of a differentiated nucleus to give rise to other cell phenotypes is provided by micromanipulation experiments in which the nucleus of an unfertilized egg is removed or inactivated and the diploid nucleus of a somatic cell is inserted in its place. The ability of the transplanted egg to undertake normal embryogenesis reflects the extent to which differentiation of the transplanted nucleus may be reversed by the egg cytoplasm.

Amphibian eggs have been used for most transplantation experiments since they are large enough to manipulate directly. With the frog Rana pipiens, the resident egg nucleus usually is removed with a glass needle and replaced by a nucleus taken from a somatic cell. The egg must be stimulated to start development by pricking with a needle. With the toad Xenopus laevis, the resident nucleus usually is inactivated with an ultraviolet microbeam, after which a new nucleus may be implanted and utilized for subsequent development without further activation (for review see Gurdon, 1964).

Injection of the donor nucleus is performed by placing donor tissue in a saline solution lacking calcium and magnesium ions; the donor cell is sucked into a micropipette of a size that breaks the cell but retains the nucleus. The nucleus is surrounded by a small amount of cytoplasm whose presence is essential if it is to survive transplantation—exposure to the medium is invariably lethal. It is more difficult to transplant nuclei from cells at progressing stages of differentiation because they have little surrounding cytoplasm compared with embryonic cells and therefore are more easily damaged.

In experiments with R. pipiens, King and Briggs (1956) found that large numbers of eggs injected with nuclei extracted from the blastula stage of embryogenesis grow to yield normal tadpoles. Successful transplantation also can be achieved with nuclei taken from different parts of the gastrula

---

result of repeated replications of the genetic material without division. One example in which not all sequences of DNA are replicated the same number of times is provided by the salivary glands of the Diptera. This results in the differential representation of sequences discussed in Chapter 16.

embryo, when cells have begun to develop specialized functions. Briggs and King (1960) later reared some of these tadpoles through metamorphosis.

Transplantations with X. laevis follow a generally similar course, although greater numbers of the transplants survive to at least the tadpole stage. Gurdon (1962a) used as donors endoderm nuclei taken from stages between the late blastula and swimming tadpole. Although the cells from which they are derived have developed differentiated functions, upon transplantation the specialization of their nuclei is reversed and many of the eggs into which they are placed develop into normal frogs. The endoderm nuclei must therefore be totipotential, possessing the ability to respond to egg cytoplasm by generating descendents that exhibit all the differentiated functions of the frog, including those that are never expressed in the endoderm cell.

As donor nuclei are extracted from cells at later stages of development, the proportion of successful transplants declines. Figure 8.1 shows that a somewhat similar decline is seen with both Rana and Xenopus nuclei, although it is more pronounced with Rana. The decline is expressed relative to the number of blastula nuclei that are able to support normal development upon transplantation, which is between one-half and one-third of the total number of transplants. Only about 10% of the nuclei transplanted from differentiating embryos and tadpoles are able to support normal development (for review see Gurdon, 1964). Poor results also are obtained

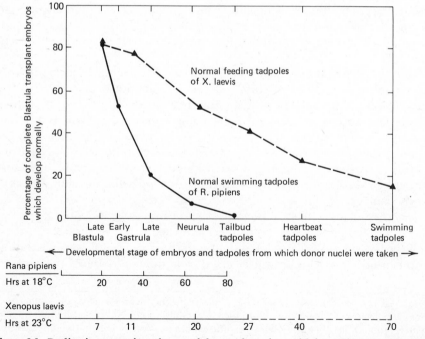

Figure 8.1. Decline in proportion of successful transplantations with increasing age of donors from which nuclei were taken. Data of Gurdon (1963).

when nuclei are transplanted from fully differentiated adult cells. Gurdon (1962c) found that when nuclei are transplanted from the intestinal epithelial cells of feeding tadpoles, some support development of normal tadpoles, but the great majority shows abnormalities that vary from lack of cleavage to the production of nearly normal tadpoles. But if the abortive cleavages—which may result when nuclei are taken from the donor at an unsuitable stage of the cell cycle—are excluded, it is possible to calculate that some 24% of the intestinal epithelial cells have a full developmental potentiality. In these cells, which represent the final state of endoderm development, differentiation is reversible.

Does the decline in successful development of transplants with donor age reflect some restriction on potentiality or is it a consequence of changes in susceptibilty to the transplantation technique itself? A hint that restrictions of developmental potential are not imposed during development is provided by observations that the same types of abnormality are seen in transplants derived from all classes of donor nuclei (Gurdon, 1960). It is reasonable to suppose that any genuine restriction of potential should be specific for the cell in which it occurs, so that cells of different specialization should possess restrictions of different genes, resulting in different and characteristic inabilities to support development after transplantation. The lack of any such specificity suggests that the decline with donor age may reflect a limitation of the technique. (Of course, it is possible that cell-specific restrictions of potential do accompany development, but that they are obscured by the random restrictions induced by the technique itself. However, the usual interpretation of these results is that all nuclei retain their full developmental potential during specialization.)

The reversibility of restrictions on the development of a transplanted nucleus can be tested by the technique of *serial transfer*, first introduced by King and Briggs (1956). A nucleus is transplanted into an egg and allowed to grow to the blastula stage. Nuclei are then extracted from this *first transfer* blastula and in turn transplanted into many more eggs; all of these *second transfer* transplants represent descendents of the single somatic nucleus used to provide the first transfer blastula. Gurdon (1963) reviewed experiments of this nature with Xenopus, which show that nuclei taken from the first transfer blastula fare no better through several generations of serial transfer than the original embryonic nuclei. Generally these experiments are interpreted to mean that when a nucleus suffers damage in transplantation, the effect is irreversible. But, of course, this does not exclude the possibility that restrictions are imposed on some nuclei in vivo and then are maintained through an indefinite number of generations in serial transfer.

The idea that induced damage may be irreversible is supported by experiments in which Gurdon (1962b) transplanted an X. laevis nucleus into the cytoplasm of an X. tropicalis egg (the two species cannot interbreed). Such transplants develop only to the late gastrula stage; when nuclei are recovered from the hybrid blastula and retransplanted into eggs of X. laevis, they are unable to support development past the neurula stage. This suggests that the X. laevis nucleus suffers irreversible damage (of an unknown nature) as the result of its passage through the X. tropicalis cytoplasm.

Even cultured cells may retain full developmental potential. When Gurdon and Laskey (1970) transplanted nuclei from cultured X. laevis epithelial cells, they found that about 75% of the transplants show abortive cleavage or fail to cleave. Most of the remaining eggs suffer partial cleavages in which part of the egg appears to consist of normal blastomeres and the remainder is uncleaved or abortively cleaved. Fewer than 5% of the transplanted nuclei cleave regularly to form blastulae and most of these die in the blastula stage. In all, less than 0.1% of the transplanted nuclei develop into normal or nearly normal tadpoles.

With donor nuclei of these cultured cells, serial transplants from a first transfer blastula show much better development than that of the original donors. Gurdon (1962c) found that serial transfers of intestinal epithelial nuclei also increase the number of tadpoles that can be reared. In these cases, as is suggested by the appearance of the blastulae, the first transfer embryos probably contain both normal and abnormal nuclei derived from the original transplanted nucleus; the serial transplantation then may allow some of the normal nuclei to develop unhampered by the damaged nuclei. This implies that the original donor nucleus is not itself damaged by transplantation, but that irreversible mistakes may be made when it divides, so that some but not all of its descendents are damaged.

Given the nature of the process, it is indeed remarkable that any transplanted nuclei are able to support the development of normal adult frogs. Success with even a single cell of a given type shows that at least some of the cells of this differentiated class retain all the genetic information of the nucleus during development. The proper development of each tissue of the adult shows further that differentiation is reversible; the genes of each specialized nucleus can revert to the pattern of expression characteristic of the embryo, and in descendants then display the usual states of gene expression characteristic of each cell type. Although the limitations of the technique prevent a formal demonstration that this is true of all cells, it is a likelier explanation to suppose that unsuccessfully transplanted nuclei have been damaged during transplantation than to postulate that some cells are not totipotential while others are.

Reactivation of Transplanted Nuclei

The ability of the nucleus of a differentiated Xenopus cell to give rise to a normal adult organism when transplanted into egg cytoplasm in itself implies that all the usual molecular controls of gene expression are exercised at the appropriate stages of development. Presumably this involves a "reprogramming" of the pattern of gene expression to fit the new environment of the nucleus. We may ask how rapidly this reprogramming is established: does it occur immediately, in the egg itself, or at some later time, when there are several descendants of the transplanted nucleus occupying the developing embryo? (This is related, of course, to the question of the extent to which the early divisions of the developing egg are independent of the expression of new genetic functions.) Answers to questions of this nature, concerning either the qualitative or quantitative control of gene expression, require direct measurements of the synthesis of specific gene products. Because of the need to manipulate individual eggs, this is difficult to accomplish. But it has been possible to follow the synthesis of certain classes of RNA (although not of individual messengers) and, more recently, to examine some of the protein gene-products. In general, nuclear activities appear to be highly responsive to the cytoplasmic environment in which the nucleus finds itself.

Replication responds rapidly upon transplantation into egg cytoplasm. Nuclei from cells that have virtually ceased DNA replication and cell division —such as adult liver, brain and blood—can recommence DNA synthesis under the stimulation of egg cytoplasm. Using autoradiography to follow the ability of transplanted nuclei to synthesize DNA, Graham, Arms, and Gurdon (1966) showed that label is incorporated within 20–40 minutes after transplantation. By 70 minutes nearly all the transplanted nuclei have completed their first mitotic division and thereafter they continue to incorporate label in the manner characteristic of the normally developing egg.

Some 10% of mouse liver nuclei are persuaded to start DNA synthesis when injected into Xenopus eggs, which argues that the signals controlling replication may be of low specificity. Nearly all Xenopus nuclei commence DNA synthesis after transplantation, so that failure to do so cannot in general be responsible for unsuccessful development. In reviewing transplantation experiments, Gurdon and Woodland (1970) observed that replication must be completed within an hour of transplantation, contrasted with a duration for S phase in the somatic cell nucleus of about 6 hours. This suggests that failure to complete replication in the time available for the first division may sometimes be responsible for the inability of transplanted nuclei to support normal development.

Distinct transitions in the pattern of RNA synthesis occur during the development of Xenopus embryos. Synthesis of mRNA and tRNA starts at late cleavage and continues at a rapid rate during gastrulation and neurulation. Synthesis of rRNA is quiescent at first and starts at the beginning of gastrulation, increasing in rate during later development. Gurdon and Brown (1965) showed that when neurala endoderm nuclei (which are therefore actively synthesizing rRNA) are transplanted into eggs, the resulting blastulae exhibit the usual failure to synthesize rRNA. But when these transplants reach the neurula stage, rRNA is synthesized at the same level as in comparable embryos derived from fertilized eggs. Transcription of rRNA genes expressed in the nucleus before its transplantation therefore is switched off early in the development of the transplant embryo and then is switched on again later at the appropriate stage. This demonstrates that reprogramming of the transplanted nucleus or its descendents includes the repression of genes previously active as well as the expression of genes previously inactive.

The nuclear activities of growing and maturing oocytes differ from those of the mature egg. Growing oocytes are engaged in RNA synthesis but do not synthesize DNA; maturing oocytes complete meiotic division and therefore possess condensed chromosomes inactive in both RNA and DNA synthesis. Comparing the responses of nuclei injected into growing or maturing oocytes therefore tests their abilities to respond to two further sets of cytoplasmic conditions. If the activities of nuclei are established by their surrounding cytoplasm, those placed in growing oocytes should synthesize RNA but not DNA; whereas those placed in maturing oocytes should lose all synthetic activities but should divide. (Nuclei transplanted into eggs, of course, immediately synthesize DNA and divide.)

The resident nucleus of an oocyte has a very large volume and possesses diffuse chromatin and multiple (up to 1500) nucleoli. Nuclei from later embryonic or adult tissues have a much smaller volume and lack multiple nucleoli. Gurdon (1968) found that when nuclei from midblastula, late gastrula, or adult brain, identified by a prior label with ^{3}H-thymidine, are injected into oocytes, they suffer a pronounced enlargement. Swelling is experienced by all injected nuclei, although its rate depends upon their previous stage of differentiation; nuclei from more highly differentiated cells swell more slowly. Blastula nuclei may enlarge more than 250-fold within three days after transplantation, to increase from 440 to 110,000 μm^3; brain nuclei may swell from a starting size of 100 μm^3 to one of 4000 μm^3. The chromatin of injected nuclei becomes more dispersed in parallel with the increase in nuclear volume. Nucleolar-like bodies develop in nuclei previously lacking them. Similar results are seen upon injection of nuclei into eggs. Active intestinal epithelial nuclei may increase in size from 160 to 4500 μm^3 within 40 minutes and in this case any nucleoli that are present disappear.

Nuclei that have enlarged after injection into growing oocytes then synthesize RNA, as shown by autoradiography. Thus midblastula nuclei, which usually synthesize little RNA but are active in DNA synthesis, are induced to commence transcription. Replication continues in the blastula nucleus for about 30 minutes after injection into growing oocytes and then ceases. This confirms that growing oocyte cytoplasm switches on RNA synthesis but switches off DNA synthesis and prevents transplanted blastula nuclei from undergoing the divisions that they usually experience. The same correlation between enlargement and reprogramming is seen in nuclei injected into eggs, where nuclei that fail to enlarge do not incorporate label into DNA and are unable to support development.

Brain nuclei divide very rarely—less than 0.1% of the nuclei in adult frog brain are in division at any time. Gurdon found that when a suspension of brain nuclei is injected into maturing oocytes—each oocyte receiving some 20–40 nuclei—they form asters, spindles and highly condensed chromosomes (although division often appears abnormal, as might be expected in an oocyte containing so many nuclei). It is not clear whether the injected nuclei enter mitosis or meiosis; but they are induced to divide by the cytoplasm of the maturing oocyte.

At each stage of development of the oocyte and egg, the cytoplasm must therefore contain information that elicits the appropriate response from any nucleus placed in it, even when that response is very different from the functions displayed by the nucleus before its transplantation. Synthesis of DNA or RNA may be either suppressed or induced according to the type of cytoplasm; and division may be repressed or induced. Nuclei therefore establish a state of activity corresponding to the surrounding cytoplasm; and nuclei of all cells appear to respond to the cytoplasmic signals in the same way, although at different rates (presumably because it takes longer to reverse changes in the state of chromatin derived from less active cells).

These assays concern the expression of classes of synthetic activities. But to what extent is the expression of structural genes reprogrammed immediately following transplantation? To investigate this question, DeRobertis and Gurdon (1977) transplanted nuclei from Xenopus cultured cells into oocytes of the newt Pleurodeles waltlii, which is only rather distantly related to X. laevis. When the proteins of X. laevis and P. waltlii oocytes are compared by two-dimensional gel electrophoresis, some are common to both species, but several proteins specific for Xenopus can be recognized. Some of these also are found in the Xenopus cultured cell line; but some are peculiar to the oocyte. Thus it is possible to assay the synthesis of some proteins specific for the oocyte and some specific for the cultured cell.

When a Xenopus cultured cell nucleus is injected into a Pleurodeles oocyte, the nucleus enlarges and by 3–7 days after injection some 6 new proteins can be seen on the 2D gels. Of these, 3 are common to Xenopus cultured cells and oocytes; and 3 are found only in the Xenopus oocyte. This implies that the Pleurodeles oocyte cytoplasm can switch on the expression of genes usually expressed in the Xenopus oocyte. Thus the nucleus of the Xenopus cultured cell can be reprogrammed to at least this extent by a new cytoplasm; and at least some of the cytoplasmic signals responsible for this control would appear to be common to organisms as distant as Xenopus and Pleurodeles.

HeLa nuclei can be injected into frog oocytes by the same techniques that are used to transplant Xenopus nuclei. Gurdon, DeRobertis, and Partington (1976) reported that such nuclei continue to synthesize RNA for at least a month (the length of time for which they have been followed). The extent of RNA synthesis—judged by autoradiography—is roughly proportional to the enlargement of the nucleus that occurs in the oocyte cytoplasm. When the total proteins synthesized by oocytes containing HeLa nuclei are compared with those of control oocytes by two dimensional gel electrophoresis, at least one additional major protein can be found at a position characteristic of a HeLa protein. This suggests HeLa messengers can be translated by the Xenopus oocyte.

This indeed is consistent with earlier observations that the cytoplasms of Xenopus oocytes and eggs can themselves support appropriate synthetic activities, as well as dictating the state of gene expression of injected nuclei. This includes the replication of DNA injected into egg cytoplasm but not oocyte cytoplasm (Gurdon, Birnsteil, and Speight, 1969; Gurdon and Speight, 1969). Heterologous messengers can be translated, so that, for example, the injection of globin mRNA is followed by the synthesis of globin protein. Indeed, this has led to the development of the Xenopus oocyte as an in vivo assay for mRNA (see Chapter 23).

We have seen that activation of an inert nucleus by transplantation depends upon its enlargement in the new cytoplasm. This must reflect the entry of proteins from the cytoplasm. These may have two roles. The physical state of the nucleus must be changed, for example, to accomplish the dispersion of its chromatin, and this may involve proteins that play a structural role. The reprogramming of gene expression, extending to individual structural genes, may depend upon the entry of specific regulator molecules.

The acumulation of proteins in injected nuclei during their enlargement can be followed by labeling the recipient eggs with radioactive amino acids prior to the transplantation. Arms (1968) followed the enlargement of Xenopus liver nuclei in cleaving eggs (which contain 2–12 cells at the time of injection) and Merriam (1969) observed the enlargement of brain nuclei in earlier eggs. Proteins begin to enter the injected nuclei even before there are any signs of enlargement, which suggests that the uptake may be needed for activation; certainly it does not appear to be a consequence of the enlargement. Nuclear uptake of proteins from the cytoplasm is not changed by the injection of puromycin together with donor nuclei, which implies that it depends upon preexisting proteins present in the cytoplasm before injection.

The proteins entering the nucleus add to rather than replace those previously present. Gurdon (1970) found that when embryos are labeled with radioactive amino acids from late blastula to neurula, nuclei extracted for transplantation retain the label during their enlargement in egg cytoplasm. There is no loss during enlargement of proteins labeled with arginine and alanine (largely histones) or of those labeled with tryptophan and phenylalanine (nonhistone proteins). This implies that it may not be necessary to remove proteins from chromatin to change its state of expression.

It is difficult to interpret experiments in which proteins of the egg are labeled before nuclear injection, because many different classes of protein are labeled and those synthesized prior to the labeling period may remain unlabeled. Gurdon therefore iodinated calf thymus histones and bovine serum albumin with ^{125}I in vitro and injected the modified proteins into eggs containing a transplanted nucleus. Histones appear to be taken up by both the transplanted nucleus and the resident nucleus (which was not removed), but the BSA enters neither nucleus.

The size of the protein appears to be an important (but not sole) determinant of its mobility. Small proteins such as lysozyme and myoglobin equilibrate rapidly between nucleus and cytoplasm, whereas large proteins such as BSA do so only very slowly (Bonner, 1975a,b). Histones accumulate in the nucleus, presumably because they are retained there after passing through the nuclear membrane, instead of remaining free to return. When nuclear contents or cytoplasm from Xenopus oocytes labeled with amino acids are reinjected into unlabeled oocytes, the proteins fall into three classes. Some are found in the nucleus and cytoplasm at similar concentrations; presumably these pass freely through the nuclear membrane. Some of the nuclear proteins accumulate in the nucleus, and this class includes some proteins that are much larger than BSA. Some of the cytoplasmic proteins accumulate in the cytoplasm. This is not surprising; since all proteins are synthesized in the cytoplasm, the ability to move to the appropriate cell compartment after synthesis must be inherent. It seems that this intrinsic ability applies to transplanted nuclei as well as to resident nuclei and may be responsible for the reprogramming of gene expression.

Fusion of Somatic Cells

Somatic cell fusion allows cells of different specialized types and from different species to be fused together to form progeny that contain the genetic information of both parental cell types. The first recent observations of somatic cell fusion concerned spontaneous events occurring in cultures containing two different lines of mouse cells (Barski, Sorieul, and Cornefert, 1961; Sorieul and Ephrussi, 1961). The conclusion that the observed polyploid cells arose by fusion and not by an increase in the ploidy of one parental line was supported by observations that they expressed antigens characteristic of both parents (Gershon and Sachs, 1963; Spencer et al., 1964).

The introduction of two further techniques strengthened the potential of this approach. Once parental lines with different deficiencies became available, it was possible to use selection to isolate hybrids by virtue of their ability to grow in conditions in which neither parent can survive. This was first used by Littlefield (1964b) to select hybrids able to grow on HAT, from parental lines one of which lacked HGPRT and the other of which lacked thymidine kinase. Of course, this approach is subject to the problem that reversion may permit parental cells to grow (see Chapter 6 and below; also Davidson and Ephrussi, 1965).

The second advance was the ability to produce hybrids at will by using inactivated Sendai virus to mediate cell fusion, instead of relying upon rare spontaneous events. This use of the Sendai virus was first noted by Okada (1962a,b) and was subsequently made the basis of a general technique for fusing together any pair of cultured cells.* Other methods for fusion since have been reported, but only the use of polyethylene glycol has proved atall productive.

The exact means of action of the Sendai virus is not clear, but a common model supposes that the virus—inactivated by ultraviolet to prevent the prosecution of infection—adsorbs to the surfaces of the cells, causing them to adhere to each other. There has been some debate on whether the cell surfaces come immediately into direct juxtaposition or for a while remain linked via the virus. At present it seems likely that the viral envelope forms the cytoplasmic bridge between the cells, after which the cells swell into this region to extend the area of contact; this is followed by a dissolution of the cell membranes over their common area. A similar series of events occurs when fusion is sponsored by polyethylene glycol, but here the initial contact must be directly between small areas of the two cell membranes. The two cytoplasms ultimately coalesce to yield a cell which contains the individual nuclei of its parents in the mixed cytoplasm of both. Fusion is not limited to two cells, for several may fuse together simultaneously or sequentially to yield multinucleate cells containing varying numbers of nuclei (for review see Harris, 1974).

The binucleate or multinucleate cells formed by fusion of parental cells of different species have been termed *heterokaryons,* in contrast to the *homokaryons* produced by fusion of like cells. Chromosome replication and nuclear mitosis may take place in these cells. In cells containing many nuclei, nuclear mitosis usually is not accompanied by formation of a spindle and division of the cell; and in such cases only some of the nuclei in the cell enter mitosis together, the others remaining in interphase. Those nuclei which enter mitosis together, however, usually are reconstituted as a single unit. Thus when all the nuclei of a cell enter mitosis simultaneously— either immediately after fusion or after intervening mitoses and reconstitutions—post-mitotic reconstitution leads to the formation of a single nucleus containing chromosomes of both parental species. Cells in which a single nucleus contains both sets of parental chromosomes sometimes are described as *synkaryons;* more often they are known simply as *somatic cell hybrids,* a term that generally is reserved for the instance when nuclei as well as cytoplasms have fused together.

In cells containing only a small number of nuclei, synchronous nuclear mitoses may be followed by formation of a spindle and division of the cell. This happens most commonly in binucleate cells, but is occasionally ob-

*References: Sendai virus—Okada and Murayama (1965), Harris and Watkins (1965), Ephrussi and Weiss (1965), Yerganian and Nell (1966); Polyethylene glycol—Pontecorvo (1975), Davidson et al. (1976); Mechanisms—Schneeberger and Harris (1966), Howe and Morgan (1969), Pontecorvo, Riddle and Hales (1977), Knutton (1979a,b); for review see Ringertz and Savage (1976).

served in cells containing more than two nuclei. In general, the success of cell division shows an inverse relationship to the number of nuclei in the cell. Following the fusion of Ehrlich ascites cells with HeLa cells, Harris et al. (1966) found that most of the multinucleate cells fail to give rise to viable progeny—often because mitoses fail to be completed—and the population of such cultures remains stationary for some days and then declines gradually. Only hybrids produced by fusion of a small number of nuclei can propagate successfully and most of the recent experiments with hybrid cells have made use of fusions between only two cells. When binucleate cells undertake a joint mitosis and division, the resulting daughter cells each have nuclei that possess a diploid complement of the chromosomes of both parents. It is such hybrid cell lines that generally are used for complementation studies, genetic mapping, and studies of the effect of one cell type upon the expression of the other phenotype.

Reactivation of Inert Nuclei

Many types of cell may be fused to generate viable progeny in which both parental genomes are active. One series of studies concerns the events during which the nucleus of a cell that was previously inert is reactivated by fusion with another cell. Another set of experiments, to which we shall come in the next chapter, concerns the effect of cell fusion upon the expression of differentiated functions.

The early experiments of Harris et al. (1966) used autoradiography to show that after fusion between HeLa cells and mouse Ehrlich ascites tumor cells—in which the two types of nucleus can be distinguished by their appearance—radioactive precursors are incorporated into RNA in both the human and mouse nuclei. Labeled leucine also is taken up and the hybrid cytoplasm synthesizes proteins. At least those genes that code for essential cellular functions must therefore be transcribed into RNA, transported to the hybrid cytoplasm and translated into proteins.

Somatic cells with differentiated properties, as well as "dedifferentiated" tissue culture lines may be used for cell fusion. To study the ability of HeLa cells to switch on different functions in partner cells, Harris et al. used rabbit macrophages, rat lymphocytes, and hen erythrocytes. Rabbit macrophages transcribe RNA but have ceased division and no longer replicate DNA. Rat lymphocytes show a similar restriction of synthetic capabilities, although they can be induced to divide in vivo by suitable antigenic stimuli. The nucleated erythrocytes of birds represent completely inert nuclei—in many animals the red blood cells in fact lack nuclei—which are rather small, contain highly condensed chromatin, and synthesize neither DNA nor RNA.

Fusion between HeLa cells and rabbit macrophages or rat lymphocytes displays the same characteristics as those seen with the Ehrlich ascites cell.

Varying numbers of cells may fuse together and the products may undergo both synchronous and asynchronous mitoses. But the differentiated features of the macrophage or lymphocyte are lost and the hybrids behave as "dedifferentiated" tissue culture cells (see Chapter 9). In both cases, the partner nuclei continue to synthesize RNA and are induced by their fusion with the HeLa cell to recommence synthesis of DNA. That this activation is not simply a random response to the cell fusion has been shown by the fusion together of rabbit macrophages and rat lymphocytes. Both nuclei in such heterokaryons continue to synthesize RNA but neither is stimulated into DNA synthesis. The characteristic changes that occur in the lymphocyte nucleus after fusion with a HeLa cell—enlargement of the nucleus, dispersion of chromatin, appearance of structures resembling nucleoli—do not take place after fusion with the macrophage.

An extensive series of studies has been made of the reactivation of avian erythrocyte nuclei by fusion with HeLa cells or mouse fibroblasts. The products of such fusion do not contain the cytoplasm of the erythrocyte; because the Sendai virus hemolyzes the erythrocyte to generate a red blood cell ghost during the fusion, its cytoplasm is lost. Fusion therefore effectively places the erythrocyte nucleus in the cytoplasm of the other partner cell, in contrast to fusions in which the two nuclei reside in a mixed cytoplasm. This makes it possible simply to study the reactivation of the erythrocyte nucleus by the new cytoplasm in which it finds itself.

[Some of the contents of a red blood cell may be transferred to a partner by cell fusion. Following demonstrations that hemoglobin can be transferred to cultured mammalian cells, Schlegel and Rechsteiner (1975) developed a method for microinjecting proteins into cultured cells by fusion. First the proteins are introduced into mammalian red blood cells by lysing the cells with a hypotonic solution; after protein has been taken up, the cells reseal when normal ionic conditions are restored. Then the red blood cells are fused with cultured somatic cells. Both thymidine kinase and bovine serum albumin have been transferred into 3T3 cells by this means. This allows the effect of introducing a given protein into a cultured cell to be determined.]

The enlargement of an erythrocyte nucleus in a dikaryon with a HeLa cell is shown in Figure 8.2. Originally the erythrocyte nucleus is compact, with deeply staining nuclear bodies. Then it begins to enlarge and the nuclear bodies disappear. The enlargement accompanies the acquisition of synthetic activities by the formerly inert nucleus, which commences synthesis of both RNA and DNA under the stimulation of HeLa cytoplasm.

In these fusions, the active partner induces appropriate synthetic activities in the dormant partner; the inactive cell does not appear to inhibit the active one. The ability of fused cells to maintain their integrity and to continue growth when the nuclei fuse together implies that the genetic mechanisms and structural organization of cells of different species are compatible; DNA may be replicated, genes expressed, and the organization of the cell

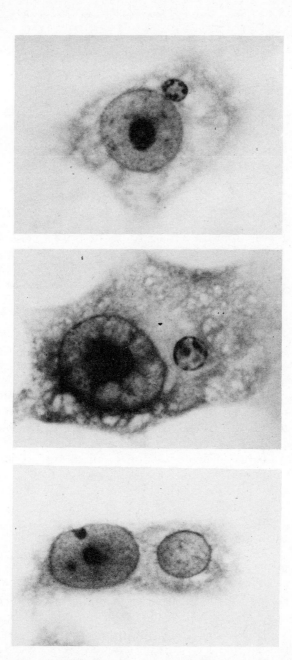

Figure 8.2. Enlargement of the erythrocyte nucleus following fusion with a HeLa cell to form a dikaryon. The upper photograph shows a product of fusion in which the erythrocyte nucleus remains compact and displays deeply staining nuclear bodies. The center photograph shows an erythrocyte nucleus that has begun to enlarge and whose nuclear bodies stain less deeply and are more diffuse. The lower photograph shows a stage of enlargement in which the erythrocyte nucleus is much less compact and has lost its nuclear bodies. Photographs kindly provided by Professor Henry Harris.

maintained. The cytoplasmic signals that activate the dormant nucleus appear to lack species specificity.

But the incorporation of precursors into nucleic acids and proteins does not prove that specific genes, rather than a random selection of sequences, have been activated. By following the reactivation of hen erythrocyte nuclei in HeLa or mouse A9 L cell cytoplasm, Harris et al. (1969) were able to identify some of the hen specific protein products. This demonstrates that in at least these instances the incorporation detected by autoradiography represents transcription and translation of meaningful sequences.

When HeLa cells are fused with hen erythrocytes, for the first 12–16 hours after fusion all the hybrids possess hen-specific surface antigens. This appears to be due to their introduction during the cell fusion. The antigens begin to disappear from the surface of the heterokaryon within 24 hours after fusion, although by this time the erythrocyte nuclei have begun to enlarge and synthesize RNA. By 3 days after fusion, 90% of the cells have lost the antigens. Their disappearance appears to be due to gradual displacement by HeLa surface antigens because the erythrocyte nucleus in the heterokaryon has not yet directed the synthesis of hen surface antigens.

By 4 days after fusion, division takes place in all the heterokaryons; and discrete erythrocyte nuclei disappear, either because irregular mitosis leads to cell death or because the erythrocyte and HeLa nuclei fuse. Mouse fibroblasts irradiated with ultraviolet, however, can grow for up to three weeks without dividing. When erythrocyte nuclei are transferred to this cytoplasm by cell fusion, hen specific antigens are lost from the hybrids more slowly than in the fusions with HeLa cells, but the loss is virtually complete within 6 days. After 8 days, however, hen-specific antigens begin to reappear on the surface of the heterokaryons, at first localized to small areas and then progressively more widespread. Figure 8.3 shows the time course of appearance of nucleoli in the reactivating erythrocyte nucleus and of the production of the surface antigens, which lags by about 4 days. Similar results are obtained with chick embryo erythrocytes, when reactivation is more rapid; nucleoli begin to appear on the second day after fusion and new chick specific antigens begin to be synthesized before those introduced in the fusion have been completely lost. The time at which new antigens appear therefore depends upon the rate of reactivation of the erythrocyte nucleus and is correlated with the time at which it forms its nucleolus.

Other gene functions as well as the surface antigens are expressed by the hen nucleus in hybrid cells. Chick embryo erythrocytes contain the enzyme HGPRT, although it is lost with the cytoplasm during cell fusion. The mouse A9 fibroblast cell line, of course, lacks this enzyme and therefore cannot utilize hypoxanthine (see Chapter 6). When Harris and Cook (1969) followed the ability of erythrocyte—A9 heterokaryons to incorporate labeled hypoxanthine, they found that for the first three days after fusion the nucleotide was not taken up. But concomitant with the development of nucleoli in the erythrocyte nucleus, at day 4, the hybrid cells began to utilize the

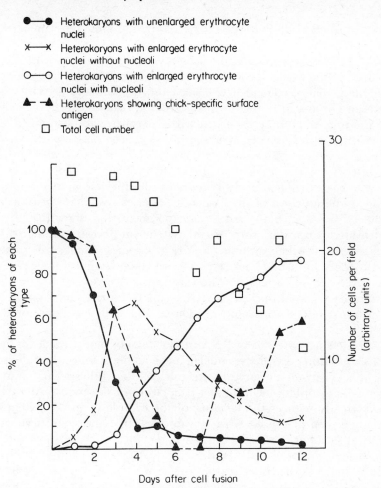

● —● Heterokaryons with unenlarged erythrocyte nuclei

×——× Heterokaryons with enlarged erythrocyte nuclei without nucleoli

○——○ Heterokaryons with enlarged erythrocyte nuclei with nucleoli

▲- -▲ Heterokaryons showing chick-specific surface antigen

□ Total cell number

Figure 8.3. Reactivation of the erythrocyte nucleus following fusion with irradiated mouse fibroblasts. The development of erythrocyte nuclei with nucleoli preceeds by about 4 days the appearance of chick specific surface antigens. Data of Harris et al. (1969).

nucleotide, their activity increasing subsequently. Cook (1970) showed that the enzyme synthesized by these cells has the electrophoretic mobility characteristic of the hen and not of the mouse, confirming that its synthesis represents an activity reactivated in the erythrocyte nucleus and not a reversion in the mouse genome (and see Chapter 9).

The kinetics of appearance of HGPRT activity are similar to those of the production of surface antigens. Since these proteins are unrelated, it seems probable that these kinetics are common to many or all genes expressed in the reactivating erythrocyte nucleus. This suggestion is supported by the observation of Clements and Subak-Sharpe (1975a) that when chick erythrocytes are fused with PyBHK cells, several chick functions are expressed at the same time.

There are two obvious interpretations for the relationship between development of the nucleolus and expression of genes in the erythrocyte nucleus. One is that of coincidence: it may simply take about the same length of time for chromatin to become dispersed and structural genes reactivated as for the nucleolus to be reconstructed. The other is to postulate a causal relationship: expression of chick genes may not be possible until a chick nucleolus is present. This might be because chick messengers cannot be transported from the nucleus in the absence of the nucleolus or because they cannot be translated in the absence of the chick ribosomes whose production is the function of the nucleolus. Species specificity in translation is unlikely, however, both because of experiments that show the efficiency of heterologous translation in several systems and because of a circularity in this argument—the synthesis of chick ribosomes would of necessity depend on translation by HeLa ribosomes of the messengers for chick ribosomal proteins. (This does not, however, formally exclude the possibility that ribosomes containing chick rRNA and HeLa ribosomal proteins are able to translate chick messengers whereas HeLa ribosomes are unable to do so.)

By suppressing synthesis of RNA in the HeLa or mouse nuclei of heterokaryons containing erythrocyte nuclei, it is possible to follow by autoradiography the ability of the erythrocyte nucleus to synthesize and transport RNA to the cytoplasm. Ultraviolet irradiation of the mouse A9 or HeLa nucleus greatly reduces the labeling of the cytoplasm in the parent cells. When they then are fused with erythrocytes, first the erythrocyte nuclei become labeled, but the hybrid cytoplasm remains unlabeled. Then after the appearance of nucleoli in the erythrocyte nuclei, label is found in the cytoplasm. This is consistent with the idea that there is some delay before a reactivating nucleus becomes able to transport messengers to the cytoplasm (for review see Harris, 1974).

The idea that the nucleolus might be involved in this transport has been suggested on the basis of experiments in which the nucleolar region of the reactivating nucleus was irradiated with ultraviolet (Sidebottom and Harris, 1969; Deak, Sidebottom and Harris, 1972). This appears to prevent transport of labeled RNA from nucleus to cytoplasm. However, it does not seem likely that this reflects any formal relationship between messenger and ribosome transport, since the synthesis of hnRNA and transport of mRNA continues when the synthesis of rRNA in the nucleolus is inhibited preferentially by low doses of actinomycin (see Chapter 25).

Molecular Reconstruction of Reactivating Nuclei

Immediately after fusion between an erythrocyte and a HeLa cell, the red blood cell nucleus displays about the same dimensions as those observed in its native state. We have seen that then the inert nucleus enlarges under the influence of the HeLa cytoplasm. Enlargement usually begins within 24 hours and by the third day unenlarged nuclei are rare. The extent of RNA synthesis in the reactivating erythrocyte nucleus correlates reasonably well with its volume (Harris, 1967). Ultraviolet irradiation of the erythrocyte nucleus before fusion does not prevent its subsequent enlargement in HeLa cytoplasm. Since the irradiation abolishes its ability to resume synthesis of RNA, this implies that enlargement is not a consequence of increased activity in RNA synthesis. This suggests that enlargement may be the cause of, or at least a prerequisite for, RNA synthesis (see also Bolund et al., 1969a).

These cytochemical studies of reactivation have been extended by Bolund, Ringertz, and Harris (1969b). The increase in volume which takes place during reactivation of the erythrocyte nucleus is accompanied by an increase in dry mass (measured by microinterferometry). The largest erythrocyte nuclei increase their mass some fivefold during the first 41 hours after fusion. Some of this increase takes place during the first 16–20 hours after fusion, while the erythrocyte nucleus remains in the G_1 phase characteristic of the red blood cell (see also Appels, Bell, and Ringertz, 1975). By 24 hours after fusion, the feulgen values of the nuclei increase as DNA is replicated; and by 41–47 hours many of the erythrocyte nuclei show G_2 values for the reaction. This supports the conclusion that autoradiographic observations of the incorporation of ^{3}H-thymidine into the nucleus reflect replication of its genetic material.

At an early stage in reactivation, the ability of the DNA of erythrocyte nuclei to bind the intercalating dye acridine orange increases. This occurs before replication and so may reflect a dispersion of the chromatin. During reactivation, the erythrocyte DNA also becomes more susceptible to denaturation, another criterion for assessing the degree of condensation. Similar changes in chromatin structure have been observed when cultures are established from epithelial kidney cells of 14-day mice, where chromatin becomes dispersed while nonhistone proteins enter the nucleus. In this case, the rate of nuclear RNA synthesis increases with the nuclear protein content rather than directly with the dispersion of chromatin, which although necessary may therefore be insufficient in itself for gene activation (Auer, 1972; Auer et al., 1973; Auer and Zetterberg, 1972). We have seen earlier, of course, that a very similar series of events occurs when nuclei from differentiated Xenopus cells are transplanted into egg cytoplasm. In all these cases, reactivation involves the enlargement of a dormant nucleus and the dispersion of its previously tightly condensed chromatin. The resumption of nucleic acid synthesis goes hand in hand with the increase in size that is accomplished by the entry into the nucleus of new proteins.

Reactivation of the erythrocyte nucleus depends upon the extent to which proteins are taken up from its new cytoplasm. Figure 8.4 summarizes the studies of Carlsson et al. (1973) on erythrocyte nuclear reactivation following fusion with rat epithelial cells. In a series of heterokaryons in which the

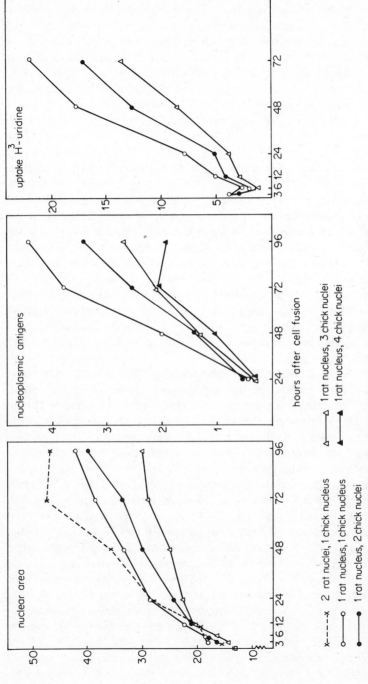

Figure 8.4. Reactivation of chick erythrocyte nuclei in heterokaryons with rat epithelial cells. The nuclear area of the reactivating chick erythrocyte nuclei, their uptake of rat nucleoplasmic antigens, and their incorporation of ⁵H-uridine into RNA all depend in a similar manner upon the ratio of chick:rat nuclei in the heterokaryon. Data of Carlsson et al. (1973).

proportion of the two types of nuclei varies from 2 rat:1 chick to 1 rat:4 chick, they found that the rate of reactivation depends upon the number of chick nuclei in the cell. The more chick nuclei, the slower is the rate of reactivation, measured by increase in nuclear area, uptake of nucleoplasmic antigens characteristic of the rat, or incorporation of ^{3}H-uridine into RNA. This suggests that the chick nuclei in a common cytoplasm compete with each other for a limiting supply of the molecules upon which reactivation depends.

In fusions with HeLa cells, mouse fibroblasts, or rat epithelial cells, the chick erythrocyte nucleus undergoes an extensive increase in dry mass during the first 48 hours after fusion. This represents the uptake of proteins from the new cytoplasm, as has been confirmed directly by following the movement of proteins that were labeled in the erythrocyte's partner before fusion (Appels et al., 1974). Probably all of the proteins entering the erythrocyte nucleus must be derived from the partner species, since at this time the reactivating chick genome has not yet started to direct the synthesis of avian proteins. The proteins responsible for enlargement and reactivation of the erythrocyte nucleus are therefore specific for neither the species nor the cell phenotype.

By isolating the erythrocyte nuclei from the heterokaryons generated by fusion with HeLa cells labeled with radioactive amino acids, it is possible to determine which proteins have entered the reactivating nucleus. By 15 hours after reactivation, both histones and nonhistone proteins have entered; the nonhistone proteins appear to enter first. Replication of DNA has not begun at this time, which again shows that the entry of proteins precedes the resumption of nucleic acid synthesis. Cytoplasmic HeLa proteins do not enter the chick nucleus, showing that the nuclear uptake is specific for those proteins concerned with reactivation of the genome (Appels, Bolund, and Ringertz, 1974; Appels et al., 1975).

The distribution of human nuclear proteins following fusion between chick erythrocytes and HeLa cells has been followed by Ringertz et al. (1971), who utilized antinuclear antibodies produced by patients with autoimmune diseases. These antibodies can be tagged with fluorescent groups, so that antibodies to human nucleoli fluoresce only over the nucleolus, those directed against the nucleoplasm react principally with this area (although also showing some response to the nucleolus); and antibodies against components of human cytoplasm reveal HeLa cells as fluorescent cytoplasms in which the nuclei appear as dark holes. Chick cells react only very weakly or not atall with these antibody preparations.

Antisera to human nucleoli and nucleoplasm react increasingly strongly with the chick nuclei as they reactivate. Cytoplasmic antigens remain unreactive. As nucleoli form in the chick nuclei, they react first with human nucleolar antigens and later during reconstruction with antibodies prepared against chick nucleoli. The HeLa nuclei of the hybrid cells at first show only a weak reaction with antigens prepared against the chick nucleolus, but as the chick nuclei grow in size and develop nucleoli, the HeLa nucleoli become as reactive as those in the erythrocyte nuclei. The intensity with which HeLa nucleoli react with antigens against chick nucleoli depends upon the ratio of HeLa to chick nuclei in the cell; the higher the number of chick nuclei, the greater is the reaction of all nuclei to chick-specific antigens. Within individual heterokaryons, the HeLa nuclei closest to the chick nuclei show the strongest reaction with chick-specific antigens.

These results imply that human nucleoplasmic and nucleolar proteins move into chick nuclei as they start to reactivate. The failure of HeLa cytoplasmic proteins to enter the chick nuclei shows that their enlargement does not depend upon a passive swelling process in which human proteins at random enter the nucleus; rather is reactivation specific, so that only appropriate (nuclear) proteins are taken up and concentrated in the corresponding regions of the chick nucleus. The observation that chick nucleoli react first with the human antigens and only later during development with the chick antigens suggests that the nucleolus in the chick erythrocyte is initially reconstructed with human nucleolar proteins. Only later, when its formation has enabled the chick genome to direct protein synthesis, are chick nucleolar proteins able to take their place.

The appearance of the chick nucleolar proteins in both chick and HeLa nuclei of the heterokaryons suggests that once synthesized in the cytoplasm, they may be utilized by any nucleus present, of either species. The effect of the relative numbers and positions of the chick and HeLa nuclei in their common cytoplasm suggests that a gradient of nucleolar proteins may be established away from the nucleus coding for them; this implies that the messenger RNAs for these proteins are translated in the cytoplasm at a location close to the nucleus from which they emanate. The structural components of the nucleoli of chick and human cells must be effectively interchangeable and able to form functional nucleoli consisting in part of components from each species.

Formation of Cytoplasts and Karyoplasts

A method that allows the preparation of enucleated cytoplasms and free nuclei from cultured cells has been developed by combining centrifugation and treatment with the drug cytochalasin B. The enucleated cytoplasms are known as *cytoplasts* and the free nuclei as *karyoplasts*. Preparations of cytoplasts and karyoplasts can be recombined by applying the protocols for Sendai virus fusion developed with somatic cells, so that it is possible to reconstruct a cell from a cytoplasm of one species and the nucleus of another. Similarly it is possible to fuse cytoplasts or karyoplasts with intact somatic cells. This makes it possible to obtain virtually any desired combination of nucleus and cytoplasm from all types of cultured cells.

Low concentrations of the drug cytochalasin B cause the disaggregation of actin microfilaments and inhibit cytokinesis and cell motility (see Chapter 3). As observed by Carter (1967) in his original report of its effects, at higher concentrations cytochalasin B causes cells to extrude their nuclei. Generally this begins to occur at a level of 1 μg/ml and becomes highly effective by 10 μg/ml, but the exact level depends upon the cell type. Figure 8.5 illustrates the technique that has been developed to use this effect for the separation of cytoplasm from nucleus. Cells attached to a coverslip disc in the form of a monolayer are centrifuged in the presence of cytochalasin B, generally at about 10 μg/ml. After centrifugation, the disc is transferred to fresh medium and the cytoplasts remain attached to it in a healthy state for some time. Karyoplasts can be recovered from the pellet at the bottom of the centrifuge tube (Prescott et al., 1972; Poste, 1972; Wright and Hayflick, 1972).

Figure 8.6 shows the process of enucleation with CHO cells, followed by Shay, Gershenbaum, and Porter (1975). After 1 minute of centrifugation in the presence of cytochalasin B, the cytoplasm has drawn up into a more compact mass, leaving filopodia (thin strands of cytoplasm) extending to the original attachment sites on the glass coverslip. By 5 minutes the nucleus of each cell has become extruded from the surface of the cytoplasm. With further time, the nucleus extends on a long stalk—essentially a cytoplasmic bridge—which eventually breaks. Attachment to the surface of the coverslip probably is necessary in order to provide the cytoplasmic adherence against which the nucleus can be pulled.

The cytoplasts consist of cytoplasms containing most of the cellular organelles—Golgi apparatus, endoplasmic reticulum, mitochondria, ribo-

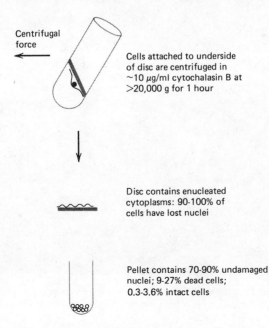

Centrifugal force

Cells attached to underside of disc are centrifuged in ~10 μg/ml cytochalasin B at >20,000 g for 1 hour

Disc contains enucleated cytoplasms: 90-100% of cells have lost nuclei

Pellet contains 70-90% undamaged nuclei; 9-27% dead cells; 0.3-3.6% intact cells

Figure 8.5. Preparation of cytoplasts and karyoplasts by centrifugation in cytochalasin B.

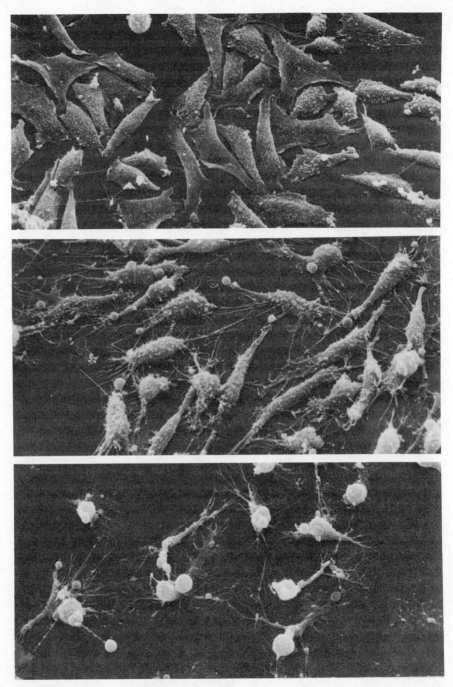

Figure 8.6 Enucleation of CHO cells. Upper: scanning electron micrograph of growing cells. Center: cells after 1 minute of centrifugation in 10 μg/ml cytochalasin B. Lower: cells after 5 minutes centrifugation. ×750. Data of Shay et al. (1975).

somes, centrioles; microtubules and microfilaments sometimes can be seen (Wise and Prescott, 1973; Shay, Porter and Prescott, 1974; Ege et al., 1974). These enucleated cytoplasms maintain normal cell shape, locomotion, and their characteristic contact inhibition or pattern of overlapping (Goldman, Pollack, and Hopkins, 1973; Pollack et al., 1974; Goldman et al., 1975). The structural network of the cytoplasm therefore appears to be present. By 24 hours after enucleation, however, the fine structure of the cytoplasm begins to suffer; it becomes vacuolated, there are changes in the Golgi apparatus, protein synthesis ceases. Presumably many of these changes are the result of the lack of a genome to direct synthesis of the messenger RNAs needed to code for protein components of the cytoplasm.

The karyoplasts consist of nuclei surrounded by a thin shell of cytoplasm and a plasma membrane; this represents about 28% of the dry mass of the cell (compared with nuclei isolated by detergent treatment, which generally have about 22% of the cellular protein). Thus when the nucleus is released, it must carry with it a portion of the plasma membrane, which is able to reseal around both the nucleus and cytoplasm. Newly isolated karyoplasts are able to incorporate labeled amino acids and precursors to both RNA and DNA. But in all of these experiments, the karyoplasts failed to regenerate new cytoplasms and died within one or two days. Small amounts of most cytoplasmic components are present in the thin shell surrounding the nucleus, which raises the possibility that the failure to regenerate may be due to the lack of a proper structural organization. In another series of experiments, however, regeneration of L cell karyoplasts has been reported. Lucas, Szekely, and Kates (1976) found that after 3 days most of the karyoplasts die but some adhere to a surface and put out short projections. These may divide to generate products that behave as daughter cells. About 10% of the initial population seems able to survive. The reason for this is not known.*

Enucleated L cells can be reconstructed by fusing cytoplasts and karyoplasts with the Sendai virus. Figure 8.7 illustrates a protocol used to distinguish the sources of the reconstituted cytoplasms and nuclei. The cytoplasms of the nuclear donors are labeled with small latex beads, while the nuclei carry the biochemical marker of ^{3}H-thymidine or the genetic marker of HGPRT$^+$. The cytoplasts were derived from cells whose cytoplasms possessed large latex beads, and whose nuclei lacked ^{3}H-label or were HGPRT$^-$, respectively (Veomett et al., 1974; Ege and Ringertz, 1975). Reconstituted cells are characterized by the presence of large latex beads and the absence of small latex beads in the cytoplasm; their nuclei should possess an ^{3}H label or be able to incorporate ^{3}H-labeled hypoxanthine, respectively. About 40% of the karyoplasts can be fused; and about 13% of these, that is, 5% of the starting material, form reconstituted cells. Lucas

---

*The only difference apparent in the protocol used to prepare the regenerating karyoplasts was that centrifugal force was applied parallel to the cell sheet instead of at an angle to it. It does not seem likely that this is a sufficient explanation for the apparent ability of karyoplasts to regenerate cytoplasms. The regenerative ability remains to be characterized.

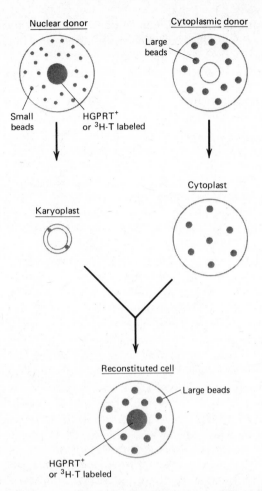

Figure 8.7. Protocols for reconstitution of cells from karyoplasts and cytoplasts.

and Kates (1976) reported that about 30% of A9-L cell reconstructions are capable of division. This offers the prospect of constructing lines that could be used to analyze nucleocytoplasmic interactions over the long term. With differentiated cells, Krondahl et al. (1977) have shown that reconstruction from L6 myotube karyoplasts and A9 cytoplasts appears to produce perfectly normal viable cells. Ringertz et al. (1978) showed further that they are able to fuse into myotubes, implying that the nuclei retain their former ability in gene expression, in spite of their new cytoplasmic environment. This contrasts, of course, with the extinction of functions often seen in somatic cell fusion (see Chapter 9).

Cytoplasts have been fused with somatic cells to effect a mixing of cytoplasms without changing the genetic complement (Poste and Reeve, 1972).

The resulting cells are described as *cybrids* (cytoplasmic hybrids). If an erythrocyte is used as the somatic cell, the loss of its cytoplasmic contents during fusion in effect places its nucleus into the cytoplast to generate a reconstructed cell. Experiments have been reported in which the effect of cybrid formation was studied on expression of differentiated phenotypes. In one set of experiments, cytoplasm of fibroblasts was able to prevent hemoglobin induction in the products of fusion with two Friend cell lines, but had little effect with two others (Gopalakrishnan et al., 1977). In other experiments, cytoplasm from L6 myoblasts had no effect upon the gene expression of embryonal carcinoma (pluripotent) cells with which it was fused (Linder et al., 1979).

Fusion between cytoplasts and somatic cells can be used also to introduce new genetic information into a cell line. Resistance to chloramphenicol is carried by the mitochondrial genome in yeast and the use of cytoplasts suggests that the same may be true of mammalian cells. Bunn, Wallace, and Eisenstadt (1974) fused cytoplasts from a mouse cell line resistant to chloramphenicol with L cells defective in thymidine kinase. Some of the progeny were able to generate descendents able to grow on medium containing BUdR and chloramphenicol, which should be possible only if a genetic recombination has occurred to give the TK⁻ line the ability to resist chloramphenicol. The frequency of such events is $1-2 \times 10^{-4}$, somewhat greater than would be expected of mutation. Similar results have been obtained with chloramphenicol resistant HeLa cells and human diploid fibroblasts (Wallace, Bunn, and Eisenstadt, 1975). Shay (1977) reported the selection of chloramphenicol-resistant cells from 3T3 karyoplasts fused with cytoplasts derived from a murine mammary tumor line. The nuclear donor is HAT^r and the cytoplasmic donor is CAP^r, which implies that the cells growing on HAT-CAP medium should have a nucleus from one parent and a cytoplasm from the other. A similar, intraspecies reconstitution has been performed with human karyoplasts and mouse cytoplasts (Nette et al., 1979). The application of selective techniques may eliminate some of the difficulties of contamination that are otherwise encountered in attempts to make use of cybrids and reconstituted cells.

Another use of the fusion techniques is to form heterokaryons by fusing karyoplasts with nucleated somatic cells. Ege, Krondahl, and Ringertz (1974) have characterized the fusion reactions of both karyoplasts and microcells with cytoplasts or somatic cells. Micronuclei can be induced by treatment with colchicine and contain only a small proportion of the cellular DNA; the amount varies widely but has an average of about 30%. They can be isolated in the form of microcells by centrifugation in the presence of cytochalasin B, and thus essentially constitute karyoplasts with a reduced genetic content. Their use in fusion experiments provides a potential method for introducing small amounts of additional nuclear genetic information into a somatic cell (see Chapter 9).

CHAPTER 9

Hybrid Cells: Chromosomes

Somatic cell hybrids are amenable to genetic manipulations that allow the relationship of the two parental genomes to change. The immediate product of fusion between two cells is a heterokaryon that contains two parental nuclei. Its successor may be a hybrid cell whose single nucleus contains both the diploid (or pseudo diploid) complements. Homospecific hybrid cells generally are stable, as are many heterospecific fusion products. But some heterospecific hybrid cells are unstable, so that their descendents possess only some of the parental chromosomes. In this chapter we shall be concerned with situations that result in the failure of some of the original chromosomes to survive in the hybrid or in which only a restricted amount of genetic material initially is transferred to a somatic cell.

Chromosomes may be lost from the complement of some heterospecific hybrid cells at a fairly steady rate following fusion. When the chromosomes of one parental genome are lost preferentially, after some generations the resulting cells have complements that consist essentially of a resident set of chromosomes from one parent with the addition of a few (varying) chromosomes from the other. It is possible then to identify the additional chromosomes and to correlate their presence with the expression of genes of this parent. Thus preferential chromosome loss provides a technique for mapping structural genes to chromosomes; and it so happens that the human genome is one for which this is a productive approach.

It is not uncommon for specialized functions expressed in a parental cell to be suppressed in a cell hybrid. In some cases, reexpression can occur following chromosome loss. This offers a potential approach for identifying putative regulator loci responsible for the suppression. And the effects of gene dosage upon expression can be studied by varying the relative numbers of chromosomes of each parent, often by comparing hybrid cells possessing diploid sets of each genome with those that have a tetraploid set of one parent combined with a diploid set of the other.

Another feature of cell fusion is that by using synchronized parental populations it is possible to fuse together cells that are in different stages of the cell cycle. This allows studies of the susceptibility of a nucleus at one stage of the cell cycle to signals inducing it to enter another stage. This is useful not only for analyzing the cell cycle, but also has the interesting characteristic that in some such fusions the chromosomes of one parent become permanently condensed and then fragmented. This may generate a hybrid cell whose genetic information is derived largely from one parent, with a small, often unstable, component from the other.

Perhaps related to this ability of small fragments of genetic material to survive in a cultured somatic cell, it is possible to introduce isolated metaphase chromosomes, only small parts of which remain in the recipient cells. Purified DNA also may be used, although with somewhat less effectiveness. Such methods offer the prospect of making possible what amounts to gene transfer between somatic cells.

Chromosome Loss in Heterospecific Hybrids

Whether chromosomes are lost from hybrid cells is a characteristic of the particular fusion. The first observation of chromosome loss was made by Weiss and Ephrussi (1966), when they noted that one month after fusion between mouse and rat cells, the number of chromosomes was much lower than the total of the two parental complements. The principal loss of chromosomes in this cross seems to take place soon after fusion, to yield stable hybrids which then retain their remaining chromosomes.

The usual pattern of loss from human × rodent hybrids is a continual disappearance of human chromosomes. When human and mouse cells are fused, hybrids which form a single joint nucleus at first appear to possess all the chromosomes of both parents. But within a few generations, many of the human chromosomes are lost, although the hybrids retain all or nearly all of the mouse chromosomes. In the first observation of this preferential loss, Weiss and Green (1967) reported that when the hybrids were first examined after 20 generations of growth, they had already lost many of their human chromosomes, with from 2–15 remaining in different clones. Further human chromosomes are gradually eliminated from the nucleus during subsequent cycles of division, so that the progeny eventually contain chromosomes derived largely or entirely from the mouse parent alone (Matsuya, Green, and Basilico, 1968; Nabholz, Miggiano, and Bodmer, 1969). Essentially these cells then contain the mouse genome with a small amount of additional human genetic information (see Figure 9.1).

Although it is usually the human chromosomes that are lost in crosses with rodent cells, this is not the inevitable direction of elimination. Instances have been reported in which guinea pig or mouse chromosomes are preferentially lost, presumably as a consequence of the particular cell types involved (Colten and Parkman, 1972; Minna and Coon, 1974; Croce, 1976). This is sometimes described as reverse chromosome segregation (for review see Ringertz and Savage, 1976).

Little is known about the mechanism responsible for chromosome loss. A reasonable speculation is that the chromosomes of one parent, for some unknown reason, fail to replicate or to interact properly with the spindle.

Given the occurrence of chromosome loss, however, it seems likely that selection acts to favor the growth of cells with reduced chromosome complements and perhaps to give an advantage to cells retaining or losing given chromosomes. In the human-mouse hybrids studied by Matsuya et al. (1968), the human parental fibroblasts grew more slowly than the mouse parental fibroblasts, whose doubling time was 15 hours. The rate of growth of the hybrids increased with the loss of human chromosomes. Hybrids with 10 human chromosomes doubled in 35–60 hours; the doubling time fell to 25–30 hours when the human chromosome complement fell below 4. Hybrids retaining only a single human chromosome doubled in 23 hours; cells which had lost all the human chromosomes doubled in 16–19 hours. This effect may therefore establish selection for hybrids that have lost increased numbers of human chromosomes.

It is difficult to establish whether the loss of human chromosomes is random, because in cases where specific human chromosomes appear to be retained more frequently than expected, it is always possible that their presence confers a selective advantage. An example is provided by an experiment in which Croce, Girardi, and Koprowski (1973) found that human chromosome 7 of SV40 transformed cells tends to be retained in hybrids with mouse cells. A possible reason is that this chromosome carries the integrated viral genome and may increase the growth rate of the cells carrying it, thus conferring a selective advantage upon them.

Preferential chromosome loss is not a characteristic of all interspecific cell fusions and it would, of course, be useful to direct the elimination of chromosomes from such hybrids so that those of either parent might be lost. Pontecorvo (1971) observed that X irradiation of either the Chinese hamster or mouse parent cell leads to the preferential loss of its chromosomes after fusion with the other cell, contrasted with the lack of any elimination in crosses between unirradiated cells. But the disadvantage of this method is that irradiated chromosomes may suffer damage, such as the induction of mutations or translocations. What is needed is a method that can be applied to established cell lines to cause chromosome loss without any effect on the chromosomes that are retained.

Genetic Mapping in Cell Hybrids

Two principal problems have stood in the way of mapping the mammalian genome with the resolution possible for procaryotes: the lack of mutants and the difficulty in obtaining recombinants.

We have seen in Chapter 6 that it is only recently, by turning to somatic cells in culture, that it has been possible to select a variety of mutants. The ability to clone somatic cells to obtain large numbers of cells of a given genotype overcomes the problem that with higher eucaryotes (especially man) it is not possible to breed sufficient progeny to isolate mutants and map them. Even so, it is true that in only a rather small number of cases has it been possible to characterize a putative mutant in detail sufficient to be sure that it does indeed represent a change in genetic information. Of course, the isolation of somatic cell mutants is restricted to those systems in which selective techniques can be applied. A possible way for widening the utility of this approach lies with the development of conditional lethal systems. Another way to obtain defined mutants is to start cell lines from individuals carrying characterized genetic disorders. This is particularly appropriate for man, where many such diseases have been identified, although the isolation of mutant cells in this way suffers from the disadvantage that they are not isogenic with control lines lacking the mutation. Of course, a restriction in all analyses depending on the use of somatic cells is that the function dealt with must be one that is expressed under the conditions of culture.

Somatic cell hybridization offers an approach that makes possible what amounts to recombination between the chromosomes of cultured cells. The fusion together of two (diploid) cells generates a tetraploid hybrid; the loss of chromosomes from this hybrid then allows different combinations of the parental chromosomes to be obtained. Preferential loss of the chromosomes of one parent allows studies of the effect of the remaining chromosomes on the other genome. Analogies might be drawn between the loss of chromosomes from cell hybrids and the reassortment of chromosomes that occurs at meiosis (although obviously the process in somatic cells is a comparatively haphazard event and does not involve recombination between homologues).

Fusions between somatic cells of the same species are useful for performing complementation analysis, to determine whether two mutations lie in the same gene (a prerequisite for mapping studies). Heterospecific fusions in which the chromosomes of one parent are lost can be used to correlate the loss or retention of a particular chromosome with the disappearance or maintenance of a given gene. A prerequisite for this analysis is the ability to distinguish the chromosomes of the two parents and to identify those remaining in any given hybrid. When hybrid cells first began to be used for chromosome mapping, this was difficult, because it was necessary to rely upon gross morphological differences such as the size of the chromosome or position of the centromere. Thus human chromosomes could be identified only as members of certain classes; and the mouse chromosomes—all of which are the same size and telocentric—could not be distinguished at all. But the techniques of chromosome "banding" with reagents such as Giemsa or quinacrine, which are discussed in Chapter 15, now make it possible to distinguish unequivocally all the chromosomes of these mammalian complements. The resolution of this technique is good enough further to identify the rearrangements responsible for some translocations, and this can be useful in determining the location of genes within as well as on chromosomes.

Because of the preferential loss of human chromosomes in hybrids with rodent cells, the hybridization technique has proved particularly useful for locating genes in the human chromosome complement. The principle of the approach, illustrated in Figure 9.1, is to fuse a human cell which possesses the gene coding for some function with a rodent cell that lacks the function. Then the hybrid cells are transferred to a medium in which they can survive only if they possess this enzyme activity. Obviously in this form the mapping technique is limited to genes for which mutations have been obtained in the rodent cells. Clearly it is necessary to be able to show directly that the hybrids surviving on selective medium do so because they possess the necessary human enzyme, and not because the rodent genome has suffered a reversion allowing production of an active rodent enzyme. The need for this is emphasized by the reports discussed in Chapter 6 of somatic cell mutants lacking the HGPRT enzyme, which were originally thought to comprise deletions because of their low rate of reversion; but revertants could be found following exposure to fusion protocols. Another possibility that must be excluded is that the human genome suppresses the defect in the rodent genome. This might take the form of correcting a nonsense or missense mutation in a rodent structural gene; or if the rodent mutation were in a control function, then the human genome might be able to cause reexpression of the rodent structural gene. This restricts the technique to analysis of genes whose products can be distinguished in the two species by some criterion such as electrophoretic mobility.

The technique establishes a selective system for cells carrying at least one copy of the gene coding for the essential human enzyme. Any hybrids which lose all their copies of the human chromosome carrying this gene therefore die. As increasing numbers of human chromosomes are lost from the hybrid cells in successive growth cycles, the survivors come to retain only the single human chromosome carrying the gene necessary for survival, or at least only a very small number of human chromosomes including this one. Identification of the human chromosome common to all survivors thus locates the gene in the human complement.

The gene coding for human thymidine kinase was the first to be located by this technique. Weiss and Green (1967) fused human diploid fibroblasts which synthesize the enzyme with a mouse cell line lacking it. Human-mouse hybrids were grown for four days in a standard medium and then transferred to HAT medium in which thymidine kinase activity is essential for growth. The parental mouse cells die in this medium and human parents grow only rather poorly; hybrid cells that synthesize thymidine kinase grow well.

In hybrid cells examined after 20 generations, all or nearly all the mouse chromosomes were present, but only a few (2-15) of the human chromosomes remained. But all the cells grown on HAT medium retained at least one human chromosome of the E group. Figure 9.2 shows the results of Matsuya, Green and Basilico (1968) in following such hybrids through a greater

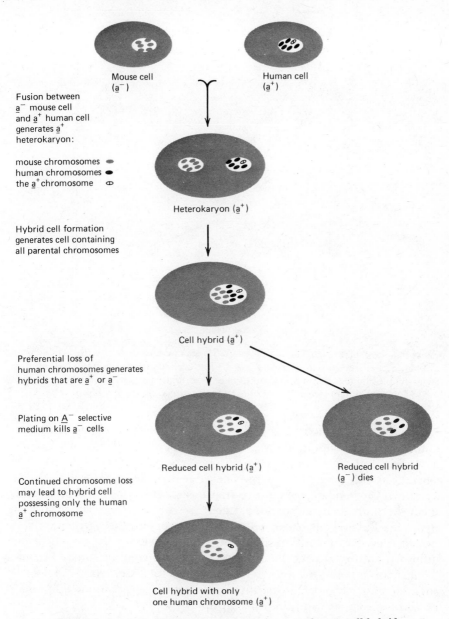

Fusion between
$\underline{a}^-$ mouse cell
and $\underline{a}^+$ human cell
generates $\underline{a}^+$
heterokaryon:

mouse chromosomes
human chromosomes
the $\underline{a}^+$ chromosome

Hybrid cell formation
generates cell containing
all parental chromosomes

Preferential loss of
human chromosomes generates
hybrids that are $\underline{a}^+$ or $\underline{a}^-$

Plating on $\underline{A}^-$ selective
medium kills $\underline{a}^-$ cells

Continued chromosome loss
may lead to hybrid cell
possessing only the human
$\underline{a}^+$ chromosome

Mouse cell
($\underline{a}^-$)

Human cell
($\underline{a}^+$)

Heterokaryon ($\underline{a}^+$)

Cell hybrid ($\underline{a}^+$)

Reduced cell hybrid ($\underline{a}^+$)

Reduced cell hybrid
($\underline{a}^-$) dies

Cell hybrid with only
one human chromosome ($\underline{a}^+$)

Figure 9.1. Survival of human chromosomes in mouse-human cell hybrids.

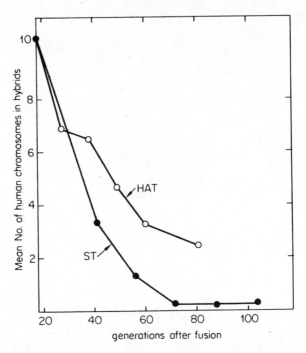

Figure 9.2. Loss of human chromosomes from human-mouse hybrid cells. On standard medium (ST) all human chromosomes may be lost over 80 generations. On HAT medium, retention of the human gene for thymidine kinase is necessary for survival; after 80 generations only 2–3 human chromosomes are present. Data of Matsuya, Green, and Basilico (1968).

number of generations. After 18 generations, when the cells were first examined, this population showed an average of 10.2 human chromosomes per hybrid cell. The human chromosomes continue to be lost and by 80 generations only 3 may be present. The value may fall to zero when the cells are grown on standard medium so that the selective pressure for cells that retain the human thymidine kinase gene is removed. Cells which lose the thymidine kinase gene occur with high frequency in each generation. Although they die in HAT medium, they can be selected by growth in medium containing BUdR, when the presence of thymidine kinase becomes lethal. Chromosomes of the E group are rare in these hybrids. These results therefore suggest that the thymidine kinase gene is located on one of the chromosomes constituting this group.

These experiments were performed before the advent of chromosome-banding techniques. Using the fluorescent banding caused by quinacrine, Miller et al. (1971c) were able to show that the E group chromosome retained in the hybrids is number 17. Green et al. (1971) found that although many of the hybrids surviving on HAT medium retain only a single copy of chromosome 17, others may have up to six, presumably as the result of nondisjunction at mitosis. Although one copy of the thymidine kinase

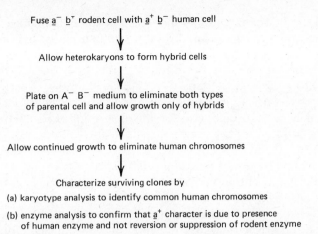

Figure 9.3. Protocol for mapping genes on human chromosomes.

gene clearly is sufficient for survival, perhaps it does not fully alleviate the selective pressure, so that it may be advantageous for cells to possess further copies. In all cases, the thymidine kinase of the hybrid cells had the electrophoretic mobility characteristic of the human enzyme, confirming that the human function on chromosome 17 is the structural gene.*

As is evident from the example of thymidine kinase, the human chromosome carrying any given gene can be identified by karyotype analysis of human-rodent hybrid cells possessing the human gene product. Figure 9.3 summarizes the procedure for chromosome mapping. A system analogous to the HAT selection is the use of alanosine to compel cells to rely upon APRT for utilization of exogenous adenylic acid (see Chapter 6). By fusing APRT⁺ human cells with APRT⁻ mouse cells, Kusano, Long, and Green (1971) demonstrated that the gene for APRT must be located on one of acrocentric human chromosomes (since identified as chromosome 16 by Tischfield and Ruddle, 1974). Similar experiments have, of course, also been performed with HAT to identify the X chromosome as the carrier for HGPRT and other enzymes (see below).

Another selective system is provided by the different susceptibilities of human and rodent cells to certain viruses. Human cells are susceptible to poliovirus whereas mouse cells are not. The human cells possess a receptor for the virus absent from the mouse cells, and by examining the susceptibility of human-mouse hybrids to poliovirus, Miller et al. (1974) were able to show that sensitive lines always possess human chromosome 19, whereas resistant lines always lack it. A related approach is possible with viruses that

---

*Cells that have lost the gene for the principal (cytosol) thymidine kinase retain about 2% of the enzyme activity; this is due to the presence of another enzyme, the mitochondrial thymidine kinase, which has been shown by similar mapping techniques to be coded by a gene on chromosome 16 of the nucleus (Willecke et al., 1977).

integrate into the human but not the rodent genome. Human cells carrying SV40 possess the T antigen; and by testing human-mouse hybrid clones for this activity, different individual human chromosomes have been identified as the site of integration in different human cell lines (Croce and Koprowski, 1974, 1975; Kucherlapati et al., 1978). Similar experiments with hybrids between African green monkey and mouse cells have identified one of the monkey chromosomes as a site for integration (Croce, Huebner, and Koprowski, 1974). Another system to which this technique has been applied is the replication of type C RNA viruses. A baboon virus is able to replicate in human but not in rodent cells; it can replicate in human-rodent hybrids that retain chromosome 6, but not in those that lack it (Lemons, Sherr, and O'Brien, 1977).

Fusion between cells of different species is most powerful as a technique for gene mapping when selective pressure can be established for the loss or retention of some enzyme activity. But correlations between chromosomes and genes can also be made for any enzyme coded by one species which is distinguishable—for example by its electrophoretic mobility—from that of the other species. By screening hybrid cell populations to identify those lacking or possessing the human enzyme, it is possible to deduce from karyotype analyses which human chromosome carries the corresponding gene.

By following the abilities of hybrid cells to synthesize two human enzymes, it is possible to establish linkage groups. If two enzymes are always retained together or lost together, they must be coded by genes that lie on the same chromosome. When a selective technique is available for one enzyme, its linkage to another can be investigated by growing hybrid cells under selective conditions and examining their retention of the second, unselected enzyme. Nabholz, Miggiano, and Bodmer (1969) used this approach to screen human-mouse hybrids, compelled to retain HGPRT by growth on HAT medium, for their ability also to synthesize glucose-6-phosphate dehydrogenase (G6PDH) and the lactate dehydrogenase subunits, LDH A and LDH B. By 40–60 days after fusion, the hybrids contained only 4–14 chromosomes in excess of the parental mouse complement. All the hybrids synthesized human G6PDH. But the activities of LDH A and LDH B varied very considerably in different clones, showing no correlation with the retention of HGPRT and G6PDH or with each other. While the common retention of HGPRT and G6PDH suggests that their genes are carried on the same chromosome, the lack of correlation with the LDH subunits suggests that they are carried on different human autosomes.

A study of the inheritance of seventeen human enzymes was performed by Ruddle et al. (1971), using starch gel electrophoresis to distinguish the human enzymes of human-mouse hybrids from their murine counterparts. During growth on HAT medium, only three human enzyme markers, HGPRT, G6PDH and PGK (phosphoglycerokinase) are invariably retained; these must all be carried on the same (the X) chromosome. The remaining enzymes present in the various clones showed no pairwise correlations in inheritance; all are independently retained or lost as the hybrids lose human chromosomes. All fourteen genes must therefore be located on different autosomes. Similar experiments have been performed by Kao and Puck (1970) with auxotrophic mutants of Chinese hamster cells.

The number of loci that can be assigned to chromosomes by the use of selective techniques obviously is somewhat limited. More sophisticated methods therefore have been introduced for correlating the presence of enzymes and chromosomes in hybrid cells that have been allowed to suffer random loss of human chromosomes. The use of the *hybrid clone panel* formalizes this approach (for review see Ruddle and Creagan, 1975). Several hybrid clones are compared, each of which contains a small number of human chromosomes that comprises a unique subset of the human complement. For example, with a panel of three clones each selected for a stable content of four human chromosomes, it is possible to identify unequivocally the markers coded by six human chromosomes if the chromosomes of the different clones overlap in such a way that two of the three clones possess and one of them always lacks a given human homologue. When these clones then are examined for the presence of human enzymes, any protein coded by one of these six homologues should be present in extracts of the appropriate two clones, and absent from the other. A number of 8–10 clones carrying suitable subsets of human chromosomes is sufficient to define all 24 human chromosomes (22 markers plus 2 sex chromosomes). An example is shown in Figure 9.4.

The early assignments of genes to human chromosomes allowed some 30 loci to be assigned to 16 homologues (for review see Ruddle, 1973). More recent analyses, together with information derived from familial studies of metabolic diseases in man, have brought the status of the human gene map to a point where at least one structural locus has been identified for each

Cell line	UMPK	Chromosome																						
		1	2	3	4	5	6	7	8	9	10	11	12	13	14	15	16	17	18	19	20	21	22	X
WA-IIa	-	-	+	+	+	-	-	-	-	-	-	+	+	-	+	-	+	-	-	-	-	-	-	+
JFA-14b	+	+	+	-	-	+	-	-	+	-	-	-	-	+	-	-	+	+	+	-	-	-	-	+
WA-Ia	+	+	+	-	+	-	-	+	+	-	+	-	+	+	-	+	+	-	+	+	+	+	+	+
J-10-H-12	+	-	+	-	-	+	-	-	+	-	-	+	+	+	-	+	+	+	+	-	+	-	+
AIM-3a	-	-	+	-	+	+	-	+	+	-	+	+	+	+	+	-	-	+	+	-	+	-	-	+
AIM-8a	-	-	-	-	-	-	+	+	-	+	-	+	+	+	+	+	+	+	+	+	+	+	+	+
AIM-11a	-	-	+	+	-	-	-	+	-	-	+	-	+	-	-	-	-	-	-	-	-	-	-	+
AIM-23a	+	+	+	+	+	+	-	+	+	-	+	-	+	-	+	+	+	-	+	-	-	+	+	+

Figure 9.4. A hybrid clone panel that allows mapping of enzymes to individual chromosomes. All cell lines retained the chromosomes indicated by (+) with a frequency of 5% or more; each chromosome is present in a unique subset of the clones. Examining all eight lines for the presence of uridine monophosphate kinase (UMPK), for example, showed an unequivocal correlation with chromosome 1. Data of Satlin, Kucherlapati, and Ruddle (1975).

homologue. More than 200 loci have been mapped to specific chromosomes, including more than 20 loci on chromosome 1 and more than 100 on the X chromosome. These include enzymes, blood group and other antigens, hormonal factors, viral and toxin susceptibilities, and some diseases whose molecular cause is not yet known (for review see Ruddle and Creagan, 1975; McKusick and Ruddle, 1977). As an example, the loci identified on the X chromosome include the enzymes PGK and G6PDH (mapped by hybrid cell studies), HGPRT and α-galactosidase (followed in hybrid cells but also identified as the deficiencies involved in Lesch-Nyhan and Fabry diseases), the X-linked surface antigen (function unknown), and classic hemophilia and color blindness (known from human familial studies).

Now that one locus is known for every chromosome, the map can be expanded more quickly by testing directly linkage between markers. Thus the joint inheritance of two enzymes, one of which already is known to be coded by a given chromosome, is evidence that their genes lie on the same homologue. Genes coded by the same chromosome sometimes are said to be *syntenic*. Although this does not abolish the need for karyotypic analysis, it does make the identification of chromosomes necessary only as a check that no translocation or other rearrangement of the genetic material has occurred.

Mapping of the Human X Chromosome by Fragmentation

Syntenic mapping, to identify the chromosomes on which genes lie, makes an important start on the construction of a genetic map for man. But the correlation of enzymes and homologues does not permit any assignment of genetic location within the chromosome. Data on the further location of genes is scanty. In most instances, they are derived from fortuitous observations of a loss of the usual relationship between the presence of an enzyme and the chromosome carrying its structural gene. When this proves to be correlated with a translocation that can be identified, it is possible to identify the chromosome region on which the gene lies. Such observations have been useful in mapping the three X-linked genes coding for the enzymes HGPRT, G6PDH and PGK. This analysis provides a good example of the technique, since it can be correlated also with the distances between these loci that have been estimated by a more general technique applicable to any set of linked genes.

This general technique relies upon the fragmentation and rearrangement of chromosomes that is induced by gamma irradiation. Goss and Harris (1975, 1977a) argued that the probability that two loci will be separated by a radiation-induced break should be proportional to their distance apart; and for any particular pair of loci the degree of separation can be controlled by altering the dose of irradiation to determine the extent of fragmentation. In these experiments, human cells were irradiated and then fused with

Chinese hamster Wg3h cells, a subline of the Don line which lacks HGPRT. Growth on HAT medium then selects for the presence of the human enzyme, and the surviving hybrid clones can be examined for the presence of other X-linked markers, such as G6PDH, PGK, and α-galactosidase. Formally this is a method for examining the linkage of these unselected markers to the selected HGPRT marker.

In controls in which the human parent was not irradiated, 95% of the hybrids containing HGPRT contained all three of the unselected markers. This corresponds to retention of an intact human X chromosome. When the human partner has been irradiated before cell fusion, there is a decline in the frequency with which the three unselected markers are found in the hybrids surviving on HAT medium. The frequency of cosurvival decreases as the dose of irradiation is increased to raise the number of chromosome breaks. At the maximum level of irradiation (4 krad), G6PDH was retained in 42% of the clones, α-galactosidase was present in 34%, and PGK remained in only 22%. If human chromosomes are lost at random from the hybrids, but only those retaining the homologue or fragment carrying the HGPRT gene are able to survive on HAT medium, the distance of each unselected marker from the HGPRT gene may be taken to be inversely dependent on its frequency of cosurvival. Thus G6PDH should be closest to HGPRT, α-galactosidase somewhat further, and PGK the most distant locus.

A quantitative estimate for the relative distance between loci may be made by considering its relationship with the probability that a break will occur between them. The frequency of cosurvival provides a measure of the proportion of cells in which no break is introduced between two loci. If a single hit is sufficient to separate the loci, and events fall on a Poisson distribution, the probability of no breaks is:

$$P_0 = e^{-t \cdot f(D)}$$

where t is the size of the target and $f(D)$ represents the appropriate function of radiation dose.

The fraction of cells showing cosurvival, F, can be taken to be equivalent to P_0 if the pattern of segregation of unlinked markers does not affect viability of the cells, that is, there is no preferential survival of hybrid clones possessing any particular human chromosome fragments. The nature of $f(D)$ can be assigned only empirically, on the basis that a dose exponent between 1 and 2 has been found for the frequency of chromosome rearrangements induced by irradiation; in these experiments it was taken as 1.6. Thus:

$$F = e^{-t \cdot D^{1.6}}$$

Solving for t, which corresponds to the distance between loci, gives:

$$t = \frac{-\ln F}{D^{1.6}}$$

Figure 9.5 shows a plot of log F versus $D^{1.6}$ for the three pairs of loci

Cosurvival of HGPRT and unselected markers

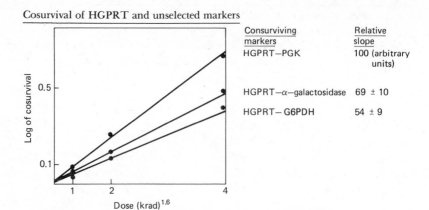

Consurviving markers	Relative slope
HGPRT—PGK	100 (arbitrary units)
HGPRT—α—galactosidase	69 ± 10
HGPRT—G6PDH	54 ± 9

Segregation of unselected markers in HGPRT$^+$ clones (at 4 krad dose)

Unselected markers	Expected frequency	Observed frequency	Observed /expected
PGK$^+$		0.22	
G6PDH$^+$		0.42	
α-gal$^+$		0.34	
PGK$^+$ G6PDH$^+$	0.22 × 0.42 = 0.09 }0.54	0.13 }0.62	1.15
PGK$^-$ G6PDH$^-$	0.78 × 0.58 = 0.45	0.49	
PGK$^+$ G6PDH$^-$	0.22 × 0.58 = 0.13 }0.46	0.09 }0.38	0.83
PGK$^-$ G6PDH$^+$	0.78 × 0.42 = 0.33	0.29	
PGK$^+$ α-gal$^+$	0.22 × 0.34 = 0.07 }0.58	0.16 }0.76	1.31
PGK$^-$ α-gal$^-$	0.78 × 0.66 = 0.51	0.60	
PGK$^+$ α-gal$^-$	0.22 × 0.66 = 0.15 }0.42	0.06 }0.24	0.57
PGK$^-$ α-gal$^+$	0.78 × 0.34 = 0.27	0.18	

Figure 9.5. Mapping X-linked loci by chromosome fragmentation. The upper part of the figure shows the cosurvival of unselected markers with HGPRT as a function of radiation dose (see text). The slope of the graph gives the relative distance from the unselected marker to the HGPRT locus. The lower part of the figure shows the cosegregation pattern of two pairs of unselected markers. The expected frequency for each pairwise combination is the product of the individual frequencies with which the markers are observed to cosurvive or fail to cosurvive with HGPRT (upper three lines). The ratios of the observed to expected frequencies for concordant and discordant segregation show that PGK and G6PDH do not differ significantly from the frequencies expected for independent segregation, while PGK and α-gal do differ significantly This conclusion is based on statistical analysis of more complete data by Goss and Harris (1975).

HGPRT-PGK, HGPRT-G6PDH, and HGPRT-α-galactosidase. The slopes of these lines give the relative distances between the loci. If the distance HGPRT-PGK is taken as 100, then the distance HGPRT-G6PDH is 54 ± 9 and the distance HGPRT-α-galactosidase is 69 ± 10.

What is the order of the four loci? The cosurvival of unselected loci should depend upon whether they lie on opposite sides or on the same side of the selected HGPRT locus. If two unselected loci lie on opposite sides of HGPRT, they should segregate independently. The probabilities that both loci will segregate with or away from HGPRT, or that one will cosurvive while the other does not, are simply the products of the appropriate individual probabilities that either locus will segregate with or away from HGPRT. Data can be tested for conformation to this expectation by a simple statistical chi-squared test. If two unselected loci lie on the same side of HGPRT, the deviation from expectation should show that whenever the locus furthest from HGPRT is present, the intervening nearer locus also should be present; whenever the nearer locus is absent, so should be the further locus. Given the occurrence of rearrangements additional to the introduction of single breaks, this is seen as an increase in extent of concordant segregation (both unselected markers present or both absent). As the data summarized in Figure 9.5 show, PGK and G6PDH segregate independently, whereas PGK and α-galactosidase show related segregation. Thus PGK and α-galactosidase must be on the same side of HGPRT, with α-galactosidase constituting the central marker since it is closer to HGPRT, and G6PDH must lie on the other side.

The data also show that local events occur not representing simple single breaks. For example, some clones do contain both PGK and HGPRT while lacking the intervening α-galactosidase. Excluding the possibility that the gene has been inactivated by mutation (this seems to occur at too low a frequency), this is equivalent to deletion of material between PGK and HGPRT. In effect this represents a localized segregational event; and if similar events take place at the other three loci, then the measured frequencies for cosurvival are distorted by the occasional removal of one locus in a manner independent of the distance between loci. This can be corrected by estimating the target size for the local event and subtracting it from the apparent distance between the loci. This gives a relative map distance between HGPRT and PGK of 66 ± 15 units, with distances between other markers of:

$$\underset{33 \pm 12}{\text{PGK}} \quad - \quad \underset{}{\alpha\text{-galactosidase}} \quad \underset{39 \pm 12}{-} \quad \underset{}{\text{HGPRT}} \quad \underset{24 \pm 11}{-} \quad \text{G6PDH}$$

Another possible distortion in the estimated distances may be caused by position relative to the centromere. Whether this is important depends on the relative chances for survival in the hybrid cell of the separated chromosome fragments possessing and lacking the centromere. If both survive equally well, perhaps because the acentric fragment usually becomes attached to some other homologue, then it will not matter whether a given break lies between HGPRT and the centromere or on the other side of HGPRT. However, if the HGPRT gene is more likely to survive when attached to its original centromere, the surviving hybrids will be biased by an increased proportion of cells in which any break has taken place on the centromere-distal side of HGPRT. Thus if PGK is the locus nearest the centromere, its closeness to HGPRT will be overestimated; correspondingly the distance from HGPRT to G6PDH will be underestimated. The extent to which such effects are significant is hard to predict.

This method can be applied to analyze the relationship of any set of syntenic markers, one of which provides the selected marker. While it is useful to be able to obtain cells carrying the selected marker by a survival method, this is not strictly necessary; any set of hybrids all of which retain a given marker can be analyzed for the presence of other, syntenic markers, irrespective of the method used to obtain the cells. Goss and Harris (1977b) have applied this method to map 8 unselected loci on human chromosome 1; their relative distances apart are in reasonable agreement with distances based on cytological mapping of individual genes to chromosome bands. This map extends across the centromere, which thus appears to have only a small effect on the segregation of genes from the irradiated chromosome.

X Chromosome Mapping by Translocation

Exceptions to the joint inheritance of syntenic genes occur whenever a chromosome break or translocation occurs spontaneously, either in the parental cell before fusion or in the hybrid cells. Examining human-mouse hybrids derived from parents in which the human cells were karyotypically normal, Miller et al. (1971d) obtained some clones possessing human HGPRT but lacking G6PDH. These presumably arose by rearrangements occurring after fusion. Another source of such rearrangements is to obtain cells from human patients with abnormal chromosome complements. By such means, several translocations in this region of the X chromosome now have been characterized. Figure 9.6 summarizes their positions relative to the standard band structure of the chromosome (see Chapter 15).

Cells with a translocation between the X chromosome and chromosome 14 have been obtained from a patient with a balanced exchange. The X chromosome is broken in its long arm, close to the centromere, with the short arm fragment forming an independent Xp chromosome (originally known as the Xq chromosome), and with the long arm becoming attached to the end of chromosome 14 to generate the large chromosome t(14q,Xq). These cells (KOP-1) possess one copy each of chromosome 14, a normal X chromosome, the Xp short chromosome, and the t(14q,Xq) translocation. Because the normal X chromosome provides the inactive sex chromatin in these cells—and is not reactivated by cell fusion—all active sex linked human genes must be located on one of the two rearranged chromosomes, Xp and t(14q,Xq).

In hybrids between the human KOP-1 cells and mouse cells, the Xp and t(14q,Xq) chromosomes may be presumed to segregate independently. In two independent series of experiments to select for HGPRT by growth on HAT medium, Grzeschik et al. (1972) and Ricciuti and Ruddle (1973a,b) found that PGK and G6PDH almost always also are retained. In the first

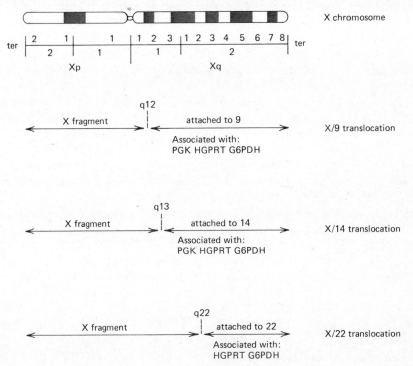

Figure 9.6. Structure of the X chromosome and some translocations. The X chromosome is divided into a short arm (Xp) and a long arm (Xq). Each arm is divided into two regions, numbered 1 and 2 from the centromere toward the terminus (ter). Within each region, the bands seen upon staining with Giemsa or fluorescence with quinacrine are numbered individually from centromere toward terminus. The first band in the long arm is part of Xq, in region 1, and constitutes band 1; it is therefore known as q11. The positions of the breaks and the fates of the two separated parts of the X chromosome are shown for translocations with autosomes 9, 14, and 22; the enzymes shown to be associated with each fragment by karyotypic analysis are described. Data of Ricciuti and Ruddle (1973b) and Shows and Brown (1975).

set of experiments, all three X-linked markers were retained in 36 of 37 human-mouse hybrid clones and in 22 of 24 human-hamster hybrid clones; in the second series, all three markers remained linked in 19 of 20 primary human-mouse hybrid clones and in 64 of 66 secondary clones derived from them. This immediately suggests that all three genes are carried by the same X chromosome fragment, that is, either all lie on Xp or all lie on t(14q,Xq). Ricciuti and Ruddle showed that in these clones, the X-linked genes also are linked to the autosomal gene for nucleoside phosphorylase. This suggests that nucleoside phosphorylase is coded by chromosome 14 and that PGK, HGPRT and G6PDH lie on the long arm of the X chromosome that is part of t(14q,Xq).

The separation of either PGK or G6PDH from HGPRT must be caused by the occurrence of a chromosome break or translocation in the hybrid cells. In the experiments of Grzeschik et al., the aberrant clones always represented the segregation of PGK from HGPRT and G6PDH, which remained linked together. Coupled with karyotypic analysis of the chromosomes remaining in these clones, this led to the suggestion that HGPRT and G6PDH lie on the short arm of the X chromosome while PGK lies on the long arm. But the idea that all three loci lie on the long X arm fits better with the overwhelming preponderance of clones in which they show common inheritance. The segregation only of PGK from HGPRT and G6PDH, observed also to be the aberrant form in the experiments of Ricciuti and Ruddle, is better explained by supposing that PGK is more distant from HGPRT than is G6PDH. This is supported by their karyotype analysis to show an association between the presence of t(14q,Xq) and the three-X-linked markers and nucleoside phosphorylase.

In the two exceptional clones in which PGK was segregated from HGPRT and G6PDH, nucleotide phosphorylase was present in the PGK$^+$ HGPRT$^-$ G6PDH$^-$ clone and absent from the PGK$^-$ HGPRT$^+$ G6PDH$^+$ clone. This suggests that PGK lies closest to the nucleoside phosphorylase gene; and since this lies on chromosome 14, PGK must lie closest to the site of the break on the long arm of the X chromosome, that is, closest to the centromere. From these data, all that can be said about the location of G6PDH is that it must lie sufficiently close to HGPRT so that breaks occur much more frequently to separate PGK from HGPRT. Contrary to some suggestions, these data do not distinguish on which side of HGPRT the G6PDH locus is to be found.

The properties of two further translocations allow the positions of these genes to be defined more closely. One involves the translocation onto chromosome 9 of the long arm of the X chromosome from band q12 to the terminus; in the other, the region of the long arm from band q22 to the terminus is translocated onto chromosome 22. Shows and Brown (1975) used cell lines carrying these translocations as the human partners for fusion with RAG or A9 rodent cells. Human HGPRT was selected for or against by growing the hybrids on HAT or 8-azaguanine medium. When the active sex linked genes were provided in the form of the X/9 translocation, all three of the enzymes HGPRT, PGK and G6PDH were retained or lost together. With the X/22 translocation, HGPRT and G6PDH always segregated together, while the segregation of PGK was independent. In both cases, the translocated chromosome carrying the long arm of the X chromosome was associated with HGPRT activity.

As summarized in Figure 9.6, this suggests that the genes for HGPRT and G6PDH lie between band q22 and the terminus of the long arm of the X chromosome; while PGK must lie between bands q13 and q22, since it is present on the X/9 translocation (carrying the region from q12-ter) and on the t(14q,Xq) translocation (carrying q13-ter), but is absent from the X/22 translocation (carrying only q22-ter). These locations are consistent with the order suggested by chromosome fragmentation mapping.

Translocations have been used also to map the locations of genes on some of the human autosomes. A translocation removing about 25% of the short arm of chromosome 1 segregates the genes for phosphoglucomutase-1 and 6-phosphogluconate DH from the gene PepC (which remains on the truncated autosome). This therefore places these two genes in the deleted region (Douglas, McAlpine and Hamerton, 1973). A human line in which part of chromosome 15 has been translocated to the X chromosome has made possible the application of HAT selection to identify genes carried on the translocated segment (Soloman et al., 1976). Other experiments using these techniques have been described by De Wit et al. (1977), Francke et al. (1977) and Chern et al. (1977).

One potential application of somatic cell genetics is to compare gene locations in different species. The genes for galactokinase and thymidine kinase both lie on the long arm of human chromosome 17; and both genes also are syntenic in the chimpanzee and African green monkey (on the chromosomes apparently homologous with human chromosome 17) and in the mouse (on chromosome 11) (Orkwiszewski et al., 1976; Chen et al., 1976; Kozak and Ruddle, 1977). Similarly the short arm of human chromosome 1 and the mouse chromosome 4 both show synteny for the group of enzyme loci, enolase, phosphogluconate DH, phosphoglucomutase, and adenylate kinase (Lalley et al., 1978). Many examples also are known, however, of loci that are syntenic in one but not another species; it is too early to say how informative such data will be about the evolution of mammalian genomes.

Expression of Differentiated Functions

Changes in phenotypic expression often take place following cell fusion. This opportunity to follow the interactions of genomes in different states of gene expression has been taken in several systems. We have seen in Chapter 8 that when cells able to replicate DNA or transcribe RNA are fused with cells that have lost either ability, the inert genome is reactivated into the appropriate nucleic acid synthesis. While this shows that an active nucleus may induce functions that are dormant in another nucleus, it does not reveal the extent to which the activation is specific for given genes, although these experiments have shown that at least some genes are properly translated as well as transcribed. These genes appear to belong to a class that is constitutively expressed; that is, they are probably expressed in all cell types. Such *constitutive* genes also have been described as *household* functions, including the enzymes of common metabolic pathways, common structural proteins, and so on. This is in contrast with *facultative* genes, also described as *luxury* functions, that code for the specialized gene products each of which is synthesized only in a certain class or classes of differentiated cell phenotypes. Obviously this distinction is far from precise, since a gene whose product is constitutively synthesized in all tissues, perhaps at a low level, may be one that is facultatively expressed at a high level in some particular cell phenotype. Nonetheless, this concept makes the point that we should look at both types of function in investigating the effect of cell fusion on gene expression.

The number of constitutive markers whose expression has been followed is not enormous, but in every instance the proteins of both parents have been found in the hybrid. Such coexpression of constitutive markers has been found in many interspecific hybrids and applies to metabolic enzymes as well as to surface antigens (for review see Ringertz and Savage, 1976). In homospecific crosses, interallelic complementation may occur, as reported for galactose-1-phosphate-uridyl-transferase (Nadler et al., 1970). In heterospecific crosses, heteropolymeric enzymes may be formed, as reported for hexosaminidase A and β-glucuronidase (Chern et al., 1976, Chern, 1977).

When differentiated cells expressing the specialized functions of the tissues from which they derive were fused with established cell lines, the first phenomenon that was noticed was the extinction of their differentiated functions. Davidson, Ephrussi, and Yamamoto (1966, 1968) and Silagi (1967) found that when hybrids are formed between Syrian hamster or mouse melanoma cells and L cell lines, the hybrids cease to make melanin or DOPA oxidase (tyrosinase), which are characteristic of the pigmented melanoma parent. Since in the interspecific fusion, other enzymes—LDH, MDH, TK— of the differentiated parent continue to be expressed, this seems to represent the operation—or failure of operation—of some regulatory system concerned with the expression of the differentiated functions.

Such extinction has been observed in several systems. Sometimes the extinguished function can be reexpressed under certain conditions. Sometimes one specialized function may be extinguished while another continues to be expressed. Data on those systems that have been the most investigated are summarized in Table 9.1.

The extinction of melanoma functions appears to be connected with a gene dosage effect. No expression of the pigment genes is seen after fusion of a diploid melanoma cell with an L cell. But expression may occur when tetraploid melanoma cells are used. Davidson (1972) found that about half of the hybrids produced by such fusion were pigmented; their stability varied, each pigmented hybrid giving rise to unpigmented cells with a characteristic frequency. Fougére, Ruiz, and Ephrussi (1972) found that all their such hybrid clones initially were unpigmented and fibroblastic in appearance, but later gave rise to pigmented cells of epithelial appearance similar to that of the melanoma parent. In both cases, these hybrids displayed chromosome complements corresponding to about 80–90% of the parental total (compared with the > 95% retention of parental chromosomes in fusions using diploid melanoma cells). The dependence of melanin expression on the ploidy of the melanoma cell suggests that a gene dosage effect is involved; and perhaps also the loss of specific chromosomes, evi-

Table 9.1 Phenotypic expression of differentiated markers in hybrid cells

Differentiated parent	"Undifferentiated" parent	Expressed markers	Extinguished markers	Conditions for reexpression of extinguished markers	Ref.
Syrian hamster melanoma	mouse L cell		melanin DOPA oxidase	4n melanoma: 2n fibroblast hybrids may express melanin	A
Rat hepatoma	mouse 3T3 or L cell or human WI38	albumin catalase complement factors	TAT aldolase B ADH	loss of human X chromosome allows TAT expression;	B
				mouse albumin may be expressed if rat:mouse chromosome ratio is high enough	C
Mouse neuroblastoma	mouse or human fibroblast	acetylcholinesterase neuronal protein 14-3-2 neuronal morphology excitable membrane	steroid sulfatase		D
Friend (mouse) erythroleukemia	human fibroblast, mouse lymphoma or bone marrow		globin	loss of X chromosome (shown in homospecific fusions)	E
"	human or Ch. hamster erythroblast	mouse globin		human or hamster globin may become DMSO-responsive	F

References: A—Davidson et al. (1966, 1968), Davidson (1972), Fougere et al. (1972); B—Schneider and Weiss (1971), Thompson and Gelehrter (1971), Croce et al. (1973); C—Bertolotti and Weiss (1972), Peterson and Weiss (1972), Malawista and Weiss (1974); D—Minna et al. (1971, 1972), Peacock et al. (1973), McMorris et al. (1974), McMorris and Ruddle (1974); E—Deisseroth et al. (1975a, 1976), Benoff and Skoultchi (1977); F—Deisseroth et al. (1975b).

dently more likely with tetraploid than with diploid melanoma parents, is pertinent. A dosage effect might take the form of requiring the presence of sufficient copies of some gene to cause synthesis of a critical amount of activator; or it might involve the presence of sufficient copies of the genome to dilute out an inhibitor. Any exact assessment of the nature of the dosage dependence is made difficult by the extensive heteroploidy of the L cell line, which means that it is impossible to say how many copies are present of any given mouse chromosome or gene.

Another specialized function that is extinguished upon cell fusion is the expression of tyrosine transaminase (TAT) in hepatoma cells. This is a liver function, present in the cells at a low level, and induced to greater levels by glucocorticoid steroids or actinomycin D. Fibroblasts have a low level of the enzyme, which is not inducible (and this may or may not represent the same enzyme as the inducible species of liver). Hybrids between fibroblasts and hepatoma cells show the characteristic activity of the fibroblast parent (Schneider and Weiss, 1971; Thompson and Gelehrter, 1971). Weiss and Chaplain (1971) reported that TAT expression suffers the same fate in rat hepatoma × rat liver epithelial cell hybrids. (Although of liver origin, the epithelial partners do not express TAT.) But descendents of the hybrid cells characterized by more rapid growth can be isolated; these have lost some of their chromosomes and among them was one clone, which although having a low level of TAT, could be induced to high levels by the steroid dexamethasone. Although difficult to interpret because the functions of the two parents cannot be distinguished, and also because both parents derive from the liver, this result raises the possibility that chromosome loss may be connected with reexpression of the extinguished function. This is borne out by the results reported by Croce, Litwak, and Koprowski (1973) with rat hepatoma × human fibroblast hybrids. Initially the hybrids cannot induce TAT. But reexpression of this function occurs following the loss of human chromosomes. When hybrids are selected in 8-azaguanine, to eliminate human HGPRT and thus the X chromosome, reexpression always occurs. Although not proving the point, since other chromosomes are lost and rearrangements can occur, the correlation between the loss of the human X chromosome and reexpression of TAT inducibility is consistent with the idea that a heterospecific repressor of this activity is sex linked in man.

Differentiated functions can be induced as well as extinguished in cell hybrids. Rat hepatoma cells produce the serum albumin characteristic of liver, whereas mouse fibroblasts produce no albumin. Following fusion of 3T3 fibroblasts with diploid or tetraploid hepatomas, Peterson and Weiss (1972) used antisera to rat and mouse serum albumins (RSA and MSA) to examine their production by the hybrids. The hybrids produced by diploid: diploid fusions all produce RSA but not MSA. Thus the rat albumin gene remains turned on and the mouse albumin gene remains turned off. This is consistent with other experiments showing that although an active nucleus may induce expression of constitutive functions in an inactive nucleus, it does not turn on facultative functions. Thus when the chick nucleus is reactivated by fusing erythrocytes with rat hepatoma or myoblast cells, the hybrids make neither chick albumin nor chick myosin, although continuing to produce the specialized proteins characteristic of the rat parent (Szpirer, 1974; Carlsson et al., 1974).

When tetraploid hepatoma cells are fused with mouse fibroblsts, the expression of facultative mouse functions depends on the karyotype of the hybrids. Cells possessing slightly more chromosomes than expected from the fusion (the excess probably being of mouse origin), produced both RSA and MSA. Cells that had lost about 15% of the rat chromosomes produced only MSA. Cells having lost 25% of the rat chromosomes produced neither albumin. This suggests that the rat genome can switch on the appropriate differentiated function in the mouse genome in a manner that depends upon gene dosage. The signal must be common to rat and mouse. Since MSA can be expressed when RSA has been lost, it seems that the regulation is not coincident with the structural gene.

Serum albumin production has been examined also in hybrids between diploid or tetraploid rat hepatoma cells and mouse leukemic lymphoblasts. Malawista and Weiss (1974) found that all the hybrid clones derived from tetraploid hepatoma parents produced both rat and mouse albumins. And 8 of 9 hybrids formed by fusion with diploid hepatoma parents produced mouse albumin; 6 of these also produced the rat protein. The reason for the difference in expression of the mouse albumin in the fusions with 3T3 fibroblasts and the leukemic lymphoblasts may lie with their chromosome complements. The lymphoblasts have about half of the number of chromosomes of the 3T3 cells, so that the same mouse:rat chromosome ratio is accomplished in diploid hepatoma × lymphoblast fusion as in tetraploid hepatoma × 3T3 fibroblast fusion. Hybrids between mouse hepatoma and human leucocyte cells also produce the albumins of both parents (Darlington et al., 1974). These results therefore again are consistent with the idea that the ability to switch on some differentiated functions is heterospecific and dose dependent. This leaves open the question of whether continued expression depends on the presence of the regulator function(s) needed for the initial switch on, or whether regulation takes the form of introducing a stable epigenetic change in the cell.

Stable changes in gene expression may occur spontaneously, such as the loss of liver functions that is seen occasionally in rat hepatoma lines. Whether these "dedifferentiated" lines are the result of genetic or epigenetic change is not known. Deschatrette and Weiss (1975) observed that when the dedifferentiated cells are fused with parental hepatoma cells, in general all the characteristic liver functions are suppressed. This suggests that dedifferentiation has led to acquisition of the ability to extinguish liver functions, not to the loss of some necessary activator.

A series of changes in the pattern of gene expression has been followed in the hybrid cells generated by fusion of mouse melanoma with rat hepatoma. Fouchére and Weiss (1978) observed that the immediate products of fusion express neither the pigment characteristic of the melanoma nor the albumin typical of the hepatoma. But upon aging of the culture, some hybrid cells express one or the other function. Subclones prepared from one hybrid line show switches in the expression of these differentiated functions: pigment of albumin may be formed, but on recloning cells appear which have reversed their state of differentiation, as judged by this criterion. In this hybrid line, the albumin that is formed is of the murine type, and thus represents activation of the mouse gene by the rat genome. Two features of this system are striking. Epigenetic reversals can be sustained over a considerable period of time (the possibility of genetic changes due to chromosome loss is not excluded, but the reversals of phenotype mean that all necessary genes must remain present). And these cells demonstrate the existence of *phenotypic exclusion*—although they retain the potential to express both types of parental function, only one can be expressed in a given cell; the other is excluded.

Globin gene expression has been followed in cell hybrids. This is a good system because molecular hybridization assays exist for determining whether globin genes are present in the cellular DNA and whether they are represented in messenger RNA (see Chapter 26). Friend erythroleukemia cells are an established mouse line in which globin synthesis is induced by the addition of DMSO (dimethylsulfoxide). Upon fusion with human fibroblasts, the hybrids retain large numbers of chromosomes of both parents and lose the ability to produce globin, although the mouse globin genes remain present (Deisseroth et al., 1975a, 1976). When the same fusion is performed using tetraploid Friend cells, the mouse globin genes are expressed, but the human genes remain uninduced (Alter et al., 1977). In another series of experiments, globin induction was found in hybrids derived by fusing fibroblasts with both diploid and tetraploid Friend cells, although at much higher frequency in the latter case. In both cases, the inducible cells displayed rounded, nonadherent morphology, whereas the noninducible cells were adherent (Axelrod et al., 1978). The nature of the linkage between morphology and hemoglobin inducibility is uncertain; but it carries the implication that in these experiments, cells should not be selected by ability to adhere, since some phenotypes then may be excluded.

Upon fusion between diploid Friend cells and human or Chinese hamster erythroblasts, the mouse complement is retained, but most of the partner chromosomes are lost. Deisseroth et al. (1975a,b) obtained a mouse-human clone in which mouse globin gene expression continued to be responsive to DMSO, but human globin was not synthesized. In another mouse-human clone, and in a mouse-Chinese hamster clone, the globins of both species were expressed upon induction with DMSO. An obvious speculation is that the continued expression of the mouse globin genes results from the erythroid nature of the partner cells (which although not yet active in synthesizing globin, are precursors to cells that will do so). The heterospecific nature of the control process again is shown by the clones in which heterologous genes have been brought under the DMSO-responsive control characteristic of the Friend cell—a control to which normally they would not respond. Whether this depends upon the particular chromosomes that have been lost in the hybrids is not known; but the mouse-hamster line has a double complement of mouse chromosomes with only 3–5 Chinese hamster chromosomes, so a dosage effect might be involved. From assays with a cDNA probe for globin messengers, it is clear that the control is exercised at a stage prior to mRNA production. Analogous results have been obtained with a line in which tetraploid Friend cells retaining only a few human chromosomes as the result of cell fusion are able to synthesize human α globin (Deisseroth and Hendrick, 1979).

Hemoglobin also is inducible by DMSO in hybrids formed between Friend cells and teratocarcinoma cells (McBurney, 1977). These hybrids lose the ability to generate many differentiated cell types that is characteristic of the teratocarcinoma, which is an undifferentiated pluripotent line. One hybrid, however, remained pluripotent; but this had lost the ability to induce hemoglobin. The hybrid cells therefore fall into either of the parental phenotypes, which seem to be mutually exclusive. Because both parents are murine, it is impossible to distinguish the parental origins of the induced hemoglobin, but there do appear to be quantitative changes from the pattern of the Friend parent which might represent expression of genes of the other parent. This is seen more clearly in similar experiments with cell lines that have different alleles for the α globin gene; the hybrids are able to induce expression of the globin gene of the teratocarcinoma genome (McBurney, Featherstone, and Kaplan, 1978).

Suppression of the response to DMSO may depend upon a locus carried by the mouse X chromosome. Benoff and Skoultchi (1977) reported that a Friend cell line lacking any active X chromosome—which was selected for HGPRT deficiency—shows the usual response to DMSO. Upon fusion with mouse lymphoma or bone marrow cells, the response to DMSO is abolished in hybrids that retain the X chromosome of the partner cell. Upon loss of the X chromosome, globin synthesis on exposure to DMSO is restored. Whether other mammalian X chromosomes carry a function(s) able to suppress the DMSO response remains to be seen.

The influence of one genome upon another in hybrid cells extends to the nucleolus as well as to the nucleoplasm. Eliceiri and Green (1969) first noted that in human-mouse hybrids segregating human chromosomes, only mouse 28S rRNA is synthesized. Even hybrids retaining several of the human chromosomes of the class that carries rRNA genes fail to synthesize human rRNA (Marshall et al., 1975). The extinction of human rRNA synthesis is associated with the absence of nucleoli at the human chromosomes where they are usually found (Miller et al., 1976a). But a different result has been found in human-mouse hybrids that lose the mouse chromosomes; Miller et al. (1976b) found that these have nucleoli associated only with the human chromosomes. Such cells synthesize only human rRNA, even though mouse chromosomes carrying rRNA genes have been retained (Croce et al., 1977). The presence of the rRNA genes of the unexpressed species since has been confirmed directly in both types of hybrid (Perry et al., 1979). Here there seems therefore to be a connection between gene expression and the direction of chromosome loss.

Division Cycles in Homokaryons

Homokaryons show an increasing tendency to establish synchrony in nuclear replication and division after their formation. Synchrony in nuclear replication begins to be imposed during the first day and by the third day is appreciable in binucleate, trinucleate, and tetranucleate cells (Johnson and Harris, 1969a,b). The imposition of synchrony can be observed also by following mitosis. At 2 hours after fusion, more than 70% of the cells which contain a mitotic nucleus are asynchronous, since only some of the nuclei in the cell have entered mitosis. By 23 and 33 hours after fusion, only 20% of the cells show asynchronous mitosis, and in the majority all nuclei divide simultaneously. By 50 hours after fusion, the proportion of cells with asynchronous nuclei has declined further to 16.5%. With continuing incubation, homokaryons therefore establish increasing synchrony in both nuclear DNA synthesis and mitosis; as nuclei continue to reside in common cytoplasm they become increasingly likely to enter simultaneous division cycles, presumably due to their exposure to common cytoplasmic control signals.

Fusion of cells in different stages of their cycles shows that the state of a nucleus may be influenced by the states of the other nuclei present in the same cytoplasm. Rao and Johnson (1970) examined these effects by performing a series of heterophasic fusions, G_1 with S phase HeLa cells, G_1 with G_2 cells, and S phase with G_2 cells. By lightly prelabeling only one of the parental cell populations, the origins of the nuclei in the homokaryons can be recognized. When a large dose of ^{3}H-thymidine is added after fusion, its labeling density is great enough to be distinct from the light prelabel.

In binucleate cells, fusion of S with G_1 nuclei induces DNA synthesis more rapidly in the G_1 nucleus. In multinucleate cells, the ratio of nuclei in advanced to early stages of the cycle determines their relative activities. With varying ratios of G_1 to S phase nuclei, the time taken for half of the G_1 nuclei to incorporate a label of ^{3}H-thymidine is:

G_1 controls	10 hours
$2G_1 : S$	3
$G_1 : S$	1.75
$G_1 : 2S$	1.5

Synthesis of DNA therefore appears to be under a positive control in which the presence of S phase nuclei may hasten the entry into replication of G_1 nuclei. Fusions between cells at different stages of G_1 produce the result expected if inducers of replication accumulate during G_1 until a critical level is reached to initiate S phase (Rao, Sunkara and Wilson, 1977). Fusions of G_2 with G_1 nuclei do not induce DNA synthesis, so that the signals which activate replication do not remain in the cell after the termination of S phase.

A compromise is seen when S phase and G_2 phase cells are fused. The S phase nucleus enters mitosis more rapidly than usual, because the period between the end of replication and the start of division is much reduced. But the G_2 nucleus is delayed in entering mitosis at least until the S phase nucleus has completed replication. There is no induction of DNA synthesis in the G_2 nucleus, even when the ratio of S phase to G_2 nucleus is high. In multinucleate homokaryons, the initiation of mitosis is dose dependent: the greater the ratio of G_2 to S phase nuclei, the sooner the S phase nuclei enter mitosis after completing replication. In fusions involving several nuclei, the time taken to reach a mitotic index of 50% is:

4S nuclei	14.2 hours
G_2 : S	13.3
$2G_2$: S	11.6
$3G_2$: S	10.7
$4G_2$	8.5

Mitosis may therefore result from the accumulation of an inducer in G_2 cells whose presence shortens the gap before the S phase nuclei enter mitosis. The presence of the S phase nuclei may dilute the effective concentration of the mitotic inducer and thus cause the delay in the time before the G_2 nucleus itself enters mitosis. This interpretation is supported by the results of fusions between G_1 and G_2 cells, in which the G_1 nucleus starts DNA synthesis at the usual time, but the entry into mitosis of the G_2 nucleus is delayed, again perhaps due to dilution of the effective concentration of mitotic inducer (see also Rao, Hittelman, and Wilson, 1975). Synchrony in cell fusion has been reviewed by Johnson and Rao (1971).

In fusions between synchronized populations of HeLa cells with mouse or Chinese hamster cells, the induction of DNA synthesis does not seem to be species specific, but the duration of S phase appears to be a characteristic of each type of nucleus. Each of the parental cell populations spends a characteristic time in G_1 and S phase. For HeLa cells G_1 is 7-11 hours; for the mouse and Chinese hamster cells it is 2-4 hours. When two mitotic HeLa cells are fused, both nuclei initiate DNA synthesis after the usual G_1 interval. But when mitotic HeLa cells are fused with mitotic mouse or Chinese hamster cells, both nuclei initiate DNA synthesis after a G_1 period which is about 4-5 hours shorter than the usual HeLa G_1. This suggests that the mouse or Chinese hamster nucleus produces the signals that induce DNA synthesis after a period only slightly longer than usual; and these signals are able to act upon the HeLa nucleus as well as upon the nucleus from which they emanate. But each nucleus then continues replication for its characteristic length of time; features intrinsic to the nucleus therefore must be responsible for determining the time required to complete duplication of all replicons.

Survival of Prematurely Condensed Chromosomes

Mitotic asynchrony occurs only rarely in G_1-S and G_1-G_2 fusions, perhaps because the nuclei usually reach an equilibrium in which division takes place at a time appropriate for both. In G_2-S fusions, however, sometimes the G_2 nucleus may enter mitosis before the nucleus in S phase is ready. This causes a *premature condensation* of the chromosomes in the S phase nucleus, in effect an atypical mitosis without a spindle. Johnson and Rao (1970) observed that synthesis of DNA is reduced but not halted completely in the condensed chromosomes, which cease transcription of RNA.

Premature chromosome condensation can be induced in all interphase cells by fusion with mitotic cells. It depends upon the ratio of mitotic to non mitotic nuclei and presumably this dosage effect reflects a titration of inducer molecules that must be produced by G_2 and M phase cells to cause condensation. The morphological changes that occur in the interphase chromatin parallel those of mitosis. Figure 9.7 shows that when the chromosomes of G_1 or G_2 nuclei are prematurely condensed by such fusion, they display single and double chromatids, respectively. Induction of premature condensation in S phase nuclei, however, yields unevenly condensed chromatin in which patches of large and small dispersed fragments separate condensed regions. No species specificity is apparent in the induction of premature chromosome condensation, which may be accomplished by fusing mitotic HeLa cells with interphase cells of the Chinese hamster, Xenopus or mosquito (Johnson, Rao, and Hughes, 1970).

Whatever factors induce chromosome condensation are produced by and remain in mitotic cells for only a short time. Rao and Johnson (1972) found that the ability of mitotic HeLa cells to induce premature chromosome condensation upon fusion with interphase cells declines when the cells are held in mitosis by incubation with colcemid for increasing periods of time. Freshly collected mitotic cells maintained in colcemid for 4 hours induce premature chromosome condensation in 88% of the resulting hybrids. But if the mitotic cells are kept in the colcemid for a further 20 hours before fusion, the proportion of hybrids with prematurely condensed chromosomes is only 10%. This suggests that the inducing factors are synthesized in late G_2 and the beginning of mitosis and are not stable. They must decay as the cell passes through division; perhaps this is necessary to allow the mitotic cells later to disperse their condensed chromosomes. The induction of premature chromosome condensation occurs in other nuclei only after breakdown of the nuclear membrane of the first nucleus to divide; and in multinucleate heterocaryons, only those nuclei nearest the dividing nucleus may suffer premature condensation (Peterson and Berns, 1979). This gradient suggests that the inducing factor(s) are released from the mitotic nucleus with the breakdown of its membrane; and either they diffuse slowly or their effect is concentration dependent.

A system in which it may be possible to investigate the nature of the

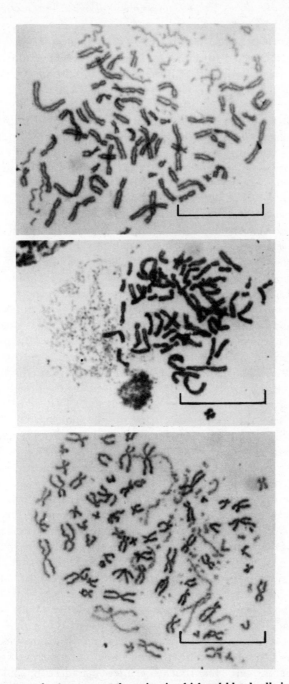

Figure 9.7. Premature chromosome condensation in chick red blood cells induced by fusion with mitotic HeLa cells. Upper: the G_1 prematurely condensed chromosomes are monovalent. Center: premature condensation in S phase generates heterogeneous elements. Lower: in G_2 all the chromosomes are condensed into bivalent structures. Data of Johnson and Mullinger (1975).

mitotic factors is provided by the observation that when extracts of mitotic HeLa cells are injected into Xenopus oocytes they cause breakdown of the nuclear membrane and condensation of the chromosomes (Sunkara, Wright, and Rao, 1979). Extracts of G_2 cells are somewhat less effective; it is not clear whether this is due to a lower level of production of the factors at this stage or to some departure from synchrony in the donor population.

By prelabeling the chromosomes of an interphase population of cells before fusion with mitotic cells, their fate can be followed during growth of the homokaryons. In the first cell cycle after fusion, the interphase chromosomes suffer premature condensation to give the appearance dictated by their state prior to fusion. Irrespective of the stage of interphase, the prematurely condensed chromosomes suffer the same fate: they fail to be integrated into the mitotic spindle and at the end of mitosis are lost from the spindle or are randomly incorporated into the reforming daughter nuclei. (In cells in which the interphase chromosome set fails to undergo premature condensation, the mitotic set of chromosomes is unable to complete division and instead forms inactive micronuclei.) In the second cell cycle after fusion, some chromosomes in the nucleus may be labeled; these must represent prematurely condensed chromosomes that survived the first mitosis and became included in the reconstituted nucleus. During subsequent cell cycles they display the same behavior as the other chromosomes of the parental set. In effect this would appear to represent the addition at random of some chromosomes from one parent nucleus to the other.

By hybridizing populations of *glyA* and *glyB* CHO mutants, cells containing both the wild type $glyA^+$ and $glyB^+$ alleles can be generated. In suitable medium, only these prototrophs can survive. When mitotic and interphase cells are fused, the rate of survival in selective medium is much lower than that resulting from fusion of two interphase populations. This implies that the formation of prematurely condensed chromosomes, with the consequent loss of one parental set from the hybrid cells, reduces the rate of survival by making unavailable one of the genes required for growth in the selective medium. But in those cells that survive the interphase-mitotic cell fusion, genes provided by both parents must be expressed. It is therefore possible for chromosomes (or parts of chromosomes) to survive premature condensation and to express their genes in subsequent cell generations. Cells whose G_1 or G_2 chromosomes are prematurely condensed have a better chance of contributing genetic markers than cells with prematurely condensed S phase chromosomes.

Genes carried by prematurely condensed chromosomes may be able to survive in cells in which there appear to be no intact chromosomes derived from their parent cell. The first experiments suggesting that chromosome fragments may be able to survive in the descendents of such heterokaryons made use of fusions between the HGPRT⁻ A9 mouse fibroblasts and chick erythrocytes. After fusion, the chick chromosome set often suffers a premature condensation when the mouse nucleus passes through mitosis.

When Schwartz, Cook, and Harris (1971) applied HAT selection to the products of this fusion, most died; but some clones resistant to HAT appeared within 2–3 weeks. These clones contained the chick enzyme HGPRT, distinguished from the mouse enzyme by its characteristic electrophoretic mobility. In similar experiments in which chick erythrocytes were fused with the HGPRT⁻ Wg3h line of Chinese hamster cells, Boyd and Harris (1973) obtained the same result. In this experiment 50 HAT^r clones were obtained per 10^6 Wg3h cells plated after fusion.

In spite of the presence of the chick enzyme, no chick chromosomes could be seen in the descendents of either set of heterokaryons. This suggests that the chick chromosomes may have been fragmented by their premature condensation, in which case fragments carrying the chick HGPRT gene may have survived either independently or by integration into chromosomes of the partner nucleus. Some light upon the state of the surviving chick genetic material can be gained by examining the stability of retention of chick HGPRT when the selective pressure is removed. After growing the cells in HAT medium for 4 months (about 75 generations), Schwartz et al. transferred them to nonselective medium for 6 weeks (about 27 generations). When the cells then were plated on medium containing 8-azaguanine to identify clones that had lost HGPRT, their frequency of occurrence was about 20% (contrasted with a frequency of about 10^{-6} survivors for parent L cells). The chick HGPRT gene therefore appears to be maintained in the hybrid cells only in an unstable state. Since no chick specific antigens could be detected on the surface of the hybrid cells, and since the genes coding for these proteins are fairly widespread in the chick complement, it seems likely that the amount of chick genetic material retained in the mouse or Chinese hamster nucleus is comparatively small. This is consistent with the idea that it takes the form of a chromosome fragment which does not undergo proper segregation at mitosis. (Fragmentation may be a consequence of the speed with which the mouse or Chinese hamster nucleus enters mitosis. When Kao (1973) fused chick erythrocytes with CHO *ade*⁻ lines, in which division can be postponed for some time by growth on medium lacking adenine, chick chromosomes carrying genes able to com-complement the *ade*⁻ defect survived in the hybrid cells and were stably maintained.)

The possibility that in at least some cases the disappearance of HGPRT activity may be due not to loss of the chick gene but to interference with its expression has been raised by experiments in which Klinger and Shin (1974) first selected erythrocyte-A9 hybrids for HGPRT, then selected against HGPRT, and finally reselected for HGPRT. The cells able to survive in the second step on medium containing 8-azaguanine lacked HGPRT activity. But they were able to survive upon the retransfer to HAT medium, when HGPRT activity was restored. These cells lacked HGPRT activity when grown on medium containing hypoxanthine and thymidine, and it appears that the enzyme activity is induced reversibly by the presence of aminopterin. This response is unique to the hybrids, since it is not shown by chick embryonic cells, in which HGPRT expression is constitutive. The implication of these results is that the chick HGPRT gene may have been stably incorporated into the mouse cells, but is expressed only in the appropriate selective conditions.

Can other genes be transferred from erythrocyte nuclei to the partner nuclei of heterokaryons? The only experiment reported is the attempt of Boyd and Harris (1973) to detect transfer of thymidine kinase to a TK⁻ line of 3T3 mouse fibroblasts. Their results are similar to those obtained with transfer of HGPRT, but upon examination, the thymidine kinase of the hybrid cells proved to be that of the mouse parent. This might mean that mutation in the 3T3 cells lies in some locus other than the structural gene for thymidine kinase, so that the chick genome could provide the corresponding function. Or it may mean that for some unknown reason the fusion technique leads to the isolation of revertants of the mouse mutation (which may or may not have gained chick genetic material). This would be akin to the experiments on reversion of HGPRT described in Chapter 6.

Small numbers of chromosomes can be added to the complement of a cell by fusion with a minicell carrying only part of the genetic information of its parent. Two techniques have been developed for obtaining such minicells. One approach is to store mitotic cells in the cold (4°C) and then to return them to normal growth conditions (37°C). Some cells fail to cleave normally and instead produce clusters of bud-like protuberances; DNA enters these buds, which then detach to form *minisegregants* (Johnson, Mullinger, and Skaer, 1975; Schor, Johnson, and Mullinger, 1975). The minisegregants form a highly heterogeneous population; and cells of a given size, that is, containing a corresponding proportion of the genome, can be isolated by centrifugation through density gradients. Presumably the formation of the minisegregants is a consequence of the disruption of the mitotic spindle in the cold followed by an abnormal reformation when the temperature is restored to normal. A second method depends on the formation of *micronuclei* that can be induced by colchicine; these again constitute compartments containing only part of the parental genetic material (Ege and Ringertz, 1974; Ege, Krondahl, and Ringertz, 1974). The preparation of microcells by centrifuging the micronucleated cells in the presence of cytochalasin B has been described in Chapter 8. Using such microcells, Fournier and Ruddle (1977a) were able to transfer mouse chromosomes into human (HeLa) cells or Chinese hamster cells. From 1–6 mouse chromosomes could be introduced in this way and then proved to be stably maintained in the hybrids. Mouse enzymes, known to be coded by given mouse chromosomes, could be identified in some hybrid lines. Similarly, using minisegregants of HeLa cells, Tourian et al. (1978) were able to show that HGPRT may be synthesized following fusion with A9 mouse cells to generate hybrids carrying only a few of the human chromosomes (including the X).

Gene Transfer with Isolated Chromosomes

One of the difficulties in conducting genetic analysis of somatic cells is the lack of a system for transferring limited amounts of genetic information. Although cells may be fused to yield hybrids in which complementation may be tested and synteny established, the presence of two complete genomes, often in a mixed cytoplasm, necessitates somewhat cumbersome procedures for mapping. In homospecific crosses the hybrid cell may exhibit gene dosage effects. In heterospecific crosses, even where the full complement of one parent is retained in company with only a single chromosome of the other, the additional content of genetic information is considerable. Even the smallest human chromosome contains 4.5×10^7 base pairs of DNA and the largest possesses more than 2×10^8 base pairs. A chromosome may therefore carry several thousand genes. It would be useful to be able to transfer small amounts of genetic information, of defined content if possible. This would allow the effects of additions or substitutions of single genes to be followed. Two methods for such transfer might in principle be envisaged. There might be viruses analogous to the transducing phages of bacteria, able to transfer small amounts of genetic material, from either random sources or specific loci; but no such eucaryotic virus is known. Or it might be possible to transfer genetic material directly into recipient cells, a process analogous to bacterial transformation.*

The source of genetic material might be chromatin, chromosomes, or DNA. No experiments have been performed with interphase chromatin. Metaphase chromosomes are somewhat more amenable to experimental handling and have been used with some success. There have been reports also of the transfer of DNA: earlier experiments were less than entirely conclusive, but more recent work has been able to utilize purified fragments of DNA coding for defined products, and thus has yielded unambiguous results.

Metaphase chromosomes may be isolated by a procedure originally developed by Maio and Schildkraut (1969). Cells are arrested in metaphase with colchicine and then homogenized in Triton detergent. Centrifugation yields a pellet containing cell debris; chromosomes are found in the supernatant. With cells whose chromosomes vary appreciably in size, it is possible to achieve a crude fractionation of the chromosomes into size classes by density gradient centrifugation.

The procedure for chromosome transfer introduced by McBride and Ozer (1973) is to add a preparation of metaphase chromosomes to a cell suspension. Generally cells are suspended at a concentration of about 2×10^6

---

*The term "transformation" is not used to describe eucaryotic gene transfer because of its prior use to describe the conversion of eucaryotic cells into alternative growth states ("transformed cells") by viruses or chemicals. The terms "transference" and "transfection)" have been introduced to describe the treatment of eucaryotic cells with isolated chromosomes or purified DNA, respectively, to generate "transferents" or "transfectants."

cells/ml and metaphase chromosomes are added to a level of about 1 diploid set per recipient cell. Then the usual procedures to detect genetic variants are followed. Cells are grown for a short period in nonselective medium before the selective conditions are imposed, after which any surviving cells are characterized.

Any system for transferring genetic material directly into recipient cells seems likely to be somewhat inefficient; and this suggests the need to use selective techniques able to isolate a small number of "transferents." An obvious restriction in attempting such transfer is the need to demonstrate that any transferents have indeed gained a donor gene rather than reverted. For this reason, the first attempts to demonstrate chromosome transfer made use of the HAT selection system to identify resistant cells among a sensitive recipient population of cells of one species treated with donor chromosomes from another. This, of course, allows the source of the HGPRT of any putative transferents to be identified by the usual criteria.

The results of several such experiments with the HAT system are summarized in Table 9.2. With donor chromosomes of human or Chinese hamster origin, and with recipient cells of mouse or Chinese hamster, the gene

Table 9.2 Frequency of HGPRT gene transfer by addition of metaphase chromosomes

Recipient cell	Source of donor chromosomes	Characterization of transferred HGPRT activity	Frequency (clones/cells plated)	Stability
Mouse A9	Chinese hamster V79	electrophoretic and chromatographic mobility (for 4 clones)	$5/6 \times 10^6 =$ 8.3×10^{-7}	1 lost at 10–20%/ division; 2 may be stable
Mouse A9	human CCRF	electrophoretic mobility (for 3 clones)	$\geqslant 9/5.5 \times 10^7 =$ $\geqslant 1.6 \times 10^{-7}$	not tested
Mouse A9	HeLa	electrophoretic mobility and immunoprecipitation	$3/1.2 \times 10^7 =$ 2.5×10^{-7}	all lost at 3%/division
Ch. hamster Wg3h (Don subline)	HeLa	electrophoretic mobility	$7/1.2 \times 10^7 =$ 1.2×10^{-6}	not tested
		lack enzyme activity	$14/1.2 \times 10^7 =$ 8.9×10^{-7}	not tested

All experiments were performed by mixing HGPRT⁻ recipient cells with metaphase chromosomes isolated from HGPRT⁺ heterospecific donor cells and plating on HAT. In the first three series of experiments there was ⩾1 diploid donor chromosome set per recipient cell; in the last series the ratio was 0.25.

References: McBride and Ozer (1973), Burch and McBride (1975), Willecke and Ruddle (1975), Wullems et al. (1975, 1976a).

for HGPRT is transferred with a frequency in the range of 10^{-6}–10^{-7} per recipient cell. The problem of identifying the transferred genetic function is excerbated by its instability. The Table therefore summarizes only those experiments in which characterization was possible; others have been reported, but the resistance to HAT was lost before the HGPRT could be characterized. The need for this characterization is emphasized by the experiments of Wullems et al. (1975, 1976a), which showed that only 7 out of 21 HAT^r clones in fact possessed HGPRT. The nature of the change in the others is unknown—it might be due to genetic transfer of some other function or to mutation in the recipient culture.

What is the efficiency of genetic transfer? This must depend on the product of the probability of uptake of a given gene and the probability that it will survive in the recipient cell. Both are unknown. To determine the frequency of transfer per donor gene by a recipient cell requires the construction of a dose response curve. But in experiments in which the ratio of donor chromosomes to recipient cells has been varied over a range of 1–10 diploid sets per recipient, no change is seen in the frequency of transfer (Burch and McBride, 1975). This suggests that the amount of donor material present is in excess of the capacity for uptake of the recipients. Indeed, if each recipient cell were able to take up only one or two chromosomes, the excess of donor material would be more than tenfold in the usual conditions employed for transfer. Thus the proportion of cells gaining a particular chromosome would be quite small. To estimate a probability of transfer would require reducing the ratio of donor chromosomes to recipient cells to obtain a dose response; but this would be difficult to measure because the frequencies of transfer would become very low.

In what condition does the ingested chromosome survive? In all but one case, karyotype analysis shows the absence of donor chromosomes. This implies that the donor material survives as a small fragment, either independently maintained or in some way associated with the host genome. As in the case of cell fusions leading to chromosome fragmentation, it is difficult to see how the donor material could be assured of continued survival without either possessing or associating with a centromere. Indeed, in most experiments the transferred character is unstable, being lost under nonselective conditions at a rate often about 0.1–0.2 per division. Instability is almost certainly due to loss of the transferred gene. In the one case in which positive revertants have been isolated from the transferents that had lost HGPRT, all proved to have regained the recipient enzyme and not the donor function (Willecke and Ruddle, 1975).

In some cases, the transferred character is inherited stably when the cells are maintained under nonselective conditions. Occasionally this behavior is displayed by an original transferent; more often a stable derivative is found among the descendents of an unstable primary clone. This prompts the obvious speculation that usually the donor fragment survives in some unstable state, but occasionally may enter a stable condition, most likely by physical integration into one of the host chromosomes. There are no reliable estimates of the frequency with which transition to the stable state may occur.

In one exceptional case, in which a human-Chinese hamster hybrid containing 3 human autosomes was used as recipient, an entire human X chromosome could be identified in a transferent clone (Wullems et al., 1976b). Several enzymes coded by the human X chromosome were present in addition to HGPRT, in contrast with their absence from transferents gaining only a chromosome fragment. It is possible that the survival of donor material might be enhanced by the presence in the recipient of chromosomes of the same species.

The size of the fragment that survives in the recipient cells can be estimated by genetic means. Willecke and Ruddle (1975) examined their three HGPRT$^+$ transferent clones for the presence of other enzymes coded by the human X and autosomes. None could be detected, including G6PDH and PGK, which lie on the X chromosome on either side of HGPRT, a distance apart corresponding to about 1% of the genome. Assuming that these constitutive genes would be expressed if present, this limits the size of the fragment carrying the HGPRT gene to a genetic content of some 2.8×10^7 base pairs of DNA. The HGPRT$^+$ transferents of Burch and McBride (1975) also lacked G6PDH and PGK, so this appears to be the general case for transfer performed by these protocols. This confirms the expectation that the fragment carrying HGPRT lacks a centromere.

A pair of closely linked genes is provided by the loci on human chromosome E17 that code for cytosol thymidine kinase and galactokinase. The distance between them is probably less than 0.1% of the human haploid genome. Both of the human enzymes can be distinguished from the mouse enzymes by electrophoresis. After selection for transfer of the human thymidine kinase gene, human galactokinase was found in 2 out of 8 and in 2 out of 12 clones in two series of experiments (Willecke et al., 1976; McBride, Burch and Ruddle, 1978). This means that about 20% linkage (very crudely estimated) is seen between loci about 3×10^6 base pairs apart. In similar experiments, Wullems et al. (1977) obtained cotransfer of the two markers in all of five clones, which suggests that under their conditions the transferred fragment may usually be longer (the reason for this is unknown).

Changes in the transfer protocols have been reported that increase the frequency of transferents. The use of calcium phosphate in the preparation of donor material is helpful; and treating recipient cells with DMSO just after transfer increases the frequency of transference (Miller and Ruddle, 1978). Together the two treatments allowed A9 transferents for HeLa HGPRT to be obtained at a frequency of 4.0×10^{-5} per recipient cell—an increase of two orders of magnitude compared with the data summarized in Table 9.2. In about half of the clones examined, human genetic material could be detected, in two cases attached to a mouse chromosome, in five cases as a small independent fragment. Both lines of the first type contained human G6PDH and PGK; in three of the second type, either G6PDH or PGK but not both were present. This suggests that the DMSO treatment in some way enhances the ability of transferred material initially to survive in the recipient, perhaps by reducing its susceptibility to degradation and thus increasing fragment size.

Somewhat similar results have been achieved by a quite different method of protecting donor material. By incorporating metaphase chromosomes from a human-mouse hybrid line into liposomes, Mukherjee et al. (1978) obtained HGPRT[+] transferents of A9 cells at a frequency of 1.0×10^{-5} per recipient. Although no report of the overall cotransfer frequency has been made, in two transferent lines both G6PDH and PGK of human origin also could be found. Karyotypic analysis has not been detailed, but there are apparently no differences detectable between the transferents and the A9 cells. This brings out one point of concern, which is that the use of a hybrid donor cell admits the possibility of rearrangements between human and mouse material before transfer.

Although there is no direct evidence on the question of whether transferred fragments may become integrated into the host genome, retransfer experiments suggest that in stable transferents the donor gene is at least closely associated with the host chromosome. In such experiments, Athwal and McBride (1977) serially transferred a human HGPRT gene first to mouse cells and then, by repeating the transfer protocol with a stable transferent as donor, to Chinese hamster recipients. The frequency of transfer in the second step was in the range of $0.3-1.8 \times 10^{-7}$, which is similar to the frequency of transfer in the first step (perhaps slightly lower). Since the method relies upon the isolation of metaphase chromosomes, the gene for human HGPRT must have been integrated into, or very closely associated with, the mouse chomosome set. As usual, karyotype analysis demonstrated no change in the recipient cell complement. Several of the second-stage transferents were themselves stable. Similar experiments have not been performed to see whether retransfer can take place from unstable transferents.

One possible way to assess the status of the human HGPRT in the first-step transferents would be to use recipient cells whose chromosomes can be fractionated into different size classes; this would allow the identification of any group with which the human genetic fragment might be associated. (This experiment cannot be performed with the mouse complement, all of whose members are much the same size.) Another approach to this question has been devised by Fournier and Ruddle (1977b), who used microcells to transfer limited numbers of chromosomes from stable human-mouse transferents into Chinese hamster recipients. Then correlations can be sought between the segregation of human HGPRT and mouse chromosomes or mouse enzymes representing different murine chromosomes. In one "tribid", the Chinese hamster recipient gained only a single mouse chromosome, identified by its banding pattern. In a second tribid, two enzymes coded by mouse chromosome 14 could be correlated with the presence of human HGPRT. Thus the human HGPRT gene appears to have become stably associated with two different mouse chromosomes in two different first-step transferents. In another case, the human gene could be associated with the presence of a recombinant mouse chromosome carrying parts of the normal homologs 14 and 15. No transfer of human HGPRT could be obtained with microcells derived from unstable first-step human-mouse transferents.

The reliance upon heterospecific systems for HGPRT transfer leaves open the question of what features may be consequences of the species difference and which are inherent in the interaction between donor chromosomes and recipient cells. Whether larger fragments or even intact chromosomes would survive homospecific transfers, for example, cannot be tested easily because of the difficulties in identifying donor genetic material and gene products. In those experiments in which homospecific transfers have been accomplished, however, the frequencies and stabilities of the transferents appear to be similar to those that have been established in the heterospecific systems. Degnen et al. (1976) tested homospecific transfer of HGPRT by obtaining donor chromosomes from a mouse cell line whose enzyme has a lower affinity for 8-azaguanine and thus is more resistant to the drug. Recipient cells that were HGPRT⁻ acquired the donor phenotype at the usual frequency; most transferents were unstable and one clone was stable. Stable derivatives appeared among unstable transferents.

The extension of chromosome transfer to dominant and codominant markers in homospecific transfers with CHO cells has been reported by Spandidos and Siminovitch (1977a,b,c). The frequencies for transfer of oua^r, mtx^{RIII}, gat^+, or $glyB^+$ all appeared similar to the results previously obtained with HGPRT; the same usual instability, with occasional stability, was found. Attempts to correlate markers with chromosomes were made by fractionating the CHO set into three size groups; each marker appeared to be transferred by one of these size classes at a frequency close to that of the unfractionated preparation, and poorly or not at all by the other size classes. However, these results since have been called into question and await independent confirmation.

Transfection with Purified DNA

Cells can be "transfected" with purified DNA at frequencies somewhat lower than those displayed in metaphase chromosome transfer. This has the advantage that it may be possible to purify fragments of DNA carrying defined genetic information. This system was first developed by transfecting the thymidine kinase gene of herpes virus DNA into cells lacking the enzyme. The protocol is to treat about 5×10^5 cells with 1–2 μg of herpes DNA in the presence of sufficient "carrier" salmon sperm DNA to bring the total to 10 μg (this improves the efficiency of transfection). The donor

DNA may be either a restriction digest or random shear product of HSV DNA (purified intact DNA cannot be used because this would lead to the infective cycle). Since the frequency of transfection is proportional to the amount of herpes DNA added, it is possible to calculate the frequency relative to the number of copies of the thymidine kinase gene made available. In restriction digests, only one fragment of HSV DNA is able to transfect, which allows the position of the thymidine kinase gene to be located on the restriction map of the viral DNA.

Transfection is equally effective with randomly sheared or enzymically digested HSV DNA (reduced in size from the 10^8 daltons of the intact genome) or with Eco RI or Hind III restriction fragments of 20 and 24×10^6 daltons. The frequency varies somewhat, but in the early experiments most commonly corresponded to about 1–3 clones of TK^+ transfectants per dish. This corresponds to about 10^{-6} transfectants per cell per μg HSV DNA, which is about 2×10^{-16} transfectants per cell per thymidine kinase gene. (This is about 2% of the frequency of lytic infection with intact HSV DNA.) The smaller fragment released by Bam I, about 2×10^6 daltons or 3400 base pairs, transfected about 20-fold more efficiently. Improvements in the transfection protocols since have brought the efficiency of the Bam I fragment to about 1 transfectant per 40 μg of Bam fragment per 10 cells, which corresponds to about 1.2×10^{-13} transfectants per donor gene per recipient cell (Bacchetti and Graham, 1977; Maitland and McDougall, 1977; Wigler et al., 1977; Minson et al., 1977).

What is the state of the herpes thymidine kinase gene in the transfected cells? In at least some cases, the gene can be perpetuated for considerable periods when the cells are grown under selective conditions. Pellicer et al. (1978) showed that the kinetics of hybridization between purified HSV TK^+ DNA and the DNA of transfected L cells correspond to very few copies of the gene, apparently not more than a single sequence in the haploid genome in several independent clones. Restriction enzyme cleavage of DNA from these lines shows that in each case the sequence originally provided in the form of the 3.4 Bam I viral DNA fragment can be recovered as a longer piece of DNA. In each case the size of the cellular sequence is different, showing that the viral fragment has been integrated into a different sequence of cellular DNA. This implies that the viral fragment may be integrated at any one of several (at least six) different sites; and there is probably a single copy per cell, stably integrated as seen in its perpetuation under selective conditions.

The integrated viral gene can be used for further transfection. Wigler et al. (1978) found that high molecular weight DNA (>40 kb) from several transfected cell lines can confer the TK^+ genotype when added to TK^- L cells. In the most efficient experiment, the addition of 20 μg cellular DNA to 10^6 recipient cells generated an average of 9.5 colonies. If there is only a single herpes thymidine kinase gene in the donor cell genome, this corresponds to a frequency of about 3×10^{-12} transfectants per donor gene per recipient cell. This is some 25-fold more efficient than the frequency achieved with the purified Bam fragment. The DNA of other first step transfectants was effective at levels from 3–20 times that of the viral Bam fragment. One possible reason for the increase in efficiency might be that the presence of cellular sequences around the viral gene facilitates integration in the second step transfectants. In one second-step transfectant, the restriction pattern of the gene suggests that it is indeed integrated in the same surrounding sequences as in the first step transfectant.

Can endogenous cellular thymidine kinase genes also be used for transfection? With DNA extracted from a variety of cells, including L (TK$^+$) cells, mouse liver, CHO cells, HeLa cells, calf thymus and chicken red blood cells, TK$^-$ L cells could be transfected with frequencies ranging from 1–8 transfectant colonies/20 μg donor DNA/10^6 recipient cells. This corresponds to 1.4×10^{-13}–1.1×10^{-12} transfectants per donor gene per recipient genome. This is only slightly below the frequency obtained with the integrated viral thymidine kinase gene. A similar frequency of transfer has been obtained in comparable experiments with the APRT gene (Wigler et al., 1979b). The frequency of successful DNA transfer is more than an order of magnitude lower than the most efficient experiments reported with metaphase chromosomes, in which a frequency of 4×10^{-5} per recipient cell treated with a diploid complement corresponds to 2×10^{-11} per donor gene per recipient cell. However, this may be an underestimate, since the usual conditions for metaphase chromosome transfer appear to represent saturation of the system (see above). It is difficult in any case to compare such frequencies, because improvements in the experimental protocols may cause large changes in the frequency of transfer. It is possible, however, that metaphase chromosomes may be better able to survive in the recipient cell; and it would be interesting to compare directly the frequency of transfer obtained with DNA and chromatin for the same gene.

The transfer of DNA appears to be independent of sequence and success may depend on the state of the recipient cells. In experiments with the herpes TK/mouse L cell system, Wigler et al. (1979a) have found that cotransformation with a mixture of viral thymidine kinase gene and other DNA such as phage ϕX174 generates TK$^+$ cells that possess large amounts of the second DNA, possibly stably integrated. While 8 out of 9 selected TK$^+$ colonies that had been cotransformed possessed ϕX174 DNA, none of 15 clones exposed to ϕX174 DNA alone had any detectable phage DNA. This suggests that eucaryotic cells may pass through a transient stage during which they are susceptible to transfection and will take up any DNA to which they are exposed; that this stage is transient and not a permanent inheritable trait is indicated by the failure of the transfected cells to be transfected in a second cycle (for another gene) at a frequency any greater than in the first cycle. It is not known whether the state of susceptibility is similar to the condition of competence through which bacteria must pass to become transformed by DNA (see Volume 3 of Gene Expression). The phage DNA appears to be integrated into cellular DNA with no specificity for site of insertion.

PART 2

Organization of the Genetic Apparatus

CHAPTER 10

Introduction: Chromosomes and Chromatin

The state of the genetic material in the eucaryotic nucleus is a central question in the analysis of gene expression. The DNA of the nucleoprotein complex carries the genetic information; but the protein components determine structure. The extent to which structure is related to, or controls, function has been an issue much debated over the past few years, and constitutes a theme that will run through this part. We have already discussed the formal analysis of the chromosome in Chapters 4 and 5 by treating mitosis and meiosis in terms of the behavior of individual chromosomes. In the chapters of this section, we shall be concerned with the structure of the genetic apparatus, and in particular with relating its visible ultrastructure to the molecular interactions of DNA and protein.

As is implied by use of the terms *chromosome* and *chromatin*, cytological observations resolve only two forms of organization of the genetic material during the cell cycle. During mitosis individual chromosomes can be seen in a condensed state. During interphase a network of chromatin can be visualized in a more diffuse condition. Although changes must occur in the structure of chromatin during interphase—most obviously during replication—this is not visible. But the cycle of condensation and dispersion now is thought of as a continuous process, with mitosis as one extreme and S phase as the other, rather than as a discontinuous transition occurring at division.

In the light microscope, individual chromosomes are distinguished chiefly by their position of the centromere and size; sometimes secondary constrictions may occur at characteristic sites. Where members of a set are similar by these criteria, it has been impossible to distinguish them by conventional cytology. More recently it has become apparent that a series of bands can be generated in each chromosome by any one of a number of treatments. These bands are useful for identifying particular chromosomes in the complement; and their existence demonstrates that a consistent ultrastructure must be generated when the chromosomes condense at mitosis (see Chapter 15). Accounting for the ultrastructure in terms of molecular structure is a difficult problem, requiring resolution of the intervening levels of organization. The precision with which interphase chromatin is organized is not known, although there appear to be no visible distinctions between the chromosomes in this state.

Differences in the degree of condensation distinguish two types of chromosome material. Heterochromatin remains highly condensed throughout the entire cell cycle. *Constitutive heterochromatin* is maintained permanently in this state and often is present in the centromeric region (see Chapters 14 and 15). *Facultative heterochromatin* enters this condition during or after certain stages of development; the best example is the inactivation of one of the two X chromosomes of female mammals, which occurs in early embryogenesis (see Chapter 15). Both types of heterochromatin appear to be genetically inactive, because they lack genes (constitutive) or because their genes are repressed (facultative). The relationship between condensed structure and genetic inactivity is of obvious interest. The leaves the *euchromatin* that undergoes cyclic condensation and dispersion as the source of the genes that are expressed in each tissue. Only a subset of the genetic complement is expressed in any given cell; this raises the question of whether there are differences in the structures of active and inactive genes within the euchromatin of each cell type, and if so how extensive the structural variations may be.

Nucleoprotein is generated by an association of DNA with roughly its own mass of histone (very basic) proteins and also with up to an equivalent amount of nonhistone (less basic) proteins. The amount of histone protein associated with DNA appears to be invariant, the cellular content increasing as DNA is replicated, so that the ratio remains the same and newly synthesized DNA possesses the same amount of histone as the parental DNA. The extent to which nonhistone protein content may change in the cell cycle, especially at mitosis, is not entirely clear, because of difficulties in excluding fortuitous association of nonchromosomal proteins with the chromosomes during preparation. The cause of mitotic condensation is not known. It is likely that additions or substitutions of nonhistone proteins are involved; it is possible also that modifications to one of the histones play a role.

There are only five types of histone, at least some of which have amino acid sequences that have been highly conserved during evolution (see Chapter 11). In view of the wide variations in the DNA sequences constituting different genomes, this suggests that histones are concerned with some fundamental aspect of structure, probably one that is independent of nucleic acid sequence. This conclusion is confirmed by the recent identification of the basic subunit of chromatin, the *nucleosome*, formed by the interaction of about 200 base pairs of DNA with a histone octamer (see Chapter 13). The octamer consists of two copies of each of four histones; a single molecule of the fifth histone is associated with each pair of nucleosomes. The nucleosome seems to be the subunit of mitotic chromosomes as well as of interphase chromatin; and both active and inactive genes are packaged in this form, although possibly with some differences.

The structure of the genetic apparatus generally has been approached from two perspectives, starting with either its DNA or protein component. The first issue concerns the form taken by the DNA of the individual chromosome. Does each chromosome contain a single duplex of DNA or does it have many individual molecules of DNA connected by (presumably) protein? This has been a principal question in cytology for some time. The second issue concerns the manner in which DNA interacts with the chromosomal proteins. The identification of the nucleosome subunit has led to a focus of attention on two questions. What is the structure of the nucleosome particle? How do nucleosomes interact with each other, in particular how are they coiled up into the higher order structures of the chromosome? This second question provides the current approach to defining the ultrastructure of the chromosome and the relationship between condensed and diffuse states (see Chapter 14).

The presence in chromatin of nonhistone proteins, which are not part of the nucleosome subunit, raises the immediate question of their function: to what extent is it structural, regulatory, or both? Are nonhistone proteins responsible for the higher order levels of chromosome structure? Are specific regulator proteins, analogous to bacterial and phage repressors (and inducers), which interact directly with DNA or with the transcription apparatus, to be found among the nonhistone proteins? Does gene expression involve changes in the structure of chromatin, to make sequences of DNA accessible (so that they can be transcribed) instead of inaccessible (when they cannot be expressed); or are all sequences in principle available to RNA polymerase, with any changes in structure depending on the act of transcription? To most of these questions there are few definite answers, but we shall come to the structure of nonhistone proteins in Chapter 12 and will take up the issue of the differences between active and inactive genes in Chapters 14 and 28.

Continuity of DNA in the Chromosome

The amount of DNA in each eucaryotic chromosome is remarkable. We have already mentioned that the DNA of a human (or any typical mammalian) nucleus would stretch for some 1.74 meters if extended as a naked duplex (see Chapter 1). In the interphase cell, a nucleus with a diameter of the order of 5 μm must contain this extraordinary length of DNA. Since individual chromosomes cannot be recognized at this stage, the length of DNA cannot be compared directly with the length of the chromosomal material. But organized into the 46 chromosomes of the diploid set, whose lengths at metaphase vary roughly from 2–10 μm, there must be some 14,000 μm of DNA in the smallest and about 73,000 μm of DNA in the largest chromosome. In its most highly condensed state, therefore, the genetic material shows a contraction in the length of DNA approaching 10,000-fold.

The variations that occur in the length of recognizable chromosomes may be considerable. In the form of the meiotic synaptonemal complex, for example, the ratio of DNA length to complex length may be of the order of up to 1000-fold (see Table 5.1). Roughly a tenfold contraction in length must therefore occur by metaphase. The highly extended lampbrush chromosomes found at diplotene during oogenesis of amphibia such as Triturus cristatus may be between 500 μm and 800 μm in length; but this contracts to some 15-20 μm later in meiosis (see Chapter 16). Another situation in which chromosomes are unusually extended is provided by the salivary gland chromosomes of Drosophila, in which DNA of length about 45,000 μm may be organized into a total length of polytene chromosomes of about 2000 μm (see Chapter 16). Any model for chromosome structure must therefore come to terms not only with the intensive packaging of DNA in mitotic chromosomes, but also with the changes that occur in the cell cycle and the existence of comparatively extended states of organization.

A fundamental question about the constitution of the chromosome is whether its genetic material takes the form of an uninterrupted duplex molecule of DNA.

Early models for chromosome structure often were based upon the supposition that the presence of very long continuous stretches of DNA would pose insuperable problems for packaging, especially for mitotic condensation and DNA replication. This led to ideas such as those of Freese (1958) that each chromosome might consist of many duplex molecules of DNA, linked end to end by a series of protein molecules. The breaks made in the covalent integrity of each duplex would allow unwinding to occur for replication; and interactions between the proteins might be responsible for mitotic condensation. An earlier model proposed by Mirsky and Ris (1951) suggested that each chromosome might consist of a longitudinal backbone of protein to which individual DNA molecules are attached laterally. More complex extensions of this model to allow for later data were proposed by Taylor (1957, 1963).

These speculations have been superseded by the view that each chromosome consists of a single fiber, constituting a deoxynucleoprotein complex, which is packaged into the characteristic ultrastructure. When chromosomes are first distinguished at prophase in the light microscope, each appears to consist of two chromatids which shorten during mitosis. Before this condensation obscures their features, each chromatid appears to comprise a single coiled thread, the *chromonema*. The question of chromosome structure then becomes the question of the constitution of this thread and the manner of its subsequent condensation.

With the greater resolution of the electron microscope, the chromonema can be seen to consist of twisted fibers in the early stages of chromosome condensation, most readily in the extended meiotic prophase studied by Ris (1956) and Ris and Chandler (1963). One early model viewed the fibers as 50-60 nm thick, consisting of two finer fibers of about 20 nm in diameter. Later observations suggested that the principal constituent of the chromosome is a fiber of about 30 nm diameter. Fibers of this order of size, with irregular diameters varying from 20 to 30 nm, have more recently been observed in a variety of preparations of both mitotic chromosomes and interphase chromatin by the technique of spreading chromosomes on a water surface before microscopy (see below).

The literature of cytogenetics contains many discussions of whether chromosomes are *unineme* (single stranded) or *polyneme* (multistranded) (for review see Prescott, 1970). These terms have two connotations.

From the perspective of structure, a unineme chromosome must possess a single duplex of DNA that is coiled up in some manner into the highly condensed condition. The high degree of packing of DNA into the chromosome implies that the basic (30 nm) fiber would consist of a DNA duplex already tightly coiled; since this fiber must itself be coiled into the ultrastructure, unineme models imply that there must be more than one level of structural organization. A polyneme chromosome, however, would consist of a fiber containing more than one duplex molecule of DNA associated in parallel. This might take the form of a single deoxynucleoprotein complex containing two or more duplexes of DNA; or it might appear as nucleoprotein subfibers, each containing a single duplex, but together associated into the basic 30 nm chromosome fiber.

From the viewpoint of genetics, a unineme chromosome corresponds to a single linkage group, in which the map distance between genes is (traditionally) proportional to their distance apart on the DNA thread (up to the limits imposed by frequency of genetic exchange). The genetic organization of a polyneme chromosome would be more complicated. If all the parallel strands were identical, there would be many copies of each gene; this repetition would have to be reconciled with the implication of mammalian genetics that there is only one copy of each gene (see Chapter 17). If the strands were different in sequence, each might constitute a conventional linkage group, but the relationship between them is not obvious to visualize. (The first situation is achieved in certain cells that become *polytene* when their chromosomes replicate many times but remain longitudinally associated instead of segregating. This multiplication to give associated chromosomes is distinct from polynemy, in which each chromosome of the haploid set would contain many strands of DNA).

Most of the arguments for polynemy were based upon observations suggesting that the chromosome consists of more than one subunit. A typical multistranded model for the chromosome was that proposed by Ris and Chandler (1963), which suggested that each 25 nm fiber consists of two fibers of diameter 10 nm wound around each other; each 10 nm fiber in turn was supposed to consist of two fibers of 4 nm diameter, each containing a duplex of DNA surrounded by protein. The early evidence taken to support such models has been reviewed by Ris (1969). However, DuPraw (1970), following the results of Barnicott (1967), has pointed out that these apparent substructures may be produced as artefacts of aggregation and separation during preparation of the chromosomes. Phase contrast and interference microscopy of living cells do not reveal parallel substructures. It is true, however, that fibers of different diameters can be resolved in interphase chromatin and mitotic chromosomes; but these appear to represent successive stages of the coiling of the basic string of nucleosomes into higher order structures (see Chapter 14). Support for the idea that each 20-30 nm fiber contains only a single duplex molecule of DNA is provided also by the observation of DuPraw (1965) that treatment with trypsin reduces the fibers to a single filament with the dimensions of the DNA duplex; treatment with DNAase degrades these filaments.

Chromosomes can be visualized as intact structures by the technique of whole mount electron microscopy, in which cells are lysed in water to float the chromosomes on the aqueous surface; they are then picked up on a grid and dried from liquid CO_2. Using this technique, DuPraw (1965, 1966) showed that each chromosome appears to consist of fibers with a diameter of 20-30 nm, rather tightly and irregularly folded into the characteristic structure of the chromosome. An example of a metaphase chromosome prepared by this means is shown in Figure 4.1. We shall come in Chapter 14 to the quantitative analysis of the constitution of the chromosome in terms of these fibers and to the question of their construction from the basic nucleosome subunits. Although it is true that changes in the appearance and dimensions of the fibers may occur during the whole mount preparation technique, the important conclusion suggested by this analysis is qualitative: the chromosome consists of a single fiber.

The absence of any other features has been taken to imply, although it does not prove, that the fiber is continuous. This in turn implies that there should be no break in the continuity of the single DNA duplex contained in the chromosome. This is supported by observations that in lampbrush chromosomes, the integrity of the axis is sensitive only to DNAase and not to RNAase or proteolytic enzymes; thus there can be no linkers of material other than DNA (see Chapter 16). Proof of continuity could in principle be provided by the isolation of DNA from a given chromosome. Unfortunately, DNA is very sensitive to breakage during preparation. Autoradiographic studies of replication have been able to identify continuous lengths of DNA of no more than 500 or 1800 μm in mammalian tissue culture cells (Cairns, 1966; Huberman and Riggs, 1968). Although some improvement over this may be expected in future studies, it is questionable whether it will be possible to identify and correlate with specific chromosomes the expected lengths of more than 10,000 μm.

Attempts to correlate the sizes of DNA molecules with chromosomes have been made with the yeast S. cerevisiae, whose genome is about 0.017-0.024 pg and appears to be divided into (probably) 17 chromosomes (see Tables 5.1 and 5.2). The average yeast chromosome then should contain about $6-8 \times 10^8$ daltons of DNA and the largest should possess about $1.3-1.8 \times 10^9$ daltons (these are only rather approximate values). This is small enough to make biophysical measurements possible. Using sedimentation analysis, Petes and Fangman (1973) obtained a range of values for extracted DNA of 5×10^7 daltons to 1.4×10^9 daltons, with a number average of 6×10^8. By electron microscopy, Petes et al. (1973) were able to identify linear molecules from 50 μm (1.2×10^8 daltons) to 355 μm (8.4×10^8 daltons), with an average of 5×10^8 daltons. This is consistent with the idea that each chromosome contains a single uninterrupted duplex of DNA.

Another technique that has been employed to measure the size of large DNA molecules depends on viscoelasticity. This gives a measurement for the largest piece of yeast DNA of 1.5–2.0×10^9 daltons (Lauer and Klotz, 1975; Lauer, Roberts, and Klotz, 1977). This again is reasonably close to the expected size of the largest molecule. It would be an over interpretation of these data to regard the discrepancy, as has been suggested, as implying that some DNA molecules may be larger than a single chromosome. Viscoelastic measurements on Drosophila DNA also have produced results not inconsistent with the idea that each chromosome contains a single duplex (Kavenoff and Zimm, 1973).

To summarize these studies, the chromosome appears to be constituted from a 30 nm fiber which contains only a single (supercoiled) duplex of DNA. This excludes simple models for polynemy, but does not prove that there are no interruptions in the covalent integrity of the duplex within a chromosome. Although structural studies are consistent with the idea that there is one DNA duplex per chromosome (the data on lampbrush chromosomes are the strongest), they fall short of a general proof of the proposition.

Semiconservative Replication of the Chromosome

One of the strongest indications that the genetic material of the eucaryotic chromosome consists of a single linear duplex of DNA is its semiconservative replication. Eucaryotic DNA is not amenable to the biochemical analysis which Meselson and Stahl (1958) used to show that bacterial DNA is replicated semiconservatively. This requires the isolation of the entire genome after each round of replication so that it can be characterized by centrifugation. However, tritiated thymidine can be incorporated into chromosomes during their replication and the segregation of the labeled daughter chromosomes then can be followed through the subsequent call divisions.

The diploid chromosome complement of the plant Vicia faba comprises twelve large chromosomes which are particularly suitable for cytological analysis. Taylor, Woods, and Hughes (1957) used Vicia faba seedlings in a protocol in which the seeds were incubated in ^{3}H thymidine for 8 hours and then transferred to unlabeled medium containing colchicine. As the treated cells accumulate in mitosis, chromosomes are found in the metaphase condition, but with sister chromatids spread apart instead of lying parallel to each other.

Because the G_2 period of these cells is about 8 hours, few if any of the nuclei that have been labeled have had time to pass through a division before treatment with colchicine. This means that virtually all the labeled nuclei are prevented from segregating their chromosomes. After transfer to the colchicine medium, cells may be allowed to grow for varying periods before they are fixed for autoradiography. The type of metaphase plate found when the cells are removed depends on the number of times the chromosome set has replicated.

When the cells have remained in colchicine for only 10 hours before fixation, all the metaphase plates contain 12 chromatid pairs. The cells have had time to replicate their chromosomes only once, in the presence of thymidine label and before their transfer to colchicine medium. Each of the chromatid pairs consists of the two sister chromatids produced by the replication, which are spread apart but remain attached at the centromere. All the chromosomes in these cells are labeled and the two sister chromatids of each pair are equally and uniformly labeled as illustrated in Figure 10.1.

When the cells are incubated for 34 hours, metaphase plates can be found containing 12, 24, or 48 pairs of sister chromatids. Those with 12 pairs usually are unlabeled. These therefore must represent cells which have replicated their DNA only once, after the end of the period of labeling with ^{3}H-

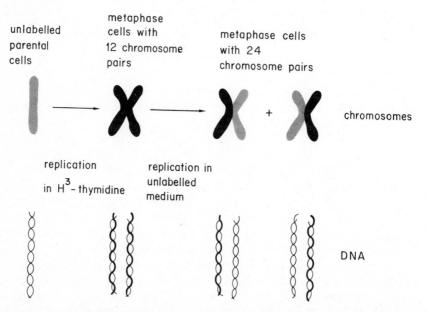

Figure 10.1. Semiconservative replication of eucaryotic chromosomes. The upper part depicts the incorporation of grains observed by autoradiography; the lower part interprets this in terms of the structure of the DNA contained in the chromosomes. At the first metaphase after replication in labeled medium, all chromosomes show an equal label (dark regions). All their DNA molecules must be hybrid, consisting of one parental (unlabeled) and one newly synthesized (labeled) strand. At the second metaphase, the labeled DNA strands segregate semiconservatively in the unlabeled medium, so that each chromosome pair has one chromatid containing hybrid DNA and one of unlabeled DNA.

thymidine. In the cells with 24 chromatid pairs, all the pairs are labeled. But only one of the two sister chromatids of each pair carries the radioactive label. These chromosomes have duplicated once in the radioactive thymidine and once later in its absence. As Figure 10.1 shows, the second replication segregates the radioactive label incorporated in the first replication as would be predicted from semiconservative synthesis of DNA. One strand of each daughter chromosome must have been labeled during the first replication; these strands separate during the second replication, in unlabeled medium, so that one of the two daughters receives a labeled strand and the other receives the unlabeled strand.

Only a few cells were found with 48 chromatid pairs, but in these cells one half of the pairs usually are labeled and the other half unlabeled. (The appearance of these cells shows some shortening in the usual 24-hour cycle of Vicia faba seedlings.) These results are consistent with a further semi conservative segregation of the labeled material.

Similar experiments since have been performed with many other cell types. The same results have been obtained in all cases. For example, Prescott and Bender (1963a,b) found that an ^{3}H-thymidine label given to human leucocytes or Chinese hamster tissue culture cells follows the predicted semi conservative segregation for the next four cell divisions. Incorporation of the label is permanent and none is lost from the chromosomes during this time. Figure 10.2 shows a second metaphase division observed by Marin and Prescott (1964), which corresponds to the stage at the far right of Figure 10.1.

When RNA is labeled during interphase, the label is lost at prophase when all the nuclear RNA is released into the cytoplasm (see Chapter 22). Nor is the segregation of labeled proteins semiconservative. Although labeled proteins are equally distributed to the two sister chromatids found at the first metaphase after addition of label, the radioactive protein is lost from the chromosomes in subsequent generations. More than one class of proteins is found in the chromosomes, of course, and these results show that that the majority of them do not behave in the same way as DNA; this does not exclude the possibility that a minor proportion does so. (The association of histones with DNA at replication is discussed in Chapter 14.) These results may be taken to show that the association of RNA with the chromosome is transient and that the protein components (or most of them) turn over, reinforcing the conclusion that the sole genetic component of the chromosome is DNA.

The simplest explanation for the semiconservative inheritance of labeled DNA is that each chromosome contains only one duplex of DNA. Models which suppose that each chromosome contains many independent molecules of DNA demand the postulation of special mechanisms to ensure that the segregation of the several replicated strands follows the observed semiconservative replication. In general, this requires the presence of some component other than DNA which also segregates semiconservatively. Although these results do not formally exclude the possibility that there is a

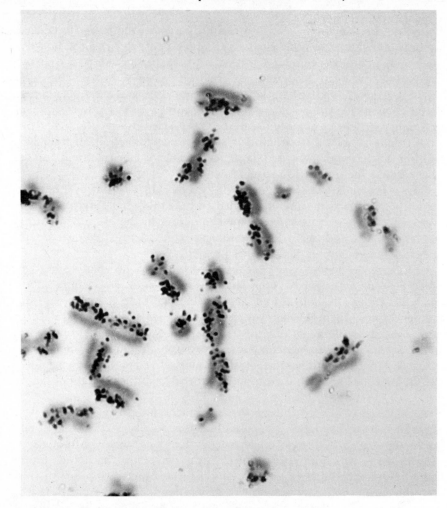

Figure 10.2. Second metaphase division after labeling DNA. All chromatid pairs possess one labeled chromatid. Sister chromatid exchanges are responsible for switches of label between the two chromatids. Data of Marin and Prescott (1964).

minor proportion of protein with such abilities, there is no evidence for its existence. These experiments therefore suggest that the sole unit of inheritance is an uninterrupted duplex of DNA present in each chromosome.

Sister Chromatid Exchange

Structural anomalies provide another approach for investigating the constitution of the chromosome. The best characterized of these is sister chromatid exchange, which was first visualized as an exchange of label between sister chromatids only one of which should carry radioactive DNA. In the experiments of Taylor, Woods, and Hughes (1957), some of the cells with 24 or 48 chromatid pairs possessed chromatids that were labeled along part of their length. But in these instances, the sister chromatid was labeled in the remaining portion. An example of such a switch can be seen in Figure 10.2. Taylor (1958) suggested that this results from a recombination event between the two sister chromatids occurring at some time after the replication that created them.

The resolution afforded by autoradiography is not great enough to allow studies of the nature of the switch with much precision. However, a more powerful technique for examining this and other anomalies has been made possible by the observation that incorporation of BUdR in place of thymidine quenches the reaction of chromosomes with the dye Hoechst 33258. This can be seen as a decrease in fluorescence; under the usual conditions it can be visualized also as a decrease in staining with Giemsa. (The use of such reagents to obtain banding patterns in metaphase chromosomes is discussed in Chapter 15). Figure 10.3 illustrates the application of this technique. Following a single round of replication in BUdR, the thymidine in one strand of each DNA duplex is replaced. This *unifilar* substitution results in a uniform reduction in the staining or fluorescence of the chromosomes as seen at metaphase. After a second round of replication in BUdR, each sister chromatid pair consists of one chromatid whose DNA is unifilarly substituted and one chromatid in which both strands possess BUdR in place of thymidine. This *bifilar* substitution results in a further decrease in staining or fluorescence. The result is that unifilarly and bifilarly substituted chromatids can be distinguished clearly (Latt, 1973, 1974; Wolff and Perry, 1974; Korenberg and Freedlender, 1974). This technique can be used to follow semiconservative replication in the same way as autoradiography and the same anomalous labeling patterns are found.

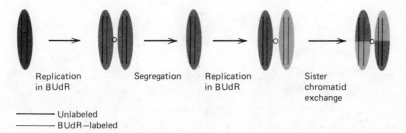

Replication in BUdR   Segregation   Replication in BUdR   Sister chromatid exchange

——— Unlabeled
——— BUdR–labeled

Figure 10.3. Chromosome segregation following incorporation of BUdR (bromodeoxyuridine). After one round of replication, each sister chromatid has one labeled and one unlabeled strand. These segregate at mitosis. Then a second round generates a sister chromatid pair, one member of which has one strand labeled, one of which has both labeled. When treated with the dye Hoechst 33258 and stained with Giemsa, the unifilar chromatid is much darker than the bifilar chromatid. A sister chromatid exchange is seen as an exchange of label between the two chromatids.

By this technique, sister chromatid exchange can be seen to represent an exact switch in label from one chromatid to its sister. An example is shown in Figure 10.4. The idea that this exchange arises by a breakage and reunion between corresponding sites of a duplex DNA in each chromatid makes some predictions about the structures that should be found. The sister chromatid exchange might take place either following the first round of replication in BUdR or following the second round (as illustrated in Figure 10.3). If cells are treated with colchicine, chromosome segregation after the first round is prevented, so that both pairs of sister chromatids resulting from an initial single chromosome are retained in the same cell. When an exchange occurs after the second round of replication, it will affect only one of the two sister chromatid pairs. When it occurs after the first round, however, it will be perpetuated by replication, so that it is seen in both pairs. Since there are twice as many chromosomes in the second round, there

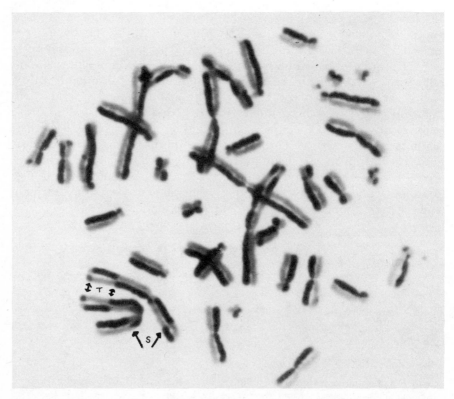

Figure 10.4. Sister chromatid exchange visualized by BUdR substitution. Tetraploid cells were produced by two rounds of replication in the presence of BudR and then were stained by a combination of fluorescence and Giemsa. The single arrows indicate sites of sister chromatid exchange (S); the double arrows indicate sites of twin exchange (T). Data of Wolff and Perry (1975).

should be twice as many "single" exchanges as "twin" exchanges. (This assumes that the frequency of exchange is the same in both division cycles, which cannot be accepted a priori because it is possible that bifilarly substituted DNA might be more active in exchange than unifilarly substituted DNA. However, Kato (1977) reported that no difference in the frequency of sister chromatid exchange is seen in the two cases.) Attempts to measure the ratio with autoradiography are difficult because of its low resolution (Taylor, 1958). But with the BUdR technique, a ratio of very close to 2:1 has been measured (Wolff and Perry, 1975). This suggests that a single (unineme) duplex of DNA constitutes the only unit that can be exchanged between chromatids. (Any exchange of single strands would generate a different pattern, including the production of totally substituted and totally unsubstituted DNA after the first round of replication, which would not otherwise be observed.)*

This technique has been used to observe sister chromatid exchange in a wide variety of cell types. In some instances, the locations of the exchanges have been reported to fall onto the Poisson distribution that would be expected if their locations were random (Wolff and Perry, 1974; Schvartzman and Cortes, 1977). In others they appear to occur preferentially in the centromeric heterochromatin (Bostock and Christie, 1976; Lin and Alfi 1976; Pera and Mattias, 1976). Yet in others the frequency is lower in constitutive heterochromatin than in euchromatin (Hsu and Pathak, 1976). No general conclusion about location can therefore be drawn.

The cause of sister chromatid exchanges has been a subject of debate for some time. With autoradiography, it has been suggested that the beta decay of tritium may cause the exchanges. With the use of BUdR, it is clear that the analogue induces exchange; the critical question then becomes whether it is responsible for all of the observed exchanges or only for increasing the spontaneous level. The frequency of sister chromatid exchange depends upon both the concentration of BUdR and exposure to light (Wolff and Perry, 1974; Latt, 1974). The dose-response curve is not linear, however, and so cannot be extrapolated to give a reliable spontaneous value for zero dose. The lowest frequencies observed were 0.15 exchanges per chromosome for CHO cells and 0.1 exchanges per human chromosome. These are upper limits for the spontaneous frequency of exchange per cycle. With D. melanogaster, however, no exchanges can be detected at the lowest BUdR

---

*After only one round of replication in BUdR, ~0.5% DNA can be detected at a buoyant density that corresponds to substitution in both strands. It has been suggested that this results from a single strand process which is involved in sister chromatid exchange (Moore and Holliday, 1976). However, in other experiments the characteristics of this DNA did not fit with the expectation that it is an intermediate in this process (Loveday and Latt, 1978). Very much larger amounts of doubly substituted DNA have been produced in other protocols, but these appear to be artefacts, probably resulting from branch migration at replicating forks, since treatment with psoralen in vivo to cross link the DNA reduces the level to about 0.3% (Tatsumi and Strauss, 1978). The origin of the low level doubly substituted material remains to be defined.

levels, which implies that their occurrence is not spontaneous, or at least is exceedingly rare, in this organism (Gatti et al., 1979).

Sister chromatid exchange can be induced by mitomycin C, a known alkylating and cross linking agent, and by caffeine, which has similar effects (Latt, 1974; Kato, 1977). This suggests a possible relationship between the exchange process and enzymes that repair damage to DNA. Caffeine causes a decrease in the frequency of exchanges involving unifilarly substituted DNA, but increases the frequency with bifilarly substituted DNA. This suggests that there may be more than a single mechanism for the induction of exchange, one of which is enhanced by caffeine. One possibility is that one involves replication and the other occurs at a different time.

The frequency of sister chromatid exchange is increased in cells derived from patients with Bloom's syndrome. This is caused by an autosomal recessive gene, which in addition to clinical effects causes instability in chromosome structure. This was first observed by increased rearrangements in the chromosomes of lymphocytes dividing in vitro (German, 1969; German, Crippa, and Bloom, 1974). Using the BUdR technique, Chaganti, Schonberg, and German (1974) showed that there is the same frequency of exchange in normal lymphocytes and in those derived from people heterozygous for Bloom's syndrome. Compared with this frequency of 7 exchanges per metaphase, a great increase to an average of 89 exchanges occurred in lymphocytes from patients with Bloom's syndrome. Exchanges also occur between nonsister chromatids of a homologue; these appear to account for the formation of the quadriradial structures originally recognized as characteristic of this disease. This suggests that the gene that causes Bloom's syndrome is concerned with recombination, since exchanges appear always to involve homologous sites. The ability of this mutation to increase the frequency of sister and nonsister chromatid exchange leaves the implication that such events may occur spontaneously at a low level as the result of the malfunction of some recombination or repair system. Cells homozygous for the mutation clearly may be useful in analyzing the system.

Another anomaly that at one time was thought to imply the presence of more than one subunit (DNA duplex) in the chromosome is *isolabeling*. Autoradiographic analysis of chromosome segregation identifies some cases where both sister chromatids are labeled for a short region, although only one of the pair is otherwise labeled. This cannot occur if the chromosome contains a single duplex of DNA, folded from one end to the other. To avoid the implication that there is more than one duplex of DNA, DuPraw (1968) suggested that this could occur if the DNA were folded longitudinally into the chromosome, in effect so that the duplex runs from one end to the other and then back again. Then an exchange could produce apparent labeling of both chromatids at the same point. However, isolabeling usually is not observed with the BUdR technique, by which it can be seen that the apparent presence of label on both chromatids is due to the occurrence of multiple exchanges over a short period (Wolff and Perry, 1974).

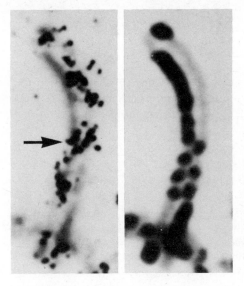

Figure 10.5. Production of apparent iso-labeling by multiple exchanges. The same chromosome is visualized in different ways on the left and right after two rounds of replication in BUdR; the first round was in ³H-labeled BUdR and the second in un-labeled BUdR. Left: autoradiography produces isolabeling at the site indicated by the arrow. Right: visualization of fluorescence and Giemsa staining shows that this site represents multiple exchanges. Data of Wolff and Perry (1974).

Figure 10.5 shows an example of a multiple exchange which would appear as isolabeling with the lower resolution of autoradiography. The corollary, of course, is that longitudinal folding does not occur; thus the distance apart of two points on the chromosome should be proportional to distance apart on the DNA thread. This implies that chromosome bands contain contiguous lengths of DNA (see Chapter 15). Another artefact that allows (genuine) isolabeling to occur is the existence of late replicating regions that manage to undergo three rounds of replication in the presence of BUdR, in contrast to the two rounds of substitution of other regions; these take the form of sister regions both of which fail to show label (Wolff, Body-cote, and Rodin, 1978).

Visible aberrations in chromosome structure can be produced by irradiation and fall into three classes. *Chromosome aberrations* affect both sister chromatids of a metaphase pair in exactly the same way. These are produced by irradiation early in interphase, before the diploid set has replicated. This causes the aberration to be perpetuated in both daughters of the damaged chromosome. *Chromatid aberrations* influence only one chromatid of a pair at metaphase. These are caused by irradiation later in interphase, after replication of the chromosome set, so that daughter chromosomes are damaged independently. This is what would be expected of a unineme chromosome.

Putative *subchromatid exchanges* provide an aberration whose inconsistency with a unineme chromosome has proved more apparent than real. Treatment with X rays during prophase or early metaphase causes aberrations which appear to be localized to half a chromatid. Figure 10.6 shows that if these apparent subchromatid exchanges are real, they should appear as

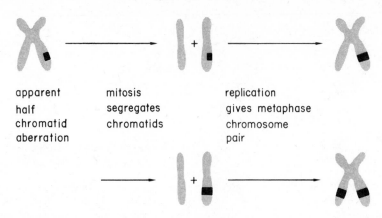

apparent	mitosis	replication
half	segregates	gives metaphase
chromatid	chromatids	chromosome
aberration		pair

Figure 10.6. Segregation of apparent half-chromatid aberrations. If the aberration represents damage to a subunit comprising half a chromatid, at the next metaphase it should appear as an aberration of one sister chromatid of a pair (upper). If the aberration really represents damage to the whole chromatid, both of the sister chromatids of the pair carrying the damage should display the aberration at the next metaphase (lower).

exchanges involving the full chromatid at the second mitosis (and should influence both sister chromatids of a pair only at the third mitosis). This prediction has been tested by Kihlman and Hartley (1967) in an analysis of breaks induced in chromosomes by 8-ethoxy-caffeine. After treatment, cells were incubated with colchicine to prevent chromosome segregation. Exchanges that had appeared at the first mitosis to constitute subchromatid aberrations were displayed at the second mitosis as full chromosome exchanges involving both sister chromatids. This implies that their appearance at the first mitosis is deceptive (for review see DuPraw, 1970; Kihlman, 1971).

CHAPTER 11

Chromatin Components: Histones

Chromatin contains proteins and RNA as well as the DNA that comprises the genetic material itself. The relative proportions of these components vary according to the tissue, organism, and method of preparation. However, the highly basic proteins constitute the greatest amount of the chromosomal protein. These are the *histones*. The remaining proteins are described as the *nonhistones;* these are less basic and at one time were known as "acidic" proteins, although this term is a misnomer since their isoelectric points are lower than those of the histones but are not necessarily acidic. There is usually a little more histone by weight than DNA; and we shall see in Chapter 13 that the nucleosome model predicts the maintenance of stoichiometry between DNA and the histones. Generally there is a somewhat lesser amount of the nonhistone proteins, whose reported proportion is more variable. The total mass of protein is therefore somewhat less than twice that of the DNA (see Bonner et al., 1968a,b). There is a much smaller amount of RNA, less than 10% of the DNA content; probably this mostly comprises nascent RNA chains that have not yet been released from the template, although there is, of course, always the possibility that this component includes molecules with a structural role.

Chromatin can be considered as a nucleoprotein particle in which the proteins interact both with DNA and with each other. The interactions involving the histones now are becoming well characterized; much less is known about the nonhistone proteins. The histones fall into five well-defined classes, which are conserved with only restricted variations across probably the entire eucaryotic domain. This conservation provides the answer to a previously long-standing controversy about their role, aroused by the suggestion of Stedman and Stedman (1950) that histones might be specific genetic repressors. The lack of either species or tissue specificity has excluded this function. The evolutionary conservation of histone structure implies that their role is structural and must be common to all eucaryotes. In this chapter we shall be concerned with the structures of the histone proteins themselves; in Chapter 13 we shall consider the interactions between them and with DNA which are responsible for constituting the nucleosome. It is, of course, the evolutionary persistence of the nucleosome which is reflected in the conservation of histone sequences.

Nonhistone proteins are not included in the nucleosome. The nonhistones consist of a much larger number of proteins than the histones and appear to show both species and tissue specificity. There are wide variations in the amount present of each nonhistone protein; but none approaches the level of a histone. The roles of nonhistone proteins remain to be established. Some may be involved in a structural capacity, concerned with the organization of nucleosomes into higher order structures. Others may represent regulators of gene expression; it is on this component of chromatin that attention now is focused as the probable residence of control molecules. The nonhistones include some proteins able to bind to DNA, among which also should be the enzymes that undertake replication and transcription. We shall come in Chapter 12 to the properties of the nonhistone proteins as such; and in Chapter 14 we shall consider the interactions between nucleosomes in which it is possible they may participate.

Separation of Histones

Early difficulties in identifying individual histones were caused by the ready degradation that occurs during extraction and by the aggregation of different histones into complexes. By comparing the histones isolated by different methods of preparation, Rasmussen, Murray, and Luck (1962) were able to show that only a comparatively small number of fractions represents genuine variations in protein sequence, although even these fractions did not comprise homogeneous proteins. The demonstration that there are only five types of histone was rendered conclusive only by the introduction of gel electrophoresis to identify the individual components of these fractions.

Histones may be isolated either from whole tissue, purified nuclei, or crude preparations of chromatin. Chromatin can be obtained by homogenizing tissue, filtering to remove intact fragments and membranes, and collecting the pellet from a low-speed centrifugation. This can be purified by centrifugation through sucrose to give a gelatinous preparation (see Fambrough and Bonner, 1966). Two general types of method have been used to isolate the histones (for review of early work see Murray, 1965; Johns, 1971). Extraction with acid, usually sulfuric acid, yields preparations of histone-salt, such as histone sulfate. These crude preparations then can be fractionated; early work relied upon elution from ion exchange columns with guanidium HCl (Bonner et al., 1968a,b). An alternative method is to use a series of extractions with salt followed by precipitation with acetone or ethanol (Johns and Butler, 1962; Johns, 1964; Phillips and Johns, 1965). Stepwise elution with acid generates fractions similar to those released by salt (Murray, 1969; Ohlenbusch et al., 1967; Kleiman and Huang, 1972).

Almost all of this work was performed with calf thymus, some with pea seedling. Five histones were eventually identified and were named according to their order of elution from the thymus in the two approaches, although the nomenclature was complicated by the further division of some of the original fractions. Correspondence between the fractions obtained by the two methods was not complete, for the peaks in each case were not homogeneous, but consisted largely of one histone contaminated with others (Starbuck et al., 1968). Similar fractions can be obtained from all sources, although not necessarily in the same order of elution. An electrophoretic separation developed by Panyim and Chalkley (1969), using acrylamide gels in the presence of urea, demonstrated that only five fractions of histone are present in virtually all eucaryotic cells. Figure 11.1 shows an analysis of calf thymus histones. On short gels, only five fractions are obtained; the same order of separation is seen with all mammalian histones; although the relative mobilities are changed in other eucaryotes, the same five fractions are always obtained. On long gels, some of the fractions are further divided into subfractions (see below).

More recently, a new nomenclature has been introduced to reconcile the previous terminologies, and their correspondence is given in Table 11.1, which summarizes the general properties of the histone classes. All the histones are small and possess large amounts of the basic amino acids. They have been classified according to their relative contents of lysine and arginine. The lysine rich histone class, H1, is the heaviest, with a molecular weight of some 23,000 and with a basicity almost completely due to its content of lysine, about 29%. Appreciably the most basic of the histones, H1 can be selectively removed from chromatin by 5% perchloric acid or trichloracetic acid. The other histones have molecular weights from 11,000 to 16,000 and about 20% of their amino acids are basic. The slightly lysine rich histones, H2A and H2B, possess more lysine than arginine, while the relationship is reversed in the arginine rich histones, H3 and H4. A comparison between cow and pea shows that the arginine rich histones are very highly conserved, while there is some variation in the slightly lysine rich histones.

The multiple bands that are seen when histones are electrophoresed on long gels are due to heterogeneity in the sequence of the protein only in the case of H1, which in calf thymus consists of several species of closely related sequences. With the other histones, this apparent heterogeneity is due to modifications made to the amino acids. These include acetylation of lysine, methylation of lysine, histidine or arginine, and phosphorylation of serine (for review see DeLange and Smith, 1971). Modification usually is incomplete at any particular site, so that histone molecules appear heterogeneous in this sense. Thus an acetylated histone may run at a different position on a gel from a nonacetylated molecule. While the proportions of the histone classes remain unchanged with different tissues and species, the proportions of different modified versions of each protein may vary.

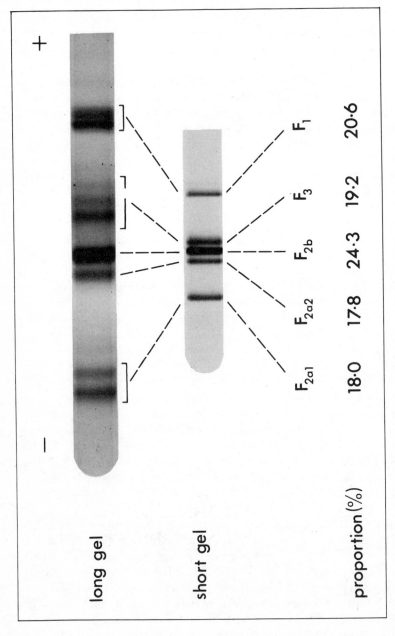

Figure 11.1. Separation of calf thymus histones by gel electrophoresis. The five bands isolated on short gels (lower) can be fractionated further on long gels (upper). The proportions of the five major bands are constant in all calf tissues; those on the sub bands may vary. The multiple bands of H1 represent proteins with very similar sequences; those of the other histones are caused by differing extents of modification, the fastest running band may represent the least modified protein species. Data of Panyim and Chalkley (1969).

Table 11.1   Classes of histones

| Histone | Species | Basic amino acids | | | Acidic amino acids | Basic acidic ratio | Apolar amino acids | Other features | Number amino acids | Molecular weight |
		Lys	Arg	Lys/Arg						
H1 (f1, I)	rabbit	29%	1.5%	21	5.5%	5.4	10%	25% Ala	213	23,000
H2A (f2a2, IIb1)	cow	11%	9.5%	1.2	15.5%	1.5	26%	13% Ala 11% Gly	129	13,960
"	pea	11%	7.5%	1.5	13%	1.5	24%	14% Ala 11% Gly		15,000/ 16,500
H2B (f2b, IIb2)	cow	16%	6.5%	2.5	13%	1.9	22.5%	10% Ala 11% Ser	125	13,774
"	pea	18%	4%	4.8	15.5%	1.5	23.5%	10% Ala 9% Ser		16,500
H3 (f3, III)	cow	10%	13.5%	0.7	13.5%	1.8	18%	1-2 Cys	135	15,342
"	pea	amino acid sequence identical to cow								
H4 (f2a1, IV)	cow	11%	14%	0.8	10%	2.7	23%	15% Gly	102	11,282
"	pea	amino acid sequence identical to cow								

Sequences of the Arginine-Rich Histones

The arginine-rich histones, H3 and H4, possess the most highly conserved sequences yet identified in the evolution of proteins. Both histones were first sequenced from calf thymus; and the entire primary structure since has been elucidated for other organisms also. Figures 11.2 and 11.3 summarize these data. The sequence of H4 is identical in several mammals—cow, pig, and rat—and shows only two changes in the pea. Both of these are conservative, replacing one apolar amino acid with another and one basic amino acid with another. Calf thymus H3 is almost identical with the sequence shared by chicken, carp and shark; the cow histone shows the interesting feature of polymorphism, with either a cysteine or a serine found at position 96, while only the serine is found in the latter three species. Pea H3 is also polymorphic at position 96, although showing either alanine or serine, and otherwise differs by three conservative substitutions elsewhere in the molecule. An indication that the conservation of sequences extends to other species is provided by the observation that the electrophoretic mobilities of H3 and H4 appear to be constant in all eucaryotes (Panyim and Chalkley, 1969; Panyim, Bilek and Chalkley, 1971). Greater variation of sequence is seen, however, in Tetrahymena H4, which can be distinguished functionally from mammalian H4. Whether this represents a separate line of evolution remains to be seen.

   The most obvious interpretation of this stringent conservation of sequences between species as distantly related as mammals, fish, and plants is that the whole sequence of each arginine-rich histone is involved in its function; and this function must be identical over a wide range, perhaps all

Calf thymus H4 sequence

                                                      Ac   Me$_2$
Ac-Ser-Gly-Arg-Gly-Lys-Gly-Gly-Lys-Gly-Leu-Gly-Lys-Gly-Gly-Ala-Lys-Arg-His-Arg-Lys-
                      10   20
  -Val-Leu-Arg-Asp-Asn-Ile-Gln-Gly-Ile-Thr-Lys-Pro-Ala-Ile-Arg-Arg-Leu-Ala-Arg-Arg-
                      30   40
  -Gly-Gly-Val-Lys-Arg-Ile-Ser-Gly-Leu-Ile-Tyr-Glu-Glu-Thr-Arg-Gly-Val-Leu-Lys-Val-
                      50   60
  -Phe-Leu-Glu-Asn-Val-Ile-Arg-Asp-Ala-Val-Thr-Tyr-Thr-Glu-His-Ala-Lys-Arg-Lys-Thr-
                      70   80
  -Val-Thr-Ala-Met-Asp-Val-Val-Tyr-Ala-Leu-Lys-Arg-Gln-Gly-Arg-Thr-Leu-Tyr-Gly-Phe-
                      90   100
  -Gly-Gly-COOH

Sequence changes in other species

Pig	none
Rat	none
Pea	val ⟶ ile @ 60
	lys ⟶ arg @ 77
Tetrahymena	15 substitutions (5 conservative) in a partial sequence of 66 amino acids

Positions of amino acid modification

Serine @ 1	terminus is acetylated
Lysine @ 16	60% acetylated in cow; also in pea
Lysine @ 20	methylated in cow; not in pea

Figure 11.2. Sequences of histone H4.   Data of DeLange et al. (1969a,b), Ogawa et al. (1969), Sautiere et al. (1970, 1971), Wilson et al. (1970), Glover and Gorovsky (1979).

higher eucaryotes. We shall discuss in Chapter 13 the evidence that there may be an interaction between H3 and H4 to generate a complex whose ability to organize DNA may be a central feature of nucleosome structure. The evolutionary restraints apparent in these histone sequences imply that the constitution and activity of this complex may be common to nucleosomes of all chromatins.

Both the arginine-rich histones are modified. In calf thymus, H4 possesses a blocked terminus, consisting of N-acetyl-serine; internal lysine residues may be acetylated or methylated. There is heterogeneity in the extent of modification; lysine 16 is acetylated in only 60% of the molecules, whereas lysine 20 is dimethylated some three times more frequently than it is monomethylated. The same modifications may occur at the same positions in H4 of other species; but their extents may be different. Histone H3 also is modified in calf thymus, and again the modification is incomplete at any given site. The same tendency is seen for modification to occur at the same sites in other species, although to characteristically different extents. Whether these differences are significant or result simply from factors such

Calf thymus H3 sequence

$$\text{Me}_{O\text{-}2} \qquad\qquad \text{Ac}$$
$$| \qquad\qquad\qquad |$$

$$\text{NH}_2\text{-Ala-Arg-Thr-Lys-Gln-Thr-Ala-Arg-Lys-Ser-Thr-Gly-Gly-Lys-Ala-Pro-Arg-Lys-Gln-Leu-}$$
$$\qquad\qquad\qquad\qquad 10 \qquad\qquad\qquad\qquad\qquad\qquad\qquad\qquad 20$$

$$\qquad\quad \text{Ac} \qquad\qquad \text{Me}_{O\text{-}2}$$
$$\qquad\quad | \qquad\qquad\qquad |$$

$$\text{-Ala-Thr-Lys-Ala-Ala-Arg-Lys-Ser-Ala-Pro-Ala-Thr-Gly-Gly-Val-Lys-Lys-Pro-His-Arg-}$$
$$\qquad\qquad\qquad\qquad 30 \qquad\qquad\qquad\qquad\qquad\qquad\qquad\qquad 40$$

$$\text{-Tyr-Arg-Pro-Gly-Thr-Val-Ala-Leu-Arg-Glu-Ile-Arg-Arg-Tyr-Gln-Lys-Ser-Thr-Glu-Leu-}$$
$$\qquad\qquad\qquad\qquad 50 \qquad\qquad\qquad\qquad\qquad\qquad\qquad\qquad 60$$

$$\text{-Leu-Ile-Arg-Lys-Leu-Pro-Phe-Gln-Arg-Leu-Val-Arg-Glu-Ile-Ala-Gln-Asn-Phe-Lys-Thr-}$$
$$\qquad\qquad\qquad\qquad 70 \qquad\qquad\qquad\qquad\qquad\qquad\qquad\qquad 80$$

$$\qquad\qquad\qquad\qquad\qquad\qquad\qquad\qquad\qquad\qquad\qquad\qquad\qquad \text{Cys}$$
$$\text{-Asp-Leu-Arg-Phe-Gln-Ser-Ser-Ala-Val-Met-Ala-Leu-Gln-Glu-Ala-}\overline{}\text{-Glu-Ala-Tyr-Leu-}$$
$$\qquad\qquad\qquad\qquad 90 \qquad\qquad\qquad\qquad\qquad\qquad \text{Ser} \qquad\quad 100$$

$$\text{-Val-Gly-Leu-Phe-Glu-Asp-Thr-Asn-Leu-Cys-Ala-Ile-His-Ala-Lys-Arg-Val-Thr-Ile-Met-}$$
$$\qquad\qquad\qquad\qquad 110 \qquad\qquad\qquad\qquad\qquad\qquad\qquad\qquad 120$$

$$\text{-Pro-Lys-Asp-Ile-Gln-Leu-Ala-Arg-Arg-Ile-Arg-Gly-Glu-Arg-Ala-COOH}$$
$$\qquad\qquad\qquad\qquad 130$$

Sequence changes in other species

Chicken, carp and shark	Cys/Ser $\longrightarrow$ Ser @ 96	
Pea	Tyr $\longrightarrow$ Phe @ 41	
	Arg $\longrightarrow$ Lys @ 53	
	Met $\longrightarrow$ Ser @ 90	
	Cys/Ser $-\longrightarrow$ Ala/Ser @ 96	
	80% 20% $\longrightarrow$ 60% 40%	
Sea urchin	Asp $\longrightarrow$ Glu @ 81	
(S. purpuratus and	Cys/Ser $\longrightarrow$ Ser @ 96	
P. miliaris)		

Positions of amino acid modification

Lysine @ 9	methylated in cow, chick and carp
Lysine @ 14	25% acetylated in cow; not in chicken or carp
Lysine @ 23	29% acetylated; not in chick or carp
Lysine @ 27	75% methylated in cow; 100% in chick; also in carp

Figure 11.3. Sequences of histone H3. Data of DeLange, Hooper and Smith (1972, 1973), Hooper et al. (1973), Patthy et al. (1973), Patthy and Smith (1975), Brandt and Von Holt (1974) and Brandt, Strickland and Von Holt (1974). For sea urchins the protein has not been sequenced, but the nucleotide sequence of the H3 coding region has been used to predict the amino acid sequence (Sures, Lowry, and Kedes, 1978; Schaffner et al., 1978; see Table 28.1).

as the growth rate of cells of the tissues from which the histones have been prepared remains to be seen.

A feature common to all histones is the clustering of basic amino acids towards the N-terminus. This is summarized below in Figure 11.8. This has prompted speculation that these regions may constitute DNA-binding sites. The affinity of the arginine-rich histones for DNA may be expected to be reduced by the elimination of basic groups; in this context, it is interesting that all the sites of modification reside in the N-terminal regions (see below). Whether the introduction of modifications plays any role in controlling the affinity of these histones for DNA is not yet known.

Sequences of the Slightly Lysine-Rich Histones

The same two slightly lysine-rich histones, H2A and H2B, can be recognized in all eucaryotic chromatins, although their sequences are not so well conserved as those of the arginine-rich histones. This was first recognized when electrophoretic analysis showed that the mobilities of H2A and H2B vary across species, in contrast with the constancy of H3 and H4. The comparison between cow and pea histones summarized in Table 11.1 shows that the sizes and amino acid compositions of H2A and H2B are different in mammals and plants, although still maintaining the same general characteristics.

The sequences that are available for these histones show less pronounced differences between mammals and fish. Figure 11.4 gives the sequence of H2A, which is almost identical in cow, rat, and chicken. The rat histone is polymorphic at two sites, at which an alternative is introduced to the amino acid present in the cow; and there is a substitution of asparagine for aspartic acid. The H2A of trout shows three sites of deletion, one insertion and four substitutions compared with calf thymus. Some of these affect the charge composition of the molecule. A greater number of changes is seen in sea urchin, where one interesting feature is the shortening of the sea urchin protein by 5 residues at the C-terminal end, possibly generated by a frameshift at codon 123. Overall, however, the same general features are maintained; there is a cluster of basic amino acids at the N-terminal end and of apolar amino acids in two central sections.

The sequences of H2B histones have been determined for cow, trout, and sea urchin. These are summarized in Figure 11.5. The difference between cow and trout is very similar in extent to that found for H2A. There is a deletion in trout H2B of two amino acids and there are seven sites of point substitution, relative to calf thymus. With one exception, all these changes lie in the N-terminal region.

A different situation, however, is found in the sea urchin Parechinus angulosus, which has three forms of H2B in sperm showing differences more widespread than those seen in the other histone polymorphisms so far characterized. There are 17 point differences between H2B(1) and H2B(2); in addition there is a difference in a repeating pentapeptide sequence, 17 amino acids long in H2B(1) and 20 amino acids long in H2B(2), which is inserted relative to the calf thymus sequence. The variant H2B(3) is very like H2B(2), but with an additional pentapeptide sequence. In most cases, the sites at which the two sea urchin forms vary also show a difference from the

Calf thymus H2A Sequence

Ac-Ser-Gly-Arg-Gly-Lys-Gln-Gly-Gly-Lys-Ala-Arg-Ala-Lys-Ala-Lys-Thr-Arg-Ser-Ser-Arg-
                                10   20

-Ala-Gly-Leu-Gln-Phe-Pro-Val-Gly-Arg-Val-His-Arg-Leu-Leu-Arg-Lys-Gly-Asp-Tyr-Ala-
                                30   40

-Glu-Arg-Val-Gly-Ala-Gly-Ala-Pro-Val-Tyr-Leu-Ala-Ala-Val-Leu-Glu-Tyr-Leu-Thr-Ala-
                                50   60

-Glu-Ile-Leu-Glu-Leu-Ala-Gly-Asn-Ala-Ala-Arg-Asp-Asn-Lys-Lys-Thr-Arg-Ile-Ile-Pro-
                                70   80

-Arg-His-Leu-Gln-Leu-Ala-Ile-Arg-Asn-Asp-Glu-Glu-Leu-Asn-Lys-Leu-Leu-Gly-Lys-Val-
                                90   100

-Thr-Ile-Ala-Gln-Gly-Gly-Val-Leu-Pro-Asn-Ile-Gln-Ala-Val-Leu-Leu-Pro-Lys-Lys-Thr-
                               110   120

-Glu-Ser-His-His-Lys-Ala-Lys-Gly-Lys-COOH

Sequence changes in other species

Trout	Glu $\longrightarrow$ Thr	@ 6
	Ile $\longrightarrow$ Val	@ 87
	Lys $\longrightarrow$ Gly	@ 99
	Gly $\longrightarrow$ Val-Ala	@ 128
	Tyr $\longrightarrow$ deleted	@ 57
	Ile $\longrightarrow$ deleted	@ 71
	Ser-His-His $\longrightarrow$ deleted	@ 122–124
Rat	Thr $\longrightarrow$ Thr/Ser	@ 16
	20% 80%	
	Lys $\longrightarrow$ Lys/Arg	@ 99
	60% 40%	
	Asp $\longrightarrow$ Asn	@ 38
Chicken	Thr $\longrightarrow$ Ser	@ 16
	Glu $\longrightarrow$ Asp	@ 121

| Sea urchin
(P. miliaris gonad) | 11 amino acid substitutions
4 amino acid deletions |
| (P. miliaris and
S. purpuratus DNA) | 20 amino acid substitutions
4 amino acid deletions |

Figure 11.4. Sequences of histone H2A. Data of Yeoman et al. (1972), Bailey and Dixon (1973), Laine et al. (1976), Wouters et al. (1978). The P. miliaris sequence is for gonad H2A; a protein in which some of the substitutions are different appears to be coded by an H2A gene that has been sequenced and is very similar in P. miliaris and S. purpuratus (see Table 28.1). The amino acid composition and molecular weight weight of pea H2A (and H2B) has been reported by Spiker and Isenberg (1977).

H2B of calf thymus. In addition, there is a large number of sites at which all sea urchin H2B molecules share an amino acid that is different from that found in calf thymus. The overall difference between calf thymus and sea urchin corresponds to changes in some 80 of the 125 amino acids, with only a quarter of these representing conservative changes. Less variation is seen when cow is compared with S. purpuratus or P. miliaris.

From these results it is clear that a greater degree of variation is permitted in H2A and H2B sequences than in those of H3 and H4. Determining the

Calf thumus H2B sequence

NH$_2$-Pro-Gln-Pro-Ala-Lys-Ser-Ala-Pro-Ala-Pro-Lys-Lys-Gly-Ser-Lys-Lys-Ala-Val-Thr-Lys-
                                10   20
  -Ala-Gln-Lys-Lys-Asp-Gly-Lys-Lys-Arg-Lys-Arg-Ser-Arg-Lys-Glu-Ser-Tyr-Ser-Val-Tyr-
                                30   40
  -Val-Tyr-Lys-Val-Leu-Lys-Gln-Val-His-Pro-Asp-Thr-Gly-Ile-Ser-Ser-Lys-Ala-Met-Gly-
                                50   60
  -Ile-Met-Asn-Ser-Phe-Val-Asn-Asp-Ile-Phe-Glu-Arg-Ile-Ala-Gly-Glu-Ala-Ser-Arg-Leu-
                                70   80
  -Ala-His-Tyr-Asn-Lys-Arg-Ser-Thr-Ile-Thr-Ser-Arg-Glu-Ile-Gln-Thr-Ala-Val-Arg-Leu-
                                90   100
  -Leu-Leu-Pro-Gly-Glu-Leu-Ala-Lys-His-Ala-Val-Ser-Glu-Gly-Thr-Lys-Ala-Val-Thr-Lys-
                                110   120
  -Tyr-Thr-Ser-Ser-Lys-COOH

Sequence changes in other species

Trout	Ala-Pro	⟶	deleted	@ 9/10
	Ala	⟶	Thr	@ 21
	Gln	⟶	Ala	@ 22
	Lys	⟶	Gly	@ 23
	Asp	⟶	Gly	@ 25
	Ser	⟶	Ala	@ 38
	Val	⟶	Ile	@ 39
	Ala	⟶	Ser	@ 77

Sea urchin   three varieties of H2B are all closely related; N-terminus is different from
(Parechinus)   cow and has variable number of repeats of penta amino acid sequence; ~35
                point changes in remainder of molecule relative to cow

(S. purpuratus)   28 amino acid substitutions, concentrated in N-terminal region and in
                central part

Figure 11.5. Sequences of histone H2B. Data of Iwai, Ishikawa, and Hayashi (1970), Kootstra and Bailey (1978) and Strickland et al. (1977a,b; 1978). The Parechinus variants are present in sperm; a gene that has been sequenced in S. purpuratus would code for an H2B protein hardly related to the Parechinus species in the N terminal region, but otherwise better related (Sures, Lowry, and Kedes, 1978). Similarly a gene for an H2B of S. purpuratus corresponds to a protein sequence more closely related to bovine H2B than to that of Parechinus (Birnstiel et al., 1978).

full extent of this variation will require sequences of these histones over a wider range of species; it would therefore be premature to come to any conclusion on what difference there may be in the relative variations permitted to the two slightly lysine-rich histones.

Modification of these histones is less common than of the arginine-rich histones. Often they run as homogenous bands on electrophoretic gels, which would imply either the absence of modification or complete modification at any given site. No modifying groups have been detected in calf thymus H2A, although modification may take place in other species (see below). Acetyl lysine has been detected within H2B, but the position of the modification is not known.

Multiple forms of both H2A and H2B have been found in the histones synthesized during early embryogenesis of the sea urchin S. purpuratus. The electrophoretic mobilities of the bands containing a pulse label in both cases show that at the 1–16 cell stage there is only a single protein species; but by gastrula, three further forms of H2A and two further forms of H2B have appeared (Cohen, Newrock, and Zweidler, 1975). Although there are changes in the histones under synthesis at each stage of early embryogenesis, the previously synthesized histones remain as constituents of chromatin; thus the result of the transitions from one form of H2A or H2B to another is to create variation in the structures of these histones present in the genetic apparatus. A similar substitution, of one form for another, is seen with H1 (see below). No transition between different forms is seen with H3 or H4. The different gel bands of H2A and H2B appear to represent proteins of different primary sequences, and not modifications of a single protein, since they are found among the proteins synthesized in vitro from mRNA of the various embryonic stages (Newrock et al., 1978).

The sequences of some of the variants of H2A and H2B now have been determined. The sequence of P. miliaris gonad H2A differs at several positions from the protein that should be coded by an H2A gene that may represent an early embryonic variant. Almost all the sites of variation are those at which there are already differences from mammalian H2A (so this scarcely affects estimates for evolutionary divergence).

In addition to the three variants of H2B present in sperm of Parechinus, at least three major variants can be distinguished during development, as well as some forms present in lower amounts (Brandt et al., 1979). Two different sequences occur early and late in embryonic development and another is present in somatic cells. This sea urchin produces a minimum of 8 variants of H2B.

Microheterogeneity of Lysine-Rich Histones

Mammalian lysine-rich histones can be fractionated into subgroups of proteins that appear to be closely related in sequence. Figure 11.6 summarizes the sequence data available for mammalian H1. Two of the five rabbit thymus subfractions and one of the four calf thymus subfractions have been analyzed. In the N-terminal 73 amino acids, 9 positions show differences between the two rabbit fractions. There are 7 differences between the cow fraction and the rabbit fraction most closely related to it.

In trout a single sequence has been determined for H1, of 194 residues compared with the 213 of rabbit. (There is some polymorphism, but it appears to be more restricted). Comparison between trout and rabbit sequences (maximizing the relationship in the C-terminal region, which has not been fully sequenced in rabbit) shows that 140 of 225 positions are identical, with 14 of the 85 changes representing conservative substitutions.

A histone gene of sea urchin can be recognized as coding for H1 only

```
Rabbit-3 Ac-Ser-Glu-Ala-Pro-Ala-Glu-Thr-Ala-Ala-Pro-Ala-Pro-Ala-Glu-Lys-Ser-Pro-Ala-Lys-Lys-
Rabbit-4 Ac-Ser-Glu-Ala-Pro-Ala-Glu-Thr-Ala-Ala-Pro-Ala-Pro-Ala-------Lys-Ser-Pro-Ala-Lys-Thr-
Calf    -1 Ac-Ser-Glu-Ala-Pro-Ala-Glu-Thr-Ala-Ala-Pro-Ala-Pro-Ala-Pro-Lys-Ser-Pro-Ala-Lys-Thr-
                                 10                                         20
           Lys- - - -Lys-Ala-Ala-Lys-Lys-Pro-Gly-Ala-Gly-Ala-Ala-Lys-Arg-Lys-Ala-Ala-Gly-Pro-
           Pro-Val-Lys-Ala-Arg-Lys-Lys-Lys-Ser-Ala-Gly-Ala-Ala-Lys-Arg-Lys-Ala-Ser-Gly-Pro-
           Pro-Val-Lys-Ala-Ala-Lys-Lys-Lys-Pro-Ala-Gly-Ala-Arg-Arg-Lys-Ala-Ser-Gly-Pro-
                                 30                                         40
           Pro-Val-Ser-Glu-Leu-Ile-Thr-Lys-Ala-Val-Ala-Ala-Ser-Lys-Glu-Arg-Asn-Gly-Leu-Ser-
           Pro-Val-Ser-Glu-Leu-Ile-Thr-Lys-Ala-Val-Ala-Ala-Ser-Lys-Glu-Arg-Ser-Gly-Val-Ser-
           Pro-Val-Ser-Glu-Leu-Ile-Thr-Lys-Ala-Val-Ala-Ala-Ser-Lys-Glu-Arg-Ser-Gly-Val-Ser-
                                 50                                         60
           Leu-Ala-Ala-Leu-Lys-Lys-Ala-Leu-Ala-Ala-Gly-Gly-Tyr-Asp-Val-Glu-Lys-Asn-Asn-Ser-
           Leu-Ala-Ala-Leu-Lys-Lys-Ala-Leu-Ala-Ala-Ala-Gly-Tyr-
           Leu-Ala-Ala-Leu-Lys-Lys-Ala-Leu-Ala-Ala-Ala-Gly-Tyr-
                                 70
Rabbit-3   Arg-16-Lys-Leu-Gly-Leu-Lys-Ser-Leu-Val-Ser-Lys-Gly-Thr-Leu-Val-Glx-Thr-Lys-Gly-
                                 90                                         100
           Thr-Gly-Ala-Ser-Gly-Ser-Phe-Lys  ⟶  213
```

Figure 11.6. Sequences of H1 aligned to show the similarities between fractions 3 and 4 of the rabbit and fraction 1 of the cow.   Cow and rabbit thymus lysine rich histones were separated into subfractions by Kinkade and Cole (1966), Bustin and Cole (1969a,b) and Sherrod, Johnson, and Chalkley (1974); the partial sequences shown above were determined by Rall and Cole (1971) and Jones, Rall, and Cole (1974). The sequence of trout H1 has been determined by Mac-Leod, Wong, and Dixon (1977). The partial sequence of a gene for sea urchin (P. miliaris) H1 corresponds to an amino acid sequence which is homologous to mammal and trout only in the 22 amino acid region from positions 65–86, which corresponds closely to the sequences at 87–108 of rabbit and 77–98 of trout (Schaffner et al., 1978).

by virtue of the presence of a highly conserved sequence of 22 amino acids; the remaining regions are not much related to rabbit or trout. But H1 proteins have not yet been directly sequenced from sea urchin.

In the sequenced proteins, clustering of related amino acids again is seen; both the N-terminal and C-terminal ends are highly basic, with an apolar region in the center of the sequence. This is consistent with the behavior of the isolated protein in solution. Acetylation and methylation have not been found in H1, but there is evidence for a cell-cyle dependent phosphorylation, although the positions that are modified generally are not known. This emphasizes the need to compare sequences rather than electrophoretic bands; thus although changes in the relative amounts (and sometimes the presence) of the H1 bands are seen when rabbit, rat or mouse sperm is compared with calf thymus, it remains to be proven that this reflects changes in gene expression rather than modification (Seyedin and Kistler, 1979a,b).

Tissue-specific variations in H1 have been identified in the sea urchin. In both the species L. pictus and S. purpuratus, the H1 histone initially synthesized forms a single band on acetic acid—urea gels; this is $H1_m$, named for the morula stage of embryogenesis, and its mobility corresponds roughly to a size of 20,000 daltons. At gastrula, its synthesis gives way to that of $H1_g$, which has a slightly more rapid electrophoretic mobility, suggesting a molecular weight of about 21,000 daltons (Ruderman and Gross, 1974; Seale and Aronson, 1973).

Is this change due to synthesis of a different H1 protein or to some difference in modification of the same protein? The same switch in electrophoretic mobility is seen when L. pictus morula or gastrula histone mRNA is used to direct protein synthesis in vitro, using a heterologous system to translate the message (Ruderman, Baglioni, and Gross, 1974; Arceci, Senger and Gross, 1976). This suggests that $H1_m$ and $H1_g$ are the products of different genes, expressed at different stages in embryogenesis. The extent of the variation between the proteins remains to be established, as does the function of the substitution.

Sequence of Avian H5

The histone constitution of chromatin is almost invariant. But there are a few instances in which changes are made. Some concern the substitution of one variant of a histone for another at a given stage of development (see above). The function of such apparently minor changes is not known. A more substantial change occurs in the nucleated erythrocytes of fish, birds and reptiles, in which the usual lysine rich histone, H1, is replaced by a new species, H5 (previously known as V or f2c). Yet a more striking change occurs during spermatogenesis in some species, when histones are replaced by protamines (see below).

Histone H5 has been found only in nucleated erythrocytes, which altogether lack H1 (Edwards and Hnilica, 1968; Sanders and McCarty, 1972). The substitution occurs during maturation of erythroblasts to erythrocytes. The total content of lysine-rich histone remains constant, while the proportion of H5 increases at the expense of H1 (Dick and Johns, 1969; Sotirov and Johns, 1972; Billeter and Hindley, 1972; Appels, Wells, and Williams, 1972; Appels and Wells, 1972). In the immediate precursor to the mature erythrocyte, H5 is the only histone to be synthesized; at this point during development the molecules turn over rapidly, but in the mature erythrocyte they are stable. All the other histones remain unchanged in amount at each stage of maturation.

Histone H5 behaves as a typical lysine-rich histone in its ability to leave chromatin in 5% perchloric acid and failure to self-aggregate (Johns and Diggle, 1969; Diggle, McVittie and Peacock, 1975). Its amino acid composition is similar to that of H1, with a lysine content almost as great (24%) but with a greater arginine content (11%); it is rich also in alanine (16%) and serine (13%). Analysis of avian, fish and amphibian erythrocytes show variations in composition, suggesting that the sequence of H5 will prove to be species specific (Greenaway and Murray, 1971).

The sequence of chicken H5 is given in Figure 11.7 and it is at once ap-

NH$_2$-Thr-Glu-Ser-Leu-Val-Leu-Ser-Pro-Ala-Pro-Ala-Lys-Pro-Lys-$\overset{\text{Gln}}{\underset{\text{Arg}}{}}$-Val-Lys-Ala-Ser
                                    10

  -Arg-Arg-Ser-Ala-Ser-His-Pro-Thr-Tyr-Ser-Glu-Met-Ile-Ala-Ala-Ile-Arg-Ala-Glu-
    20   30

  -Lys-Ser-Arg-Gly-Gly-Ser-Ser-Arg-Gln-Ser-Ile-Gln-Lys-Tyr-Ile-Lys-Ser-His-Tyr-
    40   50

  -Lys-Val-Gly-His-Asn-Ala-Asp-Leu-Gln-Ile-Lys-Leu-Ser-Ile-Arg-Arg-Leu-Leu-Ala-
      60   70

  -Ala-Gly-Val-Leu-Lys-Gln-Thr-Lys-Gly-Val-Gly-Ala-Gly-Ser-Ser-Phe-Arg-Leu-Ala-
      80   90

  -Lys-Ser-Asp-Lys-Ala-Lys-Arg-Ser-Pro-Gly-Lys-Lys-Lys-Lsy-Ala-Lys-
      100   110

-(Thr$_3$Ser$_8$Pro$_{10}$Gly$_2$Ala$_{16}$Val$_3$Lys$_{32}$Arg$_{13}$)

Figure 11.7. Sequence of chicken H5. Data of Greenaway and Murray (1971) and Sautiere et al. (1976).

parent that there is no close evolutionary relationship with H1. The most basic region of the molecule appears to be the (unsequenced) C-terminal region, which also has an appreciable content of proline, which should prevent helix formation. The molecule is close to the same size as H1, with a length of 198 amino acids. There is a polymorphism at position 15, which allows the histone to be fractionated into two species (H5a and H5b) on the basis of the charge difference between glutamine and arginine. Analysis of individual chickens has identified either H5b alone (Arg 15) or H5a and H5b (Gln 15 + Arg 15). This would suggest that the heterogeneity is due to the presence at one locus of either of two alleles, in which case it should be possible to find chickens that have only H5a. But the other histones are known to be coded by multiple genes and so it is more likely that the basis for the polymorphism will lie with the properties of a multiple gene family (see Chapter 28).

Nucleated erythrocytes are completely inactive in replication and transcription. Their nuclei are very small and the chromatin is highly condensed, a property that is acquired during maturation. A common speculation has been that the inactivation of erythrocyte chromatin may be related to the accumulation of H5. However, it is clear that the change in histone composition cannot itself be sufficient, because appreciable amounts of H5 are present in erythroblasts before the cessation of nucleic acid synthesis. One possibility is that the phosphorylation of H5 is important in this function (see below).

Evolution of Histone Sequences

The recognition of the same five classes of histones in (apparently) all eucaryotes implies that their evolution was an ancient event. It would not be unreasonable to suppose all histones evolved by divergence of duplicated sequences originally representing a single DNA binding protein. However, no remaining evidence for this has been observed in the relationship between present histone sequences.

Yet there are features common to all the histones. Figure 11.8 depicts the clustering of amino acids in each histone. In each case there is a highly basic region around the N-terminus, which provides a putative DNA bind-

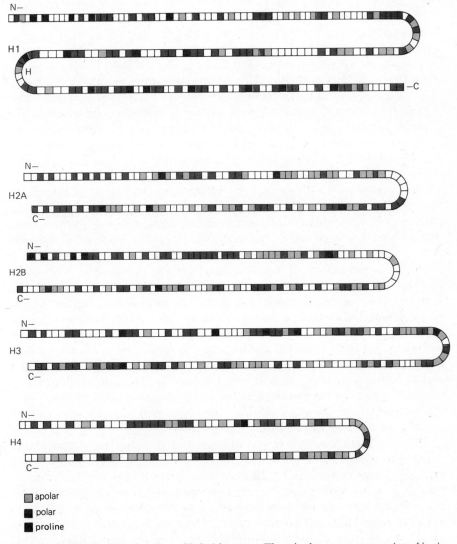

Figure 11.8. Distribution of amino acids in histones. There is always a concentration of ionic (largely basic) groups at the N-terminal end. Hydrophobic groups tend to be clustered in restricted regions, in which helix formation may be encouraged. Data are for trout H1 and for calf thymus H2A, H2B, H3, and H4.

ing site. In H1 there is also a highly basic region towards the C terminus. In the central part of each molecule there is a clustering of apolar amino acids, which would be expected to favor helix formation in this part of the polypeptide chain. This is borne out by analyses of the biophysical properties of isolated histone molecules. In each case, conformation is acquired as the salt concentration is increased, with the formation of an α helix for a distance of some 15–20 residues in the apolar region (Bradbury et al., 1967, 1972b, 1975c; Boublik et al., 1970a,b; D'Anna and Isenberg, 1972, 1974a,c; Wickett, Li, and Isenberg, 1972; Smerdon and Isenberg, 1976a; Lewis et al., 1975). In some cases it has been possible to isolate defined segments of the molecule and to show that they possess the secondary structure present in the intact molecule (Bradbury et al., 1975a,b; Chapman et al., 1976; Crane-Robinson et al., 1976, 1977; Hartman et al., 1977). This suggests a general type of structure in which there is a central region that forms a globular structure, extruding basic tails that extend to the termini. The globular regions appear to participate in interactions between the histone molecules; the N terminal tails (and sometimes the C terminal regions also) may be involved in binding DNA. The involvement of these features in the construction of the nucleosome is discussed in Chapter 13.

The persistence in diverse species of the same five types of histone in itself implies a high evolutionary stability. This is confirmed by the comparisons of the sequences of corresponding histones in different species that we have discussed above. The available data are summarized in the form of pairwise comparisons in Table 11.2; the data are not yet sufficient to allow the construction of genealogical trees for comparison with other proteins. The arginine rich histones are the most highly conserved; the striking conservation of sequence between the most distant higher eucaryotes compared, cow and pea, where there are almost no changes and those that do exist tend to be conservative, suggests that any variation in their sequences is highly inimical to function. No other case is known of such evolutionary immobility, with the possible exception of tubulin, another widely found structural protein (see Chapter 2). The slightly lysine-rich histones appear to have evolved in sequence about an order of magnitude more rapidly than the arginine rich. The lysine-rich histones seem to have evolved yet an order of magnitude more rapidly.

The same relative extents of variation are found within a given organism. Thus there seems to be no tissue-specific variation in H3 or H4, whereas multiple forms of H2A and H2B have been detected in sea urchin embryogenesis. It is interesting that the reverse relationship is seen in the modification of these histones, which is more frequent and more varied with H3 and H4 than with H2A and H2B. If variations in the electrophoretic mobility of H1 result from changes in primary sequence (as seems likely) rather representing modification (which may be cell cycle dependent), then tissue-specific variation in the lysine-rich histone(s) may be quite common. Clearly the possibility of variations in histone sequence either during embryonic

Table 11.2 Conservation of histone sequences

Histone	Species compared	Time since divergence (10^6 years)	Total amino acid changes	Nonconservative amino acid changes	Divergence/100 residues/million years
H4	cow—pig—rat	60	0	0	0
	cow—pea	1500	2	0	0.0013
	cow—Tetrahymena	2000	15(66)	10	>0.007
H3	cow—chick	300	½	½	0.0012
	cow—carp	400	½	½	0.0009
	cow—shark	1200	½	½	0.0003
	chick—carp—shark		0	0	0
	cow—sea urchin	700	1½	½	0.0016
	cow—pea	1500	3½	½	0.0017
	pea—sea urchin	1200	4½	½	0.0027
H2A	cow—rat	60	2	1	0.025
	cow—chick	300	2	0	0.005
	cow—trout	400	10	8	0.019
	cow—sea urchin	700	15/24	11/17	0.017/0.026
	trout—sea urchin		14	10	
H2B	cow—trout	400	9	7	0.018
	cow—sea urchin	700	60/28	40/20	0.07/0.03
	trout—sea urchin		32	22	
H1	cow—rabbit	60	7(/73)	4	0.16
	rabbit—trout	400	85	61	0.10
	rabbit—sea urchin	700	103(/125)	—	0.12
	trout—sea urchin		103(/125)	—	—

The total number of amino acid changes per histone is calculated from the complete protein sequences, with the exception of Tetrahymena H4 and several sources of H1, where only partial sequences are available except for trout (the length compared is indicated by the number in parentheses). When a histone is polymorphic, a difference involving only one of the two amino acids found at the site of polymorphism is counted as one-half. The number of nonconservative changes excludes those in which an amino acid is replaced by another of similar character (for example, substitutions between valine and isoleucine). The dual calculations for cow—sea urchin refer respectively to gonad/embryonic H2A and the Parechinus/S. purpuratus—P. miliaris H2B. These calculations take no account of the existence of further histone variants within a species, but simply compare the available sequences irrespective of the tissue source. All values are based on data quoted in the figures of this Chapter. A comparison between the DNA sequences of cloned sea urchin histone genes is given in Table 28.1.

stages or in adult tissues makes it more difficult to compare the sequences determined for H2A, H2B, or H1 in different species—for some of the observed variations could be characteristics of the tissue from which the histone was obtained rather than of the species. In the absence of more complete data, it has been necessary to ignore this caveat in comparing histone sequences.

On the basis of available data, Table 11.3 compares the rate of histone

Table 11.3   Rates of evolution of histone sequences

Protein	Number of positions at which no changes observed	Millions of years required to introduce 1 divergence/100 amino acids between two species
H4	100/102	770
H3	131/135	590
H2A	119/129	52
H2B	62/125	33
H1	139/213	11
Collagen (α1)		36
Glutamate dehydrogenase		55
Carbonic anhydrase		2
Cytochrome c		15
Glucagon		43
Insulin		14
Lutropin β		3
Myoglobin		6
Globin α		3.7
Globin β		3.3
Ribonuclease		2.3
Fibrinopeptide B		1.1

Data for the histones are derived from Table 11.2. Positions at which change has not been observed possess the same amino acid in all samples of the histone so far sequenced (after any necessary realignment to take account of deletions). The rate of evolution is calculated from the differences between corresponding histones of the most distant higher eucaryotic sources that have been sequenced. Data for other proteins are taken from Dayhoff (1972), Fitch (1973), and Wilson, Carlson, and White (1977).

evolution with that of other proteins. This demonstrates the difference between the rate of change in the arginine rich histones and other polypeptides. The other histones fall into the range seen with other proteins, although clearly they are located at the more slowly evolving extreme. It would be premature to draw firmer conclusions about the possible relative rates at which the slightly lysine-rich and lysine-rich histones are permitted to evolve until comparisons of sequence can be based on a larger number of species (and tissues).

A measure of the proportion of the sequence of each histone protein that is involved in functions common to all species is given by the number of positions at which no substitutions have been observed. The available data—which are rather limited—are given in the Table. From these data, it does seem that substitutions may tend to be located at corresponding sites in the histones of different organisms.

Modification of Histones

The selection of a small number of residues for modification in each histone indicates that the reaction is specific. In suitable systems it has been possible to correlate the occurrence of modification or demodification with stages of maturation of a cell line. With cells growing in tissue culture, some events have been localized at particular points in the cell cycle. Yet in spite of the evidence for specificity of location in the protein and for temporal occurrence, little is known about the function of the addition or removal of modifying groups. Nor is there much evidence on the diversity of enzymes presumably responsible for undertaking histone modification and demodification.

Modification includes acetylation, methylation, and phosphorylation. The only stable modification is the acetylation of the N-terminal amino acids of some histones. Modification to internal groups appear to be unstable; the addition of labeled modifying groups shows a rapid turnover for all histones in tissue culture cells (Wilhelm and McCarty, 1970; Balhorn et al., 1972; Jackson et al., 1975). When synchronized cells are used, most of the acetylation and methylation appears to occur during S phase, the time at which new histones are synthesized (Shepherd et al., 1971; Jackson et al., 1976). When cycloheximide is added to prevent the synthesis of new histones, acetylation and methylation continue to take place nonetheless, implying that they are not confined to newly synthesized proteins (Wilhelm and McCarty, 1970; Sung, Dixon, and Smithies, 1971; Sung et al., 1977).

The fate of newly synthesized histones can be followed by determining the electrophoretic mobility of pulse-labeled molecules. Modification probably occurs at the appropriate sites in all histone molecules immediately following their synthesis. In duck erythroid cells, for example, a 1 minute pulse of ^{3}H-leucine mostly enters the bands of histone H4 that represent modified molecules; less than 25% is unmodified. After a 1 hour chase, however, the situation is reversed, with a much increased proportion of H4 in the unmodified fraction (Ruiz-Carrillo et al., 1975). It is difficult to quantitate the modification of preexisting histones (that is, those synthesized in previous cell cycles); since even if all the molecules were modified during the course of a typical S phase (say about 6 hours), the rapid removal of modifying groups means that at any given time only those histones most recently modified would retain the acetyl or methyl groups. It is clear that preexisting histones are modified during S phase, but the quantitation of this process is obscure at present.

Modification of preexisting histones must of course occur in the nucleus. Generally it has been thought that the internal modifications of newly synthesized histones are made at the same time and place, whereas the addition of the stable N-terminal groups may occur during histone synthesis in the cytoplasm. The removal of modifying groups probably occurs only the nucleus. Some histone modifying enzyme activities have been identified in chromatin, but little is yet known about their specificity or how their action is controlled.

The general view of the acetylation and methylation of histones H2A, H2B, H3, and H4 suggested by these studies is that both preexisting and newly synthesized histones are modified during S phase; during the cell cycle the modifying groups are removed, to be replaced at the next S phase, when the cycle recurs. The obvious speculation prompted by this timing is that these modifications are involved in the replication of chromatin (see below).

A useful system for studying histone acetylation has been developed as a result of observations of the effects of butyric acid on cultured cells. Low concentrations (2–5 mM) inhibit DNA synthesis and cause histones to accumulate in a more acetylated state (Riggs et al., 1977; Hagopian et al., 1977). After 24 hours of treatment of HeLa cells, for example, 84% of the H4 molecules are acetylated instead of the usual 40%. The cause of this increase in the proportion of acetylated molecules appears to lie with an inhibition of histone deacetylase activity, an effect that is common to a large number of cell lines.* The effect is striking with H3 and H4, but may occur to a lesser extent with H2A and H2B. It offers the prospect of determining the consequences for chromatin of having an increased amount of acetylated H3 and H4 (see Chapter 14). Another situation in which acetylation is changed occurs in viral transformation with polyoma or SV40; the H3 and H4 molecules of the virions usually are highly acetylated (much more so than the histones of host cell chromatin). Certain non transforming host range viral mutants, however, generate virions with acetyl group levels similar to that of the host cell (Schaffhausen and Benjamin, 1976).

A cycle of modification and demodification is seen for H1, but its timing is distinct from the events occurring with the other histones. The lysine-rich histone(s) do not appear to be acetylated or methylated, but varying numbers of phosphate groups may be introduced. Analysis is complicated by the microheterogeneity seen in the H1 class. However, the occurrence of phosphorylation generates additional H1 bands on gel electrophoresis and, as with the other modifications, these include both the most recently synthesized and preexisting histones (Oliver et al., 1972). The timing of this phosphorylation has been unclear, with claims made both for S phase (like the other histones) and mitosis.† More recent reports suggest that a small amount of phosphorylation occurs at S phase, representing the introduction of 1–2 phosphate groups; more extensive phosphorylation, to bring the total number of groups up to as much as 6, occurs at further sites on H1 at the start of mitosis; these are removed at the end of division (Hohmann, Tobey, and Gurley, 1976; Gurley et al., 1978).

*References: Boffa et al. (1978), Candido, Reeves and Davie (1978), Cousens, Gallwitz and Alberts (1979), Sealy and Chalkley (1978a).
†References: Shepherd, Hardin, and Noland (1972), Lake and Salzman (1972), Bradbury et al. (1973), Gurley, Walters, and Tobey (1973).

At all events, it is clear that H1 is found in a phosphorylated state during mitosis of cell types as diverse as mammalian tissue culture lines and the slime mold Physarum polycephalum; it is either not phosphorylated or has fewer phosphate groups during interphase. The level of phosphate groups in H1 is determined by the net actions of a phosphokinase and phosphatase. The phosphokinase activity increases sharply around the time of the transition from G_2 to mitosis (Lake, Goidl, and Salzman, 1972; Bradbury et al., 1974a,b). When V79 cells are held in metaphase by treatment with vinblastine, the H1 remains phosphorylated (Lake, 1973). When the block is released, $\alpha^{32}P$ label is rapidly lost, as indeed is usually the case at the transition from mitosis to G_1 (Gurley, Walters, and Tobey, 1974; Balhorn et al., 1975). The timing of this H1 phosphorylation has prompted speculation that it may be involved in the mitotic condensation of chromatin. However, it is difficult to prove a causal relationship. The only evidence reported on this lies in the treatment of cells with $ZnCl_2$, which appears to prevent dephosphorylation, but does not prevent chromatin from regaining its usual condition as cells pass from mitosis into interphase (Tanphaichitr et al., 1976). This would imply that the presence of phosphate groups on H1 is not an impediment to decondensation.

The sequential phosphorylation and dephosphorylation of lysine-rich histone has been followed also in chick erythroid cells, in which H5 appears to be phosphorylated immediately upon synthesis, then gradually losing its modification (Sung et al., 1977; Sung, 1977). There are 5 sites of phosphorylation, 2 in the N-terminal region, the others possibly C-terminal (Sung and Freedlender, 1978). The proportion of phosphorylated H5 declines with the stage of maturation of these cells, which raises the possibility that it may be the amount of dephosphorylated H5, rather than the amount of H5 per se, which determines the condensation of chromatin. This is, of course, the opposite relationship between phosphorylation and condensation from that proposed with H1.

It is difficult to determine the proportion of histone molecules that suffers modification at any particular site, because of the transient association of the modifying group with the protein. Factors such as the proportions of cells in different stages of the cell cycle may therefore be responsible for the differences observed in the extent of modification at given sites in histones from different sources (see above). In view of these difficulties, it is not possible to conclude whether such differences are apparent or real and thus to decide whether histone modification plays the same role in different systems.

An extensive study of changes in histone modification has been made in the trout testis. This tissue has the advantage that cells at different stages of maturation on the spermatogenesis pathway can be isolated on the basis of size by centrifugation through density gradients. Cells become progressively smaller proceeding from primary spermatocytes to spermatozoa. The last synthesis of DNA is detected before the secondary spermatocytes enter meiosis to generate spermatids; the rate of histone synthesis (and modification) parallels that of DNA. Histones are progressively replaced by protamines as middle stage spermatids mature to spermatozoa. The protamines are small and very basic proteins, rich in arginine, with an average length of only 33 amino acids. Chromatin becomes more tightly condensed and less active as the protamines replace the histones (Dixon and Smith, 1968; Lam and Bruce, 1971).

All four of the lysine residues at positions 5, 8, 12, and 16 of H4 are acetylated. Very soon after its synthesis, H4 is acetylated twice; modification then continues at a slower rate to give forms of the protein with 3 and 4 acetyl groups. It is then slowly deacetylated to generate forms with 1 or 0 acetyl groups. The acetylation and deacetylation of a newly synthesized H4 molecule takes about 1 day (Candido and Dixon, 1971; Louie and Dixon, 1972a). A somewhat similar, although less-well characterized, series of acetylations and deacetylations takes place with H3, H2B, and H2A. The H3 molecule is acetylated principally at positions 14 and 23 and also at 9 and 18; four lysine residues can be acetylated in H2B and one in H2A (Candido and Dixon, 1972a,b).

The cycle of acetylation and deacetylation is accompanied by a similar cycle involving phosphate groups, for all histones except H3. There is a difference in the timing of events here. Histone H2A is phosphorylated soon after its synthesis; the proportion of newly synthesized histone in the phosphorylated band upon gel electrophoresis rises to about 25% during a 4 hour period and then declines as the phosphate groups are removed. Histone H4, however, suffers an appreciable lag before the introduction of phosphate groups, which then are duly removed (Dixon, 1972; Louie, Sung, and Dixon, 1973). Protamines also are phosphorylated. Each protamine gains 1–2 phosphate groups very shortly after synthesis and then more slowly gains a third. Phosphorylation is succeeded by a dephosphorylation reaction which occupies several days. Thus in spermatids that have just begun to synthesize protamines, most are found in the phosphorylated state, but with maturation the protamines become progressively dephosphorylated (Louie and Dixon, 1972b). Modification and subsequent demodification therefore appears to be a characteristic of proteins structurally associated with DNA.

Both the arginine-rich histones are methylated, but the methyl groups are stable. They are not found on either acetylated or phosphorylated H4, which suggests that they may be introduced at a later stage, after the acetyl and phosphate groups have been removed. Their lack of transience implies that the role of methylation may not be the same as that of acetylation and phosphorylation (Honda et al., 1975a,b).

What is the purpose of these modifications to histones and protamines? One possibility is that modification of the DNA-binding regions of the histones that constitute the nucleosome (H4, H3, H2B, H2A) might control their affinity for DNA. By reducing the positive charge of the protein, the modifications may reduce the affinity for negatively charged DNA (Sung and Dixon, 1970; Louie, Candido, and Dixon, 1973). The timing of most (although not all) modifications, which occur immediately upon synthesis, imply that it must be the modified forms of the proteins that bind to DNA; once binding has been accomplished, the modifying groups are removed. Yet it is not clear why such modification should be necessary, especially in view of the reconstruction of chromatin from unmodified histones that can be accomplished in vitro (see Chapter 13). Another function seems possible for the phosphorylation of lysine-rich histones (H1 and H5) and protamines; this could be concerned with the degree of chromatin condensation. The lysine-rich histones are good candidates for this role, because they are not part of the nucleosome subunit, but appear rather to be involved in the packing together of adjacent nucleosomes (see Chapter 13).

CHAPTER 12

Chromatin Components: Nonhistone Proteins

The small number of histones, their evolutionary conservation of sequence, and their universal role in the construction of the nucleosome, make it evident that they are structural components of chromatin. The four histones H4, H3, H2B and H2A form the protein constituents of the nucleosome subunit itself only H1, with an extra nucleosomal location, may be involved in the spatial connections between nucleosomes. The nonhistone proteins therefore provide the putative candidates for three functions, concerning the organization of higher orders of structure, the synthesis of nucleic acids, and the control of gene expression.

Structural components other than the histones may be concerned with the compaction of nucleosomes into interphase chromatin and with the further condensation into the typical structure of the mitotic chromosomes that occurs at division. It is necessary also to account for the difference in structure between euchromatin and heterochromatin, for the presence of regions comprising the banded ultrastructure of mitotic chromosomes, and for the formation of the specialized structures of the kinetochore and (perhaps) the telomere.

Enzymatic functions located in chromatin are responsible for replication of DNA, transcription of RNA, and modification of histones and (perhaps) nonhistone proteins. The complexity of the replication apparatus and of the transcription apparatus is not known; but it would not be surprising if several proteins were involved in the synthesis of each type of nucleic acid. The specificity with which histones are modified implies that there may be several enzymes involved. The extent to which these various activities are located in chromatin is not clear; some may be permanent constituents, whereas others associate transiently with the genetic apparatus.

The control of gene expression is presumed to involve the association of regulator molecules with recognition sites on DNA. By definition, any such species must be part of the nonhistone proteins (so long as the control molecules are protein). So little is known about the control of eucaryotic gene expression that it is not possible to estimate how many such proteins there may be or in what amount each should be present.

In this chapter we shall try to provide an assessment of the characteristics of the nonhistone proteins as such, principally concerning their diversity and the extent of possible differences between tissues and between species. Subsequently we shall consider what evidence there is on their distribution in different chromosome regions (see Chapter 14) and whether there is any significant evidence for their involvement in the control of gene expression.

Diversity of Nonhistone Proteins

The total amount of nonhistone protein reported in chromatin varies quite widely, but a typical value would be between 60 and 70% of the histone content (where the total mass of histones is not much more than the content of DNA). This corresponds to roughly 700,000 daltons of nonhistone protein for every 10^6 daltons of DNA. However, values have been reported as low as 20% of histone content or as high as 200% in various circumstances. A first question is whether this variation is apparent or real.

Nonhistone chromosomal proteins (sometimes abbreviated as NHC) are defined as the protein components remaining after the five histones have been removed from chromatin. This operational definition presents obvious difficulties. First, there is the problem of distinguishing genuine constituents of chromatin from contaminants derived from the nucleoplasm or cytoplasm. Also it may be necessary to distinguish between the proteins of chromatin itself and those that are part of its ribonucleoprotein transcripts.

Then there is the difficulty of knowing whether the proteins have been obtained in their native state, that is, without degradation; protease activities are commonly found in chromatin. Since different nonhistone proteins may have quite different features, such as affinity for DNA, any estimate of the relative amounts of individual species must be tempered by the realization that there is no reason to suppose that all nonhistones are equally well extracted. Finally, the importance of these factors may vary in chromatins derived from different sources, making it difficult to compare their nonhistone protein constitutions.

Lacking any functional assay for most nonhistone proteins, we are compelled to conclude tentatively that the mass of nonhistone protein generally is somewhat less than that of the histones, for this result is obtained with several preparative methods applied to many tissues; extensive variations probably are due to losses or to contamination during preparation. In this context, we should note also that on occasion proteins have been obtained by extraction of nuclei (rather than of chromatin); such "acidic nuclear protein" fractions may, of course, be expected to include chromatin proteins, but they may also have other components. We are here concerned with the nonhistone proteins of chromatin, which must be obtained therefore from preparations lacking other nuclear components (but even so may not be pure).

One method for preparing nonhistone proteins is to remove the histones from chromatin by extraction with sulfuric acid; then the remaining complex of DNA and protein can be dissolved in a solution containing 1% SDS, and the DNA removed by centrifugation, gel filtration or precipitation, to leave a supernatant fraction of nonhistone proteins (Elgin and Bonner, 1970). Alternatively, the nonhistones may be removed from the DNA by extraction with phenol or high ionic strength (Teng, Teng, and Allfrey, 1971). Another procedure is to dissociate chromatin completely in 2M NaCl-5M urea, after which the histones and nonhistones can be separated by fractionation, for example using ion exchange or hydroxyapatite (Graziano and Huang, 1971; Van den Broek et al., 1973; Chae and Carter, 1974; Rickwood and MacGillivray, 1975). This is often used in reconstitution experiments (see Chapter 28).

Somewhat similar results have been found in several systems by analyzing the nonhistone protein complement on SDS gels. Generally there are about 12-20 bands, ranging in molecular weight from 15,000 to 180,000 daltons (Elgin and Bonner, 1970, 1972; Shaw and Huang, 1970; Bhorjee and Pederson, 1972; MacGillivray et al., 1971, 1972; Wu, Elgin, and Hood, 1973). Usually most of the protein mass is contained in a small number of prominent bands.

There are two important drawbacks to this method of analysis. The gel separation is on the basis of molecular weight alone, so each band may represent any number of polypeptides of common molecular weight (and this may or may not be their native state since multimeric proteins would be dissociated into subunits). And only proteins representing more than about 1% of the total mass can be detected. These must be present, roughly speaking, in more than about 10^5 copies per genome (of 3×10^9 base pairs/ mammal) for an average protein of 50,000 daltons. By comparison, each of the nucleosomal histones should be present in about 3×10^7 molecules per genome.

An obvious question is the extent of differences between the nonhistone proteins of different tissues or species. It has been thought reasonable to suppose that those nonhistones with a structural role might be common to all tissues at least within a species; whereas those concerned with regulation should be tissue and species specific. On this basis, nonhistones found to vary between tissues have been considered to be candidates in the quest for regulator proteins. Two considerations greatly weaken the strength of this conclusion. First, it is not clear to what extent the pattern of gene expression varies between different tissues. Assays of messenger RNA populations suggest that only a small proportion of the expressed genes may be different when two tissues are compared; most are common to both (see Chapter 24). If this is true as a general conclusion, it would be surprising to find drastic changes in the array of regulator proteins in chromatin (assuming that it is true that genes are held in an active or inactive state by the binding of such regulator proteins). Second, accepting that there are changes to be found, it is by no means apparent that they should be detectable on SDS gels, which resolve a presumably much larger number of proteins into no more than about 20 bands; the chances of detecting changes in individual species should not be very high.

Generally similar results have been obtained in many comparisons of nonhistone proteins between related chromatins. These have involved different mammalian or avian tissues, growing and nongrowing cells, interphase and mitotic cells, virally transformed and nontransformed cells, and cells at different stages of embryonic development (Teng, Teng, and Allfrey, 1971; Bhorjee and Pederson, 1972; Elgin and Hood, 1973; Elgin et al., 1973; Platz and Hnilica, 1973; Karn et al., 1974; Hill, Maundrell, and Callan, 1974; Sanders, 1974; Yeoman et al., 1975; Krause, Kleinsmith, and Stein, 1975). In each case, quantitative scanning of the gels shows that most bands are unchanged. Usually there is a small number of changes in the amount present of some bands; sometimes a difference in the presence of one or two bands is seen. The overall impression left by these analyses is that, at the level of resolution offered by these gels, significant differences are not seen in such comparisons, since it is not possible to be sure that even those limited differences that have been noted are genuine rather than artefactual. It must be remembered that it is rare for any indication to be given of the reproducibility of the gel pattern. To conclude that a quantitative change in the amount of some band is significant requires knowledge of its variability in different preparations; standard deviations for the heights of these peaks usually have not been presented.

Identification of Nonhistone Proteins

The true number of nonhistone proteins is clearly very much larger than the number of bands seen on SDS gels. This is shown both by the isolation of subfractions of nonhistones, themselves proving to have many components (typically about 20 bands are seen on SDS gels), and by direct analysis of the total nonhistone protein fraction with methods affording greater resolution. Subfractions of the nonhistones have been isolated by elution with low concentrations of salt, by the ability to bind to DNA, or by their possession of a common type of modification. These are discussed below. Increased resolution of total nonhistones has been gained in one-dimensional SDS-Tris-glycine gels, allowing about 45 bands to be identified, and in two-dimensional SDS gel electrophoresis, allowing 50–60 spots to be distinguished, both with rat tissues (Wu, Elgin, and Hood, 1973; Yeoman et al., 1975). Again, only a small number of quantitative and qualitative differences can be seen between the proteins of different chromatins.

A technique with much greater resolution is two-dimensional gel electrophoresis using SDS in one dimension (to achieve separation on the basis of molecular weight) and isoelectric focusing in the other (where separation is on the basis of charge). Histones do not enter the gel, which affords a resolution in the range from pH 5.9 to 7.5. About 470 spots of nonhistone chromatin proteins from HeLa cells can be counted (Peterson and McConkey, 1976). Six major proteins account for almost 40% of the mass.

These proteins were obtained by preparing chromatin, digesting the DNA with S1 nuclease, and then directly applying the freed proteins to the gel. A similar analysis of nucleoplasmic proteins, taken as the supernatant from the chromatin pellet, identified 300 spots. Cytoplasmic proteins, obtained in the form of the first fraction lacking nuclei, displayed 530 spots. Some 64 proteins were common to all three fractions and accounted for about 50% of the mass. This might reflect either their wide distribution in vivo or contamination in vitro. Most of the nonhistone proteins are unique, but these represent only about 23% of the mass. Many of them are present in small amounts, of the order of 500–8000 copies per haploid genome.

Cell structural proteins are a prominent component of the nonhistone fraction. In these experiments, 3% of the mass of cytoplasmic proteins was provided by actin, which also accounted for 3.5% of the nucleoplasm and 6% of the nonhistone protein. At face value, this would imply that an appreciable proportion of the cellular actin is present in the nucleus, some at least attached to chromatin. In fact, quantitation of the actin suggests that more is present in the nuclear fractions than in the cytoplasm. This would be a surprising conclusion in view of the localization of the actin filaments described in Chapter 3. An alternative explanation is that there is extensive contamination of the nuclear fractions by each other and with cytoplasmic constituents. In this case, it becomes difficult to know what proportion of the apparent nonhistone chromatin proteins comprises the genetic apparatus.

Actin, myosin, tubulin and tropomyosin all have been identified in the nonhistone protein fraction of rat liver chromatin, where they account for almost 40% of the mass (Douvas, Harrington, and Bonner, 1975). In the chromatin of D. discoideum, actin and myosin provide some 35% of the mass of nonhistone proteins (Pederson, 1977). Again it is difficult to answer the question of whether these are genuine components of chromatin or contaminants. Certainly at least another line of evidence would be required to support a conclusion that these are structural components of the chromosomes.

Many experiments have been reported where attempts have been made to correlate changes in nonhistone proteins with switches in gene expression. Generally these have taken the form of stimulating liver with hormones, persuading resting cells to recommence growth by addition of fresh serum, or comparing virally infected with noninfected cells. Following the incorporation of labeled amino acids into nonhistone proteins has shown an increase upon stimulation (Rovera and Baserga, 1971, 1973; Rovera, Baserga, and Defendi, 1972). From this it has been concluded that nonhistone proteins are involved in the control of gene expression. But two factors render this conclusion invalid. First, only an overall change in the synthesis of nonhistones was found; the synthesis of new, individual protein species was not investigated. Second, there is no evidence that the newly synthesized proteins in any case are components of chromatin per se, let alone that they are a cause rather than effect of changes in gene expression.

The question of the true role of proteins in the nonhistone fraction is an important one. In a comparison of normal and hormone-activated rat liver, Pederson (1974b) fractionated nuclei into chromatin and ribonucleoprotein particles. Correlated with a threefold increase in RNA synthesis, there was increased synthesis of some of bands of nonhistone nuclear proteins separated on SDS gels. But all of these were part of the ribonucleoprotein fraction, not of chromatin. This immediately raises the possibility that the presence in chromatin of nascent RNA chains, not yet released from their template but already attached to protein, or contamination with ribonucleoproteins from the nucleus, may be responsible for the apparent increase in nonhistone proteins (see also Bhorjee and Pederson, 1973). These proteins therefore may be involved in packaging the gene product rather than in regulating gene expression.

HMG Proteins of Calf Thymus

The nonhistone proteins extracted from calf thymus chromatin with 0.35 M NaCl include a group of proteins soluble in 2% trichloracetic acid. These have low molecular weight, less than 30,000 daltons, and are known as the HMG nonhistones (an abbreviation for high mobility group). Gel electrophoresis identifies up to 20 bands, with two prominent components, HMG1 and HMG2, that account for a large proportion of the mass (Goodwin, Sanders, and Johns, 1973; Goodwin and Johns, 1973). These two proteins have been noticed also as contaminants of H1 isolated by perchloric acid extraction. They have molecular weights of about 27,500 and 26,000 and each is very rich in basic and acidic amino acids (which make up about 55% of the total). Thus these proteins share with the histones the characteristic of a high content of arginine and lysine, but they differ in possessing also a large amount of glutamic and aspartic acids. Each is present in about 10^5 copies per nucleus.

Both HMG1 and HMG2 bind to DNA (Shooter, Goodwin, and Johns, 1974; Goodwin, Shooter, and Johns, 1975; Yu et al., 1977). HMG1 appears to possess basic regions which bind to DNA, leaving more acidic regions free. HMG2 seems to cause DNA to take up a more compact shape and binds less strongly to denatured than to native DNA. Both have a high helical content, about 40–50% (Cary et al., 1976; Baker et al., 1976). HMG1 interacts with some, but apparently not other, subfractions of H1, while HMG2 reacts more weakly and less specifically (Smerdon and Isenberg, 1976b). No role is yet known for these two proteins, but their ready extraction in low salt, like that of H1, suggests that they are not tightly bound to DNA and may therefore be concerned with ultrastructure.

Trout testis chromatin also contains proteins extracted by low salt. Of the three major components, two are described as histone T and HMG-T and their N-terminal sequences have been determined, together with those of HMG1 and HMG2 of calf thymus (Huntley and Dixon, 1972; Watson, Peters, and Dixon, 1977). The interesting feature of these sequences is an apparent relationship between HMG-T on the one hand and HMG1 and HMG2 on the other. The size of HMG-T is 28,700 daltons, similar to that of HMG1 and HMG2; and its overall amino acid composition is very similar. Of the first 25 amino acids in each protein, 21 are identical in HMG1 and HMG2; and of these, 12 (almost half of the total sequence) are identical in HMG-T. There may therefore be an evolutionary relationship between these proteins, supporting the idea of a common structural role.

One of the other HMG proteins has been sequenced. HMG17 has 24% lysine and 4% arginine, resembling H1, but also has 24% aspartic and glutamic acids; its content of alanine, glycine, and proline is 42% (Walker et al., 1977). The molecule is 89 amino acids long, with a weight of 9247 daltons. Like the histones, the N-terminal end is very basic; 22 of the first 58 residues are lysine or arginine and only 7 are acidic. The C terminal end has an overall negative charge, with only 4 basic residues compared with 7 acidic.

DNA Binding Proteins

Nonhistone proteins that bind to DNA are generally thought to include the putative regulators as well as enzymes concerned with the use of DNA as template for nucleic acid synthesis. Proteins with a high affinity for DNA probably comprise a small minority; for example, 2 M NaCl removes about 70% of the nonhistone proteins together with the histones. But some proteins appear to be more tightly bound to DNA since they are not readily dissociated by salt.

One fraction has been isolated as a subfraction of the nonhistones not extracted by 2 M NaCl-5 M urea. These are generally low in molecular weight and represent about 8% of chromatin protein (20% of the nonhistones). When isolated, this fraction can rebind to DNA (Pederson and Bhorjee, 1975).

Another assay is to obtain a soluble rat liver chromatin preparation in 2 M NaCl-5 M urea-5 mM Tris; then when the NaCl concentration is reduced to 0.14 M, an insoluble deoxyribonucleoprotein complex forms. About 90% of the nonhistone proteins remain soluble. The remaining 10% can be recovered by adding 2 M NaCl-5 M urea. Presumably these have a high enough affinity for DNA to reassociate with it while others remain soluble. These proteins bind to phage T7 DNA about as well as to rat liver DNA, but bind well to single strand DNA and to poly(dAT) (Patel and Thomas, 1973; Thomas and Patel, 1976a,b).

The nonhistone proteins of ascites tumor cells have been fractionated into groups by their removal from chromatin when the salt concentration is increased in the presence of 5 M urea. Then the ability of each group to bind to DNA can be tested. There appears to be a wide range of binding affinities, in each case increased by the presence of histones; again there is no evidence for sequence specificity in the DNA binding (Lapeyre and Bekhor, 1976).

Another procedure is to chromatograph nonhistone proteins through a column of DNA-cellulose. Many proteins are retained by the column apparently as the result of a fairly unspecific binding. These proteins appear to have a low affinity for all DNA. A smaller number of proteins can be isolated by first running the proteins through a column of heterologous DNA to adsorb these; then the eluate is run through a column containing homologous DNA to adsorb proteins with a high affinity for specific sequences in the genome (Kleinsmith, Heidema, and Carroll, 1970; Van den Broek et al., 1973; Sevall et al., 1975). The proteins bind to DNA with varying affinities, as seen in the differences in the salt concentrations needed to elute them from the columns (Allfrey et al., 1973; Kleinsmith, 1973). A drawback in this procedure is that one might expect any protein with a high affinity for some specific DNA sequence also to have a lower affinity for all DNA (this question is taken up in Chapter 28). Thus it is not clear on what basis the fractionation proceeds; and quantitatively it becomes difficult to estimate the amount of any individual species.

Some nonhistone proteins are modified. The injection of ^{32}P is followed by incorporation of the label into nonhistones of rat liver. Most of the phosphate groups are added to serine, some to threonine. A phosphoprotein fraction can be isolated by following the label and it contains about 4–5 phosphate groups for every 100 amino acids (Kleinsmith and Allfrey, 1969; Teng, Teng, and Allfrey, 1970). The phosphate groups are not stable but turn over rapidly, with differences in stability between different phosphoprotein bands (Karn et al., 1974).

This fraction has been described as "phosphoproteins," but we should emphasize that it is heterogeneous and may include other, nonmodified species. The ^{32}P label provides a convenient way to follow its properties, but otherwise there is no reason to suppose that the constituent proteins are related in function. Among this fraction can be found proteins that bind to DNA; when applied to a DNA-cellulose column, about 1% of the label is retained. There appear to be about 20 bands when the binding fraction is analyzed on SDS gels.

The A24 Semihistone

One of the nonhistone proteins of rat liver, known as A24, has been purified and characterized. It is present in the soluble fraction extracted by 0.4 N

H_2SO_4, but is not removed by 0.35 M NaCl. Its solubility properties are therefore similar to those of the nucleosomal histones (Goldknopf et al., 1975).

Attempts to sequence the protein showed that its C-terminal region is the same as that of H2A. Two N-terminal sequences can be identified: one is that of H2A; and the other is different. The A24 protein then was shown to represent H2A with an additional polypeptide added through an isopeptide link to the ϵ-NH_2 of the lysine at position 119 (see Figure 11.4). The additional polypeptide turns out to be ubiquitin, a protein characterized for its ubiquitous presence in mammalian nuclei (Olson et al., 1977; Goldknopf and Busch, 1977; Hunt and Dayhoff, 1977).

The ubiquitin moiety is about the same size as H2A; overall it has an acidic nature and its content of glutamic and aspartic acids is sufficient to bring that of A24 overall to 20%. Although no precise quantitation has been achieved, the amount of A24 appears to be about 1% of the H2A content. What function is served by this curious modification is not known. The A24 protein is present in nucleosomes and it will be interesting to see whether these have any special location or function (Goldknopf et al., 1977).

Conclusions: Functions of Nonhistone Proteins

The difficulties inherent in demonstrating that any nonhistone protein is a genuine component of chromatin, the lack of knowledge as to the size of the replication and transcription apparatuses, and the absence of information on the expected number and diversity of regulator proteins, all mean that analysis of the nonhistone proteins in toto has as yet given little information about their biological functions. The most profitable approach may be to isolate subfractions whose members have features in common that are potentially related to function. Two approaches that have been used rely on ease of extraction from chromatin and ability to bind to DNA (hopefully in a sequence-specific manner). However, in neither case has this yet led to any understanding of the function of either the group as a whole or of any individual member.

An alternative approach is to try to work with chromatin containing (or lacking) particular nonhistones. Along these lines, attempts have been made to fractionate chromatin into transcriptionally active and inactive fractions whose nonhistones can be compared. We shall see in Chapter 28 that this has not led to the identification of any nonhistones associated with gene activity, principally because of the difficulties in achieving an effective fractionation. A more productive approach has been to prepare antigens against individual nonhistones, or groups of nonhistones, and then to attempt to locate these proteins in cases where chromosome regions can be distinguished. Work using Drosophila salivary glands and lampbrush chromosomes has led to the observation that chromosome regions are not identical in their possession of at least certain nonhistones. This is discussed in Chapter 16; but the correlation with location cannot yet be extended to identify functions with any given nonhistone. Work on the ultrastructure of the chromosome is discussed in Chapters 14 and 15; although ideas are being formulated about how the nucleosomes coil into higher orders of structure and what distinguishes different chromosome regions, this cannot yet be extended to include any discussion of what structural role might be ascribed to the nonhistones.

Perhaps the potential role of nonhistones as regulators of gene expression has attracted the most attention. It is, indeed, becoming common to see in the introduction to papers reporting research in this area statements that there is "considerable evidence" that nonhistone proteins are responsible for the specificity of gene transcription. But this is to overinterpret available data.

Two types of evidence have been taken to support the idea that nonhistones are involved in regulating gene expression. The first rests in claims that significant differences can be seen in overall comparisons of the nonhistone proteins present in chromatins derived from two sources expected to differ in their pattern of gene expression. We have seen above that it is doubtful whether these experiments should be able to detect differences, which may be smaller than can be resolved by the techniques that have been used. It is indeed dubious whether such differences have been detected; and if so, it is doubtful further whether they can be attributed to changes in proteins bound to DNA or even that are genuine components of chromatin. The second type of evidence lies in attempts to reconstitute from its components a chromatin preparation able to exhibit the pattern of transcription characteristic of its tissue of origin. We shall see in Chapter 28 that difficulties in constructing proper assays for specific transcription mean that claims to have demonstrated that this ability resides in the nonhistone fraction cannot yet be accepted. Although it remains an apparently inescapable conclusion that gene regulators reside in the nonhistone fraction, experimental evidence in support of this idea has, therefore, yet to be provided.

CHAPTER 13

The Nucleosome: Structure of the Particle

The development of the concept that the nucleosome provides the fundamental subunit of chromatin has brought about a radical change in the perspective from which the structure of the eucaryotic chromosome is analyzed. Just before the model of the nucleosome was proposed, in the first edition of this book it was possible to say in summary only that: "Our general view of the state of DNA in chromatin is of a nucleic acid duplex extensively covered in protein. The histones possess the greatest affinity for DNA and all classes of histone can bind directly to the nucleic acid; the association between DNA and histones appears to be largely responsible for ionic neutralization of the negative charges of DNA. The affinity for DNA of the nonhistone proteins is lower; they probably interact largely with each other and with histones and may bind directly to DNA to a lesser extent." While this remains true, of course, we can now proceed beyond these generalities to consider a defined structure rather than a somewhat vaguely outlined deoxyribonucleoprotein complex.

The analysis of chromatin can be focused on three general questions. What is the structure of the nucleosome? How are nucleosomes organized into chromatin and chromosomes? What structural features may distinguish regions of DNA involved in nucleic acid synthesis from those that are inactive? From the form of these questions, it is evident that discussion of both the structure and function of chromatin now is couched in terms of the nucleosome.

In this chapter and the next we shall be concerned largely with analyzing the more recent work that has been performed within this context; but where relevant we shall attempt also to reconcile this with earlier work, originally interpreted in less specific terms. Here we shall discuss the structure of the nucleosome, treating it generally as an individual particle whose existence is to be accounted for in terms of interactions between DNA and a complex of histones. In Chapter 14 we shall turn to the relationships between nucleosomes, to consider the interactions between adjacent nucleosomes, the assembly of nucleosomes and their possible alternative conformations, and the manner in which a sequence of nucleosomes may be organized into fibers of higher order structure. Analysis of the reconstitution of chromatin and its transcription in vitro will be deferred until it can be considered in Chapter 28 in the context of gene expression.

Repeating Length of DNA in Chromatin Subunits

The traditional view that the DNA of chromatin is not in general accessible to macromolecules because it is closely associated with proteins rested on three lines of evidence. First, the absence from chromatin of any appreciable amount of DNA free of protein was suggested by the elevated melting temperature of virtually all of the nucleic acid (see below). Then macromolecules have been used to probe the accessibility of DNA, including polymerases and nucleases. The extent of transcription by added RNA polymerase—or of replication by added DNA polymerase—is, as might be expected, very much less in chromatin than for free DNA (Silverman and Mirsky, 1973). Finally, the digestion of chromatin by DNAases proceeds much more slowly than the digestion of purified DNA (Mirsky, 1971; Mirsky and Silverman, 1972). The elevation of melting temperature, the reduced transcription by RNA polymerase, and the decrease in susceptibility to DNAase, all were taken to imply that little or none of the DNA in nuclei or chromatin is in a form unprotected by protein; and the protection was thought to be more or less uniform for all of the DNA. The general view derived from these experiments, although not always explicitly stated, was

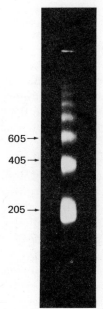

605 →
405 →

205 →

Figure 13.1. Micrococcal nuclease digestion of rat liver nuclei. The multimeric series of bands of DNA was separated by electrophoresis on 2.5% acrylamide gels (origin at the top); DNA was stained with ethidium bromide and appears fluorescent (as white bands). Data of Noll (1974a).

Figure 13.2. Micrococcal nuclease digestion of duck reticulocyte nuclei. Nuclei were digested for varying times to allow the reaction to proceed to different extents. Left: the percent of DNA entering the acid soluble condition for each digestion is indicated below each gel run. The rightmost gel is a limit digest (48% acid soluble DNA) of isolated chromatin. The acid insoluble DNA was isolated in each case and run on a 3% acrylamide gel to give the pattern shown. Black bands represent DNA that has reacted with stain. Right: a quantitative scan of a similar set of gels, showing the progressive reduction in size of the DNA. Data of Sollner-Webb and Felsenfeld (1975).

that DNA is extensively covered by protein; usually nucleohistone was thought of as a thread of DNA surrounded by protein. Given the enormous condensation of DNA into chromatin, it is, of course, true that little DNA must be exposed to the vagaries of the nuclear environment; but at the level of first interaction with protein, it now seems likely that the DNA is wrapped around the outside of a protein core to form the nucleosome.

The first evidence that DNA is organized into a regularly repeating unit in chromatin was provided by the observation of Hewish and Burgoyne (1973) that incubation of rat liver nuclei allows an endogenous enzyme to cleave the DNA into integral multiples of a unit length. The smallest is about 200 base pairs long (Burgoyne, Hewish, and Mobbs, 1974). The general significance of this observation was realized when Noll (1974a) showed that the same result is obtained upon digestion of either nuclei or isolated chromatin with the enzyme micrococcal nuclease. As shown in Figure 13.1, when the digested DNA is isolated and electrophoresed through an acrylamide gel, a series of bands is obtained. The sizes of these fragments suggest

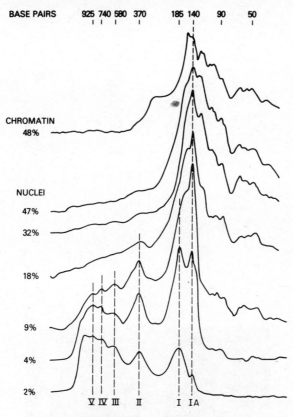

BASE PAIRS 925 740 580 370 185 140 90 50

CHROMATIN
48%

NUCLEI
47%
32%
18%
9%
4%
2%

V IV III   II   I   IA

Figure 13.2. (*Continued*)

that they comprise a set of multimers, with a monomer of 205 base pairs, and with dimers and trimers identifiable at 405 and 605 base pairs.

In earlier experiments with this enzyme, Clark and Felsenfeld (1971) monitored the digestion of calf thymus chromatin by the conversion of DNA to the acid soluble form. The reaction ceased when 54% of the DNA became acid soluble (that is, entered the form of small oligonucleotides, say <20 base pairs long). The DNA remaining after this *limit digest* took the form of small fragments. At this time, more attention was paid to the observation that over half of the DNA could be digested, since this was at apparent variance with the previous idea that DNA in chromatin is protected from such attack by its associated proteins. Since then, of course, attention has focused on the nature of the discrete fragments that are produced. These are less than 160 base pairs in length in the limit digest; but Sollner-Webb and Felsenfeld (1975) showed that the digestion of duck reticulocyte nuclei initially produces a series of multiple bands whose monomer is 185 base pairs long. Figure 13.2 shows that when the digestion is continued, the bands of dimers, trimers, and the higher multimers are converted to monomers; and then the

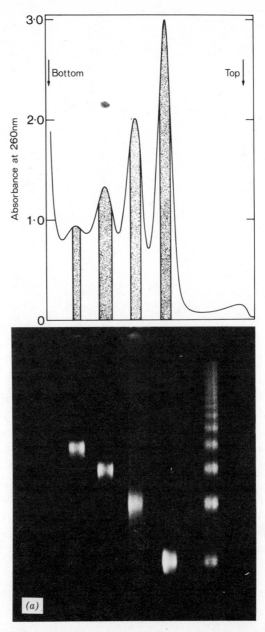

Figure 13.3. **(a) Lengths of DNA in nucleosome multimers.** The upper part of the figure shows the separation of particles by centrifugation on a sucrose gradient. Monomers are at the right and dimers, trimers, and tetramers can be identified. The DNA from the peaks indicated by the shaded areas was analyzed as shown on the gels below. Each multimer contains a corresponding length of DNA. The gel at the right shows DNA extracted from an unfractionated micrococcal nuclease digestion. (b) Electron micrographs of the four peaks, showing monomeric, dimeric, trimeric and tetrameric particles, respectively. Data of Finch, Noll, and Kornberg (1975).

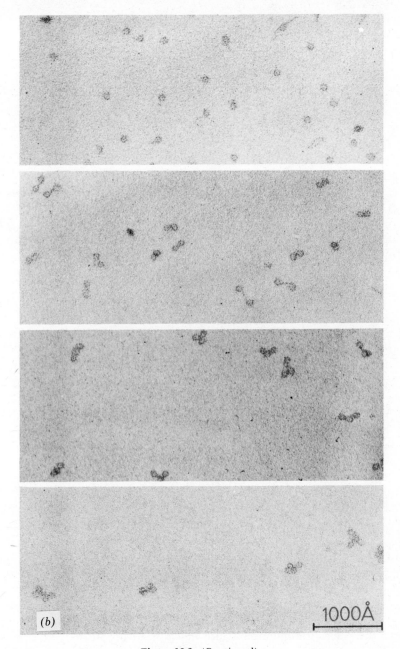

(b)

1000Å

Figure 13.3. (*Continued*)

monomers in turn are converted to the smaller fragments found in the limit digest of chromatin.

Micrococcal nuclease and the endogenous rat liver nuclease both introduce double strand breaks in DNA (see Table 13.3). The similar effects of these two different enzymes imply that the production of the multimeric series reflects some intrinsic feature of chromatin, present in nuclei and maintained during preparation, and is not due to some feature characteristic of the particular enzyme activity (and see below). The set of bands can be seen clearly when only 2% of the DNA in duck nuclei has been rendered acid soluble by micrococcal nuclease. This means that a small amount of DNA is preferentially attacked; and this must represent susceptible regions that connect less susceptible lengths of about 200 base pairs. Under conditions when all of the DNA seen on the gels enters the form of monomers and their multiples, only 13% of the rat liver DNA has become acid soluble. This implies that at least 87% of the DNA is organized into the 200 base pair repeating unit.

What is the origin of this 200 base pair length of DNA? When chromatin is digested with micrococcal nuclease and then centrifuged through a sucrose gradient, a series of particles can be found. The smallest (monomers) sediment at about 11S, the next peak (dimers) is about 15S, and peaks of larger material may be resolved with varying degrees of clarity (Noll, 1974a; Senior, Olins, and Olins, 1975).

If the smallest particles represent individual nucleosomes, they should contain the monomeric length of DNA; larger particles should contain corresponding multiples. By examining the DNA present in each peak, Lacy and Axel (1975) and Finch, Noll, and Kornberg (1975) were able to show that this is indeed the case. Figure 13.3 illustrates this relationship. Electron microscopy identified individual particles, with dimensions of about 8.5 × 10.5 nm, in the monomer peak, while dimers, trimers and tetramers of these particles are found in the larger peaks.

This suggests that chromatin consists of a series of discrete particles—nucleosomes—each containing about 200 base pairs of DNA. The DNA duplex continues from one nucleosome to the next and is especially susceptible to nuclease attack between the particles. Digestion with micrococcal nuclease thus first generates multimers, and as cutting becomes more frequent these are converted to monomers. The kinetics of reaction are consistent with an alternation of short susceptible regions (at particle junctions) and longer resistant regions (within particles) (Clark and Felsenfeld, 1974; Lohr, Kovacic, and Van Holde, 1977). More extensive digestion reduces the size of the monomers, ultimately resulting in the pattern of the limit digest seen in Figure 13.2. This is due to digestion of DNA within the 200 base pair length of the nucleosome (see below).*

---

*Chromatin extracted by methods involving shearing does not generate the 200 base pair repeating pattern upon nuclease digestion (Noll, Thomas, and Kornberg, 1975). Since this pat-

Visualization of Individual Particles

Subunits of chromatin were first visualized as discrete particles when Olins and Olins (1974) lysed interphase nuclei into water and then fixed the swollen material, which was centrifuged onto carbon grids. After staining, chromatin fibers can be seen streaming out of the ruptured nuclei, generally taking the form of roughly spherical particles, of diameter about 6-8 nm, connected by filaments of about 1.5 nm diameter (close to the width of the DNA duplex). Less stretched regions show tighter packing of the particles. (This is the technique originally developed by Miller for visualizing the transcription of genes for ribosomal RNA and is discussed in Chapter 27).

Somewhat similar results were obtained in several tissues, including rat thymus and liver and chicken erythrocytes (Olins, Carlson, and Olins, 1975). The particles were given the name ν (nu) bodies and they take the same appearance in both native chromatin and chromatin that has been reconstituted from its components. Clusters of several nu bodies that are not resolved by the staining technique may appear as thickenings in the chromatin thread with apparent fiber widths of 10-12 nm and 22-24 nm (see Chapter 14).

Figure 13.4 shows a striking example of subunit particles spilling out of a lysed nucleus, taken from a detailed study of the nature of the particles reported by Oudet, Gross-Bellard, and Chambon (1975), who introduced the term nucleosome (which we shall use here rather than the equivalent terms such as nu body). When visualized directly from lysed nuclei, the nucleosomes appear tightly packed into compact masses. Isolated chromatin proved difficult to examine because it could not be dispersed so readily. But chromatin that had been depleted of the lysine-rich histones H1 or H5 displayed nucleosomes less tightly compressed, when fine filaments sometimes could be seen to connect the particles. This gives the subunits the appearance of beads on a string, as seen in the example of Figure 13.5. That the filament connecting the particles consists of DNA is shown by its disappearance upon digestion with micrococcal nuclease.

The apparent size of the particles depends upon the method used in their preparation. When visualized directly upon a grid, nucleosomes appear roughly spherical with a diameter of 12.8 ± 1.2 nm. Fixation with 1% formaldehyde decreases the diameter to 9.6 ± 0.9 nm; while critical point drying increases it to 13.7 ± 1.0 nm. By following micrococcal nuclease digestion with sucrose gradient centrifugation, up to 80% of the DNA of chromatin could be converted into the form of individual nucleosomes. Figure 13.6 shows that these have the same appearance as those seen in the form of

tern is found upon digestion of either intact nuclei or chromatin prepared by other methods (including very brief nuclease digestion to generate rather large fragments), its existence may be taken to be diagnostic of the native state. One suggestion for the reason for its absence from sheared preparations is that this treatment may induce movement of histones (Doenecke and McCarthy, 1976).

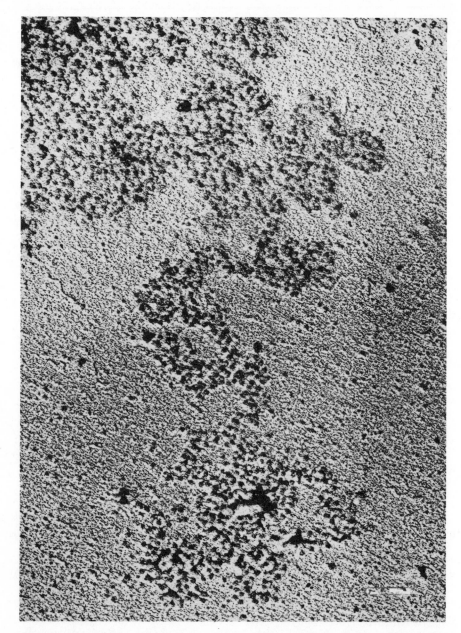

Figure 13.4. Chicken liver chromatin spilling out of nuclei lysed directly on a grid and stained with uranyl acetate. The bar is 100 nm. Data of Oudet, Gross-Bellard, and Chambon (1975).

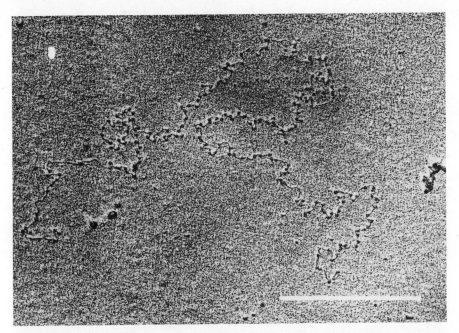

Figure 13.5. A string of nucleosomes connected by DNA after trypsin digestion to remove H1 from chicken liver chromatin. Similar results are obtained when H1 or H5 is removed by salt extraction. The bar indicates 500 nm. Data of Oudet, Gross-Bellard, and Chambon (1975).

beads on a string. Similarly to the particles characterized in other experiments (see for example Figure 13.3), these sediment at 10.5–12S. Upon digestion with proteinase, they leave DNA with a length of 66.7 ± 9.6 nm. This corresponds to about 197 ± 29 base pairs and provides an important confirmation of the size of the DNA in the nucleosome deduced from analysis of digested DNA on acrylamide gels (see Figures 13.1 and 13.2). Similar measurements, of 64 nm and 71 nm of DNA per subunit, have been made by Woodcock, Sweetman, and Frado (1976) and Senior, Olins and Olins (1975).

The presence of nucleosomes in chromatin depleted of H1 or H5 implies that the lysine-rich histones are not part of the particle, at least as it is visualized in these experiments. Two further experiments confirm this conclusion. Analysis of the histones present in isolated nucleosomes shows H2A, H2B, H3 and H4 to be present (and see below). Reconstitution experiments, in which adenovirus DNA was mixed with only these four histones, led to the formation of nucleosomes. The spacing between the nucleosomes declined as the concentration of histones was increased. Figure 13.7 shows some examples, from which it is clear that these four histones possess the ability to compact DNA into particles. The DNA of phage lambda forms the same number of nucleosomes per unit length as adenovirus DNA, which demonstrates the lack of any sequence specificity in the formation of the particle. Nucleosomes are formed in vitro only in the presence of all four

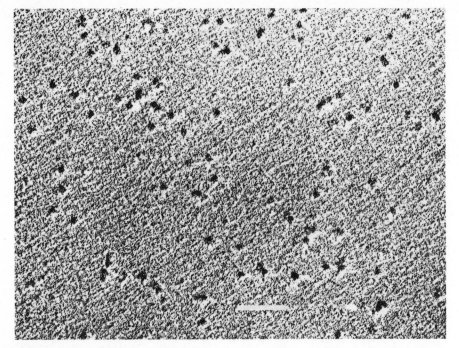

Figure 13.6. Nucleosome monomers isolated after treatment of chicken erythrocyte chromatin with micrococcal nuclease. The bar indicates 100 nm. Data of Oudet, Gross-Bellard, and Chambon (1975).

histones, H2A, H2B, H3, and H4; neither H1 nor nonhistone proteins are needed.

The compaction of DNA in the nucleosome can be expressed by the packing ratio, the length of DNA in the nucleosome divided by the length of the particle along the fiber. Measurements of the content of DNA and of the dimensions of the particle yield an approximate measure for this ratio of 5–7, corresponding to the organization of about 67 nm (200 bp) of DNA into a particle with a diameter of 10–12 nm.

Direct measurements of the packing ratio have been made from observations of the form taken by SV40 DNA. The viral DNA can be isolated, either from virions or from infected cells, as a nucleoprotein complex sometimes known as the SV40 minichromosome. This contains about equal masses of DNA and histones.* Isolated from infected nuclei in 0.15 M NaCl, the mini-

---

*Early studies showed that SV40 and polyoma virions contain proteins that electrophorese with the same mobilities as some of the histones; and similar proteins could be found as part of nucleoprotein complexes isolated from the nuclei of infected cells (Lake, Barban and Salzman, 1973; Fey and Hirt, 1974; Meinke, Hall, and Goldstein, 1975; Pett, Estes, and Pagano, 1975). This led to the view that minichromosomes in both infected nuclei and mature virions have the same structure, consisting of viral DNA and the four histones H2A, H2B, H3, and H4; histone H1 was thought to be absent in both cases. This was the model described in the papers

chromosome sediments at about 85S; treatment with 0.6 M NaCl reduces
the rate of sedimentation to about 55S, a change that might reflect the extraction of H1 (Christiansen and Griffith, 1977). Griffith (1975) found that
in 0.15 M NaCl the minichromosome takes a compact form in which it has
a width of 10.5 ± 1.0 nm and a length of 207.0 ± 12.5 nm. Since SV40 DNA
has a length of about 1500 nm under these conditions, the packing ratio
must be about 7. In these experiments, a reduction in the salt concentration
to 0.015 M NaCl converted the minichromosome to a flexible string of 21
beads, each about 11 nm in diameter, and connected by filaments of width
2 nm and average length 19 ± 8 nm (55 ± 25 base pairs); this corresponds to
a content of some 200 base pairs per bead (a little more than the 170 base
pairs originally calculated). In this form, the overall packing ratio is about
2.75.

The salt-induced change in conformation has been absent from other
experiments, in which the minichromosome of SV40 or polyoma has taken
a form depending on the number of beads. The most common structure has
20–22 beads and appears both at low and high salt as a string of particles
again connected by stretches of free DNA that are variable in length (Bellard et al., 1976; Cremisi et al., 1976). The packing ratio here was just more
than 5, with 200 base pairs of DNA present in each 12.5 nm bead. The greatest number of nucleosomes that can be found on the viral DNA in these conditions proved to be about 23–24, in which case no free internucleosomal
DNA is found. (From the 5224 base pair length of SV40 DNA, a maximum of 26 beads each containing ~200 base pairs would be predicted.
Perhaps constraints in the circular molecule limit the number). The same
results were obtained with SV40 minichromosomes extracted from virions
or from the nuclei of infected cells. The lack of a salt-dependent contraction
in structure may be due to the absence of H1 from both types of minichromosome in these experiments; since addition of H1 conferred this property
upon the complex (see Chapter 14).

An obvious question raised by these results is the nature of the free DNA
apparently connecting nucleosomes in the flexible beaded string. Is this a
genuine feature of the SV40 minichromosome (at least under conditions
when H1 is absent, as in the virion), or is it an artefact of preparation? Analysis is complicated by the lack of direct determinations of whether H1 has
been present in some of these experiments, and also by lack of knowledge
of what effect the viral capsid proteins may have upon the complex within

cited here in the text and in other early studies of the minichromosome. However, Varshavsky
et al. (1976) found that all five histones are present in the usual proportions in SV40 minichromosomes isolated with 0.15 M NaCl from the nuclei of infected CV1 monkey kidney cells. In
virions purified from the same source, there appears to be no H1, although the complex obtained by alkaline disruption may contain one of the viral capsid proteins, VP3 (Christiansen
et al., 1977). The existence of two forms of the minichromosome, possessing and lacking H1,
would constitute a useful system for examining the effect of this histone on nucleosome
structure.

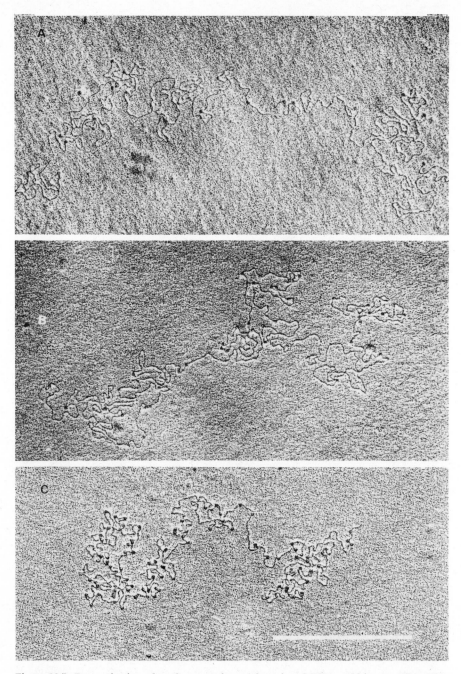

Figure 13.7. Reconstitution of nucleosomes from Adenovirus 2 DNA and histones H2A, H2B, H3, and H4 of calf thymus. The DNA was mixed with histones at varying ratios in high salt and then the salt concentration was decreased by dialysis. An increasing number of nucleosomes forms as the ratio increases from 0.4 mole histone/mole DNA in (A) to 0.8 in (B) and then to 1.0 in (C). The bar indicates 500 nm. Data of Oudet, Gross-Bellard, and Chambon (1975).

the virion. Formally the possibilities are that a variable number of nucleosomes might be assembled on SV40 DNA, leaving some DNA free when the number is insufficient to package all the genome into 200 base pair beads; or all the DNA may be nucleosomal in vivo, but some nucleosomes may disappear because their protein components are lost during preparation, perhaps in an action associated with the absence or removal of H1. Although there is no definite answer, we should remark now that there is no reason to suppose there to be stretches of free DNA in nuclear chromatin in vivo; such regions probably are generated by some unravelling of the nucleosome string in vitro. Their existence would make difficult the accomplishment of sufficient packing ratios (see Chapter 14). We shall return below to this question, but the structure of a compact mass of tightly packed nucleosomes is probably more a reflection of the native state than is the picture of beads on a string.

Histone Octamer of the Nucleosome

The original proposal that chromatin might be constituted from subunits which contain 200 base pairs of DNA and only the four histones H2A, H2B, H3, and H4 was made by Kornberg (1974), just before the biochemical and electron microscopic work that we have discussed above. The components of the subunit were inferred from the properties of histone complexes observed by Kornberg and Thomas (1974). They obtained two fractions when histones were extracted from chromatin in 2 M NaCl-50 mM NaAc and then separated on Sephadex. The first fraction contained both H1 and a tetramer of H3 and H4; their joint elution is an anomaly. The tetramer contains two molecules each of H3 and H4 according to its molecular weight and the pattern of formation of cross links with the reagent dimethyl suberimidate (which gives eight bands comprising H4, H3, $H3_2$, H3 · H4, $H4_2$, $H3_2H4$, $H3H4_2$, $H3_2H4_2$). This complex could not be formed by mixing H3 and H4 in vitro. The second fraction contained H2A and H2B in an oligomeric aggregate, whose nature was not clear, since cross linking generated a more complex pattern than would be expected of a tetramer. Mixing the H3-H4 tetramer with the H2A-H2B oligomer generated no extra bands on cross linking. But when the two histone complexes were mixed with DNA, the X-ray diffraction pattern characteristic of native chromatin, which has a prominent 10 nm repeat, was generated. Kornberg and Thomas therefore suggested that the basic unit of chromatin is assembled from an interaction of these two histone complexes with DNA.

The model for the subunit proposed by Kornberg possessed two central features: the tetramer of H3-H4; and 200 base pairs of DNA. It seemed likely (although not established as a general rule) that chromatin contained equal amounts of H2A, H2B, H3, and H4, with about half as much H1 on a molar basis.* This stoichiometry would correspond to one molecule each of the first four histones for every 100 base pairs of DNA. If the subunit contains a tetramer consisting of $H3_2H4_2$, it must therefore have 200 base pairs of DNA. This conclusion was supported by the nuclease studies, which had just shown there to be a repeating unit of this size in rat liver nuclei.

The stoichiometry between the histones suggested that two molecules each of H2A and H2B should be associated with the H3-H4 tetramer and the 200 base pairs of DNA. This idea was supported by a rough calculation that these components would have about the right mass to assemble a subunit corresponding to the well-known 10 nm repeat seen in X ray diffraction of chromatin (see below). The original model did not specify the nature of the association between the H3-H4 tetramer and the H2A-H2B complex; one possibility was that H2A-H2B might form an oligomeric thread with which discrete tetramers of H3-H3 were associated. It has since become clear, of course, that there is a discrete particle which contains two molecules each of H2A and H2B as well as of H3 and H4.

The evidence originally adduced to support this model may perhaps seem less than completely compelling. The stoichiometry between histones and with DNA even now is difficult to establish unequivocally by experiment for a given chromatin; and the isolation of the H3-H4 tetramer and the H2A-H2B oligomer does not of itself prove that it is these forms from which the repeating subunit is constituted. The length of DNA in the repeating unit is not 200 base pairs in all eucaryotes, but may vary quite widely (and in such instances one might therefore expect to see some disturbance in the ratio of histone to DNA). The calculation of the mass of the subunit needed to accommodate a 10 nm repeat can be only approximate because it involves assumptions about the density and packing together of the particles. But the model is substantiated in postulating the existence of a repeating subunit and identifying its components and size.

Nucleosomes have been isolated by sucrose gradient centrifugation of chromatin digested with micrococcal nuclease, as we have mentioned, and some of their properties are summarized in Table 13.1. Estimates for their mass range from 230,000 to 300,000 daltons (Senior, Olins, and Olins, 1975; Olins et al., 1976; also see references cited in the Table). A nucleosome in chromatin, associated with half a molecule of H1, would be expected to have a mass of 256,000 daltons; isolated nucleosomes possessing or lacking H1 should have masses of 244,000 or 268,000 daltons, respectively. Experimental values for nucleosomes tend toward the higher end of the range; the possible explanation that nonhistone proteins are associated with the nu-

---

*The exact molarity was difficult to establish because of problems resulting from differential losses during traditional methods of extraction. By overcoming this difficulty, Joffe, Keene, and Weintraub (1977) have shown for the example of chick erythroblasts that histones H2A, H2B, H3, and H4 are present in equimolar amounts.

Table 13.1 Properties of isolated nucleosomes

Dimensions	10 nm diameter, spherical or elliptical appearance by electron microscopy (range of reported diameters varies from 7 nm to 13 nm depending on method of preparation and staining)
	0.5 axial ratio from neutron scattering corresponds to dimensions of about 11 × 11 × 6 nm
	5 nm radius of gyration for DNA; 3 nm radius of gyration for protein; neutron scattering corresponds to protein-rich core surrounded by hydrated shell of DNA
Mass Content	11S peak represents monomers; 15S peak is dimer in sucrose gradient
	total mass is 250,000–300,000 daltons
	protein: DNA ratio is about 1.2–1.3 according to buoyant density in CsCl of 1.3–1.45 $g \cdot cm^{-3}$
Model Structure	nucleosome components are: 200 base pairs DNA = 135,000 daltons histone octamer = 109,000 ½ molecule H1 = 12,000 total mass = 256,000 daltons
	core particle components are: 140 base pairs DNA = 90,000 histone octamer = 109,000 total mass = 199,000 daltons

Summary of data of Olins and Olins (1974), Olins et al. (1975, 1976), Senior et al. (1975); Woodcock, Safer, and Stanchfield (1976), Woodcock, Sweetman, and Frado (1976); Oudet, Gross-Bellard and Chambon (1975), Germond et al. (1975); Noll (1974a), Finch, Noll, and Kornberg (1975), Finch et al. (1977); Honda et al. (1975); Pardon et al. (1975), Hjelm et al. (1977), Suau et al. (1977); Kornberg (1974), Kornberg and Thomas (1974).

cleosomes and increase their mass generally is excluded by experiments in which only histones have been shown to be present. But the discrepancy is not significant, since these estimates of mass usually are based upon rates of sedimentation and thus must rely on assumptions about nucleosome shape that are likely to be imprecise.

The buoyant density of nucleosomes in CsCl usually corresponds to a protein:DNA ratio of more than 1.2; although some estimates have been closer to the 0.9 ratio predicted by the model, none is actually this low. One reason for this discrepancy may be that nucleosomes are readily trimmed to "core particles," which have less DNA but retain all four histones. This would increase the ratio of protein:DNA to 1.2. (This should also reduce the mass of the particle, but there appears to be less effect upon the rate of sedimentation than might be expected, which emphasizes the insensitivity of these estimates.)

Experiments to characterize the histones present in isolated nucleosomes confirm that the four histones H2A, H2B, H3, and H4 are present in equimolar amounts (Oudet et al., 1975; Olins et al., 1976; Rall, Okinawa, and Strniste, 1977). We shall come later to the question of the location of H1, but will note now that H1 usually may be found on nucleosome oligomers, but its amount is reduced in preparations of monomers, and it is absent from core particles (see Figure 13.9).

While it is clear from such results that nucleosomes contain four histones and lack nonhistone proteins, they do not prove that every nucleosome has the identical protein content. One way to exclude the existence of heterogeneity in the nucleosome population is to apply to chromatin antigens directed against individual histones. Such experiments have shown that anti-H4, anti-H3, and anti-H2B react with the majority of nucleosomes (probably with all), but have proved less effective with anti-H2A (Goldblatt, Bustin and Sperling, 1978).

Cross linking of the proteins in nucleosomes has been used to investigate the structure of the particle (see below). From these experiments there is also evidence of the existence of the histone octamer. Thomas and Kornberg (1975) showed that dimethyl suberimidate cross links the four nucleosomal histones into a series of bands up to the size of the octamer. This represents the predominant reaction and can be equated with cross linking between the protein components within individual nucleosomes. This is suggestive evidence that much of the histone content of chromatin is organized into octamers (although, of course, it falls short of proof that all the histones are organized in this manner). In these experiments, H1 was not linked to the other histones, which provides further evidence for its absence from the nucleosomal structure. In contrast with the failure of cross links to form between nucleosomal histones and H1, some bands representing 16-mers and 24-mers were formed; these presumably represent cross linked aggregates of adjacent nucleosomes.

The nucleosome model accounts for the structure of chromatin of virtually all eucaryotic cells. Apparent exceptions have been inferred for certain lower eucaryotes, in which there have been difficulties in identifying some of the histones. But now all four nucleosomal histones have been characterized in yeast, although H1 appears absent; and all the usual histones are found in the macronucleus of Tetrahymena (Mardian and Isenberg, 1978; Glover and Gorovsky, 1978). However, histones H1 and H3 have not been found in the Tetrahymena micronucleus, which provides the only known deviation from histone ubiquity (Gorovsky and Keevert, 1975). In spite of the problems with characterization, the pairwise interactions of the lower eucaryotic histones with each other generally are similar to those of the histones of higher eucaryotes (see below).

A variation in the histone constituents of the nucleosome is seen in mammalian cells in which some of the histone H2A has been linked to an additional polypeptide to form the semihistone A24 (see Chapter 12). This may constitute up to 10% of the H2A present in chromatin and it is present in preparations of mononucleosomes (Albright, Nelson, and Garrard, 1979; Goldknopf et al., 1977; Martinson et al., 1979c). The interaction of A24 with H2B appears identical with the usual reaction displayed by unmodified H2A.

Variations in the Repeat Length of DNA

The initial cleavage of chromatin by micrococcal nuclease takes place at sites that recur at intervals of about 200 base pairs. As the reaction proceeds, two changes occur in the pattern of released DNA fragments. The proportion of multimers decreases and the proportion of monomers increases. This is due to the introduction of further scissions between nucleosomes. And the length of DNA in each fraction decreases somewhat. This decrease has been most fully followed with the monomers, but occurs also with the surviving multimers.

The reduction in length represents a "trimming" of the newly generated ends of DNA which become susceptible to the nuclease. In this section and the next we shall analyze the loss of end material from the DNA of monomeric nucleosomes; later we shall consider in detail the pattern of length reduction in multimers.

The occurrence of this decrease means that the size of the isolated monomeric DNA may underestimate the length of the repeating unit. Noll and Kornberg (1977) therefore suggested that a more accurate estimate might be obtained by comparing the lengths of DNA in the multimeric series. If all multimers suffer the same trimming of ends, at least initially, the differences in size between them should correspond to exact multiples of the repeat length. Another approach has been to divide the lengths of DNA in multimers by the relevant number of repeating units (Compton, Bellard, and Chambon, 1976). As the size of the multimer increases, the distortion caused by end trimming becomes less important.

The limitations of these approaches are discussed in Table 13.2, which summarizes data presently available on the length of the repeating unit in different organisms and tissues. Generally the repeat lengths deduced from these more sophisticated calculations are close to the initial size of the monomer. This suggests that the first action of micrococcal nuclease is indeed to make a single scission between nucleosomes; the removal of material from the newly generated ends occurs only as a second (although rapid)

Table 13.2 DNA content of the nucleosome

Source	Base pairs	Ref.
Rat liver (adult)	198	A
Rat liver (fetal)	193	B
Rat kidney	196	B
Rat bone marrow	192	B
Syrian hamster liver	196	B
Syrian hamster kidney	196	B
Rabbit liver	200	C
Rabbit cerebellar neurons	200	C
Rabbit cortical neurons	162	C
Rabbit cortical glia	197	C
Cow thymus	185	D
Chicken liver	202	J
Chicken erythrocyte	212, 207	E
Chicken oviduct	196	B
CHO cell	178	B
HeLa cell	188	B
BHK cell	190	B
Trout testis	200	F
Tobacco	194	G
Barley	195	G
Sea urchin (Arbacia lixula) gastrula	218	I
Sea urchin (Arbacia lixula) sperm	241	I
X. laevis blood	189	H
X. laevis liver	178	H
Aspergillus nidulans	154	J
Neurospora crassa	170	K
S. cerevisiae	165, 160	L

All values are based upon comparisons of the lengths of DNA present in members of a multimeric series of nucleosomes. Either the differences in length between adjacent members of the series are taken, or the repeat length for each member is obtained by dividing the total length by the number of subunits. The first method gives a value based essentially on the difference in lengths between the first and last members of the series, although the lengths of intervening members contribute to determining the standard deviation. This assumes that losses of terminal material are the same in all members of the series, which does not appear to be strictly true (see text). The second method ignores such losses, which become less significant (but not negligible) as the size of the multimer increases. Standard deviations are not quoted individually, but these measurements usually are reported to have an accuracy of ±5-6 base pairs. Measurements for the repeat length of the same chromatin in different laboratories often differ by this amount; published values for rat liver vary from about 195 to 200 base pairs and other examples are given above.

References: A—Noll (1974a), Noll and Kornberg (1977); B—Compton et al. (1976); C—Thomas and Thompson (1977); D—Sollner-Webb and Felsenfeld (1975); E—Morris (1976b), Compton et al. (1976); F—Honda et al. (1975); G—Philipps and Gigot (1977); H—Humphries et al. (1979); I—Spadafora et al. (1976); J—Morris (1976a); K—Noll (1976); L—Lohr and Van Holde (1975), Noll (1976), Thomas and Furber (1976).

step. The size of DNA in a band is generally taken to be given by the center; we shall come later to the question of the origin of the breadth of the bands, but in the case of monomers it is probably true that this depends both upon some variation in the site of initial scission and on a partial trimming.

In several mammalian cells the repeating length of DNA is very close to 200 base pairs, the value originally obtained with rat liver. In some avian tissues, fish and plants, similar values have been obtained. But different values have been found in other instances. The first to be observed was the reduction in length in three fungi, where the repeats vary from 154 to 170 base pairs. The first instance of differences within an organism was provided by rabbit brain, in which cortical neurons have a repeat length of 162, compared with the value of 200 seen in other tissues. A slight increase in repeat length is seen in chicken erythrocytes compared with liver or oviduct. Whether the reduction in repeat length in three mammalian tissue culture lines represents a feature acquired in culture or is characteristic of the original tissues remains to be seen. The longest repeat yet found is in a sea urchin sperm, 241 base pairs compared with 218 found in gastrula chromatin.

The range of variation is therefore appreciable, at present apparently from 154 to 241. Obviously these data do not represent a systematic enough survey to make possible any general conclusions on the extent and cause of the variation. Sometimes it has been suggested, implicitly or explicitly, that the repeat of 200 base pairs postulated in the original formulation of the nucleosome model is the norm, from which other values are variations. But more data will be required to say whether this is so, or whether there is in fact a range of continuous variation between organisms and even between tissues. If all the nuclear DNA is organized into nucleosomes, one implication of the variation in DNA repeat lengths is that there should be accompanying differences in the ratio of the mass of the histones to the mass of DNA.

There have, of course, been speculations on the cause and possible function of the variation (see Noll, 1976; Morris, 1976; Thomas and Thompson, 1977; Spadafora et al., 1977). One is that the presence of a long repeat may be connected with lack of gene expression; the stimulus for this proposal is the example of chicken erythrocyte and sea urchin sperm, which represent inactive cells. It is possible that such a repackaging of chromatin might occur in specialized cells in which virtually the entire genome is repressed, although obviously further evidence would be necessary to support any general conclusion. What evidence is available at present, however, argues that this is not likely to apply within a given cell, since both transcribed sequences and satellite DNA sequences display the same repeat length as that of bulk (overwhelmingly nontranscribed) DNA (Lacy and Axel, 1975; Lipchitz and Axel, 1976). Although not conclusive, this suggests that the overall repeat length of inactive and active sequences may be the same within a given cell (see Chapter 14). However, there may be differences in particular regions of the genome. The nucleosome repeat length of X. laevis oocyte-type 5S DNA, for example, is 175 bp in both blood and liver, while the average values for bulk DNA of these two tissues are 189 and 178 bp, respectively (Humphries, Young, and Carroll, 1979).

Another possibility is that the repeat length may be determined by the H1 (or H5) histone that is present. Substitutions in H1 during embryonic development, or the tissue specific differences that appear to exist, might be correlated with changes in repeat length. The extreme example, of course, is the avian erythrocyte, in which H5 is eventually substituted for H1. Here Weintraub (1978) has indeed shown that during erythropoiesis the repeat length increases gradually from 190 to 212 base pairs in parallel with an increase in H5 level from 0.2 to 1.0 molecules per nucleosome. Since the H1 level remained constant at 1.0 molecule per nucleosome, it is possible that the increase in total H1 + H5 level from 1.0 to 2.0 might be responsible for changing the repeat length, rather than the arrival of H5 per se. It is interesting that the repeat length at each intermediate stage took a single value with only the usual degree of variation and did not show a biphasic distribution.

Another possible cause of variation lies with the four major histones. The extensive sequence conservation of H3 and H4 makes it unlikely that they are involved; variations in H2A and H2B have been considered, as has been the possible influence of histone modification. However, the histones seem able to assemble the basic structure of the nucleosome, but not to determine the spacing of nucleosomes in oligomeric clusters (see below).

Is the nucleosome the subunit of mitotic chromosomes as well as of interphase chromatin? Both nuclease digestion and electron microscopic studies suggest that there is no change in the nucleosome during the condensation of chromatin. Micrococcal nuclease digestion of mitotic chromosomes isolated from CHO cells identifies DNA in the same multimeric series found with interphase chromatin, just as digestion of interphase and mitotic nuclei isolated from the slime mold Physarum polycephalum shows no difference in the repeat length of DNA (Compton et al., 1976; Hozier and Kraus, 1976; Vogt and Braun, 1976a). Electron microscopy of mitotic chromosomes of CHO and L cells shows that where the structure of the chromosome appears to unravel, particles with the appearance of nucleosomes can be seen (Rattner, Brand, and Hamkalo, 1975; Compton et al., 1976; Howze, Hsie, and Olins, 1976). In some preparations the particles appear to be connected by stretches of free DNA and there has been some discussion about whether this is a genuine feature of chromosome organization or an artefact of preparation. We shall follow the view of Compton et al. in supposing that such free DNA is probably produced by stretching during preparation; consistent with this idea, the fibers showing nucleosomes separated in this manner generally lie in a parallel orientation, as might be generated in this way. These data suggest that nucleosomes are present in mitotic chromosomes; although this falls short of proof that the structure and number of nucleosomes remains unaltered during mitosis, it seems likely that the condensation of chromatin for cell division is accomplished by a change in the packaging of nucleosomes (see Chapter 14).

Structure of the Core Particle

How does the structure of the nucleosome accommodate the variation in repeat length of DNA? When digestion with micrococcal nuclease continues beyond the initial scission, two changes occur in the monomers. The length of DNA is reduced from its initial value to about 140 base pairs; and any H1 that was associated with the monomers is lost. The resulting species are known as *core particles*.

In every case studied, irrespective of the nucleosomal repeat length, the core particle consists of 140 base pairs of DNA (reported values vary from 138 to 141 base pairs) and a histone octamer. This has led to suggestions that the DNA in the nucleosome can be considered to fall into two regions. The *core region* is always 140 base pairs.* The *linker region* comprises the rest of the repeating unit (Shaw et al., 1976; Whitlock and Simpson, 1976; Bakayev, Bakayeva, and Varshavsky, 1977; Noll and Kornberg, 1977).

The digestion of the monomer to generate core DNA proceeds in discrete steps. In the experiments of Noll (1974a) with rat liver nuclei, the monomer band formed a doublet, with peaks evident at 205 and 170 base pairs. The smaller band constitutes a partially trimmed monomer. In the limit digests characterized by Clark and Felsenfeld (1974) and Sollner-Webb and Felsenfeld (1975), there is a series of fragments ranging from 140 base pairs to as small as 20 base pairs (see Figure 13.2). Here the largest band represents core DNA and the smaller bands are derived from internal digestion of the core particle (see below). The early time course of digestion with micrococcal nuclease shows that first the 200 base pair monomer length is generated, then an intermediate 160 base pair fragment is seen, and finally this is replaced by the 140 base pair core fragment. Figure 13.8 shows the "pauses" that the enzyme makes at 160 and 140 base pairs in the early digest, before continued digestion gives the smaller fragments of the limit digest.

Not all of the DNA in the linker region is immediately susceptible to micrococcal nuclease. If any part of the 60 base pair region linking two rat liver nucleosomes could suffer the first cleavage, the width of the monomer band should range from a minimum of 140 base pairs (single core lengths) to a maximum of 260 base pairs (one core plus two full linkers). But the variation falls over the more restricted range of 160 to 210 base pairs. This suggests that only part of each linker is immediately accessible. The breadth

---

*The value of 140 bp was determined in a great many early experiments and is used here for consistency with the original reports. Actually very precise measurements suggest a length of 146 bp (as determined with rat liver and mouse tissue culture cells by Lutter, 1979). This difference does not influence any of the conclusions discussed here and its significance is analyzed later in the text.

Base pairs

← 160

← 140

|5 s   30 s   1′   2′   5′   10′

Time of digestion

Figure 13.8. Reduction in size of DNA of monomers during micrococcal nuclease digestion of rat liver nuclei. The first fragment to be released has a size of 200 base pairs; it is next reduced to 160 base pairs; and then core DNA of 140 base pairs is generated. The DNA was extracted from 11S monomers isolated on a sucrose gradient, electrophoresed on a 6% acrylamide gel, and stained with ethidium bromide. Data of Noll and Kornberg (1977).

of the monomeric band in N. crassa (where the repeat length is 170 base pairs) corresponds to only about half the variation seen in rat liver; this is consistent with the idea that it is the linker region that is smaller in the fungi and again it suggests that only part of each linker is available for the first cut (Noll, 1976). Once this cut has been made, however, the rest of the linker is rapidly removed (see Figure 13.10).

If the reduction in size from monomeric to core DNA represents trimming of accessible ends, the same total amount of DNA (60 base pairs when the repeat is 200 base pairs) should be trimmed from each multimeric fraction. We shall see later in Figure 13.16 that the size of DNA in the multimers is indeed reduced during continued digestion. The results are along the lines of, but do not exactly fit, the expectation of end trimming.

The properties of isolated core particles are consistent with the interpretation that they possess less DNA than the monomeric nucleosomes. They appear to have been first isolated by Van Holde et al. (1974) and Shaw et al. (1974), who characterized what were then described as "PS" particles from a digest of calf thymus chromatin. These appeared as globular particles of diameter about 7.4 nm. They sedimented at about 11S, corresponding to a mass average of about 180,000 daltons, with a content of DNA varying from 100–175 base pairs. In a characterization of purified cores, Olins et al. (1976) showed that their sedimentation properties are similar to nucleosomes, although their molecular weight is lower, about 215,000 daltons, which is close to the expected mass of a histone octamer bound to 140 base pairs of DNA. The shape and size of the core particle deduced from neutron scattering is little different from that of the nucleosome monomer (Hjelm et al., 1977; Suau et al., 1977).

The concept that the core particle represents the basic interaction of histones with DNA is supported by the success of Tatchell and Van Holde (1977) in reconstituting 11S particles from the four histones and DNA fragments of 140 base pairs. When the core DNA was mixed with 2 moles of each histone at high ionic strength, dialysis to a lower ionic level allowed particles to reform. The circular dichroism and denaturation curves of the reconstituted cores were the same as those of cores prepared directly from chromatin. A more penetrating assay for correct assembly is provided by the susceptibility of reconstituted material to nucleases, which for DNAase I was indistinguishable from that of cores prepared from chromatin (see below). Actually, even shorter fragments of DNA are able to form particles in which the ends seem to be similarly aligned (Tatchell and Van Holde, 1979; see Chapter 14).

We should emphasize that the characterization of two regions of nucleosomal DNA does not make any implication that there is a particular topological arrangement between nucleosomes. "Core DNA" and "linker DNA" are operational terms that describe the regions less and more accessible to nuclease, respectively. The core DNA appears to be of constant length (140 base pairs) in all nucleosomes. The linker DNA may vary from as little as 8 base pairs at each end (16 per nucleosome) to as much as 50 per end (100 per nucleosome). Its greater accessibility does not imply that linker DNA is more stretched out than core DNA or that it is extended between nucleosome cores. It may simply lie in a more accessible location at the junction of the nucleosomes, with part susceptible to the first cut and part rendered more accessible following cleavage, but with no other differences in its degree of compaction on the protein (see Figure 13.10).

The existence of linker DNA seems to depend on factors other than the four core histones. The original observations of Oudet et al. (1975), showing that particles can be assembled from DNA and the four core histones, made the point that the basic structure does not depend on H1 or nonhistone proteins. By using histones of CHO cells, calf thymus, and sea urchin sperm, Spadafora, Oudet, and Chambon (1978) have shown that the reconstituted particles do not reflect the original repeat lengths of the native chromatins (see Table 13.2). Irrespective of the source of histones, at low concentrations monomers are formed with a DNA content of about 196 base pairs. When closely packed multimers are formed, however, the repeat length is invariably 140 base pairs. Similar observations of a discrepancy between the structure of the first particle and of subsequent particles have been made with calf thymus histones by Steinmetz, Streeck, and Zachau (1978). It is as though some DNA additional to 140 base pairs can be associated at the ends of single nucleosomes; but this is compelled to become part of the next particle when more than one is assembled together. This suggests that the linker is established or maintained by the presence of additional components during assembly of adjacent nucleosomes. Addition of H1 to the in vitro system does not have any effect. In summary, the four core histones possess the ability to organize 140 base pairs of DNA into a core particle; but they do not determine the spacing of the cores.

Another indication that features additional to the histones may be important in linking adjacent core particles is provided by the properties of nucleosomes that have been depleted of H1 by extraction with 0.6 M NaCl. Micrococcal nuclease treatment of chick erythrocyte chromatin treated in this way identifies "compact oligomers," principally dimers and trimers but perhaps some tetramers (Tatchell and Van Holde, 1978). These compact dimers and trimers contain 265 ± 10 and 390 ± 20 base pairs of DNA, respectively. This is less than the 280 and 420 bp that would be expected of multimeric cores; one possible explanation is that the DNA at the ends of each core is shared with the next core. A similar series of multimeric cores with a 140 bp register has been found in calf thymus chromatin treated with 0.6 M NaCl (Steinmetz et al., 1978). This pattern is not found when H1 is removed by treatment with tRNA, which suggests that it might reflect an ability of core histone complexes to "slide" along DNA in high salt.

Location of H1 between Cores

The idea that H1 is present at the junction of nucleosomes and is not a component of the bead as such is consistent with many earlier studies of chromatin. The preferential extraction of H1 in low salt suggests that it is less tightly bound to DNA than are the other histones. Its removal does not change biophysical properties of chromatin such as optical rotatory dispersion or X ray diffraction (Tuan and Bonner, 1969; Bradbury et al., 1972a). Extraction of the lysine-rich histones from nuclei causes the chromatin fibers to take up a more dispersed appearance, which led to suggestions that H1 might be involved in condensation (Brasch, Setterfield, and Neelin, 1972). More recently it has been reported that extraction of H1 allows nucleosomes to be seen as beads on a string instead of more compact masses (Thoma and Koller, 1977; Renz, Nehls, and Hozier, 1977). This supports the earlier observation of Oudet et al. (1975) on the greater readiness with which H1-depleted chromatin can be dispersed for electron microscopy (see Chapter 14).

Isolated nucleosome dimers and high multimers contain all five histones, H1 as well as the four core proteins. Preparations of monomers may contain

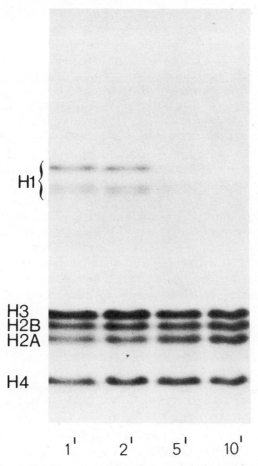

Figure 13.9. Release of H1 from nucleosome monomers during digestion with micrococcal nuclease. The proportion of H1 in the monomers is initially less than that of the other histones and it is reduced as micrococcal nuclease trims the ends of the DNA. Conditions were the same as described in Figure 13.8, except that the proteins were electrophoresed through SDS-18% acrylamide gels. Data of Noll and Kornberg (1977).

reduced amounts of H1 (or H5). The H1 present is lost when monomers are digested to core particles (Shaw et al., 1976; Varshavsky, Bakayev, and Georgiev, 1976; Noll and Kornberg, 1977). When the process of digestion is followed by isolating monomeric particles, up to five species can be distinguished by electrophoresis. These are characterized by different lengths of DNA and contents of H1 (Bakayev, Bakayeva, and Varshavsky, 1977; Todd and Garrard, 1979). Figure 13.9 shows the rate of loss of H1 from rat liver nucleosomes during digestion with micrococcal nuclease. By comparing these results with those of Figure 13.8 it can be seen that H1 is finally lost at the transition from 160 base pairs to 140 base pairs. Consistent with the earlier observations that removal of H1 renders chromatin more soluble, core particles can be solubilized more easily than monomeric nucleosomes.

These results suggest that H1 is associated with the region of linker DNA lying at the junction between nucleosomes. Part of the linker DNA constitutes the region that is susceptible to the first cleavage by nuclease. Whether H1 remains associated may depend on the position of the initial scission within the linker; at all events, removal of the linker DNA sees the loss of H1. The correlation between the location of H1 and the initial site of scission is emphasized by observations that chromatin depleted of H1 is digested more rapidly by micrococcal nuclease. In this case, the 160 base pair intermediate may be omitted and the core DNA is formed more readily (Whitlock and Simpson, 1976; Noll and Kornberg, 1977). So H1 may protect part of the linker against the initial scission, and this region becomes more susceptible in the isolated nucleosome or upon removal of H1. On the other hand, in experiments in which H1 was removed from calf thymus of HeLa nuclei by treatment at low pH, the initial size of the DNA fragments released by micrococcal nuclease was unaltered; depending on the conditions, a metastable intermediate of 168 bp may be found (Lawson and Cole, 1979; Weischet et al., 1979). Although clearly stabilized by the presence of H1, the intermediate particle may therefore be a feature of the interaction of core histones with DNA. Association of H1 with the linker regions is supported also by the observation that addition of H1 to 160 bp particles lacking it changes the melting pattern from the biphasic curve characteristic of core particles to the monophasic curve found with nucleosomes; this may result from suppression of the easier melting of the terminal regions (Simpson, 1978).

When mouse liver nuclei are digested with DNAase II, cuts are made between nucleosomes (see below). A difference is seen in the relative amounts of the peaks isolated on sucrose gradients when H1 is extracted with 0.6 M NaCl before the centrifugation. The presence of H1 is associated with a higher proportion of multimers; its removal sees an increase in the monomer peak at the expense of the larger peaks (Altenburger, Horz, and Zachau, 1976). This suggests that H1 might be holding adjacent nucleosomes together even after DNAase II has cleaved the DNA between them.

Figure 13.10 shows a diagrammatic representation of a nucleosome trimer (not implying the existence of any particular topological organization). This summarizes the salient features of the relationship between nucleosomes.

The sites at the junctions of nucleosomes that are susceptible to nucleases may also be accessible to other reagents. The existence of a 200 base pair repeat in nuclei of rat liver and Drosophila embryos is supported by the results of exposure to the cross-linking agent trimethyl psoralen. Under ultraviolet irradiation, psoralens can cross link pyrimidines across the partner strands of the duplex. The DNA of chromatin reacts to a lesser extent than free DNA; it appears to be protected by its association with proteins. Since cell and nuclear membranes are permeable to psoralens, the reaction

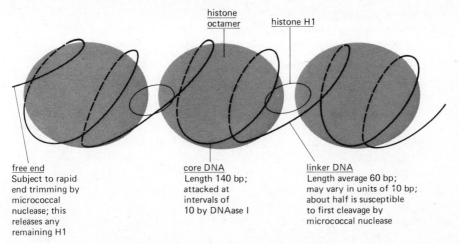

histone
octamer   histone H1

free end
Subject to rapid
end trimming by
micrococcal
nuclease; this
releases any
remaining H1

core DNA
Length 140 bp;
attacked at
intervals of
10 by DNAase I

linker DNA
Length average 60 bp;
may vary in units of 10 bp;
about half is susceptible
to first cleavage by
micrococcal nuclease

Figure 13.10. Model for structure of nucleosomes (not to scale).

can be performed in vivo. When the DNA of cross-linked Drosophila or mouse L cell nuclei is examined by electron microscopy under denaturing conditions, lengths of single-strand DNA (not cross linked) alternate with duplex regions (held together by cross links) (Hanson, Shen, and Hearst, 1976; Cech, Potter, and Pardue, 1977). The easiest pattern to interpret is the alternation of short duplex regions with longer single strand bubbles that range from 200 bases to about 900. The lengths of the bubbles show peaks with a periodicity of about 200 bases, which suggests that cross linking has occurred between nucleosomes. Also seen, however, are longer duplex lengths, again with a periodicity of 200 base pairs; whether these are genuine or an artefact (such as collapse of bubbles) remains to be seen. That the psoralens preferentially cross link the same regions that are attacked by micrococcal nuclease is suggested by the pattern of release of labeled psoralens from DNA of chromatin upon micrococcal nuclease digestion. A radioactive label in the cross link is lost more rapidly than a label in bulk DNA (Wiesehahn, Hyde, and Hearst, 1977; Cech and Pardue, 1977; Hallick et al., 1978). This suggests that cross links occur in the same linker DNA region initially attacked by the nuclease. The length of the region that is susceptible to cross linking is not known. The satellite DNAs of D. melanogaster show a different periodicity for the sites of cross linking from that of the other nuclear sequences; whether there is a difference in nucleosome structure in the satellite regions remains to be established (Shen and Hearst, 1978).

Exposure of DNA on the Nucleosome Surface

The interaction of DNA with a complex of histone proteins in vitro (see Chapter 14) and the roughly sevenfold compression of the length of DNA in the nucleosome have generally been taken to suggest that DNA is wound around the outside of the protein complex. Of itself, this evidence is suggestive rather than conclusive; for certainly it would be possible to construct models in which (for example) 140 base pairs of core DNA were surrounded by histones while 60 base pairs of linker DNA formed more exposed tails at each end. But two types of data suggest that DNA is indeed wound continuously on the surface of the nucleosomes. We shall come below to the biophysical properties of nucleosomes and core particles, which suggest that the DNA forms an outer shell while the protein forms an inner core. Here we consider the nuclease studies which suggest that both core and linker DNA have the same type of structure, in which there is a periodicity of 10 base pairs.

The first evidence for such an organization within the nucleosome was provided by the report of Noll (1974b) that DNAase I introduces single strand cuts every 10 base pairs. The properties of this endonuclease, and of other nucleases to which chromatin has been subjected, are summarized in Table 13.3. When the products of digestion with DNAase I are electrophoresed on denaturing gels, a series of bands is obtained, with the smallest of 20 bases and the largest usually in the range of 160–200 bases. An example is shown below in Figure 13.12. The bands differ by increments of 10 bases;* and there are some characteristic variations in relative intensities. The bands at 50, 80, and 110 bases are prominent; those at 60, 100, and 130 bases are weak.

The same periodicity is generated by other single-strand endonucleases, including the endogenous enzyme of rat liver and DNAase II (Simpson and Whitlock, 1976a; see below). A periodicity of 10 base pairs is seen also in the limit digest produced by micrococcal nuclease (see below). This suggests that it arises from some structural feature of chromatin.

The same pattern is obtained upon DNAase I digestion of nucleosome oligomers, monomeric nucleosomes, or core particles (Sollner-Webb et al., 1976). The coincidence of the nucleosome and core particle digests suggests that these single strand cuts are made in the region of the core DNA. Taking each cut to represent a site that is accessible to DNAase I, this suggests that core DNA may be wound on the surface of the nucleosome in an arrangement that makes sites available to the enzyme at intervals that are multiples of 10 bases. A significant feature is that the same DNAase I digestion pattern is found with chromatins of all repeat lengths, so whatever structural features are responsible for this pattern must be common to all nucleosomes.

*Core DNA is cleaved into 14 bands. The more precise length of 146 bp per core mentioned in the previous footnote assigns these an average separation of 10.4 bp, which has been confirmed by measurements on the individual bands (Prunell et al., 1979). The cutting sites therefore now are often referred to as 1, 2, 3, etc., corresponding to the bands of notional length 10, 20, 30 bases as described in the text. The original nomenclature is retained here for ease of comparison with the original research papers.

Table 13.3 Nucleases used to probe chromatin structure

Enzyme	Source	Action on purified DNA	Immediate product
Micrococcal nuclease (also known as staphylococcal nuclease) E.C. 3.1.4.7	S. aureus	hydrolyzes DNA or RNA to nucleoside 3′ phosphates and small oligonucleotides; endonucleolytic action on single strand or duplex DNA preferentially attacks Np-dT, Np-dA bonds to leave 5′-OH, 3′-P termini, with flush ends from duplex DNA; may also have exonucleolytic action in removing dN-3′-P from newly generated ends	
DNAase I (also known as pancreatic DNAase) E.C. 3.1.4.5	pancreas (usually bovine)	hydrolyzes DNA to oligonucleotides of about 4 bases; when activated by Mg^{2+} introduces single strand breaks in duplex DNA with 5′-P, 3′-OH termini; with Ca^{2+} ions also makes some duplex breaks	
DNAase II (also known as spleen DNAase) E.C. 3.1.4.6	spleen (usually porcine)	hydrolyzes DNA to oligonucleotides of about 6 bases; introduces single strand breaks in duplex DNA with 5′-OH, 3′-P termini; does not require metal ions	
Exonuclease III	E. coli	degrades duplex DNA from each 3′-OH end, releasing nucleoside 5′ phosphates, to generate short duplex region (at which reaction ceases) with two protruding single strand 5′ tails; also produces single strand breaks on 5′ side of apurinic sites; and has 3′ phosphatase activity	

References: Micrococcal nuclease—Sulkowski and Laskowski (1962), Ho and Gilham (1974), Ponder and Crawford (1977); DNAase I—Melgar and Goldthwaite (1968a,b), Price (1975); DNAase II—Koerner and Sinsheimer (1957), Young and Sinsheimer (1965), Bernardi and Sadron (1964); Exonuclease III—Richardson et al. (1964), Weiss (1976), Riley and Weintraub (1978).

A complication in interpreting these results is that the single-strand fragments cannot be located with respect to the structure of the nucleosome. Each band must result from the introduction of two breaks on the same strand of DNA, located at a distance apart which is a multiple of 10 bases. But further data are necessary to extend this to the simple conclusion that there are sites accessible to DNAase I located on both strands of DNA *every* 10 bases apart. Preferential cutting at a small number of sites could generate a 10 base ladder if some fragments came from one end of the core DNA, some from the other, and some are generated by two internal cuts. Any given denatured fragment might be derived from either strand of the duplex nucleosomal DNA, so it is not possible to say whether cutting occurs on one or on both strands. The maintenance of the same relative intensities of bands in the core particle and nucleosome digests suggests actually that cuts at the junction between core and linker DNA are common; but the remaining positions might or might not be symmetrical with respect to the two ends.

One way to locate the positions of the cuts within the core DNA is to label the ends of the DNA. By using γ^{32}P-ATP as a substrate for polynucleotide kinase, Simpson and Whitlock (1976b) were able to label the free 5′ ends of the DNA of isolated core particles. Then the pattern of labeled fragments shows the distances at which nicks are introduced from each 5′ end; while the total fragments include three species (those derived from the 5′ end, those derived from the 3′ end, and those derived from double internal nicks). Figure 13.11 compares the autoradiogram with the stained gel for the smaller fragments from DNAase I digestion. Labeled fragments occur with the usual 10 base periodicity, but with different relative frequencies. For example, the site at 80 bases from both 5′ ends is not accessible to DNAase I; this would be expected if the construction of the nucleosome were symmetrical with regard to the two ends of core DNA. A more detailed analysis by Lutter (1978) with both 5′ and 3′ labeled cores has yielded similar results, again with the implication that there may be dyad symmetry. The probabilities of cutting at individual sites that are determined from the end label studies appear able to account for the overall pattern seen in DNAase I digests, which is to say that cutting within the core is largely responsible.

The 10 base periodicity may be taken to reflect some structural feature in the winding of DNA around the nucleosomal histones. Do the relative intensities of the bands also arise from the topology of the DNA? Whitlock, Rushizky, and Simpson (1977) used two further enzymes, DNAase II and an endonuclease from Aspergillus oryzae, to analyze the susceptible sites. Although the same 10 base periodicity in denatured core DNA is obtained, there are differences in the relative intensities of the 5′ end labeled bands. The main difference is that DNAase II is able to generate an 80 base fragment whether assayed in the presence of EDTA (its usual condition) or in the presence of Mg^{2+} (the condition of assay for DNAase I), so the difference

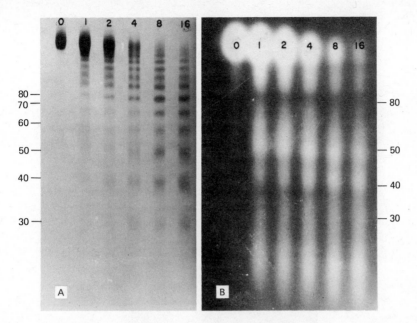

Figure 13.11. Comparison of stained gel (left) and autoradiogram (right) of single stranded DNA generated by DNAase I digestion of ^{32}**P-labeled nucleosomes.** Times of digestion with DNAase I are indicated in minutes by the numbers above the gels. A 10 base periodicity is seen in both stained and labeled gels, but band intensities vary. Data of Simpson and Whitlock (1976b).

cannot reflect an ion-induced change in nucleosome conformation. Two views have been taken of this. One is to caution that there is a 10 base periodicity, but that relative sensitivities to cleavage may differ with the nuclease used and therefore may be an unreliable guide to structure. The other has simply taken the results with DNAase I as indicative of relative susceptibilities in nucleosome structure.

An intriguing feature of the DNAase I cleavage sites is that the periodicity extends across nucleosomes. Figure 13.12 shows an experiment of Lohr, Tatchell, and Van Holde (1977) which demonstrates that the DNAase I ladder can be extended to more than 200 bases. Ladders extending sometimes to more than 300 bases have been generated from chromatin or from isolated nucleosome oligomers of HeLa cells, chick erythrocytes and yeast. The ladders of the different organisms are not exactly in phase all the way, and there are some differences in the relative intensities of the bands, but the periodicity of 10 bases is in general maintained.

Two interpretations are possible. One is to suppose that whatever structural conformation causes the 10 base cutting pattern runs continuously from nucleosome to nucleosome, with cuts occurring at 10 base intervals through linker as well as core DNA. This would imply that there is no change in structure of DNA in the linker region, compared with the core

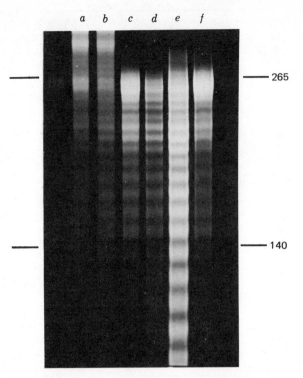

Figure 13.12. DNAase I digestion of chicken erythrocyte chromatin. Samples were run on 8% acrylamide-urea denaturing gels. Tracks *a* and *b* show the results of 15 and 30 sec digestions of isolated trimers; tracks *c–f* represent digestions of dimers for 15, 30, 60, and 15 sec, respectively. Data of Lohr, Tatchell, and Van Holde (1977).

region; indeed, it would imply further that there is no change in organization in the transition from one nucleosome to the next. Another model is to suppose that cuts occur only in the region of core DNA, so that bands larger than 140 bases are generated by cutting in adjacent cores, that is, they represent an intact linker DNA region attached to two partial core DNA lengths. This demands only that the length of linker DNA must be a multiple of 10 bases; in fact, the extension of the ladder over adjacent nucleosomes means that this must be a feature of all models.*

End labeling studies with isolated oligomers would be necessary to determine directly whether DNAase I cuts can be introduced in intact linker regions. (Monomers would not be a suitable substrate because of the possibility that micrococcal nuclease cutting in the linker might change the conformation of the regions adjacent to the newly generated ends.) In lieu

---

*This assumes that the continuous ladder is derived from typical nucleosomal arrays. If multimeric cores such as those found in chromatin treated with 0.6 M NaCl were present in vivo in any appreciable amount, the ladder might represent cutting in this subfraction lacking linkers. But probably the multimeric cores are generated in vitro (see earlier).

of such experiments, it can be said only that linker DNA is organized in a manner that restricts its length to multiples of 10 base pairs. Whether this organization is the same as that of the core DNA remains to be seen.

Another enzyme that recognizes a 10 base pair periodicity in nucleosomal DNA is exonuclease III, which digests one strand of duplex DNA from each 3'-OH end (see Table 13.3). Riley and Weintraub (1978) found that when monomers isolated by micrococcal nuclease digestion are treated with exonuclease III, the broad band of DNA of 140–170 base pairs is converted to a sharp band of 140 base pairs. This shows that linker DNA and core DNA have different susceptibilities to exonuclease III and that a sharp demarcation can be made between these regions. Because the enzyme needs a free end of DNA, this conclusion must rest only on studies with isolated monomers, where a break has already been made in linker DNA. (The isolation of a sharp 140 base pair duplex band implies also that exonuclease III has not merely digested strands terminating in 3'-OH; this would generate duplex fragments with single strand tails terminating in 5'-P. These tails must have been removed by a clipping action, a conclusion which was confirmed by showing that exonuclease III can remove a 5'-^{32}P label added to the monomers.)

Higher concentrations of exonuclease III extend the exonucleolytic (3' end trimming) reaction into core DNA, where the 5' clipping action does not seem to occur. The duplex fragments produced in this reaction are diffuse, but the single-strand fragments generated by denaturation form a 10 base ladder, which extends clearly to 100 bases and may be taken further to 70 bases. This ladder must be produced by an action which is in a sense the opposite of that of DNAase I. Instead of recognizing sites for enzyme action every 10 bases, exonuclease III must rapidly remove sequences of 10 bases but then pause at some site that represents an impediment to further progress. When dimers are treated with exonuclease III, a series of about 11 bands over the range 200–310 bases is generated. This has two important implications. Adjacent cores must have a fixed relationship that is related to their internal periodicity. Also it must be possible to digest much or all of a core DNA without destroying whatever features are responsible for maintaining the 10 base periodicity in the remaining regions.

Does exonuclease III degrade core DNA from one or from both 3' ends? All of the 140 nucleotide long core single strands can be degraded to smaller bands; and the same result is obtained with 5'-^{32}P labeled cores. This implies that the 3' strands at both ends of the core are attacked. This conclusion is supported also by the size of the material remaining after exonuclease III-digested core DNA is treated with S1 nuclease. If only one end is attacked by exonuclease III, the product should be a duplex with one single strand tail; if both ends are attacked, however, single strand tails must protrude from each end of the DNA, so that the central duplex region left after S1 treatment has removed the single strands is much smaller. The DNA formed by this combined digestion has the structure predicted by the latter model. This is consistent with the idea that the structure of the nucleosome is symmetrical as seen from either end of DNA.

Periodicity of DNA Organization

Cuts are made within the nucleosome by the enzymes DNAase I and DNAase II, which introduce single-strand nicks, and by micrococcal nuclease, which causes double-strand breaks (see Table 13.3). A periodicity of 10 is seen in each of these cutting patterns. This immediately raises several questions about the structural arrangement of DNA on the histone octamer. Are the sites recognized by all these enzymes identical? This would imply that the pattern results from a single feature of DNA organization that recurs with a periodicity of 10 and renders the DNA susceptible to attack at these sites. When single-strand nicks are made, do the susceptible sites on the two strands coincide? This is the question of whether there is a periodicity of 10 base pairs, in which case either strand may be cut as the consequence of a single recognition event, or whether different sites are independently recognized on each strand, in both cases with a periodicity of 10 bases. Is nucleosomal DNA symmetrical with regard to its ends or is there a polarity in its organization? Symmetrical organization implies that (for example) the sites at 20 base pairs from each end of the core DNA are indistinguishable; any difference between such corresponding sites implies the existence of a polarity along the DNA.

Two types of explanation have been proposed for the periodicity. The first model supposes that recognition is a feature of individual strands. Noll (1974b) suggested that DNAase I might recognize the sites at which the sugar phosphate backbone of each strand is most exposed when the duplex molecule is wound smoothly around the histone octamer. This would require a continuous deformation of structure along the axis of the DNA; theoretical calculations have shown that this is possible (Levitt, 1978; Sussman and Trifonov, 1978). The sites recognized by the enzyme then represent an inherent feature of the duplex structure itself; taking the pitch to be the conventional value of 10 base pairs per turn, the site of maximum exposure on each strand recurs every 10 bases. Figure 13.13 shows that since the two strands of the double helix are wound around each other, they should alternate on the surface of the nucleosome, with the result that their sites of maximum exposure are staggered by 6 and 4 base pairs.

The second type of model is to suppose that the periodicity reflects the recurrence of some feature every 10 (or possibly 20) base pairs along the duplex. This might be a discrete structural distortion in the DNA caused by the manner in which it is wound on the nucleosome. One proposal is that "kinks" might occur every 10 base pairs. Models for such disruptions have been presented by Crick and Klug (1975) and Sobell et al. (1976), who

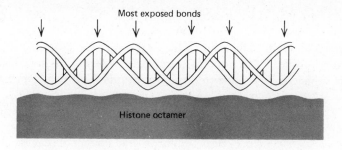

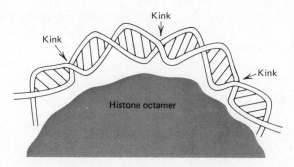

Figure 13.13. Diagrammatic projection of DNA double helix on the surface of the nucleosome. The upper part shows a length of about 30 base pairs lying in the B duplex conformation. The most exposed positions occur every 10 bases on each strand, with a stagger between adjacent sites on opposite strands of 6 and 4. Although shown here for a rod, the same sites of exposure could be maintained for a continuously distorted duplex wound around the nucleosome. The lower part shows the introduction of kinks every 10 base pairs; the double helix takes its usual conformation between the kinks, but is distorted at them. No base pairs need be broken at the kink, but stacking of adjacent pairs is disrupted.

have shown that it is possible to bend the double helix in such a way that the stacking of base pairs is interrupted while the pairing between strands remains intact. Figure 13.13 shows that in this case DNA can take its usual structure between the kinks, which represent discontinuities at which the duplex rod is bent. The responses of nucleases to kinks are not known. It is possible to postulate either that the kinks are recognized preferentially or that they inhibit recognition, confining enzyme action to base pairs located between them. Generally it is felt that the distortions are more likely to be attacked. A slightly more iconoclastic proposal is that stretches of DNA might alternate in the common B and the A form between "bends" (Selsing et al., 1979). Another possibility is that there might be a periodic interruption in the association of histones with DNA, exposing base pairs at intervals of 10. At all events, recognition of base pairs at a constant interval should cause breaks across the duplex if both strands are cut here, or cuts staggered by 10 bases if only one strand is cut at the recognition site.

The prediction that these models should be distinguished by different staggers provided the impetus for studies to determine the relationship between adjacent cuts made to opposite strands. This rests on the assumption that the enzymes cleave DNA at the sites of recognition. It turns out, however, that the cutting sites may be displaced from the recognition sites. But although this means that the models cannot at present be resolved on the basis of such evidence alone, data on the stagger reveal some interesting features of the organization of DNA on the nucleosome.

When both strands of DNA are cut at sites not too far from each other, DNAase I cleavage should release duplex fragments. The structures of the ends of these fragments should reveal the relationship between cuts on opposite strands. If there is a stagger of N bases between the sites of cutting on the two strands, each end should have a single strand protrusion of N or $(10 - N)$ bases. (Depending on the extent to which melting between cuts occurs during extraction of DNA, there may also be ends of $(10 \cdot n + N)$ and $(10 \cdot n - N)$, where n is an integer unlikely to be greater than 2.)

Double strand breaks can indeed be introduced by DNAases I and II. Cutting takes place between as well as within nucleosomes, since DNAase I or II treatment can be used to isolate 11S particles and larger multimers (Altenburger, Horz, and Zachau, 1976; Noll, 1977). Altenburger et al. found that in addition to the 200 base pair periodicity generated by digesting mouse liver nuclei with DNAase II, a 104 base pair periodicity is seen. This would correspond to the occurrence of a preferential cut half way round the nucleosome. This half nucleosome periodicity is seen only when digestion is carried out in the presence of divalent or monovalent cations (whose addition causes contraction of the chromatin). Its existence is consistent with the idea that there might be some "half nucleosome" structure, corresponding to a dyad axis of symmetry in the nucleosome (see below).

When DNA is extracted after DNAase I cutting, duplex fragments analyzed on gels form a series of bands with a periodicity of 10. The existence of these bands shows directly that both strands of DNA can be cut by the enzyme within the region of the core DNA. Sollner-Webb and Felsenfeld (1977) found that the fragments have protruding single-strand ends, for treatment with S1 nuclease (which attacks only single strands of DNA) reduces the size of each band, generating a new series with 10 base pair periodicity. The length of material removed cannot be estimated accurately, because of the uncertain relationship between size and electrophoretic mobility for DNA consisting of both duplex and single-stranded regions.

The lengths of the single-strand ends have been determined by extending the recessed 3'-OH end through the action of DNA polymerase; the number of nucleotides needed to restore a flush duplex end can be measured. Sollner-Webb and Felsenfeld (1977) used E. coli DNA polymerase II for this purpose; and Lutter (1977) used the large fragment of E. coli DNA polymerase I, which not only fills in recessed 3'-OH ends, but also removes protruding 3'-OH ends. Both series of experiments identified recessed 3'-OH ends of 8 bases; that is, the protruding 5' terminal strands are 8 bases long. The loss of material from protruding 3' strands identified lengths of 2 bases.

These results suggest that DNAase I makes the staggered cuts shown in Figure 13.14, which upon denaturation generate protuding 3' single-strand ends of 2 or 12 bases and 5' single-strand extensions of 8 (or 18) bases. The same conclusion has been reached also by Noll (1977).

Each of the bands of duplex DNA generated by DNAase I is rather broad, suggesting that its constituent fragments are heterogeneous in structure. This can be explained by the presence of the various types of protruding single-strand end. Denaturation to give single strands shows that the 50 base pair band, for example, contains lengths of both 50 and 40 bases, as well as some shorter multiples of 10 (caused presumably by the presence of internal nicks). The lengths of the flush duplex fragments generated by removal of the ends are consistent with the occurrence of the postulated protruding single-strand ends, in all possible pairwise combinations, on the original duplex fragments.

Similar studies of the products of DNAase II digestion have been carried out by Sollner-Webb, Melchior, and Felsenfeld (1978). The results show a pattern very similar to that of DNAase I, but with a different stagger: 3'-OH ends of 6 bases are recessed while 5'-P ends fall short by 4 bases (see Figure 13.14). End-labeling experiments in which ^{32}P was added to the 5' ends of DNA on core particles showed that DNAase I cuts at sites located $10 \cdot n + 2$ bases from the 5' end whereas DNAase II cuts at distances of $10 \cdot n + 3$ from 5' termini. The precision of these results, taken as ± 0.5 bases, implies that the ends of the core DNA are unique; there is virtually no variation in their location. Figure 13.15 shows the relative positions of the cuts made by DNAase I and DNAase II; the displacement of DNAase II cutting sites one base on each 3' side from the DNAase I cutting sites generates a stagger of $6 + 4$ instead of $8 + 2$.

A stagger related to these is seen also in the double strand breaks introduced by micrococcal nuclease. Although the micrococcal enzyme initially introduces scissions between nucleosomes and then produces core particles by end trimming, upon continued digestion the acid insoluble DNA is reduced to smaller fragments. The limit digest contains a series of bands ranging in size from 45 to 130 base pairs, which arise from the introduction of double-strand cuts in the DNA of the core particle (Axel et al., 1974; Axel, 1975; Sollner-Webb and Felsenfeld, 1975). The course of micrococcal nuclease digestion has been discussed above and is shown in Figures 13.2 and 13.8. Camerini-Otero et al. (1976) showed that the bands have a periodicity of 10 base pairs, although those at 150 and 80 base pairs are weak. Denaturation generates a series of fragments of lengths of exactly $10 \cdot n$ bases. When the duplex fragments are treated with S1 nuclease before

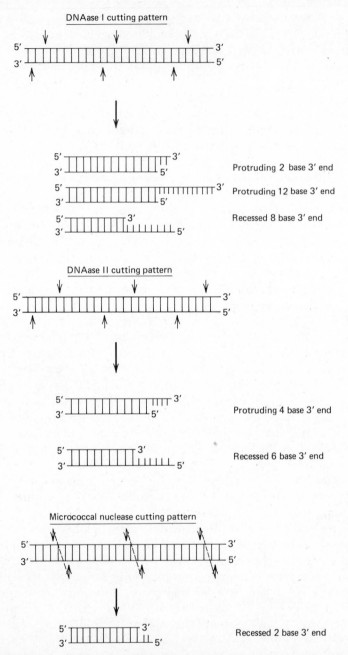

Figure 13.14. Nuclease cutting patterns of DNA in chromatin. DNAase I introduces single strand breaks with a stagger of 2 and 8; adjacent cuts on opposite strands may generate duplex fragments with the types of single strand tails illustrated. DNAase II behaves similarly, except that the stagger is 4 and 6. Micrococcal nuclease introduces only double strand breaks with a stagger of 2.

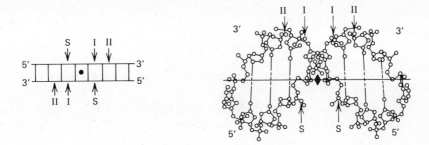

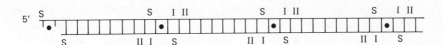

Figure 13.15. Sites cut by nucleases in the nucleosome core. The upper part shows the positions of cuts diagrammatically (left) and their locations on the two strands of the DNA duplex (right). The lower part shows a series of susceptible sites at 10 base pair intervals. The black circles indicate the simplest dyad axis for each set of cuts; an equally symmetrical set of axes exists at a displacement of 5 base pairs along the duplex. Micrococcal nuclease cuts are indicated by S; DNAase I and II cuts are indicated by I and II, respectively. After Sollner-Webb, Melchior, and Felsenfeld (1978).

denaturation, the lengths of the single strand fragments decrease to $10 \cdot n - 2$ bases. This suggests that there may be a stagger of 2 bases in the micrococcal nuclease cuts, so that one strand protrudes by two bases at each end of the duplex (and this extension is removed by S1 nuclease). Prolonged treatment with micrococcal nuclease appears to have the same effect as treatment with S1 nuclease, which suggests that the micrococcal enzyme itself removes the single strand ends at later times in the digestion. This idea is supported by the observation of Sollner-Webb et al. (1978) that a label of ^{32}P added to the 5′ ends of core particle DNA is rapidly removed by micrococcal nuclease. This suggests that it is the 5′-P ends that protrude by 2 bases. Cuts made by micrococcal nuclease within the core DNA appear to show the same behavior as the breaks at the end of core DNA, which suggests that all duplex breaks made by this enzyme take the same form, illustrated in Figure 13.14.

The removal of protruding ends by micrococcal nuclease makes it difficult to perform end-labeling experiments to investigate the sites that are cleaved within the core DNA by the micrococcal enzyme. However, in such experiments, it was possible to obtain single-strand fragments of 120 and 100 bases (these represented 15% of the total ^{32}P added to the ends). In at least these cases, therefore, internal cuts are made exactly $10 \cdot n$ nucleotides from the 5' end. If this is true of other internal cuts, all the micrococcal sites should lie 2 bases on the 5' side of each DNAase I cut, as shown in Figure 13.15. This is consistent with the existence of a micrococcal stagger of the same length of two bases seen with DNAase I, but with the orientation of the cuts reversed. As Sollner-Webb et al. point out, the arrangement shown in the figure is the only one that is consistent with the data on the locations and staggers of the cuts made by each of the three enzymes.

What do these staggers imply about the organization of DNA on the nucleosome? The locations of the cuts are inconsistent with the introduction of nicks just at the most exposed sites on each strand of a continuously wound duplex or at base pairs representing regularly occurring distortions. But both types of model can be made consistent with the data by supposing that the cutting site of each enzyme does not coincide with the site of initial recognition. However, the assumptions that are necessary in each case make some revealing comments about the structure of chromatin.

The two closest cuts made on partner strands by each enzyme have the same dyad axis, as indicated in Figure 13.15. This immediately raises the possibility that all enzymes recognize the same sites on DNA, but that in each case there is a different displacement of the cutting sites. This supposes that each cut on the two DNA strands occurs symmetrically about an axis located half way between each pair of cuts; these dyad axes recur every 10 base pairs along the duplex (as suggested by the second type of model). In this case DNAase I would cut 1 base on the 3' side of the dyad axis, DNAase II would cut 2 bases on its 3' side; and micrococcal nuclease would cut 1 base on each 5' side, presumably making cuts to both strands simultaneously in order to cause a double strand break. This model has the advantage of simplicity: each enzyme recognizes the same periodic feature in DNA.

It is possible, of course, also to generate the same cutting pattern by supposing that the most exposed sites on each strand are recognized, with cutting appropriately offset. In the cases of DNAases I and II this presents no problem: the enzymes would cut 1 base on the 5' side of the exposed peak and at the exposed peak, respectively. This makes the dyad axis for each pair of cuts a consequence of the structure of the DNA double helix and not a feature due to the mode of enzyme recognition and cutting.

But this model does not explain the action of micrococcal nuclease. The sites cut by the micrococcal enzyme lie 2 bases on the 3' side of each exposed position. But in order to make a double strand break, it is necessary to cleave at two such adjacent sites, that is, one on each strand (see Figure 13.14). If each exposed position is recognized independently, this would require two separate recognition events (each of which by itself would result in a single-strand nick). This does not accord with the introduction by micrococcal nuclease only of double strand breaks. If the breaks across both strands result from a single action, in fact, micrococcal nuclease must recognize some feature that recurs every 10 base pairs in the duplex.

It remains possible, of course, that micrococcal nuclease recognizes a regularly occurring distortion in the duplex, whereas DNAases I and II recognize most exposed bases on each strand. However, this is perhaps less attractive a model than the supposition that all enzymes have a common basis for recognition of DNA. At all events, the nature of the internal cuts made by micrococcal nuclease implies the existence of some 10 base pair periodicity in DNA, whether or not this is also the cause of the DNAase I and II patterns. Obviously any more definitive a conclusion requires information about the relationship between recognition and cutting sites. Indeed, in lieu of such knowledge, it remains formally possible that each enzyme recognizes a different site in DNA; this can be accounted for simply in terms of appropriate offsets for the cutting sites. One important conclusion can be derived unambiguously from the data, however; the sites actually cut by the three enzymes span a distance of 4 base pairs out of 10. At least these sites must be accessible to the enzyme molecules and not "protected" by histones.

For simplicity, we have discussed the cutting patterns of these nucleases in terms of the 10 base pair distance originally thought to separate adjacent susceptible sites (and this is maintained below). More recent experiments, however, have suggested that the actual distance is 10.4 bases (Prunell et al., 1979). Precise measurements of the amount of DNA in the core particle identify a length of 145.9 ± 0.3 base pairs, which corresponds to an average of 10.4 base pairs between each of the 14 DNAase I bands. Finely calibrated measurements show that individual bands vary by this amount, but it is possible that there may be some variation, with a distance between the central bands of about 10.5 bases and a distance between the more terminal bands of about 10.0 (Lutter, 1979). Whether this is a genuine feature of the nucleosome or is induced by the generation of core particles is not proven.

Taking the average distance of 10.4 bases to be typical, this refinement of length measurements does not change the basic view of nucleosome structure. Two explanations for this value have been considered. One is that the pitch of the duplex DNA actually may be 10.4 base pairs per turn and not the value of 10.0 previously assumed.* In this case, models for both continuous distortion and kinking retain exactly the same relationship to the data that we have already described. An alternative is that the periodicity of the bands does not coincide with the pitch of DNA, which might be 10.0

---

*Recent measurements using various methods including X ray diffraction, electron microscopy of DNA length, determination of the effect of inserting oligonucleotides of known length into closed circular DNA, and theoretical calculations have produced no agreement, with a range of values from 9.9 to 10.6 for the number of base pairs per turn (Griffith, 1978; Levitt, 1978; Vollenweider et al., 1978; Wang, 1979; Zimmerman and Pheiffer, 1979).

or indeed some other value. No simple explanation is forthcoming, but one possibility is that the angle of enzyme attack is related to the structure of the double helix in such a way as to change the periodicity of cleavage from that of the DNA itself. An observation consistent with the possibility that the structure or detailed susceptibility of DNA may change along the nucleosome is that the stagger between DNAase I cutting sites may deviate from the average at the ends of the core DNA, as noted above.

Phasing of Nucleosomes

Two questions may be asked about the stringency with which DNA sequences are arranged in the nucleosome. Is the content of DNA invariant in all the nucleosomes of a given chromatin? Is there any specificity with regard to nucleic acid sequence in nucleosome assembly? These are not unrelated, since the extent to which sequence specificity could be imposed must depend on the variation that is permitted in the assembly of the individual nucleosome.

The invariance of the length of core DNA, contrasted with the wide range of lengths for linker DNA in different chromatins (see Table 13.2), raises the issue of whether there might be variations in the lengths of the linkers joining a contiguous series of cores. Although the conformation in which linker DNA is arranged is not clear, a constraint on models for its organization is imposed by the DNAase I cutting pattern, which implies that its length can vary only in units of 10 base pairs (see Figure 13.12). In a chromatin with a repeat length of 200 base pairs, for example, the question then becomes whether it consists uniformly of 140 base pair cores alternating with 60 base pair linkers, or whether there are linkers of (say) 50, 60, 70 base pairs, giving an average value of 60 base pairs. (And the restriction to a particular periodicity implies that there cannot be repeat lengths of intermediate values; those that have been measured—see Table 13.2—either are slightly in error or must represent averages of different repeats each of which is a multiple of the periodic unit).

Variation in the length of linker DNA should make itself seen as heterogeneity in the length of DNA contained in monomers and small oligomeric nucleosomes. This should be reflected in two features of the micrococcal nuclease digest: the breadth of the bands and their reduction in size with time. But it is very difficult to disentangle the consequences of linker heterogeneity from the effects of cutting at different sites within the linker.

So far we have taken the general view that the breadth of the bands generated by micrococcal nuclease arises from variation in the initial site of scission. Roughly about half the linker DNA then seems to be available for micrococcal nuclease. But it is possible to argue instead that the site of cutting is fixed, although its distance from the junction with core DNA varies because the length of the linker region is not constant. Of course, models between these extremes are possible.

Different predictions are made by these models for the behavior of nucleosome multimers. If the length of linker is constant, the band width of all multimers should be the same, since they should vary only in the sites at which the terminal cuts were made. If the linker length varies, however, there may be internal heterogeneity. In this case, larger multimers should contain a greater number of linkers and so should display greater variation in width. Following this argument, Lohr, Kovacic, and Van Holde (1977) and Lohr et al. (1977) have shown that this is the case for small multimers of HeLa cells, chicken erythrocytes and yeast. The band width (taken at ¾ of peak height) for chick erythrocytes, for example, is 82 base pairs in dimers, 86 base pairs in trimers, and 106 base pairs in tetramers.

With increasing digestion, the size of each multimeric band decreases due to end trimming. Figure 13.16 shows an example of the reduction in size of bands from monomers to tetramers as a function of the extent of digestion of HeLa nuclei. A similar reduction is seen with all the chromatins in which the reaction has been followed (Shaw et al., 1976; Whitlock and Simpson, 1976; Lohr, Kovacic, and Van Holde, 1976; Gabautz and Chalkley, 1977; Martin et al., 1977). In all cases the monomer band is reduced from its

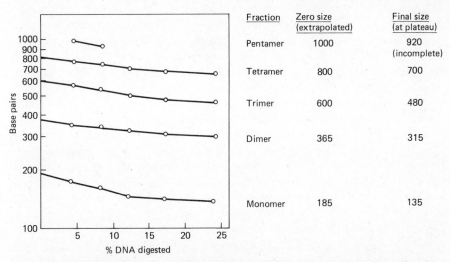

Fraction	Zero size (extrapolated)	Final size (at plateau)
Pentamer	1000	920 (incomplete)
Tetramer	800	700
Trimer	600	480
Dimer	365	315
Monomer	185	135

Figure 13.16. Decline in size of nucleosomal DNA during micrococcal nuclease digestion. The size of the DNA in each band, taken from the center position, is plotted as a function of the total percent of DNA released into acid soluble form. The table gives the initial length of DNA in each fraction, as the value suggested by extrapolation to zero digestion, and the final length after maximum end trimming (in most cases, when a plateau has been reached). These data were obtained with HeLa nuclei by Gaubatz and Chalkley (1976); similar data were obtained previously with chicken erythrocyte nuclei and HeLa cell chromatin by Shaw et al. (1976) and Whitlock and Simpson (1976).

initial size to the length of core DNA, always close to 140 base pairs. Approximately 50 base pairs are therefore usually lost by this end trimming. Trimers and tetramers always lose more DNA than this, usually 80–100 base pairs. Dimers have varied in the different experiments, sometimes showing the smaller loss characteristic of monomers, sometimes losing the larger amount of DNA characteristic of trimers and tetramers. But it is clear that in all cases the trimers and tetramers fall to a final size which is appreciably lower than would be expected if they suffered the same end trimming as the monomers.

This would not be expected if linkers were of fixed length, since in this case all fractions should lose the same two exposed linker regions at their termini. This suggests the argument that if the lengths of linker DNAs vary, multimers with shorter linkers might be less susceptible to micrococcal nuclease cutting and so might accumulate with time; while multimers with longer linkers might be preferentially removed by cutting to form monomers. In this case, one might expect the band width to decline with time, as the multimers in each fraction become more homogeneous, but it appears that there is no change in breadth with time. On the one hand this suggests that the extent of heterogeneity does not change as predicted; but on the other, it supports the idea that the length of each DNA in each fraction is not uniform after all available ends have been digested.

A possibility that is difficult to exclude is that the heterogeneity is generated by movement of the histones after cleavage of chromatin. The generation of new free ends, for example, might allow histone octamers to slide along the DNA in one direction or the other. A further complication is that these experiments do not follow the end trimming of isolated populations of oligomers, but identify the components of each fraction after digesting chromatin for increasing periods; thus new members are continually entering and leaving each population. Although these data make a case that linker DNAs may not be uniform in length, it is difficult to draw any conclusions about the range of variation or to estimate the proportions of linkers of different lengths in chromatin. It is clear that there cannot be extensive regions each containing linkers of characteristic but different lengths, since in this case there would be clear subbands in each fraction. Any linkers of differing length must therefore be interspersed, but this might be in a regularly or randomly arranged array. It might seem that in principle the best approach to determining whether linker lengths vary would be to characterize fragments of chromatin carrying defined sequences of DNA. But variations in the positions of the nucleosomes relative to particular sequences would obscure any variations in linker lengths.

It is clear that nucleosomes can be assembled without regard to DNA sequence. The DNA of phage lambda, for example, can be packaged in vitro by calf thymus histones in the same manner as adenovirus DNA (Oudet et al., 1975). But this does not prove the absence of a preferred location for nucleosomes, relative to DNA sequence, in vivo. This might be imposed, for example, during assembly, by always constructing the first nucleosome of a series at some fixed site on DNA, others being compelled by sequential assembly along the DNA to lie at specified positions. The obvious test for specificity of assembly is to see whether a given site always lies at the same location on the nucleosome or varies in position.

Sites whose positions can be investigated in this way are provided by the targets of restriction enzymes. This approach is limited, of course, to instances in which defined fragments of DNA are produced by a given restriction enzyme. Then by combined digestion with the enzyme and with micrococcal nuclease, it is possible to ask whether the site of restriction cleavage bears any particular relationship to the site of cutting by micrococcal nuclease (which is taken to define the limits of the nucleosome).

Are sites available to restriction enzymes in both core and linker DNA regions? The structure of chromatin protects restriction enzyme target sites against cleavage. Lipchitz and Axel (1976) have analyzed the susceptibility of the sites for EcoRI cleavage in bovine satellite DNA I, which consists of a tandem repetition of a sequence of 300 base pairs some 10,000 times. Due to variations in the sequence, sites recognized by EcoRI occur at intervals of 1390 base pairs. The satellite is organized into nucleosomes, since labeled fragments of purified satellite DNA hybridize with DNA extracted from nucleosomes. When nuclei are digested with EcoRI, only 2.3% of the total DNA is released in the form of 1390 base pair fragments, compared with the value of 7.3% seen with isolated DNA.

This means that only 56% of the total number of EcoRI sites is accessible to the enzyme (giving a probability that two adjacent sites will be cut of $0.56^2 = 31\% = 2.3/7.3$). But the relationship between the periodicities of 185 base pairs per calf thymus nucleosome and 1390 base pairs per EcoRI repeat means that no two adjacent restriction sites can lie in linker DNA. At least half of the restriction cuts must be made in core DNA. This conclusion cannot be made firmer because of uncertainties about the repeat lengths. For example, if the nucleosome repeat in this region was actually 198 base pairs, then either all the EcoRI sites or none of them would fall in the linker region. On the basis of random arrangement of sequences relative to nucleosome positions, however, only 24% (45/185) of the target sites should lie in linker regions. From these results it is therefore clear that EcoRI is able to cut DNA in the core region; it is not possible to calculate whether there is any difference in the relative efficiencies with which sites in core DNA and linker DNA are cut.

The availability of restriction targets in DNA and chromatin has been compared also for the α component of the African green monkey, using CV1 tissue culture cells. The enzymes EcoRI (and the less stringent activity EcoRI*), HindIII and HaeIII all cleave the DNA of this repetitive sequence into a multimeric series with a monomer of 176 base pairs. This repeating length is close to the content of the nucleosome, which has the usual broad band of DNA after micrococcal nuclease digestion, with an average size of

185 base pairs. Musich, Maio and Brown (1977) observed that EcoRI (or EcoRI*) is able to generate a 176 base pair repeating pattern upon digestion of nuclei. This could mean that the presence of nucleosomes does not influence the relative accessibility of EcoRI cleavage sites; that is, the target sites are equally available whether located in core or linker DNA regions (although they may be uniformly less accessible than in pure DNA). Or nucleosomes may occur with a constant phasing that happens to place the target sites always in accessible locations. This would demand that the nucleosome repeat length of component α chromatin coincides with that of the DNA repeat, so that it would be shorter than the average of the genome.

The HindIII and HaeIII sites lie at distances of 36 and 72 base pairs, respectively, from the EcoRI site. Neither of these is cleaved in intact nuclei, even at enzyme concentrations very much greater than would correspond to the effective concentration of the EcoRI enzyme. This situation could arise with a phased series of nucleosomes, since the distance these sites lie from the EcoRI site might place them in a constantly inaccessible position. It is possible, of course, that the relative differences in the in vitro and in vivo actions of these enzymes resides in some aspect of the enzyme functions. One way to investigate this point would be to determine whether the same difference occurs with other satellite DNAs, or whether examples can be found in which HindIII or HaeIII sites happen to be phased in accessible positions while the EcoRI sites are unavailable.

Monomeric DNA obtained by micrococcal nuclease digestion of CV1 nuclei does not appear to be cleaved by EcoRI. This implies that the target site for the enzyme lies in the region of linker DNA that is initially cleaved (or trimmed) by micrococcal nuclease. Exposure of the monomeric DNA to HindIII generates some 140 base pair fragments. This is the same length of DNA that is generated by joint cleavage of α DNA with EcoRI and HindIII (which splits the repeat unit into lengths of 140 and 36 base pairs). This implies that the ends of the monomeric DNA segments coincide with the EcoRI cleavage sites; for if the cleavage sites were located elsewhere, HindIII would generate a fragment of some other size; and if there were no phasing, HindIII would generate just a random smear. These results are therefore consistent with the idea that the EcoRI sites lie at the junction between nucleosomes. However, the identification of the fragments produced in the digestions of nuclei is based solely on size. Proof that these correspond to the expected fragments of α DNA, ultimately extending to a sequence study to determine the relatively accessible and inaccessible regions, would make a more conclusive case for the existence of phasing on this component of chromatin.

The SV40 minichromosome is especially suitable for investigating the question of whether nucleosomes occupy preferential locations or are arranged randomly with respect to sequence. The map of restriction sites on SV40 DNA is well characterized for many enzymes; and these sites can be cleaved in the minichromosome. The presence of histones confers only slight protection against restriction enzymes, since about 80% cleavage takes place with the minichromosome compared with SV40 DNA. Polisky and McCarthy (1975) observed that there is no evidence for any preferential protection of particular cleavage sites for HindIII or EcoRI. The complex used in these experiments consisted of 30% protein, 70% DNA, which suggests that only about half of the DNA is wound on nucleosomes. At face value, these results therefore suggest that under half of the nucleosomal DNA (that is 20% uncleaved/50% on nucleosomes) is protected by histones against the restriction enzymes; but no particular restriction site appears to be protected any better than any other.

These results suggest that none of the six restriction sites recognized by either HindIII or EcoRI lies in a unique location. If all SV40 genomes have the same structure, this means that each has a small number of nucleosomes and that their location is random. However, there is also the possibility that some SV40 genomes are free of histones whereas others are extensively covered. In short, the use of a complex that has less histone than the usual 1:1 mass ratio with DNA makes it difficult to draw firm conclusions about the arrangement of the nucleosomes.

With complexes of SV40 or polyoma DNA that have about 21 nucleosomes, and so are fully organized as chromatin, 15–27% of the genomes are converted from circular to linear form by EcoRI, which cleaves at a single site (Cremisi et al., 1976). Whether this degree of susceptibility corresponds with the 30% of viral DNA that lies in linker regions is not proven. Extensive protection of DNA by nucleosomes has been observed in experiments in which lambda DNA was reconstituted with varying amounts of histones (Steinmetz, Streeck, and Zachau, 1975). The degree of protection at each of five restriction sites increases with the molar ratio of histone octamer to DNA; protection becomes complete by a ratio of 1:1. At intermediate ratios some sites appear better protected than others; but overall there do not appear to be significant differences between the sites. Again this suggests that nucleosomes form at sites on DNA irrespective of nucleic acid sequence.

A more penetrating approach to analyzing the location of nucleosomes has been taken by Ponder and Crawford (1977). The polyoma minichromosome was digested first with micrococcal nuclease, which generates the same results seen with chromatin: higher multimers give way to monomers, which are digested into cores and finally subcores. The monomeric DNA can be recovered and then digested with restriction enzymes. If the restriction sites lie at particular locations relative to the ends of DNA in each nucleosome, then fragments of specific size should be generated. But if the arrangement of nucleosomes is random, the restriction sites will lie at all possible positions relative to the ends and only a smear will be obtained upon gel electrophoresis. An intermediate possibility is that the nucleosomes take up a small number of preferred positions, in which case each restriction site will lie at one of a small number of locations relative to the nucleosome ends. A restriction enzyme cutting once in the genome will then generate a series of sub bands from the 200 base pair monomer that contains this site. In the earlier experiments to see whether any site is preferentially protected, this last possibility would not be distinguished from a random arrangement.

Both the enzymes EcoRI and BamI, which cut polyoma DNA only once, generate a series of about 10 sub bands from monomer DNA. A similar result is obtained with HindIII, which cuts twice. Similar results are obtained with SV40 DNA, which gives the same results with minichromosomes isolated in vivo or reconstituted from DNA and histones in vitro. This suggests that nucleosomes may be positioned only at a small number of particular locations. Since the monomer band of DNA varies in length due to end trimming, some of the subbands may be related by loss of defined (10 base pair) segments from the ends. Others may be derived from variation in the position of the restriction site relative to the ends of the monomeric fragment that carries it. The combination of two causes for the subbands would produce a smear unless their periodicities are the same or are multiples of one another. This would therefore argue that nucleosomes take up preferred locations that vary by a number of base pairs that is a multiple of 10.

This interpretation of the results generally has been taken to assume that the position on DNA taken up by each core particle is restricted to a small number of preferred locations, probably about three or four. This might arise in vivo if the first nucleosome to be assembled (presumably at the origin of replication) were located at one of several fixed positions; if each nucleosome contains a constant amount of DNA, then the locations of all nucleosomes subsequently assembled are fixed accordingly. However, the possibility that linker DNA varies in length suggests another model. Suppose that the location of the first nucleosome is fixed at a single location. Then if the linker regions joining subsequent nucleosomes can vary in length by 10 or perhaps 20 base pairs, the location of a given site relative to a nucleosome downstream will vary accordingly. This predicts that sites close to the first nucleosome are more fixed in position than sites located further away, where there are more intervening (variable) linkers. With small systems such as the viral minichromosome, no site can be very far distant from the first nucleosome. In the replicons of eucaryotic chromosomes, however, fixed locations close to the origin would give way to effectively random locations further away. (In either case, the idea that the effect depends upon the location of the first nucleosome to be assembled after replication provides an explanation that obviates the necessity to postulate any more general specificity for sequence in nucleosomes; but unless the initial location also were preferred during assembly in vitro, it would not be possible to produce the same pattern with minichromosomes assembled outside the cell.)

This view of chromatin structure implies that the locations of nucleosomes are fixed with respect to DNA. An experiment which suggests some potential for the sliding of histone octamers along DNA has been reported by Beard (1978). When an SV40 minichromosome is cleaved at its EcoRI site and ligated to a similarly cleaved and radioactively labeled SV40 DNA, some nucleosomes can be found on the second genome. It seems likely that these have moved by sliding, rather than by dissociation from their original DNA and reassociation with a new DNA, because the reaction does not occur when the minichromosome and pure DNA simply are mixed together. The reaction is rather slow and the maximum extent of binding to labeled DNA is not achieved for some hours. Probably this means that when nucleosomes find themselves adjacent to free DNA, they are able to move along it. Whether the rate-limiting step is the initiation of movement or the mobility itself is an interesting question that might be tested by examining the structure of the minichromosome-DNA dimer at early times. If the sliding model is correct, the first regions on the minichromosome to become exposed should be those adjacent to the EcoRI cleavage site; and if movement itself is slow, the first sequences of the attached DNA to be covered should be the same. This model need not imply that mobility is a feature of chromatin in vivo, where nucleosomes are tightly packed and probably there is no free DNA.

Whatever the possible role of the origin in establishing an alignment of nucleosomes, its structure in the minichromosomes of infected cells appears different from the other regions. When SV40 minichromosomes are cleaved with micrococcal nuclease or DNAase I under conditions that introduce only a single break in the duplex, restriction mapping of the linear DNA product then shows that about 30% is preferentially cleaved in the vicinity of the origin (Scott and Wigmore, 1978). About the same proportion is cleaved at the same site by an endogenous nuclease when an extract of minichromosomes is incubated in vitro (Waldeck et al., 1978). With restriction enzymes able to cleave only at single sites within SV40 DNA, those attacking the region of the origin attack up to 95% of the minichromosomes, whereas those with target sites located elsewhere cleave only some 30%. When minichromosomes are fixed with formaldehyde before treatment with restriction enzymes, most sites become unavailable; but those covering a region of about 400 nucleotides adjacent to the origin remain susceptible (Varshavsky, Sundin and Bohn, 1978, 1979). This strongly suggests that there is a significant difference in the structure of this region of the SV40 minichromosome, which prevents formaldehyde from cross linking the DNA to any associated proteins. This region includes several control sites in addition to the origin.

Interactions between Histones and with DNA

The structure of the nucleosome can be penetrated from the perspective of either its protein or nucleic acid components. From the perspective of DNA, we may ask what form of organization accomplishes its sevenfold compression of length on the surface of the nucleosome. Overall the path followed by the DNA generates 1–2 superhelical turns upon the removal of protein; the existence of periodic sites attacked by nucleases suggests that this may be accomplished by wrapping the DNA around the protein in a regular manner. From the perspective of protein, we may ask how the histones interact with each other to generate the protein core. Do these interactions occur only in the presence of DNA or can a protein core be assembled as an independent structure? Is it a symmetrical structure, possibly reflecting the junction of two tetramers each containing one copy of each nucleosomal histone; or is it generated by the interaction of an H3-H4 tetramer with a complex of H2A-H2B? What regions of the histone molecules bind to DNA, and how do these interactions compel DNA to follow its path on the surface of the protein?

The general structural features of histones in chromatin have been known for some time. About 20% of each molecule appears to exist in a helical conformation, which is not too different from the amount of helix formed by each individual histone in solutions of appropriate ionic strength (Bradbury and Crane-Robinson, 1971). Both ionic and hydrophobic forces play a role in maintaining chromatin structure. The extraction of histones with salt shows that 0.6 M NaCl elutes H1, 1.0 M NaCl extracts H2A and H2B, and 3.0 M NaCl is necessary to remove H3 and H4 (Ohlenbusch et al., 1967; Kleiman and Huang, 1972). In the presence of 6 M urea, which reduces hydrophobic interactions, H2A and H2B dissociate at 0.15 M NaCl, whereas H3 and H4 leave chromatin at about 0.6 M NaCl (Kleiman and Huang, 1972; Bartley and Chalkley, 1972). By these criteria, the histones H3 and H4 appear more tightly bound to DNA than do H2A and H2B; for both sets, hydrophobic forces as well as ionic forces are important in maintaining their position in chromatin. The elution of H1 is less influenced by urea, consistent with the idea that its binding to DNA involves largely ionic forces; its ready removal accords with its location as an internucleosomal histone.

The properties of individual histones discussed in Chapter 11 suggest that histone-histone interactions are likely to take place in the central apolar region of each molecule, whereas binding to DNA may be a property of the N-terminal region (and with H1 also of the C-terminal region) (see Figure 11.8). This is consistent with the results of trypsin digestion, which are the same with either chromatin or isolated nucleosomes as substrate (Weintraub and Van Lente, 1974; Weintraub, 1975). First H1 (or H5) is completely degraded. Then 20–30 amino acids are removed from the N-terminal regions of H3 and H4. Finally some 20–30 amino acids are removed from H2A and H2B. When H3 and H4 are cleaved, the nucleosome is converted from an 11S to a 5S particle, which indicates that the integrity of these histones is necessary for proper folding. On the other hand, histones treated with trypsin can participate in nucleosome assembly in vitro (Whitlock and Stein, 1978).

Pairwise associations can be formed between some of the histones in solution. Interactions have been reported between H3-H4, H4-H2B, and H2B-H2A (D'Anna and Isenberg, 1973, 1974b,d). Judging from the abilities of isolated fragments of H3 and H4 to interact in solution, the central and C-terminal region of H3 is involved (amino acids 42–135) and the same is true of H4 (amino acids 38–102 are involved). Thus in both cases the N-terminal tail is not involved in the histone-histone interactions (Bohm et al., 1977). The H3-H4 complex can bind to DNA (Rubin and Moudrianakis, 1975).

Chemical cross linking of histones in chromatin identifies pairs of histones that are adjacent in the nucleosome; these include the pairwise associations found in solution. Table 13.4 summarizes the more prominent reactions that have been found. If H1 is present, it may be linked into a series of cross-linked homopolymers as the first reaction (Olins and Wright, 1973; Chalkley and Hunter, 1975). This is consistent with its location as the most superficial histone. Then the core histones become linked; these reactions may be followed also by using chromatin depleted of H1. The cross-linking pattern can be used as an assay to identify changes that occur when the structure of chromatin is unfolded or otherwise perturbed (Hardison et al., 1977; Martinson et al., 1979b; Martinson and True, 1979a). An important caveat on using these reactions to deduce the arrangement of histones with respect to one another is that the cross-linked products generally represent only a small fraction of the total histone. This makes the schematic summary of Table 13.4 useful for illustrating the interactions presently identified; but this should not be taken as a two-dimensional representation of the actual structure. One feature not accounted for on this map, for example, is the strong interaction of H3 and H4 that generates a tetramer (see below).

Another cross-linking approach that may yield information about relative organization is to cross link histones to DNA by a technique that allows scission of the DNA at the site of reaction. By using 5' labeled DNA, the length of the single strand remaining attached to the histone after denaturation identifies the location of the protein contact. Mirzabekov et al. (1978) applied this approach to obtain a linear order of histones.

A critical question is whether the histone octamer exists as a protein complex in the absence of DNA or whether its existence depends on association with the nucleic acid. Thomas and Kornberg (1975) reported that when chromatin is cross linked with dimethyl suberimidate an octamer is obtained. The same octamer is found when DNA is removed and chromatin proteins are extracted at high pH (9) and ionic strength (2). This suggests that under these conditions, the histone octamer can exist in the absence of DNA. These experiments were extended by Thomas and Butler (1977), who extracted the histones with 2 M NaCl at pH 9 from chromatin from which H1 had already been removed. Treatment with dimethylsuberimidate identified a cross-linked octamer as a major product. The cross-linked octamer sediments at 5.2S; it can also be obtained uncross-linked, sedimenting at

Table 13.4 Cross linking of histones in nucleosomes

Cross-linking agent and action	Cross-linked histones	Proportion of linked proteins	Regions in histones that are linked
Short-distance agents (<2Å)			
Formaldehyde links Lys, Arg or His within span of CH₂ bridge	H2B-H4	~25%	C-terminal part H2B—C-terminal part H4
	H2B-H4	~25%	C-terminal part H2B—central part H4
	H2B-H2A	~10%	not known
	H2A-H2A	~13%	
	H3 -H3	~6%	
Tetramitromethane links adjacent Tyr	H2B-H4		C-terminal part H2B—C-terminal part H4
Ultraviolet links adjacent Tyr and Pro	H2A-H2B	>90%	central part H2B—N-terminal part H2A
	H2B-H4	<10%	C-terminal part H2B—C-terminal part H4
Ultraviolet + tetranitro-methane	H2A-H2B-H4		H2B shows some linkages as in individual cross-linking reactions
Carbodiimide	H3 -H4		not known
Long-distance agents (~14Å)			
Dimethylsuberimidate links Lys	H2A-H2B	~ 60%	not known
	H2B-H4	~ 10%	
	H2A-H4	~ 10%	
	H3 -H3	~20%	
Methyl-4-mercaptobuty-rimidate links Lys	H2A-H2B	~ 24%	not known
	H2B-H3	~ 15%	
	H3 -H2A	~ 17%	
	H3 -H3	~ 23%	
Methylacetimidate links Lys, Arg	H2A-H2B		C-terminal part H2B

Summary of cross-linking patterns

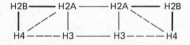

Short-distance agents cross link appropriate amino acids that are effectively adjacent (within 2Å). Long-distance agents have a substantial bridge between the two links, in the cases quoted of about 14Å. Only dimers found in proportions greater than 5% of the cross-linked species are given; since in no case does total cross linking exceed 20% of the histones present in chromatin, this is equivalent to dimeric arrangements represented with a frequency of at least 1% in toto. The regions of the histones that are linked in the dimers have been identified by determining which free peptides characteristic of the original protein are released by cyanogen bromide (cleaves at all Met residues). The summary aligns the histones to show all presently identified cross links; those seen commonly with several agents are shown as bold lines, those seen less

4.8S (the difference probably corresponds to just the addition of cross-linking reagent). The isolated octamer has an axial ratio of 2 and a molecular weight of about 107,500 ± 7700 daltons. Similar experiments have been performed by Stein, Bina-Stein, and Simpson (1977), who obtained cross-linked core particles sedimenting at 11S and cross-linked octamers at 5.3S. Isolated core DNA (of 140 base pairs) was able to interact with the cross-linked octamer to generate an 11S particle with the usual response to DNAase I. The DNA of SV40 can associate with the cross-linked octamer to give 21 ± 3 beads, with an overall contraction in length of 2.75 times. So DNA can be wrapped around the intact core protein octamer that is stabilized by cross linking. Whether the octamer exists in vivo as an intermediate in the assembly of chromatin is not established.

A main focus of discussion has been on whether the octamer is the only form in which all four histones can associate or whether a tetramer also exists. By using conditions very similar to those needed to isolate the octamer—placing histones at high (2 M) concentration in high salt (2 M NaCl)—Weintraub, Palter and Van Lente (1975) identified a tetramer. Centrifugation isolates 80% of the histone in the form of a single species which sediments at about 3.8S and has a weight of some 54,000 daltons. Complexes of all four histones whose weights suggest a tetrameric rather than octameric structure have been reported also by Campbell and Cotter (1976) and Pardon et al. (1977).

The differences in reports from different laboratories may at least in part reflect different experimental protocols. More recent experiments have shown that the concentrations of both the histones themselves and of the salt have a strong influence. Butler, Harrington, and Olins (1979) found that when the complex is isolated in 2 M NaCl it behaves as though consisting of several different types of oligomer. Among these is the octamer, which becomes the sole form when the salt concentration is increased to 4 M NaCl. Its formation also is encouraged by increase in histone concentration. This salt-dependent equilibrium has been claimed to take the form of an exchange between tetrameric and oligomeric forms by Chung, Hill, and Doty (1978). From these various analyses it now seems that under conditions of high salt and protein concentration, an octamer is the principal form of histone aggregate. Other oligomers may be found under other con-

---

frequently as narrow lines. These studies are with mammalian histones, but similar results have been obtained with UV and TNM for plant histones.

References: Formaldehyde—Van Lente, Jackson and Weintraub (1975), Jackson (1978), Martinson et al. (1979a); tetranitromethane—Martinson and McCarthy (1975, 1976); ultraviolet—Martinson et al. (1976, 1979), DeLange, Williams and Martinson (1979), Martinson and True (1979b); carbodiimide—Bonner and Pollard (1975); dimethylsuberimidate—Thomas and Kornberg (1975); methylmercaptobutyrimidate—Hardison et al. (1977); methylacetimidate—Martinson et al. (1979a).

ditions, existing in a reversible equilibrium with the octamer. The role of the tetramer in this equilibrium is not yet clear.

The "heterotypic" complexes containing all four histones may be dissociated into "homotypic" complexes that contain only the lysine-rich histones (H2A and H2B) or only the arginine-rich histones (H3 and H4). When the pH is reduced, the heterotypic tetramer of Weintraub et al. dissociates into an H3-H4 complex of 3.2S and an H2A-H2B complex of 2.4S—these are probably the forms isolated from chromatin originally by Kornberg and Thomas (1974). The H3-H4 aggregate has been characterized as a tetramer; it is a stable structure with an asymmetrical, perhaps extended, shape (Moss et al., 1976). The structure of the H2A-H2B complex has been more difficult to define; but most probably it is a dimer, perhaps capable of further aggregation. More recent studies of the dissociation process have confirmed that reductions in ionic strength, pH, or protein concentration cause the octamer to dissociate into homotypic complexes of H3-H4 and H2A-H2B (Eickbush and Moudrianakis, 1978; Ruiz-Carrillo and Jorcano, 1979). The reaction may involve the sequential dissociation of two dimers of H2A-H2B, thus proceeding through the intermediate hexamer identified by Kornberg and Thomas (1975). In contrast with earlier observations that the dissociation process is irreversible, restoration of the original conditions since has been found to allow the octamer to reform or to be generated from the individual histones. In these experiments no heterotypic tetramer was generated. It is difficult to reconcile the idea that octamers are formed by association of two heterotypic tetramers with the formation of the H3-H4 tetramer and H2A-H2B (putative) dimer; for the formation of the homotypic forms implies that the bonds between histones of like type (especially H3-H4) are much stronger than between those of different types, at least under these conditions in solution. Of course, conditions may be different in vivo, especially in the presence of DNA.

Although the structure of the octamer can be maintained in the absence of DNA and may even be generated de novo by association of histones, somewhat extreme conditions are needed in solution. This therefore leaves open the question of its origin in vivo. Nucleosomes might be assembled on the basis of an induced fit model, in which assembly of the protein core occurs only in the presence of DNA; or a protein complex, homotypic or heterotypic, tetrameric or octameric, might be formed first, around which DNA can wind. Indeed, it is possible to postulate that new nucleosomes form by using previously existing nucleosomes as a template; during replication, nucleosomes might split into "half nucleosomes," whose protein content presumably corresponds to the heterotypic tetramer, and which could assemble with another tetramer to reassemble full nucleosomes (see Chapter 14). In short, neither the form of histone aggregate involved in nucleosome assembly, nor the stage at which it associates with DNA, is known at present.

The reconstitution of chromatin in vitro takes place by a process in which the interaction of histones H3 and H4 appears to be central. The characteristic limit digest pattern produced by micrococcal nuclease (see Figure 13.2) is displayed only by chromatin reconstituted from DNA and all four nucleosomal histones. But several criteria show that "subnucleosomal" particles can be formed when H3 and H4 are mixed with DNA. Upon reconstitution with subsets of the histones, Camerini-Otero et al. (1976) found that in most cases no sharp bands are generated when the DNA of the complex is digested with micrococcal nuclease. But the combination of H3 and H4 with DNA generates the lower weight bands shown in Figure 13.17. Although H2A and H2B alone cannot generate bands, when they are added to the H3-H4-DNA complex, bands appear at higher molecular weights in

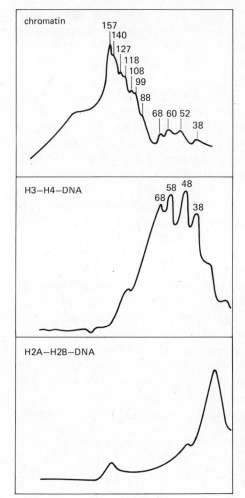

Figure 13.17. Limit digests with staphylococcal nuclease of DNA in duck erythrocyte chromatin, reconstituted H3-H4-DNA complexes, and reconstituted H2A-H2B-DNA complexes. Migration is from left to right on 6% acrylamide gels and bands have been quantitated by densitometry. The numbers indicate the length of each fragment in base pairs (determined for the upper gel by calibration against restriction fragments of known length). In chromatin, an additional fragment of 146 base pairs can be resolved on 10% gels; and a fragment of 79 base pairs can be seen in less complete digests. Data of Camerini-Otero et al. (1976).

addition to those generated by H3-H4 alone. All the bands are restored when
the H2A-H2B mass is 20% of the H3-H4 mass; their relative amounts become
normal as the proportion of H2A-H2B is increased to parity with H3-H4.*
Both H2A and H2B must be present. Of other three-way combinations, only
H4-H2A-H2B is effective, generating some of the higher molecular weight
bands (98, 118, 128, and 138 bp). Other results suggesting that nucleosomes
can be formed by sequential addition to DNA of H3-H4 followed by H2A-
H2B have been reported by Burton et al. (1978) and Jorcano and Ruiz-
Carrillo (1979).

These experiments use the nuclease digestion pattern as an assay for the
extent to which nucleosome structure has been reconstructed. A clear dependence on the H3-H4 pair has been shown by similar experiments with
DNAase I, in which Sollner-Webb et al. (1976) found that H3-H4, either
together or in combination with H2A and H2B, generates the 10 base periodicity (although the intensity of each band is inversely proportional to its
weight, instead of showing its usual value). Digestion with DNAase II,
which proceeds under different conditions (see Table 13.3), gives the same
results.

Particles formed by associating H3-H4 with DNA can be visualized by
electron microscopy as beads a little smaller than the nucleosome, with a
diameter of about 8 nm (Oudet et al., 1978). Each contains about 130 bp of
DNA, not much less than the content of a core particle. The formation of
these species imposes roughly the usual amount of supercoiling on SV40
DNA. Other results have shown that addition of H3-H4 causes a 2.75-fold
compaction in the length of SV40 DNA with the generation of some beads
(Bina-Stein and Simpson, 1977). Another criterion that has been applied to
reconstitution experiments is the reconstruction of the characteristic X-ray
diffraction pattern. Bosely et al. (1976) confirmed that all four histones can
regenerate this pattern when added to DNA; the peaks are regenerated by
subsets only when H3 and H4 are included.

These results are all consistent with the model of Camerini-Otero et al.
that the arginine-rich histones, H3 and H4, make up the "kernel" of the
nucleosome. They are capable of organizing DNA into its characteristic
path; the addition of the other histones completes the regeneration of the
normal sensitivity to nucleases. Although it is true that some reorganization has been achieved in a few experiments by histone subsets lacking H3
or H4, the general observation common to all the various studies is that the
combination of H3 with H4 always can generate at least a partial pattern.
The idea that these histones play a central role in nucleosome structure is

---

*The complex that H3-H4 forms with DNA depends on the histone:DNA ratio. At physio-
logical levels (mass ratio of 0.5) the homotypic tetramer binds to DNA as described in the text.
At greater levels, an H3-H4 octamer may be formed, causing DNA to be arranged in a more
compact shape (Simon, Camerini-Otero, and Felsenfeld, 1978). In determining the ability of
H3-H4 to organize DNA, it is therefore important to keep the histone:DNA ratio in the appropriate range.

consistent, of course, with their highly conserved sequences (see Chapter 11). But the extent to which this reflects the existence in vivo of some sub-nucleosomal structure containing H3 and H4 remains to be seen: on the one hand, H3-H4 may interact with DNA to form a particle to which H2A and H2B may be added; on the other hand a cross-linked octamer can bind directly to DNA.

Shape of the Histone Octamer and DNA Shell

The existence of a regular structure in chromatin has been recognized for some time. Early X-ray diffraction patterns showed a series of maxima at 11, 5.5, 3.5, 2.7, 2.2, and 1.8 nm (Luzzati and Nicolaieff, 1959, 1963; Wilkins et al., 1959; Pardon et al., 1967; Pardon and Wilkins, 1972). These were originally interpreted as originating from the spacing of a single structural repeat and its higher orders; and this was the basis of a model for a 10 nm regular supercoil of DNA. These data now have been reinterpreted from the perspective of nucleosome structure.

The suggestion that these peaks might originate in different components of chromatin—that is, DNA and protein—was made by Baldwin et al. (1975) on the basis of neutron-scattering studies. These experiments utilized the technique of contrast matching, which allows the contributions of each component to the scatter to be distinguished. The scatter of H_2O is different from that of D_2O; and so that of the solvent can be adjusted by varying the proportions of water and heavy water. The scatter of nucleic acid and protein in a mixture of H_2O and D_2O depends on the characteristic extent to which labile protons have been replaced by deuterium. This means that neutron scatter from DNA is "matched" at 63.5% D_2O whereas that from protein is matched at 37.5% D_2O. Thus the scatter measured at the higher D_2O level is due to the protein component; and that measured at the lower D_2O level derives from DNA.

Following the dependence of each diffraction peak upon the concentration of D_2O suggested that the 10 nm and 3.7 nm rings are derived from the histone structure. These appear to have independent origins, since they respond differently to changes in concentration; that at 3.7 nm is not affected, whereas the 10 nm ring declines when the concentration of chromatin is increased. The obvious interpretation is that the 10 nm ring depends on the spacing of nucleosomes, whereas the 3.7 nm ring is due to some feature of histone organization in the core.

The peaks at 5.5 and 2.7 nm originate with DNA. The 2.7 nm ring is found with pure DNA, that at 5.5 nm only in chromatin. This suggests that the DNA may be organized in a structure which has a pitch of 5.5 nm, perhaps as about 1.5 turns of a coil on the surface of each nucleosome.

A related approach can be used to determine the characteristics of individual particles. The general structure suggested by such analysis is of a protein core that is relatively inaccessible to water, surrounded by a hydrated shell of DNA. With nucleosomes the radius of gyration of the protein core is about 3.2 nm and that of the DNA shell is about 5.2 nm (the difference corresponds to the 2 nm diameter of the DNA double helix). Neutron scattering of cores gives very similar, although slightly lower, values, which are slightly more precise because there is less heterogeneity in the particle structure. Since the histone forms a smaller body than DNA, the nucleic acid must lie on the outside of the protein. The shape of the particle corresponds to an oblate spheroid with an axial ratio of about 0.5. Thus its overall dimensions are about 11 × 11 × 5.5 nm (Pardon et al., 1975, 1977; Hjelm et al., 1977; Suau et al., 1977).

A very similar view is suggested by the first X-ray diffraction results obtained with crystals of core particles. Finch et al. (1977) suggested a model in which the DNA takes the form of a flat superhelix of pitch about 2.8 nm. According to this model, the 3.7 nm diffraction band results from reflection off DNA in the plane of the disc, while the 2.7 nm band reflects the spacing of turns along DNA, that is, the separation between turns of the superhelix. There would be about 1.75 turns of DNA per core particle. We should note two caveats about these results. Since the data were obtained with core particles, they make no implication about the structure of linker DNA vis a vis core DNA. More important, the histones in these crystals had been proteolyzed; the extent of structural change is not known, but it is possible that removal of some regions of the histones affects the DNA conformation. Changes in conformation probably are not extensive, however, because the DNAase I pattern continued to display the usual 10 base periodicity.

Several models are consistent with the biophysical data on nucleosome structure, but all share some features, principally the shape of the protein core and the winding of DNA on the surface in a somewhat flat coil. Pardon et al. (1977) have suggested that the protein may take the form of two histone tetramers, somewhat cylindrical in shape, with a flat DNA coil lying on top of each. Another model that would preserve dyad symmetry has been proposed by Weintraub, Worcel, and Alberts (1976). Again the DNA lies in a coil on the surface of the two associated histone tetramers, which may be able to separate into "half nucleosomes" (see Chapter 14).

The properties of DNA in chromatin appear largely explicable in terms of the structure of the nucleosome particle itself. One of the early criteria that indicated the close association of DNA in chromatin with proteins was the elevation of the melting temperature of virtually all of the nucleic acid. Less than 4% of the DNA melts at the T_m of free DNA; in earlier experiments a biphasic or multiphasic melting curve was reported for the predominant components (Li and Bonner, 1971; Li, Chang, and Weiskopf, 1972, 1973; Subirana, 1973). More recently a monophasic curve has been obtained, with a T_m some 30–40°C greater than that of free DNA; and it has been suggested that the multiphasic curve may be generated from this by changes induced in chromatin during its preparation (Lawrence, Chan, and Piette, 1976). At all events, the binding of individual histones to DNA seems able to stabilize the nucleic acid, but cannot reproduce the authentic melting curve. When different classes of histones are removed from DNA, the authentic structure of chromatin naturally is lost, and the melting profiles broaden into a more generalized pattern of protection of DNA (Shih and Bonner, 1970a; Smart and Bonner, 1971a,b). Detailed denaturation curves obtained with monomeric or multimeric nucleosomes show the same pattern as native chromatin examined simultaneously (Lawrence et al., 1976; Mandel and Fasman, 1976). The melting of more free DNA is found with reduction in the size of the multimers, and this may be due to some unfolding of the array of nucleosomes, but the general conclusion suggested by these studies is that the melting of DNA in native chromatin represents the disruption of individual nucleosomes, which are principally or solely responsible for the stabilization of DNA structure.*

One of the important structural parameters of chromatin is the extent of supercoiling of the DNA. We have seen already that models for the nucleosome assume there to be between one and two turns of DNA around each particle. Such turns may generate a supercoiled structure. These structures have been characterized for small circular DNA molecules, principally the genomes of SV40 and polyoma, but also some circular plasmids. Recently the nomenclature for describing such situations has been revised (Crick, 1976). But we shall use the original terms introduced by Bauer and Vinograd (1968) and Vinograd et al. (1968), since it is in this way that most of the results with chromatin have been described. Superhelical turns are generated by twisting a closed circular duplex of DNA (for example, as a rubber band may be twisted). The number of superhelical turns, τ, is the number of times the duplex axis winds about the superhelix axis when the superhelix axis is constrained to lie in a plane. This may be quoted directly for a given molecule or calculated as the superhelix density, σ, which is the number of turns for every 10 base pairs. ($\sigma = \tau/\beta^0$, where β^0 is one tenth of the number of base pairs). Superhelicity may be positive or negative, according to the sense of the superhelical turns relative to the right hand sense of the double helix itself. Each negative turn corresponds to a denaturation or unwinding of the double helix, whereas each positive turn corresponds to an overwinding.

Direct data on the coiling of DNA in the nucleosome have been gained from analysis of the SV40 minichromosome. The circular duplex of SV40 DNA may be obtained in either a closed form or an open form that contains

---

*Core particles show a biphasic pattern in which the terminal DNA regions melt first; this is followed by a massive disruption of the major structure (Tatchell and Van Holde, 1977; Weischet et al., 1978). The 160 bp intermediate particle shows this pattern also, but the melting curve characteristic of nucleosomes is restored by the addition of H1 (Simpson, 1978a). This raises the possibility that the melting of nucleosomes may reflect more than one type of disruption of DNA-histone association.

a nick in one strand. The closed form may possess superhelical turns; these are relaxed by the nick in the open form. The superhelicity of the closed form was at first measured by intercalation studies with ethidium bromide but more recently has been assessed directly. This was made possible by the isolation of an enzyme variously described as unwinding enzyme, relaxing activity, and nicking-closing enzyme. Its effect appears to be to introduce a single strand nick in the closed molecule and then to seal the break after torsional tension in the duplex has been relieved by the loss of superhelical turns. The enzyme therefore converts closed circles, which have a high mobility upon gel electrophoresis, into relaxed circles, which have a much lower mobility. But brief treatment with the enzyme generates a series of intermediate bands on the gels. These represent partially relaxed molecules; and the number of superhelical turns can be deduced by assuming that each band differs from the next by exactly one turn. Keller and Wendel (1975) identified 21 bands, which implies the existence of -20 superhelical turns in the fully coiled closed molecule. Shure and Vinograd (1976) revised this estimate to -26 superhelical turns ($\sigma = -5.1 \times 10^{-2}$). The DNA of virions or infected cells actually is heterogeneous with respect to the number of turns, so this value corresponds to the maximum possible— some molecules have less.

When the nicking-closing activity is applied to SV40 DNA associated with histone, the extent of unwinding is restricted by the presence of nucleosomes. The number of nucleosomes can be varied by controlling the ratio of histone mass to DNA mass when the minichromosome is assembled in vitro. Then the complex is treated with nicking-closing enzyme, after which the protein is removed and the DNA is subjected to gel electrophoresis. Germond et al. (1975) found that the unwinding activity reduces the average number of superhelical turns to a value very close to the number of nucleosomes. This means that in the relaxed minichromosome the DNA must follow a path on each nucleosome which corresponds to about one superhelical turn when the nucleic acid is released by removal of the protein.

This conclusion is supported by the observation that the formation of nucleosomes on relaxed (but closed) circles of SV40 DNA induces about one superhelical turn per nucleosome (as seen by subsequent treatment with unwinding activity followed by removal of protein). When histones are assembled on to supercoiled SV40 DNA, the minichromosomes appear less twisted in the electron microscope; each nucleosome effectively removes one superhelical turn from the free DNA by utilizing it to wind the DNA onto the protein. For a nucleosome of 200 base pairs, one superhelical turn per particle would correspond to a density of $\sigma = -5 \times 10^{-2}$.

It is important to stress that this is an average value. The DNA could be evenly supercoiled along its entire length; or the 140 base pair core might contain the turn while the 60 base pair linker is free of restraint. For chromatin as a whole, the degree of supercoiling of DNA must depend on the paths followed by nucleosomes in higher order structures as well as the path of DNA on the nucleosome itself. But the overall superhelical density of DNA in chromatin, judged by experiments on the intercalation of ethidium bromide, is about the same as that seen with SV40 DNA (Benyajati and Worcel, 1976; Cook and Brazell, 1977; see Chapter 14). This may be taken as roughly one negative superhelical turn for every 200 base pairs, which corresponds to the unwinding of one turn of the double helix per nucleosome.

The Nucleosome: Organization in Chromatin

The general structure of the nucleosome is clear: it is an oblate spheroid roughly 10 nm across and comprises a histone octamer about which is wound some 200 base pairs of DNA in a topology equivalent to 1-2 negative superhelical turns. Associated with each pair of nucleosomes there is a single molecule of H1. In this chapter we shall be concerned with the relationship of this structure to the functions of DNA and with the interactions between nucleosomes that must occur in order to coil up the nucleohistone fiber into chromatin.

It has been obvious since studies of chromatin commenced that nucleic acid synthesis must involve some perturbation of structure. We can now pose this issue in more specific terms. What happens to the nucleosome during replication or transcription of DNA?

During semiconservative replication, the two strands of DNA separate to act as templates for the synthesis of daughter strands. Three models can be imagined for the behavior of the protein associated with DNA. The histone octamer might be conserved; then the daughter chromatins would contain both these "old" octamers and "new" octamers assembled from newly synthesized histones. These might be arranged in a random or nonrandom manner. Alternatively, the histone octamer might be partially conserved, perhaps splitting into half nucleosomes, each of which is restored to complete nucleosome status by the addition of a new histone tetramer. Again various models can be constructed to link the duplication of the histone octamer to the replication of DNA. Finally, the octamer might disintegrate into histone components which enter the pool containing newly synthesized histones, and from which nucleosomes are assembled de novo.

Only some of the DNA of each cell is transcribed. This has prompted questions on whether the structure of transcribed regions might differ from that of nontranscribed regions. Attempts to fractionate chromatin into active and inactive parts are discussed in Chapter 28. Here we shall be concerned with the extent to which the structure of the nucleosome is altered by transcription. A model that nucleosomes are absent from transcribed regions is rendered unlikely by observations that the corresponding DNA is found in the usual bands of a micrococcal nuclease digestion. But this does not mean that transcription of DNA must occur on the surface of the nucleosome. Nucleosomes might be temporarily displaced by RNA polymerase and there could be a reduced frequency of nucleosomes along the DNA of active genes. The essential question here concerns the relationship of the growing point of the RNA transcript to the nucleosomes. Another question is whether the structure of nucleosomes within transcribed regions is the same as that of nucleosomes elsewhere: do changes in nucleosome structure occur as a prerequisite for or consequence of transcription? Associated with these questions about the process of transcription is the issue of gene regulation. Can regulator molecules recognize specific DNA sequences on nucleosomes? Is any special structure found for sequences such as promoters or operators?

The terms in which these processes have been viewed consider replication and transcription as functions of a thread of nucleosomes. But in chromatin, of course, the nucleosomes are organized into higher orders of structure. To proceed from the packing ratio of 6–7 found in the nucleosome particle to the ratio of 10,000 or so of metaphase chromosomes would seem to require at least two higher order structures. (The network of interphase chromatin is too tangled to allow its overall length, and thus a packing ratio, to be determined; but it is not unreasonable to suppose this would be of the order of 10–20% of that seen in metaphase chromatin.)

The immediate question about the structure of the nucleosome thread is what regular structure can be discerned for the series of spheroids. Viewing the thread as a supercoil of the beads, the question becomes what interactions maintain its structure. Is H1 involved? Are particular nonhistone proteins concerned? The general nature of this structure may be amenable to electron microscopic and X-ray diffraction analysis; what components are necessary may be revealed by reconstruction in vitro. The fiber generated by this level of organization may then be supercoiled further in a regular manner in chromatin; analysis of this level of structure seems more distant.

An important issue is the level of organization at which changes in structure occur. Localized changes in organization may occur during nucleic acid synthesis; if unwinding of DNA requires nucleosomes to take up a more extended conformation, these higher order structures may be disrupted as a prerequisite for changes in the structure of the nucleosome itself. How duplicate fibers of chromatin reassemble their higher levels of organization following replication is a particularly interesting point. Related to this issue is the question of how the state of genetic expression of a given region is perpetuated in replication.

More generalized differences in organization are represented by the division of chromatin into euchromatin and heterochromatin and in the difference between interphase (eu)chromatin and mitotic chromosomes. Heterochromatin is defined by its more condensed appearance and is inert, whether containing sequences that are never transcribed (constitutive heterochromatin) or chromosome regions that are repressed in a given cell line (facultative heterochromatin). Thus euchromatin contains the active genes (although most of this fraction is not transcribed either). Both types of region appear to contain nucleosomes. Satellite DNA sequences, known to be part of constitutive heterochromatin, show the usual micrococcal nuclease digest pattern; and electron microscopy identifies beads in fibers spilling out of both. Their difference must therefore lie in the higher order structures into which the nucleosome thread is coiled. Since nuclease digestion and electron microscopy both suggest that nucleosomes remain unaltered by mitosis, the condensation of chromosomes at division may be attributed to a general change in organization occurring at an (unknown) level of higher order structure.

The concept that the thread of nucleosome beads runs continuously through the different regions of a chromosome implies that the question of what is responsible for structural changes may be translated into the form: what nonhistone proteins associate with the nucleosomes to control the higher order structures? The specificity with regard to DNA sequence of this higher order organization is an important question on which at present there is little information.

Replication of Chromatin

Two events are necessary for duplication of the eucaryotic genome: DNA must be replicated by semiconservative synthesis; and another complement of chromosomal proteins must be made available. It has been known for some time that histone synthesis usually is coordinated with DNA synthesis; it occurs during S phase to an extent sufficient to double the histone content of the cell (see Chapter 28). The timing of synthesis of nonhistone proteins is not clear; but, presumably, both structural and regulatory components of the genetic apparatus have been reproduced by the end of S phase. An interesting question is whether replication is used as an opportunity to introduce changes in the state of genetic expression of chromatin.

Replication takes place in a large number of replicons, in each of which DNA is synthesized by the bidirectional movement of two replicating forks away from an origin in the center (see Chapter 20). Different replicons are activated at different times during S phase, so that regions of DNA that have been replicated can be visualized as "eyes" upon spreading the nucleic acid for electron microscopy. The same eyes were visualized by McKnight and Miller (1977) in electron micrographs of chromatin from the D. melanogaster blastoderm (an actively replicating stage of embryogenesis). Figure 14.1 shows an example. Both sides of the replication eye (that is, the newly replicated daughter duplexes) are covered in beads, present at the same frequency that is found in nonreplicated regions. Of course, one does not know whether any given replicon has been visualized actually during replication or how long it may be since it was replicated. But given the extensive replication of blastoderm nuclei, the general implication of this analysis

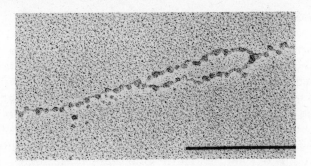

Figure 14.1. A replication eye from chromatin of the blastoderm of D. melanogaster. The string of beads on each side of the eye represents unreplicated parental chromatin; the two sides of the eye correspond to the two daughter duplexes generated by replication, each covered with beads at the same frequency as parental DNA. The bar indicates 1 μm. Data of McKnight and Miller (1977).

is that nucleosomes must be assembled fairly rapidly onto both daughter duplexes following replication: there are no extensive stretches of DNA free or partially depleted of nucleosomes.

What happens to the nucleosomes on parental DNA as the replicating fork advances? Little is known about the enzyme activities involved in synthesizing eucaryotic DNA, but replication appears to follow the same general process characterized in bacteria (see Volume 1 of Gene Expression). The parental strands must be unwound at the replicating fork to serve as templates for synthesis of daughter strands. Since DNA synthesis itself proceeds only in the direction $5' \longrightarrow 3'$, at least one strand (that synthesized overall in the direction $3' \longrightarrow 5'$) must be made by discontinuous synthesis of short (Okazaki) fragments. This implies that short single stranded regions of DNA should be present at the replicating fork, awaiting synthesis of complements. There is no direct information to show whether replication occurs on the surface of the nucleosome, so that the parental histone octamer is conserved and associates with one of the daughter duplexes as it is produced; or perhaps the nucleosome might split into half nucleosomes, each of which remains associated with one parental DNA strand. But most attention has been focused on the idea that the histone octamers may be displaced at the replication fork by a complex of replication enzymes, which undertakes parent strand separation and new strand synthesis; then nucleosomes should reform on the two daughter strands behind the replication fork.

There is evidence that DNA at the replicating fork may differ in its susceptibility to nucleases, although this does not reveal how the structure of this region is related to that of the usual organization in nucleosomes. A problem in investigating the structure of DNA at the replicating fork is to find conditions in which a label can be retrieved from this position, before it has been chased into bulk DNA. Using sea urchin blastulae, which incorporate an ³H-thymidine label very rapidly, Levy and Jakob (1978) found that newly replicated DNA is more resistant to micrococcal nuclease. A pulse label enters fractions of 150 and 305 base pairs, compared with the 140 and 410 base pairs seen for bulk monomers and dimers. The pulse label can be chased into the longer fragments. This suggests that newly replicated DNA first enters a different form of protein complex, then enters nucleosomes. From the sizes of the replicating units, it seems that the change in structure is confined to a rather short length of DNA, equivalent to only a couple of nucleosomes. Whether this represents a complex with the replication apparatus or some alternative form of the nucleosome is not shown by these results. In experiments when the movement of replication forks is followed in nuclei labeled in vitro, a pulse label enters DNA fragments shorter than usual which can be recovered from the monomeric nucleosome fraction (Seale, 1978; Schlaeger and Knippers, 1979). Under these conditions, histones are not synthesized; and so it is not clear whether this apparent association of preexisting octamers with newly synthesized DNA reflects the usual sequence of replication or is a feature of the in vitro system.

One approach to determining what happens to parental histone octamers following replication is to follow their redistribution when replication of DNA occurs in the absence of protein synthesis. This can be accomplished by using cycloheximide to inhibit protein synthesis in cells in culture. The result is that DNA synthesis continues for a short time and then ceases. Since new histone molecules are not synthesized, only preexisting histones can associate with the newly replicated DNA. The first experiment to determine the structure of the incompletely duplicated chromatin was performed by Weintraub (1973). Very short pulse doses of ³H-thymidine incorporated into chick erythroblasts in the presence of cycloheximide enter a state that is more resistant to DNAase I digestion (applied to nuclei) than is bulk chromatin. Longer pulse doses enter a state that is more susceptible to DNAase I than is bulk chromatin; but this state appears to be more resistant to degradation than expected of free DNA (although this last control involves the use of pure DNA rather than nuclei, so it may not be strictly comparable). This is consistent with the idea that DNA at the replicating fork (identified by the very short pulse label) is protected by the replication apparatus against nuclease; when the replicating fork moves on (as in the longer pulse doses), the DNA becomes more susceptible than usual because of the lack of histone synthesis upon which nucleosome assembly depends. This deficiency can be made good if protein (that is, histone) synthesis is allowed to resume before treatment with nuclease.

The same result is obtained with micrococcal nuclease: newly synthesized DNA is more sensitive than chromatin, although less sensitive than expected of free DNA, if replication has occurred in the absence of histone synthesis (Seale and Simpson, 1975). In fact, chromatin duplicated in the presence of cycloheximide is twice as sensitive to micrococcal nuclease, showing 25% resistance to degradation in a limit digest, compared with the usual 50%

(Weintraub, 1976). The normal pattern of the micrococcal nuclease digest is found in the resistant fraction, which therefore presumably consists of nucleosomes. The simplest explanation is that DNA synthesized in the absence of histone synthesis has half the usual number of nucleosomes.

Are nucleosomes arranged in a random or organized manner following this replication? The two extreme models are to suppose on the one hand that all the histone octamers lie on one of the daughter duplexes and on the other that the octamers are dispersed equally and at random on the two duplexes. The pattern of fragments released by micrococcal nuclease from histone-deficient chromatin is the same as that of bulk chromatin, at least up to the tetramer (Seale, 1976a). The kinetics of production of small oligomers are unaltered, which suggests that the relationship between adjacent nucleosomes remains the same. This does not seem to fit with the expectation of randomly arranged nucleosomes, which should consist of monomers separated by stretches of about 200 base pairs. But if nucleosomes were assembled in small groups, the observed result could be produced without implying that one daughter duplex gains all the histone octamers.

The conclusion that DNA synthesized in the absence of histone synthesis is partially depleted of nucleosomes carries the implication that there is no widespread redistribution of histone octamers on DNA—this would presumably make the newly synthesized and parental DNA structures indistinguishable. This is supported by electron microscopic observations of stretches of DNA of up to 100 kb in length that are free or highly depleted of nucleosomes (Riley and Weintraub, 1979). In some cases, two arms of a replication fork can be seen, one carrying and one lacking nucleosomes.

Contrasting results have been obtained in experiments to determine directly whether newly synthesized or old histones associate with replicated DNA. Using BUdR to identify chromatin containing replicated DNA by its increased density, with an ^{3}H-label for newly synthesized proteins, Seale (1976b) found no association between the density and radioactive labels in HeLa cells. Similar results were reported by Jackson, Granner, and Chalkley (1975, 1976) for rat hepatoma cells. These experiments used labeling periods from 30 to 180 minutes. Hancock (1978) found similar results when labeling periods of 3–30 minutes were used with a mouse cell line; with a period of 150 minutes, new histones were associated with newly synthesized DNA. Taken together, these experiments imply that newly synthesized histones associate with parental DNA, whereas newly replicated DNA associates with old histones (presumably displaced from their previous sites on parental DNA). The minimum period after which this redistribution allows new histones and recently replicated DNA to become associated remains to be established. Technical caveats in interpreting these results are that the separation of the different density fractions of chromatin often is not wide, it is not known whether incorporation of BUdR might influence the results, and differences in the presence of nonhistone proteins are ignored.

The opposite result has been found in an analysis of SV40 replication in

African green monkey cells. Histones labeled with ^{35}S-methionine are found in the fraction that contains the replicated viral genomes (identified by a label shorter in duration than one round of replication) (Cremisi, Chestier, and Yaniv, 1977). Similar results have been found with Drosophila tissue culture cells, where newly synthesized H3 and H4 appear to enter a peak of newly replicated material a little before H2A and H2B (Worcel, Han, and Wong, 1978). In these cases, therefore, there seems to be a substantial movement of new histones onto daughter duplexes of DNA.

From these various results, it is clear that a principal problem in analyzing the assembly of the nucleosome is the difficulty inherent in distinguishing the newly replicated material from the bulk of chromatin. Although this can be done with density or radioactive labels, the replicated fraction is seen as a peak not too well resolved from a high background of parental material.

Assembly of the Nucleosome

Two questions define our ignorance of the process of nucleosome assembly. Is the integrity of the histone octamer conserved when DNA is replicated or does it dissociate into either subunits or individual histones? Is nucleosome assembly achieved solely by the interaction of DNA with histones or are other proteins necessary?

The issue of whether histone octamers remain intact has been addressed by asking whether newly synthesized histones form complexes with old histones. Leffak, Grainger, and Weintraub (1977) labeled chick myoblasts with dense amino acids and ^{3}H-lysine for one hour. This is sufficient to label about one-eighth of the histones in each S phase cell. Disassembly of old octamers followed by reassembly from a pool also including new histones would require that each nucleosome should contain one heavy and seven light histone molecules. Conservation of the octamer should result in a biphasic distribution, with ⅛ of the octamers forming a dense radioactive fraction and ⅞ forming a light unlabeled fraction. If nucleosomes disassemble into half nucleosomes and then reassemble, a quarter of the nucleosomes should be half-heavy.

Using a cross-linking agent to maintain the structure of the octamer, once the DNA has been removed the protein can be banded on a CsCl density gradient. A heavy octamer then appears that remains stable for at least the 3–4 generations over which it was followed. When adjacent nucleosomes were cross linked, giving 16-mers, 32-mers, etc., there was still no shift of the radioactive label to a light density position. This suggests that the octamer may be conserved and that new histone octamers lie adjacent to other new histone octamers, rather than falling into a random arrangement with respect to the octamers of old histone.

A contrasting idea is that new histones may be associated with the newly synthesized (single) strand of DNA, whereas old histones associate with the parental strand. This has been proposed by Russev and Tsanev (1979) on the basis of analogous experiments in which histones were cross linked to DNA which was denatured before the association of protein and nucleic acid labels was examined. This model is difficult to reconcile with the conservation of the octamer, since following replication each octamer is associated with a duplex containing one old and one new strand of DNA.

Two different views of the nature of the association between DNA and the octamer are suggested by experiments that reveal unusual structures. One view rests on the observation of a change in nucleosome structure that occurs at very low ionic strength. The other depends on evidence that single stranded DNA can associate with a histone complex.

When SV40 minichromosomes are visualized under the usual conditions, they contain 20 or so nucleosomes, often separated by stretches of free DNA (see Chapter 13). Oudet, Spadafora, and Chambon (1978) showed that when the minichromosomes are incubated in the presence of EDTA to remove salt, they acquire a new form in which 40–50 beads are visible. An example is shown in Figure 14.2. The particles have diameters of about 9.3 nm compared with the 12.5 nm of full nucleosomes. From the reduction in contour length of SV40 DNA, they seem to contain about 96 base pairs of DNA. A simple interpretation is that these are "half nucleosomes," created by an unfolding into two heterotypic tetramers, each associated with half of the nucleosomal duplex of DNA. Whether this simply reflects a symmetrical organization of the nucleosome revealed by the unusual ionic conditions or mimics a transition that occurs in vivo is impossible to say. The conversion is reversible since the original nucleosomal structure is recovered when the unfolded minichromosome is incubated in 20 mm NaCl. A similar conversion can be seen with cellular chromatin, but is more difficult to visualize; it is possible that it may have occurred in experiments in which nuclei were lysed in water and this would explain some of the variation seen in such experiments to visualize nucleosomes.

When the 6000 base circles of single stranded DNA of phage fd are mixed with the four core histones (in 0.5 M NaCl), a complex sedimenting at about 45–50S replaces the peak of free fd DNA at 33S. Palter, Foe, and Alberts (1979) demonstrated that complex formation seems to occur by a sequence of events similar to those involved in nucleosome assembly in vitro (see Chapter 13). The H3-H4 pair of histones can bind first to convert fd DNA to a 39S complex; then the H2A-H2B pair can complete the reaction, although these histones are unable initially to bind to free DNA. The mass ratio of histone:DNA in the complex is not known exactly, but is in the range of 0.5–1.0. However, the histones are present in equimolar amounts. Digestion of the complex with micrococcal nuclease allows the isolation of a small amount of the input DNA (~5%) as a 9.4S peak. In addition to all four histones, this contains fragments of DNA in a size range of 50–100

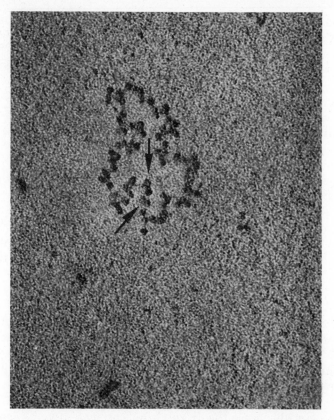

Figure 14.2. Half nucleosomes produced by removing salt from SV40 minichromosomes. The arrows point to pairs of half nucleosomes that appear closely juxtaposed and might represent the dissociation of single nucleosomes. Data of Oudet, Spadafora, and Chambon (1978).

bases. The kinetics of digestion are much more rapid than with nucleosomes and there is no obvious pause in the degradation at any particular length of DNA.

A better defined complex has been produced by reassociating short single strands of DNA with the histones. With fragments of 140–160 bases, about 30% of the DNA can be converted to the 9S peak. Fragments of DNA as small as 60 bases are effective in generating these particles. For the 140–160 base long fragments, the histone:DNA mass ratio is about 0.8, not much less than the 1.0 found when duplex DNA of this length is used for reconstitution. The particles are indistinguishable from nucleosomes by electron microscopy. Particles can also be formed from the reaction of histones with short duplex DNA fragments of less than 140 bp (Tatchell and Van Holde, 1979). Although not yet fully characterized, an interesting feature is the generation of a normal DNAase I pattern when 5′ ends are labeled with ^{32}P, in spite of heterogeneity in DNA length. This suggests that histones may have an inherent ability to organize DNA into a regular structure relative to the ends, irrespective of length.

What is the structure of the histone-single strand DNA complexes? One possibility is that a heterotypic tetramer is complexed to a molecule of single-stranded DNA. Another is that the histone aggregate is an octamer, in which case it must be associated with two single-stranded molecules of length 140–160 bases, with a corresponding increase in the number when the DNA fragment length is reduced. The general similarity of the reconstitution process with that characteristic of nucleosomes argues for the involvement of a histone octamer, although a definite proof of this will require more detailed studies, such as cross linking. A functional implication of this view is that this form of association might occur in replication or transcription, when the octamer becomes associated with only one of the two DNA strands present on the nucleosome. Palter, Foe, and Alberts show in a detailed analysis how the association of octamers with single-stranded DNA may be related to the usual structure of the nucleosome as considered in the model of Weintraub, Worcel, and Alberts (1976). The basis of these ideas is that the nucleosome consists of two heterotypic tetramers, each of which binds to only one of the two strands of duplex DNA. This allows perfect 2-fold symmetry to be achieved by the nucleosome.

The ability of DNA to reconstitute nucleosomes upon mixing with histones in vitro demonstrates that the necessary structural information is contained in the components of the particle (see Chapter 13). But since the conditions needed for assembly in vitro are somewhat extreme, involving the removal by dialysis of high (2 M) salt and (5 M) urea, it is not evident whether they mimic the process of assembly in vivo or provide a substitute for whatever other components are involved in the cell. Although not themselves components of the nucleosome, such other factors might be involved in assembly either stoichiometrically or catalytically to make possible the proper interactions under physiological conditions; precedents for this are provided, for example, by the assembly of bacteriophage head shells (see Volume 3 of Gene Expression).

A system in which assembly occurs under more physiological conditions has been developed from work with Xenopus oocytes. When SV40 DNA is injected into Xenopus oocytes, it can be assembled into minichromosomes. By injecting a sufficient excess of DNA, the capacity of the histone pool of the egg can be exhausted. Assembly then becomes dependent on added (that is, injected) histone (Laskey, Mills, and Morris, 1977). The ability of the oocyte to assemble nucleosomes has been extended into an in vitro reaction. An extract allows DNA and histones to form particles under physiological conditions of 0.15 M NaCl (Laskey et al., 1978a). The active component of the extract is an acidic protein(s) of 29,000 daltons, which appears able to form a complex (of undefined structure) with the histones. The stoichiometry of the reaction is not known (Laskey et al., 1978b). An oocyte factor able to insert supercoils into SV40 DNA in the presence of histones also has been isolated by Baldi et al. (1978).

In analogous experiments with Drosophila histones, nicking-closing enzyme has been identified as another candidate for the role of assembly protein. Upon mixing with histones and DNA, assembly of particles occurs stoichiometrically; one particle is formed for every molecule of nicking-closing enzyme (Germond et al., 1979). This has a size of 64,000 daltons, but it is not known if this is the active form (as is assumed in the calculation of stoichiometry).

In neither set of experiments have the particles formed by the assembly reaction been fully characterized. Criteria for identifying them as nucleosomes have been the formation of beads visible in the electron microscope, the insertion of an appropriate density of supercoils into DNA, and the recovery of 200 base pair fragments upon digestion with micrococcal nuclease. In lieu of experiments to demonstrate that the particles indeed contain one histone octamer with the 200 base pairs of DNA, it remains possible that they are not actually nucleosomes; experiments such as cross linking will be required to make the point unequivocal.

The idea that proteins other than histones can form particles in which DNA is wrapped around the outside is rendered plausible by experiments with some bacterial DNA-binding proteins. Liu and Wang (1978) found that E. coli DNA gyrase binds DNA to form a complex containing 140 base pairs, as seen in the release of fragments by micrococcal nuclease. At least some of the DNA appears to be on the outside, because cleavage with DNAase I generates single strand fragments varying from 46 to 96 bases with a 10 base periodicity. The enzyme consists of two subunits, whose combined mass is 140,000 daltons, and its binding of DNA is probably part of its basic action: one subunit may wrap the DNA around itself to introduce superhelical turns, while the other has the catalytic activity. The active form is a tetramer of 400,000 daltons, incidentally much larger than the histone octamer. Another bacterial protein able to form particles with DNA is the HU protein of E. coli (Rouviere-Yaniv et al., 1979).

Chromatin under Transcription

Are nucleosomes present on DNA during transcription? The two approaches that have been taken to seeking nucleosomes in the transcribed fraction of chromatin are to see if beads can be visualized by electron microscopy of DNA engaged in transcription and to determine whether sequences of transcribed DNA are cleaved into 200 base pair multiples by micrococcal nuclease.

The DNA sequences for which the most extensive electron microscopy has been performed are, of course, those coding for ribosomal RNA. The regions coding for the precursor rRNA form matrices of densely packed transcripts attached to the DNA axis; and the matrices are connected by an axis corresponding to the nontranscribed spacers (see Figure 27.4). The matrix is formed by an array of transcripts of steadily increasing length, each of which is attached to the axis by a dark spot of diameter about 12 nm; this is taken to be RNA polymerase. The similarity in size between the putative enzyme and the nucleosome points to the difficulty of using electron microscopy to distinguish the events involved in transcription.

The length of the axis of the matrix regions can be compared with the length of DNA known to code for the precursor rRNA. Complete data are available only for Xenopus laevis and Drosophila melanogaster, for which electron microscopy has been performed on the amplified (extrachromosomal) rDNA of oocytes and in early embryos, respectively; and where the lengths of the precursor rRNA coding regions are known from restriction mapping of DNA. In X. laevis, the coding length of 7400 base pairs should occupy about 2.5 μm upon spreading for electron microscopy; the matrix actually has a length of some 2.3 μm (Miller and Beatty, 1969a,b,c). In D. melanogaster, the coding length of 8000 base pairs corresponds to about 2.65 μm; the matrix has a length of 2.55 μm (McKnight and Miller, 1976). A similar relationship is found in the milkweed bug Oncopeltus fasciatus, in which the rDNA matrices take a length of 2.4 μm, compared with the (roughly) 2.8 μm expected to code for an rRNA precursor of 36S (Foe, Wilkinson, and Laird, 1976). In all of these instances, therefore, the packing ratio in actively transcribing rDNA is about 1.1-1.2. In effect, the DNA appears almost completely extended, so that there could be few, if any, nucleosomes in this region. The apparent packing, indeed, could be due to some distortion of DNA bound to RNA polymerase. When the nucleolar genes become inactive, nucleosomes appear after a lag period (Scheer, 1978).

Are the nontranscribed spacers organized in nucleosomes? For X. laevis, it is difficult to compare directly the spacer lengths separating the active matrices with the spacer distances measured in restriction mapping, because there is considerable variation in the lengths of individual spacers. Thus the spacers vary from $1.8-5.5 \times 10^6$ daltons by restriction mapping; whereas the distances between matrices vary from 0.8 to 25 μm. (Some of the very long gaps between matrices may not be genuine; for example, they could include transcription units that have failed to be activated.) In D. melanogaster there is less heterogeneity. The spacers vary in length, but occupy about 5700 ± 2000 base pairs. The average distance between matrices is 1.4 μm and is occupied by beaded structures (Laird et al., 1976). This corresponds to an overall spacer packing ratio of 1.9/1.4 = 1.35. If the beads are nucleosomes, then, either they are separated by appreciable lengths of extended DNA or they have become stretched out during preparation.

It is more difficult to perform comparable studies on nonribosomal regions, because the lengths of DNA in the transcription units are not known. However, studies of D. melanogaster and O. fasciatus suggest the same general conclusion that beads are present in transcribed regions, but at a frequency lower than in nontranscribed regions (Foe, Wilkinson, and Laird, 1976; Laird et al., 1976). To take the example of Drosophila, there are 28 ± 4 beads of diameter 13.9 nm for every μm of nontranscribed chromatin. In transcribed regions, beads can be seen between the transcript fibers that extend from the DNA axis; they occur at a frequency of 18 ± 5 per μm. There are also beads at the base of each fiber; if these include nucleosomes and do not just represent RNA polymerases, the frequency on transcribed chromatin would become 24 beads per μm. The frequency of nucleosomes in transcribed regions is therefore some 60–80% of that in nontranscribed regions, depending on the nature of the basal beads. Figure 14.3 shows an example of a nonribosomal transcription unit from Oncopeltus fasciatus.

To calculate packing ratios for these regions, it is necessary to make assumptions about the content of DNA in the beads. Depending on whether the beads are assumed to contain a core of DNA (140 base pairs) or a full nucleosome content (200 base pairs), there must be 1.9 or 2.4 μm of DNA contained in every μm of nontranscribed chromatin fiber. This is not very much less than the ratio of 2.75 found when SV40 chromatin is examined under less extended conditions (see Chapter 13). In transcribed regions the packing ratio should lie in the range 1.3–1.7.

A model system for investigating the structure of transcribed regions is presented by the SV40 minichromosome. Gariglio et al. (1979) have charac-

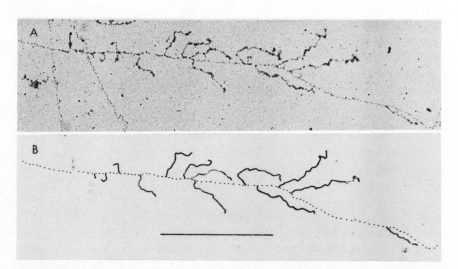

Figure 14.3. A nonribosomal transcription unit from Oncopeltus fasciatus chromatin. The electron micrograph in (A) is interpreted as a scale drawing in (B). Transcription appears to proceed from left to right, with increasing length of the chains attached to the axis. The arrow indicates the point at which initiation of transcription probably occurs, suggested by extrapolation of the chain lengths to zero. The dots on the drawing indicate the apparent positions of beads. The bar indicates 1 μm. Data of Foe, Wilkinson, and Laird (1976).

terized the transcriptional complexes that can be extracted from infected cells in a state able to continue RNA synthesis in vitro. These have the usual content of histones and display the typical beaded structure upon electron microscopy. Chains of RNA can be seen to extend from the beaded complexes. This implies that transcription in vitro can proceed around the SV40 DNA while it is wound into nucleosomes. Of course, it remains to be proven that the same state of affairs prevails in vivo.

The contrast between transcribed ribosomal DNA and other regions engaged in transcription is interesting. Perhaps it is the intensive transcription of rRNA, with about 40 polymerases per μm, that is responsible for the apparent loss of nucleosomes; whereas in nonribosomal regions, there appear to be about 6 polymerases per μm and a displacement of only some nucleosomes. In lieu of studies on specific nonribosomal transcription units, however, it is difficult to be sure whether these data are typical; for example, there must be wide variations in the frequencies with which different active genes are transcribed and these may be associated with differences in the state of the chromatin thread. The difference in packing ratios between transcribed and nontranscribed (nonribosomal) chromatin regions is interesting in suggesting that transcription may be accompanied by at least some displacement of nucleosomes. Of course, spreading for electron microscopy naturally extends chromatin fibers, and so these packing ratios cannot be taken as absolute values typical of the in vivo state.

In contrast with the absence of nucleosomes from ribosomal DNA spread for electron microscopy, the sequences coding for rRNA are found in 200 base pair multiples upon cleavage of some genomes with micrococcal nuclease. Experiments with Tetrahymena pyriformis have made use of the timing of the replication of extrachromosomal rRNA genes, which occurs at the beginning of S phase, before the synthesis of bulk macronuclear DNA. A radioactive label incorporated during this period enters the same series of multimers as bulk DNA (Mathis and Gorovsky, 1976; Piper et al., 1976). Formally this implies that newly replicated rDNA genes are organized into nucleosomes. But it is not certain what proportion of these genes is under transcription, although rRNA synthesis is active at this stage.

An approach of more general application is to use a labeled probe to assay by hybridization for the presence of complementary sequences in the multimers produced by micrococcal nuclease. With labeled rRNA of X. laevis, complementary sequences are found in monomers and small multimers isolated from chromatin of embryonic nuclei (which synthesize rRNA) and of erythrocytes (which are inactive) (Reeves and Jones, 1976; Reeves, 1977). The fraction of trimeric and tetrameric DNA complementary to rRNA is 0.049%, almost the same as the saturation value found with total DNA, 0.056%. The value drops to 0.044% for dimers and 0.041% for monomers, a decline whose reason is unknown. In animals heterozygous for a mutation removing rRNA genes, the saturation level with total DNA is 0.031%. Thus about 70% of the rRNA genes can be identified in monomeric or dimeric fragments, and about 80% can be retrieved including trimers or tetramers. Again the obvious query is whether this may correspond to the proportion of rRNA genes that is not transcribed, with the active genes lying in the 20–30% not recovered in nucleosomes. No direct estimate has been made of the number of rRNA genes engaged in transcription at given times; but in the heterozygous mutant which has only half the usual rDNA content, any further deletions of rRNA genes prove lethal. This suggests that in this case most or all of the genes may be transcribed at times of active rRNA synthesis. In this instance, 0.021% of the monomeric DNA hybridizes with rRNA, compared with a control of 0.038% for total DNA (Reeves, 1976). Just over half of the rRNA genes seem therefore to be organized in nucleosomes. If this is too many to remain untranscribed, either some rRNA genes must be transcribed while organized in nucleosomes; or 200 base pair repeats must be released from transcribed regions in spite of the absence of nucleosomes (there is no information about the effect of tightly packed RNA polymerases on micrococcal digestion). In view of these uncertainties, perhaps more significant measurements may be those on the proportion of rDNA of different oocyte stages that is represented in monomeric nucleosomes (as defined by nuclease digestion). This varies inversely with the degree of transcriptional activity characteristic of the developmental stage, from 84% to 37% (Reeves, 1978). This provides an indication that nucleosomes may be displaced for transcription, at least to some extent.

A good system for investigating the structure of transcribed rDNA is the nucleolus of Physarum polycephalum, which contains extrachromosomal rDNA arranged in pairs of genes on linear molecules (see Chapter 27). Present studies suggest that nucleosomes are present, as seen in the pattern of degradation caused by micrococcal nuclease. Attempts to correlate this result with electron microscopy have not yet produced unequivocal results (Grainger and Ogle, 1978; Stalder, Seebeck, and Braun, 1978).

Active and Inactive Nucleosomes

The question of whether nucleosomes are present on genes coding for either cellular populations of mRNA or for specific messengers can be investigated by using appropriate probes. Usually these are labeled cDNA molecules, synthesized by reverse transcription of the mRNA from its 3′ end; these therefore represent largely the promoter-distal regions of the transcription unit (see Chapter 26). With cDNA prepared from the total mRNA of rat liver, human lymphocytes, or trout testis, the same hybridization reaction is obtained with monomeric DNA (prepared by micrococcal nuclease treatment) as with DNA extracted directly from chromatin (Lacy and Axel, 1975; Kuo, Sahasrabuddhe and Saunders, 1976; Levy and Dixon, 1977b). This implies that most or all of the cell messengers are derived from genes that are present in nucleosomes. There is one caveat to this conclusion. The hybridization reaction predominantly represents mRNA species present in rather small amounts. It is therefore possible that only some of the copies of these genes are active; and it might be the inactive copies that give rise to the corresponding sequences found in nucleosomes.

A more penetrating assay for the state of transcribed genes is therefore to determine whether the sequences coding for a particular message, known to be transcribed in large amounts, also are found in nucleosomes. Lacy and Axel (1975) showed that in duck reticulocytes, globin cDNA shows an identical reaction with monomeric and total DNA, implying that the globin genes—which are actively expressed in this tissue—are present in the monomeric DNA taken to define the nucleosome. Bellard, Gannon, and Chambon (1977) found that in chick oviduct, the cDNA probes for ovalbumin (actively transcribed in this tissue) and for globin (not expressed in oviduct) both react with the multimers produced by micrococcal nuclease. Since globin and ovalbumin are the major products of reticulocytes and oviduct, respectively, most or all of the copies of these genes would seem to be engaged in extensive transcription. Thus active genes appear to be organized into repeating units of the same length as inactive genes.

But differences have been found in the organization of genes under transcription or that are potentially transcribable. The monomeric and dimeric fragments of oviduct DNA possess a higher concentration than do trimers and tetramers of sequences reacting with the ovalbumin probe. The reverse is true of the reaction of this DNA with the globin probe. As a control, there was no difference in the relative contents of ovalbumin and globin gene sequences in different members of the multimeric series derived from chick erythrocyte chromatin (which synthesizes neither protein). This suggests that the ovalbumin gene remains organized in nucleosomes in oviduct, but may be more susceptible to micrococcal nuclease in suffering more rapid cleavage to individual particles.

Two obvious explanations are that this could result from either a more extended conformation of nucleosomes (rendering more susceptible the sites of junction between them) or from a reduction in the number of nucleosomes on the gene. The removal of nucleosomes per se might be expected to result in rapid digestion of the intervening sequences, reducing the overall concentration of transcribed sequences in the digested DNA; this does not seem to happen. But presumably part of the gene is bound to RNA polymerase molecules during transcription; nucleosomes might be displaced at or around the site of RNA synthesis and the susceptibility of DNA in this region to micrococcal nuclease is difficult to assess. Also it is true that the extent of this effect must depend on the number of polymerase molecules on the gene at any time (which is difficult to determine because it is necessary to know the absolute rate of synthesis and length of the primary transcript; see Chapters 25 and 26). These problems mean that although these results seem to suggest no change in the number of nucleosomes on active genes, they do not provide unequivocal support for this conclusion. However, unless polymerase molecules are so frequent that usually they lie between adjacent nucleosomes or pairs of nucleosomes, the preferential cleavage into monomers and dimers would seem to suggest the existence of some change in conformation of the thread of nucleosomes at sites other than those of actual RNA synthesis.

Evidence for some general change in the state of organization of active genes is provided by demonstrations that they are preferentially susceptible to digestion by DNAase I. The first such experiment was performed by Weintraub and Groudine (1976), when they digested the nuclei of chick erythroblasts and then determined the ability of the undegraded DNA to react with a cDNA probe for globin message. At a point when only 10% of the DNA had been digested by DNAase I, the remaining sequences were able to react with only 50% of the cDNA probe. This contrasts with the 90% reaction displayed when the globin cDNA is reacted with digested fibroblast DNA or when ovalbumin cDNA reacts with digested erythrocyte DNA. The globin cDNA probe represented three different globin chains, so the absence of some, but not all, reacting sequences from the erythroblast DNA can be explained by the supposition that only the two genes actively expressed in this cell type have been digested; the third gene is inactive and therefore remains available to react with the corresponding sequences in the probe. (Actually this would imply about 33% protection; perhaps the level of 50% is due to some cross reaction of the globin cDNA probe with other surviving globin genes.)

But sensitivity to DNAase I is not a function solely of active transcription. The cDNA probe against adult globin mRNA reacts to only 25% with the DNA surviving DNAase I digestion of erythrocyte nuclei. This reaction appears in fact to be due to cross reaction of the probe with some embryonic globin genes that are not digested; all the three adult globin genes against which the probe is directed seem to be completely included in the first 10% of DNA that is digested by DNAase I. Since the erythrocytes have ceased to synthesize globin, this implies that it is not the act of RNA transcription that is itself responsible for the DNAase sensitivity. The erythrocytes must have retained whatever difference in conformation previously characterized the globin genes that were active in the adult red blood cells. Similarly, the globin genes of certain erythroleukemic cell lines are preferentially sensitive to DNAase I prior to as well as after the induction of globin expression. This suggests that they may be derived from an early erythroid stage in which certain globin genes have acquired the sensitive conformation although they have not yet been expressed (Miller et al., 1978).

Preferential sensitivity of transcribed sequences to DNAase I has been observed in several other instances. The kinetics of annealing between ovalbumin cDNA and the DNA remaining after DNAase I digestion of oviduct chromatin correspond to a threefold reduction in ovalbumin gene sequences (Garel and Axel, 1976). Thus the 10% of the DNA first removed by DNAase I includes some 67% of the ovalbumin gene sequences. The ovalbumin gene remains sensitive to DNAase I even after gene expression has ceased upon withdrawal of the activating hormone (Palmiter et al., 1978). The correlation with potential transcribability rather than with actual transcription is seen also in the rRNA genes of Physarum, which remain equally sensitive to DNAase I through the cell cycle, although they are not transcribed in some stages (Stalder et al., 1978). A difference in sensitivity is seen with an integrated sequence of Moloney murine leukemia virus, which is preferentially digested by DNAase I in cells that produce the virus, but not in cells that are inactive in viral RNA synthesis (Panet and Cedar, 1977).

What proportion of the genome is found in the DNAase I sensitive fraction? Weintraub and Groudine digested embryonic red blood cell nuclei with DNAase I until 20% of the DNA was acid soluble. Then the sequences remaining were tested for their ability to react with labeled whole cell DNA. Only 78% were able to do so, compared with 94% when micrococcal digest fragments were used instead of the DNAase I products. The addition of nuclear RNA to the DNAase I products increased the level of hybridization to 89%, which implies that the sequences missing after DNAase I digestion indeed correspond to those that are transcribed.

This question has been taken a stage further by Garel, Zolan, and Axel (1977), who have asked whether all genes active in the oviduct show the same level of sensitivity to DNAase I. The largest class of transcribed genes is represented by a small number of messengers per gene; that is, each gene is transcribed only infrequently, compared with the much more intense transcription of ovalbumin. When 10% of the oviduct DNA was digested by DNAase I, the reaction of a globin cDNA probe with the remaining sequences was scarcely affected. But the reaction of ovalbumin cDNA was slowed by a factor of 2, implying that only 50% of the ovalbumin sequences remained undigested. A similar result was found with cDNA corresponding to the infrequent message population. Figure 14.4 shows the dependence of the loss of these sequences on the extent of DNAase I digestion. It is clear that ovalbumin sequences are preferentially digested; the sequences corre-

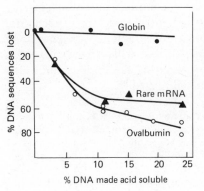

Figure 14.4. Digestion of transcribed and nontranscribed DNA sequences in chick oviduct chromatin by DNAase I. The percent of a given sequence digested is given by the reduction in the kinetics of reaction of the undigested DNA with a corresponding cDNA probe. For globin genes there is almost no reduction in frequency relative to total undigested DNA. For rare mRNA sequences and the ovalbumin gene, increasing digestion of DNA is accompanied by preferential loss of the transcribed regions. At the limit of these experiments (24% of the DNA rendered acid soluble), 60% of the rare gene sequences had been digested and 80% of the ovalbumin sequences had been degraded. Data of Garel, Zolan, and Axel (1977).

sponding to rare messengers are at first lost almost equally rapidly but then seem to reach a limit. Nonetheless, it is clear that, even if not as susceptible as the ovalbumin gene, these sequences are preferentially digested by DNAase I. The preferential digestion of DNA sequences represented in rare messages has been reported also for trout testis chromatin by Levy and Dixon (1977).

Again the difficulties of knowing the rate of transcription (as opposed to the level of mRNA) mean that the number of polymerases on each gene cannot at present be estimated. However, it seems likely that preferential susceptibility to DNAase I is a function of most or all of the genes that are subject to transcription. It is reasonable to speculate that the polymerases must be too infrequent on such genes for this effect to be due to their presence per se; thus it may reflect some conformational change introduced in active genes.

One obvious question is at what time such a change occurs. An approach to this will be to examine cells during the differentiation of a given phenotype, to see at what point in the cell lineage a given gene becomes susceptible to DNAase I, and how this correlates with the time at which it is first expressed. Clearly the same question can be asked about when the conformation returns to the inactive condition after the cessation of transcription. From the instance of the chick erythrocyte, it seems that this does not coincide with the absence of RNA synthesis.

How extensive is the region of conformational change around a transcribed gene? Flint and Weintraub (1977) have determined the sensitivity to DNAase I of the integrated viral sequences of rodent cells transformed by adenovirus 5, using two cell lines each of which contains only part of the viral genome. By using probes derived from restriction fragments corresponding to particular parts of the viral genome, it is possible to define which sequences are sensitive to DNAase I. In these cases the regions corresponding to stable mRNA are sensitive; others remain resistant. Although the limits of susceptibility are not yet fully defined, it is clear that they extend at least the equivalent of 2–3 nucleosomes downstream and several upstream from the sequences represented in mRNA. But of course the mRNA may be only a part of the primary transcript. A complication in interpreting these results is that there is more than a single integrated copy of each viral sequence; it is not known whether all are transcribed. But experiments along these lines for sequences present in single copies, where restriction fragments are available for nontranscribed surrounding regions, should allow the extent of conformational change to be delineated.*

---

*The probes used to assess DNAase I sensitivity usually have been cDNA reverse transcripts of mRNA; these may be incomplete and therefore represent preferentially the 3′ (terminal) region of the gene (see Chapter 26). Since the genes may be very much longer than the mRNA due to the presence of interruptions that are removed during the processing of the transcript, an incomplete probe may identify a genomic region that is a much smaller proportion of the whole gene than is the probe of the mRNA. Formally the available data demonstrate a difference

Given that transcribed sequences are represented in nucleosomes, an obvious question is whether the DNAase I sensitivity is a function of some change in the nucleosome itself or depends on some other structural alteration. On this point there have been conflicting reports. Weintraub and Groudine (1976) found that the same preferential susceptibility is found when erythrocyte 11S monomer particles (prepared by micrococcal nuclease action) are treated with DNAase I and assayed for globin gene sequences. This would implicate a conformational change in the nucleosome itself. On the other hand, Garel and Axel (1976) found that oviduct nucleosomes do not suffer preferential digestion of ovalbumin genes, which would argue for the implication of longer range forces or components not present on individual nucleosomes.

From all these experiments, it seems likely that some change occurs in the structure of chromatin, at the level of the nucleosome thread, in regions subject to transcription. The exact nature of this change, how extensive it is, and the timing of its introduction and cessation, remain to be established.

Constitution of Active Chromatin

What distinguishes regions of chromatin that are transcribed from those that are inactive? Can specific nonhistone proteins be detected in the company of nucleosomes representing particular regions of chromatin, such as active genes or inactive constitutive heterochromatin? Are any modifications of the histones themselves associated with a particular functional state of chromatin?

A direct approach to these questions has been to fractionate chromatin into preparations differing in some physical property; various attempts have been made to equate such fractions with particular components of the genome. Early experiments are discussed in Chapter 28; we should note here that they did not lead to the identification of any changes in chromatin structure with transcriptional state. Probably the physical parameters used to fractionate chromatin in these experiments were too crude to allow resolution of any differences.

Another approach has followed the DNAase I sensitivity of the nucleosomes of active genes. This makes it possible to ask whether the nuclease treatment has any effect on the protein composition of chromatin. Proteins preferentially released by DNAase I have been identified in sev-

---

in the conformation of (say) the 3' terminal half of the transcribed genes. It would be useful to confirm that the same sensitivity is displayed by all regions of the transcription unit; this could be accomplished by using cloned sequences representing specific regions, to allow direct assessment of their conformation. In particular it would be interesting to see whether the initiation and termination regions have the same sensitivities as internal regions.

eral systems and represent candidates for species associated specifically with active genes.

When trout testis nuclei are digested with DNAase I, the degradation of 10% of the DNA is accompanied by the release of virtually all the H6 (a histone-like protein specific to trout testis); some of the nonhistone protein HMG-T also is released. Levy, Wong, and Dixon (1977) found that micrococcal nuclease digestion rapidly releases HMG-T, suggesting that it may be located in linker regions; H6 is not preferentially released, so its removal is specific for DNAase I degradation alone. The correlation between the effects of DNAase I on active genes and on H6 and HMG-T raises the possibility that they may be associated with transcribed DNA sequences (see Chapter 28).

A similar result has been obtained with duck erythrocyte chromatin by Vidali, Boffa, and Allfrey (1977). Digestion with DNAase I causes the release into the supernatant of some of the nonhistone proteins; the specificity of this release is suggested by the observation that micrococcal nuclease preferentially releases a different set of chromosomal proteins. The proteins sensitive to DNAase I digestion fall into two groups: a low molecular weight set, 28,000–29,500 daltons, with properties similar to the HMG proteins of calf thymus, and a high molecular weight set, 57,000–59,000 and 80,000 daltons. The putative HMG proteins are present on the nucleosomes obtained by micrococcal nuclease digestion. A speculation is that they might be present on a subset of nucleosomes that is part of the transcriptionally active fraction. Similar experiments with a mouse cell line have led to a slightly different result, in which micrococcal nuclease and DNAase I release sets of proteins that overlap in large part (Defer et al., 1979).

A system in which an attempt can be made to deduce the functions of the proteins released by DNAase I is provided by D. melanogaster. Mayfield et al. (1978) prepared antibodies against five molecular weight subfractions of the proteins extracted from nuclei after digestion with DNAase I. When applied to polytene chromosomes, four of the preparations did not identify particular loci, but one reacted specifically with puffs. It is possible that this represents reaction with a protein (in the size class of about 63,000 daltons) present at active transcription sites and preferentially released by DNAase I.

A difficulty in interpreting these experiments is that they do not permit comparison of the properties of active genes with a control of inactive genes; nor is it possible to compare the structures of different active genes. A more specific assay is provided by the experiments of Weisbrod and Weintraub (1979), in which they reported that treatment of chick erythrocyte chromatin with 0.35 M NaCl abolishes the preferential sensitivity of globin genes to DNAase I. Sensitivity is regained by adding the proteins eluted with 0.35 M NaCl from either erythrocyte or brain chromatin, so the species responsible do not appear to be tissue specific. Although present data are only preliminary, the use of globin gene sensitivity as an assay may allow some characterization of the conditions needed to maintain genes in a potentially active state.

In contrast to active genes, the other region of chromatin that may be expected to have distinct properties is constitutive heterochromatin. A chromatin fraction of this sort has been prepared by treating the nuclei of African green monkey CV1 cells with the enzyme EcoRI. This releases a prominent fraction of chromatin that contains the DNA of component α, identified by its repeat length of 176 base pairs and its low buoyant density; this can be isolated by centrifugation through a sucrose gradient. Musich, Brown, and Maio (1977) found that this fraction is depleted of histone H1 and is enriched in some nonhistone proteins. The role of these proteins remains to be established.

The use of techniques that release proteins potentially involved in gene expression does not resolve the issue of their location at the level of the nucleosome. The relatively crude methods used to isolate nucleosomes originally caused the loss of most nonhistone proteins. More recently, experiments have begun with more sophisticated procedures; some nonhistone proteins now can be recognized on nucleosomes. When proteins are labeled with radioactive tryptophan (which is absent from the histones), the label found in the nucleosomes may be taken to identify nonhistone proteins. The nucleosomal complement of nonhistones is much less than that of unfractionated chromatin. Most of the nonhistone proteins are carried only by intact mononucleosomes; core particles lose most nonhistones as well as H1, but one or two nonhistones can be found on core particles or other chromatin fragments (Bakayev et al., 1978; Mathew, Goodwin, and Johns, 1979). The proteins HMG1 and HMG2 are released by micrococcal digestion of rabbit thymus and might therefore be associated with linker regions.

A long-standing question has been what effect histone modification has upon chromatin structure. Usually this is difficult to investigate because of the transience of modification. However, this problem can be circumvented by taking advantage of the observation that treatment with butyrate increases the proportion of acetylated H3 and H4 in cultured cells (see Chapter 11). Chromatin in which most of the histone molecules are acetylated is digested by DNAase I at a rate some tenfold greater than that found with control chromatin, in which only 10–20% of the H3 and H4 molecules are acetylated. Micrococcal nuclease, however, digests both acetylated and control chromatin at the same rate, although the fraction of DNA present in the monomeric form is increased with the acetylated substrate (Chang et al., 1978; Sealy and Chalkley, 1978b; Simpson, 1978b). With normal chromatin, DNAase I digestion seems to release preferentially nucleosomes containing acetylated histones (Nelson, Perry, and Chalkley, 1979; Vidali et al., 1978). Another observation is that micrococcal digestion of trout testis chromatin preferentially releases multiacetylated H4 as free histone (Levy-Wilson et al., 1979). The acetylation of histones, principally

H3 and H4, seems therefore to have some effects on chromatin similar to those that may distinguish transcribed from nontranscribed regions. Whether this coincidence indicates a role for histone modification in gene expression remains to be seen (see also Chapter 28).

Coiling of the Nucleosome Thread

Within the nucleosome, the length of DNA is compressed about 6-7 fold. Within a mitotic chromosome, the compaction factor is about 5000-10,000 fold; although clearly less in interphase chromatin, no overall value can be assigned. These figures are not particularly revealing, except to suggest that between the 6-7 fold compaction of the individual particle and the >5,000 fold compaction of the metaphase chromosome, there must lie one and probably two higher orders of organization. This is to say that nucleosomes must be organized into some higher order structure, which may itself in turn be ordered into a more complex structure.

Constituent threads of interphase chromatin and mitotic chromosomes have been visualized with various apparent attributes in cytological studies (see Chapter 10). The general concept that has been derived from electron microscopy of thin sections and whole mounts is that each chromosome consists of a continuous fiber of nucleoprotein of about 20-30 nm in diameter. In some studies this fiber appears to be generated by the coiling of a 10 nm fiber (DuPraw and Bahr, 1969). Each fiber appears to contain a single duplex thread of DNA, since treatment with trypsin generates thin patches 2.5-5 nm in diameter (DuPraw, 1965b; Abuelo and Moore, 1969). Here we are concerned with the issue of how the thread of nucleosomes is coiled into the 30 nm fiber.

The folding of the fiber appears to be random in whole mount electron micrographs, such as the example of Figure 4.1. But it is clear from the banding studies described below that each chromosome has a characteristic ultrastructure. Whether any definite organization exists in interphase chromatin, or whether the threads constituting various chromosomes are randomly tangled as they would appear to be, remains to be established. At all events, little is known about the forces governing the organization of the 30 nm fiber at the local level, although there is evidence for a regional organization within chromosomes, consisting of large loops of the nucleoprotein fiber.

Quantitative studies of the fiber have been made, but these must be considered unreliable because of the wide variation that is found in their appearance and physical parameters. There are variations over a range of some 2.5-fold from mitosis to mitosis for the dry mass found for different samples of the same chromosome, identified by its relationship to the other members of the set (DuPraw and Bahr, 1969). Indeed, the extent of this variation is emphasized by the report of Bahr and Golomb (1971) that a study of 925 human chromosomes showed a continuous spectrum of chromosome sizes rather than clusters of different size groups. Similar variation is seen in the constituent fiber; even in a study of a single nucleus, the mass varied from 2×10^8 daltons/μm to 7×10^8 daltons/μm (Golomb and Bahr, 1974a). The mass of a nucleosome is about 250,000 daltons, but it is not really possible to calculate the number of nucleosomes per μm of fiber because of uncertainty about its content of nonhistone proteins.

In the studies of DuPraw and Bahr (1969), the interphase fibers appeared to be smaller (23 nm diameter) and of lower average mass per unit length than the mitotic fibers (30 nm diameter). Whether this difference is genuine or represents an artefact remains to be established. However, it does point to the general question of whether mitotic condensation occurs only at the level of tighter packing of the 30 nm fiber into higher order structures, or includes also a change in the structure of the fiber. (Although reported diameters for the chromatin fibers vary, we shall use 10 nm and 30 nm as nominal values to indicate the apparent first two levels of organization.) The amount of material present in the chromosomes appears to increase at mitosis; but whether this involves addition of specific nonhistone proteins, or is due just to aggregation of cellular components released by dissolution of the nuclear membrane, is not clear (Huberman and Attardi, 1966; Solari, 1971; Huberman, 1973).

More recent studies to visualize nucleosomes have identified both 10 nm and 30 nm fibers in chromatin preparations. The 10 nm thread has been visualized in Drosophila interphase chromatin by Worcel and Benyajati (1977), using conditions of gentle lysis in which the whole genome is released as a folded complex (see below). The fiber appears as a thread of diameter about 10 nm; and sometimes a section can be seen as a string of beads, changing into the continuous fiber, as shown in Figure 14.5. Under these conditions of isolation, the complex has only the four core histones; it lacks H1 and nonhistone proteins. This implies that the 10 nm fiber must constitute a form of organization of nucleosomes per se.

Chromatin obtained by mild micrococcal nuclease digestion of rat liver nuclei (to avoid the distortion introduced by other methods of preparation; see Chapter 13) has been used by Finch and Klug (1976). When visualized in EDTA, fibers of 10 nm diameter, often coiled up loosely, could be seen. However, in 0.2 mM $MgCl_2$, the chromatin takes a more condensed form, with fibers varying in width over the range 30–50 nm. These fibers appear to have a helical construction; an example is shown in Figure 14.6. Presumably ionic conditions in vivo ensure the maintenance of the 30 nm fiber.

When H1 is removed, the addition of $MgCl_2$ fails to cause any condensation. This suggests that the presence of H1 (or possibly some nonhistone protein removed together with it) is necessary for the transition from the 10 nm thread into the 30 nm fiber. Another indication of such a role for H1 is provided by suggestions that the SV40 minichromosome may be able to take its more condensed form of structure only in the presence of H1 (Bellard et al., 1976; Christiansen and Griffith, 1977; see Chapter 13). In the presence of H1 and in reasonable ionic strength, the minichromosome can

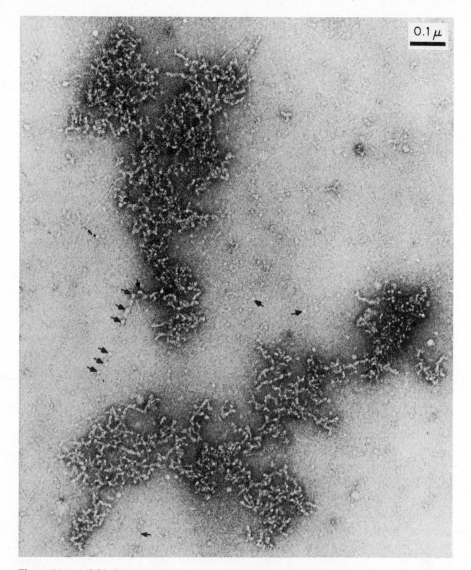

Figure 14.5. A folded Drosophila genome constituting only DNA and the four core histones.
The two large clumps of material consist of a folded 10 nm fiber; arrows point to regions in
which this has unravelled into a string of nucleosomes. Data of Worcel and Benyajati (1977).

be found in a highly compact form which is resistant to micrococcal nuclease. The minichromosome then appears as a roughly spherical particle
of diameter about 30 nm, sedimenting at 70–80S (Varshavsky et al., 1977;
Muller et al., 1978). The compact complexes appear to consist of smaller
globular subunits, whose structure is not yet clear but consists of several
nucleosomes.

Although it seems likely that H1 is involved in the first level of higher

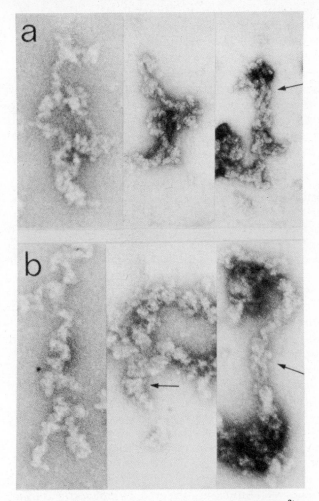

Figure 14.6. Comparison of rat liver chromatin visualized in EDTA or Mg^{2+} by negative staining. (a) in 0.2 mM EDTA the fibers are loosely twisted with a diameter of about 10 nm. (b) In 0.1 mM Mg^{2+} the fibers are thicker, in the range from 30–50 nm, apparently of helical construction. Regions of loose coiling are indicated by arrows. Data of Finch and Klug (1976).

organization of the nucleosome string, the structure of this 30 nm fiber is not yet clear. Two contrasting views have been proposed, one suggesting a continuous array of nucleosomes, the other an arrangement of specific subunits.

The peak seen at 10 nm in X-ray diffraction is concentration dependent; this has led to models based on a corresponding spacing of nucleosomes along the fiber (Carpenter et al., 1976; Carlson and Olins, 1976; see also Chapter 13). On the other hand, Finch and Klug (1976) noted the absence of this peak when chromatin is visualized in EDTA; they suggested that it derives from a higher order of coiling. Their model proposed a supercoil or solenoid with a pitch of 10 nm, which is formed upon withdrawal of water as the concentration of chromatin increases. This has about 6 nucleosomes per turn, corresponding to an overall packing ratio of roughly 40.

The principal alternative model proposes that the fiber is an organization of "superbeads," each consisting of a cluster of about 8 nucleosomes that forms a particle of about 20 nm diameter (Hozier, Renz and Nehls, 1977). Peaks consisting of 7–9-mers and ~16-mers can be generated by very mild nuclease digestion and appear as globular particles of about 20–26 nm per superbead in 100 mM NaCl (Stratling, Muller, and Zentgraf, 1978). These unfold into beads on a string at lower ionic strength.

These studies were made with interphase chromatin, but similar results are obtained with metaphase chromosomes. In mechanically lysed cells, fibers of 20–30 nm spill out of the chromosomes. These appear to represent a continuous helical coiling of the thread of nucleosomes. When the cells instead are lysed with nonionic detergents, the same fibers are seen; but they are less stable. First they break down into a series of aggregates that appear to correspond to superbeads; then these unwind to give the appearance of nucleosomes on a string (Rattner and Hamkalo, 1978a,b). One possibility therefore is that the superbeads represent a metastable intermediate in the folding of nucleosomes into chromatin.

A difficulty in all of these studies is that the great mass of chromatin remains undispersed, so that what can be visualized is a small part that becomes extended for electron microscopy. It is hard to be sure that the structures visualized are indeed typical of the organization prevailing within the mass of interphase chromatin or the mitotic chromosome.

The studies of X-ray diffraction and neutron scattering, which may provide direct structural data, have been discussed in Chapter 13. These generally have been directed toward analysis of the structure of the nucleosome itself; as yet they have not led to elucidation of the arrangement prevailing in the chromatin fiber. Sperling and Klug (1977) have reported X-ray diffraction studies of chromatin prepared by micrococcal nuclease action (rather than the shearing used in earlier studies, which may have disrupted the structure). They have discussed discrepancies in previous work and their dependence on methods of preparation. In this recent work, the 11 nm and 5.5 nm peaks both appear to represent the organization of nucleosomes, whereas the 3.8, 2.7, and possibly 2.2 nm peaks reflect the structure of the nucleosome itself.

Chromosome Integrity

In conventional cytological studies, interphase chromatin appears to be a tangled mass occupying much of the nuclear volume, while mitotic chromosomes appear as compact individual bodies. For some time it has been possible to isolate chromosomes from metaphase cells by centrifugation through sucrose gradients; and this allows their fractionation into classes of different size. Material obtained in this way is available for analyzing structure and has been used to transfer genetic information (see Chapter 9). More recently, an extension of the gentle lysis procedure originally applied to bacteria to obtain the nucleoid body as a folded genome (see Volume 1 of Gene Expression) has made it possible to isolate interphase chromatin from eucaryotic cells as a compact mass.

This protocol was used by Benyajati and Worcel (1976) to release the genome from Drosophila melanogaster tissue culture cells into sucrose gradients. When cells or nuclei are layered on top of the gradient and centrifuged through a lysis layer containing detergent, all the nuclear DNA sediments as the single broad peak of 4000–5000S shown in Figure 14.7. The gradient includes 0.9M NaCl; and the genomes contain only the four core histones: H1, nonhistone chromosomal proteins, and nuclear membrane proteins are absent. About 10% of a 10 minute label in RNA is found in this peak. The isolation of the genome in this state implies that whatever components other than DNA and the four core histones are needed to maintain the integrity of the compact structure must be minor components of chromatin. It is not clear whether there are connections between the different chromosomes, as might be inferred from the isolation of a single complex, or whether the three major chromosomes happen all to sediment at about the same rate (they are all close in size), in which case the small chromosome IV should not be part of the complex.

The structure of the complex has been investigated by its response to ethidium bromide, whose intercalation into a closed DNA molecule generates positive superhelical turns. No such effect is seen with open molecules, of course, where the strands remain free to rotate (such as in duplex circles with a nick in one strand); and this has been used to isolate closed circular molecules by centrifugation through gradients containing ethidium bromide, when the introduction of super helical turns results in an increase in the rate of sedimentation (see Volume 3 of Gene Expression). When a closed molecule possesses negative superhelical turns (as does, for example, SV40 DNA), first the intercalation causes loss of these turns, relaxing the molecule; and then it introduces positive turns, coiling it up again. The rate of sedimentation of the DNA reflects these changes, decreasing to a minimum when there are no superhelical turns and then increasing as positive turns are introduced.

Exactly this effect is seen with the compact Drosophila genome; its rate of sedimentation declines as the concentration of ethidium bromide is increased, reaching a minimum at 2.5 μg/ml; then it increases again. This level of the drug is the same as that needed to reach zero superhelical turns in SV40 DNA and in the folded E. coli genome. This suggests that the Drosophila genome has about the same concentration of negative superhelical turns as SV40 and E. coli, about 1 per 200 base pairs.

The effect of ethidium bromide is abolished by nicking with DNAase I, which relaxes the genome into a form sedimenting more slowly, at about 2500S. This relaxation does not release the histones. This suggests that

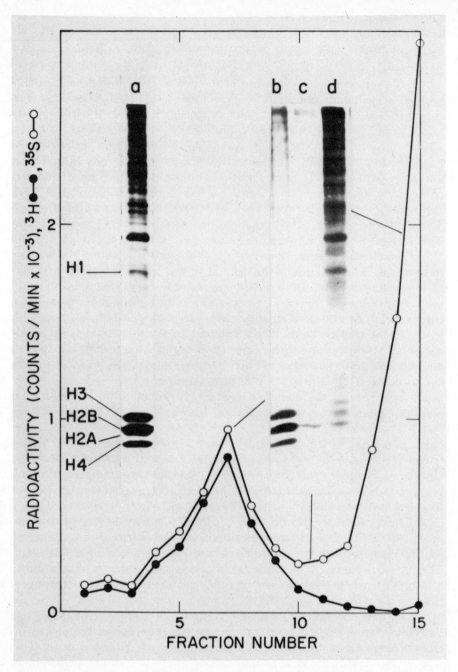

Figure 14.7. Isolation of the folded Drosophila genome by sucrose gradient centrifugation. The broad peak in the center of the gradient contains all the DNA and the four core histones, as seen by gel electrophoresis of this fraction (insert b). Histone H1 and the nonhistone proteins are found at the top of the gradient (fraction d). The gel in (a) is a control of labeled nuclear proteins. Data of Benyajati and Worcel (1976).

superhelical turns can be released by DNAase I without affecting the association of DNA with histone octamers; the implication is that supercoils are present in inter nucleosomal DNA. About 1 nick for every 85,000 base pairs (6×10^7 daltons; about 400 nucleosomes) is sufficient to achieve full relaxation. A formal interpretation would be that this corresponds to the average length of DNA that can be regarded as a closed structure; one obvious model is to suppose that there are large loops of about this size, each being independent with regard to its supercoilng. This is analogous to the model proposed from similar data for the bacterial genome (see Volume 1 of Gene Expression). By contrast with the effect of DNAase I, the intercalation of ethidium displaces histones from the genome; almost all are lost by a drug level of 2 μg/ml. One model is to suppose that histone octamers "pop out" of the DNA as the negative superhelical turns are abolished.

A difference is seen in the structure of the fibers constituting the compact 4000–5000S complex and the unfolded 2500S complex. Worcel and Benyajati (1977) found that in its compact state the complex consists of 10 nm fibers (the lack of H1, and perhaps certain nonhistone proteins, has presumably released 30 nm fibers into 10 nm threads; see above). In the unfolded state, however, the complex consists of beaded strings. Following the argument that DNAase I causes unfolding by relaxing supercoiling in internucleosomal DNA, the implication is that it is this supercoiling that is responsible for bringing together neighboring nucleosomes to generate the 10 nm fiber. A model for the 10 nm fiber and its relationship with the 20–30 nm fiber has been presented by Worcel and Beyajati and is shown in Figure 14.8. This suggests that the beaded string has −1.5 supercoils per nucleosome, while the 10 nm fiber has −1.7 to −2.3 depending on the repeat length of DNA in the nucleosome. The overall packing ratio of the beaded string is between 2 and 4 (depending on the length of DNA between adjacent core particles); that of the 10 nm fiber becomes 6.5 (compared with the value per nucleosome of 5–6). The 30 nm fiber generated by this model has a packing ratio of up to about 40. A feature of the model is that even in the 30 nm fiber, DNA remains in an exposed external position, explaining its susceptibility to nucleases.

The existence of a compact nuclear genome is not peculiar to Drosophila, since analogous results have been obtained with mammalian tissue culture cells (Cook and Brazell, 1976a,b, 1977; Cook, Brazell, and Jost, 1976). In these experiments the size of the independent units of supercoiling seemed larger, about 10^9 daltons; the concentration of supercoils was similar. It seems possible from these results that the organization of these genomes and that of Drosophila may be similar, at least qualitatively.

An approach that may lead to the definition of domains in the higher order structure of chromosomes has been developed with D. melanogaster. When nuclei are digested very mildly with DNAase I, a smear of high molecular weight DNA is obtained. Wu et al. (1979) investigated the location of particular sequences within this smear by using cloned probes repre-

Figure 14.8. Model for chromatin fiber. The lowest particle is an individual nucleosome, connected by a DNA string to three nucleosomes forming a 10 nm fiber. Then the nucleosomes coiled into a thick 30 nm fiber. The coiling of DNA on the surface of the nucleosomes leaves it exposed on the outside of the 30 nm fiber. From Worcel and Benyajati (1977).

senting sequences whose chromosomal locations are known from in situ hybridization. Each probe identifies a band of discrete but different size; the range of sizes is from about 4->22 kb. These large fragments are not found when micrococcal nuclease is used to cleave the DNA. The ends of the large DNAase I fragments may lie at specific sites, as indicated by the cleavage of each fragment into defined pieces by restriction enzymes with few recognition sites in the region. It is possible that the large fragments may define domains of the higher order structure, bordered by sites that have increased susceptibility to DNAase I.

One of the probes used in these experiments represents a gene that is expressed upon heat shock (see Chapter 28). Wu, Wong, and Elgin (1979) compared the corresponding bands generated by DNAase I under normal conditions (25°C, gene inactive) and under heat shock (34°C, gene expressed). The result of heat shock is to reduce the intensity of the large bands.

In one sense, this confirms that expressed genes are more sensitive to DNAase I. But also it raises the possibilty that the nature of the change may be extensive in influencing the higher order structure of chromatin. Assessment of this idea requires some understanding of the relationship between the DNAase I cleavage that defines the domains with the previously characterized attack on nucleosomal DNA.

Visualization of isolated mitotic chromosomes has led to the concept that each chromosome consists of a single nucleoprotein fiber, folded or coiled into its characteristic compact shape. Biochemical studies suggest that metaphase chromosomes have a greater content of nonhistone proteins than interphase chromatin; but it is not clear whether this is an artefact due to aggregation of cellular components during preparation. None of these studies has given much insight into the molecular construction of the chromosome in the sense of revealing how its typical compact structure is maintained. More recently, a procedure has been developed for removing histones from metaphase chromosomes by competition with the polyanions dextran sulfate and heparin; the resulting structures suggest for the first time that there may be a "skeleton" that maintains the structure of the chromosome.

Introducing this technique, Adolph, Cheng, and Laemmli (1977) found that HeLa metaphase chromosomes from which histones have been removed sediment at a reduced rate, in the range 4000–7000S. The histone-depleted complex can be isolated by density gradient centrifugation in metrizamide (a nonionic medium). It retains about 10% of a label in ^{35}-S-methionine; and has a buoyant density corresponding to a DNA:protein ratio of about 1:0.16, compared with the value of 1:2.2 found for intact chromosomes. The protein remaining on the complex represents about 8% of the chromosomal protein, virtually all nonhistone (99% of the histones have been removed). Gel electrophoresis resolves about 20 protein components, mostly larger than 50,000 daltons. Consistent with the lack of contribution from histones, the complex remains stable in 2 M NaCl. It is the protein components that maintain its structure: it can no longer be isolated after treatment with chymotrypsin (a procedure that is ineffective in preventing the isolation of intact metaphase chromosomes). Addition of SDS also dissociates the complex; urea unfolds it, reducing the sedimentation rate. Ribonuclease has no effect.

The histone-depleted metaphase chromosomes can be visualized by electron microscopy. Paulson and Laemmli (1977) showed that they take the form seen in Figure 14.9, with a central "scaffold" of darkly staining material surrounded by a "halo" of DNA. The scaffold seems to consist of a dense network of fibers. Threads of DNA appear to emanate from the scaffold; generally their concentration is too great to allow individual threads to be followed, but sometimes it is possible to trace a thread as a continuous loop, in which both ends are anchored close together in the central scaffold. Most of the loops are between 10 μm and 30 μm in length (30,000 to 90,000 base pairs); if compacted 40-fold in length into a 30 nm fiber, the

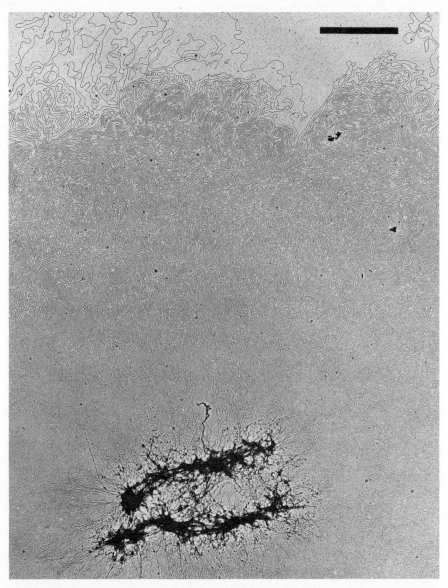

Figure 14.9. Histone-depleted metaphase chromosome from a HeLa cell. The scaffold appears to consist of a dense network of fibers. The DNA is extruded from it in the form of large loops. The bar indicates 2 μm. Data of Paulson and Laemmli (1977).

range would be roughly 0.2–1.0 μm, which is of the order of size of the diameter of the chromosome.

When chromosomes are swollen by the removal of divalent cations, their diameter is increased about 4-fold to some 2–3 μm. Thin sections allow the structure to be visualized as a series of radial loops of 10 nm fiber, held together in the central region (Marsden and Laemmli, 1979). The length of the loops varies roughly in the range 3–4 μm; if the fiber is assumed to consist of a continuous array of nucleosomes with a packing ratio of 7, this would correspond to DNA loops of 20–30 μm, in the same range as that observed in the histone-depleted chromosomes.

The scaffold can be isolated free of DNA by treating the metaphase chromosomes with micrococcal nuclease before they are depleted of histones. Adolph et al. (1977) found that 60–70% of the DNA can be rendered acid soluble; and less than 0.1–1.0% of an ^{3}H-thymidine label in DNA is associated with the isolated protein scaffold. Its properties are not altered by the removal of DNA, except for a 3–4 fold increase in the sedimentation rate.

It is at once obvious that the scaffold has the shape of a metaphase sister chromatid pair, with the sister chromatids connected at the centromere. In most cases the sister scaffolds appear to be tightly associated; sometimes they are more loosely connected by only a few fibers. The length distribution is similar to that of intact chromosomes visualized in the same conditions. The occasional single scaffold has presumably become separated from its sister. The obvious speculation is that it is this scaffold which is responsible for maintaining the structure of the metaphase chromosome; the concept of a protein core to which loops of nucleoprotein fiber are attached is familiar from early models of chromosome structure (see Chapter 10). A difference from these models, of course, is that the loops are considered to be part of a continuous fiber, not to represent individual molecules. Whether the loops seen in this structure are related to the units in which supercoiling occurs in interphase chromatin is a matter of speculation.

The nature of the attachment of the loops to the scaffold is a critical question; and in particular, the specificity with which particular nucleoprotein sequences might be attached to the protein scaffold to form the bases of the loops is an interesting issue. The extent to which it might be the scaffold that is responsible for maintaining the connections between sister chromatids is another intriguing question. An important concern is whether a scaffold exists in interphase chromatin, in which case presumably it is duplicated before mitosis to generate sister scaffolds; or whether it represents a reorganization of the nucleoprotein fiber into a structure unique for this stage of the cell cycle. Also it is interesting to ask whether a scaffold may be present in meiotic chromosomes during metaphase, how the scaffolds of sister and nonsister chromatids might be related, and whether they might have any relationship with the components of the synaptonemal complex seen earlier in meiosis.

CHAPTER 15

Euchromatin and Heterochromatin

Chromosomal material has been divided into two general classes on the basis of cytological observations of its condition during the cell cycle. The genetic material of the interphase nucleus takes the form of a dispersed network of fibers in which individual chromosomes cannot be recognized. The threads of each chromosome can first be distinguished at the beginning of mitosis, and the chromosomes acquire their characteristic condensed morphologies by metaphase. Following telophase, they return to the dispersed condition of interphase chromatin. Most of the chromosomal material comprises this *euchromatin*, which passes through a cycle of condensation and dispersion in each successive division of the cell.

Some of the genetic material does not become dispersed at the end of mitosis, but instead remains in its condensed state throughout interphase as well as division. These regions were described as *heterochromatin* by Heitz (1928), when they were first identified in the mouse; they stain densely with certain dyes and appear to represent tightly coiled regions of the chromosome.

Heterochromatin includes two types of material. *Constitutive heterochromatin* most commonly is found in the form of comparatively short regions of densely staining material; and always occurs at the same location in both members of a homologous pair (for review see Yunis and Yasmineh, 1971, 1972). The locations of constitutive heterochromatin within a chromosome set are characteristic, and any chromosome may have both heterochromatic and euchromatic regions. Constitutive heterochromatin most commonly is present at the centromeres and often at the telomeres of mitotic chromosomes; regions of heterochromatin within the chromosome arms occur, but are less common. All mammals appear to have constitutive heterochromatin at the centromeres of their chromosomes, usually on both sides of the kinetochore, but sometimes on one only (Hsu and Arrhigi, 1971). At interphase the regions of constitutive heterochromatin may aggregate to form *chromocenters*, which appear as dense clumps of genetic material, sometimes associated with the nucleolus. In the mouse the formation of chromocenters appears to vary with the cell type and stage of development (Hsu et al., 1971).

In an electron microscopic study of the interphase nuclei of salamander larvae, Hay and Revel (1963) visualized the nucleoplasm as a fine network of fibrils containing opaque aggregates. The dispersed network represents euchromatin; the dense aggregates correspond to the chromocenters generated by the heterochromatin. The dense structures appear to have the same filamentous structure as the dispersed network, but with the constituent fibers packed together more tightly. This supports the idea that heterochromatin consists of the same nucleoprotein fibers as euchromatin, organized into a more compact state.

Although constitutive heterochromatin in most mammals is confined to short regions, the chromosomes of the field vole Microtus agrestis contain extensive lengths. The two sex chromosomes are very much larger than the autosomes and thus can be identified easily in the 50 chromosome diploid complement. Constitutive heterochromatin occupies the long arm of both the X and Y chromosomes and also some one third of the short arm of the X chromosome. These regions remain heterochromatic throughout development in all cell types and vary only in the extent of condensation (Lee and Yunis, 1971a,b). This heterochromatin can be seen to consist of tightly packed fibers, which run continuously into the euchromatic regions. This reinforces the conclusion that euchromatin and heterochromatin represent different forms of organization of the same basic structure (now known to be the nucleosome string).

Constitutive heterochromatin is not transcribed into RNA. This conclusion originally was based upon the lack of incorporation of ^{3}H-uridine over heterochromatic regions as visualized by autoradiography. More recent evidence derives from the nature of the DNA sequences contained in constitutive heterochromatin. Cytological hybridization shows that satellite DNA sequences are present in constitutive heterochromatin; although the presence of other sequences is not excluded, the satellites generally are thought to constitute at least the major part of the heterochromatin. Satellite DNAs generally consist of fairly short nucleotide sequences repeated a large number of times, with some variation in the basic sequence. Their structure and organization is discussed in Chapter 19, but we may note here that the satellite sequences are not represented in the RNA population of the cell. The identification of constitutive heterochromatin with the non-transcribed repetitive satellite sequences suggests that its role in the chromosome is structural. Another property of heterochromatin is that it tends to be replicated later than euchromatin; autoradiography shows that DNA synthesis in heterochromatin commences later than that of euchromatin; and by the end of S phase, heterochromatic regions alone may be completing replication (for review see Lima de Faria, 1969). Whether this is related to some function of heterochromatin, or is simply a consequence of its more condensed condition, remains to be seen.

The only features that *facultative heterochromatin* shares with constitutive heterochromatin are the characteristic condensed morphology, inactivity in gene expression, and late replication (for review see Brown, 1966). That is to say that facultative heterochromatin does not consist of distinctive DNA sequences such as are characteristic of constitutive heterochromatin; it is generated by the condensation of chromosomes or chromosome regions that at certain times are found in the euchromatic condition. Facultative heterochromatin is produced by the inactivation of one of the two X chromosomes present in the cells of female mammals. The formation of the sex heterochromatin takes place at an early stage of embryogenesis, after which all the somatic daughter cells contain one X chromosome that appears to be euchromatic and one that is heterochromatic. The heterochromatic X chromosome remains tightly condensed through the cell cycle, is replicated late in S phase, and does not seem to be transcribed into RNA. In agreement with these biochemical properties, it behaves genetically as though inert, for the cells exhibit only those gene functions coded by the euchromatic X chromosome.

The feature common to constitutive and facultative heterochromatin is the correlation between the permanently condensed condition and its late replication and lack of genetic expression. It seems likely that the late replication may arise simply from the longer time required to make the more tightly condensed DNA available as a template. The nature of the connection between structure and lack of function is not known; but obviously there have been speculations that the condensation may at least in part be a cause of the inactivity. This is supported to some extent by the observation that translocation of genes from a euchromatic location to a position close to either constitutive or facultative heterochromatin may inhibit their expression. Whether the relationship between euchromatin and heterochromatin might be a paradigm for the structures of active and inactive genes within the euchromatin is a point on which at present it is not possible to form a conclusion.

Chromosome Banding

How specific is the organization of the mitotic chromosome? The location of the centromere is a fixed feature, of course, and so also is the position of any secondary constriction. Certain DNA sequences are associated with the centromere, with the nucleolar organizer, and perhaps with the telomere. But apart from these limited features, one might tend to either of two extreme views. The last level of folding of the nucleoprotein fiber into the chromosome ultrastructure might be more or less random, so that particular DNA sequences do not occupy any particular location (or bear any fixed relationship to one another); or the position of every sequence might be precisely specified and exactly reproducible from mitosis to mitosis.

Until recently, chromosomes could be recognized at mitosis only by their relative sizes and the shapes conferred by the positions of their centromeres (see Chapter 4). An important advance which allows individual chromosomes to be recognized, however, followed from the demonstration that each metaphase chromosome in the haploid set reacts in a characteristic way with fluorescent stains or chemical dyes. These generate a series of lateral bands, whose intensity and relationship to each other is distinct for each chromosome. This has two important consequences. As a practical matter, it allows each chromosome to be identified at metaphase; this overcomes the limitation previously restricting mapping studies with somatic cells and allows genes to be assigned to chromosomes (see Chapter 9). From a more fundamental point of view, there is the implication that the bands must originate in local and reproducible variations in chromosome ultrastructure, since different reagents are able to produce essentially the same banding pattern. It is only fair to remark that the apparent promise that this might lead to some understanding of chromosome ultrastructure has not yet been fulfilled; the structural basis of the bands remains mysterious.

The reagents usually used to generate the lateral banding patterns are the fluorochromes quinacrine and quinacrine mustard and the chemical dye Giemsa. The pattern visualized by fluorescence is known as *Q-banding* and that seen after chemical staining is described as *G-banding*. These patterns are generated after comparatively little manipulation of the chromosomes and in general are virtually identical. Under some conditions, Giemsa staining can be used to produce the *R-banding* pattern, which is exactly the reverse of this, with bands and nonbanded regions interchanged. Severe treatments of the chromosomes allow Giemsa staining to produce the *C-banding* pattern, in which the centromeres alone are intensely stained. From the point of view of chromosome structure, the most interesting question lies with the nature of the effects of the treatments that generate these various banding patterns.

Quinacrine and quinacrine mustard interact directly with fixed metaphase chromosomes to generate bands of varying but characteristic intensities. Caspersson et al. (1970, 1972) first showed that each human metaphase chromosome can be identified by its banding pattern, which is indifferent to cell type. The same is true of the mouse chromosome set, where the technique is especially useful for allowing the otherwise indistinguishable acrocentric chromosomes to be identified (Miller et al., 1971b,e). The technique has sufficient precision to allow identification of structural abnormalities involving rearrangements of material between chromosomes (Miller et al., 1971a).

A particularly intense reaction occurs with the human Y chromosome. Usually chromosomes cannot be identified in interphase cells, but after quinacrine staining the human Y chromosome (and also that of some monkeys) can be identified as an intense blob of fluorescence. Pearson, Bobrow, and Vosa (1970) showed that this reaction in effect identifies a chromatin body for the male (corresponding to the single inactivated X chromosome that forms the Barr body of the female). Variations in the appearance of the fluorescent Y body have been described (Wyandt and Hecht, 1973a,b); but its presence is an invariant feature of human male cells. Bobrow, Pearson, and Collacott (1971) were able to show by Y fluorescence that the human male chromosome is associated with the nucleolus. It is the distal part of the long arm of the Y chromosome that fluoresces; and this region appears to consist of constitutive heterochromatin.

Dense staining of the centromeres was discovered as a result of the protocols used in cytological hybridization, when chromosomal DNA is denatured in situ and a radioactive RNA or DNA probe is added to determine the location on certain sequences (see Chapter 19). When a probe corresponding to satellite DNA was visualized by autoradiography and the chromosomes were stained with Giemsa, the radioactive label coincided with dense staining at the centromeres. Then Arrhigi and Hsu (1971) and Yunis et al. (1971) found that the alkaline treatment used to denature chromosomal DNA is itself sufficient to generate the dense centromeric staining; addition of the labeled nucleic acid probe is not necessary. Thus the protocol for C banding involves exposure of fixed chromosomes to 0.07 M NaOH for 1-2 minutes followed by incubation in 2-6 × SSC (a standard saline citrate solution) at 65°C overnight. In fact, either the alkaline treatment or the prolonged incubation in SSC seems able to generate centromeric staining, but their combination produces the clearest definition (Comings et al., 1973).

The reactions of human chromosomes to Giemsa stain depend on the severity of the treatment to which they are subjected before staining. Treatment with alkali produces the C bands shown in Figure 15.1. But Sumner, Evans, and Buckland (1971) found that when chromosomes are subjected only to milder conditions of incubation by treatment with 2 × SSC for 60 minutes at 60°C, Giemsa staining generates the lateral pattern of G bands shown in Figure 15.2. In these there is little or no reaction at the centromeres (see also Schendl, 1971). Exactly the reverse pattern of banding is generated by heating at 87°C for 10 minutes followed by cooling to 70°C and Giemsa staining; the stained R bands correspond to the regions that were not banded in the G banding technique (Dutrillaux and Lejeune, 1971).

Detailed comparisons of Q bands and G bands were first made for human and mouse cells. In both cases there is virtually no difference in the locations and perhaps also the relative intensities of the bands (Sumner, Evans, and Buckland, 1971; Rowley and Bodmer, 1971; Aula and Saksela, 1972; Nesbitt and Francke, 1973). In these cells the intensity of the C bands seems to be inversely proportional to their reaction upon Q banding. The correspondence between Q bands and G bands has been reported since for other cell types. The feature responsible for generating G bands may be present in meiotic as well as mitotic chromosomes, since a G banding pattern has been generated in human pachytene bivalents (Luciani, Morazzani, and Stahl, 1975).

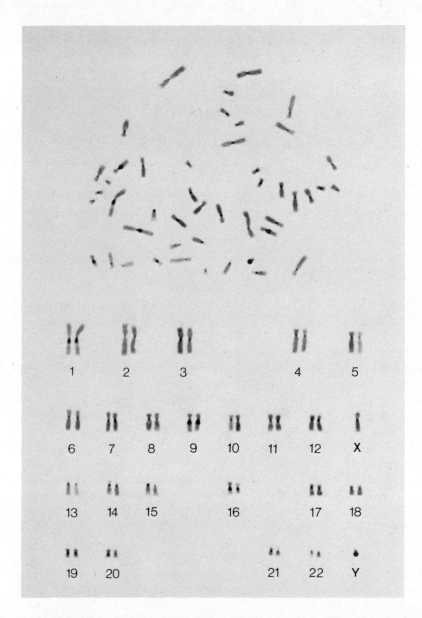

Figure 15.1. C banding of human chromosomes. Upper: metaphase plate subjected to banding. Lower: chromosomes set arranged to show individual members. Kindly provided by Mr. N. Davidson.

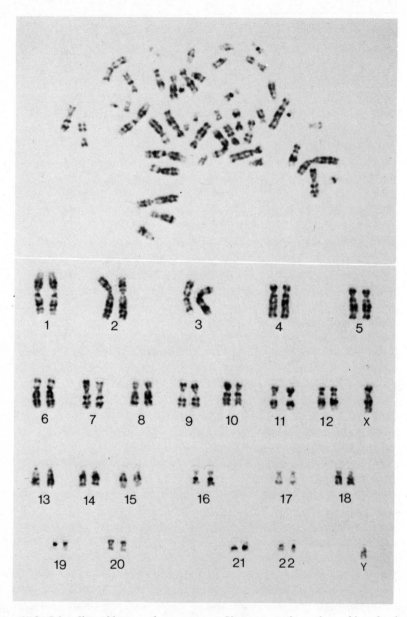

Figure 15.2. G banding of human chromosomes. Upper: metaphase plate subjected to banding. Lower: chromosome set arranged to show individual members. Kindly provided by Mr. N. Davidson.

Structure of the Bands

What is the nature of the regions that form bands in these various treatments? In the case of C banding, the intensely staining regions correspond to the constitutive heterochromatin present around centromeres. The obvious question is whether their reaction with Giemsa is due to some property of the DNA component or represents some other structural feature of this region. It was originally thought that the generation of C bands might be connected with the ability of the highly repetitive sequences of satellite DNA to renature more rapidly than other sequences of the genome. The state of DNA following treatment with alkali can be determined by staining with acridine orange, which fluoresces orange upon binding to single stranded nucleic acids but emits green with double strands. Using this technique, Stockert and Lisanti (1972) found that DNA of the centromeric regions does indeed renature more rapidly than other sequences. After 10–30 seconds of renaturation, the centromeres fluoresce green whereas other regions remain orange. But after as little as 2 minutes of renaturation, the difference in color between the centromeric and other regions decreases until they are all green. Renaturation times that allow all the DNA to reform its duplex structure do not change the relative proportions of the regions stained in C banding. Since, indeed, C bands can be generated in chromosomes in which DNA has renatured or has been maintained in a single stranded state by fixation with formaldehyde, their origin does not seem to lie with preferential renaturation of DNA sequences.

Some organisms possess large blocks of heterochromatic material that falls into the class of constitutive heterochromatin, according to its content of satellite DNA sequences. An example is M. agrestis, in which much of the material of the two sex chromosomes is of this nature (Arrhigi et al., 1970). These regions generate intense C bands and also fluoresce intensely upon Q banding (Mukherjee and Nitowsky, 1972). In several species of Drosophila, there are also intensely fluorescent regions, more like the human Y chromosome than the typical Q band. All of these correspond to sites identified as heterochromatic (although not all heterochromatin stains intensely) (Vosa, 1970). However, it is not possible to draw a general conclusion from this, because in at least one species, D. nasutoides, there are very bright fluorescent regions that do not correlate with constitutive heterochromatin as defined by C bands (Wheeler and Altenburg, 1977). From these data it is clear that C bands can be identified with constitutive heterochromatin not only in centromeric regions, and in some cases these regions may fluoresce intensely with quinacrines. But the feature(s) responsible for the generation of C bands and/or bright fluorescence remain to be identified.

The correlation between C bands and constitutive heterochromatin has inevitably led to proposals that G and Q bands also may represent heterochromatic regions, sometimes described as "intercalary heterochromatin." Before discussing the nature of these bands, however, we should note that the cause of G banding and Q banding is not yet known; and any proposal to equate these bands with heterochromatin represents an over interpretation of data presently available.

The feature of heterochromatin upon which such claims have been based is its late replication. Both the inactive X chromosome (facultative heterochromatin) and centromeric regions (constitutive heterochromatin) often replicate late in S phase (see Chapter 20). This may be a consequence of the topological difficulties presented by their apparently more condensed structure. The same correlation between heterochromatin and late replication applies in other instances, such as the large blocks of heterochromatic material around the centromeres in Samoaia leonensis, which form C bands and intense Q fluorescence and replicate late (Ellison and Barr, 1972). Thus constitutive heterochromatin can be considered to fall into at least two types, the restricted centromeric regions that do not form Q bands and the more extensive regions that show intense Q fluorescence. Both of these are late replicating.

Late replicating regions have been equated with G and Q bands in the human complement on the basis of autoradiographic experiments (Ganner and Evans, 1971; Calderon and Schnedl, 1973). But the resolution of the autoradiographic technique used to identify the late replicating regions is poor; it does not really extend to the level of individual bands. In another instance, that of Drosophila merriami, no correlation was found between late replication and the location of G and Q bands (Bostock and Christie, 1974).

A technique that allows identification of more restricted chromosome regions is provided by the incorporation of BUdR followed by staining with Hoechst 33258 (see Chapter 10). In principle, the regions replicating at any particular period during S phase can be identified by pulse labeling with BUdR during this time. Several experiments using this technique have led to suggestions that the regions replicating early in S phase correspond to the R bands (that is the regions between the G bands), while the late replicating regions correspond to the G bands; at intervening times an intermediate pattern is generated (Crossen, Pathak, and Arrhigi, 1975; Dutrillaux, 1975; Epplen, Siebers, and Vogel, 1975; Kim, Johanssmann, and Grzeschik, 1975; Latt, 1975).

Together these experiments raise the possibility that the bands could correspond to regional units into which replicons are organized (see Chapter 20). However, each band must contain many replicons, probably up to about 100, and it is not known what proportion of these must be replicated in the labeling period to make the band per se visible as a replicated segment. The implication of these results is therefore that each R band must contain a sufficient number of replicons that are activated at varying times early in S phase but must lack enough late activated replicons; the reverse must be true of the G bands. The duration of S phase is too long for there to be only two types of regional organization in the chromosomes; although it is striking that patterns reminiscent of the R and G bands can be generated via replication, the relationship between the bands and their constituent replicons remains to be established. How this relationship may be related to chromosome structure also is not yet clear.

One result that would seem to preclude a simple model is provided by observations of the behavior of the X chromosomes in female cells. Although one is euchromatic and one is condensed as facultative heterochromatin, both show the same G banding pattern (Sumner, Evans, and Buckland, 1971). Thus condensation per se does not seem to affect the formation of G bands. But the inactive X chromosome replicates later than its active partner; so that in this case the timing of replication must depend on factors other than the banded organization. An indication that there is some variability in replication times is given by the observation in some experiments that corresponding bands on homologues may sometimes replicate at different times.

Even if the G and Q bands include replicons that are active late in S phase, this is scarcely sufficient to conclude that they constitute heterochromatic material. There is no evidence for supposing that there is a similarity between G and Q bands with the C bands or the intensely fluorescent regions of constitutive heterochromatin. There has been a proposal that sequences corresponding to mRNA are found in the R band regions, which would imply that G bands do not contain active genes (Yunis, Kuo, and Saunders, 1977). However, this was based on an autoradiographic technique lacking resolution at the level of the bands.

Three types of model may be imagined to account for the origin of the bands. They might result from differences in the distribution of chromosomal material, with the bands representing concentrations of DNA produced by more compact packing of the nucleoprotein fiber. They might represent regions possessing (or lacking) certain sequences of DNA or regions showing extremes of base composition. Or they may result from variations in the association of DNA with chromosomal proteins, presumably nonhistones.

Mechanism of Banding

Untreated chromosomes stain more or less uniformly with Giemsa along the arms, although centromeres show light stain (Sumner, Evans, and Buckland, 1971). Preparations fixed with formaldehyde show a deep and uniform stain with Giemsa over all regions when subjected to the usual banding protocol (Stockert and Lisanti, 1972). This immediately suggests that the production of bands requires removal of material from the regions that are not stained; this must be prevented by formaldehyde. The implication is that the banding pattern originates with the interactions of DNA and proteins rather than with differences in the distribution of material or type of DNA.

Various protocols have been used to produce G bands. Although the original procedure comprised incubation with SSC, a large number of other treatments now has been used, including addition of proteolytic enzymes or of other agents that extract or denature proteins; these are followed by addition of Giemsa stain (Seabright, 1972; Kato and Yosida, 1972; Kato and Morikawa, 1972; Burkholder, 1975). Bands can also be visualized by staining with other reagents. The G banding pattern is produced by treatment with DNAase I followed by Feulgen staining (Burkholder and Weaver, 1977). The R banding pattern is generated by acridine, under either the usual G banding conditions or the usual R banding conditions (Bobrow, 1973). A reversal possibly similar to that seen between G and R bands can be generated by effects of different treatments on chromosomes stained with Giemsa after incorporation of BUdR; usually the presence of BUdR reduces staining, but the reverse effect can be produced (Burkholder, 1979).

The basis for reaction with both quinacrine and Giemsa appears to be binding to DNA. With isolated DNA, both quinacrine and quinacrine mustard appear to fluoresce in proportion to the content of A-T base pairs (Weisblum and De Haseth, 1972; Pachmann and Rigler, 1972; Latt, Brodie, and Munroe, 1974; Comings et al., 1975). The active component of Giemsa stain has been elusive but may comprise one molecule of eosin Y and 2 of methylene blue; it is not known to have specificity for any particular base composition (Sumner and Evans, 1973). There does not appear to be any common basis for the two reactions; the magenta compound formed by Giemsa staining is extracted by urea, whereas bound quinacrine is not. In any case, the idea that the banding patterns result from variations in base composition is a priori unlikely; each band represents a very considerable amount of DNA, far too much to expect any concentration of particular base pairs. In spite of this consideration, the idea has persisted that concentrations of A-T rich regions may be responsible for the reaction with quinacrine (Korenberg and Engels, 1978; Sahar and Latt, 1978).

Fluorescent antibodies specific for particular nucleosides can be used to generate banding patterns. Bands of the G type are generated by reaction with antiadenosine following incubation at 65°C for 1 hour in 95% formamide-SSC (Dev et al., 1972). The antibody also gives this pattern upon reaction with chromosomes that have been exposed to ultraviolet irradiation (Schreck et al., 1974). The R banding pattern is generated by using antiadenosine or anticytidine to stain chromosomes that have been exposed to methylene blue photo oxidation to destroy guanine bases (Schreck et al., 1973, 1977). Both these antibodies react only with bases in single stranded DNA; thus the generation of banding patterns depends upon the formation of regions containing single-stranded DNA rather than on any specificity with regard to sequence or composition. The implication is that the regions found in G bands are more readily denatured by heat and formamide, whereas the nonbanded regions are more accessible to methylene blue or ultraviolet.

The variety of treatments able to produce the G band pattern makes it difficult to discern any common basis for action. But in spite of differences in banding protocols, these treatments all reveal the same pattern. It is especially significant that G bands and Q bands are virtually identical. In both techniques the chromosomes are first fixed on a slide, but then for Q banding the quinacrine is applied directly, whereas for G banding the cells are incubated further (with SSC or in some other protocol). Quinacrine must presumably react to give bands in regions in which the DNA is accessible to the reagent; and these regions must therefore be distinguished from the nonbanded regions in the fixed chromosomes. It is the same lateral differentiation of structure that is responsible for the G banding pattern; but in this case, Giemsa is initially able to bind all chromosome regions and the bands are distinguished from the nonbands in the ability of the latter to lose some component necessary for staining upon incubation in SSC. This component is not the target of the dye, DNA, since the reaction with feulgen shows there to be no relationship between the amount of DNA remaining after treatment and the amount of Giemsa bound (Sumner and Evans, 1973).

Does the lateral pattern of bands exist in chromosomes in vivo? Suggestions that bands are a natural feature of chromosomes have been based upon reports that they correspond with the chromomeres seen in pachytene bivalents at meiosis and can be visualized in the absence of fixation (Okada and Comings, 1974; Comings and Okada, 1975). But the resolution with which the "natural" bands can be visualized is not high; and it is difficult to obtain a correlation good enough to be sure that they represent the same features responsible for the G bands. Another report has been that chromosomes examined by whole mount electron microscopy show a distribution of mass that corresponds with the positions of the G bands (Golomb and Bahr, 1974b); but in this case only a single chromosome was examined quantitatively. In any case, this does not fit with the demonstration that R bands can be obtained under appropriate conditions.

A feature common to all of the banding protocols is the use of fixed chromosomes. This immediately raises the question of whether the fixation may be responsible for generating the bands. Sumner, Evans, and Buckland (1973) reported that after methanol/acetic acid fixation—the method commonly used—much of the protein in the nucleus appears to have been removed. This does not prove directly that it is loss of protein that is responsible for the banding pattern, since DNA also might be lost. Indeed, according to Comings et al. (1973), the usual G banding protocols cause loss of extensive amounts of DNA; this contrasts with the lack of relationship between feulgen staining and Giemsa staining found by Sumner et al.

In summary, it is clear that the euchromatin comprising the chromosome arms can be divided into two types of material: bands and nonbands. Bands are immediately accessible to quinacrine, as seen in Q banding. Non-

bands are rendered inaccessible to Giemsa, as seen in the protocols used for
G banding. Bands are rendered inaccessible to Giemsa by the protocol for
R banding. The basis for these patterns is most likely to lie in some variation in the association of DNA with nonhistone proteins which influences
large, laterally organized blocks of material. But how this results in Q and
G bands on the one hand and in R bands on the other hand remains to be
determined.

Inactivity of Facultative Heterochromatin

Facultative heterochromatin describes large blocks of inactive genetic material, whose condition is perpetuated through indefinite cell generations,
although the cell may also contain homologous sequences in the euchromatic condition that are expressed. In the systems that we shall describe,
facultative heterochromatin represents entire chromosomes that display a
characteristically condensed appearance, replicate at the end of S phase,
and whose genes are not expressed. Although these observations do not of
themselves preclude the existence of smaller regions behaving in a similar
manner (and which might have failed to be noticed by morphological criteria), what is known about the mechanism of formation of facultative
heterochromatin would be consistent with the idea that inactivity usually
spreads through all contiguous regions of the affected material, that is, occupies the entire chromosome.

An entire chromosome complement becomes heterochromatic in the
mealy bug (Planococcus citri), in which the chromosomes that a male in-
herits from his mother are euchromatic, while the set inherited from his
father becomes heterochromatic (although it was euchromatic in the father).
Figure 15.3 illustrates this situation, in which the male has a maternal
euchromatic set of chromosomes, which remain euchromatic when passed
to a daughter but become heterochromatic when passed to a son. The paternal heterochromatic set of a male is discarded when gametes are formed,
so that all male gametes possess only the maternal genetic information. In
the female both sets remain euchromatic.

The heterochromatic chromosome set of male mealy bugs is inactive by
both genetic and biochemical criteria (for review see Brown, 1966). The pattern of inheritance shows that males express and transmit only those genes
received from their mother. The condensed heterochromatic set fails to incorporate ^{3}H-uridine into RNA and is replicated later in the division cycle
(Berlowitz, 1965).

In spite of this inactivity, the heterochromatic paternal set appears to be
necessary for the male. Irradiation of males causes lethality or sterility in
their sons; and the heterochromatic set cannot be replaced by that of a different mealy bug species following genetic crosses (Brown and Nur, 1964).
Consistent with the implication of these results, cells in some tissues of male
mealy bugs lack heterochromatic chromosomes; Nur (1967) showed that

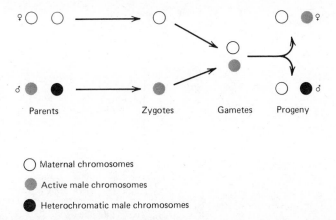

Figure 15.3. Chromosome cycle of the mealy bug. Either of the chromosome sets of a female
may enter the egg, but only the euchromatic set of the male may enter the sperm. Both euchromatic sets remain active in females; but in the male, the paternal chromosomes are inactivated.

this appears to be because the heterochromatic condition has been reversed
to generate a euchromatic paternal set. The formation of heterochromatin
may therefore be reversible in these somatic cells, although it is irreversible
in the germ line and is maintained in most somatic cells.

In eucaryotes in which the female has two X chromosomes, one of the
two homologues takes the form of heterochromatin—this is the Barr body
or sex chromatin often noted in female human cells during interphase.*
Maintaining one of the two X chromosomes in the inactive condition of
facultative heterochromatin means that both (XX) females and (XY) males
have the same number of active X-linked genes, providing a dosage compensation for the inequality of representation of the genes in males and
females.

The generation of cells in which only one X chromosome is active occurs at an early stage of mammalian development. Lyon (1961) observed
that mice heterozygous for X-linked coat color mutations have a variegated
phenotype in which some areas of the coat are wild type but others are mutant. This suggests the model illustrated in Figure 15.4 in which half the
cells have only an active paternal X chromosome while half have only an
active maternal X chromosome. The decision on which X chromosome
is to be active in a given cell line must be taken early in embryogenesis to
generate the discrete regions of the coat. Each region represents a group of
cells descended from some progenitor(s) possessing one or the other active
X chromosome and recognizable by the corresponding phenotype. Since the
pattern of variegation is random, but sees overall roughly half wild type

*Female sex chromatin was for many years thought to consist of both X chromosomes until
Ohno, Kaplan, and Kinosita (1959) showed that it comprises only one X chromosome in the rat.

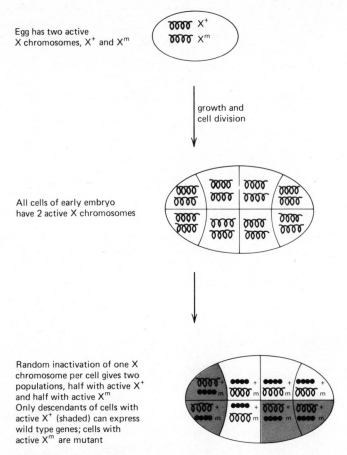

Egg has two active
X chromosomes, X⁺ and Xᵐ

growth and
cell division

All cells of early embryo
have 2 active X chromosomes

Random inactivation of one X
chromosome per cell gives two
populations, half with active X⁺
and half with active Xᵐ
Only descendants of cells with
active X⁺ (shaded) can express
wild type genes; cells with
active Xᵐ are mutant

Figure 15.4. Variegation in mice heterozygous for X linked coat color genes. Chromosome X⁺ carries a wild type gene; its homologue Xᵐ carries a mutant allele. Patches of coat color represent descendents of a common ancestor cell: those cells derived from ancestors with inactivated X⁺ show the mutant phenotype, while those descended from a line in which Xᵐ was inactivated appear wild type.

and half mutant cells, the most likely model is that at some specified time, only one X chromosome is permitted to be active in each cell of the embryo; and this state is perpetuated in all its descendants. The choice between the X chromosomes appears to be random in any given cell. That the variegated phenotype results from inactivation of the genes of one X chromosome is confirmed by the failure of autosomal coat color genes to show variegation; the coats of heterozygotes show only the dominant phenotype. Mice of the XO genotype show normal development, which confirms that one X chromosome is adequate; they do not show variegation, but express the gene carried on their single X chromosome (Lyon, 1962).

Confirmation that it is the heterochromatic X chromosome which carries inactive alleles, while the genes of the euchromatic X are active, has been obtained by correlating the phenotypes and genetic states of individual cells of mice heterozygous for an X chromosome carrying Cattanach's translocation. The translocation chromosome X^t can be distinguished cytologically from the normal chromosome, X^n, because it carries an insertion of autosomal material which makes it the longest chromosome in the mouse complement. The translocated region carries dominant autosomal genes for coat color. Ohno and Cattanach (1962) observed that in females showing variegation, the areas of the recessive coat color are composed of cells with a heterochromatic X^t; areas displaying the color specified by the autosomal genes have a heterochromatic X^n. This not only demonstrates a direct association between heterochromatin and genetic inactivity, but shows also that autosomal regions attached to the X chromosome may be inactivated together with the sex chromosome itself.

Autoradiography of cells from mice with Cattanach's translocation suggests that it is the heterochromatic X chromosome which provides the late labeling material found in many mammalian cells. Evans et al. (1965) showed that a single late labeling chromosome is found in all mouse cells; when the cells are taken from heterozygotes for Cattanach's translocation, half the cells have a late replicating X^t and half a late replicating chromosome of normal size. This implies that X^t and X^n are equally likely to provide the facultative heterochromatin. In human cells, where members of the chromosome complement can be distinguished by size, it has been somewhat easier to demonstrate that one X chromosome alone is late replicating (Morishima et al., 1962; Gilbert et al., 1965; Gavosto et al., 1968; Knight and Luzzati, 1973). Similar results have been obtained in several other systems (Lima de Faria, 1959; Evans, 1964; Hsu, 1964).

Genetic activity of the X chromosome can be inferred from the presence in the cell of proteins specified by sex linked genes. The enzyme G6PDH (glucose-6-phosphate dehydrogenase) is carried on the X chromosome of both eutherian and marsupial mammals; and has been widely used to identify active X chromosomes since different forms of the enzyme can be separated easily by gel electrophoresis. By examining the erythrocytes of human females heterozygous for a disorder in the enzyme, Beutler et al. (1962) showed that only one X chromosome is active in each cell. Although a cell population contains both isozymes, cloning produces lines displaying only one or the other form. By examining the cells of females heterozygous for two sex-linked metabolic diseases, Migeon et al. (1960) and Romeo and Migeon (1970) demonstrated a similar clonal heterogeneity for the enzymes HGPRT and α-galactosidase. The expression of only one of the two X-linked alleles in a given cell appears to occur in all cell types (for review see Lyon, 1972).

The random inactivation of one of the two X chromosomes of females which prevails in eutherian mammals such as mouse and man is replaced in marsupial mammals by a directed inactivation of the paternal X chromosome. Euros, wallaroos, and kangaroos are marsupials that can interbreed to give sterile hybrids; they possess cytologically indistinguishable X chromosomes which specify G6PDH enzymes of different electrophoretic mobilities. But female progeny always show only the electrophoretic band corresponding to the isozyme of the mother; and it is the X chromosome of the father that is late replicating (Richardson, Czuppon, and Sharman, 1971; Sharman, 1971).*

Another situation in which the two X chromosomes are not treated equally occurs in the progeny of crosses between the donkey and the horse (giving mules from a donkey male parent and hinnies from a horse male parent). Again the two X chromosomes can be distinguished, as can their G6PDH products. When fibroblasts are cloned from female mules or hinnies, any particular cell line as usual shows only one variant of G6PDH. But more lines are found with the horse enzyme and correspondingly the donkey X chromosome is the late replicating (Hamerton et al., 1971; Giannelli and Hamerton, 1971; Ray et al., 1972; Cohen and Ratazzi, 1971).

The preferential expression of the horse genes is only a tendency, as emphasized by analysis of 54 female mules performed by Hook and Brustman (1971). In two organs the horse G6PDH enzyme tends to predominate; but in others the results accord with the predictions of random inactivation, as previously reported by Mukherjee and Sinha (1964) for leucocytes. The variability in these results suggests that the cause of the preferential inactivation of donkey X-linked genes may lie in the development of the hybrids. For example, the formation of facultative heterochromatin usually may occur earlier in the donkey than in the horse; or inactivation may be random but there may be some cell selection so that the cells with inactive donkey X chromosomes have some advantage. Given the unusual circumstances prevailing in the development of such sterile hybrids, the importance of these results lies with the correlation that can be drawn between the formation of heterochromatin, genetic inactivation, and late replication, rather than with the preferential direction of the process.

Time of X Inactivation

There is no general rule for the time in the life cycle at which facultative heterochromatin is generated, although in mammals it is usually early in embryogenesis. The most data are available for the mouse, where several approaches have been used to define the stage at which female embryos come to have only one X chromosome active.

*The paternal X chromosome appears to be inactive in all tissues, with the possible exception of skeletal and cardiac muscle, in which expression of paternal PGK has been reported, although at much lower levels than the predominant maternal PGK (Vandenberg, Cooper, and Sharman, 1973). This raises the possibility that some expression of paternal X-linked genes might occur in these tissues, but the effect is small and its significance is not yet clear.

Using a variation of the chimera technique, Gardner and Lyon (1971) injected a single cell taken from a donor blastocyst of one genotype into a recipient blastocyst of another genotype. The donor cell was derived from a female embryo heterozygous for X-linked coat color genes; the recipient carried autosomal genes for a third, distinguishable coat color. If single X inactivation takes place before the blastocyst stage, each chimeric mouse should have cells expressing only one of the X-linked donor alleles (corresponding to whichever X chromosome happened to be active in the donor; half the chimeric mice should show the phenotype of one allele, and half the phenotype of the other allele). But if inactivation takes place after the blastocyst stage, patches of each X-linked coat color should be seen on each chimeric mouse; for the injected cell must have two active X chromosomes, can divide to give descendants of the same type, and only subsequently suffers random inactivation in each descendant cell. The generation of chimeric mice with both X-linked phenotypes in the coat therefore shows that X inactivation takes place after blastocyst (that is, 3.5 days of embryogenesis). Inactivation probably takes place very soon after, because biased animals often are found in which one of the two X-linked coat colors predominates, suggesting there has been chance selection from among a rather small number of cells.

Another indirect approach is to try to infer the number of cells in the embryo at the time when the decision on inactivation must have been taken. In principle, the clonal composition of embryos immediately after X inactivation should follow the binomial distribution $(p + q)^n$, where n is the number of cells and p and q are the probabilities of activation for each chromosome, each equal to 0.5 in the wild type animal. Since the cells contributing to a given phenotype in the adult (for example, the coat color) will be descended from a subpopulation of the cells of the embryo, the formula that describes the clonal composition of an adult phenotype is $(p' + q')^{n'}$, where n' is the number of progenitor cells from which the cells of the adult phenotype are derived and p' and q' are the relative frequencies of each active X chromosome in the set of progenitor cells.

In animals with random X inactivation, it is impossible to apply this formula directly because the patches of each coat color are large and each represents the descendants of an unknown number of progenitor cells. The extreme case in which the coat is entirely of one color, because inactivation occurred in the same direction in all cells, occurs too rarely to be analyzed. But such analysis becomes possible in mice carrying the X-linked O^{hv} mutation, which causes preferential activation of the X chromosome carrying it in heterozygotes where the other chromosome is wild type (see below). Then there is an appreciable frequency of occurrence of mice with coats entirely or almost entirely displaying the phenotype of the light color gene *Blo* which is carried by the O^{hv} chromosome. Ohno et al. (1974) argued that mice with completely Blo coats, or with single patches of wild-type color, provide experimental values for the frequencies of the binomial fractions $(p')^{n'}$ and $(p')^{n'-1} \cdot q'$. A value for n' of 70 can be assumed, because studies of the *Blo* mutation suggest that it acts in 35 pairs of somite clones distributed along both sides of the vertebral column. To proceed from the deduced value of p' to a value for p, it is necessary to sum several situations in which the 70 progenitor cells may have had the same active X chromosome constitution; in effect, this could have occurred in any embryo with a sufficiently biased pattern of X inactivation, not just those in which all inactivation decisions were identical. A close fit to the data is given by assuming $p = 0.8$, $n = 60$. Obviously this has uncertainties, but suggests roughly that the O^{hv} allele confers an 80% chance that its X chromosome will be active; and the number of cells in the embryo at the time of decision may be of the order of 60, which is consistent with a timing at about the blastocyst stage.

More direct studies show that the late replicating heterochromatic X chromosome cannot be detected in mouse cells until late blastocyst (see Lyon, 1972). The same is true of other mammals also. The drawback of this approach is that the stage of visible inactivation might occur sometime after the time at which genetic inactivity is established. As it happens, it would seem that the appearance of heterochromatin cannot lag far behind the decision on inactivation. It would be interesting to follow the expression of X-linked genes in early embryogenesis, but there are impediments to applying this approach to determining the time of inactivation. The expression in heterozygotes of different alleles for an X-linked product might be followed to see if both are active at first and if changes in relative expression can be found later. However, measuring the synthesis of protein products is open to the objection that cessation of protein synthesis might lag behind the cessation of transcription if mRNAs were stable for some period; it is not at present practical to assay transcription directly. In summary, it is possible to say that facultative heterochromatin first is seen at about the same time that can be inferred to be the point of decision, but the timing of repression or activation of X-linked genes remains to be established.

What is the activity of the X chromosomes in female germ cells? Ohno et al. (1961, 1962) found that in mammals the condensed Barr body characteristic of somatic cells is absent from female premeiotic germ cells. Measurements of the levels of enzymes specified by X-linked genes show that the oocytes of XO mice have half the activity of XX oocytes, contrasted with no such difference for enzymes coded by autosomal loci (Epstein, 1969, 1972; Kozak, McLean, and Eicher, 1974; Mangia et al., 1975). If the enzyme level reflects the number of active genes, then XX oocytes must have two active X chromosomes. The same is true of 1 day old embryos; but in 3 day embryos, the HGPRT activity of XO and XX genotypes is the same, suggesting that by this time only one X chromosome is active. In mature oocytes of human heterozygotes, isozymes of G6PDH display both electrophoretic bands; although 16 week fetal oocytes also have both variants, it is possible that only one may be present at 12–13 weeks (Gartler et al., 1972, 1973, 1975). These data are not conclusive, but they would be consistent with the idea that X inactivation occurs in all cells of the embryo, including the progenitors of the germ cells in which it is later reversed. The alternative would be to suppose that inactivation does not occur at all in the germ cell line.

A system for following the differentiation of mouse cells in vitro is provided by the teratocarcinoma, which may be perpetuated in the form of pluripotent "undifferentiated" embryonal carcinoma cells, but also can form bodies of differentiated cells. In a comparison of two XX teratoma lines with some XO lines, McBurney and Adamson (1976) found no difference in the levels of X-coded enzymes (with the exception of a doubling in α-galactosidase activity in one of the XX lines). No change in the enzyme levels occurred during differentiation. This suggests that the XX and XO lines possess the same number of active X-linked genes; so that presumably only one X chromosome is active in each XX line. However, this may reflect the origin of these lines in tumors initiating after the time of X inactivation, since Martin et al. (1978) have reported the existence of an XX teratoma line in which the activity of X-linked enzymes is twice that found in XO lines when the cells are in the pluripotent condition. Upon differentiation the enzyme levels are halved; for G6PDH the reduction occurs over the period of 8–12 days following the initiation of differentiation, while for HGPRT it occurs during the period of 13–15 days. Whether this reflects a difference in the time at which the two genes are inactivated or instead is a consequence of features of their expression (such as message stability) remains to be established. The correlation between inactivation and differentiation at all events is striking and should make possible analysis in vitro of the events involved in the formation of facultative heterochromatin. It is also important in showing directly that the X chromosome may indeed be active at the time of inactivation, so that gene expression is turned off and not merely preempted by the change in state.

Position Effect Variegation

Inactivation of genetic material is not confined to facultative heterochromatin itself, but can include autosomal material translocated to positions close to heterochromatin. Lewis (1950) noted that somatic mosaicism is produced in Drosophila when a gene is translocated from euchromatin to the constitutive heterochromatin of the basal region of the X chromosome. Thus heterozygotes in which the gene for white eyes is carried in this location show a variegation effect in which some of the facets of the eye are white instead of red. This cis dominant influence is known as position effect variegation (for review see Spofford, 1976). Inactivation seems to spread along the translocated autosomal segment, proving most effective at the region closest to the heterochromatin and less effective further away.

Similar effects are found in the mouse. Russell (1963) and Russell and Montgomery (1970) showed that when an autosomal gene is transferred to the X chromosome its action may be suppressed in some cells along with the inactivation of X coded functions. The effect may spread along the autosomal segment; genes closest to the sex heterochromatin may be inactivated, while those in more distant positions may remain active. The inactivation of autosomal regions also seems to depend on the point of attachment on the X chromosome; some regions of the X chromosome do not appear to inactivate attached autosomal regions (for review see Eicher, 1970). But although the inactivation of attached autosomal material may vary in this way, all regions of the X chromosome itself appear to be inactivated (Lyon, 1964).

The inactivation of autosomal loci translocated to the X chromosome has been studied in mice heterozygous for an X-autosome translocation and a normal X chromosome. Position effect variegation thus occurs only in the half of the cells in which the X–autosome is the inactivated chromosome. Although the variegated coats of such mice display the patterns that would be expected to result from inactivation of the translocated autosomal loci early in development, it is difficult to determine the effectiveness of position effect variegation because of the presence of cells in which the X-autosome remains active (Cattanach, Wolfe, and Lyon, 1962). The best resolution of the superimposed patterns of position effect variegation and X inactivation would be afforded by analysis of mice in which it was possible to follow simultaneously the expression of genes coded by the X chromosome itself and by the translocated autosomal region. This would allow the expression of the translocated autosomal genes to be studied in just those cells in which the attached X regions had been inactivated.

Cattanach's translocation X chromosome has proved useful for studies on position effect variegation. The autosomal insertion on X^t carries the wild type genes $+^p$ and $+^c$, whose recessive alleles (on autosome 7) code for pink eye (p) and chinchilla coat (c^{ch}) or albino coat (c). In mice that are heterozygous for Cattanach's translocation and a normal X chromosome (X^t/X^n), either X^t or X^n may be randomly inactivated; in cells in which X^t is inactivated, position effect variegation should sometimes inactivate $+^p$ or $+^c$, allowing the recessive autosomal alleles to be expressed.

In a study of the coats of such animals, Cattanach (1974) observed three types of region. Wild-type color represents cells in which the X^t chromosome is active so that $+^p$ and $+^c$ are expressed (and any cells in which X^t is inactive but the autosomal genes both fail to be inactivated also would have this phenotype). White regions are derived from cells in which X^t is inactive and both $+^p$ and $+^c$ have been inactivated. Fawn areas identify cells in which $+^P$ is active while $+^c$ is inactive; these cells must have an inactive X^t in which the inactivation has spread to $+^c$ but not to $+^p$. The reverse situation of patches of inactive $+^p$ and active $+^c$ does not occur. Since the $+^c$ locus is closer than the $+^p$ locus to the material of the X chromosome, this is in accord with earlier expectations that inactivation spreads sequentially into the inserted autosomal regions.

The mouse coat forms several bands each of which is derived from a clone of melanocytes. The wild type and white/fawn color regions follow the boundaries between these bands, showing that the X inactivation that permits the absence of $+^p$ and $+^c$ in the mutant bands must have occurred before each clone was established; thus each band is descended from a clone in which either X^t or X^n is inactive. The fawn regions, however, form patches within the white bands. Taking the white bands to identify clones in which X^t was inactivated, this position effect variegation of $+^c$ relative to $+^p$ implies that the difference in expression of these two loci must have occurred at some time between the decision on X inactivation (about 3.5 days) and the colonization by melanocytes that leads to coat formation (about 8 days). Thus position effect variegation is clonal and is established subsequent to X inactivation.

As these animals become older, the regions of mutant coat color become darker. The white areas first turn the fawn color characteristic of $+^p$ c cells and then turn the black (wild-type) color characteristic of $+^p$ $+^c$ cells. This suggests that the inactivating influence along the autosomal insertion gradually retreats, allowing sequential reactivation.

The production of position effect variegation in autosomal material attached to either constitutive or facultative heterochromatin suggests that a structural effect may be involved, in which the condensed condition of the heterochromatin spreads into contiguous material. The condensed structure per se cannot be the sole feature responsible for establishment and maintenance of the heterochromatic condition: constitutive heterochromatin is contiguous with euchromatin, so the condensed structure must be sequence specific, with only a limited ability to spread into adjacent regions; and the reactivation of X-translocated autosomal material shows that a distinction may be made between the sequences of the X chromosome itself (which presumably remain perpetually inactive) and the attached autosomal sequences (in which inactivation is not completely stable).

Generally it has been thought that position effect variegation depends solely upon the spread of inactivation from the adjacent heterochromatin. Given that the effect represents clones of cells in which a translocated gene either has or has not been inactivated, the implication is that at some specific developmental time a decision is made in each cell on whether the translocated material is to become heterochromatic or not; the resulting state is perpetuated in its descendants. However, the existence of reactivation at later developmental times led Cattanach to argue that perhaps this mechanism is responsible for all variegation, whether it stems from events occurring at early or late times. Thus after inactivation together with the X chromosome of a given autosomal region representing some or all of the translocated material, a progressive reactivation may take place with time, proceeding from the point most distant from the site of translocation; the rate may depend upon cell type. If reactivation at a given locus occurs at an early stage, an active gene wil be clonally inherited; athough the descendants of this cell all will possess this activity, subclones with respect to other loci may be generated as progressive reactivation proceeds further.

Formation of Facultative Heterochromatin

Two types of event may be involved in the establishment of facultative heterochromatin in a cell line. Initially the decision must be taken as to which homologue is to be inactivated. Then the chromosome must enter the heterochromatic condition and this state must be perpetuated at every replication. The formation of facultative heterochromatin usually is irreversible and may be maintained indefinitely when cells are placed in culture. Even when somatic cells are fused, a process that often reactivates inert functions (see Chapter 9), there is no change in the condition of an inactive X chromosome (Migeon, 1972).*

There has been some debate on whether the initial decision involves an activation or inactivation (for review see Cattanach, 1975). Early models generally took the view that the process takes the form of selecting one of the X chromosomes for inactivation (Russell, 1961, 1964). However, in abnormal mice with more than two X chromosomes, all but one are inactive (for review see Lyon, 1968). It is difficult to see how a cell could quantitate the number of X chromosomes to be selected for inactivation. This has led to the view that one X chromosome may be selected to be active, perhaps by protecting it against the process of inactivation that acts on all other X chromosomes in the cell. Inactivation may be accomplished by superimposing a more condensed state upon the usual structure of the X chromosome or may also involve more specific repression.

Are there discrete loci on the X chromosome at which decisions on its state are executed? Russell (1964) proposed that there might be "inactivating centers" from which the effect spreads, perhaps by a progressive condensation of material. The susceptibility of translocated autosomal regions to inactivation might then depend on their distance from the nearest center. The more recent studies of position effect variegation show that a mechanical spreading cannot be the whole explanation, since the maintenance of

---

*There has been one report of reactivation of an X-linked gene in cell culture. Kahan and DeMars (1975) found that in human-mouse hybrids carrying an active human X-autosome translocation and an inactive human X chromosome, the selection of HGPRT⁻ mutants was accompanied by loss of function of the gene on the X-A chromosome. Selection for HGPRT⁺ cells then identified hybrid cells apparently lacking X-A and containing X and an apparently active HGPRT gene. However, PGK and G6PDH remained inactive. There was therefore no general reactivation of the X chromosome; and it is possible that the active HGPRT gene could have been provided from some other source, for example, translocation of the gene from the X-A chromosome to some other location.

inactivation is not independent of the origins of sequences of X-autosome translocations. The nature of the relationship between events at the inactivating center(s) and the state of contiguous sequences is obviously an interesting question.

The existence of at least one locus on the X chromosome concerned with inactivation is suggested by the properties of Searle's translocation. This comprises a reciprocal exchange between the X chromosome and an autosome, so that heterozygotes contain the four chromosomes: X and A derived from the normal parent; and X-A and A-X derived from the translocation parent. The chromosome X-A contains most of the X material and A-X contains the remaining regions. Inactivation of the X chromosome in the translocation heterozygote is no longer random: genes carried on the normal X chromosome are always inactive while the segments of the translocated X-A and A-X chromosomes are active (Lyon et al., 1964). There are two possible explanations. One is that the inactivation is indeed random, but that cell selection takes place because cells with an inactivated X-A and/or A-X are for some reason inviable. Another is that only the normal X chromosome is susceptible to inactivation, because Searle's translocation prevents inactivation of X-A and A-X. One possible explanation for this is that the breakpoint coincides with an inactivating center and thus disrupts its function.

The position effect variegation displayed by the autosomal genes in Cattanach's translocation appears to be under the control of a locus on the X chromosome. Usually the extent of white coat color is about 30%; presumably this corresponds to the inactivation of both the translocated genes $+^p$ and $+^c$ in 60% of the 50% of the cells in which X^t is inactive. By selecting mice with greater amounts of white coat, Cattanach and Isaacson (1965) obtained a line with a level of 50%. Selection for mice with less white coat color was unsuccessful, simply retaining mice with the initial level of 30%. The high (50% white coat) and low (30% white) mouse lines must differ in some factor(s) controlling the probability that the inserted autosomal genes will be inactivated.

Does the altered probability reflect a change in the inactivation (or reactivation) of the translocated genes on inactive X^t chromosomes; or a preferential inactivation of the X^t chromosome, with a consequent increase in the frequency of inactivation of the attached material? Two loci on the X chromosome suffer the same reduction in expression as the translocated genes in the high line, which suggests that these mice differ from the low line in the probability that X^t will be inactivated (Cattanach, Pollard, and Perez, 1969).

The controlling factor is located on the X^t chromosome itself, for in crosses between the high and low lines the heterozygous X^t/X^n progeny always show the degree of variegation typical of the parent that provided the X^t chromosome (Cattanach and Isaacson, 1967). The controlling locus has been described as *Xce;* the allele *Xce^a* (originally described as *Xce^c*) is present in the high X^t chromosome, while the allele Xce^b (originally described as Xce^i) is present in the low X^t chromosome. Thus Xce^b permits random X inactivation; whereas Xce^a causes preferential inactivation of the chromosome carrying it in Xce^a/Xce^b heterozygotes. The locus appears to reside on the X chromosomal material, for it can be transferred to X^n chromosomes by recombination; the frequency in $Xce^a - X^t/Xce^b - X^n$ heterozygotes is about 3%, suggesting that Xce lies about 3 map units on the centromeric side of the autosomal insertion in X^t (Cattanach, Perez, and Pollard, 1970). What appears to be an extreme variant of this locus has been identified in the form of the allele O^{hv} of Ohno et al. (1973) and Drews et al. (1974). This exerts a cis dominant effect upon the X chromosome carrying it, which is almost always the active X in heterozygotes. Other alleles on normal X chromosomes that appear to influence inactivation have been found by Cattanach and Williams (1972).

The Xce locus appears to have an unusual instability. When males carrying an X^t chromosome with the allele Xce^b were crossed with normal females, Cattanach and Isaacson (1967) found daughters not only with the expected low white coat, but also some with the high white coat. Each type breeds true, although with a low rate of subsequent exchange between high and low coat color types. Since there can be no meiotic recombination in the X chromosome of males, this suggests that the Xce^b allele may be able to change its state to the Xce^a allele, possibly in an action analogous to the changes of state previously observed in the control elements of maize chromosomes by McClintock (1965).

Another system in which the activity of the X chromosome is controlled by a syntenic site is provided by Sciara coprophila. Here the zygote has the constitution 6A + 3X. The difference between males and females is not established until the 7th or 8th cleavage, when either one or both of the paternal X chromosomes is eliminated in somatic cell lines, to give females $(6A + X^m X^p)$ or males $(6A + X^m)$, respectively (Metz, 1938). (Whether a mother produces sons or daughters depends on the constitution of the X chromosome, which exists in the alternative forms X and X'.) The germ line of both sexes retains all chromosomes until gamete formation, when one X^p is eliminated in both cases, to give eggs with the constitution 3A + $X^{m/p}$ and sperm that are 3A + $2X^m$ (the X^m undergoes equational nondisjunction).

Each sperm contains X chromosomes of maternal origin; but in the next generation these become the paternal X chromosomes that are eliminated. Crouse (1960) suggested that when X chromosomes pass through the male and female germ lines they are given different "imprints" by which they are subsequently distinguished. (Actually it is sufficient to postulate that imprinting occurs in either of the germ lines rather than both.) The imprinting is reset in each generation; and it depends on the source of the chromosome rather than its genetic constitution. By following the fate of reciprocal translocations between the X chromosome and autosomes, Crouse localized the locus responsible for distinguishing paternal from maternal sex chromosomes to a region of heterochromatin near the centromere. In this region there must therefore be a site or sites at which the source of the chromosome can be indicated.

An analogous situation is found in mealy bugs, in which the paternal set of chromosomes becomes heterochromatic in male somatic cells and is discarded when sperm are formed (see Figure 15.3). All zygotes initially possess two euchromatic complements and it is the decision whether or not to inactivate the paternal set that determines whether a given embryo becomes female or male. Again there are two questions: how are paternal chromosomes distinguished from maternal chromosomes; and why are they inactivated in some but not all embryos? Brown and Nur (1964) proposed that the mother determines sex by producing two types of egg, one that imprints male chromosomes upon fertilization so that later they become heterochromatic, and one that lacks this action. An alternative is to suppose that all eggs possess two types of region, so that sex depends upon in which area the paternal complement finds itself (for review see Chandra and Brown, 1975). Equally likely, however, is the possibility that imprinting occurs in some but not all male gametes. Since the entire paternal complement is involved, whatever the nature of the imprinting, it must involve at least one site on each chromosome.

An even more striking example of changes in genetic activity that are independent of genetic constitution is provided by the insect Pulvinaria hydrangeae. Embryos of both sexes are produced by parthenogenesis in completely homozygous females (although males are formed they do not participate in the reproductive cycle). Here there can be no genetic difference between embryos in which one parental complement does or does not form heterochromatin (Brown and Nur, 1964).

What sort of event could be responsible for distinguishing between two homologues of identical genetic constitution? One model is to suppose that the target chromosomes each possess a site at which new genetic information can be inserted. Excision and replacement events then can account for reversal of the state in subsequent generations or for any instability too frequent to result from genetic recombination or mutation (Cooper, 1971). In the case of the mammalian X chromosome, the most likely hypothesis is that insertion of material at this receptor site would prevent the recipient chromosome from suffering inactivation. Brown and Chandra (1973) proposed that a single copy of the informational entity for insertion might be produced if it is coded by a donor locus only one of whose alleles is active in the embryo. Then all X chromosomes but the homologue gaining the insertion would be inactivated.

A form of this model has been proposed to reconcile the random inactivation of eutherians with the paternal inactivation of marsupials. In eutherians, the donor locus could be autosomal; and the allele present in the sperm might be inactivated during male gamete formation. Then the informa-

tional entity must be produced by the maternal donor locus, but may act on
either of the X chromosomes. In marsupials, the donor locus might lie on
the X chromosome itself, adjacent to the receptor locus. If its product then
is inserted only at the adjacent receptor, the inactivation of the paternal
donor during spermatogenesis would leave only the maternal X chromosome able to protect itself against inactivation. This implies that the random inactivation of eutherians may have evolved from the paternal inactivation of marsupials by translocation of the donor locus from the X
chromosome to an autosome.

There are some difficulties with both the loci postulated in this model.
Cattanach's translocation divides the X chromosome into two parts; but
both are inactivated, although not all of the autosomal insertion is inactivated. This poses a problem for models relying on the presence of a single
inactivating center in the X chromosome; yet if there were more than one,
it would not be obvious why one X chromosome should be entirely active
and its homologue entirely inactive. The idea that an informational entity
is produced in one copy by an autosomal locus predicts that the number of
active X chromosomes should equal the number of maternal autosome sets
in the cell; but it is not clear that this actually occurs in animals with more
than two chromosome complements (Lyon, 1974; Cattanach, 1975).

CHAPTER 16

Specialized Chromosomes: Chromomere Structure

Chromomeres are named for the densely staining granules formed by the chromosomes as they condense for meiosis. Pachytene chromosomes may resemble beads on a string, in which the chromomeres reflect the differential condensation of material (see Chapter 5). Whether the chromomeres have any further functional significance is not obvious. Properly speaking, the term "chromomere" may be used to describe any such unit of lateral differentiation, although in some instances "band" has been preferred. We have seen in the previous chapter that different types of bands may be generated in probably all animal chromosomes by various treatments, although their structural and functional significance remains to be clarified.

In two specialized chromosome systems it has been possible to investigate the nature of lateral differentiation on the chromosome in more detail. The term "specialized" does not imply the existence of chromosomes that are atypical with respect to common features shared by all other chromosomes— every chromosome is specialized in the sense that it contains specific nucleic acid and protein components and takes a structure appropriate for its situation. But some chromosomal systems are distinguished by features that make them amenable to analysis that elsewhere is impossible. The lampbrush and polytene chromosomes discussed in this chapter have been investigated not so much for their characteristic features per se, but in the prospect that these features may differ from those of other chromosomes in scale rather than in kind. In each system, therefore, an important question is the extent to which its properties may be a paradigm for other chromosomes.

The lampbrush chromosome represents a much extended state of organization found during meiosis. It has been most extensively characterized in the diplotene cells of amphibian oocytes, in which this state was first recognized by Flemming (1882) and Ruckert (1882), although similar structures occur in other species and at male as well as female meiosis.

Polytene chromosomes are generated by successive duplications without segregation and constitute a large number of interphase genomes organized in parallel. They were first identified in Chironomus tentans by Balbiani (1881) and since have been extensively studied principally in this species and in Drosophila melanogaster. Although characteristic of these and other

Dipteran insects, polytene chromosomes do occur in other species, most notably the ciliated protozoa.

The enlargement of both lampbrush and polytene chromosomes makes it possible to visualize their structure in the microscope. Both types of chromosome show a characteristic lateral organization, lampbrush chromosomes displaying the usual meiotic chromomeres and polytene chromosomes consisting of a series of closely spaced bands. Sites of gene expression can be recognized in both cases. The lampbrush chromosomes display prominent loops that protrude from certain chromomeres; the polytene chromosomes show swellings of certain bands that are described as puffs. Genetic activity is a characteristic of the stage of development, so that changes in gene expression can be followed in the form of the appearance or disappearance of active sites.

Two crucial questions can be investigated in these systems. What is the structure of an active segment of the chromosome vis a vis the structure of the remaining, inactive regions? What is the unit of gene expression—that is, what is its genetic content and how is this related to the visible divisions of chromosomal material? Both questions must be viewed, of course, within the general perspective of concern as to the extent to which this organization is typical of chromosomes in their more commonly found conditions.

In this chapter we shall be concerned with the structures of these two types of chromosome, taking the examples of the Triturus lampbrush chromosomes and the D. melanogaster and C. tentans polytene chromosomes. The purpose of this discussion is to define the nature of the chromomere and the structural changes associated with gene expression. In the next chapter we shall take up the issue of the genetic content of the polytene chromosome band.

Lampbrush Chromosomes

Oogenesis in amphibia may occupy a considerable length of time, and during this period the chromosomes become stretched out so that they may be observed in the light microscope. This greatly elongated state is maintained during the diplotene stage of meiosis, after which the chromosomes revert to their usual size. We may therefore take the lampbrush chromosomes to represent an unfolded version of the normal state of the chromosome.

The lampbrush chromosomes are meiotic bivalents in which the non-sister pairs are held together by chiasmata. The two sister chromatids constituting each pair are longitudinally associated with each other, forming a series of ellipsoidal chromomeres, some 1-2 μm in diameter, which are connected by a very fine thread (the chromonema). The lampbrush chromosomes of Triturus viridescens in their most extended state range in length from about 400 to 800 μm, although their length contracts into the range of 15-20 μm later in meiosis. The total length of the chromosome set may therefore reach some 5-6 mm, organized into the order of 5000 chromomeres (Gall, 1954; Callan and Lloyd, 1960).

The lampbrush chromosomes take their name from the lateral loops present in the homologues in the form of pairs of similar size and appearance. Each loop consists of an axis surrounded by a granular matrix. Figure 16.1 shows some examples of these structures, and Figure 16.2 illustrates the general appearance of the lampbrush chromosomes.

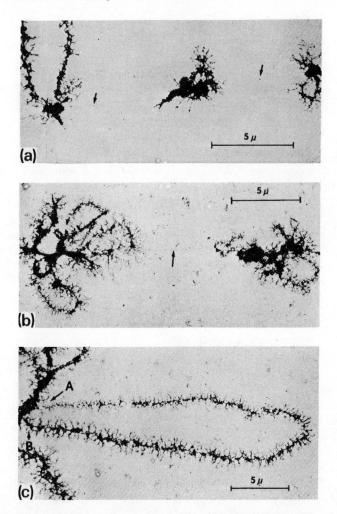

Figure 16.1. Whole mount electron microscopy of lampbrush chromosomes. (a) The main chromosome axis, indicated by the arrow, with loops protruding from the chromomeres; (b) loops connected by the main chromosome axis; (c) a loop with the typical ribonucleoprotein matrix displaying a transition from the thin end of the loop (A) to the thick end (B). Data of Miller (1965).

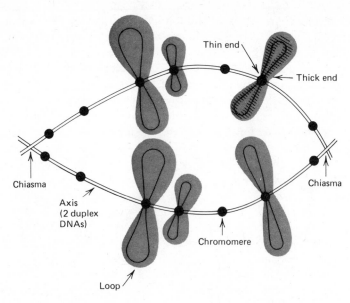

Figure 16.2. Diagrammatic representation of structure of lampbrush chromosome. Each diplotene bivalent is held together by chiasmata. The axis of each sister chromatid pair consists of two duplex threads of DNA lying close together (although drawn separately here, these cannot be distinguished under the microscope). The DNA is tightly coiled into a mass of fibers at the chromomeres. The loops which extrude from some of the chromomeres display an axis of duplex DNA surrounded by a matrix of ribonucleoprotein. The axis of the loop is continuous with the central axis of the chromosome. The matrix of ribonucleoprotein takes the form of particles, but is drawn here as the extended filaments that can be seen by spreading on water. Loops are found on pairs on each chromosome and the two homologues usually are symmetrical.

The loops take various morphological appearances and some loops can exist in alternative morphological forms. Such loops can be seen to be under genetic control, since an animal may be heterozygous or homozygous for the morphology and the appearance of the loop is transmitted in a Mendelian manner.

Early studies with stains showed that the chromomeres contain the bulk of the DNA, the loops contain RNA, and both possess basic proteins (Gall, 1954). The apparent continuity of the loops with the axis suggested that they might represent sites at which chromosomal material has been extruded to act as a template for synthesis of RNA. Callan and MacGregor (1958) confirmed that the lampbrush chromosome represents an extended thread of genetic material by demonstrating that DNAase alone destroys the integrity of the axis and also that of the lateral loops. By measuring the rate at which DNAase I introduces breaks into lampbrush chromosomes, Gall (1963) was able to show that each chromatid thread consists of a duplex of DNA. By assuming that a visible break in the chromosome is produced only when all its subunits are broken at or close to the same site, it was possible to show that two hits are needed to cause a break in a loop, whereas four hits are needed in the axis where sister chromatids are paired. Since DNAase I introduces single-strand breaks (see Table 13.3), this suggests that each chromosome consists of a single duplex of DNA, which is exposed in the loop but is tightly paired with its sister duplex in the regions constituting the central axis.

When the matrix is removed from the loops, their axial material has a diameter of some 4–10 nm; trypsin reduces these dimensions, which suggests that the axis constitutes a DNA duplex surrounded by protein (Miller, 1965). The fibers connecting the chromomeres are slightly wider, ranging from 6–20 nm, which accords with the idea that they may consist of two duplex molecules of DNA organized in parallel (although it is not possible to distinguish the two threads, except at very occasional sites where the axis appears to have split to generate an eye). The chromomeres themselves consist of a tightly compact mass of fibers, whose underlying structure has not yet been resolved (Mott and Callan, 1975).

The loops generally appear to be asymmetrical in structure, with the surrounding matrix widening along their length from a thin end to a thick end (Callan and Lloyd, 1960; Snow and Callan, 1969). The matrix consists of ribonucleoprotein particles about 30 nm in diameter (Mott and Callan, 1975). When spread on water, the matrices appear to consist of arrays of filaments, presumably generated by unpacking of the particles (Miller et al., 1970). This implies that each particle contains about 700 nm of filament. In some cases the filaments can be seen to form characteristic transcriptional matrices, in which short transcripts lie at the thin end of the loop, and their length progressively increases moving round toward the thick end (Angelier and Lacroix, 1975). In some cases the loop may represent synthesis of a single RNA product; but in others, termination and reinitiation may occur within the loop to generate more than a single species of RNA molecule.

While it is clear that the loops represent the sites of RNA synthesis, the nature of the sequences exposed in the form of their axes has been a question of some debate. Callan and Lloyd (1960) proposed that the loop may be a mobile structure, in which the axis is extruded from the chromomere at the thin end of the loop and then moves around to reenter the chromomere at the thick end. Thus at different times during development, different sequences from the chromomere would be exposed in the loop to act as templates for RNA synthesis. An alternative is to suppose that the loop is a static structure, representing the protrusion of a fixed part of the material of the chromomere, so that only a single sequence is exposed for transcription. This is consistent with the visualization of transcription matrices, in which transcription appears to proceed from a starting point at the thin end of the loop, generating ribonucleoprotein threads of increasing length around the loop. Transcription appears to proceed all round the loop, since autoradiography following the incorporation of ^{3}H-uridine shows that the label is found uniformly over each loop; although in one prominent exception, sequential labeling from thin end to thick loop end has been found (Gall and Callan, 1962).

The nature of the ribonucleoprotein transcript will be considered later, in Chapter 25. Extracted en masse, it consists of several proteins as well as RNA; and the variation in the protein components has been used to distinguish between different loops. Scott and Sommerville (1974) prepared antisera against protein fractions obtained from the RNP particles and treated lampbrush chromosomes to determine whether individual sites containing these proteins could be identified. With fluorescein-conjugated antibodies, three fractions reacted with all the chromosome loops; this implies that some proteins present in the RNP particles must be common to all loops. One fraction, however, reacted with only some 10 loops; in each case, the entire area of the loop reacted, suggesting that with regard to content of this protein the loop is a single unit. The refinement of this approach with antisera prepared against individual polypeptides of either the RNP particles or the nonhistone protein complement may allow single loops or sets of loops to be resolved; and this may prove an important approach to defining the relationship between the loop and the unit of transcription. Antisera against histones react largely with the chromomeres and not with the loops.

Polytene Chromosome Bands

The salivary glands—and also certain other tissues—of the larvae of Dipteran flies such as Drosophila contain interphase nuclei whose chromosomes are displayed in an enlarged state in which they possess both increased diameter and greater length than usual. In D. melanogaster, the total length of the giant chromosome set may be of the order of 2000 μm, consisting of about 5000 bands which stain intensely and alternate with the more lightly staining interband regions. Figure 16.3 shows a light micrograph of this set. The largest bands extend for about 0.5 μm along the axis of the chromosome; the smallest have a length of about 0.05 μm and can be distinguished only under the electron microscope. The centromeres of all four chromosomes characteristically aggregate to form a chromocenter, which in the male includes the entire Y chromosome.

The banding pattern is a characteristic for any given strain of Drosophila and the constancy in the number and linear arrangement of bands was first noted by Heitz and Bauer (1933) and Painter (1934). Rearrangements of genetic material—such as deletions, inversions, duplications—result in changes of the band order; and such observations led to the idea that the bands may be taken to represent a cytological map of the chromosome set. Cytological changes can be correlated with changes in the positions of genes on the genetic map; and Bridges (1935) first equated the linear order of bands with the linear array of genes and showed that changes at one particular

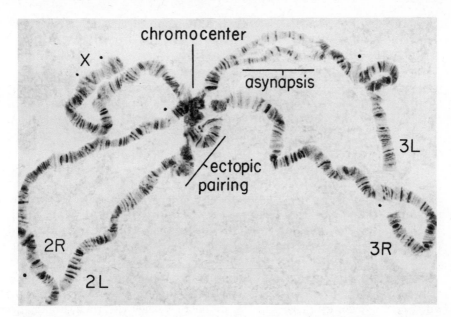

Figure 16.3. The polytene chromosome set of D. melanogaster salivary glands. The major chromosomes are indicated as X, 2 and 3 (L = left arm, R = right arm); they aggregate at the chromocenter. A region of asynapsis results where the homologues are not paired; and ectopic pairing represents the binding of nonhomologous regions. Photograph kindly provided by J. J. Bonner.

genetic locus always relate to cytological changes only in one particular band.

The high concentration of DNA in the bands is shown by their intense staining with dyes such as Feulgen. Staining for histones produces the same banding pattern as staining for DNA, which suggests that the bands represent concentrations of nucleoprotein fibers (Swift, 1964). An early controversy as to whether the interbands also contain DNA was resolved by the development of more sensitive assays, which have shown that DNA is present between the bands, although at much lower concentrations than within them (for review see Swift, 1962). Lezzi (1965) has shown that the interbands are susceptible to degradation by DNAase. Only about 50% of the total DNA is thought to lie in the interband regions.

An early experiment confirmed that the giant chromosomes represent a variation of the state of the chromosomes usually found in diploid somatic cells and not some novel reorganization of genetic material. Beerman and Pelling (1965) labeled the chromosomes of Chironomus with ^{3}H-thymidine at an early stage of embryogenesis and then grew larvae on unlabeled material to develop salivary glands. The DNA originally labeled in the normal chromosomes appeared in the giant chromosomes, exhibiting a label along the entire length of the chromosome. This suggests that the giant chromosomes are derived by the usual replication of genetic material, but represent a more extended state of organization.

The giant chromosomes are produced by many successive replications of each synapsed diploid pair; the replicas do not separate, but remain attached to one another in their unusually extended state. By using a quantitative Feulgen assay to follow the increase in DNA content of salivary gland nuclei during the development of D. melanogaster larvae, Swift (1962) found that the nuclei start with the usual diploid content, 2C. The amount of DNA then increases in roughly geometric steps to give cells containing 4C, 8C, 16C etc., up to a maximum of 1024C, which corresponds to 9 successive doublings of the diploid DNA content. Chemical analysis of the DNA content of polytene chromosomes of C. tentans suggests that here the maximum number of doublings may be 13 (Daneholt and Edstrom, 1967). This suggests that each giant chromosome consists of hundreds of identical copies of its genetic material, all of which are replicated in each successive cycle.

Each giant chromosome consists of many parallel fibers running longitudinally along its axis; an early estimate for the number of fibers based on thin section techniques was in the range 1000–2000 (Beerman and Bahr, 1954). In whole mount preparations, these fibers take the appearance characteristic of the constituent fibers of interphase chromatin, with numbers per Drosophila chromosome in the range 1000–4000 (Rae, 1966; DuPraw and Rae, 1966). Fibers appear to run continuously through the bands, condensed into compact masses within bands, and more extended between them. Each of these fibers presumably represents one copy of the nucleoprotein which constitutes the normal chromosome, that is, it contains a single duplex of DNA (for review see Swift, 1962). The state of these chromosomes is therefore described as *polytene* (contrasted with a *polyneme* chromosome, whose basic fiber would contain many duplex molecules of DNA; see Chapter 10).

The exact number of bands in the polytene chromosomes of D. melanogaster salivary glands is not clear, principally because of a problem in deciding whether doublets are counted as one band or two. In the original count of Bridges (1938), using the light microscope, each instance in which a band(s) took a double structure was included as two bands (presumed to be derived by duplication at some past time of a single ancestral band). This gives a total number of 5059 bands. On the other hand, Berendes (1970) found that many doublets could not be resolved in the electron microscope, where they appeared as single bands. A detailed map for part of the X chromosome gave a total band number of 116, compared with the previous value of 174 counted by Bridges. If the doublets seen in the light microscope are in many cases artefacts of fixation, the total band number may therefore be as much as one third below the commonly accepted value of the order of 5000. Yet in some cases, very fine bands can be resolved by electron microscopy that have not been distinguished in the light microscope. From such analysis, it seems likely that the number of bands is indeed somewhat less than that estimated by Bridges, but at present it is difficult to give an accurate value for the total (for review see Lefevre, 1976).

These uncertainties about the precise band number, coupled with the possibilities for error in the estimated genome size of D. melanogaster (see Appendix 1) pose a problem in assessing the DNA content of the bands. A rough estimate is to suppose that there are about 5000 bands, which contain some 73% of the haploid genome of 10^{11} daltons (the remainder is present in the chromocenter). This corresponds to an average band content of 1.6×10^7 daltons (or 25,000 base pairs) of DNA per haploid genome. In C. tentans, a genome probably slightly larger than that of D. melanogaster is organized into some 2000 polytene bands, corresponding to an average haploid genome content of about 7×10^7 daltons (100,000 base pairs) per band. Details of these calculations are given in Table 17.1.

Even these approximate values for the average band DNA content serve to show that a band contains a considerable amount of potential genetic information (some 8000 amino acids, or almost 10^6 daltons of protein, per D. melanogaster band). Clearly this is much too large to represent a single coding sequence; either each band must contain several coding sequences or most of the DNA must serve some function other than coding for protein. We shall discuss the relationship between bands and genes in the next chapter.

Replication of Polytene Chromosomes

Polytene chromosomes often provide systems in which there is uneven replication of different parts of the genome. Over replication occurs in species of the Sciaridae (a Dipteran family), in which normal polytene chromosomes form during larval development; but then in late larval stages, certain bands suffer additional replication cycles to form *DNA puffs*. Preferential replication occurs in the hypotrichous ciliates, in which a macronucleus is formed by conversion of some, but not all, chromosomes into the polytene state; then the polytene chromosomes are broken into segments corresponding in size to the bands. Under replication is found in Drosophila, where the full extent of polytenization occurs with the euchromatic regions that form the banded chromosomes, while the heterochromatin aggregated into the chromocenter replicates a smaller number of times. It is not uncommon, therefore, for the occurrence of polytene chromosomes to be associated with changes in the proportions of different DNA sequences.

The first evidence that the D. melanogaster genome is not uniformly replicated when polytene chromosomes are formed was provided by the demonstration that the amount of DNA in polytene nuclei is not an exact geometric multiple of the DNA content of the diploid chromosome complement. By comparing the response to feulgen of the DNA of polytene nuclei of salivary glands and ganglion cells with that of diploid cells, Rudkin (1969) found that a progressive increase in DNA values follows the sequence: 1.0, 1.6, 3.5, 6.7, 12.6. The best fit of this series to the replication of Drosophila DNA is given by assuming that some of the genome replicates a much smaller number of times than the rest. The heterochromatin contained in the chromocenter constitutes about 26% of the genome (the exact value depends on the sex of the cell); and the observed result would be generated if it replicated 3–4 times, while the euchromatin replicated 8–9 times (Rodman, 1967).

Since the chromocenter is a fairly compact structure, its content of DNA can be measured as a function of its apparent mass. A comparison of the proportion of heterochromatin in polytene and diploid nuclei of ganglion cells of D. hydei showed that the polytene cells possess relatively less of this heterochromatin (Berendes and Keyl, 1967). The reduction corresponds to a cessation of replication in the chromocenter after 2–3 doublings.

Another approach is to compare the proportion of total DNA extracted from polytene and diploid cells that is occupied by the satellite DNA sequences present in the constitutive heterochromatin (see Chapter 19). The hybridization kinetics of total DNA show that probably all the euchromatic sequences are replicated equally well during polytenization; but the content of satellite DNA sequences is very much reduced, corresponding to either no or very few replications in D. melanogaster and D. hydei (Dickson et al., 1971; Gall, Cohen, and Polan, 1971; Hennig, 1972a,b; Laird et al., 1973).

Cytological distinctions sometimes have been drawn between the material of the chromocenter per se, denoted α-*heterochromatin*, and the condensed regions immediately adjacent, denoted β-*heterochromatin*, which connect the chromocenter to the chromosome arms. It is necessary, therefore, to emphasize that the lack of replication applies only to the α-heterochromatin; the β-heterochromatin, which has a banded structure (though admittedly more difficult to visualize than the euchromatic arms), appears to replicate with the euchromatin and also to be transcribed (Lakhotia and Jacob, 1974; Lakhotia, 1974). This makes it questionable whether the "β-heterochromatin" should in fact be regarded as heterochromatin; it may constitute euchromatin, polytenized to the normal degree, whose band structure is affected by the close presence of the "α-heterochromatin." For the purposes of discussion here, therefore, we shall refer to the heterochromatin of the chromocenter in the understanding that this includes only the regions usually described as α-heterochromatin.

What is the structure of the junction between the more and less replicated sequences on an individual chromosome? The additional rounds of replication in euchromatin may be visualized as proceeding throughout each chromosome arm, until the replication forks in the replicons adjacent to heterochromatin cease movement at the junction. The large discrepancy between the numbers of rounds of replication occurring in the two regions means that in fact there should be a series of replication forks within replication forks at the junction. One possibility is that the forks remain in position to connect the differentially replicated sequences (Laird et al., 1973). The structure of such static replication forks would be of some interest. An alternative is that a break is made in the integrity of the chromosome thread, abolishing the forks, so that the more extensively replicated sequences represent molecules shorter than the usual duplex DNA of the chromosome.

In addition to the under replication of heterochromatin, the cistrons coding for ribosomal RNA also achieve only a limited degree of polyteny. In hybridization measurements of the frequencies of these sequences, Henning and Meer (1971) observed that the rRNA genes are appreciably under represented in the polytene nuclei of D. hydei. By determining saturation values for hybridization between rRNA and polytene or diploid DNA of D. melanogaster, Spear and Gall (1973) observed that the proportion of rDNA in salivary gland DNA is about 0.08% in flies of either XX or XO genotype. In diploid cells, XX flies have about 0.47% rDNA, whereas XO flies have about 0.26%, consistent with the location of most rRNA genes on the X chromosome. Compared with the 9 doublings of euchromatin, the rRNA genes appear to pass through only 6–7 doublings, to reach a final level that is the same whether one nucleolar organizer (in XO cells) or two organizers (in XX cells) were originally present. The multiplication of rDNA in this and other situations is discussed in Chapter 27.

In summary, there are therefore at least three controls over the replication of DNA to form polytene chromosomes. The greatest number of doublings is experienced by the euchromatic arms of the chromosomes. That it is the number of doublings which is controlled, rather than the final amount of DNA, is shown by an early experiment demonstrating that in XY cells the polytene X chromosome has half as much DNA as in XX cells (Aaronson, Rudkin, and Schultz, 1954). The X chromosome therefore undertakes the same number of replications, irrespective of whether the initial species is a single chromosome, or constitutes two synapsed chromosomes like the autosome pairs. By contrast, the number of replications of the rRNA genes may be regulated in terms of the final gene number; but this type of control may be specific for rDNA, rather than for the process of polytenization. The DNA sequences comprising the heterochromatin of the chromocenter show the least amount of replication (possibly none); whether this is a function of the structure of the chromocenter per se or of the constituent sequences remains to be established.

The entire DNA content of the euchromatic regions of the polytene chromosomes is replicated in several successive and separate cycles. The discrete geometric series obtained when total DNA content is measured for nuclei at different stages of polytenization suggests that one cycle must be completed before the next commences (see above). Autoradiographic studies show that at any given time only some nuclei are labeled, suggesting the existence of gaps between S phases.

The absence of division in polytene cells leaves no external reference point for establishing the start of an S phase. To distinguish events occurring at different times within a round of replication, Keyl and Pelling (1963) used a double label protocol. The principle is that two different labels are given to larvae within an interval shorter than the period required to complete a round of replication. Then any chromosomes possessing both labels must have incorporated the first early in S phase and have taken up the second at a later time in the same cycle. (This depends on the existence of a gap between successive rounds of replication that is longer than the interval between the two times of labeling.) These experiments suggested that replication of all parts of Chironomus chromosomes begins within about 30 minutes of the start of the cycle, to produce a continuous label along the chromosome axis, in which each region is labeled in proportion to its DNA content. A label incorporated at later times, however, shows a discontinuous pattern, in which local regions, perhaps corresponding to bands, are labeled. Pelling (1966) therefore suggested that all bands may start replication at the beginning of an S phase; but the larger bands require longer to complete DNA synthesis and so continue to incorporate label after the smaller bands have terminated replication. This argues that the band can be equated with the unit of replication, since if each band contained many replicons, all active simultaneously or at least overlapping in time, all bands should be able to complete replication in the same period of time. This rests on the assumption that all the polytene copies of the DNA in each band replicate synchronously.

Some subsequent studies have been consistent with the model that bands start replication together but end at different times, although others have identified differences in the times at which replication is initiated in different regions (Plaut, Nash, and Fanning, 1966; Plaut, 1969; Arcos-Teran, 1972; Hagele and Kalisch, 1974; Kalisch and Hagele, 1976). The control of polytene chromosome replication thus remains to be elucidated; present data, although not proving the point, remain consistent with the idea that the band is a unit for replication. The amounts of DNA in the haploid component of the bands are within the range of the sizes of replicons seen in autoradiography of replicating DNA (see Chapter 20).

The idea that the band may be the unit of replication was first proposed by Keyl (1965) on the basis of a comparison of band sizes between two related subspecies of Chironomus thummi. Some parts of the homologous chromosomes can be distinguished from each other in the hybrid; and the amounts of DNA in corresponding bands then can be determined. The bands of C. th. thummi always contain 2, 4, 8, or 16 times the amount of DNA in the homologous bands of C. thummi piger. Since the total ratio of DNA in the two species is the same in salivary glands and spermatocytes, this multiplication does not appear to be caused by the selective polytenization of some bands. Rather would duplication appear to have doubled the length of DNA in the haploid component of these bands at some time in the evolutionary development of C. th. thummi. Since unequal crossing over should produce triplet ratios in addition to doublings, recombination alone cannot explain the origin of the geometric series. This suggests that replication may be responsible, with some aberrant event doubling the length of a band; this implies that the bands are replicated as units. Whether or not this mechanism is responsible, the implication remains that the band is treated as a unit in evolution (see Chapter 17).

Puffing of Bands

Some of the bands of polytene chromosomes are found in an expanded or puffed state, in which chromosomal material is extruded from the axis of the chromosome. A characteristic pattern of puffing is observed in different tissues and at different times during larval development. Any given band may undergo a characteristic cycle of activity, in which it changes from the quiescent state to display a puff during some developmental period, until a specific time when the puff regresses. The parallel with the extrusion of loops from the chromomeres of lampbrush chromosomes is obvious.

Thin-section studies suggest that the puffs represent sites where chromosome fibers unwind, although the fibers remain continuous with those packed into the axis of the chromosome (Beerman and Bahr, 1954). Puffs usually appear to emanate from single bands, occasionally from pairs of bands (Beerman, 1965). Large puffs, often described as Balbiani rings, may represent such extensive swelling, however, that the underlying band pattern is obscured and it becomes difficult to determine which of the several bands at the site of puffing is in fact the active locus. Figure 16.4 shows an example of some prominent puffs in the salivary gland of C. tentans.

The correspondence between the puffs and the sites of RNA synthesis was first shown by the autoradiographic studies of Pelling (1964). Puffs stain with reagents specific for RNA, whereas unpuffed bands fail to do so. Autoradiography of salivary gland cells previously exposed to labeled RNA precursors reveals a pattern of labeling that corresponds to the puffing pattern; the extent to which each puff is labeled is related to its size. In addition to visible puffs, other bands may be labeled by ^{3}H-uridine; these may be less active sites of transcription (Zhimula and Belyaeva, 1975; Bonner and Pardue, 1976). Beerman (1964) found that actinomycin, which inhibits transcription, prevents puff formation and causes the regression of existing puffs. This suggests that each puff represents a site of transcription, the size of the puff providing some measure of the activity of the band from which it derives.

The dependence of the puffing pattern upon both tissue and stage of development was first shown by Beerman (1952); and four categories of puff

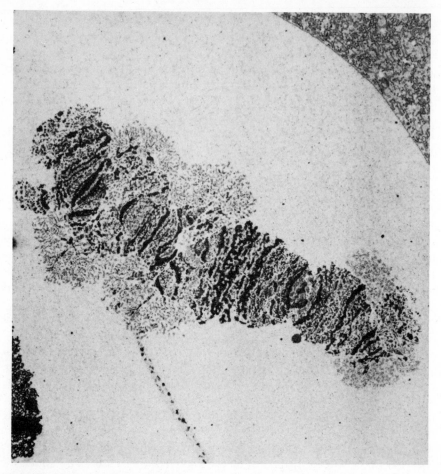

Figure 16.4. Chromosome IV in a salivary gland of C. tentans. Three large Balbiani rings are present. Data of Daneholt (1975).

now may be distinguished. Some puffs are restricted to particular stages of development; in particular, the major periods of puff activity precede larval moults. Some puffs are tissue specific and are found in only one or some of the types of polytene cell (Berendes, 1965a,b; 1966). A third class of puff is neither tissue specific nor restricted to particular stages of development. The most likely role for these puffs is to represent the expression of genes that code for functions needed in all the polytene tissues at all larval stages. In reviewing the appearance of puffs in salivary glands, Ashburner (1970a) noted that in Drosphila most of the puffs (more than 80%) are specific in appearing or regressing at particular stages of development; but in C. tentans the situation is reversed and only a minority of the puffs shows developmental specificity. A final class of puffs is induced by environmental treatments such as temperature shock or oxygen deprivation (Berendes, 1968; Leenders and Berendes, 1972; Beerman, 1973); these are not related to the usual pattern of puffing found in larval development (see Chapter 28).

The pattern of puffing in salivary glands of D. melanogaster during the last 10 hours of the third larval instar and the subsequent 12 hour prepupal period has been defined by Becker (1959) and Ashburner (1967, 1969a, 1973). Puffs develop only at certain sites; and their formation and regression occurs in a well-defined sequence. Some 108 autosomal loci form puffs during one or both of these developmental periods; the timing or size of 83 puffs depends on the stage of development. The X chromosome has a further 21 loci which are active during these periods, bringing the total number of active sites to 129. This is almost certainly much less than the number of genes expressed in these cells; measurements of the number of different mRNA species found in Drosophila cells or larvae (although not specifically made in salivary glands) yield much greater values (see Chapter 24). Either each band must synthesize more than one mRNA species; or the visible puffs may represent only the more active sites, with the remaining mRNAs synthesized by bands whose much lower activity does not result in visible perturbation of structure.

The same puffs are found in different strains of D. melanogaster, but the extent of puffing may vary. When Ashburner (1969b) compared the Oregon and vg-6 strains of D. melanogaster, only 64% of the puffs proved to be identical during the third instar and prepupal stages. Some 12% differ in size, while 19% differ in time of activity; 5% differ in both size and timing. It would be interesting to know the extent to which the visible changes in the degree and timing of puff development are accompanied by changes in the mRNA population.

There is only one exception to the invariance of the loci involved in puffing. This is a puff that is active only in vg-6 and which segregates as a single Mendelian locus. In heterozygotes of Oregon and vg-6, this puff develops in only the vg-6 homologue when the homologues are asynapsed; but it develops in both homologues when they are synapsed. This implies that puffing at this locus is controlled by interactions at the level of the strands of the polytene chromosomes.

A striking increase in puff activity occurs before the larval moults and prior to metamorphosis, when sudden and multiple changes take place in the puffing pattern (Berendes, 1967). This resetting of the pattern of puffs coincides with the time when the moulting hormone ecdysone is released from the prothoracic gland. Two types of experiment show that ecdysone controls puff development.

Larvae can be ligatured so that the salivary gland is divided into two regions, the anterior part lying in the same half of the larva as the prothoracic gland and the posterior region separated from it. Becker (1962a) found that when larvae are ligatured before the last larval moult, the anterior region alone develops the puffing pattern characteristic of the puparium, whereas the posterior region remains at the intermoult stage. Tying the ligature at different times shows that the hormone is released about 3-5 hours before the moult. That the puffing pattern of the salivary gland chromosomes is established by the developmental stage is also suggested by the transplantation experiments of Becker (1962b), which showed that a transplanted gland acquires the puffing pattern characteristic of the stage of development of its host.

Injection of ecdysone into larvae, first performed by Clever and Karlson (1960), confirms the role of the hormone. When ecdysone is injected into fourth instar intermoult larvae, the two puffs that appear are located at the sites that are usually the first to puff before moulting. Clever (1966) noted that these two puffs have different threshold levels for their hormone response, corresponding to about 10 and 100 molecules of ecdysone per haploid genome. This dose must be increased very considerably—by more than two orders of magnitude—for maximum reaction. In addition to the two puffs that respond immediately to the injection, other puffs respond some hours later. Berendes (1967) showed that injection of ecdysone into third instar larvae of D. hydei causes the same overall pattern of changes as that observed in normal development before pupation.

The puffs that respond rapidly—within 15-20 minutes of injection—are presumably activated by the interaction of ecdysone with the cell. The puffs that respond only after a delay of 4-6 hours may represent activation by some of the gene products of the puffs that respond immediately. The same biphasic pattern is seen in vivo or upon injection of ecdysone into the larvae or incubation of salivary glands in ecdysone (Ashburner, 1972a). The usual appearance of a series of puffs in developmental sequence may therefore depend upon a cascade of interactions, in which the hormone acts directly on only a small number of loci, some of whose gene products are responsible for the activation of further loci.

Puffs and Gene Products

Puffs represent sites at which gene expression can be visualized. Two questions can be asked directly concerning the nature of the puff. What is the mechanism by which a puff is generated—what changes occur in the structure of the band? And what is the unit of transcription? This second question concerns the issue of whether each band contains only a single gene, in which case a puff at a given site always produces the same single transcript; whereas if the genetic content of a band codes for more than one protein, a single puff either may be responsible for the production of several gene products, or at different times might be producing different gene products. Chapter 17 discusses the relationship between bands and genetic units; here we are concerned with the structure of the puff and with the correspondence between sites of puffing and gene products. In Chapter 28 we shall consider the genetic constitution and control of expression of certain bands whose puffing has been investigated in detail.

Initial attempts to equate protein products with particular chromosomal sites focused upon the prominent puffs of the salivary gland (see Beerman, 1961). In Chironomus, the salivary glands synthesize proteins that are secreted for the larva to use in spinning a fine thread. Grossbach (1969, 1973) identified six major proteins by electrophoresis of the polypeptides secreted by the salivary gland of C. pallidivittatus. A seventh major polypeptide is synthesized by some, but not most, salivary gland cells. This last protein, and one of the generally present six proteins, are not synthesized in C. tentans. Genetic crosses locate the genes for both these proteins on chromosome IV, in each case in a region of a few bands that contains a single prominent Balbiani ring. In the instance of the seventh protein, those cells synthesizing it possess an additional Balbiani ring that is absent from the other salivary gland cells. While this is not yet conclusive, further work may be able to identify these two protein products with particular bands that puff in the salivary gland. Indirect evidence that another Balbiani ring on the fourth chromosome is responsible for synthesizing another of the proteins has been provided by the demonstration that a reduction in size of the puff can be correlated with reduced synthesis of the protein. A similar correlation was demonstrated previously in the midge Acricotopus, in which regression of one of the Balbiani rings was associated with the cessation of synthesis of a secretory protein identified by its content of hydroxyproline (Baudisch and Panitz, 1968). The conclusion suggested by these results is that the major sites of activity in the Chironomus salivary gland may be prominent Balbiani rings which are responsible for directing synthesis of the secretory proteins that are produced in large amounts.

Similar secretions appear to be the function of the larval salivary glands in Drosophila, where Korge (1975) has isolated secretional variants that have been used to map some of the proteins. Different strains of Drosophila vary in the electrophoretic mobilities of three of the four major secretory proteins. Genetic mapping shows that one of these proteins (protein 4) is coded within a region of six bands on the X chromosome. Within this region, there is a prominent puff in salivary gland cells that is absent from other tissues. This puff is absent also from a strain lacking protein 4 (Korge, 1977). This makes a strong case that the puff at band 3C codes for protein 4.

Another of the secretory proteins (protein 3) similarly was located, although with less precision, to a region of chromosome 3 that contains a prominent salivary gland puff (68C). By examining flies with genetic deficiencies in this region, Akam et al. (1978) showed that the structural gene for protein 3 must lie within a region of 8 bands including the puff in 68C. Similar techniques located the structural gene for larval serum protein 2, which is synthesized by the fat body, to one of two bands (68E3 and 68E4), although so far neither of these has been shown to puff in the fat body polytene cells.

These results with Drosophila do not yet prove that these proteins are coded by individual bands, although this is in principle a matter of achieving greater genetic resolution in the cytogenetic mapping. The presence of puffs in the appropriate regions, however, and more strongly the correlation between the presence of some puffs and the synthesis of certain proteins in both Drosophila and Chironomus, make a good case that the puffs represent active structural genes. This does not answer the question, of course, of whether a given band codes for only the protein so far identified with it or may also specify other proteins.

The synthesis of RNA has been associated with the sites of puffing by showing that ribonucleoprotein particles accumulate at puffs and that inhibition of RNA synthesis is inimical to the puffs. Induction of puffs can be prevented by several inhibitors of transcription, including actinomycin, α-amanitin, and 3'-deoxyadenosine, although regression of existing puffs is not always complete (Beerman, 1964, 1971; Ellgaard and Clever, 1971; Ashburner, 1972b). Protein synthesis may remain unaffected, which suggests that mRNA may be stable in the salivary glands, at least at later stages of development (Clever, Storbeck, and Romball, 1969; Clever, 1969; Doyle and Lauffer, 1969; Clever and Storbeck, 1970; Rubinstein and Clever, 1972).

The RNA transcribed at active bands is packaged into ribonucleoprotein particles that resemble similar bodies found in diploid nuclei (Beermann and Bahr, 1954; Swift et al., 1964; Stevens and Swift, 1966). Generally these are about 30 nm in diameter, although in some puffs a different morphology is displayed (Derksen, Berendes, and Willart, 1973). The presence of such particles also has been reported in interbands in cases where the adjacent bands are apparently too thin and well defined to have contributed the particles by spreading (Skaer, 1977). Morphological criteria alone, however, do not provide sufficient evidence to locate the site of transcription and it remains to be seen whether interbands are indeed transcribed.

The origins of specific RNA molecules can be identified by hybridizing the labeled RNA with polytene chromosomes and locating the label by autoradiography. This is possible, of course, whether or not the RNA was synthesized in the polytene cells. In other words, the polytene chromosomes provide a cytological assay for identifying the bands that contain DNA complementary to a given nucleotide sequence. This assay has been used to identify the loci corresponding to the mRNAs produced when Drosophila cells are heat shocked; and these coincide with the bands at which puffs develop upon elevation of temperature (see Chapter 28). This demonstrates directly that these puffs represent sites of transcription. A similar analysis has been possible with one of the prominent Balbiani rings of the Chironomus salivary gland, where the structure of the transcript is being defined (Daneholt, 1975). In these instances it should become possible to identify the genes coding for specific proteins and to determine their relationship to the band.

Transcription at Puffs

The appearance of puffs suggests that they represent sites at which the nucleoprotein fiber has expanded from its usually condensed state within the band. This may be necessary to relieve topological restraints that would prevent the extensive transcription of DNA within a band. Bands that synthesize RNA without generating visible puffs may represent less active sites of transcription.

What is responsible for the dispersion of band structure? Holt (1971) observed by interferometric measurements that the formation of puffs is accompanied by an increase in protein mass at a puffed site. This does not reflect changes in histone content; for the early studies of Gorovsky and Woodard (1967) showed no change in quantitative staining with alkaline fast green, and more recent studies with indirect immunofluorescence show no concentration of histones at puffs (Silver and Elgin, 1976; Alfagame, Rudkin, and Cohen, 1976; Jamrich et al., 1977). However, Holt (1970) found a local increase in the capacity of puffs to bind naphthol yellow S, which reacts with the indole group of tryptophan. Since tryptophan is largely absent from histones, this suggests that nonhistone proteins accumulate at the puffs. The order of events in development of a puff suggested by Berendes (1968) is: accumulation of nonhistone protein, dispersion of the nucleoprotein, and lastly transcription of RNA. Since inhibition of transcription prevents formation of new puffs (see above), the stage of visible dispersion may be relatively late.

How specific is the accumulation of nonhistone proteins—to what extent are the same nonhistones involved in dispersing all bands for puffing or are different proteins involved with different puffs? This question can be approached by examining the locations of particular proteins within the polytene set by immune reactions. The most convenient technique is indirect immunofluorescence (see Chapter 2) and the only modification that this requires to the usual techniques for handling polytene chromosomes is that fixation with formaldehyde or glutaraldehyde must be introduced, since the usual method of preparing squashes in acetic acid leads to the extraction and perhaps also movement of chromosomal proteins. A limitation inherent in this approach is that there are no particular proteins to use as probes as, for example, isolated messenger RNAs can be used to determine the location of corresponding sequences in the genome. With the exception of RNA polymerase, it is necessary therefore to isolate fractions of nonhistone proteins against which antisera can be prepared for testing against the polytene chromosomes; this means that the antisera are likely to react with proteins present in larger amounts in the nonhistone complement, so that the probability of obtaining a reaction against a regulator protein present in small amount is not high.

Indirect immunofluorescence directed against RNA polymerase II shows a correlation with puffing. No fluorescence is found over the chromocenter or on metaphase chromosomes (of diploid cells); but with salivary glands a large number of loci fluoresce, among which are the puffs (Plagens, Greenleaf, and Bautz, 1976; Jamrich et al., 1977). A particularly good example of the correlation between RNA polymerase and puffing is provided by two of the heat shock loci, at which RNA polymerase II accumulates upon temperature shock (Greenleaf et al., 1978). On the other hand, RNA polymerase II is found at many sites not thought to be active in transcription and the analysis of Jamrich, Greenleaf, and Bautz (1977) suggested that it may be located in interbands. In fact, the pattern of indirect immunofluorescence against RNA polymerase II amounted to a negative of the usual banding pattern, with interbands reacting and the bands inert. A difficulty in interpreting this observation lies in the unknown details of the immune reaction. Is it possible that the reactive component of the interbands is not RNA polymerase—perhaps a subunit or some other factor than the complete enzyme? How many protein molecules are needed to generate fluorescence? And to what extent may the sites at which RNA polymerase is located fail to coincide with those that are transcribed? It is, at all events, difficult to take the view that transcribed genes lie in all the interbands: autoradiography identifies grains over the bands; less DNA (thought to be about 5% of the total) resides in the interbands than probably is transcribed; but if the interband were the unit of transcription, it would be surprising to find every single gene transcribed in the salivary gland. Clearly it is intriguing that interbands can be positively identified by such experiments; but the significance of this result remains to be seen.

Indirect immunofluorescence shows that antibodies prepared against the complete nonhistone protein complement stain all the bands of the polytene set. With antisera directed against subfractions of the nonhistone proteins, more specific results may be obtained (Silver and Elgin, 1976, 1977, 1978). Of antisera against three individual proteins, one reacts with all bands and two with a restricted distribution; the first may be involved in chromatin structure, the other with specific sites. Reaction against a group of nonhistones characterized by a molecular weight range of 80,000—110,000 daltons results in a selective staining which includes all the sites at which puffs are present and also those at which puffs will be induced later in development. Upon heat shock, fluorescence is seen at the new puff sites. This suggests that among this group of nonhistones (termed the rho proteins), there may be proteins whose presence is necessary for puffing. Whether the specificity of reaction reflects accumulation at these sites or changes in accessibility is not known; nor is it known how many proteins are included in the immune reaction. But in principle, refinement of these experiments should lead to definition of the relationships between puffs and given proteins. Other experiments along these lines have identified a small number of sites that react to antisera against the D1 nonhistone protein (apparently similar to the HMG proteins of calf thymus; see Chapter 12); these coincide with the sites highly stained by quinacrine and it may be interesting to see whether the D1 protein is involved in this response (Alfagame, Rudkin, and Cohen, 1976).

The potential of the polytene chromosomes as an assay system for the location and function of nonhistone proteins is indicated by the reaction of salivary glands with antisera prepared against proteins released by DNAase I digestion of Drosophila embryonic cells. Mayfield et al. (1978) divided the proteins released by 5–10% digestion into five molecular weight fractions. Antisera against four of these fractions produced general staining; the corresponding proteins are likely therefore to play some common structural role in the chromosome. But the antiserum against a molecular weight fraction of about 63,000 daltons showed a pattern similar to that of the anti-rho serum, staining not only puffs but also sites expected to puff at other stages. The active nonhistone protein(s) of this fraction again might be component(s) associated with gene expression.

PART 3

Organization of the Eucaryotic Genome

CHAPTER 17

Introduction:
Genes and Gene Number

The critical issues in approaching the function of the eucaryotic genome lie with the nature and number of its constituent genes. Is the gene a unit identical with that defined in procaryotes or does it take a more complex form? What is the unit of gene expression—does it include information coding for a single gene product or for more than one protein? How many genes are present in the genome of a given eucaryote—indeed, what proportion of the eucaryotic genome is accounted for by genes?

The importance of each gene is an unresolved question that stands in the way of making genetic estimates for genome complexity. Is the function of every gene critical at some point or are there genes whose deletion is without serious effect on the organism? In the former case, it should be possible in principle to obtain lethal mutations to identify every gene; but if many genes are superfluous, in the sense that the absence of their protein products can be compensated by other gene products, then the number of lethal loci may be very much less than the number of functional genes. In short, it is unknown what proportion of the genes of a eucaryote can be identified by lethal or at least detectable mutations.

It is necessary here to distinguish between gene and gene product. If every gene is unique and codes a distinct protein product, the question becomes whether every gene product is necessary for viability or whether some gene products can be substituted by other gene products. But if some proteins are coded by genes present in more than one identical copy per haploid genome, the deletion of one or even several copies of the gene might not prevent synthesis of an adequate supply of the gene product. This poses the question of to what extent genes may be repeated within the genome and of how such copies are to be detected. This is to ask whether there is a discrepancy between the number of genes and the number of gene products.

Taking the gene to contain the nucleotide sequences required to code for a single polypeptide chain, the question of the number of different genes becomes that of how many different proteins are synthesized in toto during the development of a given organism (including all embryonic stages as well as the adult state). A direct answer is precluded by the impracticality of counting the number of proteins. An estimate for the number of genes active in a given tissue or cell type can be provided by the number of different messenger RNA species; but to determine the total gene number it would be necessary to assay every single cell type or developmental stage and to determine the extent of overlap between the messenger populations. The isolation of messengers, however, permits determination of the number of sequences in the genome corresponding to each gene.

The context within which each coding sequence lies is the critical issue in defining the unit of gene expression. Is each coding sequence organized into a single unit; or may more than one coding sequence reside in a single unit? Clearly the unit of gene expression must be larger than the coding sequence in order to include necessary regulator elements. But what is the proportion of regulator to coding sequences? Does the situation mimic that of the procaryotic genome, where comparatively short regulator elements control the expression of much longer adjacent coding sequences? Or may the eucaryotic regulator sequences be much longer, perhaps even longer than the coding sequences, so that much less of the genome has the function of coding for protein. Are all the noncoding sequences contained within a unit of gene expression critical, or could some be unimportant?

The constitution of the unit of gene expression must determine its susceptibility to mutation. If most identifiable mutations occur within coding sequences, each complementation group can be taken to identify a gene. But if regulator sequences comprised the major part of each unit of gene expression, most mutations presumably would inactivate the unit by interfering with its expression; this would prevent the function of however many genes were part of the unit.

What proportion of the eucaryotic genome is organized into units of gene expression? Are other sequences present and what may be their functions? What is the effect of mutation of such sequences? These questions underline some of the fundamental issues that remain to be resolved about the components and organization of the eucaryotic genome.

Genes and Complementation Groups

Like the question of how many angels can dance on the head of a pin, the question "what is a gene?" often seems incapable of resolution. Yet the definition of "gene" is not trivial. While it is straightforward to say that a gene is a nucleic acid sequence coding for a polypeptide chain, there is no single experimental test that identifies a gene. Genes usually have been defined either in terms of a genetic complementation test or by virtue of the gene product, whether protein or RNA (see Lewin, 1970). But neither of these approaches as such now affords a satisfactory definition that encompasses the forms that genes may take. Without indulging in semantic discussions, however, we may define the gene in operational terms that allow us to consider its organization and expression within the genome.

The cis/trans complementation test was devised to determine whether two mutations lie in the same or in different functional genetic units (see Volumes 1 and 3 of Gene Expression). If two mutations lie in different units, the double mutant of a heterozygous diploid will have wild phenotype irrespective of whether the mutations lie in cis on the same chromosome or in trans on homologous chromosomes; for in both cases, the cell possesses one mutant and one wild-type copy of each gene. If two mutations lie in the same unit, they will give wild phenotype only when lying in the cis arrangement, when one gene copy has two mutations and one has none; in the trans arrangement, each gene has one of the mutations, so neither is functional. The practical form of this complementation assay is to form the trans heterozygote, when mutations lying in the same gene do not complement and the mutant phenotype reveals the absence of a functional gene. A *complementation group* consists of all those mutants that fail to complement in pairwise trans arrays. The cis arrangement of double mutations then may be used as a control.

Can the complementation group be equated with the gene? The inability of the complementation assay to distinguish between a gene and an adjacent cis dominant regulator site was recognized by Jacob and Monod (1961) in their formulation of the operon model. The operon includes control elements, adjacent to the structural genes coding the proteins; these elements do not code for diffusible products but are recognized by RNA polymerase or regulator proteins. Since this recognition controls the expression only of the contiguous structural genes and does not influence those present on other molecules of DNA, the control elements are said to be *cis dominant*. A cis dominant element cannot be distinguished by genetic means from the gene(s) that it controls; thus in the cis/trans test it fails to complement mutations in any of the contiguous structural genes.

A mutation in the control element adjacent to a structural gene therefore behaves in the complementation test as part of the same complementation group as structural gene mutations. To distinguish mutations in the control element from those in the structural gene, it is necessary to determine whether they lie within the nucleotide sequence coding for the protein. It is necessary to demonstrate that control mutations do not affect the structure of the protein; but this is not sufficient since mutations in the structural gene may be silent in this respect. In extreme cases, sequence studies may be necessary to locate mutant sites with respect to the structural gene.

The complementation test therefore in principle distinguishes genes coding for diffusible products from each other, leaving cis dominant control elements unresolved from the contiguous structural genes. Such elements sometimes have been described as "genes" but here we shall reserve the use of the term for sequences that give rise to diffusible products. The control elements will show the same genetic behavior whether they are not transcribed or are transcribed as part of the same RNA molecule that carries the coding sequence. The elements can be defined by mutations that map outside the limits of the structural gene and by characterization of the sequences that bind proteins (although this latter criterion identifies only the stable binding sites; when one site is recognized by a protein that then moves immediately to a second, adjacent site, it may be possible to identify the first site only by mutation).

Delineation of the limits of the structural gene turns on the definition of gene products. *Cistron* was introduced to describe the sequence of DNA that codes for a single polypeptide chain (Benzer, 1959, 1961). The implication of the cis/trans test is that all the nucleotide sequences coding for a polypeptide must be arranged in linear order on a single DNA molecule; "cistron" has been used interchangeably with the earlier term "gene." Since the messenger RNA from which a protein is translated may be somewhat longer than the protein-coding sequence, the full extent of the DNA region used for producing the protein can be determined only by comparing the messenger sequence with that of the DNA. The noncoding sequences that surround the coding sequence may or may not be considered to comprise part of the "gene," depending on semantics.

This argument extends further, since messenger RNA itself need not be the immediate product of transcription; the message may be derived from the primary transcript by processing steps that involve the removal of some of the noncoding sequences of the primary transcript. Definition of the unit of transcription therefore requires comparison of the primary transcript with DNA. If the unit of transcription is monocistronic, including information coding for only a single protein, it will correspond to a single complementation group. Mutations in nontranscribed cis dominant elements on DNA, mutations in the noncoding sequences of the primary transcript or mature message, and mutations in the coding sequence all will fall into the same group. The relative proportions of mutations of each type will depend on the relative lengths and susceptibilities of these regions to mutation; it is possible that the coding sequence may represent only a small part of the unit of transcription, but in lieu of information on the functions of the noncoding sequences, it is impossible to predict the likelihood that their function will be impaired by mutation.

When the unit of transcription contains more than one cistron, each cistron represents a complementation group. Several possibilities for the stage at which gene products become separated are illustrated in Figure 17.1. In the classical operon, a single polycistronic messenger is translated into individual proteins. In some phages, the primary transcript is polycistronic, but it is processed into individual messengers for translation. In some eucaryotic viruses, a polycistronic mRNA is translated into a polyprotein, which is then cleaved into individual proteins. The genetic consequences are the same insofar as all those mutations lying in a region that gives rise to an independently diffusible element will form a complementation group. As with the bacterial operon, mutations affecting the unit as a whole, or at least more than one of its independent products, will fail to complement with mutations of more than one of the groups.

In the concept that a gene is a sequence of DNA coding for a single polypeptide chain, there has been implicit the understanding that the DNA constitutes an uninterrupted series of nucleotides, which read in triplets from one strand correspond to the amino acid sequence. But this need not be true. Recently it has become apparent that at least some eucaryotic genes contain "intervening" sequences (see Chapter 26). These are stretches of nucleotides present within the coding sequence of DNA that are not represented within the amino acid sequence of the protein. The intervening sequences may be transcribed into RNA, then to be removed from the transcript during the processing steps that produce the mature mRNA. It seems likely, as implied in Figure 17.1, that the removal of the intervening sequences is an intramolecular reaction, retaining a situation in which those sequences on one molecule of DNA give rise to a protein chain of corresponding sequence. Essentially the parts of the coding sequence are "spliced" together on the RNA. Although the parts of the coding sequence lie in linear array on the DNA, that is, in the order in which the corresponding sequences fall in the protein, they can no longer be assumed to be contiguous.

The presence of the intervening sequences implies that genetic map distances between mutations will not correspond to distance apart of corresponding sites in the protein. Point mutations affecting the protein structure should fall into clusters representing each contiguous segment of the coding sequence, separated by distances corresponding to the intervening sequences. The function of the intervening sequences is not known; so it is impossible to assess their susceptibility to mutation. As with other noncoding sequences, if they constitute part of the primary transcript that is degraded during messenger processing, any mutations will lie in the same complementation group as the coding sequence; only if such sequences give rise to stable RNA molecules that are independently diffusible would they represent different complementation groups. Thus so long as splicing is an intramolecular reaction, each coding sequence forms a single complementation group; the corollory is that if intermolecular joining of different parts of the coding sequence were possible, the cis/trans test would identify each such part as an independent cistron. If an intervening sequence were to be represented as an independently diffusible molecule (whether in RNA or translated into protein), the corresponding complementation group might therefore by located within another complementation group.

A second concept of genetics recently disproven is that a given sequence of DNA can code for no more than one protein. Nucleic acid sequences represented in more than one polypeptide now have been identified in both procaryotic and eucaryotic viruses; the forms that they take are illustrated in Figure 17.1.

In the simplest case, one gene represents part of another; a "full length"

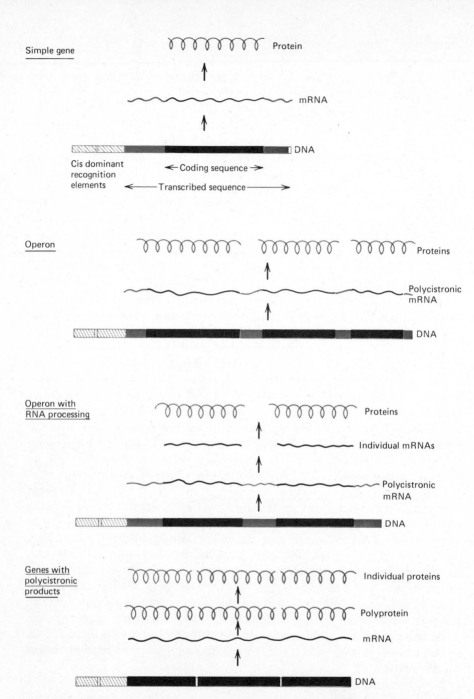

Figure 17.1. Relationships between genes and gene products. In a *simple gene* a single coding sequence is transcribed into a messenger RNA that is translated to give a polypeptide product. The transcript may be longer than the coding sequence, with noncoding regions at either or both ends. Transcription of the message may be controlled by cis dominant elements that are recognized by regulator proteins or RNA polymerase.

In an *operon* the unit of transcription contains more than one gene. The polycistronic tran-

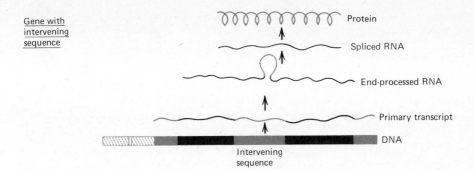

Gene with
intervening
sequence

Protein

Spliced RNA

End-processed RNA

Primary transcript

DNA

Intervening
sequence

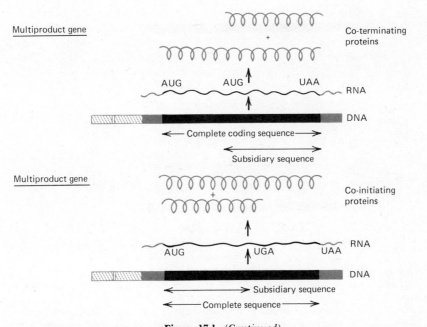

Multiproduct gene

Co-terminating
proteins

AUG   AUG   UAA

RNA

DNA

◄— Complete coding sequence —►

◄————————►
Subsidiary sequence

Multiproduct gene

Co-initiating
proteins

AUG   UGA   UAA

RNA

DNA

◄————————► Subsidiary sequence

◄——— Complete sequence ———►

Figure 17.1. (*Continued*)

script may or may not include noncoding sequences at the ends or between adjacent coding sequences. In the classical model the polycistronic transcript is translated as such; but it may also be processed into unicistronic messengers, with or without loss of some material. When a *polycistronic protein* is translated from adjacent genes, it is its cleavage into individual proteins that separates the gene products. This may or may not be accompanied by loss of some amino acid sequences.

In a gene with an *intervening sequence*, the coding sequence is arranged into parts that are in linear order on the DNA but are interrupted by noncoding sequences. Any intervening sequences must be spliced out of the RNA to generate a messenger in which the entire coding sequence is uninterrupted. Processing of the ends of the primary transcript to remove some or all of the terminal noncoding sequences also may occur.

Multiproduct genes generate more than one protein from a single coding sequence. Additional initiations may occur within the coding sequence, to generate partial C terminal proteins as well as full length proteins; or a termination codon within the coding sequence may be partially suppressed to generate full length polypeptides as well as the partial N-terminal polypeptides. In both cases, the smaller protein represents part of the sequence of the full length

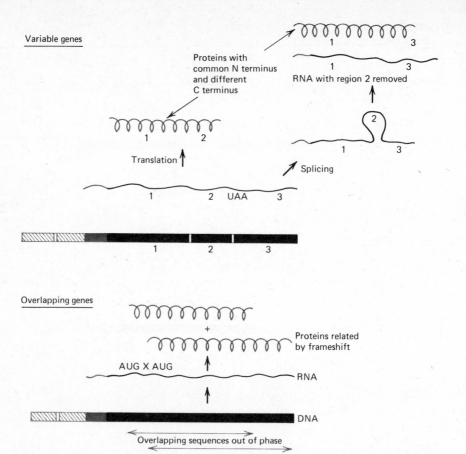

Figure 17.1. (*Continued*)

protein. The relative proportions of the two products depend on the relative efficiencies of the initial and internal initiation codons or the efficiency of readthrough past the internal termination codon.

Variable genes represent the situation in which alternative sequences are joined together. A simple case is illustrated in which the transcript may follow two pathways: if translated as such, it generates a protein corresponding to regions 1 and 2 (a termination codon occurs at the end of 2); if spliced, region 2 is removed and the protein corresponds to regions 1 and 3. The relative amounts of the proteins depend on the efficiency of splicing. More complex processing events would allow splicing to join a given sequence to more than one other sequence, to generate proteins in which a common region is joined to one of several alternative regions, which might be independent or overlapping.

Overlapping genes are related by a shift in phase. Thus the same nucleotide sequence is read in two different phases to generate two proteins whose amino acid sequences are unrelated per se. Both multiproduct and overlapping genes are shown here to result from translation of single messengers; but it is equally possible that changes in the sites at which transcription is initiated and/or terminated could occur.

protein is translated from the complete gene, but a shorter protein representing only part of this sequence also is produced, by using initiation or termination sites that lie within the gene (all the punctuation signals lying in the same phase). This situation presents no particular genetic problem; it is in principle no different from the partial production of a smaller protein by degradation of part of the sequence of a precursor polypeptide. A well-characterized example of partial termination is provided by the coat protein gene of phage Qβ, where most of the proteins terminate at a UGA codon to generate a product of 14,000 daltons; but some readthrough occurs to generate a longer protein of 36,000 daltons, as discussed in Volume 3 of Gene Expression (Weiner and Weber, 1971, 1973). A similar partial suppression of termination may occur in certain mammalian C-type RNA viruses (Philipson et al., 1978; Weiss et al., 1978). An example of internal initiation is seen in the A gene of phage φX 174 (Linney and Hayashi, 1973).

A more complex situation arises when variations in splicing allow the same initial sequence to be joined to more than one terminal sequence, to generate proteins with common N-terminal regions and different C-terminal regions (or vice versa). This situation has been identified in the production of T antigen by SV40 (see Chapter 26). The genetics of such situations are difficult to predict, since the abilities of mutations in the spliced regions to complement one another may depend on the functional relationships between the proteins. In the simplest case, in which a common sequence is joined to alternative sequences, mutations in the common sequence may fail to complement mutations in any of the alternative sequences, whereas mutations in the different alternative sequences may complement each other. This would be akin to the relationship between a cis dominant control element and its contiguous structural genes. Although the consequences of these events are the same as the rearrangement of DNA that may occur to link constant to variable immunoglobulin regions, in the immunological system there is no genetic consequence for the organism, because the rearrangement occurs only in somatic cells and each cell line expresses a different rearrangement (see Chapter 26).

Another difficult situation arises when a single nucleotide sequence is translated in more than one reading frame. Then a mutation may affect either or both proteins depending on where the changes lie in the overlapping triplets. Roughly one-third of the mutations will affect the sequence of both proteins and one-third each will affect the sequence of only one protein. This should produce a complicated complementation pattern in which mutations affecting two different proteins are intermingled. Examples of overlapping reading frames have been found in phage φX174 and are discussed in Volume 3 of Gene Expression (Barrell, Air, and Hutchinson, 1976; Smith et al., 1977; Weisbeek et al., 1977).

If these situations have evolved in response to the special need of viral genomes for compact organization, they may fail to occur in cellular genomes.

The C Value Paradox

The large amount of genetic material in the eucaryotic genome, and very wide variations in genome size within the eucaryotes, raise the question of the function of the DNA.* How much of the haploid DNA codes for proteins? What other functions reside in nucleotide sequences? To what extent do evolutionary restraints on change of sequence vary among different parts of a eucaryotic genome? The C value paradox arises from the belief that the DNA content of eucaryotic genomes (known as the C value) may be excessive in the sense that there is far more DNA in the genome than would be required to code for all the proteins that it specifies.

A simple view would be to expect the C value of an organism to be proportional to its complexity, for example, to the number of different cell types in the animal. But there is no clear correlation of this type. As Figure 17.2 shows, most classes of eucaryotes comprise a range of species with variations in genome size of about 10 fold. Some of the better substantiated data concern mammals and amphibia. Mammals have genomes that fall into a particularly small range of DNA contents, with a C value usually of 2-3 picograms ($2-3 \times 10^9$ base pairs). Amphibia, by contrast, vary very much more widely, from less than 1 picogram to almost 100 picograms. Even closely related amphibia may have greatly different contents of DNA in the haploid genome. It is difficult to believe that there could be accompanying variation in the number of genes.

The minimum size of genome reported for each class of eucaryotes increases with evolutionary development as shown in Figure 17.3. This implies that a certain increase in content of genetic information must accompany evolutionary progression. A possible corollory is that all (or at least more) of the genome fills essential functions in the species with the minimum genome size for their class. But many species have a much greater content of DNA than the minimum content for the class: what function is served by the increase in genome size?

If the number of protein products remains similar in related species, two extreme models describe the possible functions of the increased DNA content. One is to suppose that only part of the genome codes for proteins and that this part remains of similar size in the related organisms. The proportion of the genome coding for protein then must decline with increasing C value in this class; the increased content must represent the addition of sequences serving some other purpose. Since it seems reasonable to assume that regulatory circuits will be roughly similar in size for related organisms with similar gene numbers, the increased DNA content cannot readily be attributed simply to increased numbers of control elements. An alternative

---

*It is true that absolute values for DNA content are difficult to obtain as outlined in Appendix 1. But comparative studies using assays such as feulgen make it clear that there is indeed very wide relative variation in the size of the haploid genome.

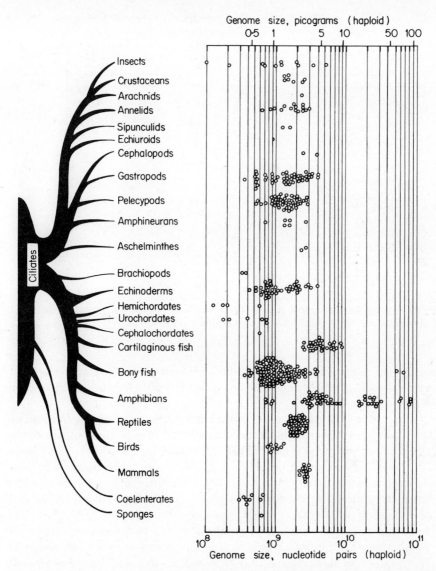

Figure 17.2 Genome sizes of the eucaryotes. Data compiled by Britten and Davidson (1969).

model is to suppose that there may be more than one copy of a given gene. Then the number of copies could be increased in species with larger genomes. To reconcile this situation with the genetic identification of single loci responsible for coding particular proteins, "master-slave" hypotheses have been advanced (Callan, 1967; Whitehouse, 1967). These postulate that one copy of the gene is a master, which corresponds to the complementa-

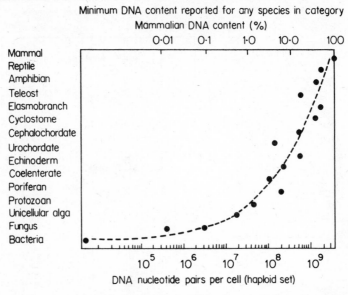

Figure 17.3. Minimum haploid genome size for each class of species.   Based on the data in
Figure 17.2 compiled by Britten and Davidson (1969).

tion group identified by mutation; the sequences of the slave copies are sub-
servient to this, so that mutations in them affect only the individual and are
not inherited.

The issue of how many copies of each gene are present in the haploid genome can be considered in terms of the extent of repetition of nucleotide
sequences within the genome. Hybridization analysis shows that only some
of the DNA is unique in sequence: this *nonrepetitive* fraction often comprises the majority (see Chapter 18). The remaining sequences are repeated
within the genome in the form of families: in the repetitive fraction each
family consists of members whose sequences are related but not identical.
The *moderately repetitive* component shows a small degree of repetition
(see Chapter 18); the *highly repetitive* component comprises satellite DNA
sequences, often rather short and repeated a large number of times (see
Chapter 19).

The presence of nonrepetitive DNA in the high C value species as well
as in the minimum C value species of a given class precludes the possibility
that the increase in DNA content represents a multiplication of the number
of copies of every gene. Such a model would predict that the sequence complexity of the genome should remain the same in spite of increases in DNA
content; as the genome size increases, the nonrepetitive component should
disappear, to be replaced by an increasing number of identical repetitions.
But this does not happen: nonrepetitive DNA continues to be found in the
genome and the degree of repetition of the repetitive component does not

necessarily increase. One example is provided by three related amphibia whose genomes span a 7-fold range of DNA content. All have a large component of nonrepetitive DNA (Straus, 1971).

If each protein is coded only by a single gene, then the coding sequences must constitute part of the nonrepetitive fraction. The location of coding sequences can be determined by identifying the sequences in the genome complementary to messenger RNA. The messenger populations of given cell types or tissues invariably prove to be derived for the most part from nonrepetitive DNA (see Chapter 24). With individual messengers representing particular proteins, the corresponding sequences can be shown to be present in only single copies in each haploid genome. Most coding sequences therefore appear to be unique in the genome.

Since nonrepetitive DNA represents a proportion of the genome that varies widely among eucaryotes, it is perhaps the complexity of this component that should be compared between species and with the apparent number of genes. But this offers no relief from the C value paradox. There is no relationship apparent between nonrepetitive DNA complexity and evolutionary development; nor does it seem likely that the entire nonrepetitive component can represent coding sequences. The proportion of nonrepetitive DNA represented in the messenger population of a given cell may be quite small; and overlaps between mRNA populations of different cell types may be extensive. The number of different messengers found in a mammalian cell population, for example, generally is of the order of 10,000; and only a small proportion of these may be specific for the cell type (see Chapter 24). But if all its sequences are assumed able to code for proteins, the capacity of mammalian nonrepetitive DNA must be about 10^6 genes. Again assuming that the number of genes in related species is similar, irrespective of variations in genome size, the presence of nonrepetitive components in high as well as low C value species makes the point that the proportion of unique DNA that represents coding sequences may vary considerably.

This leaves an unsatisfactory situation. If much of the DNA does not code for protein, what is its function? How is it able to change so greatly in amount? A further question is to what extent the total amount of DNA may affect the expression of the coding sequences; for example, an increase in genome size might influence the ability of a regulator protein to bind to a specific control element (see Chapter 28).

There appear to be two general types of solution to the paradoxical relationship between DNA complexity and gene number. One is to suppose that the genes represent only a small part of the genome; the remaining sequences, including nonrepetitive as well as repetitive, must serve some other function. The extreme view of this model supposes that this function, at present entirely unknown, is not connected with the expression of the genes per se. The alternative is to suppose that "genes" have a much more complex organization in eucaryotes than that familiar from procaryotes. This is to imply that each unit of genetic expression contains much more DNA than is involved simply in coding for the protein it specifies. Such sequences might take the form of lengthy control elements that are cis dominant to adjacent genes; or they might be flanking sequences that are transcribed but are not represented in messenger RNA. The discovery of intervening sequences has provided an impetus for models that suppose the unit of transcription to be much larger than the gene itself. The extreme form of this model takes the view that the number of genetic units will be similar in related organisms, and that differences in genome size will be accommodated by corresponding changes in the size of the units (see Chapter 26).

Although most genes are unique, some are indeed present in multiple copies in the genome. Two instances in which there are clusters of identical genes arranged in tandem array are presented by the ribosomal RNA and histone genes. Virtually no polymorphism has been detected in ribosomal RNA; and it is therefore assumed that all copies of the genes are identical, although intervening sequences are present in some (see Chapter 27). Similarly the high conservation of sequence of at least histones H3 and H4 suggests that all the copies of their genes may be identical (see Chapter 28). An obvious problem, analogous to that which the master-slave hypothesis tried to answer for a more general case, is how selection can be applied when there are many copies of a gene.

Other genes also may be represented in more than one copy in the genome, since a small part of cellular mRNA populations often finds complementary sequences in the moderately repetitive component. But coding sequences do not seem to be typical of the repetitive DNA and comprise only a small part of it. The general behavior of this fraction is of sequences related to, but not identical with, each other; often it is present in the form of rather short sequences, too short to represent genes, interspersed with much longer nonrepetitive sequences.

Gene Number in Drosophila

The C value paradox implies that the amount of DNA in the genome cannot be taken as an indication of the number of genes of an organism. Two approaches have been taken to determining the gene number. One is to estimate the number of gene products in the form of the complexity of mRNA; as we have remarked, this allows the number of active genes in any given cell or tissue to be estimated, but does not lead directly to estimates for total gene number because of difficulties in determining the degree of overlap in gene expression of different cell types. Also it leaves open the possibility that there might be genes that are expressed only occasionally or transiently and whose messengers are present at levels too low to estimate accurately. Another approach is to attempt to identify every gene directly by mutation. This is subject to two critical reservations. The mutable unit, as defined by the complementation test, might not coincide with a single protein coding sequence. And not all genes may be identifiable by mutation, given the criteria that are available for isolating mutants; since the individual gene products are unknown, selection must be applied to identify mutants by virtue of lethality or visible defects. Immediately this raises the question of whether every gene is essential for survival of the organism: in how many cases would the deletion of a gene be lethal, cause some lesser deleterious effect, or be without effect?

Given a large number of lethal mutants, the formation of heterozygotes can be used to assign mutations to complementation groups. For any eucaryotic organism, the task of mapping the entire genome in this manner is too extensive to be practical. A more limited alternative, however, is provided by the Diptera, in which a cytological map of the genome is provided by the polytene chromosomes. Then it becomes possible to isolate lethal mutations corresponding to a given chromosome region and to determine the number of complementation groups into which they fall. This can be correlated with the visible structure of the region; and relying only upon the assumption that the relationship between band and genetic function is similar throughout the genome, the number of complementation groups corresponding to a given number of bands can be extrapolated to give the total number of groups comprising the genome.

The idea that the bands of polytene chromosomes might in fact represent functional units was suggested when the equivalence of the genetic and cytogenetic maps was demonstrated in D. melanogaster (Bridges, 1935; Painter, 1944; Muller and Prokofyeva, 1935). Several visible mutations have been mapped to individual bands by cytogenetic means (largely the use of deletions); and in none of these cases have mutations lying in different complementation groups been mapped to the same band. (Although there are certain "complex loci" at which mutations have pleiotropic effects yet to be explained in terms of single genes.) In most cases, however, the mutants are identified by phenotype and the molecular nature of the mutation remains to be established.

A more systematic attempt to equate genetic loci with bands has been to saturate by mutagenesis all the mutable loci in a given chromosome region. Then mutants are recovered, placed into complementation groups, and mapped within the region by determining their ability to complement with deletions removing defined series of bands. If the distribution of mutations to complementation groups falls into a Poisson distribution, it is possible further to estimate how many groups remain in which no mutations have been found.

A protocol used for such experiments is illustrated in Figure 17.4. This has been applied by Judd, Shen, and Kaufman (1972) to the 3A2-3C2 region of the X chromosome of D. melanogaster, which coincides with the region of the genetic map lying between the loci *zeste* and *white*. This occupies 0.75 map units of the total length of 66 map units; and extends for about

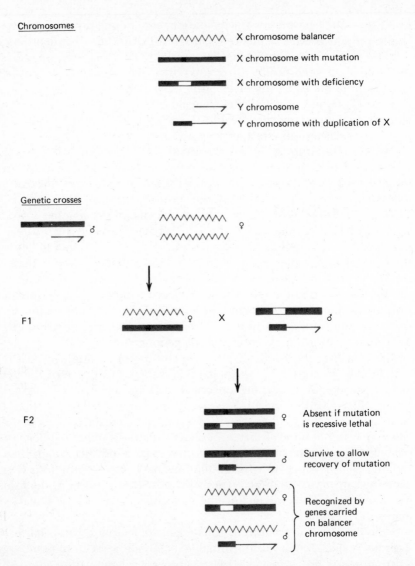

Figure 17.4. Protocol for isolation of recessive lethals on X chromosome of D. melanogaster. Males are mutagenized and then crossed with females homozygous for a balancer X chromosome. The daughters carry the mutant X chromosome and the balancer X chromosome, which suppresses crossing over in the sex pair. These F1 females are crossed with males whose X chromosome has a deficiency in the region being examined and whose Y chromosome carries additional material that covers this region. In the F2, females carrying the mutant chromosome do not survive if the mutation is a recessive lethal in the region covered by this deletion; but the males survive because of the material attached to the Y. The nature of the mutation is therefore defined by the absence of the females; and the mutant chromosome can be recovered from the males. The males and females carrying the balancer X chromosome are distinguished as such by the presence of genes identifying this chromosome. After Judd, Shen, and Kaufman (1972).

16 bands of the total of 1012 on this chromosome. The region therefore represents just over 1% of the X chromosome on either the genetic or cytogenetic map.

The protocol identifies recessive lethal mutations and a total of 121 independent mutants behaved as single-site mutations which could be mapped into 14 complementation groups. (Since the mutations are recessive lethals, the complementation test asks whether the heterozygote *m1/m2* is wild type (flies are present) or mutant (flies are absent)). The complementation groups mapped into a linear array, 12 between *z* and *w*, and 2 to the left of *z*. Subsequent experiments to seek further complementaton groups have so far identified only one clear additional lethal group (Judd and Young, 1973; Lim and Snyder, 1974; Liu and Lim, 1975). The distribution of mutations to groups does not fit the Poisson curve, so it is not possible to estimate by this means whether saturation has been reached.

The number of bands in this region is not certain, because of the presence of some rather fine bands in 3B that are difficult to resolve. The present estimate is that there are 16 chromomeres (including doublets) to be seen in the light microscope, with up to three more visible in the electron microscope (Kaufman et al., 1975). The relationship between the bands and complementation groups is shown in Figure 17.5. In addition to the 15 recessive lethal groups, several groups defined by nonlethal mutations are present.

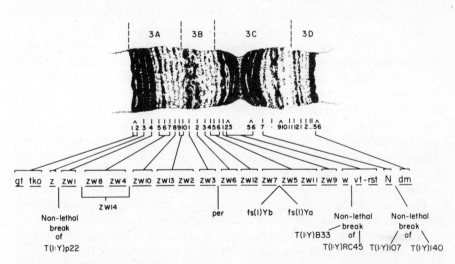

Figure 17.5. Correlation of bands in the 3A-D polytene map region of D. melanogaster with complementation groups in the zeste (z)—white (w) region. Lethal complementation groups are numbered *zw1-14*. Other markers are *per* (locus determining circadian rhythm), *fs(l)Ya* and *fs(l)Yb* (produce sterility in females); to the left of *z*, markers *gt* and *tko* identify loci producing giant larvae and adults sensitive to shock by blows, respectively; to the right of *w*, markers *vt-rst*, *N* and *dm* affect the eyes, wings, and bristles. Nonlethal break points identify sites at which it is possible to survive breakage of the chromosome. Data of Judd, Shen, and Kaufmann (1972), Judd and Young (1973), and Young and Judd (1978).

The total number of lethal complementation groups is close to the number of bands; and most of these groups can be localized to a specific band by deficiency mapping. This implies that each band may contain no more than a single lethal complementation group. It is consistent with the conclusion that saturation of loci able to generate recessive lethals has been reached or is at least very close.

Similar conclusions have been reached for other regions of the D. melanogaster genome. In earlier work, Lifschytz and Falk (1968, 1969) mapped 105 lethal mutations into 34 complementation groups, which is close to the number of bands found in the corresponding region of the X chromosome. In attempting to identify all mutable loci in the small chromosome IV, Hochman (1971, 1973) has now brought to 43 the total number of lethal complementation groups. In this case a Poisson distribution can be applied and suggests that the total number of such groups may be 60. Again this is not far from the total band number of 45.

These results with lethal mutations have lent credence to the idea that the band represents a unit of genetic function, or at least that it contains no more than one essential unit. In relating this genetic analysis to the structure of the genome, however, it is necessary to know the nature of the lethal mutation, and in what proportion of the genetic units such mutations can occur.

A critical question about the lethal complementation group is whether each includes only a single genetic function. An indication that this may be so is that all the mutants within a given group appear to have similar defects, at least in so far as can be judged by the time of developmental block and the similarity of any morphological defects (Shannon et al., 1972). A more stringent examination of this point was made by Shearn (1974) for third chromosome late larval lethals, many of which have defects in imaginal disc development that can be characterized morphologically. Noncomplementing mutants almost always showed the same or very similar defects; differences are generally quantitative rather than qualitative. Also consistent with the idea that each band represents a single functional unit is the observation of Lefevre (1973) that all rearrangements of material that can be seen to affect a given band show the same lethal defect.

The situation becomes more complex when the existence of complementation groups not identified by lethal mutations is considered. At present, such nonessential mutations have been systematically related to the organization of lethal groups only in the 3A2-3C2 region, where the two classes of complementation group appear distinct (Young and Judd, 1978). The existence of the nonessential groups z and w has been known for some time and none of the lethal mutations falls into either group; and indeed it is known that deletion of the w gene does not impair viability (Lefevre and Wilkins, 1966). Other nonessential functions in this region are identified by the locus per and by two groups (which may possibly really be one) that cause female sterility (fs). Additional nonessential sites are identified by the points at which nonlethal breaks may occur in the chromosome. To the right of *w* are some further nonlethal complementation groups. At least some of these nonessential loci appear to represent distinct complementation groups that lie in the same band as a lethal complementation group. This is not consistent with the hypothesis that there is only one gene for every band.

It is not possible to exclude the existence of unknown numbers of complementation groups in which recessive lethal mutation does not occur; and which either do not happen to have been identified previously by visible mutations or in which mutation is in fact undetectable (see below). An indication that a reasonable proportion of the genome may be occupied by such functions is given by the observation that many chromosome rearrangements are not lethal; this means that there are many sites at which breakage can occur without interrupting an essential function. The *z* and *w* loci have become the classical examples of functions that appear unidentifiable by the technique of lethal mutation and which occupy bands to which genetic functions could not be ascribed in terms of lethal complementation groups.

On the other hand, virtually every other band in the 3A2-3C2 regions appears to be occupied by one of the identified lethal complementation groups; so that any further complementation groups must share these bands (as seems to be true for the example of *per*). This would represent a situation in which the genome consists of both lethal and nonlethal complementation groups, but organized in such a way that each band contains no more than one lethal group, accompanied by an unknown number of nonlethal groups. It is not apparent why there should be a systematic relationship of this nature between the lethal and putative nonlethal groups. The total number of lethal complementation groups appears to be close to the number of bands, which implies that a minimum number of about 5000 functions is necessary for the survival of D. melanogaster. What proportion this represents of the total number of genetic units remains unknown.

What is the constitution of the lethal complementation group—do the lethal mutations lie in coding sequences or control elements? The simplest model is to suppose that most mutations lie in coding sequences, in which case each complementation group identifies a single gene together with any adjacent control elements. In this case, the D. melanogaster genome contains about 5000 genes coding for proteins whose absence is lethal. The absence of the proteins coded by any further, nonessential genes cannot be lethal or severely detrimental to the organism. Any other sequences in the genome must have some function other than coding for protein. In this case, either their function is not detectable by lethal mutation; or if mutation is lethal, it falls into the same complementation group as an essential coding sequence.

This last model is related to the general idea that most of the genome might be devoted to control rather than to coding functions. An extreme model following from this is to suppose that most mutations lie in the predominant component, that is, in the control elements. In this case, each lethal complementation group would correspond to such an element together with however many genes it controlled. But in contrast with this conclusion is the argument that, irrespective of the relative proportions of control elements and structural genes, when saturation is achieved mutations should have occurred to block the function of every structural gene. Even if most mutations were to lie in control elements, there should nevertheless be complementation groups corresponding to each individual gene. This makes it unlikely that the lethal complementation group includes more than a single gene.

Yet the Drosophila genome clearly contains far more DNA than can be accounted for by 5000 coding sequences. Table 17.1 summarizes data on three Dipteran genomes; in spite of the uncertainties in these estimates, in each case the haploid DNA content of the average band greatly exceeds the 10^6 daltons or so that might be expected to represent a single polypeptide. It would be interesting to know whether the difference in band numbers between D. melanogaster and D. hydei does indeed represent a more than twofold variation in the number of genes between genomes of very similar

Table 17.1   DNA content of Dipteran polytene chromosome bands

Species	Chromosome number	Haploid DNA (daltons)	Haploid DNA in bands (daltons)	Band number	Average DNA/band (daltons)
D. melanogaster	4	1.1×10^{11}	8.4×10^{10}	5059	16×10^6
D. hydei	6	1.2×10^{11}	9.6×10^{10}	1988	48×10^6
C. tentans	4	1.3×10^{11}	1.3×10^{11}	1900	68×10^7

These values are only approximate because there are uncertainties about both the content of DNA in the haploid genome and the number of bands. The genome size of D. melanogaster is based on a feulgen reaction using chick erythrocytes as standard (see Appendix 1); the D. hydei value is based on a hybridization analysis comparing complexity with that of B. subtilis (itself not certain as noted in Appendix 1); and the C. tentans DNA content is estimated from measurements of the increasing DNA content of cells of successive degrees of polyteny. The uncertainty is indicated by hybridization analysis of D. melanogaster, which suggests that D. hydei has a genome 1.3–1.6 times larger, rather than the 1.1 times indicated here. The haploid DNA contained in the bands is calculated by subtracting the amount of DNA apparently present in the constitutive heterochromatin of the chromocenter from the total genome. The number of bands is based on light microscope estimates; and, at least for D. melanogaster, this may be too high. The range of band sizes, judged from widths, is about tenfold. Even with these reservations, however, the difference in band size between D. melanogaster and D. hydei or C. tentans would appear to be appreciable.

References: D. melanogaster—Bridges (1938), Lefevre (1976), Rasch et al. (1971); D. hydei—Berendes (1963), Dickson et al. (1971); C. tentans—Beerman (1952), Daneholt and Edstrom (1967).

size; or whether the equivalence of bands and lethal complementation groups is not the same in the two species. Assuming the same equivalence, the proportion of the genome presumably not engaged in coding for essential proteins varies from very roughly 90% in D. melanogaster to 98% in C. tentans. To summarize the two questions we have discussed about its function, the first is to what extent it codes for nonessential proteins, and the second is to what extent noncoding sequences form elements that are part of the same complementation groups as the essential structural genes.

Essential and Nonessential Genes

Attempts to determine directly the number of genes in the eucaryotic genome necessarily have relied upon the assumption that all genes are subject to mutations whose effects can be detected by virtue of the altered properties of the protein product. Of course, some mutations will be without effect on the phenotype, either because they fail to change the amino acid sequence of the protein or because the change does not impair its function. But if a gene product is essential for development of the organism, at least some mutations must have deleterious effects.

Early attempts to deduce total gene numbers were based upon comparing the rate at which recessive lethals are generated at specific loci with that for the entire X chromosome of D. melanogaster. Using X irradiation to induce mutations, Demerec (1933, 1934) found that the average rate of visible and lethal mutation at 22 loci was 2.46×10^{-4} per locus; X chromosomes bearing deleterious mutations were generated at a rate of 0.125. This would correspond to the presence of about 500 genes capable of mutation to give deleterious alleles. Other experiments using similar approaches suggested totals of around this value (Muller, 1927; Muller et al., 1949; Dobzhansky and Wright, 1941; Herskowitz, 1950). The more detailed analysis of the 3A2-3C2 region now suggests, of course, that the total number of lethal groups should be close to the number of bands, about 1000, corresponding to some 5000 groups in the entire genome.

In most organisms it is not possible to isolate sex-linked lethals with the same facility as with Drosophila. However, a systematic attempt to isolate mutants of the nematode C. elegans has been reported by Brenner (1974). Following treatment with the mutagen EMS, visible mutants were generated at a rate of 5×10^{-4} per gene. The frequency of 0.15 with which lethal X chromosomes were induced would therefore correspond to 300 genes in toto, so long as lethal mutations are induced at the same rate as visible mutations. C. elegans has 6 chromosomes; an exact comparison of sizes cannot be accomplished as with the Drosophila polytene set, but the X chromosome represents 15% of the total genetic map (which is admittedly incomplete, based on some 100 loci). This suggests a total content of about 2000 loci in which recessive lethal mutations can occur. The genome of C.

elegans is 8×10^7 base pairs, of which some 6.7×10^7 base pairs is nonrepetitive. This corresponds to about 30,000 base pairs (2×10^7 daltons) for every lethal complementation group, a value close to that seen with D. melanogaster (Table 17.1).

It has often been assumed that most mutations that cause a change in amino acid sequence will prove to be deleterious to the protein function. This is tantamount to saying that observed mutations represent most (around 70%) of the alterations that arise in essential coding sequences by spontaneous or induced mutation. The assumption that mutations generally are deleterious has been the basis for calculations to estimate the maximum number of loci that a eucaryotic genome could contain without accumulating an unacceptable genetic load due to mutation. This is to say that if too many recessive lethal mutations arise in each generation, a species will rapidly become extinct.

The number of recessive lethals that is generated is the product of the deleterious mutation rate and the number of susceptible genes. The rate of mutation in higher eucaryotes is generally taken to lie in the range 10^{-5}–10^{-6} per locus per generation.* This appears to be valid for visible and lethal mutation at many individual loci in D. melanogaster (Muller, 1927; Muller et al., 1949). Estimates for the spontaneous rate at which recessive lethals

---

*The mutation rate per generation may depend on mutations occurring both as errors during replication of the genome and as time-dependent random events in nonreplicating DNA. In this case the number of mutations per gamete will be a function of the error rate and the number of replications and of the period of time available for mutations to accumulate. A simple model is to consider the mutation rate per gamete in terms of the observed rate of mutation per cell division and the number of cell divisions that separate successive generations of gametes. This predicts that the rate per gamete should be correspondingly greater than the rate per division. Thus in D. melanogaster, if 40 cell generations separate successive formations of gametes, a mutation rate per locus of 1×10^{-6} per gamete would correspond to a rate of about 2×10^{-8} per cell cycle (following Drake, 1969). This is only approximate, since different stages in gamete formation may display different susceptibilities to mutation, due to varying efficiencies of repair systems. For example, the ability of mouse oocytes to repair damage to the resident genome is well characterized; more recently it has been shown also that damage to the male genome can be repaired following fertilization, at a frequency strongly dependent on genotype (Russell, Russell, and Kelly, 1958; Russell, 1967; Generoso et al. 1979). The data on mutation rates per mammalian cell cycle summarized in Table 6.6 generally fall into a range of 10^{-7}–10^{-8}; but they are insufficient to say whether this is a generally applicable rate, and the rate per gamete of 10^{-5}–10^{-6} quoted in the text is not widely enough based to determine whether there is indeed a discrepancy of some 100-fold. The rate for bacteria is in the range 10^{-6}–10^{-7} per locus per replication cycle; but this cannot be compared directly with eucaryotic rates since the target size of the eucaryotic locus is not known. It has been suggested on the basis of sensitivities to X irradiation that the induced mutation rate per locus increases with C value, which would imply that the target size depends on genome size (Abrahamson et al., 1973). Since the error rate, efficiency of repair processes, pertinent number of cell generations, and period between gamete formations, all may be expected to take different values in different species, there is no a priori reason to expect similar mutation rates per locus per gamete, unless there is selection for some overall rate of evolutionary change (which would imply an unlikely degree of teleology).

are generated in the X chromosome depend on genotype, but generally lie in the range 1-3×10^{-3} per generation (Demerec, 1929). Reversing the usual argument about gene numbers, if there are 1000 susceptible loci on this chromosome, the average rate of mutation per locus must be about 1-3×10^{-6} per generation. In mammals, data are available principally for mice, in which a rate of $\leqslant 10^{-5}$ per gamete has been suggested (Russell et al., 1958).

The overall rate of molecular evolution seems to be consistent with this rate of mutation. Proteins evolve at characteristically different rates (see Table 11.3), but a general rate of 10^{-8}-10^{-9} amino acid changes per generation has been suggested (Ohta and Kimura, 1971). This is similar to estimates of the rate at which nucleic acid sequences have evolved, which in comparisons of man and monkey DNA or of rat and mouse DNA appear to be about 10^{-8}-10^{-9} nucleotides per generation (Kohne, 1970; Rosbash; Campo, and Gummerson, 1975). In the latter case, the rate at which sequences represented in mRNA change appears to be about half of that displayed by other nonrepetitive sequences. It should be noted, however, that these calculations assume a generation time of 1 year. In any case, these molecular rates would generate changes per locus in the range of 10^{-5}-10^{-6} per generation.

If the mammalian genome of 3×10^{9} base pairs comprised 2×10^{6} genes of average length 1500 base pairs each, the number of loci mutated in each generation would be in the range of 2-20 per genome. Ohno (1971) suggested that this would constitute an unacceptable genetic load if all these mutations were deleterious to coding sequences. The implication is that the number of susceptible loci must be much less, which means that only a small part of the genome can code for essential proteins. This is consistent with the earlier suggestions of Muller (1950, 1956, 1967) that the number of recessive lethals actually arising each generation in man is of the order of 0.1-0.5 per genome. (But we should note that this estimate was based principally on work with Drosophila.)

An alternative way out of the dilemma apparently posed by the mutation rate is to suppose that not all mutations are deleterious. This is to follow the view of evolutionary change which supposes that most mutations are selectively neutral, so that there are bound to be polymorphisms at neutral sites; and in which case there is no need to suppose that deleterious mutations constitute a proportion of the total that is high enough to impose any restriction on the number of coding sequences in the genome (Kimura and Ohta, 1971).

To summarize these arguments, given the range of mutation rates over an order of magnitude, uncertainties about the proportion of mutations that is deleterious, and the lack of information on what would constitute an unacceptable genetic load, it seems doubtful whether this evidence is adequate to support any conclusion about the number of genes.

Even if an unknown proportion of mutations in a given gene is not deleterious because the protein product can tolerate certain changes in sequence,

it has been assumed that if the protein is necessary for viability, some mutations must be lethal (such as the extreme case of deletion). But in addition
to such "essential" genes, are there also genes whose products are dispensable and whose absence therefore does not impair viability? The only satisfactory criterion for considering a gene to be dispensable is indeed to demonstrate that deletion is without effect; the isolation of innumerable point
mutations within a gene that are not lethal does not prove that others with
lethal effects cannot be found. The w gene of the D. melanogaster 3A2-3C2
region is an example that can be deleted; and the failure of any of the large
number of lethal mutations of this region to fall into the z complementation group suggests that this locus also may be incapable of generating
lethals. Both of these loci therefore stand identified only by virtue of their
visible effects on eye color. This prompts the question of how many other
loci there may be at which lethal mutations do not occur; and of whether
some of these may in fact be without discernable effect on the phenotype.

Only a comparatively small number of the genetic loci of D. melanogaster can be identified with a gene product. O'Brien (1973) pointed out
that of the then roughly 30 such loci, 13 had been shown to possess "null"
alleles which completely eliminate the protein product, but which are not
lethal when homozygous (for a more recent summary see MacIntyre and
O'Brien, 1973). Only 4 of these loci have alleles that produce visible recessive phenotypes, leaving the implication that the other 9 loci would not
have been discovered had there not been screening to identify the genes of
their specific enzymes. While it would be rash to make quantitative estimates for the proportion of such loci from these limited data, they do raise
the possibility that the number of genes may be greater than can be identified by lethal and visible mutations.

It is difficult to see how such loci might be maintained in the genome,
since if the gene product is truly dispensable, there should be no selective
pressure against the accumulation of mutations abolishing enzymatic activity, and indeed eliminating the protein altogether. Perhaps these loci are
dispensable under the conditions in which the null alleles were isolated,
but do have selective value elsewhere, for example in flies of other genotypes; they could constitute parts of duplicate pathways in which more than
one locus must be inactivated to obtain a detectable change in phenotype.
Perhaps a small amount of protein product remains (this must be less than
5%), but this is adequate to supply the organism's needs. Perhaps the loci
really are dispensable, but constitute a small, atypical part of the genome.
In any case, to account for the full amount of potential coding DNA in D.
melanogaster that does not represent genes, the ratio of dispensable to essential loci would have to be about 20. Although there is evidence for the existence of null alleles, it does not suggest an effect of this magnitude sufficient
to account for the apparent excess of DNA.

CHAPTER 18

DNA Sequence Organization: Nonrepetitive and Repetitive DNA

The constitution of the eucaryotic genome is one of the most rapidly advancing topics discussed in this book. When the first edition was written in 1973, the organization of DNA sequences could be analyzed only in somewhat general terms; it was possible to define different sequence components in the genome and to characterize the products of transcription at stages subsequent to the act of RNA synthesis, but detailed analysis of the organization and expression of individual genes could not be accomplished. However, recent advances have made it possible now to approach the structure of the genome directly at the molecular level, so that in this volume we may discuss the organization of coding sequences per se and the context within which they reside. As well as analyzing sequences within a given genome, it is possible to compare the organization of corresponding sequences in the genomes of related organisms, so that evolutionary change can be followed at the molecular level at which it occurs.

This striking progress rests upon the combination of developments in several techniques. Nucleic acid hybridization not only allows the sequence components of the genome to be characterized, but means that individual genes can be recognized through the identification of their messenger RNA products. Cleaving DNA with restriction enzymes allows a genome to be divided into defined fragments; and those corresponding to particular regions can be inserted into bacterial plasmids or phages and amplified. In fact, it has become possible to obtain the DNA corresponding to any structural gene whose messenger RNA can be isolated. By using reverse transcription to prepare a duplex DNA copy of the mRNA (or any precursor to mRNA), the same cloning technique can be applied to any RNA product of the gene, so that the structural gene obtained from the genome can be compared directly with its products. Indeed, recent advances in sequencing DNA mean that it is practical to determine the nucleotide sequences of genes and their products from the cloned fragments.

The development of nucleic acid hybridization into a variety of techniques for analyzing DNA and RNA sequences has taken place over the past decade. The ability of complementary single-stranded nucleic acid molecules to form duplexes has been applied both to the genome itself and to the products of transcription. In this chapter we shall discuss the reassociation of previously separated DNA strands, which defines three general components in the eucaryotic genome. Here we are concerned in detail with the properties of nonrepetitive DNA, which represents sequences unique in the haploid genome, and moderately repetitive DNA, which occurs as a multiplicity of related though not identical sequences. Principally we shall discuss the extent to which genomes consist of these components and the organization of their constituent sequences. In the next chapter we shall analyze the structure and organization of the highly repetitive sequences, which occur as blocks of tandemly repeated variations of usually rather short sequences.

We shall discuss the general principles of DNA reassociation in this chapter; and we shall turn in Chapter 24 to the techniques that have been developed for following the hybridization of RNA with DNA.* Applied to mRNA populations, these allow the nonrepetitive component of the genome to be identified as the residence of most structural genes; and they allow the total number of such genes that is expressed in any cell to be determined. When individual messengers are available, corresponding fragments of the genome may be isolated by hybridization. We shall discuss the characterization of such fragments in Chapter 26.

Denaturation and Renaturation of DNA

It has been known for many years that when a solution of DNA is heated, striking changes occur in many of its physical properties, such as viscosity, light scattering, and optical density. This *denaturation* or *melting* of DNA occurs over a narrow temperature range and represents the disruption of the double helix into its complementary single strands (see Marmur, Rownd, and Schildkraut, 1963). The process usually is characterized by the temperature of the midpoint of transition, T_m, which is sometimes described as the *melting temperature*.

The denaturation of DNA originally was followed by optical density. The heterocyclic rings of the nucleotides adsorb ultraviolet light strongly, with a maximum around 260 nm; but the absorption of DNA itself is some 40% less than would be expected from a mixture of free nucleotides of the same composition. This is described as the *hypochromic effect;* although

---

*Reactions involving DNA usually are referred to as *reassociation* or *renaturation*, since the single strands that are reacting are reforming the duplex state in which they originally existed. Reactions between DNA and RNA usually are referred to as *hybridization*, since the product consists of two types of nucleic acid. Apart from a possible difference in the rates of DNA-DNA and DNA-RNA reactions, there is no difference in the governing principles.

the cause is not completely understood, it appears to result from mutual interactions of the electron systems of the bases and is thus at its greatest when they are stacked in the parallel array of the double helix. The degree of hypochromocity is a sensitive measure of the physical state of DNA, since any departure from the ordered configuration of the double helix is reflected by a loss of hypochromocity—or increase of *hyperchromicity*—that is, an increase in optical density.

When DNA which has been denatured by heating is cooled, the ultraviolet absorbance decreases again, typically to a level some 12% above that of the original solution. Doty et al. (1960) found that when this solution is in turn reheated, the optical density increases again, but in a manner that depends upon the speed with which the denatured DNA was cooled. After a quick cooling, reheating results in a gradual increase in optical density with no sudden transition. This suggests that although complementary regions of DNA have associated, they have formed a variety of probably rather short duplex regions, which melt over a wide range of individual transition temperatures to generate the gradual increase in optical density. But when the denatured solution is cooled more slowly, on reheating there is a characteristic transition close to the original T_m. This suggests that slow cooling allows a substantial proportion of the original duplex molecules to reform to give a DNA that more closely resembles native DNA; it is this process that is usually described as *renaturation*.

Early experiments in which renaturation was followed by buoyant density also supported the idea that fast cooling produces a random aggregation whereas slow cooling permits formation of longer duplex regions. Single stranded and duplex DNAs have different characteristic buoyant densities, in both cases depending on base composition. For duplex DNA, buoyant density increases more or less in proportion to the content of G-C base pairs, from 1.700 g-cm^{-3} at 42% G-C to 1.718 g-cm^{-3} at 59% G-C. The empirical formula

$$\rho = 1.660 + 0.00098 \ (\% \ \text{G-C}) \tag{1}$$

often is used to deduce the base composition of duplex DNA from its buoyant density. The buoyant density of single strands is greater, so that denaturation can be followed by the generation of peaks of increased density; in the example of E. coli, the change is from 1.710 g-cm^{-3} for duplex DNA to 1.725 g-cm^{-3} for single strands (Meselson, Stahl, and Vinograd, 1957). At all events, molecules renatured slowly regain a buoyant density close to that of the native DNA before denaturation, whereas rapidly cooled molecules have a higher density, such as might arise from the random aggregation of complementary parts of several different strands along one given partner strand.

A technique more recently developed to separate single stranded and duplex DNA is chromatography on hydroxyapatite. The ability of hydroxyapatite to retain DNA depends on the phosphate ion concentration. At very low concentrations (~0.001 M phosphate buffer) both duplex molecules and single strands are bound. At the level of 0.12 M phosphate buffer, hydroxyapatite retains only duplex DNA, while single strands are eluted. The denaturation of DNA can therefore be followed by applying duplex molecules to a hydroxyapatite column and following the release of single strands as the temperature is increased; this is known as *thermal elution*. The melting temperature determined on hydroxyapatite is described as the T_i; it is greater than the T_m by about 2°C, because a molecule must be denatured in its entirety before the single strands are released.

Reassociation of a solution of complementary single strands of DNA can be assayed by the amount of DNA retained by hydroxyapatite under appropriate buffer conditions. Hydroxyapatite chromatography offers the advantage that both single stranded and duplex molecules may be isolated from a given preparation of DNA; for the duplex molecules retained in lower salt can be eluted by increasing the concentration to the range of 0.27 M phosphate buffer; this is known as *salt elution*.

Figure 18.1 shows the melting curves of several DNAs followed by increase in optical density; all display the same general melting profile, although each has a characteristic T_m. Figure 18.2 shows that the difference is correlated with the G-C base pair content of the DNA. The increase in duplex stability with G-C content depends partly, although not entirely, on the difference between the three hydrogen bonds of the G-C pair and the

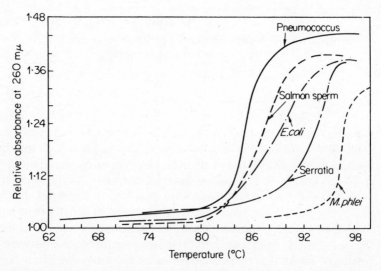

Figure 18.1. Melting curves of DNA from various sources. Each DNA has a characteristic T_m, but all the melting curves take the same general form. Data of Marmur and Doty (1959).

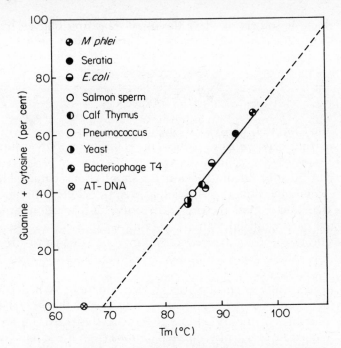

Figure 18.2. Dependence of melting temperature on base composition of DNA. Stability increases with G-C content. Data of Marmur and Doty (1959).

two hydrogen bonds of the A-T base pair. The effect is linear and the T_m increases about 0.4°C for every percent increase in G-C content. The melting temperature also is influenced by ionic strength and increases some 16.6°C for every tenfold increase in monovalent cation concentration. Taking these two effects into account, we can write the general equation:

$$T_m = 0.41 \ (\% \ \text{G-C}) + 16.6 \log M + 81.5 \tag{2}$$

where M is the concentration of monovalent cations between 0.0001 M and 0.2 M. The T_m is influenced also by organic solvents that assist denaturation; the most commonly used of these has been formamide. The T_m is reduced by 0.6°C for every 1% formamide. At physiological ionic strength, the T_m usually lies in the region of 80–90°C; the most common experimental condition is the use of 0.12 M phosphate buffer, which provides a monovalent Na^+ concentration of 0.18 M. However, it is possible to reduce the T_m to as low as 40°C by the use of high concentrations of formamide.

Another parameter that influences thermal stability is the length of the DNA. The melting temperature is decreased by a reduction in length (because less energy is required to separate the partner strands), but this effect is smaller than the others we have discussed. Shearing native DNA to a standard length of 450 nucleotides reduces its stability by only 1–2°C, according to the formula:

$$\Delta T_m = \frac{650}{L} \tag{3}$$

where L is the length of the sheared fragments.

Any mispairing between bases reduces the stability of a DNA duplex. In a duplex of native DNA, comprising two strands that are perfectly paired, every base pair must be disrupted before the strands can separate. The presence of any nonmatched pairs of bases that are not held together by hydrogen bonds therefore reduces the energy required for separation. Since G-C and A-T base pairs contribute different stabilities to a duplex, the magnitude of the effect depends on base composition; for a DNA consisting of 40% G-C pairs, it is usually reckoned that every 1.0% mispairing reduces the T_m by about 1°C (Bonner et al., 1973).

Mispairing generally is encountered in two situations. When DNA has been denatured and renatured, the presence of sequences that are related but not identical results in the formation of imperfectly paired duplex molecules. The reduction in T_m of such preparations from the value characteristic of native DNA provides a measure of the extent of mispairing that has occurred during renaturation; and from this, the degree of relationship between the mispaired sequences can be inferred. Similarly, when two heterologous DNAs are reassociated together, the reduction in T_m provides a measure of their relatedness, a technique that can be applied to the genomes of different organisms.

Complexity of Nonrepetitive DNA

The reassociation of two complementary sequences of DNA depends upon random collision and follows second-order kinetics (Wetmur and Davidson, 1968). The rate of reaction is therefore governed by the equation:

$$\frac{dC}{dt} = -kC^2 \tag{4}$$

where C is the concentration of DNA that is single stranded at time t and k is the reassociation rate constant. Integrating this equation between the limits of the initial concentration of DNA, C_0, at time $t = 0$, and the concentration C remaining single stranded after some time t, gives:

$$\frac{C}{C_0} = \frac{1}{1 + k \cdot C_0 t} \tag{5}$$

When the reaction is half complete,

$$\frac{C}{C_0} = \frac{1}{2} = \frac{1}{1 + k \cdot C_0 t_{1/2}}$$

so that:

$$C_0 t_{1/2} = \frac{1}{k} \tag{6}$$

The reassociation of any particular DNA may therefore be described by the rate constant k, given in units of liters-nucleotide moles^{-1}-sec^{-1}, or in the form of its reciprocal, $C_0 t_{1/2}$, given in nucleotide moles $\times$ seconds/liter.

The colloquial expression "Cot" often is used to describe the progress of the reaction in terms of the $C_0 t$; and similarly the value required for half reassociation is referred to as the Cot$_{1/2}$. Since this conveniently expresses the concept that the parameter controlling the reaction is the product of DNA concentration and time of incubation, we shall use this form here (in general omitting its units for simplicity). Reassociation of DNA usually is followed in the form of the *Cot curve* introduced by Britten and Kohne (1968). This plots the fraction of DNA that remains single stranded (C/C_0) or the fraction that has reassociated into duplex molecules ($1 - C/C_0$) against the log of the Cot.

From the several Cot curves shown in Figure 18.3 it is immediately obvious that the Cot required for DNA reassociation depends upon the size of the genome. This occurs because as the genome becomes more complex, the number of copies of any particular sequence within a given mass of DNA is reduced. For example, comparing a bacterial genome of 0.004 pg with a haploid eucaryotic genome of 3 pg, it is clear that 12 pg of DNA will contain 3000 copies of each sequence in the bacterial genome, but will have only 4 copies of each eucaryotic genome. Thus in a solution containing the same absolute concentration of DNA measured in moles of nucleotides per liter, each bacterial sequence actually will be present in a concentration 750 times greater (3000/4) than each eucaryotic sequence. Put another way, to obtain the same concentration of each nucleotide sequence in these two examples, it is necessary to have an absolute concentration of eucaryotic DNA that is 750 times greater than that of the bacterial DNA. Thus the Cot$_{1/2}$ for the eucaryotic reaction will be 750 times the Cot$_{1/2}$ of the bacterial reaction.

Implicit in this example is the concept that it is the concentration of each reassociating sequence that determines its rate of renaturation. Conversely, the Cot$_{1/2}$ of a reaction indicates the total length of different sequences present. This is described as the *complexity* and is usually given in base pairs, but can be expressed in daltons or any other mass unit. The complexity is independent of the number of copies of each sequence present. Thus if a given genome consists of 1 molecule of a sequence a nucleotides long, 10 copies of a sequence b nucleotides long, and 100 copies of a sequence of

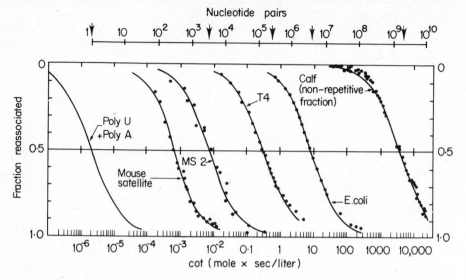

Figure 18.3. Cot curves for reassociation of DNAs of various complexities. The correlation between Cot and sequence complexity is indicated by the axis showing nucleotide pairs, on which arrows mark the size of each genome; ideally these points should coincide with the $Cot_{1/2}$ values shown by the points where the Cot curves cross the 0.5 reassociation line. These data were plotted by Britten and Kohne (1968) by following the decrease in optical absorbance as renaturation proceeds in 0.12 M phosphate buffer.

c nucleotides, the complexity is $a + b + c$. The *repetition frequencies* in the genome of sequences a, b, and c are 1, 10, and 100, respectively.

When the renaturation of two DNA sequences is compared, each should therefore display a $Cot_{1/2}$ proportional to its complexity. This means that the complexity of any DNA can be determined by comparing its $Cot_{1/2}$ with the $Cot_{1/2}$ of a standard of known complexity. Usually the standard used is E. coli DNA, whose complexity is taken to be identical with the length of the genome (that is to say that every sequence in the E. coli genome of 4.2×10^6 base pairs is assumed to be unique). Thus we can write:

$$\frac{\text{Cot}_{1/2} \text{ (any DNA)}}{\text{Cot}_{1/2} \text{ (E. coli DNA)}} = \frac{\text{complexity (any DNA)}}{4.2 \times 10^6} \tag{7}$$

Figure 18.4 compares the renaturation of calf thymus DNA with E. coli DNA. It is apparent that the bovine reaction takes place over a much wider range of Cot values than the bacterial reaction. This indicates that bovine DNA contains more than one sequence component, since a single kinetic component must always renature over a range of Cot values spanning no more than two orders of magnitude between 10% and 90% reaction. (The ratio of the Cot values needed to drive the reaction to roughly 10% and 90% completion can be seen to be about 100 by substituting appropriate values in the second-order equation—see curves in Figure 18.3).

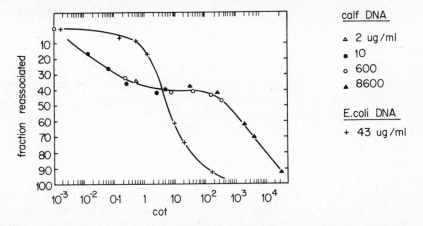

Figure 18.4. Reassociation of denatured calf thymus DNA. The reaction has (at least) two components; and in order to follow reassociation over the necessary range of Cot values, different concentrations of DNA are combined with appropriate periods of incubation; the symbols indicate the DNA concentrations used for each Cot point. Data of Britten and Kohne (1968).

Three reassociating components can be distinguished in the bovine curve of Figure 18.4. The *slow component* renatures between Cot values of about 10^2 and 10^4. This represents 60% of the DNA and its span of little more than two orders of magnitude suggests that it is kinetically homogeneous. Little renaturation appears to take place over the plateau between Cot values ranging from 10 to 10^2. An *intermediate component* renatures over a range from Cot 10^{-2} to 10; probably some of this component renatures before the first point at Cot 10^{-2}, so that it appears to span too wide a range of Cot values to be kinetically homogeneous. This component must occupy some or all of the first 40% of the DNA to reassociate. However, there may also be a *fast component*, since even at the first point of Cot 10^{-2} almost 20% of the DNA has renatured; it would be necessary to extend the Cot curve to much lower values to resolve the number of components in this fraction.

The slow component has a $Cot_{1/2}$ of about 4000, much greater than that of E. coli DNA, which is about 6.5 under these conditions. Equation (7) gives the complexity for any single kinetic component and can be applied directly to this example after compensating for the presence of other components in the Cot curve. This gives a complexity for the slow component of 1.6×10^9 base pairs. The bovine genome appears to comprise some 3.2×10^9 base pairs by chemical analysis, so a component representing 60% of it should correspond to 1.9×10^9 base pairs. The close agreement between the complexity measured by reassociation and the physical length implied by the genome size suggests that the slow component comprises sequences that are unique in the haploid genome.

The intermediate component has a $Cot_{1/2}$ of about 0.03, much lower than that of E. coli DNA, and therefore corresponding to a less complex sequence. Assuming for the purposes of calculation that this component occupies the remaining 40% of the genome, its complexity is about 8×10^3 base pairs. But 40% of the genome should represent a length of 1.3×10^9 base pairs. This implies that the intermediate component must consist of a sequence of some 8×10^3 base pairs repeated some 160,000 times in each haploid genome. Given the apparent heterogeneity of this component, these figures are at best only averages, which is to say that at face value this analysis identifies sequences whose total length is 8×10^3 base pairs in the genome and which on average are repeated 1.6×10^5 times. These values are not in fact accurate, since the Cot curve starts at 20% reassociation; but they illustrate the nature of calculation of the properties of intermediate components.

Formally the same analysis may be applied to the slow component, which would then be considered to consist of a sequence of 1.6×10^9 base pairs repeated an average of 1.2 times in order to occupy its length in the genome of 1.9×10^9 base pairs. Obviously the lowest degree of repetition actually possible in the genome is 2; and thus the value of 1.2 times repetition may be considered to fall within the experimental error expected for the 1.0 repetition of sequences that are unique. The discrepancy between the complexities of the slow and intermediate components is so great that for practical purposes the complexity of the eucaryotic genome may be equated with that of the slowest reassociating component.

Proceeding to a more formal analysis, suppose a genome of size G base pairs contains a slow component whose complexity of N_{nr} base pairs consists of sequences present in only one copy per haploid genome. Then if this *nonrepetitive* or *single copy* component occupies a proportion α of the genome,

$$N_{nr} = \alpha \cdot G \tag{8}$$

If this genome contains an intermediate (or fast) component which consists of a sequence of complexity N_r base pairs repeated F times in each haploid genome, and if this *repetitive* component occupies a proportion β of the genome,

$$N_r = \frac{\beta \cdot G}{F} \tag{9}$$

This equation may be applied to each repetitive component in a genome containing more than one.

Values for N_{nr} and N_r can be determined experimentally from the $Cot_{1/2}$ values characterizing their respective sequence components. The $Cot_{1/2}$ as measured for each component during renaturation of the whole genome is determined by the C_0, which is calculated from the total amount of DNA. Thus the actual concentration of sequences concerned with the reaction of each independent component is correspondingly less; in fact, if each component were purified, its reassociation would occur at a lower Cot, because all the sequences included in the C_0 value would participate in the reaction. Thus to obtain the $\text{Cot}_{1/2}$ of a pure component for comparison with a standard, the $\text{Cot}_{1/2}$ observed in the renaturation of whole DNA is multiplied by the proportion of the reassociation reaction represented by the component. For nonrepetitive and repetitive components:

$$N_{nr} = \alpha \cdot \text{Cot}_{1/2}^{nr} \text{ (obs. whole DNA)} \cdot \frac{4.2 \times 10^6}{\text{Cot}_{1/2} \text{ (E. coli)}} \tag{10}$$

$$N_r = \beta \cdot \text{Cot}_{1/2}^{r} \text{ (obs. whole DNA)} \cdot \frac{4.2 \times 10^6}{\text{Cot}_{1/2} \text{ (E. coli)}} \tag{11}$$

Comparing equations 8 and 10 for N_{nr}, we see that:

$$G = \text{Cot}_{1/2}^{nr} \text{(obs. whole DNA)} \cdot \frac{4.2 \times 10^6}{\text{Cot}_{1/2} \text{ (E. coli)}} \tag{12}$$

This means that the size of any genome can be determined directly by comparing the $\text{Cot}_{1/2}$ of its nonrepetitive component with that of the bacterial standard.

Comparing equations 9 and 11 for N_r, we see that:

$$\frac{G}{F} = \text{Cot}_{1/2}^{r} \text{(obs. whole DNA)} \cdot \frac{4.2 \times 10^6}{\text{Cot}_{1/2} \text{ (E. coli)}} \tag{13}$$

Now dividing equation 12 by 13,

$$F = \frac{\text{Cot}_{1/2}^{nr} \text{ (obs. whole DNA)}}{\text{Cot}_{1/2}^{r} \text{ (obs. whole DNA)}} \tag{14}$$

This means that the repetition frequency of any repetitive fraction may be obtained directly by comparing the $\text{Cot}_{1/2}$ values determined for nonrepetitive and repetitive DNA in the same renaturation curve. In effect, this uses the nonrepetitive component as an internal standard, assuming that it represents the reassociation of unique sequences (formally, its repetition frequency = 1.0).

Applying equation 12 to the example of Figure 18.4 gives a size for the bovine genome of 2.6×10^9 base pairs, reasonably close to the chemical value of 3.2×10^9. Equation 14 suggests a repetition frequency for the intermediate component of 130,000 times (which is slightly less than the estimate of 160,000 obtained above by proceeding through equation 11, because the latter is calculated on the same basis that sets a frequency for the nonrepetitive fraction of 1.2 rather than 1.0).

Appendix 2 summarizes available data to compare the sizes of eucaryotic genomes estimated by reassociation kinetics with the values determined by chemical analysis. In almost every case there is close agreement, confirming that the $\text{Cot}_{1/2}$ of any single kinetic component of DNA is indeed proportional to its complexity, and demonstrating that the slow components of these genomes consist of sequences that are unique in the haploid genome. The concordance of kinetic and chemical data is plotted in Figure 18.5.

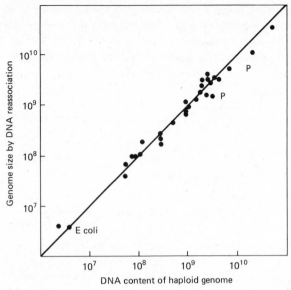

Figure 18.5. Relationship between genome size estimated by reassociation kinetics and the haploid DNA content determined by chemical analysis. Polyploid genomes are indicated by a P. From the data of Appendix 2.

Kinetics of DNA Reassociation

For the Cot values of two different preparations of DNA to be compared, they must be measured under the same conditions of ionic strength, temperature, and fragment size. If two preparations are renatured under different conditions, their $Cot_{1/2}$ values can be compared only after a suitable correction is made to allow for any differences in renaturation rate resulting from the conditions of incubation. The standard conditions for renaturation now comprise incubation of single strands of average length 450 nucleotides in 0.12 M phosphate buffer (corresponding to 0.18 M Na^+) at 60°C.

When the reaction is performed under different ionic conditions, the observed $Cot_{1/2}$ is multiplied by an empirical correction factor, γ, which is the ratio of the reassociation rate under the experimental conditions relative to that under the standard conditions (Britten et al., 1974). The resulting value is described as the *equivalent Cot* (occasionally referred to as ECot) and reassociation reactions usually are quoted in terms of equivalent Cot values. With increasing ionic strength, renaturation proceeds more rapidly, so that at (for example) 0.24 or 0.48 M phosphate buffer, the equivalent Cot is obtained by multiplying the observed Cot by 2.9 or 5.6 times, respectively.

Reassociation of two complementary single strands to give a duplex takes place in two stages. First a comparatively short length of duplex is formed by a nucleation reaction. This appears to be rate limiting. Then a rapid zipper-like reaction brings the remaining sequences into the base paired conformation. Given this model, the rate of reassociation should be proportional to the lengths of the reassociating single strands. However, the observed dependence is on $\sqrt{L}$ (Wetmur and Davidson, 1968; Hutton and Wetmur, 1973a). This may reflect an excluded volume effect, in which increasing length of a single strand makes it more difficult for a second strand to interpenetrate to find complementary sites. To convert Cot values determined at one fragment length to another, the formula used is therefore:

$$\text{Cot}_{1/2}^{L1} = \text{Cot}_{1/2}^{L2} \cdot \sqrt{\frac{L2}{L1}} \tag{15}$$

Temperature controls the *stringency* of the hybridization reaction. The higher the temperature, the more precisely matched two sequences must be to anneal with each other. By lowering the temperature at which a reassociation reaction is performed, it is therefore possible to allow duplexes to form with greater degrees of mispairing. For unique sequences, the rate of reassociation is at a maximum in a range between 15 and 30°C below the T_m, as shown in Figure 18.6. The difference between the T_m and the temperature of incubation is sometimes referred to as the *criterion*; and this is usually set at 25°C, which in practice means that reassociation reactions are performed at 60°C.

The high Cot values required for reassociation of nonrepetitive eucaryotic DNA often require incubation for extensive periods of time, when high

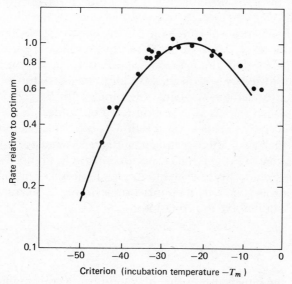

Figure 18.6. Dependence of renaturation rate on temperature of reassociation. Data obtained for T4 and T6 DNA by Bonner et al. (1973).

temperatures may tend to cause chain breakages and depurination reactions in the DNA. This can be avoided by performing the reaction in the presence of formamide to lower the T_m. Thus in 48% formamide, renaturation at 37°C lies within the optimum temperature range (McConaughy, Laird, and McCarthy, 1969).

The effect of the proportion of G-C base pairs upon the reassociation of DNA is not clear. By correlating G-C content with the discrepancy between sequence complexity and genome size for several DNAs, Wetmur and Davidson (1968) suggested that increasing G-C content accelerates the rate of reassociation. An apparent linear relationship for some time has been used in the form of the equation:

$$\text{Cot}_{\text{corrected}} = \text{Cot}_{\text{observed}} \times [1 - 0.018 \, (51 - \% \, \text{G-C})] \qquad (16)$$

taking E. coli DNA with 51% G-C as the standard to which Cot values are corrected. However, this rests on the reassociation of DNAs of widely varying complexities and a more satisfactory demonstration of such a relationship would require the use of a better defined series of DNAs. Some subsequent work has failed to detect an effect of G-C content on reassociation (Gillis and De Ley, 1975); and here we shall not introduce a compensation for changes in G-C content. Indeed, the data of Appendix 2 are plotted without reference to G-C content; the correlation between reassociation rate and genome size shown in Figure 18.5 therefore suggests that any effect may be small.

The renaturation reaction may be followed either by the decrease in optical density as duplex regions reform or by using hydroxyapatite. When hydroxyapatite is used, all molecules containing duplex regions at a given Cot are retained, even though only part of their length may be double stranded. This retention of unrenatured single-strand tails along with the reassociated duplex regions means that the reaction appears to proceed more rapidly. Generally the reaction appears twice as fast on hydroxyapatite as when followed by optical density. Thus the $\text{Cot}_{1/2}$ for E. coli DNA of 4.0 that is seen under the standard conditions with hydroxyapatite is increased to 8.0 when the reaction is monitored optically.

Another procedure that is used to follow reassociation is to measure duplex regions surviving after treatment with the S1 nuclease derived from Aspergillus oryzae. This enzyme is specific for single-stranded DNA; it will remove single-strand tails attached to duplex molecules without degrading the duplex regions. The kinetics of reaction followed by S1 nuclease are not second order, but observe the equation:

$$\frac{S}{C_0} = \frac{1}{(1 + k\text{Cot})^{0.45}} \qquad (17)$$

where k takes the same value as that of equation 5. S corresponds to the proportion of unpaired nucleotides, as opposed to C in equation 5, which refers to the proportion of fragments lacking any duplex region. This means that the reaction followed by S1 nuclease proceeds more slowly than that observed on hydroxyapatite (Smith et al., 1975; Britten and Davidson, 1976). The retardation appears to result from a reduction in the rate of reaction of the single-strand tails of partially duplex molecules, compared with the rate of reaction displayed by free single strands. Partly this is due to a reduction in the length of available single stranded regions as they enter duplexes; partly it represents a particle effect caused by the proximity of duplex regions to the available single strands. Indeed, a similar effect is seen also in the reaction monitored on hydroxyapatite, but it is less effective because the single-strand tails are removed from the reaction by the presence of the duplex region adjoining, whereas in the S1 reaction they must become duplex before they are recorded as having reacted.

Families of Repetitive Sequences

The reassociation of a eucaryotic DNA commonly occurs over a range of Cot values spanning up to eight orders of magnitude. The individual components of the reaction usually are not well separated; to resolve them a computer fitted program is used to generate a set of Cot curves whose sum fits the experimental data. Individual curves represent single kinetic components and are fitted to the second-order equation by least-squares analysis of the data; and the program usually is run on the basis of finding the minimum number of curves able to fit the data. An example is shown in Figure 18.7, in which reassociation of the DNA of the sea urchin S. pur-

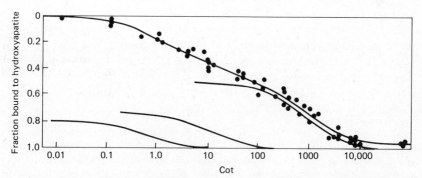

Figure 18.7. Reassociation of DNA of the sea urchin Strongylocentrotus purpuratus. The reaction was performed with 450 nucleotide long fragments at 60°C in 0.12 M phosphate buffer. The curve through the data is the best least squares solution for three second order components. The lower curves show the individual components: 50% with a $Cot_{1/2}$ of 870; 27% with a $Cot_{1/2}$ of 8.3; 19% with a $Cot_{1/2}$ of 0.53. Other solutions are possible; for example, another classification identifies the four components: 38% with a $Cot_{1/2}$ of 800; 25% with a $Cot_{1/2}$ of 43; 27% with a $Cot_{1/2}$ of 3.3; 10% with a $Cot_{1/2}$ of 0.1. Data of Graham et al. (1974).

puratus is resolved in three components. While the slowest component represents nonrepetitive DNA whose properties are clear, the two repetitive components represent averages of the behavior of their constituent sequences. These are convenient for indicating the approximate repetition that occurs, but do not necessarily correspond to the actual situation, in which presumably there are somewhat larger numbers of individual components with varying repetition frequencies. Thus it is possible that the repetitive DNA of S. purpuratus consists of a continuum of components reassociating over a range of from about 20 to 10,000 times the nonrepetitive rate, rather than the two discrete components that correspond to repetition frequencies of about 10 and 160 (Graham et al., 1974).*

Appendix 3 summarizes data on the repetitive components of eucaryotic genomes. Either two or three components are adequate to define the range of repetitive sequences; for each component the complexity represents the total length of a series of individual sequences whose repetition frequencies have the average value indicated by the $Cot_{1/2}$. The proportion of the genome occupied by the repetitive components varies very widely. This variation means that the complexity of the genome as seen in the total length of the nonrepetitive component tends to increase with overall genome size, but is not tightly related to it. For lower eucaryotes, most of the DNA usually is nonrepetitive, with as little as 10–20% falling into one or more moderately repetitive components. In animal cells, often up to half of the DNA is repetitive, including highly as well as moderately repeated sequences. In plants and amphibia, the repetitive proportion may be even greater, sometimes comprising the vast majority of the genome. The range of repetition frequencies of individual members of the moderately repetitive components usually is substantial, often from as little as 10 to as much as several thousand. Crab genomes are characterized by the apparent absence of moderately repetitive DNA. In plants that have become polyploid comparatively recently, there may be no unique sequences, so that the entire genome consists of sequences repeated from 2 or 3 to many more times.

The different components of eucaryotic DNA can be isolated by using hydroxyapatite columns to separate the duplex molecules renatured at any given Cot from the remaining single strands. Renaturation to a very low

---

*Indeed, usually it is possible to provide alternative sets of component Cot curves that fit the data all almost equally well. These may involve slight changes in the proportions and $Cot_{1/2}$ values of the same number of components; or may represent the introduction of additional components. Generally curves have been plotted with an "unrestricted fit," in which all the components are allowed to vary to best fit the data, giving the characteristics for the nonrepetitive fractions summarized in Appendix 2. More recently, "restricted fits" have been used, in which the curve for the nonrepetitive component is fixed to the rate expected for its genome size (that is, its repetition frequency is established as 1.0 as an internal control) and only the rates of the other components are varied (see Goldberg et al., 1975; Pearson et al., 1978). Broadly similar but not identical fractions are resolved by each type of analysis. This emphasizes again that the components resolved in this approach do not represent actual kinetically pure fractions, but serve rather to indicate the general properties of repetitive DNA.

Cot, generally to the order of 10^{-2}, allows only the fast fraction to form duplexes. The single strands recovered from hydroxyapatite then should contain the intermediate and slow fractions. The intermediate components can be obtained by reassociation of these single strands to a low Cot, for most DNAs in the range of 10–50. Then the duplex molecules identify the intermediate fraction; and the single strands should contain only nonrepetitive DNA. Of course, the DNA reassociating in each reaction is to some extent contaminated by other components, so that successive denaturations and reassociations may be necessary to purify each fraction.

Reassociated nonrepetitive DNA generally has a thermal profile resembling that of native DNA; it melts sharply at a T_m only slightly below that of native DNA sheared to the same size. This shows that strand association is accurate; each unique sequence anneals with its exact complement. As Figure 18.8 shows, however, reassociated repetitive DNA may melt gradually over a wide temperature range (Britten and Kohne, 1968; McConaughy and McCarthy, 1970).

The low stability of reassociated repetitive DNA shows that there must be considerable mispairing in the renatured duplex molecules. Thus the reaction does not represent perfect matching between each strand and its original complement. Rather must duplex formation take place by the association of strands that are derived from related, but not identical, sequences. This suggests the concept of Britten and Kohne (1968) that repetitive components consist of *families* of related sequences. The members of each family comprise a set of nucleotide sequences that have sufficient similarity with each other to renature, but are not identical.

A general effect of mismatching is to reduce the rate of reassociation be-

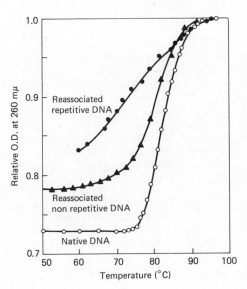

Figure 18.8. Thermal denaturation of bovine DNA fractions. Native DNA has a sharp T_m at 82°C (measured in 0.08 M phosphate buffer). Nonrepetitive DNA was prepared by removing sequences that renature before a Cot of 100; the remaining single strands were reassociated to an equivalent Cot of 3000. The renatured duplex molecules melt with a sharp T_m at 80°C; the reduction from the value of native DNA is due in part to the reduced size of the reassociated nonrepetitive DNA (which was sheared to 3400 base lengths before reassociation). Repetitive DNA was prepared by reassociation to low Cot; the renatured duplex molecules melt gradually over a range from 60 to 90°C. Data of Britten and Kohne (1968).

tween two sequences, so that a higher Cot is required for strand association. This means that the repetitive fractions observed to renature at a given Cot are less complex (and more repeated) than implied by the direct comparison with bacterial DNA. Values for the complexity and degree of repetition of these fractions, such as those shown in Appendix 3, are therefore only approximate, since the $Cot_{1/2}$ values may be overestimated by factors of 2 or more (Bonner et al., 1973).

The wide melting range of reassociated repetitive DNA implies that it contains duplexes whose degree of mismatching varies widely. Reassociation between the most closely related members of a family gives duplex molecules with extensive base pairing, which melt at a high temperature. Reassociation of the least related members of a family generates the less-well-paired duplex molecules that melt at the lowest temperature of the curve. Thermal fractionation of reassociated repetitive DNA on hydroxyapatite supports this idea that sequences of increasingly close relationship form the duplex molecules that melt at increasing temperatures. When re-associated repetitive DNA bound to hydroxyapatite is incubated at increasing temperatures, the single strands eluted at each step can be recovered. When each set of single strands then is reassociated, the resulting duplex molecules in turn have average T_m values that reflect their origins. The single strands originally eluted at 60°C, for example, reassociate to give a duplex fraction that melts at a lower T_m than that regenerated from the strands originally eluted at 70°C (Britten and Kohne, 1968). Data from a related experiment using membrane filters are shown in Figure 18.9. (This follows the earlier studies of the importance of the temperature of renaturation performed by Martin and Hoyer, 1966 and by McCarthy, 1967 and McCarthy and McConaughy, 1968.)

This suggests that each set of fragments represents the reassociation of

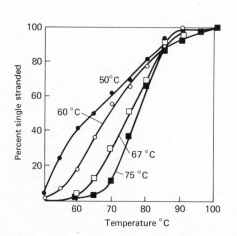

Figure 18.9. Stability of repetitive mouse DNA fractions reassociated under conditions of different stringency. Preparations of single-strand DNa were reassociated on membrane filters at the temperatures between 50 and 75°C indicated on the left. These conditions allow the renaturation only of repetitive eucaryotic sequences (see Chapter 24). The melting curve on the right shows the stability of each fraction. The higher the original temperature of reassociation, the higher the T_m of the renatured DNA. With B. subtilis DNA, reassociated fractions have the same stability irrespective of the temperature of reassociation, showing that they do not have the option of selecting appropriately related sequences, but are restricted to finding exact complements. Data of McCarthy (1967).

strands derived from sequences with characteristic degrees of divergence. From this it is clear that the apparent size of a family of repetitive sequences may be controlled by the stringency of the conditions used for reassociation. As the stringency is decreased, usually by reduction of temperature, the proportion of perfectly paired bases required to form a duplex is decreased; thus sequences increasingly less well related are allowed to reassociate. Conversely, with increasing stringency, ultimately it becomes possible for sequences to reassociate only with exact complements and not with any other members of the family. Thus reassociation of bovine DNA at a temperature only 8°C below the T_m identifies less than 10% of the genome as repetitive; whereas at 36°C below the T_m, 55% of the genome reassociates at the low Cot values characteristic of repetitive components (Britten and Davidson, 1971). Similar results are found with other organisms (Laird and McCarthy, 1968, 1969; Bendich and McCarthy, 1970; McCarthy and Church, 1970). The proportions of the genome in each sequence component generally are taken as those detected at 25°C below the T_m, the optimum rate for reassociation of nonrepetitive DNA; but this is an arbitrary condition used for convenience, so these data cannot be taken to represent any absolute partition of the genome.

However, there may also be some well-defined repetitive families, whose members diverge to a more limited extent, and which therefore are not extended further by lowering the criterion for reassociation. An example of a genome in which such families may be quite prominent is provided by S. purpuratus, in which the kinetics of reassociation are scarcely altered by a reduction in temperature from 60 to 50°C (Graham et al., 1974). In this instance, therefore, there may be comparatively few repetitive families that possess a range of divergence sufficient to allow increased reaction at lower temperature. Repetitive families that appear relatively homogeneous in the degree of divergence between their members also have been identified in D. melanogaster and in a variety of plants (Bendich and Anderson, 1977; Bouchard and Swift, 1977; Wensink, 1978).

The kinetics of renaturation reveal only the total length of all the repeated sequences within the intermediate fraction (or within each component resolved) and their average degree of repetition. However, this makes no implication about the lengths of individual repetitive sequences or their distribution within the genome. Definition of the organization of the repetitive components requires characterization of the individual families: how long are the repeating units, what is the degree of repetition and extent of divergence, and where are the various members located in the genome?

Individual repetitive families can be identified by following the reaction of a single member with the other sequences. This has been made possible by the use of cloning techniques that allow individual sequences, in this case isolated at random, to be amplified by incorporation into a bacterial plasmid (see Chapter 26); then large amounts of the cloned sequence can be recovered from the plasmid. A series of such sequences has been cloned from the DNA of S. purpuratus by Klein et al. (1978). Their repetition frequencies vary from 3 to 12,500, and the corresponding families can be classified into three groups according to their extents of divergence.

When reassociated with bulk DNA at either 55 or 45°C, the three examples comprising the first group form duplexes whose T_m is no different from that of reassociated nonrepetitive DNA, with a ΔT_m relative to native DNA of about 4°C. Presumably the families to which these sequences belong have members that are all identical or very closely related. The second group is defined by the failure of a reduction in the criterion for reassociation from 55 to 45°C to effect any difference in the melting profiles of the renatured duplexes, which have T_m values less than that of renatured nonrepetitive DNA. The seven examples investigated possessed ΔT_m values between 5.7 and 10.8°C. These are therefore moderately diverged families that do not appear to possess the more highly diverged members who would be permitted to anneal at lower temperature. The third class of families has such members: when the criterion for reassociation is reduced to 45°C, further members of the family are recognized in addition to those detected at 55°C; and the T_m of the duplexes renatured at 45°C is lower than that of duplexes formed at 55°C. The ΔT_m values for the five families identified varied from ~20 to ~25°C.

Assignment of membership to these groups is arbitrary in the sense that different criteria, for example 60 and 50°C, would alter some of the relationships; but it serves to make the point that repetitive families may vary from those in which there is in fact little divergence (which allows a reasonably precise definition of the extent of the family) to those in which there are members with at least as much as 25% divergence in sequence.

The extent of divergence between the genomes of different species also can be followed by characterizing the melting of the heteroduplex formed by annealing their DNAs. Usually the reaction is performed with an excess of DNA from one species and a small amount of labeled DNA from the other, to provide the "tracer" conditions described in the next section in which the labeled species only forms heterologous (and not homologous) duplexes. With bulk DNA, this reaction may be difficult to interpret for repetitive components because of the divergence present to begin with among the members of each family; a more powerful technique for investigating the evolution of repetitive DNA is provided by the ability to compare the reaction of a cloned repetitive sequence with the members of its family in its own and in other species. For nonrepetitive DNA the reaction may be performed directly; generally the divergence between the nonrepetitive DNAs of different species fits with their known evolutionary separation (Kohne, 1970). There is some evidence that the nonrepetitive DNA sequences represented in mRNA may diverge more slowly between species than the overall rate (Rosbash, Campo, and Gummerson, 1975; Angerer, Davidson, and Britten, 1976). The technique has been extended to investigate the degree of polymorphism in populations by comparing the renaturation of nonrepetitive DNA derived from a single (diploid) sea urchin with the reaction between the DNAs of different individuals. The ΔT_m due to reaction of different genomes appears to be about 4°C, which suggests a polymorphism of some 4% (Britten, Cetta, and Davidson, 1978).

Interspersion of Repetitive and Nonrepetitive Sequences

The ubiquity of repetitive sequences in eucaryotic genomes poses the question of what common functions they might fulfill. The highly repetitive sequences of satellite DNAs appear to have some structural role, since they are not transcribed and are clustered in distinct regions of the chromosomes (see Chapter 19). By contrast, members of intermediate components are located in all regions of the genome and may be transcribed. The characteristics of the reaction between repetitive sequences during the reassociation of DNA suggest that these sequences are interspersed with nonrepetitive sequences, in a pattern that has been found in many species. Usually the repetitive sequences are rather short, while the nonrepetitive sequences are longer.

When the reassociation of DNA is followed by retention of duplex molecules on hydroxyapatite, the progress of the reaction depends not only on the conditions of incubation, but also is controlled by the length of the DNA fragments that are used. To follow this effect, the reaction usually is performed by reassociating a radioactively labeled *tracer DNA* with a much larger amount of unlabeled *driver DNA* (or *carrier DNA*). The reaction is governed by the reassociation between single strands of driver DNA, with the tracer essentially participating at the appropriate Cot established by the driver DNA. When the tracer represents a single purified component, it reacts over only one part of the reassociation curve. For example, with a tracer of purified nonrepetitive DNA, the radioactive reaction coincides with the driver reaction over the last part of the curve. Because a given tracer fragment may renature with a single-strand region representing either an unreacted single-strand fragment or part of a fragment that has become duplex elsewhere, the reaction is governed by equation 17 rather than by equation 5.

When the length of the tracer DNA is longer than that of the driver, the rate of reaction for nonrepetitive DNA is increased over that between driver DNA fragments by the ratio of lengths, implying that all regions of the long tracer single strands are available for nucleation (Hinnebusch et al., 1977). Thus:

$$\text{Cot}_{1/2}^{\text{tracer}} = \text{Cot}_{1/2}^{\text{driver}} \cdot \frac{L^{\text{driver}}}{L^{\text{tracer}}} \tag{18}$$

When the length relationship is reversed, so that the driver is longer than the tracer, this equation no longer holds, because the reaction of the tracer is retarded by the presence of long driver fragments (this effect is probably due to interference with nucleation). An empirical equation that applies in this circumstance is:

$$\text{Cot}_{1/2}^{\text{tracer}} = \text{Cot}_{1/2}^{\text{same}} \sqrt{\frac{L^{\text{driver}}}{L^{\text{tracer}}}} \tag{19}$$

where $\text{Cot}_{1/2}^{\text{same}}$ is the $\text{Cot}_{1/2}$ that the tracer displays when annealed with driver fragments of its own length (Chamberlin et al., 1978).

Systematic studies of the effect of changes in tracer fragment length on the extent of reaction were introduced with the protocols first applied to the DNAs of X. laevis and S purpuratus by Davidson et al. (1973) and Graham et al. (1974). Figure 18.10 shows the results of an experiment in which tracers were provided by shearing labeled X. laevis DNA to give preparations of various defined lengths. After denaturation, each preparation was reassociated with an excess of a standard preparation of unlabeled X. laevis DNA of length 450 nucleotides. The reaction was followed at low Cot values at which only repetitive DNA can reassociate.

By a Cot of 50, only 40% of the driver fragments have reassociated. This

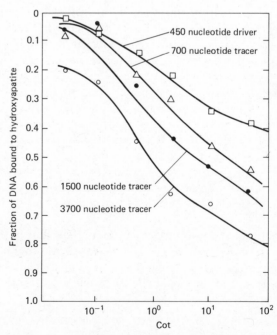

Figure 18.10. Reassociation of labeled tracer DNA of different lengths with 450 nucleotide long driver DNA of X. laevis. The reaction proceeds more rapidly (has a lower $\text{Cot}_{1/2}$) as the length of the tracer is increased. Data of Davidson et al. (1973).

corresponds to the reaction of repetitive components as followed in a standard renaturation curve and usually would be taken to define the content of intermediate repetitive DNA (compare with the two components listed in Appendix 3). But 55% of the 700 nucleotide tracer, 65% of the 1500 nucleotide tracer, and 75% of the 3700 nucleotide tracer, is retained on hydroxyapatite at this low Cot. The increase in the proportion of fragments retained at longer tracer lengths implies that the reassociating repetitive sequences are interspersed with nonreacting nonrepetitive sequences.

The total amount of repetitive duplex formed in each reaction must be the same, irrespective of tracer length. Since the criterion for retention on hydroxyapatite is the presence of some duplex region on each molecule, the amount of tracer bound at low Cot can be greater than its content of repetitive sequence if each bound tracer possesses two types of sequence: its repetitive sequence forms a duplex with the driver DNA and is responsible for binding; while a nonrepetitive region does not react and remains as a single-stranded tail. The increase in the number of fragments bound with longer tracer lengths suggests that increases in fragment length raise the probability that each tracer molecule will include a repetitive sequence.

This interpretation is supported by the results of an experiment in which the 3700 nucleotide fragments retained on the hydroxyapatite were recovered, sheared to a length of 450 nucleotides, denatured, and then renatured with the standard unlabeled (450 nucleotide) preparation. Only 46% of the now shorter labeled fragments formed duplex regions, demonstrating that the repetitive and nonrepetitive sequences originally present on the same molecule of 3700 nucleotides may be separated by shearing to a length of 450 nucleotides.

The length of the repetitive sequences can be estimated by isolating the renatured duplex fragments. After renaturing DNA fragments of length 2000–2500 nucleotides to low Cot, Chamberlin, Britten, and Davidson (1975) showed by electron microscopy that the product comprises short duplex regions (the repetitive elements) with up to four single-stranded tails each (the nonrenaturing nonrepetitive elements). By using the S1 nuclease to degrade all the unpaired sequences in such molecules, the lengths of the purified duplexes remaining can be measured by their mobilities on agarose gels. Both approaches suggest that most (60–75%) of the repetitive regions are rather short, averaging about 300 nucleotides in length.

The average distance between the short repetitive sequences can be determined from the relationship between the length of the labeled fragments and the proportion that contains repetitive sequences (that is, binds to hydroxyapatite after renaturation to low Cot). The principle underlying this analysis is illustrated in Figure 18.11. At short fragment lengths close to the length of the repetitive elements, most of the nonrepetitive DNA will exist as fragments lacking repetitive sequences and so is not retained at low Cot; most of the renatured molecules will consist principally of renatured duplex regions, with only short tails of unrenatured nonrepetitive DNA. As

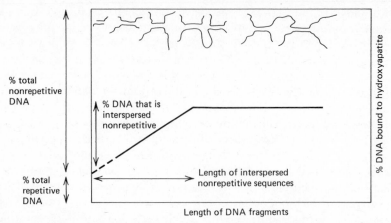

Figure 18.11. Interspersion plot of relationship between fragment length and hydroxyapatite binding. Tracer DNA of different lengths is reassociated with driver DNA of standard length to a low Cot that allows only repetitive sequences to renature. The graph shows the dependence of the percent DNA bound to hydroxyapatite on the length of the tracer DNA. As illustrated across the top of the figure, at low fragment length the renatured molecules consist largely of duplex regions. The increase in retention with fragment length is due to the presence of unrenatured single-strand tails of increasing length. This reaches a plateau when the tails become long enough to extend to the next repetitive sequence. Thus the length of the interspersed nonrepetitive sequences is given by the size of tracer at which the transition occurs. The total percent of repetitive DNA is given by extrapolating the curve to zero fragment length; correspondingly, the total nonrepetitive DNA represents the remaining fraction, and the amount of nonrepetitive DNA that is interspersed is given as a fraction of the total genome by the height of the initial rise in the curve.

the fragment length is increased, an increasing proportion of the nonrepetitive sequences will be included in fragments that also possess repetitive DNA; thus the renatured molecules will represent duplex regions of the same length, but with increasingly long single-strand tails of nonrepetitive DNA. This increase in the proportion of DNA retained on hydroxyapatite will cease when the fragments become longer than the distance occupied by the nonrepetitive sequences that separate successive repetitive regions, for by then all fragments should contain a repetitive sequence and so should be retained.

Even at short fragment lengths, not all the DNA retained on hydroxyapatite will consist of duplex regions; because the genome is randomly sheared, there must always be some molecules consisting in part of repetitive and in part of nonrepetitive sequences. This implies that when DNA reassociation is followed on hydroxyapatite, the equation of DNA bound at low Cot with repetitive sequences provides an overestimate of their representation in the genome. The extent of the overestimate decreases as the fragment length is reduced. Thus the true proportion of repetitive DNA may be taken by extrapolating the length dependence curve to zero fragment length.

The reassociation of X. laevis DNA shown in Figure 18.12 at first displays

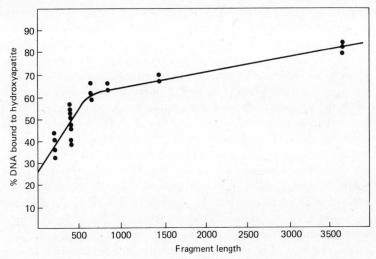

Figure 18.12. Relationship between fragment length and hydroxyapatite binding for DNA of X. laevis. Labeled tracer DNA of various lengths was reassociated with an excess of 450 nucleotide long driver DNA to a Cot of 50 under standard conditions. Data of Davidson et al. (1973).

a rapid increase in the amount of DNA retained at low Cot as the fragment length is increased; and then, instead of a plateau, there is a more gradual increase with fragment length as far as the curve has been followed. The change in slope occurs at a length of 700–800 nucleotides. Up to this point, the curve represents roughly half of the X. laevis genome, apparently organized in a pattern of interspersion in which repetitive elements of about 300 nucleotides alternate with nonrepetitive sequences of about 750 nucleotides. The slow rate of rise of the remaining part of the curve suggests that much of the remaining DNA shows a similar, but longer, repeating pattern, in which repetitive elements (of undefined length) are separated by nonrepetitive sequences that must be longer than 4000 nucleotides (the farthest point on the curve: to determine the actual value, it would be necessary to proceed to greater fragment lengths). In addition to these two modes of interspersion, about 6% of X. laevis DNA forms large duplex structures upon low Cot reassociation. These are excluded from the interspersion analysis and appear to represent clusters of repetitive sequences, which may represent up to about a quarter of the repetitive DNA.

To perform this analysis, DNA is renatured to a Cot at which all repetitive sequences will have reassociated, while all nonrepetitive DNA remains single stranded. In cases where more than one intermediate repetitive component can be resolved, the interspersion of each component could in principle be determined by renaturation to an appropriate Cot. However, since the reassociation curve usually is continuous, so that any such components are not clearly resolved, it is difficult to achieve conditions under which there would be significant enough reassociation of one, but not another, intermediate repetitive component. This contrasts with the greater ease with which highly repetitive components may be removed by very brief renaturation; usually these are excluded from the interspersion analysis.

Table 18.1 summarizes studies that show that the *short period* interspersion characteristic of the predominant part of the X. laevis genome is found in a wide range of other species. The length of the nonrepetitive elements generally is in the range of 800–1500 base pairs, but may be longer; the repetitive elements usually are rather short, around 250–350 base pairs. This organization seems to prevail for the majority of nonrepetitive sequences. Many of the remaining nonrepetitive sequences are organized in an interspersion pattern in which much longer lengths of nonrepetitive DNA separate repetitive sequences of unknown length. Usually most of the intermediate repetitive DNA appears to be interspersed with nonrepetitive DNA, since only a small proportion can be identified in the form of long duplex molecules when renatured at low Cot.

The total amount of intermediate repetitive DNA identified by the interspersion plot, or assayed by its resistance to S1 nuclease after low Cot reassociation, clearly is always less than implied by the kinetics of reassociation as followed on hydroxyapatite. Inaccuracies in determining the intercept on the ordinate of the interspersion plot arise from the scatter of points in this region; usually there are several almost equally good lines that can be drawn. Partly this results from the difficulty of shearing preparations to homogeneous fragment lengths. The proportion of DNA resistant to S1 nuclease may underestimate the repetitive component, because the enzyme may attack mispaired regions resulting from reassociation of different members of a family. Hyperchromocity can be used to assay the exact proportion of paired bases, although again this ignores mispairing. For S. purpuratus, these three approaches yield estimates of 31, 25, and 25%, respectively; overall the best estimate is that some $27 \pm 3\%$ of the sea urchin DNA represents repetitive sequences (defined by the ability to reassociate by Cot 20). This contrasts with the value of 46% obtained in a standard reassociation curve of 450 nucleotide fragments on hydroxyapatite; and even with shorter fragments of 330 nucleotides, the apparent repetitive component is 34%. Comparison of Table 18.1 with Appendix 3 shows that an effect of this magnitude is common.

Genomes that lack short period interspersion are described in Table 18.1. The electron microscopic analysis of Manning, Schmid, and Davidson (1975) first showed that in D. melanogaster, interspersion occurs in a long-period pattern, in which repetitive sequences of average length 5,600 base pairs are separated by nonrepetitive sequences of length greater than 13,000 base pairs. This has been confirmed by reassociation analysis, which has also shown that short period interspersion is absent from the genome of the bee, Apis mellifera (Crain et al., 1976a,b). This situation is not a characteristic of the insects, since the housefly, M. domestica, shows the usual short period pattern; so the organization of the genome has changed in the

Table 18.1 Interspersion patterns in eucaryotic genomes

Species	Short period interspersion Nonrepetitive Per cent	Short period interspersion Nonrepetitive Average length	Short period interspersion Repetitive Average length	Long period Nonrepetitive Per cent	Long period Nonrepetitive Average length	Intermediate repetitive component Total percent by Interspersion plot	Intermediate repetitive component Total percent by S1 nuclease	Percent intermediate DNA not interspersed
D. discoideum	60	1500 bp	350 bp	>10	>3700 bp	22	25	50
S. purpuratus	40	1050 bp	350 bp	>30	>3400 bp	31	25	20
A. pernyi	60	800 ?	250 bp	>25	>2800 bp	15	22	25
A. californica	80	1500 bp	300 bp	<10	>4000 bp	28	41	30
G. lateralis		850 bp	300 bp			36	44	
R. norvegicus	60	2700 bp	250 bp & >1500 bp	>10	>7000 bp	25	—	
H. sapiens	50	2250	600 bp	>20	>20,000 bp	20	—	15
X. laevis	50	800 bp	300 bp	>30	>4000 bp	25	—	20
N. tabacum	50	1400 bp	300 bp	<5	>4000 bp	65	57	
G. hirsutum	50	1800 bp	800 bp	>15	>4000	35	—	20
G. domesticus	35	4500 bp	~2000 bp	>10	>17,500 bp	12	17	
D. melanogaster	>13,000 bp nonrepetitive DNA interspersed with 5,600 bp repetitive DNA							

Data are based on interspersion plots of the form shown in Figure 18.11. For the short-period interspersion, the content and length of nonrepetitive DNA are taken from the initial part of the curve; the length of repetitive DNA segments usually has been determined by analysis of the duplex fragments remaining after renaturation to low Cot followed by S1 nuclease digestion. For the long-period interspersion, the data give the minimum content of nonrepetitive DNA and average length, based upon the furthest fragment length at which the continuation of the second slope has been measured; cases where a maximum content is shown identify situations where the second slope may in fact be a plateau. The total content of intermediate repetitive DNA is given according to the intercept in the interspersion plot and as directly measured by resistance to S1 nuclease at low Cot (where measured). The proportion of noninterspersed intermediate DNA gives the amount that can be identified in the form of long clusters of repetitive sequences. These data concern only intermediate repetitive and nonrepetitive DNA components; zero time binding DNA usually is excluded from the data before plotting (see Chapter 19). Note that in crabs there is little moderately repetitive DNA and the component described approaches the highly repetitive. References are as given in Appendix 2 except for D. discoideum (Firtel and Kindle, 1975). In the example of Gossypium hirsutum (cotton), the plant is tetraploid with a DNA complexity corresponding to a haploid genome size of about 8×10^8 bp, of which 60% is nonrepetitive and 27% intermediate repetitive by hydroxyapatite reassociation analysis (Walbot and Dure, 1976). The predominant interspersion pattern of the chicken, although summarized under the short-period heading, can be seen to be intermediate between the usual short- and long-period categories. In addition to the data summarized in this table, less detailed analyses have been performed for five marine organisms, C. virginica, S. solidissima, L. polyphemus, C. lacteus, and A. aurita, in which comparison of reassociation kinetics for short and long tracers shows the presence of interspersion; the repetitive sequences in each case are about 300 bp long, suggesting the occurrence of a predominantly short period pattern (Goldberg et al., 1975). The same is true of the housefly, M. domestica; so the absence of short period interspersion seen in the bee, Apis mellifera, where 2200 long fragments reassociate at the same rate as short fragments, and in D. melanogaster, which has a long period interspersion pattern, is not a characteristic of all insects (Crain et al., 1976a,b; Manning et al., 1975). Interspersion has not been detected in some lower eucaryotes, including A. bisexualis (tested to a length of 135,000 bp by Hudspeth et al., 1977) and the nematode Panagrellus silusiae (tested to 2000 bp by Beauchamp et al., 1979).

evolution of the Diptera. The chick genome has a pattern intermediate between the short period and long period. Some lower eucaryotes lack interspersion.

Interspersion is probably a general phenomenon in higher eucaryotic genomes. In addition to the detailed studies summarized in Table 18.1, indirect evidence of its occurrence is provided by earlier studies of DNA reassociation for other species, which showed that the hyperchromicity of reassociated fractions is less than that of native DNA, presumably due to the presence of unpaired nonrepetitive single-strand regions interspersed with the repetitive sequences (Britten and Kohne, 1968). Of course, this does not reveal the pattern of the interspersion.

What is the significance of this pattern of interspersion? Its occurrence in genomes so separated in evolution suggests that it may have a persistent and basic function. Since mRNA is transcribed predominantly from nonrepetitive DNA, an obvious speculation has been that the repetitive sequences may represent control elements interspersed with nonrepetitive structural genes (see Chapter 28). However, since only a small proportion of the nonrepetitive DNA usually is represented in mRNA, this is not a compelling argument; indeed, the coding fraction may be small enough to permit structural genes to be located in regions of nonrepetitive DNA other than those displaying the predominant form of organization. At all events, the pattern of short-period interspersion appears to correspond to too large a proportion of the genome to represent a simple alternation of control elements and structural genes. Yet in S. purpuratus, mRNA hybridizes preferentially with nonrepetitive sequences of DNA that are closely linked to repetitive sequences, which at least in this case would support the idea that the interspersed repetitive sequences are connected with the expression of nonrepetitive structural genes (see Chapter 24). But another argument against a simple model is that, at least in mammals, it is clear that transcription units must be much larger than mRNAs, due to the presence of sequences both within and at the ends of genes that are transcribed but later removed from the RNA: the transcription units would correspond to several of the interspersed arrays. Although this does not mean that the interspersed repetitive elements are not control elements, it implies that their function might be mediated in RNA rather than in DNA, and that successive nonrepetitive sequences are unlikely to be structural genes, if there is only one gene in each transcriptional unit. A difficulty in discussing the role of the interspersed repetitive sequences is that the pattern represents an average of many repetitive families. If the members of a given family do indeed provide a set of related controls, then ultimately the significance of interspersion may be found by identifying the functions of the regions adjacent to the different members of individual repetitive families.

CHAPTER 19

DNA Sequence Organization: Inverted and Tandem Repeats

All eucaryotic genomes contain a component whose slow reassociation corresponds to a complex nonrepetitive sequence. Interspersed with the nonrepetitive DNA are the sequences of the moderately repetitive DNA, reassociating at intermediate Cot values. Together these components usually, although not always, account for the major part of the genome. The low Cot at which the most rapidly reassociating component reacts corresponds to a sequence of low overall complexity repeated very many times in each genome. All of these components show the bimolecular reactions discussed in the last chapter and thus can be characterized by their $Cot_{1/2}$ values. Reassociating even more rapidly, there is a fraction of the genome that is found in duplex form even at the lowest Cot values examined. The values given for the last two (rapidly reassociating) classes in Appendix 3 are only rather approximate, since usually they comprise only a small proportion of the genome, and reassociation of purified fractions is necessary to determine their properties with any degree of accuracy.

Considering these classes in the order of their reassociation from single-stranded DNA, the first fraction to react sometimes has been described as the *zero-time binding DNA*, because it comprises the duplex molecules retained on hydroxyapatite after even the shortest periods allowed for reassociation. (When reassociation of other genome components is studied, often the zero-time binding fraction is removed first.) The reassociation of this fraction is unimolecular, implying that each pair of reassociating complements lies on a single strand of DNA. Thus the reaction is an intramolecular reassociation of sequences that must be repeated in reverse orientation close together. These are known as *inverted repeats;* sometimes they are described as *palindromes* or the fraction is referred to as *foldback* or *snapback DNA*. The inverted repeats seem to be widely dispersed throughout the genome and appear to include members of all sequence classes. At least some of these sequences are transcribed.

The most rapidly reassociating of the components following the usual second-order kinetics has properties quite distinct from those of other sequence components. It consists of rather short sequences repeated in many tandem copies, not interspersed with the intermediate repetitive or nonrepetitive DNA. A consequence of this form of organization is that the highly repetitive components often may be isolated as "satellites" of the main peak when DNA is fractionated according to buoyant density. With appropriate variations in the density fractionation procedures, most of the highly repetitive components seem to form satellite peaks; but, of course, this does not mean that every highly repetitive tandem sequence does so. Taking *satellite DNA* to be synonymous with highly repetitive DNA, such components that are not detected as distinct fractions sometimes are referred to as "cryptic satellites."

The extent of sequence variation among the members of a highly repetitive family differs widely. Originally it was thought that, like the more moderately repeated sequences, there would be appreciable variation among the copies of each highly repeated sequence. This indeed occurs in many cases, in particular in those mammalian satellites whose sequences have been determined; although, of course, the tandem repetition of the members of each family means that its bounds may be subject to definition and not indefinitely extensible by reductions in the criteria for reassociation. (Reassociation shows the usual dependence on criterion, with reduction in temperature allowing more distantly related members of the family to react; but this should not affect the total number of sequences in the satellite fraction). In some instances, notably in the Arthropods (including Drosophila), satellites consist of a many-fold repetition of essentially identical sequences; variations on the predominant sequence may occur in as few as 5% of the copies. Thus divergence of sequence is not necessarily a feature of highly repeated families.

The large blocks of satellite DNA are located in constitutive heterochromatin, generally in centromeric regions, but also in other loci identified by their ability to form C bands (see Chapter 15). As might be expected from its simple sequence constitution, satellite DNA is not transcribed. Although its role may be presumed to be structural, therefore, there is at present little evidence on its nature (in spite of many theories).

Distribution of Inverted Repeats

The reassociation of foldback DNA occurs at very low Cot values, below 10^{-5}-10^{-4}, in a unimolecular reaction. This implies that the reassociating single-strand sequences must be in very close proximity. This might be accomplished in either of two ways. Cross links might exist between the two strands of DNA, so that they are held together at these sites upon denaturation. Or complementary sequences might exist on the denatured single strands themselves. These must take the form of inverted repeats, such as:

$$5'\quad A\ T\ C\ G\ \cdots\ C\ G\ A\ T\quad 3'$$
$$3'\quad T\ A\ G\ C\ \cdots\ G\ C\ T\ A\quad 5'$$

in which the sequence read in the same direction (say 5′ to 3′) on each strand is the same.

These possibilities can be distinguished by the form taken by the reassociated DNA, as illustrated in Figure 19.1. Cross-linked DNA spread under denaturing conditions should give 4-ended molecules, held together at the cross link; upon renaturation, the molecules should return to the linear state, with the length of duplex determined simply by the starting size of the DNA preparation. The inverted repeats should exist on linear single strands in the denatured state; but upon renaturation the complementary sequences on each single strand should fold back to form a hairpin. When the inverted repeats are immediately contiguous, the hairpin should be seen as a simple duplex protrusion from the single strands, its length being determined by the length of the complementary sequences. When the inverted repeats are separated by some other sequence, the hairpin should have a loop of single strand DNA corresponding in length to the distance apart of the complementary sequences.

Electron microscopy of the zero time binding DNA shows the foldback structures expected for both immediately adjacent and separated inverted repeats; there is no evidence for any cross-linked DNA (Cech and Hearst,

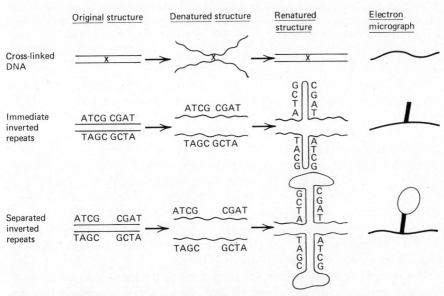

Figure 19.1. Models for zero time binding DNA. Cross links in DNA would maintain the association between denatured strands and lead to reassociation of linear duplex DNA. Inverted repeats allow duplex regions to form by intramolecular reassociation. When the repeats are very close, a duplex hairpin is formed, with single-strand ends. When the repeats are separated, the structure takes the form of a duplex stem with a single-stranded loop. The last column shows the appearance the reassociated structures take in the electron microscope, when the duplex regions are distinguished from the single-strand regions by their greater width.

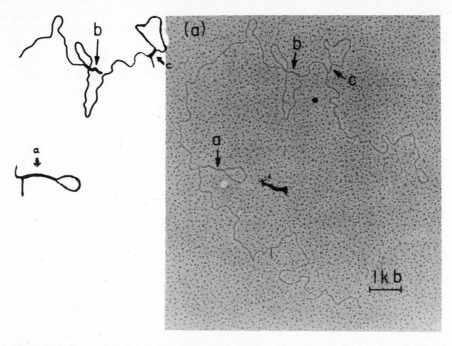

Figure 19.2. Inverted repeats in D. melanogaster DNA. Three repeats lie on one molecule of DNA; *a* and *c* are separated inverted repeats; *b* is a simple hairpin. The structures are illustrated in the drawing, where the dimensions are slightly exaggerated for clarity. The arrows point to the renatured duplex regions of the repeats; in *a* and *c* single strand loops represent the material separating the complementary sequences. Data of Schmid, Manning, and Davidson (1975).

1975; Schmid, Manning, and Davidson, 1975). Some examples are shown in Figure 19.2. The proportion of the foldback sequences that has interruptions between the inverted repeats depends on the genome; available data are summarized in Table 19.1.

Another prediction made by the existence of cross links would be that denatured DNA of a given length containing cross links should be twice the weight of single strands not cross linked. Similarly, following one round of density labeling with BUdR, cross-linked DNA should take an intermediate position, contrasted with the heavy and light non-cross-linked single strands. Neither prediction is fulfilled by the zero time binding DNA (Perlman, Phillips, and Bishop, 1976).

The content of the foldback fraction depends on the length of the fragments of denatured DNA. There are two effects, one depending on the organization of the inverted repeats, the other reflecting their dispersion in the genome.

When there is an interruption between the inverted repeats, the rate of reassociation depends upon their distance apart. In fact, by allowing only a very short time for reassociation, immediately adjacent inverted repeats may

Table 19.1 Distribution of foldback sequences in eucaryotic genomes

Species	Proportion of genome, %	Average spacing of foci, bp	Number of foci	Average duplex length, bp	Proportion and average length of loops
D. melanogaster	3.0	40,000	3.5×10^3	1300	80% of 3400 bp
H. sapiens (HeLa)	5.0	55,000	5.4×10^4	250	64% of 1600 bp
M. musculus (L)	3.0	75,000	4.0×10^4	1000	60% of 3000 bp
T. aestivum	1.7	25,000	6.8×10^5	300	20% of 1000 bp

The proportion of the genome occupied by foldback DNA is the percent resistant to digestion by S1 nuclease after intramolecular reassociation. The average spacing between foci is determined from the dependence on fragment length of the percent DNA retained on hydroxyapatite after very low Cot reassociation (spacing is given by the reciprocal of the slope of the curve). This corresponds to the long spacing between foldback foci; but each focus may possess more than one inverted repeat sequence. The number of foci is calculated by dividing the genome size by the average spacing. In all the instances summarized here, clusters of at least 2–3 inverted repeats are common (the spacing between individual foldback sequences generally is of the order of 3000 bp). The average length of duplex formed by an inverted repeat, the proportion of these repeats that have a single-stranded loop (as opposed to simple hairpins), and the length of the loops are largely derived from electron microscopic studies.

References: D. melanogaster—Schmid et al. (1975); H. sapiens—Wilson and Thomas (1974), Schmid and Deininger (1975), Deininger and Schmid (1976), Dott et al. (1976); M. musculus—Cech et al. (1973), Cech and Hearst (1975); T. aestivum—Bazetoux et al. (1978). Other studies have been reported on the cotton, Physarum and P. silusiae (nematode) genomes by Walbot and Dure (1976), Hardman and Lack (1977, 1978), Hardman et al. (1979), and Beauchamp et al. (1979).

be able to renature while separated inverted repeats are unable to do so. Thus Wilson and Thomas (1974) obtained a fraction consisting of only simple hairpins by quenching denatured DNA directly into cold phosphate buffer. But even when more conventional procedures are used, an inverted repeat will be retained in the foldback fraction only if both complementary sequences remain on the same single strand of DNA. Those inverted repeats that are separated thus will be excluded from the foldback fraction as the length of DNA is reduced below the distance that separates the complementary sequences. Ultimately, only immediately adjacent inverted repeats should remain.

Interruptions between inverted repeats can be detected also when the foldback DNA fraction isolated from long fragments of DNA is treated with S1 nuclease to remove all unpaired single strands. When the treated foldback DNA is denatured and then again reassociated, the proportion of sequences surviving this "second cycle" is always less than 100%. The lost inverted repeats must have been interrupted by a sequence that formed a single strand loop at the end of the foldback hairpin, and which attracted the attention of S1 nuclease. Upon denaturation, the inverted repeats then would fall onto separate molecules. Correspondingly, the surviving repeats must have had loops absent or short enough to be resistant to the nuclease.

The second length-dependent effect arises from the interspersion of the foldback DNA with other sequences in the genome. This is similar to the length dependence caused by interspersion of intermediate repetitive and nonrepetitive DNA (see Chapter 18). As the fragment length is increased, an increasing proportion of the genome is retained as single-strand tails attached to the reassociated inverted repeats. When the proportion of DNA retained on hydroxyapatite is plotted against fragment length, often the line is straight, with no plateau even at long fragment lengths; this has been taken to imply that the foldback sequences are interspersed through most or all of the genome (Perlman, Phillips, and Bishop, 1976).

The reciprocal of the slope of this curve (that is, fragment length divided by proportion) gives the distance apart between reassociating foci. Such foci always are far apart, typically of the order of 50,000 base pairs (see Table 19.1). However, each focus may contain more than one inverted repeat. This should be seen by a steep slope to the early part of the length dependence curve (reflecting the separation of different inverted repeats within the cluster); and at fragment lengths longer than the cluster, the slope should be shallower, reflecting the distance apart of clusters (termed foci in Table 19.1). The detailed shape of the curve has been considered by Wilson and Thomas (1974) and Schmid, Manning, and Davidson (1975); but usually the data do not clearly resolve the curves predicted by models for random or clustered distributions of inverted repeats.

Electron microscopy of the foldback fraction, however, shows the presence of appreciable numbers of clustered inverted repeats. Often two or three hairpins are connected by a stretch of single-stranded DNA. The individual inverted repeats usually are rather short, as seen either by electron microscopy or by isolation of the duplex fragments surviving S1 nuclease treatment. The thermal stability of the hairpins generally is high, suggesting that the complementary sequences comprising each inverted repeat are closely related if not identical.

Which sequence components are represented in the inverted repeats? When a labeled tracer of foldback DNA is reassociated with an excess of nuclear DNA, the radioactive curve follows the bulk curve (sometimes with a slight enrichment of sequences reacting at intermediate Cot). Thus the inverted repeats include both repetitive and nonrepetitive sequences. Similar results are obtained whether the foldback fraction is used directly, in which the inverted repeats presumably continue to foldback, so that the tracer reaction follows the flanking sequences; or whether S1 treatment is used to isolate the foldback duplexes, in which case the proportion of inverted repeat sequences per se is increased. Obviously a difficulty in this analysis is that some inverted repeats, most particularly the class forming simple hairpins without loops, will continue to reassociate by intramolecular reaction. However, it seems clear that inverted repeats are found in contexts of both repetitive and nonrepetitive DNA and may themselves include both types of sequence component.

Are the inverted repeats located at specific or random sites in DNA? If they lie at certain loci, the flanking sequences should constitute a specific fraction of the genome, with a corresponding complexity; if, for example, there are some 10^5 foci in a genome of 3×10^9 base pairs, then at a fragment length of (say) 2000 base pairs, about 10% of the complexity of the genome should be present in the foldback fraction. Two experiments have been addressed to this question. Wilson and Thomas (1974) fractionated mouse L cell DNA into a foldback fraction (about 20% at 7500 nucleotide length) and a nonfoldback fraction. Immobilized on filters, each fraction showed a 2-3 fold preference for hybridizing with added DNA of its own, compared with the opposite, source. Only repetitive sequences react in filter hybridizations, so this implies that inverted repeats located within repetitive DNA must reside in particular families.

A more extensive attempt to characterize the sequences surrounding the inverted repeats has been reported by Perlman, Phillips, and Bishop (1976). The foldback fraction of X. laevis DNA was isolated at a fragment length of 3500 base pairs, sonicated to an average length of 600 base pairs, and then used in excess to drive a reaction with a labeled tracer of nonrepetitive DNA. Only 10% of the original foldback fraction remains able to reassociate by intramolecular reaction after the reduction in length; so the reaction is driven by the 90% of the preparation constituting surrounding sequences and any inverted repeats that have been separated. The nonrepetitive tracer reacts to the same extent when driven by the foldback-derived fraction as when driven by unfractionated total DNA. This implies that the foldback fraction contains no fewer sequences than are present in total DNA. The failure to isolate a specific set of nonrepetitive sequences representing the regions within 3500 base pairs of the inverted repeats implies that the foldback sequences reside at random locations. This surprising result would seem to suggest that the inverted repeats might be located at different sites in different genomes; a polymorphism this extensive would imply further that the sites of inverted repeats may change with great speed on the evolutionary time scale. It would be interesting to determine whether the same result is obtained with DNA derived from a single animal, contrasted with a population.

Little is known about the function of inverted repeats. Many appear to be transcribed, since foldback sequences are found in nuclear RNA (see Chapter 25). Several clones of phage lambda carrying randomly selected EcoRI restriction fragments of CHO cell DNA have been selected for the presence of transcribed inverted repeats, by virtue of their ability to hybridize with nuclear RNA containing duplex regions (Jelinek, 1978). Through the isolation of individual inverted repeats by this and more direct means, it may become possible to define their structure and function.

The only instance in which a clear function in the eucaryotic genome can be assigned to inverted repeats is provided by the unusual case of the ciliated protozoan Oxytricha. Ciliates contain two types of nuclei: the micronucleus is diploid and is perpetuated as the germ line through mitotic and meiotic divisions; the macronucleus exists only during vegetative growth and is generated from a diploid nucleus by the formation of polytene chromosomes which are then fragmented into individual bands. Most of these sequences are degraded, so that finally a small part of the genome is represented in the form of a large number of copies of DNA molecules of average length about 3000 base pairs. Upon denaturation and renaturation, these fragments form single-stranded circles with short duplex stems. The stems must represent inverted repeats. The length of the inverted sequence is 26 base pairs and it appears to be the same in most or all of the fragments (Lawn, 1977; Herrick and Wesley, 1978). This suggests that the repeats may be involved in the process of forming macronuclear DNA.

Fractionation of DNA by Buoyant Density

The ability of centrifugation through a CsCl density gradient to separate distinct components of the genome was first realized with mouse DNA. Kit (1961) observed that in addition to the main band—which sediments at about 1.701 $g\text{-}cm^{-3}$, corresponding to its G-C content of 42%—about 8% of the DNA is found as a satellite band of buoyant density about 1.691 $g\text{-}cm^{-3}$. Satellite DNA bands since have been separated from the main band of DNA in a variety of eucaryotes; they may be either heavier or lighter than the main band and it is relatively unusual for their proportion of the total DNA to exceed about 10%.

Centrifugation through gradients of Cs_2SO_4 which contain silver ions improves the resolution of satellite DNA bands. As Figure 19.3 shows, the mouse satellite is more clearly separated from the main band under these conditions. In many cases, centrifugation through Ag^+ Cs_2SO_4 gradients reveals satellites that are not separated from the main band in CsCl. In the guinea pig, for example, the single satellite band found on CsCl as a shoulder to the main band is clearly resolved as a heavy satellite on Ag^+ Cs_2SO_4; and two further satellites are found on the light side of the main band (see Figure 19.3). Since the buoyant density of each band is altered by the presence of silver ions, it is customary to characterize the satellites isolated on Ag^+ Cs_2SO_4 by the buoyant density that they show when recovered and run again on CsCl (for examples see Corneo et al., 1968, 1970, 1972; Yasmineh and Yunis, 1971). Other DNA binding agents, such as Hg^+ ions or actinomycin, may be used with similar effect.

The behavior of satellite DNA on both CsCl and Ag^+ Cs_2SO_4 gradients is anomalous. The base composition of a satellite rarely corresponds with the value predicted from its buoyant density on CsCl. Examples from both insects and mammals are given in Tables 19.2–19.4; in some cases, two satel-

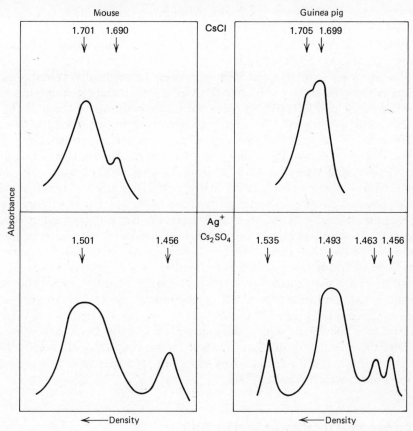

Figure 19.3. Separation of satellite DNAs from main band DNA. The resolution achieved on CsCl gradients (upper) is much improved by the addition of silver ions (lower). With mouse DNA (left) a single satellite is found and its separation from the main peak is increased on $Ag^+Cs_2SO_4$. With guinea pig (right) the single satellite present on CsCl as a shoulder to the main peak is resolved as a separate heavy band on $Ag^+Cs_2SO_4$ and two additional satellites are resolved on the light side of the main band. Data of Corneo et al. (1968, 1970).

lites may even have the same buoyant density on CsCl, but quite different contents of G-C base pairs. The basis for the improved resolution of satellites in the presence of silver ions is not clear. There is again no correlation between base composition and buoyant density. Presumably the tandem repetition of a simple sequence is in some way responsible for the anomalous behavior on CsCl and the very different abilities to bind Ag^+ ions.

On alkaline gradients many satellite DNAs can be separated into constituent heavy (H) and light (L) strands. This reflects an asymmetry in the base composition of the single strands, whose buoyant density increases with the content of T and G bases. This is a common feature of satellite DNAs, although there are exceptions; and this behavior is not found with main band

DNA, where the buoyant density increases upon denaturation but remains in the form of a single peak. The separation of satellite strands in alkali is a useful property, because it allows the sequence of each strand to be determined separately.

The main band of DNA can itself be resolved into subsidiary components. Usually some 60–65% of the total DNA of a mammalian genome lies in a density range of 1.697–1.700 g-cm^{-3}, with the remainder of the main band comprising another 20–30% at slightly greater densities, about 1.704 and 1.709 g-cm^{-3}. The distribution of densities within the main band does not seem to be continuous; there may be concentrations at some densities (Filipski, Thiery, and Bernardi, 1973; Thiery, Macaya, and Bernardi, 1976; Macaya, Thiery, and Bernardi, 1976). Whether this has any functional significance seems doubtful, since the proportions of the apparent components of the main band are not invariant with regard to tissue. Up to about 10% of the DNA is found as minor components, of density either greater or lower than that of the main band. These include satellite DNAs and also other sequences, such, for example, as mitochondrial and ribosomal RNA. Somewhat similar results qualitatively are found with other vertebrates, but amphibian, fish, and insect DNAs all appear to have symmetrical main bands not suggesting the presence of subsidiary components. There are, of course, exceptional species in which the content of satellite DNA is very much greater than seen in most cases; we shall in due course discuss the satellites of Drosophila virilis, which occupy 40% of the genome, and those of Dipodomys ordii, which represent 50%.

Chromosomal Location of Satellite DNA

Reassociation kinetics show that satellite DNA consists of rather short sequences repeated very many times. The isolation of satellites at quite large DNA fragment sizes implies that there must be tandem repeats at least to the length of several hundred base pairs. Where are these simple sequence clusters located in the genome? Are they interspersed with other sequence components?

An extension of nucleic acid hybridization techniques allows the location of satellite sequences to be determined directly in the chromosome complement. In the protocols for *in situ* or *cytological* hybridization, chromosomal DNA is denatured in situ and then hybridized with a solution containing radioactively labeled DNA or RNA. The procedure is to squash cells beneath a cover slip, freeze in dry ice, wash with ethanol and dry; then the slides are dipped in an agar solution, the agar is allowed to gel, and the chromosomal DNA is denatured by treatment with alkali. After incubation at 60°C for several hours with the solution of radioactive nucleic acid, the slide is washed to remove any unreacted DNA or RNA; then the location of the labeled nucleic acid that has been retained can be determined by autoradiography (Gall and Pardue, 1969; Pardue and Gall, 1969; John, Birnstiel, and Jones, 1969). This technique corresponds to reassociation at fairly low Cot values; and it may therefore be used to follow the reaction of repetitive but not nonrepetitive DNA.

Using labeled mouse satellite DNA, Pardue and Gall (1970) found that the satellite sequences are confined to the constitutive heterochromatin found at the centromeres of mitotic chromosomes. An example is shown in Figure 19.4. All mouse chromosomes except the Y have satellite DNA at the centromere. Usually metaphase plates are used for cytological hybridization, so that individual chromosomes can be identified. However, when interphase nuclei are used, the satellite sequences bind to the chromocenters which appear to be formed by aggregation of the centromeric heterochromatin. Similar results were obtained by Jones (1970), using a labeled RNA transcript (obtained by using satellite DNA as a template in vitro for E. coli RNA polymerase). If main band DNA or its transcript is used for cytological hybridization, the grains detected by autoradiography are located over all regions of the genome of either mitotic or interphase cells, as would be expected from the dispersion of intermediate repetitive sequences throughout the genome.

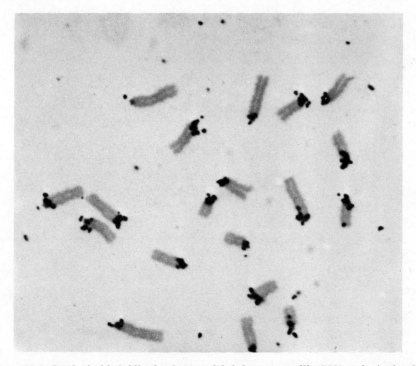

Figure 19.4. Cytological hybridization between labeled mouse satellite DNA and mitotic cells. The label is located at the (acrocentric) centromeres of the chromosomes. Data of Pardue and Gall.

The correlation between satellite DNA and constitutive heterochromatin appears to be a general one (for review see Walker, 1971a,b). Indeed, all regions able to form C bands with Giemsa, as described in Chapter 15, seem to contain satellite DNA. Thus some mouse cell lines possess C bands in some chromosome arms, as well as at the centromeres, and these are able to hybridize with satellite DNA (White, Pasztor, and Hu, 1975). From the results obtained with many species, it is clear that satellite DNA is located only within constitutive heterochromatin;* of course, this does not reveal whether other sequences also are present in the heterochromatic regions.

When there is more than one satellite DNA in the genome, is each present on a different set of chromosomes; and may a given centromere possess more than one satellite sequence? In the human genome there are four satellites; and cytological hybridization suggests that they reside at sets of overlapping but not identical centromeres (Gosden et al., 1975). Only 13 of the chromosomes show hybridization with these satellites; and some features of the overlapping pattern—for example every satellite fraction hybridizes strongly with chromosome 9—raise the possibility that some hybridization may involve contaminating sequences. The precise distribution of satellites remains to be resolved; one approach in use is to determine which satellites can hybridize to hybrid cell lines retaining only a single human chromosome (Beauchamp et al., 1979). This should allow the investigation of differences in components of a given satellite that are present on different chromosomes, for example by restriction mapping. The four satellites appear to represent all of the highly repetitive human sequences, since the rapidly renaturing DNA can be separated into buoyant density peaks that coincide with the native satellites (Marx, Allen, and Hearst, 1976a,b).

Evidence that the evolution of some of the human satellites occurred before speciation is provided by cross-hybridization reactions with the cells of the chimpanzee, gorilla, and orangutan. All of the human satellites hybridize with at least some of the monkey species. In some cases, the reaction involves chromosomes thought to be homologous on the basis of their banding patterns. In all the monkeys, as in the human genome, the Y chromosome is a major site of reaction (Gosden et al., 1977; Mitchell et al., 1977). These reactions are subject to the same reservations that we have noted for the homologous human reaction. In at least the case of the chimpanzee, the genome contains a satellite whose properties are similar to a human satellite (III); there is some cross hybridization between them (Prosser et al., 1973).

In the cow, four satellites can be resolved (Kurnit, Shafit and Maio, 1973). One appears to be present at the centromeres of all autosomes; the others each appear to be present at some, but not all, centromeres. This implies

---

*Examples of satellites located in heterochromatin have been described by Yasmineh and Yunis (1969), Arrhigi et al. (1970), Arrhigi and Hsu (1971), Jones and Robertson (1970), Jones and Corneo (1971), and Corneo et al. (1971).

that more than one satellite may be present at a given centromere. Like the mouse, none of the satellites appears to reside on the Y chromosome (or on the X), which lack centromeric heterochromatin.

In contrast with the cow and mouse, simple sequence DNA occurs on the Y chromosome of the human and on the sex determining chromosome of other species. Comparisons of the simple sequences released by restriction enzyme cleavage from male and female human DNA have raised the possibility that there may be a male-specific sequence located on the Y chromosome: its relationship to the common satellites remains to be established (Cooke, 1976; Cooke and McKay, 1978; Kunkel et al., 1977; Bostock, Gosden, and Mitchell, 1978; Manuelides, 1978). Snakes and birds have sex-determining systems in which males are ZZ and females are ZW; in both classes, satellite DNA is present which hybridizes strongly to the heterochromatic W chromosome (Stefos and Arrhigi, 1974; Singh, Purdom, and Jones, 1976, 1979).

A good system in which to examine the distribution of satellite sequences is provided by Drosophila. The satellites of D. melanogaster have been analyzed in detail; and the polytene chromosome map is available for locating sequences in the genome. A difficulty, of course, is that the satellite sequences are under replicated in the polytene chromosomes (see Chapter 16 and below). However, there is evidence both for differential distribution of the different satellites in D. melanogaster chromosomes and for the existence of some regions in which different satellites may be contiguous. It has also been possible to identify two cases in which a copy of a repetitive gene resides adjacent to a satellite DNA (Carlson and Brutlag, 1978a,b). One of these genes is the transposable *copia* unit (see Chapter 26).

Defined by its ability to form distinct buoyant density peaks, satellite DNA invariably consists of long stretches of DNA from which other sequences are absent. This is in fact more than a definition arising from the method of isolation, because usually most if not all of the highly repeated DNA sequences of a genome can be identified in the form of satellites. It seems likely that any nonclustered highly repetitive sequences are in a small minority, since reports that such sequences can be found interspersed with other sequence components are not common; so far they have been confined to mouse and to D. melanogaster DNA (Kram, Botchan, and Hearst, 1972; Cech, Rosenfeld, and Hearst, 1973; Cech and Hearst, 1976).

Satellite DNAs of Arthropods

The sequences of satellite DNAs have been determined for the genomes of some mammals, notably small rodents, and for some insects and crabs. In this section we shall discuss the sequence of crab and Drosophila satellites, both in the phylum of Arthropods. In the next section we shall describe the sequences of rodent satellites. Finally, we shall consider models for the evolution of satellite DNA.

At present, the Arthropod satellites whose sequences have been determined all appear fairly homogeneous; a single short repeating unit can be identified which may represent about 90% of the copies constituting the satellite. However, the rodent satellites whose sequences have been determined all show appreciable divergence; usually the predominant sequence that can be identified represents no more than 25% of the copies. The others are related to the predominant sequence by a variety of base substitutions and deletions or insertions. Whether this is significant or simply fortuitous remains to be decided when further sequences have been determined.

Determining the sequence of a satellite DNA is complicated by its short repeating structure, especially when there are variations in the repeating units. A common approach has been to separate the two strands of the satellite so that each may be transcribed into a labeled RNA. Then the RNA is sequenced by conventional techniques, relying upon cleavage with RNAases of known specificity to generate a series of small oligonucleotides. For example, the use of RNAases cleaving on the 3' side of G residues essentially reduces the satellite to a series of repeats defined by the location of guanine. Coupled with nearest-neighbor analysis, this approach can identify the major repeating fragments at a small size level. A similar approach has been to subject satellite DNA itself to pyrimidine tract analysis; after treatment with diphenylamine to degrade purines, the pyrimidine oligonucleotides remaining in each strand can be identified. For the Arthropod satellites, this approach has been able to define the unit of the tandem repeats that represents almost all of the satellite.

The three satellites of Drosophila virilis occupy some 40% of the haploid genome and are interesting for two features: each satellite is highly homogeneous; and each is related to the others by simple base substitutions. The sequences of these satellites have been determined by Gall and Atherton (1974), from the labeled RNA transcripts synthesized from isolated satellite strands by E. coli RNA polymerase, and these data are summarized in Table 19.2. A high degree of homogeneity is indicated by the small ΔT_m of reassociated satellite DNA ($< 1.5°C$) and by the occurrence of almost all of the radioactivity in the pancreatic RNAase spots predicted by simple repetition of the predominant sequence. There must, however, be some variations in sequence; for example, although there is no G in the predominant sequence of the light strand of satellite I, some labeled GTP can be incorporated into the transcript of the heavy strand.

The three satellite sequences are closely related. A single base substitution is sufficient to generate the sequence of either satellite II or III from the predominant heptanucleotide of satellite I. Indeed, these sequences provide a good illustration of the abnormal biophysical properties of satellite DNAs. Although satellites II and III actually have the same base composition in the simple repeat (only the position of a G-C pair is different),

Table 19.2 Satellite DNAs of *D. virilis*

Satel-lite	Buoyant density	Percent G-C pairs	Predominant sequence			Number of copies	Propor-tion of genome, %
I	1.692 (33%)	29	5′ A C A A A C T 3′ T G T T T G A	(L) (H)		1.1×10^7	25
II	1.688 (29%)	14	5′ A T A A A C T 3′ T A T T T G A	(L) (H)		3.6×10^6	8
III	1.671 (11%)	14	5′ A C A A A T T 3′ T G T T T A A	(L) (H)		3.6×10^6	8

The satellites are numbered according to buoyant density proceeding away from the main peak at 1.700 g-cm^{-3}. The proportion of G-C base pairs expected on the basis of the buoyant density is shown in brackets. The predominant sequence in each case accounts for >95% of the labeled oligonucleotides found in the cRNA transcripts of the satellite DNA. The number of copies is calculated for each sequence from the observed proportion of the genome, which is taken to be 3.2×10^8 base pairs. Data of Gall and Atherton (1974).

they display different buoyant densities. Since the buoyant density of each satellite is unaltered even at high molecular weight (up to 10^5 base pairs, or about 12,000 repeats), each seems to be organized as a long array of tandem repeats.

The satellites can cross hybridize with each other in a reaction characterized by Blumenfeld, Fox, and Forrest (1973). Unfortunately, this has the consequence that it is impossible to determine the relative locations of the satellites on the chromosomes. One or more satellites is present on each chromosome, except possibly the Y. All are located in the centromeric heterochromatin of mitotic chromosomes.

The relationship between the satellites immediately suggests that they are derived from a common ancestor. The three species of Drosophila most closely related to D. virilis all have satellite I peaks whose amount and buoyant density is similar to that of D. virilis. In D. a. americana, this satellite has been shown to have the same heptanucleotide sequence. Satellites II and III are not present in these other species. This implies that satellite I must have been present before separation of the species. Either satellites II and III have been lost in the others since speciation, or (more likely) they have evolved more recently in D. virilis. (The other Drosophila species assigned to the D. virilis group, but which are more distantly related to D. virilis itself, lack satellite I, which therefore must have arisen since the separation from them of D. virilis and its three closer relatives).

All the satellites are found in reduced proportions in polytene nuclei. We have already noted in Chapter 16 that in salivary gland cells the chromocenter contains far less satellite DNA than would be expected from the degree of polyploidy. This is caused by the failure of satellite DNA to replicate while the other chromosomal sequences are duplicated. In D. virilis the extent of the under replication of the satellites varies in different polyploid tissues, perhaps corresponding to the cessation of satellite replication at different times during the polyploidization. In pupal ovaries there may even be a slight increase in satellite content; this disappears by the time of adult emergence, and is replaced by a reduction within 5–6 days (Blumenfeld and Forrest, 1972; Endow and Gall, 1975; Renkawitz-Pohl and Kunz, 1975). There may be differences in the extent to which each satellite is underreplicated in the various polytene cell types. In lieu of information on their chromosomal locations, it is impossible to say whether this reflects differences in the control of replication on each chromosome or of each individual satellite.

The contrast between the differences in the physical properties of the three D. virilis satellites and the similarities of their sequences makes the point that it may be misleading to assess the relationship between satellites until sequence information is available. Thus the three related species D. hydei, D. neohydei, and D. pseudoneohydei all have satellites whose buoyant densities correspond to high G-C content and which vary in amount from 3.0 to 6.5% of the genome. When transcribed into RNA, the satellites of D. neohydei and D. pseudoneohydei each hybridizes well with its parent DNA but only 5–10 times less effectively with the DNA of the other species (Hennig, Hennig, and Stein, 1970). Since these Drosophila species are interfertile, this might seem to imply that evolution of satellite DNA is very rapid, but it would be premature to form a firm conclusion in lieu of sequence data.

D. melanogaster has four satellites, three of which have been investigated in some detail, as can be seen from the summary of Table 19.3.* All exist as large blocks of DNA; there is almost no reduction in the amount of satellites I, II, and IV when the fragment length is increased from 4500 to 225,000 base pairs. Relative to the number of copies of each satellite indicated in the Table, this implies that the number of different sites at which each satel-

---

*Evidence for tandem repetition of sequences in the genome has been gained by forming circles from duplex molecules treated with exonuclease III to generate single-stranded ends. The ability of the tails to anneal to form circular structures implies that complementary sequences are repeated a short distance further along the genome (Thomas et al., 1970). By using DNA extracted from polytene chromosomes of Drosophila species, Lee and Thomas (1973) attempted to identify circles from the main component of DNA. The contribution of satellite DNA to circle formation should be small, because of its underrepresentation in polytene chromosomes; the successful formation of circles was therefore taken to indicate the presence of serial repetition within the main components of the genome. However, Schachat and Hogness (1973) showed directly that most of the circles formed from D. melanogaster DNA in fact represent satellite sequences; and Peacock et al. (1973) showed that the removal of satellite DNA reduces circle formation in the remaining fraction by some 75%. The conclusion that there is little tandem repetition in the principal part of the genome, with most circles derived from satellite DNA, has been confirmed by Hutton and Thomas (1975).

lite may occur must be rather limited, of the order of say no more than 10. In some cases, arrays of different satellites may be adjacent. Brutlag et al. (1977) found a skew in the peak of satellite IV; judged by hybridization with other satellites, this may be due to the presence of sequences of satellites I and II. This suggests that there may be covalent linkage between different satellite clusters. Two different sequences that are transcribed have been found in DNA containing satellite III, although since these are repetitive it is not known whether the copy adjacent to the satellite is expressed (Carlson and Brutlag, 1978a,b).

Cytological hybridization with ^{3}H-RNA transcribed from each satellite suggests that they reside at a restricted number of loci. All are concentrated in heterochromatic regions. The Y chromosome contains large amounts of every satellite; other chromosomes have characteristic contents of the various satellites. In addition to the predominant localizations shown in the Table, small amounts of each satellite may be present on the other chromosomes. All the satellites appear to anneal to a single band in the polytene set (in the 21 C-D region of chromosome 2), but the significance of this is not clear.

All the satellites display renaturation kinetics corresponding to rather short repeating units. The exact reassociating length is difficult to determine for such short units, especially given the uncertain effects of mismatching (see next section); however, it is very low for satellites II and IV, somewhat longer than the unit shown in Table 19.3 for satellite I, and more complex for satellite III than for any other (Brutlag et al., 1977).

Satellite I has a low density; and nearest-neighbor analysis shows that some 84% of its sequence must be a simple alternation of A and T (Fansler et al., 1970). The disparity between the single strands that permits their isolation on alkaline density gradients shows that there must be another feature, however, and this takes the form of additional A-T pairs present in fixed orientation between the strands (Gall and Atherton, 1974; Peacock et al., 1973). Two possible forms of the repeating unit are summarized in the Table, each a variant of $(AAT)_n (AT)_m$.

A repeating sequence of 10 base pairs has been derived for satellite II; and satellite IV consists of two related repeating units, organized in separate clusters (Endow, Polan, and Gall, 1975; Brutlag et al., 1978). Polypyrimidine sequences identified as a cryptic satellite may be part of satellite IV (Birnboim and Sederoff, 1975; Sederoff, Lowenstein, and Birnboim, 1975). All the sequenced satellites fit the general structure $(AAN)_n (AN)_m$, where the variable positions N may be occupied by a G or a T, or by a combination of G, T, and C. Thus the families may share a common ancestor or have arisen in stages from each other. The three satellites that have been sequenced are highly homogeneous.

None of the three sequenced satellites has been shown to possess any repeating unit longer than that identified directly from the sequence studies. Satellites I and II are cut by the restriction enzyme HaeIII into heterogene-

Table 19.3   Satellite DNAs of D. melanogaster

Satel-lite	Buoyant density	G-C base pairs, %	Predominant sequence	Number of copies	Proportion of genome, %	Chromosomal distribution
I	1.673 (13%)	7	generally $(AAT)_n(AT)_m$ may be either 5′ A A T A A T A T A T A T 3′ T T A T T A T A T A T A or both 5′ A A T A T (60%) 3′ T T A T A and 5′ A A T A T A T (40%) 3′ T T A T A T A	6×10^5	5.2	68% on Y 22% on *4*
II	1.686 (27%)	22	5′ A A C A T A G A A T 3′ T T G T A T C T T A	4×10^5	2.6	50% on Y 24% on *2* 18% on *3*
III	1.688		has 378 bp repeat with considerable variation		4	48% on X 44% on Y
IV	1.705 (46%)	38	contains AAG, AG, G in ratio 3:4:1 90% is: 10% is: 5′ A A G A G A A G A G A G 3′ T T C T C T T C T C T C	1×10^6	3.8	52% on Y 33% on *2*

In addition to the satellites described above, a peak at 1.680 g-cm^{-3} corresponds to mitochondrial DNA. Further satellite sequences have been identified in the peak at 1.697 g-cm^{-3} that contains ribosomal DNA and in a presently uncharacterized peak at 1.690 g-cm^{-3}. Homogeneity in a satellite sequence is indicated by the recovery of > 90% of a radioactive label in the spots of the in vitro RNA transcripts corresponding to the predominant sequence. Satellite I consists of AAT and AT sequences and alternative arrangements have been proposed; these may be resolved by cloning satellite fragments for direct sequencing. Satellite II may have an additional AAT sequence every fourth repeat. Satellite IV consists of two related sequences organized into discrete clusters as shown by cloning. The distribution of the satellites on chromosomes is calculated from grain counts following in situ hybridization of labeled satellites with mitotic chromosomes of diploid cells.

References: Fansler et al. (1970), Botchan et al. (1971), Blumenfeld and Forrest (1971), Gall and Atherton (1974), Endow, Polan, and Gall (1975), Peacock et al. (1973, 1978), Brutlag et al. (1977, 1978).

ous patterns. This means that the sites recognized by the enzyme must have arisen independently in random repeating units. Nor can a regular pattern be detected in satellite IV, which is not cut at all. In contrast, however, satellite III is partly (13%) cut by HaeIII or Hinf into a regular series of bands whose monomer is about 350 base pairs (Manteuil, Hamer, and Thomas, 1975; Shen, Wiesehahn, and Hearst, 1976). This may correspond to the repeating unit of the entire satellite according to the suggestion of Brutlag et al. (1978) that the loss of restriction sites in many of the repeating units is responsible for their resistance to the enzyme.

The satellites characterized in D. virilis and in D. melanogaster are not obviously related. However, a possible link between them has been found in the form of a minor, cryptic satellite of D. virilis. This has the repeating structure $(AATATAG)_n$ (Mullins and Blumenfeld, 1979). It is related to satellite II of D. virilis by two base changes; it could be related to satellite I of D. melanogaster by one base change.

The satellite DNAs of crabs first attracted attention because of the identification of a satellite of very low buoyant density (1.683 g-cm^{-3}), which it was thought might comprise a simple alternation of bases in the form of poly-dAT (Sueoka, 1961; Swartz, Trautner, and Kornberg, 1962). Originally identified in Cancer borealis, in which this component represents 30% of the genome, A-T rich satellites since have been found in many Crustacea (and, as we have seen, in D. melanogaster, a much more distantly related Arthropod). However, this satellite may be present in lesser amounts, or may be absent altogether; and there is no correlation between the amount of satellite and evolution in the Crustacea (Skinner, 1967; Beattie and Skinner, 1972).

Satellites rich in G-C also may be found in crabs; and two minor satellites, representing 0.2 and 0.7% of the genome, have been sequenced in the hermit crab Pagurus pollicaris (Skinner et al., 1974; Chambers, Schell, and Skinner, 1978). As with the other Arthropod satellites, there is little variation in the sequences, which are:

$$5' \quad \text{T A G G} \quad \text{(H)} \qquad 5' \quad \text{A G T G C A G} \left(\begin{matrix} \text{C T G} \\ \text{G A C} \end{matrix} \right)_{3\text{-}12}$$
$$3' \quad \text{A T C C} \quad \text{(L)} \qquad 3' \quad \text{T C A C G T C}$$

and

The lack of any apparent relationship between these predominant sequences contrasts with the situation in Drosophila and suggests that these satellites may be of independent origin. (The sequence of the first constitutes part of that of the guinea pig satellite I, described below, but this is presumably coincidental.)

Mammalian Satellite DNAs

The combination of several approaches is necessary to reconstruct the hierarchy of periodicities characteristic of a satellite DNA whose repeating units have diverged in sequence. Analysis of the oligonucleotide products of ribonuclease digestion (of labeled transcripts) or pyrimidine tracts (of satellite DNA itself) identifies several sequences present in varying molar yields. Usually a small number of sequences, related to each other by base substitutions or deletions, is predominant, with other related variants present in smaller amounts. This variation between repeating units means that such sequence analysis is limited to determining the principal short sequence(s) from which the satellite has evolved; this is not sufficient to define the overall constitution of the satellite, since it does not reveal the pattern of relationships between adjacent variants of the repeating unit.

With more recent sequencing techniques, it is possible to obtain directly a sequence of a hundred or so nucleotides from any defined fragment of

DNA (usually produced by cleavage with a restriction enzyme) (see Chapter 26). The application of such techniques to satellite DNA is difficult, however, because variation in the repeating units provides substantial heterogeneity at many positions; restriction fragments therefore may not have unique sequences, so it may be impossible to derive an unequivocal sequence over the range of distances of 100–200 base pairs.

Although some difficulties have been encountered in cloning satellite DNAs in bacterial plasmids, this may be the best approach to determining how adjacent short repeats are related. The limitation of this approach, of course, is that each clone is useful for sequencing only a single stretch of DNA (~200 bases). Over longer distances, the same problems of heterogeneity that afflict direct sequencing of satellite DNA may make it difficult to obtain a defined order of fragments for sequencing. A large number of independently derived clones may give a wider view of sequence organization at this level, but with the drawback that the locations of the individual sequences within the satellite are unknown.

Repeating units longer than the short oligonucleotides prominent in direct sequence analysis can be recognized by reassociation analysis. The highly repetitive nature of satellite DNA was first recognized by the low $Cot_{1/2}$ characterizing the kinetics of reassociation. But the length of the repeating unit corresponding to these kinetics is usually much greater than that revealed by sequence studies. It is difficult to determine the exact length of the reassociating unit when the renaturing sequences are related rather than identical, because mismatching may influence the rate of reassociation. Although the reaction seems to be slowed by mismatching, efforts to quantitate the effect have produced estimates varying from almost zero to about 3-fold (Sutton and McCallum, 1971; Southern, 1971; Bonner et al., 1973). Thus the observed $Cot_{1/2}$ is likely to be too great, but by an undetermined amount.

In spite of this uncertainty, however, it is clear that the reassociating unit often is of the order of >100 bases or so, compared with the smallest sequence unit of the order of 10 base pairs. The reassociation kinetics represent annealing of the shortest sequence able regularly to recognize its complement. Since this appears to consist of an array of several of the repeating units identified by sequence analysis, the presumption must be that these represent variants sufficiently different to compel reassociation to occur in only one phase. The reassociating unit may, of course, itself be organized into longer regular arrays; but the variation between adjacent reassociating units must be small enough to allow them to renature.

A technique that yields information about long-range periodicities is the use of restriction enzymes. When an entire eucaryotic chromosome is digested with a restriction enzyme, the sites that are attacked should occur with a spacing that is determined statistically by the length and base composition of the recognition sequence. In effect, each restriction enzyme selects a different set of randomly spaced sites. Over distances as great as the eucaryotic genome, the absence of any regular arrangement of sites has the result that the cleaved DNA forms an extended smear when analyzed for fragment size on acrylamide gels. An exception is provided by the tandem repetitions constituting satellite DNA. Here a given restriction site may be repeated regularly; so its cleavage generates many fragments of the same size. These may sometimes be recognized as distinct bands in digests of whole DNA; and, of course, can be generated by cleavage of an isolated satellite DNA. These bands may identify a repeating unit longer than that responsible for reassociation.

The first evidence for a periodicity in a mammalian satellite shorter than that identified by reassociation was provided by an analysis of satellite I of the guinea pig. After separating the complementary strands in alkaline CsCl, Southern (1970) subjected each to degradation with diphenylamine to generate pyrimidine tracts. The most frequently occurring fragment in the L strand is the sequence CCCT. And all except 10% of the total tracts isolated from this strand can be derived from CCCT by a single base substitution, such as CCTT, CTCT, etc. In the H strand the most frequent sequence is TT; and T, TTT, and T_2C are also common. This is consistent with the occurrence of a basic repeating unit of the structure:

$$5' \quad T \; T \; A \; G \; G \quad (H)$$
$$3' \quad A \; A \; T \; C \; C \; C \quad (L)$$

which together with related sequences makes up some 90% of the satellite. Southern therefore suggested that the satellite evolved by the amplification of this short sequence, which must have been modified by the accumulation of mutations during evolution.

The kinetics of reassociation identify a repeating unit that is much longer, in the range of 100–300 base pairs, depending on the correction made for mismatching. The discrepancy between the length of the basic 6 base pair repeat and that of the unit of reassociation suggests that the evolution of satellite DNA may have occurred in stages. Southern's model thus suggests that initially there may have been a multiplication of the basic sequence to produce (say) 10–20 copies. The introduction of mutations into this longer sequence could have produced the simple variants. A further amplification of a unit consisting of these variants then could generate the repeat identified by reassociation kinetics. Further mutations since may have been introduced to render the reassociating units nonidentical.

Similar internal heterogeneity is seen in the three satellites of the kangaroo rat, Dipodomys ordii, which together occupy about half of the genome. Table 19.4 summarizes the general properties of these satellites and Figure 19.5 shows the relationship between the major sequence variants. It is clear that most of the satellite is occupied by variants that are related to the predominant sequence by simple changes, usually single base substitutions or deletions or insertions. Sometimes part of the repeating unit may vary in length.

Members of the genus Dipodomys all appear to possess satellites at the same densities, suggesting that they may have originated before this line of

Table 19.4 Satellite DNAs of dipodomys ordii

Satellite	Buoyant density	G-C base pairs, %	Predominant sequence	Number of copies	Proportion of genome, %
MS	1.707 (47%)	39	5′ G C T $\left(\text{T C T}\right.$ (H) 3′ C G A $\left.\text{A G A}\right)_{1\text{-}2}$ (L)	1.5×10^8	22
HSα	1.713 (54%)	43	5′ T T A G G G (H) 3′ A A T C C C (L)	1.5×10^8	19
HSβ	1.713 (54%)	66	5′ C C C G C T G T G T (H) 3′ G G G C G A C A C A (L)	5.0×10^7	11

All satellites are heavier than the main peak of DNA and the G-C content predicted from the buoyant density (given in parentheses) is different from that actually found. In HSβ the content of 5-methyl cytosine is 6.7%, corresponding to methylation of 20% of the C residues, at unknown positions. The number of copies is calculated from the proportion of the genome occupied by each satellite, the length of the predominant sequence, and the assumption of a haploid genome size of about 5×10^9 bp. Data on the general properties of the satellites are taken from Hatch and Mazrimas (1974) and Hatch et al. (1976); data on the sequences from Figure 19.5.

evolution commenced (Mazrimas and Hatch, 1977). This suggestion needs confirmation by sequence studies. There are wide variations in the amount of satellite DNA in the genome of each species, possibly correlated with the number of metacentric (versus acrocentric) chromosomes (Hatch et al., 1976). This is consistent with the location of satellite DNA in centromeric heterochromatin; it seems reasonable to suppose that changes in satellite DNA amounts might occur during rearrangements involving the centromere. There is evidence that HSα sequences are located at the centromeres of all but 2–3 pairs of the D. ordii chromosome set; the other satellites may be located in short chromosome arms (Prescott et al., 1973). At least some extracentromeric locations might be expected from the large amount of satellite DNA in the D. ordii genome.

The sequences of the D. ordii satellites show two especially interesting features. First, the predominant short sequences of the different satellites do not appear to be related. This suggests that the satellites may be of independent origin; more formally, any common ancestral sequence would have to be very distant in view of the degree of divergence. Second, the predominant sequence of satellite HSα is the same as that of guinea pig satellite I. This satellite may therefore antecede the separation of guinea pig and kangaroo rat.

This satellite indeed appears to exist in other rodents. Fry and Salser (1977) found that similar fingerprints are generated from the in vivo transcripts of corresponding satellites in two other rodents, the pocket gopher (Thomomys bottae) and the antelope ground squirrel (Ammispermophilus leucurus). This satellite is therefore found in all three suborders of the rodents and must have evolved before their separation (40–50 million years

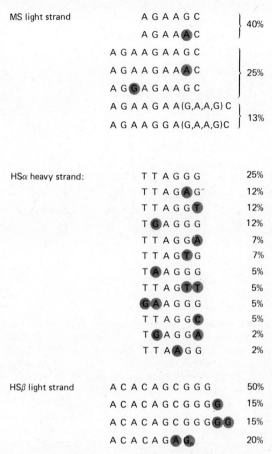

Figure 19.5. Probable variations on the predominant sequences of D. ordii satellites. The most commonly found longer sequences, derived from cleavages at C in MS and at G in HSα and HSβ, are given; positions of base changes relative to the predominant sequence are indicated by shading. Data of Fry et al. (1973), Salser et al. (1976), and Fry and Salser (1977).

ago). The similarity of fingerprints suggests that the major variants of the predominant sequence may prove to be the same in each case, which would imply that the present structure of the satellites has been maintained over a long evolutionary period (see next section).

The most extensive sequence analysis now has been performed with mouse satellite DNA, in which several periodic levels of organization have been identified. The pyrimidine tracts of H strand DNA take the form T_nC (Harbers et al., 1974). When RNA transcribed from the H strand is digested with T1 ribonuclease, about half takes the form of fragments with the structure A_nTGA, with some variation. The analysis of these fragments performed by Biro et al. (1975) is summarized in Figure 19.6. Again it is clear that much of the satellite represents a set of short sequences, related by base substitutions and changes in overall length.

5′   3′

G	A	A	A	A	A	T	G	A				~4%
G	A	A	A	A	–	T	G	A				~4%
G	A	A	A	A	C	T	G	A				~4%

T
C G A A A T G   ~4%
GA

G	T	A	A	A	A	T	T	A	(G)			~1½%
G	–	A	A	T	A	T	G	(G)				~1½%
G	C	A	A	A	–	T	G	C				~1½%
G	C	A	A	A	A	T	C	A	T	G	(G)	~1%
G	–	A	A	A	T	C	A	C	G	(G)		~1%
G	–	A	A	A	–	T	A	T	G	G	A	~1%
G	A	A	A	A	A	(G)						~1%
G	A	A	A	–	C	(G)						~1%
G	T	A	C	–	–	T	G	A				~1%
G	T	A	T	G	(G)							~1%
G	C	A	A	(G)								~1%

Figure 19.6 The most common oligonucleotide sequences of the light strand of mouse satellite DNA. The nucleotides were generated by ribonuclease T1 digestion of labeled cRNA transcribed from a template of mouse satellite heavy strand DNA. Only the larger of the common oligonucleotides are given here, since the smaller species cannot be aligned unequivocally with the sequence. The proportions indicated for each sequence are based on total radioactivity and are only rather approximate; they represent about 35% of the total, and the omitted smaller oligonucleotides account for about 60%.

Since RNAase T1 cleaves on the 3′ side of G residues, each fragment is presumed to commence with a G. Those shown as terminating in A represent instances where this was shown to be the next nucleotide by nearest neighbor analysis; those terminating in G are shown with the last base in brackets, since it might also be the first base of the next repeating unit. Actually the proportion of free G residues, derived from GG or GGG sequences, is high. This method of analysis is biased against showing positions in which G has been generated within the repeating unit (as aligned here), because the fragments produced by RNAase T1 cleavage of such units are small and therefore have been omitted as noted above. T is shown in place of the U actually detected in the RNA product. Spaces have been introduced as indicated by the dashes to maximize homology between the sequences; in the runs of A residues, the exact locations of putative deletions are arbitrary. Calculated from data of Biro et al. (1975) and Reis and Biro (1978).

The sequence of part of a short cloned fragment of the satellite has been obtained by Reis and Biro (1978); as can be seen from Figure 19.7, this consists of a series of tandem repeats related to the predominant sequences identified in the oligonucleotide fragments. This unique sequence was obtained by cloning a 240 base pair fragment released from the satellite DNA by the enzyme EcoRI; the total oligonucleotides released from a transcript of this sequence by ribonuclease are identical in constitution, but different in relative amounts, from those generated from total satellite DNA. The cloned sequence may therefore be taken to represent a particular component

5′   G-A-A- T -A—T-G-C-G-A-
      G-A-A- A -A-C-T-
      G-A-A- A -A—T-C-A-
      G-A-A- A —C—G-A-
      G-A-A-[A]—C-T-C-A-C-T-[G]
      G-A-C- G————A-C-T-
      G-A-A- A————T-G-A-C-
      G-A-A——T-C-T-
      G-A-A- A -A-A————C-G-T-
      G-A$_n$-   3′

Figure 19.7. Sequence of a cloned fragment of mouse satellite DNA, aligned to show homologies between probable repeating units. Uncertain bases shown in brackets. Data of Reis and Biro (1978).

of the satellite. The 78 nucleotide sequence that has been determined represents one end of the cloned fragment. This sequence is related to a sequence obtained from total satellite DNA by priming synthesis of labeled DNA products from EcoRII sites; the adjoining sequences must be heterogeneous as shown by uncertainties in the sequence of the labeled product. In spite of the ambiguity found in bases at many sites, this sequence generally can be aligned with the cloned sequence. [Some variation in the regular sequence organization must exist, since isolated H or L strands are able to support self reassociation to a limited but reproducible extent—about 15% of either strand self-anneals (Flamm, Walker, and McCallum, 1969). This implies that each strand must contain short sequences that are complementary rather than homologous.]

Evidence for longer periodicities in mouse satellite DNA has been obtained by cleavage with restriction enzymes and by reassociation analysis. Southern (1975a) showed that the products of cleavage with EcoRII form a series of bands on agarose gel electrophoresis. Figure 19.8 shows that after partial digestion these constitute an arithmetic series, representing multiples of up to at least 26 times the length of the 240 base pair monomer. Also seen more faintly are bands corresponding to ½, 1½, 2½, and 3½ of these repeats. Thus the sequence recognized by EcoRII, $\frac{GGACC}{CCTGG}$, must occur at regular intervals; this is a feature of much of the satellite DNA, since a large proportion can be cleaved into these bands. The width of the monomeric band is sufficient to correspond to a heterogeneity of about 20 base pairs between the longest and shortest of these repeating units. Even after complete digestion, the satellite is not reduced completely to a single monomeric band, but some longer bands remain intact. These probably result from mutations that have eliminated the restriction sites in some repeats. These data suggest that the short fragments identified by sequence analysis are organized into longer repeats. The repeating unit consists of a tandem array of variants of the short sequence; the variation in its length from 230–240 base pairs, and the disappearance of some of the EcoRII recognition sites, show that the copies of this unit have suffered divergence since it first arose.

A repeating unit of about the same length can be identified by reassociation kinetics. Renaturation of the EcoRII bands occurs with a $Cot_{1/2}$ of

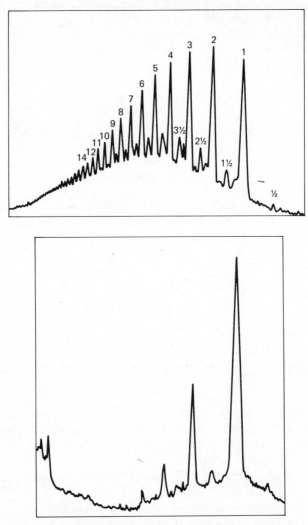

Figure 19.8. Digestion of mouse satellite DNA with EcoRII. The upper pattern is a partial digest analyzed on a 1% agarose gel; the lower pattern is a limit digest analyzed on 2% agarose. Data of Southern (1975).

about $3\text{-}5 \times 10^{-4}$.* Allowing for the uncertain effect of mismatching on the rate of reassociation, this corresponds to a reassociating length for satellite DNA in the range of 150–300 base pairs.

If the reassociating unit coincided with the EcoRII monomer, its length

*Actually the renaturation of the monomeric, dimeric and trimeric EcoRII bands gives $\text{Cot}_{1/2}$ values of 2.9×10^{-4}, 3.8×10^{-4}, and 5.0×10^{-4}, respectively. This relationship between length and the rate of reassociation, as observed previously by Hutton and Wetmur (1973b), is the reverse of that shown for nonrepetitive DNA (see equation 15 of Chapter 18).

should be 240 base pairs; and the reassociated molecules should be in register, that is, should be in the duplex state along their entire length (apart from any mismatching).* But if shorter lengths of DNA can reassociate, there should be some stagger, so that on occasion only part of each complementing single strand anneals, to form duplex molecules with single strand tails. These tails should in turn be able to reassociate with other single strands, generating longer molecules. Southern showed that such molecules are formed, with the fraction associating in register declining from 56% with the EcoRII monomer to 25% and 17% for the dimer and trimer.

This is what would be expected if the unit for reassociation were half of the monomeric length; then there would be two reassociating units per monomer, four per dimer, and six per trimer, so that random association of the first complements on each molecule to renature would produce ½, ¼, $\frac{1}{6}$ in register with the increase from monomer to dimer to trimer. There must therefore be a periodicity in mouse satellite DNA of less than the EcoRII monomeric length; and this may be equal to half the monomer. Taking each periodic unit to represent a stage in the evolution of satellite DNA, this suggests that there may have been at least two stages in the amplification of the original short repeat, with somewhat different degrees of divergence accumulating between them (see below).

Two types of restriction pattern in mouse satellite DNA have been distinguished by Horz and Zachau (1977). One is the EcoRII pattern, which in a limit digest generates a high proportion of monomers and extends only so far as the pentamer. A pattern with the same periodicity is generated by the enzymes HindII, Bsu, Alu, and EcoRI; but even in a limit digest, only a small proportion of the satellite DNA is cut, and the band pattern extends at least as far as the decamer. The presence of a large amount of undigested DNA makes it difficult to quantitate this band pattern; but upon double digestion with EcoRII and either HindII or Bsu, 0.6–1.4% of the repeat units are stated to contain two restriction enzyme sites. If the relative arrangement of EcoRII sites versus HindII and Bsu sites is random, the total number of repeats possessing HindII or Bsu sites must be about 0.7–1.5% (since 80% of the repeats have an EcoRII site).†

If the sites for HindII or Bsu were randomly distributed throughout the entire satellite, the probability of their occurrence in adjacent or closely lo-

---

*This model applies because the reassociating DNA is derived by restriction cleavage of the satellite and therefore has defined end points. This contrasts with the random end points generated when DNA is sheared, when differences in the lengths of reassociating complements would always generate tails.

†Verification of this assumption requires an experiment in which satellite DNA is digested with EcoRII after which the insusceptible fraction as well as the susceptible fraction is digested with HindII or Bsu. The conclusion quoted in the text rests upon a joint digestion experiment in which the number of repeat units suffering two cleavages was estimated on the basis of the amounts of the resulting intermediate size bands. This gives the number of HindII or Bsu sites existing in the 80% of the satellite that is cleaved by EcoRII, but leaves open the possibility that the proportion might be different in the undigested 20% fraction. For the purposes of the text calculation, the proportion is assumed to be the same.

cated repeats would be rather small. To generate the observed band pattern, these sites must therefore be clustered in distinct domains of the satellite. These two enzymes appear to identify different domains, since the DNA resistant to cleavage by Bsu is still cleaved into the usual pattern by HindII. The same may be true of the other enzymes, since joint digestions seem to generate larger amounts of the standard series of bands. Each domain may represent a region of some 5–10% of the satellite, in which the sites recognized by the enzyme occur with separations varying from adjacent to more than 15 repeats, with an average spacing of about 8 repeats. Whether each domain is a contiguous region or consists of several smaller clusters is an interesting question that may bear on the evolution of this organization (see below).

Comparable digestion patterns are given when other mammalian satellites are treated with restriction enzymes. Bovine satellite I is digested into a series of bands representing multiples of a 1400 base pair monomer; a half length series also is found (Botchan, 1974). The α satellite of the African green monkey is digested by HindIII into multiples of a 176 base pair repeat; contrasted with the 88% of the satellite cleaved by this enzyme, only 25–30% is cleaved into a similar pattern by the enzymes EcoRI or Bsu (Fittler, 1977). This again suggests the existence of domains within the satellite. With some enzymes, guinea pig satellite I gives a random smear or a series of bands with no simple repeating pattern; but with Bsu there is a series based on a 215 base pair monomer, showing the bands, 1, 1¼, 1¾, 2, etc. (Horz, Hess, and Zachau, 1974; Altenberger, Horz and Zachau, 1977). The Bsu enzyme appears to cleave twice within the repeating unit, at sites 55 base pairs (¼ length) apart. This raises the possibility that evolution has proceeded through successive doublings of the 55 base pair unit.

An interesting feature with implications for restriction enzyme studies has been found in the bovine satellite I. Gautier, Bunemann, and Grotjahn (1977) observed that when the 1400 base pair EcoRI repeat is cloned in E. coli, three sites within the repeat become available for cutting by the enzyme SmaI. This suggests that the $\frac{CCCGGG}{GGGCCC}$ sequence recognized by SmaI may be methylated in the native satellite, rendering it immune to the enzyme; but it fails to be methylated in E. coli. The same result is given by the enzyme HapII, which recognizes the sequence $\frac{CCGG}{GGCC}$; but the enzyme BsuI cuts the native satellite and cloned DNA equally well, although its action is inhibited by methylation of its recognition sequence $\frac{GGCC}{CCGG}$. This suggests that methylation in the satellite occurs only in the sequence $\frac{5'\ CG\ 3'}{3'\ GC\ 5'}$. About 7% of the C residues in the satellite are methylated. The existence of this methylation may not only influence the pattern of restriction cutting, but also means that attempts to obtain direct sequences may encounter the difficulties resulting from base modification.

Returning to rodent satellites, there seem to be quite wide variations in the proportions and buoyant densities of satellites even among quite closely related species (Hennig and Walker, 1970). However, related satellites which appear to have diverged from a common ancestor are found in the genomes of some Mus species. The Asian mice M. caroli, M. famalus, M. cervicolor, and M. castaneus all have satellites on the light side of the main band DNA; that of M. castaneus actually has the same density as that of M. musculus (to which it is closely related as shown by their interfertility). When separated ^{32}P-labeled strands of M. musculus satellite DNA are reassociated with the complementary strands of the other satellites, reassociation occurs to generate somewhat poorly matched duplexes, according to their low T_m (Sutton and McCallum, 1972; Rice and Straus, 1973).

The nature of the relationship between the satellites of M. musculus and M. caroli has been defined in more detail by Sutton and McCallum. The cross reassociation reaction occurs more slowly than either homologous reaction (both have about the same $Cot_{1/2}$). At 60°C, the heterologous $Cot_{1/2}$ is about 60 times the homologous $Cot_{1/2}$; when the criterion is lowered to 50°C, the discrepancy is reduced to about 3-fold. Qualitatively this confirms that the satellites are related, but suggests that the divergence is great enough so that the increase in criterion from 50 to 60°C makes it difficult for sufficiently homologous sequences to find each other. The kinetics are consistent with the relationship seen in melting studies of the cross-reassociated satellites; the divergence between satellites of different species is greater than the divergence within each satellite.

Less detailed studies of five species of mice of the genus Apodemus suggest that they too may possess a satellite originating before speciation. Digestion of total DNA with EcoRI in each case generates a band of 430 base pairs, as well as some larger bands whose patterns overlap but are not identical in all species. HindIII produces bands corresponding to 370 base pairs and dimers and trimers in three of the species (Cooke, 1975). The nature of these tandem repeats remains to be established; but they raise the possibility that related sequences may be present in each species.

Evolution of Tandem Repeats

The existence of tandem repeats in the genome poses several intriguing questions. From the perspective of the present arrangement of sequences, an immediate concern is the specificity of the sequence. To what extent is sequence necessary for function? This cannot be answered directly, since there is almost no information about the functions of satellite DNA. The question is tantamount to asking what selective pressure applies to the sequence. An indication that the answer may not always be the same is provided by the contrast between the constancy of sequence within the Arthropod satellites and the divergence in mammals. Yet it is difficult to see how selection can affect the individual members of so large a family. Even if sequence were critical, presumably it would be possible for mutations to occur in a large number of repeating units before the number with the necessary sequence became insufficient. Thus an accumulation of mutations might be expected before selection became effective. This concern is related to the question of how uniformity of sequence is maintained in repetitive gene families, an issue we shall take up in Chapter 27.

The evolution of tandem repeats is perplexing. These sequence clusters display their history in a sense that nonrepetitive sequences do not; for within the cluster can be seen many descendants of an original sequence, not just one. More precisely, the extant clusters show the structure of successful lines; there is, of course, no evidence about any unsuccessful variants. It is clear from the repetitive nature of the sequences comprising the cluster that satellite DNA has arisen by amplification of short ancestral sequences, sometimes with the acquisition of longer periodicities, sometimes not. Two types of process have been proposed to account for this amplification: saltatory replication; and unequal recombination.

The concept of *saltatory replication* was introduced by Britten and Kohne (1968) to describe a sudden event in which a particular sequence might suffer replication to produce a large number of copies. Because base substitutions of the putative ancestral sequences of diverged satellites do not seem to have occurred with equal frequencies at all positions, Walker (1971a) suggested that there might have been a series of successive replications separated in time by the introduction of mutations. The presence of periodic hierarchies, in which each repeating unit consists of a number of variants of a smaller (less well related) unit, led Southern (1975) to suggest that each successive replication may have involved a unit of larger size.

This model is illustrated in Figure 19.9. Suppose an initial short sequence *a* is replicated several times to produce a tandem repeat. After this sudden event, there is a period of some generations during which mutations occur to produce variants *b*, *c*, etc., closely related to the original sequence. This divergence may take the form of simple base substitutions, deletions, and insertions. Then another saltatory replication occurs, this time to amplify a sequence of the length of several ancestral units, and which includes more than one variant. Again mutations accumulate; and again there may later be another saltatory replication to amplify part of the then current sequence. Because divergence occurs among the members of the cluster between each saltatory replication, successive periodicities are generated. The amount of time that passed between successive amplifications can be recognized by the degree of divergence found between successive components in the periodic hierarchy. The frequency with which any sequence occurs reflects the time at which it was amplified; in terms of the shortest components, the ancestral sequence itself should be the most common, variants introduced before the first saltatory replication the next common, and so on. The nature of the event responsible for a saltatory replication is unspecified; several possible mechanisms have been proposed, but there is no evidence on which (if any) actually may have occurred.

The diverged satellite for which the most data are at present available is that of the mouse M. musculus. The putative ancestral sequence is GA_5TGA, which is one of the predominant short sequences, and from which most of the other prominent sequences shown in Figure 19.6 can be derived by simple base substitutions or deletions or insertions. The four most prominent sequences are closely related; the second and third can be derived from the first by a single change (an $A \rightarrow C$ base substitution and an $A_5 \rightarrow A_4$ deletion, respectively); and the fourth can be derived by a deletion and a substitution. Thus a good candidate for the first unit of saltatory replication would be a sequence of four or more ancestral unit lengths, its components showing about 12% divergence (differences at 4 out of 32 total positions).

This divergence is fairly close to that seen between the satellites of different Mus species. The ΔT_m for reassociated M. musculus satellite DNA is about 7°C; that for the cross reassociation reaction with M. caroli, M. cervicolor, or M. famalus is about 23°C (Sutton and McCallum, 1972). The 16°C difference would correspond to an interspecies divergence of roughly 16–20%. The similarity suggests that this amplification step might have occurred at about the same time as the divergence of the species.

The next stage of amplification might have involved a unit of about 120 base pairs. The yields of the minor oligonucleotides shown in Figure 19.6 suggest that roughly a further 10% divergence occurred before this sequence was amplified.

Only a rather small degree of further divergence seems to have been introduced before a 240 base pair unit was taken for the most recent saltatory replication. The ΔT_m of 7°C for reassociated satellite DNA compared with native DNA suggests that there is only low divergence between the reassociating units. Actually, since this reaction represents reassociation in both register and half register, this is an average (as explained in Figure 19.9). However, the melting curve does not show two components, but is broader than that of native DNA by an amount suggesting that the two registers differ in T_m by about 2°C (Southern, 1975). This suggests that only about 2% divergence occurred between the last two saltatory replications.

Divergence in the longest (240 base pair) register can be estimated from the restriction data. If every repeat unit originally possessed an EcoRII site, each instance in which a higher multiple is present on a gel identifies the loss of one (or more) sites by independent mutation. The fraction of EcoRII sites that has been lost appears to be about 14%. If a single mutation in any one of the 5 base pairs of the site is sufficient to prevent cleavage, then there has been roughly 3% divergence (i.e., 14%/5) since the last saltatory replication.

If there has been no selection for particular sequences in the mouse satellite, and if mutations accumulated at the same rate as in other mouse DNA sequences, the rate of divergence would be about 1% per million years (Southern, 1975). Then the saltatory replication events should have occurred at times of roughly 3×10^6, 5×10^6, 10^7, and 2×10^7 years ago. However, the assumptions on which this calculation is based can at best be only ap-

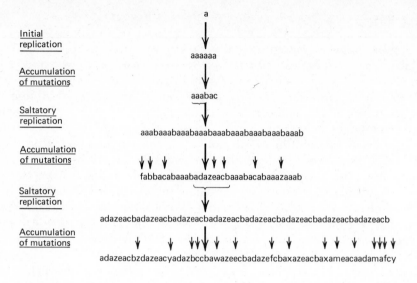

Initial replication

a

aaaaaa

Accumulation of mutations

aaabac

Saltatory replication

aaabaaabaaabaaabaaabaaabaaabaaabaaab

Accumulation of mutations

fabbacabaaabadazeacbaaabacabaaazaaab

Saltatory replication

adazeacbadazeacbadazeacbadazeacbadazeacbadazeacbadazeacbadazeacb

Accumulation of mutations

adazeacbzdazeacyadazbccbawazeecbadazefcbaxazeacbaxameacaadamafcy

Restriction maps for sites in:

b	10	3	8	8	8

c	8	7	1	8	8	8	8	8

d	8	8	16	24

e	8	16	1	7	8	8

z	5	3	8	8	8	8

Figure 19.9. Model for evolution of satellite DNA by successive saltatory replications. A sequence *a*, probably consisting of 5–10 nucleotides, is replicated to produce several copies. Mutations accumulate at the sites indicated by arrows, generating components *b*, *c*, etc, related to *a* by single base changes. A longer part of the sequence, *aaab*, then is selected for saltatory replication; again mutations accumulate in the amplified sequence. Mutations of the components *b*, *c*, etc. are designated *z*, *y*, etc.; these are related to *a* by double base changes. In the next saltatory replication the sequence *adazeacb* is amplified; and mutations accumulate in the product to generate the present sequence. Some of these mutations are the same as those occurring in previous steps (e.g. *a* ⟶ *b*); others are new (e.g. *a* ⟶ *f*; *z* ⟶ *m*).

If the mutations generating the sequences *b,c,d,e,z* created sites recognized by restriction enzymes, the final sequence would be cleaved into the fragments shown. In each case, the predominant product is a monomer consisting of a length of 8 ancestral units; dimers and trimers may also be present, as well as a small number of other fragments resulting from mutations occurring since the last amplification.

Direct analysis of sequence would reveal short components to be present in proportions depending on the time at which they originally arose in the lineage of the present sequence. The ancestral sequence *a* is the most common; sequences related to it by 1 substitution should be more common than those related by 2 or 3 substitutions, which on average should have occurred later. But this need not be true of any individual sequence; thus *z* happened to be contained in the second unit of saltatory replication and is common, compared with *y*, *x*, and *w* which have arisen since then by mutation; and *f* originated only recently, since it happened not to be included in the early units of saltatory replication.

Ignoring the mutations occurring since the last saltatory replication, reassociation in register with the unit of 8 ancestral sequence lengths would occur without mismatching. Reassociation

562

Sequence analysis shows constitution:

Ancestral sequence:	a̲	21 copies	33%
Sequences related by 1 substitution:	b̲	6	
	c̲	9	
	d̲	5	45%
	e̲	7	
	f̲	2	
Sequences related by 2 substitutions:	z̲	7	
	y̲	2	19%
	x̲	2	
	w̲	1	
Sequences related by 3 substitutions:	m̲	2	3%

Hybridization analysis shows possible
matches in last unit of saltatory replication

In register	a d a z e a c b ȧ d̊ ȧ ż e̊ å ċ b̊	0 mismatches
In half register	a d a z e a c b e̊ å ċ b̊ å d̊ å ż	1 mismatch/unit
In quarter register	a d a z e a c b å ż e̊ å ċ b̊ å d̊	1.5 mismatch/unit
	a d a z e a c d ċ b̊ å d̊ å ż e̊ å	1.5 mismatch/unit

Figure 19.9. *(Continued)*

in half register, corresponding to a unit of 4 ancestral lengths (that of the first saltatory replication) would occur with only 1 mismatch/unit since each reassociating element is related to its complement by a single change. In quarter register the number of mismatches is increased, because many of the corresponding units would be related by two or more changes. In eighth register, corresponding to reassociation of units of ancestral length, the mismatching/unit length may be greater again. The discrepancy between matching in register and in half register may be small enough to allow both to occur; for example, if the ancestral unit were 10 base pairs long, in this instance the mismatching would be 10%. This would identify a repeating unit of 4 ancestral unit lengths, with an average mismatching of 5%. Taking into account mutations occurring since the last saltatory replication, there would be few sequences able to anneal without some mismatching. The spectrum of sequences might mean that the difference between register and half-register would not be significant; while the difference in the quarter register might prevent reassociation.

proximate, because the other rodent sequences probably have been subject to selection; at all events, there is no reason to suppose that the selective pressure on satellite DNA is the same as that on other sequence components.

If genetic recombination occurs in satellite DNA, unequal crossing over should happen whenever the two homologues become aligned out of register. In the mouse satellite, this is likely to occur in half register because of the low divergence between the two components of the longest repeating unit. The result should be to generate equal numbers of ½-mers and 1½-mers in the restriction pattern, as illustrated in Figure 19.10. Higher multiples containing half length repeats (2½-mers, 3½-mers, etc.) should occur whenever mutation removes an EcoRII site adjacent to a half repeat, or when the recombination event occurs near a missing restriction site. These longer bands are present at the expense of shorter bands in proportions about those that would be expected. The frequency of unequal crossover is about 1 for every 15 repeats of the 240 base pair unit. In addition to the ½-mer band, there are some other rather minor bands smaller than the monomer; these could be produced by less frequent crossing over in other, shorter registers. The presence of intermediate bands is a common feature in restriction maps of satellite DNA and may be taken as general evidence for the existence of subregisters able to support unequal recombination.

Unequal recombination provides an alternative mechanism that might account for the evolution of satellite DNA. The theory of this model has been developed by Smith (1973, 1976) and rests on the assumption that crossing over at unequal sites occurs frequently between sister chromatids. The result of a series of such crossovers is illustrated in Figure 19.11. Computer simulation shows that the result of successive random events is to impose uniformity on the region involved. Its size may increase or decrease, depending on the position of the crossover, but eventually the entire region comes to consist of repeats of a single sequence. This process has been described as *crossover fixation* and can be characterized by the number of generations required for uniformity to be achieved.

This model makes two important implications about the evolution of satellite DNA. First it predicts that short tandem repeats will evolve to occupy any region of DNA that is not subject to selective pressure. Second it predicts that uniformity will be achieved within a region of tandem repeats, because any new variants created by mutation either will be eliminated or will expand to occupy the entire array.

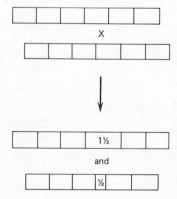

Figure 19.10. Generation of new repeating units by unequal crossover.   If two clusters are aligned out of register, a recombination event generates one unit larger than, and one unit smaller than, the monomer. In the illustration, the two clusters are aligned in half-register, generating one ½-mer and one 1½-mer. These may give rise to 2½-mers and higher multiples when the adjacent restriction sites are lost by recombination. (This order of events could be reversed.)

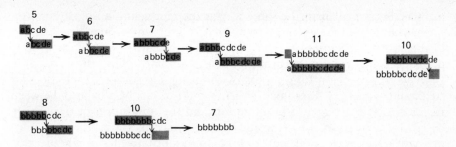

Figure 19.11. Model for evolution of satellite DNA by unequal crossover. When two recombining sequences are misaligned, a genetic crossover may change the size and constitution of the unit. Although reciprocal products are generated by each event, the fate of only one is shown (indicated by the shaded area). Crossovers within the unit generate duplications and deletions. Thus the very first event shown generates *abbcde* (as drawn, with a duplication of *b*) and also *acde* (not shown, which has lost the sequence *b*). The sequences at the very ends (*a* and *e*) can be lost only by terminal crossovers; anomalous sequences may therefore be retained in these positions. The size of the unit may either increase or decrease, as indicated by the numbers which give the total length in each generation.

Sequences of simple repetition could arise from crossovers between sister chromatids as shown in Figure 19.11, where each unequal event results in reciprocal deletions and duplications. These would be lethal if the sequence were subject to selective pressure; but in the absence of selection, some sequences survive while others by chance are lost. Eventually one sequence will predominate in the population. The sites of crossover might be determined by the locations of rather short sequences which fortuitously are homologous. With the assumption that recombination can occur at matching sites 4 base pairs long, a sequence of 500 base pairs acquires a short predominant repeat length, ranging in different computer runs from 5 to 39 base pairs, in 200 cycles. Each cycle consists of the introduction of 1 random mutation and 500 "attempted crossovers," which represent alignments of randomly chosen sites in the sister chromatids, which lead to recombination only when the 4 base pair criterion for matching of sequence is met. Another limitation was that the overall size of the array should always remain between 450 and 550 base pairs. Whether these conditions correspond at all well with those prevailing in nature naturally is dubious; but the same general result might be expected to occur over a time span determined by the actual values taken by the controlling parameters.

When a sequence has become at least partially repetitious, most recombination events will occur with the repeats in register, although they may involve units at different positions in the two recombining arrays. Such events result in duplications and deletions that expand or contract the size of the array. The computer simulation shows further that uniformity is imposed on the repeating unit. Thus repeats will diverge whenever mutation creates a new member of the set. But the number of copies of the new member is controlled in the same way as the representation of original se-quences of a random starting array. That is, the number may be expanded by duplication, eventually coming to occupy the entire population, or may be reduced by deletion, eventually being eliminated. This allows satellite DNA to consist of uniform repeats without the imposition of selection for particular sequences. Presumably there would in fact be a balance between the creation of divergence by mutation and the imposition of uniformity by crossover fixation, which would be determined by the relative speeds of the two processes. It is difficult to know in quantitative terms whether this model can account for the evolution of satellite DNA. Qualitatively, how-ever, it seems that if unequal crossovers were to occur sufficiently frequently relative to mutation, they could first generate and then maintain arrays of tandem repeats.

Can this model account for the existence of hierarchical periodicities? Longer repeats might be generated from smaller repeats by crossovers that fix variants of a repeating unit in some particular order. Two examples are shown in Figure 19.12. There are some difficulties, however, in this aspect of the model. The principal problem is that the variants of the original re-peat (x and y in the figure) must diverge sufficiently from each other to be treated as different sequences in recombination. If they remain similar enough to recombine in register, they will simply be treated as variants sub-ject to crossover fixation as shown in Figure 19.11. In this case, one or the other would predominate, instead of an array of both. This will be prevented only if the variants diverge sufficiently to make alignment for recombina-

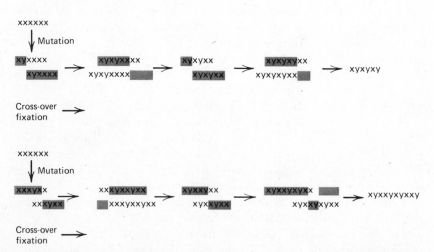

Figure 19.12. Generation of higher order repeats by crossover fixation. An initial sequence consists of tandem repeats of a unit x, one member of which is converted to the related sequence y by mutation. In the upper example a new repeating unit of xy is generated; in the lower ex-ample a new repeat of $xyxxy$ results.

tion more likely in the longer register. This would seem to require frequent mutation relative to crossover frequency; and it is not clear whether the relative frequencies needed for crossover fixation per se would be consistent with those needed for evolution of longer repeats from shorter repeats. Generally, it would be expected that whenever there is substantial internal repetition within a repeat, a smaller unit is likely to be generated from the longer by crossover fixation.

How is the existence of domains in satellite DNAs to be explained? To account for these by divergence due to mutation requires the postulate that different levels of mutation occur in different regions. Unequal recombination here provides a more likely model, for when a new mutation arises, the initial process of crossover fixation will spread its sequence initially through copies close to the original. Thus domains could represent regions in which a mutation has spread from its original site of occurrence, but has not yet filled the entire satellite. Another possibility is that the domains might represent regions of the satellite that have become independent (see below).

In some mammalian satellites there appears to be a connection between the repeat length of restriction fragments and the length of DNA present in the nucleosome. Variations of both the saltatory replication and cross-over fixation models in which events are more likely to occur between nucleosomes have been considered by Maio et al. (1977).

A critical feature of the crossover fixation theory is that the sequence of satellite DNA is not subject to selective pressure. It would be possible to apply pressure for general features, such as the overall length of the blocks of tandem repeats or the average base composition, but not for a precise sequence; otherwise the deletions and duplications produced by unequal crossovers would be lethal. Indeed, one question about this model is the effect of the genetic load that would be generated by lethal changes in other (specific) sequences if the events responsible for crossover fixation in satellite DNA occurred at the same frequency throughout the rest of the genome. Of course, this requires evidence on whether the frequency of sister chromatid exchange varies with the chromosome region. With a repetitive family of sequences, some number of variants could presumably accumulate as the result of mutation before any selective pressure became effective; but eventually selection would dictate the constitution of the cluster.

Another prediction of the model is that uniformity can spread only through a single linkage group. This implies that satellite sequences on different chromosomes, which presumably are unable to recombine with one another, must evolve randomly with respect to crossover fixation. If descended from a common ancestral sequence, the model would suggest that different variants are likely to be fixed by unequal crossover in each independent cluster. Uniformity could spread from one centromeric cluster to another only if recombination could occur between as well as within chromosomes. Except for satellites located at telomeres, this would disturb the arrangement of linkage groups and seems improbable. Testing this prediction of the model requires more information about the relationship between satellite sequences in different locations than is presently available (the sites occupied by domains would be interesting); also it would be necessary to know for how long the centromeric clusters have been separated onto their present chromosomes.

The conservation of very similar or identical sequences in the satellites of guinea pig, kangaroo rat, and probably some other rodents raises a question about the importance of specific sequence components. The rodent satellites must have originated before the separation of these species, roughly 40–50 million years ago. Their persistence over this evolutionary period makes it seem unlikely that sequence is unimportant. This is supported by the apparent conservation of the minor sequence variants; if this indeed proves to be extensive, the implication will be that there are quite precise demands for specificity of even variations in the sequence. The stability of these satellites essentially excludes all models requiring that satellite evolution is independent of sequence; the crossover fixation model, in particular, would predict that by chance different variations should have been established in each species, even if descended from the same ancestral sequence.

The Arthropod satellites which consist of a repetition of essentially identical copies of a short sequence could be explained either by a single saltatory replication of a short sequence or by crossover fixation proceeding to completion. The existence of three related satellites in D. virilis implies that a saltatory replication must have amplified different members of an original family containing these repeats, or must have amplified some variants of the first family at later times. The latter model is rendered more likely by the existence of the first satellite in the related species D. americana; and in this case, the unit for saltatory replication must have remained identical with the repeating unit of the first family. Since then the sequences have been maintained, with few mutations allowed to persist. This implies that there has been selection for specific sequences. An alternative view is that the D. virilis satellites represent clusters in which crossover fixation has spread different variants through each, effectively to the stage of uniformity; but the existence of one satellite in D. americana argues against the absence of selective pressure.

It may seem out of place to leave mention of satellite DNA function until the conclusion of this chapter, but so little is known about its role that any discussion must be largely speculative. Its ubiquitous occurrence and identification with centromeric (and other constitutive) heterochromatin regions argues that satellite DNA plays some fundamental role in chromosome structure. Beyond this its role is nebulous. Possibilities that have been raised include a function in chromosome mechanics (perhaps a bulk of centromeric DNA helps resist damage that might be induced during mitotic and meiotic movement); or in recombination (involving chromosome pairing); or simply in the recognition of homologous centromeres at meiosis. Certainly it is true that large deletions in the constitutive heterochromatin of the X chromosome of D. melanogaster have an effect on meiosis (see Chapter 5); but the molecular basis of the defect has not been correlated with the properties of satellite DNA.

CHAPTER 20

DNA Replication

Replication of eucaryotic DNA presents some imposing topological problems. During S phase, the single strands of the duplex must separate to act as templates for synthesis of daughter strands. This must involve considerable rearrangement of the parental nucleosome thread, in which DNA is highly packaged via several levels of organization. As the daughter duplexes are synthesized, the chromatin structure must reassemble. The process of disassembly, reproduction, and reassembly, must occur continually through S phase, passing successively through each DNA sequence. All this must be accomplished within the confines of the nuclear membrane, so there cannot be any extensive length of unfolded free DNA. Later in the cell cycle, as division approaches, the material of the daughter chromosomes must become disentangled, so that they can be segregated into the daughter cells.

In this chapter we shall be concerned with the replication of DNA itself. The genetic material of each eucaryotic chromosome is replicated semiconservatively (see Chapter 10); but the enormous length of DNA in each chromosome means that replication cannot proceed sequentially along the DNA molecule; instead it must involve the simultaneous action of many replicating forks. The unit in which replication occurs, the *replicon*, possesses an origin from which DNA synthesis proceeds in both directions. The process of replication may be described in terms of the replicon: how long are replicons, how many are there in the haploid genome, is the organization of replicons invariant or does it depend on the cell type, what governs the initiation of replication in each unit, and are the replicons activated in a particular sequence?

The mechanism of DNA synthesis takes the familiar form of semiconservative replication and appears to be accomplished by a series of events similar to those characterized in bacteria. The details of this process, involving the separation of DNA strands, discontinuous synthesis of at least one (and possibly both) complements, by extension of DNA chains from RNA primers, and joining of the discontinuous fragments, are discussed in Volume 1 of Gene Expression. We have already discussed in Chapter 14 the events that may be involved in the disassembly and reassembly of the nucleosome as the replicating fork passes along DNA. Here the evidence is consistent with a rather rapid process, in which the amount of single-stranded DNA awaiting replication, or free duplex DNA just synthesized, is rather small. Although the evidence is not yet decisive, a reasonable view for the present is that only a few nucleosomes on either side of the replicating fork are absent at any time. How the disruption of nucleosome structure is related to the process of DNA synthesis, and whether these events are identical on both strands of DNA, remains to be seen. Also, of course, little is known about how the higher orders of chromatin organization are affected by the movement of the replicating fork.

Organization of Replicons

Most of the evidence about the structure of the replicon and the relationship between replicons is derived from autoradiographic studies. The usual technique is to label cells with ^{3}H-thymidine, after which they are lysed to release DNA, which is allowed to adhere to filters or glass slides. These are exposed to film; and the labeled regions of DNA generate tracks of silver grains that identify regions replicated during the period of labeling (Cairns, 1966). The initial study of Huberman and Riggs (1968) showed that the replicated regions occur in tandem units, which would be generated if each origin initiated replication simultaneously in both directions, as illustrated in Figure 20.1. Then a short period of labeling should detect the two symmetrical replicating forks on either side of the origin; while in longer periods entire replicons may be labeled and adjacent replicons may have fused. Unambiguous evidence that replication is bidirectional is provided by the successive use of two thymidine precursors, each labeled with ^{3}H to a different level. Then a change in activity at some time during the labeling period produces a switch in the tracks between intense and faint labeling; the two members of each tandem array thus can be seen to be proceeding in opposite directions (Huberman and Tsai, 1973). Some results are shown in Figure 20.2. (The tracks seen in autoradiograms usually are single; it is rare to see separated sister chromatids until rather long continuous stretches of DNA are labeled.)

The length of the replicon cannot be estimated directly from the autoradiographic tracks, since termini cannot be identified by this method. If a fork arrives at a terminus, labeling should stop during the period of exposure to ^{3}H-thymidine, generating a short track. The distance between the termini of the tandem repeats would correspond to the length of the replicon. These termini would not be seen, however, if adjacent replicons were active at the same time so that they fused at or soon after the forks reached the terminus joining them. It is not yet clear whether in fact each replicon has definite termini, or whether each replicating fork continues to move until it meets one progressing in the opposite direction, in which case the only important parameter governing replication would be the use of the origins (McFarlane and Callan, 1973; Callan, 1973; Hand, 1975).

The average size of the replicon therefore is usually estimated by the mean

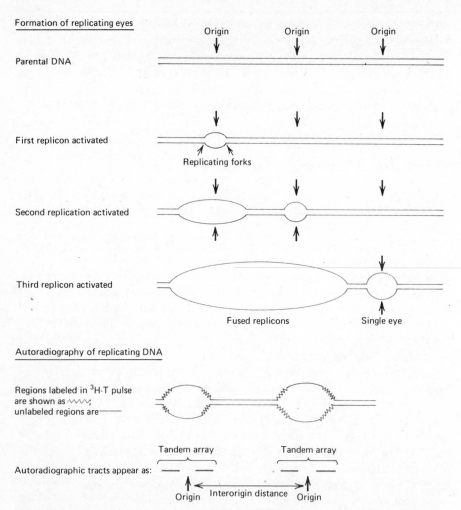

Formation of replicating eyes

Parental DNA

First replicon activated

Second replication activated

Third replicon activated

Autoradiography of replicating DNA

Regions labeled in ³H-T pulse are shown as ⌁⌁⌁; unlabeled regions are——

Autoradiographic tracts appear as:

Figure 20.1. Replication of eucaryotic DNA. The parental DNA contains several replicons, each of which is activated at a different time. Two replicating forks proceed in opposite directions away from the origin. When the replicating forks of adjacent replicons meet, the two smaller replication eyes become fused into one longer eye. Each replicon appears on autoradiography as a pair of tracks of equal length; the origin is taken to be the midpoint between them, and the size of the replicons is estimated by means of the interorigin distance between two adjacent active replicons.

distance between adjacent origins (each identified as the midpoint between two members of a tandem array). This requires at least two adjacent replicons to be active during the labeling period. This seems to be a common occurrence; indeed, clusters of active replicons often can be found, suggesting that all the origins may be activated at about the same time in regions of DNA containing several replicons (Hand, 1975). A caveat in this approach is

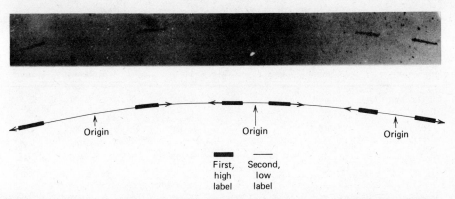

Figure 20.2. **Autoradiography of replicating DNA from HeLa cells.** The tracks identify three replicons. Each growing point is identified by a reduction in the intensity of the track; labeling was performed for 15 minutes with high specific activity ^{3}H-thymidine followed by 15 minutes with low specific activity label. The inferred locations of the origins are given in the drawing. The bar is 50 μm. Data of Huberman and Tsai (1973).

that the measurement of replicon size might not be typical if replicons of one class were organized in clusters, while there was also another class, activated singly, with different properties.

The units of replicating DNA observed in mammalian somatic cells generally range from replicons of 15 μm to about 120 μm, with a mean of 50–60 μm. Reasonably similar measurements have been made in avian and amphibian cells and in plant cells. Available data, converted to kilobases, are summarized in Table 20.1.

A complication in estimating the rate of replication is that a given fork may have either started or ceased movement during the period of labeling. The rate therefore is usually taken to be defined by the longest tracks formed during the labeling period. This usually turns out to suggest a rate of about 30–60 μm/hour. Thus an average replicon of 60 μm, with two replicating forks proceeding bidirectionally at 30 μm/hour, will be completely replicated in 1 hour. To determine the rate of replication, the labeling period must be determined empirically for each system; but usually is in the range of 30 minutes, so that not too many replicons are able to complete replication. During longer labeling periods, increasing numbers of replicons will fuse in the replicated state, so that the rate of replication will be underestimated because of the premature cessation of fork movement.

The number of replicons can be estimated from their average size and the haploid content of the genome. In mammalian cultured cells this seems to be about 20,000. If the average replicon takes some 30–60 minutes to complete replication, then at any time during a 6 hour S phase there must be roughly about 2500 active replicons in each haploid genome. This calculation is only very approximate, because it assumes that replication proceeds

Table 20.1 Characteristics of eucaryotic replicons

Cell type and genome size	Method of synchrony	Length of S phase	Temp., °C	Replicon lengths, kb	Mean replicon length, kb	No. of (haploid) replicons	Rate of fork movement, bp/min
CHO culture (3×10^9 bp)	FUdR	7 hr	37	50–400	100	30,000	2700
L culture (3×10^9 bp)	FUdR		37	30–300	145	20,000	2200
chick somatic primary culture (1.2×10^9 bp)	FUdR	7½ hr	37	80–475	200	6,000	1600
X. laevis somatic culture (3.1×10^9 bp)	FUdR	13 hr	25	65–325	200	15,000	500
T. cristatus somatic culture (2.2×10^{10})	FUdR	2 days	25	> 300	> 300	< 75,000	1100
T. vulgaris neurula	none	4 hr	18	~130	130		300
D. melanogaster embryonic culture line (1.4×10^8 bp)	none	10 hr	25	28–57	40	3,500	⩾ 2600
D. melanogaster cleavage nuclei	none	4 min	25	3–16	8	17,500	2600
Vicia sativum root meristem (2.4×10^9 bp)	FUdR	8 hr	23	30–350	175	14,000	900
Vicia hirsuta root meristem (4.5×10^9 bp)	FUdR	8 hr	23	30–400	250	18,000	1300
Vicia faba root meristem (1.1×10^{10} bp)	FUdR	8 hr	23	30–500	300	35,000	1600
Pisum sativum root meristem (3.9×10^9 bp)	none	6 hr	23	65–450	175	22,000	1600
Arabidopsis thaliana seed (1.9×10^8 bp)	none	3 hr	22	50–130	78	2,500	300
Physarum polycephalum (2.9×10^8 bp)	none	3 hr	28	25–50	30	9,500	650

These data have been converted from the lengths of DNA measured as μm on autoradiographs by using the approximation 1 μm = 3250 base pairs (= 3.25 kb).

References: CHO—Huberman and Riggs (1968); L—Hand and Tamm (1973, 1974), Cohen, Jasny and Tamm (1979); chick—McFarlane and Callan (1973); Xenopus and Triturus—Callan (1972, 1973); D. melanogaster—Blumenthal et al. (1973); Vicia—Gaddipati and Sen (1978); Pisum—Van't Hof (1975); A. thaliana—Van't Hof et al. (1978); P. polycephalum—Funderud et al. (1979).

evenly throughout S phase, which is not true, but it serves to indicate the magnitude of the problem of regulation of replicon activation. The precision with which their activation is controlled is not known. The identification of clusters of several active units in reasonably short labeling periods suggests that there may be some regional organization, but its nature is unknown.

The rate of eucaryotic DNA replication is very much slower than that of bacteria. The chromosome of E. coli constitutes a single replicon of 4.2×10^3 kilobases; at 37°C, the two replicating forks complete replication in 40 minutes, which corresponds to a rate of 50,000 base pairs per minute (see Volume 1 of Gene Expression. The bacterial replicon is thus very much longer than the 200 kb units of eucaryotic cells, and the rate of replication is 25 times greater than an average eucaryotic rate of 2000 base pairs/minute. Presumably at least part of the reason for the slower replication of eucaryotic DNA lies with its more complex state of organization, but there are not yet enough data to permit conclusions about factors that may influence the rate. Temperature has an effect, of course, and so the growth temperature of each cell type is noted in Table 20.1. The only comparative study appears to be that of mature cells of the plant Pisum sativum, where S phase varies from 22 hours at 10°C to 4 hours at 35°; the variation in rate of replication is twofold, from 300 to 600 base pairs/minute (Van't Hof et al., 1978).

The rate of replication is not constant throughout S phase, but increases after the first hour or so in the two cell types in which different periods have been examined. In P. sativum, it appears initially to be about 500 base pairs/minute, eventually reaching the rate of 1600 base pairs/minute found through most of S phase (Van't Hof, 1976a,b, 1977). In CHO cells also, roughly a threefold variation has been found, from an early value of 650 base pairs/minute to a later rate of about 2000 base pairs/minute (Housman and Huberman, 1973). Although absolute rates are difficult to determine accurately, the relative variation within S phase seems clear. Probably this reflects a slow beginning to the synthetic period during which some components of the replication apparatus are in short supply, limiting the rate of replication.

In many experiments, cells have been synchronized with FUdR in order to obtain replicating DNA. This has been noted in Table 20.1, since the effects of FUdR are controversial and it is possible that this treatment affects the characteristics of the subsequent replication cycle; one report is that the interorigin distance is altered as well as the rate of replication (Ockey and Saffhill, 1976).

Is the control of replication inherent in the genome or does it vary with the cell phenotype? Detailed studies of whether a given replicon is always activated at the same time, and of whether the two copies in the diploid nucleus are activated at the same time, are not at present possible. However, there is general evidence that there may be rather extensive changes in the patttern of replicon organization between embryonic and adult cell types in some organisms. Studies of D. melanogaster show very different patterns in tissue culture cells and the nuclei of cleavage embryos. The tissue culture cells grow with a fairly lengthy S phase of 10 hours, in contrast to the very rapid synthesis of DNA in cleavage nuclei, which is completed within about 4 minutes. Electron microscopy of the DNA of cleavage nuclei identifies long series of replication eyes that must represent individual or fused replicons. The distance between adjacent midpoints has been estimated as 8–12 kb; actually the eyes may be multiples of 3.4 kb, and if this is the size of the replicons, the number would be correspondingly increased from that given in Table 20.1 (Wolstenholme, 1973; Kriegstein and Hogness, 1974). The contrasts with the estimate of 40 kb determined by autoradiography of the DNA of cultured cells (Blumenthal et al., 1973). Thus cleavage nuclei appear to contain a large number of short replicons, up to ten times the number of correspondingly longer replicons in the cultured cell line.

The rate of replication fork movement probably is not very different, and it seems that the rapid completion of DNA synthesis in the cleavage nuclei is accomplished not only by increasing the number of replicons, but also by activating all the origins simultaneously instead of sequentially. These results have the interesting implication that there must be a large number of origins for the short replicons that are recognized in cleavage nuclei but are not utilized in the cultured cells; whether these origins include as a subset those of the culture line or are completely different is not known.

Comparable differences in Triturus vulgaris embryonic cells and spermatocytes have been reported by Callan (1973), where the shorter S phase of embryogenesis appears to be accomplished by a reduction in the length and increase in number of replicons, without change in the rate of fork movement.

Other methods as well as autoradiography have been used to follow the replication of DNA. Usually these involve following density shifts as DNA replicates; this has the advantage that the entire DNA population can be analyzed (instead of the small number of samples taken in autoradiography). However, these methods are less reliable quantitatively, and although they are broadly in agreement with the results of autoradiographic analysis, too much reliance should not be placed upon them (Painter et al., 1966; Painter and Schaeffer, 1969; for review see Edenberg and Huberman, 1975).

Regional Patterns of Replication

The problem in investigating the control of replication at the level of the chromosome is that it is impossible to identify chromosome regions at interphase when replication occurs. Instead it is necessary to incorporate a label during S phase and then to examine the chromosomes at the subsequent mitosis. By then, of course, the chromosomes are condensed, so that only comparatively large regions can be distinguished.

Little is known about the detailed sequence of events during S phase. Evidence on the question of whether a given replicon is always activated at the same time is sketchy. In some early experiments, Mueller and Kajiwara (1966a) labeled synchronized HeLa cells with ^{3}H-thymidine during the first two hours of S phase, allowed the cells to grow for some generations, and after resynchronizing them, demonstrated that BUdR is incorporated into the labeled regions at the beginning of the next S phase. These experiments used amethopterin to achieve cell synchrony. Later experiments with CHO cells synchronized by mitotic selection showed that a region labeled early or late during one S phase tends to be replicated at the same time in the next S phase; however, synchronization with FUdR reduced very greatly the proportion of material replicating at the same time in the two S phases (Adegoke and Taylor, 1977). This adds concern to the problem of the effects of FUdR, although suggesting that there may usually be some assignment of replicons to be activated at specific periods of S phase.

The recent technique of using incorporation of BUdR to quench the bright staining of Hoechst 33258 suggests that there is a difference in the locations of the replicons activated at the beginning and end of S phase. We have already discussed the technique in Chapter 10 and the nature of the banding patterns it produces in Chapter 15; here we need only reiterate that a pattern similar to that of the G bands is generated by incorporation of BUdR at late times, with the reverse pattern generated by incorporation at early times. Since each band and interband probably contain of the order of 100 replicons, and these structures divide the chromosome into only two types of region, they are presumably too large to be single units of replication; the relationship between bands and replicons remains to be established. However, this work does provide evidence for some preferential time of activation in different regions of the chromosome. This supports earlier studies, made with the lower resolution of autoradiography, which suggested that each chromsome may replicate its regions in a specified order (Hsu, 1964). The fact that homologues usually replicate together, but sometimes behave differently, shows that variations can occur. In human-mouse hybrid cells, the remaining human chromosomes may replicate in a different sequence from that found in the parental cell (Farber and Davidson, 1978). The nature of this perturbation is not known. Nor in general is it yet possible to resolve the question of what variations in the sequence of replication may exist between different cell types.

A prominent feature of the order of replication is that heterochromatic chromosome regions usually are the last to complete replication. These include both the constitutive heterochromatin often present at the centromeres and the facultative heterochromatin that may occupy whole chromosomes, such as one X homologue in female mammals. We have discussed this feature of heterochromatin in Chapter 15; here it is necessary only to add that there is no evidence from these studies for any difference in the rate of replication of euchromatin and heterochromatin. The heterochromatic regions start replication later, perhaps because of their more condensed structure, and thus complete synthesis last (Comings, 1967; Gavosto et al., 1968; Knight and Luzzatti, 1973).

The late labeling of centromeric heterochromatin should represent the late replication of satellite DNA. There have been few experiments to examine directly the synthesis of satellite sequences. The mouse satellite does indeed appear to be replicated during the middle and at the end of S phase (Flamm, Bernheimer, and Brubacker, 1971). In the genome of D. ordii, which consists of about 50% satellite sequences (see Table 19.4), each satellite appears to be replicated principally at one time, although there is a low level of replication of each throughout S phase. The main band DNA is replicated predominantly during the first half of S phase; then there is a switch to satellite DNA replication, each satellite achieving its maximum rate of synthesis at a different time (Bostock, Prescott, and Hatch, 1972; Bostock et al., 1976). During the second part of S phase, a new component is seen in the autoradiographic tracks, consisting of long series of tandem arrays that correspond to replicons all activated around the same time (Hori and Lark, 1974). The average spacing of origins is the same as in main band DNA, about 100 kb. These appear to represent a fairly synchronous initiation of satellite DNA synthesis.

The Replication Apparatus

An array of enzyme activities analogous to some of those involved in bacterial replication has been found in a variety of eucaryotic cells. In addition to DNA polymerases, this includes ligase (presumably for joining discontinuous fragments), nicking-closing enzyme (characterized by its ability to remove supercoils from circular duplex DNA), single-strand binding protein (possibly analogous to the gene 32 protein of the T4 replication apparatus), several types of DNAase (although it is not known which, if any, are involved in replication and repair), ribonuclease H (able to degrade the RNA strand of DNA-RNA hybrids, and a candidate for the role of removing RNA primers). None of the known RNA polymerases (see Chapter 22) has been implicated in the role of synthesizing the RNA primer. Although these activities may be among those needed for replication, they are unlikely to be adequate; there is little information on which may be involved in replication and nothing is known about the nature of interactions between the replication apparatus and the chromatin thread (for review see Edenberg and Huberman, 1975).

DNA polymerases have been investigated mostly in mammalian cells, in particular calf thymus, rat tissues, and human lymphocytes (for review see Weissbach, 1975). Our discussion here is directed particularly to the mammalian enzymes. A DNA polymerase in calf thymus was identified as long ago as 1960 by Bollum; and only more recently has it become apparent that there are several different DNA polymerases. Their properties are described in Table 20.2. Most of the cellular activity resides in the first enzyme to be identified, DNA polymerase α (using the present nomenclature). For

Table 20.2 Mammalian DNA polymerases

Enzyme	DNA polymerase α	DNA polymerase β	DNA polymerase γ
Location	nucleus and cytoplasm (by leakage)	nucleus	nucleus (~80%) and mitochondrion (~20%)
Proportion of total activity	~80%	10-15%	2-15%
Activity in cell cycle	increases in S phase and with resting $\longrightarrow$ growing transition	generally invariant	not known
Size	110-220,000 daltons; subunit composition not clear	45,000 daltons; single polypeptide chain (in calf thymus & human KB cells)	60,000 daltons
Cationic requirements	5-10 mM Mg^{++} or 0.3 mM Mn^{++}	15-25 mM Mg^{++} or 1 mM Mn^{++}	5-10 mM Mg^{++} or 0.5 mM Mn^{++}
Salt requirements	50-100 mM KCl or 60 mM KPO_4	60-100 mM KCl or 40-50 mM KPO_4	150 mM KCl or 50 mM KPO_4
Involvement of SH groups	yes	no	yes
Exonuclease activities	none	none	none
Synthetic activities	extends 3'-OH DNA or RNA primer on gapped DNA; cannot use RNA templates	extends 3'-OH DNA primer on gapped DNA; about half as active with poly(A)/oligo(dT)	uses (poly(A) template from oligo(dT) primer 5 × faster than gapped DNA

None of the enzymes has been purified completely to homogeneity, so details of size and ionic or salt requirements are based on preparations of varying purity. All the enzymes form aggregates that obscure their molecular nature; with the possible exception of DNA polymerase β, the active form has not been identified. The sizes given here are the smallest values found for active preparations. There are slight differences in the ionic and salt requirements of the enzymes; but these vary in different systems. It is not known whether variations in reported properties for each enzyme are due to genuine differences between the corresponding enzymes of different cell types or to artefacts of the in vitro systems depending on assay conditions and degree of purity.

References: DNA polymerase α—Bollum (1960), Chang and Bollum (1973), Spadari and Weissbach (1974a, 1975), Holmes et al. (1974, 1976), Hesslewood et al. (1978), Sedwick et al. (1975), Fisher and Korn (1977), Fisher, Wang and Korn (1979); DNA polymerase β—Weissbach et al. (1971), Spadari and Weissbach (1974a, 1975), Chang and Bollum (1972), Chang (1973), Wang et al. (1974, 1975); DNA polymerase γ—Fridlender et al. (1972), Tibbetts and Vinograd (1973amb), Knopf et al. (1976), Bolden et al. (1977), Bertazzoni et al. (1977), Hubscher et al. (1977). Apart from the mammals, the best investigated eucaryotic DNA polymerases are from chicken, D. melanogaster and yeast: see Stavrianopoulos et al. (1972), Karkas et al. (1975), Brakel and Blumenthal (1978), Chang (1977), Wintersberger (1978).

some time, most of this activity appeared to be located in the cytoplasm, with only a small proportion remaining in isolated nuclei. However, this distribution appears to have been due to leakage of the enzyme in aqueous isolation medium; with a nonaqueous procedure, the enzyme activity is found solely in the nucleus (Foster and Gurney, 1976). When cells are enucleated with cytochalasin B, the karyoplasts (nuclei) have >85% of the DNA polymerase α activity (Herrick et al., 1976).

The enzyme is large in size, although its tendency to aggregate (shared by the other DNA polymerases), and the difficulty of obtaining assay systems representative of in vivo conditions, mean that its subunit composition and active form have yet to be determined. Different forms of the enzyme appear to be generated by proteolysis. Like other known DNA polymerases, it cannot initiate synthesis of DNA chains, but is able to extend a primer from a 3'-OH end along a complementary template. The usual assay for this enzyme is its ability to fill in gaps on DNA containing 3'-OH termini adjacent to long single strand stretches in otherwise duplex molecules. This is the only one of the known eucaryotic enzymes able to extend an RNA as well as DNA primer. Like the other eucaryotic enzymes, but in contrast with the bacterial DNA polymerases, there appears to be no nuclease activity associated with the synthetic activity.

DNA polymerase β was originally identified as a second enzyme obtained by salt extraction of HeLa cells (Weissbach et al., 1971). Its location is nuclear, which is to say that it does not share the propensity of DNA polymerase α to leak out during nuclear isolation. Generally it represents <20% of the activity of DNA polymerase α. It is smaller in size and has slightly different ionic requirements for its synthetic activities. It shares with DNA polymerase α the ability to extend 3'-OH primers on template strands, although only from a DNA priming sequence. It can also use a template consisting of a poly(A) strand to which an oligo(dT) sequence has been annealed to act as primer, although this action is less efficient. It cannot use natural RNA as a template for DNA synthesis.

DNA polymerase γ was isolated originally by virtue of its ability to copy the poly(A)/oligo(dT) substrate efficiently; this assay was used in an attempt to identify cellular enzymes with RNA-dependent DNA polymerase activities analogous to those of the C type RNA tumor virus reverse transcriptase activities (Fridlender et al., 1972). Although the enzyme is active in this assay, this does not appear to be its role in vivo, since it does not utilize RNA templates. It has the usual ability to extend a 3'-OH primer on gapped DNA.

A separate DNA polymerase in mitochondria was reported at one time, but it has since been shown that this is an artefact. Instead, the enzyme present in mitochondria has been identified as DNA polymerase γ. The two enzymes appear identical by biophysical criteria; some rather slight differences in synthetic activity cannot be ascribed to different enzymes, but could be due to the impure state of the extracts.

In none of these cases has the active form of the enzyme been determined unequivocally; this will depend upon the development of assay systems also containing the other components of the replication apparatus. Attempts to determine which enzyme(s) undertake DNA replication have led to the suggestion that in the nucleus this may be DNA polymerase α. This is the only enzyme whose level increases during S phase; and there is a much larger increase in its activity upon the transition from resting to growing state than that seen with DNA polymerase β (Chang, Brown, and Bollum, 1973; Spadari and Weissbach, 1974b; Bertazzoni et al., 1976). Because of the impure state of the enzymes, and the lack of assay systems that distinguish sufficiently well between their activities, these data at present provide only tentative support for this suggestion. The conclusion would imply that DNA polymerase β has a repair function or possibly some subsidiary role in replication. Presumably DNA polymerase γ undertakes replication in the mitochondrion; but its nuclear role is unknown.

The most promising attempts to develop assay systems for the complete replication apparatus have made use of rather crude systems, in which all the components probably are retained. Systems in which HeLa lysates or nuclei are able to continue replication have been developed; some of the necessary components appear to be present in nuclear extracts, which may therefore be fractionated on the basis of their ability to support the active systems. The DNA synthetic activity appears to be confined to completing replication of activated replicons; initiation of new replicons does not seem to occur. However, the continued movement of the preexisting replication forks seems to include initiation, extension and joining of Okazaki fragments, not just polynucleotide chain elongation (Fraser and Huberman, 1977; Planck and Mueller, 1977).

Synthesis of DNA by the HeLa lysate system is resistant to high concentrations of the analogue 2'3'-dideoxy thymidine-5'-triphosphate (Waqar, Evans and Huberman, 1978). Since DNA polymerases β and γ are sensitive to the analogue, whereas the α enzyme is resistant in vitro, this suggests that DNA polymerase α may be the only polymerase involved in the synthesis of new polynucleotide chains. This supports the observation that DNA polymerase α is the only polymerase present in replicating complexes of SV40 DNA, which are able to continue DNA synthesis in vitro and which also are resistant to the dideoxythymidine analogue (Edenberg, Anderson and DePamphlis, 1978).

The HeLa nuclear system is stimulated by the addition of ribonucleoside triphosphates and forms 4S Okazaki fragments in an action that is resistant to α-amanitin (Brun and Weissbach, 1978). This suggests that priming of Okazaki fragments with RNA may be occurring, and that the primase enzyme is either RNA polymerase I or a new RNA polymerase activity. The priming reaction is inhibited by antibodies against RNA polymerase I, which would favor the first possibility.

A replication system in which it is practical to fractionate the active components is provided by X. laevis eggs. When injected into eggs, the nuclei of nondividing cells are induced to commence replication (see Chapter 8). A cytoplasmic extract induces the appearance of replication eyes in added DNA. Using the ability of the extract to induce θ structures in a circular template (consisting of part of the Xenopus rRNA genes inserted in a bacterial plasmid), the extract has been fractionated into 9 (crude) components (Benbow and Ford, 1975; Benbow, Krauss, and Reeder, 1978).

There is no evidence for any localization of the replication apparatus in a particular area of the cell nucleus. At one time it was thought that replication might occur at the nuclear membrane. However, the cells used for these studies were synchronized by the use of amithopterin and it turns out that one effect of this drug is to induce peripheral labeling; following this the DNA is degraded, so in fact this is a stage in a process of drug-induced cell death (Ockey, 1972; Ockey and Allen, 1975).

Our general view of DNA replication at present, therefore, is of a rather complex replication apparatus, containing several enzyme activities coordinated in an unknown manner, which is associated with chromatin. Replication occurs in the interphase nucleus through the sequential activation of a large number of replicons, each of which possesses two replicating forks (that is, two copies of the replication apparatus), which create a probably transient perturbation of chromatin structure as they proceed.

CHAPTER 21

Organelle Genomes

The residence of some genetic information outside the nucleus accomplishes a dispersion of loci concerned with the elaboration of mitochondria and chloroplasts. These organelles contain genes that specify some of the components of the energy conversion apparatus. Other components are coded by nuclear genes, so that organelle function rests upon the expression of genetic information from each cell compartment. In this chapter we shall be concerned with the nature of the genetic information carried by the organelles; and with the relationship between the expression of these genes and those in the nucleus that code for organelle functions.

The first evidence for the occurrence of extranuclear hereditary events was provided by the genetic studies initiated by Correns and by Baur in 1909, in which non-Mendelian inheritance was observed in plants. Several characters showed two unusual features in crosses. Abnormal segregation ratios sometimes represented uniparental (maternal) inheritance, sometimes a predominance of the character contributed by one parent. In either case, segregation of the mutant and wild phenotypes occurred during somatic growth. No recombination occurred between markers, restricting these studies to equating the changes with nonnuclear events, and leaving open the question of the relationship between different markers (for review see Sager, 1972). Indeed, the subsequent early studies were characterized by discussion as to whether these events represented the presence of extra chromosomal genetic information or might be due to epigenetic changes (for review see Caspari, 1948).

Non-Mendelian genetics are a feature of certain mutations in some unicellular eucaryotes, in particular yeast, as well as in plant cells, although they have not been observed in animal cells. In yeast, the site of these mutations can be identified unequivocally with the mitochondrion. The mutants are deficient in respiratory activities; and their isolation is made possible by the alternate life styles of yeast, which allow it to survive aerobically (when mitochondrial activities are essential) or anaerobically (when the mitochondrion is dispensable). Since animal cells lack this second option, mutations abolishing mitochondrial activity may fail to be inherited because they are invariably lethal.

Plant cells possess both mitochondria and chloroplasts, as do some unicellular eucaryotes such as the algae. In plants, non-Mendelian inheritance is displayed by a wide variety of phenotypic characters, including not only chloroplast development, but also pollen sterility, size and shape of the leaf, and differentiation of the flower (for review see Kirk and Tilney-Bassett, 1967; Sager, 1972). Which of these characters is carried by the chloroplast or mitochondrion (or even some other extranuclear determinant) is not known, although it is likely that at least some of the mutations visibly affecting chloroplast development reside in this organelle.

Extra nuclear genetic information is an invariable constituent of the eucaryotic cell; but in the absence of methods for inducing, selecting and mapping mutations, it is difficult to define its nature. Two systems have been developed in which these difficulties have been overcome. In genetic crosses involving either the yeast Saccharomyces or the alga Chlamydomonas, recombination occurs between markers affecting the mitochondrion and chloroplast, respectively. This has allowed a linkage group for the organelle markers to be constructed in each case. Then the genetic map can be correlated with the physical form of the genome. In systems lacking this facility, direct analysis of the genome must constitute the sole approach.

Cytochemical studies first suggested that DNA might be present in mitochondria and chloroplasts, although this work was far from definitive (Ris and Plaut, 1962; Nass and Nass, 1963a,b). The use of density gradient centrifugation then demonstrated that particular DNA fractions can be identified with organelles (Chun et al., 1963; Sager and Ishida, 1963; Luck and Reich, 1964). These early experiments used unicellular eucaryotes in which the organelle DNA conveniently forms a satellite distinguished from the single peak of nuclear DNA by its buoyant density. Subsequent experiments have been performed also with systems in which there is no density difference, when it becomes necessary to purify the organelles more extensively in order to extract their DNA.

The presence of genetic information in each organelle accounts for the features of non-Mendelian inheritance. First, the representation of organelle genetic information in a zygote will depend upon the number and fate of the copies of the organelle contributed by each parent. Contrasted with the contribution of a haploid set of nuclear chromosomes by each parent, organelles may be provided largely or entirely by one parent because of a differential contribution of cytoplasm to the zygote. Or the organelle genomes of the two parents may have different probabilities of surviving in the zygote. Second, the segregation of markers during somatic growth suggests the presence of multiple copies of the organelle genome in the cell. During the cell cycle, each organelle generates progeny of its particular genetic disposition; when there is heterozygosity for some organelle gene, the distribution of the progeny into daughter cells at each division eventually generates cells of only one or the other organelle genotype.

What is the nature of the organelle genome? The DNA molecules present in the mitochondrial or chloroplast population of a cell might in principle all be identical or might consist of a variety of different sequences. In fact, only a single type of DNA is to be found in each class of organelle. This almost invariably takes the form of a circular DNA of constant contour length (although some circular dimers may be found and linear molecules presumably derived by breakage are common). Investigations of the complexity of DNA by renaturation kinetics or restriction mapping suggest that the circular molecule represents a unique sequence. It is important to know how many copies of this molecule are present in each individual mitochondrion or chloroplast; in other words, what is the ploidy of the organelle? And how many organelles are present in the cell, and what governs their distribution at division? In cases where recombination occurs, we may ask how the juxtaposition and distribution of organelle genomes is accomplished.

Turning to the genetic content of the organelle genome, we must know how many (and which) genes are carried on organelle DNA and how they are expressed. Both mitochondria and chloroplasts are able to perform protein synthesis; indeed, each contains a complete apparatus for replication, transcription, and translation (and presumably in some instances, recombination). Some of this apparatus, in particular the RNA components, is coded largely or entirely by the organelle genome; other components, including most of the proteins, are coded by nuclear genes. There is dispersion of the structural genes coding for the proteins that comprise the enzyme complexes which fulfill organelle functions; usually some are located in the nucleus and some in the organelle genome. Thus in no case does an organelle appear to be self-sufficient; there is always an intricate relationship between expression of organelle and nuclear genes. One interesting question about this relationship is whether organelle genes are expressed solely within the organelle and nuclear genes solely within the cytoplasm; or whether mRNA (and tRNA) may be transported between organelle and cytoplasm for use in translation.*

Chloroplast and Mitochondrial DNA

Organelle genomes are identified as duplex molecules of DNA present in purified chloroplasts and mitochondria. Because the organelle DNA is small, it forms a sharp buoyant density peak; and, when its density is sufficiently distinct from that of nuclear DNA, may be isolated as a satellite. Tables 21.1 and 21.2 show that this option is not available in every case; in particular, the chloroplast DNA of higher plants has a density too close to that of nuclear DNA. In some cases there has been difficulty in correlating satellite

*Molecules or events located within an organelle here are described as such; so we shall use the terms *organelle DNA, organelle protein synthesis.* On occasion, organelle DNA has been described as "cytoplasmic DNA" and its genetics as "cytoplasmic inheritance." But we shall not use these terms because they are ambiguous in view of the need to distinguish translational events occurring within the organelle from those within the surrounding cytoplasm. Thus *cytoplasmic protein synthesis* here describes the translation of mRNA in the general cytoplasm.

Table 21.1   Complexity

Species	Nuclear DNA density	Ct DNA density	Percent of total DNA
Chlamydomonas reinhardii	1.724	1.694	7
Euglena gracilis	1.707	1.685	1 (dark); 3 (light); 5 (max. growth)
Higher plants (e.g. Zea mays, Spinacia oleraceae, Nicotinia tabacum)	1.691–1.703	1.697	not known

Values given are present best estimates; other estimates can be found in the cited literature. The renaturation complexities of Chlamydomonas and Euglena appear to be too high, but the reason is not known. Several higher plant chloroplast genomes contain a sequence of about 15% of genome length that is present in two copies arranged as an inverted repeat; this explains the excess of contour length over complexity.

References: Chlamydomonas—Chun et al. (1963), Sager and Ishida (1963), Chiang and Suekoa (1967), Sueoka et al. (1967), Bastia et al. (1971), Wells and Sager (1971),

DNAs with their source; in particular with D. melanogaster, in which the mitochondrial satellite was confused with nuclear satellite DNA, and with higher plants, in which mitochondrial contamination of chloroplast fractions led to misassignment of buoyant density fractions. The situation is now clear, however, and the tables summarize available data on the systems that have been the principal targets for investigation.

The organelle genome appears almost always to take the form of a circular DNA duplex, with both closed (twisted) and open (relaxed) circles visualized by electron microscopy. The qualification is due to difficulties in obtaining intact organelle DNA; often most of the DNA is linear, but so long as some circles can be found, it is assumed that this is in the vivo form, with the linear molecules resulting from breakage, usually random to generate a fairly dispersed distribution of linear lengths. In those cases in which most of the molecules remain circular, this property may be used to separate intact mitochondrial DNA, for example by centrifugation in the presence of ethidium bromide. There are few cases in which no circles have been found; and these seem more likely to represent unusual difficulties in obtaining unbroken molecules than to represent the existence of a linear genome, although the point is hard to prove.

of chloroplast DNA

| Chloroplast genome size | | | |
Electron microscopy	Restriction analysis, kb	Renaturation complexity, kb	Number of Ct DNA molecules per organelle
62 μm = 200 kb circle	190	380 (for 90% of DNA; rest faster)	1 chloroplast/gamete should contain ~7000 kb = 35 × 200 kb
44 μm = 140 kb circle	130	270	mean 15 chloroplasts/cell; 200 − 1000 Ct DNA (as 1–5% total) = 13 − 70/chloroplast
45 μm = 146 kb circle	132	136	not known

Howell and Walker (1976), Behn and Herrmann (1977), Whiteway and Lee (1977), Rochaix (1979); Euglena—Brawerman and Eisenstadt (1964), Ray and Hanawalt (1964), Stutz (1970), Klein et al. (1972), Manning and Richards (1972), Rawson and Boerma (1976a), Gray and Hallick (1977), Kopecka et al. (1977); higher plants—Whitfield and Spencer (1968), Herrmann et al. (1975), Kolodner and Tewari (1975a, 1979), Bedbrook and Bogorad (1976).

The coincidence of three types of estimate for genome length suggests that the organelle DNA represents a single population of molecules all identical in sequence. The contour length of the circles observed in electron microscopy identifies the size of the individual molecules. The kinetics of reassociation of organelle DNA usually correspond to a complexity very close to the circular length, suggesting that the circle is a unique sequence. Restriction enzymes cleave organelle DNA into fragments that can be separated by gel electrophoresis. The equimolar yields of these fragments, and the equivalence of the sum of their lengths with the circular contour, suggest that the circle comprises a single type of sequence. Exceptions to a close coincidence of all three size estimates are rare; probably they are due to artefacts characteristic of the individual system.

In cases in which circles form a minor part of the population, which largely consists of smaller linear molecules, the correlation between circular contour length and the reassociation or restriction complexity suggests that the circle is the usual form of the genome. In the cases of the Saccharomyces mitochondrial DNA and the Chlamydomonas chloroplast DNA, this is supported by the genetic map, which is circular (see below).

Table 21.1 summarizes the general features of chloroplast DNA. In higher plants, the genome is usually about 140,000 base pairs, a little larger than a large bacteriophage such as T4 (110,000 base pairs). In lower eucaryotes the

Table 21.2 Complexity

Species	Nuclear DNA density	Mt DNA density	Percent of total DNA
S. cerevisiae	1.698	1.684	18
N. crassa	1.712	1.701	<1
C. reinhardii	1.720	1.706	0.6
D. melanogaster	1.701	1.680	not known
X. laevis	—	—	not known
G. domesticus	1.698	1.707	not known
M. musculus (L cell)	1.703	1.698	<0.2
Human (HeLa cell)	—	—	—
P. sativum (pea)	1.698	1.706	not known

*l*alues given are present best estimates. Where there have been several reports of electron microscopy, the range of contour lengths observed in different laboratories is given. This is based solely on measurements of circles, which are usually a small, sometimes very small, proportion of total. The length is converted into kilobases by the approximation 1 μm = 3.25 kb, except when internal standards were provided for correlation. The restriction analysis gives the sum of the lengths of the restriction fragments determined by gel electrophoresis. Renaturation complexities have been determined by comparison of kinetic rates with a phage T4 or E. coli standard. The same basis for calculations was used in Table 21.1.

References: S. cerevisiae—Mounolou et al. (1966), Moustacchi and Williamson (1966), Williamson (1970), Hollenberg et al. (1970), Lazowska et al. (1974), Michel et

genome may be larger yet, about 200,000 base pairs in Chlamydomonas. Whether the larger genome codes for additional functions is not yet known. Of course, even these (relatively) large organelle genomes still represent only a small part of the total DNA complexity of the cell.

There appears to be more than a single chloroplast DNA in the individual organelle in Chlamydomonas and Euglena. In C. reinhardii, there is only a single chloroplast in the cell; as seen by feulgen staining, the DNA is confined to discrete areas, often appearing as two bodies. Yet the total amount

of mitochondrial DNA

Mitochondrial genome size			
Electron microscopy	Restriction analysis, kb	Renaturation complexity, kb	Number of Mt DNA molecules per organelle
26 μm = 84 kb	67–76 depending on strain	108	22 mitochondria/diploid cell contain average ~4 genomes each
20 μm = 66 kb	69	not known	not known
4.6 μm = 15 kb	not known	15.7	~660 kb/cell = 42 mt genomes; no. of mitochondria unknown
6 μm = 18.4 kb	17.3	17.8	not known
5.4 μm = 18.4 kb	17.2	not known	not known
5–5.6 μm = 17.2 kb	not known	not known	not known
4.7–5 μm = 15.8 kb	15.8	16.0	1100/cell corresponds to ~2/organelle on basis of ~500 mitochondria
4.8 μm = 15.6 kb	16.6	not known	8800/cell corresponds to ~10/organelle if ~800 mitochondria
30 μm = 110 kb	116	not known	not known

al. (1974), Sanders et al. (1975a), Morimoto et al. (1977); N. crassa—Luck and Reich (1964), Schafer et al. (1971) Bernard et al. (1975a,b), Terpstra et al. (1977); C. reinhardii—Ryan et al. (1978); D. melanogaster—Bultman and Laird (1973), Polan et al. (1973), Wolstenholme (1973), Wolstenholme and Fauron (1976), Fauron and Wolstenholme (1976), Klukas and Dawid (1976), Goldring and Peacock (1977); X. laevis—Dawid and Wolstenholme (1967), Ramirez and Dawid (1978); G. domesticus—Rabinowitz et al. (1965), Sinclair et al. (1967), Van Bruggen et al. (1968); L and HeLa cells—Sinclair et al. (1967), Radloff et al. (1967), Nass (1969c), Brown and Vinograd (1974), Robberson et al. (1974), Bogenhagen and Clayton (1974), Potter et al. (1975); P. sativum—Kolodner and Tewari (1972).

of chloroplast DNA corresponds to about 35 times the apparent size of the individual genome. How such multiple copies may be distributed remains to be seen. In Euglena, the number of chloroplasts varies from about 10 to 30 per cell; and comparison with the total amount of chloroplast DNA again suggests the presence of multiple copies. An interesting feature of this system is that when Euglena is grown in the dark its chloroplasts fail to develop; instead it possesses proplastids, essentially an immature form of the organelle. Upon exposure to light, the proplastids are converted to chloroplasts; without any change in the total number of organelles, there is a substantial increase in the amount of chloroplast DNA. The conversion of the proplastid into the chloroplast therefore seems to involve not only morphological changes in size and differentiation, but also sees an increase in the number of DNA molecules per organelle.

Table 21.2 shows that mitochondrial DNA is smaller than chloroplast DNA. In animal cells the genome is about 17,000 base pairs long. A similar size is found in avian, amphibian, and insect mitochondrial DNA. The coding capacity of such a genome obviously is rather limited, not much more than 10 genes. Mitochondrial DNA of unicellular eucaryotes sometimes is somewhat larger. Yeast has been well characterized as about 80,000 base pairs; Tetrahymena (although not visualized in circular form) is about 55,000 base pairs. One interesting question is to what extent the change in size is associated with the acquisition of additional genetic functions. Some of the yeast mitochondrial genome is occupied by AT-rich sequences, in the form of short regions probably lacking coding functions. Some rather lengthy regions of yeast mitochondrial DNA code for single proteins (see below). A comparison of the genes present in yeast and in animal mitochondrial DNA may therefore reveal fewer differences than would be suggested by the discrepancy in size. Certainly the number of transcripts of the mitochondrial genome is not very different (see below). Drosophila mitochondrial DNA also has an A-T rich sequence, representing about 20% of the genome; changes in its proportion seem to be largely responsible for variations in size between the mitochondrial genomes of different Drosophila species. There is no evidence for comparable putative noncoding sequences in the other, smaller mitochondrial DNAs. Because so much attention has been focused on the chloroplast in plant cells, less is known about their mitochondrial DNA. The pea, however, contains the largest mitochondrial genome yet reported, approaching the size of chloroplast DNA, although restriction studies will be necessary to characterize it fully.

In animal cells, mitochondrial DNA occupies a small proportion of the cellular content, certainly less than 1%. Its contribution to the overall genetic complexity is, of course, minute. The exact number of mitochondrial DNA molecules in the cell has been difficult to determine because of the unknown extent of losses during procedures to separate mitochondrial from nuclear DNA. The preferential labeling of mitochondrial DNA in L cells and HeLa cells that have lost the nuclear thymidine kinase activity but retain a mitochondrial activity has made it possible to label the organelle preferentially. This shows that the number of genomes is subject to some variation, over a twofold range in two different lines of L cells and by almost an order of magnitude compared with HeLa cells (Bogenhagen and Clayton, 1974).

The proportion of the HeLa cell occupied by mitochondria appears to be greater than that of the L cell, but unfortunately there seems to have been no exact comparison of the number of mitochondria in these lines. Generally the number of organelles in the cellular cytoplasm is thought to lie in the range 500–800, which would correspond to the presence of from 2 to 10 mitochondrial genomes per organelle. This is consistent with a body of earlier work in which somewhat approximate measurements suggested that there is roughly 0.5 mg DNA per gram of mitochondrion protein, with about 0.1 pg protein per mitochondrion (Nass, 1969a; Borst and Kroon, 1969). These values correspond to about 50 kb of DNA per mitochondrion, say about 3 genomes. The electron microscopic observations of Nass (1969b) showed that there are usually 2 or 3 nucleoids of DNA in the L cell mitochondrion, but may be up to 6; however, there is no evidence on how many genomes might constitute a nucleoid.

In yeast the proportion of organelle DNA is much higher, varying with the stage of growth, but generally is in a range of up to 18%. The proportion is the same in haploid and diploid cells, so that a doubling of the mitochondrial DNA content accompanies the nuclear doubling. This seems to reflect an increase in the number of mitochondria, from a mean of 10 per cell to a mean of 22; the effect of this increase is to maintain the mitochondrial proportion of cytoplasmic volume at about 14%. Under the conditions of these measurements mitochondrial DNA constituted 13% of total DNA, which corresponds to about 45 genomes per haploid cell and 85 genomes per diploid cell. This suggests an average value of about 4 mitochondrial DNA molecules per mitochondrion (Grimes, Mahler, and Perlman, 1974). This is a little greater than an earlier estimate of 1–2 genomes per organelle (Williamson, 1970).

In the context of this discussion, it may be misleading to think of the mitochondrion as an invariable structure; mitochondria may vary substantially in size and in the complexity of their cristae in a single animal or yeast cell, and it may be that this is related to the number of genomes. Indeed, the concept of the individual organelle has been questioned in yeast, where it has been suggested that many or all mitochondria may be joined in a branched network (Hoffman and Avers, 1973). This contrasts with the identification of individual mitochondria by Grimes et al. (1974). At all events, however, it is possible that the number of genomes is flexible in terms of the organelle and is determined by some smaller unit of mass or volume. This would be consistent with studies to visualize DNA in plant chloroplasts, where larger chloroplasts appear to have greater amounts of DNA (Herrmann, 1970; Kowallik and Herrmann, 1972). This consideration is interesting in view of the question of how organelle genomes from different parents may mix and be segregated if mitochondria or chloroplasts fuse following mating. Here we shall incline to the general view that the mitochondria form a fluid network, with constant fusions and separations, forming neither a single fixed huge structure nor many smaller truly independent units.

Inheritance of the Chlamydomonas Chloroplast

Organelle genes are identified by their non-Mendelian behavior, as seen in the failure to generate expected ratios in crosses, differences in the results of reciprocal crosses, the occurrence of somatic segregation during vegetative (clonal) growth, and independent assortment from nuclear genes. Early studies revealed the existence of such loci, but were not able to map them onto linkage group(s). Whether recombination occurs between organelle genes in higher plant and animal cells is indeed still a matter for conjecture; although in cases where there is strict uniparental inheritance, of course, there would usually seem to be no opportunity for recombination to occur.

A linkage group corresponding to the chloroplast genome now has been described in the unicellular alga Chlamydomonas reinhardii. The only other organelle genome mapped in detail is the mitochondrial DNA of the yeast Saccharomyces cerevisiae, although the essential prerequisite of recombination between organelle genes has been found for the mitochondria of other fungi also. Extensive reviews of organelle genetics have been made by Sager (1972) and Gillham (1978).

C. reinhardii offers a particular advantage for the analysis of the chloroplast genome since it possesses only a single chloroplast per cell, which occupies about half of the volume. The chloroplast genetics therefore concern events occurring in an individual organelle rather than a population. Another potential advantage is that mitochondria are also present, so that their genomes can be analyzed in the same system, to permit definition of relationships between the operations of chloroplasts and mitochondria.

The cells of C. reinhardii are haploid. The alga has a sexual life cycle, illustrated in Figure 21.1, in which the alternate alleles of a single nuclear gene ensure that cells are either mt^+ or mt^-. Fusion occurs only between cells of opposite mating type to give mt^+ mt^- diploid zygotes, which suffer meiosis to generate two mt^+ and two mt^- offspring from each cross. The existence of nonnuclear genetic loci was revealed when Sager (1954) isolated a streptomycin-resistant mutant with an unusual pattern of inheritance. Instead of the 2:2 Mendelian ratio, this showed 4:0 inheritance, so that in the cross str^r $mt^+ \times str^s$ mt^-, *all* the progeny are str^r. In the reciprocal cross the ratio is 0:4, since str^s $mt^+ \times str^r$ mt^- generates progeny *none* of which is str^r. Thus whichever allele is carried by the mt^+ parent is transmitted to all the progeny, in contrast with a Mendelian gene, which would be inherited by half the progeny. This is the classic pattern of maternal inheritance; all other nonnuclear mutations since have been defined by the same characteristic inheritance.*

How is this uniparental inheritance accomplished? C. reinhardii is isogamous: both parents contribute equal amounts of cytoplasm, excluding the possibility that only one chloroplast genome enters the zygote. This

---

*In the context of maternal inheritance, it is interesting to note that there has been a report that in C. reinhardii it may be the mt^+ parent that produces a fertilization tubule, and thus corresponds to the male sex (Friedman et al., 1968). Usually the mt^+ mating type has been equated with the female parent on the basis of the direction of uniparental inheritance. To avoid confusion about the nature of the events involved in uniparental inheritance, the mating types here are described only as mt^+ and mt^-, with no sexual assignment.

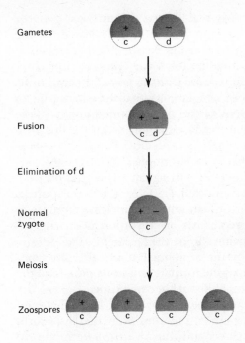

Gametes

Fusion

Elimination of d

Normal zygote

Meiosis

Zoospores

Figure 21.1. Inheritance of chloroplast geno-type in C. reinhardii. Two haploid gametes of mating type (+ and −) fuse to form a diploid zygote. The chloroplasts of both parents (*c* and *d*) enter the zygote. Shaded areas indicate the nucleus, white areas the chloroplast. The chloroplast genome of the mt^- parent is eliminated. The haploid zoospores produced by meiosis have a 4:0 ratio for chloroplast genotype, all retaining that characteristic of the mt^+ parent (*c*). Nuclear genes all are inherited with a Mendelian 2:2 ratio, as indicated by the example of the mating type. (Recombination may have occurred between nuclear genes, which fall into several linkage groups.)

problem was realized in the initial studies and implies that the chloroplast genomes of the two parents may be treated differently in the zygote, so that only that derived from the mt^+ parent is inherited, although both are present. This contrasts with the apparent origin of maternal inheritance in plants, which generally has been thought to result from the contribution of more cytoplasm (and hence more or the only copies of the organelle) by the female gamete. However, Sager (1972) has proposed that in some plant crosses the situation may in fact be analogous to that seen with C. reinhardii, with preferential inheritance due to differences in survival rather than provision of organelle genomes. Certainly in view of the existence of uniparental inheritance in an isogamous system, unilateral contribution of genetic information cannot be assumed to be responsible in any system unless the absence of alleles from one parent is indeed proven.*

---

*In animals it is assumed that mitochondrial inheritance is maternal because of the great excess of mitochondria in the egg compared with the sperm. The lack of mutants means that no genetic evidence is available on whether paternal organelle genes may be inherited. It is not known whether paternal mitochondrial DNA is completely excluded or may enter the egg in small amounts. The only direct evidence has been reported in Xenopus, in which there is maternal inheritance of mitochondrial DNA in crosses between X. laevis and X. mulleri (Dawid and Blackler, 1972). This involves measuring the ability of mitochondrial DNA of hybrid progeny to hybridize with mitochondrial RNA of either parent; it does not exclude the presence of a small amount of paternal mitochondrial DNA. Strict maternal inheritance also has been claimed in donkeys, horses, and rats, but the evidence is not satisfactory (Hutchinson et al., 1974; Buzzo et al., 1978).

What is responsible for the exclusion of the mt^- chloroplast genome from the progeny of a sexual cross? Sager and Lane (1972) followed the fate of the chloroplast DNA, identified by its buoyant density, in direct labeling experiments. In control crosses in which the DNA was unlabeled, the chloroplast DNA of the zygote was found to band at 1.690 g-cm^{-3} instead of its usual 1.694 g-cm^{-3}. When the DNA of one parent was distinguished by an ^{15}N-density label, the chloroplast DNA of the zygote formed only a single peak, instead of the double peak that would be produced by the differing densities of the two parental organelle DNAs. This peak bands at a density somewhat less than that of the mt^+ parental chloroplast DNA, the shift being variable but usually in the range 0.005–0.010 g-cm^{-3}. There is no peak corresponding to the density of the chloroplast DNA of the mt^- parent. The density shift is not affected by treatment with ribonuclease, pronase or α-amylase, which suggests that it represents an alteration occurring actually in the chloroplast DNA provided by the mt^+ parent. The density shift in the mt^+ chloroplast DNA occurs at about 6 hours after the commencement of mating, a time when zygote formation is at its peak, and the peak of mt^- parental DNA disappears at about the same time.

The density shift could be accounted for by methylation; 1% methylation causes a reduction of about 0.001 g-cm^{-3}. This has been the basis for a model, elaborated in some detail by Sager and Kitchin (1975), for modification and restriction of the chloroplast DNA during mating. The essential features of this model are that the mt^+ chloroplast DNA becomes methylated, presumably at specific sites, but the mt^- chloroplast DNA is not modified. Then a restriction activity degrades unprotected (mt^-) DNA in the zygotes. Modification and restriction must be linked specifically to mating, since the density shift occurs only at this time, and the surviving DNA later returns to the usual density. Support for the first part of this model has been provided by the observation that up to roughly half of the cytosine in mt^+ DNA is converted to 5-methyl cytosine in zygotes analyzed at 6 or 24 hours after mating. The methylation is specific for genotype since mt^- DNA analyzed at the same time contains little methyl cytosine ($< 10\%$ of the C residues). No methyl cytosine is found in vegetative cells or gametes of either mating type (Burton, Grabowy, and Sager, 1979).

Functions of both parents appear to be involved in the events responsible for preferential survival of the mt^+ chloroplast genes. Both the genetic constitution of the parents and the conditions of mating may influence the outcome of a cross. This is revealed by their effects upon the occurrence of *exceptional zygotes*, which transmit chloroplast genes from both parents to the progeny. These zygotes thus represent the survival of the mt^- as well as the mt^+ chloroplast genes. Their genetic properties have been studied in some detail, although biochemical experiments to identify the nature of the events involved in the persistence of mt^- chloroplast DNA have not yet been reported.

Two mutations that may identify components of the system involved in the inheritance of chloroplast DNA have been identified by Sager and Ramanis (1974). The mutant *mat-1* is linked to mt^- and increases the frequency of exceptional zygotes; the mutant *mat-2* is linked to mt^+ and decreases their spontaneous frequency. Environmental effects that may increase the survival of the mt^- chloroplast genes are seen in the treatment of mt^+ parents with ultraviolet irradiation or ethidium bromide before mating; some effects of inhibiting protein synthesis in either parent also have been noted. From these results it is clear that the mt^+ parent plays a critical role in ensuring the loss of mt^- chloroplast genes from the zygote; it is not yet clear what is the contribution of the mt^- parent to this process.

Linkage Map of Chlamydomonas Chloroplast DNA

Exceptional zygotes were first noted as rare occurrences in which both parental chloroplast alleles are transmitted to the progeny. The spontaneous rate at which they arise varies widely, but often is less than 1% (Sager and Ramanis, 1963; Gillham, 1963). The progeny of a single cross display 8 different genotypes: each of the 4 nuclear genotypes (2 parental and 2 recombinant) generates subclones carrying the alternate chloroplast alleles. Thus the zoospores (the first haploid products of meiosis) must carry the organelle alleles of both parents, as shown in Figure 21.2. Segregation of the parental alleles occurs rapidly during the succeeding mitotic divisions, generating cells that are homozygous for chloroplast genotype. This is complete within about 10 mitotic divisions, when no heterozygotes remain. This suggests that the number of segregating units is low.

In normal progeny possessing markers derived only from the mt^+ parent, the absence of mt^- parental markers precludes any opportunity for recombination. But recombination does occur between chloroplast markers present in the exceptional zygotes (Sager and Ramanis, 1965). Segregation of recombinant genotypes occurs during growth of the progeny, until eventually all cells are homozygous. Recombinants continue to arise in each generation in which heterozygotes remain.

An advance in using this system to map the chloroplast genome was made possible by the discovery that ultraviolet irradiation of the mt^+ parent immediately before mating leads to the recovery of 100% exceptional zygotes (Sager and Ramanis, 1967). These zygotes always are able to transmit the allele contributed by the mt^- parent; they may or may not be able to transmit alleles of the mt^+ parent, any deficiency apparently arising from damage to the mt^+ genome during irradiation. This indeed provided the first suggestion that, in a normal mating, the mt^+ parent is responsible for preventing the zygote from transmitting the mt^- chloroplast alleles to the progeny. Ultraviolet irradiation inactivates whatever function of the mt^+ parent is responsible, so that the mt^- genome always survives. By controlling the conditions, it is possible to obtain largely biparental exceptional

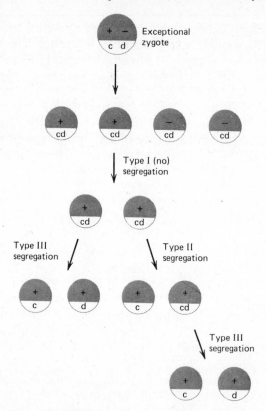

Figure 21.2. Behavior of exceptional zygotes in C. reinhardii. Both chloroplast genotypes remain and the zoospores usually all are heterozygous (possess both *c* and *d*). The heterozygosity may be maintained for some subsequent mitotic divisions (these are type I divisions); or segregation of the heterozygous markers may occur to generate cells that are homozygous for either chloroplast genotype. Segregation takes two forms: in type II segregation one of the daughter cells remains heterozygous and the other is homozygous for one of the parental marker alleles; in type III segregation one daughter cell is homozygous for one allele and the other daughter cell is homozygous for the other allele. The progeny of one zoospore are followed in the figure until no heterozygous cells remain. There is no somatic segregation of nuclear markers.

zygotes, in which both chloroplast genomes are present and recombination between them can be followed.

Relying upon biparental exceptional zygotes, pedigree analysis was introduced to allow events occurring in the zygote to be distinguished from those taking place in later generations. This procedure involves allowing the zoospores to pass through one mitotic division, to generate 8 octospores, which are then respread on a plate and allowed to divide again; the sixteen colonies resulting from the next division all are classified for the chloroplast markers. This effectively allows the fate of each of the 8 single DNA strands produced by meiosis to be followed. An important result of this approach was the demonstration that recombination occurs at a four-strand stage, analogous to that of conventional meiosis. This is true of recombination occurring in somatic cells as well as at meiosis (Sager and Ramanis, 1968, 1970; for review see Sager, 1972).

At each division, the segregation of chloroplast alleles falls into one of the three types of event illustrated in Figure 21.2. The first is lack of segregation: both progeny remain heterozygous. The second is the generation of homozygosity for (either) one of the parental markers in one of the daughter cells; the other cell remains heterozygous. The third is the disappearance of heterozygosity, with the generation of daughter cells each homozygous for one of the parental alleles. It is at once obvious that two types of event might be responsible for the segregation: sorting out and recombination. Models of these types have been discussed by Gillham (1974), Gillham et al. (1974), and Sager (1977); for review see Gillham (1978).

Sorting out would occur if there were two or more copies of the chloroplast genome and if the daughter chromosomes were randomly assigned to the progeny cells. The rate of segregation and the allelic segregation ratio (the relative proportions of homozygous cells of each allelic genotype) would depend on the total number of genome copies and on whether equal or unequal numbers were contributed to the biparental zygote by each parent.

If recombination is responsible for segregation, none of the homozygous segregants should have the parental genotype when scrutinized for additional markers. All should have recombinant genotypes. If segregation occurs only as a consequence of recombination, the reason for this effect must be that the genetic exchange alters the distribution of alleles to daughter cells. This implies that there must be a specific pattern for the distribution of the daughter chromosomes at division, controlled by some feature of the chromosome analogous to a centromere. This is akin to the phenomenon of recombination relative to the centromere that is used for (nuclear) genetic mapping in some fungi.

Extensive genetic data have been marshalled to show that the chloroplast genome behaves as though diploid, with an ordered distribution of daughter chromosomes (Sager and Ramanis, 1976a,b; Singer, Sager, and Ramanis, 1976). Three types of information have been used to construct a genetic map from the pedigree data. Conventional recombination analysis relies upon the relative frequencies with which recombinant genotypes arise between pairs of loci. Segregation analysis follows the relative rates at which heterozygosity at different loci is converted to homozygosity. Cosegregation analysis reveals the frequency at which segregation occurs simultaneously at a pair of loci rather than just at one.

To determine linkage between markers from the fraction of recombinants, it is necessary to analyze cells within a couple of generations after meiosis. Thus linkage can be found when recombinant proportions are measured in the total population after two (post meiotic) divisions. The linkage disappears when the population is examined after six generations; this reflects the continuation of recombination in the cells remaining heterozygous in each generation. The linkage data allow markers to be ordered in a single linear array. Thus all the known chloroplast genes fall onto a single linkage group (Sager and Ramanis, 1970).

When segregation at individual loci is followed, the heterozygosity shows an exponential decay with growth. This means that there is a constant rate of segregation per doubling. The rate of segregation, however, as measured by the slope of the decay curve, is different for each locus. Segregation of either type II or III usually involves only part of the chloroplast genome. Thus when segregants at one locus are examined for other markers, usually some of these remain heterozygous. This suggests that segregation does not depend on reassortment of whole genomes, which would generate homozygosity simultaneously at all loci. Nor can homozygosity be achieved by reduction to haploidy.

When the two types of segregation are distinguished by pedigree analysis, the difference in rate at each locus proves to be due entirely to type III (reciprocal) segregations; the frequency of type II (nonreciprocal) segregation is the same at each locus. The simplest explanation for the type III segregation is that it represents reciprocal recombination. Implicit in this model is the concept that the chloroplast genome is diploid, which is sustained by the 1:1 allelic ratios characteristic of type III events. Figure 21.3 shows that the variations in rates can be explained by supposing that there is an ordered distribution of chloroplast genomes at each division, governed by a centromere-like region, so that each daughter cell gains one centromere of mt^+ origin and one of mt^- origin. Loci then can be ordered on the basis that the farther they lie from the centromere, the more likely is their transfer by recombination to a centromere of opposite type, and hence the greater is the rate of segregation (Singer, Sager, and Ramanis, 1976).

This model proposes that the type II segregations represent nonreciprocal recombination due to gene conversion.* Although individual events may result in deviations from 1:1 allelic ratios, the nature of these deviations depends upon the timing of the recombinational event. In zygotes, the deviations are greater and occur more often at certain loci. In zoospores, a lower level of bias appears to have an allele-specific basis. It is possible therefore that gene conversion is initiated in two ways. A general mechanism involving events similar to those occurring in recombination in this

---

*Gene conversion can be visualized as occurring by a series of events in which recombination is initiated by formation of a heteroduplex between two DNA molecules, but the process is completed by rejoining of original strands on either side of the hybrid region, instead of the strands of opposite parents as seen in reciprocal recombination (see Volume 1 of Gene Expression). Any heterozygosity within the hybrid DNA must segregate with the separation of strands at the next replication; or it may be removed by repair enzymes that correct mispaired bases. Correction may be biased in one direction or the other for any particular mispair, so that heterozygosity is removed predominantly in favor of one parental type, resulting in a deviation from 1:1 allelic ratios.

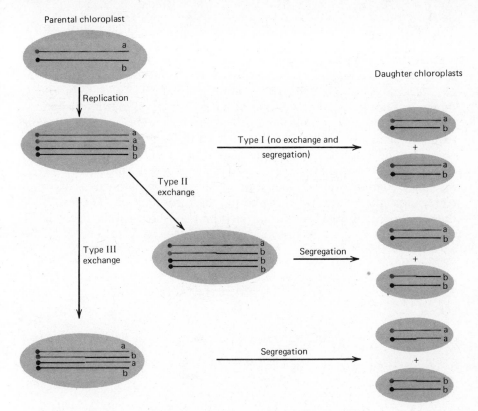

Figure 21.3. Model for diploid inheritance of chloroplast genes. Each genome is shown as a linear molecule with the centromere (●) at one end (to conform with the actually circular genome, it is necessary only to make all genetic exchanges occur in even numbers). When no recombination occurs, a type I event produces two daughter cells each of which remains heterozygous for the alleles *a* and *b* because each daughter always gains one centromere of each parental origin. Type II segregation is produced by nonreciprocal recombination and is independent of location on the map; one daughter cell is heterozygous and one is homozygous for one of the alleles. Type III segregation is produced by reciprocal recombination between a locus and the centromere, generating daughter cells each homozygous for one of the parental alleles.

and other systems may be present in both cell types; in addition, another mechanism may exist in the zygote, possibly related to the system for destruction of the *mt⁻* genome. By contrast, type III segregation shows the same characteristics in either zygotes or somatic cells.

The involvement of only restricted regions of the chloroplast genome in segregation of either type II or type III allows cosegregation to be used to map loci. Most segregation events involve only a single locus. Thus although homozygosity is achieved in the population as a whole rather rapidly, this is due to a series of independent events rather than to wholesale reassortment. The frequencies with which pairs of loci cosegregate may therefore be taken as a measure of their closeness; and similar results are obtained with either type II or type III events (Sager and Ramanis, 1976a,b).

Each method of mapping allows all markers to be arranged into one linkage group. This suggests that the chloroplast genome corresponds to a single DNA molecule. The sequence of markers obtained by each technique is shown in Figure 21.4. The segregation map gives distance from the centromere; and so may be arranged with the centromere either at the end of the sequence or in the middle. The biarmed arrangement represents a permutation of the linear cosegregation map. This suggests that the genome actually may be circular. A problem in achieving a definitive sequence is that the recombination map does not agree exactly with the maps based on segregation; the marker *ery* is placed in a different location. The reason for this discrepancy is not clear, but the circular map shown in the figure is the

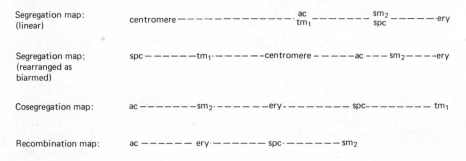

Segregation map:
(linear)

centromere $-----------\genfrac{}{}{0pt}{}{\text{ac}}{\text{tm}_1}-------\genfrac{}{}{0pt}{}{\text{sm}_2}{\text{spc}}-----$ery

Segregation map:
(rearranged as
biarmed)

spc $------$tm$_1\cdot------$centromere$-----$ac$---$sm$_2----$ery

Cosegregation map:

ac $-------$sm$_2\cdot-------$ery$\cdot--------$ spc$--------$ tm$_1$

Recombination map:

ac $------$ ery$\cdot------$ spc$\cdot------$sm$_2$

Circular composite map:

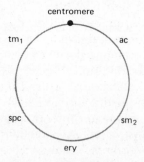

Figure 21.4. Maps of the C. reinhardii chloroplast genome. The segregation map gives distance from the centromere and may be arranged as a linear array (centromere at one end) or bilinear array (centromere in middle). The cosegregation map gives the relationship between pairs of markers. These two maps are related by a permutation of sequence which can be explained if the actual form is circular (Sager and Ramanis, 1976a,b; Singer et al., 1976). The recombination map also is linear, with the same order determined by Sager and Ramanis (1970) and Harris et al. (1977). This can be construed as a linear sequence drawn from the circular map, except for the position of *ery*.

present best estimate for the structure of the genome. It would be useful to demonstrate by a single method of mapping that the map is circular; as can be seen from the figure, this would require following pairs of markers on either side of the centromere. Since there is no information on what properties the feature corresponding to the centromere may have in an organelle, there may be some difficulty in this. At all events, it would be interesting to relate the map to the physically circular structure of chloroplast DNA, in particular to establish what is responsible for the centromeric properties.

Two crucial points are raised by the identification of a single linkage group. How many copies of the genome are present in each chloroplast? What governs their distribution at division? Before we consider these issues, however, we should take up the question of proof that this linkage group is indeed located in the chloroplast (rather than the possible alternative of the mitochondrion).

Two arguments have been made to identify the linkage group with the chloroplast. The first concerns the nature of the chloroplast DNA, which is inherited only from the mt^+ parent in a normal mating, in parallel with the uniparental inheritance of markers. This is not strong, because there is no evidence to say that mitochondrial DNA does not show the same behavior. The second derives from the nature of the mutations. Most of the mutations have been induced by streptomycin, which acts as a mutagen in this system, a capacity not demonstrated in any other system. The mutations fall into two types. Auxotrophic markers clearly are probably located in the chloroplast, since they affect specific functions such as photosynthesis. Drug-resistant markers could in principle be located in either chloroplast or mitochondrion, since they reflect the ability to continue protein synthesis in the presence of antibiotics that are effective in inhibiting bacterial, mitochondrial, or chloroplast protein synthesis. In this system, one such mutation has been shown to affect a chloroplast ribosomal protein; the presence of the other mutations all on the single linkage map implies that they too confer antibiotic resistance upon chloroplast protein synthesis. This is puzzling: why among the many drug-resistant markers has none been found in the mitochondrion? There seem to be three possibilities. Chlamydomonas mitochondria may differ from those of other organisms and may lack sensitivity to the antibiotics. Or they may be sensitive, but may lack genes able to mutate to confer resistance (unlike other mitochondria). But then they must be able to function in the drug-resistant cells. This would imply that resistance is conferred by the mutations in the chloroplast genome; in other words, there must be a functional relationship between the protein synthetic systems of the chloroplast and mitochondrion (see Bogorad, 1975; Surzycki and Gillham, 1971). Finally, mitochondrial mutations may occur (and must be present in the drug-resistant cells in addition to the chloroplast mutations), but they may escape detection for some reason that we do not understand.

It is evident from Figure 21.4 that the range of markers that has been available for analyzing the Chlamydomonas chloroplast genome has been very restricted. A significant extension of mutant genotypes has been made possible more recently by the observation that treatment with FUdR increases the frequency of chloroplast gene mutation (Wurtz et al., 1979). Whether this is related to its effect in reducing the amount of DNA per chloroplast is not known (Wurtz, Boynton, and Gillham, 1977). Mutants that are incapable of photosynthesis can be selected (rather inefficiently) by their ability to grow on arsenate; they are characterized further by their dependence on acetate. Shepherd, Boynton, and Gillham (1979) have classified 16 chloroplast mutants into 9 loci, according to their ability to recombine to generate photosynthetically competent progeny. In lieu of a complementation assay for this system, it is not possible to determine whether mutants with similar phenotypes lie in the same or different genes. Two of the mutants are defective in ribosome production; three lack the CP1 component of the thylakoid membrane, three lack polypeptides (*4.1–4.2*) that are thought to be part of the CF_1-ATPase complex, and one has no discernible defect in these functions. The recovery of these mutants doubles the number of markers available, will allow the construction of a linkage map containing more diverse functions, and should lead to unequivocal definition of which chloroplast functions are coded by the organelle genome and which by the nucleus.

The genetic crosses of chloroplast markers have been interpreted in terms of two contrasting models. Both must explain the demonstration that segregation occurs only by recombination and not by sorting out, which carries the implication that there is an ordered, not random, distribution of parental genomes at each division (as shown in Figure 21.3). This is perhaps surprising. In the normal course of events, there would be no possibility for such a situation, since the zygote would transmit only the mt^+ chloroplast genome. To take this point further, the existence of biparental zygotes raises the question of their relationship to the usual uniparental zygote. Is there a difference in ploidy of the chloroplast genomes; or is this rectified by a reduction in the ploidy of the exceptional zygote to that of the normal level in the progeny?

The model proposed by Singer, Sager, and Ramanis (1976) regards the organelle genome as diploid, in contrast with the model of Gillham, Boynton, and Lee (1974), which proposes that it is multicopy. The events involved in type II segregation are not entirely clear, and may even differ in zygote and zoospore; and part of the impetus for the two models has been a difference in results concerning this process. We have already described the data suggesting that type II segregation involves 1:1 allelic ratios, interpreted by Sager and Ramanis (1976a,b) in terms of gene conversion between pairs of alleles. The data obtained by Gillham et al. (1974) under different conditions, however, suggested a skew in the ratio in favor of maternal alleles. Both sets of data agree on the events involved in type III segregation, which takes the form of a reciprocal recombination, generating a 1:1 allelic ratio that is consistent only with a diploid genome. A balanced review of this discussion has been achieved by Gillham (1978).

The proposal that the chloroplast genome is genetically diploid offers a paradox in view of the presence of many DNA molecules per chloroplast (see Table 21.1). None of the attempted reconciliations seems plausible. It is hard to accept the view of Sager (1972) that the chloroplast DNA at present characterized is a repetitive component that does not carry the genetic markers that have been followed, which are supposed to lie on an unidentified, much larger, DNA. This flies in the face of available evidence on organelle genomes. But nor do simple multicopy models, based on parallels with phage population genetics, explain the 1:1 allelic segregation ratios; also they would predict some segregation by reassortment. Yet treatment with FUdR to change the amount of chloroplast DNA has a parallel effect on parental gene transmission (Wurtz, Boynton, and Gillham, 1977). The number of genomes followed in inheritance can be reduced from the total number of DNA molecules by models derived from those for plasmid incompatibility, in which the genomes in effect compete for attachment sites that are necessary for replication and passage to progeny. But while these can explain ordered distribution and a reduction in the number of segregating units from the total number of genomes present, it is necessary to introduce rather severe, perhaps unlikely, restraints in order to reduce the number of effective genetic units to only two. It is a matter for speculation whether the discrepancy between the genetic results and physical structure of chloroplast DNA is related to the apparent organization of many DNA molecules in the much smaller number of nucleoids that is visible morphologically. To summarize what seems to be a somewhat unsatisfactory situation, at least some of the genetic evidence calls for diploidy, whereas the biochemical evidence makes it clear that there must be far more DNA molecules than two. A model reconciling all the data has yet to be proposed; a reasonable suspicion is that the properties of the centromerelike region may prove to be critical.

If recombination occurs between chloroplast DNAs of the two parents forming an exceptional zygote, the two genomes must have come into juxtaposition. This implies that there must be a fusion between the chloroplasts analogous to that occurring between the nuclei; or there must be transfer of DNA from one chloroplast to the other. Evidence for genetic recombination has been gained in lieu of information on organelle fusion; and there is still little knowledge of how organelles may recombine or reproduce. In Chlamydomonas, chloroplast fusion may have been observed, since their juxtaposition with possible breakdown of membranes has been reported (Cavalier-Smith, 1970). It would be interesting to know whether recombination also occurs between mitochondrial genes in C. reinhardii.

The Petite Mutation of Yeast

Extranuclear mutations conferring a variety of phenotypes have been found in several fungi. In some cases, most notably Aspergillus and Podospora, some phenotypic changes bear no obvious relationship to mitochondrial functions, and it remains to be seen whether the mutations reside in the mitochondrial genome or there is other extranuclear genetic information (for review see Sager, 1972). In Neurospora and Saccharomyces there is clear evidence to show that the mitochondrion is the site of mutation. Indeed, similarities between these two systems suggest that the general form of organization of their mitochondrial genomes may be the same. This may be paralleled in a broad sense by the organization of plant and algal chloroplast genomes; perhaps also by that of vertebrate mitochondrial genomes, although here there is no direct information. In fungal mitochondria and in chloroplasts, mutations affecting organelle functions may display Mendelian inheritance (are located in the nucleus) or non-Mendelian inheritance (located in the organelle). This argues for an intimate relationship between nuclear and organelle genetic functions, which is borne out by studies of the assembly of organelle enzyme complexes.

Extranuclear inheritance in Neurospora was discovered with the isolation of the *poky* mutant of N. crassa. Now known as *mi-1*, this arose spontaneously, causes slow growth, is inherited maternally, and is associated with the absence of cytochromes *a* and *b* and an excess of *c* (Mitchell and Mitchell, 1952; Mitchell et al., 1953). Since then some 10–15 additional genes causing slow growth have been isolated; they may be distinguished by their phenotypes and (sometimes) by complementation in heterokaryons. Which proteins the *mi* genes code for is not yet known, but evidence is accumulating for various disruptions of cytochromes *a* and *b* (Bertrand and Werner, 1977). Nuclear mutations conferring the poky phenotype also have been found, as have nuclear genes that suppress the phenotype. In heterokaryons, mutant mitochondrial genotypes may suppress the wild type; and complementation may occur between mutants (Gowdridge, 1956; Pittinger, 1956; Bertrand and Pittinger, 1972). The occurrence of complementation means that there must be either diffusion of mitochondrial enzymes or products through the cytoplasm or fusion to bring different genomes into the same compartment. Mitochondrial fusion, of course, would create an opportunity for recombination to occur.

Interest in the mitochondrial genetics of yeast was aroused when Ephrussi et al. (1949a,b) discovered petite mutations, which render the cell able to live only in the anaerobic mode. Saccharomyces cerevisiae usually has the ability to grow under either aerobic or anaerobic conditions. Growth in the aerobic mode depends on mitochondrial respiratory functions; but because the cells can survive under anaerobiosis, these functions in effect are dispensable. Petite mutant strains are unable to adjust to aerobic conditions because they are deficient in respiration; but they can live in the anaerobic mode if given a fermentable sugar. This provides a conditional system, in which the anaerobic mode allows retrieval of mutations that would be lethal if the cells were constrained to grow in the aerobic mode. Mutations may therefore be obtained in any mitochondrial function.

The genetic versatility of yeast allows genetic manipulations to be performed that have led to a detailed definition of the mitochondrial linkage group. S. cerevisiae can exist as either haploid or diploid, or even triploid, cells. Haploid cells possess either the *a* or *α* mating type allele; cells of opposite mating type may fuse to form a diploid zygote, which can divide mitotically to generate a diploid clone. Diploid cells may sporulate; the meiosis produces a tetrad of haploid ascospores with 2:2 ratios for alleles of nuclear genes. Each ascospore divides to give a haploid clone. A diploid gamete obtains mitochondria from both parents; and subsequent growth through mitotic or meiotic divisions is accompanied by segregation and recombination of parental markers.

Petite mutations originally were generated by treatment with acriflavine, but since have been induced by a variety of treatments including ultraviolet irradiation and the addition of ethidium bromide. In some yeast strains they occur at high spontaneous rates, generating frequencies of up to 1%. Susceptibility to induction is the same for haploid and diploid cells; this independence of nuclear gene dosage was one of the early indications that the mutations might be non-Mendelian. When petite mutants of either mating type are crossed with wild type cells, the diploid progeny are normal. When they sporulate, all their haploid progeny are normal. The petite mutation has effectively disappeared and cannot be retrieved by backcrosses, a property quite inconsistent with Mendelian genetics. All the early petite mutants were of this type, known as the neutral or recessive petite.

A second class of non-Mendelian petite, the suppressive petite, was isolated by Ephrussi et al. (1955). When crossed with wild type cells these give entirely petite diploids; and sporulation generates petite haploid colonies. Thus the petite mutation has suppressed the wild type. Actually the degree of suppressiveness is not 100%, but is itself a characteristic of the strain. All these mutations show non-Mendelian ratios in crosses (Ephrussi and Grandchamp, 1965; Ephrussi et al., 1966). Both the neutral and suppressive petites represent irreversible genetic changes, since neither ever reverts to wild type. The suppressive petite genotype is inherited independently of nuclear markers (Wright and Lederberg, 1957).

The petite phenotype also can be generated by nuclear mutations. At least 20 nuclear genes are susceptible and all the mutations show normal Mendelian inheritance (Chen et al., 1950). These are able to revert to wild type.

The idea that non-Mendelian petite mutations represent gross changes in the mitochondrial genome originated with early work to characterize yeast mitochondrial DNA (Tewari et al., 1965; Mounolou et al., 1966; Moustacchi and Williamson, 1966). Actually some of this early work was in error; but its conclusion is well taken. The mitochondrial DNAs of suppressive petite strains may exhibit buoyant densities that are different from each other and from that of wild type; and components of the melting curve, which has several discrete peaks, may disappear (Bernardi et al., 1970). This suggests the occurrence of large changes in base composition. The change seen in neutral petites is more extreme: mitochondrial DNA is entirely absent (Nagley and Linnane, 1972).

The terminology now used to describe the nature of petite mutations was introduced with the postulate that these changes occur in a discrete entity, the *rho factor*. Wild type, or *grande*, strains, are described as rho^+. Suppressive petites possess an altered, inactive factor; and are described as rho^-. Neutral petites lack the factor and are described as rho^o. From this description it is evident that the rho factor may be equated with mitochondrial DNA. Although suppressiveness depends upon the presence of some (altered) mitochondrial DNA, the degree of suppressiveness does not seem to be correlated with the extent of the change in DNA (Nagley and Linnane, 1972). An exception to this, of course, is the property of zero suppressiveness, that is neutral petite mutation, which is conferred only by the absence of mitochondrial DNA.

What is the nature of the change in the mitochondrial DNA of rho^- petites? The reduced complexity of their melting curve is accompanied by more rapid kinetics of renaturation, suggesting that the length of unique nucleotide sequences has been reduced, sometimes by the loss of regions of distinct base composition (Bernardi et al., 1970). The kinetic complexities of rho^- strains relative to wild type vary from 0.2 to 36% (Michel et al., 1974). Hybridization between the mitochondrial DNA of rho^- strains and that of grande strains shows that the mutants always have lost some of the wild-type sequences; but all of the mutant sequences are derived from the wild-type mitochondrial genome. The proportion of the wild-type genome that is retained may vary from < 1 to 50%. (Lazowska et al., 1974). Hybridization between the mitochondrial DNAs of different rho^- strains may show any degree of relationship from none to virtually complete (Hollenberg et al., 1972; Faye et al., 1973; Sanders et al., 1973). This suggests that rho^- mutants arise by the deletion of mitochondrial DNA sequences selected at random from any part of the genome and varying from about half to the whole genome in extent.*

The reduction in sequence complexity naturally is accompanied by a loss of genetic markers. The complete absence of mitochondrial DNA from rho^o cells explains the loss of all mitochondrial genetic markers; formally, the rho^o mutation is pleiotropic for all mitochondrial genes. The retention of a random part of the mitochondrial genome in each rho^- petite means that any given marker may or may not be present. A formal interpretation of the retention of the ery^r marker in some rho^- petites at one time led to the incorrect conclusion that *ery* and *rho* were independent nonnuclear genetic determinants (Gingold et al., 1969)! The interpretation that there are many

---

*The ability to generate petite mutants has been most fully characterized in S. cerevisiae, but is found also in certain other "petite positive" yeasts, which similarly may suffer loss or deletion of the mitochondrial genome. By contrast, "petite negative" yeasts are unable to form respiratory deficient mutants by loss of mitochondrial DNA; although nuclear mutations conferring petite phenotypes have been found in some cases. The basis for the difference between the types of yeast is not known. Possibilities are that mitochondrial DNA of petite positive yeasts may have some peculiar feature allowing loss or deletion; or that mitochondria of petite negative yeasts may carry genes that are essential for viability.

types of *rho⁻* mutation, each characterized by the loss of a different part of the mitochondrial genome, was suggested by Nagley and Linnane (1972). A more complete characterization showed that two *rho⁻* mutants retaining some marker(s) in common always show some cross hybridization of their DNAs (Faye et al., 1973; Lazowska et al., 1974). Other similarities may be found between the mitochondrial DNAs of *rho⁻* petites showing common genetic markers; characteristic buoyant densities are associated with the retention of the *ery^r* and *chl^r* markers.

In spite of the loss of sequences from the mitochondrial genome in *rho⁻* strains, there does not appear to be a reduction in total amount of mitochondrial DNA, which varies from 12–21%, compared with the 18% of grande strains (Nagley and Linnane, 1972). This implies that the reduction in size of the genome must be associated with an increase in the number of copies of the surviving sequence. Such amplification might be accomplished by increasing the number of individual molecules, each containing one copy of the surviving sequence, or by increasing the number of copies in each molecule. Both seem to happen. The mechanism is obscure; indeed, the apparent ability of any random sequence of yeast mitochondrial DNA to survive in a petite implies that there can be no unique origin necessary for replication. At all events, however, the yeast mitochondrial genome appears to show a unique fluidity in its ability to undertake rearrangements of sequence in petite strains. This indeed may be an important feature in the phenomenon of suppression. Two mechanisms can be proposed for the ability of *rho⁻* petites to suppress the wild type in crosses: the truncated petite genomes may be preferentially replicated, thus excluding the wild type genomes from the progeny; or the petite genotype may be spread through the mitochondrial DNA population by recombination with the wild type genomes. The complete absence of mitochondrial DNA from *rho°* petites immediately explains their genetic disappearance in crosses with wild type.

When examined by electron microscopy, the mitochondrial DNA of *rho⁻* petites can be visualized in the form of rather small circular molecules (Hollenberg et al., 1972; Sanders et al., 1973). Denaturation mapping shows that in each *rho⁻* mutant the circular DNA forms a multimeric series. The multimers constitute a set of either even or odd numbers of tandem repeats or of even numbers of inverted repeats (Locker, Rabinowitz, and Getz, 1974a,b; Lazowska and Slonimski, 1976). The kinetic complexity of the petite DNA is related to the size of the monomeric circle, which thus appears to represent a single copy of the sequence retained from the wild-type mitochondrial genome. Restriction analysis of *rho⁻* petite mitochondrial DNA usually generates a series of fragments that constitute a subset of the grande pattern, but sometimes the array is more complex and cannot be seen directly to correspond to a simple part of the grande mitochondrial genome (Morimoto et al., 1975).

The larger circles in the multimeric petite series clearly are constituted from integral repeats of the sequence present in the monomer. This suggests that they may arise by recombination between homologous sequences. A single genetic exchange between two circles fuses them into one circle; an alternative origin would be by recombinational closure of a circle from the tail of a replicative rolling circle. Up to at least the tetramer, the frequency of an n-meric circle seems to be $1/n$ times the frequency of the monomer, so that on a mass basis each class represents the same fraction of total DNA. While some rho^- mutants contain simple multimeric series of DNA, others have multiples of different monomeric series. These may form joint circles that contain both types of repeating sequence. One such petite contains 73% of its DNA in the form of a 0.45 μm monomer and its higher multiples, 20% in a form based on a 1.18 μm circle, and the remainder as 1.53 μm circles representing the sum of the 0.45 μm and 1.18 μm monomers. This suggests that the origin of at least the joint circle lies in recombination.

To summarize, rho^- petite mutants contain mitochondrial DNA in which large deletions of random parts of the genome have occurred. Actually, some markers appear to be lost more frequently when petites are formed; the reason for such preference is not known (Fukuhara et al., 1978). However, any sequence appears able to survive in a petite and may then be amplified to increase the total content of DNA, both by multiplication of individual molecules and by the generation of repetitive genomes. Exactly how the presence of these sequences suppresses the wild-type mitochondrial DNA in crosses with grande strains is not known; there is no relationship between the complexity of petite DNA and suppressiveness, although there is some connection with the presence of large rearrangements (Heyting et al., 1979). The rho^0 petite mutants have lost all mitochondrial DNA and thus retain no mitochondrial genetic markers; fusion with a wild-type strain restores normal mitochondria to the mixed cytoplasm. The nature of the events responsible for generating petite mutations is not entirely clear, although it seems to reside in interference with mitochondrial DNA replication. Ethidium bromide may generate either rho^- or rho^0 mutants depending on the severity of treatment (Goldring et al., 1970). Presumably its preferential inhibition of mitochondrial DNA replication rests upon the effect of its intercalation in circular DNA. However, calculations of the number of targets for ethidium bromide in induction of petites have been inconclusive (Deutsch et al., 1974). Acridine preferentially causes the complete loss of mitochondrial DNA, explaining why only rho^0 petites were isolated in the original studies (Mattick and Nagley, 1977).

Recombination in Yeast Mitochondrial DNA

Several antibiotics inhibit mitochondrial protein synthesis, while not influencing the action of ribosomes in the cytoplasm. Their effect upon yeast cells is to produce phenocopies of the petite type. Mutants resistant to these antibiotics can be isolated; and may be located in the mitochondrial genome, when they show non-Mendelian genetics. The first such marker to be identified was resistance to erythromycin. Crosses of $ery^r \times ery^s$ haploids to give diploid progeny always produce mixed clones, consisting of some ery^r and some ery^s cells, in relative proportions that vary widely. Subclones are completely resistant or completely sensitive; thus segregation of the parental markers is completed at an early stage and the mixed clones include homozygous progeny of both types rather than heterozygotes. When the diploid progeny are sporulated, their haploid progeny show non-Mendelian ratios of 4:0 for resistance and 0:4 for sensitivity (Thomas and Wilkie, 1968a; Linnane et al., 1968). Thus the zygotes of a cross possess a mixed population of mitochondria, derived by fusion of the parental cytoplasms; but these segregate very rapidly during growth of the diploid progeny to generate cells that are homozygous for mitochondrial genotype.

Recombination was first detected when haploid parents bearing different drug resistance markers were crossed to generate diploid progeny with new combinations of the markers (Thomas and Wilkie, 1968b). Obviously the first concern here is to see whether the "recombination" involves a genetic exchange or some consistent reassortment of intact parental genomes. Following crosses between strains resistant to erythromycin and chloramphenicol (chl^r), Coen et al. (1970) showed that the recombinants do not segregate their markers either during vegetative growth or when haploid progeny are generated by meiosis. This suggests the formation of new mitochondrial genotypes.

A complication in analyzing the progeny of a yeast cross is that the zygote reproduces by budding instead of sequential fission of both progeny. Thus the zygote undergoes a series of divisions, each of which produces a bud that itself divides to generate a progeny clone; early buds may give rise to further progeny even while late buds continue to be produced by the zygote. Since recombinants continue to arise during formation of the first few buds, the representation of any genotype in the progeny population depends upon the time at which it is generated. Thus the frequencies of genotypes in each zygote clone may be quite different.

Two approaches have been used to analyze the results of such crosses. One is to use large enough populations to compensate for the fluctuations in individual events. Another is to take advantage of the opportunity to distinguish events occurring at early and late times by microdissection of buds (Lukins et al., 1973; Wilkie and Thomas, 1973). This shows that each of the early buds may be heterozygous for mitochondrial genotype, and so may be the next generation. Homozygosity is established very rapidly by segregation, however, since subsequent cells give clones of homozygous genotype. Recombinants arise frequently. It is not clear whether recombination is actually restricted to the zygote itself or may occur also in any immediate progeny which remain heterozygous.

All possible recombinants may emerge from any one zygote (although not simultaneously), but there is no reciprocity evident in their production. Indeed, there is no particular relationship between the genotypes produced in successive buds. The extreme case of this is the genotype of late buds, which remains constant but may be either parental or recombinant. The former situation is interesting in suggesting that a parental genome would seem to have been preserved in spite of the previous production of recombinants in the earlier buds.

The details of the recombination of yeast mitochondrial DNA are not yet defined, but the general nature of the system is considered to be a population cross. Thus the zygote gains a number of mitochondrial genomes from each parent and there is intrinsically a random nature to the formation of new genotypes from among the population. This is akin to a bacteriophage cross, although the system cannot be so readily manipulated (Dujon, Slonimski, and Weill, 1974). A central feature of this model is the proposal that the mitochondrial genomes of the zygote form a single "panmictic" pool, in which any mitochondrial DNA may recombine with any other. A constraint on the detailed analysis of recombination may be that panmixis is established slowly (in contrast with the immediacy of a phage cross) and this may induce a bias into the results, as has been shown with Schizosaccharomyces pombe (Seitz-Mayr et al., 1978; Wolf and Seitz-Mayr, 1978). Viewing yeast mitochondrial recombination in the general terms of this model, important questions that remain to be resolved concern the number of copies of each mitochondrial genome present in the zygote, the number contributed to each bud and the manner of distribution, the proportion of genomes that participates in genetic exchange, whether the mitochondrial genomes indeed form one population or several for the purposes of mating, the reciprocity of individual recombination events, and the number of rounds of recombination.

The standard genetic crosses used for mapping involve mating two haploid strains and collecting a large number of diploid cells that represent a random sample of progeny from many zygotes. In such crosses there may be a bias in the proportions of different genotypes found in the progeny. Two factors may influence the bias: relative input and mitochondrial genotype.

Using conditions in which the relative input of parental genomes was adjusted by treatments that discourage the transmission of genes from one parent, Birky et al. (1978) observed a tendency toward uniparental inheritance of the organelle markers provided by the predominant parent. This is not absolute, since recombination occurs, but it seems as though the input bias may be amplified in the progeny, possibly because of some preferential selection of the majority type of genome. The basis for such an effect is not clear.

The representation of parental alleles in the population is strongly influenced by a single mitochondrial locus. Bolotin et al. (1971) defined two types of yeast cell with respect to recombination. Polar matings result from crosses of $\omega^+ \times \omega^-$ and are recognized by the preferential transmission of markers close to the ω^+ allele. This establishes a gradient of polarity that can be used to order markers. Nonpolar crosses lack this bias and are produced by the matings $\omega^+ \times \omega^+$ or $\omega^- \times \omega^-$. This suggests that some difference between the ω^+ and ω^- alleles may cause gene conversion to proceed preferentially in crosses between opposed genotypes (Dujon et al., 1974). Another allele at this locus since has been found; ω^n shows no polarity in crosses with either ω^+ or ω^- (Dujon et al., 1976). The ω^n genotype may be derived from the ω^- but not from the ω^+.

The suggestion that polarity results from a gross difference in the structure of mitochondrial DNA at the ω locus has been borne out with the observation that ω^+ strains possess an insertion of about 1150 base pairs that is absent from the mitochondrial DNA of ω^- strains. This corresponds to the intervening sequence that is present about 400 bp from one end of the gene for large ribosomal RNA as noted in Table 26.5 (Faye et al., 1979; Heyting and Menke, 1979). The basis for the polar effect cannot be explained simply as preferential repair of a single stranded sequence resulting from heteroduplex formation because of the lack of polarity in crosses of ω^+ with ω^n (whose detailed structure is not known, but which lacks the 1150 bp insertion). This suggests that some special property of ω^- might be responsible; for example, this could take the form of a specific sequence whose integrity is interrupted by the insertion in ω^+ and which has been deleted or mutated in ω^n. Asymmetric gene conversion has been observed between alleles of slightly different size that are thought to code for the *var-1* protein; these events generate new alleles (Strausberg et al., 1978). Whether the same events are involved at both loci remains to be seen, but it is possible that these situations represent the same feature of the yeast system for mitochondrial recombination. The persistence of ω^- genotypes in spite of the preferential conversion to ω^+ in polar matings raises the possibility that lack of the insertion may confer some selective advantage.

Recombination Mapping with Petites

Conventional mapping experiments involve crosses between grande strains bearing drug-resistance markers. Because the rate of recombination is high, this technique is limited in its ability to map markers that are not very closely linked. An important extension of mapping techniques was made possible by the discovery that petites may recombine with grande strains. In fact, testing whether a marker can be rescued from petite mitochondrial DNA in a cross with a grande strain provides the only way to determine the genetic constitution of a petite. Direct determination of the markers retained in a petite is impossible, of course, because the phenotype is recognized by its pleiotropic loss of all mitochondrial functions, precluding examination of individual markers.

The initial report of Coen et al. (1970) demonstrated that in petite × grande crosses, functions absent from the grande can be provided by the petite. This left open the question of whether this was due to recovery of markers through genetic exchange or to complementation via cytoplasmic diffusion. The first would imply the creation of a new (homozygous) genotype, the second the perpetuation of a stable heterozygosity. The nature of the events involved was made clear by the demonstration that genetic recombination can occur between the mitochondrial genomes of petite strains.

Again the occurrence of recombination cannot be assessed directly because of the pleiotropy of the petite mutations. Thus the cross is performed in two steps. First two haploid petite strains (carrying markers previously defined by crosses with grande strains) are crossed to give diploid petites. These in turn are crossed with a haploid grande strain to give triploids. Cells are selected for the simultaneous recovery of two markers absent from the grande mitochondrial genome, each of which must be derived from a different one of the original (haploid) petite genomes. Then the diploid petite clone that was able to provide both these markers is examined.

Evidence for physical recombination was gained by Michaelis et al. (1973), who used two petite strains, one carrying ery^r and with mitochondrial DNA of buoyant density 1.687 g-cm^{-3}, the other carrying chl^r with a density of 1.681 g-cm^{-3}. The diploid petites able to confer drug resistance to an ery^s chl^s grande strain had mitochondrial DNA of intermediate density, 1.683 g-cm^{-3}. Comparison of the kinetic complexity and melting profile of the mitochondrial DNA of the diploid petite with the properties of its two haploid parents suggests that recombination is achieved by a junction of the parental petite DNAs in equal ratios (Michaelis et al., 1976).

An electron microscopic analysis of the mitochondrial DNAs of these strains shows that the genomes take several forms, related by simple repetition of a basic sequence (Lazowska and Slonimski, 1977). The ery^r petite contains a series of circles with an 0.6 μm monomer; the predominant form is a 1.1 μm dimer which comprises an inverted repeat. The chl^r contains a predominant circular form of 3.38 μm, shown by denaturation mapping also to comprise an inverted repeat, of a 1.75 μm monomer. One recombinant contains circles of 4.48 μm contour, consisting of two inverted sequences of length 2.35 μm. Each of these contains one of the parental repeats, that is, units of 1.75 + 0.6 μm. Thus if the two parents have the mitochondrial genomes E.E' and C.C', the recombinant has the genome E.E' .C.C' (expressed in linear form and with a prime indicating an inversion of sequence). Another recombinant is very similar, but with a slightly different repeating length. An ery^r chl^r petite selected in a single step from a grande strain has a similar structure, although with a genome about 5% longer; where the 5% length has been lost in the reconstruction from separate petites is not known. The restoration by recombination of what appears to be an original sequence suggests that the mechanism may involve a site specific exchange, depending on sequences present at the junctions of the inverted repetitions. At all events, the mechanism cannot involve extensive homology, because the parental sequences show scarcely any similarity; there is less than 2% cross hybridization. The apparently precise location of the rejoining would argue against recombination by chance homology of short sequences.

Two points of general interest are evident in these results. First, the occurrence of recombination between petites means that the system responsible must be coded by the nucleus; like the replication apparatus, its presence in the mitochondrion must be independent of mitochondrial functions. Second, the recombination between petite and grande strains must involve some form of marker rescue; either the events are nonreciprocal or reciprocal genetic exchanges must occur in even numbers to replace a part of the grande circle with a part of the petite circle. In fact, of course, the recovery of recombinants is only one possible outcome of the cross; the other is suppression of the wild genotype. Naturally it is necessary, therefore, to use petite strains of low suppressiveness if recombinants are to be obtained. Whether recombination also is responsible for suppression is not known; but if this is so, it may be that it is the nature of the genotypes produced by each recombination event that determines whether the progeny are to be grande or petite.

A practical consequence of the ability of petites to recombine with grande strains is that they may be used for deletion mapping. The only proviso necessary for the success of this procedure is that the generation of a petite must involve the retention of contiguous regions of the grande mitochondrial genome; there must not be extensive rearrangement of the relationship between markers. That this condition is at least approximately fulfilled is suggested by the ability to construct a consistent genetic map and by the direct analysis of the DNA fragments retained in petites. Detailed examinations of the structures of petites have been made to identify secondary alterations that may arise after the initial deletion event and have emphasized that for really fine structure mapping it is necessary to use petites whose molecular structures are defined (Heyting et al., 1979; Lewin et al., 1978).

The likelihood that two markers will be retained in the same petite is a genetic measure of their closeness (Molloy, Linnane, and Lukins, 1975; Bolotin-Fukuhara and Fukuhara, 1976). A circular genetic map can be obtained on the basis of frequencies of coretention. The use of petites for deletion mapping (or more accurately, retention mapping) allows the physical location of markers on mitochondrial DNA to be determined directly (Sriprakash et al., 1976). The position of any mutation can be defined within the sequence held in common between two petites that retain the marker. Originally defined by electron microscopy or hybridization with grande

DNA, such overlaps now are routinely determined by using a library of petites that have been the subject for detailed restriction mapping.*

The physical map constructed from restriction analysis of grande and petite strains is comparable to the genetic map, although there are some interesting discrepancies between physical and genetic distances (see below). Poly(dAT) regions, which account for about half of the genome, are concentrated in one quadrant (Prunell and Bernardi, 1974, 1977; Sanders and Borst, 1977). A comparison of the restriction maps of two strains of S. cerevisiae and the related yeast S. carlsbergensis shows that the major differences represent sequences inserted in this region. There are also many smaller differences, although the overall organization of the map seems to remain similar (Sanders et al., 1977).

Complementation Mapping of *mit⁻*Mutations

Petite mutants have proved invaluable for mapping the yeast mitochondrial genome, but cannot be used for investigating the functions of individual loci because they are multisite and pleiotropic. The recent dissection of the mitochondrial genetic apparatus has drawn upon two types of mutant with more limited defects, representing point mutations or perhaps rather small deletions.

Mutants of the *mit⁻* type are defective in some specific respiratory function. They are distinguished from petite mutants by their ability to support protein synthesis (Tzagoloff, Akai, and Needleman, 1975b). They can be mapped into seven complementation groups by determining the ability to respire of diploids formed by mating two haploid *mit⁻* parents (Foury and Tzagoloff, 1978). Their map locations are shown in Figure 21.5; the organization of some of these groups appears to be complex and their relationship with the mitochondrial gene products is discussed below.

Mutants of the *syn⁻* type are defective in some specific function of mitochondrial protein synthesis. These have to be conditional mutants, since the failure of mitochondrial protein synthesis induces the petite state. Temperature-sensitive mutants of this type have been isolated by Bolotin-Fukuhara et al. (1977).

In addition to these classes of mutants, there are also the previously characterized drug resistance markers. These are of two types. Resistance to inhibitors of protein synthesis, such as erythromycin and chloramphenicol,

---

*In earlier work using hybridization as the criterion for establishing overlaps, a problem in using petites was the possibility that some sequences might be differentially amplified relative to others, or even that new sequences might be added. This was overcome by hybridizing the petite DNA to labeled grande mitochondrial DNA; the reacting labeled sequences all are part of the wild-type mitochondrial genome and are present in equimolar amounts. These then may be used as a probe for comparison with a second petite genome.

is acquired by virtue of alterations rendering the ribosomes immune to inhibition. Resistance to inhibitors of energy production, notably oligomycin, is conferred by mutations that lie in the appropriate *mit* loci.

Mutations in individual loci may be characterized by recombination and by complementation. Recombination mapping between closely linked markers or between markers and petites locates groups of mutations on the genetic and physical maps. To determine whether each cluster corresponds to a gene requires complementation analysis. Until recently it was thought that the rapid rate of recombination and segregation of genotypes would make complementation studies impossible.

However, Foury and Tzagoloff (1978) found that a cross between two *mit⁻* mutants allows complemention mapping on the basis of whether respiration is restored in the immediate diploid cells produced by fusion. The first zygotes appear at about 3 hours after mating of the haploid parents. When respiration rises to 20–40% of wild type levels at 4 hours and to 70–80% at 10 hours, the *mit* mutations can be assigned to different complementation groups. When respiration is not restored within this period, but is found only later in a minority of (recombinant) cells, the *mit* mutations are assigned to the same complementation group.

Evidence that complementation and not recombination is responsible for the observed activity is inferential rather than direct, since the inevitable occurrence of recombination at later times precludes any demonstration that it has not occurred in the zygotes. But the respiratory activity at the time when the zygotes are analyzed is either high or absent, determining whether the mutations are assigned to different or the same complementation groups. Later only a small proportion of cells can respire; these indeed correspond to recombinants and the frequencies may be equally high for mutations lying in different groups and mutations at the distant ends of the same group. Other experiments argue against biochemical complementation (which would be achieved by mixing of wild-type subunits present in each parental strain) or interallelic complementation (when the different defects in the same protein would compensate upon mixing to form a homomultimer). The results of the complementation analysis are consistent in showing that clusters of mutations in different loci fall into different groups; and especially important is the demonstration that some adjacent clusters of mutations fall into the same group. This provides critical evidence for determining the extent of each gene.

A similar technique has been used by Strausberg and Butow (1977) to attempt zygotic gene rescue of markers in crosses between petite and grande strains. These experiments followed the expression of alleles of the *var-1* locus, which generate proteins of different electrophoretic mobilities. An allele carried by the petite is first expressed at about 4 hours after mating; by 5 hours the intensity of this protein band may be equivalent to that of the grande allele product. This is greater than would be expected from the 20%

issue of recombinants; it could be due to the repetition of sequences in the petite DNA, causing extensive transcription and translation upon exposure to the expression apparatus of the grande cell. Neither biochemical nor interallelic complementation can be involved, of course, since there is no mitochondrial protein synthesis in the petite parent.

In both systems the evidence for complementation is essentially quantitative. But the only way to achieve a more satisfactory basis for determining the level of the effect would be to show its existence under conditions when recombination cannot occur. At present this is not possible; although an interesting speculation is that there may be nuclear genes whose mutation would prevent recombination in mitochondrial DNA.

How might complementation occur? Formally the only requirement is that the protein products of the two parents are able to act in the same cell, so that a complete metabolic pathway is present. But the particular features of the mitochondrion would seem to impose a more stringent demand. The products of mitochondrial protein synthesis are located on the inner membrane of the organelle; there is no evidence of any ability to enter the cytoplasm, which would argue against diffusion of the enzymes. Diffusion of the products of enzyme action is unlikely; in some cases they too are large and in others the complementation groups represent different subunits of the same protein complex, all of which must be in one location for function. This suggests that the two parental DNA must be in the same cell compartment, so that the products of translation can mix directly. This implies the existence of mitochondrial fusion, or the transfer of DNA between mitochondria. Of course, this is the same requirement imposed by the occurrence of recombination; there is at present no information on what may be its mechanism.

Protein Synthesis in Organelles

Organelles undertake both transcription and translation. Expression of the organelle genome is interesting for both its mechanism and function. In terms of mechanism, how are the processes of RNA and protein synthesis related to the corresponding events in the nucleus and cytoplasm? And to what extent is the nucleic acid and protein synthetic apparatus of the organelle specified by its own genome? From the perspective of function, what is the form of cooperation between nuclear and organelle gene products in the assembly of organelle protein complexes? Which organelle proteins are coded by the nucleus and which by the organelle itself; are their messengers translated exclusively on ribosomes of the same origin?

Both chloroplasts and mitochondria contain a complete apparatus for protein synthesis which is distinct from that of the surrounding cytoplasm. Purified preparations of organelles may be able to undertake protein synthesis as seen from the incorporation of labeled amino acids into completed products. Within the cell, events occurring in the organelle and in the surrounding cytoplasm can be distinguished by the use of drugs. Cycloheximide inhibits cytoplasmic protein synthesis but leaves organelle translation unaffected; many antibiotics inhibit organelle protein synthesis without influencing events in the cytoplasm. The organelle protein synthetic apparatus in some respects is similar to that of bacteria (in particular in its sensitivity to antibiotics); and this has been taken to support the theory of an endosymbiotic origin in some ancestral fusion of eucaryotic cells with a procaryotic progenitor to the organelle. However, there are also some striking differences.

The relative sensitivities of nuclear and organelle transcription to inhibition are less well defined. But there is no evidence for susceptibility of organelle RNA synthesis to α-amanitin, an inhibitor of animal nuclear RNA polymerases at varying levels of sensitivity (see Chapter 22). Camptothecin has been used to suppress the appearance of labeled mRNA in the cytoplasm while leaving mitochondrial RNA synthesis intact (Hirsch, Spradling, and Penman, 1974). Susceptibility of organelle transcription to rifampicin, an inhibitor of bacterial RNA polymerase, varies according to published reports; but the lack of sufficiently purified systems precludes a definite evaluation (for review see Borst, 1972). Ethidium bromide inhibits the replication of mitochondrial DNA, an effect particularly well characterized in yeast, of course, and also may be used to inhibit RNA synthesis (Zylber, Vesco, and Penman, 1969).

The components identified in the best characterized organelle protein synthetic systems are summarized in Table 21.3. Ribosomes, tRNAs and aminoacyl-tRNA synthetases have been found; but little is known about detailed functions such as the operation of initiation or elongation factors. Organelle ribosomes usually are smaller than those of the eucaryotic cytoplasm, which are typified as 80S, with subunits of 60S and 40S. The only organelle in which the ribosomes approach the size of the cytoplasmic is the higher plant mitochondrion. The smaller organelle ribosomes fall into two general classes. Ribosomes of chloroplasts and the mitochondria of lower eucaryotes are fairly close in size to the bacterial 70S ribosome, with subunits often about 50S and 30S. The tendency of the subunits to dissociate may be somewhat different from those of bacteria. The rRNAs are about the same size as bacterial rRNAs; they seem to suffer a much lower level of methylation, as seen in yeast. A 5S rRNA is found in plant chloroplasts and mitochondria, but seems to be absent from animal and yeast mitochondria. Mitochondria of vertebrates have smaller ribosomes, sometimes assigned a value of 55S and known as miniribosomes. Their rRNAs are rather small. (Indeed, because of the small size of the miniribosome, there was for some time a certain amount of confusion as to whether these particles actually constituted complete ribosomes or subunits.)

The ability of organelles to synthesize proteins implies that they must possess a full complement of tRNAs. Minimum numbers of tRNAs have

Table 21.3   Protein synthetic apparatus of organelles

System	Ribosome size, S	Number ribosomal proteins	Sizes of rRNAs, bp	Number of tRNAs	Other identified components
Yeast mitochondrion	74 = 50 + 37	38 + 37	3900 2100	>19 αα	RNA polymerase tightly bound to membrane; difficult to purify
Neurospora mitochondrion	73 = 57 + 35	34 + 25	3000 2200	>17 αα ~26 tRNA	>15 αα-tRNA synthetases; RNA polymerase of 64,000 daltons
Drosophila mitochondrion			1500 860		
Xenopus mitochondrion	60 = 43 + 32	44+ + 40	1650 900		RNA polymerase of 46,000 daltons
HeLa mitochondrion	60 = 45 + 35		1650 1050	>16 αα >17 tRNA	>14 αα-tRNA synthetases
Maize mitochondrion	78 = 60 + 44		3800 2300		
Maize chloroplast	70 = 50 + 30		3200 1700	>16 αα	RNA polymerase of 180,000 and 140,000 dalton subunits
Euglena chloroplast	70 = 50 + 30		3300 1700		
Chlamydomonas chloroplast	68 = 33 + 28	22 + 26	3200 1600		

The sizes of ribosomes and their subunits are only approximate in view of fluctuations in sedimentation coefficients with conditions. The number of ribosomal proteins is based on two-dimensional gel electrophoresis; the size of rRNAs on gel electrophoresis. The number of tRNAs is shown in terms of the minimum number of different accepting activities found for amino acids (αα); where found, the number of separated tRNA molecules is shown. Estimates for the number of tRNA genes in organelle DNAs are given in Table 21.4. The minimum number of aminoacyl-tRNA synthetases is given where identified. Where RNA polymerases have been found, they do not appear to coincide with any of the nuclear enzymes.

References: Yeast—Grivell et al. (1971); Reijnders et al. (1973), Tsai et al. (1971), Faye and Sor (1977); Neurospora—Barnett et al. (1967); Barnett and Brown (1967), Rifkin et al. (1967), Kuntzel and Noll (1967), Kuntzel and Shafer (1971), DeVries et al. (1979), Lambowitz et al. (1979); Drosophila—Klukas and Dawid (1976); Xenopus—Swanson and Dawid (1970), Dawid and Chase (1972), Wu and Dawid (1972), Leister and Dawid (1974); HeLa—Attardi and Ojala (1971), Robberson et al. (1971), Lynch and Attardi (1976); maize—Leaver and Harmey (1973), Pring (1973), Cunningham et al. (1976), Lyttleton (1962), Clark et al. (1964), Boardman et al. (1966), Stutz and Noll (1967), Smith and Bogorad (1974), Haff and Bogorad (1976); Euglena— Rawson and Stutz (1969); Chlamydomonas—Hoober and Blobel (1969), Hanson et al. (1974).

been determined by identifying the ability of organelle extracts to place labeled amino acids onto tRNA. Although the number is less than 20 in each case, probably this is just due to technical difficulties in charging some tRNAs. The number of tRNA species is more difficult to determine directly. Isoaccepting tRNAs have been identified in one or two instances, but it has not yet been possible to fractionate the individual species. The best estimates for total tRNAs are provided by assays for the number of tRNA genes in the organelle DNA. These are performed by hybridizing tRNA charged with a labeled amino acid with organelle DNA; the proportion of organelle DNA that hybridizes at saturation can be used to calculate the total number of tRNA genes. As can be seen from Table 21.4, this number is larger than the number of different accepting activities. Probably these values are underestimates, due to some of the same difficulties encountered in isolating certain charged aminoacyl-tRNA species.

Estimates for the total number of tRNA species coded by each organelle genome have been steadily rising and may yet prove to reach the minimum number of 32 required by the wobble hypothesis for recognition of all 61 codons (see Volume 1 of Gene Expression). But in lieu of demonstrations that this is so, three explanations have been considered for any discrepancy. The simplest is that tRNA species additional to those coded by the organelle genome may be coded by the nucleus and will be imported from the cytoplasm; there is at present no evidence on whether all tRNAs found in organelles are coded by the organelle genome. Another idea is that not all codons may be used in the organelle; this would impose a unique restriction on the sequence of the genome and seems unlikely. Or the recognition between codon and anticodon in the organelle may escape the restraints of the wobble hypothesis, perhaps by using unusual bases in the anticodon that are able to recognize codons irrespective of irrelevant third bases; individual examples of such tRNAs have been found in bacteria, but this would require the construction of an entire set.

One feature of organelle protein synthesis that parallels the procaryotic rather than eucaryotic cytoplasm is the use of formyl-methionyl-tRNA as initiator. It will be interesting to make detailed comparisons of sequences of the organelle and cytoplasmic tRNA sets, in particular in organisms that possess both chloroplasts and mitochondria. Charging of tRNAs occurs within the organelle, which presumably must possess a full complement of aminoacyl-tRNA synthetases. This has been confirmed experimentally in two cases. Again a comparison of their properties with those of the cytoplasmic enzymes may be interesting with regard to the question of how tRNA features are distinguished.

What part of the protein synthetic apparatus is coded by the organelle genome? Available data on the RNA components are summarized in Table 21.4. At least a large proportion of the tRNAs is coded by the organelle genome. The ribosomal RNAs appear always to be coded by the organelle. Mitochondrial genomes at present characterized appear to have one copy

Table 21.4 Genes in organelle DNA

System	Number tRNA genes	Number and organization of large and small rRNA genes	Number of discrete mRNAs or protein products
Yeast mitochondrion	22	1 copy of each; not linked	8 poly(A)$^+$ RNAs; 5 major proteins of 45 + 29 + 21 + 12 + 8 = 115 K
Neurospora mitochondrion		1 copy of each; 5–6 kb apart	
Drosophila mitochondrion		1 copy of each; ~160 bp apart	8 poly(A)$^+$ RNAs total length ~12 kb
Xenopus mitochondrion	21	1 copy of each; ~140 bp apart	
HeLa mitochondrion	19	1 copy of each; <500 bp apart	8 poly(A)$^+$ RNAs total length 7–12 kb
L cell mitochondrion		1 copy of each; <200 bp apart; possible joint precursor	7 poly(A)$^+$ RNAs; total length ~6 kb; 7 proteins, 10–48 K, made on mit ribosomes
Maize chloroplast	23	2 copies of each and 2 5S genes; may be joint precursor of 8 kb; 2 rDNA sets form inverted repeat	
Euglena chloroplast	26	3 copies of each in tandem repeat of 5.6 kb; single precursors of 3.5 & 1.9 kb	~50% of genome transcribed as seen in hybridization with total cell RNA

The number of tRNA genes is determined by hybridizing labeled aminoacyl-tRNA with organelle DNA to saturation. Due to loss of some aminoacyl-tRNA species, these values are probably under estimates. The number of rRNA genes initially was determined by saturation hybridization; this has been confirmed by studies of organization using restriction fragments. The number of mRNAs is determined by the predominent discrete bands of poly(A)$^+$ RNA separated on gel electrophoresis; protein products usually have not been separated with the same resolution.

References: Yeast—Nagley et al. (1974), Faye et al. (1975), Padmanaban et al. (1975), Sanders et al. (1975b), Schneller et al. (1975), Van Ommen et al. (1977), Martin et al. (1977), Martin and Rabinowitz (1978); Neurospora—Schafer and Kuntzel (1972), Kuriyama and Luck (1973), Heckman and Raj Bhandary (1979); Drosophila—Hirsch et al. (1974), Klukas and Dawid (1976); Xenopus—Dawid (1972), Ohi et al. (1978), Ramirez and Dawid (1978); HeLa—Aloni and Attardi (1971), Hirsch and Penman (1973), Ojala and Attardi (1974), Angerer et al. (1976); L cell—Lansman and Clayton (1975), Battey and Clayton (1978); maize (and other higher plants)—Hartley and

of each gene. Usually these are located close together and sometimes there is tentative evidence for the existence of a common precursor, analogous to that of bacteria and the nucleus. Yeast is an exception in which the two rRNA genes are far apart and must be transcribed separately. In chloroplast genomes there may be more than one copy of the set of rRNA genes, with the multiple sets organized in tandem or inverted repeats. The repetition of these genes contrasts with the general view of the organelle genome as restricted in coding capacity to the point where many functions must be specified by nuclear genes.

Polyadenylated RNA is found in mitochondria, although the presence of poly(A) on yeast mitochondrial mRNA is controversial. The number of discrete mRNA species resolved on acrylamide gels is low, usually about 8. It is interesting that the number seems to be no greater in the yeast mitochondrion than in the mitochondria of Drosophila and mammals where the organelle genome is much smaller. The number of messengers is close to the number of proteins that can be identified as products of organelle protein synthesis. Each single poly(A)$^+$ RNA probably represents a discrete monocistronic messenger. It may be translated by organelle ribosomes and arrays of polysomes have been found in mitochondria. An interesting question is whether the (approximately) eight structural genes of the mitochondrial genome code for the same protein products in the mitochondria of different species.

The locations of structural genes can be mapped by hybridizing restriction fragments of mitochondrial DNA with the RNA products. Actually this also allows the transcripts to be defined in more detail. Direct isolation of mitochondrial mRNAs is difficult because of their small amounts; but by using a mitochondrial DNA restriction fragment as probe, it is possible to isolate the corresponding messengers as well as any precursors. The direction of transcription of each gene can be inferred from the ability of the RNA to hybridize with the separated strands of the genome; animal mitochondrial DNAs can be separated into their constituent strands by alkaline density centrifugation and these are known as the H (heavy) and L (light) strands (the same nomenclature applied to satellite DNAs; see Chapter 19).

A clear result has been obtained with the L cell mitochondrion, where all 7 poly(A)$^+$ transcripts hybridize with the H strand, as do the two rRNAs (Battey and Clayton, 1978). These regions together account for about 60% of the genome. Larger, less abundant, transcripts map in the same regions; this suggests the possibility that there are precursors which are processed to give the mature mRNAs. No transcripts of the L strand can be found. Similarly

Ellis (1973), Thomas and Tewari (1974), Haff and Bogorad (1976), Bedbrook et al. (1977), Whitfield, Herrmann and Bottomly (1978); Euglena—Chelm and Hallick (1976), Rawson and Boerma (1976b), Schwartzbach et al. (1976), Scott (1976), Gray and Hallick (1978), Jenni and Stutz (1978).

10 poly(A)$^+$ mRNA species and the two rRNAs have been mapped on the X. laevis mitochondrial genome; all hybridize with the H strand and in every case the sequence of the gene appears to be continuous with the sequence of mRNA (Rastl and Dawid, 1979).

The situation has been less clear in HeLa cells. Early work showed that the H strand is transcribed. Saturation hybridization with RNA suggested that virtually all of this strand is represented in the transcripts, which appeared heterogeneous in size (Attardi et al., 1970; Aloni and Attardi, 1971). Subsequent work identified the 8 discrete poly(A)$^+$ RNA species described in Table 21.4, which may be taken to identify the structural genes (Hirsch and Penman, 1973). A larger number of RNA species complementary to the H strand since has been reported (Amalric et al., 1978). Presumably these represent primary transcripts and any intermediate precursors produced by processing; as with the L cell, it will be interesting to see to what extent sequences present in the precursors but absent from the mature RNAs might account for the apparently nonutilized part of the genome. The relationship of the putative precursors to the mature HeLa mitochondrial RNAs has yet to be determined; however, the dispersion of tRNA genes raises the possibility that processing may involve not just the removal of additional sequences but perhaps the production of tRNAs and mRNAs from the same precursor. Definition of the processing pathway will require mapping with restriction fragments; until such data are available, it is premature to assess the suggestion that there may be a sole primary transcript of the H strand which represents a complete turn of the genome. The ability of L strand DNA to hybridize with DNA has been reported; and this is the basis of the suggestion that transcription may be symmetrical, using both DNA strands as templates (Murphy et al., 1975). The absence of L strand transcripts from the mature mRNA population implies that any such transcripts must lack coding functions; their concentration is not known but may be low. Such transcripts might be rapidly degraded.

It is likely that the mitochondrial mRNAs are the only messengers translated in the organelle, although there is no definitive evidence to exclude the possibility that some mRNA is obtained from the cytoplasm. In a yeast mutant that shuts down nuclear RNA synthesis at high temperature, mitochondrial protein synthesis continues as normal (Mahler and Dawidowicz, 1973). If it is true that no tRNAs are added to the complement coded by the organelle genome, and that the only genes expressed within the organelle are those residing in its genome, then there may be no import of any RNA into the organelle. Similarly it seems unlikely that there is any export of RNA for use in the cytoplasm. The organelle may therefore be self-sufficient for RNA components.

But it is at once obvious by comparing the numbers of proteins needed to constitute the protein synthetic apparatus (Table 21.3) with the complexities of organelle genomes (Tables 21.1 and 21.2) that the organelle DNA cannot code for these proteins, let alone all the enzyme functions associated with the organelle. On the question of the origin of aminoacyl-tRNA synthetases, RNA polymerase, DNA polymerase, and other synthetic activities, there is little evidence, although usually it is assumed that all of these are coded by the nucleus. What is known of organelle RNA polymerases suggests that they are distinct from nuclear enzymes (see Chapter 22). Mitochondrial DNA polymerase of mammalian and avian cells appears to be identical with the nuclear enzyme DNA polymerase γ (see Chapter 20).

The locations of the genes coding for ribosomal proteins have been approached through the existence of drug-resistant mutants. Antibiotics that inhibit bacterial ribosome function usually act upon proteins of either the large or small subunit; less often the object of mutation is concerned with the rRNA, but this involves methylation, not a change in sequence (see Volume 1 of Gene Expression). The nature of the mutations conferring drug resistance upon organelles has been investigated in Chlamydomonas chloroplasts and Saccharomyces mitochondria.

In C. reinhardii the chloroplast mutations to streptomycin and spectinomycin resistance alter the electrophoretic mobility of proteins of the small ribosome subunit (Boynton et al., 1973; Conde et al., 1975; Ohta, Sager, and Inouye, 1975). Mutation in any one of three genes, two nuclear and one chloroplastic, can confer resistance to erythromycin; each of these causes a change in the electrophoretic mobility of a different protein of the large subunit (Mets and Bogorad, 1972; Davidson et al., 1974; Bogorad, 1975). Subject, of course, to the caveat that the conclusive demonstration that a gene codes for protein is an amino acid charge upon mutation, this suggests that the chloroplast genome contains genes for both small and large subunit proteins; other genes, probably the majority, must reside in the nucleus. In each of these cases, the mutant protein resides in the same ribosomal subunit upon which the drug acts in bacteria.

In yeast mitochondria, mutants resistant to erythromycin, lincomycin, spiramycin, or chloramphenicol have ribosomes that are resistant to inhibition in vitro. Resistance is associated with the large subunit, as it is in bacteria (Grivell et al., 1973). But no changes in ribosomal proteins can be detected by the criterion of electrophoretic mobility. This is not definitive, of course, but fits with the genetic map location of the three resistant loci, which appear to lie close to or within the gene for the subunit rRNA (Faye, Kujiwa, and Fukuhara, 1974; for review see Borst and Grivell, 1978). The loci map in the order *rib2-rib1-ω-rib3* and at least *rib1* and *rib3* seem likely to reside within the rRNA gene by restriction mapping (Morimoto et al., 1978; Heyting and Menke, 1979). Confirmation must await the nucleotide sequences of wild-type and mutant alleles.

The existence of a mitochondrial drug resistance mutation residing in rRNA would contrast with the lack of such mutations in bacteria. One possible explanation is that the repetition of genes for rRNA in bacteria is sufficient to preclude the isolation of mutants, so that resistance is conferred only by changes in ribosomal proteins or in enzymes that methylate the rRNA. The existence of the latter class shows that the properties of bacterial ribosomal RNA may control drug resistance.

Whether all the genes for yeast mitochondrial ribosomal proteins reside in the nucleus is not proven; but it is clear from the present genetic map that there can be room for very few, if any, on the organelle genome. Most, probably all, mitochondrial ribosomal proteins must therefore be synthesized in the cytoplasm. The question of how they are distinguished for transport to the organelle remains open.

Organelle and Nuclear Genes and Products

Which mitochondrial functions are coded by the organelle genome? The most detailed information is available in yeast where it has been known from early studies that petites always lack cytochromes *a*, *a3*, and *b*, but retain cytochrome *c* (Sherman and Slonimski, 1963). [Cytochrome *c* is synthesized on cytoplasmic ribosomes and is coded in the nucleus (Sherman et al., 1966; Montgomery et al., 1978; for review see Sherman and Stewart, 1972).] The major protein complexes of the yeast mitochondrion are located on the inner membrane and each is constructed from subunits some coded in the organelle and some in the nucleus (for review see Schatz and Mason, 1974). The assembly of some subunits may depend upon the presence of others, so that a single mutation removing one subunit may prevent the others from associating with the membrane. This pleiotropy affects function rather than synthesis, since the other subunits often may be recovered in unassembled form, but it has made more difficult the problem of correlating mutations with gene products. Table 21.5 summarizes data on the three yeast complexes to which both organelle and cytoplasm contribute components.

The oligomycin-sensitive ATPase is a large complex of nine subunits, which may be dissociated into two components. The F1 ATPase is a soluble complex of four subunits, which is not itself sensitive to oligomycin. They may be distinguished from subunit 7, also known as the OSCP, which appears to link the F1 ATPase to the membrane factor rendering the ATPase activity sensitive to oligomycin. These subunits are synthesized in the cytoplasm and transported to the mitochondrion. Three nuclear genes involved in their synthesis have been identified; but none has been equated with an individual subunit and some or all of them may code for other functions concerned with assembly of the complex but not comprising part of it. The membrane factor consists of four subunits synthesized in the mitochondrion; these are located as a complex of the membrane.

Mitochondrial oligomycin-resistant mutants fall into three complementation groups. Their pleiotropic effects result in the absence of more than a single protein subunit in mutant cells of each type; this makes it necessary to rely on stringent criteria for equating genes with proteins. Demonstrations that different mutations in the gene cause different changes in electrophoretic mobility have been used; direct sequencing of the wild-type and mutant proteins provides the ultimate proof. The locus *oli-1* codes for subunit 9; the other *oli* loci remain to be assigned products. Two *pho* loci also have mutants in the ATPase; *pho-2* is the same as *oli-1*, while *pho-1* is uncharacterized.

The function of the *oli-1* locus in coding for the ATPase subunit 9 has been confirmed directly by determining the sequence of the DNA present in appropriate petite mutants. This includes an uninterrupted sequence of 230 bp corresponding to the amino acid sequence of the protein except at one position.* The coding sequence is rich in G-C base pairs and is flanked on either side by sequences almost exclusively of A-T base pairs (Hensgens et al., 1979; Macino and Tzagoloff, 1979).

The cytochrome *c* oxidase consists of 7 subunits, 3 coded by organelle genes and 4 by the nucleus. Several nuclear mutations abolish cytochrome *c* oxidase activity; some of these may be located in structural genes, but others must represent proteins that participate in its assembly in some other capacity. Two of the nuclear mutants in fact have all the cytoplasmic components in an unassembled state, but lack particular mitochondrial subunits. This suggests the possible existence of rather specific interactions in assembly, influenced by nuclear gene products that are not part of the complex itself. The three mitochondrial *oxi* loci presumably correspond to the three mitochondrial subunits, although one of these has not been formally assigned.

A similar situation is seen in the synthesis of the cytochrome bc_1 complex. Here the only protein to be coded by the mitochondrial genome is the cytochrome *b* apoprotein. In its absence, some of the nuclear-coded subunits are not found in the mitochondrion. Again more nuclear loci have been shown to influence complex formation, some of them affecting assembly of the apoprotein, then can be accounted for by structural genes.

There is a close correlation between the location of structural genes and the sites of protein synthesis. Although products have not yet been equated with every mitochondrial complementation group, it seems likely that all the organelle genes will prove to code for proteins synthesized within the organelle. Whether any other proteins are synthesized in the mitochondrion is more difficult to determine. Up to 20 proteins were shown to fit the

---

*The discrepancy between nucleotide and amino acid sequence takes the form of a codon CUA (which corresponds to leucine) at a position where threonine (coded by four codons ACX) is found in the protein. This may be explained by the presence of an unusual tRNA in the mitochondrion; a gene coding for a tRNA accepting threonine has been sequenced and has the predicted anticodon sequence 3'-GAU-5', which is expected to respond to CU_G^A (barring base modifications in tRNA) (Li and Tzagoloff, 1979). The full implications of this deviation from the genetic code are not yet clear, since it is not known how all other CT_G^A codons are read. However, most Leu residues seem to be coded by UUA (including all the others in the subunit 9 gene).

Table 21.5   Organization of genes

Protein complex	Components synthesized in organelle	Components synthesized in cytoplasm
Yeast mitochondrial oligomycin-sensitive ATPase (340 K)	membrane factor: subunit 5 = ~29 K subunit 6 = ~22 K subunit 8 = ~12 K subunit 9 = ~8 K (sub 9 = DCCD)	F1 ATPase is soluble complex: subunit 1 = ~58 K subunit 2 = ~54 K subunit 3 = ~38 K subunit 4 = ~31 K subunit 7 = ~18 K (sub 7 = OSCP)
Yeast mitochondrial cytochrome *c* oxidase	subunit 1 = ~40 K subunit 2 = ~27 K subunit 3 = ~23 K	subunit 4 = ~14 K subunit 5 = ~13 K subunit 6 = ~10 K subunit 7 = ~9½ K
Yeast mitochondrial cytochrome bc_1 complex	cytochrome *b* apoprotein = ~32 K	subunit 1 = ~44 K subunit 2 = ~40 K subunit 3 = ~32 K subunit 4 = ~17 K subunit 5 = ~14 K subunit 6 = ~11 K
Maize chloroplast RuBP carboxylase (8 LS + 8 SS)	large subunit, LS = ~55 K	small subunit, SS = ~15 K

The site of synthesis of components of each protein complex has been identified by resistance or sensitivity to inhibitors of organelle or cytoplasmic protein synthesis. Components are identified by molecular weight (K = 1000 daltons); the values indicated are only approximate. The genes that have been identified in organelle genomes all have been located on the genetic and restriction map of DNA. Genes are equated with particular proteins when mutation causes the absence and/or specific change in the individual product only. Other genes are identified by mutations abolishing activity, but have not yet been shown to reside in particular subunits. The numbers of nuclear genes involved in each complex are based upon placing mutants lacking activity into complementation groups; this is therefore a minimum estimate, although often higher than the number of cytoplasmic proteins. No yeast nuclear genes have been equated with individual cytoplasmic protein products;

criteria for mitochondrial products in the study of Douglas and Butow (1976); but many are present in small amounts and could be subsidiary products of the major proteins. Data on the components synthesized in the cytoplasm are less well advanced; structural genes have yet to be identified. To summarize, there is at present no evidence to support the possibility that messenger RNA may be transported in either direction across the

coding for organelle functions

Genes identified in organelle genome	Genes identified in nuclear genome
oli-1 (*pho*-2) codes subunit 9	2 genes: lack Fl ATPase
oli-2 product not known	1 gene: has unassembled Fl ATPase and may lack one of *mit* products
oli-3 product not known	
pho-1 product not known	*aur-1* mutant makes subunit 2 that fails to bind and is resistant to aurovertin
oxi-1 codes subunit 2	10 genes lack or have much reduced activity
oxi-2 product not proven	1 gene: lacks *mit* subunit 3 ⎤ but have all subunits
oxi-3 codes subunit 1	1 gene: lacks *mit* subunit 2 ⎦ 4–7 unassembled
cob (*box*) codes apoprotein	9 genes: some lack apoprotein b
1 gene codes LS	1 gene codes SS

some lack specific mitochondrial products as well as total activity; some appear to possess all known cytoplasmic products and therefore do not seem to represent structural genes for the complex.

References: Yeast ATPase—Tzagoloff and Meagher (1972), Tzagoloff et al. (1972, 1975c,d, 1976b), Perlman et al. (1977), Douglas et al. (1979); yeast cytochrome oxidase—Mason and Schatz (1973), Rubin and Tzagoloff (1973), Poyton and Groot (1975), Tzagoloff et al. (1975a,c), Slonimski and Tzagoloff (1976), Cabral et al. (1978), Cabral and Schatz (1978), Eccleshall et al. (1978); yeast cytochrome bc_1—Tzagoloff et al. (1975a, 1976a), Katan et al. (1976); maize (and other plants and Chlamydomonas) RuBP carboxylase—Chan and Wildman (1972), Coen et al. (1977), Gelvin et al. (1977).

mitochondrial membrane; probably each mRNA is translated in the compartment corresponding to the location of its gene.

The dispersion of genes coding for the components of organelle protein complexes is common to chloroplasts as well as to mitochondria. The enzyme ribulose biphosphate (RuBP) carboxylase is a major chloroplast component, which consists of two types of subunit. The large subunit is coded by the chloroplast DNA and synthesized in the organelle; the small subunit is coded by the nucleus and synthesized in the cytoplasm, where its mRNA is an appreciable component of total cytoplasmic poly(A)$^+$ mRNA. This poses an interesting problem in the coordination of gene expression, since hybridization kinetics show there to be only one (or very few) copies of the nuclear structural gene per haploid genome, whereas the large number of chloroplast genomes per cell means that the gene for the large subunit must be present in many more copies (Cashmore, 1979).

How are proteins coded by nuclear genes distinguished for transport from the cytoplasm into the organelle? The small RuBP carboxylase subunit synthesized on cytoplasmic ribosomes of either higher plants or Chlamydomonas first appears as a precursor of about 20,000 daltons (Dobberstein, Blobel, and Chau, 1977; Highfield and Ellis, 1978). When the plant precursor synthesized in vitro is added to intact chloroplasts, it enters the organelle where it is found in the mature form of about 15,000 daltons (Chua and Schmidt, 1978). A processing enzyme is probably located on the chloroplast envelope. A similar series of events is seen with the yeast mitochondrion, where the three largest subunits of the F1 ATPase synthetized on in vitro translation appear as precursors some 4000–6000 daltons larger than the mature polypeptides. On exposure to mitochondria, these are incorporated into the organelle with an accompanying size reduction (Maccecchini et al., 1978). A significant feature of the transport system of both types of organelle is that the substrate for processing is the completed precursor polypeptide (for review see Chua and Schmidt, 1979). This may contrast with the secretion of proteins through cellular membranes, where transport appears concomitant with translation, and it is the nascent polypeptide on the ribosome that is the substrate for processing (see Chapter 23).

To what extent is the pattern of gene dispersion seen for yeast mitochondrial protein complexes conserved in other species? In Neurospora and in HeLa mitochondria the complexes have generally similar constitutions to those shown for yeast in Table 21.5. The ATPase is the best investigated and appears invariably to consist of a membrane activity and soluble enzyme activity (for review see Sebald, 1977). Not enough data are available yet in other systems to see how extensive is the overall correspondence with yeast.

One of the components of the ATPase membrane factor, however, can be isolated specifically by virtue of its proteolipid nature. This is subunit 9 in the yeast mitochondrial complex, coded by the mitochondrial locus *oli-1* (Tzagoloff, Akai and Foury, 1976b). A similar small proteolipid component of the ATPase of Aspergillus nidulans has an altered amino acid composition in an extranuclear oligomycin-resistant mutant, *or-11* (Marahiel et al., 1977). In N. crassa, however, it has been impossible to detect synthesis of the small proteolipid in the mitochondria, although it is present in the complex; a proteolipid of the right size is synthesized in the cytoplasm, raising the possibility that this component may be coded by a nuclear gene (Jackl and Sebald, 1975; Kuntzel et al., 1975). Such a difference in location would presumably have taken place by transfer of the gene between an an-

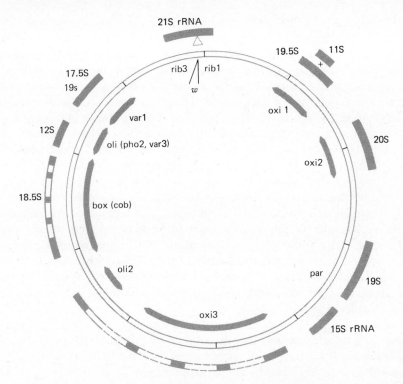

Figure 21.5. Map of yeast mitochondrial DNA. The circular map is divided into 100 units, each of about 0.75 kb. Inside the circle are shown the positions of genetic loci as defined by petite deletion mapping of complementation groups. Mutants *oxi* affect cytochrome oxidase, *oli* are resistant to oligomycin, *box* lie in cytochrome b, *var* are polymorphic for size of a protein. Synonyms for loci are noted in parenthesis. Outside the circle are shown the approximate locations of RNAs; the 21S and 15S rRNAs have been mapped in the most detail. Multiple products of a given locus are indicated by additional bars; the sizes indicate the S value of the RNA. The 19S RNA counterclockwise from the small rRNA gene is of unknown function; its location is not certain and may extend into the rRNA gene. The *box* and *oxi3* loci are very much larger than would correspond to the protein products and represent interrupted genes; sequences that are represented in mRNA are shaded and the white regions between represent intervening sequences that are transcribed but removed during processing. For *oxi3* this is only very approximate indeed. The organization of *box* is described in Figure 26.9. Further details of the functions coded at each locus are given in Table 21.5.

cestral mitochondrion and the nucleus, although the mechanism can be a matter only for speculation. Comparison of the protein sequences may be interesting.

The locations of the yeast mitochondrial complementation groups are shown in Figure 21.5. In most cases the extent of the locus is within the range of what would be expected from the size of the probable or known protein product. There are two interesting exceptions. Mutations affecting the cytochrome *b* apoprotein map into several clusters, known as *cob* or *box*, that belong to a single complementation group (Foury and Tzagoloff,

1978; Slonimski et al., 1978b). These represent regions that code for the protein, separated by intervening sequences (in some of which mutations also may occur to prevent protein synthesis). This genetic analysis is supported by the restriction map, which shows that the region occupies about 7000 base pairs altogether, much larger than the 875 bp that would be needed to code for the cytochrome *b* apoprotein. This organization is discussed in detail in Chapter 26. The discrepancy between genome length and protein size is comparable in the *oxi3* complementation group, which extends over some 6400 base pairs in S. cerevisiae, compared with the 1100 bp needed to code subunit 1 of cytochrome oxidase. In the related yeast S. carlsbergensis, this region contains inserts relative to S. cerevisiae, bringing its total length to 11,400 bp (for review see Borst and Grivell, 1978). Again the structure of the gene probably is mosaic, with coding regions separated by intervening sequences. Because of the restricted size of its genome, and the ability to analyze the entire apparatus, the yeast mitochondrion provides a good system in which to investigate the nature of intervening sequences and their possible functions. In fact, at present it constitutes the only system in which it has been possible to analyze some of the genetic relationships predicted for interrupted genes in Chapter 17.

A map of the most abundant transcripts found in the mitochondrion of S. cerevisiae is summarized in Figure 21.5. In this analysis, performed by testing the ability of restriction fragments of the DNA to hybridize with messengers that were separated by gel electrophoresis, Van Ommen et al. (1979) identified transcripts representing all the known genes. In several cases multiple transcripts, overlapping in sequence, correspond to a single region. Probably these represent mature and precursor RNAs. Thus RNAs of 17.5S and 19S represent *var*, while RNA species of 11S and 19.5S represent *oxi1*. The RNA species representing the *box* and *oxi3* loci show a complex organization. For *box*, at least four intervening sequences appear to be removed from the RNA; in addition to the 18.5S RNA whose organization is summarized in the Figure, there are other species that may represent processing intermediates. For *oxi3* the situation is complicated. It is not yet clear whether a single transcript extends from *oli2* through *oxi3*, but there is some evidence for a relationship like this. The regions of *oxi3* that are represented in mRNA are dispersed, but their exact arrangement is not yet known. The insertions that are found in one strain of yeast relative to another appear to be transcribed; presumably they are removed during processing to generate a functional mRNA.

Intervening sequences have been found in the large rRNA genes of several mitochondria and chloroplasts. The summary of Table 26.5 shows that an interruption is present in the mitochondrial genome of N. crassa, may be present in the S. cerevisiae mitochondrion (the polymorphism for presence and absence corresponds to the ω^+ and ω^- genotypes), is present in both copies of the large rRNA gene in the Chlamydomonas chloroplast, but is absent from the maize chloroplast. It is not yet known to what extent the presence of intervening sequences may explain the wide variations in mitochondrial genome size that do not appear to be related to genetic content. The presence of intervening sequences in the organelles, however, could mean that the origins of their genomes may be more complex than would be apparent from the view that they are more closely related to procaryotes than to the surrounding cytoplasm and nucleus. This difficulty could be overcome, however, by supposing that the endosymbiotic event occurred before procaryotes lost their ability to remove intervening sequences. This still leaves unanswered the question of the relationship between small and large mitochondrial genomes.

Is the expression of genetic information in organelles invariant or subject to regulation by the environment? Developmental regulation of chloroplast gene expression is well established in plants. The etioplasts of seedlings grown in the dark are much less active in RNA synthesis than the chloroplasts into which they are converted by exposure to light. The increase in transcription includes not only a greater rate but also the expression of new regions of the genome. The new RNA species can direct the translation in vitro of the LS subunit of RuBP carboxylase and a 34 K protein of presently unidentified function (Bedbrook et al., 1978).

The expression of the LS gene indeed depends upon the state of differentiation of the cell in which the chloroplast finds itself. Bundle sheath cells and mesophyll cells differ in their CO_2-fixation functions and the enzyme is found only within the chloroplast of the former. The mRNA of the LS gene is synthesized only in the bundle sheath chloroplasts (Link et al., 1978). Thus the transcription of the chloroplast genome is controlled by the state of differentiation of the surrounding cytoplasm. The chloroplast genome appears identical in chloroplasts from both types of cell (Walbot, 1977).

Replication of Organelle DNA

The mechanism by which mitochondria and chloroplasts reproduce has been a matter of controversy for some years, between the opposing viewpoints that assembly of a new organelle must take place de novo and that it must occur by fission of a parental organelle. An analogous discussion has taken place about the origin of the centriole (see Chapter 4). The discovery that genetic information is present in mitochondria and chloroplasts naturally strengthened the view that the assembly of daughter organelles must depend upon a parental structure. The replication of mitochondrial or chloroplast DNA clearly must precede or accompany the formation of new organelles. This raises intriguing questions on how daughter genomes are distributed to daughter organelles, how the initiation of organelle DNA replication is linked to production of new organelle structures, and how the number of organelles is related to cell division.

The number of mitochondria is presumably a feature of each cell type and is perpetuated through division. Little is known about any developmental processes that may influence the number of mitochondria. The replication of mitochondrial genomes has been studied largely in tissue culture cells, in which it is assumed that the number of mitochondria doubles during the cell cycle, to be reduced by segregation into the daughter cells. The essential problem in following mitochondrial DNA replication is to distinguish the small amount of organelle DNA from the nuclear DNA. Early experiments in which mitochondrial DNA was obtained either by purification of mitochondria or by centrifugation in the presence of ethidium bromide led to confusing conclusions (Pica-Mattoccia and Attardi, 1972; Ley and Murphy, 1973). In some cases, mitochondrial DNA replication appeared to occur at the same time as S phase; in others it appeared continuous through the cell cycle. Possible artefacts include contamination of the mitochondrial DNA with nuclear DNA and damage induced by the cell synchronization techniques.

These problems were overcome by Bogenhagen and Clayton (1977) by a technique that investigates the ability of mitochondrial DNA to replicate twice within a period appreciably shorter than the cell cycle. Mitochondrial DNA is labeled with ^{3}H-thymidine and then chased with unlabeled precursor before labeling with BUdR. Any shift of the radioactive label into higher density DNA identifies molecules that have replicated in both the first and second labeling periods. The extent of the shift is measured as a function of the length of the chase period. Two extreme models are that mitochondrial DNA replication is continuous and random (in which case the extent of shift should be independent of chase length in excess of the time required to complete one replication); or that it occurs during only one phase of the cell cycle (in which case the extent of shift should show a peak corresponding to this time). At all chase times, the same 9% of labeled mitochondrial DNA was shifted to higher density.

This suggests that DNA molecules are selected at random for replication from a pool containing the mitochondrial genomes of the cell; and replication must be independent of the stage of the cell cycle. This is analogous to the mode of replication of some plasmids (see Volume 3 of Gene Expression). But an important difference is that the plasmid DNAs all replicate in what is presumably a single cell compartment; if mitochondrial DNAs are segregated into individual mitochondria, the analogy would raise the question if whatever is responsible for controlling initiation of mitochondrial DNA replication is a factor freely diffusible between mitochondria.

In yeast the replication of mitochondrial DNA may be influenced by both environmental and genetic factors. The amount of mitochondrial DNA may be estimated readily by the content of the low buoyant density peak. There has been a report that in S. lactis the proportion fluctuates sharply, corresponding to replication during a discrete period of the cell cycle which does not coincide with S phase (Smith et al., 1968). On the other hand, with S. cerevisiae, precisely the opposite behavior has been found, with mitochondrial DNA replication free to occur at any stage of the cell cycle (Williamson, 1970). The proportion of mitochondrial DNA may be changed by treatments that inhibit nuclear DNA synthesis but leave mitochondria able to support replication for at least some period. These include treatment with mating factor and inhibition of cytoplasmic protein synthesis (Cryer et al., 1973). Temperature-sensitive mutants that are defective in what appears to be the initiation of nuclear DNA synthesis may be able to continue to replicate mitochondrial DNA; but two mutants apparently defective in nuclear DNA chain elongation also are temperature sensitive for mitochondrial replication (Cottrell et al., 1973; Newlon and Fangman, 1975). This shows that the level of mitochondrial DNA is not adjusted in direct response to the level of nuclear DNA, but responds to some other signal, which is independent of at least some of the functions needed to initiate nuclear DNA synthesis. At least some of the enzymes needed for nuclear DNA synthesis may be involved in mitochondrial DNA synthesis also.

The organelle replication apparatus is not well defined, apart from the characterization of DNA polymerase-γ as the likely replicase of animal and avian mitochondria (see Chapter 20). The inability of this enzyme, like other procaryotic and eucaryotic DNA polymerases, to initiate DNA synthesis implies that there must be separate enzyme activities responsible for synthesizing a primer. The circularity of the organelle genome implies further that there must also be an activity such as the nicking-closing enzyme to provide a swivel mechanism for unwinding.

It has generally been assumed that organelle DNA synthesis occurs by conventional semiconservative replication; although experiments to prove the point directly have for the most part been less than fully convincing because of labeling problems (Reich and Luck, 1966; Chiang and Sueoka, 1967; Parsons and Rustad, 1968). Electron microscopy of animal cell mitochondrial DNA identifies both mature circles and circles engaged in replication. The mature circles are predominantly of one length but include a small proportion (about 5%) of dimeric length (Nass, 1969b). Denaturation mapping shows that dimers found in human mitochondrial DNA and in pea chloroplast DNA represent two monomers in tandem repeat (Clayton et al., 1970; Kolodner and Tewari, 1975b). These could arise by either replication or recombination; the latter may seem more likely because of the absence of any replicative intermediates likely to generate dimers. The structure of the circles engaged in replication shows that mitochondrial and chloroplast DNA duplication is initiated by a novel mechanism.

The *displacement loop* or *D loop* was first identified in L cell mitochondrial DNA as a short region in the molecule where the two strands of DNA have been separated; one of them is bound to a short complement which has displaced the original partner strand in the form of a single stranded loop (Kasamatsu, Robberson, and Vinograd, 1971). The length of the D loop is about 3–3.5% of the circle, roughly 500 bases. Denaturation of molecules containing D loops (which can be separated from simple circles by their reduced sedimentation velocity) generates a single stranded DNA molecule of some 7S, about 450 bases. When annealed with the separated strands of mitochondrial DNA, it hybridizes solely with the light strand. In dimeric circles there are two D loops diametrically opposed, which suggests that their formation occurs at a unique site.

Expanded D loops are present in mitochondrial DNA in proportions lower than the frequency of the smallest D loops. By organizing such molecules into a series, Robberson, Kasamatsu, and Vinograd (1972) constructed the model for mitochondrial DNA replication summarized in Figure 21.6. The formation of the D loop may be a rate-limiting step in replication, since it is found in as many as 25% of the mitochondrial genomes. The length of the single-stranded loop may be expanded from its initial 3% to as much as 60% of the circle. Then with any further increase in size, duplex regions of somewhat varying length appear on the displaced strand. The obvious explanation is that an origin for this synthesis is exposed when a region about 60% round the circle from the position of the D loop becomes single stranded. There may also be subsidiary origins in the region between 60 and 100%, since sometimes duplex regions appear to be located here. An alternative mechanism, that some feedback prevents the second initiation until a certain amount of synthesis has occurred from the D loop, requires a more complicated control system. Synthesis on each template strand appears to occur unidirectionally from a single origin. The discrepancy in timing of the two initiations means that one strand will complete synthesis of its complement while the other is partly duplex and partly single stranded. This explains the occurrence of gapped circles, which are largely duplex but may have single-stranded regions occupying up to 10% of the length or more.

A single D loop similar in size to that of mammalian mitochondria is found in avian mitochondrial DNA (Ter Schegget et al., 1971). In X. laevis the D loop is somewhat larger, about 1500 bases (Hallberg, 1974). In chloroplast DNA of higher plants there are two D loops, each of the order of 800 bases in length, and located about 7000 base pairs apart. Unidirectional replication expands the D loops toward each other; after they have passed, the replicating DNA shows the usual pattern seen in bidirectional replication of circles (Kolodner and Tewari, 1975c). Tetrahymena mitochondrial DNA has been visualized only in linear form (Suyama, 1966). The molecules may contain up to 6 D loops; these are not present at fixed positions on the linear map, but their locations become consistent if the molecules are regarded as random permutations of a circle (Arnberg et al., 1972). The only exception at present reported to this form of replication is Drosophila mitochondrial DNA, where no D loops can be found (Rubenstein, Brutlag, and Clayton, 1977).

When L cell or HeLa cell mitochondrial DNA is cleaved with restriction enzymes, one end of the D loop is found at a constant location while the other moves unidirectionally away when D loops are expanded (Robberson, Clayton and Morrow, 1974). The first end identifies the presumptive origin;

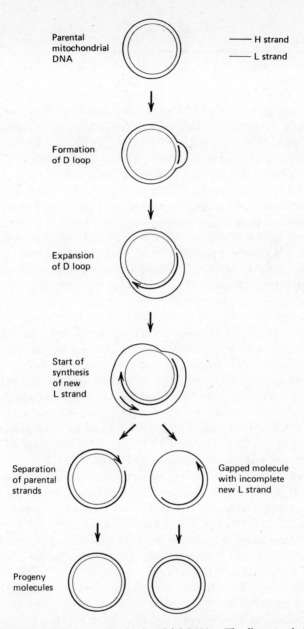

Parental mitochondrial DNA

————— H strand
————— L strand

Formation of D loop

Expansion of D loop

Start of synthesis of new L strand

Separation of parental strands

Gapped molecule with incomplete new L strand

Progeny molecules

Figure 21.6. Model for replication of mitochondrial DNA. The first step is formation of a short primer H strand, which displaces the parental H strand in this region to form a D loop. The displaced loop of single strand parental H DNA is then expanded by unidirectional synthesis of the new H strand. When this is about 60% complete, synthesis of a new L strand commences in the opposite direction. When displacement is complete, the parental strands are separated; this may leave a short gap in one product if synthesis of the new H strand need not be quite complete for separation; it must leave a long gap in the other product, since synthesis of the new L strand starts so late. The progeny molecules are closed duplex circles.

the second is the point at which synthesis of the D loop ceases, until expansion occurs, presumably by a different enzyme. Subsequent detailed studies have shown that there may be microheterogeneity in the locations of the ends of the D loop; but no intermediates in its synthesis have been identified.

The form of the microheterogeneity depends upon the origin of the mitochondrial DNA. When the isolated 7S DNA is fractionated on acrylamide gels, it forms a family of bands. Mammalian mitochondria give characteristic series such as those of L cell (500, 525, 540, 580 bases) or KB cells (555, 585, 615); the larger D loops of Xenopus oocyte mitochondrial DNA form bands at 1350 and 1510 bases. Digestion with restriction enzymes to separate the 5' and 3' ends of the molecules shows that in L cells the variation occurs entirely at the 3' end, in human placental cells it is entirely at the 5' end, and in KB cells it may be at both ends (Gillum and Clayton, 1978; Brown, Shine, and Goodman, 1978). There are two possible explanations. Initiation and termination of D loop DNA may occur at fixed sequences; but the populations of mitochondrial DNA molecules may be polymorphic due to variations in length between these two points. These could take the form of small insertions/deletions altering the distance of the ends from the sites recognized by restriction enzymes. Or there may be variation in the exact locations at which initiation and/or termination occur. Variation in the sequences present at the immediate 5' end suggests that initiation may not always start with the same base; but this does not reveal whether the variation is sufficient to account for the length differences. Sequence studies of the D loop DNA will be able to resolve the cause of the heterogeneity. The 5' end of one of the 7S DNA species of HeLa cells has been sequenced and thus located precisely on the genome (Crews et al., 1979).

The D loop functions as primer for synthesis of one strand of mitochondrial DNA. It is not known whether the other strand also is initiated by formation of a comparably discrete primer or instead by a more continuous mechanism. The 7S DNA of L cell mitochondrial D loops appears to turn over rapidly; the large proportion of molecules with D loops seems to reflect a situation in which D loops are rapidly synthesized, then to be removed and resynthesized (Bogenhagen and Clayton, 1978a). Whether this occurs by degradation or displacement is not known, although no free 7S DNA can be found. This means that most of the D loops are not used to prime initiation, but may be removed first. Whether they have any function other than that of primer remains an open question.

The parental strands of the expanded D loop circles remain closed, which implies that replication must proceed via a continuous nicking and closing (Robberson and Clayton, 1972; Berk and Clayton, 1974). Two forms of L cell mitochondrial DNA varying in the extent of supercoiling can be distinguished at stages following replication. The earliest closed circular product has zero or few supercoils; a label can be chased from it into a form with an average of about 100 turns per molecule (Bogenhagen and Clayton, 1978b). Then the formation of a D loop in the next cycle of replication relaxes the superhelical turns. The same two types of mitochondrial DNA molecule, with zero and 100 turns, can be found in D. melanogaster oocytes; but here both are stable and there is no apparent relationship with replication. In tissue culture cells, however, mitochondria contain almost exclusively the low density form (Rubenstein, Brutlag, and Clayton, 1977). The function of these two forms, and the reason for the differences in their occurrence in the two cell states are not known, but the predominance (75%) of high density forms in eggs may be related to the probable lack of mitochondrial DNA replication.

A novel feature of mitochondrial DNA is the presence of ribonucleotides in mature genomes. Their presence was revealed when it was discovered that high pH or RNAase H introduces single-strand scissions into mammalian mitochondrial DNA (Grossman, Watson, and Vinograd, 1973; Miyaki, Koida, and Ono, 1973; Wong-Staal et al., 1973). They appear to be present in both low-density and high-density forms, which suggests that they do not represent an intermediate prior to removal. There have been suggestions that they occur at specific locations, but analysis of the RNAase-susceptible locations relative to restriction enzyme cleavage sites has yet to be reported. Alkali-labile sites occur also in chloroplast DNA (Kolodner et al., 1975). One possible explanation of their presence is that RNA is used to initiate DNA synthesis; an alternative is that mitochondrial DNA polymerase fails to exclude ribonucleotides rigorously. The alkaline lability is lost upon cloning mitochondrial DNA in a plasmid in E. coli (Chang et al., 1975; Brown et al., 1976).

Apart from the genetic studies showing that chloroplast and mitochondrial DNAs of different parents are rapidly segregated following mating of Chlamydomonas or yeast, nothing is known about mechanisms for the distribution of organelle genomes to organelle progeny. The Chlamydomonas genetic map possesses a centromere-like feature; no such feature is evident in the map of the yeast mitochondrial genome, but when the DNA is released from mitochondria by treatment with Triton, it appears to be attached to a protein-membrane structure at or near the origin for replication (Albring, Griffith, and Attardi, 1977). Further studies of the site of attachment on the DNA and the components attached to it should reveal whether this is fortuitous or plays a specific role in replication or distribution.

PART 4

Expression of
Genetic Information

CHAPTER 22

Introduction:
The Transcription Apparatus

The invariance of genetic information in each somatic cell of an organism implies that each cellular phenotype results from the expression of a particular subset of the genes. This conclusion is supported by observations that during development there is no loss of unexpressed genes or amplification of active genes. The most compelling evidence for the retention of a full set of genetic information in each cell is inferential, provided by the demonstrations of totipotentiality through amphibian nuclear transfer, which have been discussed in Chapter 8. Data on the lack of gene amplification in cells expressing a given function are less general; but in those cases in which the number of copies of a gene has been determined (by hybridization with the corresponding mRNA) there is no increase associated with gene expression. More general experiments on the constancy of DNA sequences fail to identify any changes in cells of different phenotypes; but these are rather limited in resolution and simply exclude any large addition or loss of sequences (see Davidson, 1976). One subsidiary point needs emphasis. Although these data generally are taken to imply the absence of changes in genetic content during development, they do not prove that the arrangement of sequences in the genome is invariant; the possibility remains that transpositions of genetic material might occur, although the transplantation experiments suggest that they would have to be reversible.

The concept of differential gene expression is supported by examination of the sequences present in the mRNA populations of different cell types. Each cell phenotype is characterized by the presence of a particular set of messengers. These experiments rest largely on the use of nucleic acid hybridization to characterize the sequences present in cellular RNA populations. The most striking demonstrations of specific gene expression are provided by comparing cells that synthesize one predominant protein with those inactive in this function; the corresponding messenger invariably is found only in the active cells (see Chapter 23). More general comparisons between mRNA populations show that two different cell types may possess messenger sequences that overlap in large part but show some differences, presumably representing the specialized functions of each cell (see Chapter 24).

Gene expression proceeds through several stages, at any of which regulation might occur. Initially a precursor RNA is *transcribed;* then this molecule is *processed,* by the removal of some sequences, to form the mature messenger, which is *transported* to the cytoplasm; finally the mRNA is *translated* into protein. It has been clear for some time that there is transcriptional control in the sense that only a part of the genome is transcribed. What is becoming clear only as the result of recent work, however, is the extent to which changes in the pattern of transcription are involved in producing changes in the phenotype.

Early work showed that nuclear RNA is much larger in size than messenger RNA and that much of the mass of the RNA transcribed in the nucleus is degraded there, never to be transported to the cytoplasm; but only more recently has it been possible to determine the extent to which this degradation is specific for particular sequences (see Chapter 25). Comparing the hybridization of nuclear RNA and cytoplasmic RNA with DNA shows that there is a large reduction in sequence complexity proceeding across the nuclear membrane; it is important now to show whether this reduction simply represents loss of nonmessage sequences from the long precursors, or whether some potential coding sequences are degraded.

The processing of RNA occurs exclusively in the nucleus, and involves three known types of activity. The discovery that coding sequences in the genome may be interrupted by sequences that are absent from the mature messenger RNA implies that these intervening sequences must be removed from the primary transcript (see Chapter 26). This *RNA splicing* involves cutting and resealing the ribonucleotide chain at both ends of each intervening sequence. This in itself accomplishes a reduction in the size of the transcript that may be considerable. Whether further reductions in the size of spliced molecules result from removal of sequences from the ends is not clear. The proportion of structural genes that contains intervening sequences, and thus the proportion of primary transcripts that must suffer size reduction by splicing, is not known.

The other modifications concern the ends of the RNA molecule: the 5′ end of all messengers is marked by the presence of a methylated *cap;* the 3′ end may carry a sequence of poly(A). Both these modifications are accomplished within the nucleus and may be detected on nuclear RNA precursors to mRNA. Originally such modifications, in particular polyadenylation, were thought to be involved in the removal from one end of the precursor of a continuous length of RNA to serve as messenger. What proportion of the primary transcripts represents transcription units that are longer than the mature messengers simply because additional sequences are present at one or the other (or both) ends is not known. To what extent the terminal modifications represent signals indicating that the molecule should be transported to the cytoplasm has been an issue of some concern.

The discrepancy in size between the primary transcripts and the mature messengers implies that there must be sequences whose function is to be transcribed but not translated. In considering the question of what proportion of the genome is expressed, it is therefore necessary to view the unit of expression as a longer segment of DNA than would be expected from the messenger RNA or the protein product. Whether the discrepancy is large enough to contribute to a solution of the C value paradox is hard to say; this would require that the size of the transcription unit be related to the size of the genome. Data on the sizes and complexities of nuclear RNA populations in different species are not yet complete enough to reveal how the number and/or size of units changes with genome size; but changes in the number of genes represented in messenger RNA seem to be much less than changes in genome size.

The mechanisms involved in RNA processing now are being defined in increasing detail, but the critical question remains the nature of their roles in selecting specific sequences for expression in the cytoplasm. In other words, to what extent are changes at the stages of transcription, processing, or translation responsible for accomplishing the differences in protein synthesis associated with changes in cell phenotype? To anticipate the results discussed in this last section of the book, it is clear that there is indeed transcriptional control. This may be seen both in the specific production of certain primary transcripts only in cells in which the mature messenger is to be translated; and in the changes in sensitivity to DNAase suffered by genes that are expressed (see Chapter 14). But transcriptional control is not sufficient. Processing does indeed appear to involve selection of sequences for maturation and nucleocytoplasmic transport; in some cases, the same primary transcripts are found in more than one cell type, but give rise to mature messengers only in some. Control at the level of translation is the least-well-substantiated mechanism for adult somatic cells; but is utilized in early embryogenesis, where the egg possesses a store of nontranslated (maternal) messengers, whose use as templates for protein synthesis occurs at varying times after fertilization. At present it is difficult to generalize about the use of these controls in situations other than those in which they have been characterized. No overall picture of the extent to which an organism relies on each level of control during genetic development can yet be provided.

The primary function in the apparatus responsible for production of messenger RNA is the enzyme RNA polymerase that synthesizes the primary transcript. Eucaryotic cells contain three classes of nuclear RNA polymerase, responsible for synthesizing different classes of RNA. Little is known about the mechanisms involved in the initiation of RNA synthesis, that is, about the selection of DNA sequences for transcription. Some of the enzymes involved in splicing, capping and polyadenylation have been identified; but again there is little information on what signals govern their activities.

In this chapter we shall be concerned with the constitution of the transcription apparatus. The structures of the three RNA polymerases now are being resolved; but as yet there are few instances in which their actions can be defined in vitro. Present research has not reached the stage of defining the active forms of the enzyme molecules. More is known about the products of transcription. Following the definition of mRNA and nuclear RNA structure and complexity in the next three chapters, we shall be able in Chapters 25 and 26 to consider the mechanisms for processing RNA and the role that they may play in the selection of sequences for expression. Ribosomal RNA also is produced by processing of mature molecules from larger precursors; the less complex mechanism of this pathway is considered in Chapter 27. Finally, we shall consider model systems for investigating the control of gene expression, including the largely unsuccessful attempts to reproduce differential gene expression in vitro.

Functions of RNA Polymerases

Nuclear transcription occurs in two locations: the nucleolus and nucleoplasm. The genes coding for ribosomal RNA are associated with the nucleolus, which is devoted to the synthesis, maturation and transport of rRNA. The nucleolar enzyme RNA polymerase I is responsible for the transcription of the major ribosomal RNAs. Usually the activity of this enzyme accounts for more than half of the total nuclear transcription. A second major activity is provided by RNA polymerase II, much of which is associated with chromatin; this can be characterized by its presence in nucleoplasmic fractions after removal of the nucleolus. This enzyme is responsible for the synthesis of the heterogeneous nuclear RNA, which provides the precursors for mRNA. Generally its activity represents somewhat less than half of the total transcription. A third enzyme, RNA polymerase III, accounts for the smaller proportion of RNA synthesis, roughly some 10% of total, represented by the production of certain small RNAs, including 5S RNA, tRNA, and some nuclear species. Although the most detailed results have been obtained with animal cells, a similar division of labors may be common to all eucaryotes, since comparable enzyme activities have been found in plant, insect, and fungal cells.

The existence of at least two types of enzyme system was first indicated by experiments that showed that the nature of the RNA product depends upon the conditions under which nuclei are incubated. Widnell and Tata (1966) found that in the presence of ammonium sulfate, transcription is most active when Mn^{2+} is provided as the necessary divalent cation. But in the absence of the salt, the reaction works best with Mg^{2+}. Under the first set of conditions the reaction is less sensitive to inhibition by actinomycin (which preferentially inhibits rRNA synthesis) than under the second protocol, which sees the production of RNA similar in base composition to rRNA. Maul and Hamilton (1967) correlated these ionic conditions with the sites of transcription by showing that in the first case ^{3}H-uridine is incorporated preferentially over the chromatin; but in the second the grains lie at the nucleolus.

One explanation would be that the same enzyme is involved, but works optimally under different conditions in different locations. However, the idea that there are several enzymes, two stimulated at higher $(NH_4)_2SO_4$ and by Mn^{2+}, and the other stimulated by lower $(NH_4)_2SO_4$ and by Mg^{2+}, was supported by the demonstration of Roeder and Rutter (1969) that different enzyme activities can be fractionated by DEAE-Sephadex chromatography. This indeed has become the standard method for separating different RNA polymerase activities, which are named I, II, and III according to their order of elution by increasing ionic strength of ammonium sulfate. This is effective in almost all cases, although other methods of fractionation have been necessary on occasion (for review see Roeder, 1976).

The activities of the isolated enzyme preparations correlate well with the behavior of isolated nuclei under different conditions. Table 22.1 summarizes the properties of the enzymes. RNA polymerase I functions best at low ionic strength and either Mg^{2+} or Mn^{2+} is equally effective as the divalent

Table 22.1   General classification of nuclear RNA polymerases

Property	RNA polymerase I	RNA polymerase II	RNA polymerase III
Location	nucleolar	nucleoplasmic	nucleoplasmic
Proportion of cellular activity	50–70%	20–40%	10%
Elution on DEAE-Sephadex at $(NH_4)_2SO_4$	~0.1 M	~0.2 M	0.2–0.3 M
Optimum activity in $(NH_4)_2SO_4$	~0.05 M	~0.09 M	0.05–0.17 M
Mn^{++}/Mg^{++} ratio of optimum activities	1–1½	2–5	2
α amanitin concn. for 50% of inhibition			
mammals	>400 $\mu g/ml$	0.025 $\mu g/ml$	20 $\mu g/ml$
insects		0.03 $\mu g/ml$	>1000 $\mu g/ml$
yeast	600 $\mu g/ml$	1.0 $\mu g/ml$	>2000 $\mu g/ml$

These are typical values determined in calf thymus, rat liver, and MOPC 315 plasmacytoma cells, except where noted for insects (B. mori and D. melanogaster) and yeast (S. cerevisiae). The elution properties and ionic optima appear to be similar for other species, including amphibia and plants.

References: Roeder and Rutter (1969), Roeder et al. (1970), Kedinger et al. (1972), Roeder (1974), Schwartz et al. (1974), Weil and Blatti (1975), Schultz and Hall (1976), Hager, Holland and Rutter (1977); for further references see Chambon (1975) and Roeder (1976).

cation. RNA polymerase II functions best at higher ionic strength and shows a preference for Mn^{2+}, although the extent of this varies with the system. RNA polymerase III shows a broad biphasic peak of ionic dependence, often being active at somewhat higher levels than the other enzymes, and has a slight preference for Mn^{2+}.

Fractionation of nuclei shows that RNA polymerase I is associated with the nucleolus, while RNA polymerases II and III are present in the nucleoplasm (sometimes RNA polymerase III is found in the cytoplasmic fractions, presumably by leakage). First suggested by assays of the polymerase activities of different fractions under various ionic conditions, of necessity a somewhat unsatisfactory protocol, this conclusion since has been confirmed by studies of preferential drug sensitivity with isolated enzymes (Roeder and Rutter, 1970; Jacob et al., 1970; Lindell et al., 1970).

The three RNA polymerases are distinguished by their sensitivities to inhibition by α-amanitin, a bicyclic octapeptide isolated from the poisonous mushroom, Amanita phalloides. In animal, plant, and insect cells, RNA polymerase II is highly sensitive to the drug; in yeast cells it remains the most susceptible enzyme, but at somewhat higher drug levels. In animal cells, RNA polymerase I is essentially refractory to the drug. The mammalian enzyme RNA polymerase III is sensitive to an intermediate level; but in insects and yeast, RNA polymerase III is the most resistant of the enzymes. For this reason, an alternative nomenclature for the RNA polymerases based on sensitivity to α-amanitin, in which enzymes A, B and C correspond to animal cell I, II, and III, now generally is less preferred than the original description, which seems valid for at least a wider range of eucaryotes.

The crude enzyme activities all appear to be proteins of rather large molecular weight, generally about 500,000 daltons (for review see Chambon, 1975; Roeder, 1976). Electrophoresis through SDS acrylamide gels has been used to identify the subunit structures of the enzymes. Table 22.2 summarizes data that show that in each case the enzyme consists of two large subunits, one generally about 200,000 daltons and one about 140,000 daltons, and a number of smaller subunits ranging from 90,000–10,000 daltons. The constitutions of enzymes analyzed separately cannot be compared directly because there may be systematic differences in estimating molecular weights. But in those cases in which enzymes have been analyzed on parallel gels, usually there are some subunits apparently identical in molecular weight between the different enzymes of one cell type or the homologous enzymes of different cell types. More discriminating assays, involving two-dimensional electrophoretic separation, and ultimately tryptic peptide mapping, will be required to determine whether some subunits are common to all classes of enzyme and to what extent there has been evolutionary conservation of structure. In the case of yeast, tryptic maps suggest that three subunits of RNA polymerases I and II are indeed identical (Buhler et al., 1976).

Table 22.2 Subunit structures of some nuclear RNA polymerases

RNA Polymerase I			RNA Polymerase II			RNA Polymerase III		
calf thymus	MOPC 315	yeast	calf thymus	MOPC 315	yeast	X. laevis	MOPC 315	yeast
				240				
				or				
			214	205				
			or	or				
197	195	185	180	170	170	155	155	160
126	117	137	140	140	140	137	138	128
51/-	60/-	41	34	41	41	92	89	82
44	50	35	34	30	33	68	70	
25	27	28	25	25	28	52	53	53
25	16	28	25	22	24	42	49	41
16		24	20	20	18		41	37
16		20	16	20	14	33	33/32	34
		16	16	16	12	29	29	28
		14	16		12	19	19	24
		12					19	20
								14
								11
500	466	540	506	554	497	627	676	632
or	or		or	or				
449	407		472	519				
				or				
				484				

Subunit molecular weights are given in kilodaltons, determined by fractionating the RNA polymerase on SDS acrylamide gels. These are present in approximately equimolar amounts; in cases where a molecular weight band is present in roughly dimolar or trimolar amounts, this is shown as more than one subunit of the appropriate weight. Multiple forms of animal RNA polymerase I are found in which the third largest subunit may be present (weight shown) or absent (indicated by /-). Multiple forms of animal RNA polymerase II are related by differences in the size of the largest subunit. Generally similar subunit structures have been reported for RNA polymerase II of chick (almost identical with calf thymus), and higher plants (little variation among several plant species), and D. melanogaster. The subunits of RNA polymerase III from three species are arranged so as to maximize possible homologies. Two forms of the MOPC enzyme differ in the weight of one subunit. A similar subunit composition has been reported for the enzyme of Bombyx mori. Total molecular weights for each enzyme are given at the bottom of the table.

References: Gissinger and Chambon (1972), Kedinger and Chambon (1972), Kedinger et al. (1974), Krebs and Chambon (1976), Schwartz and Roeder (1974, 1975), Sklar and Roeder (1976), Sklar et al. (1976), Goldberg, Perriard, and Rutter (1977), Greenleaf and Bautz (1975), Jendriask and Guilfoyle (1978).

In several cases, an RNA polymerase activity has been resolved into more than one enzyme by chromatographic separation. Thus two RNA polymerase I enzymes and either two or three RNA polymerase II enzymes have been distinguished in some animal cells. The subunit structures of the different enzymes within a class are closely related. With RNA polymerase I, one of the subunits is present in one subclass and absent from the other. With RNA polymerase II, the alternative forms of the enzyme appear to be related by changes in the size of the largest subunit. The variants within a class do not show any differences in activity. It therefore remains to be established whether these multiple forms represent a genuine polymorphism or are simply artefacts of the isolation and characterization procedures.

The structure of the "real" RNA polymerase, indeed, is not yet clear in any instance. Two critical problems have been the lack of in vitro systems in which defined templates can be transcribed by initiation at promoter sites and the inability to reconstitute active enzymes from subunits. In lieu of such evidence, it is not possible to say whether the isolated enzyme preparations represent basic enzymes, which need assistance from further proteins for physiological activity, or whether they include extraneous proteins that are not part of the RNA polymerase per se. (The subunit constitution of the enzymes usually is assigned on the basis of identifying electrophoretic bands that remain in equimolar quantities through successive stages of purification. Obviously this is only a tentative criterion and must in due course be superseded by systems in which reconstitution of enzyme activity from defined subunits is possible.)

Because of these uncertainties, we shall not delve into the large number of putative factors involved in RNA synthesis; nor shall we be concerned with speculations on the possible roles of the multiple forms of each polymerase. From the general properties of each class, however, it is possible to see in broad terms what function each has. A principal tool in distinguishing the enzyme functions has been to correlate the susceptibilty of cellular RNA synthesis to α-amanitin with that of the purified enzymes. The drug acts on the polymerase enzyme directly to inhibit RNA chain elongation; mutant cells resistant to α-amanitin possess an altered RNA polymerase II enzyme that is not inhibited by the drug in vitro (Chan, Whitmore, and Siminovitch, 1973).

The first activity to be inhibited in animal cells is that of synthesis of heterogeneous nuclear RNA; the least affected is the synthesis of ribosomal RNA. In cells infected by adenovirus, the synthesis of viral mRNA is sensitive to treatment with α-amanitin. Thus it is clear that the bulk of the transcription of hnRNA (and mRNA) is accomplished by RNA polymerase II, whereas it is RNA polymerase I that is responsible for rRNA synthesis (Blatti et al., 1970; Reeder and Roeder, 1972; Bitter and Roeder, 1978).

The first suggestion that small RNAs might be synthesized by a different enzyme was made on the basis of their continued synthesis when the principal nucleoplasmic and nucleolar activities are inhibited (Zylber and Penman, 1971a). The correspondence of this activity with RNA polymerase III for cellular transcripts was shown by Weinman and Roeder (1974). More detailed studies showed that RNA polymerase III is responsible also for synthesizing certain small RNAs from the adenovirus genome (Price and Penman, 1972; Weinman et al., 1976). Although this enzyme is responsible for synthesizing tRNA, 5S RNA, viral small RNA, and some small nuclear RNAs, it does not synthesize all the cellular small nuclear species; some appear to be transcribed by RNA polymerase I (Benecke and Penman, 1977; Zieve et al., 1977).

It is not known whether the specificity of the class I and II enzymes relies largely upon their localization in the nucleolus and nucleoplasm, respectively, or whether they recognize different classes of promoters. Attempts to demonstrate template specificity with in vitro systems, in spite of claims for partial success, do not yet support any general conclusion that the enzymes as presently characterized possess intrinsic abilities to recognize different classes of transcription units.* In the case of RNA polymerase III, promoters specifically recognized by the enzyme may exist at least for the transcription of 5S genes, which appear to be transcribed in vitro with proper orientation by this enzyme but not by the other enzymes (see Chapter 27).

There is little information on whether changes in the transcription of structural genes are accomplished by mechanisms extrinsic to the synthesis of RNA per se, or may involve changes in the RNA polymerase enzyme itself. Resolution of this question will require definition of functions of the putative structural subunits of the RNA polymerases. Changes in the relative amounts of RNA polymerases I and II may sometimes be correlated with the extent of synthesis of their respective types of product, so that the level of RNA polymerase could be the first stage at which transcription is controlled. However, the crudity of RNA polymerase preparations precludes any precise measurement of the amount of each enzyme per cell. Probably it will be necessary to develop specific immunological assays to follow the control of synthesis of the enzymes.

RNA Synthesis in the Cell Cycle

Transcription occurs throughout interphase, but ceases or is much reduced at mitosis. There is no satisfactory evidence for any relationship between the extent or nature of RNA synthesized and the stage of interphase. The transcriptional apparatus is disrupted by the reorganization of cellular struc-

*References: Butterworth, Cox, and Chesterton (1971), Smuckler and Tata (1972), Flint et al. (1974), Beebee and Butterworth (1974, 1975), Holland, Hager, and Rutter (1977).

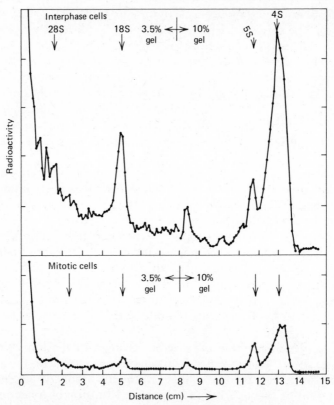

Figure 22.1. RNA synthesis in interphase and metaphase-arrested HeLa cells. In interphase cells all RNAs are synthesized; but in mitotic cells only 5S and 4S RNAs are produced. Data of Zylber and Penman (1971b).

ture at mitosis. Entry into mitosis is accompanied not only by the cessation of transcription, but also sees a reduction in the translation of messengers synthesized during the preceding interphase.*

Autoradiography shows an incorporation of an ^{3}H-uridine label declines in early prophase and virtually ceases by midprophase (Prescott and Bender, 1962; Terasima and Tolmach, 1963; Davidson, 1964). Transcription thus ceases before the disintegration of the nucleolus and nuclear membrane; and it resumes in telophase, before reformation of nuclear structures. The relative timing of these events therefore implies the existence of a specific control, rather than a mechanical response to chromosome condensation.

Figure 22.1 shows that the synthesis of ribosomal RNA and large nucleoplasmic RNA is almost completely abolished at mitosis, but the synthesis of 5S and 4S RNAs continues. This implies that RNA polymerase III re-

*It is only the transcription and translation of nuclear genes that is inhibited at division; mitochondrial synthetic activities continue unabated (Fan and Penman, 1970a).

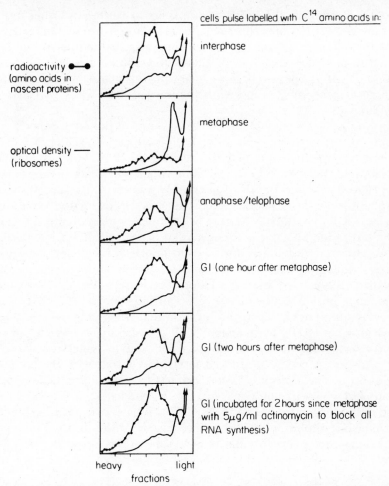

Figure 22.2. Protein synthesis in interphase and mitotic Chinese hamster cells. The ^{14}C amino acid pulse identifies nascent proteins on the polysomes; the optical density trace follows the state of the ribosomes. Interphase cells possess a large peak of active polysomes. In metaphase cells protein synthesis is reduced to 25% of the interphase level and the ribosomes are present as monomers. Reformation of polysomes begins in telophase as incorporation of amino acids resumes. The interphase state is restored in early G1. Since the same results are obtained in cells incubated with actinomycin to block new RNA synthesis, the resumption of protein synthesis must rely upon RNA templates synthesized in the previous cell cycle. Data of Steward et al. (1968).

mains able to transcribe these products even from the condensed mitotic chromosomes, although the activities of RNA polymerases I and II cease.

What happens to the existing nuclear and cytoplasmic RNA molecules during mitosis? Autoradiography suggests that any segregation between nucleus and cytoplasm is lost during division; but it appears to be regained at the beginning of interphase (Prescott, 1963). At least in the case of the nucleolar transcripts, the status quo appears to be frozen during mitosis. Thus the large rRNA precursors remain stable in cells arrested in metaphase for up to 6 hours, in contrast with their short average lifetimes in interphase cells (Fan and Penman, 1971). While their maturation is halted, the precursor molecules appear to be associated with chromatin. Whether a similar cessation of processing occurs with the transcripts of RNA polymerase II is not known.

During division of mammalian cells, the rate of protein synthesis drops to some 20–40% of the interphase level. This appears to be due to a disruption of polysome structure (Scharff and Robbins, 1966; Steward, Schaeffer, and Humphrey, 1968). Figure 22.2 shows that polysomes disaggregate to be replaced by monomeric ribosomes during mitosis. At the start of the next interphase, polysomes reform; and since this reconstruction occurs even when new RNA synthesis is blocked by treatment with actinomycin, it must be the utilization of stable messengers that is controlled. Since the drop in the proportion of polysomes during mitosis exactly parallels the decline in protein synthesis, but the total number of messengers and ribosomes remains constant, it seems likely that translation here is controlled by the ability of ribosomes to initiate on the messengers (Fan and Penman, 1970b). This conclusion is supported by the demonstration that when polysomes are maintained in mitotic cells by treatment with cycloheximide (which preferentially prevents movement of ribosomes along the message), the ribosomes remain able to function at reduced levels. In contrast with the decline in translation during mitosis, cell-free extracts from mitotic cells are able to initiate and synthesize proteins virtually as efficiently as extracts from S phase cells (Tarnowka and Baglioni, 1979). Whatever factor is responsible for inhibiting inhibition must therefore have been lost during preparation of the extract; the possibility has been raised that this factor may be a small nuclear RNA that is released into the cell by the dissolution of the nuclear membrane.

CHAPTER 23

Structure of Messenger RNA

A debate on whether messenger RNA exists in eucaryotic cells as a counterpart to bacterial mRNA was settled some years ago with the isolation of the first individual messengers (coding for α- and β-globin of reticulocytes). Much of the difficulty in isolating eucaryotic messengers originally arose from the need to isolate mRNA from a polysomal fraction in which it is only a minor component of the RNA mass. Direct extraction of RNA yields a fraction in which ribosomal RNA is present in overwhelming excess. Techniques to distinguish the mRNA from the rRNA by preferential labeling do not exclude contaminating RNA fractions that also gain the label. The observation that messenger RNA may be released from the polysomes in the form of a ribonucleoprotein particle which is distinguished from the contaminating RNP species was necessary to lead to the first characterization of cellular mRNA. These difficulties since have been overcome by the discovery that most messengers carry a sequence of polyadenylic acid at the 3′ end. This allows their isolation by virtue of the reaction of the poly(A) "tail" with oligo(dT) or oligo(U) linked to cellulose or sepharose. The poly(A)$^+$ fraction may be isolated directly from whole cell preparations or (preferably) from polysomes; generally it corresponds to some 1–2% of the mass of cellular RNA (about 90% of which is ribosomal RNA).

From such fractions the general properties of the messenger RNA population have become clear. Each molecule possesses a "cap" of methylated nucleotides at the 5′ end as well as a tail of poly(A) at the 3′ end. Both these modifications are found in the nuclear precursors to mRNA and are introduced posttranscriptionally by enzyme activities independent of RNA polymerase. Each messenger usually codes for only a single polypeptide, although it is somewhat longer than is necessary just for the coding function, due to the presence of untranslated regions at both the 5′ and 3′ ends.

The amount of a particular messenger RNA present in a cell is related to the extent of synthesis of its product. Thus messengers for particular proteins may be present in relatively large proportions in specialized cells in which these proteins are the predominant products. Purified messengers coding for individual proteins may be characterized by demonstrating that the appropriate product is synthesized in vitro in a heterologous system, that is, with purified ribosomes, factors and tRNAs from another source (usually wheat germ or reticulocyte). Except in the most favorable cases, however, even these mRNA preparations may be contaminated with messengers coding for other proteins.

The whole problem of messenger purity now may be circumvented, however, by the use of DNA cloning. A given messenger or mRNA population is reverse transcribed into cDNA copies, which are converted into duplex form and inserted into bacterial plasmids. Cloning the chimeric plasmids in bacteria provides an exacting method for obtaining only single sequences, usually characterized by their ability to hybridize back to the individual messengers. The sequence that has been cloned in a plasmid may be retrieved and used directly for sequence studies or to provide a probe to isolate and quantitate the corresponding message for structural or translational studies.

In this chapter we shall describe the structure of messenger RNA determined from studies of cellular populations and individual species. We shall be concerned to demonstrate the ubiquity of 5′ methyl capping and to characterize two classes of mRNA according to the presence or absence of 3′ poly(A) tails. We shall defer until the discussion of nuclear RNA in Chapter 25 the question of the origins of these modifications and the issue of whether they play any role in the selection of sequences for transport to the cytoplasm. From the properties of isolated individual messengers, we can formulate a view of their general sequence organization and ribonucleoprotein structure and of the factors that influence their ability to be translated. We shall postpone until Chapter 25 our discussion of the derivation of mRNA from the primary transcript, and shall come in Chapter 26 to the question of its relationship with the transcription unit in the genome.

Recognition of mRNA and mRNP in Polysomes

A small proportion of the RNA in the cytoplasm of HeLa cells was first identified as mRNA because it is associated with polysomes, has sedimentation properties corresponding to the size expected to code for proteins, and has a base composition generally resembling that of DNA (Girard et al., 1965; Latham and Darnell, 1965). Techniques for specifically labeling the mRNA were based upon differences in its synthesis compared with ribosomal RNA. One method is to use short pulse labels, which reach cytoplasmic mRNA within 5–10 minutes, but do not enter rRNA for 25–40 minutes (Perry and Kelley, 1968). Another is to suppress rRNA production by treatment with low doses of actinomycin, which preferentially inhibit nucleolar RNA synthesis (Penman, Vesco, and Penman, 1968).

Messenger RNA labeled by these means is found in the form of ribonucleoprotein particles associated with ribosomes. The polysomes sediment as a dispersed fraction at rates > 80S (the size of the monomeric ribosome) on sucrose density gradients. When fixed with formaldehyde and banded on CsCl, however, two peaks are found, one at the buoyant density of 1.54 g-cm^{-3} characteristic of mRNA-ribosome complexes, and one at lower val-

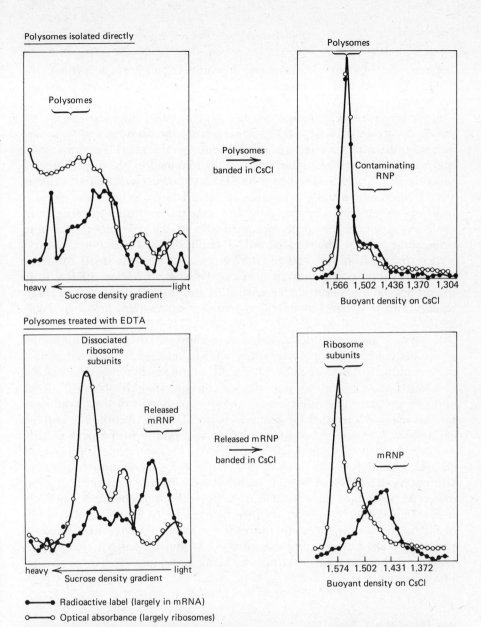

Polysomes isolated directly

Polysomes

heavy ← Sucrose density gradient → light

Polysomes

Polysomes
→
banded in CsCl

Contaminating RNP

1,566 1,502 1,436 1,370 1,304
Buoyant density on CsCl

Polysomes treated with EDTA

Dissociated ribosome subunits

Released mRNP

heavy ← Sucrose density gradient → light

Released mRNP
→
banded in CsCl

Ribosome subunits

mRNP

1.574 1.502 1.431 1.372
Buoyant density on CsCl

●——● Radioactive label (largely in mRNA)
○——○ Optical absorbance (largely ribosomes)

Figure 23.1. Release of messenger ribonucleoproteins from L cell polysomes by EDTA. The radioactive label is used to follow the mRNA (and contaminating RNA) while the optical density identifies the ribosomes.

The upper profiles show polysomes extracted by a detergent method in which they form a dispersed (> 200S) distribution on sucrose gradient sedimentation. Banding in CsCl shows that in addition to the polysomes (density about 1.54 g-cm^{-3}) there is also a contaminating component of lower density.

The lower profiles show that treatment with EDTA dissociates the polysomes into ribosomal subunits (60S and 40S) and a peak of ribonucleoprotein (< 20S). Banding in CsCl shows that the released RNP fraction sediments in a single broad peak of about 1.45 g-cm^{-3}, separated from the ribosome subunits of 1.50 and 1.57 g-cm^{-3}. Data of Perry and Kelley (1968).

655

ues around 1.40 g-cm^{-3}, representing contaminating ribonucleoprotein. To isolate the mRNP, it is necessary to separate it from the contaminating fraction.

Messenger ribonucleoprotein may be distinguished from the other component by treatment with EDTA, which causes the ribosomes to dissociate into subunits with the release of mRNA, but does not affect the contaminating material. Figure 23.1 shows the results obtained by the application of this protocol. The mRNP released by EDTA sediments with a mean value of 16–18S (the range usually is roughly 12–60S), relative to the ribosome subunits of 60S and 40S. The contaminating material continues to sediment in the heavy position previously occupied also by the polysomes, extending up to some >200S. The released mRNP bands on CsCl at 1.40–1.45 g-cm^{-3}, well separated from the 1.50 and 1.57 g-cm^{-3} bands of the small and large ribosome subunits. Puromycin also can be used to release mRNA from polysomes. The general characteristics of the mRNP defined by EDTA-release from polysomes in several systems are summarized below in Table 23.1.

Eucaryotic messenger RNA therefore comprises fairly small molecules which are contained in polysomes and may be released from the ribosomes and contaminating material by EDTA. All of the newly synthesized mRNA is released in the form of a particle that contains roughly about 40% RNA and 60% protein. The mRNA and protein components of the ribonucleoprotein may be separated by treatment with SDS and phenol. The mRNA extracted in this way generates a polydispersed profile with a peak usually around 10S, extending to a tail of up to 40S. This corresponds roughly to an average length of the order of 2000 nucleotides. The nature of the contaminating RNP is not entirely clear. It may in part constitute nuclear material, but might also include messenger-like molecules not presently under translation. However, its presence precludes the direct extraction of mRNA from polysomes and makes it necessary to use the EDTA-release procedure. The advantage of this protocol is that the mRNA corresponds exactly to the spectrum of sequences being translated, that is, to the expressed genes.

Poly(A) at the 3' Terminus

Most eucaryotic messengers possess a sequence of polyadenylic acid at the 3' end. The poly(A) tail is not coded in the genome but is added to the transcript in the nucleus; inhibition of polyadenylation prevents the appearance of the mRNA in the cytoplasm (see Chapter 25). However, poly(A) sequences are found not only on nuclear RNA and cytoplasmic RNA, but also on the messengers synthesized by some viruses which reproduce within the cytoplasm and in mitochondrial messengers. This suggests that, if poly(A) has a function in events preceding the appearance of mRNA in the cytoplasm, this cannot be its sole purpose. The function of poly(A) in mRNA has indeed been mysterious. Since messengers are translated from 5′ to 3′, termination of protein synthesis must occur before ribosomes reach the poly(A), which renders unlikely any function in translation per se. Other possible functions concern the stability of mRNA or the attachment of messengers to membranes. However, the demonstration that some cellular messengers lack poly(A) implies that its function cannot be obligatory for successful translation, although it does also raise the possibility of finding differences between poly(A)⁺ and poly(A)⁻ mRNA that might be related to the function of the poly(A).

One of the first observations of the existence of poly(A) was made by Edmonds and Caramela (1969) when they isolated RNA from Ehrlich ascites cells grown in mice injected with ^{32}P. After extraction with phenol, about 1% of the RNA is retained by poly(dT)-cellulose, due to the presence of poly(A) sequences able to hybridize with the poly(dT). By measuring the poly(A) content of RNA fractions isolated from HeLa nucleus and cytoplasm, Edmonds, Vaughan, and Nakazoto (1971) showed that poly(A) is absent from the nucleolar (rRNA) fractions, but is present in the RNAs of both nucleoplasm and cytoplasm. Other methods since have been used to isolate RNA by virtue of its content of poly(A); the most reliable have proved to be the use of oligo(dT) or oligo(U) immobilized on cellulose or sepharose, or poly(U) in solution.

The poly(A) can be isolated by virtue of its resistance to degradation by pancreatic ribonuclease. Figure 23.2 shows that when mRNA is labeled with ^{3}H-uridine, ribonuclease is able to degrade the label completely. But when the labeled precursor is ^{3}H-adenosine, a peak of resistant material is obtained (Darnell et al., 1971; Lee, Mendecki, and Brawerman, 1971). This is poly(A).

The location of the poly(A) in mRNA was identified by experiments with a purified exoribonuclease specific for 3′-OH termini. If poly(A) is located at the 3′ terminus of mRNA, it should be degraded first. Molloy et al. (1972) found that this enzyme degrades the poly(A) of HeLa or L cell mRNA very rapidly relative to other regions. Some 90% of the poly(A) segments isolated by resistance to pancreatic RNAase treatment of mRNA can be degraded by the enzyme. This implies that they contain free 3′-OH ends derived from the 3′ messenger terminus, rather than the 3′-P termini that would be derived by cleavage from internal parts of the molecule. Probably the remaining 10% of the poly(A) also is 3′ terminal, but has lost its original 3′-OH termini as the result of internal cleavage during extraction.

The poly(A) fraction of HeLa cell mRNA is about 2.5–5.0% of the nucleotide mass. An inverse relationship between the proportion of poly(A) and the size of the messenger fraction suggests that all messengers, irrespective of their overall size, contain the same length of poly(A). This is supported

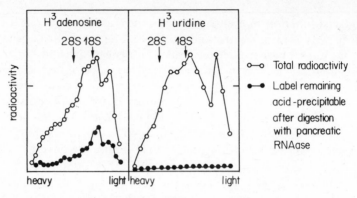

Figure 23.2. Identification of poly(A) in messenger RNA. HeLa cells were labeled with ^{3}H-adenosine (left) or ^{3}H-uridine (right). The mRNA fraction extracted from polysomes was sedimented through a sucrose gradient. The total radioactivity in each fraction was determined after acid precipitation of RNA. The RNAase-resistant label was determined as the counts remaining acid precipitable (that is, in polynucleotide form) after digestion with pancreatic ribonuclease. The distribution of total messenger is the same with either labeled adenosine or uridine. But material resistant to RNAase is found only when adenosine is labeled. This identifies the poly(A). Data of Darnell, Wall, and Tushinski (1971).

by experiments in which mRNAs of different sizes were isolated and digested with pancreatic ribonuclease; the product sediments at a constant rate, about 10% faster than the 4S of tRNA (Edmonds, Vaughn, and Nakazoto, 1971; Darnell, Wall, and Tushinski, 1971). This corresponds to a size of less than 200 nucleotides and suggests that each messenger contains only one sequence of poly(A).

Few if any bases other than adenine are present in the poly(A) sequence. By degrading the poly(A) with alkali, Lee et al. (1971) detected one 3'-OH terminus for every 200 adenine residues. Working with isolated HeLa nuclear or cytoplasmic poly(A), Molloy and Darnell (1973) identified only 2 UMP and 1–2 CMP residues for every 195 residues of AMP. By electrophoresing the poly(A) recovered from newly synthesized HeLa messengers, Sheiness and Darnell (1973) obtained a peak corresponding to 190 ± 20 nucleotides, with a tail corresponding to shorter poly(A) segments.

The length of the poly(A) decreases with the age of its messenger, as shown in Figure 23.3. In these experiments, HeLa cells were labeled with ^{3}H-adenosine in the presence of a low dose of actinomycin to suppress rRNA synthesis. Then further synthesis of mRNA was blocked by the addition of a high dose of actinomycin. After chases of 90 and 180 minutes, the length of the poly(A) has decreased considerably. This shortening does not seem to be related to the translation of the messenger since it continues at the same rate in cells in which protein synthesis is inhibited. The rate at which

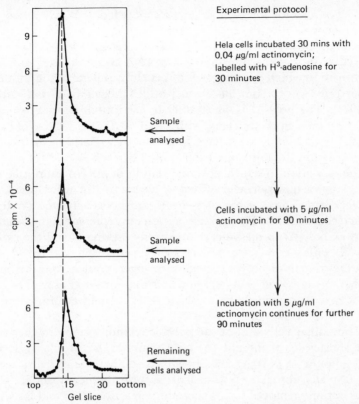

Experimental protocol

Hela cells incubated 30 mins with
0.04 μg/ml actinomycin;
labelled with H^3-adenosine for
30 minutes

Cells incubated with 5 μg/ml
actinomycin for 90 minutes

Incubation with 5 μg/ml
actinomycin continues for further
90 minutes

Figure 23.3. Shortening with age of poly(A). HeLa cells were labeled with ^{3}H-adenosine and chased in the presence of actinomycin to prevent further synthesis of mRNA. The polysomal RNA was extracted immediately after labeling and after two chase periods; poly(A) was isolated as the material resistant to pancreatic and T1 RNAase. Its mobility when electrophoresed on acrylamide gels decreases during the chase, corresponding to a reduction in length. Data of Sheiness and Darnell (1973).

poly(A) is shortened is not constant for all messengers; for two adenovirus mRNAs that turn over relatively rapidly in rat cells, the poly(A) is shortened at rates somewhat faster than that of the bulk cellular message population (Wilson et al., 1978).

The reduction in length has been characterized for globin messengers in both mouse and rabbit cells. When newly synthesized globin mRNA is obtained from young erythroid cells, its poly(A) segment is about 150 nucleotides in length. In mature blood cells, however, shorter poly(A) segments are present on the globin mRNA, corresponding to discrete size classes of roughly 100, 60 and 40 bases (Gorski et al., 1974; Merkel et al., 1975; Nokin et al., 1976). There is indirect evidence that length reduction may occur in discrete steps also for the two adenovirus messengers. Whether shortening occurs in steps also for the cellular population is not known.*

Mitochondrial protein synthesis is distinct from that of the cytoplasm, in particular in that the translation of mRNA occurs in the same cellular compartment in which it is transcribed. Yet HeLa cell mitochondrial mRNA possesses poly(A) of about 50–60 nucleotides in length (Hirsch and Penman, 1974). The poly(A) is added after transcription, although somewhat more slowly than in the nucleus, with a half life for the process of about 60 minutes. Presumably it must have some function related to the expression of mRNA within the mitochondrion.

Another situation in which polyadenylation of mRNA cannot be related to processing or transport is provided by the sea urchin egg. Upon fertilization, preexisting cytoplasmic messengers are polyadenylated (Wilt, 1973; Slater and Slater, 1974). The reaction appears to represent both the addition de novo of poly(A) to messengers previously lacking it and the extension of poly(A) tails on messengers previously possessing shorter sequences. Overall the content of poly(A) in egg messengers doubles, but this includes the replacement of some poly(A) sequences that turn over as well as the synthesis of new or extended chains (Wilt, 1977). Again the function of the polyadenylation is obscure.

In mammalian cells almost all polyadenylation occurs in the nucleus; and the shortening of the poly(A) tails occurs in the cytoplasm. However, a small amount of poly(A), less than 10 bases, may be added in the cytoplasm; this addition is only transient because the newly added terminal bases are rapidly removed in the shortening reaction (Diez and Brawerman, 1974). The significance of the reaction is not known.

Poly(A)$^+$ and Poly(A)$^-$ mRNA

The proportion of cytoplasmic messenger RNA that can be isolated by its possession of poly(A) varies with the cell system and isolation technique, but usually is substantial, sometimes apparently approaching 90% (Greenberg and Perry, 1972). The 9S histone mRNA fraction has been known for some time to lack poly(A), but the failure of other material to react with oligo(dT) or oligo(U) was thought to be an artefact. Because under most conditions some of the poly(A) fails to bind to the column of complementary nucleotides, and because these assays are highly susceptible to the gen-

---

*A length of up to 200 nucleotides of poly(A) is typical for mammalian cells, on which most work has been done. Somewhat shorter 3' terminal regions of poly(A) have been found in other species. The mRNA of silkmoth follicular cells has poly(A) that is about 50 nucleotides when newly synthesized, but shortens with age to about 30 nucleotides (Vournakis, Gelinas, and Kafatos, 1974). The general behavior of the poly(A) therefore is similar, although it is shorter.

eration by internal cleavage of message fragments lacking poly(A), all mRNA was assumed to possess poly(A).

The presence of a messenger fraction in HeLa cells that does not contain poly(A), however, was first demonstrated by Milcarek, Price, and Penman (1974). Some 30% of the mRNA failed to bind to oligo(dT) cellulose columns, under conditions in which 96% of the poly(A)$^+$ mRNA isolated on the column is able to rebind to it. When ribonuclease resistant tracts of ^{3}H-adenosine were prepared from the poly(A)$^+$ and poly(A)$^-$ fractions, the former had about 11% of its label in this form, the latter only < 0.4%. Thus the poly(A)$^-$ mRNA has less than 5% of the poly(A) content of the poly(A)$^+$ preparation. Without characterizing the 3' termini of the messengers in detail, it is not possible to exclude the possibility that the poly(A)$^-$ mRNA has some polyadenylic acid, but this could not be longer on average than 10 nucleotides. Similar observations have been made in L cells, also with about 30% poly(A)$^-$ mRNA, and in sea urchin embryos, where the proportion of poly(A)$^-$ mRNA is about 50% (Nemer et al., 1974, 1975; Greenberg, 1976).

About one third of the poly(A)$^-$ mRNA usually represents histone mRNA. The remaining fraction does not seem to be derived by cleavage of poly(A)$^+$ messengers since its sedimentation pattern is essentially identical with the poly(A)$^+$ mRNA (with an average of 16S in the mammalian cells, 22S in the sea urchin). Exposure of polysomes to puromycin releases both fractions of mRNA; thus both poly(A)$^+$ and poly(A)$^-$ mRNA seem able to serve as template for protein synthesis. The stabilities of both classes of mRNA are the same; so is the delay before they become labeled after addition of a radioactive precursor. The only difference so far demonstrated in their metabolism is that synthesis of the poly(A)$^+$ mRNA is 95% inhibited by deoxyadenosine (cordycepin), whereas production of poly(A)$^-$ mRNA is only 60% inhibited. Virtually no functional differences are therefore apparent between the two populations.

Do poly(A)$^+$ and poly(A)$^-$ messengers represent different sequences? Originally this appeared to be so, because a cDNA probe prepared against poly(A)$^+$ mRNA reacted only very poorly with the poly(A)$^-$ mRNA. But in a more detailed analysis Kaufman et al. (1977) found that the poly(A)$^-$ mRNA represents a subset of the sequences present in the poly(A)$^+$ preparation. Comparison on 2D gels of the products obtained in translating the messenger populations in vitro shows that most of the proteins coded by poly(A)$^-$ mRNA also are made by the poly(A)$^+$ mRNA. The poly(A)$^+$ mRNA appears to code for a greater number of proteins, including many that are not specified by the poly(A)$^-$ mRNA. This comparison implies that the most abundant messengers in the poly(A)$^-$ population also are abundant in the poly(A)$^+$ sequences. Correspondingly, about 10% of a cDNA probe representing the 20% most abundant sequences in the poly(A)$^+$ fraction was able to hybridize with the poly(A)$^-$ mRNA rather quickly; the remaining abundant poly(A)$^+$ sequences hybridized with poly(A)$^-$ mRNA but more slowly. This suggests that some of the abundant poly(A)$^+$ sequences also are abundant in poly(A)$^-$ mRNA; others are present in the poly(A)$^-$ mRNA, but at much lower concentrations. What proportion these sequences represent of the total poly(A)$^-$ mRNA is not known; to determine this would require an experiment in which the poly(A)$^-$ mRNA was labeled and tested for its ability to hybridize with the poly(A)$^+$ cDNA. Thus at present it remains unknown whether some sequences are present in poly(A)$^-$ mRNA but not in poly(A)$^+$ mRNA. The major product of both poly(A)$^+$ and poly(A)$^-$ mRNA is actin; this identity reinforces the question as to what significance the presence or absence of poly(A) may have. Another system in which poly(A)$^+$ and poly(A)$^-$ messengers have been characterized for the same protein is provided by the synthesis of protamine in trout testis (Iatrou and Dixon, 1977). Again the functional significance is not clear, although a greater proportion of the poly(A)$^-$ protamine mRNA appears to be under translation.

The existence of both poly(A)$^+$ and poly(A)$^-$ mRNA on polysomes, and the demonstration that at least some of the sequences in the two populations are the same, emphasizes the difficulty of defining what may be the function of poly(A). Direct attempts to search for a role of polyadenylation in messenger function have concentrated on possible influences of poly(A) on messenger translatability or stability. Translation of messengers with different lengths of poly(A) has been followed in vivo by Bard et al. (1974) by identifying "old" messengers with a pulse label followed by a chase with unlabeled precursors, while "new" messengers are identified by their possession of a very recent pulse label. No difference can be seen in the proportions of the two age classes in polysomes; and when cells are raised to 42°C to disaggregate polysomes, subsequent reduction of the temperature allows both age classes to enter polysomes again with equal efficiency. Similar experiments have been performed with vaccinia virus RNA by Nevins and Joklik (1975). Thus for both cellular and viral messengers there appears to be no correlation between the age-dependent reduction in length of the poly(A) and the ability of the messengers to initiate and elongate protein chains.

The poly(A) tail can be removed from isolated mRNA by treatment with an exoribonuclease specific for 3′-OH ends; and the deadenylated messengers then can be tested for ability to support translation in vitro. With L cell, HeLa cell, or purified globin messengers, the deadenylation has no effect (Bard et al., 1974; Munoz and Darnell, 1974; Williamson et al., 1974). The nature of the system used for translation may influence the result, however, since deadenylated ovalbumin mRNA is translated as efficiently as the (normal) poly(A)$^+$ message in a wheat germ in vitro system, but less efficiently in a reticulocyte system (Doel and Carey, 1976). A difficulty in interpreting the results of in vitro translation, however, is that none of the available systems matches the efficiency of protein synthesis in vivo. Thus each messenger may suffer only a few rounds of translation; and it is possible that the limited lifetime of the system may be insufficient for any effect of poly(A) to become apparent. While it is clear that the lack of poly(A) per se does not prevent the use of mRNA as template, these results do not exclude a possible role for poly(A) in some subsidiary role in vivo.

The possibility that the lack of poly(A) may indeed lead to a more rapid decline in the ability of globin message to be translated has been raised by the results of Huez et al. (1974, 1975). With an in vitro cell-free system, the translation of poly(A)$^-$ globin mRNA declined during longer incubation periods (90 minutes) much more than that of the poly(A)$^+$ globin message. A striking difference is evident in the ability of the poly(A)$^+$ and poly(A)$^-$ globin mRNAs to be translated in Xenopus oocytes: initially their activities are the same, but by 20 hours of incubation the poly(A)$^-$ preparation has become inactive while the poly(A)$^+$ mRNA can still be translated. A control in which readenylation of the poly(A)$^-$ preparation restores its ability to sustain translation for more protracted periods shows that the decline is caused by the removal of poly(A) and not by any random damage suffered by the mRNA. An obvious cause for this effect would be a reduction in stability of the mRNA resulting from the removal of the poly(A). In a preliminary report, Marbaix et al. (1975) have indeed reported that a hybridization assay with a cDNA probe against globin finds a disappearance of message sequences when poly(A)$^-$ but not poly(A)$^+$ mRNA is injected into oocytes. However, a full assessment of this possibility must await the publication of more complete data.

A contrary report has been that removal of poly(A) tails from interferon mRNA has no effect upon protein synthesis in the oocyte over the 48 hour period for which this message remains active (Sehgal, Soreq, and Tamm, 1978). Here it would be useful, however, to have confirmation that the deadenylated message lacks poly(A) by some technique other than inability to bind to poly(U); also it would be useful to show that the oocyte is not able to readenylate the message.

Messenger Ribonucleoprotein Particles

Both messenger RNA in the cytoplasm and its precursor in the nucleus have been extracted from many cell types of various species in the form of ribonucleoprotein particles. It is possible that such particles might form as artefacts during preparation of mRNA by a nonspecific association between the negatively charged RNA and basic proteins. But the binding between mRNA and proteins in the particle seems to be more stable than that of randomly formed complexes; and in some instances specific proteins have been identified in the RNP particle. One possible function for this structure is to transport mRNA from the nucleus to cytoplasm; another is to control its translation.

Early experiments which used sucrose gradient sedimentation alone to characterize the rapidly labeled RNA showed only that it sediments more rapidly than free mRNA, without revealing the nature of the components associated with it. For example Joklik and Becker (1965) and McConkey and Hopkins (1965) found that rapidly labeled RNA first appears in the cytoplasm in a form sedimenting at 45S.* Subsequently these particles have been characterized following EDTA release from polysomes, which identifies them with messengers as shown in Figure 23.1. Generally they sediment with a heterogeneous distribution that spans a range of S values of about twice those shown by the isolated mRNA. Their buoyant density in CsCl corresponds to a protein content of half or more as seen in the summary of Table 23.1.

Do the mRNP particles represent the condition of mRNA in vivo or are they artefacts of preparation? Ribonucleoprotein particles can indeed be formed during extraction of mRNA (Baltimore and Huang, 1970). But the bound proteins are released from the RNA by washing with 0.5 M salt, whereas the mRNP particles released from rat liver polysomes are stable under such conditions (Lee and Brawerman, 1971). Since this treatment suffices to remove initiation factors from polysomes, the proteins remaining bound to the mRNA would not appear to represent persistent translation factors.

In addition to the mRNP particles released from polysomes by EDTA, ribonucleoproteins are found in the form of the material contaminating the polysome peak. The heavier RNP particles which sediment with the polysomes probably represent nuclear material, but other, lighter ribonucleoproteins have been found in the cytoplasm. The role of the free cytoplasmic RNP particles is not clear, but in some cases they may contain messenger RNA not presently under translation. This is well substantiated for certain embryonic systems and similar suggestions have been made for some cultured cells.

The eggs of several marine or amphibian species contain mRNA synthesized during oogenesis, but which is not translated until following fertilization. These maternal messengers may be used by the early embryo up to about the stage of gastrulation. Early work focused on the loach Misgurnis fossilis, in which mRNP can be found in the form of particles sedimenting in the range from 20–110S and with a buoyant density of 1.39 g-cm^{-3} (Spirin, 1969, 1972). Similar particles have been found in the sea urchin (Infante and Nemer, 1968; Dworkin et al., 1977). They have been described as "informosomes" in a model in which they contain mRNA that is to be used for translation later. Newly synthesized RNA enters these particles as well as polysomes, but can be translated in vitro, consistent with the idea that they represent a storage form of mRNA.

*This led to suggestions that the complex might represent a 40S ribosome subunit bound to mRNA, perhaps as a means to transport the messenger from nucleus to cytoplasm. However, centrifugation on CsCl shows that the complex is a separate mRNP particle, distinguished from ribosome subunits by a much lower buoyant density (Henshaw, 1968; Parsons and McCarty, 1968; Henshaw and Loebenstein, 1970; Kumar and Lindberg, 1972).

In HeLa cells, some of the newly synthesized RNA enters cytoplasmic RNP particles rather than polysomes (Spohr et al., 1970). The possibility that there might be a precursor-product relationship between these particles and the polysomal mRNP was raised by Schochetman and Perry (1972) on the basis of temperature shock experiments. Upon increasing the temperature of incubation of L cells, the polysomes disaggregate and yield mRNP complexes that are inactive in protein synthesis. When the temperature is then reduced to normal, the polysomes reform at the expense of the free mRNP peak. Whether the behavior of the mRNP released from the polysomes in these conditions can be equated with the RNP particles present in the cells at normal temperature is not proven. A similar situation occurs when mouse sarcoma cells are starved for amino acids: the initiation of protein synthesis is blocked, polysomes are converted to 80S ribosomes, and mRNA is freed in the form of ribonucleoprotein. Restoration of amino acids allows the ribosomes to resume translation of the mRNP (Sonenshein and Brawerman, 1977).

One of the problems in investigating the relationship of the nonpolysomal RNP particles with the polysomal mRNP has been the use of cellular RNA populations presumably containing many different sequences. It would be interesting to know whether the same or different sequences are represented in the polysomal and nonpolysomal RNP particles, a point that should be tested by in vitro translation or by hybridization with cDNA probes. But any demonstration of a precursor-product relationship may best rest upon experiments to follow defined sequences. There have, of course, been many experiments to attempt to identify nontranslated messengers in cells; but usually these have not included assays of the cellular location of the putative stored messenger. In some cases sequences apparently able to code for particular proteins have been identified in nonpolysomal RNPs (Bag and Sarkar, 1975; Buckingham et al., 1976). But there has yet to be a full kinetic study in which newly synthesized mRNA is followed into nontranslated RNP form and then is chased into polysomal mRNP. In lieu of such data, it is difficult to say whether the nonpolysomal RNP in any particular case includes coding sequences and is an intermediate to translation or a dead end.

Whatever the role of free RNP particles, the mRNA under translation by the ribosomes appears to be maintained as a ribonucleoprotein. This raises several questions. Where are the protein components located on the mRNA and what is their influence on translation? Are the same proteins associated with all cellular messengers or are there different proteins with specificity for different sequences?

The number of proteins associated with mRNA in the ribonucleoprotein particles released from polysomes by EDTA is rather small. With cellular messenger populations, the total number of proteins in the isolated mRNP complex usually is not more than ten. Some of these proteins are present in minor amounts and it is not clear whether they are genuine components of

Table 23.1　Properties of messenger ribonucleoprotein particles

Source of complex	mRNP size	mRNP density	mRNA size	Sizes of major proteins										Ref.
Rabbit reticulocyte globin	20S		9S			52		78					(<3 minor)	A
Duck erythroblast globin	15S		9S			49		73					(~6 minor)	B
KB cell	10-100S	1.38	10-40S			56	68	78		130			(~5 minor)	C
HeLa cell	15-60S	1.40	8-30S			52		76		120			(~10 minor)	D
Ehrlich ascites	heterogeneous	1.46	10-30S			52	68	78					(~3 minor)	E
L cell	heterogeneous	1.45	10-40S	32	36			76	87	103			(no minor)	F
Mouse kidney	10-80S	1.43	5-40S				69	75	80	84	90	118	(~5 minor)	G
Trout testis protamine	14-16S		6S		29			73					(no minor)	H

In all cases mRNP was prepared by EDTA-dissociation of polysomes, usually followed by fractionation on oligo(dT) cellulose and a low salt (0.5 M NaCl) wash to remove loosely bound proteins. The sizes of the major proteins present are given in kilodaltons; the number of minor proteins is indicated. Proteins were separated by size through electrophoresis on SDS gels. In some fractions a 34K protein which is bound to 5S RNA was present as a contaminant in addition to the noted major proteins.

References: A—Blobel (1972), Bryan and Hayashi (1973); B—Morel et al. (1973), Mueller et al. (1977); C—Lindberg and Sundquist (1974); D—Kumar and Pederson (1975); E—Barrieux et al. (1975); F—Greenberg (1977); G—Irwin et al. (1975); H—Gedamu et al. (1977).

the complex or contaminants. If they are indeed associated with mRNA in vivo, only some messengers can possess them. Usually less than five proteins comprise the great part of the protein mass; these could be distributed equally on all messengers. Cleaner results might be expected with individual messengers and with globin it is clear that there are only two principal proteins associated with the mRNA (including both α and β messages). These have sizes of about 52,000 and 78,000 daltons; and if they were present in equal mass with the mRNA, there would be only 1-2 molecules of each protein with each mRNA molecule.

Available data on the sizes of the proteins present in mRNP preparations from various sources are summarized in Table 23.1. One feature that has often been remarked upon is the presence in almost all preparations of two proteins with the same sizes as those found on the globin mRNA (49,000-56,000 and 73,000-78,000). It has been suggested that these may be identical or at least very similar proteins serving a function common to perhaps all mammalian cells. But of course size alone is an inadequate criterion for such judgment; further data, such as immunological cross reaction or the existence of common tryptic fingerprints, would be necessary to confirm that these proteins are common components in packaging mRNA. Proteins of this size are present in both polysomal mRNP and free RNP of chick embryonic muscles, although the other proteins differ in both amount and type (Jain and Sarkar, 1979).

Some of the protein components of the messenger ribonucleoprotein particle may be bound to the poly(A) tail. The first indication of this was provided when Kwan and Brawerman (1972) found that ribonuclease releases poly(A) from mRNP in the form of a 12-15S ribonucleoprotein. In a more thorough study, Blobel (1973) found that ribonuclease treatment of reticulocyte or L cell polysomes releases the 78K and 52K proteins. The 78K protein remains attached to a fragment of RNA somewhat heterogeneous in size, as judged by its rate of sedimentation (4-14S), while the 52K protein shows more variable behavior, sometimes released free of RNA, sometimes attached to it. After labeling with ^{3}H-uridine or ^{3}H-adenosine, the material attached to the released 78K protein appears to comprise at least 80% adenine bases. Similar results have been obtained with HeLa cells, in which a 75K protein is associated with the longest (newly synthesized) tracts of poly(A); the smaller protein, measured variably as 52K or 62K, also is released with poly(A), but apparently is associated with shorter (older) tracts (Kish and Pederson, 1976; Schwartz and Darnell, 1976). Since the 75K protein also is associated with the poly(A) of nuclear RNA (see Chapter 25), this suggests that it may be involved in some transport function as well as constituting a major component of the cytoplasmic mRNP. The role of the smaller major protein, and of any minor components, is less clear, but these too may be located on the poly(A).

The possibility has been raised that poly(A) may have a role in maintaining secondary structure in mRNA. Jeffery and Brawerman (1975) reported that when poly(A) is isolated by ribonuclease treatment of mRNA, it may be associated with uridine-rich material from which it is released by heating. This material appears to be somewhat longer than the poly(A) itself, which would argue against the possibility that it comprises the oligo(U) sequences characterized in mRNA by Korwek et al. (1976). These appear to be roughly 30 bases long and to be present at about one fifth of the number of poly(A) tracts. Whether poly(A) is associated in vivo with other nucleotide sequences, by inter- or intramolecular reaction, remains to be proven, however, since it is difficult to exclude the possibility of an adventitious association.

Another possible role for poly(A) was suggested by the observation of Milcarek and Penman (1974) that it is associated with membrane fractions. Whether this is a function of the poly(A) itself or of the proteins associated with it remains to be seen.

A useful comparison in examining the locations and functions of the proteins associated with poly(A) would be to determine whether they are also present on poly(A)$^-$ messengers. Similarly the ability of the poly(A)$^-$ polysomes to attach to membranes might be compared with that of poly(A)$^+$ polysomes, as might the location and secondary structure of oligo(U) putative complements.

To summarize the uses and abuses of poly(A), it is clear that most cellular messengers contain a 3' terminal tract of poly(A) which is associated with at least some of the protein components of the messenger ribonucleoprotein. This location explains why the proteins do not interfere with translation of the messenger by the ribosomes. The presence of the poly(A) tail has made possible the development of protocols to isolate poly(A)$^+$ mRNA with high specificity. But two important reservations about this technique must be noted. First, the presence of poly(A)$^-$ messengers means that some of the cellular messengers are not included; it remains possible that these might contain some sequences different from those of the poly(A)$^+$ population. Second, the isolation of RNA through its poly(A) tail sometimes has been applied to the cellular or cytoplasmic fraction directly. While the specificity of this technique may appear to make redundant the need for an EDTA release procedure, such an omission does not distinguish translated mRNA from any nontranslated messengers; thus the fraction will include any poly(A)$^+$ RNA that may not have been attached to ribosomes. This is especially important in cells in which there are likely to be stored messengers, such as those of early embryos. In spite of much information about the structure of poly(A) and the evidence for its ubiquity, little is known about its function. It does not appear to be necessary for translation per se or to be involved in transport. The significance of its effects upon stability in the oocyte remain to be extrapolated to its natural habitat. In all of this work, the lack of a clearly demonstrable function for poly(A) raises the problem that it is impossible to define whether there may be a minimum length of poly(A) necessary for its role. Thus it is possible that a reduction in the length of poly(A) below some critical level might mean that messengers apparently possessing poly(A) in fact are functionally indistinguishable from poly(A)$^-$ messengers. Conversely, some messengers apparently lacking poly(A) might in fact have a short tract long enough to sustain whatever function it has. In short, it is difficult at present to assess the significance of the presence of poly(A) in most messengers and its absence in the others.

Methyl Capping at the 5′ Terminus

The discovery that methylation occurs in eucaryotic mRNA has led to the identification of a novel type of modification to RNA, in which the 5′ terminus is blocked by the formation of a 5′-5′ pyrophosphate linkage and a cluster of methyl groups is added to this structure. This field was opened by the observation of Perry and Kelley (1974) that a methyl label enters purified L cell poly(A)$^+$ mRNA at a level about 17% of that seen with rRNA. Prior to this experiment, the high level at which methyl groups enter rRNA had obscured the label in mRNA, because the impure messenger fractions were too highly contaminated with rRNA.

The presence of methyl groups within ribosomal and transfer RNA of both procaryotes and eucaryotes has been known for many years. In eucaryotic rRNA, most of the methyl groups are found on the 2′-O position of ribose and only some 20% are attached to the bases. Since methylation occurs largely on the precursor rRNA and the methyl groups are conserved, their function may be concerned with the selection of sequences to comprise mature ribosomal RNA (see Chapter 27). With tRNA there is, of course, a large number of diverse base modifications, whose functions are presumably concerned with the various recognition reactions of the molecule (see Volume 1 of Gene Expression). In residing in a new type of structure at the messenger terminus, the function of the major mRNA methylation appears distinct from the purpose of the modifications to rRNA and tRNA. Its most likely function is to influence translation. A minor proportion of the mRNA methyl groups is found internally and may serve a different (unknown) purpose from the terminal modification. Similar methylated sequences are found in hnRNA as in mRNA, although at lower levels (see Chapter 25).

The structure of the methylated termini was first deduced from work with viral RNA. Purified vaccinia and reovirus (and other) virions contain RNA methylase activities able to transfer CH_3 groups from S-adenosyl methionine to the viral mRNA (Shatkin, 1974; Wei and Moss, 1974). This makes it possible to use labeled SAM to identify the methylated nucleotides in digests of the RNA. Upon treatment with alkali, the label is recovered largely in the form of a structure behaving on DEAE-cellulose chromatography as though containing about 5 negatively charged groups (compared with the lower charge of the mononucleotide products).

The methylated structure is located at the 5′ terminus of the RNA. When the viral RNA is synthesized under conditions preventing modification, the 5′ terminal nucleotide can be isolated after alkaline hydrolysis by its possession of more than one phosphate group. When methylation is permitted, this product disappears, to be replaced by the highly negatively charged structure. This has no free 5′ end: after treatment with alkaline phosphatase, which removes only 2 negative charges from the 3′ phosphate, the product cannot be phosphorylated by polynucleotide kinase, showing that the 5′ end must be inaccessible.

The components of the methylated structure have been defined by conventional means—degradation with nucleases of various specificities followed by characterization of the products of chromatography. In vaccinia, for example, there is always an amount of 7-methyl-guanosine which is equal to the total of the amount of 2′-O-ribose methylated adenosine and guanosine. By providing a ^{32}P label in various positions of the triphosphate precursors, it is possible to show that only ATP and GTP contribute to the structure and that it contains phosphate groups not released by treatment with penicillium nuclease and alkaline phosphatase, which usually reduces RNA to 5′ mononucleotides. However, these phosphate groups can be released by nucleotide pyrophosphatase. Such results suggest that the terminal structure consists of 7-methyl guanosine linked by a 5′-5′ pyrophosphate bridge to either 2′-O methyl G or 2′-O methyl A. The structure

$$^{\text{7-methyl}}G^{5'}ppp^{5'}G/A^{2'\text{O-methyl}}pNp$$

is resistant to alkaline cleavage because the orientation of the terminal G

Table 23.2   Sequences of methylated

Cell	Cap 0		Cap 1		
S. cerevisiae	^{7Me}G ppp Pu		absent		
D. discoideum	^{7Me}G ppp Pu	75%	^{7Me}G ppp A^{Me}p	^A_U	20%
D. melanogaster	^{7Me}G ppp X	5%	^{7Me}G ppp X^{Me}p	Y	85%
A. salina	absent		^{7Me}G ppp A^{Me}p	G	75%
			^{7Me}G ppp G^{Me}p	G	25%
S. purpuratus	absent		^{7Me}G ppp Pu^{Me}p	Y	92%
			^{7Me}G ppp Py^{Me}p	Y	8%
Animal cells	absent		^{7Me}G ppp X^{Me}p	Y	65%
(Hela, L, myeloma)			^{7Me}G ppp A^{Me_2}p	Y	20%

Cellular messenger populations usually are the poly(A)$^+$ class, but when poly(A)$^-$ mRNA has been analyzed the results have been the same. In the structures shown, X and Y indicate that all bases are found at the position, Pu and Py indicate purine and pyrimidine (but usually 80–90% of Pu is A rather than G), N^{Me} indicates 2'-O ribose methylation, and A^{Me_2} indicates methylation at 2'-O ribose and at N^6. The proportion of each type of cap is given in percent.

Cap structures were first worked out for viral messengers of vaccinia, reovirus and VSV by Abraham et al. (1975a), Furuichi et al. (1975a,b), Moss and Koczot (1976),

is reversed from usual and the 2'-O methyl group prevents cleavage at the second base position. It has no free 5' end. But both the 2'-OH and 3'-OH groups of the terminal base are free, so that it is susceptible to β elimination upon treatment with periodate oxidation and aniline; this confirms that it must be the 5' group of the terminal G that is linked to the original 5' terminus of the primary transcript, which thus occupies the second position after modification.

The modification of the 5' end therefore forms a *cap* which renders the original terminal base inaccessible. Quantitation of the methyl groups originally suggested that there is about 1 cap for every 2200 bases in HeLa mRNA. This implies that all messengers in the animal cell possess caps; the same now appears to be true for several other species, including lower eucaryotes.

Available data on the form taken by the caps in cellular messengers are summarized in Table 23.2. Three types of cap have been recognized according to the number of positions at which methylation occurs.

In yeast and in the slime mold, the most prominent type is the *cap O*, in which 7-methyl-G is added to the 5' terminus of the messenger, but no further methylation occurs. A small amount of this cap type is found also in Drosophila. It is absent from animal cells.

The type of cap identified in the viral messengers is described as *cap 1* and is methylated also at the second position. Apart from yeast, in which it is

groups in cellular messengers

Cap 2		Internal methylation	Ref.
absent		absent (<0.1/cap)	A
absent		absent (<0.2/cap)	B
^{7Me}G ppp X^{Me}p Y^{Me}	10%	not known	C
absent		~2 A^{N6Me}/cap	D
absent		~2 A^{N6Me}/cap	E
^{7Me}G ppp X^{Me}p Y^{Me}	15%	~1 A^{N6Me}/1000 bases occurs as $GA^{Me}C$ and $AA^{Me}C$	F

Moyer et al. (1975), Rhodes et al. (1974), Wei and Moss (1975). The first exception to be found was poliovirus, which lacks modification (Nomoto et al., 1976). Data above are from A—Sripati et al. (1976); B—Dottin et al. (1976); C—Levis and Penman (1978a); D—Groner, Grosfield and Littauer (1976); E—Faust et al. (1976), Surrey and Nemer (1976); F—Cory and Adams (1975), Furuichi et al. (1975c), Perry et al. (1975), Salditt-Georgieff et al. (1976), Wei et al. (1975, 1976), Wei and Moss (1977a), Schibler, Kelley, and Perry (1977).

absent, and the fruit fly, in which it is minor, this is the predominant type of cap. Most commonly the second position is occupied by adenine (corresponding to the use of ATP as the most common nucleotide to initiate transcription). But guanine also is found; and in animal cells pyrimidines too may occupy this location. Methylation always is at the 2'-O ribose position. However, when the base is adenine, a second methylation may occur in animal cells at the N^6 position to generate dimethyl ($2'$-O; N^6) adenosine. The structure of a cap methylated in this way is illustrated in Figure 23.4.

In some cases methylation occurs not only at the first and second positions, but also at the third position. This is always a 2'-O-ribose methylation and creates the *cap 2* type. This is found as a minor component in animal cells. Its presence also in Drosophila makes this the only organism at present known to have all three types of cap.

In addition to these methylations, some N^6 methyl adenine may be found within the messenger. This is a feature of animal cells and sea urchin, but not the lower eucaryotes. The frequency of internal methylation is great enough for every messenger to have at least one N^6 methyl adenine, but it is not known if this in fact is the situation.

Are the different types of caps found on different messenger sequences or do they occur as differing extents of modification on the same messengers? To demonstrate the specificity of the capping reaction, individual messengers have been examined. Available data are summarized in Table 23.3. It seems likely at present that each messenger has a specific cap structure,

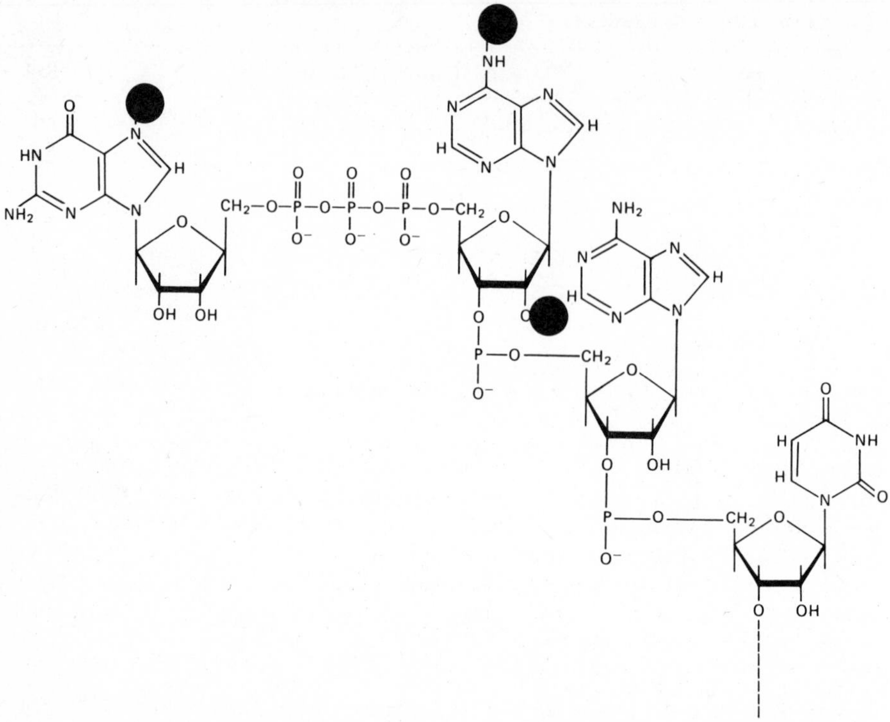

Figure 23.4. Structure of the 5′ cap of mRNA for a molecule terminated by $^{7Me}G(5')ppp(5')\cdot A^{2'OMe}_{N_6Me}pU\text{-}$.

Table 23.3 Sequences of methylated groups in individual messengers

Messenger	Cap sequence	Internal methylation	Ref.
Rabbit α globin	^{7Me}G ppp A^{Me_2}p $C^{Me?}$p	absent (<0.2/molecule)	A
Rabbit β globin	same	same	A
MPC-11 immunoglobulin H gamma chain	^{7Me}G ppp A^{Me_2}p A^{Me} p C p	probably 1 A^{N_6Me}/molecule	B
MPC-11 immunoglobulin L kappa chain	^{7Me}G ppp G^{Me}p A^{Me} p A p	perhaps 1 A^{N_6Me}/molecule	B
B. mori silk fibroin	^{7Me}G ppp A^{Me}p U^{Me} p C	absent (<0.4/molecule)	C
HeLa histone ($\geqslant 5$ mRNA species)	^{7Me}G ppp Pu^{Me} p Y p ^{7Me}G ppp Pu^{Me} p Y^{Me}p	absent (<0.3/molecule)	D

Cap sequences are indicated as in Table 23.2 except that Me? is used to indicate partial methylation. In the case of the histones, Pu^{Me} may be A^{Me}, A^{Me_2}, or G^{Me}; it is not known whether the variation reflects differences between or within messengers. In the case of immunoglobulin mRNA, doubt on internal methylation is due to possible contamination.

References: A—Cheng and Kazazian (1977), Heckle et al. (1977), Lockard and RajBhandary (1976); B—Marcu et al. (1978b); C—Yang et al. (1976); D—Moss et al. (1977), Stein et al. (1977).

although the apparent partial methylation of the third position of rabbit globin mRNA leaves the possibility that some molecules have a cap 1 and some a cap 2.

Capping occurs after the initiation of transcription and the reactions involved have been studied mostly in two systems, vaccinia virus and the HeLa cell (for review see Shatkin, 1976). In both cases cap formation occurs by the sequential series of events summarized in Figure 23.5. First a guanylyl transfer occurs in which the 5′-5′ pyrophosphate bridge is formed by the addition of the blocking residue. Then the terminal guanine is methylated at the N^7 position. The next step is methylation of the adjacent base in the

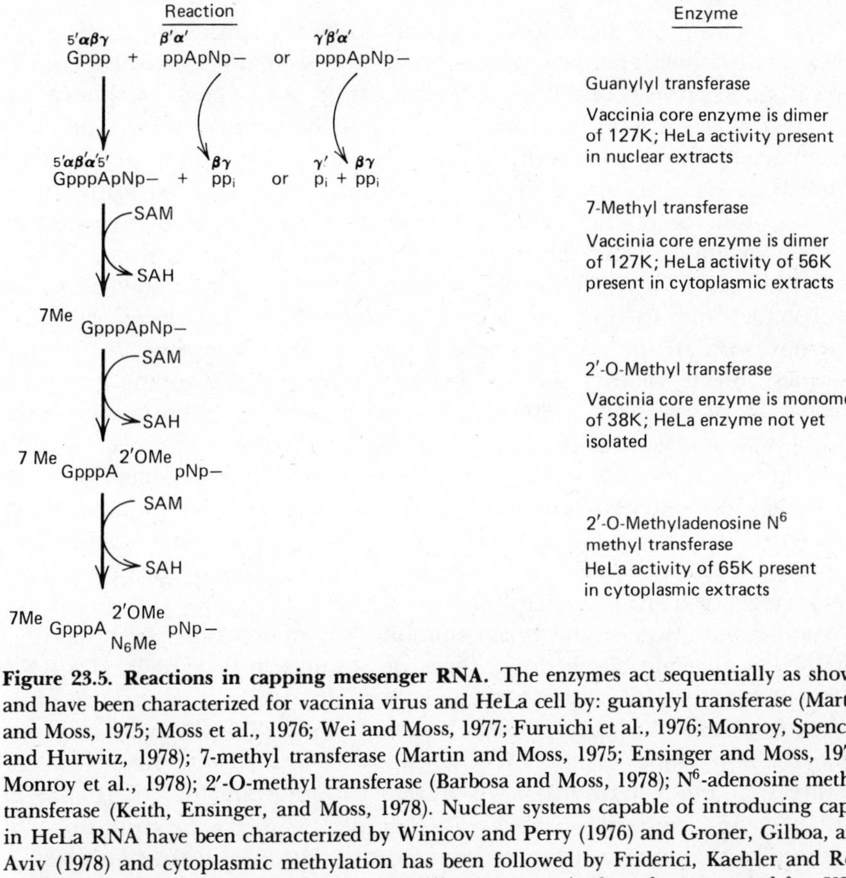

Figure 23.5. Reactions in capping messenger RNA. The enzymes act sequentially as shown and have been characterized for vaccinia virus and HeLa cell by: guanylyl transferase (Martin and Moss, 1975; Moss et al., 1976; Wei and Moss, 1977; Furuichi et al., 1976; Monroy, Spencer, and Hurwitz, 1978); 7-methyl transferase (Martin and Moss, 1975; Ensinger and Moss, 1976; Monroy et al., 1978); 2′-O-methyl transferase (Barbosa and Moss, 1978); N^6-adenosine methyl transferase (Keith, Ensinger, and Moss, 1978). Nuclear systems capable of introducing cap 1 in HeLa RNA have been characterized by Winicov and Perry (1976) and Groner, Gilboa, and Aviv (1978) and cytoplasmic methylation has been followed by Friderici, Kaehler and Rotman (1976). Alternative viral systems with different properties have been reported for: VSV, where a guanylyl transferase links the GDP moiety of GTP to a 5′ mono phosphate terminus (Abraham, Rhodes, and Bannerjee, 1975b); and cytoplasmic polyhedrosis virus, in which a 5′ pNHppA terminus prevents not only capping but also RNA synthesis (Furuichi, 1978). Enzymes that may be involved in changing the number of 5′ phosphate groups in vaccinia are nucleotide phosphohydrolase and 5′ phosphokinase (Paoletti and Moss, 1974; Spencer et al., 1978).

2'-O position on ribose.` Finally any further modifications occur to generate dimethyl adenosine or the cap 2 structure with an additional 2'-O ribose methylation.

The reaction catalyzed by guanylyl transferase links a GMP moiety provided by GTP to the terminus of the RNA. The use of ^{32}P labels in different positions of the GTP shows that the β and γ phosphates are released as pyrophosphate; only the α phosphate of the GTP enters the 5'-5' pyrophosphate bridge. There have been alternative reports on the nature of the 5' terminus on the RNA substrate. Originally it was suggested that the best substrate is a 5' diphosphate; subsequently the reaction has been reported to occur with a 5' triphosphate, in which case its terminal γ phosphate group must be released.

The substrate for the reaction appears to be the initial sequence of the RNA chain, since guanylyl transfer does not occur to individual nucleotides (that is, prior to initiation of transcription). Addition to a triphosphate terminus would allow direct modification of the primary transcript at the first base. A demand for a diphosphate terminus would require the prior removal of the γ phosphate from the first nucleotide after the initiation of transcription. In neither case, of course, would it be possible to cap the product of an internal cleavage of RNA, which would terminate in a 5' monophosphate; it would be necessary to introduce at least one additional phosphate group to generate the proper substrate. Vaccinia virus cores contain not only all the enzymes needed for capping per se, but also possess enzymes able to modify the number of 5' terminal phosphate groups; a nucleotide phosphohydrolase can remove 5' terminal phosphates while a 5' phosphokinase is able to add them.

A slightly different mechanism has been reported for the guanylyl transferase of VSV (vesicular stomatitis virus). Here a GDP moiety appears to be transferred from GTP to a 5' monophosphate terminus. This would allow capping of ends generated by internal cleavage; although no such activity has been reported in uninfected cells.

Another variation on the most common series of events has been reported for the cytoplasmic polyhedrosis virus of Bombyx mori. The viral mRNA is initiated with ATP, which is then capped. However, the use of AppNHp, whose terminal phosphate cannot be removed by phosphohydrolase, prevents the appearance of mRNA. The implication is that the presence of the analogue at the 5' terminus may prevent capping; and this deficiency in turn prevents synthesis of RNA. This would mean that capping must be involved in the initiation or immediate continuation of RNA synthesis.

In vaccinia the guanylyl transferase and 7 methyl transferase activities appear to reside in the same enzyme, probably coded by the virus. With extracts from HeLa cells, these activities appear to reside in different enzymes. In both cases, however, the guanylyl transfer must precede the methylation, which is specific for guanine nucleotide linked in the 5'-5' pyrophosphate bridge.

The further methylation events all are catalyzed independently. Vaccinia undertakes only one further reaction, the 2'-O methylation of the second nucleotide in the cap. HeLa cells may add a second methyl group at the N^6 position to 2'-O-methyl adenosine occupying this site. The enzyme that undertakes this reaction does not appear able to generate internal N^6-methyl adenine, which must therefore presumably be the product of yet another enzyme reaction. The formation of cap 1 structures appears to occur in the HeLa nucleus as judged both by their presence in nuclear RNA and the ability of nuclear extracts to undertake the reaction. Cap 2 formation may occur in the cytoplasm, but it is not known whether the enzyme responsible for adding the further 2'-O-methyl group is the same as the nuclear enzyme performing this reaction for cap 1. When viral mRNAs possess a cap 2 structure, the last methyl group must be added by cellular enzyme, in contrast with the formation of cap 1 by solely viral enzymes.

When messenger RNA is degraded, what happens to the cap? HeLa cell cytoplasmic extracts contain an enzyme activity able to hydrolyze the 5'-5' pyrophosphate linkage, probably with substrates of $^{7Me}GpppN^{Me}(Np)_{<7}$ rather than with intact RNA (Nuss and Furuichi, 1977). This would correspond to the return to the nucleotide pool of the 5' terminal fragments left by message degradation. Removal of the methyl group from the oligonucleotide prevents the pyrophosphorolysis. Caps thus appear as stable as the messenger molecule; by contrast, internal N^6-methyl adenine turns over more rapidly than mRNA (Sommer, Lavi, and Darnell, 1978).

Capping and the Initiation of Transcription

Caps appear to be present on all messenger RNA molecules that are translated in the eucaryotic cytoplasm, although they have not been found in mitochondrial mRNA. Two possible functions seem obvious: to protect the messenger against degradation from the 5' end; and to provide a feature for recognition by ribosomes. As with poly(A), however, the occurrence of the principal modification within the nucleus leaves open the possibility that the cap may be involved in selection or transport of mRNA sequences.

The role of the cap in protein synthesis has been tested by comparing the ability of messengers possessing and lacking caps to be translated in vitro. Noncapped messengers have been isolated in two ways. The terminal guanosine may be removed from the RNA by treatment with periodate and aniline to leave a triphosphate terminus. We shall describe this as "decapped" mRNA. Its use has the disadvantage that the chemical treatment might introduce damage within the molecule. An alternative that can be applied to some viral messengers is to synthesize RNA in vitro under conditions that inhibit capping. With reovirus, the inclusion of SAH results in the synthesis of molecules with a mixture of ends, 70% ppG and 30% GpppG. We shall describe this as "uncapped" RNA. This has the disadvantage that while most molecules may be uncapped, there are always some that are capped although unmethylated, even though conditions may be adjusted to render the latter proportion reasonably small.

Two types of experiment have suggested that caps are necessary for efficient translation. Decapped or uncapped messengers are translated less efficiently than capped messengers in vitro in a variety of systems. And the addition of analogues such as [7Me]GMP inhibits the translation of capped messengers. But this is not true of every messenger; there are some exceptions, comprising messengers that are not capped in vivo, and which are translated efficiently in vitro without suffering inhibition by [7Me]GMP.

The cell-free protein synthetic systems possess the ability to cap noncapped mRNA; and so in order to test the dependence of translation upon capping, SAH usually is included in the in vitro system to prevent remethylation. When the wheat germ system is operated under these conditions, decapped reovirus or VSV RNA is translated only at 10–15% of the level of activity of capped mRNA. Cellular messengers, such as reticulocyte mRNA, show a similar drop, to about 20% of the capped activity (Muthukrishnan et al., 1975a,b). In these experiments, the addition of SAM only partially restored the lost activity, suggesting that some damage might have occurred during decapping. But in other experiments comparing uncapped and capped reovirus mRNA, recapping completely restored translational activity (Both et al., 1975).

The difference in the translational capacity of capped and noncapped messengers results from the effect of the cap upon the ability of ribosomes to bind to the mRNA. Comparing the ability of wheat germ ribosome subunits to form 40S-mRNA complexes with capped and uncapped reovirus mRNA, Both et al. (1975) found that virtually no activity remained with the uncapped messengers. Decapping of reovirus RNA results in a 4-fold reduction in ability to bind wheat germ ribosomes; but when the bound messenger sequences are recovered, most prove to possess capped termini (Muthukrishnan et al., 1976). Thus the residual binding capacity results mostly but not completely (about 70%) from incomplete β elimination.

The analogue [7Me]GMP inhibits the translation of all tested capped RNAs by wheat germ extract; with increasing concentration it inhibits the binding of labeled reovirus RNA to ribosomes. However, it does not block the translation of STNV viral RNA in wheat germ extracts or EMC viral RNA in HeLa extracts (Hickey, Weber, and Baglioni, 1976; Weber et al., 1976; Shafritz et al., 1976). Both of the viral messengers naturally lack caps. This implies that messengers that are capped in vivo rely upon the cap structure for recognition by ribosomes; this recognition may therefore be inhibited by competition with [7Me]GMP (although the analogue must be present at a concentration 2–3 orders of magnitude greater than that of the mRNA). By contrast, a different recognition mechanism must exist for messengers that are not capped in vivo; this is not inhibited by analogues of the cap structure. The addition of [7Me]GMP also may fail to inhibit some or all of the translation of uncapped vaccinia RNA, which suggests that the low extent to which these messengers are recognized by ribosomes reflects the use of features independent of the cap structure (Weber et al., 1977).

Different systems may rely upon recognition of the cap structure to differing degrees. Rose and Lodish (1976) found that a reticulocyte protein synthetic system showed only a reduction of 50% in ribosome binding to mRNA upon decapping. The greater level of a reduction to 25% in overall translation was therefore taken to be due to internal damage caused in messengers during the β elimination procedure. The comparatively high level of binding of the decapped mRNA led to questions about the involvement of the cap in mRNA recognition by reticulocyte ribosomes. Comparisons of the abilities of wheat germ and reticulocyte systems to translate the same messengers indeed show that decapped or uncapped messengers are translated about 3 times relatively more efficiently by reticulocyte ribosomes (Muthukrishnan et al., 1976; Lodish and Rose, 1977). Typically the wheat germ system may translate noncapped messengers at less than 10% of the level of capped messengers, whereas reticulocyte ribosomes show up to 25% or more activity. The difference in the inhibitory effect of ^{7Me}G is more pronounced. Whether this reflects a difference prevailing in vivo or occurs because the reticulocyte system has a more lax demand for the capped structure under the conditions of in vitro operation is hard to say. However, the effect may be more apparent than real, since experimental conditions do influence the dependence on the cap. As the K^+ concentration is decreased in both systems, the relative translation of noncapped messengers increases (Weber et al., 1977; Bergmann and Lodish, 1979). The effect is more pronounced with the wheat germ system, which usually is operated at its optimum of about 100 mM K^+ compared with the 56 mM K^+ optimum of the reticulocyte.

Which features in the cap are responsible for ribosome recognition? The blocked but unmethylated GpppNp termini present in uncapped viral preparations do not seem to be recognized any better than the ppNpNp termini. Although ^{7Me}GMP is an effective inhibitory analogue, GpppG is ineffective. Terminal methylation as well as blocking to generate the 5′-′5 pyrophosphate bond therefore seems necessary. The 2′-0 ribose methylation at the second position probably is not of primary importance but does make a contribution. Its presence does not confer recognition ability upon decapped mRNA in which the terminal ^{7Me}G has been removed to leave $pppN^{Me}pN$. However, it does increase ribosome binding to analogue messengers such as $^{7Me}GpppG$-poly(U) (Both et al., 1976; Muthukrishnan et al., 1976). The ability of 40S subunits to bind to capped poly(U), poly(U,C), or poly(A,C) [though not to poly(A)] implies that the cap structure may to some extent be recognized independently of the adjacent sequence. However, 60S subunits may join to generate mRNA-80S ribosome complexes only with capped poly(A,U) or poly(A,U,G), which implies some need for specific sequences at least at this stage (see below).

The process of messenger binding by eucaryotic ribosomes is not yet sufficiently well defined to resolve questions about the relative extents to which ribosome subunits or initiation factors are responsible for recognizing the salient features on mRNA. However, by using more purified systems for protein synthesis, attempts have been made to identify initiation factors that may be involved in cap recognition. One factor that may be implicated is reticulocyte eIF-4B (previously known as IF-M3), which is involved in assisting mRNA binding to ribosomes. Its binding to capped mRNA is reduced to 10–30% of the previous level by the addition of ^{7Me}GMP (Shafritz et al., 1976). Since its ability to bind the naturally uncapped EMC RNA is unaffected, its use of caps only where naturally present mimics the overall reaction.

Another approach is to seek directly factors able to bind the ^{7Me}GpppG analogue. A protein generated by salt wash of Artemia salina ribosomes is able to do so, in contrast with the lack of reaction by reticulocyte initiation factors (Filipowicz et al., 1976). However, this protein is equally well bound by GTP, which suggests that it cannot (at least yet) be taken to identify a cap binding function.

An alternative is to identify proteins able to react with caps that have been rendered reactive by chemical modification. After periodate oxidation has converted the terminal ribose to a reactive aldehyde, an 80S initiation complex can be formed with wheat germ ribosomes. Reduction with $NaBH_3CN$ should cross link the aldehyde groups to amino groups of neighboring proteins. Four proteins have been identified by such cross linking with ribosomes; 2 are found after formation of a 40S complex (Sonnenberg and Shatkin, 1977). The same approach has been used to see whether purified initiation factors can bind to reactive caps. A single 24,000 dalton protein of reticulocyte or mouse ascites factors is cross linked; it may be a minor component or contaminant of eIF-3 (Sonnenberg et al., 1978).

Although none of these approaches yet has defined the proteins responsible for cap recognition, it seems likely that one or more of the initiation factors will prove to be necessary. At present there is no evidence to implicate proteins of the ribosome itself.

Another system in which the effect of the cap can be investigated is the Xenopus oocyte. The removal of the ^{7Me}G causes a 90% reduction in the translation of injected globin mRNA (Lockard and Lane, 1978). Reovirus mRNAs that are blocked (by GpppG) or fully capped (with ^{7Me}GpppG^{Me}) appear to be more stable in both the oocyte and cell free translation systems (Furuichi, LaFiandra, and Shatkin, 1977). In this regard, the terminal methylation has little effect, since either ^{7Me}GpppG or GpppG termini survive 2–4 times better than unblocked mRNA. The possibility that the 5′-5′ pyrophosphate bridge confers resistance to nucleases is supported also by the observation that removal of ^{7Me}G from CPV mRNA with pyrophosphatase is followed by rapid degradation of the RNA in wheat germ extract (Shimotohno et al., 1977). There is a contrary report that globin mRNA survives well in wheat germ after T4 kinase has been used to split the pyrophosphate bond (Abraham and Pihl, 1977).

Is the cap part of the ribosome-binding site at which translation is initiated? The sequences of mRNA at which stable ribosome binding occurs can be defined by recovering the fragments protected against degradation with ribonucleases. This can be done separately for either 40S subunits or complete 80S ribosomes. The 80S binding site usually comprises a shorter part of the region covered by the 40S subunit, perhaps because some conformational contraction takes place upon junction with the 60S subunit (Legon, 1976; Kozak and Shatkin, 1977a,b; Rose, 1978). The binding site always includes the AUG codon at which protein synthesis initiates. Generally the 80S ribosome protects about 30–40 bases, with the AUG codon often but not inevitably in the center. The additional sequences protected by the 40S subunit alone seem sometimes to extend further in the 5′ than in the 3′ direction.

Available data are summarized in Table 23.4, which gives the 5′ terminal sequences of messengers from the cap to the initiation codon. Where ribosome binding sites have not been characterized by experiment, it may be assumed that they represent the region immediately surrounding the AUG initiation codon. The distance from the message terminus to the initiation codon varies very widely. At present the known extremes are 10 bases and 241 bases. For many messengers the AUG codon lies close enough to the cap for both to be included in the 40S binding site; sometimes both lie within the region protected by 80S ribosomes. In other cases the distance is much too great for both to be simultaneously bound to the ribosome.

Most experiments on ribosome binding have been performed with the VSV and reovirus messengers. Here the cap is always part of the sequence bound by the 40S subunit, whose stable binding site may therefore coincide with the initial recognition region. Rebinding experiments with the reovirus RNA fragments protected by ribosomes show that ribosomal recognition is reduced as sequences are removed from either end (Both et al., 1975; Kozak and Shatkin, 1978). Full capped fragments rebind with about 90% efficiency; removal of the cap reduces the level to about 10%, but the same sequences are bound by ribosomes as with capped messengers. Removal of the 3′ terminal sequences of the protected fragments steadily reduces rebinding; and the level falls to zero as soon as the AUG codon is lost. This suggests that both the cap and the AUG triplet are essential features for formation of a ribosome-mRNA complex.

This does not answer the question of whether the cap and AUG (and any other pertinent features) are recognized simultaneously or sequentially. It is clear that 80S ribosomes may be stably bound to a sequence of RNA that does not include the cap, although the cap may have been recognized initially in 40S subunit binding. It is therefore possible that the 40S subunit first must recognize the cap, but then binds by recognizing the adjacent sequence in an action now independent of the presence of the cap. Since in almost every case there are no AUG triplets between the 5′ terminus and the initiation codon, one model is to suppose that 40S subunits initially recognize the terminus and then move along the message until halted by the first AUG sequence (for review see Kozak, 1978). The idea that ribosomes directly recognize only the 5′ terminus of the mRNA is consistent with the

Table 23.4   5′ Terminal and initiation sequences

Messenger	Nontranslated length	Sequence
α globin (rabbit)	37 bases	m m₂ GpppACACUUCUG-
α globin (human)	38	m GpppACUCUUCUG-
α globin (mouse)	33	m m₂ GpppACUUCUG-
β globin (rabbit)	54	m m₂ GpppACACUUGCUUUUGACACAACUGU-
β globin (human)	51	m GpppACAUUUGCUUCUGACACAAC-
β globin (mouse)	50	m m₂ GpppACAUUUGCUUCUGACAUAGUUG-
β globin (mouse)	50	m m₂ GpppACGUUUGCUUCUGAUUCUGUUG-
ovalbumin (chick)	62	m GpppACAUACAGCUAGAAAGCUGUAUUGCCUUUAGCAG-
insulin (rat)	57	AGCCCUAAGUGACCAGCUACAGUCGGA-
reovirus small 54	32	m m GpppGC-
reovirus small 45	28	m m GpppG-
reovirus small 46	19	
reovirus medium 52	30	m m GpppG-
reovirus medium 44	19	
reovirus medium 30	14	
VSV N	14	
VSV NS	11	
VSV L	11	
VSV M		
VSV G		
BMV coat	10	
AMV coat	37	m GpppGUUUUUA-
TMV genome	68	m GpppGUAUUUUUACAACAAUUACCAACAACAACAAACAACA-
STNV coat	29	
Ad2 fiber	202 or 383	m GpppACUC-
Ad 2 hexon	241	m GpppACUCUCUUCCGCA-
SV40 16S	239	m m₂ GpppAUUUCAGGCC*AUGG*..46b..*CAUGG*..27b..**GCCUUU**..17b..**GCUUUUGC**UGCAA-

hypothetical complement to 18S rRNA
3′ terminal sequence of 18S rRNA

Sequences have been determined directly from mRNA except for globin, ovalbumin, SV40 16S, and Ad 2 hexon whose cDNA reverse transcripts have been sequenced. The AUG initiation codon is indicated in italic, as are any other AUG sequences. Sequences of at least 4 bases that are complementary to the 3′ end of 18S rRNA are indicated in bold type; these include possible G-U pairs. Ribosome binding sites are indicated by the boxes: sequences protected by 80S ribosomes are shaded and the additional 5′ sequences protected by 40S subunits alone (where determined) are indicated by outline. Sequences beyond the initiation codon are given only where ribosome protection sites have been identified. The sequences are continuous and are broken here only to fit on the page.

Sequence	Ref.
GUCCAGU**CCG**ACUGAGAAGGAACCACC*AUG*	A
GUCCCCACAGACUCAGAGAGAACCC*ACCAUG*	
AUUCUGACAGACUCAGGAAGAAACC*AUG*	
GUUUACUUGCAA**UCCCCC**AAAACAGACAGA *AUG*GUGCAUCUGUCCAGUGAGGAGAAGUC	
UGUGUUCACUAGCAACCUCAAACAGACACC*AUG*	
UGUUGACUCACAACCCC*A*GAAACAGACAUC*AUG*	
UGUUGACUUGCAACCUCAGAAACAGACAUC*AUG*	
UCAAGCUCGAAAGACAACUCAGAGUUCACC*AUG*	B
AA**CCAUCAGCA**AGCAGGUCAUUGUUCCAAC*AUG*	C
UAUUUUGCCU**CUUCCC**AGACGUUGUC**GGCA***AUG*GAGGUGUGCUUGC	D
CUAAAGUCACG**CCUG**UCGUCGUCACU*AUG*GCUUCCUCACUCAG	
m m GpppGCUAU**UCG**CUGGUCAGUU*AUG*GCUCGCUGCGCGU	
CUAAU**CUG**CUGACCGUU*A*CUCUGCAAAG*AUG*GGGAACGCUCU	
m m GpppGCUAAAGU**GACCGUG**GUC*AUG*GCUUCAUUCAAG	
m m GpppGC**UA**UUCGCGGUC*AUG*GCUUACAUCGCAG	
m m₂ GpppAA**CA**GUAAUCAAA*AUG*UCUGUUACAGUCAAG	E
m m₇ GpppAA**CA**GAUAUC*AUG*GAUAAUCUCACAAAAG	
m m₇ GpppAA**CA**GCAAUC*AUG*GAAGUCCACGAUUUG	
m GpppG........UUA**UCCC**AAUCCAUUCAUC*AUG*AGUUCCUUAAAGAAG	
m GpppG.............UUU**CCUUG**ACACU*AUG*GAAGUGCCUUUUGUACUUAG	
m GpppGUAUUAAUA*AUG*UCGACUUCAGGAACUGGUAAGAUGACUCGCGCGCAGCGUCG	F
UUUUUAAUUUU**CUUUC**AAAU*A*CUUCC*AUC**AUG*	G
AACAACAUUACAAUUACUAUUUACAAUUACA*AUG*	H
ppAGUAAAGACAGGAAACUUUACUGACUAAC*AUG*	I
UCUUCCGCAUCG.......184 or 365..........AG*AUG*	J
UCG....213 bases.....CC**GCUUUCC**AAG*AUG*	
UUUUGU..90b..**ACUUCUG**CUCUAAAAGCUU*AUG*AAG*AUG*	K

AUCCUUCCGCA
3′HO-AUUCCUCCACUAGGAAGGCGUCC......5′

References: A—Legon (1976), Baralle (1977a,b), Chang et al. (1977), Efstratiadis et al. (1977), Baralle and Brownlee (1978); B—Mc-Reynolds et al. (1978); C—Lomedico et al. (1979); D—Kozak (1977), Kozak and Shatkin (1977a,b); E—Rose (1978); F—Dasgupta et al. (1975); G—Koper-Zwarthoff et al. (1975); H—Richards et al. (1978); I—Leung et al. (1979); J—Akusjarvi and Pettersson (1979), Zain et al. (1979); K—Ghosh et al. (1978).

monocistronic nature of most eucaryotic messengers. It is supported by observations in several systems that only the first coding sequence is translated in a viral multicistronic messenger; the subsequent coding sequences appear to be translated only after a cleavage has generated a smaller molecule in which they lie immediately subsequent to the 5′ terminus. This suggests that initiation may be unable to occur internally.

There are some possible exceptions to this, but the only clear case at present is the 16S mRNA of SV40 which codes for the major capsid protein VP1. Here it seems clear that there are AUG codons a considerable distance before the initiation sequence. It would be interesting to determine directly for messengers such as this or the Ad 2 hexon or fiber, where a great distance separates the cap from the AUG initiation codon, whether the cap is needed for binding. If it is necessary, then either the 5′ terminus must constitute an initial recognition site from which the ribosome moves to a stable binding site at the initiation sequence; or the cap must be brought into juxtaposition with the initiation sequence by the formation of secondary structure.

What other features may be part of the stable binding site? In E. coli, the ribosome binding sites of mRNA include a short sequence complementary to part of the 3′ terminal sequence of the 16S rRNA (Shine and Dalgarno, 1974). The rRNA sequence is conserved in B. stearothermophilus (Sprague et al., 1977). The idea that base pairing occurs between the mRNA and rRNA is supported by the isolation of base-paired rRNA-mRNA fragments (Steitz and Jakes, 1975). An analogous proposal has been made for eucaryotic systems by Hagenbuchle et al. (1978) following the demonstration of extensive conservation of the 3′ terminal sequence in eucaryotic 18S rRNA. In mouse, wheat, and B. mori, the 3′ terminus is the same apart from a change in the terminal base; it is related to the E. coli sequence by a substitution at 2 positions and by the deletion of the 5 base sequence thought to be involved in pairing with bacterial messengers. The sequence in the lower eucaryotes yeast and D. discoideum is intermediate between the bacterial and higher eucaryotic sequences, which are

mammalian

$A^{Me_2}A^{Me_2}$C C *U G C G G A A G G A U*C A U U A–OH

bacterial

$A^{Me_2}A^{Me_2}$C C U G C G G U U G G A U C A *C C U C C* U U A–OH

The italicized sequences indicate the regions proposed as the complements to sequences in the mRNA initiation regions.

Table 23.4 shows all possible complements in regions between the 5′ terminus and the initiation codon for each message. Most messengers have a prospective binding sequence, although at somewhat variable distances from the initiation codon. The strength of the potential interaction varies somewhat, depending on the length of the complementary sequence and the number of G-U pairs that must be included to achieve pairing. An interesting feature of several mRNAs has been noticed by Lomedico et al.

(1979). Within about 10 bp of the AUG initiation codon, a sequence of up to 12 bp is complementary to a sequence lying roughly 20 bp closer to the 5' end. If these complements base pair to form a helical structure, the intervening 20 base pairs will be extruded as a single stranded loop. The potential complement to 18S rRNA always lies within this loop.

Some viral messengers entirely lack any possible complement, however; and so if pairing with the 16S rRNA terminus is involved in initiating translation, it cannot be obligatory. Like the utilization of the cap, there must be an alternative. Whether the mechanism is used at all, of course, remains to be established. If this mechanism is used, however, it should be available to bacterial ribosomes since they possess the appropriate complement (although it is not used as such in bacteria). Thus bacterial ribosomes should be able to translate eucaryotic messengers (presumably ignoring the cap structure). But eucaryotic ribosomes lack the bacterial complement and so should be unable to function with bacterial or phage messengers. Apart from the potential pairing sequence, there are no common sequences in the 5' terminal regions of the eucaryotic mRNAs that might serve as ribosome recognition features.

Two procaryotic messengers have been translated in vitro by a wheat germ system and activity depends upon the addition of a cap; only the first (5') coding sequence of a polycistronic messenger is translated (Paterson and Rosenberg, 1979; Rosenberg and Paterson, 1979). Neither messenger has an adequate complement to eucaryotic 18S rRNA in the short region between the 5' terminus and the AUG codon, which would argue that the cap and the initiation codon may be sufficient for ribosome binding.

Characterization of Individual Messengers

The isolation of individual messenger RNA species until recently was possible only for proteins synthesized in large quantity by some particular cell type. Relying upon the presence of large amounts of a single messenger in the polysomes, it was possible to obtain somewhat purified preparations from which the mRNA could be characterized. The paradigm for this approach is globin mRNA, which comprises a 9–11S fraction representing some 2% of the total RNA of the avian or mammalian reticulocyte in which globin is by far the predominant product of protein synthesis (for review see Chantrenne, Burny, and Marbaix, 1967). Another approach, which can be applied to messenger RNAs present in reasonably large but not such overwhelming amounts, is to precipitate polysomes engaged in synthesis of some protein by reaction with antibody against the protein. From these polysomes it is then possible to isolate the mRNA, an approach first used successfully with immunoglobulin mRNA in myeloma cells and albumin in rat liver (Schechter, 1974; Taylor and Schimke, 1974). The more recent development of cloning technology since has made it possible to isolate any mRNA by virtue of its reaction with a cloned sequence (see Chapter 26).

The ability to segregate any number of sequences by the cloning of plasmids containing a random representation of complements to any mRNA population means that a probe containing any sequence can be obtained in large amounts. This can then be used to purify specifically the corresponding messenger by hybridization. However, it remains true that it is necessary to be able to apply some selective technique to isolate the plasmid carrying the desired sequence from among the others. Essentially this requires some prior purification of the mRNA. Combinations of these techniques may be productive. For example, it is possible to use immunoprecipitation of polysomes to obtain a preparation of mRNA sufficiently purified to allow cloning techniques then to be applied to obtain complete purity (Strair, Yap, and Shafritz, 1977). The essential criterion for messenger purity has been the synthesis of the appropriate protein as the sole product in heterologous systems for in vitro translation. The development of techniques for rapid sequencing of DNA now has made it possible to determine the nucleotide sequence of DNA reverse transcribed from the mRNA; and, of course, the correspondence of coding sequence with amino acid sequence then may be shown directly (see Chapter 26).

The characterization of globin mRNA provided the first convincing evidence for the existence of eucaryotic mRNA. When reticulocyte polysomes are dissolved in solution containing 0.5% SDS and centrifuged through a sucrose gradient, peaks of RNA are found at 23S and 16S (the rRNAs), 4S (tRNA), and 9–11S. The precise sedimentation coefficient of the last peak depends upon both the species of reticulocyte and method of isolation. The equation of this peak with messenger RNA was indicated by its susceptibility to ribonuclease; when polysomes are treated with RNAase before extraction of RNA, only the 23S, 16S, and 4S peaks remain—there is no 9S RNA. Dissociation of the polysomes with EDTA releases the 9S fraction in the form of a ribonucleoprotein complex (see Table 23.1). Quantitation shows that there is one molecule of 9S RNA in each polysome (Evans and Lingrel, 1969a,b).

The 9S RNA fraction contains messengers for both α and β globin; both proteins are synthesized when 9S RNA is provided as template for in vitro systems (Laycock and Hunt, 1969; Lockard and Lingrel, 1969). The two types of polysome can be distinguished by their response to certain treatments and by a slight difference in size; β-globin often is produced in greater amounts than α-globin, apparently due to a somewhat greater ability to bind ribosomes exhibited by β-globin mRNA (Lodish and Jacobsen, 1972; Temple and Housman, 1972). The cell-free systems show no specificity for species of either ribosomes or initiation factors and can translate the globin message in the form of either free mRNA or as mRNP (Matthews, 1972; Sampson et al., 1972; Schreier and Staehelin, 1973).

Separating the two globin messengers proved difficult for some time. Electrophoresis on acrylamide gels generates two bands under denaturing conditions, contrasted with the single band found in the absence of reagents such as formamide. The sizes of the messengers can be estimated from their electrophoretic mobilities. The range determined for rabbit reticulocyte globin mRNAs by this and other techniques is given in Table 23.5. The sizes are rather similar, but β-globin mRNA appears to be slightly larger than α-globin mRNA. Thus the failure of the two mRNAs to separate under nondenaturing conditions implies that their mobilities then do not reflect simply the molecular size, but must be influenced by secondary structure. This is generally true of mRNA and means that reasonable size estimates cannot be obtained unless the messenger is maintained in a denatured state lacking secondary structure. Another cause of the diffuse nature of the bands is a size variation in each mRNA created by the presence of poly(A) tails of varying lengths. This can be overcome by removing the poly(A); then it becomes possible to isolate two separate and sharp bands, the α-globin mRNA displaying a greater mobility (smaller size) than the β-globin mRNA (Maniatis et al., 1976).

The properties of globin mRNA are typical of the properties of the many other messengers since isolated. Available data are summarized in Table 23.5. The imprecision of the size estimates for mRNA is indicated by the range of values reported for each messenger. Usually this is due to the differences between techniques, which include electrophoresis, centrifugation, electron microscopy, and complexity determined by hybridization. Probably this is due to the difference in extent of secondary structure and its resistance to denaturation in each case.

Even within the limitations of these data, some general features of mRNA are apparent. The messenger length appears always to be greater than would be needed just to code for the protein [after allowing for the presence of a possibly variable length of poly(A)]. Yet the length excess is not great enough to code for a second protein. Eucaryotic cellular messengers therefore appear to be monocistronic. Even the polypeptide components of multimeric proteins, such as hemoglobin, are translated on different messengers.

Each messenger can be divided into a coding sequence and the additional, nontranslated regions. Table 23.6 shows that for those messengers whose complete sequences are known, nontranslated regions are present at both the 5′ and 3′ termini. This conclusion appears to be generally true of mRNA, since in several other cases partial sequences have been obtained that identify appreciable lengths of noncoding material at one or the other terminus. In globin and ovalbumin mRNAs, the length of the 5′ nontranslated region is somewhat less than that of the 3′ nontranslated region. Part of the 5′ region may be involved in ribosome building (see Table 23.4); but apart from the putative ribosome binding features, there are no sequences common to the termini even of α- and β-globin mRNAs. The role of the 3′ nontranslated region is unknown. Part might perhaps be involved in protein binding in the formation of mRNP particles. In the globin and ovalbumin mRNAs, and in an immunoglobulin mRNA, the hexanucleotide AAUAAA occurs about 20 bases before the 3′ terminus (Proudfoot and Brownlee, 1976). It is possible that this might be a signal connected with some 3′ terminal function, although the only such function known at present is the addition

Table 23.5 Lengths of isolated messengers estimated by biophysical techniques

Cell type	Protein	Coding length	Apparent messenger length	Poly(A) length	Proportion base paired	Ref.
Rabbit reticulocyte	α-globin	429	550–630	40–50	55–60%	A
Rabbit reticulocyte	β-globin	444	550–710	40–50	—	A
Chick oviduct	ovalbumin	1164	1600–2640	45–90	41%	B
Chick oviduct	(pre)ovomucoid	627	850–900	+	—	C
Chick oviduct	conalbumin	~1800	~2750	+	—	C
Chick oviduct	lysozyme	441	625–660	+	—	C
Rat liver	(pre)albumin	1925	1800–2700	+	—	D
Rat liver	α-fetoprotein	~1800	~2700	+	—	E
Rabbit uterus	(pre)uteroglobin	270	~600	+	—	F
Dog pancreas	(pre)amylase	1575	~2100	+	—	G
Bovine parathyroid	(pre)parathyroid hormone	345	600–775	60	—	H
Chick liver	vitellogenin	~6300	~6950	+	—	I
X. laevis liver	vitellogenin	~5500	5500–6600	+	—	J
Chick calvaria	α-procollagen	~4000	5150–5450	140	49%	K
Chick lens	δ-crystallin	1260	~2000	+	—	L
Calf lens	B2 crystallin	525	~750	50	—	L
Calf lens	α-A2 crystallin	522	~1460	50–200	—	L
Rat lens	α-A^{Ins} crystallin	588	~1460	+	—	L
B. mori silk gland	fibroin	~14,000	~16,100	100	—	M
Mouse myeloma MOPC41	light Ig (κ)	~660	1200–1300	200	—	N
Mouse myeloma MPC11	light Ig (κ)	~730	~1200	+	—	O
Mouse myeloma RPC20	light Ig (λ)	~650	~1150	+	—	P
Mouse myeloma MPC11/31C	heavy Ig (γ)	~1450	1800/2200	+	—	O
Mouse myeloma H2020/J558	heavy Ig (α)	~1450	1800/2200	+	—	O
Mouse myeloma MOPC3741/104E	heavy Ig (μ)	~1800	2150/2800	+	—	O
Maize kernal	zein (22K)	~600	~1120	+	—	Q
Maize kernal	zein (19K)	~520	~950	+	—	Q
S. purpuratus	histone H4	310	~400	0	—	R
S. purpuratus	histine H2B	375	~520	0	—	R
S. purpuratus	histone H2A	387	~470	0	—	R
S. purpuratus	histone H3	405	~520	0	—	R
S. purpuratus	histone H1	621	~670	0	—	R
Trout testis	protamines	96–99	215–270	0–60	—	S
Trout testis	protamines	96–99	235–290	0–60	—	S
Trout testis	protamines	96–99	250–310	0–60	—	S
Trout testis	protamines	96–99	275–330	0–60	—	S

Coding lengths are derived from the amino acid sequence of the protein where available (or the protein precursor, indicated by "pre," where appropriate); coding lengths derived from overall protein size are shown as approximate. Messenger lengths are based upon sucrose gradient centrifugation, acrylamide gel electrophoresis (usually under denaturing conditions, probably the most accurate method), and occasionally electron microscopy. Where several techniques have been used, the range of estimated values is given. Where a single value is given, this does not indicate greater accuracy, but usually the application of only a single technique. Poly(A) lengths are given when determined for the steady-state population; the presence of poly(A) of undetermined length is indicated by (+) and usually this will be <100 bases. The absence of poly(A) is indicated by (0). The proportion of bases involved in formation of secondary structure is given where determined; a dash indicates lack of experimental data.

References: A—Labrie (1969), Gaskill and Kabat (1971), Hunt (1973), Hamlyn and Gould (1975), Holder and Lingrel (1975); B—Haines et al. (1974), Shapiro and Schimke (1975), Woo et al. (1975), Van et al. (1977); C—Groner et al. (1977), Buell et al. (1978); D—Taylor and Tse (1976), Strair et al. (1977); E—Innis and Miller (1977); F—Atger and Milgrom (1977); G—MacDonald et al. (1977); H—Stolarsky and Kemper (1978); I—Deeley et al. (1977); J—Shapiro and Baker (1977); K—Boedtker et al. (1976), Brentani et al. (1977); L—Berns et al. (1974), Zelenka and Piatigorsky (1974), Cohen et al. (1976, 1978), Chen and Spector (1977); M—Lizardi et al. (1975); N—Brownlee et al. (1973), Honjo et al. (1974), Stavnezer et al. (1974); O—Marcu et al. (1978a), Faust, Heim and Moore (1979); P—Honjo et al. (1976); Q—Weinand and Fox (1978); R—Kedes et al. (1975), Sures et al. (1978); S—Iatrou and Dixon (1977), Gedamu et al. (1977).

Table 23.6 Regions of individual messengers from complete sequences

Messenger	Total length	5' Nontranslated region	Coding region	3' Nontranslated region	Ref.
α-globin (rabbit)	551	36	429	86	A
α-globin (human)	575	37	429	109	B
β-globin (rabbit)	589	53	444	92	C
β-globin (human)	628	50	446	132	D
Ovalbumin (chick)	1859	64	1164	631	E
Insulin (rat)	443	57	333	53	F

Lengths are based upon the sequences of the cDNA reverse transcripts determined for the entire distance of each messenger. The posttranscriptional modifications of cap and poly(A) are not included. The coding region includes both the initiation and termination codons.

References: A—Proudfoot et al. (1977), Baralle and Brownlee (1978); Heindell et al. (1978); B—Baralle (1977b), Chang et al. (1977), Proudfoot et al. (1977), Wilson et al. (1977); C—Baralle (1977a), Efstratiadis et al. (1977), Proudfoot (1977); D—Baralle (1977b), Chang et al. (1977), Marotta et al. (1977); E—McReynolds et al. (1978); F—Cordell et al. (1979), Lomedico et al. (1979).

of poly(A). It is noticeable, however, that the evolutionary conservation of sequences extends to the nontranslated as well as the coding sequences in the α- or β-globin mRNAs of different mammals. We shall discuss this, together with the structure of the intervening sequences, in Chapter 26.

Data on individual messenger RNAs at present concern species that are abundant in the cell. Are these typical of the entire cellular message population? A comparison of the lengths of the mRNA population of HeLa cells with the distribution of protein sizes suggests that most messengers are indeed longer than their coding regions (see Davidson and Britten, 1973). The average size of mRNA is about 2200 bases, corresponding to a length of 2000–2100 bases excluding poly(A). The median length for the mRNA population is about 1400 bases, that is, 1200–1300 excluding poly(A).

Products of Messenger Translation

Isolated messenger RNA may be characterized by translation in two types of system. Reconstituted cell-free systems, comprising ribosomes, protein synthetic factors, and tRNAs, can be obtained from various sources. The most common are wheat germ, rabbit reticulocyte, and mouse ascites, but many others have been used (Roberts and Paterson, 1973; Pelham and Jackson, 1976). The degree of purity—and hence the extent of background due to translation of endogenous messengers—depends on the system. Usually the cell-free systems work relatively inefficiently; each messenger is translated only a few times before activity ceases, generally within 90–120 minutes. The Xenopus oocyte provides an alternative in which the in vivo condition of the protein synthetic apparatus is assured. The system is more efficient and can undertake translation of injected messengers up to the point at which it is saturated. Usually translation continues for 24–48 hours, during which there is repeated initiation and completion of protein synthesis. First characterized for globin mRNA, the oocyte system since has been used with many other messengers; and more recently it has been possible to follow coupled transcription and translation of injected DNA (Lane, Marbaix, and Gurdon, 1971; Berns et al., 1972; Marbaix and Lane, 1972; De Robertis and Mertz, 1977).

In neither type of system does there appear to be any tissue or species specificity. Messengers from a wide variety of eucaryotic systems have been successfully translated; the coding functions of many of the messengers summarized in Table 23.5 have been confirmed in this way. This in itself demonstrates the general lack of translational controls in the form of initiation factors able to discriminate between different messengers. Probably the protein synthetic apparatus functions just to translate the templates provided for it, apparently with little species dependence in the mechanism of initiation. Of course, this does not preclude translational control at earlier stages; indeed, one test for the presence of nontranslated mRNA in a cell is to show that mRNA templates for a given protein may be translated in vitro or in oocyte, even though they are inactive in the cell itself. This is prima facie evidence for seeking features that may be involved in preventing the mRNA from reaching the ribosomes in vivo.

Another important use of the heterologous translation systems is to identify precursors that may be unstable in vivo but may accumulate due to lack of processing in vitro. The oocyte test system is not suitable for this purpose, since in at least some cases it is able to process precursor proteins; examples include vitellogenin coded by Xenopus liver mRNA and the polyproteins coded by RNA viruses (Berridge and Lane, 1976; Laskey, Gurdon, and Crawford, 1972; Ghysdael et al., 1977). The source of the protease(s) responsible remains to be established; but there is the possibility that at least some processing reactions are common to Xenopus oocytes and other tissues and species. On the other hand, the oocyte has been used to characterize the precursor to an insect protein toxin, which is not processed (Kindas-Mugge et al., 1974).

With cell-free protein synthetic systems, messengers coding for hormones or secreted proteins may be translated into products larger than any found in vivo. Several examples for which such "pre" forms have been found are mentioned in Table 23.5. (Often these are precursors to "pro-proteins" that are stable precursors to the mature form.) The additional molecular weight in each pre-protein is accounted for by an extra N terminal sequence, ranging in various examples from 16 to 29 amino acids (Burstein and Schechter, 1977; Strauss et al., 1977). No conserved amino acid sequence is apparent in the additional regions of the immunoglobulins and other secreted proteins whose precursors have been characterized, although a common feature is the presence of a high proportion of hydrophobic amino acids.

The function of the additional material is concerned with the secretion of the protein through the cell membrane. A *signal hypothesis* linking the synthesis of secreted proteins to the secretory process has been proposed by Blobel and Dobberstein (1975a). The initial impetus for the model was the observation that the mRNA for a light chain immunoglobulin of a myeloma cell is synthesized exclusively by ribosomes bound to membranes. This contrasts with the synthesis of nonsecreted proteins on free polysomes.

The membrane bound polysomes synthesize only the mature form of the protein. However, upon detachment from the membrane, the polysomes generate two products: one corresponds to the mature immunoglobulin; but the other is a larger protein. Under conditions when the detached ribosomes complete protein synthesis but cannot initiate new chains, the first products are of mature size; and only the later products are of larger size. This suggests that an additional sequence is present only on nascent polypeptides attached to ribosomes at the beginning of the message (which are the last to complete translation). On the ribosomes that have progressed further before detachment, the precursor has already been processed, so that the first ribosomes to complete translation release only the mature protein. Probably about 80 amino acids must be incorporated into the nascent protein chain before removal of the additional material occurs.

The signal hypothesis postulates that the additional N terminal material constitutes a *signal sequence*, whose presence distinguishes secretory proteins from other proteins. The signal sequence is recognized by membrane receptors, perhaps by virtue of its hydrophobic content. This serves to insert the precursor protein into the membrane, probably as soon as the signal sequence and perhaps a few additional amino acids have been incorporated into the growing chain. Soon after, however, when the chain is only a little longer, the signal sequence is cleaved by a protease of the membrane. Presumably the protein is now properly orientated in the membrane; and once released from the ribosomes upon completion, can pass through the membrane. Since the precursor processing is completed while the protein chain is still nascent, the only product observed in vivo is the secreted form of the protein.

The full form of the model takes the role of the signal sequence further in proposing that it is the appearance of the N terminal region of the nascent polypeptide chain that is responsible for attaching the ribosome to the membrane. This is to argue that neither mRNA nor the ribosome itself attaches to the membrane. This implies that translation of all messengers is initiated in the same way, by binding of free ribosomes. Only subsequently are those ribosomes engaged in synthesis of secretory proteins bound to the membrane via the signal sequence. The evidence for this aspect of the hypothesis, however, is indirect, and it is not proven whether the ability of ribosomes themselves to bind to stripped microsomal membranes is indeed irrelevant. The ribosome-membrane binding interaction has been characterized by Borgese et al. (1974), and the system has been taken further by Blobel and Dobberstein (1975b) in experiments showing that reconstituted systems are able to segregate newly synthesized immunoglobulin into a protease-resistant form (presumably membrane bound). This activity is associated with processing and does not happen when globin mRNA is translated in the same system.

In summary, it is clear that the signal sequence is a feature of many secreted proteins that is necessary for their membrane transport. The function of the sequence is exercised while the protein is in a nascent condition. Points that remain to be established concern the nature of the interaction between ribosome and membrane, especially the question of whether the signal sequence alone is indeed responsible for polysomal attachment. Be this as it may, however, the process of membrane transport clearly involves transfer of proteins from the ribosomes to the membrane. Analogous models have been proposed for the secretion of bacterial proteins (Smith et al., 1977, 1978; Randall and Hardy, 1977). An exception is provided by the example of ovalbumin, which is translated as a primary sequence identical with the sequence secreted in chick oviduct; no N terminal region corresponding to a signal sequence can be identified (Gagnon et al., 1978; Palmiter et al., 1978).

Another type of transport is involved in the entry of proteins synthesized in the cytoplasm into organelles. Components that are transferred to mitochondria or chloroplasts are synthesized and released from free cytoplasmic ribosomes in a precursor form. The additional material may be removed during passage of the completed protein through the membrane into the organelle in which the mature form is found (see Chapter 21). Thus although recognition and cleavage of a special sequence is involved, the reaction is independent of protein synthesis.

What is the specificity of the signals for secretion? When messengers are injected into oocytes, the location of the newly synthesized protein may depend on its nature. Zehavi-Willner and Lane (1977) found that globin mRNA directs the synthesis of protein that is found in the supernatant fraction of a homogenate, corresponding to the free cytoplasm. But liver mRNA directs the synthesis of albumin, which is found in a membrane fraction consisting largely of smooth vesicles. The proteins synthesized by guinea pig mammary gland mRNA are resistant to proteases unless detergent is applied, which suggests that they may be secreted in vesicles. This implies that there is neither tissue nor species specificity in the signals for secretion that are recognized by the oocyte. The fate of the heterologous proteins may follow the division of locations found for newly synthesized oocyte proteins. Whether the segregation involves cleavage of the secretory signal is not known.

Stability of Messenger RNA

The level of messenger RNA in the cytoplasm depends upon both the rate of production and transport from the nucleus and the rate of degradation. Changes in the production of mRNA clearly are implicated in the control of gene expression. To what extent are changes in the stability of messengers important in regulating translation?

Messenger stability may be assessed by two types of criterion: physical and functional. Direct measurements of mRNA [usually of poly(A)$^+$ mRNA], or the use of probes representing mRNA populations or individual messengers, essentially allows the nucleotide mass of the mRNA(s) to be measured. However, this does not reveal whether the molecules remain functional or may have been inactivated by small changes not detected in the assays. Assay of the translation of mRNA determines the amount of functional message. For this to correspond to the state of the mRNA, the translation must be performed in an in vitro system; since when translation is followed in vivo, failure to synthesize protein might be due to causes other than degradation. A corollory is that translation in vitro does not reveal the physical condition of material that fails to be translated. Ideally the two approaches should be considered complementary, since together such data allow the stability and translatability of mRNA to be fully defined.

Measurements of messenger lifetimes have been made in two general situations. The rate of decay of the cellular message population has been followed in various systems; for technical reasons this approach is restricted largely to cells in culture. Usually no single value can be given for the stability of mRNA; but the decay curve can be resolved into components each characterized by a particular half-life. These range from the order of an hour to a day. The rate of decay of individual messengers has been followed in certain specialized cells, usually the end products of differentiation devoted principally to synthesis of one or only a few proteins. Usually their messengers resemble the most stable cellular components, with half lives of several hours or even a couple of days.

Early experiments on the stability of mRNA made use of protocols in which actinomycin was added to block synthesis of new messengers, so that the decay of preexisting mRNA could be followed. Rather short lives, in general of about 3 hours, were inferred from the rate of decline of protein synthesis. A critical assumption upon which this technique relies is that actinomycin inhibits only the synthesis of messenger and not its translation. However, Singer and Penman (1972) showed that addition of actinomycin inhibits the attachment of ribosomes to the preexisting messengers, so that the ability of cells to synthesize proteins declines due to reduced translation as well as because of degradation.

When the fate of an ^{3}H-uridine pulse label is followed in mRNA over several days, the longer lived components of a cellular population may be distinguished (Singer and Penman, 1973). With HeLa cells the decline of radioactivity in mRNA does not follow simple exponential kinetics, but displays a more complex curve which can be fitted to the exponential decline of two components. These now appear to have half lives of about 4 hours and 24 hours. (A somewhat shorter lifetime for the less stable component has been assigned from experiments with very short pulse labels by Puckett and Darnell, 1975.) About 60% of the mRNA is found in the more rapidly decaying component of HeLa cells identified by pulse labels. When cells are labeled to reach steady state, over a period of about 6 days, the result is different, of course, with the less stable component forming a smaller proportion of the population upon subsequent chase (about 30%). Two components also are found in cultured insect cells, with half-lives of about 1 hour and 20 hours (comparable to the generation time) (Spradling, Hui, and Penman, 1975). Further details are given in Table 25.2.

Another technique is to follow the approach to steady-state labeling as ^{3}H-uridine is incorporated. This identifies longer lived components and shows that the half-life is some hours in L cells (Greenberg, 1972; Perry and Kelley, 1973). In cells growing at 37°C with a doubling time of about 15 hours, the mRNA displays a half-life of about 10 hours, corresponding to an average messenger lifetime of about 15 hours. Growth at 30°C slows the cell doubling time to 41 hours, and the mean mRNA lifetime is increased to about 42 hours. Decay is stochastic: this implies that all mRNAs, old and new, have an equal chance of decaying at any given moment. Any given molecule has a 63% chance of decaying within a period equal to the mean lifetime. The coincidence of the messenger lifetime with cell generation time may be fortuitous; for example, it might indicate that nucleases responsible for degradation respond to temperature in a manner similar to overall cell growth rate.

The presence of messenger components with very different stabilities implies that within the cell the relative expression of genes is not influenced only by their production of mRNA. In the next chapter, we shall consider the question of whether any function can be discerned for the different decay components; there is evidence that rapid decay is associated with messengers present at lower levels per cell. What mechanism may be responsible for generating the different components is unknown. One possibility is that degradation involves discrimination between different stability classes. Another is that factors such as messenger length may be important—longer messengers may be more susceptible to nucleases because they provide larger targets.

Although the messenger decay curves can be resolved into discrete components, this does not mean, of course, that any particular mRNA must be characterized by one of these values. An example is provided by adenovirus-infected rat cells, in which particular viral mRNAs turn over more rapidly than any component detectable in the bulk population (Wilson et al., 1978). The range of individual decay rates may therefore be somewhat wider than indicated by the resolution of just a short-lived and a long-lived component for cellular mRNA.

In erythroid cells, in the form of either mouse spleen cells or Friend cells, the mRNA population falls into a major (70%) class with a half-life of about 36 hours and a minor class decaying with a half-life of 3–6 hours. Globin mRNA constitutes about 10% of the population and has a half-life of about 17 hours (Aviv et al., 1976). In reticulocytes, the end product of erythroid differentiation, globin mRNA may account for more than 90% of the mRNA; but has the same half-life of 17 hours (Bastos, Vollack, and Aviv, 1977). If this is constant, the accumulation of globin mRNA to such a high proportion of the cellular message population can be accounted for only if there is a decrease in the stability of the long-lived messenger component (Bastos and Aviv, 1977a). On the other hand, in Friend cells the induction of globin is succeeded initially by a rapid accumulation of globin mRNA, then by a modest decline. For the first 48 hours after induction, globin mRNA displays a half-life of 50 hours; this declines to 17 hours for the second 48 hours, corresponding to the value previously measured (Lowenhaupt and Lingrel, 1978). The stability of other mRNA classes does not change during induction. These results suggest the possibility that differentiation of erythroid cells is accompanied by changes in the stability of mRNA. The specificity of these changes and the extent to which they are responsible for the maturation of blood cells is not yet fully established.

Relatively stable mRNAs coding for the proteins that are the predominant products of synthesis in specialized cells have been reported in several instances, including the BR2 gene of C. tentans salivary glands, fibroin in the silk gland of Bombyx mori, and crystallins of lens. A system in which synthesis of a specialized product may be switched off is provided by chick oviduct, in which ovalbumin mRNA is synthesized upon induction with estrogen, but ceases to be produced when the hormone is withdrawn. During induction, ovalbumin mRNA has a half-life of about 24 hours. When assayed by translation of mRNA in a cell-free system, hormone withdrawal is followed by a rapid decline in the survival of ovalbumin messenger. This does not follow first-order kinetics, but accelerates in rate with time, reaching a value corresponding to a half-life of about 3 hours (Palmiter and Carey, 1974). This implies that the hormone control of ovalbumin production involves not only control of mRNA synthesis but also control of stability. Another situation in which enzyme synthesis ceases upon hormone withdrawal is provided by response of hepatoma cells to steroids. The enzymes tyrosine and alanine aminotransferase both are induced by steroids; but upon removal of steroid, AAT synthesis in vivo declines with a half-life of 13 hours, while TAT synthesis declines with a 2 hour half-life (Stiles, Lee, and Kenney, 1976). Whether this represents degradation or the operation of some other mechanism rendering the mRNA unavailable to ribosomes is not clear in lieu of studies on mRNA levels or translatability in vitro. All the same, it is interesting that different decay rates for enzyme production are seen.

CHAPTER 24

Complexity of mRNA Populations

As the final intermediate between structural gene and protein, the messenger RNA present in the polysomes defines the spectrum of genes that is expressed in any cell. In this chapter we shall be concerned with analyzing the nature and number of structural genes through the constitution of mRNA. By utilizing the ability of the RNA to hybridize with the sequences in DNA which it represents, it is possible to approach some of the issues that are central to the definition of eucaryotic gene expression. Thus we may ask how many copies of each gene are present in the haploid genome? How many genes are expressed in each cell type? What overlap is there between the sets of genes expressed in different cells?

As soon as the renaturation reaction between complementary strands of DNA was characterized, it was realized that denatured DNA should be able to anneal also with complementary sequences of RNA. The nature of the product of this reaction, a hybrid with one strand of DNA and one of RNA, gave rise to the description of the process as *hybridization*. It is necessary, however, to modify the protocol for renaturation of DNA in order to follow the reaction of cellular DNA with its transcripts, because the hybridizing RNA must compete for the complementary DNA sequence with the original DNA partner strand. Thus a way must be found to allow hybridization to proceed to completion in the presence of renaturation.

The first attempts to accomplish this relied on an approach in which the denatured single strands of DNA are immobilized so that they are unable to renature, although they remain free to react with any added nucleic acid (single strand DNA or RNA). The first immobilization technique relied upon incorporating denatured DNA into a gel of agar; this can be used in the form of a column to which complementary sequences bind specifically (Bolton and McCarthy, 1962). This then was superseded by the more convenient method of immobilizing DNA on membrane filters, which could be incubated with solutions of single-strand nucleic acids to test the presence of complementary sequences able to adsorb (Nygaard and Hall, 1963, 1964; Gillespie and Spiegelman, 1965; Denhardt, 1966). We should note also that the protocol of in situ hybridization, described in Chapter 19, in which

DNA is denatured within the cell, can be considered an immobilization technique.

It was realized only subsequently, with the characterization of different repetitive sequence classes of DNA, that these techniques permit hybridization only with repeated DNA sequences. The rate of reaction is much reduced when one reacting strand is immobilized; the use of agar or filters in effect allows the reaction to proceed only to rather low Cot values in terms of the reaction in solution. Thus the reaction terminates before reaching hybridization with nonrepetitive DNA sequences. This precludes its general application to the reaction of messenger RNA transcribed from structural genes present in one or a few copies.

A more recent approach is to perform hybridization reactions with a great excess of one component. The excess must be sufficient to ensure that this component drives the reaction to completion with the minor component. Then any competition between renaturation and hybridization becomes irrelevant. Two protocols were originally developed to follow the reaction between mRNA and cellular DNA. The common feature in these and other such techniques is that the minor component bears a radioactive label and the parameter that is followed depends on the entry of the label into hybrid form.

If a small amount of labeled RNA (or DNA) is included in a standard renaturation reaction, it behaves as a *tracer* which does not influence the course of the reaction. Because the cellular DNA is present in great excess, the amount withdrawn into RNA-DNA hybrid does not change the concentration of DNA sequences, which is governed by the renaturation of strands of DNA. This is described as a "DNA-driven" reaction. The tracer RNA (or DNA) participates in the reaction as though it were part of the genome component(s) that it represents. At completion all of the tracer should be hybridized. The parameter followed is the *rate* of reaction; this is a kinetic technique. Thus the tracer is characterized by the Cot values at which it enters duplex form, which correspond to the appropriate degrees of repetition of the corresponding cellular DNA sequences. The $Cot_{1/2}$ of each hybridizing component gives the number of copies in the haploid genome representing each messenger in the component. The application of this technique shows that most, though not all, messengers are derived from genes residing in nonrepetitive DNA.

In saturation analysis, a great excess of RNA is hybridized with a small amount of labeled DNA. This constitutes an "RNA-driven" reaction in which the excess of the driver ensures that any complementary sequences in the DNA hybridize to RNA and do not have the opportunity to renature with their DNA complements. RNA-driven reactions are controlled by the product of RNA concentration and time, the *Rot*, which is strictly analogous to the Cot by which DNA-driven reactions are characterized. Provided that the reaction is followed to a great enough Rot value, a plateau of hybridization is reached at which all DNA sequences complementary to RNA have reacted. Only a small proportion of the RNA is hybridized. The parameter that is followed is the *extent* of reaction. Thus the proportion of DNA that is hybridized to the RNA should correspond to the template sequences from which the RNA is derived. This reaction can be used only to characterize the nonrepetitive genes. Transcripts of repetitive sequences, like the DNA from which they are derived, are able to hybridize with DNA sequences related to the original template as well as with the exact complement. This may increase the saturation value many fold, making it impossible to estimate the number of sequences in the genome from which the mRNA was derived. To estimate the number of genes represented in a population of transcripts, it is therefore necessary to follow the saturation of purified nonrepetitive DNA. Invariably only a small proportion is hybridized at saturation with the mRNA of any given cell type or tissue.

A kinetic technique also has been developed to assay the complexity of a messenger population and to distinguish different classes within it. This relies upon preparing a labeled cDNA by reverse transcription of mRNA. When the cDNA is hybridized with an excess of the mRNA, the rate of reaction depends upon the mRNA complexity. The reaction for an individual cDNA/mRNA reaction, such as for globin message, can be characterized by the $Rot_{1/2}$ value. The reaction for a population will have a $Rot_{1/2}$ value that is higher in proportion with its greater complexity. With most cellular populations, the reaction occurs over several orders of magnitude of Rot; and as with renaturation reactions displaying such kinetics, it is possible to resolve several components. Each component may be characterized by a $Rot_{1/2}$, which gives its complexity when compared with the $Rot_{1/2}$ of an individual control of known complexity. If the total amount of mRNA in the cell is known, the total mass of each component can be calculated; from this it is possible to deduce the number of copies of each messenger per cell. Such analysis shows that the cellular message population consists of a large number of sequences present in rather a small number of copies per cell, with a much smaller number of sequences represented in many more copies per cell. This technique is applicable only to RNA for which a reverse transcript can be obtained, which in effect limits its application to the $poly(A)^+$ population.

These techniques allow the number of genes expressed in a given cell to be estimated. Three further types of protocol have been developed to determine what differences there may be between the messenger populations of different cells or tissues.

The simplest technique is to perform a double saturation. A labeled DNA is hybridized with an excess of RNA derived from two tissues; the saturation level is compared with that obtained with the RNA of either tissue alone. Any reduction below the sum of the two separate saturation levels must be due to sequences that are present in both RNA preparations; any increase in the double saturation above the single level of either tissue identifies sequences present only in the RNA of the other tissue. The disadvantage of this technique is that it assays only the total difference between tissues. Because it is performed in excess of RNA, sequences that are present in one tissue at low concentration but in the other tissue at high concentration will simply appear as common transcripts.

Another technique is to perform an exhaustion analysis, in which labeled DNA is reacted with excess RNA to obtain one preparation depleted of sequences represented in the RNA (this is the nonhybridized DNA that remains single stranded) and a second preparation corresponding to the transcribed sequences (this is retrieved by denaturing the hybrids). These may be separately tested for their abilities to hybridize with RNA from a second source. This is more sensitive since the exhaustion procedures can be carried to specified Rot values, allowing some distinction to be made between sequences represented abundantly or scarcely in RNA.

Competition techniques directly measure the homology between two RNA preparations. These were originally developed for use with membrane filters and follow a protocol in which bound denatured DNA is tested for its ability to hybridize with one, labeled RNA preparation in the presence of increasing amounts of a second, unlabeled preparation. A plateau is reached when the second preparation is at high enough concentration to prevent reaction of any of its sequences that also are represented in the labeled preparation. By reversing which tissue provides the labeled RNA, it is possible to assess how many sequences in each tissue also are found in the other. The technique has been used successfully with procaryotic systems (see Volume 3 of Gene Expression).

The general conclusion suggested by experiments to determine homology is that the overlap between RNA sequences of different cells may be considerable. In several situations, including the viral transformation of cells, transition from resting to growing state, and differentiation of a precursor cell type into a specialized cell, many of the RNA sequences may be common to both cell situations or types. Although there are changes in the sequences expressed and their frequency of representation in RNA, the proportion of change is small enough to raise the possibility that an appreciable proportion of the total number of genes may in fact be expressed in any one cell.

Reiteration Frequencies of Structural Genes

By defining the genome sequence components that are represented in mRNA populations, it is possible to characterize the general nature of the structural genes that are expressed in a cell. When individual messenger RNAs are available, it is possible to determine how many copies of the corresponding gene may be present in the haploid genome. The technique of including a small amount of labeled tracer mRNA or cDNA derived from it in a reaction driven by excess cellular DNA has been applied to both situations. In addition a technique has been developed for characterizing individual genes in which the cDNA is present in moderate excess over the cellular DNA, so that the number of complements in the genome is calculated from the extent of competition between cDNA and cellular DNA sequences.

Tracer reactions generally follow the kinetics that we have described in Chapter 18; generally equation **17** is used, subject to the correction for short tracer lengths that is given in equation **19**. The potential of this approach for characterizing the sequences represented in mRNA was first recognized by Greenberg and Perry (1971). A complication in using an RNA tracer is that there has been some doubt about the relative rate of RNA-DNA hybridization compared with the DNA-DNA reaction. This has been variously reported to vary from almost identical to a 3–4 fold retardation with RNA (Melli et al., 1971; Hutton and Wetmur, 1973a; Davidson et al., 1975). In a comparison of RNA-driven and DNA-driven reactions, there appeared to be very little ($< 25\%$) retardation when RNA is in excess, but a substantial (4-fold) retardation when DNA is in excess (Galau et al., 1977a,b). The reason for this is obscure and, of course, it is not the behavior predicted of a random collision process. The consequence, however, is that RNA tracers may hybridize more slowly than would the same sequences in the form of a DNA tracer. The observed $Cot_{1/2}$ values for hybridization may therefore be too high; in other words a slightly repetitive sequence transcript ($f < 4$) might appear to be part of the nonrepetitive DNA fraction unless compensation is introduced for the retardation. Because the effect is empirical, the basis for correction is not sure, and the $Cot_{1/2}$ values determined for RNA tracers generally are regarded as having an uncertainty of a factor of about 2. One way to avoid this difficulty is to use a tracer of labeled cDNA prepared by reverse transcription of the mRNA, which also solves the problem of introducing the radioactive label into the tracer. A minor disadvantage in this approach is that the cDNA usually is not full length and represents only the $3'$ terminal regions of the mRNA; it is therefore necessary to assume that this is typical of the entire messenger. A more serious assumption is that the cDNA population evenly represents the mRNA population; this will not be true if some sequences are reverse transcribed more or less readily than others.

A typical curve for the hybridization of mammalian tracer mRNA with DNA is shown in Figure 24.1. About 15% of the input mRNA forms a component hybridizing with a $Cot_{1/2}$ of about 10; a further 25% of the input RNA hybridizes over a higher Cot range, from 10^2–10^4. The reaction terminates at this point, with only 40% of the input RNA hybridized. Since the second transition does not reach a plateau, an accurate $Cot_{1/2}$ value cannot be assigned, but the apparent midpoint of the experimental data is close to that

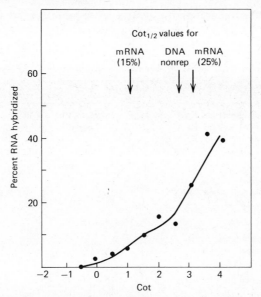

Figure 24.1. Hybridization of rat myoblast mRNA with excess DNA. The $Cot_{1/2}$ values observed for the first and second RNA transitions and for the control of nonrepetitive DNA are indicated by the arrows. The second RNA transition is only approximate since the reaction has not proceeded to completion. Data of Spradling et al. (1974).

displayed by nonrepetitive DNA. Thus the second transition may be taken to represent RNA sequences transcribed from unique DNA; by comparison the first transition represents the transcripts of sequences repeated 10^2-10^3 times in the genome.

Table 24.1 summarizes data from this and other systems. The source of the mRNA is usually polysomal or cytoplasmic poly(A)$^+$ material; in one case where both have been used, the same result is obtained. The comparison of HeLa poly(A)$^+$ and poly(A)$^-$ RNA sequences discussed in Chapter 23 suggests that their overlap is substantial enough to make measurement of the poly(A)$^+$ RNA complexity an accurate estimate of total messenger complexity (see Kaufman et al., 1977). Also those experiments performed with total messenger released from polysomes give results similar to those obtained with the poly(A)$^+$ RNA population.

Two quite different interpretations have been offered for the problem caused by the termination of mRNA hybridization at such low input levels, as typified by the different conclusions arrived at by Spradling et al. (1974) and Klein et al. (1974) from similar data on HeLa mRNA. The question is the nature of the unhybridized RNA. One view is that all sequence components hybridize only partially, so that the unhybridized material represents all classes equally well. In this case, the true content of repetitive transcripts

Table 24.1   Reiteration frequencies of sequences represented in mRNA populations

Cell	Source of mRNA	Nonrepetitive reaction (%)	Repetitive %	Repetitive Frequency	Total Reaction RNA	Total Reaction DNA	DNA Excess	Ref.
HeLa	polysomal poly(A)$^+$ RNA	25	15	~10^3	40	90	>10^5	A
	polysomal poly(A)$^+$ RNA	25	6	~10^3	31	81	5 × 10^3	B
	polysomal poly(A)$^+$ RNA	29	7	~10^3	36	81	5 × 10^4	B
	cDNA (500 b)	70	10	~10^3	80	90	>10^5	C
Rat myoblast	polysomal poly(A)$^+$ RNA	25	15	~10^3	40	90	>10^5	A
	polysomal poly(A)$^+$ RNA cytoplasmic poly(A)$^+$ RNA	32	18	~10^3	50	85	4 × 10^4	D
	cDNA (500 b)	60	10	~10^2	70	85	10^6	D
Mouse Friend	cDNA (300 b)	65	8	~10^2	73	80		E
S. purpuratus	polysomal puromycin	38	<3	—	38	95	10^4	F
	polysomal puromycin	45	<3	—	45	95	10^6	F
	polysomal puromycin	65	<3	—	65	95	5 × 10^4	G
Aedes culture	polysomal poly(A)$^+$ RNA	45	8	~10^2	53	85	>10^5	A
D. discoideum	cytoplasmic poly(A)$^+$	63	10	~10^2	73	95		H

All experiments were performed by hybridization of labeled mRNA or cDNA derived from it in the presence of an excess of cellular DNA. The source of the mRNA indicates whether the preparation was obtained from isolated polysomes or unfractionated cytoplasm, and whether it was purified by poly(A) content or by puromycin release. Preparations of cDNA represent only the 3′ ends of mRNA populations for the average distances indicated in bases. Transcripts of nonrepetitive DNA hybridize at the high Cot values characteristic of unique sequences; transcripts of repetitive DNA sequences hybridize at lower Cot values, the reiteration frequency of the DNA being indicated very roughly by the $Cot_{1/2}$ and given in the table as an order of magnitude. The extent of reaction is given as the percent of total RNA for both nonrepetitive and repetitive transcripts; the total extent of reaction of the RNA is compared with that of the (control) total DNA reaction. The excess of DNA is indicated on a mass basis; it is much less in terms of reacting sequences.

References: A—Spradling et al. (1974); B—Klein et al. (1974); C—Bishop et al. (1974); D—Campo and Bishop (1974); E—Birnie et al. (1974); F—Goldberg et al. (1973); G—Davidson et al. (1975); H—Lodish et al. (1973).

is given by dividing the proportion of the repetitive component by the proportion of total hybridizing RNA.* The other view is that all the repetitive sequence transcripts have hybridized, but that a large proportion of the nonrepetitive transcripts fails to hybridize. Then the proportion of repetitive sequence transcripts is given by the plateau of the first transition; all the remaining material is assumed to represent nonrepetitive sequence transcripts. The idea that most of the repetitive sequence transcripts have indeed reacted is supported by the experiments with cDNA tracers, which

*In the data of Figure 24.1 this would be 15%/40%, leaving 62.5% in the nonrepetitive transcript fraction. In support of this view, certainly it is true that reactions with RNA often terminate before those with DNA; but usually the discrepancy is not so great, say 10% or so.

proceed to higher plateau levels of 60–70%, but in which only the hybridization of nonrepetitive transcripts increases. The most likely situation is therefore that the values given in Table 24.1 underestimate the proportion of repetitive sequence transcripts only by perhaps 10%, due to incomplete reaction; while the remaining sequences are transcripts of nonrepetitive DNA. The most obvious reason for the failure of one RNA component to hybridize in these reactions would be if the excess of DNA were insufficient. However, this is likely to account for the failure to react of only 20% or so.* Thus it must be admitted that the basis for the failure of most sequences to hybridize is obscure, leaving the interpretation problematical rather than definite.

If the conclusion is correct, more than 80% of the transcripts of mammalian cells are derived from nonrepetitive sequences of the genome. The proportion may be greater in other eucaryotes. In particular, in sea urchin the tracer experiment identifies no repetitive component, setting an upper limit for it of 3%. In an experiment in which the DNA hybridized by the mRNA was retrieved and characterized, it proved to be entirely nonrepetitive (Galau et al., 1974).

The experiments summarized in Table 24.1 identify the total amount of material in the RNA population that is derived from the repetitive component of the genome. But this does not reveal its organization vis à vis the nonrepetitive sequence transcripts. Do the two types of sequence comprise separate sets of RNA molecules or are the repetitive elements represented in the form of sequences covalently attached to the nonrepetitive transcripts?

In mammalian cells it is possible to fractionate the mRNA population into transcripts derived solely from repetitive or nonrepetitive DNA. Indications that the two types of sequence transcript are not linked are provided by experiments in which treatments of the RNA do not alter the pattern of hybridization. Thus the amount of RNA retained in the HeLa repetitive

---

*A sufficient excess is considered to be 10^2 to 10^3 copies of a sequence in DNA for every copy in RNA. Thus if the total complexity of the mRNA population is a minimum of 1% of the genome complexity, an excess mass ratio of DNA:RNA of 10^4 to 10^5 is adequate (this is 10^2 to $10^3 \div 0.01$). However, this assumes that all sequences are equally abundant in the mRNA population. To anticipate the results of Table 24.4, in many cases a small proportion of the population (~20% of the mass) comprises a small number of sequences represented perhaps 10^3 times more frequently than the majority of sequences. The concentration of these sequences in DNA may only be equal to, or in just slight excess over, the concentration in RNA, reducing the hybridization. If insufficient excess is the cause of any failure to hybridize, an increase in the excess ratio should increase the hybridization of the relevant component. Spradling et al. (1974) found no increase in hybridization over an order of magnitude increase in the excess ratio. Goldberg et al. (1973) and Klein et al. (1974) found the increases summarized for HeLa and sea urchin in Table 24.1. In both cases these affected the nonrepetitive and not the repetitive component, suggesting that available repetitive sequence transcripts all have hybridized. Yet increases in the excess ratio do not produce major changes in the hybridization plateau, so that insufficient excess seems likely to account for the failure to react of only the small proportion of the mRNA sequences present in great abundance.

mRNA-DNA hybrids is not reduced by treatment with ribonuclease, implying the absence of any covalently attached single strand tails of nonrepetitive sequence transcripts (Klein et al., 1974). When poly(A)$^+$ mRNA of rat
myoblasts is treated with alkali to introduce breaks, neither the 3′ ends
(identified by binding to oligo-dT) nor the 5′ ends (which fail to bind to
oligo-dT) show any change in the proportion of repetitive sequences (Campo
and Bishop, 1974). Thus repetitive sequences are not clustered at either end
of mRNA molecules that are otherwise derived from nonrepetitive sequences.

The rat myoblast messengers have been separated into repetitive and nonrepetitive sequence transcripts. A preparation of unlabeled repetitive DNA
was denatured and linked to Sepharose. About 13–15% of a labeled mRNA
population binds to the DNA preparation. Figure 24.2 shows that when
recovered and rehybridized with fresh (total) DNA, the bound fraction shows
a single transition with a $Cot_{1/2}$ of about 10. The unbound fraction of
mRNA which did not bind to the DNA-Sepharose also displays a single
transition when hybridized in excess DNA, but at the high $Cot_{1/2}$ of about
800 typical of nonrepetitive DNA. Thus the repetitive sequence transcripts
form a distinct set of mRNA molecules, which must represent genes present
in multiple copies or at least genes for which related sequences are present
elsewhere in the genome. The nonrepetitive sequence transcripts represent
genes that are unique or at least are present in very few copies in the haploid genome.

An alternative model for mRNA, which postulates a structure in which
most of the molecule is derived from nonrepetitive DNA, but a short sequence at the 5′ end represents a repetitive sequence, has been proposed in
two systems. Crippa et al. (1973) reported that repetitive "tags" of 50–60
nucleotides are present at the 5′ ends of Xenopus neurula messengers, whose
remaining sequences are entirely nonrepetitive in origin. This conclusion
was based on the retention on DNA-membrane filters of a greater proportion of the RNA than could be accounted for by the repetitive component
characterized in solution (a protocol analogous to the interspersion experiments described in Chapter 18). However, the messenger RNA was extracted
by phenol from polysomes with no other purification; there may therefore
have been contamination with other RNA and the conclusion cannot be
accepted in lieu of experiments with better purified preparations.

In Dictyostelium about 12% of the mRNA sequences are derived from
repetitive DNA. Yet about 50% of the molecules are retained on hydroxyapatite after hybridization to low Cot, compared with the 90% retention at
high Cot (Firtel and Lodish, 1973). This suggests that about 60% (i.e., 50%/
90%) of the messengers may be derived largely from nonrepetitive DNA, but
contain short regions corresponding to repetitive genome sequences. In
spite of a speculation that these might be located principally at the 5′ end,
alkaline cleavage generates 5′ and 3′ halves of molecules which possess the
same repetitive content as the parental molecules. This contrasts with the
preferential location of repetitive sequence elements in the 5′ regions of

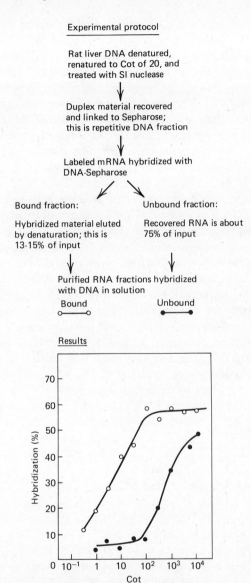

Figure 24.2. **Fractionation of rat myoblast poly(A)$^+$ mRNA into repetitive and nonrepetitive sequence transcripts.** Data of Campo and Bishop (1974).

the nuclear precursors, which are only about 20% longer than the cytoplasmic messengers. One cloned sequence of D. discoideum DNA, however, carries a complement to an mRNA which consists of a repetitive sequence at the 5′ end, the remainder being nonrepetitive. The repetitive sequence hybridizes with mRNA of a wide range of sizes, representing about 1% of the total mRNA population (Kimmel and Firtel, 1979; Kindle and Firtel, 1979).

A similar reaction is seen with another cloned sequence, so possession of a common repetitive element may define families of mRNAs whose remaining sequences are unique. Data to validate the suggestion that these repetitive sequences may always lie at the 5′ end remain to be gained.

Most mammalian, and probably other eucaryotic, messengers are derived from nonrepetitive DNA. The presence of repetitive tags in the mammalian systems and in sea urchin can be excluded by the stringency of the hybridization conditions that were used. This is all the more noteworthy since in DNA most of the nonrepetitive sequences are interspersed with repetitive sequences as described in Chapter 18. These include the transcribed sequences, which can be isolated by retrieving the DNA hybridized by mRNA.

The nonrepetitive sequences adjacent to repetitive sequences of DNA were isolated by Davidson et al. (1975) by a protocol in which fragments of average length 1800 nucleotides were renatured to low Cot. Those containing a duplex region (of minimum length 50 base pairs) were retained on hydroxyapatite. The unbound fraction represents nonrepetitive sequences; the bound molecules, which take the form of four tailed duplexes, include those nonrepetitive sequences that lie within 1750 bases of the repetitive sequence responsible for duplex formation. After performing a second cycle to isolate all these sequences, the bound fraction was sheared to a length of 450 nucleotides and renatured to low Cot to remove the repetitive sequences. Repeating this step a second time sees elution from hydroxyapatite of the "repeat-contiguous" fraction of DNA, that is, the nonrepetitive sequences that reside close to repetitive sequences.

The proportion of labeled mRNA that is hybridized by an excess of the repeat-contiguous DNA is 50%; this compares with the 65% hybridized by total nonrepetitive DNA (the third sea urchin experiment shown in Table 24.1). Thus 50/65 = 77% of the hybridizing mRNA is represented in the repeat-contiguous sequences. This is close to the level that would be observed if all mRNA sequences were adjacent to a repetitive sequence at one end.

The repeat contiguous DNA fraction largely represents about one third of the sequence complexity of the sea urchin genome. Yet the total proportion of the genome that appears to be represented in mRNA is only about 4% (see below). Thus although most or all mRNAs are derived from nonrepetitive sequences adjacent to repetitive sequences, only a small part of the repeat contiguous regions is involved in transcription. While certainly it is possible that some of the short-period interspersion of repetitive and nonrepetitive sequences found in eucaryotic genomes may reflect an alternation of control regions with message regions, such an organization does not seem likely to represent the major part of the interspersion. Also we should note that since mRNA initially is transcribed in the form of larger precursors, the role of sequences adjacent to the mRNA sequences may be exercised during processing of mRNA rather than in transcription (see Chapter 25).

Identification of Individual Genes

Knowledge of the number of copies of the gene representing a given protein naturally is a prerequisite for analyzing the control of gene expression. Two general techniques now are available for determining the number of sequences in DNA corresponding to a particular mRNA, which must be available in purified form or as a purified cDNA reverse transcript. The first approach was to use hybridization techniques to assay the number of DNA complements for a probe by either saturation or kinetics. More recently restriction enzyme analysis has been used to determine how many genome fragments can be obtained that correspond to a given probe; when there is more than one copy of a gene, each surrounded by different sequences, each generates a unique fragment.*

Here we shall discuss the results that have been obtained with various hybridization techniques; and in Chapter 26 we shall consider the detailed definition of individual gene maps. In most cases most of the present data have been obtained with the same systems, representing messengers that are available in large amounts as the predominant products of specialized cells. But as we have noted previously, the use of cloning technology in principle allows the extension of these approaches to any mRNA.

Individual gene frequencies may be measured by the same techniques applied to mRNA populations. Saturation analysis measures the amount of DNA in the genome complementary to an mRNA or cDNA probe; division by the size of the probe gives the number of copies. The problem with the general application of this approach in its direct form is that the proportion of cellular DNA corresponding to a single gene is so small that it is very difficult to measure accurately. Kinetic analysis may be used in the same way that we have described for mRNA populations: a tracer is included in the DNA renaturation reaction and its $Cot_{1/2}$ of hybridization identifies the repetition frequency.

Table 24.2 summarizes available data on hybridization experiments performed with isolated messengers or their cDNA reverse transcripts. The approach of kinetic analysis is perhaps the most generally used. In preparations of duck or mouse globin mRNA, containing both α and β sequences, the $Cot_{1/2}$ of reaction of a cDNA probe with excess DNA is close to that of nonrepetitive DNA; the same result is found with ovalbumin mRNA and a crystallin mRNA. The genes coding for these messengers thus either are unique or, given the uncertainty of the reactions, at least are repeated no more than 2-3 times. (This does not exclude the presence of further, more distantly related sequences, with which hybridization would be much slower.)

*This is true for nonrepeated genes; restriction analysis provides a sensitive assay for determining whether more than one copy is present when a gene appears to be unique by hybridization analysis. For the tandemly repeated members of a multiple gene family, however, the complexity of the restriction pattern in principle depends on the variation between individual copies; often one pattern of identical repeats is predominant, although the number may be hard to quantitate. Here hybridization analysis may be a better guide to total gene number.

Table 24.2　Reiteration frequencies of genes for isolated messengers

Cell	Messenger	Reaction components	Reaction parameter	Gene number	Ref.
Duck erythroblast	$\alpha + \beta$	excess cDNA × DNA	analyt. complexity = 633 physical complexity = 249	2–3 for $\alpha + \beta$	A
Duck erythroblast	$\alpha + \beta$	cDNA × excess DNA	$Cot_{1/2}$ = 200 nonrepetitive $Cot_{1/2}$ = 350	1–2 α 1–2 β	B
Mouse reticulocyte	$\alpha + \beta$	cDNA × excess DNA	$Cot_{1/2}$ = 800 nonrepetitive $Cot_{1/2}$ = 800	~1 α ~1 β	C
Human reticulocyte	α globin	excess cDNA × DNA	analyt. complexity = 1480 kinetic complexity = 700	~2	D
Human reticulocyte	β globin	excess cDNA × DNA	analyt. complexity = 390 kinetic complexity = 280	~1½ (1β, 1δ?)	E
Human reticulocyte	γ globin	excess cDNA × DNA	analyt. complexity = 440 kinetic complexity = 290	~1½	F
Chick oviduct	ovalbumin	cDNA × excess DNA	$Cot_{1/2}$ = 480 nonrepetitive $Cot_{1/2}$ = 660	~1	G
Chick embryonic lens	δ crystallin	cDNA × excess DNA	$Cot_{1/2}$ = 1740 nonrepetitive $Cot_{1/2}$ = 1135	≤1	H
B. mori silk gland	fibroin	excess mRNA × DNA	saturation at 0.0017% or 0.0022% DNA	<2–3	I
Chick erythrocyte	histone H5	mRNA × excess DNA	$Cot_{1/2}$ = 110 nonrepetitive $Cot_{1/2}$ = 1300	~12	J
Human (HeLa)	histones H2A, H2B, H3, H4	mRNA × excess DNA	$Cot_{1/2}$ = 180 rRNA $Cot_{1/2}$ = 40 (f = 180)	~40	K
Sea urchin (P. miliaris)	histones	mRNA × excess DNA	$Cot_{1/2}$ = 2.5 E. coli $Cot_{1/2}$ = 15.9	~1200	L

References: A—Bishop and Freeman (1973); B—Bishop and Rosbash (1973); C—Harrison et al. (1974); D—Tolstoshev et al. (1977); E—Old et al. (1976), Ottolenghi et al. (1975); F—Old et al. (1976); G—Harris et al. 1973); H—Zelenka and Piatigorsky (1976); I—Suzuki et al. (1972), Gage and Manning (1976); J—Scott and Wells (1976); K—Wilson and Melli (1977); L—Weinberg et al. (1972).

In one case in which saturation analysis has been successful, the number of genes is clearly very low, probably unity for silk fibroin.

With histone mRNAs the same approach identifies a degree of repetition that varies with the species. With human and sea urchin there is some uncertainty caused by the use of a control other than the cellular nonrepetitive DNA reaction; but it is clear that there are up to 100 copies of each histone gene in the HeLa (estimated) haploid genome, with rather more in sea urchin. We should note that, as with the mRNA population studies, the $Cot_{1/2}$ of the tracer reaction gives the average repetition frequency of all messengers in the probe.

The most detailed data are available with the globin system. Here there is the advantage that genetic analysis of human globins is possible with the large number of known variants. These data suggest that for the major adult chains there may be two copies of the α globin gene and a single copy of the β globin gene. For other genes, there is probably a single copy of the δ gene and may be two copies of the γ gene. The nucleotide sequences of the multiple genes need not necessarily be identical (see Chapter 26). Some of the genes resemble each other well enough to show some cross hybridization, which increases the apparent gene copy number seen by hybridization.

Early experiments in which globin gene frequencies were measured by saturation analysis did not prove very satisfactory, giving rather high results that appear to be spurious. The first kinetic experiments identified somewhat lower values, roughly 2-3 copies of each gene, and these since have been refined further, especially with the introduction of a hybridization technique developed by Bishop and Freeman (1973). A moderate (roughly tenfold) excess of cDNA is hybridized with cellular DNA and the proportion of cDNA that enters duplex form is measured. Because its excess is only moderate, the cDNA must compete with the natural DNA strand for its complement. Presumably when the reaction is complete the complement will be completely bound to cDNA and its natural partner in proportion to their relative concentrations. Thus

$$\frac{D_0^c}{n_i} = \frac{xD_0^c}{n_c - xD_0^c} \tag{1}$$

where D_0^c is the amount of input cDNA, n is the amount of either the identical or complementary sequence in the natural DNA, indicated as n_i and n_c to explain the derivation of the equation, and x is the proportion of the cDNA that has become duplex at completion.

Since n_i and n_c are equal, equation 1 may be reduced to:

$$n = \frac{xD_0^c}{1 - x} \tag{2}$$

The amount of natural complementary sequence per unit of input cellular DNA is n_c/D_0, where D_0 is the total amount of cellular DNA. The total length of the natural complement is then given by setting n_c as equal to n of equation 2 to give:

$$l' = \frac{Cn_c}{D_0} = \frac{CxD_0^c}{(1 - x)D_0} \tag{3}$$

The parameter C is the complexity of the cellular DNA: if this is provided in daltons, then l' is given in the same units. Since the cDNA is a reverse transcript of the mRNA, the natural identical and complementary sequences, n_i and n_c, represent the nontranscribed and transcribed strands, respectively.

All the values in equation 3 are measured in the moderate cDNA excess $\times$ DNA reaction. The length of the natural complement (transcribed sequence) given by l' is denoted the *analytical complexity*. This is the total length of material in the genome complementary to the cDNA. Thus dividing the analytical complexity by the physical length of the cDNA gives the number of copies of the gene representing it. This is strictly true only if the cDNA represents a single message; if the cDNA represents more than one message, for example both α- and β-globin, then the division gives the number of genes representing both messengers together.

The analytical complexity is given by l' to distinguish it from the *kinetic complexity*, l, which may be calculated from the rate of reaction in the usual way. For a moderate excess reaction, the rate constant is given by:

$$k = \frac{\ln (2 - x)}{D_0^c t_{1/2}} \tag{4}$$

The value determined for k may be converted into the corresponding complexity in the usual way, by comparison with the rate constant for a renaturation reaction of known complexity, usually that of E. coli.

This gives the technique greater versatility. The ratio of kinetic complexity to the size of the cDNA gives the number of different sequences represented in the cDNA. Thus for a pure sequence, the ratio should be unity; for a mixture of α- and β-globin sequences it should be 2. Comparison of the kinetic complexity and cDNA size is therefore a useful check on purity; given the difficulty of accurately determining cDNA sizes on gels, the kinetic complexity is sometimes used for comparison instead of the physical size when the ratio is indeed close to unity. Formally the ratio l'/l gives the number of copies in the genome of each separate sequence represented in the cDNA probe.

The results summarized in Table 24.2 show that there is a low number of genes for α- and β-globins in the duck. In the human genome, there appear to be about 2 α-globin genes, more than one β-globin gene (which may be due to cross reaction of the β-globin cDNA with the δ-globin gene), and about 1-2 γ-globin genes. We should note that these numbers may be taken to be only approximate; the technique is influenced strongly by the necessity to reach the right degree of excess—at higher or lower excesses, the results are not accurate. In some, but not in all cases, this has been established by using various excesses and following the reaction as a function of excess. The results summarized in the table represent an attempt to use the most reliable values, but still should be regarded as accurate within perhaps a factor of about 2.

To summarize the results available with individual messengers, two types of result have been obtained. Most specialized messengers are represented by genes present in single or very few copies in the genome, as seen by hybridization analysis. We shall see in Chapter 26 that this conclusion is borne out by restriction analysis of the same genes. The histones are an example of a multiple gene family; again, restriction and cloning data confirm the presence of multiple genes, and show further that they lie in a tandem array.

Number of Expressed Structural Genes

Reaction of DNA with an excess of RNA should drive into hybrid form all the genome sequences represented in the RNA population. The reaction with nonrepetitive DNA requires the same conditions that are necessary for renaturation of unique sequences: because each hybridizing sequence of DNA represents only a small proportion of the input material, a high Rot is needed to drive the reaction to completion. The reaction may be characterized either in terms of the $Rot_{1/2}$ or by the rate constant. Because the RNA is in vast excess over the DNA, its concentration essentially remains unchanged throughout the reaction, which therefore follows pseudo-first-order kinetics. Thus

$$\frac{R}{R_0} = e^{-k \cdot Rot} \tag{5}$$

so that when the reaction is half complete and $R/R_0 = 0.5$,

$$k = \frac{\ln 2}{Rot_{1/2}} \tag{6}$$

(Note the difference from the second-order relationship of equation **6** of Chapter 18.)

Termination of the reaction is established by following the increase in percent hybridized DNA with increasing Rot until a saturation plateau is reached. The complexity of the expressed sequences then is given by multiplying twice the proportion of hybridized DNA by the complexity of the nonrepetitive DNA component. The factor of 2 is necessary because the asymmetry of transcription means that only one strand of any expressed DNA sequence is represented in the mRNA, so that only half of the DNA in principle is available for reaction. Often the saturation level is determined from a double reciprocal plot of 1/% reaction against 1/Cot; this gives a straight line whose extrapolation to the ordinate yields the value that should be obtained at infinite Cot. Sometimes this has been used as a substitute for following the reaction to true termination, which is more desirable.

An example of a saturation reaction is shown in Figure 24.3. Galau, Britten, and Davidson (1974) used two ways to plot the reaction of nonrepetitive DNA with the messengers released from sea urchin gastrula polysomes by puromycin. The upper curve shows the approach to saturation as hybridization is plotted against linear Rot. The lower curve shows the familiar kinetic plot of hybridization against log Rot. Confirmation that the reaction has proceeded to saturation was obtained by recovering the unhybridized DNA and reacting this preparation with an excess of fresh mRNA; there was virtually no reaction. An important control is to show that only nonrepetitive DNA has reacted with mRNA, because the apparent saturation value would be highly susceptible to inflation upon contamination with repetitive sequence hybrids. This was performed by using the recovered DNA as a tracer in a renaturation reaction; it reacted as a single component with the $Cot_{1/2}$ characteristic of nonrepetitive DNA. This is known as a "playback" experiment.

At saturation, 1.01% of the ^{3}H-labeled nonrepetitive DNA is hybridized to mRNA. Since a control DNA renaturation proceeded only to a completion of 75% under these conditions, this means that 1.35% of the available

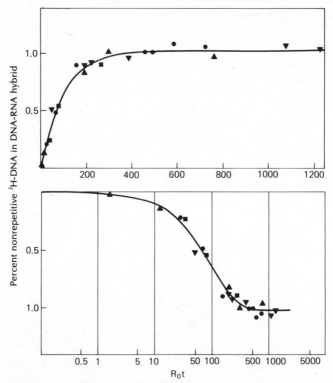

Figure 24.3. Reaction of excess mRNA of S. purpuratus gastrula polysomes with ³H-labeled nonrepetitive DNA. The per cent of DNA hybridized is plotted against the Rot on a linear scale (upper) or a log scale (lower). The reaction has a $Rot_{1/2}$ of 70 and is complete by a Rot of 300. It was performed at an RNA:DNA mass excess of 3–40 fold; but since only some 1% of the DNA is complementary to the mRNA, the excess ratio on the basis of sequence becomes 300–4000 fold. Data of Galau, Britten and Davidson (1974).

nonrepetitive DNA is represented in the mRNA. Assuming that transcription is asymmetrical, this corresponds to the expression of 2.7% of the nonrepetitive DNA. Since nonrepetitive DNA represents 75% of total cellular DNA, 2% of the haploid genome of 8.1×10^8 bp is transcribed in the gastrula. This corresponds to a complexity of 1.7×10^7 bp; if the average messenger is 2000 bases in length, the number of genes expressed in toto in the gastrula is about 8000. Some of the messengers may have been synthesized in the oocyte and stored for later utilization; others may be those synthesized since fertilization. The importance of this complexity value, however, is that it represents the number of genes that is translated into protein at the gastrula stage.

Most of the mass of the messenger population may consist of a small number of sequences represented a large number of times. Correspondingly, most of the sequences present in the mRNA are represented in only a small number of copies. Since a saturation reaction essentially depends on driving the reaction until these scarce sequences have saturated all complements on DNA, only a small proportion of the mRNA mass is involved in pressing the reaction to its conclusion; most of the mass rapidly saturates the small number of corresponding DNA sequences and then essentially plays no further part in the reaction. In other words, the Rot calculated from the total mass of RNA is much higher than the effective Rot. This results in a discrepancy between the observed and expected $Rot_{1/2}$ values, which can be used to calculate the proportion of mRNA driving the reaction. The $Rot_{1/2}$ expected of an RNA preparation with a complexity of 1.7×10^7 bp is 5.6.* The ratio $Rot_{1/2}$-expected/$Rot_{1/2}$-observed gives the proportion of RNA that contains most of the complexity; this is therefore $5.6/70 = 8\%$. The kinetic curve contains only the single transition corresponding to this component; there is no transition identifiable for a fast component. Thus the remaining 92% of the mRNA molecules must contain a relatively small number of sequences, with a complexity only 5–10% of that seen for the total population; each of these sequences must be represented a large number of times.

The total amount of mRNA in the gastrula embryo is about 7×10^{10} nucleotides, which corresponds to about 3.5×10^7 molecules of 2000 bases. If 8% of these molecules represent 8000 genes, then there must be some 350 messengers representing each gene in the polysomes. The embryo is thought to consist of about 600 cells at this stage, so that this complex class of messengers is present on average in less than 1 copy/cell. This must imply that there are different messenger sequences present in different cells. These data do not support any calculation of the complexity and hence number of messengers representing the major (92%) class. But presumably these abundant messengers are common to many or all cell types of the embryo. The concept that messengers vary greatly in their abundance of representation in the population is consistent with the data obtained by hybridization of mRNA in large DNA excess, which show that the proportion of hybridizing RNA can be increased by increasing the excess ratio (see Table 24.1 and earlier footnote). In the next section we shall see that the abundance classes may be defined in more detail by the hybridization reaction between mRNA and cDNA.

Table 24.3 summarizes available data on total mRNA complexities. For

---

*The expected $Rot_{1/2}$ can be calculated in two ways. One is to rely on the $Cot_{1/2}$ expected for a reaction of the same complexity. Because RNA-excess reactions drive hybridization at a rate very close to that of the DNA-DNA reaction, there is no need to compensate for the use of RNA. The expected $Rot_{1/2}$ is therefore simply half of the corresponding $Cot_{1/2}$; the factor of 2 represents the effective doubling of any RNA sequence relative to DNA because of the asymmetry of transcription. For standard conditions in which the E. coli renaturation reaction has a $Cot_{1/2}$ of 4.0, the expected $Rot_{1/2}$ would be 0.5 per 10^6 bases ($2.0/4.2 \times 10^6$). Alternatively the $Rot_{1/2}$ measured for a known RNA-driven reaction under the same conditions can be used. The most commonly used standard is the reaction of globin mRNA with its cDNA; under the usual conditions this has a $Rot_{1/2}$ of 0.0006 for a complexity of 1200 bases (see next section). This also corresponds to 0.5 per 10^6 bases. Obviously the $Rot_{1/2}$ expected for any particular reaction depends on the experimental conditions used in the way that has been described for Cot values in Chapter 18.

convenience, the percent saturation of available nonrepetitive DNA is converted into percent expression of the total genome and then into complexity. The number of different mRNA species has been calculated by dividing the complexity by the estimated average size of mRNA. This provides an estimate for the number of genes expressed. In the somatic tissues of higher eucaryotes, this is only a small proportion of the genome; the number of expressed genes is between 10,000 and 20,000. It is similar for plant tissue

Table 24.3   Total sequence complexity of mRNA populations

Cell and source of mRNA	Driver proportion, %	Saturation of available nonrepetitive DNA, %	Genome proportion expressed, %	Complexity, kb	Number of mRNAs
Yeast total poly(A)$^+$	38	20.0	40.0	6,000	4,000
Achlya ambisexualis total poly(A)$^+$	—	5.8	9.5	3,000	2,600
S. purpuratus					
intestine polysomal puromycin	10	0.5	0.75	6,000	3,000
Gastrula polysomal puromycin	8	1.35	2.0	17,000	8,000
Blastula polysomal puromycin	—	2.10	3.1	26,000	13,000
Oocyte total RNA	8	3.0	4.5	37,000	18,500
Xenopus oocyte total RNA	—	0.75	1.0	30,000	15,000
Tobacco leaf polysomal poly(A)$^+$	20	2.6	1.7	26,000	18,000
Chick					
oviduct polysomal poly(A)$^+$	23	1.8	2.5	30,000	15,000
liver polysomal poly(A)$^+$	23	2.05	2.8	34,000	17,000

All experiments are based upon the saturation of purified nonrepetitive DNA with excess mRNA. The percent saturation given has been corrected for the incomplete extent of reaction when nonrepetitive DNA is renatured. The percent of genome expression takes a value for the proportion of nonrepetitive DNA from Appendix 3 and assumed that transcription is asymmetrical so that only half of the DNA is available for reaction with mRNA. All calculations refer to the haploid genome. The driver proportion is calculated by comparing the rate of reaction with that expected for a preparation of the observed complexity.

    Complexity in kilobases is converted to the number of mRNAs by assuming a number average molecular weight for mRNA of 2000 bases as determined for HeLa cells (Bishop et al., 1974). Exceptions are A. ambisexualis (average 1150 bases), yeast (average 1500 bases) and tobacco (average 1450 bases). The use of the number average follows from the argument that the complexity corresponds to $\Sigma\ N_i M_i$, where N_i is the number and M_i the length of the mRNA molecules in each size class. Since the number average molecular weight is defined as $\Sigma\ N_i M_i / \Sigma\ N_i$, the division of complexity by this value gives $\Sigma\ N_i$, the total number of different messengers. In nontechnical terms, the use of the number average takes into account the greater contribution made to the sequence complexity by larger molecules. However, Davidson and Britten (1973) have used an alternative parameter, the median molecular weight, which is the messenger length compared with which 50% of the molecules are longer and 50% shorter. This is about 1200 bases for animal cells and its use explains why the estimates for messenger number in their papers are greater than those calculated here.

References: Yeast—Hereford and Rosbash (1977); A. ambisexualis—Rozek et al. (1978); S. purpuratus—Galau et al. (1974, 1976), Wold et al. (1978); Xenopus—Davidson and Hough (1971); tobacco—Goldberg et al. (1978); chick—Axel et al. (1976).

and for vertebrate tissue. Most of the expressed genes are represented by few mRNA copies; their proportion of the total mRNA mass is small and is given by the driver proportion. The only substantial exception to this range of values has been reported for mammalian brain (see footnote in next section).

The number of expressed genes is similar for X. laevis and S. purpuratus oocytes, in spite of a 4-fold difference in genome size; this argues that larger genomes do not necessarily contain larger numbers of structural genes (at least as represented in the oocyte). The S. purpuratus oocyte has a larger number of expressed genes than the gastrula embryo or the adult intestine; presumably many are represented in stored but not translated mRNA. It is noteworthy that with increasing specialization of these sea urchin tissues, the number of active genes is reduced.

In lower eucaryotes, the number of active genes is less; but this represents an appreciably greater proportion of the genome. The number for Achlya is not much greater than the value generally assumed for the number of bacterial genes, around 2000.

Abundance Classes of mRNA

In any hybridization or renaturation reaction, the kinetics are established by the nucleotide complexity of the reacting sequences. In RNA-driven reactions, the $Rot_{1/2}$ should reflect the complexity of the RNA population. However, we have seen that when an excess of RNA is hybridized with total cellular DNA, the variation in abundance of different mRNA sequences means that only some of the RNA drives most of the reaction. A corollary of the discrepancy between observed and expected $Rot_{1/2}$ values that is seen when individual components are not resolved is that the $Rot_{1/2}$ of the reaction cannot be used to determine the complexity of the RNA population. (This would yield much too high a level since the Rot values needed to complete the reaction are greatly increased by the presence of a large amount of RNA that contains few sequences.)

Conditions under which the kinetics of the hybridization reaction can be used to determine complexity, however, have been established with the introduction of another technique by Bishop et al. (1974). Excess mRNA is hybridized with labeled cDNA previously prepared from it by reverse transcription. All of the mRNA sequences should be represented in the cDNA population, with a frequency corresponding to their abundance in mRNA. Since all of the cDNA sequences are derived from the mRNA, all of the labeled cDNA should be driven into hybrid form. For a single component, the $Rot_{1/2}$ of reaction is proportional to complexity in the same way that $Cot_{1/2}$ is determined by complexity in a renaturation reaction. Thus the complexity of an unknown RNA population may be determined by comparing its $Rot_{1/2}$ with the $Rot_{1/2}$ of a standard reaction. Usually the standard taken is the reaction of excess globin mRNA with its cDNA. If the globin mRNA contains both α and β sequences, its complexity is 1.2 kb, so that:

$$\frac{\text{complexity (any RNA)}}{1.2\ \text{kb}} = \frac{\text{Rot}_{1/2}\ (\text{any RNA})}{\text{Rot}_{1/2}\ (\text{globin mRNA})} \tag{7}$$

This is exactly analogous to equation 7 of Chapter 18 for determining the complexity of unknown DNA preparations by comparing the $\text{Cot}_{1/2}$ with that of a standard. In the same way, the calculated complexity gives the total length of different sequences without any implication about organization.

A single component displays a Rot transition that occupies about 2.5 decades. When reaction occurs over a greater Rot range, more components must be involved; usually they are resolved by computer simulations that fit a curve to the experimental data by summing several independent components each behaving with ideal kinetics. As with the comparable analysis of renaturation curves, this is arbitrary in the sense that it resolves the minimum number of discrete components; often equally good fits may be achieved using more components (see Quinlan et al., 1978). Complexities can be determined from equation 7 only when the $\text{Rot}_{1/2}$ values of single components are compared. Thus when a reaction is resolved into several components, the $\text{Rot}_{1/2}$ measured for each component is converted to the value that would be shown if it were purified:

$$\text{Rot}_{1/2}\ (\text{corrected}) = \text{Rot}_{1/2}\ (\text{observed}) \cdot \alpha \tag{8}$$

where α is the proportion of the reaction occupied by the component. Essentially the observed $\text{Rot}_{1/2}$ is too great because the value used for RNA concentration includes the RNA present in all components, whereas only the mass of each individual component drives its reaction.

Figure 24.4 shows an example of the reaction between excess mRNA and cDNA of chick oviduct, which occurs over about six orders of Rot magnitude and is resolved into three components. The first component has a $\text{Rot}_{1/2}$ of 0.0015 and represents 45% of the input cDNA. Since the reaction terminates at 90% of input, this component corresponds to 50% of the hybridizing cDNA. Its corrected $\text{Rot}_{1/2}$ is therefore 0.00075. This may be compared with the control $\text{Rot}_{1/2}$, which in this experiment was 0.008 for the reaction of ovalbumin mRNA with its cDNA. The identity of the $\text{Rot}_{1/2}$ values immediately suggests that this component is in fact ovalbumin mRNA, which is the predominant product of gene expression in oviduct and is known to correspond to about half of the messenger mass. The next component has an observed $\text{Rot}_{1/2}$ of 0.04, which may be converted to a corrected value of 0.006 by multiplying by its proportion of the reaction (0.15). Comparison with the control shows that its complexity is 0.006/0.0008 = 7.5 times that of ovalbumin, which is about 14 kb if ovalbumin mRNA is 1.9 kb in length. This would correspond to 7 messengers of average length of 2000 bases. Similarly the last component has a complexity of $0.35 \times 30/0.0008 \times 1.9 =$ 25,000 kb or about 12,500 mRNA species of average length.

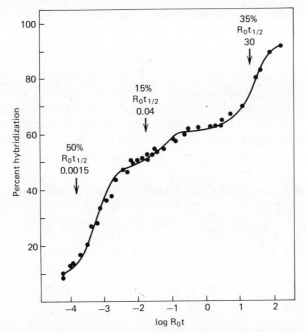

Figure 24.4. Hybridization of labeled cDNA with excess mRNA from chick oviduct. The proportion of the reaction and the $Rot_{1/2}$ values of the three resolved components are indicated by arrows. Data of Axel et al. (1976).

If the total amount of mRNA in the cell is known, the average number of copies of the messengers in each component can be calculated. This is known as the *abundance* or *representation*. Since for each component:

total mRNA mass = abundance of mRNA × complexity of mRNA

it is evident that

$$\text{abundance} = \frac{\text{gms of mRNA in cell} \times \text{fraction in component} \times 6 \times 10^{23}}{\text{complexity of component in daltons}}$$

$$(9)$$

The equation is usually expressed in this form since total RNA is measured in picograms while complexity is given in bases or daltons.

While equation **9** may readily be used to calculate abundance from the complexity and total mass, this procedure may be circumvented by calculating abundance directly from the observed $Rot_{1/2}$. Since the complexity of the component is calculated from equations **7** and **8** through the use of terms also present in equation **9**, these may be cancelled by substitution to give:

$$\text{abundance} = \frac{\text{total daltons mRNA per cell}}{\text{observed } Rot_{1/2}} \times \frac{Rot_{1/2} \text{ standard}}{\text{daltons in standard}} \quad (10)$$

Applied to the data in Figure 24.4, this means that there are roughly 100,000 copies of the first component (ovalbumin mRNA) per cell, while the second component's 7 messengers each are represented about 4000 times in the cell; but the last component consists of the much larger number of 12,500 genes each represented only about 5 times in mRNA.

Table 24.4 summarizes available data from a variety of systems. A common feature is the presence of more than one component; often three are resolved, occasionally two will suffice. Usually the first component or first two components contain of the order of half the mass of mRNA in the form of less than a few hundred active genes represented in many copies each.

Table 24.4 Complexity and abundance classes of mRNA populations

Cell and source of mRNA	Reaction, %	Transitions, %	Corrected $Rot_{1/2}$	Complexity, kb	mRNA number	Abundance	Total mRNA /cell, pg
		Total cDNA					
Yeast total poly(A)$^+$	75	23	0.025	28	20	200	0.01
		51	0.56	645	430	20	
		26	3.2	3,700	2,500	1½	
		globin =	0.0013				
A. ambisexualis total poly(A)$^+$	78	32	0.03	33	30	1,000	0.06
		32	0.23	250	200	140	
		36	3.1	3,400	3,000	11	
		globin =	0.0011				
Schneider Drosophila culture: cytopl. poly(A)$^+$	64	6.5	0.004	8	4		
		59	0.17	350	175		
		34	6.4	13,000	6,500		
		globin =	0.0006				
Tobacco leaf polysomal poly(A)$^+$	84	9	0.006	120	80	4,500	0.38
		52	0.53	9,500	6,500	340	
		39	7.9	14,000	10,000	17	
		globin =	0.0007				
Chick oviduct polysomal poly(A)$^+$	91	50	0.00075	2	1	100,000	0.275
		15	0.006	14	7	4,000	
		35	10.5	25,000	12,500	5	
		ovalbumin =	0.0008				
Chick liver polysomal poly(A)$^+$	88	16	0.00043	1	<1	33,000	0.15
		40	0.08	210	100	750	
		44	8.8	22,000	11,000	7	
		ovalbumin =	0.0008				
HeLa cell cytoplasm poly(A)$^+$	77	67	0.4	800	400		
		33	9.2	18,000	9,000		
		globin =	0.0006				
L cell polysomal poly(A)$^+$	90	4.5	0.002	4	2	6,000	0.31
		40	0.25	500	250	450	
		45	6.1	12,000	6,000	20	
		globin =	0.0006				
Mouse embryo polysomal poly(A)$^+$	75	56	2.8	840	420		
		44	66	20,000	10,000		
		globin =	0.004				

Table 24.4 (*Continued*)

Cell and source of mRNA	Total cDNA			Com-plexity, kb	mRNA number	Abundance	Total mRNA /cell, pg
	Reac-tion, %	Transi-tions, %	Corrected $Rot_{1/2}$				
Mouse liver polysomal poly(A)⁺	75	50	1.5	450	225		
		50	70	21,000	10,500		
		globin =	0.004				
Mouse liver cytoplasmic poly(A)⁺	75	22	0.0086	17	9	12,000	0.54
		41	0.67	1,400	700	300	
		37	11.4	23,000	11,500	15	
		globin =	0.0005				
Mouse kidney cytoplasmic poly(A)⁺	77	10	0.003	6	3	12,000	0.38
		45	0.52	1,050	500	300	
		45	10	20,000	10,000	15	
		globin =	0.0005				

For each cell type a cDNA population prepared by reverse transcription of the mRNA was hybridized back with an excess of the RNA. The total amount of the cDNA that has hybridized at completion is given in the first percent column. Hybridization takes place over several orders of magnitude of Rot and the next column gives the proportion of the hybridizing cDNA assigned to individual components of the reaction. This assumes that the reaction has proceeded to completion so that the nonhybridizing cDNA belongs equally to all classes. Control reactions of globin mRNA x cDNA usually terminate at about 90%, so the assumption is in doubt only for apparent termination values somewhat lower than this. If the nonreacting sequences in fact are part of the most complex (last hybridizing) class, the number of genes in this class will be slightly greater (about 10% more) than estimated; and the number of genes in the less complex classes will be slightly less than given.

Individual components are resolved by curve fitting to give the minimum number able to fit the data; equally good fits sometimes may be obtained by postulating greater numbers of components. The observed $Rot_{1/2}$ for each component has been converted into the value that would be displayed if the component were pure by multiplication by the proportion it occupies of the reaction. The corrected $Rot_{1/2}$ is converted into a complexity measure by division by the $Rot_{1/2}$ for a standard of known length (globin = 1.2 kb; ovalbu-min = 1.9 kb). The standard used in each analysis is given as the last Rot value in the series.

The complexity is converted into messenger number as described in Table 24.3. The abundance of the sequences in each class can be calculated if the total cellular content of mRNA is known by comparing the mass of each component with its complexity. Note that in two different analyses of mouse liver the total complexity remains the same, although one relies on two components and one identifies three. However, the abundance of the more common species would be difficult to estimate in the first case.

References: yeast, achlya, tobacco, chick—see Table 24.3; Drosophila—Levy and McCarthy (1975); L cell—Ryffel and McCarthy (1975); mouse tissues—Young, Birnie and Paul (1975), Hastie and Bishop (1976); HeLa—Williams and Penman (1975). Note that the values originally given for HeLa by Bishop et al. (1974) appear to be somewhat too great. Additional data to those given in the table have been obtained by: AKR mouse cells—Getz et al. (1976); 3T6 cells—Williams and Penman (1975); trout liver and testis—Levy and Dixon (1977a).

Where this *abundant* material has been resolved into two classes, the first component represents a very small number of genes, often of the order of 10, represented several thousand times per cell. In some cases some of the highly abundant sequences can be equated with specialized functions of the cell, such as the example of ovalbumin in oviduct, or actin in growing HeLa cells.

In all cases less than half of the mass of the mRNA consists of a large number of sequences, of the order of 10,000, each represented by a small number of copies of mRNA per cell, of the order of 10. The actual values shown for abundances in the table are only approximate, since they rest on determinations of mRNA content that vary in accuracy; but the range of representation clearly varies over 1000 fold or more between the most and least abundant messengers. The slowest reacting component identifies the *scarce* or *complex* messenger class. Given the possibility that unreacted cDNA sequences might preferentially reside in this class, the number of scarcely represented genes determined by this technique is likely to be a lower limit.

An important check can be made by comparing the total complexity of the mRNA population determined by the saturation and kinetic techniques. Since it should be the members of the complex class that drive the saturation reaction, the driver proportion in the reaction of excess mRNA × celular DNA should correspond with the proportion of the mRNA occupied by the slowest component in the excess mRNA × cDNA reactio Where both techniques have been applied to the same system, the resu₋ts are in reasonable agreement, as summarized in Table 24.5. The difference between the complexities estimated by the two techniques emphasizes that the results identify a range of values for the number of active structural genes; usually each technique is considered to be accurate within some 20% or so. As might be expected, the total complexity determined by the kinetic technique is less than that determined by saturation. The proportion of the scarce component, and thus its apparent abundance, shows a twofold variation between the techniques.

The general conclusion suggested by these results is that the somatic tissues of higher eucaryotes, or individual cell lines perpetuated in culture, have of the order of 10,000–20,000 active structural genes.* For tissues the actual mRNA complexity and copy number per cell will depend on the extent of heterogeneity in the cellular components of the tissue.

What is responsible for establishing the abundance of a messenger RNA? Clearly the amount of any messenger in the cell is the result of the balance between its production and turnover. Rates of production may be influenced by both transcription and processing (see Chapter 25); rates of turnover fall into the two general classes of short lived and long lived that we have described in Chapter 23 (and see Table 25.2). Is either production or turnover correlated in a simple way with abundance? In cultured Drosophila cells, abundance appears to be related to turnover, since the abundant

---

*A possible exception to these levels of gene expression is provided by mammalian brain. Recent estimates based on saturation and kinetic analyses have produced results ranging from levels similar to those obtained with other tissues to much greater values, the present extremes being about 12,000 to 70,000 genes expressed (Bantle and Hahn, 1976; Hastie and Bishop, 1976; Young, Birnie, and Paul, 1976). Until the reasons for this variation are resolved, it will be difficult to form any firm conclusion on the number of genes that is active in brain. It should be noted that the complexity of nuclear RNA also appears to be greater in brain than in other tissues (see Chapter 25).

Table 24.5 Comparison of active gene numbers estimated by saturation and kinetic techniques

Cell	Saturation		Kinetic	
	Complexity as mRNA no.	Driver proportion, %	Total complexity	Proportion of slow component, %
Chick oviduct	15,000	23	12,500	35
Chick liver	17,000	23	11,000	44
Tobacco leaves	18,000	20	17,000	39
Yeast	4,000	38	3,000	26

Data are taken from Tables 24.3 and 24.4 and show the total complexity determined by saturation or by summing all kinetic classes. Proportions are taken from the driver in saturation and the slow component in kinetics.

messengers belong to the long-lived class, whereas the scarce messengers are part of the rapidly turning over set (Lenk, Herman, and Penman, 1978). In HeLa, however, no such simple relationship could be found; each abundance class appeared to contain both short- and long-lived messengers. On the other hand, in L cells the abundant class appears not only to correlate with the long-lived component but also to consist of shorter molecules than the scarce, shorter lived class (Meyuhas and Perry, 1979).* In sea urchin gastrula, however, only a single stability class has been distinguished, which contains both abundant and scarce messenger components (Galau et al., 1977c). Studies on further cell types will be needed to determine whether there is any common basis for abundance differences.

Overlaps between Messenger Populations

An important question is the extent to which different tissues share the same transcripts. Some similarities are expected, of course, since all cells have certain common features. Differences corresponding at least to the known specialized functions of each tissue also are to be expected. But what proportion of the expressed structural genes is the same in different tissues; and from this, is it possible to make any estimate of the total number of structural genes utilized by the organism? In attempting to determine structural gene number, it is necessary, of course, to take account of embryonic as well

*The equation of smaller messengers with the more abundant class implies that complexity measurements may be distorted by taking the same average size value for all members of the cellular mRNA population. First the average size will be determined from the mass distribution in which the abundant messengers are predominant. Second the use of this average will (slightly) underestimate the number of species in the abundant class and will (more seriously) over estimate the number of species in the scarce classes. Abundancies will be affected in the converse manner.

as adult stages. Thus comparisons of messenger complexities have focused on successive states of embryonic development as well as on adult tissues. In both cases, the subject of comparison contains an unknown number of different cell types; so it is useful also to know the nature of the genes expressed in a single (cultured) cell type. The comparison that can be made with the greatest facility, of course, concerns the appearance or disappearance of the highly abundant mRNAs coding for predominant products of specialized cells; this offers the best prospect for unravelling the sequence of events involved in messenger production and turnover.

The messengers translated during stages of sea urchin embryogenesis have been compared in a series of experiments by Galau et al. (1976). As can be seen in Figure 24.3, gastrula mRNA is able to saturate 1.35% of the non-repetitive DNA, corresponding to a complexity of 17,000 kb (see Table 24.3). The protocol in Figure 24.5 shows the use of this reaction to divide the non-repetitive DNA into two fractions, one represented in gastrula messengers (gastrula mDNA), and one unrepresented (null mDNA). Then polysomal mRNA extracted from different embryonic stages or adult tissues can be hybridized with both fractions to determine the amount of DNA complementary to the expressed sequences. Any sequences of gastrula mDNA that react must be expressed in both gastrula and the second tissue; sequences of gastrula mDNA that do not react are represented in gastrula polysomes but are not expressed at the second stage. Sequences of null DNA that react are expressed at the second stage but not in gastrula.

The results shown in Figure 24.5 reveal that there is a steady decline in the total complexity of the sequences under translation as development proceeds from the oocyte to the adult tissues. What is especially significant is that the sequences expressed in the gastrula constitute a large part of the messenger population at each stage. In particular, all the sequences expressed in the gastrula are present in the oocyte. Measurements of synthesis and turnover rates suggest that by the time of gastrulation most or all mRNA molecules are the products of embryonic gene expression, not species physically inherited from the oocyte (Galau et al., 1977c). But the important point is that no new genes are turned on at gastrula, although some genes have ceased to be represented in the polysomes. Some of the gastrula sequences were not expressed at the previous stage of blastula, although they were present in the oocyte.

A similar pattern is revealed by comparisons of the RNA population of the oocyte with the subsequent embryonic stages (Hough-Evans et al., 1977). A decline occurs in the representation of oocyte sequences on polysomes; relative to the reaction of the 37,000 kb complex oocyte RNA with an oDNA preparation; the polysomal mRNA of the 16 cell stage showed 73% reaction, equivalent to a complexity of 27,000 kb. The mRNA population of blastula and gastrula showed 49 and 53% reactions, equivalent to about 18,000 and 19,000 kb complexities. For blastula, whose total mRNA complexity is about 26,000 kb, this shows that some 70% of the expressed sequences are identical with those previously expressed in the oocyte; for

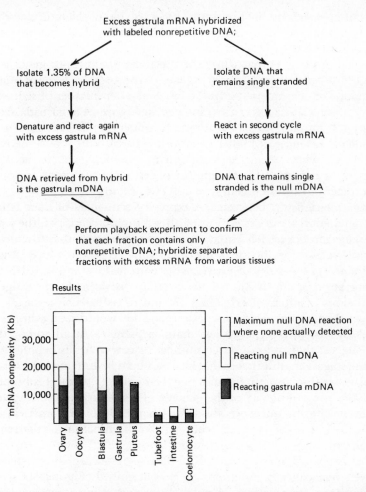

Figure 24.5. Comparison of structural genes active in sea urchin embryos and adult tissues. Nonrepetitive DNA was divided into the gastrula mDNA fraction represented in gastrula mRNA and the null DNA fraction not represented. The proportions of these preparations able to hybridize with mRNA extracted from various tissues is converted to total base complexity. In the control reaction of gastrula mDNA with gastrula mRNA, 63% of the DNA is hybridized at saturation; this is taken to represent the complexity of 17,000 kb previously characterized for gastrula messengers. Null mDNA does not react with gastrula mRNA; 86% of it reacts with unfractionated DNA and this is used as the control for calculating complexity in reactions with mRNA. Data of Galau et al. (1976).

gastrula this confirms the lack of any new sequences relative to the oocyte. This demonstrates that some new genes are expressed at blastula but are turned off again by gastrula. (It is noteworthy also that the overall complexity of total cytoplasmic RNA, and the proportion of oocyte sequences represented in it, is greater than is found for polysomal mRNA. This may imply that some maternal sequences are present in the cytoplasm, although they fail selectively to be translated; however, the problem of possible contamination of total cytoplasmic RNA with nuclear RNA precludes a firm conclusion.)

The main point is that while the spectrum of expressed genes is clearly specific for each stage of embryogenesis, the overlap between messenger populations is considerable. In the much smaller number of sequences expressed in the adult tissues investigated in Figure 24.5, gastrula sequences occupy a high proportion.* It should be noted that these tissues at present provide the example of higher eucaryotic cells with the lowest mRNA complexity. To what extent the reduction in complexity of the adult tissues represents a reduction in the number of cell types per tissue as opposed to the number of expressed genes per cell is not known.

If the relationship between the expressed sequences of these embryonic stages and adult tissues is typical of all sea urchin cell types, the sequences represented in oocyte RNA may correspond to most of the structural genes that are expressed in the life of the organism. (To be tested fully, this would require a detailed series of experiments with mDNA and null DNA representing appropriate tissues.) If the overlaps between the messenger RNA populations of different cell types are indeed as extensive and closely related as they appear from these examples, this would set a number for the expressed structural genes of less than, say, 25,000. At all events, these results suggest that unless a large number of new genes is expressed in embryonic stages or adult tissues that have not been investigated, the total number of sea urchin structural genes may be unable to account for more than, say, 5% of the complexity of the genome. This reinforces the C value paradox by setting the questions raised in Chapter 17 on a numerical basis: if this value is roughly correct, what can be the function of the remaining 95% of sequences of the sea urchin genome?

Additive saturation with mRNA from two sources allows the overlap between messenger populations to be determined. Figure 24.6 compares the sequence complexities of chick liver and oviduct messengers. With mRNA derived from the individual tissues, saturation in excess RNA was achieved at 2.05 and 1.8% of DNA, respectively (as quoted in Table 24.3). With combined RNA, saturation is achieved at 2.4%, a level only slightly

*As might be expected from the relationships shown in Figure 24.5 and described in the text, similar overlaps are found in other comparisons. Thus 15% of the available blastula mDNA can be driven into hybrid form by intestine mRNA. This overlap corresponds to 3900 kb, or to 65% of the total complexity of 6000 kb of the intestine (Wold et al., 1978). One note of caution, however, is that only in the experiments with gastrula mRNA has both an mDNA and null DNA preparation been obtained. In all these experiments, the reaction with the mDNA preparation has been characterized relative to the control reaction of this preparation when driven by its homologous mRNA. With blastula mDNA this reaction proceeds to 78%, with oocyte to 58–77% depending on the preparation, with gastrula to 58%. The need to make such a large correction for incomplete reaction makes it useful to confirm the total overall complexity of the message population for each stage by saturation of total nonrepetitive DNA. This has been done for oocyte and gastrula (of course) and for intestine and blastula (see Table 24.3).

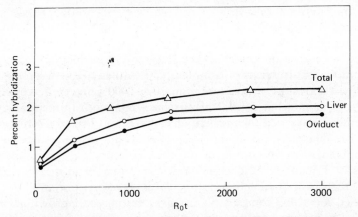

Figure 24.6. Comparison of liver and oviduct mRNA of chick. ^{125}I-labeled nonrepetitive DNA was hybridized with an excess of mRNA derived from liver, from oviduct, or from both combined, until a saturation plateau was reached. Data of Axel et al. (1976).

greater than that of the individual tissues and very much less than the 3.85% that would result if the two sequence sets were entirely different. This suggests that about 75% of the sequences expressed in liver and oviduct are the same, although since this is a saturation experiment there is no information on whether they are present in the same or different abundances in the two tissue types.

The reaction of excess mRNA × cDNA can be used to yield information about abundance differences, although its measurement of complexity differences is perhaps less sensitive. In a comparison of mouse liver, kidney, and brain, Hastie and Bishop (1976) found that all three tissues have about 12,000 expressed genes, divided into three abundance classes (see Table 24.4). In a heterologous reaction, brain cDNA or kidney cDNA was hybridized with excess liver mRNA. Both give a curve with two transitions identical with the two higher transitions of the homologous liver reaction. Thus the first transition of highly abundant sequences is absent. This means that the highly abundant sequences of liver are not highly abundant in kidney or brain, but must be specialized for the tissue.

The abundant sequences can be isolated by retrieving the cDNA that reacts at low Rot in the first transition of the homologous reaction. When the highly abundant liver cDNA is hybridized with kidney or brain mRNA, it reacts as though part of the last (scarce) transition. This means that the sequences that are highly abundant in liver also are expressed in kidney and brain, but are represented scarcely instead of abundantly. Similarly, the abudant kidney cDNA hybridizes to brain or liver mRNA as though part of the middle abundance class.

Considered in terms of mass of mRNA, most of the differences between these tissues are accounted for by changes in the 10 or less sequences comprising the highly abundant class in each case. Heterologous reactions between middle abundance and scarce cDNA with mRNA show that liver and kidney appear to have identical sequences in the second transition, whereas about 90% of the sequences in the final, scarce transition are common. This would imply that in terms of complexity, the difference between the tissues lies in the expression of no more than 1000–2000 genes represented in the scarce class. Somewhat similar results have been seen in comparisons between embryonal carcinoma (pluripotent) cells and Friend (erythropoietic) cells or myoblasts (Affara et al., 1977). Most of the sequences are the same, with differences in about 10% of the total complexity. The abundant sequences of Friend cell and embryonal carcinoma are different, although those of the pluripotent cell and the myoblast appear similar.

Another revealing comparison involves resting and growing cultured cells. The transition from resting state to active growth is accompanied by an increase in the production of mRNA, both as a proportion of total cellular RNA (about 2–3 fold) and absolutely (about 3–5 fold). The increase is characteristic of the cell type and its control is discussed in Chapter 25. When the total complexities of the mRNA populations are determined in the two conditions, there is little difference in 3T6 cells or mouse embryo AKR cells. The same abundance classes therefore are present with the same relationship, although the absolute abundance in each case is increased upon the transition into active growth.

The extent to which different sequences may be present has been measured in two ways. Williams and Penman (1975) used an exhaustion procedure with cDNA of 3T6 cells, in which the cDNA of resting cells was hybridized through two cycles with excess mRNA from growing cells. This should remove from the cDNA the sequences common to resting and growing cells, leaving the sequences specific for the resting state. When the exhausted cDNA preparation was hybridized with resting mRNA, 19% reacted, compared with an 8% reaction with a control of growing mRNA. These values correspond to 5.7 and 2.4% of the input cDNA; the difference of 3.6% identifies the proportion of the cDNA specific for the resting cell. This would be about 500 sequences. The same result is found when the reaction is performed in the reverse direction. Although quantitatively this cannot be considered too reliable, since it depends on the difference between larger numbers after several hybridization steps, qualitatively it does suggest that resting and growing cells each have only a comparatively small number of sequences specific for the growth state. A similar conclusion is suggested by experiments in which Getz et al. (1976) followed the ability of cDNA from resting or growing AKR cells to be hybridized with excess mRNA from cells in the alternate growth state. Resting cDNA is completely hybridized by growing mRNA, whereas growing cDNA is hybridized to more than 90% by resting cDNA. This would correspond to the presence in growing cells of up to about 1000 additional sequences, with no sequences present in resting cells that are not present in growing cells. Again this suggests that changes in only a small proportion of the number of expressed structural genes are necessary for the resting-growing transition. Generally similar results have been obtained in comparison of transformed with non-transformed cells (Williams, Hoffman, and Penman, 1977; Getz et al., 1977).

Changes in the scarce messengers or those of middle abundance are difficult to follow in detail. To investigate the control of gene expression, it is necessary to use systems in which specific genes may be activated; essentially this is to focus attention on members of the highly abundant class. The concentration of an individual mRNA may be measured by titration, using hybridization of a cDNA probe with excess mRNA to determine the proportion of the complement in the overall RNA population. For a cDNA probe representing a single sequence, the $Rot_{1/2}$ is inversely proportional to the concentration of its complement in the RNA, so that:

$$\text{proportion of sequence} = \frac{Rot_{1/2} \text{ (control)}}{Rot_{1/2} \text{ (unknown preparation)}} \qquad (11)$$

where the control is the reaction between cDNA and its pure complementary mRNA. This can be used with either total RNA or poly(A)$^+$ RNA from cellular or polysomal preparations; all that is needed to calculate the amount of the mRNA per cell is the total amount per cell of the RNA source.

Four systems that have been used are summarized in Table 24.6. In each case the conversion of an inactive cell into an active cell—either as the result of hormone treatment or by time-dependent cell differentiation—is accompanied by the appearance in the polysomes of large amounts of new mRNA species. The table gives data on the most prominent of the new messengers; but others may appear similarly, subject to coordinate control. The induction factor may be very considerable, with virtually no stable cytoplasmic mRNA detectable before induction, but with 10^4–10^5 mRNA molecules per cell present in the active state. In these and in the other instances discussed in the table, gene expression is controlled by the production of messenger RNA.

What other changes occur in the population of messengers? In the case of oviduct, other proteins such as ovomucoid and lysozyme show the same response as the major induction of ovalbumin. Their messengers presumably reside in the middle abundance class. But comparisons of the total populations by the excess mRNA × cDNA reaction show no substantial change in complexity (Cox, 1977; Hynes et al., 1977). Similar conclusions have been suggested for the response of rat prostate to androgens and for chick myoblasts during fusion (Parker and Mainwaring, 1977; Paterson and Bishop, 1977). Differentiation along a single pathway from one cell type to its immediate product thus appears to be accompanied by great changes in the abundance distribution, with the appearance of large amounts of a small number of messengers coding for the prominent proteins characteristic of the new cell phenotype. However, there is little change in the total number of structural genes expressed.

Table 24.6 Induction of messenger RNA synthesis

System	Condition	Amount of protein	Measure of mRNA	Total mRNA content
Globin in Friend cells	uninduced	$<4 \times 10^6$ globin (0.1 pg)/cell	$<7.5 \times 10^{-5}\%$ of total cell RNA	<30 mRNA/cell
	induced by DMSO or hemin	8×10^7 globin/cell	0.017% of total cell RNA	3000 mRNA/cell
Ovalbumin in chick oviduct	unstimulated	no detectable ovalbumin	$<7 \times 10^{-6}\%$ of total cell RNA	<1 mRNA/cell
	stimulated by estrogen	2.5×10^9 ovalbumin/ tubular gland cell	0.06% of total cell RNA; or 50% of total mRNA	100,000 mRNA/cell
	estrogen withdrawn	ovalbumin synthesis ceases	0.011% of total mRNA	<30 mRNA/cell
Vitellogenin in X. laevis liver	unstimulated	no protein	$<5 \times 10^{-6}\%$ of total cell RNA	<0.2 mRNA/cell
	stimulated by estrogen	vitellogenin synthesized and processed	0.04% of total cell RNA	36,000 mRNA/cell
Myosin (heavy) in myoblast	replicating myoblast	6×10^6 chains/ cell	$<1.4 \times 10^{-3}\%$ of total cell RNA	<37 mRNA/nucleus
	fused myoblast	16×10^6 chains/ myotube	0.08% of total cell RNA	1400 mRNA/nucleus

The amount of protein (where available) and the steady-state number of its mRNA molecules per cell are given for both the inactive and active condition. The first mRNA column shows how the particular mRNA was quantitated and the second converts this to the number of messengers for this protein per cell. Limits of detection are indicated as $<$ amount detectable. With Friend cells, the values depend on the clone used; uninduced cells may have undetectable levels of globin or levels at some percent of the induced level. Induced globin mRNA levels as great at 13,000 molecules per cell have been reported. Globin includes both α and β. For ovalbumin, the per cell measure depends on assumptions about the proportion of oviduct occupied by the tubular gland cells in which ovalbumin is synthesized. Fully induced levels of ovalbumin mRNA have been reported in the range 85,000–150,000 mRNA per cell. Less detailed data have been obtained in other systems to show that increase in protein content is correlated with increase in mRNA (assayed directly with a cDNA probe or indirectly by translation in vitro or in oocyte). These include the production of tryptophan oxygenase or tyrosine aminotransferase in rat liver in response to glucocorticoids, synthesis of α_{2u}-globulin in response to glucocorticoids and other agents, and changes in liver albumin with the diabetic condition. A result in contrast is that α_{2u}-globulin message may be induced in the absence of growth hormone but is translated only upon addition of the hormone (Kurtz, Chan, and Feigelson, 1978).

References: Globin—Ross et al. (1974), Ross and Sautner (1976), Lowenhaupt and Lingrel (1978); ovalbumin—Harris et al. (1975), McKnight et al. (1975), Palmiter et al. (1976), Hynes et al. (1977); vitellogenin—Baker and Shapiro (1977): myosin—John et al. (1977); liver enzymes—Schutz et al. (1975), Kurtz et al. (1976), Kurtz and Feigelson (1977), Roewekamp, Hofer and Sekeris (1976), Nickol, Lee and Kenney (1978), Peavy, Taylor, and Jefferson (1978).

Two general views have been advanced to account for the existence of most of the cellular mRNA in the form of a large number of messengers present in a small number of copies (see Lewin, 1975b). One is that the control of gene expression may be "leaky," so that although many fewer genes are necessary for maintenance of cell type, most of the structural genes of the organism are expressed at a low level because of incomplete repression. This view originated with the first observations of messenger complexity in cultured cells, representing reluctance to believe that HeLa or L cells genuinely might need the expression of as many as 10,000 genes. This would predict that much of the scarce class should comprise genes that are expressed at high levels in various tissues. Leakiness might be characteristic of cultured cells only; or it might be a general feature of eucaryotic control. The latter interpretation would be implied by the subsequent demonstrations that similar numbers of active genes are expressed in the same abundance classes in various tissues. But the demonstrations of in vivo gene expression render more likely the alternative explanation that the scarce class represents genes whose products are needed for sustenance of all cell types. These might be equated with the constitutive or housekeeping genes mentioned in Chapter 9. Although generally about 10,000 in mammalian, avian, and plant cells, the number clearly can be lower elsewhere, as shown by the range 2000–3000 for the expressed gene number in certain adult sea urchin tissues. The highly abundant, and perhaps at least in part the middle abundant, messengers then would correspond to the expression of the specialized or luxury genes. There is as yet no definitive answer to the question of the extent to which genes that are highly expressed in one tissue may be scarcely expressed in others. Assays for globin and ovalbumin mRNA, however, suggest that repression can be effective in nonexpressed tissues to levels below that characterizing the scarce mRNA class. Accepting this general division of genes, the total number will correspond to the 10,000 or so common housekeeping genes plus the specialized functions of each cell type. It seems unlikely that this could be great enough to correspond to the major part of the genome complexity of mammals and other higher eucaryotes.

CHAPTER 25

Heterogeneous Nuclear RNA

The structure of heterogeneous nuclear RNA and its relationship with messenger RNA has been a subject of intense debate over the past decade. Production of RNA in the nucleus falls into two principal categories: transcription of ribosomal RNA precursors and their maturation in the nucleolus; and transcription of heterogeneous nuclear RNA (hnRNA) and its degradation in the nucleoplasm. Although taking place in morphologically distinct parts of the nuclear structure and catalyzed by the different enzymes that we have discussed in Chapter 22, both types of event have in common the transcription of a nuclear precursor much larger than the mature RNA found in the cytoplasm. (This contrasts with the production of small RNA species by the third RNA polymerase.) In the nucleolus the large transcripts comprise precursors from which the sequences of the mature rRNAs are conserved, the remaining sequences degraded. We shall discuss the genetic organization and expression of rRNA genes in Chapter 27. The question of whether a similar relationship exists between hnRNA and mRNA has aroused partisanship, with reports that mRNA is derived by cleavage of much larger nuclear precursors opposed to claims that there is no evidence for any such relationship (for review see Lewin, 1975a).

Evidence on the relationship of hnRNA and mRNA is essentially of two types: concerning the physical nature of the molecules; or comparing the sequences represented in them. The similarities between mRNA and hnRNA immediately suggest the possibility of a product-precursor relationship: both contain 5′ methyl caps and 3′ poly(A). Since these modifications are introduced only in the nucleus but are found also in the cytoplasm, the obvious inference is that mRNA is derived from hnRNA. But there is a critical difference between the nuclear and cytoplasmic molecules: the size of the hnRNA is several fold greater than that of the mRNA.

This presents a paradox. How can both ends of an hnRNA precursor molecule be preserved in a much shorter messenger product? Two solutions offer themselves and are illustrated in Figure 25.1. Both take account of the fact that much of a radioactive label incorporated into hnRNA turns over in the nucleus and so cannot give rise to mRNA; the ability to do so therefore must be the prerogative of no more than part of the mass of hnRNA.

The solutions differ in the relationship that they provide between the sequences of hnRNA that give rise to mRNA and those that do not.

The first depends on the location of the 5′ and 3′ terminal modifications in mRNA with respect to the primary transcript and turns on the question of whether all modified sequences are conserved. If the terminal modifications present in hnRNA are not necessarily conserved, then the large nuclear molecules may be cleaved at some distance from one or the other end so that appropriate modification can be introduced at either the new 5′ or new 3′ end group. This is to say that modification may occur at ends created posttranscriptionally as well as at those representing initiation and termination sites on the primary transcript. In this case, one of the products of cleavage may be preserved to act as messenger RNA, while the other (including an originally modified end) is degraded. Naturally this bears on the question of whether the capping or polyadenylation reactions are involved in selecting sequences for transport to the cytoplasm, since if either caps or poly(A) turn over in the nucleus their addition per se cannot be sufficient to indicate a message sequence.

The second solution is in a sense more direct: both ends of the molecule may be conserved but sequences from within may be lost. This proposal would have seemed bizarre before the discovery of interruptions in genes and of the existence of RNA splicing. This is the subject of Chapter 26; but in anticipation we may say now that it is clear that it is possible for internal sequences to be lost from nuclear RNA precursors. Since the splicing reaction has been characterized only for certain individual molecules, this does not prove the question of whether splicing or simple cleavage of terminal (or internal) regions is responsible for the derivation of the majority of mRNA molecules from nuclear precursors.

The sequence complexity of hnRNA is very much greater than that of messenger RNA. This is seen in two ways. The hnRNA population contains a greater proportion of sequences derived from repetitive elements; and the complexity of its nonrepetitive sequence components is much greater. Comparison with mRNA suggests that hnRNA contains all of the sequences found in the cytoplasm. Thus the message sequences must be selected for transport to the cytoplasm, the nonselected sequences suffering degradation within the nucleus.

Important questions on the processing mechanism that need to be answered are whether each hnRNA species gives rise to only a single or to more than one mRNA species; and what proportion of the hnRNA molecules serves as precursor to mRNA. Concerning the selection of sequences for conservation, it is important to know whether some hnRNA molecules are degraded in their entirety or all contain (at least) one mRNA sequence. Comparisons between hnRNA populations of different cells can be used to ask whether differences in messenger sequence content result from changes in transcription or from the choice of different sequences for conservation from a constant hnRNA pool.

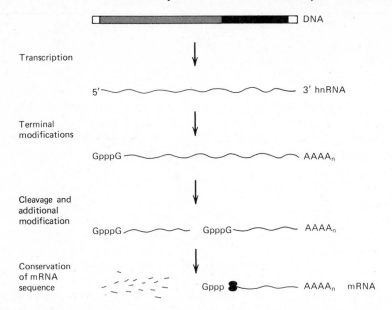

Figure 25.1. Models for derivation of mRNA from hnRNA. The DNA represents a single transcription unit in which black indicates the regions represented in mRNA and shading denotes regions that are transcribed but not translated. The primary transcript is capped at the 5′ end and polyadenylated at the 3′ end. In the first model, the hnRNA is cleaved and a cap is added to the new 5′ end; the polyadenylated end of the molecule is conserved as mRNA. The cap on the original 5′ terminus is degraded with this part of the molecule. Similar events could result in opposite orientation if the 3′ end generated by cleavage were polyadenylated and this molecule were conserved; then the original polyadenylated sequence would turn over. In the second model, the message regions in DNA are separated by an intervening sequence which is spliced out from the hnRNA and then degraded. This means that the original 5′ and 3′ termini of the primary transcript may be conserved.

Fate of Rapidly Labeled hnRNA

Much of a radioactive pulse label incorporated into nuclear RNA appears to turn over in the nucleus without entering the cytoplasm; and Harris (1963) first suggested that a large proportion of the nucleoplasmic RNA must be degraded in the nucleus. Characterizing the nuclear RNA presents the same problem that was encountered in distinguishing mRNA from rRNA: most of a radioactive label enters 45S precursor rRNA and matures through a series of precursors to give the stable rRNAs. A pulse label enters nucleoplasmic RNA more rapidly than nucleolar RNA; for the first 5–15 minutes there is little labeling in nucleolar RNA, but then it becomes predominant (Penman, 1966; Warner et al., 1966a). Characterized by the early fate of a pulse label, the nucleoplasmic RNA displays a heterodispersed distribution ranging in size from much larger than the 45S rRNA to less than that of mature mRNA. This gave rise to its description as heterogeneous nuclear RNA. Once label has begun to appear in the 45S rRNA precursor, it becomes impossible to distinguish the hnRNA sedimenting at less than 45S

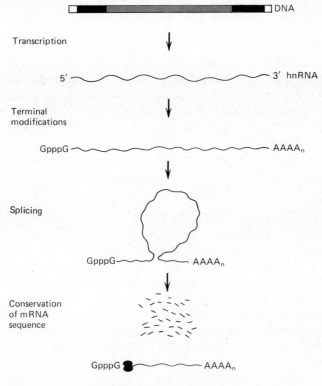

Figure 25.1. (*Continued*)

from the rRNA molecules. Some experiments therefore have followed the behavior of the >45S hnRNA molecules on the assumption that this will be typical of the entire population. The kinetics of hnRNA synthesis also can be followed by using low doses of actinomycin, which selectively block rRNA synthesis but allow nucleoplasmic RNA synthesis to continue. The label in hnRNA can be chased by applying a high dose of actinomycin to block all RNA synthesis (Penman, Vesco, and Penman, 1968; Greenberg and Perry, 1971).

When HeLa cells are incubated in low actinomycin, all of a radioactive label enters hnRNA; after the addition of high actinomycin, the label decays rapidly but does not appreciably enter the cytoplasm. Most of the hnRNA therefore turns over rapidly within the nucleus. More sophisticated labeling procedures have been introduced to control pool sizes and permit measurements of decay rates. The instability of hnRNA is a feature common to all cell types that have been investigated, including HeLa and L cells, cultured insect cells, and sea urchin blastulae. There may be one or more decay components; half lives are summarized in Table 25.2 and range from 20 minutes for either the entire population or the short-lived component to as much as 180 minutes for a longer lived component.

The decay within the nucleus of most of a radioactive label incorporated into hnRNA has made it difficult to chase the label into the cytoplasm. However, two types of experiment suggest that mRNA is indeed derived from hnRNA, although it can represent no more than a small part of its mass. The synthesis of mRNA shows the same sensitivity to actinomycin as hnRNA transcription, some hundredfold less than the sensitivity of rRNA synthesis (Penman, Vesco, and Penman, 1968; Perry and Kelly, 1970). And the terminal modifications at the 5′ and 3′ termini of hnRNA can be chased into mRNA. Of the polyadenylated 3′ ends, most can be chased into mRNA but some may turn over within the nucleus. At least a large proportion, and possibly all, of the 5′ methyl caps are conserved for transport to the cytoplasm.

Size and Secondary Structure of hnRNA

Early experiments to characterize hnRNA used fractionation on gradients or gels in the absence of denaturing agents (Warner et al., 1966a; Greenberg and Perry, 1971). Under these conditions the largest molecules are very large indeed, apparently up to about 100S, with this maximum size corresponding to perhaps 50,000 bases. The content of duplex material is about 5%. Although some hnRNA molecules are as small as mature messengers, these constitute only a small proportion of the population; most of the hnRNA appears to lie in a size range of from 10 to 100 times the length of mRNA, which for animal cells has an average of about 2000 bases. But under denaturing conditions the apparent range of sizes of hnRNA is somewhat reduced, as seen by a reduction in the sedimentation rate which is greater than can be accounted for simply by the unravelling of secondary structure. The reduction in size requires fairly stringent denaturing conditions; a concentration of 80% DMSO reduces the sedimentation range so that most molecules reside between 28S and 45S, typically with a duplex content of about 1.5% (Federoff, Wellauer, and Wall, 1977).

Two views have been taken of this effect. One is that the large molecules represent genuine primary transcripts, but that nicks have been introduced as steps in processing; however, the molecule has remained held together by the existence of intramolecular duplex regions. The effect of denaturation then is to allow the molecule to separate into the different processed regions (Derman and Darnell, 1974). The implication is that hnRNA must be characterized under nondenaturing conditions to determine the true size of the primary transcript. The other view is that the rapidly sedimenting species are artefacts due to aggregation. This must depend on the formation of intermolecular duplexes. In this case the true nature of the nuclear transcripts can be established only under denaturing conditions that prevent the formation of aggregates. A corollory is that it becomes difficult to determine the true proportion of intramolecular duplex regions that may be a genuine feature of hnRNA, since they will respond to denaturation in a manner similar to that of the artefactual intermolecular associations.

The demonstration that many of the duplex regions in hnRNA result from intermolecular reaction suggests that the formation of very large molecules is mostly artefactual (Federoff, Wellauer, and Wall, 1977). When labeled hnRNA was mixed with excess unlabeled hnRNA under denaturing conditions and then returned to nondenaturing conditions, the labeled material entered high molecular weight aggregates over a 20 minute period. This time dependence suggests that association is intermolecular (compare with the almost instantaneous snapback of intramolecular complements in DNA discussed in Chapter 19.) When the duplex regions are recovered by degrading the single-stranded RNA with ribonuclease, upon hybridization with excess DNA they appear to be transcribed from repetitive sequences. (The general absence of these sequences from messengers explains why the same problem is not encountered with mRNA). The intermolecular aggregates have been visualized directly by spreading for electron microscopy, when the large species can be seen to contain several molecules linked by duplex regions of average length about 300 base pairs. This implies that the true size distribution of primary transcripts lies largely in the range 5000–14,000 bases as seen in the denatured condition. The number average size is about 9000–10,000 bases.

Another attempt to define the true condition of the primary transcript has made use of very short pulse labels to identify nascent molecules (Derman, Goldberg, and Darnell, 1976). Under these conditions processing does not seem to have occurred to labeled molecules, so their lengths should reflect the sizes of the transcription units. This shows that the median size is about 5000 bases; only 10% of the molecules are longer than 20,000 bases. These lengths were estimated under denaturing conditions, but are only slightly longer in the absence of denaturation.

The question of whether the observed distribution of hnRNA molecules represents the primary transcripts, or is biased by the inclusion of cleavage products at least to some extent, is hard to resolve. Ultraviolet sensitivity mapping has been applied to compare the relative sizes of the transcripts with the lengths of their transcription units in DNA. The technique measures the sensitivity to inhibition of transcription by increasing doses of ultraviolet. Since the irradiation acts by introducing pyrimidine dimers in DNA which terminate transcription, the sensitivity of any particular transcript depends upon its length. The sensitivities of different size classes of HeLa hnRNA are proportional to their physical sizes, that is, their synthesis is inhibited at the same rate per unit length (Giorno and Sauerbier, 1976). This suggests that each size class corresponds to primary transcripts.

Messenger RNA displays a sensitivity between that of 18S and 28S ribosomal RNAs (Goldberg, Schwartz, and Darnell, 1977). The sensitivities of hnRNA and rRNA precursors appear to be similar; and if the rRNA is used as a calibration, this would place the sensitivity of the mRNA in the range 5000–10,000 bases. The discrepancy between this estimate and the physical length of mRNA suggests that messengers are derived from longer primary transcripts, whose length distribution is close to that observed for hnRNA.

Using this calibration assumes that RNA polymerases I and II must be equally impeded by pyrimidine dimers. A better calibration would be to use an individual mRNA precursor of known length; but this is not yet possible.

Formally the parameter determining sensitivity to ultraviolet is distance from the promoter. When a molecule is derived by cleavage of a larger precursor, its sensitivity will depend on its location and on the demands of the processing system. If an mRNA sequence is located at the 3' end of a longer primary transcript, its sensitivity should coincide with the size of the transcript rather than with the messenger length. If it lies at the 5' end its sensitivity will coincide with its own size if cleavage can occur as soon as its synthesis per se is complete; but the sensitivity will be set by the length of the primary transcript if processing requires completion of the precursor. The increased sensitivity of mRNA compared with its physical length therefore implies that it is derived from longer precursor molecules but does not necessarily mean that the message sequences must lie at the 3' end.

The general burden of these results is that mRNA is derived from longer primary transcripts, but that the discrepancy between the sizes of mRNA and hnRNA is not so great as originally thought. Usually the messenger lengths have an average of 2000–2100 bases in animal cells, compared with the average length of denatured hnRNA of 8000–10,000 bases. Data for several systems are summarized in Table 25.1. Of course, individual genes may show considerable variation around these average values. However, the general conclusion that there is often a substantial length discrepancy between hnRNA and the corresponding mRNA is supported by more detailed characterization of particular sequences. Cloned DNA sequences complementary to 9 mRNAs of CHO cells varying from 1300 to 3500 bases in length hybridize with nuclear RNA molecules ranging from 2400 to 13,400 bases long; the ratios of nuclear/cytoplasmic length for the individual species range from 2 to 6 (Harpold et al., 1979). Characterization of several avian and mammalian interrupted genes shows a similar range (see Table 26.6).

Although some duplex regions form in hnRNA by intermolecular aggregation, intramolecular pairing occurs also, creating duplex regions that presumably are involved in maintaining secondary structure. These can be found in hnRNP as well as in isolated hnRNA. Under these conditions they provide about 3% of the nucleotide mass (Calvet and Pederson, 1977). This is perhaps the best estimate for native duplex content, since duplex regions probably are less likely to form as artefacts when the hnRNA is maintained as ribonucleoprotein.

When hnRNA is allowed to renature under conditions that should permit only intramolecular duplex formation, a large part of the duplex material is derived from repetitive sequence elements (Jelinek and Darnell, 1972; Jelinek et al., 1974). Up to some 4% of the mass of HeLa hnRNA forms a relatively simple series of spots upon ribonuclease fingerprinting, with a total complexity probably less than 1000 bases (Robertson, Dickson, and

Jelinek, 1977). When the inverted repeat fraction isolated from DNA (see Chapter 19) is transcribed in vitro with E. coli RNA polymerase, the product has a fingerprint pattern very similar to that of the hnRNA duplex material (Jelinek, 1977). This suggests that some of the intramolecular duplexes formed in hnRNA represent simple or complex hairpins derived from the inverted repeats of DNA. The possibility that these regions might be involved in processing has been in mind ever since their discovery. Whether it is relevant that they appear to be less tightly bound to protein than the single stranded regions is hard to say at present (Calvet and Pederson, 1978). Their existence in vivo has been inferred from the ability of the nucleic acid cross linking agent psoralen to increase the proportion of foldback material in hnRNA extracted subsequently (Calvet and Pederson, 1979).

Although hnRNA is found in the form of ribonucleoprotein, this has been more difficult to characterize than mRNA, partly because of the lack of a functional assay comparable to release from polysomes (see Chapter 23). But this form of organization is apparent in a wide variety of organisms, including animal cells, marine species, and even lower eucaryotes. The array of proteins complexed with hnRNA is more complex than that found with mRNA, covering a wide range of molecular sizes from 40,000 to 180,000 daltons. Fairly wide variations have been reported in the spectrum of associated proteins in different systems; whether this is genuine or reflects artefacts of extraction is hard to say. In many cases, however, about 6 predominant proteins in the molecular weight range 25,000–45,000 daltons provide some 75–90% of the mass.* The proteins are associated with hnRNA with different degrees of firmness, as seen in their progressive release by salt. Some of them are modified, for example, by phosphorylation.

Particles of discrete size have been isolated from hnRNP by treatment with ribonuclease. The mass of hnRNP then is converted to species sedimenting at a size variously reported as 30S to 40S; the poly(A) segment is found in a 15S ribonucleoprotein particle. An early model suggesting that the structure of hnRNP represents stable protein aggregates connected by RNA has given way to the view that these represent beads of RNP on a string, in which the hnRNA periodically is coiled into particles by association with protein, the particles being connected by the thread of RNA. However, the possibility that breakage of RNA induces rearrangement of proteins, possibly forming particles as artefacts, has been raised; one approach to exclude such occurrences would be to use cross linking agents before ribonuclease treatment analogous to experiments with nucleosomes (see Chapter 13). Newly synthesized RNA in Triturus or Drosophila has been visualized by electron microscopy in the form of 20 nm ribonucleoprotein beads on a string; the contraction in length compared with free

*References: Georgiev and Samarina (1971), Niessing and Sekeris (1971), Olsnes (1971), Lukanidin et al. (1972), Pederson (1974a), Firtel and Pederson (1975). Augenlicht and Lipkin (1976), Beyer et al. (1977), Karn et al. (1977), Quinlan et al. (1977), Stevenin et al. (1977, 1978, 1979), Blanchard et al. (1978).

RNA fits well with the physical characterization of the hnRNP particles (Malcom and Sommerville, 1974, 1977; McKnight and Miller, 1976). In animal cells, the buoyant density of hnRNP is about 1.4 g-cm^{-3}, which corresponds to a protein: RNA ratio of about 4 by mass. If the isolated 40S particle is taken to have a mass of about 1.5×10^6 daltons, it could consist of about 40 protein chains each of 30,000 daltons associated with some 900 bases of RNA.

The isolated poly(A) segments are associated with only two protein species in HeLa cells, a major species of 74,000 daltons and a minor protein of 86,000 daltons (Kish and Pederson, 1975). Comparison with Table 23.1 suggests that the same major protein may be bound to poly(A) in hnRNA and mRNA.

The poly(A)-protein complexes isolated from either hnRNP or mRNP contain an appreciable proportion of UMP residues (about 20%). In hnRNP the U can be recovered in the form of oligo(U) stretches of about 20 bases, which correspond roughly with tracts previously identified as a component of hnRNA (Molloy et al., 1972). In mRNP the nature of the U residues is less clear (see Chapter 23). The oligo(U) of hnRNP can be released from the poly(A) by thermal denaturation. Kish and Pederson (1977) suggested that the oligo(U)-poly(A) duplex regions might provide binding sites for proteins.

The hnRNP may be attached to the nuclear lamina as a network of particles. Almost all (>96%) of the DNA can be removed from isolated nuclei by treatment with DNAase and ammonium sulfate; but half of the hnRNA is retained. The same result is obtained with either a steady state or pulse label, the latter implying that hnRNP must associate very rapidly with the nuclear ultrastructure (Herman, Weymouth, and Penman, 1978; Miller, Huang, and Pogo, 1978a). A network of RNP fibers attached to the lamina can be seen in the electron microscope, in agreement with earlier studies (Narayan et al., 1967; Manneron and Bernhard, 1969; Faiferman and Bernhard, 1969; Faiferman and Pogo, 1975). Digestion with RNAase destroys the network; many of the hnRNP proteins then are lost from the nuclear fraction. Prominent among the sequences comprising the surviving 2% of hnRNA mass are duplex regions and poly(A). These results suggest that newly synthesized hnRNA is packaged as a ribonucleoprotein and then attached to some structural component of the nucleus upon its release from chromatin. While surviving in the nucleus, hnRNA remains in this condition. It is possible that this may be important in the movement of messenger sequences to the cytoplasm.

Nuclear Polyadenylation and hnRNA Turnover

Sequences of poly(A) are found in nucleoplasmic hnRNA and have been characterized by the same methods applied to mRNA (see Chapter 23). The proportion of poly(A) in hnRNA lies in the range 0.5–1.0% and decreases with the length of RNA (Edmonds, Vaughan, and Nakazoto, 1971; Derman and Darnell, 1974). The length of HeLa nuclear poly(A) is about 230 nucleotides, a little longer than is seen when messenger RNA first appears in the cytoplasm (Sawicki, Jelinek, and Darnell, 1977). Almost all synthesis of poly(A) occurs in the nucleus; only a small amount (about 15 nucleotides) of terminal addition may occur in the cytoplasm (Brawerman and Diez, 1975). The nuclear site of polyadenylation, the similarity in length of poly(A) in hnRNA and mRNA, the identical location at the 3′ terminus, and the tenfold discrepancy in proportion, have been taken to suggest that the single tract of poly(A) on an hnRNA molecule together with the adjacent sequence may constitute the messenger that is conserved for transport to the cytoplasm.

Poly(A) is added to nuclear RNA after transcription and can be chased into cytoplasmic mRNA. The first experiments demonstrating that poly(A) is not coded in the template DNA made use of adenovirus, whose denatured DNA does not hybridize with poly(A). Yet apparently all adenovirus nuclear transcripts possess a length of poly(A) [actually up to 80% can be retained by virtue of poly(A); given the susceptibility of the isolation procedure to cleavage, this probably indicates 100% polyadenylation] (Philipson et al., 1971; Wall, Philipson, and Darnell, 1972). Similar results have been obtained for vaccinia virus (Sheldon and Kates, 1974).

The absence of sequences coding for poly(A) in cellular genomes cannot be demonstrated so directly, although there do appear to be far fewer poly(dA:dT) tracts than would be necessary if poly(A) were derived by transcription (Bishop, Rosbash, and Evans, 1974). But other lines of evidence indicate strongly that polyadenylation occurs after transcription.

The synthesis of poly(A) has characteristics distinct from that of hnRNA. Concentrations of actinomycin that inhibit hnRNA synthesis by 96% inhibit polyadenylation only by 60% (Darnell et al., 1971). More striking is the effect of cordycepin (3′ deoxyadenosine), which inhibits nucleolar RNA synthesis by causing production of molecules shorter than the 45S precursor (Penman, Rosbash, and Penman, 1970). The analogue 3′-deoxycytidine has the same effect (Abelson and Penman, 1972). Presumably the 3′-deoxynucleotides inhibit the nucleolar but not the nucleoplasmic RNA polymerase. But in spite of its failure to inhibit hnRNA synthesis, 3′-deoxyadenosine prevents the appearance of a label in mRNA. Figure 25.2 shows that this effect is peculiar to it and is not shown by 3′-deoxycytidine. This observation is open to two interpretations: hnRNA may not provide the precursor to mRNA; or if hnRNA is the precursor, 3′-deoxyadenosine may interfere with the processing of mRNA from hnRNA. The latter conclusion is suggested by subsequent observations that cordycepin prevents the synthesis of poly(A); of course, this explains the failure of 3′-deoxycytidine to have the same effect (Darnell et al., 1973).

Support for the idea that actinomycin and cordycepin affect hnRNA syn-

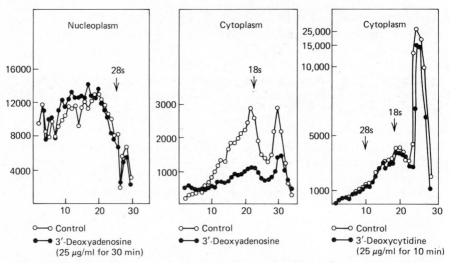

Figure 25.2. Involvement of polyadenylation in processing mRNA from hnRNA. Left: treatment with 3'-deoxyadenosine (cordycepin) does not inhibit synthesis of hnRNA. (Similar results, not shown, are obtained with 3'-deoxycytidine.) Center: the treatment with 3'-deoxyadenosine greatly reduces the appearance of mRNA in the cytoplasm. Right: 3'-deoxycytidine does not prevent appearance of cytoplasmic mRNA. Data of Penman, Rosbash, and Penman (1970) and Abelson and Penman (1972).

thesis and poly(A) synthesis differently because different enzyme processes are involved, and not because of some differential effect on adenosine metabolism, is provided by the experiments of Nakazoto, Edmonds, and Kopp (1974). HeLa cell hnRNA contains short internal sequences of oligo(A) that are transcribed as part of the molecule. Comparing the effects of the two drugs on oligo(A) synthesis and poly(A) synthesis showed that oligo(A) is inhibited like hnRNA by actinomycin and is not inhibited by cordycepin; whereas poly(A) synthesis is less inhibited by actinomycin and largely inhibited by cordycepin.

Polyadenylation takes place one base at a time; there is no pool of free poly(A). A poly(A) polymerase activity has been identified in both vaccinia virions and infected cells (Moss, Rosenblum, and Paoletti, 1973; Brakel and Kates, 1974a,b). It is not clear whether the same enzyme is involved in both cases. The addition of poly(A) to both newly synthesized and preexisting RNA molecules also has been accomplished by an enzyme present in the poliovirus replication complex (Spector and Baltimore, 1975). Multiple enzyme activities able to synthesize poly(A) have been reported in rat liver nuclei, yeast nuclei, wheat germ nuclei and chloroplasts, and hepatoma mitochondria.* The reality of these variations, which in each case rest

References: Edmonds and Abrams (1960), Winters and Edmonds (1973a,b), Burkard and Keller (1974), Haff and Keller (1975), Niessing (1975), Rose, Morris, and Jacob (1975).

on the isolation of different chromatographic peaks with enzyme activity, remains to be established by purification of the enzyme proteins.

Several observations suggest that hnRNA and mRNA have a precursor-product relationship in which addition of poly(A) to hnRNA plays an essential role. Labeled adenosine first is incorporated into the poly(A) of hnRNA but then can be chased into polysomal mRNA (Darnell et al., 1971; Mendecki, Lee, and Brawerman, 1972). When nuclear poly(A) is labeled with ^{3}H-adenosine and further synthesis is prevented by addition of cordycepin, the cellular redistribution of the label can be followed. During the first 30 minutes of the chase, the label of poly(A) in hnRNA declines and the label in mRNA increases (Jelinek et al., 1973a,b). Although complete equality could not be demonstrated, in these experiments most of the poly(A) added in the nucleus was conserved and transported to the cytoplasm. About 70% of the cytoplasmic poly(A) is present on polysomal mRNA and about 30% is found on RNA that is not associated with the ribosomes.

Polyadenylation and transport to the cytoplasm appear to be slow stages in the maturation of mRNA. The time required to synthesize an hnRNA molecule probably is fairly brief, say of the order of 1 minute for every 5000 nucleotides. Labeling for only 45 seconds with ^{3}H-adenosine, a time shorter than is needed for synthesis of most hnRNA molecules, sees incorporation of radioactive poly(A) into molecules of *all* size classes. Following treatment with actinomycin to inhibit hnRNA synthesis, exposure to ^{3}H-adenosine sees continued incorporation of label into poly(A). This again suggests that synthesis of poly(A) is independent of transcription of hnRNA and that it occurs after synthesis of the polynucleotide chain is completed.

By labeling cells with ^{3}H-uridine for 20 minutes before incubation with high levels of actinomycin (to inhibit hnRNA synthesis) and cordycepin (to inhibit polyadenylation), Penman et al. (1970) found that some of the label can enter the cytoplasm. Thus hnRNA molecules labeled during the 20 minutes can be processed to yield messengers, implying that the addition of poly(A) occupies less than 20 minutes. By shortening the labeling time in such experiments to 7.5 minutes, Adesnik et al. (1972) obtained a 70% reduction in the appearance of label in mRNA. This implies that many messengers which could be polyadenylated to become available for transport to the cytoplasm within a 20 minute period cannot mature in this way during the briefer 7.5 minute period. In experiments with L cells, Perry, Kelley, and LaTorre (1974) showed that the delay between transcription and the appearance of mRNA in the cytoplasm is about 20 minutes; but polyadenylation of the mRNA sequences occurs about 7 minutes before it enters polysomes. These results suggest that following transcription there may be a delay before polyadenylation occurs. An incidental observation is that when poly(A) synthesis is inhibited, those molecules that escape to the cytoplasm have shorter poly(A) tails than usual; thus the full length of poly(A) is not necessary.

Poly(A) may be added to hnRNA molecules in more than one state. Derman and Darnell (1974) have examined the addition of poly(A) to hnRNA molecules of different size classes, following very brief labels with ^{3}H-adenosine or ^{3}H-uridine. Large (>32S) molecules possess poly(A) that is added very soon after transcription, as judged by their high content of uridine label, which can only have been incorporated immediately before the completion of transcription. Small (<32S) molecules possess poly(A) that is added after some delay, as judged by their low content of uridine label, which implies that they must have been synthesized at some time preceding the addition of the labeled uridine. One possible explanation is that the smaller molecules are derived by cleavage of longer primary transcripts, and that the process of freeing a new 3′ end for polyadenylation takes some time.

Early models for messenger maturation postulated that addition of poly(A) to hnRNA takes place only at the 3′ terminus of the primary transcript. In none of these data, however, is there any evidence to exclude the possibility that polyadenylation occurs also at 3′ ends created by internal cleavage. The occurrence of such events has been identified in adenovirus infection by Nevins and Darnell (1978). The region *16–99* of adenovirus DNA constitutes a single transcriptional unit from which several different mRNAs may be derived. All transcriptional events involve the continuation of RNA polymerase to the end of the unit, as seen in the equimolar production of RNA for all parts. However, poly(A) may be added to 3′ ends representing several internal sites. Two models can be distinguished: a single polyadenylation event might occur for each molecule after its completion, involving the selection of a particular site for addition; or poly(A) might be added to an internal end after RNA polymerase has passed, but before transcription has been completed. The first model predicts that a pulse label should first enter polyadenylated molecules with 3′ ends corresponding with the termination point, then working back through polyadenylated molecules with ends successively farther from the terminus. But the results fit with the prediction of the second model, that all ends are labeled simultaneously. Processing may therefore precede completion of transcription.

There has been much discussion about whether poly(A) added to hnRNA in the nucleus is conserved entirely for transport to the cytoplasm. Conservation or near conservation has been reported by Jelinek et al. (1973a) and Nevins and Darnell (1978); turnover of appreciable amounts of poly(A) in the nucleus has been found by Perry, Kelley, and LaTorre (1974). This bears on the question of whether addition of poly(A) is sufficient to preserve the adjacent sequence for transport to the cytoplasm (for review see Lewin, 1975b). Whether or not addition of poly(A) is a signal for conservation, however, it is clear that its addition is essential for the maturation and appearance in the cytoplasm of poly(A)$^+$ messengers. But this cannot be the sole mechanism for selection of messengers, since histone mRNA and other poly(A)$^-$ mRNAs do not need polyadenylation. Although histone mRNAs appear in the cytoplasm after a much shorter lag than that found for poly(A)$^+$ mRNA synthesis, the other poly(A)$^-$ messengers show delays similar to those of the poly(A)$^+$ class (Milcarek, Price, and Penman, 1974). Here the difference cannot be due to the time required for polyadenylation, the lack of which originally had been thought to be responsible for the more rapid appearance of histone mRNA.

Is the extensive turnover of hnRNA the result simply of the degradation of those parts of long precursors that are not represented in mRNA, or are some hnRNA molecules degraded in their entirety and not used at all for messenger synthesis? Recent developments in pulse chase techniques have made it possible to determine the proportion of the mass of hnRNA that is converted into poly(A)$^+$ mRNA. Table 25.1 summarizes data for several systems on the relative sizes of mRNA and hnRNA and on the proportions of the molecules that are polyadenylated. Usually less than 30% of the hnRNA bears poly(A), compared with more than 60% for mRNA. The hnRNA turns over more rapidly than the mRNA, but may have a component with a half life of up to 3 hours, somewhat longer than originally thought. Table 25.2 shows that generally less than 10% of the mass of the hnRNA of higher eucaryotes is converted into poly(A)$^+$ mRNA. This can be used to calculate the proportion of molecules giving rise to mRNA sequences, on the assumption that each hnRNA contains the sequence of no more than one mRNA. The proportion generally is less than 30%; even allowing for the existence of poly(A)$^-$ mRNA, this implies that at least half of the hnRNA molecules must be degraded in their entirety within the nucleus. Often the proportion of hnRNA molecules that gives rise to poly(A)$^+$ mRNA is similar to the proportion that is polyadenylated; this would be consistent with the idea that at least most poly(A) sequences are conserved for transport to the cytoplasm.

Only a subfraction of hnRNA therefore gives rise to mRNA. Can this be distinguished from the molecules that suffer total degradation? There have been two reports of putative mRNA-precursor subfractions in the nucleus. One is the observation of Price, Ransom, and Penman (1974) that about 10% of HeLa hnRNA is eluted from a pellet of chromatin with ~500 mM ammonium sulfate. It takes the form of ribonucleoprotein particles shorter than the total population, with hnRNA of average size about 28S, say three times larger than mRNA. This fraction contains about 40% of the nuclear poly(A); in the presence of cordycepin this may be completely chased out, in contrast with the poly(A) in the predominant (90%) hnRNA fraction, more than half of which remains in the nucleus after such treatment. Whether the short subfraction consists of primary transcripts or processing intermediates is not known. In contrast with expectation, it does not seem to be enriched in message sequences (Herman, 1979).

Another subfraction has been identified by the use of DRB, an inhibitor

Table 25.1   Size and polyadenylation of hnRNA and mRNA

System	Average size		Proportion polyadenylated	
	hnRNA	mRNA	hnRNA, %	mRNA, %
Dictyostelium	1500	1200	30–60	>75
Drosophila tissue culture	4200	2100	~20	~60
Aedes tissue culture	8400	2100	13–17	—
Sea urchin blastula	8800	2100	12–20	~50
HeLa cell	10000	2100	20–30	~70

The sizes of hnRNA and mRNA are number averages determined from sedimentation rates and include any poly(A) tails. Probably they are accurate only within 20% or so. Note that somewhat smaller values have been obtained in some experiments, with lengths of about 5000 bases for hnRNA and 1500 bases for mRNA in mammalian cells. The proportion of polyadenylated molecules is determined by reaction of poly(A) with oligo(U) or oligo(dT) after correction for any internal tracts of oligo(A).

References and details of other systems: Dictyostelium—Firtel and Lodish (1973); other lower eucaryotes also have small RNA, including Achlya ambisexualis in which both hnRNA and mRNA are about 1150 bases (Timberlake et al., 1977) and yeast in which mRNA is 1500 bases (Hereford and Rosbash, 1977). Drosophila and Aedes—Lengyel and Penman (1975). Sea urchin—Nemer et al. (1974, 1975), Dubroff and Nemer (1975, 1976); HeLa—Derman, Goldberg and Darnell (1976), Federoff, Wellauer, and Wall (1977), Jelinek et al. (1973a,b), Milcarek, Price, and Penman (1974). Lower estimates for hnRNA size (4500 bases) and mRNA size (1400 bases) have been made in mouse brain by Bantle and Hahn (1976); but a length for mRNA of about 1950 bases has been obtained in several mouse tissues by Hastie and Bishop (1976). In chick oviduct mRNA is about 2200 bases long (Hynes et al., 1977).

of RNA synthesis. Under conditions when treatment with DRB inhibits the synthesis of total hnRNA by about 70%, the production of mRNA is inhibited by about 95% (Sehgal, Darnell, and Tamm, 1976). This might imply that there exist two classes of hnRNA, sensitive transcripts, which are mRNA precursors, and insensitive transcripts, which are not. On the other hand, the effect of DRB is not yet fully characterized; originally thought to act at or soon after initiation, now it appears to terminate chain growth prematurely, allowing perhaps a few hundred base pairs of chain elongation (Sehgal, Fraser, and Darnell, 1979; Fraser, Sehgal, and Darnell, 1979). It is hard to exclude the possibility that DRB has a second effect, one specific to the processing of mRNA from hnRNA. Synthesis of the minor (10%) hnRNA fraction is more sensitive to DRB than is that of the predominant fraction (Herman, 1979).

Table 25.2 Approximate kinetic characteristics of hnRNA and mRNA

System	Decay rates ($T_{1/2}$) hnRNA	mRNA	Proportion of hnRNA converted to poly(A)$^+$ mRNA mass, %	number, %
Dictyostelium	not known	not known	~75	~100

Drosophila	poly(A)$^-$: 12 min
	poly(A)$^+$:
	  70% 20 min
	  30% 180 min
		80% 50 min
		20% 4 hr
			14–20	28–40

Aedes	not known	80% 60 min 20% 20 hr	~3	~13
Sea urchin	total: 23 min	not known	4–7	17–29

Hela	total: 70 min
	poly(A)$^+$:
	  40% 60 min
	  60% 150 min
		60% 4 hr
		40% 24 hr
			3–6	15–30

The decay rates are determined by following the disappearance of label from RNA when further synthesis is prevented. The proportions given for components of different stability refer to pulse label experiments and thus describe the turnover of newly synthesized hnRNA or mRNA. The steady state population contains a much greater proportion of the long lived component, usually with about 70% of the mRNA population, for example, representing the more stable component. The mass conversion of hnRNA into mRNA gives the proportion of nucleotides incorporated into hnRNA that enters cytoplasmic poly(A)$^+$ mRNA. This is used to calculate the proportion of hnRNA molecules that gives rise to mRNA molecules by relying on the size discrepancy noted in Table 25.1 and assuming that each hnRNA molecule can give rise to no more than 1 mRNA molecule. The proportion would be correspondingly reduced if the size of hnRNA were smaller than given. Allowing for the presence of poly(A)$^-$ mRNA, the conversion factors would be up to about 30% greater than given above.

References: Dictyostelium—Firtel and Lodish (1973); Drosophila—Lengyel and Penman (1975, 1977), Levis and Penman (1977); Aedes—Lengyel and Penman (1975), Spradling, Hui, and Penman (1975); sea urchin—Brandhorst and Humphries (1972), Davidson (1976), Galau et al. (1977c), Grainger and Wilt (1976); HeLa—Brandhorst and McConkey (1974), Herman and Penman (1977), Singer and Penman (1973), Scherrer et al. (1970), Soiero et al. (1968).

Capping and Internal Methylation

Both methylated caps and internal N^6-methyl adenine are found in hnRNA. On average there are about 0.4 m^6A residues per 1000 nucleotides of HeLa or L cell hnRNA, compared with about 1.0 m^6A residues per 1000 bases of mRNA (Salditt-Georgieff et al., 1976; Schibler and Perry, 1976, 1977). The content of m^6A in poly(A)$^+$ hnRNA is greater than that of total hnRNA, close to the value seen with mRNA (Lavi et al., 1977). There are about 0.1 caps per 1000 bases of hnRNA, compared with about 0.5 caps per 1000 nucleotides in mRNA (see Table 23.2). Similar results on cap ratios are obtained with poly(A)$^-$ and poly(A)$^+$ hnRNA, although reports have differed on the question of whether small and large hnRNA molecules have the same proportion of caps.

Direct measurement of the proportions of different termini in hnRNA show that caps represent about 20% of the 5′ ends after a 30 minute period of labeling, but this increases to 50% after 2 hours of labeling. This implies that caps are metabolically more stable than the other 5′ ends, which take the form of 20% tri- and diphosphate and 30% monophosphate after the 2 hour label. In molecules labeled for short periods, the proportion of triphosphate termini is greater in smaller molecules than in large hnRNA. This may mean that some hnRNA molecules are capped at the 5′ terminus before chain growth has been completed.

The caps in hnRNA are of type 1 and the proportion of different bases in the methylated position is similar to that found in mRNA (see Table 23.2) (Perry et al., 1975; Salditt-Georgieff et al., 1976). The proportion of purine bases is greater in hnRNA (79%) than in mRNA (64%). If transcription is taken to initiate exclusively with purines, as reflected by the presence only of pppA and pppG at 5′-triphosphate ends, this may reflect the presence in mRNA of an increased number of molecules with 5′ termini derived by internal cleavage of the primary transcript. In fact, this would set a minimum proportion of 36% for the number of mRNA molecules not starting with the first base of the primary transcript, and therefore presumably not derived from the 5′ end. Although both C and U are found as methylated pyrimidines, in 5′-monophosphate termini the most common sequence is pU, implying some specificity in cleavage (Schibler and Perry, 1976, 1977). Little is known about the substrate used in vivo for the capping reaction, in particular whether it is the same for 5′ termini of primary transcripts (originally triphosphate) and for 5′ termini derived by cleavage (originally monophosphate) (see Chapter 23).

Cap 2 structures are found only in mRNA and the second methylation may occur in the cytoplasm (see Figure 23.5). The proportion of bases found in the first methylated (X^m) position differs significantly from the average constitution of mRNA and hnRNA, which suggest that a specific subset of the cap 1 sequences may be selected for further modification to yield cap 2 (Perry and Kelley, 1976). The second methylated position (Y^m) is most commonly occupied by a pyrimidine (U = 57%, C = 28%, Pu = 15%). Another difference between the caps of hnRNA and mRNA is that the proportion of dimethyl adenosine is much greater in the cytoplasm than in the nucleus.

Are methyl groups in hnRNA conserved in the formation of mRNA or do they turn over in the nucleus? Pulse chase experiments show that label in the caps can be followed from hnRNA to mRNA; this supports the inference drawn from the similarity of base compositions of hnRNA and mRNA caps that there may be a precursor-product relationship (Perry and Kelley, 1976). The results are consistent with, although they do not prove, the conservation of cap structures in processing. The internal m^6A also may be conserved. It occurs in trinucleotides of the same composition in HeLa and L cell mRNA, a point perhaps of evolutionary significance, and takes the same form in L cell hnRNA: 75% is found as $GA^{Me}C$ and 25% as $AA^{Me}C$ (Wei and Moss, 1977a; Schibler, Perry and Kelley, 1977). Again the similarity suggests the possibility that the internally methylated bases in hnRNA are precursors to those in mRNA. This is supported further by the report of Chen-Kiang et al. (1979) that during adenovirus infection the kinetics of m^6A labeling in nuclear and cytoplasmic RNA are consistent with conservation of the methyl groups. Since the methylation occurs on large nuclear molecules before their processing to mRNA, this could mean that it is involved in providing some sort of signal for processing. This is consistent with the increase in number of m^6A residues/1000 bases proceeding from large hnRNA to small hnRNA to mRNA. This role would be similar to that postulated for the conserved methyl groups in ribosomal RNA processing (see Chapter 27). Internal methylation cannot be essential for the generation of mRNA, however, since some messengers—such as globin or the histones— do not possess m^6A (see Table 23.3).

In summary, the modification of hnRNA appears to involve sequences that may be destined to become messengers. At least a large part of the caps or poly(A) added to 5′ and 3′ ends in the nucleus may find its way into cytoplasmic mRNA. Capping might occur at 5′ ends generated by internal cleavage as well as the terminus of the primary transcript; the same clearly is true for polyadenylation of adenovirus nuclear RNA and may apply also to cellular species. But this provides only some relief from the size discrepancy between hnRNA and mRNA; since large molecules of hnRNA may contain both cap and poly(A), in these cases either internal sequences must be removed or one or the other of the modifications must be lost by cleavage as described in Figure 25.1. This is to assume that each long hnRNA molecule contains the sequence of no more than one potential messenger; although it is possible that some hnRNA molecules could contain multiple messenger sequences, in those cases in which the precursors for individual mRNA species have been characterized, they do not appear to include additional stable sequences (see Chapter 26). Even with this assumption, only some of the degradation of hnRNA can be attributed to loss of non-message regions of long precursors; an appreciable proportion of the hnRNA of several species appears to represent molecules that are entirely degraded, that is, without ever giving rise to a messenger for cytoplasmic export. Perhaps there are several different pathways for the processing of mRNA from hnRNA, reflecting variations in the location of the message sequence within the transcription unit; it is too early yet to say to what extent each might be used in the nucleus.

Sequence Complexity of hnRNA

To answer the question of what proportion of the genome is transcribed, it is necessary to determine the nucleotide complexity of hnRNA. This is much greater than that of mRNA, the difference identifying sequences that must be degraded within the nucleus. Do these sequences represent only the nonconserved parts of precursors to mRNA; or do some molecules represent sequences destined to remain entirely within the nucleus? This is to ask whether the processing of hnRNA involves a quantitative or qualitative selection of sequences to act as mRNA. We have seen that only some hnRNA molecules appear to contribute messengers to the cytoplasm while others are entirely degraded. The question is whether these two sets of molecules represent the same or different sequences. In the first case all transcripts must possess a potential message region; each copy will have a certain probability of preserving it. The alternative is that only some molecules possess potential messages, while others possess no such region (or possess a region that is utilized with zero probability in the given cell type). Difficulties in working with hnRNA, in particular in obtaining a preparation of primary transcripts rather than partially processed molecules, mean that present information is equivocal; but what direct data are available, and the inferences that may be drawn from comparing hnRNA populations in different cells, suggest that the genome may contain transcription units whose products do not give rise to messengers in certain cells although they are transcribed.

The use of labeled hnRNA as a tracer in a renaturation reaction of excess cellular DNA gives results similar to those obtained with mRNA and subject to the same limitations. Table 25.3 summarizes available data. Comparison with the characteristics of mRNA given in Table 24.1 suggests that there is perhaps a tendency toward the inclusion of a greater proportion of repetitive component in hnRNA; but in most cases it is difficult to make firm conclusions because of the failure of the reaction to proceed to completion. The repetitive sequences represented in hnRNA occupy a wide range of repetition frequencies, in HeLa cells from about 200 to some 10,000 (Darnell et al., 1970; Pagoulatos and Darnell, 1970).

A clear difference between hnRNA and mRNA is provided by the sea urchin, in which the presence of repetitive sequence transcripts in hnRNA contrasts with their virtual absence from mRNA. Most of the hnRNA mole-

Table 25.3   Reiteration frequencies of sequences represented in hnRNA populations

Cell	Source of labeled hnRNA	Proportion hnRNA		Extent of reaction		
		Nonrep., %	Repetitive, %	RNA, %	DNA, %	DNA excess
Hela cell	actinomycin	25	15	40	90	$>10^5$
L cell	actinomycin	27	13	40	100	200
Friend cell	cDNA = 400 bases from 3′ terminus	48	10	58	80	4×10^5
Rat myoblast	actinomycin $>50S$	25	15	40	90	$>10^5$
Rat ascites	$>80S$	30	25	55	90	4×10^5
S. purpuratus gastrula	pulse label $>40S$	19 25	8 8	27 33	90 90	300 1.5×10^4
Aedes culture	actinomycin	38	15	53	85	$>10^5$

Details are similar to the reactions described in Table 24.1 and represent the hybridization of hnRNA in a great excess of DNA. The source of the hnRNA indicates whether specific labeling was achieved by suppression of rRNA synthesis with actinomycin or by use of a pulse label and whether a specific size class of hnRNA was utilized. Because of the incomplete nature of the reaction, estimates for the proportion of each component are only approximate; the repetition frequency of the repetitive component cannot be determined, but usually appears to correspond to 10^2–10^3.

References: HeLa, rat myoblast, and Aedes—Spradling et al. (1974); L cell—Greenberg and Perry (1971); Friend cell—Getz et al. (1975); Rat ascites—Melli et al. (1971); S. purpuratus—Smith et al. (1974).

cules of sea urchin gastrula contain a repetitive sequence. In hnRNA sheared to a length of 1100 nucleotides, some 23% of the molecules can react at low Cot, although only 8% of the mass is resistant to ribonuclease (Smith et al., 1974). This discrepancy is akin to the classical demonstration that repetitive and nonrepetitive sequences are interspersed in the genome (see Chapter 18). The average length of the repetitive sequences in hnRNA is about 300 bases (the same as the major interspersed repetitive component of the genome). Assuming that the distribution of repetitive sequences in hnRNA is random, at least one such sequence must be present in 70% of the hnRNA molecules present in the initial population before shearing. Contrasted with the sole representation of nonrepetitive sequences in mRNA, this suggests that the repetitive regions are among those lost when mRNA is processed from hnRNA.

A similar interspersion of repetitive and nonrepetitive sequences has been reported in rat hnRNA. When molecules sheared to different sizes were hybridized to low Cot, the reaction falls from 80% at a length of 10,000 bases to 40% at a length of 1200 bases; below this there is a steep fall as repetitive sequences are separated from the nonrepetitive sequences (Holmes and Bonner, 1974). Indirect data suggesting that repetitive sequences are more common in hnRNA than in mRNA are provided by demonstrations that hnRNA molecules aggregate by intermolecular duplex association involving repetitive elements (see above). Thus although in most cases the total difference in apparent representation of repetitive sequences in mRNA and hnRNA is not striking, there is an important difference in organization. The repetitive sequences present in hnRNA are covalently linked to the nonrepetitive sequences, in contrast with the sequestration of the two sequence types in different molecules in mRNA. Thus nonrepetitive messengers must be divorced from any repetitive sequences to which they were linked in hnRNA.

The reaction of excess hnRNA with purified nonrepetitive DNA is the most satisfactory way to assay the complexity of the nuclear transcripts. As in the reaction with mRNA, it is necessary to exclude any possible reaction involving repetitive sequences. Their interspersion with nonrepetitive sequences, and difficulties in handling hnRNA, make the kinetics of reaction with a reverse cDNA transcript less suitable. Table 25.4 summarizes available saturation data. It is evident that there is a wide spread in reported

Table 25.4 Sequence complexity of hnRNA

Cell	Source of hnRNA	Saturation of nonrepetitive DNA with hnRNA, %	Complexity of hnRNA, kb
Mouse tissues: nonrepetitive DNA of complexity 1.8 × 10^9 bp			
Friend cell	nuclear RNA >35S	5.9	210,000
Friend cell	poly(A)⁺ nuclear RNA	5.4	194,000
Embryonal carcinoma	poly(A)⁺ nuclear RNA	3.0	108,000
Liver	total cell RNA	1.6	58,000
		3.0	108,000
		4.8	173,000
Kidney	total cell RNA	1.8	65,000
		3.0	108,000
		3.9	140,000
Brain	total cell RNA	7.8	281,000
		9.0	324,000
Brain	total nuclear RNA	21.2	763,000
Brain	poly(A)⁺ nuclear RNA	13.5	486,000
Rat tissues: nonrepetitive DNA of complexity 2.0 × 10^9 bp			
Brain	total nuclear RNA	15.6	590,000
Liver	total nuclear RNA	10.9	435,000
Kidney	total nuclear RNA	5.3	210,000

Table 25.4   (*Continued*)

Cell	Source of hnRNA	Saturation of nonrepetitive DNA with hnRNA, %	Complexity of hnRNA, kb
Sea urchin (S. purpuratus): nonrepetitive DNA of complexity 6.2 × 10⁸ bp			
Gastrula	nuclear RNA > 40S	14.0	174,000
Adult intestine	total nuclear RNA	17.9	223,000
Sea urchin (T. gratilla): nonrepetitive DNA of complexity 5.1 × 10⁸ bp			
Blastula	total nuclear RNA	16.6	169,000
Pluteus	total nuclear RNA	17.3	176,000

Experiments were performed by reacting labeled nonrepetitive DNA with an excess of RNA as described in Table 24.3. The RNA sequence complexity is calculated from the saturation value by assuming the nonrepetitive DNA complexity given in the table; note that values used in published calculations vary over about a 2-fold range, causing some variation in the conversion to RNA complexity. Another factor causing variation is the proportion of cellular DNA actually retained in the isolated nonrepetitive fraction. Values for poly(A)$^+$ mRNA complexity have been obtained in the same work for Friend cell (42,000 kb), mouse brain (144,000 kb) and S. purpuratus gastrula (17,000 kb). In two instances kinetic estimates for poly(A)$^+$ mRNA complexity have been determined from the $Rot_{1/2}$ for excess reaction with cDNA. For Friend cells this gives 90,000 kb for hnRNA and 19,000 kb for mRNA, a discrepancy of about 2.5-fold with the corresponding saturation estimates. For embryonal carcinoma this gives 102,000 kb for hnRNA and 16,000 kb for mRNA. In both cases the kinetic approach resolves two components in hnRNA, one abundant and one scarce. Note that this method suffers from problems posed by the large discrepancy between the long length of hnRNA and the short length of the region that is reverse transcribed into cDNA from the 3′ terminus.

References: Friend cell—Birnie et al. (1974), Getz et al. (1975), Kleiman et al. (1977); Embryonal carcinoma—Jacquet et al. (1978), Jacquet and Gros (1979); Mouse tissues—with total cell RNA the different values represent the experiments of Brown and Church (1971), Hahn and Laird (1971), Grouse, Chilton, and McCarthy (1972) and with brain nuclear RNA the results are from Bantle and Hahn (1976); Rat tissues—Chikaraishi et al. (1978); S. purpuratus—Hough et al. (1975), Wold et al. (1978); T. gratilla—Kleene and Humphreys (1977).

estimates but that they are always substantially greater than the complexity levels generally displayed by mRNA (see Tables 24.3 and 24.4). The lack of a second method to check the data means that these estimates must be treated with some caution.

Early data using total cell RNA from mouse tissues identified saturation levels of 2–3%, except for brain where the level was higher. With purified nuclear RNA somewhat greater levels have been obtained in rat tissues and mouse brain, although even in the greatest estimates only a small proportion of the nonrepetitive DNA appears to be transcribed. The increased levels of saturation in more recent experiments may be due in part to improvements in the handling of RNA and in part to the ability to drive the reaction to greater Rot values. A feature common to all these preparations is that (as with mRNA) only a small part of the mass drives the reaction. Thus a small number of sequences must be present in high abundance, with most of the sequences represented very much more poorly. Calculations of abundance can be made, but are of less obvious significance than with mRNA because of the rapid turnover of hnRNA; abundance will be biased in favor of the more stable components, as is true of any steady-state population compared with the spectrum of molecules upon synthesis. The range of abundances poses the usual problem in driving the reaction to completion; and this may be exacerbated by the use of total cell RNA in which abundant mRNAs are present together with the nuclear RNAs. The estimates obtained with purified hnRNA fractions therefore should be more reliable.

Comparison of total nuclear RNA with polyadenylated hnRNA in the two cases of Friend cell and mouse brain suggest that some transcription units represented in hnRNA fail to gain poly(A). If it were true that the total hnRNA population represented primary transcripts, this would mean that some fail to contribute messenger sequences. However, if some of the poly(A)$^-$ hnRNA molecules are products of processing that have been split from the poly(A) but have not yet been degraded, this would not be true. But the complexity of the poly(A)$^+$ hnRNA population remains much greater than that of mRNA, about 4-fold in Friend cell or mouse brain. Thus only about 25% of the sequences attached to poly(A) find their way into mRNA. Whether this represents cleavage of nonmessenger regions from the poly(A) or nuclear turnover of some polyadenylated sequences is not evident from these data. The mammalian data show such a wide spread, and in particular the discrepancy between saturation and kinetic analysis in Friend cell, and between different estimates for brain hnRNA and mRNA complexity, means that these conclusions can only be tentative. Perhaps the most thorough characterization is that of the gastrula of the sea urchin S. purpuratus, in which the difference between the complexities of total hnRNA and mRNA is 11-fold. It is difficult to make such clear quantitation in the mammalian systems, but as a general conclusion it is evident that the discrepancy between hnRNA and mRNA complexities is great enough to imply that more nonrepetitive sequences turn over exclusively in the nucleus than are transported to the cytoplasm. Also it is clear that in no tissue, even in the case of brain, is a major proportion of the nonrepetitive DNA transcribed; taking into account the asymmetry of transcription and the proportion of nonrepetitive DNA, transcription in most of the mammalian tissues does not represent a great enough genome proportion to offer any relief from the C value paradox. The discrepancy between sequence complexities of hnRNA and mRNA cannot be compared with the size discrepancy, because most of the sequences are rather rare, while the size is determined by the small number of sequences that is abundant.

Sequence Relationship of mRNA and hnRNA

The direct approach to the question of the relationship of mRNA and hnRNA is to determine to what extent the sequences of mRNA are found in hnRNA; and then the linkage between messenger and nonmessenger sequences in the nuclear RNA can be defined (for review see Lewin, 1975b). The representation of messenger sequences in hnRNA can be quantitated by hybridization experiments in which probes representing mRNA are reacted with hnRNA (and vice versa). With populations, this allows the proportion of hnRNA sequences found in mRNA to be determined. With individual message species, it further permits the characterization of discrete precursors (see Chapter 26).

Early experiments introduced the idea that all the sequences of mRNA are found in the nucleus, but that nuclear RNA in addition contains many sequences not represented in the cytoplasm. These mostly made use of competition hybridization experiments with filters and therefore concern repetitive sequences, although some experiments were performed in solution.* More precise measurements have become possible through the availability of DNA probes complementary to the mRNA; these have been prepared either by reverse transcription to give the usual cDNA probe, or by hybridization with cellular nonrepetitive DNA to isolate an mDNA preparation. The reaction of such probes with hnRNA can be used to assess its degree of homology with mRNA. The principal technical difficulty lies with the problem of contamination. Suppose, for example, that the mRNA population used to prepare the probe is contaminated by 5% of mass with hnRNA. Then the sequences represented in the DNA probe will include these; and even if they are rare, any reaction of excess DNA × hnRNA may give a high spurious level of saturation (spurious in the sense that the hnRNA sequences are in fact reacting with DNA representing themselves rather than mRNA). Some reassurance on this question might be obtained by determining the complexity of the probe. Similar problems are encountered by messenger contamination of hnRNA when the reactions are performed in reverse orientation, that is, with probes against hnRNA used to react with mRNA.

The presence of messenger sequences in the hnRNA population of L cells was demonstrated by Hames and Perry (1977) with an mDNA probe. This was obtained by hybridizing excess mRNA with isolated nonrepetitive DNA, removing unhybridized single strands of DNA with S1 nuclease, and then isolating the surviving DNA. When an excess of this DNA was hybridized with the mRNA preparation, the reaction could be followed to 53% hybridization of the RNA. In the reactions of excess DNA with hnRNA preparations, poly(A)⁻ hnRNA was about 13% hybridized, large poly(A)⁺ hnRNA was about 15% hybridized, and small poly(A)⁺ hnRNA was about

---

References: Church and McCarthy (1967a,b), Shearer and McCarthy (1967, 1970), Soiero and Darnell (1969, 1970), Scherrer et al. (1970).

29% hybridized. Setting the level of 53% reaction with mRNA as completion (a necessary if not entirely satisfactory procedure), it is possible to calculate the proportion of hnRNA sequences that is represented in mRNA. With poly(A)$^-$ hnRNA this is independent of size at about 25%. With poly(A)$^+$ hnRNA, about 30% of the mass of large molecules ($>$28S) corresponds to message sequences, but with small molecules ($<$28S) this increases to about 56%. These results not only demonstrate directly that both large and small poly(A)$^+$ hnRNA sequences are precursors to poly(A)$^+$ mRNA, but show that poly(A)$^-$ hnRNA molecules contain messenger sequences. Because the content of message sequences is determined on a mass basis, this does not demonstrate what proportion of the hnRNA molecules represents precursors to mRNA.

Another approach has been used to set limits on the proportion of poly(A)$^+$ hnRNA molecules bearing message sequences. Herman, Williams, and Penman (1976) found that a cDNA probe prepared against HeLa cell hnRNA was 67% hybridized by excess hnRNA but only 47% hybridized by mRNA. Thus only 47/67 = 70% of the poly(A)$^+$ hnRNA molecules carry sequences represented in mRNA. This demonstrates directly that polyadenylation does not distinguish message from nonmessage sequences; for some nonmessage sequences gain poly(A) but nevertheless then must turn over within the nucleus. All the poly(A)$^+$ messenger sequences are present in the poly(A)$^+$ hnRNA, since a cDNA probe prepared against mRNA was equally well hybridized (70%) with excess mRNA or hnRNA. Experiments with fractionated hnRNA preparations showed that large as well as small hnRNA molecules act as messenger precursors.

The frequencies with which messenger sequences are represented in hnRNA are somewhat different from mRNA. In these experiments, the cDNA probe against hnRNA reacted with mRNA with kinetics showing that the same message sequences are relatively more abundant in both nucleus and cytoplasm. But the abundant proportion occupies only 35% of hnRNA compared with 67% of the mRNA. Also there is less difference between the relative representation of abundant and scarce message sequences in the nucleus; and the absolute number of molecules is much less than in the cytoplasm. Similar results have been obtained in rat liver and in Drosophila cultured cells (Sippel et al., 1977a,b; Levy and McCarthy, 1976). The implication is that messenger abundance reflects qualitatively the relative concentration of sequences in the nucleus, but that this alone is not responsible for the frequency with which a sequence is represented in mRNA. Variations in the relative abundance of sequences in hnRNA and mRNA have been characterized also for 9 cloned CHO messages (Harpold et al., 1979). Since the hnRNA used in these experiments represents the steady-state population, this does not say whether changes in the relative abundance of nuclear sequences occur following transcription.

To summarize these results and those of the previous section, the complexity of hnRNA is much greater than that of mRNA, say usually by 4–10 fold as seen in the context of nonrepetitive transcripts. All the sequences of mRNA are present in the hnRNA population; the difference in complexity must represent sequences that turn over within the nucleus. The differences in the sequences represented in mRNA populations of different tissues appear at least in some cases to represent a relatively small proportion of the total expressed complexity. This immediately directs our attention to the question of whether changes in mRNA content result from control of transcription or changes in processing. This is to ask whether the sequences that are confined to the nucleus in one cell type may represent precursors to mRNAs that are processed only in some other cell type. Direct evidence in principle can be obtained by hybridizing cDNA or mDNA representing mRNA specific to one cell type with hnRNA from another. Technically this is at the limit of present techniques, as is made clear by the comparisons of mRNA populations (see Chapter 24). The experiment is simply too difficult to perform when two cell types show very substantial overlap of their messenger populations. Although difficult with populations, such experiments can be performed, of course, with probes representing individual messengers; this is discussed in Chapter 26, but in anticipation we may note that genes such as globin and ovalbumin appear to be transcribed only in those tissues in which mature mRNA is found.

A different result has been obtained in one case in which it has been possible to compare different cell types for the presence of nuclear precursors to messenger sequences found only in one of them. This is provided by the sea urchin S. purpuratus, in which the complexity of the mRNA population of adult tissues is much less than that found in embryonic tissues (see Figure 24.5). Thus intestine mRNA has a complexity of 6000 kb compared with the 26,000 kb of blastula. Wold et al. (1978) prepared a blastula mDNA preparation, which contains the 2.1% of nonrepetitive DNA that is complementary to blastula mRNA. About 12% of the blastula mDNA is hybridized by excess mRNA of adult intestine, compared with the control reaction of 78% driven by blastula mRNA. Thus the great majority (85% = 100 − 12/78) of blastula mRNA sequences are not present in the intestine mRNA population (see footnote in Chapter 24). But the blastula mDNA probe is hybridized to a saturation value of 76% by nuclear RNA from adult intestine. Thus virtually all the blastula message sequences are represented in the adult intestine nucleus, in spite of their absence from the cytoplasm. This implies that the expression of these genes is controlled by whether or not their transcripts are selected to provide sequences for transport to the cytoplasm. Similar results are obtained when blastula mDNA is reacted with adult coelomocyte or with embryonic gastrula nuclear RNA. The total complexities of the nuclear RNA populations of intestine and gastrula are broadly similar, 230,000 and 174,000 kb, respectively, and bear no relationship to the mRNA complexities. Taken together, this series of results raises the possibility that control of sea urchin gene expression may occur principally after transcription.

A contrasting result is obtained when the presence of repetitive sequence transcripts is followed. Most repetitive families are represented in oocyte RNA, since excess RNA can drive 80% of the repetitive DNA into hybrid form. The repetitive DNA forms two components, one long and isolated, the other short and interspersed with nonrepetitive DNA. When 9 cloned repetitive sequences were examined, 2 probably representing the first class and 7 the second, all proved to be represented in both oocyte RNA and gastrula and intestine hnRNA (Constantini et al., 1978; Scheller et al., 1978). The relative concentrations with which the different clones are present vary over roughly a 100-fold range; but these amounts are greatly different between the three tissues (and are not related to the reiteration frequencies of the families). All the sequences are present in long RNA molecules whose other sequences presumably are nonrepetitive. Thus these repetitive sequences appear to reside in transcription units whose expression is under tissue specific control. The repetitive sequences are confined to the nucleus, since no hybridization is detected with mRNA; the function of the transcripts containing them is not known.

The complexities of hnRNA populations of different cells can be compared in much the same way that mRNA sequences are compared. Early experiments using saturation with total cell RNA showed that the combination of two tissues increases the complexity level less than additively. Mouse liver plus kidney gave a level of about 7.5%, and liver plus spleen a level of about 5.5%, compared with individual saturations of 4.8, 3.9, and 5.0% for liver, kidney, and spleen, respectively (Gelderman, Rake, and Britten, 1971; Brown and Church, 1972; Grouse, Chilton, and McCarthy, 1972). This shows that there are some differences in transcription between these tissues; but it does not address the question of whether they are related to differences in mRNA sequences. More recent experiments with nuclear RNA isolated from rat tissues suggest that the overlap is considerable, with differences representing only a small proportion of the total complexity. Thus brain and liver nuclear RNAs together saturate 16.5% of nonrepetitive DNA compared with 15.6% for brain RNA alone; liver and kidney RNA saturate 8.6% compared with individual controls of 8.4 and 4.6%, respectively (Chikraishi et al., 1978). Thus 92% of the liver nuclear RNA sequences are found in brain; and 96% of the kidney sequences are found in liver. However, additivity experiments are comparatively insensitive; while the experiments raise the possibility that overlaps in primary transcript sequences may be very extensive, they do not prove the case.

A more detailed study has been made of the nuclear RNAs of blastula and pluteus stages of the sea urchin T. gratilla. Complete overlap is suggested by additivity experiments, in which the kinetics and saturation level of reaction of both nuclear RNAs together is essentially the same as the RNA of either stage alone (Kleene and Humphreys, 1977). A more sensitive experiment is to fractionate nonrepetitive DNA into components complementary to, or devoid of sequences complementary to, the nuclear RNA of either stage. Then the two DNA fractions are tested for ability to be hybridized by excess nuclear RNA from the other stage. (This is analogous to the experiments with sea urchin mRNA summarized in Figure 24.5.) The same results are obtained with the experiment in either orientation: blastula⁻ DNA and pluteus⁻ DNA do not hybridize with pluteus RNA or blastula RNA, respectively; whereas blastula⁺ DNA and pluteus⁺ DNA hybridize equally well with the heterologous as with the homologous RNA. This suggests that there is no difference in the sequences represented in nuclear RNA at the two stages. But in a different sea urchin, S. purpuratus, the difference in complexity of messenger RNA populations is almost 2-fold. If this is true also of T. gratilla, the inference is that this change in mRNA complexity is brought about by processing twice as many sequences to be messengers from among the same population of transcripts.

A comparison yielding a different result has been made between S. purpuratus gastrula and intestine hnRNAs. An excess of intestine hnRNA drives 6% of a gastrula⁻ DNA preparation into hybrid form; whereas a gastrula hnRNA control hybridizes 2.4% of the gastrula⁻ DNA (Ernst et al., 1979). This implies that 3.6% of the mass of the gastrula⁻ DNA preparation is represented in intestine but not gastrula hnRNA; this corresponds to about 36,000 kb, or roughly 15% of the total complexity of intestine hnRNA. Since the sequences corresponding to mRNA do not appear to be differentially represented in nuclear RNA of different tissues, presumably these intestine-specific nuclear RNAs have some function other than coding for mRNA. If typically interspersed with repetitive sequences, these nonrepetitive sequences could represent sufficient transcripts to account for the tissue-specific variation in repetitive sequence transcripts.

The extent to which gene expression generally is controlled at the level of processing mRNA from hnRNA vis a vis the production of primary transcripts is not known. Given the comparatively large overlaps between many mRNA populations that are suggested by the (admittedly still limited) data discussed in Chapter 24, even relatively small changes in the complexity of nuclear RNA would be sufficient to accommodate the difference. On the other hand, there is the possibility that changes in transcription may be minor or absent and that the mRNA population may be changed by the selection of different sequences for transport to the cytoplasm to act as messengers. It is clear that messenger production involves the selection of a minor proportion of hnRNA sequences, but the extent to which this involves selection of different primary transcripts versus selection of messenger regions of longer precursors is not known. However, it is clear that polyadenylation cannot be the sole selective mechanism, since some poly(A)⁺ hnRNA sequences (and by inference primary transcripts) do not pass to the cytoplasm. Whether the hnRNA sequences that are degraded in one tissue include messages that are conserved in another tissue is a critical question about the selectivity of the processing mechanism; this appears to occur in the sea urchin, but data are not yet available for other species.

Control of RNA Processing

Although there is little direct evidence on the relationship between the mechanism of processing and the selection of sequences destined to become messengers, quantitative changes in the proportion of hnRNA that is converted into mRNA have been implicated in the control of messenger levels in the transition from the resting to the growing state. It has been known for a long time that growing cells have an increased content of RNA, associated with their increased activity in protein synthesis (this indeed was one of the early clues that RNA is part of the translation apparatus). Measurements of the relative amounts of different classes of RNA show that when 3T3 (or 3T6) cells are compared in growing and resting states, they have 5.6 (or 3.2) times more mRNA per cell and 3.9 (or 2.2) times more rRNA (Johnson et al., 1974). (The increase is less if expressed relative to the content of DNA, which on average is about 1.4 times greater in growing than resting cells.) Thus amounts of both mRNA and rRNA increase when cells are growing, by an amount characteristic of the cell type, but in each case mRNA increases about 1.5-fold more than rRNA. Thus in the 3T3 (or 3T6) systems, mRNA represents 1.1% (or 1.8%) of rRNA content in the resting state, but increases to 1.6% (or 2.7%) in growing cells. At least during the first few hours of transition, mRNA and rRNA systhesis are regulated independently, since inhibition of rRNA synthesis (by treatment with fluorouridine) does not alter the accumulation of mRNA (Johnson, Penman, and Green, 1976).

The rate of transcription of hnRNA is not altered during the resting-growing transition, in contrast with the transcription of rRNA which is increased. Correspondingly there is no increase in the activity of RNA polymerase II, but an increase in RNA polymerase I is observed (Mauck and Green, 1973; Mauck, 1977). The stability of poly(A)$^+$ mRNA is the same in both resting and growing cells, so that some posttranscriptional change in production must be solely responsible for the increase in content (Abelson et al., 1974). The stability of rRNA increases in growing cells, where virtually no turnover is seen. Thus the increase in rRNA level results from both increased production and stability. The amount of tRNA increases broadly in parallel with that of rRNA, but shows transient discrepancies; because in addition to a regulatory increase in transcription, tRNA production depends on the doubling of gene number at DNA replication (Mauck and Green, 1974; Willis, Baseman, and Amos, 1974). The rate of protein synthesis increases in parallel with the production of mRNA, suggesting that provision of templates may be the rate-limiting step in translation.

The fraction of hnRNA that is polyadenylated in growing and resting cells is the same; the poly(A) comprises about 2% of the 3T6 hnRNA in either state. The total nuclear content of poly(A) also is about the same, confirming that there is no change in the level of hnRNA in response to the resting-growing transition (Johnson et al., 1975). If the production of poly(A)$^+$ hnRNA and the degradation of poly(A)$^+$ mRNA remain unchanged in the two growth states, the increase in messenger content must occur at some intermediate stage. When the ratio of labeled cytoplasmic to nuclear poly(A) was followed during continuous labeling for periods varying from 15 to 260 minutes, the cytoplasmic proportion of poly(A) was always greater in the growing cells. A maximum ratio was reached after 60 minutes, when the ratio in growing 3T6 cells was 1.8 compared with the ratio of 0.8 in resting cells. Thus about 2.25 times more poly(A) is transported from nucleus to cytoplasm in growing cells. When resting cells are stimulated into growth by addition of serum, the ratio increases to reach the growing level about 2-3 hours after the transition is initiated. An analogous result was obtained in pulse chase experiments, in which a brief exposure to ^{3}H-adenosine was followed by inhibition of further poly(A) synthesis by treatment with cordycepin; the increase in cytoplasmic poly(A) during the chase in growing cells is twice that seen in resting cells. Another approach has made use of techniques that allow label to be chased from hnRNA into mRNA; by comparing the initial rates of labeling of hnRNA and mRNA, Johnson et al. (1976) showed that 3-4% of the mass of labeled hnRNA enters the cytoplasm of growing cells, compared with 1-2% in resting cells. (The growing conversion ratio is comparable to that seen with HeLa cells; see Table 25.2).

The simplest interpretation of these experiments is that a greater proportion of the poly(A)$^+$ hnRNA molecules is converted into poly(A)$^+$ mRNA in growing cells compared with resting cells. This means that the efficiency with which nuclear polyadenylated sequences are converted into mRNA must be controlled in response to the conditions of growth. The implication, of course, is that in the resting state some of the polyadenylated sequences must turn over within the nucleus. These results do not reveal whether the resting-growing transition is accompanied just by an increase in the probability with which a polyadenylated nuclear sequence will give rise to poly(A)$^+$ mRNA or whether there is some change in the selection of sequences for nuclear export; but from the comparisons of messenger complexity discussed in Chapter 24 it seems clear that the major part of the change is quantitative rather than qualitative. There are two possible objections to these conclusions. One is that only poly(A)$^+$ mRNA has been followed; so it would not be true that there has been an increase in the total content of mRNA if the rise in poly(A)$^+$ mRNA were at the expense of poly(A)$^-$ mRNA. This would mean rather that there had been a change in the relative export of polyadenylated and nonpolyadenylated sequences from the nucleus. This alternative is rendered unlikely by the observation that the relative proportions of poly(A)$^+$ and poly(A)$^-$ mRNA are the same in resting and growing CHO cells (Levis, McReynolds, and Penman, 1977). The second objection is more difficult to exclude: mRNA might be the product of a subfraction of hnRNA whose synthesis is controlled separately from the bulk of the nucleoplasmic RNA. The transcription of this subfraction might be specifically increased by the resting-growing transition, while the major part of hnRNA synthesis remains unaltered. Since the measurements of hnRNA transcription apply to the entire nuclear population, this increase in a minor fraction would not be observed. In other words, the conclusion that the efficiency of conversion is increased depends on the assumption that the synthesis of hnRNA molecules does not differ according to whether they are or are not utilized as message precursors.

Small Nuclear RNA

Eucaryotic nuclei contain a fraction of RNA that sediments in the range of 6-8S and which can be resolved by gel electrophoresis into roughly 10 or so bands. First discovered by Weinberg and Penman (1968), these species are relatively stable; and the various bands appear to be present in amounts varying from about 10^4-10^6 molecules per cell. The lengths of the molecules vary from about 100-200 nucleotides, they contain methylated and other modified bases, and may have 5'-terminal cap-like structures. They have been most fully characterized in animal cells, where there appears to be extensive conservation of structure; there is tissue-dependent variation (Rein and Penman, 1969). Comparable snRNA molecules are to be found in a wide variety of other eucaryotic nuclei, but their structures are not known.

Table 25.5 summarizes the properties of mammalian snRNAs. The preparations include nucleolar and ribosomal RNAs; SnA is found in nucleolar (but not cytoplasmic) 32S-28S RNA, SnE/ScE, and SnG/ScG are components of the ribosome that are found in both nucleolus and cytoplasm. The locations of the remaining molecules have been investigated by Zieve and Penman (1976), by identifying the species present in different nuclear fractions. The snRNAs are assigned to the nuclear skeleton if they are present in the fraction remaining after most of the chromatin has been digested by DNAase. They are assigned to chromatin if they are released from the nuclei by this treatment; and to the nucleoplasm if they are associated neither with chromatin nor with the skeleton. In comparable experiments, Miller, Huang, and Pogo (1978b) have suggested that all the snRNAs of rat liver nuclei are associated with the skeleton.

Both SnC and SnD (the most abundant snRNA) are found in the nuclear skeleton with this fractionation procedure. However, Pederson and Bhorjee (1979) identified some small RNA molecules that appear to be covalently linked to duplex DNA. These are identical with SnC, SnD, and SnG'. In the case of SnC and SnD, less than 10% of the total nuclear content takes this form. The nature of their association with DNA, and the relationship of these molecules with those in the predominant location of the nuclear skeleton, are topics for further definition. These two molecules, and presumably also other snRNAs, are obtained as ribonucleoprotein complexes upon extraction; this could be their form in vivo (Raj, Ro-Choi, and Busch, 1975).

Two small RNA molecules were found in the cytoplasm by Eliceiri (1974),

Table 25.5 Structure and location of small nuclear RNA

Species of RNA in Hela	in Novikoff	Size	Abundance in HeLa	Half-life in HeLa	Location
SnA	U3	~180	2×10^5	stable	nucleolar (with 32S-28S rRNA)
SnB	—	~210	1×10^4	~20 hr	nuclear skeleton
SnC	U2	196	1×10^5	~25 hr	nuclear skeleton/chromatin (10%)
SnD	U1B	171	1×10^6	stable	nuclear skeleton/chromatin (10%)
SnE/ScE	5.8S --------------------		-----5.8S rRNA	---------------------------------	
SnF	U1A	~125	3×10^4		nucleoplasmic
SnG/ScG	5S I and II -------------		--5S rRNA	---------------------------------	
SnG′	5S III	~120	2×10^5	stable	nucleoplasm/chromatin
SnH	4.5S I	96	1×10^5	~30 hr	nuclear skeleton
SnH	4.5S II	~96			
SnH	4.5S III	~96			
SnI	----------------------		--- tRNA	---------------------------------	
SnK	—	~260	2×10^4	stable	nuclear/cytoplasmic
SnP	—	~130		stable	nuclear
ScL (viral 7S)		~260		stable	cytoplasmic/membranes
ScM (A)		~180		7 min	cytoplasmic
ScD (B)		~180		10 min	cytoplasmic (precursor to SnD)

Small nuclear RNAs have been separated by gel electrophoresis in HeLa cells (Weinberg and Penman, 1968, 1969) and Novikoff hepatoma cells (Hodnett and Busch, 1968, Ro-Choi et al., 1970). Small RNAs present also or instead in the cytoplasm are designated scRNA. Where sizes are shown as approximate they are based on gel mobility; the three exact measurements are the lengths of the Novikoff snRNAs that have been sequenced (Ro-Choi et al., 1972; Reddy et al., 1974; Shibata et al., 1975). Abundance and stability estimates are from Weinberg and Penman. Locations have been determined by Prestayko, Tonato and Busch (1970) for U3, Eliceiri (1974) for (A) and (B) which are identical with ScM and ScD, Zieve and Penman (1976) for all the HeLa snRNAs, and Pederson and Bhorjee (1979) for the HeLa snRNAs associated with chromatin. Among the snRNAs identified as such on gels are 5.8S rRNA, 5S rRNA, and tRNA, as indicated.

where they are unstable. One of these, originally named (B) and now re-named ScD, appears to be a precursor to SnD (Zieve and Penman, 1976). This implies that the molecule must move from its site of synthesis in the nucleus into the cytoplasm and back again.

A fraction of small RNA molecules that is hydrogen bonded to both nuclear and cytoplasmic poly(A)$^+$ RNA has been isolated by Jelinek and Leinwand (1978) from CHO cells. Their concentration is greater in the nuclear RNA. They appear to constitute a family of related sequences. It is not known whether this group of RNAs coincides with the snRNAs characterized in other studies; on the basis of size, these could be equated with the SnH fraction.

The question of the synthesis of the snRNAs has been a little confused.

Originally it was thought that all were the products of action by RNA polymerase III, together with its synthesis of 5S rRNA and tRNA. However, Zieve, Benecke, and Penman (1977) found a difference in the synthesis of 5S rRNA, tRNA, SnK, and SnL, which in isolated nuclei appear to be transcribed by RNA polymerase III according to the characteristics of the reaction, while SnA, C, D, F, H are not transcribed in nuclei in vitro, but are inhibited in vivo by conditions that preferentially prevent nucleolar transcription. Yet Benecke and Penman (1977) have characterized the small nuclear products of polymerase I action (as defined by synthesis in the presence of high α-amanitin) and have shown that they include none of the snRNAs described in the table; instead there appears to be a group of about 30 quite different RNA products. However, their synthesis is resistant to levels of actinomycin that prevent the major rRNA synthesis. They therefore constitute a second set of small nuclear RNAs, less well characterized than the major set of snRNAs.

Almost nothing is known of the function of snRNA. The low complexity of the species described in the table suggests that their role is not likely to be one of regulation; an obvious speculation is that they may have a structural role, concerned for example with the organization of the nuclear matrix or its association with chromatin. Another possibility is that they are involved in the processing of hnRNA.

CHAPTER 26

Interrupted Genes and RNA Splicing

More than a hundred years of work has led to the concept that the gene is a contiguous region of DNA that is colinear with its protein product. This definition encompasses the complementation test, which formally defines the gene as a unit all of whose parts must be present on one chromosome; and it has found fulfillment in the detailed molecular studies of the past two decades demonstrating that DNA first is transcribed into RNA and then is translated into protein by reading the triplets of the genetic code (see Volume 1 of Gene Expression. Colinearity appeared to be preserved through the stages of gene expression.

The recent discovery that eucaryotic genes contain intervening sequences that are not represented in the mRNA from which the protein product is synthesized has caused this concept now to be discarded. We have already discussed in Chapter 17 the implications of these interruptions for our view of the constitution of the genetic complementation group. This has made it difficult, if not impossible, to arrive at a single satisfactory definition of the gene. In this chapter we shall be concerned with the molecular nature of the intervening sequences and with the processes by which they are removed from RNA so that the protein product can be synthesized. Also we shall consider the variations that may occur in this RNA splicing, rendering a single sequence of DNA able to code for multiple, overlapping products. In short, we shall attempt to achieve by example what amounts to an operational definition of the eucaryotic gene, taking this in the broader sense to include the entire unit of gene expression and not just the genome sequences representing the protein product.

The techniques that have made possible the resolution of eucaryotic gene structure at the molecular level follow from the discovery of the properties of restriction enzymes. Of procaryotic origin, there are many such enzymes, each with the ability to recognize and cleave duplex DNA at a characteristic, usually rather short, sequence of base pairs. The direct application of restriction cleavage is to obtain a restriction map of any particular DNA, representing a linear sequence of the sites recognized by various enzymes. A related use is to obtain fragments representing some particular sequence. With eucaryotic genomes, any restriction fragment representing a unique sequence is present in amounts too small to determine directly; but the combination of recent "blotting" techniques with nucleic acid hybridization allows such fragments to be recovered.

One of the most striking applications of restriction cleavage is seen in the reconstruction of DNA from restriction fragments. Given defined fragments of DNA, it is possible by various means to link them covalently. In particular, a restriction fragment of eucaryotic DNA may be inserted into a bacterial plasmid or phage. Such chimeric plasmids (or phages) may be used to clone eucaryotic DNA sequences, so that large amounts of a unique sequence may be recovered. This may be useful as a probe or to study directly. Of some practical concern, as well as of interest for the light it casts on evolution of the apparatus for gene expression, is the ability of cloned eucaryotic genes to be expressed in bacteria.

Restriction Mapping of DNA

Originally identified as part of host modification and restriction systems that appear to serve to protect bacteria against foreign DNA, restriction enzymes now have been identified in many bacteria, sometimes specified by resident plasmids.* Their characterization and mode of operation is discussed in Volume 1 of Gene Expression. Each restriction enzyme has a specific target, a sequence of DNA usually in the range of 4–8 base pairs. In a complete digest, a given DNA will be cut at every occurrence of the target sequence. For a long enough sequence of DNA, targets occur with a statistical likelihood that depends on their length and base constitution relative to the overall base composition of the DNA. By appropriate choice of restriction enzymes, it is therefore possible to cleave a given DNA into a larger or smaller number of fragments.

The fragments into which a defined DNA is cut by a restriction enzyme can be separated on the basis of size by agarose gel electrophoresis. (It is necessary to quantitate the bands to check whether each represents a unique fragment or more than one of the same size). The technique of restriction mapping originated with the observation of Danna and Nathans (1971) that a mixture of Hin enzymes cleaves SV40 DNA into 11 equimolar pieces, the sum of whose mass corresponds with that of the intact genome. These were ordered into a cleavage map by the use of partial digests with Hin and double digests with Hin and HpaII (Danna, Sack, and Nathans, 1973; Sharp, Sugden, and Sambrook, 1973).

When a partial digest is made, usually by reducing the amount of en-

---

*Restriction enzymes conventionally are named by an abbreviation that indicates their origin, with a (Roman) numeral where more than one has been isolated from the same source. Thus EcoRI is obtained from E. coli carrying certain R factors, HindIII is obtained from Hemophilus influenzae, HpaI is derived from Hemophilus parainfluenzae, and so on. Restriction fragments of any DNA usually are named for the enzyme and indicated as A, B, C, etc. in order of decreasing size (See Danna, Sack, and Nathans, 1973).

zyme or period of incubation, larger fragments are obtained, each consisting of two or more of the smaller fragments generated by complete digestion. Redigestion of the partial digest products leads to cleavage of their intact restriction sites, releasing the appropriate complete digest products. Thus adjacent fragments can be recognized by their linkage in the partial products. Double digests allow the cleavage fragments produced by two different enzymes to be ordered with respect to each other. Some information can be obtained purely on the basis of defining the sites cleaved by each enzyme in the fragments produced by the other. More detailed definition of overlaps may be produced by hybridization between denatured fragments. A useful variation of the mapping technique is to use end-labeled DNA and then to identify the hierarchy of sizes of fragments produced by partial digestion (Smith and Birnstiel, 1976).

The locations of individual fragments may be determined also by using electron microscopy to identify the positions of the heteroduplex regions formed by hybridization with denatured intact DNA. For these experiments it is necessary to have a point of reference. With linear DNA molecules this is provided by the ends. With circular genomes, an arbitrary origin for the map is defined by the single cleavage site recognized by an enzyme that happens to have only one target. In the example of SV40, this was provided by the early observation that EcoRI cleaves only once, converting the circular molecule into a linear molecule of the same length (Morrow and Berg, 1972; Mulder and Delius, 1972).

Restriction mapping as such allows any defined DNA to be recognized as a linear series of cleavage sites. A critical step comes in relating this map to the genetic map. The simplest method is to compare the restriction map of the wild-type genome with the patterns of fragments generated by mutants that have discernible physical effects, viz., deletions. This also serves the purpose of defining the exact size and location of deletions otherwise characterized only by their multisite genetic effects. This has been used successfully with plasmids, phages, and eucaryotic viruses. A variety of other sophisticated methods, often peculiar to the system, has been developed to identify the genetic information carried by restriction fragments. Here we shall be concerned more with the application of restriction mapping to fragments of eucaryotic DNA. Usually genetic markers are not available and the restriction map must be related to the location of the gene by determining the DNA sequence. Given the base pair sequence, of course, the location of restriction targets can be predicted; and this much increases the power of restriction techniques by allowing the use of appropriate enzymes to obtain exactly defined regions of DNA.

Nucleic Acid Cloning

The use of restriction enzymes to cleave duplex DNA into defined fragments has been accompanied by the development of techniques for linking such fragments together. When a fragment of DNA from some foreign source is linked to a bacterial plasmid or phage, it becomes part of the genome of this genetic element and is replicated as part of it. (This mimics the insertion into such elements of sequences of the bacterial genome that occurs naturally.) The *hybrid* or *chimeric plasmids* (or phages) formed by reconstruction with restriction fragments may be obtained in large amounts by growth in bacteria; and the foreign region may be recovered by restriction cleavage. This is useful for amplifying individual sequences that have been placed on plasmids or phages; and when a large variety of sequences, representing perhaps a genome or population of messengers, is used to form chimeric plasmids, individual sequences may be isolated by cloning the bacterial hosts. Identification of any particular sequence requires the use of a selection technique.

The essential feature of any plasmid or phage that is used as a *cloning vector* is that there should be only a single cleavage site available to the enzyme that creates the break where foreign DNA is to be inserted. This must lie at a point where cleavage and insertion does not disrupt the replication of the element. For a circular plasmid genome, the break generates a linear molecule, whose circularity must be restored by joining foreign DNA to the two free ends. For a linear phage genome, the two fragments resulting from the break must be rejoined via the ends of the foreign DNA.

Some restriction enzymes introduce staggered cuts in DNA, generating single-stranded ends that are complementary and may be used directly to accomplish the reconstruction reaction. The first such activity to be discovered was EcoRI, which cleaves the two strands of DNA four bases apart within its target site. This generates the "sticky end" sequences:

$$3' \quad N\ N\ N\ N\ T\ T\ A\ A \qquad\qquad N\ N\ N\ \ 5'$$
$$5' \quad N\ N\ N\ N \qquad\qquad A\ A\ T\ T\ N\ N\ N\ \ 3'$$

which can reanneal by base pairing. When two different EcoRI fragments are mixed together, cross-wise annealing may take place. When a single EcoRI break is introduced in a plasmid DNA, the resulting linear molecule has sticky ends at both termini; these may react with the sticky ends on any foreign DNA fragment that is added (Cohen et al., 1973; Hershfield et al., 1974; Morrow et al., 1974; Hamer and Thomas, 1976). As illustrated in Figure 26.1 this reaction regenerates a circular molecule, with the covalent integrity of each strand interrupted by nicks on either side of each junction. These may be sealed in vitro by the action of DNA ligase or in vivo following infection of E. coli.

An advantage of this technique is that the annealing of sticky ends regenerates the restriction target sites. Thus the foreign DNA may be recovered as a restriction fragment by cleaving the chimeric plasmid with the enzyme used in the construction reaction. A disadvantage is that any sticky end can anneal with any other sticky end. Thus some circular plasmids will reform by annealing of the ends of the linear molecule, some may gain a

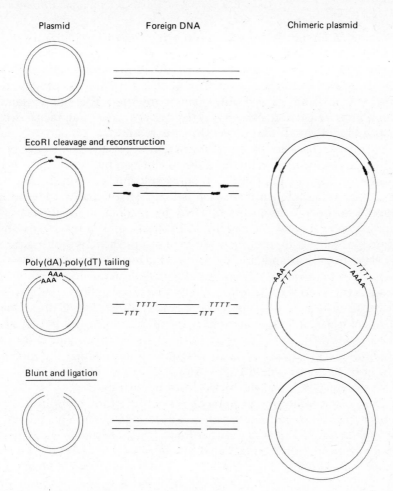

Plasmid   Foreign DNA   Chimeric plasmid

EcoRI cleavage and reconstruction

Poly(dA)·poly(dT) tailing

Blunt and ligation

Figure 26.1. Methods for constructing chimeric plasmids. For joining at sites of restriction, both plasmid and foreign DNA are cleaved by EcoRI or another enzyme (PstI or BamHI for example) that generates sticky ends. Mixing allows duplex formation to occur between complementary single stranded ends; chimeric plasmids have a nick in the integrity of each strand at the EcoRI (or other) sites that are reconstituted at each hybrid junction. For poly(A):poly(dT) tailing, the plasmid is cleaved at a suitable site with a restriction enzyme and free 3′ ends are polyadenylated; the foreign DNA is cleaved by shearing to generate random ends that are polythymidylated. Mixing allows chimeric plasmids to form with junctions of A:T base pairs; gaps may exist when the dA_n and dT_n lengths are not identical. For blunt end ligation, the plasmid is cleaved with a restriction enzyme generating blunt ends or with an enzyme generating sticky ends that then are filled in. Phage T4 ligase then catalyzes joining to foreign DNA that has blunt ends. The chimeric plasmid has no breaks in the integrity of its strands.

single insert of foreign DNA, and in some there may be several units of foreign DNA linked end to end via a series of EcoRI sites. It is therefore useful to be able to apply a selection technique to isolate plasmids that have gained insertions. And it is necessary to determine that any chimeric plasmid used for further studies has gained only a single insertion; this can be done by checking that restriction cleavage generates only two fragments, whose combined length equals that of the chimeric plasmid.

A general drawback to the use of restriction fragments as a source of eucaryotic DNA for cloning is that in order to clone an intact gene, it is necessary to use an enzyme that cleaves on either side but not within the region. This may be difficult when the region is long. An alternative to restriction cleavage is to obtain fragments of DNA by random shear; their average size then is controlled by the conditions of shearing. To use these for insertion into plasmids requires the generation of complementary single-stranded ends on the fragments and cloning vector. One method for generating reactive termini is to use terminal deoxynucleoside transferase to add a poly(dA) segment to the 3′ ends of the plasmid DNA (at a convenient site cleaved by a restriction nuclease), while poly(dT) is added to the 3′ ends of randomly sheared DNA (Jackson, Symons, and Berg, 1972; Lobban and Kaiser, 1973). Figure 26.1 shows that reaction between the complementary ends of the plasmid and other DNA generates a circular molecule in which the junctions consist of duplexes of poly(dA:dT). Any gaps in the circle can be filled in by DNA polymerase I; and the nicks closed by ligation.

Of course, the technique of poly(dA:dT) tailing can be applied also to the ends generated by restriction enzymes in foreign DNA. This allows use of the much larger number of enzymes that becomes available when there is no need to generate sticky ends. An advantage of the method is that reaction can occur only between the plasmid DNA and foreign DNA, so that only a single fragment of eucaryotic DNA can be inserted in each chimeric plasmid. This is useful when a large number of random fragments is to be cloned. A disadvantage is that the inserted fragment cannot be recovered by cleavage with restriction enzymes, because the formation of the poly-(dA:dT) abolishes any restriction sites present at the junctions.*

---

*Several methods have been developed for recovering the DNA inserted by poly(dA:dT) tailing, but none is completely satisfactory. Hofstetter et al. (1976) found that S1 nuclease preferentially cleaves the poly(dA:dT) junctions when the chimeric plasmid is subjected to moderate denaturation. However, degradation at other sites also may occur. Goff and Berg (1978) converted the chimeric plasmid to linear form by introducing a single cleaveage in the vector; denaturation followed by brief renaturation then causes the inserted sequence to form a single stranded loop joined by a poly(dA:dT) duplex region to single stranded tails of vector material. When the tails are removed by single strand exonuclease and the inserted region is renatured to form a duplex, about 30–50% of the cloned DNA is recovered. The problem is that many manipulations are involved and recovery is not achieved in as satisfactory a manner as with restriction cleavage. A condition of the technique also is the need for a restriction enzyme that cleaves the vector and not the inserted region. Another approach has been developed by Rougeon and Mach (1977). Using a plasmid cleaved with EcoRI, the single-stranded ends were

The range of manipulations possible with DNA fragments has been increased by the development of blunt end ligation. It has been known for some time that the T4 DNA ligase has the ability to join together two DNA duplex molecules at their ends in what appears to be a relatively inefficient side reaction of its ability to seal single strand nicks in duplex DNA (Sgaramella, Van de Sande, and Khorana, 1970). The establishment of conditions that permit more efficient reaction have made possible its use in constructing chimeric plasmids (Sugino et al., 1977). The reaction requires duplex substrates with blunt ends, and so may be applied directly to the products of digestion with restriction enzymes that cleave both strands at the same position. The single-stranded ends created by enzymes that make staggered cuts may be filled in with DNA polymerase to generate suitable ends, in which case the restriction target site remains available at the junction after blunt end ligation.

The advantage of this technique is that any pair of ends may be joined together, irrespective of sequence. Thus the ends made by one restriction enzyme may be joined to those made by another. It is especially important when fragments of defined sequence are to be joined without introducing any additional sequence at the junction; this makes possible sophisticated reconstructions of sequence using particular fragments created (for example) by double restriction digests (that is, with different sequences at each end). The reaction has been used in chemical synthesis of DNA fragments of particular sequences; one particular use in this context is to join chemically synthesized EcoRI target sites to the ends of some DNA fragment, thus allowing it to be inserted into, and retrieved from, chimeric plasmids by restriction cleavage (Marians et al., 1976; Heyneker et al., 1976; Scheller et al., 1977). A disadvantage of the technique is that there is no control over which ends are joined; any pair of blunt ends in the reaction mixture may be ligated. This makes it necessary to characterize the product in detail.

In all of these techniques for inserting DNA into plasmids, the foreign DNA fragment may be inserted in either orientation into the plasmid (that is, with either of its ends joined to either of the plasmid ends). This will not be important when the purpose of cloning is simply to amplify a given sequence; but it may influence the ability of the inserted sequence to be expressed by readthrough from the plasmid. Which orientation is found in any particular chimeric plasmid can be determined from the restriction map of the hybrid DNA. Sequences may be inserted in particular orientations by generating different sticky ends at the termini through digestion with two restriction enzymes. Thus when the sites of two such enzymes lie

---

filled in by a repair reaction and then joined to foreign DNA by poly(dC:dG) tailing. The reconstructed EcoRI sites then should be available on the plasmid sides of the poly(dC:dG) junctions. But the inserted DNA could be recovered by EcoRI cleavage from only one of several chimeric plasmids. The reasons for this low success rate are not known; similar fill-in reactions appear to have been used successfully with blunt end ligation (see text).

close together on a plasmid, double digestion removes a small region, leaving (for example) one EcoRI sticky end and one BamHI sticky end. If a fragment of DNA for cloning also is provided with one sticky end of each type, insertion can occur in only one orientation; and reformation of the original plasmid is not possible. Some of these manipulations are indicated in Table 26.3 by the note of restriction sites present at junctions in chimeric plasmids.

A variety of cloning vectors has been developed to serve different purposes. The techniques involved in modifying plasmids or phages so that restriction sites are available for insertion at convenient locations are discussed in Volume 3 of Gene Expression. Table 26.1 summarizes the derivation of the cloning vectors that at present are in most common use. These share the feature that vectors gaining foreign DNA may be selected from those that have reformed without insertion; then the chimeric plasmid population may be screened further.

With plasmid vectors, drug resistance markers may be used to select bacteria carrying plasmids and to distinguish chimeric plasmids. If the plasmid DNA carries two drug-resistance markers, one can be used simply to select bacteria that have gained a resident plasmid following exposure to the DNA. If the site for insertion of foreign DNA lies within the other drug resistance locus, formation of a chimeric plasmid will cause loss of this resistance; this can be used to distinguish chimeric plasmids from others. Plasmid vectors offer the advantage that virtually unlimited lengths of inserted DNA can be cloned, but it is necessary to isolate the DNA directly from bacteria (usually by relying on the properties of circular molecules).

Plasmid vectors usually are constructed from plasmids that have "relaxed" replication control, which means that they exist in multiple copies (about 10–20) per cell; a further accumulation occurs in the case of Col El and related plasmids, when treating cells with chloramphenicol to prevent bacterial DNA replication allows the plasmid number to increase to 1000–3000 per cell. A series of popular cloning vectors has been derived from the Col El factor by linking useful drug resistance markers to this replication control system. Some of these rearrangements have involved reconstruction through restriction cleavage and rejoining; others have utilized the ability of transposon (Tn) elements carrying drug resistance markers to move into new sites. Some of the reconstructions have involved "scrambling" of genomes, in which random mixtures of restriction fragments have been selected to isolate chimeric circles with the desired properties. Others have been reconstructed with specific sequences in order to obtain expression of inserted DNA; these are discussed in the next section.

Phage vectors have been derived from manipulations of lambda that create a shortened genome with only a single restriction cleavage site. The cloning vector per se is too short to be packaged into the phage head; this establishes selection for insertions long enough (but not too long) to permit packaging. Although the use of phages limits the length of DNA that

Table 26.1 Origins and characteristics of cloning vectors

Vector	Derivation	Size	Markers	Single restriction sites
Col El	natural	6.4 kb	colicin El production and immunity	EcoRI in nonessential location
pSF2124	amp^r gene on TnA transposed into Col El from R1	11.0 kb	as above + ampicillin resistance	as above
pMB1	natural; identical with Col El but also has genes for EcoRI	8.3 kb	colicin El production and immunity; produces EcoRI	as above
pMB9	restriction digest of pMB1 scrambled with tet^r fragment of pSC101	5.3 kb	colicin El immunity; tetracycline resistance	EcoRI in nonessential location; BamI, SalI, HindIII in tet^r
pBR312 and pBR313	amp^r gene on TnA transposed into pMB9 from PSF2124	9.5 kb 8.6 kb	as above + ampicillin resistance	as above + further BamI site
pBR322	rearrangement of pBR312	4.4 kb	tetracycline and ampicillin resistances	nonessential EcoRI; BamI, SalI, HindIII in tet^r; PstI in amp^r
pSC101	unknown; possibly a natural conjugative R factor	1.0 kb	tetracycline resistance	nonessential EcoRI
λgtWES	lambda modified by deletions to have only 2 EcoRI sites	~30 kb	small fragment lost on EcoRI cleavage must be replaced by DNA of 1–15 kb for viability; similar cleavage sites for SstI in same region	EcoRI at chimeric junctions; same when SstI is used
pJC703	cos and rif^r added to Col El	25 kb	colicin El immunity; rifampicin resistance; amp^r present in some further variants	variants have single sites in colicin immunity or drug resistance loci; packaged in vitro

All vectors derived from Col El have relaxed replication control and are present in 10–20 copies per bacterial chromosome, rising to 1000–3000 upon addition of chloramphenicol. pSC101 has stringent replication control and is present in single copies per bacterial chromosome. Phage lambda may be perpetuated by lysogeny or released through the lytic cycle.

References: Col El—Hershfield et al. (1974), Clarke and Carbon (1975); pSF2124—So et al. (1975); pMB and pBR series—Betlach et al. (1976), Rodriguez et al. (1976), Bolivar et al. (1977a,b), Sutcliffe (1978b); pSC101—Cohen and Chang (1977); lambda derivatives—Murray and Murray (1974), Rambach and Tiollais (1974), Thomas, Cameron and Davis (1974), Cameron et al. (1975); Leder et al. (1977); see also Blattner et al. (1977); pJC708—Collins and Hohn (1978).

can be cloned, the packaging reaction purifies the cloned DNA from bacterial DNA. The utility of these vectors has been increased by the development of systems for packaging in vitro. An attempt to gain advantages of both plasmid and phage vectors has led to the construction of "cosmids," which are plasmids into which have been inserted the phage lambda cohesive (cos) sites on which packaging depends. These vectors may be grown as plasmids, but packaged in vitro as phages (although still subject to the length limitation on how much DNA can be contained in a phage head).

When particular sequences of eucaryotic DNA are obtained for cloning, either by selecting fragments of the genome or by reverse transcription of specific messenger RNAs, chimeric plasmids carrying the desired sequence constitute a large proportion of the hybrid population and may be purified with according ease. When no selection is imposed before insertion of DNA, it is necessary to screen a population of chimeric plasmids carrying randomly derived fragments in order to obtain the desired sequence. The fragmentation of a eucaryotic genome into random pieces for cloning has been described as a "shotgun" experiment. It is important that the fragments should be long enough to contain intact genetic units and this means that their ends must be generated by cleavage at random sites, whose distance apart can be controlled by the conditions. This is accomplished either by shearing to a given average size or by making partial digests with restriction enzymes whose target sites are short (usually 4 base pairs) and therefore approach a random distribution. Generally a fragment length of about 15–20 kb is considered adequate. The number of random fragments that must be cloned to have a high probability that every sequence of the genome is represented at least once in a chimeric plasmid decreases with the fragment size and increases with the genome size and the probability.* For a probability level of 99%, almost 1500 cloned fragments are needed with the E. coli genome, 4,600 with yeast, 48,000 with Drosophila, and 800,000 with mammals. "Libraries" of cloned fragments achieving this probability have been established with all these species (Wensink et al., 1974; Clarke and Carbon, 1976; Maniatis et al., 1978).

The frequency with which any particular genome sequence will be represented in a cloned library is inversely proportional to the size of the library. With more complex genomes, a selective technique of considerable resolving power is therefore necessary to isolate individual sequences. In the technique of colony hybridization, bacterial colonies carrying chimeric plasmids are formed and then lysed on nitrocellulose filters; then their DNA is denatured in situ and fixed on the filter. The filter is hybridized with a labeled RNA or cDNA probe that represents the desired sequence and then any colonies in which it is present are visualized by autoradiography. Bacteria

---

*The number of transformants, N, is given by the expression $N = \ln(1 - P)/\ln(1 - L/G)$, where P is the desired probability, L is the average fragment length, and G is the genome length. This assumes that the cloned length L is sufficiently longer than the length of the genetic unit to make the occurrence of random breaks within the unit rather rare (Clarke and Carbon, 1976).

carrying the corresponding chimeric plasmids can be recovered from the reference plates that were made by replica plating before lysis of the colonies on the filter (Grunstein and Hogness, 1975). A related method using a lambda vector has been developed in which plaques are formed by lysis of the bacteria, phages are allowed to adsorb to nitrocellulose filters by contact, and those possessing a given sequence are visualized by hybridization with a labeled probe followed by autoradiography (Benton and Davis, 1977). This allows a greater number of chimeric vectors to be screened quickly.

A critical question is the fidelity with which a cloned eucaryotic sequence is perpetuated in bacteria. The best indication that changes in sequence do not occur is given by the cloning of regions that code for proteins of known sequence. In each case so far characterized, the nucleic acid sequence of the cloned coding region remains exactly related to the amino acid sequence by the triplet genetic code (see below). A further assumption in the construction of libraries is that all genome sequences are cloned with equal probability and fidelity. There is no reason to doubt the assumption for nonrepetitive and for dispersed repetitive regions; but what is the fate of tandem repeats? Difficulties have been encountered in cloning satellite DNA and these are likely to be due to the frequent opportunities that exist for recombination in the inserted segment (Brutlag et al., 1977b). Some loss of cloned segments is common with all inserted sequences; but with highly repetitive tandem elements, the extent of loss and the difficulties of excluding the possibility that rearrangement has occurred seem at present to limit the application of cloning.

Isolation of Eucaryotic Genes

Two approaches have been taken to isolating eucaryotic genes. One is to obtain the appropriate sequences directly from the genome. Only in the case of repetitive gene clusters can enough material be obtained in this way for further studies; in other cases the genes must be amplified by cloning. A second technique is to prepare a DNA copy of the isolated messenger RNA; again there are quantitative difficulties that can be overcome by cloning. When these techniques were conceived, it was thought that they would identify essentially the same sequences, although the use of mRNA has the disadvantage of ignoring any sequences in DNA that are not transcribed or that are discarded from the ends of the primary transcript. Now that the existence of interruptions within the unit of transcription is known, these approaches prove to be complementary, since it is necessary to compare the different sequences obtained in each case to determine the relationship between genome and messenger.

When eucaryotic DNA is digested with a restriction enzyme and fractionated by gel electrophoresis, a few prominent bands are formed from cleavage of tandemly repeated sequences (see Chapter 19) but most of the material forms a continuous smear. A method for detecting specific sequences in the restriction smear was developed by Southern (1975b) and has been widely used, together with some subsequent variations. The DNA fractions are transferred ("blotted") from an agarose gel to strips of nitrocellulose. Essentially this is accomplished by placing the gel on a filter paper soaked in very concentrated (20 ×) SSC solution. The nitrocellulose strip is placed on top of the gel and is in contact with dry filter paper. The SSC passes through the gel, drawn to the dry filter paper, and carries the DNA, which becomes trapped in the nitrocellulose. On the nitrocellulose, the DNA can be hybridized directly with labeled RNA and visualized in the form of bands by autoradiography. (Hybridization cannot be performed with DNA on the agarose gel; and cutting the gel, eluting the DNA, and then hybridizing, leads to unacceptable losses of material and lack of resolution). The bands present on the nitrocellulose strip can be compared with the DNA present on the agarose gel as visualized by fluorescence after ethidium bromide treatment.

The technique can be applied to either one- or two-dimensional gels; and the material transferred to the nitrocellulose filter often is referred to as "Southern blots." An example is shown below in Figure 26.6, in which labeled β-globin sequences are used as a probe to identify the corresponding restriction fragments of the genome. Combinations of approaches to isolate eucaryotic genes also may be useful. For example, a partial purification of globin gene fragments by Southern blotting has been followed by formation of a cloned library of the isolated fragments, from which chimeric phages carrying the globin genes were selected (Tilghman et al., 1977). Also it should be noted that the blotting technique can be used with RNA,* for example to isolate precursors for a particular message from among the nuclear RNA population. This is sometimes described colloquially as "Northern blotting" and an example is shown below in Figure 26.17.

Duplex DNA also may be prepared by using purified mRNA as a template for DNA synthesis. The first step is reverse transcription of the mRNA using the reverse transcriptase enzyme obtained from RNA tumor viruses. This requires a primer, which usually is provided by annealing oligo(dT) to the poly(A) tail of the mRNA. In this form the technique is generally applicable to poly(A)$^+$mRNA and hnRNA; and the product is a single-stranded cDNA, which is complementary to sequences extending from the 3' end of the RNA. The only problem at this stage is to obtain a full length reverse transcript of the messenger; naturally this is more severe with longer templates. In the first experiments with globin mRNA, for example, with which the technique was worked out, cDNA of size close or similar to the mRNA was obtained, but constituted only a proportion of reverse transcripts (Verma et al., 1972; Kacian et al., 1972; Ross et al., 1972; Marrotta et

---

*Because RNA does not bind to nitrocellulose, the method must be adapted by using an alternative medium. Diazobenzyloxymethyl (DBM) paper has been introduced by Alwine, Kemp, and Stark (1977).

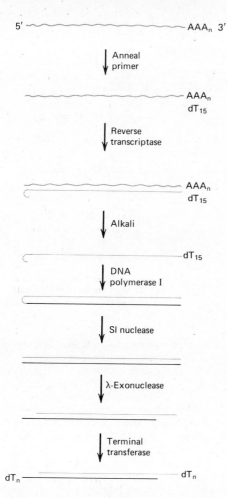

5′ ~~~~~~~~~~~~~~~~~~~~~ AAA_n 3′

↓ Anneal
  primer

~~~~~~~~~~~~~~~~~~~~ AAA_n
                    dT_15

↓ Reverse
  transcriptase

~~~~~~~~~~~~~~~~~~~~ AAA_n
                    dT_15

↓ Alkali

~~~~~~~~~~~~~~~~~~~~ dT_15

↓ DNA
  polymerase I

↓ SI nuclease

↓ λ-Exonuclease

↓ Terminal
  transferase

dT_n ―――――――――― dT_n

Figure 26.2. Method for copying mRNA sequences into duplex DNA. The first five steps generate a duplex copy of the messenger sequence, extending from the 3′ end to as far as the reverse transcriptase proceeded toward the 5′ end. The DNA is prepared for insertion into a plasmid by treatment with lambda exonuclease to generate suitable ends to which dT tails may be added.

al., 1974). With mRNA populations, often the cDNA population is only 300–400 bases long on average, as we have noted in discussing its use in hybridization in Chapter 24.

The reason for the production of reverse transcripts shorter than the template probably lies with premature termination at discrete sites, rather than cessation of reverse transcription continually with increasing time. Thus the reverse transcripts of globin mRNA contain a discrete set of bands when electrophoresed on acrylamide gels (Efstratiadis et al., 1975). Whether the discrete sites correspond to particular sequences or to regions of secondary structure is not known; but a similar feature is found also in the action in vitro of E. coli DNA polymerase I or RNA polymerase and has been useful for ordering fragments in sequencing studies (Gilbert, Maizels, and Maxam, 1973; Sanger et al., 1973). The longest cDNA molecules obtained from globin mRNA are close to the apparent size of the mRNA; their proportion can

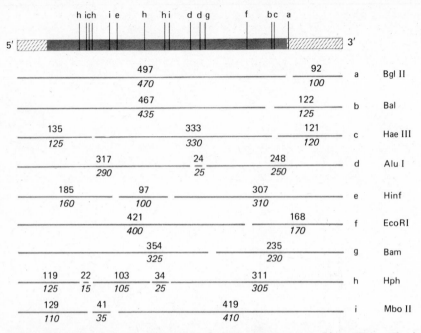

Figure 26.3. Restriction map of the rabbit β globin gene represented in mRNA. The duplex DNA sequence is conventionally oriented so that the left end corresponds to the 5′ end of mRNA. The sequence is 589 nucleotides long, with 444 bases in the coding region (shaded) and 53 and 98 bases in the 5′ and 3′ nontranslated regions (striped), respectively. Positions cleaved by a series of restriction enzymes are indicated to scale as predicted by the protein sequence and confirmed by the nucleic acid sequence. The fragments released by cleavage with each individual enzyme are depicted below the map. The bold number above each line corresponds to the predicted length of the fragment; the italic number below the line shows the size determined by gel electrophoresis of fragments generated by cleavage of a duplex DNA copy of the gene as shown in Figure 26.4. The left most fragment of each enzyme may be slightly smaller than predicted because reverse transcription may fall slightly short of the 5′ end (for example, by 12 bases in one cloned "full length" duplex copy). The right most fragment of each enzyme may be slightly too long because of the presence of poly(A):oligo(dT) sequences derived from the primer. The relationships between restriction sites are very close to expectation. Data of Maniatis et al. (1976) and Efstratiadis et al. (1976).

be increased by changes in the conditions of reaction. This is not always necessary, for with some other mRNA templates there have not been the same difficulties in obtaining full length reverse transcripts (Buell et al., 1978).

The steps in converting the cDNA into a duplex DNA are illustrated in Figure 26.2. After removal of the mRNA template by alkaline hydrolysis, the full-length reverse transcripts prove to contain hairpin structures, as seen in a level of 2–8% resistance to S1 nuclease. The hairpins are small, about 13–22 bases, and appear to lie at the 3′ end of the cDNA (corresponding to the 5′ end of the mRNA). Apparently they form because the reverse transcriptase loops back on itself for a very brief period. Their occurrence is

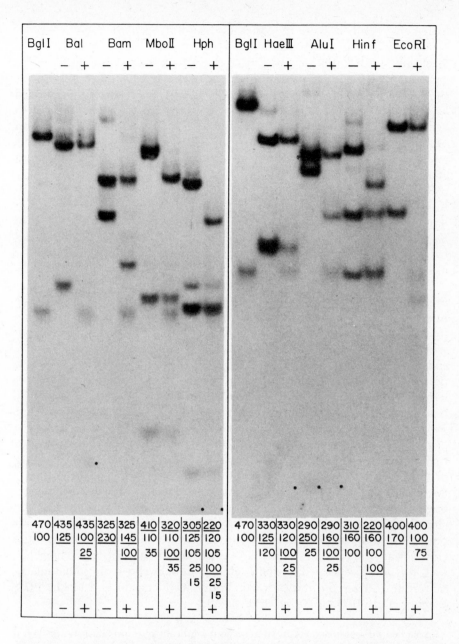

The table below each gel:

BglI	Bal		Bam		MboII		Hph		BglI	HaeⅢ		AluI		Hinf		EcoRI	
470	435	435	325	325	410	320	305	220	470	330	330	290	290	310	220	400	400
100	125	100	230	145	110	110	125	120	100	125	120	250	160	160	160	170	100
		25		100	35	100	105	105			100	25	100	100	100		75
						35	25	100			25		25		100		
							15	25									
								15									
–	+	–	+	–	+	–	+	–	+	–	+	–	+	–	+

Figure 26.4. Restriction fragments of duplex DNA copy of rabbit β-globin mRNA. The BglII and all the (–) columns represent analysis of the fragments cleaved by the single enzyme indicated. All the (+) columns represent double digestion by BglII and the other enzyme indicated. The sizes of the fragments resolved are indicated below each gel; those that are changed between the (–) and (+) columns are underlined, to identify where the BglII site falls in relation to the other restriction sites. The experiments were performed in the two gel runs indicated by the two panels, with a BglII included in each. Data of Maniatis et al. (1976).

775

extremely useful, because they provide "natural" primers for DNA polymerase I, which then can synthesize a complement to the cDNA by chain extension. With β-globin, this can be characterized as a covalent molecule of about 1200 bases when electrophoresed under denaturing conditions. In the duplex form, this consists of about 600 base pairs, the two strands linked by the hairpin at one end. This can be cut by treatment with S1 nuclease to generate a conventional duplex (Efstratiadis et al., 1976). The formation of hairpins at the end of the reverse transcripts is common to other systems and now generally is used for obtaining a duplex copy of mRNA (Seeburg et al., 1977; Buell et al., 1978).

Restriction mapping confirms the structure of the duplex DNA, as illustrated in the example of Figures 26.3 and 26.4. The restriction map can be related to the sequence of the gene predicted by the amino acid sequence of β globin (assuming appropriate nucleotides to be present at flexible third-base positions). This allows the restriction fragments to be placed in order more readily.

Using the technique of poly(dA:dT) tailing, a duplex copy of the rabbit β-globin mRNA was inserted into a plasmid (Maniatis et al., 1976). The inserted sequence can be characterized by its restriction map. Usually the inserted DNA is cleaved from the chimeric plasmid by using a restriction enzyme that cuts at the junctions (which is possible when restriction sticky ends have been used for insertion), or close to them (which is necessary when poly(dA:dT) tailing has been used). Then the released fragment is characterized by cleavage with further enzymes; when excision has occurred by cleavage within the vector, the fragments on either end will include vector sequences and the hybrid junctions. The congruence of the restriction map of the inserted sequence with that predicted from the protein and/or with that of the isolated duplex DNA demonstrates the fidelity of the cloning process, which is further confirmed by the determination of the sequence for the inserted DNA (see below). Similar experiments in which a less thorough characterization of the inserted sequence was achieved, and where the cloned segment may not necessarily have been a full length copy, were reported by Rougeon and Mach (1976), Higuchi et al. (1976), and Rabbitts (1976). Many other copies of mRNAs since have been cloned and characterized, including separated α- and β-globin, human chorionic somatomammotropin and rat pituitary growth hormone (Rougeon and Mach, 1977; Seeburg et al., 1977a,b; Shine et al., 1977).*

---

*A point that emerges from these initial studies is the difficulty of knowing whether the cloned segment corresponds to the entire mRNA. Since the copying process is initiated by reverse transcription at the 3′ terminus of mRNA, the inclusion of this end usually is assured. Whether copying has proceeded past the N terminus of the protein can be determined by comparing the nucleotide sequence of this end of the cloned segment with the amino acid sequence of the protein; but even if the coding region is intact, this does not prove whether the cloned segment extends to the 5′ terminus of the mRNA. This must be determined by direct comparison of the two sequences.

Bacterial Translation of Eucaryotic Genes

The universality of the genetic code implies that any given coding sequence will have the same meaning in all cells. The only formal demand that is made for translation per se is that tRNAs should be present to read all codons and that punctuation signals should be recognized. Differences have been found in the extents to which particular synonym codons for any amino acid are used in different systems. But their significance has been somewhat overrated: there is no evidence that any particular cell actually is unable to to read any particular codon(s). Punctuation signals are constant in the simplest sense: AUG is the usual initiation codon; and UAA, UAG, and UGA all indicate termination. In a wider sense, however, the extent to which heterologous initiation of translation may occur within bacteria is more doubtful; there is insufficient evidence on whether ribosome binding sites will be recognized efficiently and on the stability of heterologous messengers. A hint that this might not be a problem is provided by the success of heterologous systems for translation in vitro (see Chapter 23). There are therefore two practical problems to be resolved in the expression of cloned eucaryotic genes in bacteria: the cloned segment must be placed in a context in which it is transcribed into mRNA; and then the messenger must be translated by bacterial ribosomes. A subsequent difficulty is that the protein product may be unstable in the bacterial cell.

Transcription of a cloned segment of DNA can take place either by initiation within the eucaryotic region or by readthrough from a promoter within the adjacent plasmid sequence. When a segment of the genome has been cloned, transcription might initiate within it, either by heterologous recognition of the natural promoter, or because of the fortuitous occurrence in the right location of a sequence resembling a bacterial promoter. Insufficient knowledge is available about bacterial promoter sequences to predict the likelihood of their occurrence in any particular cloned sequence or to construct synthetic regions for linkage to cloned DNA. When cloned DNA is derived by copying mRNA, there is in any case little probability that transcription will be initiated, since the cloned segment lacks any genomic control regions. But when the site of insertion in a cloning vector lies within a transcription unit, the cloned sequence may be read by extension from the vector DNA. Presumably the RNA polymerase continues until it meets a termination sequence, most likely at the usual point on the other side of the cloned segment. An impediment to expressing cloned DNA from eucaryotic genomes is presented by the occurrence of intervening sequences, which the bacterial host may be unable to remove from the transcript. This has compelled most experiments to obtain expression of eucaryotic genes in bacteria to rely on the insertion of cloned copies of mRNA at sites downstream from a promoter on the vector. More precisely, the cloned segment often is a copy of the coding region, since there is no particular evident need for it to include the 5′ nontranslated regions of the messenger.

Table 26.2 Complementation of bacterial

Bacterial mutation	Deficient enzyme	Source of chimeric plasmid
hisB deletion	imidazole glycerol P dehydratase	S. cerevisiae DNA cleaved with Eco RI and inserted in λ*gt*.
hisB deletion	imidazole glycerol P dehydratase	S. cerevisiae DNA randomly sheared and inserted in Col El
leuB3 point mutations and deletion	β-isopropyl malate DH	S. cerevisiae DNA randomly sheared and inserted in Col El
leuB6 single point mutation	β-isopropyl malate DH	S. cerevisiae DNA randomly sheared and inserted in Col El
argH deletion	arginosuccinate lyase	S. cerevisiae DNA randomly sheared and inserted in Col El
aroD mutation not defined	5 dehydroquinate hydrolase	N. crassa DNA cleaved with Hind III and inserted in pBR322

The first two columns give the locus and type of bacterial mutation and the enzyme activity that is deficient in the auxotroph. E. coli is the bacterial host in each case except the *leuB* deletion, which is present in S. typhimurium. In each case a library of vectors carrying random parts of a fungal genome was tested for the presence of chimeric plasmids able to complement the bacterial deficiency by isolating infected bacteria able to grow in the absence of the appropriate nutrient. The plasmid pres-

A general approach to determining the ability of segments of a eucaryotic genome to be expressed in bacteria is to test the ability of a cloned library to complement bacterial mutations. Auxotrophic bacteria are infected with chimeric plasmids and tested for their ability to grow in the absence of the appropriate nutrient. The occurrence of a clone should indicate that some function carried by the plasmid is able to substitute for the function that has been mutated in the bacterial host. To make a negative result significant, it is necessary to be sure that the cloned library does indeed represent virtually all the segments of the eucaryotic genome.

To exclude the possibility of reversion or suppression of the bacterial mutation, it is necessary to use deletions. The need for the plasmid to support growth can be demonstrated by removing it through acridine curing. The sequence carried by the chimeric plasmid must be characterized as part of the cloned foreign genome in order to exclude the possibility that some bacterial function has been acquired during its perpetuation in prior hosts.

mutants by chimeric plasmids

Size of insert in active plasmid	Nature of complementation
10.3 kb (λgt-Sc2601)	Probably expression of IGP dehydratase structural gene of yeast as *his3⁻* yeast mutants cannot provide complementing plasmids
10.7 kb (pYe*his* 1,2,3)	no evidence
13.5 kb (pYe*leu* 10)	Possibly expression of structural gene as plasmid may carry DNA from yeast chromosome III on which *leu2* locus resides
8.9–13.2 kb (pYe*leu* 11,12,13) 11.1–11.9 kb (pYe*arg* 1-4) and pYe*leu* 116)	Probably suppression since different sequences found on several pYe*leu* and pYe*arg* plasmids all suppress *leuB6* but not other *leuB* point mutations
11.1–11.9 kb (pYe*arg* 1-4 and pYe*leu* 118)	plasmids carry variants of same sequence; action on *argH* lies in different part of DNA from action on *leuB6*; basis is unknown
6.2 kb (pVK55)	Probably expression of Neurospora structural gene since enzyme activity has molecular weight of N. crassa species

ent in each such clone is recovered and characterized. The name given to the complementing plasmid(s) is noted in parenthesis after the size of the insert(s).

References: Vector λgt—Struhl, Cameron and Davis (1976), Struhl and Davis (1977); Col El vector—Ratzkin and Carbon (1977), Clarke and Carbon (1978); pBR322 vector—Vapnek et al. (1977).

This can be done by hybridization with the original eucaryotic genome; and usually the size of the inserted DNA is determined by restriction mapping. Ultimately it is necessary to purify the protein in which the complementing enzyme activity resides; but as a first step often it is possible to show that it has the size or some other feature characteristic of the corresponding (or some other) eucaryotic protein.

Available data are summarized in Table 26.2 and in several cases there is prima facie evidence for supposing that the vector is expressing a eucaryotic gene. Three bacterial deletions, in *hisB*, *leuB*, and *argH* have been complemented by chimeric vectors present in libraries representing the S. cerevisiae genome. In these cases it seems likely that transcription and translation have been initiated within the yeast DNA because the same segment of cloned DNA is equally active in either orientation relative to the vector. In the examples of *hisB* and *leuB* there is indirect genetic evidence to suggest that the complementing function is the corresponding gene of yeast. In *argH* it

is unknown; although as with *leuB* and *hisB*, all the complementing plasmids carry the same yeast DNA sequence, implying that at least only a single
eucaryotic function is involved.

In one case, suppression rather than complementation appears to be responsible, because several plasmids, carrying different parts of the yeast genome, are able to suppress the bacterial *leuB6* point mutation, but are ineffective with other point mutations in this gene. It turns out also that all the
plasmids selected for ability to complement a bacterial deletion in *argH*
are able to suppress *leuB6*, which may mean that a copy of whatever function is involved lies near the yeast counterpart of the *argH* gene. A sequence
derived from another fungus, Neurospora crassa, also is able to complement
a bacterial deletion, probably by expression of the eucaryotic gene corresponding to *aroD* of E. coli. This system makes a useful point that the ability to complement may depend on the bacterial genotype, since complementation is prevented by some host bacterial strains while occurring in others.

No successful attempts to express cloned genomic DNA have been reported with any other species, notably the higher eucaryotes. Possible reasons for the negative results include the inability to recognize transcription
and/or translation initiation sites and the presence of intervening sequences.
A corollary is that if the complementation described in Table 26.2 indeed
represents expression of fungal genes, these must lack intervening sequences
(this is to exclude the possibility that bacteria may remove them from mRNA).
The synthesis of protein by chimeric plasmids carrying segments of Drosophila DNA has been reported in several cases, but in none is there any evidence that the proteins represent natural products (although it has been
argued that long polypeptides are unlikely to arise by chance translation
because of the probable occurrence of termination codons in nontranslated
regions; it is possible that some of these represent parts of natural products)
(Meagher et al., 1977; Miller et al., 1977; Rambach and Hogness, 1977).
Cloned mouse mitochondrial DNA appears to suffer transcription of the
proper strand in E. coli, but the protein products are smaller than those
seen in vivo (Chang et al.,1975).

The problems of initiating transcription and translation in cloned DNA
and of expressing genes containing intervening sequences can be circumvented by inserting coding regions (derived by copying part or all of mRNA)
into a vector site that lies within a transcription unit. Either the cloned segment must be inserted only in the appropriate orientation for expression;
or following insertion in random direction, half of the chimeric plasmids
should have the active orientation. Whether the cloned segment is mean-
ingfully translated depends on its location with respect to the gene into
which it is inserted. So long as the site of insertion lies within a coding region, the eucaryotic segment will be translated as well as transcribed by
readthrough from the vector. If the inserted coding region is in the same
phase as the vector coding region, this will generate a hybrid protein that
consists of an N terminal sequence of the vector protein, replaced from the

point of insertion by the eucaryotic protein sequence. Termination will occur at the usual site at the end of the eucaryotic coding region. If the two coding regions are out of phase, translation will continue into the eucaryotic DNA to generate a meaningless product, terminating at the first chance nonsense codon that occurs in the phase of reading.

Several systems have been used to translate eucaryotic coding regions by linkage to vector proteins; and available results are summarized in Table 26.3. For these purposes it is useful to have the eucaryotic insert start as close as possible to the first amino acid of the eucaryotic protein. This may be accomplished by finding an appropriate restriction site in a cloned copy of the mRNA that includes the 5' nontranslated region or by using reverse transcripts that are selected for a terminus close to the initiation codon.

One useful system is provided by the plasmid pBR322, which contains an amp^r gene that codes for β-lactamase (penicillinase). This protein is synthesized in the form of a precursor containing an additional 23 N terminal amino acids, which are removed when it is secreted into the periplasmic space (Ambler and Scott, 1978; Sutcliffe, 1978a). A PstI cleavage site exists at the position of amino acid 182 of the protein. Several eucaryotic sequences have been inserted at this site, either by using the PstI sticky ends or by poly(dC:dG) tailing. In the first case, it is necessary for there to be by chance a PstI site in proper phase with the subsequent eucaryotic coding region; in the second case, the plasmids should fall into three classes, depending on the length of the dC:dG linkers, one of which should contain both coding regions in the same phase. The active chimeric plasmids should produce a hybrid protein that contains 182 N-terminal amino acids of β-lactamase (including the precursor signal for secretion), joined either directly or through a linker represented in amino acids, to the eucaryotic protein.

With this and with other cloning systems, active products have been detected in bacteria carrying chimeric plasmids by radioimmunoassay for the protein. Characterization of the active proteins is not yet complete enough to say whether the expectations of the system are fulfilled, but there are indications that this may be so. The chimeric pBR322 plasmids carrying eucaryotic coding regions that appear to be active have been sequenced directly. In the case of insulin, a sequence starting with the fourth codon of preproinsulin is linked by an integral number of codons of linker material to the β-lactamase gene. In the case of growth hormone, PstI cleavage directly links the vector to a codon within the prehormone that lies 23 amino acids before the start of the mature protein. In the latter case a protein of about the right size to be a hybrid product is synthesized when the chimeric plasmid is carried by minicells. A chimeric plasmid carrying the DHFR gene, however, has a nonintegral number of codons separating the eucaryotic coding region from the β-lactamase. The production of DHFR therefore cannot occur by simple readthrough. One possibility is that initiation may take place at the eucaryotic AUG, perhaps because the poly(G) linker in mRNA mimics a ribosome binding site. Another possibility is that slippage

Table 26.3   Translation of

Eucaryotic gene	Structure of chimeric plasmid at insertion
Duplex DNA copy of rat pre-proinsulin mRNA	pBR322—β-lactamase—**182 dG$_{17}$A** + 4-preproinsulin—>*300*-dG$_n$—pBR322 *Pst I* *Pst I*
Cloned DNA copy of rat growth hormone (GH) mRNA	pBR322—β-lactamase—**182 -23**—pre GH—*800*—pBR322 *Pst I* *Pst I*
Duplex DNA copy of mouse DHFR mRNA	pBR322—β-lactamase—**182 dG$_{11}$AUG +1**—DHFR—*1500* dG$_n$—pBR322 *Pst I* *Pst I*
Cloned DNA copy of chick ovalbumin mRNA	pBR322—2.85kb-β-gal—**7 +5**—ovalbumin—*1700*—pBR322 *Hind III* *Eco RI/blunt* *Hha/blunt*
Cloned DNA copy of chick ovalbumin mRNA	pMB9—203bp-β-gal—**7 dN$_6$-8**—ovalbumin—*2200*—pMB9 *Hae III* *Eco RI* *Eco RI*
Synthetic 60bp gene for somatostatin	pBR322—203bp-β-gal—**7 dN$_6$ AUG +1** somatostatin—*60*—pBR322 *Eco RI* *Eco RI* *Bam HI*
Synthetic 60 bp gene for somatostatin	pBR322—6.5kb-β-gal-**1005 AUG +1** somatostatin—*60*—pBR322 *Eco RI* *Eco RI* *Bam HI*
Cloned synthetic genes for human insulin A (77bp) and insulin B(104bp)	pBR322—6.5kb-β-gal-**1005 AUG +1**—insulin—*77* or *104*—pBR322 *Eco RI* *Eco RI* *Bam HI*

The structure of the chimeric plasmid shows the cloning vector and the gene into which the foreign DNA is inserted. In some experiments this is the *amp*^{r} of pBR322 itself, which codes for β-lactamase (penicillinase). In the others a fragment of the lactose operon, including the control region and part of *lacZ*, has been introduced into the cloning vector; and the foreign DNA then is inserted into the β-galactosidase gene. The site of insertion relative to the beginning of the gene in which it lies is indicated by the first bold face number; this gives the position of cleavage in terms of the number of amino acids from the initiation site of the plasmid protein. Any linker sequences are shown; and then the last boldface number gives the amino acid position with which the insert starts. (+) indicates position after the usual startpoint, i.e., within the protein; (-) indicates preceding positions, which in growth hormone lie within the amino acid sequence of the precursor but for ovalbumin lie in the untranslated region. Italic

eucaryotic genes in bacteria

Characterization of product	Amount
insulin activity detected by radioimmunoassay; size of protein not known; secretion may occur	~100 molecules/cell
GH activity detected by radioimmunoassay; 46,000 dalton protein is synthesized in minicells; GH activity not secreted	~24,000 molecules/cell
DHFR activity detected in 22,000 dalton protein by radio-immunoassay; DHFR activity shows response to inhibitors characteristic of mammals	~0.01% of soluble protein
ovalbumin activity detected in 43,000 dalton protein by radioimmunoassay; not secreted	~40,000 molecules/cell
ovalbumin activity detected in protein larger than native ovalbumin by radioimmunoassay; 50% secretion into periplasmic space	~1.5% of total protein = 100,000 molecules/cell
no somatostatin activity detectable	
somatostatin activity detected by radioimmunoassay in cell extracts treated with cyanogen bromide to cleave hormone from β-galactosidase; original product not analyzed	<0.03% of total protein
insulin A or B chain activity detected by radioimmunoassay in products cleaved by cyanogen bromide from protein of size of β galactosidase	not assayed directly, but putative hybrid protein is 20% of total protein

numbers indicate the total length of each insert in nucleotides, which in all cases appears to be long enough to include the end of the coding region. The sequences at the first junction have been confirmed by experiment in all cases except ovalbumin. Restriction sites used to form the junctions of the chimeric plasmids are indicated (blunt indicates removal of single strand ends followed by blunt end ligation); linkers that were used are indicated in the sequence. DNA from cloned genes was obtained by restriction enzyme cleavage of the carrier plasmid. All estimates of the amount of eucaryotic protein synthesized are only very approximate.

References: Insulin—Villa-Komaroff et al. (1978); growth hormone—Seeburg et al. (1978); DHFR—Chang et al. (1978); ovalbumin—Mercereau-Puijalon et al. (1978), Fraser and Bruce (1978); somatostatin—Itakura et al. (1977); synthetic insulin—Goeddel et al. (1979).

occurs in the homopolymer linker to allow correction of the phase. In either case there is likely to be some loss of efficiency in translation of the eucaryotic segment. Independent initiation seems more likely because the active product has a size close to that of mammalian DHFR. The extent to which these hybrid proteins are secreted by virtue of their N-terminal leader sequence is not yet entirely clear.

Another approach is to insert into the cloning vector a region representing the start of the lactose operon. A fragment that is particularly useful for this is a 203 base pair restriction product that carries the promoter, operator, and part of *lacZ*; but other fragments have been used in the same way. An advantage of the lactose system is that promoter mutants are available which increase the rate of initiation of transcription. There is an EcoRI site at the seventh codon of β-galactosidase that can be used to insert foreign DNA; this then should be expressed under control of the lactose repressor to produce a hybrid protein that consists of only the first 7 amino acids of β-galactosidase linked to the eucaryotic sequence. The procaryotic sequence is indeed so short that it may fail to contribute significantly to the properties of the protein product. Ovalbumin has been synthesized in this way, using chimeric plasmids in which the protein starts with either the fifth amino acid of the usual ovalbumin sequence or with the eighty preceding codon in the 5'-nontranslated region.

With short eucaryotic proteins it is possible to synthesize chemically a "gene" that codes for the function. Three such sequences have been formed, with EcoRI sticky tails at one end and BamHI sticky tails at the other. This allows insertion only in the proper orientation in the pBR322/*lactose* vector cleaved jointly with EcoRI and BamHI. In each synthetic gene an AUG codon precedes the codon for the first amino acid, not to be used for initiation, but to ensure that the procaryotic and eucaryotic protein sequences are linked by a methionine residue. Then cleavage of the hybrid protein with cyanogen bromide releases the two parts; of course, this approach cannot be used for eucaryotic proteins that contain methionine. The synthetic gene terminates with two nonsense codons to ensure that no additional material is present at the C terminal end of the protein product. When a sequence coding for somatostatin was linked to the seventh codon of β galactosidase, no active product could be detected. An indication that the cause might be instability of the product is given by the successful production of somatostatin or insulin activity upon insertion of a synthetic gene at position 1005 of β-galactosidase. Perhaps the longer bacterial protein sequence protects the eucaryotic protein sequence; in any case, the presence of preceding amino acid sequences in the hybrid protein is irrelevant to the eucaryotic region, since they are removed before assay.

In most of these experiments the yield of eucaryotic protein is comparatively small. This seems less likely to be due to problems in gene expression than to result from instability of the hybrid product, presumably because of proteolysis of the foreign amino acid sequence. When fully induced, a single

β galactosidase gene synthesizes enough enzyme to provide about 2% of total cellular protein. With a multicopy plasmid such as pBR322, about 10 times this level should be achieved, but this has been approached in only one report.

DNA Sequence Determination

The conventional approach to nucleic acid sequencing has involved cleavage of the nucleic acid into small fragments whose sequence is determined by direct analysis. By analogy with the sequencing of proteins, the main problems are the lack of sufficient diversity in base constitution and the difficulty of deriving overlaps. In any case, this approach has been more successful with RNA than with DNA, because of the existence of ribonucleases with specificity for particular bases. The modest amount of information gained with this method emphasizes the progress represented by the revolution in sequencing that has resulted from the development of techniques that allow sequences of reasonable length to be determined directly. These were made possible by the observation that DNA molecules of the same sequence and with one common end, but differing in length at the other end by only 1 nucleotide, can be separated by electrophoresis on acrylamide gels. Bands identifying single stranded DNA molecules of increasing length can be followed over a range of about 20–200 nucleotides. While several protocols now have been developed to obtain such series of molecules, they share in common the principle that the bands are divided into sets according to their terminal nucleotide. This is accomplished by running several gels in parallel, each of which contains the bands of a given set. By following successive bands across the sets, the base responsible for each single increase in length can be identified, and thus the DNA sequence can be "read" directly off the gels. These techniques now are so powerful that a coding region of DNA may be sequenced more rapidly than its protein product.

Two types of approach have been used to obtain sets of bands that identify each possible terminal base. One is to use chemical reactions that lead to cleavage of DNA strands at individual bases. The other is to synthesize DNA in vitro by copying a template strand in such a way that the reaction products terminate specifically with given bases.

Chemical modifications with appropriate specificities have been made the basis of a sequencing technique by Maxam and Gilbert (1977). DNA in either single stranded or duplex form is labeled at either the 5′ or 3′ terminus with ^{32}P. With a substrate of duplex DNA, both ends will be labeled, and so they must be separated before analysis. This can be done by denaturing the DNA and isolating the individual single strands. An alternative is to cleave the duplex DNA with a restriction enzyme, so that the two labeled ends fall onto different duplex molecules which can be separated before the single strands are isolated for analysis. The DNA is treated in two steps:

first a reagent is added which excises a base from the polynucleotide chain; then the deoxyribose moiety that is exposed by its loss is removed by alkali treatment or β elimination, creating a break in the chain. The first reaction must be performed under conditions when modification is partial, occurring in any given molecule at 1-2% of the available sites. The second reaction must be complete, so that every modified site is cleaved. This generates a population of molecules each broken at random at a site occupied by the base that is subject to modification. When single strands are electrophoresed, the lengths of the labeled fragments can be determined by autoradiography. They comprise a set of molecules that should represent the distance from labeled terminus to each position at which the modified base was present.

Purines can be recognized by their reaction with dimethyl sulfate, which methylates the N7 position of guanine about 5 times more effectively than the N3 position of adenine. Upon heating the methylated base is lost. When end labeled fragments are identified by autoradiography of gels, a dark band is generated for every position at which guanine was present, while a light band results from the less frequent breakage of adenine. The pattern can be reversed by releasing the methylated bases with acid instead of by heat, when adenine is lost more efficiently to form a dark band and the less effective loss of guanine forms a light band. Pyrimidines are analyzed by treatment with hydrazine, which causes cytosine and thymine to be lost equally well, thus generating bands of the same intensity for both bases. These may be resolved by performing the reaction in the presence of high salt, when only cytosine is able to react to form bands.

After these four reactions have been performed, the products are analyzed in parallel on a set of gels. Autoradiography reveals a series of bands on each gel, each representing the occurrence of the appropriate base(s) at a different position in the DNA. Taking all four gels together, a band should be present for every nucleotide position in the DNA. Reading up from the bottom of the gels, successive bands therefore represent additions of one nucleotide to the distance from the labeled terminus. Which base is present at each position is revealed by the gel(s) on which the band is found. The sequence may therefore be read up the gel until the bands become too close together for positions to be distinguished. An example is shown in Figure 26.5.

The first copying method for sequencing was developed by Sanger and Coulson (1975). This uses a single stranded DNA as template for synthesis of a complement by E. coli DNA polymerase I; the reaction starts at a primer annealed to the template and relies upon a ^{32}P labeled dNTP precursor to identify the product. If the reaction is allowed to proceed in an asynchronous manner, synthesis of the complementary DNA strand should cease at random locations, generating a population of labeled products that include molecules of all possible lengths between the primer and the longest chains. These are isolated still annealed to the template strand and are used as substrates for two further types of reaction. In the "minus" reaction, the mixture is reincubated in four separate incubates, each of which lacks a different one of the four dNTP precursors. Synthesis proceeds in each case as far as possible before encountering a position that must be filled by the absent nucleotide. Given the nature of the starting population, each of the four reactions therefore generates a set of molecules terminating one base short of every position that should have been filled by the missing base. The "plus" reaction relies on the ability of T4 DNA polymerase to degrade duplex DNA from the 3' end, but stopping at any position for which a triphosphate is present. Four reaction mixtures are set up, each containing a different dNTP precursor. The result in each case is a set of products terminating at every position filled by the appropriate base. After the single-strand products have been recovered, they are separated by gel electrophoresis and identified by autoradiography. The plus and minus gels for each nucleotide should have a series of bands that differ in length by one base; except when a run of the same nucleotide occurs, when the distance between bands corresponds to the number of nucleotides in the run. This is a weakness in the method, since sometimes the distance may be difficult to determine accurately.

A simpler copying method for sequencing is the dideoxy technique developed by Sanger, Nicklen, and Coulson (1977). Dideoxynucleotides inhibit DNA synthesis when incorporated into the growing chain, because no 3'-OH group is available for further extension. The frequency with which termination occurs can be controlled by the proportion of ddNTP to dNTP in the reaction mixture. For sequencing purposes, a primer is annealed to a single-strand DNA template and four separate reactions are set up in which DNA polymerase I synthesizes DNA in the presence of one of the dideoxynucleotides and all four deoxynucleotides, one of which carries a ^{32}P label. Under suitable conditions, termination has a low probability of occurring at each position occupied by the appropriate base; and this generates the now familiar series of products, extending from the primer to each base position. Comparison of the four sequencing gels allows the nucleotide sequence to be read off from the bands. The method is simple to use because few manipulations are involved, requiring only a single copying reaction.

Another sequencing method involves partial ribosubstitution, relying upon the ability of DNA polymerase I to incorporate ribonucleotides into DNA. In the protocol of Barnes (1978), a primer is briefly extended with DNA polymerase I to allow the incorporation of ^{32}P-labeled deoxynucleotides; and then the reaction is continued in four separate incubates, in each of which about 2% ribosubstitution of a different base occurs. Alkali is used to cleave the DNA at each ribosubstituted position. As previously, the sequence is determined by comparing the four parallel gels.

Yet another option is provided by the reaction of nick translation. The method of Maat and Smith (1978) introduces nicks at random sites in duplex DNA by the action of DNAase I. The 3'-OH groups exposed by each nick serve as primers for chain extension by DNA polymerase I; four exten-

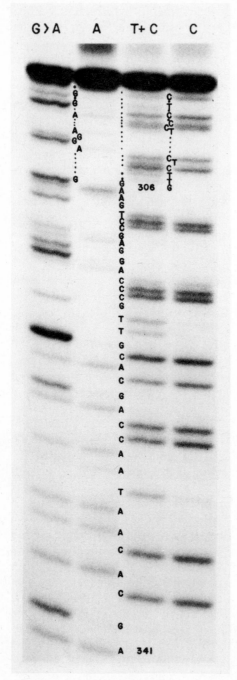

Figure 26.5. Sequence analysis by the Maxam-Gilbert method. The sequence is read 5′ to 3′ from bottom to top of the gel. Data of Efstratiadis et al. (1977).

sion reactions are performed, each in the presence of one dideoxynucleotide to cause random partial termination. The DNA substrate carries a 5′ terminal label at one end only, which is achieved by labeling both 5′ termini and then cleaving off one end of the duplex with a restriction enzyme before applying the sequencing protocol.

All of these methods allow sequences of DNA to be determined subject only to the limits of resolution of the gels that are used to separate the fragments. Probably the maximum distance that can be read in one experiment is in the range of 300–400 bases, but more often only 200 or so are read. Characteristic difficulties are encountered in each method so that there may be a few positions in any sequence whose base assignment is equivocal. An advantage of the Maxam-Gilbert and nick translation methods is that they may be used with duplex DNA; the former in fact was developed as the result of experiments to characterize regions on duplex DNA that are protected from modification when bound by protein and this is a useful feature. The copying methods relying on primer extension require the provision of purified single-strand templates and suitable primers; usually the primer is provided by annealing a denatured restriction fragment to the template strand. All the methods are effective with DNA, but some have been adapted for use with RNA. The Maxam-Gilbert method may be applied to the cDNA synthesized by reverse transcription of mRNA, although this is limited to the sequences extending immediately from the primer. The plus and minus method has been adapted for use with reverse transcriptase, but again is subject to the limitation of proximity to the primer (Brownlee and Cartwright, 1977). The dideoxy method has been applied to RNA replication by $Q\beta$-replicase (Kramer and Mills, 1978). For the most part, however, RNA has been sequenced by using a DNA copy amplified by cloning. More recently a set of chemical reactions has been developed to allow direct sequencing of RNA by a procedure similar to the Maxam-Gilbert DNA sequencing method (Peattie, 1979).

The protocols that are available allow any DNA sequence to be determined from a fixed point. By using a restriction enzyme to establish this position, virtually any length of DNA can be sequenced once its restriction map is established. Restriction fragments of suitable size are chosen from the map and sequenced individually. To ensure accuracy at the junctions of restriction fragments, overlapping sets of fragments are used; this is necessary especially to ensure that small fragments, unnoticed on the restriction map, do not lie between the larger restriction segments. Generally the accuracy of a sequence of a single strand of DNA may be taken to be >95%. This may be increased appreciably, to say 99%, by separately determining the sequence of both strands of a duplex; it is rare for equivocal assignments to lie at the same position of both strands. Another check is possible in coding regions by comparing the nucleotide sequence with the amino acid sequence of the protein.

Occurrence of Intervening Sequences: Globin Genes

Two independent lines of research converged to provide the first demonstrations of the existence of intervening sequences in DNA and their absence from mature RNA.

The first observation of intervening sequences, in fact, was made in a system in which it is not clear whether the interrupted genes are expressed. Restriction mapping showed that some of the 28S rRNA genes of D. melanogaster are longer than others; electron microscopy of the hybrids formed with the rRNA showed that the longer genes contain an interruption of variable length that is not represented in the rRNA (Glover and Hogness, 1977; White and Hogness, 1977; Wellauer and Dawid, 1977; Pellegrini, Manning, and Davidson, 1977). Because of the multiplicity of rRNA genes, it is not possible to say whether any particular copy, or class of copies, is expressed.

But similar interruptions then were observed in unique genes, for which no other, uninterrupted copies exist. Observed first with rabbit and mouse β-globin, and then with chick ovalbumin, as well as with a yeast suppressor tRNA, intervening sequences not represented in the mature RNA are found within the coding region (Jeffreys and Flavell, 1977b; Tilghman et al., 1978a; Breathnach, Mandel, and Chambon, 1977; Goodman, Olson, and Hall, 1977). Here there is no escape from the conclusion that the intervening sequences are present in the same genes whose transcripts lack the interruptions.

At the same time, analysis of the structure of late messenger RNAs of adenovirus and SV40 showed that the 5' regions represent sites on the viral DNA that are distant from the regions corresponding to the coding sequences of the messengers (Berget, Moore, and Sharp, 1977; Chow et al., 1977; Klessig, 1977; Dunn and Hassell, 1977; Aloni et al., 1977). The adenovirus 5' "leader" region is an amalgam of several short RNA sequences that represent widely separated sites in the viral genome; the same leader is common to all late mRNAs, being joined to a variety of coding sequences that correspond to the different late genes.

The simultaneous discovery of intervening sequences in cellular genomes and of common leaders that are joined to disparate viral messengers immediately suggested the existence of a common mechanism: transcriptional units may include internal sequences that separate regions that are contiguous in the messenger RNA. This form of genetic organization is found in lower as well as higher eucaryotic genomes and in viruses; and it occurs in all classes of genes, representing mRNA, rRNA, and tRNA.

The juxtaposition in mRNA of sequences that are not adjacent in DNA might be accomplished in several ways, the most likely of which seemed to be an adjustment in transcription so that RNA polymerase would "skip" the intervening sequences, or the removal from the primary transcript of these interruptions, by cleavage and rejoining of the RNA chain. The presence of intervening sequences in primary transcripts indicates that their absence from mRNA is caused by a rearrangement of the RNA sequence; and this now is known as RNA splicing.

Comparison of the restriction maps of a cloned copy of a messenger RNA and of the corresponding regions of cellular DNA shows whether the sequences of the genome and its mature transcript coincide. The simplest approach is to cleave cellular DNA with restriction enzymes that do not find target sites in the cloned duplex DNA copy of the message. If the gene is present as a single uninterrupted sequence, a single fragment should be released from the genome. This result may be taken to demonstrate that a gene is unique; although it does not exclude the possibility that it contains intervening sequences lacking the recognition site. If two (or more) fragments released from cellular DNA are able to hybridize with a probe for the messenger sequence, there are two possible explanations. The gene may be present in more than one copy per haploid genome, each copy surrounded by a different flanking sequence in which the restriction sites lie at different distances from the structural gene. Or there may be an interruption within the gene that contains the enzyme recognition site. These possibilities may be distinguished by using probes specific for the 5′ (left) or 3′ (right) ends of the mRNA. If the fragments represent multiple copies of the gene, each will hybridize with both probes. If they represent different regions of the gene separated by cleavage in an intervening sequence, each will hybridize with a different probe. (The difficulties in distinguishing duplicate genes that lie close to one another from a single gene containing intervening sequence(s) are illustrated in the example of human α-globin shown in Figure 28.3.)

The location of intervening sequence(s) may be characterized by determining the relationship between restriction sites within the gene. Usually double digests are used so that distances may be calculated relative to a cleavage site in the coding sequence, since the sizes of terminal fragments cannot be compared directly because the flanking sequences are different in the cloned copy and genome sequence. When an intervening sequence contains new target sites, its existence is revealed by the presence of additional fragments. When an intervening sequence does not contain a new target site, its presence increases the distance between two sites lying in the coding sequence, as revealed by the increased size of the appropriate restriction fragment. To identify the sizes of intervening sequences, a complete map must be constructed by cleavage with several enzymes, since the method of blotting and identifying fragments by ability to hybridize with a probe for the message sequence does not detect any fragments that are produced by double cleavage within an intervening sequence.

The β-globin genes of both rabbit and mouse possess an intervening sequence located about 200–250 bases from the end of the gene. The rabbit gene has been characterized by applying Southern blotting to cellular DNA; after cleavage with one or more restriction enzymes, the DNA is elec-

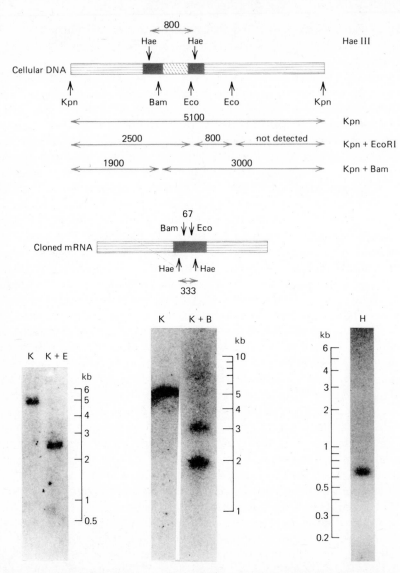

Figure 26.6. Restriction mapping of an intervening sequence in rabbit β-globin. The upper map shows the locations of restriction cleavage sites in a 5100 bp fragment containing the globin gene that is released from cellular DNA with the enzyme Kpn. EcoRI cleaves at two sites within this fragment, generating a 2500 bp fragment containing the left (5′) end of the gene and an 800 bp fragment containing the right (3′) end; the remaining fragment is not detected because it contains no globin sequences. BamHI cleaves at one site within the Kpn fragment, releasing a left fragment of 1900 bp and a right fragment of 3000 bp. This places the EcoRI and BamHI cleavage sites about 600 bp apart. The lower map of sites in the cloned copy of mRNA is taken from Figure 26.3 and shows that here the BamHI and EcoRI sites are only 67 bp pairs apart. Similarly the HaeIII sites in the cloned message copy are 333 bp apart; but HaeIII cleavage of cellular DNA releases a fragment of about 800 bp (this contains internal sequences of the globin gene but the fragments carrying the terminal regions presumably are not detected

trophoresed on agarose, denatured and transferred to nitrocellulose, and detected by reaction with a labeled probe carrying β-globin sequences. Two mouse genes have been characterized in the form of sequences isolated by such means and then cloned for characterization (which gives somewhat cleaner results). The similarity in the two sets of results relieves any apprehension that artefacts might be introduced during cloning.

A single sequence containing the β-globin gene is released from rabbit DNA by cleavage with the enzymes Kpn or PstI, implying that there is only one copy of the gene (although some minor fragments reacting to a lesser degree also are found in the PstI digest; and when EcoRI digestion produces two fragments by cutting within the gene, an additional two, minor fragments are found). The presence of an intervening sequence was indicated by the observation that about 600 extra base pairs are found between the BamHI and EcoRI sites in cellular DNA compared with the cloned copy of mRNA (Jeffreys and Flavell, 1977a,b). An example of the data is shown in Figure 26.6. The intervening sequence is not related to the globin coding sequences, since it does not hybridize with the probe for β-globin (this means that it cannot be detected directly in a BamHI/EcoRI double digest, since the released fragment must contain only 67 base pairs of globin sequence, which is not enough for reaction with the probe). The same restriction digest pattern is found in all tissues, including somatic cells in which the globin gene is or is not expressed and in sperm cells. Thus the intervening sequence is a constant feature of the genome, a conclusion that applies also to the other intervening sequences that have been characterized.

Cleavage of mouse DNA with EcoRI releases fragments of 7 and 14 kb that hybridize with a β-globin probe. Both have been cloned and characterized by restriction mapping. Each contains an entire β-globin gene, but has different flanking sequences. Subsequent studies have shown that they correspond to two nonallelic β-globin genes that are closely linked, the 7 kb fragment representing the β^{maj} gene and the 14 kb fragment representing the β^{min} gene.* Both the β-globin genes possess an intervening sequence that lies just to the right of the BamHI site (which is in the same position

*The two genes β^{maj} and β^{min} characterized in these studies are present in mice homozygous for the Hbb^d haplotype, which synthesize the two types of β-globin chain. Mice homozygous for the alternative haplotype Hbb^s synthesize only a single type of β globin chain, but appear nonetheless to have two β globin genes which are found as two different fragments of about the same size (10 kb) in EcoRI digests of the DNA (Weaver et al., 1979). The relationship of the Hbb^s genes to the Hbb^d gene structures described in the text remains to be established.

because their content of globin sequences is too low to react efficiently with the probe). This is only a small sample of the data by which the fragments of cellular DNA have been ordered and shown to contain an additional 500–600 base pairs between the BamHI and EcoRI sites (indicated by the striped region in the upper map). Below the two maps are shown Southern blots of rabbit DNA restricted with the indicated enzymes and identified by autoradiography following hybridization with a labeled cloned copy of β-globin mRNA. Data of Jeffreys and Flavell (1977b).

relative to the protein sequence as the BamHI site of rabbit; although there is no EcoRI site within the mouse coding sequence).

Electron microscopy of a hybrid formed between β-globin mRNA and the cloned β-globin genes directly identified the location and size of this intervening sequence. This depends on the technique of R loop mapping introduced by White and Hogness (1977), in which RNA is annealed with duplex DNA at elevated temperature in the presence of a high concentration of formamide. Under these conditions RNA-DNA hybrid duplexes are more stable than DNA-DNA double helices. Thus the RNA pairs with its complement in the DNA to form an RNA-DNA hybrid whose position is revealed by the displacement of the homologous DNA strand as a single stranded *R loop*. Using this technique, Tilghman et al. (1978b) showed that the DNA paired with mRNA in the hybrid duplex formed with the β^{maj} globin clone is not a continuous length, but represents two separated regions; as can be seen from Figure 26.7, its lack of complementary sequences in mRNA means that the intervening sequence cannot participate in hybrid formation, and thus is compelled to form a loop of DNA that is extruded

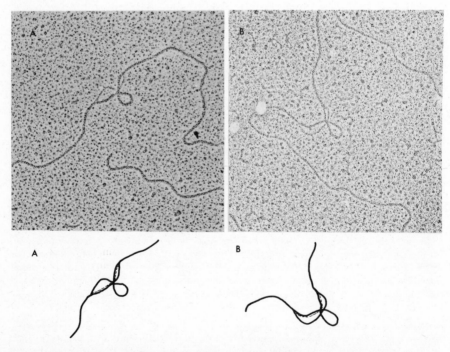

Figure 26.7. R loop mapping of the mouse β-globin genes. The hybrids formed between mRNA and the cloned copies of the β^{min} gene (A) or the β^{maj} gene (B) each have a principal intervening sequence at about the same site. The heavy line in the drawing indicates DNA and the dashed line shows RNA. The R loop is formed by the displacement of one DNA strand in the region corresponding to the mRNA. The intervening sequence is shown by the loop of duplex DNA extruded from the middle of each R loop, where RNA has not displaced one strand of the DNA. Data of Tiemeier et al. (1978).

from the RNA-DNA hybrid. This is about 550 base pairs long and lies some 250 base pairs from the 3′ end of the gene. Sometimes a very small loop is observed near the 5′ end of the gene, as little more than a knot in the hybrid duplex, accompanied by a slight increase in length of the displaced R loop relative to the hybrid length. This represents a small intervening sequence present earlier in the gene.

Figure 26.7 shows that the R loop formed by annealing globin mRNA with the β^{min} gene is almost the same as that formed with the β^{maj} gene. Tiemeier et al. (1978) found that the intervening sequence appears as a loop of about 585 base pairs some 210 bp before the 3′ terminus. Heteroduplexes formed between the two cloned genes show two regions of homology, about 850 bp at the 5′ end and about 230 bp at the 3′ end, separated by a distance of about 310 bp of nonhomologous sequences. The nonhomologous region corresponds to about half of the large intervening sequence; the other half has remained the same in both the β^{maj} and β^{min} genes. The lack of non-homology elsewhere is consistent with the idea that the small intervening region is present also in the β^{min} gene, although the resolution of electron microscopic mapping is not sufficient to determine the extent of homology.

The complete sequence of the mouse β^{maj} globin gene shows that the two interruptions occur at amino acid positions 30 and 104 (Konkel, Tilghman, and Leder, 1978). Neither intervening sequence is likely to code for protein, since termination codons occur frequently in all three reading frames. Failure to remove either intervening sequence would result in termination of translation by ribosomes reading through from the coding region. In the β^{min} gene of the mouse, the intervening sequences lie at the same sites as in the β^{maj} gene (see Table 26.4). The smaller intervening sequence has indeed been conserved, while the larger shows conservation at the ends and in the 5′ region; the 3′ region has drifted extensively to produce the nonhomology seen by electron microscopy. Perhaps this difference in behavior of the two intervening sequences is a function of their length; for example, some minimum length around the junction might be conserved more stringently. Sequencing of the junction between coding regions and intervening sequences shows that both interruptions occur at exactly the same sites in the amino acid sequence of the rabbit β-globin gene (Van den Berg et al., 1978). The sequences at the ends of each interruption appear to have been highly conserved between mouse and rabbit.

Restriction data suggest that at least the larger intervening sequence is present also in the β-, δ-, and both γ-globin genes in man. Figure 28.4 shows that the BamHI and EcoRI sites in each of these genes lie some 800–900 bp apart instead of the 67 expected from the amino acid sequences. Direct sequencing of the appropriate parts of the β and δ genes confirms that the interruption is located between amino acid positions 104 and 105 in each case. Two chicken β globin genes have the same general interrupted structure as seen by electron microscopy of R loop hybrids (Dodgson, Strommer and Engel, 1979).

The mouse α-globin gene also has two intervening sequences, one small and one large, although the larger is only about 150 bp, compared with the 600 bp of the β genes (Leder et al., 1978). Sequencing studies show that these interruptions occur at the amino acid sites homologous to those where the intervening sequences are located in the β genes (these are at α amino acids 30 and 99). The occurrence of the large intervening sequence may therefore prove to antedate the evolution of the mammalian globins and even to precede the separation of mammals and birds; insufficient evidence is yet available to say whether this is true also of the smaller intervening sequence. Although the size and sequence of the large interruption has diverged considerably between species and between different globin genes, its removal must be inadvisable.

Duplicate Insulin Genes

An interesting relationship has been found between the two genes that code for insulin in the rat; unlike most mammals, which synthesize only one type of insulin, two are found in the rat and in the mouse (and also in certain fish). The two rat proinsulins differ in 4 amino acids; both proteins are synthesized together and their genes appear to be nonallelic. The two genes have been characterized by isolating the restriction fragments that hybridize with a probe obtained from insulin mRNA. In several rat strains the insulin II gene is obtained as two EcoRI fragments, of 3.7 and 0.8 kb; this is the result of EcoRI cleavage within an intervening sequence. The insulin I gene is obtained as a single fragment of 9.4 kb in one strain, 7.2 kb in another strain, with both fragments present in a third strain (Cordell et al., 1979). The nature of the polymorphism is not yet known.

Both the genes have been completely sequenced and their structures can be compared with the sequence of mRNA. Lomedico et al. (1979) found that they share the common feature of an intervening sequence of 119 bp located within the leader region of the mRNA (see Table 26.4). A striking difference is seen in the two coding regions; insulin I is continuous, but insulin II has an intervening sequence of 499 bp. Apart from this feature the two gene sequences are closely related; their homology extends from a point about 500 bp preceding the 5' end of the gene, but terminates at the site corresponding to the 3' end of the mRNA. There are three amino acid differences in the sequences of the two pre (signal) regions, bringing the total divergence of amino acids to 7 in preproinsulin. Excluding the nucleotide changes involved in this divergence (which corresponds to 3% change in sequence), there is 22% divergence between the two genes in silent sites (mostly third bases) of the coding region. Except for the conserved terminal 10 nucleotides at each end of the small intervening sequence (see Figure 26.11), about the same divergence is seen in this region as in the silent coding sites. This suggests that they are equally free to drift. A somewhat lower rate of about 10% divergence is found in the 5' leader region and the preceding sequence, and in the 3' trailer region.

The two insulins of the mouse are identical with those of the rat, but only one insulin is found in other rodents such as the spiny mouse and Syrian hamster. This suggests that a gene duplication may have occurred after the divergence of the latter two species, but before separation of rat and mouse. This is consistent with a separation between the two insulin genes that occurred about 20–35 million years ago; the drift between the nonconserved positions of rabbit and mouse β-globin is about 68%, which corresponds with a rabbit/mouse divergence about 85 million years ago. It is not yet known whether the two rat insulin genes are linked together; the same doubt exists about the two mouse β-globin genes.

Both rat insulin genes appear to be expressed equally well, so the interruption in the coding sequence does not appear to influence the formation of mRNA. The presence of the intervening sequence in one but not the other gene does not in itself reveal whether the sequence arose by insertion in the first gene or by deletion from the second. However, if the other mammals that synthesize only a single insulin protein possess only the expected single gene, comparison with its structure may reveal whether the absence or presence of the intervening sequence corresponds to the ancestral form prior to gene duplication.

Ovalbumin and its Relatives

A highly complex pattern of intervening sequences is found in the chick ovalbumin gene. There are no EcoRI cleavage sites in cloned DNA representing the ovalbumin message; but three or four fragments are found when chicken DNA is restricted with EcoRI and ovalbumin sequences are isolated by Southern blotting (Breathnach, Mandel, and Chambon, 1977; Lai et al., 1978; Weinstock et al., 1978). When probes representing the left or right halves of the gene are used to identify fragments, bands of 2.35 and 1.8 or 1.3 kb react with the 5' probe, while a band of 9.2 kb reacts with the 3' probe. This implies that there are at least two intervening sequences.

The EcoRI fragments have been cloned and characterized further by detailed restriction mapping and R loop electron microscopy (Dugaiczyk et al., 1978; Garapin et al., 1978a,b). The 3' terminal 9.2 kb fragment contains a continuous region of 1032 bp represented in the mRNA. The 2.35 kb 5' terminal fragment contains several intervening sequences. Further intervening sequences are present in the 1.8 and 1.3 kb fragments, which are allelic and represent the central region of the gene. Chickens may be homozygous for either genotype or may be heterozygotes that possess both fragments. The difference between the types is caused by the presence of an additional EcoRI site within the 1.8 kb region (Mandel et al., 1978; Weinstock et al., 1978; Lai et al., 1979a). This polymorphism occurs within one of the intervening sequences; the 0.5 kb fragment that should also be released by the additional cleavage is not detected in blotting experiments, because it consists entirely of intervening material that does not react with ovalbumin probe (LePennec et al., 1978). This emphasizes the need to construct a complete restriction map to characterize the intervening sequences.

Altogether seven intervening sequences have been identified in the ovalbumin gene, whose coding sequences consist of seven short regions (varying from 45 bp to 188 bp) that constitute the 5' half of the mRNA and are interrupted by the much longer intervening sequences, with the 3' half of the mRNA constituting a single long sequence in DNA (see Table 26.4). Much of the gene has been directly sequenced and the positions of the intervening sequences relative to the coding sequences can be assigned by comparison with the known sequence of mRNA (see below). The results of this detailed analysis introduce the note of caution that distances determined by restriction mapping or electron microscopy may prove to be only approximate when the nucleotide sequence is known; however, the overall structure assigned to the gene is confirmed. The problem is more acute with short regions, since a minimum length of coding sequence is needed for reaction with the probe; and small intervening sequences may be missed in restriction mapping or electron microscopy. While such difficulties can be avoided by working with cloned copies of the intact cellular gene, this may be hard to isolate for very long units.

A particular limitation of the mapping techniques is indicated by the absence of the extreme 5' end of the mRNA sequence from the cloned EcoRI ovalbumin fragments. This is indicated by the absence from the 2.35 kb fragment of a Taq restriction site that occurs at nucleotide 41 of the mRNA; and has been confirmed by sequencing studies that show that the ovalbumin sequence in the 2.35 kb fragment starts with nucleotide 46 of the mRNA (Dugaiczyk et al., 1978; Breathnach et al., 1978). The missing 45 nucleotides should in theory form a single stranded tail when mRNA is annealed with DNA; but in practice this would be too short to observe, so that electron microscopic mapping is able only to say that any such sequence must be absent or shorter than 50–100 bases. Also, of course, this is too short a sequence to detect by hybridizing blotted genome restriction fragments with a probe; for practical purposes it is about at the limits of detection. This can be overcome by using another restriction enzyme to generate a 5' terminal fragment that is not cleaved in the first intervening sequence, since this will carry a greater amount of ovalbumin sequence. Using a chick DNA library containing long fragments of randomly terminated DNA, a sequence containing the 45 bp leader has been identified; it is separated from the next region of the gene by an intervening sequence of about 1550 bp (Dugaiczyk et al., 1979; Gannon et al., 1979).

In this context, the problem in locating 5' terminal regions that are separated from the remainder of the gene is exacerbated by the frequent absence from cloned message copies of just these regions. (This is caused by the premature cessation of reverse transcription; see footnote above). One ovalbumin chimeric plasmid used as a probe, for example, lacks the 13 5' terminal nucleotides. This is unlikely to cause practical difficulties when the missing terminal sequence is so short, but in some cases it extends for a hundred or so nucleotides, which much increases the chance that the 5' termini in the genome may fail to be detected.

Most eucaryotic genes have been characterized in isolation. Even where related genes are known to exist, in examples such as mouse globin or rat insulin it has not yet been possible to demonstrate whether they are linked. The context within which a gene resides, in the sense of its relationship to adjacent genes, is difficult to determine unless there is some knowledge a priori of a relationship. Thus on the basis of information about mutants and gene products, it has been possible to seek duplicate human globin genes; in the human genome there are probably two α globin genes about 3 kb apart (see Figure 28.3); and there is a cluster of the γ^G, γ^A, δ, and β genes in a span of about 40 kb (see Figure 28.4). The ovalbumin gene also is not alone. When a cloned 46 kb chicken DNA fragment carrying the ovalbumin gene is hybridized with total poly(A)$^+$ mRNA from oviduct, two genes transcribed in the same direction are found on the 5' side (Royal et al., 1979). About 11.5 kb upstream is the Y gene, which has a similar structure to ovalbumin itself, with seven intervening sequences evident by electron microscopy; the organization of the coding sequence includes what is probably a short leader and also there is a long continuous 3' terminal region. The total length of the gene is 2000 bp and it extends over a total genome length of 6.5 kb. A further 5.5 kb upstream is the X gene, only part of which is present on the cloned fragment; this also has intervening sequences. Both X and Y genes are partly related to ovalbumin, since some cross hybridization occurs between their sequences and probes for ovalbumin. (This explains the weakly hybridizing fragments previously detected in genomic restriction digests.) The exact relationship of the three genes has yet to be worked out; but it seems likely that they represent a triplication of some ancestral gene. Their functions may be related, since all three mRNAs are found in oviduct only following stimulation by estrogen. So like the globin gene clusters, these provide an opportunity to seek relationships in the control elements. The problem of determining the distance to the next, unrelated gene remains; without an idea of what probe to use, it is difficult to perform such experiments to define the limits of the region concerned with expression of a certain function(s). However, the existence of further members of the same family could be investigated by using cloned X or Y genes as probes.

Occurrence of Intervening Sequences: Higher and Lower Eucaryotes

Available data on the presence of intervening sequences in higher eucaryotic structural genes coding for protein are summarized in Table 26.4. Mammalian globin and insulin genes have only one or two intervening sequences, while the genes expressed in chick oviduct have a more highly mosaic organization. Intervening sequences are present also in immunoglobulin genes, which are deferred to Figures 26.22 and 26.23. A mosaic structure also can be inferred from genetic data on the cytochrome b gene of the yeast mito-

Table 26.4 Intervening sequences

Gene	Organization of structural regions
Rabbit β globin	143 126 222
Mouse β globin[maj]	142 116 222
Mouse β globin[min]	? ? 222
Mouse α globin[(1)]	126 100 207
Rat insulin I	*40* 119 403
Rat insulin II	*40* 119 203
Chick ovalbumin	*45* 1550 188 246 53
Chick ovomucoid	100 1000 20 700 100 500 50
Chick lysozyme	200 1000 440 1250
Yeast cytochrome c	
Sea urchin histones (H1, H2A, H2B, H3, H4)	
Drosophila hsp70 genes	

The shaded sequences identify regions represented in mRNA and the intervening sequences are striped. Distances are indicated in base pairs; in most cases these are approximate, based on R loop mapping and restriction digests, although nucleotide sequences have been determined for mouse β[maj] globin and ovalbumin. In all cases except chick lysozyme, for which the probe lacks the 5′ terminal 90 bp of the mRNA, the structural sequences identified in the genome correspond to the full length of the mRNA. The continuous lengths of uninterrupted genes are noted.

chondrion (see Figure 26.9). However, a nuclear cytochrome c gene has been sequenced and has no interruptions; and the ability of cloned yeast genome fragments to complement bacterial mutations may mean that this is true of at least these genes also (see Table 26.2). Interruptions are absent also from the histone gene clusters of sea urchins and probably from certain heat shock genes of D. melanogaster.

In each case, although the coding regions are separated in the genome, they are present in the same order as in messenger RNA. The genes thus are split rather than dispersed: expression requires omission of the intervening sequences from mRNA, but does not involve any rearrangement of the order of segments or any assembly of sequences from different molecules of DNA. Almost all of the intervening sequences lie within the protein-coding region of the gene; occasionally they are present in nontranslated 5′ leaders or 3′ trailers (which are indicated by italics in the table).

in eucaryotic structural genes

| 580 | 224 |

| 646 | 256 |

| 585 | 210 |

| 150 | 250 |

| | |

| 499 | 200 |

| 576 | 132 | 398 | 118 | 860 | 142 | 370 | 155 | 1625 | 1032 |

| 200 | 100 | 700 | 100 | 1100 | 100 | 600 | 150 |

| 350 | 400 | 100 |

continuous 330 bp

continuous 400 bp

continuous 2200 bp

References: globin—Konkel, Tilghman, and Leder (1978), Leder et al. (1978), Tiemeier et al. (1978), Van den Berg et al. (1978); insulin—Cordell et al. (1979), Lomedico et al. (1979); ovalbumin—Breathnach et al. (1977, 1978), Garapin et al. (1978a,b), LePennec et al. (1978), Mandel et al. (1978), Gannon et al. (1979), Lai et al. (1978), Dugaiczyk et al. (1978, 1979), Catterall et al. (1978), Robertson et al. (1979), Weinstock et al. (1978); ovomucoid—Catterall et al. (1979), Lai et al. (1977b); lysozyme—Ngyuyen-Huu et al. (1979), Baldacci et al. (1979); cytochrome—Smith et al. (1979); histones and hsp70—see Chapter 28.

Since the initial observation with D. melanogaster, intervening sequences have been observed in the genes coding for the larger ribosomal RNA in a wide variety of eucaryotes. Their frequency of occurrence in tandem clusters of higher eucaryotes is not yet established. Available data are summarized in Table 26.5, although the organization of ribosomal DNA is discussed in more detail in Chapter 27. When an interruption occurs, almost always it takes the form of a single intervening sequence; no interruptions have been found in any gene for the smaller rRNA. The intervening sequence typically is present at a site about two thirds of the distance along the gene; whether this is actually a homologous location is not known. Only in Physarum is there more than this one interruption.

The interrupted rRNA genes of D. melanogaster differ from interrupted unique genes in two ways. First is the variation in the intervening sequence,

Table 26.5 Interruptions of genes for large ribosomal RNA

Organism	State and number of rRNA genes	Proportion in class	Structure of large rRNA gene
D. melanogaster	chromosomal (250 repeats)	50%	2500 | 500, 1000 or 5000 | 1200
		15%	2500 | 3000–4000 | 1200
		35%	3700
D. virilis	chromosomal (250 repeats)	75%	2450 | 9600 | 1000
		25%	?
T. pyriformis strain 6UM	extrachromosomal (30,000 repeats)	>96%	2200 | 390 | 820
strain 8ALP	extrachromosomal (30,000 repeats)	all	3000
D. discoideum	extrachromosomal (180 repeats)	all	4100
P. polycephalum	extrachromosomal (150 repeats)	all	2400 | 680 | 700 | 1200 | 540
S. carlsbergensis	chromosomal (110 repeats)	all	3450
S. cerevisiae	chromosomal (110 repeats)	all	3360
C. reinhardii	chloroplast (2 repeats)	both	2105 | 940 | 270
Z. mays	chloroplast (2 repeats)	both	3100
N. crassa	mitochondrion (1 copy)	only	2700 | 2300 | 500
S. cerevisiae	mitochondrion (1 copy)	only	3200 | 1160 | 510

The structure of the larger rRNA gene is based on R loop mapping between the rRNA and isolated genomic or cloned rDNA. Coding sequences are shaded; and the intervening sequences seen as loops of DNA within the R loop are striped. The length of each segment is indicated in base pairs. In almost every instance these assignments have been confirmed by restriction mapping. Where different types of rRNA gene are found, the proportions are determined by quantitating the appropriate classes of restriction fragments. In the yeast mitochondrion the intervening sequence is present in some strains but absent in others; this corresponds to the ω insertion (see Chapter 21). In the C. reinhardii chloroplast the sequences of the junctions of the intervening sequence have been determined as:

Table 26.5 (*Continued*)

A A A A C G T A A A T A ⋯   ⋯ T C A T G C G T G A

The length of the coding sequence alone is given in cases in which no intervening sequence has been detected; formally these exclude the occurrence of interruptions longer than about 50 bp. Interruptions are absent also from the transcription unit of X. laevis rDNA, which is discussed in detail in Chapter 27.

References: D. melanogaster—Glover and Hogness (1977), White and Hogness (1977), Pellegrini, Manning, and Davidson (1977), Glover (1977), Wellauer and Dawid (1977, 1978); D. virilis—Barnett and Rae (1979); T. pyriformis—Wild and Gall (1979); D. discoideum—Frankel et al. (1977); P. polycephalum—Campbell et al. (1979); S. carlsbergensis—Meyerink et al. (1978); S. cerevisiae—Philippsen et al. (1978); C. reinhardii—Rochaix and Malnoe (1978), Allet and Rochaix (1979); Z. mays—Bedbrook et al. (1977); N. crassa—Hahn et al. (1979), Heckman and RajBhandary (1979); S. cerevisiae—Bos et al. (1978), Faye et al. (1979),. Heyting and Mencke (1979).

both in length and in type. There are two major classes of intervening sequence; the sequences within each are related but vary in length; the two types of sequence are unrelated. The individuality of the two types suggests an independent origin or extensive evolution since establishment; the variation could correspond to the consequences of lack of selection for a specific sequence. Second is the occurrence elsewhere in the genome of sequences related to the intervening sequences (Dawid and Botchan, 1977). In D. virilis the intervening sequence is longer and more homogeneous; it is related to the predominant type of intervening sequence in D. melanogaster. Again it is found elsewhere in the genome. This contrasts with the nonrepetitive nature of the intervening sequences in globin and ovalbumin genes.

About one third of the large rRNA genes in D. melanogaster lack any interruption detectable by restriction mapping or electron microscopy. These generally are thought to represent intact genes, but formally the possibility is not excluded that there might be very short (<30 bp) interruptions. As with the structural genes coding for proteins, ultimately sequencing is necessary to resolve the integrity of coding regions. This is more difficult to perform with repeated than with unique genes. But accepting the absence of intervening sequences from these genes, immediately the question becomes whether the interrupted rRNA genes are expressed. Ribosomal RNA appears homogeneous in sequence; and no product corresponding either to a transcript containing the additional sequences or truncated by premature termination within the intervening sequence has been detected. It is possible that the interrupted genes do not function; but it is possible also that their products are spliced rapidly to yield the mature rRNA. In D. virilis, it is not yet clear whether the minor proportion of rRNA genes of the expected length actually lack intervening sequences or contain short interruptions.

An interesting situation is presented by the ciliate Tetrahymena pyriformis, where some strains may possess an intervening sequence in all of the copies of the gene for the larger rRNA in the amplified extrachromosomal rDNA of the macronucleus. Since these genes are actively expressed, the intervening sequence must be spliced out from the initial transcript. The presence of the interruption depends on the strain; it is entirely absent in some. No difference is discernible between the strains, which raises the question of whether the intervening sequence has any function.

In contrast with the absence of intervening sequences from the nuclear rRNA genes of the lower eucaryotes Saccharomyces and Dictyostelium, interruptions are found in the larger rRNA genes of chloroplasts and mitochondria from several sources, including yeast. Their occurrence is not inevitable, as shown by the absence of interruptions in the rRNA gene of the maize chloroplast. However, the existence of intervening sequences in organelle genomes contrasts with the resemblence that has been noted between organelle and bacterial protein synthetic systems. In this respect, the organelles do not conform to what would be predicted from the theory that their genetic systems resemble procaryotes because of their descent from endosymbiotic bacteria, unless this occurred prior to the loss of splicing ability for procaryotes.

Intervening sequences have been found in the yeast nuclear genome as rather short interruptions in transfer RNA genes. The three of the eight tyrosine tRNA genes that have been characterized possess an insertion of 14 base pairs in the DNA at the site immediately on the 3′ side of the anticodon (see Figure 26.8). At least one of these genes must be expressed, since a single base pair change in the anticodon sequence enables it to give rise to the mutant tRNA in cells able to suppress ochre mutations (Goodman, Olson, and Hall, 1977). Three of the (roughly) ten phenylalanine genes of yeast have been sequenced; and again an intervening sequence is present, located one or two base pairs from the anticodon position and extending for 18 or 19 base pairs (Valenzuela et al., 1978). The intervening sequences in different copies of the gene are closely related (though in one case the flanking sequences are different). An intervening sequence of similar size is present in a gene for one tRNA^{Ser} species; but intervening sequences may be absent from the genes that code for the other tRNA^{Ser} species (Etcheverry, Colby, and Guthrie, 1979). A longer intervening sequence is found in tRNA^{Trp}.

From the sequences given in Figure 26.8 it is evident that in all four types of gene the interruptions lie at almost identical sites, just on the 3′ side of the anticodon, although their sequences are unrelated. The intervening sequences present in the tRNA^{Tyr}, tRNA^{Phe}, and tRNA^{Trp} genes share with those of higher eucaryotic genes the feature that a short sequence repetition prevents exact definition of their ends. But this is not true of tRNA^{Ser}, where the ends of the intervening sequence can be uniquely defined.

Whether the intervening sequence serves any function in determining the use of tRNA, or is an inevitable but transient stage in the production of active tRNA, is not known. But as Figure 26.8 shows, a common feature of all three structures is that the intervening sequence may cause the molecule to take up a new conformation for the anticodon arm, in which

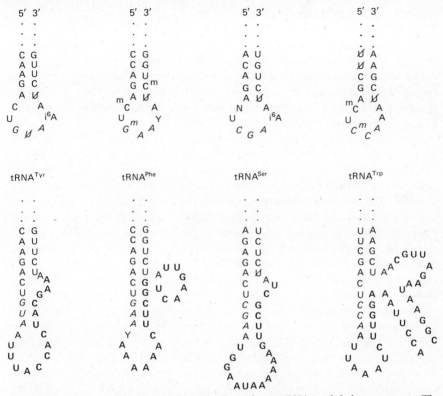

Figure 26.8. Sequences surrounding the anticodons of yeast tRNAs and their precursors. The intervening sequences indicated in bold face occupy 14 bases in tRNA^{Tyr}, 18 or 19 in different genes for tRNA^{Phe}, 19 in tRNA^{Ser}, 34 in tRNA^{Trp} in each case located at or close to the 3′ side of the anticodon (indicated by italics). In tRNA^{Tyr}, tRNA^{Phe}, and tRNA^{Trp} one end of the intervening sequence repeats the base(s) present at the coding side of the other end (*AA* in tRNA^{Tyr}, *A* in tRNA^{Phe} and tRNA^{Trp}), so that the sequence does not define the ends of the interruption unequivocally. The ends assigned to the interruption in tRNA^{Phe} and tRNA^{Trp} therefore are arbitrary. Splicing of tRNA^{Tyr} in vitro excises a unique intervening sequence, defining the ends as drawn. In tRNA^{Ser} there is no base repetition so the ends of the intervening sequence are unambiguous. The sequences of the three interruptions are related neither to each other nor to the sequences summarized in Figure 26.12. Purified precursors containing the intervening sequences lack the base modifications found in tRNA. Data of Goodman, Olson and Hall (1977), O'Farrell et al. (1978), Valenzuela et al. (1978), Etcheverry, Colby and Guthrie (1979), Knapp et al. (1979), Ogden et al. (1979).

the anticodon is base paired instead of available at the end of a single stranded loop. That this conformation actually may be achieved is suggested by the nuclease sensitivities of the isolated precursors to tRNA^{Tyr} and tRNA^{Ser}. The presence of the intervening sequence prevents the precursor from being aminoacylated in vitro. The isolation and processing of the precursors in vitro demonstrates directly that the interrupted genes are expressed (see below).

Genetics of Mosaic Genes

In contrast with the prospective absence of intervening sequences from nuclear genes coding for yeast proteins, some genes of the yeast mitochondrion have a mosaic structure. Other loci in the organelle appear to represent uninterrupted genes, so both types of organization may be found in the same DNA. The interruption in the larger rRNA gene is polymorphic, being present in ω^+ and absent in ω^- strains.

The *box* yeast mitochondrial mutants comprise a set of 22 *mit⁻* mutations that lack the cytochrome b apoprotein and as a result do not assemble the cytochrome bc_1 complex (see Table 21.5). Of the mutants, 2 are multisite and the remaining 20 appear to be point mutations; 16 of them have been reverted. Their genetic properties are unprecedented; indeed, we have referred in Chapter 17 to the implications that the existence of intervening sequences have for genetic mapping and the yeast mitochondrion provides a rare example of a system in which the genetics of mosaic genes can be analyzed.

The *box* mutations map into six clusters. Each cluster consists of two or more mutations that are very closely linked; but the clusters are less tightly linked. A possible seventh cluster is at present identified by only a single mutation. The summary given in Figure 26.9 shows that the genetic map agrees well with a restriction map on which the *box* clusters have been located by deletion mapping. It is at once evident that this region of some 8000 base pairs is about ten times too long to code for the cytochrome b apoprotein of roughly 30,000 daltons. This excludes the possibility that the mutations represent regions of a continuous coding sequence that for some reason are more susceptible to mutation.

The complementation properties of these clusters are unusual. In an innovative analysis, Slonimski et al. (1978a) found that although *box1*, *box2*, *box4*, and *box6* are widely separated, no complementation occurs between mutations in these clusters. Formally this identifies cis related elements, all of which are needed for cytochrome b synthesis, but which extend over a length of some 10% of the yeast mitochondrial genome. Even the explanation that the clusters represent the sequential coding regions of an interrupted gene is inadequate in view of the ability of *box3* mutants to complement the other *box* clusters. This means that *box3* is trans-acting with regard to *box1, 2, 4,* and *6;* thus *box3* must be responsible for the synthesis of one diffusible product, while the surrounding clusters of *box1, 2, 4,* and *6* are responsible for the synthesis of another product. Similar but less detailed results have been reported in which mutations, classed as *cobB*, probably equivalent to *box3*, reside in a separate complementation group within other clusters that form a single complementation group (Haid et al., 1979).

Three of the *box* clusters can clearly be identified as components of the structural gene. Drug resistant loci that are thought to represent changes in cytochrome b are allelic to *box1*, *box4* and *box6*. Claisse et al. (1978) showed that at each of these loci, some *box* mutants lack the 30,000 dalton cytochrome b protein and display instead a protein of lower molecular weight.

The idea that this is due to premature termination of translation is supported by an increase in fragment size as the map positions of the mutations proceed from left to right. Each of these mutations can be reverted to generate mitochondria that possess the 30,000 dalton protein and lack the smaller fragment. In *box4* and in *box6* there are also mutants that retain the 30,000 dalton protein; these are putative missense mutations.

Again similar results have been obtained in a less detailed study, with mutants *cobA*, *cobC*, and *cobE* probably equivalent to *box4*, *box1*, and *box6*, respectively (Haid et al., 1979). The relationship of these proteins to cytochrome b was confirmed by demonstrating that they show similar tryptic fingerprints. One mutation in *cobE* causes the appearance of a protein related to cytochrome b but of greater molecular weight; this may result from readthrough past the termination codon. The proteins synthesized by three mutants whose defects have been mapped in the left part of the *cob* region, in or beyond *box4*, have been studied in detail (Hanson et al., 1979). In all three cases, cytochrome b activity is lost together with cytochrome oxidase activity. One mutant is an object lesson; it proves to have two mutations, one in *cob* and one in *oxi3*. The dual effects of the other two, however, appear to result from single *cob* mutations, one of which causes cytochrome b to be replaced with a related protein of 15,000 daltons, the other of which causes the production of a new protein of 45,000 daltons, apparently unrelated to wild-type cytochrome b or cytochrome oxidase. The locations of these mutations, described here as *cob* to distinguish them from the series defining the *box* clusters, are not yet known with the regard to the latter; but their properties are consistent with the view that this region has an intricate organization connected in some presently unknown way with the *oxi3* locus.

Less complete information is avilable on the properties of *box2* and *box5* (Claisse et al., 1978). Several new minor proteins appear in all *box2* mutants; and these continue to be synthesized in addition to the 30,000 dalton cytochrome b in revertants. On the other hand, some of the major proteins characteristic of individual *box2* mutants may be premature termination fragments of cytochrome b; others may result from readthrough into a noncoding region. Present biochemical evidence, together with the complementation data, therefore suggests that *box2* represents a region coding for part of the cytochrome b protein. It is not yet clear whether *box5* is an independent region or part of the same cluster as *box4*. It has the property unexpected of a coding region that mutants cause loss of cytochrome oxidase subunit I as well as of cytochrome b; but the disappearance of the two proteins is not accompanied by appearance of any other product, which leaves the nature of the effect to be explained.

The ability of *box3* mutants to complement the *box1, 2, 4,* and *6* mutants implies that *box3* cannot be part of the sequence coding for cytochrome b.*

---

*Formally a possibility consistent with the genetic results is that *box3* mutations do lie in the cytochrome b coding region, but that unlike all other such mutations they can be overcome by intermolecular splicing to generate functional mRNA. This is excluded by other data.

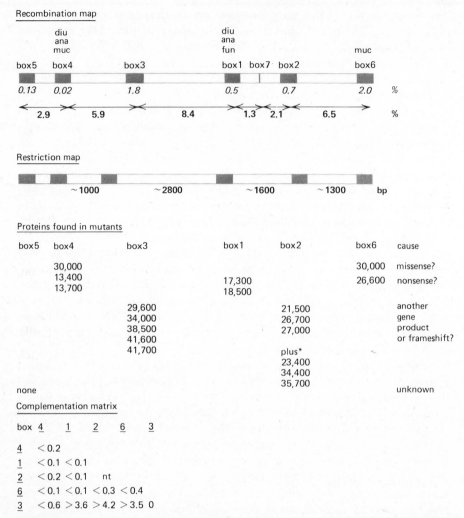

Figure 26.9. The _box_ locus of the yeast mitochondrial genome. Twenty _box_ mutants that appear to result from point mutation fall into seven groups, the members of each group showing tight linkage (<2.0% recombination) as indicated in italics. The distance between groups is greater (>4.0% recombination) as indicated in bold face. Two exceptions are _box4-box5_, which are separated by a distance only slightly greater than the span within the group (and might in fact constitute a single group), and _box7_ which is identified by only a single mutation. Loci at which mutation to drug resistance can occur are indicated by the abbreviations diu, ana, muc, and fun.

The absolute recombination frequencies given in the map are not an accurate representation of distance on DNA. For example, they are subject to the classical phenomenon of map expansion. Different distances between the groups are obtained by mapping the frequency of separation, but the same order and approximate relationship is obtained. Deletion mapping with _rho⁻_ mutants has been correlated with the restriction fragments produced by petites to generate the restriction map, which shows the approximate distance on DNA between groups.

Complementation between mutants has been tested by following the abilities of zygotes to

This cluster appears also to be directly involved in the synthesis of subunit I of cytochrome oxidase. In *box3* mutants the 40,000 dalton subunit I is replaced by proteins of varying molecular weight. One possibility is that *box3* actually is part of an interrupted gene coding for the cytochrome oxidase subunit, the rest of which is identified by the *oxi3* locus. This would mean that the genes for cytochrome b and cytochrome oxidase I are interspersed (see Figure 21.5). Perhaps a more likely possibility is that *box3* codes for some function that is involved in the production specifically of both cytochrome b and cytochrome oxidase I. This could be either a protein or RNA sequence. A protein coding function seemed to be indicated by the isolation of *box3* mutants that are thermo- or cryo-sensitive; but a hint that this is not so is provided by the ability of petites (which are deficient in protein synthesis) to complement *box3* mutations. Mutants in *box7* have yet to be fully characterized, but present indications are that this locus also serves some function other than coding for cytochrome b; its effect on cytochrome oxidase I synthesis is similar to that of *box3* mutants and minor proteins are produced similar to those resulting from *box2* mutations.

To summarize these data, it is clear that the gene coding for cytochrome b is a highly mosaic structure containing several intervening sequences. The coding regions include the clusters *box5/4-1-2-6*, since these belong to a single complementation group, are the sites of drug resistant mutations (*box4-1-6*), and generate protein products that could correspond to non-sense and missense mutants. It is clear that *box3* belongs to a different complementation group; together with *box7* it is involved in the synthesis of subunit I of cytochrome c oxidase, whose structural mutations have been identified at the distant *oxi3* locus. Thus *box3* (and *box7*) represent intervening sequences in which mutation prevents both the synthesis of the product of the surrounding coding regions and the synthesis of another gene product. The ability of *box3* mutations to complement mutations in the coding clusters shows that at least this intervening sequence exercises a function involving synthesis of a diffusible product. A likely function for

respire shortly after mating. The values shown in the complementation matrix are a measure of oxygen uptake per zygote. Mutants crossed with others in the same group do not complement (values < 0.4) and nor do crosses between *box1, 2, 4,* and *6*. Crosses with *box3* complement well for *box1, 2,* and *6* (10-fold increase in respiration) and to a lesser degree for *box4*.

Wild type mitochondria synthesize a 30,000 dalton protein thought to be cytochrome b (see Table 21.5). A protein of this size is present in one *box6* and one *box4* mutant, both of which retain the cyt b spectral absorption bands, although lacking enzyme activity. Neither the spectral band nor protein of this size is present when the proteins synthesized by the other mutants are analyzed by gel electrophoresis. The new proteins that appear on the gels are indicated by their molecular weights; each value represents the protein produced by a different *box* mutant, except for the three bands indicated as plus* which are present as minor proteins in all *box2* and *box7* mutants. Some of the major *box2* proteins may be fragments of cytochrome b or read-through products. Data of Slonimski et al. (1978a), Claisse et al. (1978), Colson and Slonimski (1979).

this is to mediate the splicing of the transcripts for cytochrome b and cyto-chrome oxidase subunit I.

This system provides the first genetic markers in intervening sequences. The relationship between the genetic and physical maps shown in Figure 26.9 demonstrates that the *box* mutations identify only small, specific sites within the locus. As further mutations are characterized, it seems likely that the genetic organization of the *box* locus will prove to comprise an alterna-tion of clusters of mutations that represent coding and intervening regions. At saturation each coding region should be of a size roughly proportional to the length of the polypeptide chain that it specifies (subject to the usual limitations on the proportion of mutated sites that prevent protein func-tion and are thus detectable). Each intervening cluster, however, may repre-sent only a small part of the noncoding region, so that the two types of clus-ter may be separated by apparently "silent" regions. Further data will be needed to see whether these have any function. Since the *box* mutations are isolated by virtue of their inability to produce cytochrome b, only those mutations in the intervening sequences that prevent synthesis of this pro-tein will be detected; these are likely, for example, to lie in the restricted sites involved in the splicing out of the intervening sequence. In view of the ex-ample of *box3*, it is possible that other functions may be carried by the in-tervening sequences; but these will be detectable by mutation only when appropriate selective techniques are applied that rest upon the loss of this other function. The size of the intervening sequences may vary, since a sub-stantial change in the length of this region is seen in some yeast strains, which have an insertion of about 3000 base pairs between *box4* and *box1;* what function this may have is unknown.

The most remarkable feature of the organization of the *box* mutants is the interspersion of sites representing at least two different complementa-tion groups. Summarizing their work which revealed this situation, Slo-nimski et al. (1978b) have shown that simpler explanations cannot be sus-tained. It remains to be seen whether RNA splicing here follows the pattern established in other cellular genomes of joining together sequentially sep-arated sequences, although with at least one additional function for a sequence intervening in the structural gene; or whether the pattern of or-ganization is more complex, possibilities for which are intermolecular RNA splicing (recombination), variations in the splicing of a single tran-script to generate different messengers, or interspersion of structural genes so that the coding sequences of each gene are spliced out of the other.

Does this situation occur elsewhere in the yeast mitochondrial genome? Another locus at which it is likely that there are interruptions is *oxi3*, which varies from 6,400 bp in S. carlsbergensis to 11,400 bp in a strain of S. cere-visiae (see Borst and Grivell, 1978). This is 5–10 times longer than the 1100 bp needed to code for the 40,000 dalton subunit I of cytochrome c oxidase. Other loci may possess intervening sequences that alter the total length less dramatically.

One of the curiosities of mitochondrial genomes has been the large discrepancy in size between the length of the DNA of yeast and fungi compared with the mitochondrial DNA of higher eucaryotes (see Table 21.2). An ironical point is that the yeast mitochondrial genome cannot be said to resemble the nuclear genome, which probably does not possess extensive intervening sequences. But in higher eucaryotes, the small mitochondrial genomes are accompanied by the presence of very long intervening sequences in nuclear DNA. The difference in organelle genetic organization contrasts with an apparent similarity in the proteins coded by the genomes (although they are not identical since, for example, variations in which cell compartments code for components of the ATPase complex have been reported). The presence of intervening sequences in yeast mitochondrial DNA could account for its extra length; and it will be interesting to know to what extent this is accompanied by additional genetic functions. It is not yet known whether interruptions are altogether absent or merely very much shorter in the mitochondrial DNA of higher eucaryotes.

Evolution of Interrupted Genes

The presence of intervening sequences in a wide range of eucaryotic species and in all types of genes clearly suggests that their origin is ancient. A question at present unanswerable is whether the interrupted genes have arisen by insertion of intervening sequences into genes that previously were continuous; or whether the interrupted genes originally evolved in this form and have been maintained in it ever since.

The first view supposes that the uninterrupted genes of procaryotes, which may also be present in the nuclei of at least certain lower eucaryotes, represent the original form of the coding unit. A problem in accounting for the evolution of interrupted genes from ancestral continuous genes is that insertions in functional genes would presumably be lethal in the absence of ability to reconstitute the coding sequence during gene expression. But would such an ability be present in cells lacking interrupted genes?

At this point it is necessary, of course, to note that there is no direct evidence that procaryotes are unable to perform splicing reactions. This inability is suggested, however, by the form of fine structure genetic maps, in which a more or less continuous series of mutations defines a given gene, and by present data on the physical structure and sequences of coding units. The instability of bacterial mRNA precludes any direct comparisons of product with template. In lower eucaryotes such as yeast and Dictyostelium, the size of nuclear RNA is not much different from that of messenger RNA; in the latter case, the average lengths are 1500 and 1200 nucleotides, respectively (see Table 25.2). This does not exclude the existence of intervening sequences, but suggests that highly mosaic genes are unlikely to form any appreciable proportion of the population. And we have seen that at least some yeast genes can be expressed in E. coli (see Table 26.2). Yet yeast, a simple unicellular eucaryote, has the capacity to remove intervening sequences from both nuclear and mitochondrial genes.

The view that interrupted genes originated as such raises the question of whether these are typical of some or of all genetic functions of higher eucaryotes. If the ability to express such genes was acquired subsequent to the development of a genetic apparatus consisting of uninterrupted genes, the present-day genetic apparatus might contain two classes of genes: those of more ancient origin that remain uninterrupted; and those of more recent origin that have been assembled in mosaic form. The paucity of present information allows no answer; but perhaps it is worth noting that (for technical reasons) all the genes whose organization so far has been determined might be characterized as "specialized" functions. It will be interesting to see whether genes coding for common "household" functions, such as general metabolic enzymes, share the same form of organization. (An entertaining question is whether the genes coding for splicing enzymes are themselves interrupted.)

If all higher eucaryotic genes are interrupted, however, this is tantamount to excluding any ancestor for the eucaryotes whose organization was similar to that of the present procaryotes, that is, with continuously organized genes. To take this argument to its logical conclusion, indeed, one might wonder whether the common ancestor of procaryotes and eucaryotes had interrupted genes; but the interruptions have been sloughed off during evolution of the present procaryotes, which no longer have the flexibility to eliminate intervening sequences during gene expression. As we have remarked, at all events it seems necessary to suppose that the ability to splice RNA sequences must have been present in the cell before the first split genes became organized as such. A speculation is that if splicing originated very early in evolution, the same enzyme activity might initially have undertaken rearrangement of both DNA and RNA in the primitive cell. Splicing then might be viewed as developing from the separation of activities for DNA and RNA, which allowed the cell to produce controlled deletions in the RNA without suffering the damaging instability that might result from continued rearrangement of DNA at high frequencies. As the genome became more stable, the DNA-dependent activities might have evolved into more precisely acting recombination enzymes.

This is all to argue that the genetic apparatus of primitive cells may have been organized in a wasteful manner, with procaryotes since having come under selective pressure to achieve greater efficiency in gene expression, and accomplishing this by eliminating the intervening sequences and thus reducing the size of the genome. Presumably this has not been demanded during evolution of eucaryotes, whose cell cycles are much longer. In this context, it is interesting that both certain bacteriophages and some eucaryotic viruses seem to be under pressure to carry more genetic information than can be accommodated by continuous readout of their genomes (whose length may be limited by the capacity of the virion). Yet although some phages resort to an organization of overlapping genes, read in different phases from the same nucleotide sequence, they do not display the variable splicing utilized by the eucaryotic viruses; this may be another indication that the host bacteria lack the apparatus for expression of interrupted genes.

Arguments on the evolutionary advantages of particular forms of genetic organization tend to have a teleological bent, but it is possible that interrupted genes might present a way to facilitate the evolutionary development of more complex proteins. Two formerly separate protein functions might be brought together by placing their coding sequences within the same transcriptional unit. Even rather imprecise splicing might yield adequate amounts of a bifunctional protein with sufficient selective advantage for subsequent evolution further to improve the efficiency of its production. This contrasts with the much closer in-phase reconstruction of two DNA coding sequences that would be required to achieve such linkage if RNA splicing did not exist. Perhaps the ability to recombine parts of proteins rather imprecisely, and to change the proteins coded by a segment of DNA by altering the splicing pattern, may have been important in allowing the eucaryotic genome to evolve more rapidly than would otherwise have been possible. Another potential advantage of splicing is that it might allow the cell to try out new combinations of proteins without losing the original, separate functions, at least until the bifunctional species becomes established. And possibly the increased size of the interrupted gene allows a greater rate of recombination between the coding segments; this argument is not compelling in view of the ability of organisms to adjust recombination rates, but perhaps there might be an advantage in allowing recombination to occur in the noncoding regions so that heteroduplex formation is avoided in the coding segments.

This model views present proteins as originating with the junction of several ancestral proteins, probably sequentially rather than simultaneously, and with further evolution presumably acting to bring the different functions into concert. To the extent that this form of evolution remains reflected in the present function of the protein, the separated coding regions might correspond to different functional domains of the polypeptide. This appears indeed to be the case for the immunoglobulin genes, where the individual coding regions appear to correspond to particular domains in the antibody protein (see below); whether this is a special case remains to be seen, since it is not yet clear if such separation of function is discernible among the products of most interrupted genes. In the case of globin, the central coding region of the gene may specify the heme binding segment of the polypeptide; its functional independence can be tested by preparing the appropriate part of the protein. A more obvious separation of functions might be sought in genes coding for enzymes that catalyze multiple steps in a given metabolic pathway.

Once a protein has evolved by assembly of several domains into a split gene, what advantage is there in retaining this form of organization? This is tantamount to asking whether the intervening sequences play any active role in genetic organization. If they serve no purpose other than to be eliminated when the gene is expressed, we might expect to see the accumulation of mutations and deletions in the intervening sequences, with no selective pressure on the nucleotide sequence, except at the junctions where splicing occurs. In this case, some polymorphism might be expected within an intervening sequence; and ultimately it might even be lost. Its persistence therefore implies some function. There are two apparent possibilities. Its presence (if not its exact sequence) may be necessary for expression of the gene in which it resides; for example, because the splicing reaction has become an essential step in the mechanism by which mRNA sequences are selected from their precursors for conservation. Or an intervening sequence might have another function, unconnected with the function of its flanking coding sequences. The nearest approach to this seems to be presented by the yeast mitochondrial *box* locus, where at least one of the sequences that intervenes in the region coding for cytochrome b may be needed for expression of cytochrome oxidase.

Individual intervening sequences within the globin and ovalbumin genes appear unrelated to each other, to the coding sequences, or to sequences found elsewhere in the genome; they behave as part of the nonrepetitive component together with the surrounding coding regions (Jeffreys and Flavell, 1978b; Roop et al., 1978). Insufficient data are available at present to make a general comparison of their rate of sequence divergence with that of other noncoding regions or with sites at which mutation is silent. Polymorphism of restriction sites in the intervening sequences has been detected in the chick ovalbumin gene, where one EcoRI site and two HaeIII sites are present in one allelic form but absent in another (Lai et al., 1979a). In the case of the EcoRI site, this is not due to variable methylation, because the pattern is retained upon cloning in E. coli (when methylation does not occur in an appropriate host).

Variable methylation has been found, however, at a site in the rabbit β-globin gene. Waalwijk and Flavell (1978) used two restriction enzymes that recognize the same sequence, CCGG, which occurs in the longer intervening sequence, about 200 bp to the left of the EcoRI site. The difference between the two enzymes is that HapII does not cleave DNA in which 5-methyl cytosine is present at its target site, but MspI is indifferent to the methylation. The degree of methylation thus is determined by a reduction in cleavage by HapII, relative to a control in which complete cleavage by MspI confirms the presence of the site and excludes polymorphism. The results are tissue specific. In liver, spleen, bone marrow, and blood there is 50% digestion with HapII; but sperm shows zero digestion (complete methylation) while brain shows only 20% digestion (80% methylation). It is not known whether the variation is allele specific or if what is observed is a general average, possibly with different degrees of methylation existing in different cells in a tissue.

In contrast with the nonrepetitive genes coding for proteins, in genes coding for ribosomal or transfer RNA, versions both with and without intervening sequences are known. Here it is tempting to speculate that the intervening sequences may have arisen as insertions after the evolution of the (intact) gene, rather than to suppose that the intact gene has arisen by precise deletion of the interruptions. Whether the interruptions present in some genes of a cluster also containing uninterrupted copies should be regarded in the same light as intervening sequences in unique genes is not clear; after all, it is not known whether the interrupted genes are transcribed. The possibility that the intervening sequences may be present at homologous positions in the larger rRNA of many species would argue that there is at least a propensity for the two parts of the gene to be separated. With tRNA the genes are not clustered, but as with rRNA there is the paradox that intact tRNA genes are known, but that different intervening sequences are found at the same position in the different genes. Certainly it would be hard to think in terms of the intervening sequence bringing together two parts of the tRNA gene that previously were unconnected. When more yeast tRNA genes have been characterized, and it is evident also what proportion is intact or interrupted, it may be possible to estimate the stage of evolution at which the insertion occurred.

Sequences of Splicing Junctions

The sequences at the junctions between coding regions and intervening segments now are known for several nuclear and viral genes; in some cases the sequence of the entire intervening region has been determined. A striking feature is that in no case can a unique site be assigned for the splicing reaction from the sequence data. A sequence of at least 1, and of up to 4, nucleotides is repeated at the left and right junctions of coding and intervening DNA. Figure 26.10 illustrates the example of mouse and rabbit β-globin, in which splicing to remove the longer intervening sequence could

	gly	*arg*	*leu*
Messenger sequence	G G C	A G G	C T G

Genome sequence	G G C A G G T G ...... T T T A G G C T G
Possible splicing frames	G G C .... A G G C T G
	G G C A .... G G C T G
	G G C A G .... G C T G
	G G C A G G .... C T G

Figure 26.10. Possible splicing frames for the second intervening sequence of mouse and rabbit β-globin. The trinucleotide AGG is repeated at each end of the intervening sequence, so that splicing between any pair of corresponding phosphodiester bonds in each repeat can generate the same messenger sequence.

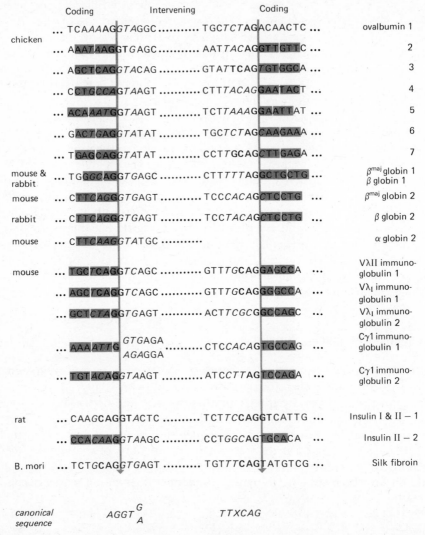

Figure 26.11. Nucleotides sequences at junctions between coding regions and intervening sequences. The sequence shown is for the DNA strand homologous with mRNA and intervening sequences are numbered according to location in the gene from left to right. The coding sequence is read in triplets as indicated. In no case can the exact point of splicing be identified, because from 1-4 of the last bases in the left coding region are repeated at the right end of the intervening sequence. These are shown in bold face. The junction sequences are aligned to maximize homologies with splicing occurring at the points indicated by the vertical lines. The actual splice point can be shifted at both junctions to any other pair of corresponding positions within the bold region. (The number of possible splice points is $n + 1$, where n is the number of bases repeated at each junction.) The repeated sequence, and thus the possible splice points, may lie to either left or right or on both sides of the hypothetical cutting site. Common elements of sequence are apparent at both junctions and define the possible recognition region shown in italic. The canonical sequence indicated for left and right junctions gives the base

Left site							Right site					

$$T_9 \quad A_{13} \quad A_{21} \quad G_{27}^{\downarrow} G_{28} \quad T_{28} \quad A_{18} \qquad T_{15} \quad T_{19} \quad A_9 \quad C_{18} \quad A_{26} \quad G_{27}^{\downarrow} G_{12}$$

$$A_6 \quad C_{10} \quad G_3 \quad A \qquad\qquad\qquad G_7 \qquad\qquad C_9 \quad C_5 \quad G_6 \quad T_7 \quad C \qquad C_5$$

$$C_6 \quad G_4 \quad T_2 \qquad\qquad\qquad\qquad C_2 \qquad\qquad G \quad G_2 \quad C_6 \quad G \qquad\qquad T_6$$

$$G_7 \quad T \quad C_2 \qquad\qquad\qquad\qquad T \qquad\qquad A \quad A \quad T_6 \quad A \qquad\qquad A_4$$

Figure 26.12. Base frequencies at splicing junctions. The left trinucleotide GGT and the right dinucleotide AG are strongly conserved; other positions show more variation, although at some the conservation remains high in terms of purine versus pyrimidine. The derivation of canonical sequences on this basis treats all the junctions as belonging to a common class; it is possible that longer canonical sequences may characterize particular groups of junctions, but further data will be needed to draw such distinctions. Arrows indicate the hypothetical cleavage sites used to align the sequences. Data from Figures 26.11 and 26.20.

occur between any one of four pairs of sites to give the same coding sequence. The only difference between the four reactions would lie in the terminal bases of the intervening sequence that is released. Whether the splicing reaction actually permits the use of alternative sites, or relies upon a unique pathway, is not known. As the reaction is characterized in terms of the breakage and rejoining of RNA, it may become clear whether this repetition plays any role in reducing errors in the formation of the coding sequence.

Available data on globin, ovalbumin, insulin, and immunoglobin systems are summarized in Figure 26.11. Comparable data for viral splicing junctions are summarized below in Figure 26.20. These show that all of the sequences may be aligned around a hypothetical splice point in such a way as to reveal homologies in the presumptive recognition sequence. This suggests the canonical sequences for the left and right junctions:

$$A \quad G^{\downarrow} G \quad T \quad A \qquad \text{and} \qquad T \quad X \quad C \quad A \quad G^{\downarrow}$$

All of the junctions are closely related to these sequences. Some positions are more subject to conservation than others; and at some one pyrimidine or purine may be able to substitute for the other. The frequencies with which different bases are present in each position for the viral and cellular systems are summarized in Figure 26.12.

The intervening sequences present in three yeast tRNA genes are unre-

---

that is predominant at each position. The alternatives given for the left splicing junction of the first γ1 intervening sequence were determined in different laboratories for newborn mouse and myeloma, respectively. They differ by the presence of GTG on the right of the junction. It is not known yet whether this is genuine or one sequence is in error. This sequence is therefore omitted from subsequent comparisons. Data of Breathnach et al. (1978), Catterall et al. (1978), Konkel, Tilghman and Leder (1978), Leder et al. (1978), Van den Berg et al. (1978), Tonegawa et al. (1978a), Bernard, Hozumi and Tonegawa (1978), Sakano et al. (1979a), Honjo et al. (1979), Lomedico et al. (1979), Tsujimoto and Suzuki (1979).

lated to these canonical sequences (see Figure 26.8). Nor is there any relationship with the sequences at the junctions of the intervening sequence of the 23S ribosomal RNA genes of the C. reinhardii chloroplast (see Table 26.5). In yeast tRNA^{Ser} and in the chloroplast rRNA, there is no terminal repetition and the intervening sequences are uniquely defined.

At present the intervening sequences characterized in the higher eucaryotes represent a biased selection of only a small number of systems. More data will be required to support any firm conclusions on the features recognized by the splicing enzymes. But some points are apparent already; and we may consider their implications for the models of splicing depicted in Figure 26.13. It is clear that splicing does not demand complementary base pairing between the ends of the intervening sequence that is to be removed. The lack of complementary sequences thus excludes simple models in which the intervening sequence forms a loop in the RNA, held together by a base paired stem, across which breakage and reunion occurs. Nor is there any very extensive sequence homology that identifies all the left or all the right junctions. The canonical sequences seem to be insufficient to serve this purpose since they are both too short (and therefore occur elsewhere by chance) and show too little variation (and thus do not distinguish one left or right junction from another).

Models to account for the selection of the correct splices from among other possible combinations of left and right junctions (or with entirely different sites, for that matter) can be considered in two general classes. One is to suppose that each proper pair of left and right junctions is specifically recognized by virtue of features additional to the canonical sequences. It seems hard to believe that there is a hierarchy of splicing enzymes, each able directly to recognize an appropriate pair of junctions. A more likely alternative is that there might be an RNA complementary to the sequences that are to be spliced. This could bring the junctions into juxtaposition by base pairing with either the termini of the coding regions that are to be joined or with the ends of the intervening sequence (or both). If the RNA were a small diffusible molecule, it would represent a trans-active regulator of splicing; it is possible to take this model further and suggest that the synthesis of this molecule would determine whether any particular cell was able to splice out the mature sequence from a given primary transcript. Another possibility is that the RNA sequence might be derived from another part of the primary transcript, possibly at one or the other end, possibly elsewhere. In this case the splicing reaction would depend on intramolecular base pairing and would be cis active. Combinations of the models also are possible: for example, a necessary complementary sequence might be present in an intervening region and could be able to react rapidly in an intramolecular reaction, or more slowly as an independent molecule able to sponsor intermolecular reaction after its splicing out.*

---

*My formulation of these models has been much influenced by the ideas of P. Berg on SV40 splicing and of P. Slonimski on expression of the yeast mitochondrial *box* locus.

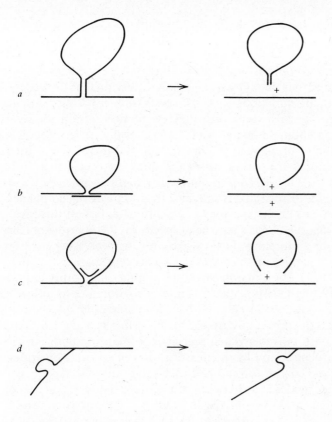

e ⌣⌣⌣⌣⌣⌣ A_n → ⌣⌣⌣⌣⌣ A_n → ⌣⌣⌣ A_n

Figure 26.13. Models for splicing out intervening sequences. Three models involve the formation of base paired RNA duplex regions, which might be recognized by an enzyme able to perform cleavage and ligation. In a the ends of the intervening sequence are complementary and pair directly; this is excluded by the lack of any inverted sequence repeats at the boundaries of the intervening sequences. In b and c another RNA sequence pairs with either the ends of the coding region or with the ends of the intervening sequence, bringing the sequences that are to be spliced together into juxtaposition for cleavage and ligation. The pairing RNA might be the small molecule depicted or might be part of another region of the primary transcript. Two models involve sequential splicing along the transcript, which precludes reaction between incorrect left and right splicing junctions. This might allow a protein to undertake the reaction directly, without mediation of RNA. In d splicing follows rapidly upon transcription, so that only one pair of junctions is exposed at any time. This is excluded by the isolation of full-length polyadenylated molecules containing up to all of the intervening sequences. In e splicing is obliged to wait until the molecule has been completed and polyadenylated; it might then proceed in the $3'$–$5'$ direction, but present evidence suggests that this is not the case. However, it is possible that the initiation of splicing does require a complete primary transcript, even if the subsequent events do not follow an obligatory pathway (see text).

The sequences that have been determined in globin and ovalbumin show no such complements able to bring together the splicing junctions. However, this is not a compelling argument against intramolecular duplex formation, because a complication in attempting to make deductions about the splicing reaction from the known sequences is that there is no basis for supposing the splicing junctions to be the only sites involved. Splicing might involve successive reactions utilizing several sites within the intervening sequence. Without knowing what criteria apply, it is difficult to search for appropriate complements. The concept that there may be intermolecular RNA duplex formation raises the question of whether intermolecular splicing (in effect RNA recombination) might occur, even if only occasionally. Evidence against this is provided by the failure in complementation of different *box* clusters specifying cytochrome b; but an interesting possibility is that the role of *box3* may be to specify an RNA sequence complementary to splicing junctions that is needed for the processing reaction.

Another type of model is to reduce the dependence on sequence specificity in selection of left and right junctions by restricting the availability of junctions at any given time. One possibility would be for splicing to follow so rapidly upon transcription that enzymes have available only the ends of a single intervening sequence at any time. However, this is excluded by the isolation of full length polyadenylated precursors containing both the globin intervening sequences, and by the identification of a variety of ovalbumin precursors, some containing most or all of the intervening sequences (see below). Another model is to impose a demand for sequential action on the enzyme(s), which might be obliged to proceed strictly from one end to the other; however, there is no precedent for such action among processing as opposed to synthetic enzymes. What evidence is available suggests that intervening sequences may be removed in a preferred, but not obligatory order from the precursor. Splicing appears to follow the completion of transcription and polyadenylation and this might be explained in several ways. One is that the reactions involved might simply occur rather slowly. Another is that splicing might occur at a different location from transcription, for example, at some site on the nuclear membrane that represents a stage in nucleocytoplasmic transport. Or there may be a more formal link between the 3′ end and the initiation of splicing, such as an involvement in permitting some particular splice to occur.

Finally it is perhaps worth noting that there is no evidence on the accuracy of splicing; it is not known how often erroneous splicing occurs or what would be the fate of the resulting molecules. Might they account for part of the RNA that turns over within the nucleus; or would they be sent to the cytoplasm to be translated into aberrant proteins?

Sequences that may be involved in controlling the synthesis of the primary transcript or in the processing of flanking regions have been sought in the form of homologies at the 5′ and 3′ ends of the genes. A heptanucleotide sequence TATAAAA has been noticed about 30 bp before the start of one sea

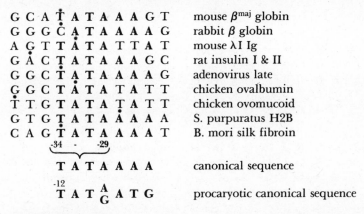

G C A T̊ A T A A A G T   mouse β^{maj} globin
G G G C̊ A T A A A A G   rabbit β globin
A G T T Å T A T T A T   mouse λI Ig
G Å C T A T A A A G C   rat insulin I & II
G G C T̊ A T A A A A G   adenovirus late
G̊ G C T Å T A T A T T   chicken ovalbumin
T̊ T G T A T A T̊ A T T   chicken ovomucoid
G T G T̊ A T A Å A A A   S. purpuratus H2B
C A G T̊ A T A A A A T   B. mori silk fibroin
     -34 - -29

T A T A A A A   canonical sequence

-12
T A T ^A_G A T G   procaryotic canonical sequence

*indicates base -30

Figure 26.14. Putative promoter sequences preceding the messenger startpoint. A conserved sequence rich in A-T base pairs is found about 30 bp preceding the messenger startpoint. This sequence was originally identified by D. Hogness in the Drosophila histone gene cluster. Bases conforming with the canonical sequence are shown in bold type. The data above are taken from sequences determined by Levy, Sures and Kedes (1979), Konkel, Tilghman, and Leder (1978), Bernard, Hozumi, and Tonegawa (1978), Tonegawa et al. (1978a), Cordell et al. (1979), Lomedico et al. (1979), Ziff and Evans (1978), Gannon et al. (1979), Lai et al. (1979b), Tsujimoto and Suzuki (1979). An analogous sequence has been identified as a component of eucaryotic promoters by Pribnow (1975).

urchin histone mRNA; it is present about 80 bp before the initiation codons of other histone genes, where the start of the message is not known (see Chapter 28). Identical or closely related sequences have been observed in similar locations in insect, avian, mammalian, and viral transcription units, as summarized in Figure 26.14. These short A-T rich sequences tend to be surrounded by G-C rich sequences; this may be a feature that distinguishes them from chance sequences of similar constitution that occur elsewhere. The putative promoter is related to a canonical sequence found in RNA polymerase recognition sites in E. coli, although this is located starting 12 or 13 base pairs from the point where transcription is initiated (see Volume 3 of Gene Expression).

Another short conserved sequence is found in mammalian transcription units at the site corresponding to the start of the mRNA (as identified by the capped 5' terminus). This has the canonical sequence GTTGCTCCTXAC as can be seen from the summary in Figure 26.15. It is noteworthy that this sequence is not clearly present in the chick ovalbumin gene and is absent from the ovomucoid gene. No trace of it is evident in silk fibroin or the sea urchin histone genes (which have another common sequence before the putative 5' ends of the mRNAs; see Chapter 28). The extent of the ubiquity of this element therefore remains to be established.

The interspecies ubiquity of the TATAAA box suggests that it might play a fundamental role in transcription, whereas the capping sequence

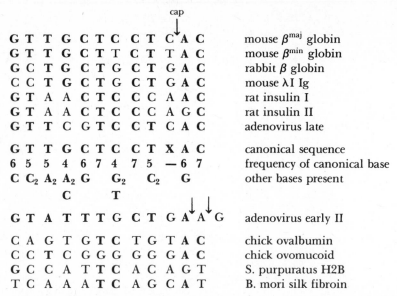

Figure 26.15. Sequence adjacent to the site of capping. A canonical sequence can be deduced for the mammalian genes and adenovirus. The messenger startpoints are known exactly except for insulin, which is established within one base (and which is assigned above to give maximum homology with the putative recognition element; but capping could occur one base downstream). The adenovirus early region has a sequence preceding the capping site that can be aligned with the canonical sequence as shown; but capping actually occurs at two locations, as indicated by the arrows, one and two residues downstream. The two chicken gene sequences, the sea urchin histone gene, and silk fibroin do not conform with the canonical sequence and are shown for comparison. Bases conforming with the canonical sequence are shown in bold type. References as in Figure 26.14.

may be specific to mammals (it has yet to be seen whether a counterpart is present elsewhere). One interesting contrast with this situation is provided by early region II of adenovirus. Baker et al. (1979) have shown that there are two sites of capping, possibly resulting from the initiation of transcription at either of adjacent bases. The sequence preceding the capping site in DNA can be aligned with the canonical sequence if an adjustment is made in the location of the capped bases so that they lie downstream from the usual point of alignment. But no TATAAA sequence can be found upstream. The same situation occurs in the late transcripts of polyoma and SV40, where the startpoints of transcription appear to be heterogeneous and TATAAA sequences are absent. In these cases the capping sequences also are absent (see below). This shows that transcription can be initiated in the absence of this sequence, although it would be interesting to know the relative efficiencies of comparable transcription units possessing and missing the sequence. It prompts the speculation that the TATAAA sequence might be concerned with aligning RNA polymerase; so that if it is absent, the initiation event takes place with less precision.

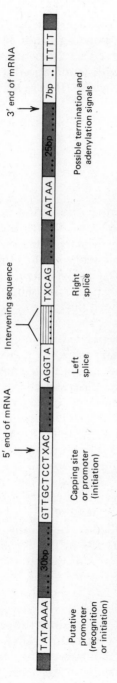

Figure 26.16. Mammalian transcription unit showing canonical sequences possibly involved in control of transcription or processing. Two models have been envisaged for the relationship between the putative components of the promoter. The TATA box could comprise the initial recognition site for RNA polymerase binding, although the enzyme then actually initiates RNA synthesis at the capping site. This means that the 5' end of the primary transcript coincides with that of RNA. Or the TATA box could be the site for binding and initiation, in which case the cappng sequence would be recognized in RNA as a signal for cleavage and capping. In this case the initial 5' region of the primary transcript must be degraded extremely rapidly after its synthesis, since no transcription of any 5' flanking region has yet been detected. Characterization of the initiation reaction in vitro may be required to distinguish these models, although the first is generally favored at present.

823

Some 25 bases before the 3' end of the mRNA, all the messengers so far sequenced have the pentanucleotide AATAA; this might be a signal for the termination of transcription or for cleavage and polyadenylation downstream. Just after the position corresponding to the last base of mRNA, there is a tetranucleotide sequence of TTTT (sometimes it is longer); this again could be a signal concerned with termination. This is a feature of transcription units read by RNA polymerase III as well as RNA polymerase II.

An idealized transcription unit possessing the various canonical sequences is illustrated in Figure 26.16.

Splicing of Nuclear Transcripts

The long debate about the size of primary transcripts has been brought to an end by the discovery of interrupted genes. If the intervening sequences are removed after transcription, the primary transcripts must be correspondingly longer than the mature messengers. The minimum size for the transcription unit is given in Table 26.6 as the distance between the points in the genome corresponding to the 5' and 3' ends of the messenger. For several genes expressed in the chick oviduct, the range is from 4 to 6 times the length of the mRNA; for β-globins it is about twice the length of the cytoplasmic mRNA; and only for α globin and rat insulin I is it not much longer than the message. Obviously the primary transcript may be longer than this if the flanking sequences in the genome are represented in the RNA; of course, this is the question that we have discussed in Chapter 25 of whether the 5' and 3' ends of the mRNA coincide with the termini of the primary transcript or are derived by cleavage of terminal regions. The overall size discrepancy between the lengths of hnRNA and mRNA summarized in Table 25.2 lie in the range of up to 4–5 fold for mammals, decreasing as the genome size becomes less for Aedes and Drosophila.

Available data are insufficient to decide to what extent the size discrepancy between nuclear RNA and messenger RNA can be accounted for by the elimination of intervening sequences. Taking this point further, a question that has been raised in the past is whether the C value paradox might be accommodated by supposing that the size of the genome is proportional more to the length of the transcription unit than to the number of genes. Now in terms of the structure of interrupted genes, the question becomes whether any such increase might be reflected in the length of transcribed flanking regions or intervening sequences. The actual length of transcriptional units cannot be resolved without isolating the primary transcripts; but the possibility that the amount of intervening DNA increases with the C value, because the numbers or lengths of intervening sequences are greater, can, of course, be tested by characterizing nuclear genes.

Irrespective of whether any additional sequences are flanking or inter-

Table 26.6   Overall lengths of mRNA and structural genes

Gene	mRNA size	Minimum transcript length	Ratio
Rabbit β globin	589	1295	2.2
Mouse β^{maj} globin	620	1382	2.2
Mouse β^{min} globin	575	1275	2.2
Mouse α globin	585	850	1.5
Rat insulin I	443	562	1.3
Rat insulin II	443	1061	2.4
Chick ovalbumin	1859	7500	4.0
Chick ovomucoid	883	5600	6.3
Chick lysozyme	620	3700	6.0

In most cases the length of the mRNA has been determined from mapping and sequencing studies of cloned reverse transcripts. The minimum size for the primary transcript is given by the distance in the genome that separates the first and last bases corresponding to the mRNA. Based on data of Table 26.4.

vening, however, any model that relates the size of the genome to the size of the transcription unit would predict that the complexity of nuclear RNA should be a relatively constant proportion of genome complexity. As can be seen from the summary of Table 25.4, insufficient data are available to support any conclusion; but in no case would even a majority of the genome sequences appear to be transcribed. Yet it is necessary to remember that the complexity measurements are made with steady-state populations of nuclear RNA, in which the concentration of any transient precursors may be much reduced (see below). To the extent that the nuclear RNA population represents partially processed molecules instead of primary transcripts, its complexity will therefore be underestimated.

The use of cDNA probes representing the 3' terminal ends of the cytoplasmic mRNA population has shown that these sequences can be found in both larger and smaller molecules of nuclear RNA; but this does not answer the question of the relationship between the coding and the additional sequences, in particular the issue of whether the reacting nuclear molecules are primary transcripts or partially processed precursors. Results suggesting diametrically opposed conclusions on precursor size have been obtained with probes representing individual messengers. Globin cDNA apparently reacts with nuclear molecules of a variety of sizes widely ranging from more than 30,000 bases to less than 3000 (Imaizumi, Diggelman, and Scherrer, 1973; Spohr, Imaizumi, and Scherrer, 1974). Contrast this with the apparent failure of ovalbumin cDNA to react with any nuclear molecules longer than the mature mRNA (McKnight and Schimke, 1974). The complicating factor in all these experiments has been that the reaction was followed by using a labeled probe, which hybridizes with available complements in the steady-state population; precursors may be present in very low concentrations, and of course the reaction is susceptible to being overwhelmed by hybridization with the greater amounts of (processed) mRNA that may be present. Two approaches have been taken to overcome this problem. One is to use a pulse label to identify the precursor(s), which are isolated by reaction with an excess of the cDNA probe. Another is to use probes of cloned sequences, which offer the technical advantage of possessing much higher levels of radioactive label; also these make it possible to use sequences representing not mRNA but individual intervening sequences; these should be able to react only with precursors and so do not have any background of reaction with mature mRNA.

Several techniques have been used to identify the pulse labeled molecules able to react with globin cDNA. Ross (1976) found that a sucrose gradient of total RNA from mouse erythroid cells contains a pulse labeled peak at about 15S which forms RNAase resistant material on hybridization with unlabeled cDNA. The mature mRNA could be identified in the same gradient at about 10S by its ability to render a labeled cDNA resistant to S1 nuclease. In a pulse chase experiment, the labeled RNA moved from the 15S position to the 10S fraction. Another approach in which pulse labeled and steady-state molecules were analyzed simultaneously on the same gradient was developed by Curtis and Weissmann (1976). Friend cells were induced with DMSO to synthesize globin and then were labeled for 16 hours with ^{3}H-uridine followed by 20 minutes exposure to ^{32}P-phosphate. When fractions of a sucrose gradient were analyzed for globin sequences by reaction with cDNA, the ^{3}H-label was found at the 10S position, while the ^{32}P label was found in two peaks, a major one at 10S and a smaller one at 15S. Again using denaturing gradients to fractionate pulse-labeled material, several experiments in which excess cDNA has been bound to cellulose have led to the identification of 15S and 10S peaks which have a precursor-product relationship, in Friend cells, mouse cells, and duck cells (Bastos and Aviv, 1977b; Kwan, Wood, and Lingrel, 1977; Strair, Skoultchi, and Shafritz, 1977). In all cases the half life of the precursor is rather short, of the order of a few minutes.

The cDNA probes used in these experiments were obtained by reverse transcription of globin mRNA preparations containing both α and β sequences. The 15S peak, however, contains only β sequences. Fingerprinting the ribonuclease digest of this material shows that only the pattern of β mRNA is included; the ribonuclease fragments typical of α globin mRNA are found in a fraction on the heavy side of the 10S peak (Curtis et al., 1977). A better resolution of the precursors has been accomplished by electrophoresis in the presence of formamide; Ross and Knecht (1978) found that a pulse label enters two precursor peaks, one of about 1800 bases (corresponding to the 15S sucrose gradient fraction) and one of about 850 bases (only a little larger than the mature messenger of 600). Hybridization experiments showed that the 1800 base peak contains only β sequences, while the 850 base peak contains both β and α sequences. The concentration of these two peaks at steady state is low, the larger representing less than 0.03% of the total globin mRNA and the smaller amounting to no more than 0.6%. In lieu of any larger precursor, the 15S sequence is taken to be the primary transcript for β globin mRNA*; while the 850 base pair sequence may be the primary transcript for α globin. The nature of the β sequences present in the 850 base size fraction remains to be established; one possibility is that these represent partially processed molecules (which must have lost part or all of the longer intervening sequence).

The putative primary transcripts are around the sizes expected from the structures of the interrupted genes; they are polyadenylated, with tails of up to some 150 bases in length. The β precursor appears to be perhaps longer than the minimum transcript lengths of 1382 and 1275 bases for the major and minor genes, respectively. The α precursor is the same size as the minimum transcript, 850 bases. Hybridization experiments suggest that the 15S fraction contains precursors for both the major and minor genes; the α precursor has not yet been characterized.

When hybridized with cloned β^{maj} or β^{min} genes, the 15S RNA forms a continuous R loop, with no extrusion of intervening sequences (Tilghman et al., 1978b). Although this should mean that both genes are represented in the precursor population, presumably there should be some cross hybridization to generate small extrusions representing the nonhomologous regions of the intervening sequence. Characterization of the precursor may provide an easier assay for determining the extent to which the two types of β gene are transcribed than would be possible from working with isolated mRNAs (which will be more similar).

Another line of experiment is to hybridize the precursor with full length globin cDNA. The hybrid formed between 15S RNA and a cloned copy of the β messenger forms an R loop of about 420 bp, corresponding to the longer intervening sequence. The individual regions of the precursor can be identified by treating hybrids of 15S RNA and full length cDNA with ribonucleases (Kinniburgh, Mertz, and Ross, 1978; Smith and Lingrel, 1978). Upon treatment with RNAase A, any nonhybridized RNA regions will be destroyed. The fragments recovered should represent the different segments of the coding region. With a control of globin mRNA, a single fragment is recovered. With the precursor, three fragments are recovered, corresponding in size to the lengths of the coding segments in the genome. Upon treatment with RNAase H, RNA hybridized to the cDNA is destroyed, leaving intact the intervening sequences. These take the form of two fragments,

---

*In some of the earlier experiments, a 27S peak was obtained in addition to the 15S peak. Haynes et al. (1978) showed that while this may be found upon hybridization with β cDNA, it is not present when cloned β globin sequences are used as probe. It appears therefore to represent reaction with a contaminant present in the cDNA preparation.

again reasonably close to the expected sizes. In the two sets of experiments, the total length of the fragments was 1750 and 1500 bases. Using this approach, the characterization has begun of intermediate precursors which may be partially processed. One β molecule may contain only part of the large intervening sequence, which would imply the use of two splicing steps in its removal (Kinniburgh and Ross, 1979).

Are globin genes transcribed only in erythroid cells? It is clear from experiments with cDNA probes directed against adult globin mRNAs that in the usual course of erythroid development the messengers are found only in cells expressing their proteins. Several attempts have been made to determine whether nuclear sequences representing globin genes are found in cells from which the messenger RNA is absent. These experiments have encountered the difficulties involved in assessing precursors in steady state populations, as is emphasized by the calculation of Ross and Knecht (1978) that in mouse fetal liver, there may be only some 20–50 large precursor β globin RNA molecules per cell, with about 2000 in the smaller α and β peak, compared with about 30,000 messengers. However, the reaction of labeled cDNA with steady-state RNA from erythroid cells shows that the amount of globin RNA increases with erythroid development. In chick cells, for example, a cDNA probe prepared against adult α and β globin mRNA identifies 0.057% of the total nuclear RNA from primitive erythroblasts as globin, compared with 0.163% in adult erythroid cells (Groudine et al., 1974). In muscle cells, fibroblasts and blastoderm, no reaction is detectable, setting an upper limit for nuclear globin sequences of $3.4 \times 10^{-6}\%$ of the steady-state population. Thus there is an increase in steady-state nuclear globin sequences of erythroid cells of at least 20,000 times. Of course, such measurements do not show to what extent de novo transcription or reduced degradation may be responsible for the increase in amount. Somewhat different results with nonerythroid cells have been obtained by Humphries, Windass, and Williamson (1976), who reported that cultured cells (lymphoma, L cells, 3T3 cells) may contain nuclear globin sequences. In conditions in which fetal mouse liver had 4600 reacting sequences per nucleus, these cells possessed from 2 to 70 molecules per nucleus able to react with $\alpha + \beta$ adult globin cDNA, with somewhat smaller amounts of reacting material in the cytoplasm. It is difficult to make any firm conclusion in lieu of data on the physical state of the reacting molecules; if they were indeed globin precursors, this might imply some degree of leakiness in the gene control of the cultured cells, allowing low-level expression of specialized genes. It would be useful to perform such experiments with cloned globin sequences and when the precursors are identified by pulse labels rather than in the steady state.

A series of putative precursors for ovalbumin mRNA has been identified by the reaction of oviduct nuclear RNA with probes derived from cloned regions of the ovalbumin gene. A clear demonstration that ovalbumin gene expression is controlled at or prior to the level of nuclear RNA is provided by hybridization of nuclear RNA with the cloned intact copy of mRNA. Using the titration protocol described in equation 11 of Chapter 24, Roop et al. (1978) found that there are roughly 3000 copies of the ovalbumin message sequence per nucleus after estrogen stimulation, but less than 2 in the absence of hormone. The EcoRI 2.4 and 1.8 kb fragments of the ovalbumin gene contain largely intervening sequences from the 5′ end and middle, respectively, and react with nuclear RNA from stimulated but not withdrawn oviduct. Excess nuclear RNA is able to drive the reaction to completion, implying that these intervening sequences are transcribed. Titration with these probes suggests that the number of copies at steady state is about 250 per nucleus. Formally two conclusions are possible: the intervening sequences may be transcribed about ten times less efficiently than the structural sequences; or they may be degraded much more efficiently, reducing their steady-state concentration. Following the second interpretation, this emphasizes the need to use pulse label conditions to identify levels of control.

The more rapid turnover of intervening sequences is what would be expected if mature messengers are processed by splicing the precursor. Blotting experiments with the cloned intact mRNA sequence identify the series of longer nuclear molecules shown in Figure 26.17. There appear to be up to seven size classes, ranging from 2600 to >7800 bases in length. These bands are present only in oviduct tissue stimulated by estrogen. When the cloned EcoRI 2.4 and 1.8 kb fragments are used to identify complements in nuclear RNA, the first is represented only in the two largest bands, while the second reacts with all bands. This is difficult to interpret clearly since the cloned fragments do contain some structural regions as well as intervening sequences, but it might mean that most of the intervening sequences present in the 2.4 kb fragment have been spliced out from all but the two largest bands. The existence of a discrete series of molecules suggests that splicing may be an ordered process in which some intervening sequences are removed to leave intermediates that are stable enough to be represented in the steady-state population. More detailed experiments with probes representing individual intervening sequences will be needed to see whether they are removed in a given sequence with the same or with different efficiencies. Pulse chase experiments will be necessary to confirm that these bands are precursors to ovalbumin mRNA and to follow the fate of individual segments.

Similar experiments have been performed with probes representing ovomucoid mRNA (Nordstrom et al., 1979). The mature mRNA forms a band at 980 bases and the putative precursors form a series of seven bands ranging from 1650 to 5800 bases. With this gene as with ovalbumin and globin, the possibility exists of defining a pathway for processing precursors into messengers.

A principal target since the discovery of splicing has been to isolate the enzymes responsible for breaking and rejoining RNA. An ingenious system in which this may be possible is provided by yeast, where a temperature-

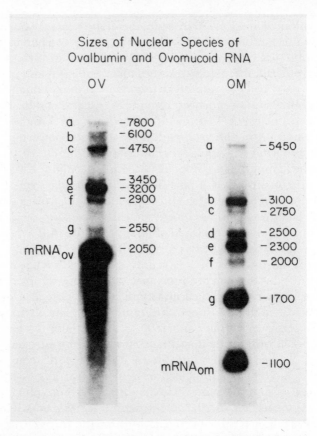

Figure 26.17. Identification of nuclear RNA containing ovalbumin or ovomucoid message sequences. Oviduct nuclear RNA was electrophoresed on agarose gels in the presence of the denaturing agent methyl mercury hydroxide. The RNA was transferred by Northern blotting from the gel to DBM paper and hybridized with ^{32}P-labeled coding sequences isolated from a cloned copy of ovalbumin mRNA (left) or ovomucoid mRNA (right). The putative precursor bands are labeled according to size. Data of Roop et al. (1978) and Nordstrom et al. (1979).

sensitive mutant has been known for some time to accumulate RNA in the nucleus. Originally believed to possess a defect in nucleocytoplasmic transport of RNA, the mutant more recently has been shown to accumulate tRNA precursors in the nucleus (Hopper et al., 1978). Among these molecules are species that contain the mature 5′ and 3′ ends but which have retained the short intervening sequence. The full consequences of the mutation have not yet been determined, but from these results it is clear that it does not impede the processing of flanking sequences from the tRNA ends; nor does it prevent the usual delivery to the cytoplasm of tRNA processed from transcripts of genes that do not possess intervening sequences. This raises the possibility that the mutation prevents the splicing reaction and that splicing is necessary for nucleocytoplasmic transport of molecules whose precursors contain intervening sequences; although it is not possible to exclude the model that the mutation directly affects transport and indirectly influences splicing.

By using the tRNA^{Tyr} and tRNA^{Phe} precursors obtained from the mutant as substrates for an in vitro preparation obtained from wild-type cells, Knapp et al. (1978) and O'Farrell et al. (1978) have been able to splice out the intervening sequence. The assay is simple: after treatment with the extract, tRNA is electrophoresed to identify molecules that have been reduced in size by the length of the intervening sequence. The structures of both the precursors and products have been confirmed by RNA sequencing. (A similar assay could be used to identify splicing enzymes in other systems.)

Two models come immediately to mind as possibilities for the mechanism of splicing. A single breakage and reunion event involving the ends of the intervening sequence would correspond to an RNA recombination. Free termini would not be released at any point; the intervening sequence would be released as a circle. Presumably the reaction would be undertaken by a single enzyme activity. An alternative is to separate the cleavage and rejoining reactions, so that first the intervening sequence is cleaved from the precursor; and then the ends of the two fragments representing the mature RNA would be ligated. This could involve separate enzyme activities.

Two steps in the splicing reaction of tRNA have been distinguished in the vitro system by their different dependence on ATP. In the absence of ATP, cleavage occurs to release two "half tRNA" molecules and a linear intervening sequence; the presence of ATP is required for a second reaction that links the half molecules into a mature tRNA by phosphodiester bond formation. The presence of excess mature tRNA inhibits the joining reaction and also causes the accumulation of cleaved intermediates (Peebles et al., 1979). Knapp et al. (1979) have characterized the cleavage intermediates, which have the unusual termini of 5'-OH and 3'-phosphate. Removal of the 3' phosphate prevents ligation of the two tRNA halves. This contrasts with all other known endonucleases, which cleave RNA to generate 5' phosphate termini. Among DNA ligases, the common requirement is for such 5' phosphate termini (although a 3' phosphate is involved in the reaction of eucaryotic DNA-relaxing enzymes). The intervening sequence excised from the tRNA^{Tyr} precursor has unique ends, defining the interruption in tRNA as shown in Figure 26.8.

It seems likely that this pathway is common for the splicing of other interrupted yeast tRNAs; and a sensible speculation is that the cleavage and ligation depends on recognition of secondary structure. The persistence of the intervening sequences near the anticodon in the three characterized examples makes it possible that the illustrated conformation of the anticodon region in the precursor is sufficient for recognition. The ligation stage may rely upon recognition of the structure of the molecule formed by base pairing between the two tRNA halves; this would be consistent with the ability of mature tRNA to inhibit the reaction.

Whether this mechanism is similar to that involved in other systems is not yet known. If analogous features are characteristic of the reactions in higher eucaryotes, a speculation would be that the use of complementary RNA to establish a structure for the splicing junctions, as illustrated in Figure 26.13, might provide a counterpart to the secondary structure that is an inherent feature of tRNA. The pertinent features of yeast tRNA precursors can be recognized in Xenopus oocytes; when cloned tRNA^{Tyr} genes are injected, they are transcribed into precursors, which are correctly spliced (De-Robertis and Olson, 1979). This has been taken further with the development of an extract of oocyte nuclei that can both transcribe and process (including splicing) the yeast tRNA^{Trp} gene into mature tRNA (Ogden et al., 1979).

Another in vitro system has been derived from the inability of isolated HeLa nuclei to splice transcripts of adenovirus DNA. Blanchard et al. (1978) found that when infected cells are labeled before nuclei are prepared, they contain a transcript of 5000 bases that represents the adenovirus 72,000 dalton protein. Under normal conditions, the transcript is spliced to yield a 2000 base mRNA that lacks two intervening sequences. In incubated nuclei, the 5000 base molecule is stable. Addition of the postnuclear supernatant sees a reduction in size of the molecule to that of mRNA. This is accompanied by the production of an RNA of about 1.5–2.0 kb that is a candidate for the spliced out fragment representing the longer intervening sequence. The cellular location of the splicing activity is not certain: the nuclei were prepared by swelling cells in hypotonic buffer, so that some nuclear enzymes may have leaked into the supernatant.

A potential mutation in the splicing apparatus has been found in the form of a temperature-sensitive nuclear mutant of Neurospora which is defective in the synthesis of mitochondrial ribosomes. Instead of the 25S large ribosomal RNA, a 35S RNA accumulates at high temperature; this may represent a primary transcript from which the intervening sequence has not been removed (Mannella et al., 1979).

Alternative Splicing Pathways in Adenovirus

The splicing of adenovirus transcripts was revealed by the observation that the 5′ segments of late messengers correspond to regions of the genome quite distant from those that code for the protein products. Both electron microscopic and restriction mapping have shown that the 5′ nontranslated "leader" of these messengers is a common structure, assembled from three different regions on the genome.

Adenovirus 2 DNA consists of 35,000 base pairs divided into 100 units numbered from left to right. Early genes are present in several regions, some transcribed rightward (map regions *1.5* →*11.5; 76* →*86*), and some transcribed leftward (map regions *61* ← *75; 91* ←*99*). Most late genes are transcribed rightward in the region $16 \rightarrow 92$, but some leftward transcription occurs from $11 \leftarrow 15$. Thus some of the early regions lie within the 85% of the genome in which the major late transcription occurs ($16 \rightarrow end$).

By R loop mapping the hexon mRNA, a late rightward transcript, Berget, Moore, and Sharp (1977) showed that the main body of the mRNA hybridizes to the region at coordinate $51.7 \rightarrow 61.3$ which is contained within a HindIII restriction fragment. The 3' terminal region forms an unhybridized tail of poly(A). And the 5' region also forms a nonhybridizing tail, about 160 bases long. When hybridized with the denatured DNA of EcoRI fragment A, which contains regions 0–60, the 3' end of the hexon mRNA is identified by its unhybridized tail. The 5'-terminal region hybridizes with some widely separated segments of the DNA, forming short hybrids of about 80, 110, and 110 base pairs with DNA at map coordinates 16.8, 19.8, and 26.9, respectively, with continuous hybridization from 51.7 to the end of the DNA fragment (60). The DNA between the hybrid regions is not represented in the mRNA and forms excluded single-strand loops of 1010, 2350, and 8060 bases; in other words, it represents intervening sequences.

The tripartite loops are not common and are difficult to measure precisely; so Chow et al. (1977) used another method to identify the DNA segments represented in the nonhybridized 5' tails that they observed when many late messengers were hybridized with adenovirus DNA. After forming R loops between the DNA and the main region of the mRNA, denatured restriction fragments representing different parts of the genome were tested for ability to hybridize with the single stranded 5' RNA tails. The location of the corresponding segment in DNA is given by the point of contact with the restriction fragment. This identified short sequences at map coordinates 16.6, 19.6 and 26.6 which are present on all late messengers, whose coding segments are derived from distant parts of the genome between 30 and 92. Similar splicing between leader segment and coding region is found for early messengers, but the structures of the mRNAs from different early regions are of course different (Kitchingman, Lai, and Westphal, 1977).

A related approach is to perform two-part hybridizations with filters. In such "sandwich" hybridization, a DNA fragment representing a given coding region is bound on a filter and hybridized to its mRNA; then labeled restriction fragments from other DNA regions are tested for their ability to bind to any exposed tails. This shows that RNA from the Ad2$^+$ND1 hybrid virus has regions that are continuous in the RNA but are separated on the genome (Dunn and Hassell, 1977). A further indication of such structures is given by identifying the products of cell-free translation of mRNAs that have been isolated by their ability to hybridize with different regions of the genome. This shows that RNAs representing genes that lie in the right-hand part of the adenovirus genome can be isolated by virtue of their ability to hybridize with fragments derived from sites distant on their left (Lewis, Anderson, and Atkins, 1977).

Restriction mapping of the 5' ends of late mRNAs confirms that all initiate with the same capped sequence, representing the point at map coordinate *16.4*.* The first hint of this was provided by the observation that a single undecanucleotide capped fragment is very prominent in RNAase digests of late message preparations (Gelinas and Roberts, 1977). This terminal fragment can be released by mild RNAase treatment after the mRNAs for two late proteins, 100K and fiber, have been hybridized to their coding regions on DNA, which start at the distant locations of *66.1* and *86.2*, respectively. By contrast, a region between *14.7* and *17.0* is able to protect the undecanucleotide from RNAase (Klessig, 1977).

The entire sequence of the leader has been determined for both the fiber and hexon mRNAs (Akusjarvi and Pettersson, 1979a; Zain et al., 1979). There is only one base difference, which may be due to variation in adenovirus strains. Comparison with the known sequences of the DNA shows that the leader consists of three segments of lengths 41, 71, and 88 nucleotides. The sequences of the splicing junctions are summarized below in Figure 26.20. The first segment of the leader starts with a sequence that is found in the genome at map coordinate *16.4* and which coincides with the site at which transcription appears to be initiated (Ziff and Evans, 1978). This conclusion is supported by the observation of a sequence related to the canonical promoter just 31 base pairs upstream (see above).

Variations in splicing have been observed. The most common form of the fiber message starts with the tripartite leader, but up to three additional short leader segments may be present, derived from locations closer to the start of the gene (*76.9–77.3; 78.6–79.1; 84.7–85.1*). It is possible that these are in fact precursors to the native tripartite structure. The most common variant is a quadripartite leader consisting of the usual tripartite leader joined to the second of the three supplementary segments (Chow and Broker, 1978). This has been sequenced; the additional segment is 181 nucleotides long and its presence therefore roughly doubles the total length of the leader (Zain et al., 1979). It appears to be derived from a region that is transcribed and translated in the same direction as part of an early mRNA, which is consistent with the presence of one open reading frame. This suggests that the splicing of at least this region changes from early to late times; possibly it is determined by the condition of the primary transcript in which the sequence finds itself.

Neither the tripartite nor quadripartite leader has any AUG codons. The leader carries a putative ribosome binding site close to the 5' terminus, as shown in Table 23.4, but otherwise is devoid of such signals. The only such sites in the additional segment of the quadripartite leader are more than 150 bases from the final splice. Together with the lack of homology between the third and fourth segments of the leaders, this suggests that their func-

---

*The map coordinates quoted for each experiment are those originally reported by the authors. The minor discrepancies between different authors are due to slight variations that depend on the conditions of experiment. These are insignificant and well within the standard error of deviation. The same is true for the discussion of SV40 (see below).

tion is not involved with events occurring immediately around the site of initiation. Messengers of the hybrid virus Ad2$^+$ND1 that start with tripartite and quadripartite leaders are translated equally well in vitro (Dunn et al., 1978). The distance from the last splice to the initiation codon in the body of the message may vary; it is 39 nucleotides in hexon mRNA but only 2 nucleotides in fiber message. Although the hexon body carries a potential ribosome binding site just before the AUG codon, there is of course no room for this in the fiber body.

A large number of positions exists at which the tripartite leader may be spliced to the body of a messenger. Electron microscopy and restriction mapping show that the late messengers fall into the five groups depicted in Figure 26.18. The members of each group have a common 3' end (within the resolution of the mapping techniques). They differ in the total length of messenger, that is, in the distance between the 3' end and the map coordinate at which splicing to the leader has occurred (Chow and Broker, 1978; McGrogan and Raskas, 1978; Nevins and Darnell, 1978a; Ziff and Evans, 1978).

The members of each group thus carry overlapping information, with identical sequences in their 3' regions, but with some members carrying additional sequences in the 5' regions. The total number of RNA species is greater than the number of proteins identified with the late region; and it is not clear whether all these RNAs are translated. The relative abundance of the members of a group varies widely, presumably due to the efficiencies with which the differing splicing joins are made. Some of the larger species may be precursors, perhaps involving splices at intermediate locations that are later removed; some might even be errors in splicing. However, some coding regions may be represented in more than one active messenger, but may be translated only in one. If it is true that ribosomes cannot initiate translation at internal coding regions (see Chapter 23), each member of a group may function as template only for the coding region that is adjacent to the leader, while coding regions toward the 3' end remain untranslated; these will become functional in the smaller members of the group, in which the splice point has brought the leader into juxtaposition with the appropriate initiation codon. The map locations of the relevant coding regions are not yet well enough defined to determine whether this model applies, but it would seem to be consistent with the available data on protein products and coding locations summarized in Figure 26.18.

How is the same tripartite leader joined to different coding segments? There are two broad possibilities. The leader segments might be synthesized separately and joined to transcripts initiating at different locations within the late region; this would require intermolecular splicing. Or transcription may start only at or prior to the first leader segment, generating a primary transcript in which an intramolecular reaction splices out the sequences between the third leader and the start of the body of the message. The form of this reaction will depend on whether there is any repetition of the leader se-

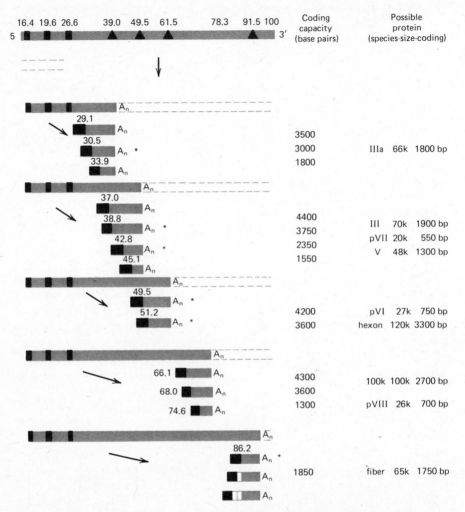

Figure 26.18. The rightward late transcription unit of adenovirus 2. Adenovirus DNA consists of 35,000 bp divided into percent units on the restriction map. Late transcription is initiated at *16.4* and continues to coordinate *100*, except for some molecules that terminate prematurely before coordinate *25* (indicated by the short dashed lines). Five potential sites for addition of poly(A) are indicated on the genome map by triangles. Shortly after RNA polymerase has passed one of these, cleavage may occur, to be followed by addition of poly(A); the rest of the transcript is synthesized but does not give rise to mRNA, as indicated by the dashed lines. A sequence contiguous to the polyadenylated 3′ end is spliced to the tripartite leader segment at the 5′ end, which is assembled from the sequences indicated by the black squares. For each of the 3′ termini, there are several points at which the contiguous sequence may be attached to the leader; their coordinates are indicated. For the fiber message, additional leader segments may be found between the common tripartite leader and the message body; these are derived from points at map coordinates *76.9*, *78.6*, and *84.7* and are indicated by the white squares. The asterisks indicate messengers present in larger amounts. The potential coding capacity of each message is calculated as the distance between the junction with the leader and the 3′ end. The

quences in the genome. If there is more than one copy of each leader segment, leaders might be spliced onto more than one messenger body from each transcript molecule. If there is only one copy of each leader segment, each transcript can undergo only one series of splicing events, in which the leader is joined to one of the several possible messenger bodies; this appears to be the case.

Nascent RNA chains start only with the capped sequence that corresponds to map coordinate *16.4* (Evans et al., 1977; Ziff and Evans, 1978). This site appears to be the only promoter for the late rightward transcription. Studies of ultraviolet sensitivity originally suggested that the various individual messengers might have the same dose response as the distal regions of the primary transcript, which would imply that completion of transcription might be obligatory for processing of earlier regions (Goldberg, Nevins and Darnell, 1978). But a dependence on distance from the promoter since has been reported; and this is consistent with the conclusion discussed in Chapter 25 that polyadenylation involves cleavage at any one of a number of sites prior to the completion of transcription (Nevins and Darnell, 1978b; Fraser et al., 1979).

The termination site for the RNA polymerase appears to lie beyond any of the regions represented in late mRNA, since transcription between coordinates *25* and *99* is equimolar.* This implies that the 3' termini of all five groups of messengers are generated by cleavage of the transcript rather than by termination of transcription at the different sites. (This may therefore be a good system in which to distinguish signals for cleavage and polyadenylation from those for termination of transcription.) The cleavage event appears to occur shortly after passage of the polymerase, which nonetheless is compelled to continue to the end of the transcription unit. Each primary transcript therefore gives rise to a single message, which contains the common leader and one of the several possible bodies. The sequences intervening between the components of the leader, and between the leader and the body, as well as the sequences distal to the site of cleavage presumably are degraded. This model is supported by measurements that show virtually all the poly(A) to be conserved, whereas only 20–25% of the sequences

*Some premature termination of transcription appears to take place in the initial region, ceasing before the third leader segment, since synthesis of RNA representing map units *16–25* occurs at levels 3–6 times greater than that of subsequent regions (Evans et al., 1979). Discrete bands are found when this RNA is analyzed on acrylamide gels, suggesting the existence of a large number of termination sites. The production of the short RNAs is enhanced by the inhibitor DRB (see Chapter 25). The prematurely terminated transcripts are capped (Fraser, Sehgal, and Darnell, 1979).

---

proteins that lie in the appropriate map regions are indicated, although not all can be identified with individual messengers. Protein sizes are given in kilodaltons (k) and the length of the necessary coding sequence in base pairs. Data of Chow and Broker (1978), McGrogan and Raskas (1978), Goldberg, Nevins, and Darnell (1978), Nevins and Darnell (1978a,b), Ziff and Evans (1978), Ziff and Fraser (1978), Evans et al. (1979), Fraser et al. (1979).

transcribed in any region find their way into mRNA (Nevins and Darnell, 1978a).

How is the selection of a point for splicing the body to the leader coordinated with the cleavage of a particular 3′ terminus? In pulse label studies, the first poly(A)$^+$ transcripts to be detected contain continuous sequences from the *16.4* initiator to the 3′ terminus (Nevins and Darnell, 1978b). This implies that cleavage and polyadenylation of the 3′ end precedes splicing out of intervening sequences. The conclusion that splicing is slow compared with polyadenylation is consistent with the isolation of precursors for cellular messengers that contain intervening sequences and are polyadenylated (see above). It is supported also by the isolation of discrete precursors for the early adenovirus mRNA representing regions *61–75*. Goldenberg and Raskas (1979) found that the primary transcript is a 28S nuclear RNA that is polyadenylated and contains two intervening sequences. This is processed to a 23S RNA that lacks the first intervening sequence; then the second intervening sequence is removed to generate a 20S mRNA with a dipartite leader. The two intervening sequences therefore appear to be removed solely, or at least predominantly, in order from the 5′ end.

For each of the five possible polyadenylated precursors derived from the late adenovirus transcription unit, a restricted number of splicing sites is available for leader-body junction; there are up to four members of each group shown in Figure 26.18, although these do not necessarily identify all possibilities. As cleavage sites more distant from the promoter are used, each precursor has a longer intervening sequence that contains the splicing junctions used by the shorter precursors as well as those that it may use itself. There is no evidence on how the splicing system is compelled to use the appropriate set of sites. One possibility is that splicing proceeds progressively from the 5′ end, recognizing at least one member of each group, but proceeding further if another group is available. However, this is hard to reconcile with the existence of different members in each group, which means that the final splice does not always take place at the recognition site closest to the 3′ end. Another possibility is that the 3′ terminus in some way controls secondary structure so as to expose only a given range of possible splicing junctions. Yet another option is to suppose that some action of the transcriptional apparatus, possibly dictated at the time of initiation, ensures that splicing and cleavage are coordinated. To summarize, splicing in the late adenovirus transcription unit has two consequences: it joins a common leader to all the late messengers; and it selects from the primary transcript a variety of possible bodies.

Similar events occur in the splicing of early adenovirus messengers. The mature mRNAs have been mapped on the genome by a procedure developed by Berk and Sharp (1978b), in which an RNA is annealed with labeled single-stranded DNA representing the region from which it is transcribed. After any single-stranded ends of the DNA not represented in the RNA have been removed by exonuclease, the DNA can be recovered as a labeled fragment representing the region between the beginning and end of the mRNA on the genome. If the hybrid is treated with S1 nuclease, any intervening sequences in the DNA that have formed single-strand loops will be digested. Separate fragments of DNA then are recovered for each genome region that is represented in the RNA. The two DNA preparations, containing and lacking intervening sequences, can be characterized by gel electrophoresis and hybridization with restriction fragments of the genome. This identifies the regions represented in RNA and the interruptions between them. Each of the four early regions of the adenovirus genome generates RNAs that are spliced together by junction of leaders (which may be continuous or segmented) with messenger bodies. In each case, there are alternative splicing pathways in which the leader is joined to a different site on the body; again the various RNAs differ in frequency of occurrence, but their total number is greater than needed to account for the known polypeptides coded by the early regions.

The existence of splicing in the transcripts of C type RNA tumor viruses is implied by the observation that the same 5′ terminal sequence, derived from the region adjacent to the 5′ end of the viral RNA, is present on all size classes of mRNA present in infected cells (Krzyzek et al., 1978; Panet et al., 1978; Rothenberg, Donoghue, and Baltimore, 1978). The relationship between the size classes is akin to that between the members of each late adenovirus group: all messengers contain sequences from the 3′ end, and as the size class increases, the point of splicing to the leader has come closer to the 5′ end. It may be that in each size class only the coding sequence adjacent to the 5′ end is translated.

Variable Splicing of SV40 Transcripts

Probably all of the messengers of the DNA viruses SV40 and polyoma are spliced. Both have circular genomes: SV40 consists of 5226 base pairs of known sequence; polyoma is about the same size and has been partially sequenced. The sequences of the two viral genomes are not generally related, but their overall form of genetic organization is similar. In both the genome is about equally divided into early and late regions that are transcribed in opposite directions. The early region codes for proteins identified as tumor antigens; the late region codes for components of the virion. In both regions in each virus, there are several genes that overlap in nucleotide sequence and are expressed by the production of appropriate mRNAs through variations in the splicing pathway.

The initial observation that SV40 mRNAs represent noncontiguous regions of the viral DNA was made with late messengers. There are three late genes coding for virion proteins. VP3 is part of VP2; it represents the C terminal region of the longer gene and is read in the same phase, although apparently by independent initiation of translation at an AUG within VP2.

The initiation site for VP1 lies prior to the termination codon for VP2/VP3, but in a different phase (Contreras et al., 1977). Together the coding regions occupy some 2100 bp of the 2600 bp of the late region (see Figure 26.19). The late messengers fall into two size classes, 16S (coding for VP1) and 19S (coding for VP2; although not yet identified, coding capacity for VP3 may lie in this class). Both size classes are able to hybridize with a region of DNA between map coordinates *670–760*, which is distant from the main coding segment (Aloni et al., 1977; Bratosin et al., 1978; Lavi and Groner, 1977).

More detailed studies of messenger structure have led to the results summarized in Figure 26.19. The 16S RNA is formed by splicing a leader of 202 base pairs to a messenger body that starts 43 nucleotides before the VP1 initiation codon. The 19S RNAs are formed by splicing a messenger body that contains all the coding sequences to a variety of leaders derived from the same region as the 16S leader; the junction with the messenger body is located 5 bases before the VP2 initiation codon. Both ends of the leader sequence are variable in location. The 5′ end may be located at any one of several sites over a region of some 250 bases; the same heterogeneity is seen in nuclear RNA, which makes it possible that it is due to variation in the site of initiation rather than to variable cleavage. The point at which the other end of the leader is spliced to the body of the messenger also varies; it may lie at any one of at least three sites separated by a distance of some 250 bases. The total length of the leader therefore varies from roughly 250–500 bases. In poly(A)⁻ nuclear RNA representing the late region, the 3′ end is variable in position; but in poly(A)⁺ RNA it is constant, which suggests that (as with adenovirus) the 3′ terminus of the mRNA may be generated by cleavage and polyadenylation. It will be interesting to see if the same is true of cellular mRNAs.

From the point of view of coding purposes, the difference in splicing between 16S and 19S RNAs is that each type of event brings a different initiation codon, for VP1 or VP2, respectively, into juxtaposition with the leader. It is this event that allows the region of overlap between the two genes to be read in one phase as the terminal sequence of VP2/VP3 and in another phase as the initial sequence of VP1. No splice to any junction close to the initiation codon of VP3 yet has been reported, although small amounts of an 18.5S nonspliced RNA starting at map position *780* have been found.

The 16S RNA functions as a messenger for VP1 and Table 23.4 gives the salient features of the 5′ region, from which it is clear that AUG codons and potential ribosome binding sites both are present in the leader segment, although none is close to the splice points. The 19S species vary in abundance and it is not known which of them serve as templates for synthesis of VP2. The most abundant is the middle of the three shown in Figure 26.19. It is possible that the outer RNA species is a precursor to this; while the inner 19S RNA, which has the same leader as the 16S RNA, might even represent an error in splicing. The function(s) of the leaders are unknown; it is inter-

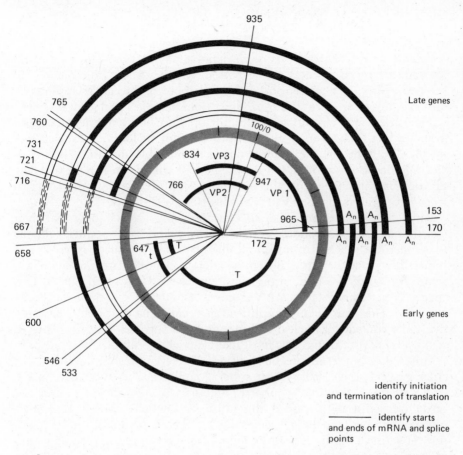

Labels within figure:
935
Late genes
765
760
731
721
716
667
658
600
546
533
100/0
834 VP3
766
VP2
947
VP1
965
A_n A_n
A_n A_n A_n A_n
153
170
647
t   T
172
T
Early genes

identify initiation
and termination of translation

———— identify starts
and ends of mRNA and splice
points

Figure 26.19. Transcription and translation map of SV40. The shaded circle represents the DNA of SV40 divided into tenths and numbered so that 100 units = 523 base pairs. Within this circle are shown the positions of the coding sequences; AUG initiation codons are indicated by tails and nonsense termination sites by arrowheads. Outside the circle are shown the principal cytoplasmic RNAs, with 3′ termini indicated by A_n. Black indicates conserved regions; white indicates sequences that are removed by splicing. Both early mRNAs start at map position *658* and differ only in whether the regions *600 → 533* (for T) or *546 → 533* (for t) have been spliced out. The late RNAs fall into two classes. The 16S mRNA for VP1 has the leader *721 → 760* spliced to the body *935 → 170*. The 19S RNAs, which include those coding for VP2 and possibly VP3, have leaders whose 5′ ends appear to be heterogeneous, as indicated by the stippled regions between *667 → 721*. These are spliced to the messenger body *765 → 153*. The coordinates given here are as adjusted by Ghosh et al. (1978b); although the entire nucleotide sequence of the DNA is known, different numbering systems have been used for the nucleotides and so these are not used here, although these data have been used to assign the coding sequences (Fiers et al., 1978; Reddy et al., 1978). Late gene data are from Lai, Dhar and Khoury (1978) and Ghosh et al. (1978a,b); early gene data are from Berk and Sharp (1978a), Crawford et al. (1978), Paucha et al. (1978), Paucha and Smith (1978), Volckaert et al. (1978).

841

esting that mutations at map positions *700* and *750* alter the splicing patterns of the late RNAs, but do not render the mutants nonviable (Villarreal, White, and Berg, 1979).

A similar series of events may occur in polyoma infection. The three virion proteins are coded by messenger RNAs in which a leader from one region has been spliced to a messenger body coded at one of three, overlapping distant regions (Horowitz, Bratosin, and Aloni, 1978). The late RNAs start with a variety of 5′ termini, as seen in the presence of seven different capped dinucleotides (Flavell et al., 1979). All the termini are found in each class of late RNAs. Whether the termini represent multiple initiation points or cleavage sites is not proven; but the presence of only purines in the first base position is consistent with the possibility of initiation. Similar observations have been made for SV40 (Canaani et al., 1979).

Ribonuclease fingerprints of the late polyoma leaders show that their constituent oligonucleotides are present in equimolar amounts with the remainder of the RNA in unspliced nuclear transcripts made in vivo or in transcripts made in vitro. But these oligonucleotides are present at increased molarity, 3–5 fold, in the mature mRNAs (Legon et al., 1979). This suggests that some of the mRNAs contain multiple copies of the leader sequence. How might this amplification occur? The discrepancy between the leader contents of nuclear and mature RNA suggests that it is achieved at the level of processing. The most likely model is to suppose that late transcription does not terminate, but continues for several revolutions of the circular genome; then leader-leader splicing could produce multiple tandem copies of the 5′ terminal sequence. Of course, if the remaining sequences of the transcript are discarded, this implies a low mass conservation of sequences representing the messenger bodies. The structure of the leaders is not yet defined, but there are hints that it may be more complex than simple tandem repeats of a given sequence.

The early region of SV40 codes for tumor antigens that are synthesized from overlapping genes that are expressed by means of variations in the splicing pathways. The large T antigen has been known for some time as a protein of 90,000–100,000 daltons. This requires a coding capacity equivalent to virtually the entire early region of 2600 base pairs, lying between map positions *658* and *160*. Yet more recently the small t antigen has been identified as a viral product of 15,000–20,000 daltons which is immunologically related to large T and shares with it some peptides (Prives et al., 1977). The unusual relationship of the sequences coding for two proteins was first revealed by observations that mutations and deletions affecting T map between *430* and *200*, whereas deletions in regions *590–540* alter or abolish t, but do not affect T (Crawford et al., 1978). But the two proteins have the same amino terminal sequence, which is coded by the DNA sequence at *650* map units (Paucha et al., 1978; Volckaert, Van de Voorde, and Fiers, 1978).

This suggests the model shown in Figure 26.19. Small t is coded by a continuous sequence read from an AUG at *658* to a termination codon 174 triplets later at position *533*. This has been confirmed by examining the tryptic peptides present in the altered t protein produced by deletion mutants lacking part of the region *590-540* (Paucha and Smith, 1978). Large T is coded by two noncontiguous regions, *650-590* (shared with t) and *540-200* (the C terminus). Thus the two proteins have the same N terminal sequence but different C terminal sequences. It is possible that further proteins may be coded by other variations in the splicing pattern.

A pattern of this sort is seen in the polyoma early proteins. Three tumor antigens, of 90-100K (large T), 55-60K (middle T) and 15-22K (little t), all share the same N terminal tryptic peptides. Middle T and small t share some further peptides that are absent from large T (Hutchinson, Hunter, and Eckhart, 1978; Smart and Ito, 1978). Although the locations of the coding sequences remain to be fully defined, it seems likely that alternative pathways of splicing are responsible for seeing that a common N terminal coding sequence is succeeded by three different C terminal sequences. It is probable that changes in reading frame are involved.*

Messenger RNAs corresponding to the SV40 T and t proteins have been analyzed by Berk and Sharp (1978a), using their mapping technique. A 2200 base mRNA contains a 5′ region of 330 bases (*670-600*) joined to a 1900 base region (*540-140*). The sequence of the 5′ region of the mRNA has been determined; comparison with SV40 DNA shows that 346 bases have been spliced out (Thompson, Radonovitch, and Salzman, 1979). This codes for T; and the splicing event has removed the termination codon at the end of the t sequence. A 2500 base mRNA consists of a 630 base region (*670-540*) joined to the same 1900 base region (*540-140*). The sequence that has been spliced out is rather short, about 65 bases. Whether this mRNA has any function other than translation of its 5′ end to code for t is not known. Consistent with the assignment of the mRNAs to these functions, deletions between *590* and *540* reduce the size of the larger mRNA coding for t, but do not affect the mRNA coding for T (May, Kress, and May, 1978).

The use of deletion strains of SV40 and of variants constructed in vitro has cast some light on the necessity of splicing for gene expression. Deletions in either the early or late region that remove a splicing junction prevent appearance of the corresponding mRNA. This contrasts with the appearance of shortened mRNAs when the deletions lie entirely within the coding regions or of normal mRNA when they lie within the intervening sequence. When a deletion removes the splicing junctions of one mRNA, but lies within a nonspliced region of an overlapping mRNA, only the latter is found (Lai and Khoury, 1979; Gruss et al., 1979; Khoury et al., 1979).

---

*One implication of this sort of arrangement is that it is not possible to tell by examining DNA for AUG and UAA, UAG, UGA codons where genes commence and end. Nonsense codons that provide apparent blocks to reading in the frame in which translation is initiated may be spliced out or rendered ineffective by splices that change the reading frame. The presence of DNA sequences that contain termination codons in all three reading frames thus cannot be taken to imply a lack of coding function.

Extensive sequences around the splicing junctions cannot be generally necessary for the removal of intervening sequences; deletions as close as 11 bp to the T splice sites leave unaltered the production of T mRNA. Yet an indication that sequences additional to the splicing junctions may be involved in the reaction is provided by the reduction that is seen in the amount of the shortened mRNAs with deletions located elsewhere; some small deletions located 200 bp upstream from the small t splice site may reduce the amount of the t mRNA, presumably because indirectly or directly they alter the conformation of the molecule.

When transcription complexes are prepared from infected cells and incubated under conditions that allow chain extension in vitro, label is incorporated into the sequences that are absent from mRNA. Although not precisely quantitated, the level of synthesis seems to be roughly similar in wild type and mutant infections. This suggests that the deletion of the splicing junctions does not influence transcription, but exerts its effects at some subsequent stage. Lack of splicing may therefore mean that the transcript becomes unstable. This would accord with a model in which a transcript must suffer one of two fates: it is spliced and transported to the cytoplasm; or it is degraded. The alternatives of remaining stable for some period in the nucleus or entering the cytoplasm unspliced do not seem to be available.

This implies that splicing is an essential step in gene expression, which is to take the view that the presence of an intervening sequence in an interrupted gene is necessary for its function. But this does not account for the presence of several intervening sequences: the analysis of viral gene expression would suggest that one intervening sequence is sufficient for this function. Another question is whether uninterrupted genes are present in genomes also possessing interrupted genes; presumably the expression of continuous genes must involve some particular alternative to splicing to allow transition through these stages of gene expression.

Accepting these views, any sequence that is attached to proper splicing instructions probably can be processed and exported to the cytoplasm. Cloned DNA copies of the rabbit β-globin mRNA can be transcribed and translated when monkey cells are infected with SV40 in which the globin gene has been inserted in place of either the VP1 or VP2 gene (Mulligan, Howard, and Berg, 1979; Hamer et al., 1979). The production of a hybrid mRNA in which the SV40 leader is spliced to the globin message implies that virtually the entire late coding region is not necessary for splicing to occur. And at least when attached to the SV40 leader, removal of the intervening sequences in the globin gene cannot be essential for expression.

If the reconstructed vector lacks the splicing junctions of SV40, however, the hybrid mRNA is not found; again this appears to be due to degradation of the transcripts, which can be detected in vivo with short pulse doses, but then soon decline in size. When part of the mouse β^{maj} globin gene including the principal intervening sequence was inserted into SV40 in place of the VP2 coding region, the globin intervening sequence was removed during the formation of hybrid mRNA (Hamer and Leder, 1979). This implies that mouse splicing signals can be recognized in monkey cells. It would be interesting to know whether the presence of the globin intervening sequence would allow a hybrid mRNA to be produced from a vector lacking the SV40 splicing signals.

A hint that splicing ability may depend on cell phenotype is provided by the demonstration that SV40 early RNA is not spliced in mouse teratoma cells (which are not successfully infected by SV40), although early mRNA and t/T antigens can be synthesized upon infection of differentiated mouse embryo or 3T3 cells (where there remains a subsequent block to SV40 production). Segal, Levine, and Khoury (1979) found that the SV40 DNA can be transcribed in the teratoma cells, as seen in the ability of complexes to continue transcription in vitro. Small amounts of unspliced transcripts can be detected in the infected teratoma cells by hybridization with SV40 restriction fragments. These unspliced molecules can be found in extracts of both nuclear and cytoplasmic RNA. This contrasts with the presence of abundant amounts of properly spliced early mRNA in differentiated mouse cells.

The splicing junctions that have been sequenced in the viral systems are summarized in Figure 26.20. They show the same general features as those characterized in cellular genes: there is always a repetition of 2–4 nucleotides between the ends of the mature sequence and the intervening sequence so that the splice points cannot be read unambiguously; but if aligned in the same way as the cellular splicing junctions, they show the same canonical sequences (see Figure 26.12).* As with the multiple splices that occur in cellular genes, the similarities of the sites offer no hint as to how the splicing systems may distinguish between them. What controls the utilization of alternative splicing pathways is unknown.

What is to be made of the contrast between the uses to which splicing is put in the cellular and viral systems? Splicing of the transcripts of nuclear genes is responsible for assembling a continuous coding sequence from regions that may be spread out over large distances in the genome. The alternative pathways for splicing of viral transcripts allow an intensive use to be made of a given region, to produce multiple products with shared sequences. Are these different situations or does the viral system provide a paradigm for a cellular system capable of even more complicated reactions than the reconstruction of intact coding units?

A critical question about the significance of interrupted cellular genes is therefore whether this organization is just part of the intricate processes by which eucaryotic genes are expressed or whether it is involved in regulation.

---

*No sequences corresponding to the canonical promoter sites have yet been identified for either early or late SV40 genes. The putative capping sequence is absent at the sites corresponding to the 5′ ends of late 16S mRNA and early T/t mRNA. TATAAA boxes also are absent. Either could be involved in the existence of multiple startpoints (see text above).

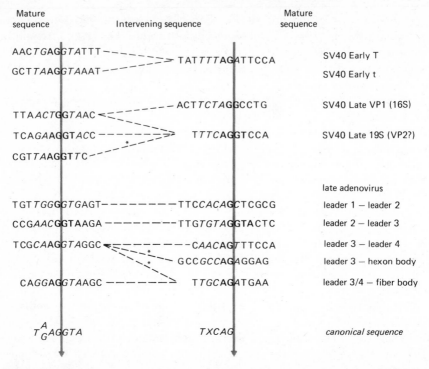

Figure 26.20. Splicing junctions in SV40 and adenovirus. Data are arranged as described in Figure 26.11. Possible splicing connections are indicated by dashed lines; alternative routes that seem to be followed most often are indicated by the asterisks. Data of Ghosh et al. (1978a,b), Bina-Stein et al. (1979), Thompson, Radonovich, and Salzman (1979), Akusjarvi and Pettersson (1979a,b), Zain et al. (1979).

Regulatory functions might be quantitative or qualitative. In the first case, the splicing of a given transcript might be the step deciding whether mature sequences are to be exported to the nucleus. In other words, splicing could constitute an on/off switch for control of gene expression at the level of nuclear RNA. A more complex possibility is to suppose that alternative splicing pathways exist also for cellular genes. There is no evidence to support any idea that this might be true of the eucaryotic genes we have discussed so far, but such concepts can be applied to complex gene systems, such as immunoglobulins.

It is tempting to speculate that the relationship between the overlapping early or late gene sets of SV40 and polyoma may have evolved in response to the need of the virus to produce more proteins than can be coded by continuous readout of the length of DNA that can be packaged in the virion. Another factor might be that the viruses need to generate proteins with partially overlapping sequences, but that repetition of the coding sequence in the genome would lead to unacceptable consequences, such as unequal crossing over. A similar argument can be made for the repetition of nontranslated leader sequences in adenovirus. In this sense, one might suspect that the viruses have subverted the cellular splicing systems to their own advantage (a conclusion that may be applied also to bacterial viruses, although with the difference that splicing does not seem to be among the options available in procaryotes).

Diversity of Immunoglobulin Genes

The generation of antibody diversity has been one of the most provocative issues in molecular biology. The ability of the organism to produce an apparently innumerable number of antibodies in the immune response (generally reckoned for practical purposes to be of the order of 10^6 different immunoglobulins) poses the immediate question of whether this diversity is coded in the genome of the gametes or is generated in somatic cells. The structure of the polypeptides comprising each antibody complex raises the complementary issue of the nature of the relationship between gene and protein.

As illustrated in Figure 26.21, each immunoglobulin is a tetramer consisting of two (identical) *light (L) chains* of about 216 amino acids each and two (identical) *heavy (H) chains* of about 450 amino acids each. Each type of chain consists of two regions: the N terminal part comprises the *variable (V) region;* and the C terminal part is the *constant (C) region*. The combining site of the antibody (which determines its specificity for antigen) is formed by the association of the light and heavy chain variable regions (V_L

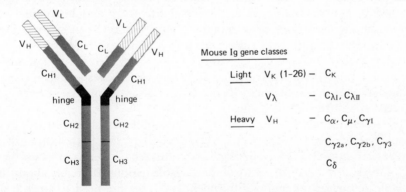

Mouse Ig gene classes

Light	V_K (1–26)	–	C_K
	$V\lambda$	–	$C_{\lambda I}$, $C_{\lambda II}$
Heavy	V_H	–	C_α, C_μ, $C_{\gamma I}$
			$C_{\gamma 2a}$, $C_{\gamma 2b}$, $C_{\gamma 3}$
			C_δ

Figure 26.21. Structure of immunoglobulins. The tetramer antibody contains domains formed by the reactions V_H-V_L, C_{H1}-C_L, C_{H2}-C_{H2}, C_{H3}-C_{H3}; the hinge regions separate the first two from the second two domains. Mouse light chains may be of the kappa or lambda type; at least 26 variable subgroups of the kappa type exist, but there is only a single subgroup of the lambda type. There is no evidence for more than one type of constant region in the kappa type; two types of constant lambda region have been found. In humans (and other mammals), the V_λ type may show much greater diversity. At least seven types of C_H region exist; some subgroups exist among the V_H types, but these have not yet been classified.

and V_H to form the *V domain*. The association of the constant region of the light chain (C_L) with part of the constant region of the heavy chain (C_{H1}) forms one constant domain; two further constant domains are formed by the association of the corresponding parts (C_{H2}, C_{H3}) of the heavy constant region. The constant domains mediate the effector functions responsible for the immune response.

The variable and constant regions take their names from the pattern of amino acid sequences seen when different light or heavy chains are compared. Relatively few differences are found in the constant regions, but extensive changes exist in the variable sequences of different immunoglobulin chains. Thus the region for determining antibody specificity shows versatility; but the region involved in effecting the immune response is conserved. Within the variable region, two types of subregion may be defined by the relative extent of variation in their constituent amino acid sequences; the *framework* regions are variable, but the *hypervariable* regions show much greater differences between chains of the same type.

Sequence homologies have been used to assign both light and heavy chains into different classes. Most data are available in mouse and in man; here we shall be concerned largely with murine immunoglobulins, since only their genes have been characterized at present. There are two types of light chain, kappa and lambda, recognized by the types of variable region (V_κ and V_λ). Extensive sequence diversity is found among variable regions of the kappa type; but little diversity exists in the murine lambda chains. This is the only substantial difference between mouse and man, since human lambda chains show high sequence diversity. In mouse there appears to be only one type of constant region in kappa chains; two have been distinguished in lambda chains. In the heavy chains there may be distinct subgroups of the variable region; but these have not yet been systematically analyzed. Several different heavy constant regions have been found, classified into the major types C_α, C_μ, and C_γ; the complete antibody tetramer is named for the type of C_H region as IgA, IgM or IgG, respectively. Genetic studies suggest that the genes for each family (kappa, lambda, or heavy) are located on a different autosome.

The existence of large numbers of immunoglobulin chains in which the variable region is different while the constant region is the same distinguishes two questions about their genes. In this secton we shall discuss the issue of the source of the diversity of the variable regions. Probably about 1000 variable sequences of both light and heavy types are needed, since if any light chain may associate with any heavy chain, this would be sufficient to generate 10^6 different antibodies. This (proposed) ability is known as *combinatorial association*. In the next section we shall consider the issue of the events responsible for linking different variable regions to the same constant region. Here an important consideration is that each cell line usually produces only a single type of antibody. Discussion of models to account for antibody production has focused on whether genes coding for the immunoglobulin chains are present in the gamete genome or are generated during somatic differentiation.*

The *germ line theory* proposes that all (or at least most) of the immunoglobulin genes are inherited. An extreme form of the model would be to suppose that there are 1000 different V genes, each accompanied by a copy of the same (or of one of a few) C genes. More favored models have suggested that there may be no repetition of C genes, so that synthesis of each different immunoglobulin requires one of the 1000 different V genes to be joined to one of few (less than 10) C genes. In both cases the full panoply of diversity is carried by the gamete genome; although the first would require only the expression of a chosen V + C gene, whereas the second would require its construction from component V and C genes.

Somatic theories propose by contrast that variation is generated during lymphocyte differentiation. This might occur by *somatic mutation* to generate new sequences from one or very few germ line genes. In the extreme model there need be no more than one copy of a progenitor V gene. An alternative is that there might be a small number of germ line V genes from which new variants are created by *somatic recombination*. In neither model is there any need to postulate repetition of C genes.

Current thinking favors intermediate models, in which there are many germ line V genes and few C genes. But the number of V genes may be insufficient to account for the full diversity of variable regions and may therefore be increased by the introduction of further variants during somatic differentiation. Hybridization data to analyze the number of V and C genes are consistent with this view, but do not prove it. Rearrangement of V and C genes is indeed required to join one of the many variable sequences to one of the few constant sequences; this is clear from restriction mapping of V and C genes in embryonic and immunoglobulin-producing cells.

The heterogeneous array of immunoglobulins produced by an individual makes analysis difficult or impossible with lymphocyte populations. Most work to identify antibody proteins and characterize immunoglobulin genes has made use of myelomas—tumors that produce immunoglobulins. A single myeloma generally secretes a single immunoglobulin; the messenger RNA may be present in large amounts, up to 10% of the total poly(A)$^+$ mRNA. Immunoglobulin messengers have been isolated from many myelomas; the properties of a typical array are summarized in Table 23.5. Each mRNA codes for a single polypeptide chain, heavy or light; this demonstrates that whatever the properties of the V and C genes, the immunoglobulin chain is translated in the conventional way from a template coding for the whole molecule. The messengers display the usual property of a length

*For the purposes of this discussion we shall take "gene" to identify a region of the genome coding for a discrete part of the final immunoglobulin polypeptide. Thus "V genes" code for variable regions and "C genes" code for constant regions, although neither type is expressed independently. In this formulation, two "genes" code for one polypeptide chain, a concept first proposed by Dreyer and Bennett (1965).

greater than needed to code for the protein product and can be considered to possess four regions: the 5' nontranslated region, the variable coding region, the constant coding region, the 3' nontranslated region. Since they are polyadenylated, they can be used as templates for the synthesis of cDNA by reverse transcription. Their structure implies that a short cDNA reverse transcript may extend only into the C region, while longer cDNAs may represent both C and V regions. This makes detailed characterization of such cDNA probes imperative. Because of the difficulties of obtaining mRNA or cDNA representing precisely the C or V regions, hybridization studies have become more effective with the more recent availability of restriction fragments obtained from cloned V and C coding sequences.

Hybridization experiments with appropriate probes should encounter no difficulty in determining the number of C genes: there are few types of constant region and each should hybridize with either a single gene (or with a very small number) if these are not extensively repeated. But if the C genes have a repetition frequency that matches the diversity of the V genes, there would be a large number of identical C genes able to react with a single C probe. With V genes such resolution is much more difficult to achieve. If the diversity of variable sequences results from the presence in the genome of multiple V genes, these will be related but not identical in sequence; a probe representing a given V region should be able to hybridize with some of the other V genes in addition to its own, but how many will depend on the degree of relationship. Even if only a few V genes react with a given probe, there might be many more present in the genome that are not close enough in sequence to react.

Early experiments with mRNA or cDNA probes produced confusing and inconclusive results. When short cDNA reverse transcripts were used as tracers in DNA renaturation, they followed the unique component (repetition frequency <3) (Faust, Diggelmann, and Mach, 1974; Honjo et al., 1974; Stavnezer et al., 1974). If these are taken to represent the C region, this would imply little repetition of C genes; but this conclusion cannot be firm because of difficulties in knowing precisely how much of the C region was included in each probe. Experiments with mRNA labeled with ^{125}I at first yielded results in which either the $Cot_{1/2}$ corresponded to repetitive DNA, or in which the kinetics were biphasic, with both repetitive (about 200–300 fold) and nonrepetitive components (Delovitch and Baglioni, 1973; Premkumar, Shoyab, and Williamson, 1974). These experiments are misleading, however, because a critical problem was the lack of purity of the mRNA; there was no evidence to exclude the possibility of contamination with other mRNA species or to show that the repetitive and nonrepetitive components were covalently linked. Different results have been obtained with mRNA purified by electrophoresis, when a single component reacts at high Cot, with a repetition frequency of <4 (Tonegawa et al., 1974; Tonegawa, 1976; Farace et al., 1976; Rabbitts, Jarvis, and Milstein, 1975). Repetitive components thus appear to be contaminants. These results certainly seem to exclude the possibility that there is any substantial repetition of identical C genes. They do not, however, resolve the status of the V genes. Although in some cases several related V sequences are known, which might be expected to cross hybridize with the probe, the discrepancy is not great enough to conclude that additional diversity exists that is not present in the germ line. Several problems are inherent in these reactions, among which one important difficulty is the probably reduced rate of reaction between related but not identical V gene sequences; the reaction of the tracer may be unable to compete effectively with perfect renaturation of the original DNA strands (see Smith, 1977).

One approach to define the pattern of diversity of variable regions has focused on the relationship between their amino acid sequences. In the mouse kappa group, the known sequences have been classified into 26 V_κ subgroups or "isotypes" on the basis of amino acid homologies (McKean, Bell, and Potter, 1978). Probably there will prove ultimately to be about 50 such groups on this basis. A geneological tree shows that the relationship *within* each subgroup is such that members might be derived from a basic germ line gene by different somatic mutations; these are considered to be "allotypes," which represent variation between individuals not necessarily inherited in the germ line. But for members of *different* subgroups to be derived from the same germ line gene would require extensive parallel mutation (or parallel function of whatever other system might generate somatic diversity) (Hood et al., 1977). Taking such parallel action to be unlikely has been the basis for suggesting that each isotype must be represented by at least one germ line gene. As more sequences accumulate, subgroups may be further divided; within the $V_\kappa 21$ subgroup, diversity has been taken to be sufficient to imply the presence of at least six germ line genes (Weigert et al., 1978). If this is typical of other subgroups, the minimum number of germ line V_κ genes would be about 200. This approach has not been applied so extensively to other variable types, but human V_κ is probably similar to mouse V_κ; mouse V_λ sequences fall into a single subgroup, whereas human V_λ sequences show subgroup diversity; and V_H sequences in the mouse probably have a diversity between the extremes of kappa and lambda (Berstad et al., 1974; Weigert and Riblet, 1977).

This model suggests that the germ line genes identified by each isotype have evolved through duplication and sequence separation from primordial V_κ, V_λ, and V_H genes. A caveat in this analysis is that it depends on the assumption that whatever process might be responsible for generating somatic diversity is unlikely to produce the same multiple changes on independent occasions; but the nature of the process is completely unknown. The model does not resolve the source of further, allotypic variation, except to say that the members of each isotype are closely related enough to support the possibility that their diversity is generated by somatic changes of one or few genes (again relying on expectations about the nature of the system generating diversity). This does not exclude the possibility that the variation is due to a multiplicity of closely related germ line genes (with the members of each isotype having evolved more recently than the isotypes themselves). In some cases the germ line gene itself may be expressed; several mouse lambda chains have identical sequences, whose lack of diversity is taken to imply that it is the germ line sequence (λ_o) that has been expressed. One model for the much reduced diversity of mouse lambda sequences compared with mouse kappa or human lambda is to suppose that at some time the mouse has suffered a deletion of most of its V_λ germ line genes. This is to ascribe the primary responsibility for V gene diversity to the germ line. This accords with present views, which may be summarized in the model that at least each cluster of related sequences represents a different germ line V gene; the extent to which further diversity is generated in somatic cells or is inherited remains controversial.

Attempts to determine the number of variable coding sequences directly have become more refined with the use of individual nucleic acid probes representing protein chains that are members of well-defined isotypes. This allows the number of germ line genes to be compared with the number of known allotypes; and this should reveal to what extent additional diversity is generated during somatic differentiation. Data at present available have been taken to suggest that much, but not all, of the diversity is inherited. However, although these clearly demonstrate that there may be several V genes in a given isotype (which excludes extreme somatic mutation or recombination models), the resolution of the hybridization data is not adequate to support an unequivocal conclusion that the number of proteins is greater than the number of genes.

The mouse $V_\kappa 21$ subgroup has been investigated by Valbuena et al. (1978), who suggested that there are 2–3 C_κ genes and 4–6 $V_\kappa 21$ genes; in another subgroup there seemed to be 8–11 $V_\kappa 19$ genes. Since there are more than 6 members of the $V_\kappa 21$ subgroup, at first sight this result would imply that the additional members have been generated by somatic variation. However, problems with this saturation analysis make the gene number only approximate and leave open the possibility of an underestimate sufficient for there to be full coding capacity for the isotype in the germ line.* In another case, Seidman et al. (1978a) have shown that at least six EcoRI frag-

---

*In these experiments a cDNA prepared against mRNA of one $V_\kappa 21$ type appears able to hybridize virtually completely with the mRNAs synthesized by other members of the subgroup. It should be noted, however, that this is seen by 90% hybridization compared with the 60–70% reaction that occurs with V_κ mRNA of a different isotype, due to cross hybridization of the C_κ regions. When mouse embryo DNA was hybridized with an excess of full length $V_\kappa 21$ cDNA, no saturation could be detected, probably because of 5–10% contamination with other sequences. Compensation for this, effectively to subtract out the contaminating reaction, suggested the plateau values given in the text. The reliability of this procedure is reduced further by the use of short cDNA to represent C_κ genes and long DNA to represent $C_\kappa + V_\kappa$ so that another subtraction is involved to calculate the V_κ gene number. An indication, however, that the result may be of the correct magnitude is given by the use of a cloned sequence, but this represented $C_\kappa + V_\kappa$; separate fragments will give more reliable results.

ments from the DNA of the myeloma MOPC149 hybridize with a cloned probe representing the V_κ subgroup expressed in these cells. Two of these fragments have been sequenced; and comparison with the amino acid sequence of the expressed immunoglobulin suggests that they are indeed other members of the subgroup. In this case the number of genes is close to the present number of protein members of the subgroup, suggesting that most and possibly all of these V genes are coded in the genome.*

Although any attempt to define V gene number suffers from the difficulties involving reactions of related but not identical sequences that we have discussed above, the hybridization of cloned fragments representing particular C or V types with genome fragments obtained by Southern blotting (perhaps at varying criteria), should allow the number of V genes related to any mRNA to be determined. However, a caveat in comparing this with the number of protein members of the allotype is that these are determined by sequencing myeloma immunoglobulins. These may not represent the necessary random selection of antibodies. The V_κ chain sequences of myelomas derived from two different inbred strains of mice (Balb-c and NZB) form two distinct populations (Loh et al., 1979). Whether this is because different genes have been fixed in the two strains, because there are other genetic differences causing different genes to be expressed or generated, or because different lymphocyte subpopulations are transformed, is not known. At all events, this leaves open the possibility that the number of types may be greater than presently estimated. To summarize the status quo, it is clear that mouse DNA may contain several V_κ genes of a given type; these appear sufficient to code for most, although it is not yet clear whether all, of the presently known protein diversity; however, it is difficult to assess the accuracy of the estimate of protein diversity.

Rearrangement of V and C Genes

Although each light or heavy immunoglobulin chain comprises a covalently continuous polypeptide which is translated from a single messenger RNA, the V and C genes lie at distant sites in the embryonic genome. Two successive events are involved in bringing the V and C regions into juxtaposition in the messenger template. First a transposition in DNA occurs to bring a V gene into proximity with a C gene. This event may occur on a

---

*Note that the source of "genome" DNA was the myeloma. It is preferable to use gamete or early embryo DNA to represent the germ line, since there is always the possibility that somatic changes occur not only to allow the expression of one gene, but also in the number of nonexpressed genes. With the EcoRI fragments no difference has been reported between the myeloma and embryonic patterns; possibly this may mean that any rearrangement occurs close to the EcoRI sites (see next section). With BamHI a rearrangement is reported, but this does not affect the total number of genes. It should however be noted that the relevant bands are faint and hard to discern in published data (Seidman and Leder, 1978; Seidman et al., 1978b).

single chromosome. This forms a transcription unit that possesses intervening sequences that then must be removed from the precursor during production of mRNA.

Evidence for somatic rearrangement of genome sequences coding for immunoglobulins has been provided by comparing the restriction maps of embryo and myeloma DNA. Early experiments used mRNA probes to detect the corresponding restriction fragments. Hozumi and Tonegawa (1976) used a full length κ mRNA to identify V and C fragments, while a 3′ mRNA fragment was used to identify C regions. With embryo DNA, two BamHI fragments of 9 and 6 kb hybridize with the intact mRNA, whereas only the 9 kb hybridizes with the 3′ half. This suggests that in the germ line the 9 kb fragment contains the C gene and the 6 kb fragment carries the V gene. The possibility that their separation results from the fortuitous occurrence of a BamHI site between adjacent V and C regions is excluded by the observation that they continue to lie on different restriction fragments when other enzymes are used. The sizes of the separate fragments suggest that the V and C genes must be at least a considerable distance apart. But in myeloma DNA, both RNA probes hybridize with a single 3.5 kb BamHI fragment. This must have resulted from a transposition bringing the sequences together. An alternative is that changes have occurred just in the BamHI sites; but first this would require a more complicated change than just a single mutation, and second it is again excluded by the reproducibility of such results with other enzymes. In this case the germ line pattern could not be detected in the myeloma; which means that, if the myeloma genome remains diploid, both alleles must have been rearranged. Similar rearrangements have been detected for other kappa chains, in one case for one very closely related to the first, where one pattern can be seen in embryonic DNA (or in one experiment in kidney DNA which was taken as a paradigm for the germ line inheritance), while another is generated in a myeloma; in these cases the myeloma retained the germ line pattern as well as gaining the new pattern, suggesting that one of two alleles may have been rearranged (Tonegawa et al., 1978b; Rabbitts and Forster, 1978). Cloned DNAs representing several kappa sequences have been tested in such experiments; these identify restriction fragments of embryo DNA that do not carry C genes, but again in one myeloma DNA a different fragment is found that also carries a C gene (Lenhard-Schuller et al., 1978). When probes are used that represent different intact kappa chains, the fragments detected appear to identify different V genes but the same C gene (Tonegawa et al., 1978b; Seidman et al., 1978b).

The isolation and cloning of immunoglobulin coding regions shows in more detail how somatic rearrangement occurs. Three EcoRI fragments have been identified in mouse embryo DNA which code for λ genes: a $V_{\lambda I}$ gene is carried on a 3.5 kb fragment, a nearly identical $V_{\lambda II}$ gene on a 4.8 kb fragment, and an 8.6 kb fragment carries $C_{\lambda I}$; the $C_{\lambda II}$ gene has not yet been identified. The cloned 4.8 kb fragment hybridizes with about half of the length of the λ Ig mRNA; this identifies the V_λ gene as a sequence of about 400 bp in the middle of the fragment (Tonegawa et al., 1977). The sequence of the fragment has been determined directly and shows that it carries a $V_{\lambda II}$ gene divided into two parts (Tonegawa et al., 1978a). A leader consists of (presumably) some untranslated sequences followed by the sequence coding for the first 15 amino acids of the light chain precursor (−19 to −4)*; then there is an intervening sequence of 93 bp which is followed by the region coding for amino acids −4 to +98. The next 1250 base pairs is unrelated to immunoglobulin sequences.

The same organization is found in the $V_{\lambda I}$ gene, which has been sequenced both in the form of an embryonic gene (as a cloned 3.5 kb fragment) and as part of a myeloma gene, where it is present on a 7.4 kb fragment that also carries the $C_{\lambda I}$ gene. The 7.4 kb fragment is not present in a kappa myeloma and therefore represents a specific rearrangement connected with the expression of the λ_I light chain (Brack et al., 1978). The $C_{\lambda I}$ gene also is found as an immunoglobulin coding sequence divided by an intervening sequence. In the embryo DNA, the 8.6 kb EcoRI fragment carries a short "J" sequence coding for amino acids 98–110, separated by an intervening sequence of 1250 bp from the $C_{\lambda I}$ gene which codes for the sequence from amino acid 110 to the end of the gene. The J sequence represents the terminal part of the V region; this is one of the framework sequences. Figure 26.22 illustrates the rearrangement that must take place in myeloma DNA, where Bernard, Hozumi, and Tonegawa (1978) have shown by direct sequencing that the V gene is transposed to a position contiguous with the J segment. This leaves the gene in three fragments, coding for amino acids −19 to −4, −4 to +110, and +110 to the C terminus. The first (93 bp) intervening sequence separates most of the signal sequence from the variable region, which is now intact since the major V segment is contiguous with the J segment. The second intervening sequence (1250 bp) is derived from the region that separates the J segment and C gene in the embryo DNA. This explains the prior observation of electron microscopy that a 1250 bp sequence separates the V and C genes in a cloned myeloma fragment (Brack and Tonegawa, 1977). The V region has the same sequence in the embryo and myeloma DNA; it appears to be a member of the λ_o class.

In the mouse kappa chain system, the J region provides an interesting source of diversity. By sequencing the appropriate region, Sakano et al. (1979b) and Max, Seidman, and Leder (1979) have shown that it contains five J segments, each separated from the next by a distance of 300–350 bp. Four of these J segments correspond to the amino acid sequences present in various kappa light chains; the other (J3) is identified as a J segment by

---

*Immunoglobulin chains are synthesized initially as precursors carrying an N-terminal signal sequence as described in Chapter 23. This is removed during secretion and the amino acids are given negative numbers proceeding from the cleavage point to the N terminus. The first acid of the mature polypeptide identifies position +1.

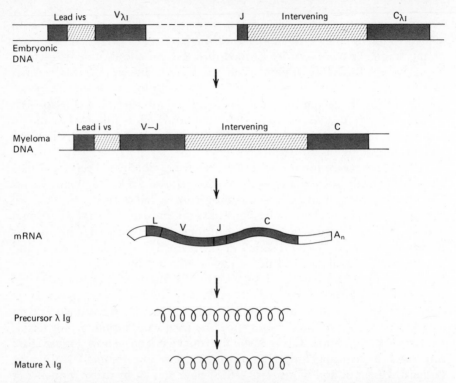

Figure 26.22. Expression of mouse λ_I immunoglobulin genes. Embryonic DNA contains two different regions. One carries an 81 bp leader sequence consisting of 36 bp corresponding to the nontranslated 5′ terminus and the 45 bp corresponding to amino acids −19 to −4 of the protein, separated by an intervening sequence (ivs) of 93 bp from the V segment of 282 bp which codes for amino acids −4 to +97. The second region has a 39 bp region coding for the J segment of amino acids 98 to 110 separated by an intervening sequence of 1250 bp from the constant region coding for amino acids 110 to the C terminus. In myeloma DNA the end of the V segment is joined to the start of the J segment, making a continuous V region coding for amino acids −4 to +110. Both intervening sequences are excised during the production of mRNA; nontranslated sequences remain at both the 5′ and 3′ termini. The messenger is translated into a λ_I light chain precursor, starting at amino acid −19 (methionine). Then the first 19 amino acids are cleaved from the polypeptide to generate the mature immunoglobulin chain. Based on sequence data of Bernard, Hozumi, and Tonegawa (1978).

its homology with the four expressed J sequences, but is not itself represented in any known immunoglobulin chain. However, there are 10 amino acid sequences known for the J region. But each of the additional protein sequences that is not coded by one of the identified J segments differs only in the first amino acid. This could be explained by shifting the position of recombination between V and J to alter the sequence of the first codon in the J region. This would be a source of somatic diversity.

Each J segment has two features whose recognition is part of the process of generating immunoglobulin mRNA. Within the 40 bp on the left side,

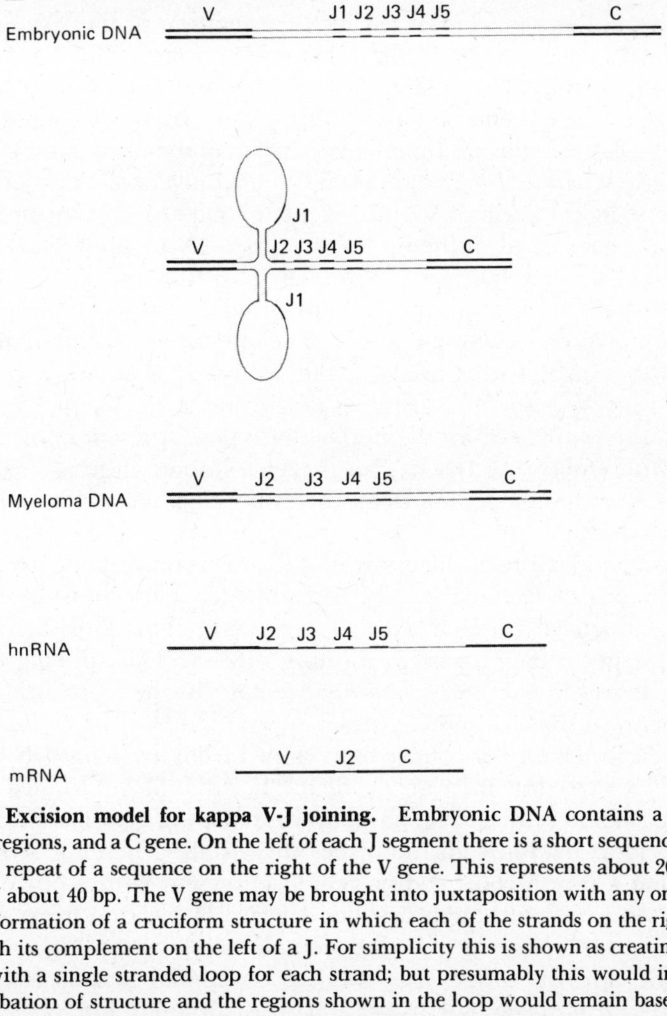

Figure 26.23. Excision model for kappa V-J joining. Embryonic DNA contains a V gene, cluster of 5 J regions, and a C gene. On the left of each J segment there is a short sequence which is an inverted repeat of a sequence on the right of the V gene. This represents about 20 bp out of a region of about 40 bp. The V gene may be brought into juxtaposition with any one of the J regions by formation of a cruciform structure in which each of the strands on the right of V base pairs with its complement on the left of a J. For simplicity this is shown as creating a base paired stem with a single stranded loop for each strand; but presumably this would in fact be a local perturbation of structure and the regions shown in the loop would remain base paired. Cutting and rejoining in the region at the base of the stem would create a continuous V-J gene and excise the region between V and J. Recombination is shown for V and J2. Any J regions on the left of the recombining J region are deleted as part of the loop. All J regions to the right of the recombining J remain in the myeloma DNA and are transcribed into hnRNA. They must be removed by splicing between the right end of the V-J gene and the left end of the C gene region to create Ig mRNA.

there are about 20 bp that represent an inverted repeat of a sequence present on the right side of the V gene. This could represent a signal for V-J recombination as illustrated in the model of Figure 26.23. If recombination takes place by cleavage and rejoining at the base of the duplex stem formed by the pairing of the inverted complements, the sequence between V and J should be deleted. This prediction is difficult to test directly, because if only one of two alleles is reorganized in the formation of the V-J gene, the sequence will be present on the other chromosome. However, using two probes that contain either the right end of V and the downstream flanking sequence or the left end of J and its upstream flanking sequence, Sakano et al. (1979) showed that no fragments other than the recombined V-J and the embryonic separate V genes are found in a λI myeloma. This demonstrates that the reorganization of V and J is not accompanied by the incorporation of the sequence between them into any other DNA region, since this would be detected by new fragments hybridizing with the probe (although these would be hard to see on the gel reproductions). The most likely explanation is that the region between V and J is deleted when recombination occurs. One prediction that this model makes is that if V genes are organized in clusters, any V genes lying between the recombining V and J genes will be deleted, this would reduce the number of V genes present in myeloma compared with embryonic DNA. This makes it important to use germ line (or at least nonimmunogenic) DNA for estimating total gene numbers (see footnote above).

The second feature of the multiple J segments of the kappa system is that the right side of each except J3 has a sequence that conforms with the left splicing junctions summarized in Figure 26.11. This must be used to join the V-J gene to the C gene in forming mRNA. This splicing event must delete the additional J segments that remain in the reorganized myeloma DNA between the V-J and C genes.

Two heavy chain genes have been cloned from myeloma DNA and characterized by electron microscopy of the heteroduplexes formed with H Ig mRNA. Sakano et al. (1979a) have shown that a $C_{\gamma 1}$ gene has a mosaic structure. R loop mapping distinguishes three coding regions; sequence data show that a fourth short coding region is present within one of the intervening sequences, although it is not detected by electron microscopy. Each coding region represents a distinct domain; as shown in Figure 26.24 these represent C_{H1}, the hinge, C_{H2} and C_{H3}. A similar arrangement has been detected in a C_{α} gene by Early et al. (1979); although not analyzed in such detail, the dimensions are such that each coding region may correspond to a domain. In this case an intervening sequence of 6.8 kb separates the C_{α} gene from the adjacent V_H gene. This form of C gene organization explains the observation of mutant immunoglobulins in which a single domain has been deleted from the constant region.

This analysis of both light and heavy chains reveals a striking correlation between coding regions and the protein domains summarized in Figure 26.21. In myeloma DNA, the light chain coding regions represent the three functional segments of the polypeptide: the signal, variable region, and constant region. (The discrepancy between the location of the intervening sequence at −4 and the cutting of the polypeptide at −1/+1 might be explained if the position of cleavage were offset from the position of recognition.) The correlation is the stronger for comparing the organiza-

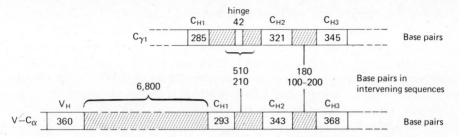

Figure 26.24. Organization of C$_H$ genes. Electron microscopy shows that three regions homologous to mRNA in the constant region are separated by two intervening sequences. In the α gene an intervening sequence of 6.8 kb separates the V$_H$ region from the start of C$_H$. Possible J segments and the hinge coding region are too short to be detected by electron microscopy. Sequence studies of the C$_\gamma$ gene show directly that each coding region corresponds to the amino acid sequences thought to constitute each domain; this also identifies the hinge sequence as a short coding region. Probably the same relationship between coding regions and domains applies in the α gene. Data of Sakano et al. (1979) and Early et al. (1979).

tion of C$_L$ and C$_H$ genes: the light constant region is a single domain, whose coding region is uninterrupted in DNA; the heavy constant region contains several domains, each of which is represented by a different coding region. The variable region of the heavy chain gene has yet to be analyzed in detail. Tonegawa et al. (1978a) have suggested the terminology "intron" to identify the intervening sequence and "exon" to identify the sequences represented in mRNA. These terms have been applied also to other interrupted genes. Whether the functional relationship evident between coding regions and protein domains is a feature peculiar to the immunoglobulins—whose genetic organization after all is distinct from other known genes—or reflects the general mode of evolution of proteins remains to be seen (see above).

It seems very probable, but has not yet been proven directly, that the intervening sequences in the rearranged myeloma immunoglobulin genes are spliced out of a precursor nuclear RNA to generate mRNA. Large precursors have been found for both heavy and light chains, of about 11 kb and 5.4 kb, respectively, in one myeloma, but their functional organization has not yet been determined (Schibler, Marcu, and Perry, 1978). In another myeloma a cloned kappa gene has been used to select nuclear molecules; a pulse label passes from a 40S precursor through a 24S intermediate into the mature 13S mRNA (Gilmore-Hebert and Wall, 1978). The nature of the processing events again remains to be defined. With this myeloma, Rabbitts (1978) has shown that when cDNA of 700 nucleotides representing both V + C regions is hybridized to a 27S nuclear precursor, only a region of 295 bases is protected against digestion with S1 nuclease. This is presumed to be the C region, which thus does not lie adjacent to the V region in the precursor. Since this is the same pattern obtained when the cDNA is hybridized with myeloma DNA, whereas the full 700 bases of cDNA may be protected upon hybridization with mRNA, the inference has been drawn that the precursor retains the interrupted genome organization. However, this leaves the critical question of why no second fragment corresponding to the V region was isolated in the reaction with nuclear precursor or DNA, as would be expected from this procedure (which is related to the Berk-Sharp mapping technique). A firm conclusion on the structure of the precursor seems premature until this question is resolved.

The organization of V and C genes in embryo DNA and their reconstruction in the myeloma allows many old questions about immunoglobulin gene structure and expression to be reformulated in more specific terms, susceptible to molecular answer. We have seen that the fundamental issue of the basis of V gene diversity has yet to be definitively resolved; but already it is clear that there may be multiple V genes of a given light chain type, each capable of linkage in the myeloma to the same constant gene. This immediately implies that the choice between alternative recombination pathways is one level at which a specific immunoglobulin gene is constructed for expression. This may extend to the construction of V genes from V segments and J segments; and of course to the selection of different C_H genes. Reconstruction is specific for the gene under expression; other immunoglobulin genes continue to display the embryonic form of organization. Whether any control of immunoglobulin expression occurs at the level of splicing is not known; but it is possible that there may be clusters of V or C genes, so that reconstruction generates a transcriptional unit containing more than a single V and C gene. Such models have been invoked to explain cases in which one cell lineage successively produces different immunoglobulins; another possible model is the occurrence of sequential rearrangements.

Allelic exclusion is another problem. This describes the ability of cells apparently to express only one of the two alleles for any immunoglobulin gene. One possibility is that this may result from rearrangement of the genes on only one chromosome; this is consistent with the structure of DNA in some myelomas, that both generate a new pattern and retain the embryonic pattern. However, others have been reported in which only the new pattern is found. This could represent the loss of the other alleles, since myelomas do not remain diploid, or could represent the imposition of homozygosity by some means; at all events it is not obvious how the same rearrangement might be coordinated in the homologues.

The nature of the process responsible for the somatic recombination remains entirely unknown; it is scarcely necessary to point out that the joining of the V and J segments into the intact V region is accomplished with some precision. In this context it is interesting that neither does the junction leave an intervening sequence between the juxtaposed coding segments, but nor is the occasion taken to remove the intervening sequences that exist in the V gene, between V and C, or within C. What the function of the intervening sequences may be is unclear, as it is with other cellular genes. Whether the recombination event involves a transposition of one gene from its embryonic location to another, or a simple recombination event that deletes the sequences between the genes, or takes some other form, is not certain.

Mobile Elements in the Genome

With the exception of the directed rearrangement of the immunoglobulin genes, our discussion of genome organization has been conducted in terms of an invariant structure. In cases in which the restriction maps of non-repetitive genes have been examined, this is borne out by the data, which show the same fragments to be present in all tissues investigated. But two striking cases involving the transposition of repetitive sequences have been reported in S. cerevisiae and in D. melanogaster. The generality of such occurrences is not yet clear.

In yeast there is a family of repetitive sequences described as *Ty1*, present in about 35 copies of length 5.6 kb. The repeats are dispersed, although there are some tandem clusters of two or more. The various copies of *Ty1* are related in sequence but not identical; each has a repetition of the same short element, a sequence of about 250 bp described as δ, at each end. There are actually about 100 copies of the δ sequence in the genome, so that about 70 are associated with *Ty1* sequences and about 30 are located elsewhere. The function of *Ty1* is not known; it was isolated as a cloned EcoRI fragment of the genome which hybridizes not just with one corresponding fragment in genomic digests, but with about 35 bands.

The *Ty1* sequences do not appear to be fixed in location. When different isolates of S. cerevisiae are compared, there is variation in the pattern of genomic EcoRI fragments that hybridize with the cloned probe. The extent of the variation is hard to quantitate, but contrasts with the lack of change in the patterns of nonrepetitive genes. Cameron, Loh, and Davis (1979) provided further evidence for changes in *Ty1* at a specific location, by identifying two strains that differ in the presence/absence of *Ty1* in the *sup4* region. Also, when two strains were cultured for a month, each showed one change in the *Ty1* band pattern. The mechanism of transposition is unknown, but the direct repetition of δ sequences at the ends of *Ty1* resembles the structure of some bacterial transposons.

In D. melanogaster, several families of dispersed genes have been identified. These are defined as genes by their representation in poly(A)$^+$ RNA; indeed, they were originally isolated as cloned genome sequences able to hybridize with the RNA. Three such families share the feature that a direct repetition of a short sequence is present at each end of the unit. The length of *copia* is about 5 kb, with direct repeats of about 300 bp; the *412* sequence is about 7.3 kb long with direct repeats of 500 bp; and the sequence of *297* is less well defined. Potter et al. (1979) found that the total number of each element shows variation; in the Oregon R strain, hybridization kinetics suggest the presence of 60 copies of *copia*, 40 copies of *412*, and 30 copies of *297*. These are close to the number of sites identified by in situ hybridization, suggesting that the elements occur singly. In two cultured cell lines, however, there are 90 and 170 copies of *copia*, 120 and 40 copies of *412*, and 110 and 170 copies of *297*. So the number of copies of each can vary widely and independently.

Does this change reflect the transposition of these elements to new sites? Restriction mapping suggests that the increase in the number of copies of *copia* in cultured cells represents the presence of additional copies at new sites (not a tandem duplication at previously occupied sites). A comparison of the location of the 40 *412* elements present in Oregon and a cultured cell suggest that some are at different sites. Again there is an instance in which a specific site can be identified in two different D. melanogaster strains, in one occupied by *copia* and in the other lacking it. In this case, the sequence into which *copia* is inserted appears to have weak homology with the element. Comparisons by restriction mapping and in situ hybridization suggest that all three sequences have different distributions in each of four different strains of D. melanogaster (Strobel et al., 1979). This suggests that these genes change in location, although the rate of transposition is hard to estimate. Another sequence that may be of this type, *225*, has been identified by Tchurikov et al. (1978).

Modification of DNA

The presence of 5 methyl cytosine in eucaryotic genomes has been known for some time. The modified bases occur largely or entirely in CpG doublets; their function is mysterious, and inevitably this has encouraged speculations that it might be concerned with the regulation of gene expression. The procaryotic counterpart lies in the existence of host restriction and modification enzymes, which may either methylate a target sequence or, if it is not methylated, cleave it. Since enzymes may differ in their target sites, in effect this provides a system for distinguishing the source of a DNA as native to a given bacterium or foreign (see Volume 1 of Gene Expression).

Some of the features of the eucaryotic methylation system have been described for the ribosomal RNA genes of X. laevis by Bird and Southern (1978). In these genes about 13% of the C residues are methylated in chromosomal DNA; but modification does not occur in the amplified oocyte nuclear ribosomal genes. This allows the restriction digest patterns of methylated and unmethylated DNA to be compared directly. Enzymes whose target sites include CpG may either be unable to cleave the site when the C is methylated; or they may be unimpeded by the modification. In the first case a difference is seen in the restriction digests of methylated and unmethylated DNA; the methylation reduces the number of cleavage sites, or, if complete, abolishes restriction entirely.

With several such enzymes that together recognize a total of about 27% of the estimated CpG doublets in the region, cleavage is suppressed by the methylation. The lack of a series of bands from methylated DNA shows that the CpG sites almost all are heavily methylated; approximately 10–20 sites would seem to be unmethylated per 1000. There is therefore a 98–99% probability that a given site will be methylated. With two exceptions, the pattern of nonmethylated sites is random. The exceptions are seen by the formation of specific bands upon restriction, which implies that the same site is not methylated in all, or at least many, of the DNA molecules. One of these sites appears to be unique; the other may comprise several non-methylated targets in the same region.

Since CpG is a self-complementary sequence, in duplex DNA every doublet has the structure $^{CpG}_{GpC}$. Restriction enzymes that are impeded by methylation probably are unable to cleave a target site when the C residue on either strand is unmethylated (this has not been proven directly for every case, but is true for those examined). Thus the methylated sites in the ribosomal genes might contain 5-methyl-cytosine on one, the other, or both strands. These possibilities can be distinguished by denaturing the methylated DNA and renaturing it with a large excess of X. laevis ribosomal DNA that has been cloned in E. coli, so that methyl groups are absent. Every methyl group in the native DNA strand then should face a nonmethylated C residue in the cloned strand. If the methylated sites in the duplex DNA all are modified on both strands, every target site should continue to have a methyl group in the native/cloned hybrid duplex. Then there should be no change in the restriction pattern. If some or all of the sites in native DNA were methylated only on one strand, however, these now should be found in the native/cloned hybrid half as completely unmethylated and half as methylated on one strand. This should increase the susceptibility to restriction enzyme. The lack of such an effect therefore implies that the methylated sites are modified on both strands of DNA (Bird, 1978). A further test of this is to denature and then renature the native DNA; since the unmethylated sites occur at random (apart from the two prominent exceptions), the background of cleavage should be much reduced, since the probability of any site remaining unmethylated is reduced to $1 \times 1\% = 0.01\%$. This also is found.

These results suggest a model in which a given site is methylated on both strands of DNA. At replication the strands separate; and synthesis of the daughter strands generates two duplex molecules each of which is methylated only on the parental strand. This could be the recognition signal for a methylase to add a methyl group on the daughter strand. This is supported by the demonstration that methyl groups are added only to the newly synthesized strand at S phase. If methylases are able only to complete the methylation of the half methylated sites created by replication, this provides a system in which the pattern of methylation is perpetuated through cell division. This raises the interesting question of how changes in the pattern are introduced, as for example must be necessary to explain the tissue specific methylations of the rabbit β globin gene found by Waalwijk and Flavell (1978) (see above).

The pattern of methylation in the entire sea urchin genome has been investigated by separating the DNA of Echinus esculentus into fractions that are degraded by or survive restriction with enzymes impeded by methylation of their target sites (Bird, Taggart, and Smith, 1979). The resistant fraction corresponds to about 40% of the genome and cross hybridizes at a reduced rate with the susceptible fraction. This suggests the existence of compartments of methylated and nonmethylated DNA; and these appear to be grossly similar in all tissues examined.

CHAPTER 27

Ribosomal Gene Clusters

Constituting some 80 or 90% of the total amount of cellular RNA, the components of the ribosome comprise the predominant products of transcription. To accomplish this high output, the genes coding for rRNA are repeated in many identical (or nearly identical) copies in the haploid genome. But although a higher eucaryotic genome may contain more than 200 copies of each gene (so that there are 400 in the diploid nucleus), the ribosomal DNA still accounts for less than 1% of the haploid complement.* Intense transcription of these genes is necessary to produce sufficient ribosomal RNA; and this is the function of the nucleolus.

The genes for the major rRNA species of higher eucaryotes are organized as clusters of tandem repeats that are transcribed by the nucleolar RNA polymerase I (see Chapter 22). Each repeating unit is transcribed into a single precursor RNA molecule that is then processed to give the mature rRNA species, 18S (present in the 40S ribosome subunit) and 5.8S and 28S (present in the 60S ribosome subunit). The 5S rRNA of the large ribosome subunit is transcribed independently by RNA polymerase III, which also transcribes other small RNAs. This is true whether the 5S genes are organized as a separate cluster as is usually the case, or interspersed with the major rRNA genes as is found in some lower eucaryotes. All four rRNA species are found in all eucaryotes, contrasted with the three species of 16S, 23S, and 5S rRNA that are synthesized by expression of a common operon in procaryotes.

The high concentration of active RNA polymerases along each major transcription unit produces the characteristic appearance of a Christmas tree, with transcripts increasing continuously in length from the newly initiated to the just completed. Methylation of the regions of the primary transcript that are destined to be conserved as one of the three mature ribosomal RNAs distinguishes them from the nonribosomal regions, which are degraded. These *transcribed spacers* represent as little as 20% of the primary transcript of lower eucaryotes, but as much as 50% in mammals. We should stress that the transcribed spacers simply separate the various regions of

---

*Ribosomal DNA (rDNA) is used here exclusively to describe the genes coding for ribosomal RNA and the associated nontranscribed regions. In eucaryotes nothing is known about the copy number or organization of the genes coding for the (roughly) 70 ribosomal proteins and protein synthesis factors. In bacteria these genes are not repeated to match the repetition of rRNA genes. Of course, the expression of genes coding for protein can be amplified via the level of messenger RNA, whereas the products of the rRNA and tRNA genes must be used directly.

the primary transcript that are to be conserved, and which may be released by endonucleolytic cuts. Usually each such region is a continuous sequence corresponding to the mature ribosomal RNA. In some cases an intervening sequence is present in the gene coding for the large ribosomal RNA as summarized in Table 26.5; and if this copy of the gene is transcribed, the intervening sequence must be removed by splicing, so that ligation as well as cutting is necessary. The presence of an intervening sequence naturally reduces the proportion of the primary transcript that is conserved.

In higher eucaryotes there may be one or more clusters of the major transcription unit, but each cluster takes the form of a linear succession of repeating units. The repeating unit of DNA is longer than the transcription unit; and when the tandem array is visualized in transcription by electron microscopy, each Christmas tree is separated from the next by a *nontranscribed spacer*. The transcription units all are the same length (except for any diversity that may result from the presence of intervening sequences). In some organisms the repeating units are homogeneous in length, but in others they vary widely. This results from heterogeneity solely in the length of the nontranscribed spacers. Such spacers are present also in the clusters of 5S genes, where they may show the same property of varying in length while the transcription units remain fixed.

The tandem clusters of major transcription units are present on the chromosomes that are associated with the nucleolus. In special situations, the best known example of which is presented by the amphibian oocyte, the gene copy number is increased by the *amplification* of some repeating units to generate large amounts of extrachromosomal rDNA. In certain lower eucaryotes, notably the Protists, the micronucleus may have integrated genes for ribosomal RNA, but the macronucleus may have a larger number of genes in the form of many individual extrachromosomal molecules of DNA. The extreme case of such an arrangement appears to be that of the slime molds, where all the ribosomal RNA genes may be extrachromosomal, again organized as a large number of molecules of characteristic size.

Synthesis of Ribosomal RNA

There has been evidence for many years to suggest that synthesis of ribosomal RNA and assembly of the ribosome takes place within the nucleolus (for review see Perry, 1967). Indirect evidence was that rapid protein synthesis—which demands the production of many ribosomes—is correlated with an increase in the size of the nucleolus; the composition of nucleolar RNA resembles that of ribosomal RNA and the nucleolus contains particles that resemble ribosomes (Birnstiel et al., 1966; Vaughan et al., 1967). Cytological studies suggest that ribosomal RNA is first transcribed at the fibrillar core of the nucleolus; and then in the form of precursor particles it moves to form the granular cortex (Das et al., 1970; Bachellerie et al., 1977).

The primary transcript of the rRNA genes is the predominant species identified by a radioactive label in ribonucleotides. Table 27.1 compares the size of this initial precursor with the mature ribosomal RNAs. The available data suggest that the transcripts fall into roughly two classes. Lower eucaryotes, amphibia and marine organisms have a transcription unit of 7000–8000 base pairs, only about 25% larger than the procaryotic transcription unit. The mature ribosomal RNAs are of the order of 4000 and 2000 nucleotides long, so that the proportion of the precursor which is conserved

Table 27.1   Size of the primary transcript and mature ribosomal RNAs

Species	Primary transcript		Ribosomal RNA length		Proportion conserved
	S value	Length (bases)	Large (26–28S)	Small (18S)	
E. coli		6000	3000	1541	80%
S. cerevisiae	37S	7200	3750	2000	80%
D. discoideum		7350	4075	1800	80%
P. polycephalum	40S	8625	3700	1950	65%
T. pyriformis	35S	6900	3725	1975	83%
D. melanogaster	34S	7750	4100	2000	78%
S. gairdneri		7750	4450	1875	81%
X. laevis	40S	7875	4475	1925	79%
I. iguana		7875	4350	1775	78%
N. tabaccum		7900	3700	1900	71%
L. variegatus		7500	3200	1800	67%
G. domesticus	45S	11250	4626	1800	57%
P. tridactylis	45S	11750	4750	1900	56%
M. musculus	45S	13400	5000	1950	52%
H. sapiens	45S	12900	4850	1740	52%

The S values given for the longest precursors observed (assumed to be primary transcripts) are "nominal" and may vary with actual conditions. Length estimates are based on electron microscopic and gel electrophoretic data where available; estimates based on sedimentation data are less reliable. For practical purposes variations of ~200 bases in the quoted values should not be regarded as significant. The proportion conserved gives the sum of the lengths of mature rRNAs divided by the length of the primary transcript.

References: E. coli—Brosius et al. (1978), Morgan et al. (1977, 1978), Young, Macklis, and Steitz (1979); Saccharomyces—Uden and Warner (1972), Klootwijk, De Jonge, and Planta (1979); Dictyostelium—Batts-Young et al. (1977), Batts-Young and Lodish (1978), Frankel et al. (1977); Physarum—Jacobson and Holt (1973); Tetrahymena—Niles (1978); Drosophila—Perry et al. (1970), Levis and Penman (1979); Salmo—Perry et al. (1970); Xenopus—Loening, Jones, and Birnstiel (1969), Wellauer and Dawid (1974); Iguana—Perry et al. (1970); Nicotiana—Perry et al. (1970); Lytechinus—Wilson et al. (1978); Gallus—Perry et al. (1970); Potorous—Perry et al. (1970); Mus (L cell)—Perry et al. (1970), Wellauer et al. (1974a); Homo (HeLa cell)—Loening, Jones, and Birnstiel (1969), Wellauer and Dawid (1973).

approaches the procaryotic value of 80%. In avia and mammals, however, the transcription unit is much larger, of the order of 12,000–13,000 base pairs. There is a slight increase in the length only of the larger rRNA, towards 5000 bases, and so the overall proportion of the primary transcript that is conserved falls to little more than 50%.

The outline of the processing pathway was made clear by following the pattern of methylation of rRNA precursors in HeLa cells. Methylation takes place entirely on the nucleolar 45S precursor, except for one secondary methylation event that later generates two adjacent dimethyl adenines in 18S rRNA (Greenberg and Penman, 1966; Zimmerman and Holler, 1967; Zimmerman, 1968). Labeled methionine first enters the 45S precursor, then passes through a 32S nucleolar RNA, and finally enters cytoplasmic 28S and 18S rRNA. By using gel electrophoresis to separate the nucleolar RNAs, Weinberg and Penman (1970) were able to refine this scheme to include further intermediates. Figure 27.1 shows the resolution of these species and summarizes their properties. As the 45S precursor is cleaved to successively smaller species, there is an increase in the ratio of the ^{3}H label in methyl groups to a ^{32}P label in nucleotides; this implies that the regions lacking methyl groups are lost. The total of methyl groups in the mature 18S and 28S rRNAs equals that of the 45S precursor, which suggests that conservation of methylated sequences may be part of the processing mechanism. The base composition of the 45S and 32S precursors is much richer in G-C nucleotides than is mature rRNA, so that the discarded sequences appear to be very rich in G-C bases.

Two processing schemes were considered when the large precursors were first identified. There might be a single precursor molecule, containing the sequences of both 18S and 28S ribosomal RNAs as well as sequences that are discarded. Or there might be two precursor species of about the same size, one carrying the 18S and the other the 28S sequence. The first experiments to indicate that there is only a single precursor showed that ribonuclease treatment of purified 45S RNA releases the oligonucleotides characteristic of 18S and 28S rRNA as well as additional fragments corresponding to the nonribosomal regions (Jeanteur, Amaldi, and Attardi, 1968; Birnboim and Coakley, 1971). Using labeled methyl groups to identify the oligonucleotides, Maden, Salim, and Summers (1972) showed that the 45S pattern is virtually identical to the sum of the 18S and 28S patterns. This indicates further that virtually none of the discarded regions is methylated. The pattern of the 41S RNA is indistinguishable from that of the 45S RNA, which shows that this conversion must be accomplished by cleavage of one end of the primary transcript. The 32S RNA has the same pattern as 28S rRNA, whereas the 20S RNA has the same pattern as 18S RNA. This suggests that the 41S RNA is cleaved to give the 32S and 20S RNAs, which are the immediate precursors to the mature ribosomal RNAs.

The processing pathway has been followed in more detail by electron microscopic visualization of the precursors. Wellauer and Dawid (1973) prepared secondary structure maps of the precursors and mature ribosomal

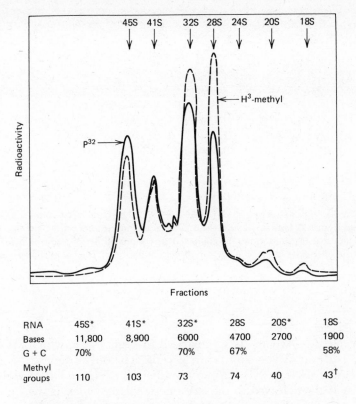

RNA	45S*	41S*	32S*	28S	20S*	18S
Bases	11,800	8,900	6000	4700	2700	1900
G + C	70%		70%	67%		58%
Methyl groups	110	103	73	74	40	43[†]

*indicates nucleolar RNA.

[†]number includes 4 additional groups added in cytoplasm.

Figure 27.1. Precursors to ribosomal RNAs in HeLa cells. The upper curve shows an electrophoretic separation of precursor and rRNAs, identifying molecules by a ^{32}P label in the nucleotides and an ^{3}H label in methyl groups. The proportion of methyl groups: unit length (i.e. $^3H/^{32}P$) increases as the precursors are processed to lose additional nonmethylated sequences. The lower table shows the approximate length, base composition, and methyl group content of the RNA molecules. The total number of methyl groups added in the nucleolus is conserved (110 in 45S RNA, 113 in 28S + 18S rRNAs). Data of Weinberg and Penman (1970), with additional information from Amaldi and Attardi (1968), Jeanteur and Attardi (1969), Maden and Salim (1974), Willems et al. (1968).

RNAs by spreading the molecules to allow their duplex hairpin regions to be distinguished from the single stranded regions. An example is shown in Figure 27.2. The patterns of both the 18S and 28S rRNAs are present in both the 45S and 41S molecules. The only difference between the 45S and 41S species is the lack of one end, which is found as a 24S RNA. The 18S pattern lies near one end of the 45S and 41S RNAs; it is cleaved as a 20S RNA, containing additional terminal sequences that are removed as the last stage in maturation. The 28S pattern is found at the other end of the 45S and 41S RNAs and in the 32S RNA. A 36S molecule may identify a minor alternative processing pathway or an aberration, since it appears to be the sequen-

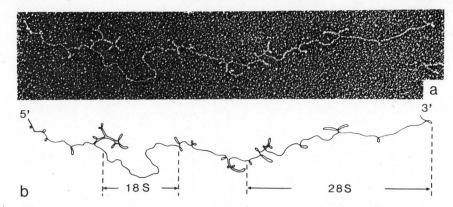

Figure 27.2. Secondary structure of HeLa 45S RNA. The electron micrograph shows a molecule spread from 80% formamide and 4M urea. The tracing illustrates the hairpins that form the secondary structure and shows the characteristic pattern of the 28S segment and of the 18S segment (the only extended region actually devoid of hairpins). Data of Wellauer and Dawid (1973).

quences remaining when the end of the 45S molecule containing the 28S sequence is removed.

The processing scheme is summarized in Figure 27.3, which also shows that the pathway is very similar in L cells. The only difference is that the 41S precursor is cleaved directly to give the 18S rRNA and a 36S RNA. The 36S RNA is then cleaved to the 32S intermediate, which is the immediate precursor to 28S rRNA. The sole difference between the HeLa and L cell pathways is therefore a reversal in the order of cleavage at two sites (Wellauer et al., 1974a).

The maturation of X. laevis rRNA from its somewhat smaller primary transcript has been followed by the same techniques (Wellauer and Dawid, 1974). The 40S primary transcript contains an 18S sequence close to one end and a 28S sequence at the other end. First one end of the molecule is cleaved to generate a 38S molecule; then internal cleavage generates 18S rRNA and 34S RNA. The 34S RNA is cleaved at one end to release the 30S RNA that is the immediate precursor to 28S rRNA. The same sequence of events occurs in D. melanogaster, although the molecules are slightly smaller (Levis and Penman, 1978b).

In early experiments the ends of the precursor molecules were distinguished by following the action of an ascites exonuclease thought to be specific for attack on the 3′ end. This identified the 28S sequence as comprising the 5′ end of the molecule. This assignment has since been shown to be in error. The impetus for more detailed analysis of the structure was provided by observations that the synthesis of 18S rRNA is more sensitive to inhibition by ultraviolet irradiation than is 28S rRNA synthesis (Hackett and Sauerbier, 1975). This implies that the 18S sequence lies nearer to the promoter of the common transcription unit (see Chapter 25). Actually this con-

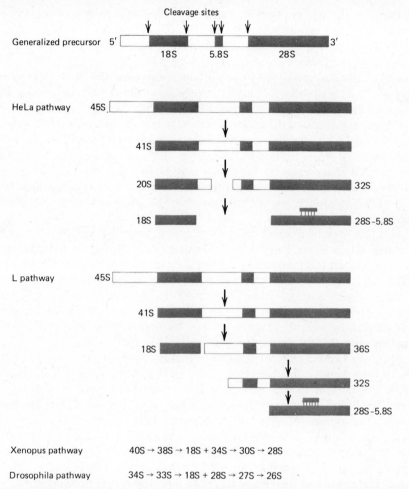

Figure 27.3. Processing pathways for ribosomal RNA. The typical eucaryotic precursor has a transcribed spacer at the 5′ end, followed by the 18S rRNA sequence; a 5.8S sequence lies within the internal transcribed spacer; and the 28S sequence lies at the 3′ end of the molecule. The HeLa and L cell initial precursors are the same size but there is a difference in the order of cleavage at the two sites on the 3′ side of the 18S gene. The pathways in Xenopus and Drosophila are the same as in L cells, except that the overall sizes of the precursors are smaller, as indicated.

clusion was anticipated in the analogous observation that premature termination of nucleolar transcription by cordycepin has the same effect; partially completed 45S molecules in HeLa cells can give rise to 18S but not to 28S rRNA (Siev, Weinberg, and Penman, 1969). Further evidence for this order was provided by the observation that when transcription proceeds in vitro, the methyl groups present on partially completed chains first are found on 18S sequences (Liau and Hurlbert, 1975b). This conclusion was confirmed by direct structural analysis by examining the relationship of the 45S precursor to EcoRI fragments of X. laevis DNA representing defined parts of the 18S and 28S genes (Dawid and Wellauer, 1976; Reeder et al., 1976a; Schibler et al., 1976).

The general arrangement of sequences is therefore the same in both procaryotes and eucaryotes: a 5′ leader sequence is followed by the 18S sequence, which is succeeded by a transcribed spacer, with the 28S sequence lying at or close to the 3′ end of the molecule. In bacterial rRNA operons, tRNA genes may lie in the transcribed spacer between the 18S and 28S rRNA sequences; and the 5S rRNA is transcribed as part of the same unit, distal to the large (23S) rRNA. Further genes for tRNA may be found beyond the 5S gene. In eucaryotes, the only other sequence present in the transcription unit is the 5.8S rRNA, which is part of the transcribed spacer separating the 18S and 28S genes.

The presence of a small rRNA associated with the 28S rRNA was first revealed when Pene, Knight, and Darnell (1968) found that denaturation releases a small fragment from the 28S species. This is the 5.8S rRNA, about 160 bases long, which is noncovalently bound to 28S rRNA. Its sequence is highly conserved (much more so than 5S rRNA), with only 5–6 differences among HeLa, Xenopus, trout, and turtle; the yeast 5.8S rRNA has about 75% homology with these species.* The secondary structure is not yet fully resolved, but there appears to be extensive base pairing.

Synthesis of 5.8S rRNA is linked to that of the larger rRNAs. The oligonucleotides produced when HeLa 5.8S rRNA is digested with T1 ribonuclease are present in 32S precursor RNA, whether or not it is denatured. They are found in 28S rRNA that has not been denatured, but are absent after denaturation (Maden and Robertson, 1974; Nazar et al., 1975). The single methylated nucleotide present in 5.8S rRNA is found in both the 45S and 32S precursors. A more exact location for the 5.8S sequence in the transcribed spacer has been determined by restriction mapping studies (see below). The timing of its release may vary with the species; in vertebrates presumably this is accomplished when 32S RNA is cleaved to generate 28S rRNA. In D. melanogaster, it may occur later, since a cytoplasmic 26S RNA can be obtained in which the 5.8S sequence (which is smaller than vertebrate 5.8S RNA, about 130 nucleotides) is still covalently linked to the 26S rRNA sequence (Jordan, Jourdan, and Jacq, 1976). No small precursor to 5.8S rRNA has been recognized, except in yeast where there is a 7S RNA that contains an additional 150 nucleotides at the 3′ end (De Jonge et al., 1978).

The timing of the association between 5.8S and 28S rRNA is not yet known. The preferred model is that the 32S precursor acquires the appropriate conformation, in which the two sequences are base paired and a loop of nonribosomal sequences between them is removed. An alternative is that the

---

*The 5.8S rRNA was originally described as 7S rRNA. Its sequence has been determined for several species by Rubin (1973, 1974), Nazar and Roy (1976, 1978), Nazar, Sitz, and Busch (1976), Ford and Mathieson (1978), Selker and Yanofsky (1979).

cleavage events are made separately to release the two molecules, which then associate. The complex between 5.8S rRNA and 28S rRNA can be reconstituted under appropriate conditions in vitro. By annealing ^{32}P labeled 5.8S rRNA with unlabeled 28S rRNA, it is possible to search for a sequence in the 5.8S rRNA that is protected from ribonuclease in the complex but not in the free molecule. This identifies a possible reactive sequence at the 3' end of the 5.8S sequence (Pace, Walker, and Schroeder, 1977). This sequence is complementary to the 5' end of the 5.8S rRNA; and in models for the secondary structure of the free RNA, the 3' and 5' ends have been base paired. These regions appear to be able to interact between molecules to form multimers; perhaps this may indicate that actually they are free in the individual molecule (Sitz, Kuo, and Nazar, 1978).

Maturation of rRNA Sequences

The conservation of methyl groups during rRNA processing suggests that this may be the mechanism by which conserved sequences are distinguished from those that are degraded. The process of maturation appears to take place in ribonucleoprotein particles that are precursors to the mature ribosomal subunits; possibly the modification and processing enzymes may be part of the particles.

Methylation of ribosomal RNA appears to be equimolar: all modified sites probably are methylated in all molecules, with no or few partially methylated rRNAs. In HeLa cells all except 5 of the total 110 methyl groups are present on ribose moieties and most lie in different oligonucleotide sequences (Maden and Salim, 1974). Methylation of the ribose groups follows rapidly upon transcription as seen in vitro (Liau and Hurlbert, 1975a). Base methylations occur more slowly. Roughly the same number of methyl groups seems to be present in all vertebrate rRNAs; in yeast and in D. melanogaster the total is only about 60, but the same addition to the primary transcript and subsequent conservation appears to be the rule (Klootwijk and Planta, 1973; Levis and Penman, 1978b).

Some methylated trinucleotides are common to the large rRNA of all eucaryotes so far examined (Khan and Maden, 1976). Whether this is coincidence or represents common signals for processing or function remains to be seen. One highly methylated sequence ($Gm_2^6Am_2^6$) occurs in a highly conserved sequence near the 3' end of 18S rRNA (Alberty, Raba, and Gross, 1978). This would be a candidate for some role in ribosome function. When rRNA is degraded with S1 nuclease, 50% of the methylated nucleotides are released rapidly; but the remaining half are protected (Khan and Maden, 1978). One suggestion is that this might indicate a change in their conformation since methylation occurred (presumably to accessible sequences), but the use of techniques other than S1 susceptibility would be needed to confirm this.

Early experiments suggested that methylation might be essential for processing, since starvation for methionine prevents ribosomal RNA synthesis (Vaughan et al., 1967). But there are complications in this interpretation, since the effects of amino acid starvation per se may be significant. Cycloleucine has been used specifically to inhibit methylation of 45S RNA in CHO cells; its effect on rRNA synthesis seems to be to reduce the rate of transcription and processing, but transfer of unmethylated rRNA to the cytoplasm continues (Caboche and Bachallerie, 1977). Unless unlabeled methyl groups have escaped into the rRNA during the period of inhibition, this would argue against a rigorous link between methylation and processing.

The processing of ribosomal RNA could be accomplished by a discrete number of endonucleolytic cuts as indicated in Figure 27.3. However, formally only two cuts are necessary, to release the 5.8S rRNA with separation of the 18S and 28S sequences; the remaining reduction in size could be accomplished by exonucleolytic action (although this would have to be very rapid to explain the lack of molecules of intermediate size). Some indirect evidence for end trimming is provided by the heterogeneity of the 5' termini of the 45S, 32S, and 28S RNAs of rat hepatoma; by contrast the 18S and 5.8S rRNAs have homogeneous termini (Kominami and Muramatsu, 1977; Kominami et al., 1978). The extent of the trimming is not known. In contrast with the single (processed) phosphate group found at the 5' terminus of mammalian 45S RNA, the initial precursors of Drosophila, Dictyostelium, Tetrahymena and yeast all bear triphosphate termini, and the terminus of X. laevis 40S RNA also has more than a single phosphate, identifying them as untrimmed primary transcripts.

Precursor particles that contain rRNA sequences and some of the ribosomal proteins can be identified in the nucleolus. The proteins of the ribosome must be synthesized on cytoplasmic polysomes, after which they enter the nucleolus to associate with rRNA. HeLa cells contain a limited pool of ribosomal proteins, since inhibition of protein synthesis with cycloheximide does not immediately prevent ribosome assembly (Warner at al., 1966b).

The inhibition of protein synthesis, either suddenly by addition of cycloheximide or gradually by incubation in hypertonic medium, reduces the flow of rRNA sequences from 45S RNA (Pederson and Kumar, 1971). The half life of the 45S molecule is increased and it accumulates in ribonucleoprotein particles of about 80S. Under normal conditions, a label passes through the 80S particles into 55S particles. Continued transcription may be necessary for the late stages of maturation of the 55S particles (Kumar and Wu, 1973).

The processing of rRNA seems to take place entirely in these particles; the precursor RNAs are not found in free form. The 80S particles contain 45S RNA, 5S RNA, and some of the proteins that are found on both the 60S and 40S ribosomal subunits (Yoshikawa-Fukada, 1967; Kumar and Warner, 1972; Shepherd and Maden, 1972; Prestayko et al., 1974). This suggests that some of the proteins of each mature subunit associate with the appropriate regions of the 45S precursor before it is cleaved into precursors to the individual ribosomal RNAs. Nonribosomal proteins also are found on the 80S particle; these seem to be restricted to the nucleolus and may be maturation proteins involved in the assembly of successive ribosomes.

The only product of the 80S particle that has been identified is the 55S particle, which appears to be a direct precursor of the 60S subunit. It contains 32S RNA, 5S rRNA, and most of the proteins present on the 60S subunit. Again some nonribosomal proteins also are present. No intermediate precursor for the small ribosomal subunit is known; either its existence is brief or a 40S subunit is produced directly from the 80S precursor particle.

Virtually nothing is known of the enzymes responsible for accomplishing rRNA processing. None of several candidates for this activity yet has been shown to be involved in vivo. Endoribonucleases have been found in precursor ribonucleoprotein particles or in isolated nucleoli (Prestayko et al., 1972; Winicov and Perry, 1974). An exoribonuclease active on 3′ ends and inhibited by methyl groups has been found in mouse nuclei (Perry and Kelley, 1972). An enzyme activity localized in the chick and mouse nucleolus is active on double-stranded RNA (Grummt, Hall, and Crouch, 1979). By analogy with the role of RNAase III, which acts on duplex regions and is involved in processing rRNA in E. coli, this has been suggested as a processing enzyme. This activity is present on the appropriate precursor particles.

Repetition of Ribosomal Genes

The genes coding for ribosomal RNA are redundant even in bacteria and in eucaryotes there are typically 200 or more copies of each transcription unit (18S-5.8S-28S) per haploid genome. Table 27.2 summarizes the number of repeat units based on the proportion of nuclear DNA that is saturated by hybridization with 18S and 28S ribosomal RNAs. In higher eucaryotes these genes are clustered at a few chromosomal locations which can be recognized by their ability to organize a nucleolus. A characteristic number of nucleoli can be recognized as dark staining regions of the nucleus of the interphase cell; and the chromosomal regions responsible for their organization are seen as secondary constrictions at metaphase (Heitz, 1931; McClintock, 1934).

In mammals several chromosomes may contain nucleolar organizers; in humans and chimpanzee the number is five and in mouse it is six. A recent technique to visualize nucleoli uses staining with ammoniacal silver, which presumably reacts specifically with nucleolar proteins. The stained regions correspond exactly with the sites that hybridize with ribosomal RNA in situ.* The nucleolar organizers are not all the same size and may show differences in different strains of the organism. On average each must have

References: McConkey and Hopkins (1964), Huberman and Attardi (1967), Evans, Pardue, and Buckland (1974), Henderson et al. (1974), Goodpasture and Bloom (1975), Hsu, Spirito, and Pardue (1975), Tantravahi et al. (1976), Dev et al. (1977).

Table 27.2   Number of rRNA and tRNA genes in eucaryotic genomes

Species	Number of 18S/28S genes	Major rDNA/ total DNA, %	Number of 5S genes	Number of tRNA genes
E. coli	7	1.0	7	60
S. cerevisiae	140	5.5	140	250
D. discoideum	180	17	180	not known
P. polycephalum	150	1.0	not known	not known
T. pyriformis				
micronucleus	1	0.04	325	800
macronucleus	200	0.49	325	800
L. donovani	160	2.0	not known	not known
D. melanogaster (X)	250	1.3	165	860
(Y)	150			
C. tentans	40	0.3	not known	not known
X. laevis	450	0.18	24,000	1150
HeLa	280	0.4	2,000	1300

The number of major transcription units per haploid genome is calculated from the level of hybridization obtained when DNA is saturated with an excess of 18S + 28S rRNA. The proportion of this rDNA per genome is calculated to include transcribed and nontranscribed spacers as well as the sequences coding the mature rRNAs, i.e., it is the total proportion of the genome concerned with production of 18S, 5.8S, and 28S rRNAs. These values are averages that do not take account of polymorphism in the number of units per individual genome. The numbers of 5S and tRNA genes also are calculated from saturation hybridization data; the tRNA gene numbers tend to be underestimates.

References: E. coli—Nomura, Morgan and Jaskunas (1977); Saccharomyces—Schweizer et al. (1969), Rubin and Sulston (1973); Dictyostelium—Firtel et al. (1976); Physarum—Newlon, Sonenshein, and Holt (1973), Hall, Turnock and Cox (1975); Tetrahymena—Engberg and Perlman (1972), Yao, Kimmel and Gorovsky (1974), Kimmel and Gorovsky (1976), Yao and Gall (1977); Leishmania donovani (hemo-flagellate)—Leon, Fouts and Manning (1978); Drosophila—Tartof and Perry (1970), Tartof (1973), Spear and Gall (1973), Spear (1974); Chironomus—Lambert et al. (1973), Hollenberg (1976); Xenopus—Dawid, Brown, and Reeder (1970), Brown and Weber (1968a), Brown, Wensink, and Jordan (1971); HeLa—Hatlen and Attardi (1971); values for various amphibia have been given by Vlad (1977).

about 40 transcription units. An interesting question is whether and how sequence identity is maintained between the separate clusters.

The values shown for the number of genes in Table 27.2 may represent only averages, for in some cases there is evidence for polymorphism in the size of the gene clusters. Variations in the size of the secondary constriction representing a given nucleolar organizer have been known for some time, but of course such observations left open the question of whether the cause lies with variations in the number of genes or in their activity. This was resolved when Miller and Brown (1969) showed that about 40% of the toads, Bufo marinus, collected from natural populations, showed nucleoli and secondary constrictions that varied in size. The variation is correlated with the proportion of the genome that hybridizes with labeled rRNA. Similar later studies using in situ hybridization with salamanders suggest the same conclusion (MacGregor, Vlad, and Barnett, 1977). In P. cinereus one population showed a 7.5-fold variation in the amount of rDNA per individual animal.

More precise investigations are possible in species in which all the rRNA genes (excluding 5S rRNA) are present in a single cluster. This is the situation in both X. laevis and D. melanogaster, in which mutants with deletions of some or all of the genes are available. The *anucleolate* mutant of X. laevis is a recessive lethal that lacks a nucleolus altogether; heterozygotes (nu^+/nu^-) have only one nucleolus compared with the two of normal diploid animals. The mutation prevents synthesis of 18S and 28S rRNA; and DNA from homozygous mutants anneals poorly to rRNA, while DNA from heterozygotes shows a level about half that of the wild type (Brown and Gurdon, 1964; Wallace and Birnstiel, 1966). This suggests that all the rRNA genes are carried in a single cluster on one homologue. Mutants that have partial deletions of this region and as a result form smaller nucleoli have since been isolated; their properties suggest that the minimum number of rRNA genes necessary for viability is between 25 and 50% of the wild-type level (Knowland and Miller, 1970; Miller and Knowland, 1970).

A similar effect is seen with *bobbed* mutants of D. melanogaster, which have partial deletions located in the X chromosome. Different mutants remove different proportions of the number of rRNA genes. By using a range of mutant flies, Ritossa and Spiegelman (1965) showed that the amount of DNA able to hybridize with rRNA is proportional to the number of doses of the nucleolar region. As with Xenopus, this confirms directly the earlier inferences that nucleoli form at the chromosomal sites carrying the genes for rRNA. The properties of the Drosophila mutants suggest the same conclusion supported by the results with Xenopus: the number of rRNA genes in the wild type is greater than needed for viability and the mutant phenotype is displayed when the level falls below about one third of wild type (Ritossa, Atwood, and Spiegelman, 1966). It turns out further that different strains of D. melanogaster possess different numbers of rRNA genes, so the exact size of the cluster cannot be critical (Ritossa and Scala, 1969; Spear and Gall, 1973).* In any case there is significant variation between females and males, since the nucleolar organizer on the X chromo-

---

*The total number of genes detected by hybridization may be greater than the active number, since typically about 35% of the D. melanogaster rRNA genes possess an intervening sequence (see Table 26.5 and below). If these genes are not expressed, their presence may be irrelevant to considering the number necessary for survival. In this context, it is interesting that the amount of DNA hybridizing with rRNA in the six species of the melanogaster subgroup of Drosophila varies from 0.14 to 0.43%, corresponding to from 80 to 250 rRNA genes per X chromosome (Tartof, 1979). The proportion of restriction fragments presumed to correspond to interrupted gene copies declines with reduction in total gene number, so the number of active (uninterrupted) genes may show less variation than the apparent total number.

some carries (approximately) 250 rRNA genes, but the organizer on the Y chromosome carries somewhat less, about 150 (Spear, 1974). The relationship between the two clusters is interesting in view of the lack of genetic recombination in male meiosis in D. melanogaster.

In yeast at least 90% of the rDNA is carried on a single chromosome, which now appears to be chromosome XII as seen by the linkage of rDNA to markers and the correlation of its total amount with chromosome doses (Petes, 1979). Two types of rDNA gene cluster can be distinguished by their restriction maps, one having six sites for cleavage by EcoRI and the other having an additional site. Diploids may have either or both types of unit; when both are present, their amounts are about equal. Haploids have only one type of unit, so the variation is due to a polymorphism in the type of cluster rather than within the cluster (Petes and Botstein, 1977; Petes, Hereford, and Skryabin, 1978). This means that the difference can be used to follow the inheritance of the clusters from heterozygotes. In crosses of the two types, usually there is 2:2 segregation of the two parental clusters as expected of a single Mendelian factor (formally this demonstrates that all the rRNA genes lie on one chromosome). But since the rDNA comprises about 5% of the genome, the level of recombination (70 crossovers per meiosis) should correspond to some 4 exchanges within the cluster. The lack of any meiotic exchange implies that recombination is suppressed within this region, relative to other regions. Unusual segregants that have both rDNA patterns occur in 2:2 reciprocal recombinant ratios that correspond to crossing over at the mitosis preceding the meiosis.

Organization of Transcription Units

Two techniques have been used to analyze the arrangement of ribosomal transcription units. The gene clusters can be visualized directly by electron microscopy of ribosomal DNA, either engaged in transcription as isolated from the nucleolus, or following hybridization of rRNA with denatured single strands of rDNA. This approach requires long stretches of rDNA and therefore is applied comfortably to situations in which the rDNA may be separated readily from other nuclear DNA sequences. The techniques for restriction mapping and cloning that we have discussed in Chapter 26 may be applied to rDNA with some facility, since the redundancy of the genes increases their proportion of the genome to a point at which gene isolation becomes practical, either by reliance on gross differences such as buoyant density of rDNA, or by restriction cleavage and blotting.

Electron micrographs of ribosomal DNA in the throes of transcription suggest that the sequences coding for precursor RNA alternate with inactive spacer sequences. In developing the visualization technique, Miller and Beatty (1969a,b,c) made use of the extrachromosomal DNA in amphibian oocytes that is engaged in intensive production of rRNA (see below). This is present as a large number of nucleoli, taking the form of compact fibrous cores (containing DNA, RNA, and protein) surrounded by the granular cortex (containing only RNA and protein). In solution of low ionic strength, the core and cortex can be separated and the core DNA dispersed for electron microscopy.

When the cores are unwound, each consists of a thin axial fiber some 10-30 nm in diameter, periodically coated with a matrix as revealed in Figure 27.4. The axial fiber of each core forms a circle; treatment with DNAase breaks the axis and treatment with trypsin reduces its diameter to about 3 nm. The axis thus consists of DNA and the length covered by one matrix segment is 2-3 μm, which should correspond to about 7200 bp, close to the size of the amphibian precursor RNA measured directly (see Table 27.1). Each matrix appears as a "Christmas tree" of about a hundred thin fibrils connected at one end to the core axis, and increasing in length from the "thin" to the "thick" end of the unit. The fibrils are the RNA precursors and they reach a maximum length (at the "thick" end) of 0.5 μm, which corresponds to roughly a twelvefold reduction from the length expected of a free RNA molecule. This presumably is due to the association of the protein that can be detected on the fibrils by appropriate staining. Each fibril terminates upon the axial core in a spherical granule of about 12.5 nm diameter, which is presumed to be RNA polymerase. This means that about a third of the total length of the gene is covered in enzyme molecules, a much greater concentration than is seen with nonribosomal genes (see Chapter 14).

Each matrix segment represents a transcription unit engaged in the synthesis of the 40S precursor and is separated from the next by an inactive stretch of axial core. The nontranscribed spacers on average are about two thirds of the length of the transcription unit. The presence of nontranscribed spacers can be inferred also from the denaturation mapping of Xenopus rDNA. When DNA is partially denatured, single strand loops are generated in A-T rich regions, forming a characteristic repeating pattern. Successive repeats lie head to tail, that is, all in the same orientation (Wensink and Brown, 1971; Henikoff, Heywood, and Meselson, 1974). The average length of the repeating unit is 5.4 μm, which is about 13,000 bp, substantially longer than the transcription unit that resides within it. The arrangement of transcription units is shown in more detail by comparing the secondary structure map of single-stranded rDNA with that of the 40S precursor (see Figure 27.2). This shows that transcription units alternate with nontranscribed spacers, making up a repeating unit of 11,700 bp of which only 7850 bp are transcribed (Wellauer and Dawid, 1974). A similar conclusion is suggested by hybridizing 18S and 28S rRNA to denatured Xenopus rDNA; short (18S) and long (28S) hybrid regions alternate, with the two types of hybrid separated by a distance corresponding to the transcribed spacer, and each pair of genes separated from the next by a greater

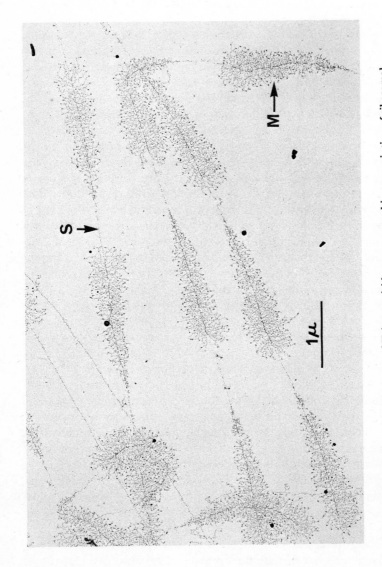

Figure 27.4. Nucleolar genes of oocytes of Triturus viridescens engaged in transcription of ribosomal precursor RNA. Each axis consists of deoxyribonucleoprotein periodically coated with a matrix (M) which consists of ribonucleoprotein fibrils 5–10 nm in diameter and up to 0.5 μm long, attached to the axis at spherical granules of 12.5 nm diameter. The length of the matrix is 2–3 μm. Each matrix is separated from the next by an inactive length of spacer (S). Photograph kindly provided by O. L. Miller, Jr.

distance that corresponds to the nontranscribed spacer (Forsheit, Davidson, and Brown, 1974).

The first work to characterize biochemically the arrangement of genes coding for rRNA took advantage of the greater G-C content of the rDNA of Xenopus laevis (67%) compared with the genome average (40%). After fragmenting the chromosomal DNA, an rDNA fraction can be isolated directly as a satellite of distinct buoyant density (Brown and Weber, 1968a,b). The satellite is absent from DNA of the anucleolate mutant. By fragmenting the rDNA to smaller sizes, it is possible to test whether the 18S and 28S genes are interspersed (both present on molecules of about the size of the genes) or separate (both present on molecules large enough to contain two blocks of genes). Since the linkage between the genes is lost only when DNA is fractionated to a size less than that of the genes, these results suggested an alternation of the two types of gene sequence. The work was succeeded by similar experiments to show that the genes for 5.8S rRNA are linked to the 18S–28S gene pairs (Spiers and Birnstiel, 1974).

Detailed restriction maps now have been made for the ribosomal DNA of many eucaryotes. The existence of tandem head to tail arrays is shown by the formal circularity of the map (this occurs because the fragments connecting repeating units greatly outnumber terminal fragments that connect the array to adjacent nonribosomal DNA. In fact, the ends have not been identified for any such gene cluster). This shows that each transcription unit consists of sequences for the 18S, 5.8S, and 28S rRNAs, separated by the transcribed spacer regions; usually there is also an external transcribed spacer, which precedes the start of the 18S sequence. The sequences represented in rRNA usually are continuous; but in some cases intervening sequences are present in some or all of the 28S genes. Table 26.5 describes the interruptions that have been characterized, in D. melanogaster, Tetrahymena and Physarum. In D. melanogaster there is another quirk in the map, which is the existence of a gap in the 28S gene; about 140 bp of DNA does not hybridize with the large rRNA. This appears to correspond to a hidden break in the polynucleotide chain, which is revealed by denaturation (Jordan, Jourdan, and Jacq, 1976). Probably some material is removed from the rRNA at the site of the break, explaining its inability to hybridize with the DNA.

Available data are summarized in Table 27.3, which compares the length of the repeating unit with the transcription unit (as inferred from the size of the largest rRNA precursor). In all cases that have been studied, a nontranscribed spacer separates adjacent transcription units; there appears usually to be variation in the length of the spacer. In mammals the transcription unit is the longest known, about 13,000 bp, but the nontranscribed spacer is 2–3 times this length, making the total repeating unit longer than 40,000 bp. As with all rDNAs, the restriction fragments corresponding to the transcription unit appear invariant, suggesting that there is little if any variation between the multiple copies. Heterogeneity is seen in the sizes

Table 27.3 Organization of ribosomal RNA genes

Species	Length of repeat	Length of transcript	Nontranscribed spacer	Arrangement and location
S. cerevisiae	8950	7200	1750	tandem repeats on chromosome XII
N. crassa	9100	not known	not known	not known
D. discoideum	42000	7350	34250	extrachromosomal palindromic dimers
P. polycephalum	28000	8625	19375	extrachromosomal palindromic dimers
T. pyriformis micronucleus	—	—	not known	single chromosomal gene
macronucleus	9550	6900	1650	extrachromosomal palindromic dimers
P. tetraurelia macronucleus	7875–8350	not known	not known	extrachromosomal tandem repeats
L. donovani	13,500	not known	not known	chromosomal tandem repeats
D. melanogaster	11500–14200	7750	3750–6450	tandem repeats on chromosomes X and Y
C. tentans	8400	not known	not known	chromosomal tandem repeats
L. variegatus	12100	7500	4600	chromosomal tandem repeats
X. laevis	10500–13500	7875	2300–5300	tandem repeats on single chromosome
M. musculus	44,000	13400	30,000	tandem repeats on six chromosomes
H. sapiens	43,650	12,900	30,750	tandem repeats on five chromosomes

The length of the repeating unit is based on restriction mapping in most cases, sometimes also using data from denaturation mapping and electron microscopy of rDNA hybridized with rRNA. Transcript lengths are based on analysis of the largest precursor (see Table 27.1) or its hybrid with rDNA. The nontranscribed spacer is the distance between the 3′ end of one transcript and the 5′ end of the next. Where variations in length have been characterized, the range from the smallest to largest repeating units is given. For palindromic dimers the repeat length is half the total size of the dimer. Extrachromosomal locations are indicated by the isolation of multiple molecules of constant size, except in the case of Paramecium where the macronucleus contains entirely extrachromosomal DNA amplified from the micronuclear chromosomes. The number of chromosomes carrying rDNA repeats is indicated by genetic analysis or by the number of nucleolar organizers, as described in the text (references in the footnote).

References: Saccharomyces—Bell et al. (1977), Nath and Bollon (1977), Kramer, Phillipsen and Davis (1978), Skryabin et al. (1978); Neurospora—Free, Rice, and Metzenberg (1978); Dictyostelium—Cockburn, Newkirk, and Firtel (1976), Cockburn, Taylor, and Firtel (1978), Maizels (1976); Physarum—Molgaard, Matthews, and Bradbury (1976), Vogt and Braun (1976b); Tetra-

of the fragments corresponding to the nontranscribed spacer; the relative amounts of each of four recognized size classes vary among humans (Arnheim and Southern, 1977; Krystal and Arnheim, 1978). When hybrid cells are formed, heterogeneity is not reduced with the reduction of number of human nucleolar organizers, which suggests that each cluster of ribosomal genes may show heterogeneity in spacer lengths.

The best characterized gene clusters are those of Xenopus and Drosophila, in which repeating units are of the order of 11,000 to 14,000 bp, compared with the 7800 bp transcription unit. The size variation is due entirely to the existence of nontranscribed spacers of different lengths. The values in the table are given for repeating units that lack intervening sequences within the 28S rRNA gene; the presence and variation in length of these interruptions further increases the heterogeneity of the repeating units of Drosophila. The tandem repeats of the sea urchins also are about 12,000 bp; in the lower eucaryotes such as fungi and ciliates, the length usually is about 7000 bp. We should emphasize that restriction mapping per se simply demonstrates that the rDNA is organized as tandem repeats, and identifies the source of any heterogeneity, but does not reveal the total length of the individual cluster or prove that it resides on a chromosome. The chromosomal location is proven by the identification of nucleolar organizers, whose number per haploid genome gives the best indication of the number of separate clusters of rDNA.

Some interesting variations on the usual organization of rDNA genes have been found in lower eucaryotes. In several cases the ribosomal genes are arranged as palindromic dimers. The rDNA is found as extrachromosomal molecules, identified as such because they have a constant size. Denaturation and restriction mapping shows that each consists of an inverted repeat; and this can be seen more dramatically by the formation of hairpins when the molecules are denatured and intrastrand renaturation occurs. Gentler conditions of denaturation allow cruciforms to be produced. The two rRNA transcription units may lie at various positions relative to the inverted repeat, as illustrated in Figure 27.5.

Palindromic dimers were first characterized in the polyploid macronucleus of Tetrahymena. About 100 molecules, each containing two rDNA transcription units, are derived from the single rDNA transcription unit that is integrated in the chromosomal DNA of the micronucleus. How the gene is duplicated, the copies placed in reverse orientation, and amplified in the macronucleus, is not yet known. A short repeating sequence,

hymena—Engberg et al. (1976), Karrer and Gall (1976), Yao and Gall (1977); Paramecium—Findly and Gall (1978); Leishmania—Leon et al. (1978); Drosophila—Glover and Hogness (1977), Pellegrini, Manning, and Davidson (1977), Wellauer and Dawid (1977); C. tentans—Degelmann et al. (1979); Lytechinus—Wilson et al. (1976), Blin et al. (1978); Xenopus—Wellauer and Dawid (1974), Boseley et al. (1978); Mus—Cory and Adams (1977); Human—Wellauer and Dawid (1979).

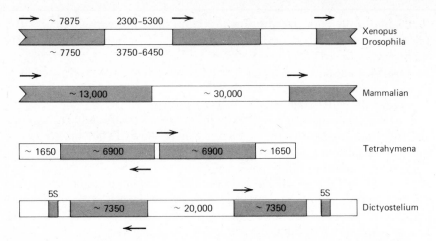

Figure 27.5. Organization of rDNA repeating units. Transcription units are indicated by shaded regions, nontranscribed spacers are indicated in outline. Arrows show the direction of transcription. Tandem chromosomal repeats, such as Xenopus, Drosophila, mammalian, are arranged head to tail; extrachromosomal palindromes, of definite size, are transcribed in reverse orientation. For details see Table 27.3.

$\left[\begin{smallmatrix} CCCCAA \\ GGGGTT \end{smallmatrix}\right]_n$ has been found close to the ends of the molecule (Blackburn and Gall, 1978). Within this repeating unit there are short discontinuities involving single-strand breaks, whose origin and function are unknown. The hexanucleotide repeats occupy about 400 base pairs at each end of the palindrome; variation in the exact number of repeats in different extrachromosomal molecules could explain why the terminal restriction fragments are not precisely uniform in length. There has been some difficulty in identifying the 5′ and 3′ termini, which leaves open the possibility that a hairpin is formed (perhaps involving the repeating sequences) or that a protein is attached. In other cases, macronuclear amplification results in the production of a tandem array of genes, such as in Paramecium.

In the slime molds Dictyostelium and Physarum, rDNA takes the form of extrachromosomal palindromic dimers that are inherited as such. Replication of the Physarum rDNA is not directly coordinated with the duplication of chromosomal DNA at S phase, since it starts in the later part of S but continues through G2. Although the total number of rDNA molecules is maintained from division to division, any individual molecule has the same probability of being selected for replication at a given time, irrespective of the time since its last replication (Vogt and Braun, 1977). This is reminiscent of the control of mitochondrial DNA replication (see Chapter 21) and of multicopy plasmid replication (see Volume 3 of Gene Expression). Replication is bidirectional from the central portion of the molecule toward the ends, but may involve more than one origin. Similar bidirectional replication, from a central origin, occurs with the macronuclear rDNA of Tetrahymena (Truett and Gall, 1977).

Structure of Nontranscribed Spacers

Extensive studies of the structure of rDNA repeats have been carried out on the chromosomal and amplified sequences of X. laevis. We have seen that there are about 450 repeats located on a single chromosome, which are amplified some 1000–4000 fold as circles of extrachromosomal rDNA in the oocyte. When the rDNA is cleaved with EcoRI, two types of fragment are released (Wellauer et al., 1974b). The most prominent is about 4500 bp long and contains most of the 28S gene. The remainder of the repeating unit, comprising the 3′ terminal region of the 28S gene, the entire nontranscribed spacer, and most of the 18S gene, is found in the form of a series of larger bands, each present in lower molar amounts. These vary in size from about 6000–9000 bp. The different bands represent nontranscribed spacers of different sizes (see Figure 27.5).

The structure of the nontranscribed spacers has been investigated by cloning the larger EcoRI fragment class. Wellauer et al. (1976a) found that reassociation of fragments denatured from the same clone leads mostly to the formation of perfect homoduplexes. But some molecules are formed in which there are pairs of short single-strand loops. This can be explained if each nontranscribed spacer is internally repetitious, so that denatured single strands can reassociate in the wrong register; each strand then forms a single-strand loop of unpaired sequences. The variable positions of the loops indicate the extent of the internal repetition. (An analogous phenomenon has been investigated from a different perspective in satellite DNA; see Chapter 19).

Loops also are seen when heteroduplexes are formed by reassociating DNA from different nontranscribed spacer elements. Then single loops are found as well as the pairs resulting from alignment out of register. The single loops represent variations in length between the nontranscribed spacers in different repeats. They occur in the same regions that form the pairs of loops, and thus appear to represent differences in the number of internal repeats present rather than substitutions of sequence. From these experiments the nontranscribed spacer falls into four regions. Proceeding from the 3′ end of one transcription unit towards the 5′ end of the next unit: first there is a region A devoid of loops, which appears to be maintained between spacers; then there is a region B that is internally repetitious and variable in length; followed by another conserved region, C; and finally another variable repetitious region, D.

The arrangement of repeats in chromosomal and amplified rDNA of the same individual frog has been compared by Wellauer et al. (1976b). Both show an array of sizes of nontranscribed spacers, although the relative abundances of the size classes may be different. This suggests that amplification does not produce new rDNA size classes, but selects among those present in the chromosomal rDNA. There is a tendency for siblings to amplify the same size classes.

The pattern of repeats in long single strands of rDNA can be examined by hybridizing a given sequence—such as a particular cloned spacer or transcribed sequence—and examining the lengths of contiguous repeats. With a cloned spacer, adjacent repeats are seen to differ in size in chromosomal rDNA, but to be the same size in amplified rDNA. This suggests that amplification starts with a single repeat, which might be amplified (for example) by a rolling circle (see below). Whether a single oocyte has one or more such extrachromosomal molecules is not known, since individual frogs have thousands of oocytes. Thus the overall heterogeneity in size classes in amplified rDNA could be due to the selection of different repeats for amplification in different oocytes or could represent the presence of different amplified repeats in each oocyte.

The occurrence of variation in repeat lengths between adjacent units excludes models that maintain sequence identity between the multiple copies by any form of sudden correction, such as the master-slave hypothesis (see Chapter 17). We should note, however, that there has been some confusion about the relationship between adjacent repeats. Early electron microscopy visualized arrays of repeating units all apparently of similar size. This conclusion has been suggested also by Buongiorno-Nardelli et al. (1977), who found that long strands of rDNA either renature perfectly or form displaced loops of constant size, which would be expected only if each strand contains a series of identical repeats. This contrasts with the results of Wellauer et al. (1976), showing identical lengths in only amplified and not chromosomal rDNA.

Comparison of the structure of rDNA from X. laevis and the related species X. borealis (previously known as X. mulleri), with which it forms an infertile hybrid, shows that their organization is the same head to tail array of tandem repeats. The 18S and 28S sequences are indistinguishable and cross hybridize perfectly between the species. The denaturation map shows similarity in the (presumed) transcription unit, but differences in the nontranscribed spacer (Brown, Wensink, and Jordan, 1972; Wellauer and Reeder, 1975). Two cuts are made by EcoRI in the same two locations in X. borealis as are cleaved in X. laevis; and the nontranscribed spacer varies in length in a similar way. This means that while selection is able to impose uniformity on the rRNA coding sequences in the two species, their nontranscribed spacers are different. How selective pressure is made effective on multiple gene copies remains to be established, but the general similarity of variation in nontranscribed spacer length in spite of sequence difference implies that this may be involved.

The structure of the nontranscribed spacer of X. laevis has been characterized in more detail by restriction mapping and direct sequencing of cloned segments. Restriction mapping of two cloned repeats shows its four regions take the structures: *A* is about 500 bp long and has no internally repetitive organization; *B* is internally repetitive and is cut by SmaI into many small (~100 bp) fragments; *C* lacks a repetitive structure; and *D* is cleaved by BamHI into large (660 bp) fragments and by HaeIII into small fragments (Botchan, Reeder, and Dawid, 1977).

This general division of the nontranscribed spacer into four regions is supported approximately by the sequences determined from a cloned repeat, although the lengths of the regions are different from what had been expected on the basis of electron microscopy and the repetitious pattern is a little more complicated. Figure 27.6 summarizes the results of Boseley et al. (1979). Region *B* consists of a repeat of 97 bp, rich in alternating blocks of C and G; this is defined as repetitious region *1*. Region *C* is a nonrepetitious sequence of 320 bp, in the middle of which is a BamHI site and which gives rise to its description as a "Bam island." Region *D* corresponds to a structure in which one repetitious region, *2*, is separated by a Bam island from another repetitious region, *3*. Repetitious regions *2* and *3* have the same internal repeating structure, consisting of alternating canons of 60 bp and 81 bp; these differ only in the absence/presence of a 21 bp segment. Occasionally one of these units is missing, so that two of the same length are adjacent. The two Bam islands that separate repetitious region *1* from *2* and then *2* from *3* are very similar, in large part identical. Without detailing an extensive analysis, it is clear that this structure could be derived by duplications, partial saltatory amplification, and deletions.

The initiation site of the transcription unit has been identified by locating the point corresponding to the 5′ end of the 40S precursor. This has been identified with the startpoint by Reeder et al. (1977), who applied a technique in which the complex of capping enzymes from vaccinia virus is used to distinguish 5′-tri- or diphosphate termini (which are substrates) from 5′-monophosphate termini (which it cannot use). The incorporation of a labeled cap thus is taken to indicate the presence of the first transcribed base (assuming the absence of any enzyme activity converting monophosphate to polyphosphate termini). Although only 25% of the X. laevis 40S RNA molecules are labeled, their 5′ terminal sequence is the same as that of bulk 40S RNA, which is therefore assumed to represent the primary transcript.

The use of terminally labeled 40S RNA then allows the corresponding restriction fragments to be isolated from the rDNA. The initiation site has

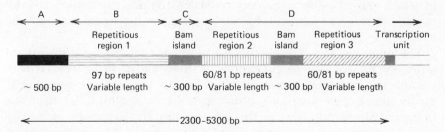

Figure 27.6. Nontranscribed spacer region of X. laevis rDNA. This organization is based on the sequence of a single cloned repeat and the earlier microscopic analysis of length variation. Region A and the Bam islands are constant in length; the lengths of the repetitious regions may vary according to the number of repeats present in the individual repeat. Data of Boseley et al. (1979).

```
-40                  -30                -20              -10            +1 ─────────►
  |                   |                  |                |             |
CTCCATGCTACGCTTTTTTGGCATGTGCGGGCAGGAAGGTAGGGGA
```

Figure 27.7. Sequence at the initiation site of X. laevis 40S RNA. Data of Sollner-Webb and Reeder (1979).

been located on the nucleotide sequence of the region by hybridizing the restriction fragment with rRNA, removing nonhybridized single-stranded material with S1, and analyzing the surviving DNA, whose 3' end corresponds to the 5' terminus of the RNA. The same result can be obtained by hybridizing a DNA fragment to a downstream segment of 40S RNA and reverse transcribing up to the 5' end of the RNA. These experiments identify the startpoint exactly; and this has been confirmed by sequence analysis directly of the first few nucleotides at the 5' end of the 40S RNA (Sollner-Webb and Reeder, 1979). Figure 27.7 shows the initiation sequence; it is evident that RNA polymerase I does not have as its recognition site any sequence corresponding to the regions that are conserved in the promoters for RNA polymerase II either at the (presumed) initiation site or at the Hogness box 30 bp upstream (see Figures 26.14 and 26.15).

A sequence resembling that at positions -127 to $+4$ preceding and at the initiation site is repeated at the first Bam island upstream (positions -1147 to -1017). Presumably this is present also at the other Bam island. Whether these sites are used to initiate transcription is not known, but a low-level activity could explain the occasional starts of transcription that have been seen in the nontranscribed spacers in electron micrographs. Definition of the sequences comprising the RNA polymerase I promoter may require more direct studies of polymerase binding and initiation, because it is not possible to compare a variety of transcriptional units all recognized by the enzyme, such as occurs with RNA polymerase II. The sequences of three rDNA initiation regions have been determined in different clones and are virtually identical.

Similar experiments to those on the initiation site have been used to locate the 3' terminus of the 40S RNA, which should identify the region in which transcription terminates on rDNA. Again it is not yet possible to determine which structural features comprise the recognition site for termination.

The inheritance of Xenopus rDNA has been analyzed by following the restriction pattern from parent to progeny (Reeder et al., 1976b). This is made possible by performing crosses in which at least one parent is heterozygous for the anucleolate mutation, so that the inheritance of a single nucleolar organizer can be followed in the nu^+/nu^- progeny of crosses $nu_1^+/nu_2^+ \times nu^+/nu^-$ or $nu^+/nu^- \times nu^+/nu^-$. Usually no change is seen in the chromosomal rDNA and the spacer pattern is transmitted as a simple Mendelian unit. Two deviations from this rule have been seen in 50 matings. In one case, a new EcoRI band appeared in the rDNA of one nucleolar organizer, which had previously been present in its allele (that is, it was transferred from nu_1 to nu_2 in the progeny of the first cross given above). In the other instance, the bands remained the same, but one was greatly increased in abundance; this band also was amplified in the oocyte. These changes are not easily explained by genetic crossover between homologues, which would require the size classes to be clustered in blocks rather than dispersed. A model that may explain such evolution of band patterns is to suppose that a block of the amplified rDNA is inserted into the chromosomal rDNA; presumably the individual members of this block subsequently must be scattered to achieve a general arrangement of dispersed size classes. The dispersed arrangement means that it is not possible to tell from genetic crosses how often recombination occurs in the rDNA cluster; and so at present it is unknown what contribution this may make to maintaining the structure of the cluster.

The general structure of the rDNA repeats of D. melanogaster appears similar to that of X. laevis, but with the additional complication that intervening sequences are present in some of the 28S genes. The nontranscribed spacers seem to vary in length but not in sequence, since heteroduplexes show deletion loops and not substitution loops. There may be internal repetition comparable to that of Xenopus. As summarized in Table 26.5, the intervening sequences fall into two principal classes. Type 1 have lengths about 500, 1000, and 5000 bp and are related in sequence. Type 2 have lengths mostly about 3000–4000 bp representing a different sequence class.

The structures of the rDNA gene clusters on the X and Y chromosomes can be distinguished by comparing the patterns of female (XX) and male (XY in which the X is bobbed) flies. The coding sequences appear to be the same, as seen in the pattern of fingerprints of the rRNAs (Maden and Tartof, 1974). The nontranscribed spacers all belong to the same class. Different patterns are seen in the restriction fragments corresponding to the intervening sequences in the 28S genes. In the X chromosomal rDNA about 49% of the repeats have type 1 insertions, while 16% have (mostly) type 2 insertions. But on the Y chromosome only 16% of the 28S genes are interrupted, probably all by type 2 insertions (Tartof and Dawid, 1976; Wellauer, Dawid, and Tartof, 1978). The total number of rDNA repeats on the Y chromosome also is less; and it is as though the class of genes with type 1 insertions had been deleted at some time.

A question with interesting implications for the evolution of the rDNA gene cluster is whether exchange occurs between the X and Y nucleolar organizers in spite of the absence of genetic recombination in D. melanogaster males. The identity of the coding sequences and nontranscribed spacers on the two chromosomes suggests that some form of exchange should occur, especially since there is no evidence to support the existence of selective pressure that might maintain the sequence of the spacers. But the clear difference in the patterns of intervening sequences suggests that the type 1 units cannot be transferred to the Y chromosome. Whether this means that exchange does not occur or that these units are excluded in some way is not known. One model to reconcile the apparent need for exchange with the existence of differences would be to suppose that the process involves extrachromosomal rDNA representing only certain repeating units.

The nontranscribed spacers have been compared with the rRNA coding regions of the melanogaster subgroup of six Drosophila species by following the hybridization of cloned sequences of D. melanogaster with the rDNA of the other species (Tartof, 1979). The ΔT_m of the heteroduplexes relative to the homoduplexes is always low (0–3°C); and in no case is it any lower for the spacer than for the coding region. This implies that spacer as well as coding regions have been well conserved, a contrast with the different behavior of the two regions seen in Xenopus evolution.

Multiplication of Ribosomal Genes

Several situations are known in which the control of rDNA content is different from that of the other sequences comprising the genome. At one extreme is the example of the extrachromosomal palindromic dimers of the slime molds, which may be regarded as episomal components of the genome. In amphibia and in many other species, rDNA may be amplified to produce the extrachromosomal nucleoli found in the oocyte; this is a feature of the cell phenotype and the additional sequences are not inherited (unless inserted into the chromosomal rDNA by genetic exchange). The control of rDNA replication in D. melanogaster is seen to be distinct in polytene cells, in which it undergoes fewer rounds of replication than the majority of genome sequences (see Chapter 16). A reverse effect is seen in *bobbed* mutants in which the amount of rDNA may be increased to overcome the deficiency resulting from the deletion of some of the chromosomal sequences. This increase occurs in two ways, one heritable and one not. Whether common mechanisms exist in these situations is somewhat doubtful; but the variety of instances in which adjustments are made in the rDNA content argues that the importance to the cell of being able to synthesize sufficient rRNA may have led to the development of these mechanisms for introducing sudden changes in rDNA content.

Ribosomal gene *amplification* occurs in the oocytes of many species, although by far the best described are the amphibia, in particular X. laevis. The presence of multiple nucleoli was observed in very early cytological studies of oocytes, but sensible interpretation became possible only after identification of the constituent DNA. The mistaken interpretations that preceded modern work have been discussed by Gall (1978). The first studies to make clear the real situation were those of Brachet (1940) and Painter and Taylor (1942), in which Feulgen staining was used to show that a "cap" of DNA is formed in the pachytene cell, on the side of the nucleus opposite the attachment of the chromosomes. The granular cap breaks up and spreads out to form multiple nucleoli. The significance of these observations was made apparent when it was discovered that the nucleoli contain newly synthesized ribosomal DNA, whose production represents an increase in the cellular content relative to its usual proportion of the genome (Brown and Dawid, 1968; Evans and Birnstiel, 1968; Gall, 1968). Similar work has been performed on insects but it took longer for its significance to be appreciated.

Actually there are two distinct periods in amphibian oogenesis when the number of rDNA genes is increased. The primordial germ cells of Xenopus do not have any amplified rDNA, which first appears with the onset of sexual differentiation and mitosis in the germ cells of the tadpole. Kalt and Gall (1974) found that both oogonia and spermatogonia have a low level of rDNA amplification (<50-fold) as seen in the presence of multiple nucleoli and the extent of in situ labeling with ^{3}H-rRNA. At the onset of meiosis the amplified copies are lost in both sexes, reappearing in much greater amounts, ~2500-fold, only in oocytes. Whether the first set of amplified copies is lost completely in the female or is involved in the meiotic amplification in oocytes is not yet known.

Two general types of origin seem plausible for the amplified rDNA. It might be inherited in an extrachromosomal form which is amplified only at the oocyte stage. Or the amplified genes of the oocyte may be copied from chromosomal rDNA. Brown and Blackler (1972) distinguished the two models by examining the amplified rDNA of hybrids generated by crossing X. laevis and X. borealis, whose rDNAs can be distinguished by their buoyant density (the result of G-C differences in the nontranscribed spacer). An episome model would predict strict maternal inheritance (subject to the assumptions discussed in Chapter 21). But the hybrids predominantly or exclusively amplify X. laevis DNA, irrespective of the direction of the cross. This suggests copying of chromosomal sequences, with X. laevis preferred to X. borealis for unknown reasons.

We have seen above that the structure of the amplified rDNA is essentially the same as that of the chromosomal rDNA, but varies in its pattern of nontranscribed spacers in two significant ways. The lengths of the spacers may be a subset of those seen in chromosomal rDNA or may vary in abundance; and adjacent spacers tend to be similar in length rather than different. This suggests that the unit that is amplified is rather short, of the order of size of the individual repeat. The mechanism by which it is amplified is thought to be a rolling circle. (The features of such structures are discussed in Volume 3 of Gene Expression, in brief this consists of a circular template, around which a replicating fork moves many times to produce a single-stranded tail of indefinite length. The tail must be rendered duplex by synthesis of a complementary strand by a second DNA polymerase complex.)

The rDNA fraction of X. laevis oocytes—isolated as a buoyant density satellite—contains up to 9% of the molecules as circles varying in size from 12-210 kb (1-20 rDNA repeats). Hourcade, Dressler, and Wolfson (1973) found that at low sizes these fall into a discrete multimeric series (variations in the sizes of individual repeats obscure any periodicity at greater lengths). The size of the circle corresponds to the number of originally contiguous units that are amplified to form each extrachromosomal molecule. About one in six of the circles have the long tails characteristic of rolling circles. Denaturation mapping confirms that both simple circles and rolling circles consist of rDNA repeats.

As is always true of structures identified in low proportion by electron microscopy, these results do not prove that the rolling circles are either replication intermediates or the source of amplified rDNA. Rochaix, Bird, and Bakken (1974) therefore tried to identify replicating DNA directly. After a 30–120 minute pulse of ^{3}H-thymidine, the proportion of rolling circles in the material identified by autoradiography increases to 18%. A chase with unlabeled material reduces this to a level of 1%. If these structures are indeed rolling circles engaged in replication, a radioactive label should move continuously along the tail away from the circle. The movement of the labeled segment should continue during a chase period. These expectations are fulfilled. Free circles also become labeled, but these appear to represent rolling circles whose tails have been broken off; the labeling pattern follows the predictions for this origin rather than for alternatives such as generation by recombination.

Many questions remain to be answered. Is each rolling circle responsible for the formation of one nucleolus? What controls the selection of sequences for amplification in this manner? How are the circular templates derived from the chromosomal rDNA? These issues concern the amplification process itself, which is responsible for allowing the oocyte to achieve the massive synthesis of ribosomal RNA that is necessary at this stage of development. This is a transient feature of the life cycle, essentially a somatic phenomenon. An evolutionary issue with implications for the genetics of the organism is the frequency with which amplified sequences are inserted as large blocks into chromosomal rDNA. If this were to occur with sufficient frequency, amplification might be involved directly in the operation of selective forces to maintain the proper rRNA coding sequence. Otherwise the amplified sequences will be involved in selection only to the extent that amplification of mutant sequences may impair viability of the developing organism.

The *bobbed* mutants of D. melanogaster are recessive lethals with pleiotropic properties originally recognized as defects in bristle length and other morphological features. The mutations may be located on either the X or Y chromosome; as mapped on the X they lie in the heterochromatic region near or at the nucleolar organizer. We have mentioned that hybridization of *bobbed* DNA with rRNA showed that the mutants result from deletions of varying amounts of the ribosomal DNA (Ritossa, Atwood and Spiegelman, 1966). Complete deletion of a nucleolar organizer is described as bb^0 and the partial deletions are denoted bb; wild type is bb^+, but the actual number of repeats may vary (see above). A reduction in the number of rDNA genes below about one third of the wild type level causes a decrease in the synthesis of rRNA which is presumably responsible for the mutant defects by inhibiting protein synthesis (Mohan and Ritossa, 1970).

A remarkable feature of the *bobbed* locus is the ability of D. melanogaster to compensate for the deficiency in rDNA. This appears to be accomplished by increasing the number of rDNA repeating units and occurs in two situations, described respectively as *compensation* and *magnification*.

In flies that carry only a single nucleolar organizer—either X/O males or X^+/Xbb^0 females—the number of repeats comprising the single tandem array may increase from the initial 250 to about 400 (Tartof, 1971, 1973). This compensation occurs only on the X chromosome, since a wild type Y chromosome maintained in a stock with an $X\ bb^0$ chromosome shows no such increase. The extent of compensation appears to depend on the initial number of rDNA genes, since the longer is the single nucleolar organizer, the greater the number of extra genes. From 20-60% of the number of genes present in the nucleolar organizer may be added. Compensation occurs in the first generation of flies to possess the genotype $X\ bb/X\ bb^0$, which means that it must therefore be achieved by an increase in the amount of rDNA in somatic cells. The state of the additional rDNA is questionable, for two reasons. It is doubtful whether it is inherited, since although flies continue to have the increased amount of DNA associated with the *bb* nucleolar organizer so long as it has no partner, the amount of DNA returns to the starting value when the X chromosome is transferred into a Y bb^+ strain. Also there is no further increase in the level in subsequent generations, although a bb^+ nucleolar organizer with the increased amount of DNA would show an increase initially. A point that is more difficult to substantiate because it rests on assessment of the phenotype is that flies may continue to display the bobbed phenotype even after compensation has increased the amount of rDNA; which raises questions about whether the additional genes are functional.

A change in genotype appears to be accomplished by magnification of rDNA genes on an X *bb* chromosome when it is perpetuated in the company of the Y bb^- chromosome (whose properties have been somewhat controversial, but which now appears a deletion of all but about 40 of the rDNA genes). One protocol used is to cross females carrying an attached $\widehat{XX}$ chromosome and a Y bb^- chromosome to males that are X $bb/Y\ bb^-$; the male progeny have the constitution X $bb/Y\ bb^-$ and are back crossed in each generation to the $\widehat{XX}$- Y bb$^-$ females. Thus the same X bb chromosome is effectively passaged for many generations with Y bb^- (Ritossa, 1968, 1972; Ritossa et al., 1971). The male progeny lose the bobbed phenotype in as little as two generations; and an increase in their total rDNA occurs, presumably by increase in the number of genes on the X bb nucleolar organizer. Again the state of the magnified genes is not obvious, since the amount of rDNA appears to increase before the bobbed phenotype is ameliorated; it is possible that this represents a "premagnification" in which the genes are amplified but not yet active. It was originally thought that magnification was unstable, but this appears to have been due to the presence of another locus in the genotype which causes a phenotype similar to bobbed; actually the magnified number of genes may be retained when the appropriate chromosome (denoted X bb^m) is placed in company with a bb^+ chromosome (Tartof, 1973, 1974). However, a *reduction* in the number of repeats present at any bb^+ locus may occur when it is placed opposite Y bb^-. Reduction has the same features as magnification, but operates in the opposite direction. Two features therefore distinguish magnification/reduction from compensation: the change in rDNA content may occur over more than one generation; and it is heritable.

Compensation and magnification might be related if compensation were a necessary first step preceding magnification; or the two processes may occur by quite independent mechanisms. The definition of genetic conditions that prevent compensation but do not affect magnification therefore argues for independent origins for the two processes. Procunier and Tartof (1978) found that the occurrence of compensation depends on the location of the deletions removing rDNA from the deficient homologue. With deletions (presumed to be) entirely within the cluster, or extending from it towards the centromere, compensation does not occur. With deletions extending from the cluster to loci that are distal from the centromere, compensation occurs as described above. This suggests that compensation on the normal homologue may result from deletion of some locus distal to the rDNA cluster on the other homologue. The presence of this locus is necessary on the normal homologue in a contiguous cis location, since inversion of the region distal to the rDNA cluster prevents compensation from occurring. Describing the locus as *cr*, in formal terminology this means that compensation occurs only to increase the number of rDNA genes closely linked in cis to the cr^+ allele in cr^+/cr^- heterozygotes. It is as though the cr^+ allele "senses" the absence of a partner in trans, and acts in cis to compensate for this by increasing the number of rDNA genes from the cluster contiguous with it. But this function is not involved in magnification, since the inversion X chromosomes that cannot compensate are able to magnify as usual when placed in conjunction with Y bb^-.

Two types of mechanism might be involved in increasing the amount of rDNA: increased replication of part of the tandem array of genes; and unequal crossing over between sister chromatids. Additional replication could take the form either of extra rounds of replication within the rDNA cluster, creating replication eyes; or extrachromosomal rDNA might be produced, to survive as an episome. Either can explain the lack of heritability following compensation. An episomal location for the additional rDNA could also account for inefficient gene expression by supposing that these genes fail to associate properly with the nucleolus. The episomal mechanism can be invoked to explain magnification by adding the postulate that the presence of the Y bb^- chromosome allows some of the additional rDNA to become integrated (but only in the X chromosome). To explain reduction it is necessary then to postulate excision of some rDNA genes. Unequal crossing over has the difficulty in explaining compensation that it predicts neither the lack of heritability nor incompetence of expression; it is more attractive for explaining magnification/reduction, which should be reciprocal results of the same exchange event; successive crossing over events can account for the steady increase in rDNA content over some generations.

These two mechanisms have provided the basis for more detailed models to account for the occurrence of magnification. Ritossa (1973) has proposed that extrachromosomal rDNA is produced in the form of circles that are then integrated into the chromosome. The only direct evidence for this proposal is the isolation of circles of DNA from an additional peak of material hybridizing with rRNA, which is present in the testes of males undergoing magnification, but is not found in wild-type males or males that have completed magnification (Graziani, Caizzi, and Gargano, 1977). The circles are small, only 5–10 μm in length, which corresponds to only 16–32 kb, or just 1–2 rDNA repeats. It is not clear whether sufficient material is present in this peak to account for the magnification; and it has yet to be characterized fully as rDNA by denaturation or hybridization mapping. If circles of rDNA were involved in magnification, however, the mechanism might be similar to that envisaged for amplification in Xenopus.

Unequal crossing over at mitosis has been implicated by Tartof (1974), who found that magnified progeny appear in clusters from matings involving X bb/Y bb^- fathers; this suggests that magnification is premeiotic, occurring in germ cells during mitotic division. This generates frequencies of magnification in single males that are much too high to be accounted for by meiotic recombination. Unequal crossover between sister chromatids should produce as reciprocal recombinants one magnified and one reduced chromosome. Whether this actually occurs is not clear: reduced flies are found as progeny of the same crosses generating magnified flies, but occur in low, not equal, numbers. This may be due to selective effects, but precludes a clear demonstration of reciprocity. The best evidence that crossing over may be involved in magnification is provided by an experiment in which its occurrence was prevented by the construction of a ring X chromosome. A reduction in the frequency of magnification would be expected because this imposes a demand for an even number of crossovers if intact ring chromosomes (necessary for survival) are to be recovered. At all events, it is hard to see why an integration model should be affected by the status of the X chromosome.

From this discussion it is clear that many features of the magnification system are not at all well understood. Apart from the questions of whether episomes or unequal sister chromatid exchanges are involved, a particular puzzle is why only the rDNA cluster of the X chromosome is magnified. If episomes are formed, why can they not integrate into the Y bb^- rDNA cluster as well as into the X *bb* cluster? If mitotic recombination occurs, why cannot sister chromatid exchange occur in the Y rDNA and also homologue exchange between the Y rDNA and X rDNA?

Structure of 5S Gene Clusters

In most eucaryotes the organization of 5S genes is distinct from that of the other rRNA genes. The total number of genes may be different, their location(s) are separate from the major rRNA clusters, and 5S rRNA is synthesized by RNA polymerase III (see Table 27.2). The only exceptions known at present are in Saccharomyces and Dictyostelium, where one 5S gene is present in each major repeat unit. But as seen in yeast, its expression remains independent of the other rRNA genes. In all cases, equimolar amounts of 5S rRNA and the other rRNAs are required for ribosome assembly, but it is not clear how the activities of two different RNA polymerases on different gene clusters are coordinated; in Xenopus, in fact, regulation is discoordinate at any particular time, although overall the appropriate proportions of each transcript are produced during early development (Ford, 1971; Miller, 1973, 1974).

There is no common pattern in the organization of 5S genes. Always they are arranged in clusters, but there may be only one or very many in a given species. In D. melanogaster a single cluster is identified by in situ hybridization in region 56EF of chromosome 2, whereas in D. virilis there are two clusters (Wimber and Steffensen, 1970; Cohen, 1976b). The lampbrush chromosomes of Triturus vulgaris have a single 5S locus, those of Triturus viridescens have several loci (Barsacchi-Pilone et al., 1977; Hutchison and Pardue, 1975). In chimp, gorilla, and man, an evolutionary conservation of the position of the cluster is seen, since the major site of in situ hybridization lies in a subterminal position on one arm of chromosome 1 (Henderson et al., 1976). An interesting arrangement of clusters is seen in X. laevis, where most or all of the 18 chromosomes have 5S genes present at or very close to the telomere of the long arm (Pardue, Brown, and Birnstiel, 1973).

One question worthy of pursuit is whether or how sequence identity is maintained between different clusters of repeated gene sequences. The telomeric location of the X. laevis 5S genes raises the possibility that genetic exchange could occur without deleterious effects, presumably at zygotene when the chromosome ends all are associated. In D. funebris, ectopic pairing can be seen (in salivary glands) between two of the three 5S gene clusters (Cohen, 1976a). But it is clear that the occurrence of single genetic exchanges between clusters located within chromosomes would lead to an instability of the genome that would be highly deleterious.

A single 5S gene cluster that is being analyzed in some detail is that of D. melanogaster. Genetic analysis has been made possible by the isolation of mutants that have deletions for up to about half of the 165 genes. Procunier and Tartof (1975) showed that these *min* mutants have phenotypes comparable to that of *bobbed* when hemizygous (that is, when the other homologue has a deletion for the entire cluster). Homozygous *min* mutants appear wild type, which has the same implication made previously for the major cluster that less than half of the full number of genes is needed to sustain normal viability. The *min* deletions are derived from a translocation (*L62*) which appears to break the cluster in the middle; formally these results show that either half of the cluster can function independently. This excludes the existence of cis dominant control sites at either end.

Several plasmids have been obtained carrying up to 32 repeats of the 5S gene cluster (Artavanis-Tsakanos et al., 1977; Hershey et al., 1977). Partial denaturation mapping shows that A-T rich regions (nontranscribed spacers) alternate with G-C rich regions (which include the genes). The total length of the repeating unit seen either by electron microscopy or by restriction mapping is some 370–390 bp. A small amount of length variation, up to about 20 bp, is seen in the 250–270 bp spacer; all the copies of the 120 bp gene are presumably identical in sequence. The plasmids hybridize to the expected chromosome region, possibly occupying two adjacent bands.

Cleavage with EcoRI generates a single chromosomal fragment containing all the 5S genes. Its exact length is difficult to determine but appears to be >50 kb. This is reduced in size by the *min* deletions to about 21 kb (Procunier and Dunn, 1978). This implies that the cluster is probably a continuous tandem array of spacers and genes, although the possibility of some interruptions or unusually longer spacers cannot be formally excluded. [This contrasts with the earlier suggestion of Procunier and Tartof (1976) that the cluster might consist of two separate arrays, separated by a substantial distance. This was based upon a mistakenly low estimate for the size of the EcoRI fragment (23 kb) and its apparent failure to be reduced in size by *min* deletions.]

Unlike the single type of 5S rRNA identified in Drosophila, several sequences have been detected in Xenopus. There are two principal types of 5S RNA in both X. laevis and X. borealis; one is found exclusively in oocytes and the other is synthesized in somatic cells. The oocyte and somatic 5S rRNAs of X. laevis differ in 6 bases. In X. borealis, the two species differ from their counterparts in X. laevis by 1 and 4 residues, respectively (Brownlee et al., 1972; Wegnez et al., 1972; Ford and Southern, 1973; Ford and Brown, 1976). This means that the 5S RNA of each type is more closely related to its counterpart in the other species than it is to the 5S rRNA of the other type in its own species. This suggests that the two types had evolved before the species separation of X. laevis and X. borealis; and reinforces the idea that the sequences are specific for function and are not some chance consequence of the evolution of repeated genes. From the minor variants present in each type, the rate of change appears to be about 4 times greater in the oocyte type than the somatic type (which may be due to the greater number of genes).

Most of the 5S genes of X. laevis can be isolated as a buoyant density satellite on Ag^+-CsCl (Brown, Wensink, and Jordan, 1971). The cluster is organized in the familiar form of genes alternating with spacers. Denaturation mapping shows that the repeating unit is some 750 bp long. The gene sequences in the satellite are largely or entirely those of the major oocyte 5S rRNA, which accounts for more than half of the 5S RNA at this stage of development (Brownlee, Cartwright and Brown, 1974). This means that most of the 5S genes of the organism code for 5S rRNA that is synthesized only in the oocyte. Presumably this is a counterpart to the amplification of the major rDNA in oocytes, making possible the devotion of these cells to ribosome synthesis.

Another cluster that corresponds to a minor oocyte 5S rRNA species, called trace 5S rRNA and accounting for about 10% of the total, has been obtained as a satellite by actinomycin binding (Brown, Carroll, and Brown, 1977). This also consists of a tandem array of spacers and genes, although the repeat length is only 350 bp and the spacer appears unrelated to that of the oocyte cluster. The trace 5S rRNA has a sequence in which 2 of the variable positions have the bases found in somatic 5S rRNA, 3 are in common with the oocyte sequence, and 3 others are different from either.

The number of somatic 5S genes is not known; nor is their location, although it is generally assumed to constitute a separate cluster. It would be interesting to know whether the different types of genes are present on different chromosomes. Probes of 5S rRNA cannot be used to test this because of cross hybridization, but if the spacers are different for each type of gene, these could be used as probes.

In the oocyte 5S gene cluster, a sequence rich in A-T base pairs of about 400 bp alternates with a sequence rich in G-C base pairs of about 300 bp (which includes the 120 bp of the gene). The length of the spacer is heterogeneous and denaturation mapping shows that this resides in the A-T rich region. In cloned multiple repeats, adjacent units may be different in length (Carroll and Brown, 1976a,b). Restriction mapping supports this conclusion in more detail. HindIII cleaves each repeat once and generates a series of bands with a separation of about 14 bp. HaeIII makes three cuts, releasing two fragments of constant size; the third forms a series of bands of size separation again about 14 bp. This is consistent with earlier studies showing that the major oligonucleotides derived from cRNA transcripts of 5S DNA made in vitro constitute A-T rich species of 15 nucleotides, related but not identical in sequence. This suggests that the length of the A-T rich region varies because it contains internal repetitions of a variable number of 15 bp repeats, analogous to the organization of a satellite DNA.

The sequence of the oocyte 5S repeating unit now is known in detail from direct sequence determination of several cloned repeats as well as bulk DNA. Federoff and Brown (1978) showed that the A-T rich part of the spacer comprises a repeat of 24–25 nucleotides. Figure 27.8 shows that three regions can be recognized: A1 is an invariant 40 bp sequence; A2 consists of a variable

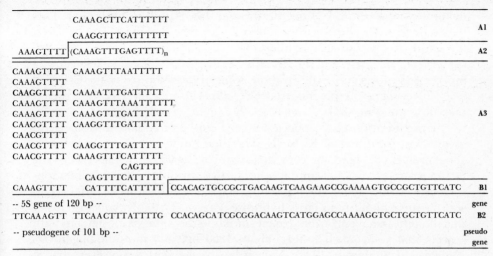

CAAAGCTTCATTTTTT		
CAAGGTTTGATTTTTT		**A1**
AAAGTTTT (CAAAGTTTGAGTTTT)ₙ		**A2**

CAAAGTTTT CAAAGTTTAATTTTT		
CAAAGTTTT		
CAAGGTTTT CAAAATTTGATTTTT		
CAAAGTTTT CAAAGTTTAAATTTTTT		
CAAAGTTTT CAAAGTTTGATTTTT		**A3**
CAACGTTTT CAAGGTTTGATTTTT		
CAACGTTTT		
CAACGTTTT CAAGGTTTGATTTTT		
CAACGTTTT CAAAGTTTCATTTTT		
CAGTTTT		
CAGTTTTCATTTTT		
CAAAGTTTT CATTTTCATTTTT CCACAGTGCCGCTGACAAGTCAAGAAGCCGAAAAGTGCCGCTGTTCATC	**B1**

-- 5S gene of 120 bp --		**gene**
TTCAAAGTT TTCAACTTTATTTTG CCACAGCATCGCGGACAAGTCATGGAGCCAAAAGGTGCTGCTGTTCATC	**B2**
-- pseudogene of 101 bp --		**pseudogene**

Figure 27.8. The repeating unit of X. laevis 5S DNA. The sequence is written to show the nature of the internal repeats. The A-T rich part of the spacer consists of 24–25 bp repeats of the sequence

$$CAA_A^C GTTTTCAA_{GG}^{AA} TTTGA_T^{C} \ {}_A^{G} TTTT(T).$$

This is internally repetitious and may have arisen by a duplication. Region A1 consists of 2½ repeats, A2 consists of a variable number of repeats of part of the sequence, and A3 consists of about 12 repeats, including some half repeats. The B1 part of the G-C rich spacer joins this sequence to the 5S gene. One of the A-T units is then repeated, followed by B2 which is closely related to B1, and the sequence based on the first 101 nucleotides of the gene. The sequence as written represents the anticoding strand of DNA (same sequence as the transcript). Data of Federoff and Brown (1978) and Miller et al. (1978).

number of repeats of a 15 bp sequence, most often about 6 but up to 16; and A3 is about 170 bp long, probably able also to vary in length. The relative ages of the internal repeats are indicated by the greater heterogeneity in the units of A1 and A3, which suggests an older origin; compared with the lesser heterogeneity in A2, which argues for more recent repetition of a 15 bp part of the 24–25 bp repeat. The variation in length of A2 suggests that this could be a preferential site for crossing over out of register. The A-T rich (A1–A3) region makes up most of the spacer and is separated from the gene by a 49 bp G-C rich region.

The G-C rich part of the repeating unit includes the gene and comprises a duplication. This was first recognized in the presence of a lengthy sequence adjacent to the 5S gene itself which is very similar to the coding sequence. Jacq, Miller, and Brownlee (1977) showed that this *pseudogene* contains residues 1–101 of the gene with some 9 base changes. The lack of any detectable heterogeneity in the entire G-C rich segment of the gene cluster argues that its sequence is conserved along with the gene itself. Although not detected in the earlier studies that identified only 1 5S gene per repeat, the pseudogene is able to hybridize with the gene under less stringent conditions. The sequence obtained by Miller et al. (1978) shows that, in fact, the entire region from −73 to +101 around the 5S itself is duplicated. This means that the 5S gene is followed by a short segment of the A-T rich spacer, the entire 49 bp sequence of the G-C rich spacer that separates the A-T spacer from the gene, and the first 101 nucleotides of the gene. There are 23 mismatches in the entire duplication, some of them located in the G-C rich region preceding the pseudogene (B2) which mimics the G-C rich region preceding the gene (B1).

The organization of the oocyte 5S gene cluster of X. borealis appears to be different from that of X. laevis. The total number of 5S genes is lower, about 9000 in X. borealis, compared with the 24,000 of X. laevis; but the total amount of rDNA is similar at some 0.7%. In spite of the close relationship between the genes, the noncoding sequences appear to be different, as seen in their inability to cross hybridize. Originally the X. borealis cluster was thought to have an arrangement similar to the X. laevis cluster, but with spacers that were much longer, about 1800 bp, and more homogeneous (Brown and Sugimoto, 1973). Now it has been reported, however, that at least part of the X. borealis satellite instead consists of clusters containing variable numbers of 5S genes about 80 bp apart, each cluster separated from the next by spacers of heterogeneous length (unpublished results quoted in Korn and Brown, 1978). The sequence of a cloned segment consisting of three oocyte genes of X. borealis shows that the cluster is preceded by a tandemly repeated sequence of 21 bp; variable numbers of these repeats may account for the heterogeneous spacing between clusters. The 80 bp sequences separating the genes (including the sequence preceding the first and following the last) are very similar. Point variations occur in the 11 bp immediately preceding the start of the gene; positions −12 to −40 are invariant; and point variations occur in the remaining half (−40 to −80). Of the genes themselves, the first two differ by 2 base substitutions; one of these corresponds to the major oocyte 5S RNA sequence. The third gene differs at 15 positions; and its 5′ half can be found as a restriction fragment of bulk 5S DNA. This implies that an appreciable proportion of the X. borealis oocyte 5S genes differ from the rest in this way. It is possible that this gene is like the pseudogene in X. laevis and does not function in vivo. All three genes can be transcribed in vitro, which implies that the gene and 80 bp sequence are sufficient to support initiation and termination of transcription.

The sequences immediately preceding the 5S oocyte genes of X. laevis and X. borealis and also a tRNA gene of X. laevis as well as the VA gene of adenovirus are summarized in Figure 27.9. The significance of the short sequences that represent the only possible homologies is somewhat doubtful. Since there is little similarity of sequences around the genes, this implies that RNA polymerase III may not function by recognition of a consensus sequence flanking the 5′ side of the gene. None of these higher eucaryotic sequences shows any homology with the sequence preceding the 5S gene of S. cerevisiae.

The unavailability of promoter mutants in eucaryotic genes emphasizes

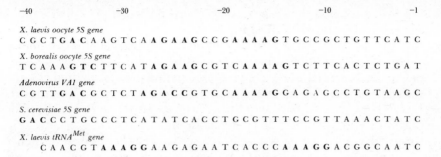

Figure 27.9. Sequences preceding the starpoints of genes transcribed by RNA polymerase III in various species. The only homologies are rather short as indicated in bold face. Data of Korn and Brown (1978), Miller et al. (1978), Pan, Celma and Weissman (1977), Valenzuela et al. (1977), Telford et al. (1979).

the need to develop systems for in vitro transcription in which binding and initiation sites can be determined and on which the effect of mutation can be judged. Progress toward this aim has been slow with purified preparations of either RNA polymerase I or II, but has been more rapid with RNA polymerase III and with crude systems containing it. When 5S DNA of either X. laevis or X. borealis is injected into the germinal vesicle (nucleus) of X. laevis oocytes, 5S RNA is transcribed (Brown and Gurdon, 1977, 1978). The ratio of coding: noncoding strand transcripts is ~10; some transcription of spacer regions occurs in addition to the gene sequences. A plasmid carrying a single 5S repeat of X. laevis can be transcribed. Most such plasmids give 5S RNA; one generates a longer molecule probably resulting from readthrough past the termination site, and this may resemble molecules of this sort that has been detected in vivo (Denis and Wegnez, 1973). Formally these experiments prove the existence of initiation and termination signals in each repeat unit.

An important extension of this system has been made possible by the demonstration that a crude nuclear extract from the oocytes can transcribe the cloned 5S genes accurately in vitro (Birkenmeier, Brown, and Jordan, 1978; Ng, Parker, and Roeder, 1979). Molecules of 5S RNA constitute at least half of the products; some longer molecules again may result from readthrough past termination signals. The sensitivity to α amanitin and other characteristics of the reaction indicate that the active enzyme is RNA polymerase III. Progress in fractionating the system into the enzyme and other active components has been made by Mattoccia et al. (1979). This work parallels the successful development of systems in which purified RNA polymerase III transcribes 5S chromatin or DNA (see Chapter 28). Together the two approaches offer the prospect of defining the action of the enzyme and other components of the transcription apparatus in initiating and terminating transcription.

In X. laevis, Federoff (1979a) has obtained deletions in the A-T rich region of the spacers of cloned 5S repeats propagated in E. coli. The ability of the deleted repeats to be transcribed in vitro shows that expression is independent of the A-T spacer and must depend on the 49 bp G-C rich spacer region B1 and/or the gene itself. Similarly a cloned tRNA^{Met} gene of X. laevis DNA can be expressed when a sequence including only 22 bp on the 5′ side of the gene itself is present on a plasmid injected into oocytes (Telford et al., 1979). It is clear, therefore, that there is no extensive 5′ flanking region needed for recognition by RNA polymerase III.

Using the oocyte extract system, this approach has been taken further in an elegant series of experiments by Sakonju, Bogenhagen, and Brown (1980) and Bogenhagen, Sakonju, and Brown (1980). The location of the promoter recognized in a single repeating unit of the X. borealis 5S gene carried on a plasmid was defined by the use of deletions extending into the gene from either direction. The 5S RNA or an RNA of very similar size continues to be synthesized when the 5′ flanking sequence and up to the first 50 nucleotides of the gene are deleted and replaced by plasmid DNA. But there is no transcription when the deletion exceeds 55 base pairs. This suggests that the 5′ boundary of the promoter lies between residues 50 and 55 of the gene; more precisely, it appears that RNA polymerase III recognizes a sequence on the 3′ side of this location, but then starts transcription upstream. In the normal gene, presumably this occurs at the first base of the 5S sequence; in the deleted genes, initiation occurs at the same distance from the promoter, starting at the A residue closest to this point if one is not present at the exact location. This generates a molecule of the same size as 5S RNA, but with a 5′ region consisting of plasmid sequences in place of the usual sequences. This model is supported by the construction of "maxigenes" in which an additional sequence is inserted between the normal 5′ end of the 5S gene and the start of the promoter; again initiation occurs an appropriate distance before the promoter, rather than at the usual 5′ sequence.

The 3′ boundary of the promoter was defined by analogous experiments in which deletions entered the gene from the far end. Deleted genes that retain the first 83 base pairs appear to initiate transcription normally. Deletions that leave less than 80 nucleotides prevent transcription. This locates the boundary of the promoter between bases 80 and 83. Supporting the conclusion that the promoter is an internal region of the gene, occupying roughly 30 base pairs from positions 50/55 to 80/83, an internal gene fragment representing the region from bases 41 to 87 can direct the specific initiation of transcription at a site a corresponding distance upstream in a plasmid.

The simplest model that this suggests for initiation is to suppose that the 50/55–80/83 region of the 5S gene constitutes the site that is initially recognized by RNA polymerase III. The enzyme may be large enough to contact DNA simultaneously at the site upstream where initiation occurs. This is the most straightforward explanation for the apparent ability of the enzyme to count backwards from the promoter. The sequence of bases present at the initiation site may influence the exact startpoint for transcription, within certain topological restraints. Possibly the homologies indicated in Figure 27.9 are important in this context, possibly not. At all events, it is clear that initiation can occur at approximately the appropriate location even when this bears no sequence relationship to the usual region. Since the "promoter" usually is considered to constitute the site that is essential for initiation of transcription, it is a moot point whether the initiation region should be considered part of it, or whether the recognition region alone can be considered to comprise the promoter.

One intriguing implication of these results is that RNA polymerase III transcribes the tRNA genes as well as the 5S genes. It seems unlikely that there will prove to be any extensive conservation of sequence in the corresponding positions of all these genes. If RNA polymerase III acts by the same mechanism in all its transcription units, and if there is no variation in the structure of the enzyme, this would imply that it has the capacity to recognize many different promoter sequences. An alternative is to suppose that the structure of the enzyme varies; the subunit(s) responsible for recognition could constitute dissociable factors, or there might be a population of enzymes with different (permanent) subunits.

The assumption that the 5' end of 5S RNA coincides with the startpoint for transcription is based upon the presence of a triphosphate. The location of the 3' end is less certain, an ambiguity that is reinforced by the apparent readiness with which different 3' termini are generated in vitro and possibly in vivo. The sequences at or just beyond the 3' end of the mature 5S RNA molecule have clusters of 4-6 T residues (read on the anticoding strand), which are thought to be possible terminators.

In D. melanogaster it appears that 5S RNA is processed from a longer precursor. Rubin and Hogness (1975) discovered that heat shock of cultured cells causes the accumulation of a new form of 5S RNA. Described as $5S^+$ RNA, this has an additional sequence at its 3' end. The possibility that this represents a precursor to 5S rRNA whose processing is inhibited by the heat shock was explored by Jacq, Jourdan, and Jordan (1977). A chase following reduction of temperature from 37 to 25°C shows that $5S^+$ rRNA can indeed be converted to 5S rRNA; a small amount of this material also can be found under normal conditions. The 15-16 extra nucleotides on the precursor terminate in $(Up)_6$-OH, which could be a control signal.

Evolution of Tandem Gene Clusters

The feature common to all tandem gene clusters is that they code for products that all cells must synthesize in vast amounts. Thus every cell must rely upon the synthesis of ribosomes to maintain its protein synthetic apparatus and upon the synthesis of histones to organize its genetic apparatus. To match the synthesis of ribosomal RNA to that of ribosomal proteins (about whose genes nothing is known, but which presumably are transcribed into abundant messengers), there is always repetition of the major rDNA transcription units associated with the nucleolus; and also of the 5S genes organized in independent clusters. To match the production of histones to the replication of DNA in every cell cycle, presumably the usual amplification through high transcription of mRNA would be inadequate, for these constitute the one well-characterized example of protein-coding genes that are highly repeated and clustered together (see Chapter 28). (This is to except the immunoglobulins, which may have a repetitive organization, but which are not expressed independently; see Chapter 26).

The same problem is encountered by all these gene clusters. Usually the question is posed in terms of how identity of sequence is maintained between the individual members of the cluster. This practical issue follows from the general question of how selection can be imposed to prevent the accumulation of deleterious mutations. As with any gene system, the generation of mutant copies will be balanced by their removal by adverse selection. But the effect of any single mutation will be diluted by the large number of unmutated copies. Whereas selection may be expected at once to eliminate a genome whose only copy of some essential gene has been mutated, when there is a large number of copies it seems unlikely that every one must be wild type for viability; mutant copies might therefore reach some appreciable proportion of the population before selection eliminates the genome. Lethality in fact will be quantitative. This point is made all the more striking by the demonstrations that more than half of the units of the rDNA clusters of Xenopus or Drosophila may be deleted without immediate effect on the viability of the organism. This should imply that a similar proportion of the genes can accrue deleterious mutations before selection becomes effective.

But this conclusion poses a dilemma: if mutations accumulate to the point of collective lethality, sooner or later this must be the fate of all genomes. This implies that to counteract the accumulation of mutations there must be some mechanism for regenerating clusters of wild-type genes, or at least for increasing the number of wild-type gene copies at the expense of mutant copies. More precisely there must be some mechanism that allows a given variant to be scrutinized by evolution: this may be the current wild-type sequence, a new version of it that proves to have an advantage, or a deleterious version that is eliminated. The nature of this mechanism has been the subject of much speculation, but two classes of model have been developed on the assumption that the general feature must be the construction of genomes in which one particular gene sequence accounts for all or most of the tandem copies and is therefore subject directly to selective scrutiny.

Sudden correction models postulate that at some point the entire gene cluster is replaced by a new set of copies that represent one or a very few of the copies present previously. This might be achieved by the multiplication of the new sequences which then physically replace the old; or the same effect could be achieved as in the master/slave hypothesis by correcting all the copies to comply with the sequence present in one of them (Callan, 1967; Whitehouse, 1967; Buongiorno-Nardelli et al., 1972; see also Chapter 17). Usually the correction event is postulated to occur every generation but formally it could occur less frequently than this (although an irregular periodicity would be harder to explain). This model predicts that the cluster should have a homogeneous structure, since all the repeating units should have been recently derived from a common sequence. This is at odds with the heterogeneity of spacer lengths seen in clusters such as the principal Xenopus rDNA and the 5S oocyte DNA.

An alternative model relies upon the occurrence of unequal crossing over; if the total size of a tandem cluster is restricted, unequal crossing over will tend to maintain sequence identity, by expanding and contracting the representation of particular elements (Smith, 1973, 1976; see also Chapter 19). This generates clusters in which most or all of the members are derived (with additional mutations) from as few as one member of an ancestral cluster, which is therefore exposed to selection. The difference between the heterogeneity of spacers and the homogeneity of genes can be explained by supposing that deleterious mutations in the genes are eliminated (when concentrated by unequal crossover), whereas changes in the spacer sequences are less subject to selective forces and may therefore accumulate. Certainly the variations between spacers show the sort of relationship that would be expected to result from crossovers that are not only unequal with regard to the ends of the cluster but also with respect to individual repeating units. The organization of the spacers is reminiscent of that of some satellite DNAs; and might be thought in the same way to represent the condition achieved by DNA whose exact sequence is not critical for its function.

One central feature of the maintenance of sequence identity by unequal crossing over is the need for all members of the cluster to be available for the genetic exchange. Although this presents no problem for situations such as the single nucleolar organizer of Xenopus, there are several clusters in which it is not clear whether this condition can be fulfilled. One is the rDNA cluster of D. melanogaster, where recombination does not occur in the male; we have already discussed features of the X and Y rDNA gene clusters in this connection. Another is presented by situations in which there are several tandem clusters in different locations. The example of the X. laevis 5S DNA may escape difficulties if recombination occurs at the telomeres, but there are other cases of less favorable locations.

Since the process of crossover fixation may be relatively slow (compared with sudden correction mechanisms), at any given time there should be polymorphism with regard to mutant elements that are being spread through the population, but have not yet reached a concentration in the cluster sufficient to force its elimination. Actually it is difficult to estimate the extent of any such variation from sequence data on bulk DNA, since (for example) a large number of different variations each occurring below the 1% level would not be detected. While clustered gene sequences are indistinguishable by available criteria, it may be necessary to determine the sequences of a sufficient number of individual clones to assess the extent of variation at the nucleotide level. At present it is not possible to decide what proportion of a cluster might be occupied by nonactive, mutant units.

Mutations may take two forms, of course, either influencing the activity of the gene product or occurring in a site that has no effect (such as the third base of a codon). Little is known about the sequences of eucaryotic rRNA, but the evolutionary conservation of the histone H3 and H4 sequences is so stringent as to imply that probably any change in amino acid sequence has a high chance of inactivating the protein. It may therefore be illuminating to examine the extent of silent third base substitutions in individual repeats of a histone gene cluster (see Chapter 28); this should reveal whether selection is imposed solely on the protein sequence or whether homogeneity is generated by control of other features of the gene organization or expression.

What is the role of spacers in the tandem gene clusters? One view is that they play some function at present not understood, but which is independent either of sequence or at least of the exact constitution of the sequence. Another is that the spacers need be assigned no particular function in the operation of the gene cluster, but are an inevitable consequence of whatever mechanisms produce and maintain tandem gene families (for review see Federoff, 1979b). The dissimilarity of the spacers in the three identified 5S gene clusters (X. laevis oocyte and trace, X. borealis oocyte) at least suggests that any sequences with common functions may constitute only a small part of the spacer. Looked at the other way, the argument becomes that any (short) conserved sequences are likely candidates for control elements.

CHAPTER 28

Gene Families

The relationship of a gene to the other constituents of the genome might be considered the pivot of the analysis of gene expression. How many copies of each (unique) gene sequence are present? Are there other genes closely related in sequence? In what cells is the gene expressed and how is the control of its expression coordinated with the regulation of other genes whose products contribute to the cell phenotype? The molecular approach to these questions is to define the sequences of the gene itself and the context within which it resides. In this chapter we shall be concerned with attempts to define families of genes, related either structurally or functionally, which represent systems for seeking features of common control.

Most genes are unique, or at least are present in the haploid genome in very few copies (see Chapter 24). Only in the case of proteins that are produced abundantly in some particular cell phenotype is it possible to define a functional family of unique genes. The chick oviduct proteins stimulated by estrogen provide one such example, some of whose members have been identified. All are mosaic genes and the 5' flanking sequences will be examined for homologies that might represent common control sites (in addition to the short homologous elements found also in other genes that identify the possible promoter sequences discussed in Chapter 26). Muscle proteins constitute another family, expressed in high amounts upon fusion of myoblasts into myotubes, but a complication in seeking features of coordinate control is the presence of the same or related proteins as structural components probably of all cell types. The best characterized gene family codes for the globin proteins; the relationships between the different globin proteins are well established, in mammals the proteins are expressed in a well-characterized developmental pathway, mutations are available in the form of human thalassemias, and the wild-type and mutant gene organization is susceptible to molecular analysis. An extreme example of a functional family of genes is provided by the heat shock loci of D. melanogaster, whose functions in fact are not known, but which have in common their expression when cells (apparently of any phenotype) are incubated at 37°C instead of 25°C.

Structural gene families may have any number of members from a simple duplication to a lengthy tandem cluster. Little is known about moderately repetitive gene families, but in the cases of Dictyostelium actin, Xenopus vitellogenin, mouse major urinary proteins, and mammalian keratin there is evidence that there may be a number of genes coding for related but not identical proteins; in the case of actin at least some of the genes are clustered together (Kindle and Firtel, 1978; McKeown et al., 1978; Wahli et al., 1979; Hastie, Held, and Toole, 1979; Fuchs and Green, 1979). Other such genes may provide the templates for the minority of the cellular message population that is derived from repetitive DNA (see Chapter 24). The repetition and rearrangement of immunoglobulin genes has been discussed in Chapter 26.

The existence of histone genes as a repetitive cluster has been known for some time. This takes the usual form of a tandem alternation of coding sequences and spacers. There must be more than one type of coding sequence at least for the histones (H2A and H2B) that are found in the form of different variants at certain developmental stages (in sea urchin), as well as for H1, which exists as several closely related sequences in many species. The spacers of the sea urchin histone gene cluster do not seem to have the internal repetition characteristic of the ribosomal gene clusters. The lack of variation in histone H3 and H4 proteins can be contrasted with the variations in the coding sequences found within and between species.

Little is known about the control of transcription. Putative promoter sequences can be identified by short homologies between different genes expressed via a common RNA polymerase, although this approach is not practical for RNA polymerase II because its template comprises a single type of transcription unit. Additional homologies between members of the gene families recognized by RNA polymerase I may identify regulator elements. But no clear role for any such site can be assigned until recognition by RNA polymerase and/or regulator proteins can be demonstrated in vitro. Then it will be possible not only to identify the recognition and binding sites but also to determine the effects of introducing mutations in them in vitro. The lack of purified enzymes that initiate properly in vitro has impeded the development of transcription systems of defined components; at present, for example, it is hard to say whether regulator elements function independently or might perhaps comprise additional polypeptides of the RNA polymerase complex. Another approach has been to try to obtain preparations of chromatin which reproduce in vitro the pattern of transcription previously prevailing in vivo. Claims have been made that the characteristic features of such preparations can be retained through the dissociation and reconstitution of chromatin, which would, of course, offer the prospect of identifying the responsible components; but these are fraught with difficulties.

The Histone Gene Cluster: Organization

Repetition of the histone genes as a tandem cluster has been found in Drosophila, sea urchins, avian, and mammalian cells; it is reasonable to suspect that this will be a feature of all higher eucaryotic genomes. Since there are five classes of histones, H1 represented in chromatin in half molar amounts, and H2A, H2B, H3 and H4 occurring in equimolar proportion, the first question to resolve is whether the genes for each type of histone occur in separate blocks or are intermingled in a common repeating unit. Then it is possible to ask how genes coding for the minor variants of H2A and H2B, or for the microheterogeneous species of H1, are organized relative to the predominant types.

The first step in isolating histone genes is to obtain histone mRNAs to use as probes. Sea urchins have been particularly well investigated, because histone synthesis occupies a large part of the protein synthetic activity of the early embryo, where the histone messengers can be found as an abundant peak of small polysomes (Moav and Nemer, 1971). The mRNA of these polysomes represents a large proportion of the poly(A)⁻ mRNA population (see Chapter 23). In D. melanogaster, histone messengers have been purified by taking advantage of the observation that they comprise a larger part of the total mRNA population present after heat shock (Burkhardt and Birnstiel, 1978). In other cell types, including mammalian cultured cells, advantage has been taken of the appearance of histone synthetic activity specifically during S phase, again in the form of small polysomes (see below).

In all these cases the histone mRNAs are small, although apparently longer than necessary just to provide coding sequences (for example see Table 23.5). In all cases the hybridization of the histone mRNAs with DNA occurs at a $Cot_{1/2}$ indicating repetition of the corresponding sequences in the genome (for example see Table 24.2). In sea urchins the repetition frequency of the histone genes is high; the original estimates for Psammechinus miliaris are for 400–1200 copies of each gene (Kedes and Birnstiel, 1971; Weinberg et al., 1972). Lower degrees of repetition appear to be characteristic of the mammals, with estimates of about 25 for mouse and 40 for HeLa (Jacob, 1976; Wilson and Melli, 1977). Xenopus appears to have a similar repetition of about 40; and the chick may be even lower at about 10 (Jacob, Malacinski, and Birnstiel, 1976; Crawford et al., 1979). None of these estimates is accurate; all rest on the use of histone mRNA as a hybridization probe, and since the messengers generally are not polyadenylated, it is hard to achieve an adequate degree of purification. Better estimates could now be obtained by using cloned genomic probes.

Histone mRNA usually sediments at about 9S and can be translated in vitro into all the histone proteins. The individual messengers have proved hard to separate, due to their erratic behavior during gel electrophoresis (Levy et al., 1975, Gross et al., 1976a; Burkhardt and Birnstiel, 1978). By adjusting the conditions of fractionation, the histone mRNAs of sea urchin or Drosophila have been separated into groups, each of which codes in vitro for a single type of histone; a single group, however, may consist of more than one mRNA species when fractionated under appropriate conditions. For example, the H4 mRNA present at early blastula in sea urchins can be fractionated into three size classes (Grunstein and Schedl, 1976; Grunstein,

Schedl, and Kedes, 1976). It is not known whether these represent different genes or are mRNAs of different lengths representing the same genes. At all events, the groups have been used as probes to identify the corresponding coding sequences by hybridization. Using cloned gene sequences as probes, the procedure can be reversed to identify particular messengers in a population. This is also effective with cross-species hybridization, at least for the most highly conserved histones. For example, it may be necessary to turn to this approach to identify individual mammalian histone mRNAs (or genes), since the messengers have not yet been resolved by direct fractionation (Borun et al., 1977).

Cytological data show that histone genes are organized in clusters, of which there may be one or more. The reaction between sea urchin histone mRNA and Drosophila DNA has been used to identify the locus coding for histones. A single region, consisting of more than one salivary chromosome band, hybridizes in situ; each histone mRNA appears to cover the same breadth of the chromosome, suggesting that the genes are intermingled (Pardue et al., 1977). In Triturus, however, lampbrush loops on several different chromosomes react with a sea urchin mRNA probe (Old, Callan, and Gross, 1977).

The histone genes of the sea urchin P. miliaris can be isolated as a buoyant density satellite on CsCl-actinomycin gradients; this peak corresponds to about 0.5–0.8% of the DNA (Birnstiel et al., 1974). This shows directly that the genes are clustered; and the buoyant density of the satellite suggests that the G-C rich coding sequences must be interspersed with A-T rich spacers. Cleavage with EcoRI or HindIII generates a single fragment of about 6 kb from this satellite, which suggests that there is a single predominant type of repeating unit (Schaffner et al., 1976). Denaturation mapping of a cloned 6 kb repeating unit shows an alternation of A-T rich sequences (spacers) with G-C rich sequences (genes) (Portmann, Schaffner, and Birnstiel, 1976).

Histone genes have been isolated directly from the unfractionated DNA of two other sea urchins, Lytechinus pictus and Strongylocentrotus purpuratus. Fragments obtained by cleavage with EcoRI have been cloned to form a library from which plasmids carrying the histone genes were isolated by hybridization with the mRNA (Kedes et al., 1975). Here two cuts are made by the enzyme, so the repeating unit is found as two fragments on different plasmids. Complete repeating units have been obtained from S. purpuratus by cleavage with HindIII (Overton and Weinberg, 1978).

Restriction mapping and electron microscopy have identified a similar basic repeating unit in all three sea urchin species; its organization is summarized in Figure 28.1. Assignment of the individual gene sequences depended initially on hybridization of subfragments with separated mRNAs.

All five histone genes are arranged in the same order and orientation; although each is thought to be transcribed separately, the arrangement of the genes leaves open the possibility of a common control of transcription. Although the overall size of the cluster is different in each species, the rela-

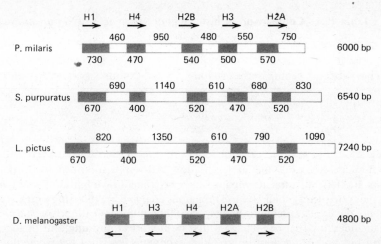

Figure 28.1. Organization of histone gene clusters. The repeating units of the tandem arrays of three sea urchins have been defined in cloned fragments by restriction mapping and electron microscopy of the sequences corresponding to individual histone mRNAs. The general arrangement is the same in each, although the precise sizes of genes and spacers may differ, as indicated by the lengths given in base pairs. The direction of transcription is indicated by the arrows. A different arrangement has been reported for D. melanogaster, although no data have been presented. References: P. miliaris—Gross et al. (1976b), Schaffner et al. (1976, 1978), Birnstiel et al. (1977); S. purpuratus—Kedes et al. (1975), Weinberg et al. (1975), Cohn et al. (1976), Sures et al. (1976, 1978), Wu et al. (1976); L. pictus—Cohn and Kedes (1979a,b); D. melanogaster—Lifton et al. (1978).

tive separation of the genes is the same; for example, the spacers between H4-H2B and between H2A-H1 are always the largest. However, there appears to be no sequence relationship between the spacers of different species, as judged by cross hybridization. No intervening sequences are found in any of the histone clusters. According to the cross hybridization between mRNA and DNA, the coding sequences of the different species may be expected to have diverged more extensively than the amino acid sequences.

Nucleotide sequences have been determined for much of two cloned repeating units from P. miliaris and S. purpuratus (Schaffner et al., 1978; Sures, Lowry, and Kedes, 1978). The coding sequences are likely to correspond to the histone variants found in early embryogenesis in each case. Table 28.1 summarizes the differences between the four core histone genes (H1 has not been determined in both species). Apart, of course, from the differences needed to accommodate the amino acid changes in H2A and H2B, almost all the substitutions occur in third base positions that do not affect codon meaning. There are a few instances where synonym codons differing in two bases can be found, presumably as the result of sequential mutations at the two base positions. Overall the variation in sequence corresponds to about 10–15% divergence, much what would be expected from the ΔT_m of cross hybrids.

Table 28.1   Comparison of histone gene sequences of S. purpuratus and P. miliaris

Histone	Total codons	Codons compared	Identical codons	Third base substitutions	Other synonyms	Coding differences
H4	104	53	33 (62%)	19	1	0
H3	137	137	96 (70%)	41	0	0
H2A	125	125	78 (62%)	42	3	2
H2B	124	102	62 (50%)	30	0	10

The number of codons includes initiation and termination codons. Initiation is always at AUG; all three terminators have been identified, but UAA and UAG appear more frequently than UGA. The codons analyzed are those for which complete sequence data are available in both species. The number of third base substitutions also includes any first base changes that do not affect meaning, but these are very few (in Leu and Arg codons only). Other synonyms describe cases where the meaning is not changed although there is more than one substitution in the codon. Although the use of codons in each gene may be nonrandom, these do not appear biased in any particular direction; for example, in H2A two particular Leu codons are TTG and CTA in one species and are CTC and TTG in the other species. There are no differences in the amino acid sequences of H4 and H3 between the species, slight differences in H2A, and more substantial change in H2B. Based on sequence data of Grunstein and Grunstein (1978), Schaffner et al. (1978). Sures et al. (1978).

Although the coding regions can be identified by translating the triplet sequence into amino acids, the precise extents of the flanking regions that are represented in the nontranslated leaders and trailers of mRNA usually are not known. The region corresponding to the mRNA has been identified for H4 of S. purpuratus by comparing the oligonucleotide fingerprint of the mRNA with the sequence of the genome (Grunstein and Grunstein, 1978). This suggests that the mRNA may start at base position –63; and it is interesting that the homology between S. purpuratus and P. miliaris extends before the H4 gene to position –119. The region represented in H2B of the S. purpuratus repeat has been defined by determining the sequence formed upon using the mRNA as template for reverse transcriptase after annealing to a short restriction fragment to form a primer (Levy, Sures, and Kedes, 1979). This locates the 5′ end of the mRNA at position –71. Probably the leader lengths are similar for all the histone mRNAs; and this assumption has been made in seeking sequences that may be involved in the control of expression (see below). Some short homologies precede the AUG codons of the different genes in each cluster; these are candidates for such elements. The limits of each spacer region cannot yet be delineated precisely; but there seems to be no internal repeating structure or relationship between the spacers separating different genes.

The only repeat apart from the sea urchins that has been investigated is that of D. melanogaster, which has been reported to take a different form.

This again contains all five histone genes, but they appear in a different order and orientation. At least two rearrangements would be needed to interconvert the sea urchin and Drosophila types of repeating unit. Whether one represents an ancestral arrangement from which the other has been derived, or whether both differ from a common ancestor, is not known.

The heterogeneity in the histone gene repeats is less than that seen with (for example) rDNA, as judged by the cleavage of bulk histone DNA into fragments relatively invariant in size. Since the detailed results summarized in Figure 28.1 are derived from analysis of individual cloned repeats from each species, they give no information on the existence of variant repeat units in the population. But the existence of different histone variants implies that not all repeating units can have identical coding functions. The early embryonic types coded by the repeat units of P. miliaris and S. purpuratus that have been sequenced appear in each case to be the predominant component of the cluster. But it is important to know to what extent there may be differences in the genes and spacers of individual repeating units, within and between different genomes. This has provided the impetus for a more detailed examination of possible heterogeneity.

Two plasmids carrying different complete repeating units of S. purpuratus have been mapped in detail by Overton and Weinberg (1978). One appears to represent the major type of repeating unit. Two differences are seen in restriction sites that are present in one plasmid, absent in the other; these could result from single base substitutions. Two slight changes in the sizes of restriction fragments could result from insertions/deletions of about 20–30 bp. When the histone DNA of individual sea urchins is examined instead of using pooled material, differences are found in the size of the major repeating unit; in some individuals multiple bands can be resolved. It seems therefore that each individual contains a large number of copies of a limited number of types, but that there may be polymorphism in the structure of the repeat.

Two major types of repeating unit can be resolved in L. pictus. Each has the same length and arrangement of genes, but they differ in their susceptibility to restriction enzymes. Cohn and Kedes (1979a,b) found that one is cleaved by EcoRI into fragments of 6.0 and 1.2 kb, the other into fragments of 4.1 and 3.1 kb. The isolation of these fragments from bulk DNA implies that the two types of repeating unit are not intermingled but must be located in different clustered groups. Both types of cluster are present in all individuals, which suggests that they are not alternative alleles. Heteroduplex mapping shows that every spacer has regions of nonhomology between the two types of unit; neither has any homology with the spacer sequences of S. purpuratus. Both types of gene cluster form R loops with histone mRNAs of the early embryonic type, so it is likely that they code for the same sequences (the difference between the early and late embryonic histone mRNAs is great enough to imply that cross hybrids would not form under the conditions used for R looping).

The existence of developmental variants of histones H1, H2A, and H2B in the sea urchin has been shown by direct analysis of the proteins; this is reproduced when mRNA extracted from the appropriate stage is translated in vitro, which suggests that the variation is coded in the form of different genes (see Chapter 10). Actually the sea urchin embryo derives its vast amount of histone protein from two sources. Histone mRNA comprises one of the major types stored in the maternal mRNA of the unfertilized egg (Skoultchi and Gross, 1973; Gross et al., 1973; Arceci, Senger and Gross, 1976; Lifton and Kedes, 1976). These mRNAs probably are the same types as the messages synthesized in high amounts at the onset of embryogenesis. The combination of stored mRNA and high synthesis of new mRNA ensures production of enough histone at a stage of very rapid cell division. These histones define the early embryonic type, whose synthesis is superseded by the late embryonic type as changes occur in the mRNAs transcribed at blastula and gastrula. An obvious question is whether the early and late genes are randomly arranged in repeating units or whether they reside in particular clusters. In the first case each gene must be controlled independently, and some members of a given repeat might be active and other inactive at one developmental stage; but in the second situation, entire repeats might be activated at the appropriate developmental stage.

The early and late embryonic mRNA sequences of S. purpuratus have been compared directly. Kunkel and Weinberg (1978) hybridized the histone mRNAs synthesized at early and late stages to a plasmid carrying the repeating unit described above, all of whose members are early genes. A reduction in the T_m of about 10°C is seen when late mRNA is hybridized to the early genes. With bulk histone DNA, the early mRNA forms a hybrid of which 80% melts accurately at high T_m, while 20% melts sooner as expected for the cross reaction with late genes. Probably these proportions correspond approximately to the relative numbers of each type of gene. A significant feature of the reaction is that when subfragments of the cloned repeat were used, the expected T_m depression was seen not only with the developmentally regulated histones H1, H2A, H2B, but also with H3 and H4.

The implication that there may be more than one type of gene coding even for histones whose sequence is invariant has been substantiated for H4 of L. pictus. Grunstein (1978) found that the H4 mRNAs of early and late stages vary in length as do the other histone messengers; and there are differences in the oligonucleotide fingerprint patterns of the two H4 mRNAs. This raises the possibility that the developmental switch is accomplished at the level of the repeating unit, and that early and late units have evolved a separate relationship between amino acid and nucleotide sequence. One interesting possibility is that this might be accomplished if early and late units were organized in separate groups, so that sequence identity could be maintained within the groups but not across the entire histone gene cluster. In this context, there is of course no evidence on whether the sea urchin histone genes exist as one cluster or several smaller clusters, so these groups might be part of a single locus or located at different chromosomal sites.

The difference in T_m between early and late sequences in S. purpuratus suggests a sequence divergence in the two types of gene of about 15%. This is equal to or greater than the divergence between the corresponding early genes of S. purpuratus and P. miliaris. This suggests that the division between early and late histone genes may have preceded the speciation of the sea urchins. A possible guess on timing would be that if S. purpuratus and P. miliaris separated 60×10^6 years ago, the early and late sequences diverged within the preceding 40×10^6 years. Overall the rate of substitution in the sea urchin genes seems to be about 3×10^{-9} substitutions per codon per year.

The Histone Gene Cluster: Expression

Several features distinguish the expression of histone genes from other nuclear genes. Histone proteins appear to be synthesized only during S phase, coordinately with the replication of DNA, in contrast to the expression of almost all other genes independently of the cell cycle (see Chapter 7). The kinetics of synthesis and degradation of the small histone messengers differ from those of the mRNA population as a whole, probably due to differences in processing and nucleocytoplasmic transport. The most striking structural feature of the histone mRNAs vis a vis other mRNAs is the absence or reduction of poly(A) content.

Early observations showed that the chromosome as a whole, not just its genetic material, is reproduced in S phase. The ratio of a Feulgen stain for DNA to a fast green stain for basic proteins stays constant during the division cycle of plant or animal cells (Alfert, 1958; Bloch and Goodman, 1955). The progress of chromosome reproduction can be followed visually in the macronucleus of Euplotes, where histone duplication can be seen to accompany DNA replication (Gall, 1959; Prescott, 1966).

Several lines of evidence suggest that histones actually are synthesized at the same time as DNA (not withdrawn from a pool of stored protein). By incubating onion root cells with labeled arginine or lysine, or instead with labeled thymidine, Bloch et al. (1967) showed that autoradiography of metaphase plates produces the same parameters for the cell cycle with either type of label. This implies that both components of the chromosome are synthesized at the same time.

A more direct approach is to measure the ability of the cytoplasm to synthesize histones in synchronized cultures. A large increase in the production of histone occurs at the start of S phase and persists so long as DNA synthesis continues (Robbins and Borun, 1967; Gallwitz and Mueller, 1969). Synthesis of the histones occurs on a peak of small polysomes, whose mRNA is about 9S and which can be translated in vitro into histones (Borun, Scharff, and Robbins, 1967; Adesnik and Darnell, 1967; Jacobs-Lorena et al., 1972; Breindl and Gallwitz, 1973). According to the in vitro translation assay, active histone mRNA is present only during S phase. This is true of yeast as well as animal cells (Moll and Wintersberger, 1976). Treatment of cultured animal cells with cytosine arabinoside to inhibit DNA synthesis appears to prevent histone synthesis by causing the histone polysomes to disappear, but there is a high background in such experiments (Borun et al., 1975). This conclusion is intriguing because histone mRNAs appear to be moderately stable, with a half life of some hours (Perry and Kelley, 1973). The concept that histone synthesis is tightly coupled to DNA synthesis has been challenged for cultured HTC cells, where inhibition of DNA synthesis by hydroxyurea is accompanied by only a rather modest decline in histone synthesis; the decline is greater (to about 30% of the usual level) in HeLa cells (in both cases measured overall on unsynchronized cultures) (Nadeau, Oliver, and Chalkley, 1978). It is possible, therefore, that the relationship between histone synthesis and DNA replication may depend upon the cell type. Experiments using a cloned probe carrying a sea urchin histone gene repeat have shown that corresponding sequences can be detected in cellular RNA throughout the cell cycle, and after treatment with cytosine arabinoside, although cytoplasmic mRNA is present only in cells synthesizing DNA (Melli, Spinelli, and Arnold, 1977). This raises the possibility that the control of histone gene expression may lie posttranscriptionally. One general feature of these experiments that should be noted is that all have been performed with cell lines adapted to grow in culture.

The demand for synthesis of equivalent amounts of histone and DNA need not necessarily be met by exact temporal coordination. Unfertilized and newly fertilized eggs of Xenopus and Arbacia (sea urchin) synthesize large amounts of histone, much in excess of the amount of DNA synthesized during the first two divisions, in anticipation of the very rapid replication of DNA that commences subsequently. During this second period the synthesis of DNA outstrips the synthesis of histones (Adamson and Woodland, 1974, 1977; Arceci and Gross, 1977). Chromatin reproduction presumably rests on the prior accumulation of adequate supplies of histone to form a pool for later use. Probably this histone synthesis is accomplished by translation of maternal messengers.

These experiments all were performed by using as an assay the translation of mRNA into histone proteins in vitro; now, of course, it is possible to make more direct measurements of mRNA levels by using cloned probes. The histone mRNAs were at first thought to constitute the sole messenger species lacking poly(A), a feature to which was attributed the more rapid exit of newly synthesized molecules from the nucleus (Adesnik and Darnell, 1972). Ironically, however, as other nonpolyadenylated mRNAs have been discovered, so has it been reported that Xenopus and Triturus histone mRNAs (and presumably also the histone messengers of other species) contain up to a third or so polyadenylated molecules (Levenson and Marcu, 1976; Ruderman and Pardue, 1978). Since the major class of poly(A)$^-$ mRNAs has kinetics of synthesis indistinguishable from that of the poly(A)$^+$ mRNAs, it does not now seem that any difference in polyadenylation accounts for the distinct kinetics of histone messenger synthesis.

One feature that distinguishes the histone genes is the apparent absence of any intervening sequences from the clusters in sea urchins and D. melanogaster. It remains to be seen to what extent the features of histone mRNA synthesis are influenced by the omission of splicing. No definite evidence is yet available on whether histones are transcribed into precursor RNAs from which the histone messengers are cleaved or whether the mRNAs may in fact be the primary transcripts. In HeLa cells, the production of each histone mRNA, measured by the synthesis of histone proteins in vivo or in vitro, is equally sensitive to inhibition by ultraviolet irradiation (Hackett, Traub, and Gallwitz, 1978). This would suggest that each gene lies the same distance from its promoter. Since the arrangement of histone genes is not known in HeLa cells, this cannot be fully interpreted, but would argue against the existence of a precursor RNA containing the sequences of all or several histone genes, which would place them at differing distances from the promoter. On the other hand, a cloned probe of sea urchin histone genes hybridizes with pulse-labeled HeLa nuclear RNA (Melli et al., 1977). The nuclear RNA is of high molecular weight (> 4000 kb), is not disaggregated by denaturing treatments, and is effectively competed by the 9S histone mRNA fraction. These results do not give any information about the structure of the large nuclear RNA, but the most obvious model is to suppose that it carries all or at least several of the histone message sequences. But further characterization of the putative precursor will be needed before it can be concluded that it represents an intermediate in the synthesis of histone mRNA. In sea urchin early embryos, the only RNA hybridizing with a cloned histone gene repeat is of the same size as the mature mRNAs. On nondenaturing gradients a high molecular weight peak is evident; but denaturation reduces its size to 9S. Kunkel, Hemminki, and Weinberg (1978) found, however, that later in development, after the switch to synthesis of late embryonic histones, a hybridizing peak of higher molecular weight material can be found even under denaturing conditions. This material occupies a rather broad size range. These results raise the possibility that the switch in histone gene expression is accompanied by a change in the pathway for synthesis of histone mRNA, caused by either the formation of a large percursor(s) or perhaps by a reduction in the rapidity with which such species are processed.

The disadvantage of using cloned probes representing the entire histone gene repeat, of course, is that the mRNAs and putative precursors for all histones should be equally well detected. Probably the existence of precursors will not be clarified until cloned probes for individual genes are used. In the sea urchin, the head to tail arrangement of all five histone genes opens the possibility that a common precursor might exist. This could be tested the most effectively by using cloned probes representing the spacer regions, an approach analogous to the use of cloned intervening sequences to detect and characterize the precursors to messengers representing mosaic genes (see Chapter 26). In contrast with the idea of a common transcript, in the S. purpuratus cluster the putative promoter sequence found 30 bp before the probable transcription initiation site of other genes (see Figure 26.14), is found about this distance before the point corresponding to the start of the H2B mRNA; it is also present about 100 bp before the initiation codon for H3. In P. miliaris it is present about 100 bp before the H2A coding sequence. Further data will be needed to see whether it precedes the other genes. Its presence would be consistent with the occurrence of independent initiation of transcription for each gene. At the point corresponding to the start of the mRNAs for H2B and H4, a sequence PyCATTCPu is found (Levy, Sures, and Kedes, 1979). This is present 50-70 bp before the initiation codon for all the sequenced S. purpuratus histone genes; it might in principle indicate the startpoint of transcription or the site of cleavage from a precursor. It is noteworthy that it is not related to the sequence found at the corresponding positions of the unique genes summarized in Figure 26.15. In contrast with the arrangement of sea urchin genes, in D. melanogaster the inverted arrangement of histone genes means that they must be transcribed from different strands of the DNA and therefore cannot be part of a single common transcription unit. The most obvious model is to suppose that each gene is independently transcribed. If any other model is adopted, it becomes necessary to suppose that a gene may be transcribed on both strands, one during the formation of mRNA, and the other as a sequence that is present in a precursor and is discarded during formation of mRNA for other gene(s).

Globin Gene Relationships: Gene Numbers

The structure and genetics of hemoglobin has been the subject of intensive study since the genesis of molecular biology; the more recent application of techniques for characterizing gene organization directly has made it possible to test and extend by molecular analysis the inferences that had been drawn previously. Figure 28.2 summarizes the relationship between the human globin gene family and hemoglobin proteins. There is both genetic and structural evidence on the number and organization of fetal and adult genes (α, β, δ, γ); the embryonic ζ and ϵ genes have not yet been characterized.

The existence of two α genes per haploid genome was suggested by the presence of normal HbA in individuals homozygous for some α chain vari-

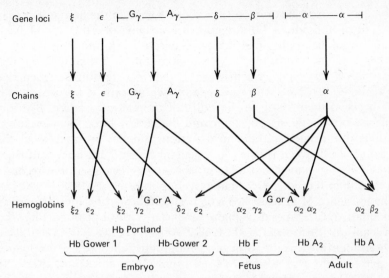

Figure 28.2. The human globin gene family. All hemoglobins appear to be tetramers consisting of two types of chain. Related to the adult alpha and beta chains, delta differs from beta only by 10 residues, gamma has many differences from the beta chain, epsilon appears to have features in common with both beta and gamma, and zeta has similarities to alpha. The first chains to be synthesized in the embryo are the zeta and epsilon polypeptides, followed by alpha and gamma. Thus Hb Gower 1 and 2 and Hb Portland all are found in embryos up to 8 weeks, although their exact temporal relationship is not yet defined. From 8 weeks to term the predominant hemoglobin is HbF. Then this is replaced by adult hemoglobin; HbA constitutes about 97%, HbA_2 provides about 2%, and up to about 1% HbF persists. For review see Weatherall and Clegg (1976, 1979).

ant (Brimhall et al., 1970; Lie-Injo et al., 1974).* Formally the genetic result means that there is more than one α gene. The basis for setting the number at 2 lay in observations of the proportion of HbA in individuals homozygous or heterozygous for α chain variants (Lehmann and Carrell, 1968). This is rather weak evidence at best, however, in view of the many factors that may intervene in the representation of a gene in protein. The hybridization of α mRNA or cDNA probes with the DNA of human-mouse hybrid cell lines carrying different human chromosomes shows that the presence only of chromosome 11 correlates with the human α gene sequences (Deisseroth et al., 1977). The α gene copies therefore must be present on the same chromosome.

*There are some data on individuals who are homozygous for α variants and lack normal HbA. At first these were taken to suggest that these individuals might represent a population with only one α gene in the haploid genome. This would imply that the human genome is polymorphic for the number of α genes. However, these individuals are found only in populations carrying α thalassemia, so it is now thought that probably the second α gene has been inactivated (for review see Weatherall and Clegg, 1979). The present view is therefore that probably all human populations have two α genes in the wild-type haploid genome.

There appears to be only one β and one δ gene in the human haploid genome, since individuals homozygous for variants of either type have only the abnormal hemoglobin. Linkage between the β and δ genes was first inferred from analysis of the Lepore hemoglobin, in which the non-α chain consists of N terminal sequences of δ fused to C terminal sequences of β (Baglioni, 1962). More than one type of Lepore now is known, the differences between them lying in the point of transition from δ to β sequence. Their structure suggests that the δ and β genes lie close together, are transcribed and translated in the same direction, and may be fused by deletion of the sequences between them. Close genetic linkage has been confirmed by studies showing that β and δ chain variants segregate without recombination (Weatherall et al., 1976; Stamatoyanopoulos et al., 1978).

The existence of at least two γ genes is indicated by the presence in all individuals of two types of γ chain; γ^G has glycine at position 136, whereas γ^A has alanine (Schroeder et al., 1968).* A fusion between the γ and β chains has been identified in the form of Hemoglobin Kenya; the γ residues are N terminal and the β residues are C terminal (Huisman et al., 1972b; Smith et al., 1973). Since normal γ^G is found in individuals with this variant, presumably γ^A is involved in the fusion. This suggests that there is a cluster of at least three globin genes in the order $\gamma^A - \delta - \beta$. Also the γ^G locus is assumed to be part of the same cluster. Presumably this set of genes has evolved by duplication followed by accumulation of differences by mutation. Studies of the ability of DNA from hybrid cell lines to hybridize with β or γ probes suggest that both types of gene lie on human chromosome 16 (Deisseroth et al., 1978). Since there is good cross hybridization between β and δ gene sequences, a further inference is that δ also must lie on this chromosome.

Two forms of direct gene analysis have been applied to quantitate the globin genes. Measurements of the hybridization of mRNA with cellular DNA are discussed in Chapter 24. As can be seen from Table 24.2, the results show clearly that there is a low number of copies of each globin gene. The imprecision of the method, however, does not resolve the exact number. It is fair to say perhaps that there seem to be between 1 and 3 copies each of the α, $\beta + \delta$, and γ genes. Claims to have verified the gene numbers predicted by genetic studies are therefore premature. The lack of greater resolution is in part inherent in the method and in part results from the use of probes that are not completely purified. Some improvement could therefore be obtained by using cloned sequences, but greater precision is in any case available from the alternative approach of characterizing the genomic

*There have been suggestions that there may be more than two γ genes in the human genome. Levels of the abnormal hemoglobin in variants are highly variable, but these data are not too reliable as an indicator of gene number (Huisman et al., 1972a). Another variant, γ^T has threonine at position 75 instead of isoleucine; this is a form of γ^A since it has alanine in position 136 (Ricco et al., 1976; Saglio et al., 1979). It is thought that γ^T is probably an allele of γ^A, rather than a duplicate locus, but direct proof of this requires genetic studies.

restriction fragments that possess globin sequences. This allows detailed maps to be constructed of the genes (see below).

Human thalassemias are of great interest because they result from mutations that prevent globin synthesis, some of which may reside in control elements. Thalassemias may reside in either α or β chain synthesis and are divided into two general classes. The types α^0 and β^0 have no detectable synthesis of the named adult chain type, in contrast with the types α^+ and β^+ that have reduced levels of the affected chain. The major thalassemias are described in Table 28.2. Both the hybridization and blotting approaches have been applied to the thalassemias to resolve the question of whether the deficiency in synthesis of a given globin chain results from a gross deletion or from a lesser mutation. Here the hybridization method is concerned not with determining absolute values for gene copy number, but with comparing the wild type and thalassemic DNAs to see whether a large loss of hybridizing material can be found. Available data are summarized in Table 28.3. The corresponding approach with restriction fragments is to see whether fragments blotted from the wild type genome are absent or reduced in size in thalassemic genomes.

The α^0 thalassemias occur in individuals homozygous for what is described as the *α-thalassemia-1* genotype. This is thought to result from the deletion of both α genes, but as can be seen from Table 28.3 the hybridization data are sufficient only to conclude that there has been a gross deletion of α gene sequences; they do not prove unequivocally the absence of all α genes. Less severe thalassemic syndromes are associated with the *α-thalassemia-2* genotype, either homozygous or heterozygous with α-thal-1. The α-thal-2 genome is thought to result from the deletion of one of the α genes; it is not clear whether different α-thal-2 genomes should be expected to have deletions of the same or alternative α genes. Hybridization with the DNA of an α-thal-1/α-thal-2 heterozygote suggests that it contains α genes at a lower concentration than wild type. This would be consistent with the proposal that one genome has no α genes and the other has 1 α gene, giving 1 α gene per diploid cell instead of the expected 4. But the data are not sufficiently precise to prove the point. The numbers of active α genes given for the various α thalassemias described in Table 28.2 are calculated on the assumption that the α-thal-1 genome has no α genes, whereas the α-thal-2 genome has 1. A further type of α chain defect is presented by the example of Hb Constant Spring, whose α chain is extended due to mutation of the termination codon. This behaves as an α-thal-2 type, as do other such termination mutants, because the output of the mutant gene is eliminated or much reduced (Clegg et al., 1971; Weatherall and Clegg, 1975).

The β thalassemias are heterogeneous in type and appear to result from a variety of defects at the molecular level. A feature common to all β^0 and β^+ types that have been tested is the absence of any detectable reduction in β-globin gene sequences. As can be seen from Table 28.3, however, this con-

Table 28.2   Characteristics of human thalassemias

Type	Description	Genotype	Clinical effects	Probable diploid gene constitution
α^0	α thalassemia-1	α-thal-1/α-thal-1	Hb Bart's hydrops fetalis	0 α genes
α^+	α thalassemia	α-thal-1/α-thal-2 or α-th-1/Const Spr	HbH hemolytic anemia (HbH = β_4 tetramers)	1 α gene 2 α genes
α^+	α thalassemia-2	α-thal-2/α-thal-2	minor thalassemic symptoms	2 α genes
α^+	heterozygous α thalassemia-1	α-thal-1/wild type	minor thalassemic symptoms	2 α genes
α^+	silent carrier	α-thal-2/wild type	no effects	3 α genes
α^+	Hb Constant Spring	Hb-CS/Hb-CS	minor thalassemic symptoms	4 α genes
β^0	types 1, 2, 3	homozygous	Cooley's anemia; 98% HbF & A_2	2 β genes
β^0	types 1, 2, 3	heterozygous	minor thalassemic symptoms	2 β genes
β^+	type 1	homozygous	Cooley's anemia: 80% HbF	2 β genes
β^+	type 2 (Negro)/ type 3 (Silent)	homozygous	intermediate thalassemia: 30–50% HbF/15–30% HbF	2 β genes
$\delta\beta^0$	$\gamma^G \gamma^A$ type of $\delta\beta$ thalassemia	homozygous	Cooley's anemia: HbF has both γ types	no β or δ genes
HPFH	Negro $\gamma^G \gamma^A$	homozygous	deficiency in δ and β chains compensated by persistent synthesis of γ chains	no β or δ genes

These principal types of thalassemia have been characterized at the molecular level. Details of further, less well-characterized syndromes can be found in Weatherall and Clegg (1976, 1979), from whom this summary is taken. The number of diploid genes indicates the expected total of wild type *and* mutant gene sequences present, that is, it is reduced only where total or major deletions are thought to remove gene sequences.

clusion rests on the use of an impure cDNA$_\beta$ probe that cross reacts with δ gene sequences. Thus although a complete reduction of the β gene should be detectable as a 50% reduction in hybridization, these results do not formally exclude the possibility of a smaller deletion in part (for example the 5' end) of the gene. Of course, in the β^+ thalassemias the production of β globin mRNA indicates the presence of the gene.

     The syndromes of $\delta\beta^0$ thalassemia and HPFH (hereditary persistence of fetal hemoglobin) appear to have a similar basis in the joint deletion of the

Table 28.3 Globin gene frequencies in human thalassemias

Thalassemia	Assay	Results
α^0 (α-1/α-1)	cDNA$_\alpha$ × excess DNA	DNA of α-thal-1 saturates only 20% or 40% of cDNA probe compared with levels of 50% and 70%, respectively, reached with wild type DNA; indicates deletion of at least half of the cDNA complements
α^+ (α-1/α-2)	cDNA$_\alpha$ × excess DNA	thalassemic DNA drives probe with $Cot_{1/2}$ of 3 × 10^3 compared with wild type $Cot_{1/2}$ of 10^3; indicates roughly 3 fold reduction in concentration of cDNA$_\alpha$ complements in thalassemia. Probe is 60% reacted by thalassemic DNA, 75% reacted by wild type DNA; indicates presence in thalassemic DNA of most or all sequences represented in cDNA$_\alpha$
β^0 and β^+	cDNA$_\beta$ × excess DNA	reactions of cDNA$_\beta$ tracer with wild type or thalassemic DNA are indistinguishable; indicates no detectable deletion of cDNA$_\beta$ complements in thalassemia
HPFH	cDNA$_\beta$ × excess DNA	DNA of HPFH saturates only 20% of cDNA$_\beta$ probe compared with 75% level reached with wild type DNA; indicates deletion of at least 75% of sequences complementary to cDNA$_\beta$ (includes both β and δ genes)
$\delta\beta^0$	moderate cDNA$_\beta$ excess × DNA	cDNA$_\beta$ reacts 1/3 as well with thalassemic DNA as with wild type; indicates deletion of 67% of cellular DNA sequences complementary to cDNA$_\beta$ probe

All cDNA preparations represent probes prepared by reverse transcription of mRNA; "DNA" indicates the use of total cellular DNA. The cDNA × excess DNA reactions use the labeled probe as a tracer to determine either its $Cot_{1/2}$ or extent of reaction when renaturation of cellular DNA is complete. A deletion of the entire sequence corresponding to the probe should result in no hybridization; partial deletion should reduce final extent of reaction of probe, but does not affect $Cot_{1/2}$; reduction in number of copies should increase $Cot_{1/2}$ of reaction, but does not affect terminal level. All cDNA probes used here were enriched for α or β sequences appropriately, either by exhausting the other type of sequence from cDNA by reaction with mRNA from thalassemia, or through preparing by reverse transcription of mRNA of opposing type of thalassemia. Probably only about 75% purity is achieved. This means partial reactions may indicate either that only some of the probe sequences have been deleted; or more sequences could be deleted, with reaction due to contaminating sequences in probe. Extent of deletion is probably therefore underestimated. However, all probes are rather short and therefore detect the sequences complementary largely to the 3′ regions of the messenger. The cDNA$_\alpha$ probe may react to some extent with zeta genes as well as α genes; cDNA$_\beta$ will react with both β and δ genes.

References: α^0—Ottolenghi et al. (1974), Taylor et al. (1974); α^+—Kan et al. (1975a); β—Tolstoshev et al. (1976), Ramirez et al. (1975); HPFH—Forget et al. (1976); $\delta\beta^0$—Ottolenghi et al. (1976).

linked δ and β genes. The $\delta\beta$ fusion in Hb Lepore has the same effect as the $\delta\beta^0$ type. The difference between $\delta\beta^0$ thalassemias and HPFH lies in whether γ chains continue to be produced in adult life; one possibility is that this may be determined by the length of the deletion, possibly by whether it removes particular control sites. Because both of the sequences with which the cDNA$_\beta$ probe reacts are removed in these genotypes, the hybridization data here offer the clearest evidence that a gross deletion has occurred.

Globin Gene Relationships: Mutant Genes

The human α globin gene(s) have been mapped by isolating the restriction fragments able to hybridize with an α cDNA probe. The only corresponding sequences present in an EcoRI digest reside on a 22.5 kb fragment (Orkin, 1978a). If this is a unique fragment, all α-globin genes must lie within its span; the alternative that this fragment is represented more than once in the genome seems unlikely because it would demand that the repeated sequence is > 22 kb in length.

The difficulties of distinguishing duplicated genes from interrupted genes solely on the basis of restriction mapping are illustrated in Figure 28.3, which shows two possible gene organizations for the 22 kb fragment. Double cleavage with EcoRI and HpaI yields two fragments carrying α-globin sequences, but this can be equally well explained by duplication or interruption. The enzyme HindIII is known to cut within the middle of the coding sequence, but the fragments that are produced could represent either cleavage in each of two gene copies or cleavage in both an intervening sequence and coding sequence of one gene copy. The four fragments produced by cleavage using both HpaI and HindIII make it necessary to suppose that if there is only one gene copy, it has two intervening sequences, one with an HpaI site and one with a HindIII site.

The two models can be distinguished by using a probe representing one end of the gene. With cDNA enriched for short reverse transcripts, the 3′ end of the gene alone should react. This cDNA reacts with both HpaI fragments, as would be expected if each contains one copy of the gene; but with HindIII the two smaller fragments react just as well as with longer cDNA probes and it is hard to tell whether there is a reduction in reaction with the largest fragment (the prospective 5′ end). The difficulty with short cDNA probes is their lack of purity; cleaner results might be obtained with cloned fragments, which would also allow the necessary reciprocal experiments to be performed for both 5′ and 3′ ends of the gene. This evidence therefore favors but does not rigorously prove a two gene model in which both genes lie in the same orientation (left to right as drawn in Figure 28.3).

If there are two genes, they can be placed about 3.7 kb apart if it is assumed that there are no intervening sequences in either gene. Actually these data say little about the possible presence of intervening sequences in the pro-

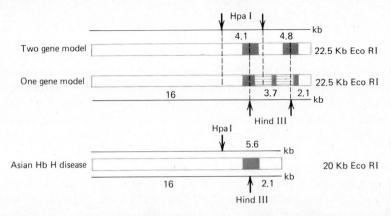

Figure 28.3. Restriction map of human α globin gene(s). Sites of cleavage have been mapped by restriction of total DNA followed by agarose gel electrophoresis, blotting onto nitrocellulose, and identification of fragments able to hybridize to a labeled cDNA probe for α-globin. Two models are shown to account for the location of sites on the 22.5 kb EcoRI fragment. Both generate two hybridizing fragments on cleavage with HpaI, three fragments on cleavage with HindIII, and four fragments on cleavage with both enzymes. The restriction map of the 20 kb EcoRI fragment from Asian HbH disease is most readily interpreted as the presence of a single α-globin gene, generated by removal of the sequences lying between the two HindIII sites in the two gene model. Shaded segments indicate regions represented in mRNA, striped regions are intervening sequences. Data of Orkin (1978a) and Orkin et al. (1979).

spective two genes; except that (reversing the arguments on the location of the restriction sites for the possible single interrupted gene) any such interruptions in the two gene structure may be expected to lack HpaI and HindIII sites. The presence of intervening sequences in the two genes must bring them closer together if both lie in the same orientation.

The complications in analyzing a duplicated gene structure in which there may be also prove to be intervening sequences emphasize the need for a second, independent line of evidence. This may best be provided by using electron microscopy to characterize the hybrids formed with α-globin mRNA in cloned 22.5 kb fragments. This should show unequivocally whether there are two genes, how far apart they lie, and whether large intervening sequences are present. The fine structure may be determined then by restriction mapping of cloned subfragments and ultimately by sequencing.

A variety of restriction fragments has been found to contain α globin sequences in the genomes of individuals suffering from α thalassemias. The severe case of hydrops fetalis is accompanied by a loss of material hybridizing to a cDNA$_\alpha$ probe; presumably this corresponds to the anticipated complete deletion of α globin sequences from both chromosomes. Individuals displaying HbH disease are expected to have only one active α globin gene per cell; but in contrast with the expectation that the others all have been deleted, several different restriction patterns are found in the DNAs of different patients (Orkin et al., 1979a).

In an example of HbH disease in Asians, a single EcoRI fragment of 20 kb is found. As Figure 28.3 shows, double cleavage with EcoRI/HpaI releases a single α fragment instead of the wild type two fragments; double cleavage with EcoRI/HindIII releases the same two fragments assigned to the ends, but the central fragment is absent. The simplest explanation of these results is that in wild type DNA there are two α globin genes and that in the thalassemia these have been fused by deletion extending from a point in one gene to the corresponding point in the next. This could occur by unequal crossing over (the reciprocal event should be seen in a chromosome with 3 α globin genes). It is the synthesis of some α globin chains in HbH disease that implies the existence of an active gene on the 20 kb fragment, which in turn supports the two gene rather than one gene model for the wild type. To explain the severity of the thalassemia, it is necessary to assume that the 20 kb fragment represents the α gene of one chromosome only; its homologue is expected to have a complete deletion. This is the "classical" HbH disease described above.

Other restriction patterns have been seen in HbH patients of Mediterranean origin. In one type an EcoRI fragment of the normal 22.5 kb size is present together with a fragment of 2.6 kb; in another a 20 kb fragment is present together with the 2.6 kb band. The 22.5 kb and 20 kb fragments present in these types show the same responses to further restriction with other enzymes as shown in Figure 28.3 for wild type and Asian HbH, respectively. A single example has been reported for another type in which only a 22.5 kb fragment is present. The simplest interpretation is to assume that the 22.5 kb, 20 kb, and 2.6 kb fragments seen in these thalassemias each are homogeneous and represent the same defects in different individuals. To arrive at the same value of 1 active α gene per cell, then it is necessary to suppose that the 22.5 kb fragment actually carries two gene sequences, but only one is active. In the individual with only this fragment, the other chromosome must be completely deleted for α genes. It follows that the 2.6 kb fragment must carry α gene sequences that are not expressed, so that the two EcoRI fragments present in the 22.5/2.6 kb and 20/2.6 kb patterns in each case represent the sequences present on the homologues. This predicts that there might be a 2.6/2.6 kb pattern with a defect similar to hydrops fetalis.

To summarize present data, the simplest interpretation is to follow the two gene model, but this is not yet rigorously proven. Obviously it will be necessary to construct more detailed maps to define the basis of the defects in the α thalassemias. The most immediate question is to determine the nature of the α sequences present in the hybridizing fragments. Reductions in size do not necessarily imply loss of coding sequences, but could be derived from changes in flanking or intervening sequences. Measurements of total gene number are not sufficiently reliable to resolve the point, since mostly they use cDNA probes biased in representation of 3' regions. But these results suggest that there may be heterogeneity in the types of mutation responsible for reducing α gene expression; different mutant loci may be predominent in different human populations. An especially important point is the demonstration that some α thalassemic genomes do not have gross deletions detectable by present restriction mapping. Like the case of the β thalassemias, this raises the possibility that there may be mutations to be found in control regions.

A detailed restriction map for the region containing the γ, δ, and β human globin genes has been constructed for both wild type and various mutant DNAs. A complication in identifying the δ and β sequences is their cross reaction, but this has been approached by hybridizing cloned β sequences with restricted DNA under conditions of differing stringency. An increase in stringency reduces the reaction of some fragments; these are seen as fainter bands and are assigned to the δ locus. With genomic DNA fragments the best evidence for gene assignments is provided by comparing the wild type and Lepore restriction patterns (Flavell et al., 1978; Mears et al., 1978). Direct linkage of the δ and β genes is demonstrated by the existence of only a single XbaI fragment of ~12 kb that reacts with the β probe; this is reduced to ~4 kb in the Lepore genotype.

More detailed analysis generates the map summarized in Figure 28.4, in which the δ and β genes lie some 7 kb apart. Sequencing of the relevant parts of a cloned (randomly cleaved) fragment of wild type human DNA confirms the identity of the two genes and shows further that the large intervening sequence present in each lies between codons 104 and 105 (Lawn et al., 1978). A smaller intervening sequence also is present earlier in at least the β gene. The appropriate restriction fragments corresponding to the interruptions in the δ and β genes do not hybridize, which suggests that their sequences are largely different. The exact point of fusion between the genes in the Lepore deletion is not known, but must lie between amino acids 87 and 116 in the example (Lepore Boston) that has been mapped; this means that the fusion may have occurred just on either side of or within the intervening sequence. The most likely origin for the fusion is the occurrence of unequal crossing over between the δ and β genes on meiotic homologues. The reciprocal of this event should be the anti-Lepore gene, identified in the form of abnormal globin chains carrying N terminal β and C terminal δ sequences; its origin could be confirmed by demonstrating that the mutant genome has a fused $\beta\delta$ gene lying between normal δ and β genes.

The two fetal γ-globin genes of man have been shown to be adjacent by the isolation of a single genome fragment that hybridizes with a cloned γ sequence (Little et al., 1979; Tuan et al., 1979). A more detailed map shows that both genes have a large intervening sequence and lie no more than 3.5 kb apart. The γ^G and γ^A genes can be distinguished because their difference in sequence means that a PstI site is present only in the latter. Close linkage between the two γ genes and the δ and β genes has been shown by demon-

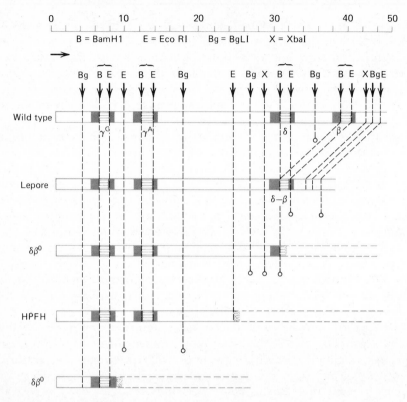

Figure 28.4. Restriction map of the human β-like globin gene cluster. In the wild type, the two γ genes are separated by 3.5 kb, the δ gene is about 16 kb distant and is separated from the β gene by 7 kb. Transcription occurs from left to right. All four genes have the large intervening sequence as shown by the separation of the internal BamHI and EcoRI sites (indicated by the bracketed B-E pairs). A smaller intervening sequence may also be present closer to the 5′ ends, but has not been detected yet by the restriction patterns. In the Lepore mutation the 5′ δ sequence has been fused to the 3′ β sequence by deletion of about 8.4 kb. The exact point of fusion is not known, so the internal BamHI and EcoRI sites in the fused gene may be derived from either the δ or β wild-type gene. In the cases of δβ° thalassemia and HPFH, deletions are found with left ends in the γ-δ gene region and with right ends that lie beyond the EcoRI site to the right of the β gene. In one case of δβ° thalassemia, the left end of deletion lies in the large intervening sequence in the δ gene; in the other it lies just to the right of the γ^G gene (which must be intact since γ chains can be produced). In two independent cases of HPFH the same deletion was found, starting just on the right of the EcoRI site lying between the γ^A and δ genes. For simplicity only a few of the many mapped restriction sites are shown. The presence of a site in a given DNA is indicated by lines passing through the genome; its absence is indicated by the termination of the line in a circle above the genome. The δ-β loci and the Lepore deletion have been mapped by Flavell et al. (1978), Lawn et al. (1978), Mears et al. (1978); the γ genes and the δβ° and HPFH deletions have been mapped by Fritsch, Lawn, and Maniatis (1979), Little et al. (1979), Tuan et al. (1979).

strating that a short subfragment of the cloned δ-β gene complex can hybridize with fragments carrying the γ sequences. Linkage between globin genes of like type also has been found in the chick genome, where a fragment has been found carrying both an adult and an embryonic β gene (Dodgson, Strommer, and Engel, 1979).

Restriction mapping of the region between the human γ genes and the δ-β genes has been used to locate deletions that cause $\delta\beta^0$ thalassemia or hereditary persistence of fetal hemoglobin (Tuan et al., 1979; Fritsch, Lawn, and Maniatis, 1979). The deletion in HPFH takes the expected form of removing completely all δ and β sequences and some of the region on their left. One form of $\delta\beta^0$ thalassemia has a deletion whose left end is within the large intervening sequence in the δ gene. The other $\delta\beta^0$ thalassemia has the most extensive deletion, removing all sequences to the right of the γ^G gene. These data do not fit the model in which the HPFH syndrome is less severe than $\delta\beta^0$ thalassemia because the HPFH deletion is longer than the $\delta\beta^0$ deletion and removes some site usually responsible for switching off the γ genes in adult life. However, the clinical distinction between HPFH and $\delta\beta^0$ thalassemia is not straightforward; and more detailed analysis of the hemoglobins produced by these individuals may be necessary to resolve the pattern of gene expression. It remains to be seen, therefore, precisely how γ gene expression is influenced by deletions extending from the $\delta\beta$ gene region. If a regulator site for adult γ expression does exist, however, it may be expected to lie in the region just to the left of the δ gene. There is evidence that the control of γ gene expression in $\delta\beta$ deletions is cis dominant, that is, only γ genes contiguous with the deletion are active. This would therefore imply that a cis dominant site exists on the 3' side of the gene.

The β^0 and β^+ thalassemias are highly heterogeneous. Less attention has been paid to the β^+ type, in which reduced amounts of globin mRNA are present; the basis of the defects is not known, but the presence of β mRNA implies that the gene is present. Attention on the β^0 thalassemias has been focused on the belief that the β globin gene is present but fails to be expressed. Actually the evidence from hybridization studies is not decisive; certainly β globin sequences seem to be present in β^0 thalassemic DNA, but this does not prove the presence of an intact gene (see Table 28.3). More recently a series of β^0 thalassemic DNAs has been examined by restriction mapping; and in most instances no change is seen in the sizes of the fragments generated by several enzymes (Flavell et al., 1979; Orkin et al., 1979b). But in three patients of Indian origin, cleavage with PstI generated two fragments containing the β globin gene, one the usual size and the other 600 bp shorter. The first fragment is cleaved as expected by EcoRI; but the second fragment is not. Both fragments appear to have the internal BamHI site. A complication in interpreting these results is the presence of the normal pattern as well as the appearance of a new pattern in the Indian β^0 thalassemia. A plausible interpretation is that the patients are heterozygous for two different β^0 thalassemic defects. One chromosome carries a β gene whose gross structure is normal, as is apparently common of most β^0 thalassemias. The homologous chromosome carries a β globin gene that has suffered a deletion of about 600 bp, which includes the internal EcoRI site. As with other studies, the use of cloned fragments that can be mapped in detail will be needed to confirm the inferences that are drawn from mapping of genomic DNA.

The first measurements of β mRNA levels in β^0 thalassemias led to some confusion, because in some studies no β mRNA could be detected, whereas in others it was possible to detect material hybridizing with a $cDNA_\beta$ probe (Forget et al., 1974; Tolstoshev et al., 1976; Kan et al., 1975b). The situation has been clarified by the demonstration that the heterogeneity is genuine and that at least two general categories of β^0 thalassemia can be distinguished by the level of β mRNA (Benz et al., 1978; Old et al., 1978).

In the first type no β-globin mRNA can be detected by hybridization with cDNA probes enriched or specific for β sequences (although it should be noted that there is a high background in these experiments). In a comparison of two patients of this type, one showed a concentration of β sequences in the nucleus no greater than the background in the cytoplasm, while the other possessed appreciable reacting nuclear RNA (Comi et al., 1977). This would suggest that the absence of β mRNA may result from either transcriptional or post transcriptional defects.

The second type of β^0 thalassemia displays β globin sequences in mRNA, but the amount of mRNA may be much reduced or it may hybridize only incompletely to the cDNA probe. It is likely that the defect in β-globin chain synthesis takes several different forms. In most cases the mRNA is unable to act as a template for protein synthesis; in different cases probably a different defect is involved. In the unusual case of Ferrara β^0 thalassemia, it is possible that the β mRNA may be able to function upon addition of supernatant factors from wild-type reticulocytes (Conconi et al., 1972).

The molecular defect has been identified in one case of β^0 thalassemia. The globin mRNA present in reticulocytes of a Chinese patient includes the oligonucleotides characteristic of the β message (Temple, Chang and Kan, 1977). Chang and Kan (1979) have determined the sequence of part of this mRNA by applying the Maxam-Gilbert method to a fragment of cDNA. This suggests that the AAG codon for lysine at position 17 has been mutated to a UAG amber codon. Although this thalassemia appears therefore to be due to a nonsense mutation, a further interesting feature is the small amount of β globin mRNA that can be detected. This raises the possibility that the premature termination of globin chain synthesis may lead to degradation of the mRNA. This type of mutation is the reverse of that seen in some α thalassemias, such as Hb Constant Spring, in which the usual chain termination codon is mutated to a sense codon, allowing synthesis of an elongated α chain (but which is produced only at very low levels).

Drosophila Heat Shock Genes

The response of Drosophila cells to heat shock defines a family of genes, dispersed at various chromosomal locations, whose expression is under common control. The existence of a response to heat shock in polytene cells has been known for some time; when Drosophila larvae or tissues are incubated at 37°C, a set of puffs is immediately induced (within 1 minute) at 9 polytene bands in D. melanogaster, 6 bands in D. hydei. Within an hour, the bands that were previously active at normal temperature (25°C) regress, leaving only the heat shock loci active; the size of the heat shock puffs depends upon the temperature and duration of heat shock (Ritossa, 1964; Berendes, van Breugel and Holt, 1965; Ashburner, 1970b). Although known as the heat shock puffs, these sites are activated by many other agents, many of which act on mitochondrial electron transport and oxidative phosphorylation (for review see Ashburner and Bonner, 1979). The purpose of the response, in the teleological sense, is unknown; but may comprise a homeostatic mechanism, possibly a response to anaerobiosis.

Present studies of the heat shock genes were initiated by the observations of Tissieres, Mitchell, and Tracy (1974) on protein synthesis in D. melanogaster following heat shock. The increase in temperature causes a dramatic change in the bands resolved by gel electrophoresis when newly synthesized proteins are labeled. The large number of bands previously present is replaced by about 7 bands; and the pattern shown in heat shock is the same in both polytene and diploid larval tissues and in adult tissues. Addition of actinomycin D or α amanitin to inhibit the induction of puffs is able to prevent synthesis of the heat shock proteins (Lewis, Helmsing, and Ashburner, 1975).

This work extends the cytological observations in two important directions. First by demonstrating that the changes in puffing patterns of polytene chromosomes do indeed reflect changes in gene expression, this opens the possibility of correlating specific proteins and genes with particular genetic loci. This aim has been hard to fulfill without the specific probes made possible by the heat shock system (see Chapter 16). Their common response to heat shock makes these genes a suitable system in which to seek the control elements that define their membership of the family. A second point of general interest is that by turning to the criterion of protein synthesis, it is clear that the heat shock response is not a feature just of the larval cells in which it can be seen cytologically; it appears to be common probably to all cells and stages of development. It may become possible to analyze the mechanism of induction through the observation of Compton and McCarthy (1978) that cytoplasm from heat shocked cells can induce puffing at the heat shock loci in salivary glands incubated at normal temperature.

The heat shock response is displayed by tissue culture cells, which has made it possible to identify the messengers transcribed from the heat shock genes (McKenzie, Henikoff, and Meselson, 1975; Spradling, Penman, and Pardue, 1975). When polysomes are extracted from cultured cells grown at 25°C, their mRNAs show the usual polydispersed array of sizes and hybridize in situ with a large number of polytene bands. But upon heat shock these messengers disappear from the polysomes, to be replaced by heat shock messengers. This loss of normal messengers does not appear either to be a simple consequence of the cessation of transcription at the corresponding puff sites or to result from displacement by new messenger species. Polysome disaggregation occurs more rapidly than puff regression; and it happens even when actinomycin is added to prevent synthesis of the heat shock messengers. The normal mRNAs may remain intact in the cell following displacement from the polysomes, as judged from the ability of nonpolysomal poly(A)$^+$ RNA to support translation in vitro of the proteins characteristic of normal temperatures (Mirault et al., 1978). This suggests that the switch in gene expression upon heat shock may involve translational as well as transcriptional controls.

The poly(A)$^+$ mRNA of heat shock cells forms two peaks at 20S and 12S on denaturing gradients; and these are further separated into three bands each by gel electrophoresis. The abilities of these separated fractions to hybridize in situ with polytene chromosomes are summarized in Table 28.4. The products of these mRNAs upon translation in vitro identify the loci corresponding to several of the heat shock proteins. The messengers and proteins corresponding to the three minor puff sites have not yet been identified. One of the major puff sites, 93D, does not appear to be represented at all in the poly(A)$^+$ mRNA; and this site also differs from the others in the characteristics of its induction. One protein has been identified as the product of 63C; another is coded by 95D. Two proteins that are not related in sequence appear to be the products of 67B. The best investigated situation is the production of what may be several closely related proteins of about 70,000 daltons by a messenger RNA fraction that hybridizes with both 87A and 87C. In addition to these heat shock specific proteins, histones continue to be synthesized following heat shock.

A series of experiments has now begun to characterize the DNA sequences present at the heat shock loci. The common approach is to screen a library of clones carrying D. melanogaster DNA to isolate sequences complementary to the heat shock mRNAs. Then the clones can be equated with particular loci by in situ hybridization with polytene chromosomes. The first to be analyzed correspond to the A2 and A4 messengers.

Three plasmids carrying sequences found at 87A and 87C have been mapped by restriction cleavage and R loop formation (Lis, Prestidge and Hogness, 1978). Heteroduplex formation shows that they possess internal repetitions and these are due to the presence of two types of repeats, designated $\alpha\beta$ and $\alpha\gamma$. The α sequence is about 500 bp long and the β sequence about 1100 bp; a single HindIII cleavage site exists in α and releases 1.5 kb fragments from tandem arrays of the form $\alpha\beta\alpha\beta$. . . . These sequences hy-

Table 28.4 Heat shock gene expression in D. melanogaster

Locus	Puff size	Relative mRNA production	Messenger RNA size and abundance	Protein size and abundance	
33B	minor	3%	minor poly(A)$^+$	not identified	
63C	major	6%	A1 (20S) major	82K	
64F	minor	3%	minor poly(A)$^+$	not identified	
67B	major	5%	A5 (12S) minor	23K and 26K	
70A	minor	2%	minor poly(A)$^+$	not identified	
87A	major	10%	A2 (20S) major }	70K {	50% of total
87C	major	34%	{ A2 (20S) major J { A4 (12S) minor		heat shock protein none
93D	major	35%	no poly(A)$^+$ RNA	not identified	
95D	major	4%	A3 (20S) major	68K	

Puffs are divided into major and minor classes on the basis of the size seen in salivary glands. Production of mRNA is assessed by the number of grain counts observed at each locus when total labeled heat shock RNA of imaginal discs is hybridized in situ (Bonner and Pardue, 1976). The 93D puff is unlike the others in showing wide variation in expression, subject to experimental conditions, according to both criteria. The heat shock mRNAs are either 20S (about 2.6–2.9 kb) or 12S (about 1.5–1.9 kb) in size; gel electrophoresis fractionates them into species A1-A6, decreasing in size. Five heat shock loci are hybridized in situ by individual mRNA fractions; other loci are noted for hybridizing or not with total poly(A)$^+$ heat shock RNA (Spradling, Pardue, and Penman, 1977). The proteins equated with individual message fractions have been identified on the basis of size and by comparison with in vivo heat shock proteins by tryptic peptide mapping. Proteins of 22K and 27K have not yet been assigned to mRNAs (Henikoff and Meselson, 1977; McKenzie and Meselson, 1977; Mirault et al., 1978). Neither the assignment of mRNAs to polytene bands nor the attribution of proteins to mRNAs is absolutely certain because of the difficulties of obtaining mRNA fractions uncontaminated by the other mRNAs. Preliminary results with cloned sequences confirm these assignments, however.

bridize in situ with the 87C locus, which appears to have upto some 12 tandem repetitions of the $\alpha\beta$ repeat judged by the intensity of the HindIII 1.5 kb band produced by digestion of Drosophila DNA. The presence of the tandem repeat only at this locus is indicated by the absence of the 1.5 kb band from flies homozygous for a deletion that removes 87C. The same result is found for the $\alpha\gamma$ repeat, which is released as a 1.3 kb band by HindIII, and which is present in about 6 copies. Although the alternating repeats are found only at 87C, other copies of the individual elements α, β, and γ may be present elsewhere in the genome, as shown by the presence of corresponding restriction fragments of other sizes. In particular, the γ sequence is present at 87A, as seen by the ability of a suitable probe to hybridize in situ.

The A4 RNA species is about 1800 bp long and corresponds to the sequence $\alpha\beta\alpha$. This means that it has a repetition of all or part of the α sequence at each end. Which elements in the tandem $\alpha\beta$ repeating arrays actually are transcribed into the RNA is not yet known. No protein coded by this RNA has been identified. Table 28.4 notes that a problem in assigning proteins to mRNAs is the impurity of the messenger fractions; but the more precise technique of hybrid arrested translation has been applied to this sequence. This involves the addition of a cloned sequence to the mRNA preparation before translation in vitro. Any mRNA corresponding to this sequence should be removed from the active fraction by hybridization; this is reflected by disappearance of the protein it codes for. But the $\alpha\beta$ sequence has no discernible effect on the pattern of protein synthesis (Livak et al., 1978). The role of the poly(A)$^+$ A4 message is therefore mysterious: if it codes for protein, it must be translated extremely poorly, so it has been suggested that it may have some other function, possibly regulatory.

The absence of a protein-coding function for the A4 mRNA is supported by knowledge that deletion of the 87C locus does not remove any heat shock proteins (Ish-Horowicz et al., 1977, 1979). The same lack of effect is seen in deletions of 87A. This implies that either the sequences present at 87A or 87C do not code for heat shock proteins; or that, if they do, the coding sequences must be duplicated at more than one of the heat shock loci. While A4 appears to be an example of the first type, the A2 mRNA corresponds to the second. All the genes corresponding to the A2 mRNA are contained within the 87A and 87C loci, since the deletion of both prevents synthesis of the 70,000 dalton heat shock protein coded by the mRNA.

Several clones have been obtained in various laboratories that carry sequences corresponding to the A2 mRNA. The use of these clones either to purify the mRNA or in hybrid arrested translation has confirmed that it codes for the 70,000 dalton heat shock protein. Probably the proteins comprise several related but not identical polypeptides, which may vary in size from 70,000 to 72,000 daltons. The same variants appear to be coded by both loci, since with only one exception the deletion of either locus alone does not alter the pattern; a single tryptic peptide appears to be unique to 87C, which may therefore code for one variant not coded at 87A. The A2 mRNA hybridizes in situ to both 87A and 87C, as do the cloned sequences representing it. This shows directly that copies of this gene are present at both loci; it does not reveal the extent of differences between the copies at each locus.

Restriction mapping of the cloned sequences corresponding to A2 suggests the general organization illustrated in Figure 28.5. The coding sequences are organized in tandem arrays; the longest yet identified contains 3 copies of the gene, separated by two spacers that are identical, and with flanking sequences at each end of the cluster that are different. The coding region of each of these genes is preceded by a short region that may have homology with the γ sequence (Craig, McCarthy, and Wadsworth, 1979). Other plasmids have one or two copies of the gene and a detailed comparison shows that the repeating unit is about 3000 bp, somewhat longer than the 2200 bp mRNA. The region of homology between the repeats is longer than the coding sequence; this is described as the 2500 bp z region. Preceding this region are two short regions, x and y, that can be distinguished by the pattern of their variation in the different repeats (Artavanis-Tsakanos

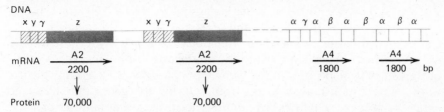

Figure 28.5. Schematic representation of sequences present at the 87C heat shock locus. Several repeating units represent the A2 mRNA that codes for the 70K heat shock protein(s). Two are shown. The mRNA corresponds to part of the sequence *z* that has the same restriction map in adjacent repeats; but differences may be found between different cloned copies, probably representing point mutations. The length of the coding region is 2200 bp by mapping, although the mRNA has been measured as 2600 bp. The coding region is preceded by regions *x* and *y* that may differ between different repeats. The initial region of *z* may be related to or may actually comprise a γ sequence. The locus also contains αγ alternating repeats and a larger number of αβ repeats; the sequence αβα corresponds to the A4 poly(A)$^+$ RNA, which does not appear to be translated. The relationship between the A2 repeats and the A4 repeats is not known, as indicated by the broken lines. This map combines features reported in several cloned repeats whose relationship is not yet defined; the A2 repeating unit is found also at 87A, but the A4 repeats are confined to 87C. Based on data of Lis, Prestidge, and Hogness (1978), Livak et al. (1978), Schedl et al. (1978), Artavanis-Tsakonas et al. (1979), Craig et al. (1979), Moran et al. (1979).

et al., 1979; Moran et al., 1979). Which of these cloned repeats resides at 87A and which at 87C is not yet known; nor have possible differences in the proteins that they represent yet been identified. It is clear, however, that the 70,000 dalton heat shock proteins are represented by multiple genes, probably organized in tandem clusters. An obvious focus of research is to identify the control elements that are assumed to reside at the 5′ ends of the genes. The only link between the two different types of gene present at 87C is the possible presence of a γ sequence in both clusters; its significance is not yet known.

Already this analysis of the heat shock loci suggests two important conclusions. A single band (87C) may possess sequences coding for two different RNAs (A2 and A4). And a given gene (A2) may be present at more than one band (87A and 87C). Whether this organization is typical of other loci remains to be seen. The cloning of genes from the other heat shock loci should allow the search for common features of control to be extended; and in due course this may make it possible to identify regulator sequences that are responsible for ensuring that these genes respond to heat shock.

Chromatin in vitro: Transcription

Perhaps the broadest definition of a gene family is to encompass within a single class the entire set of active genes whose products generate the phenotype of the cell. Complexity studies of messenger RNA populations suggest that often the number of active genes will be about 10,000 (see Chapter 24). This large set will include both the class of "household" genes, possibly concerned with essential metabolic functions and active in all cells, and sets of "specialized" genes, involved in the production of proteins characteristic of the cell type. A prime aim for many years has been to determine the features that distinguish active genes from both inactive genes and other non-expressed sequences, particularly in the hope of isolating families of active genes. In this section and the next we shall pursue the question of the extent to which this approach has succeeded or failed. To conclude the chapter we shall consider the types of model that have been proposed to explain how a given gene may be included in overlapping families of active genes, as defined by its contribution to different cell phenotypes.

The rationale of experiments to identify active genes in vitro has been that in the process of transcription the organization of chromatin may be altered sufficiently to create structural differences that can be recognized at the biochemical level, either as such or in the ability of the gene to be preferentially transcribed in vitro. The impetus for these initial efforts was provided by the view that DNA must lie within a package of histones that would have to be removed for nucleic acid synthesis. This might both allow the DNA to be recognized more readily by RNA polymerase as well as creating gross structural differences between active and inactive regions. Now as described in Chapter 14 it seems that DNA may be transcribed while it remains on the surface of the nucleosome; nonetheless, it is likely that some relaxation in the packing of nucleosomes into higher order structures may be necessary to provide a congenial environment for RNA synthesis. Possibly this may resemble the puffing of Dipteran polytene chromosome bands (see Chapter 16).

We shall be concerned here with assessing the validity of two complementary approaches on the nature of active chromatin. First there have been attempts to fractionate chromatin into active and inactive regions on the basis of structural differences. The criterion for success is the relative content of sequences corresponding to expressed genes in the two fractions; if a successful fractionation were achieved, it might be possible to determine the nature of the protein components involved in the control of transcription (see below). Then there have been attempts to accomplish the transcription in vitro of the same sequences that are expressed in vivo. This is assayed either by comparing the populations of transcripts made in vivo and in vitro or by following the synthesis in vitro of RNA corresponding to an individual gene that is expressed in the cells from which the chromatin was prepared. If specific transcription were achieved, it might be possible to isolate regulator molecules from chromatin by dissociating and reconstituting active genes from the components of DNA and protein.

A critical assumption that pervades all of this work is that gene expression is regulated at the level of transcription, so that the selection of transcribed sequences corresponds to the direct isolation of the class of active genes. It is clear already that in some cases this may not be true; in Chapter

25 we have seen that some comparisons of sea urchin tissues have led to the conclusion that nuclear RNA populations may be the same, while selection in processing results in different messenger RNA populations. Even in such cases, however, the complexities of nuclear RNA correspond to transcription of only a minority of the sequences of the genome; it remains important to determine their distinguishing features, although it should be recognized that this may represent only the first stage of control. Since most of these sequences may be transcribed but not represented in the cytoplasm (whether because of the size discrepancy between hnRNA and mRNA or because of the restriction of some transcripts altogether to the nucleus) the proper comparison for in vitro transcription is with nuclear rather than cytoplasmic RNA.

Asking whether there is specific transcription of chromatin in vitro is akin to enquiring whether there is life after death: many claims have been made, but good evidence is lacking. It would not be profitable to review in detail the large body of work on in vitro transcription. Briefly, however, we may note that the impetus for this work was the observation that isolated chromatin is much less active as a template for added RNA polymerase than is free DNA. Difficulties in obtaining eucaryotic RNA polymerases have meant that most of these experiments have relied upon the bacterial enzyme. The only basis for supposing that its pattern of transcription might have biological significance would be the belief that the parameter controlling transcription is simply access to DNA. This is to argue that nonexpressed DNA sequences are inaccessible to the endogenous enzyme and similarly to added enzyme from any source, while any exposed sequences are members of the active class of genes and may be as well recognized by bacterial RNA polymerase as by the natural enzyme. This view that the source of the enzyme is irrelevant for the specificity of transcription implies a somewhat simple minded model in which gene expression is determined solely or at least largely by the exposure of active sequences, with no subsequent control of the initiation event itself. This is to dismiss the alternative that eucaryotic promoters have sufficient features in common with bacterial promoters to be recognized specifically by E. coli RNA polymerase; also it ignores the possibility that bacterial enzyme may show selection for nonrandomly located sites which fortuitously resemble bacterial promoters. Both these complications can in any case be avoided by using core instead of complete bacterial RNA polymerase to minimize the specificity of initiation.

A large number of experiments in which chromatin from various sources was transcribed in vitro with bacterial RNA polymerase led many authors to conclude that the pattern of transcription resembled that of the tissue from which the chromatin had been obtained.* The main basis for this con-

References: Allfrey and Mirsky (1963), Huang, Bonner, and Murray (1964), Paul and Gilmour (1968), Gilmour and Paul (1969), Smith, Church, and McCarthy (1969), Tan and Miyagi (1970), Smart and Bonner (1971c), Spelsberg and Hnilica (1969, 1971a,b), Spelsberg, Hnilica, and Ansevin (1971), Teng, Teng, and Allfrey (1971).

clusion lay with the ability of RNA extracted from the tissue to compete effectively with the in vitro products when hybridized to DNA. A control to show that this was not due to the presence of endogenous RNA was to extract RNA directly from chromatin that had not been incubated with RNA polymerase. A control to show that the restriction of transcription was specific relied upon comparison with the transcripts obtained with pure DNA as template, which should be not only quantitatively more numerous but also represent all sequences equally (and thus are less well competed by the in vivo RNA). The main flaw in these experiments was the use of a filter assay to follow the hybridization, which restricts the reaction to transcripts of repeated sequences (see Chapter 24). These constitute a minor proportion of the population of transcripts; and in any case are particularly difficult to assay by competition protocols. A demonstration that the observed competition may be illusory was provided by the work of Reeder (1973), which revealed the extremely low sensitivity of the competition assay. The only difference in the transcripts of chromatin and free DNA actually lay in the use of satellite DNA as template in the latter but not in the former condition; but the failure of satellite sequences in chromatin to be transcribed allowed the competition assay to suggest the existence of a spurious close resemblance between the in vivo and in vitro transcripts. Satellite DNA constitutes an appreciable proportion of repetitive DNA; its lack of activity in the chromatin template has been confirmed by Gjerset and McCarthy (1977).

A further caveat about the in vitro experiments is that in these cases chromatin was not prepared by the present (relatively) gentle methods, but by earlier protocols that may have led to a rearrangement of structure (see Chapter 13). Thus while it is clear that less transcription occurs with chromatin than with free DNA, it is by no means evident that the reduction represents any specific repression of nonexpressed genes. Largely it is probably due simply to the reduced accessibility of DNA in chromatin.* The exact nature of the effect with bacterial RNA polymerase is perhaps unimportant in view of the basic inadequacies of the system; but it is interesting that with eucaryotic RNA polymerase, the effect is specifically to reduce the number of initiation sites (Cedar, 1975).

A major drawback of this approach is that in competition or titration protocols it is not possible sufficiently to exclude contamination of the apparent in vitro product with endogenous RNA (controls notwithstanding). The apparent success in mimicking the authentic pattern of transcription with bacterial polymerase in vitro, however, led to experiments to recon-

---

*Although there was at one time some question on whether the reduction in transcription could be an artefact of reduced solubility, this is not now considered to be a problem. A continuing discussion about chromatin solubility of early preparations may be followed through Zubay and Doty (1959), Sonnenberg and Zubay (1965), Roy and Zubay (1966), Barr and Butler (1963), Butler and Chipperfield (1967), Bonner and Huang (1966), Johns and Forrester (1969), Johns and Hoare (1971).

stitute chromatin with DNA, histones and nonhistone proteins from different sources. These resulted in two types of claim for specificity. One was that the source of the nonhistone proteins determined the pattern of transcription in vitro; in these and subsequent reports, probably the cause of the apparent specificity was the copurification of endogenous RNA with the nonhistone proteins. The other proposal was that there exist small chromosomal RNA molecules that control transcription; the evidence for their reality was not satisfactory. In any case, it goes without saying that in view of our criticisms of the assay for detecting specific transcription, these conclusions may be dismissed.*

Many of the intrinsic difficulties in analyzing the entire population of transcripts can be averted by following the ability of chromatin to act as template for synthesis of a particular RNA. The most intensive experiments have been performed with globin and ovalbumin. Using a labeled cDNA probe allows the proportion of complementary (that is, message) sequences among the in vitro transcripts to be determined by titration (see equation 24.11). Typical results show that complements can be detected among the RNA transcribed in vitro from chromatin of erythroid cells, but are absent when chromatin of other cells or free DNA is used as template.† The proportion of globin transcripts varies, but one reasonable value is 0.007% of the total products. Similar results have been obtained with ovalbumin, where a cDNA probe detects complements that provide 0.01% of the total products. In some experiments, mammalian or plant RNA polymerases have been used as well as bacterial RNA polymerase, generally with rather similar results. Although this might be interpreted to show that the bacterial enzyme is indeed a faithful mimic of the eucaryotic enzyme, a more likely interpretation is that the eucaryotic enzyme is liable to the same artefacts as the bacterial enzyme and is failing to reflect the natural state of affairs.

*The citations in the previous footnote include claims for regulation by nonhistone proteins. Additional reports of specific reconstitution assayed with probes against individual mRNA species include Barrett et al. (1974) and Stein et al. (1975a). Another claim for faithful chromatin reconstitution based on inadequate evidence is from Stein et al. (1975b). There has even been a claim that dephosphorylation of nonhistones alters the pattern of transcription of reconstituted chromatin (Kleinsmith, Stein, and Stein, 1976). The inadequacy of this approach is highlighted by the outré report of Gadski and Chae (1977) that proteolysis of chromatin proteins does not alter the pattern of transcription upon reconstitution! The involvement of chromosomal RNA was claimed in a long series of papers from Huang and Bonner (1965), Huang and Huang (1969), Bekhor, Kung, and Bonner (1969), Bekhor, Bonner, and Dahmus (1969), Dahmus and McConnell (1969), Jacobson and Bonner (1971), Mayfield and Bonner (1971, 1972), Sivolap and Bonner (1971), Holmes et al. (1972). Probably the cause of the artefact is the breakdown of larger RNA molecules (Artman and Roth, 1971; Heyden and Zachau, 1971).

†Erythroid chromatin has been used by Axel, Cedar, and Felsenfeld (1973), Gilmour and Paul (1973), Steggles et al. (1974), Wilson, Steggles, and Nienhuis (1975), Barrett et al. (1974); oviduct chromatin by Tsai et al. (1975, 1976), Buller et al. (1976), Harris et al. (1976); myeloma chromatin by Smith and Huang (1976).

There are some obvious reasons why these experiments cannot be taken to indicate proper transcription, as well as more subtle technical artefacts. In those cases where the symmetry of strand transcription has been followed, both strands of the DNA appear to be transcribed equally well. This is not what would be expected of specific transcription. All these experiments involved the use of cDNA probes prepared by reverse transcription of mRNA. Although data have not always been given, the probes probably are rather short, say of less than one half of the length of the mRNA. In view of the mosaic structure of the genes, this means that what has been assayed is the transcription in vitro of the sequences at the 3' end of the long transcription units. The products of transcription rarely have been characterized; but probably they are rather short—certainly there is no evidence for the synthesis of transcripts even remotely approaching the length that would be necessary to account for proper initiation (see Table 26.6). This means that the most that can be said from these experiments is that the terminal regions of transcription units might be preferentially accessible to both bacterial and eucaryotic RNA polymerases in such a way that both strands are available for random initiation and termination of transcription. Even this conclusion is not secure, however, because the hybridization protocols used to detect the transcripts involve the reaction of labeled cDNA with the in vitro products; this makes it difficult to distinguish newly synthesized RNA from endogenous RNA or to exclude the possibility that extension of existing chains rather than de novo initiation has occurred in vitro.

To avoid the complications presented by the presence in chromatin of endogenous RNA chains whose synthesis was initiated in vivo, and which may be present in amounts much greater than the products of in vitro transcription, the in vitro reaction has been performed with mercurated nucleotides. When one of the four nucleoside triphosphates has been replaced by a mercurated precursor, Hg-UTP (with mercury covalently linked at the 5 position of the base), its incorporation into RNA allows the nucleic acid to be isolated on a sulfhydryl column (Dale and Ward, 1975). It was expected that this would allow an unequivocal separation of newly synthesized RNA from preexisting chains; and indeed it was soon reported that with this technique, specific transcription in vitro with bacterial RNA polymerase could be obtained for globin, ovalbumin, or Drosophila heat shock genes, with chromatin of appropriate sources.*

But these conclusions may be as illusory as those based on the earlier work. An interesting artefact in the technique was discovered by Zasloff and Felsenfeld (1977), who found that the E. coli RNA polymerase is able to use endogenous RNA in chromatin as a template. The result is a duplex RNA, one strand representing the endogenous message, the other com-

References: Crouse, Fodor, and Doty (1976), Reff and Davidson (1979), Towle et al. (1977), Beissman et al. (1976, 1978).

prising newly synthesized material containing the mercury label. The duplex RNA is purified as such on thiol agarose; and so when it is used for hybridization with labeled cDNA, the reaction is due principally to the contaminating endogenous message. The nature of the material isolated by affinity for thiol was checked by denaturing the RNA isolated from the in vitro reaction. This reduces the amount of material that is able to bind to the thiol column and is complementary to globin cDNA to less than 4% of the previous value. Thus 96% of the original hybridizing sequences are contaminants of endogenous globin mRNA that are retained on the column because they are not separated from the newly synthesized antiglobin RNA that carries the mercury label. The predominance of anti-globin RNA sequences in the mercurated material was confirmed by hybridization with an anti-globin cDNA probe (that is, cDNA homologous with mRNA). Similar experiments have been performed by Konkel and Ingram (1977, 1978), who have noted another artefact; endogenous RNA may be retained by thiol columns as a result of nonspecific aggregation with the mercuri-substituted products. This problem, as well as the formation of sense/antisense RNA duplexes, can be circumvented by isolating the mercurated RNA on thiol-agarose at high temperature. In these experiments, a very low degree of apparently specific transcription was observed: with erythroid chromatin as template about 0.001% of the mercurated transcripts are complementary to globin cDNA, compared with the level of 0.00025% displayed by transcripts of brain chromatin. Two features of the reaction are significant. At apparent saturation no more than 15–20% of the labeled cDNA probe is hybridized, compared with the usual value of 85–90%; this implies that the in vitro transcripts do not correspond to the entire region of the cDNA probe. And the fourfold increase in representation of globin sequences is lost upon dissociation and reconstitution of chromatin by the usual methods.

To summarize the results of these series of experiments, the maximum level of possibly specific transcription of chromatin in vitro is so low as to cast doubt on its biological significance. It is possible that this might arise entirely from short extension of endogenous RNA chains by the bacterial enzyme (this activity has not been demonstrated but must be considered as likely as any other explanation). What criteria therefore should be satisfied in order to demonstrate the specific transcription of chromatin in vitro? It is necessary initially to demonstrate that the material is indeed newly synthesized (for which the Hg/SH technique is adequate with proper modifications), it is essential to show that its synthesis has been initiated in vitro (which requires the incorporation of a γ-^{32}P-label in ATP or GTP; and of course the absence of a capping system that might remove it), and it is important to show that transcription is asymmetrical (with sequences present that correspond only to the mRNA and not to its complement). Knowledge of the size of the in vitro transcript is a prerequisite for determining the number of initiation events that occurs in vitro; together with the absolute amount of transcript synthesized per template and per polymerase.

If an increase in the proportion of specific transcription of a given gene is obtained, it is necessary to know the total amount of transcription as well as the proportion represented by this gene in order to decide whether the increase is due to enhanced expression of the one gene or to reduced expression of the others.

In none of the experiments that have been reported is there any evidence along these lines to suggest that transcription in vitro at all resembles the pattern of gene expression in vivo. It would indeed be surprising if bacterial RNA polymerase were able to mimic eucaryotic transcription; as we have noted, this might be expected only on the naive view that function is determined exclusively by structure. With this realization, claims for the potential use of E. coli RNA polymerase to transcribe chromatin have become more muted: it may be suggested that it is useful as a probe for differential accessibility of transcribed versus nontranscribed regions, without other implications for function. (This is to abandon all thought of preferential recognition of the initiation region, which in any case is excluded by the use generally of incomplete reverse transcripts that correspond only to the terminal regions of long interrupted genes.) In this sense the use of the bacterial enzyme becomes analogous to the application of any other probe for chromatin conformation; for example, the use of DNAase I to distinguish by preferential degradation the different regions of chromatin. But even this limited objective has not been met with in vitro transcription by bacterial RNA polymerase: nor is it likely that success would be so illuminating as to justify the extensive efforts that have been devoted to what may be a chimera.

The principal aim of work with in vitro systems for RNA synthesis has been to obtain specific initiation, which would allow studies of the factors that control transcription. Little progress has been made until recently. The ability of nuclei to use mercurated nucleotides has been used to develop systems in which incubated nuclei continue to synthesize specific transcripts of globin or ovalbumin genes, probably by elongation of nascent chains rather than by de novo initiation (Orkin, 1978b; Nguyen-Huu et al., 1978). Early attempts to use eucaryotic RNA polymerase II with chromatin showed that the characteristics of the reaction differ from those of the bacterial enzyme; eucaryotic RNA polymerase II initiates at fewer and different sites, but the specificity of their selection remains uncertain.[*] The first in vitro system to initiate properly has been developed by Weil et al. (1979a), using a crude extract of KB (human) cells with purified KB RNA polymerase II to transcribe the late transcription unit of adenovirus DNA. Initiation occurs at the late promoter; since restriction fragments lacking the distal part of the transcription unit have been used as template, it is not yet known whether termination occurs as usual or prematurely. It is significant that an enzyme system from uninfected cells can recognize adeno-

---

[*]*References:* Butterworth, Cox, and Chesterton (1971), Keshgegian and Furth (1972), Meilhac and Chambon (1973), Cedar (1975).

virus DNA directly; the structure of chromatin does not appear to be essential. This system has obvious potential for defining the events involved in the initiation of transcription of nucleoplasmic genes.

The most favorable system for obtaining accurate transcription in vitro has been presented by the action of RNA polymerase III upon the 5S genes. The synthesis of 5S RNA represents an appreciable proportion of the reaction when chromatin from X. laevis oocytes is incubated to allow transcription by the endogenous enzyme. The product is identified as authentic (or close to authentic) 5S RNA by its mobility on gel electrophoresis; and it may be quantitated by hybridization with 5S DNA. Addition of exogenous enzyme stimulates transcription of the 5S genes by up to some 10-20 fold (Parker and Roeder, 1977; Yamamoto, Jones, and Seifart, 1977). The specificity of strand selection was examined by comparing hybridization of the labeled product with denatured 5S DNA (both sense and antisense strand transcripts should hybridize) with the extent of hybridization when excess unlabeled 5S rRNA is added (which competes with the reaction only of the 5S products, leaving the antisense transcript to hybridize). This is not in itself decisive because a labeled 5S RNA product might hybridize with unlabeled 5S rRNA; but no such formation of RNA-RNA duplexes could be found. With this assay there appears to be reasonable strand asymmetry, with a selection ratio of about 10:1. Appreciable transcription of the spacer regions also occurs, as measured by the difference in hybridization of the product with 5S X. laevis DNA (products react with genes and spacers) and 5S X. borealis DNA (products react only with genes). From the extended duration of reaction and the proper size of the product, it seems likely that the enzyme is continuously initiating and terminating, at least at approximately correct sites. A minor caveat is that the data do not prove directly that the exogenous enzyme undertakes transcription; formally it is possible that it might instead stimulate the activity of the endogenous enzyme, a point that is consistent with the retention of roughly the same ratio of gene: spacer transcription in both the endogenous and stimulated reactions. RNA polymerases I and II do not cause specific transcription of the 5S genes; an incidental point is that E. coli RNA polymerase is quite ineffectual.

A complementary approach has made use of nuclei to provide the template for transcription. Isolated nuclei synthesize 5S rRNA and tRNA. When nuclei isolated from a mouse plasmacytoma are incubated with exogenous RNA polymerase III, the extent of 5S gene transcription is increased some 3-4 fold by the homologous enzyme and some 2-fold by RNA polymerase III of X. laevis oocytes. The sense strand seems to be selected with high preference (Sklar and Roeder, 1977). It is not clear directly whether spacers are transcribed, but the same degree of stimulation is obtained when 5S products are assayed by hybridization with X. borealis 5S DNA, in which presumably only the genes are available for reaction. A yeast system also has been developed along these lines (Tekamp et al., 1979).

An extension of this system to another template has been accomplished by using nuclei of KB cells infected with adenovirus. The 5.5S VA RNA of adenovirus is transcribed by RNA polymerase III; a 3–4 fold stimulation is seen upon addition of homologous RNA polymerase III or the Xenopus enzyme. In this case the endogenous reaction appears to show poor strand selection (as assayed by the competitive technique), but the stimulated reaction shows about 2:1 strand selection (Jaehning and Roeder, 1977). In this system the endogenous enzyme may be inactivated by N-ethylmaleimide; the exogenous enzyme then remains able to produce 5S RNA, a demonstration that its role is indeed that of transcription.

Attempts to use the (relatively) purified RNA polymerase III preparations to transcribe either 5S DNA or adenovirus DNA were unsuccessful. This cannot be due to any deficiency in the purified template, since upon injection into Xenopus oocytes either genomic or cloned 5S X. laevis DNA is transcribed as described in Chapter 27 (Brown and Gurdon, 1977, 1978). This suggests that factor(s) additional to the enzyme itself that are components of chromatin or nuclei also are necessary for transcription; the ability of crude extracts of the oocyte to transcribe purified 5S DNA opens the way to purify these components (Birkenmeier, Brown, and Jordan, 1978; Ng, Parker, and Roeder, 1979). In fact, a soluble post chromatin supernatant is able to confer upon the RNA polymerase III preparations the ability to transcribe the 5S genes specifically. A similar system from KB cells transcribes 5.5S VA RNA from purified adenovirus DNA in vitro (Wu, 1978). There are differences between the human and amphibian extract systems in their preference for transcribing particular genes; the amphibian extract transcribes only one of two VA genes, while the human extract transcribes both; the amphibian extract transcribes the dominant type of oocyte 5S gene, whereas the human extract transcribes at least two different types (Weil et al., 1979b). This suggests that although there may be general features common to the initiation and termination signals of genes recognized by RNA polymerase III within and between species, additional controls (not necessarily present in a given in vitro system) may regulate the utilization of these genes in particular cells. The authentic nature of both the 5S and VA RNA products has been demonstrated by ribonuclease fingerprint analysis; in some cases, there appears to be readthrough by chain extension for a small distance beyond the apparent natural termination site. This analysis of the products demonstrates directly that the enzyme is functioning in vitro very close to its usual manner in vivo. It does not exclude the possibility that in addition there is transcription of spacer regions or antisense strands; the use of cloned probes and separated strands will be necessary to define the extent of discrimination against aberrant initiation.

Chromatin in vitro: Fractionation

An approach that seems to have had persistent attraction in spite of limited successes is the attempt to separate chromatin into "active" and "inactive" fractions. The rationale of such experiments has been the belief that structural differences are likely to distinguish regions that are transcribed from those that are not expressed and that these differences should be reflected in differential sensitivities to fractionation procedures. The basis of early work lay in the view that active regions would be characterized by a less condensed structure than inactive regions; essentially this is in microcosm the traditional view of the difference between euchromatin and heterochromatin. A variety of approaches has been used to capitalize upon these putative differences.* Thermal denaturation has been used to identify sequences that may be less well bound to protein, possibly because of transcriptional activity. Centrifugation of sheared chromatin has been used to separate rapidly sedimenting (purportedly inactive, tightly condensed) material from more slowly sedimenting (prospectively unfolded, active) material. A related technique is the use of ion exchange chromatography with Ectham-cellulose to separate early and late eluting fractions.

Various criteria have been applied to assess the success of these fractionation procedures. One approach has been to measure the ability of the different fractions to support transcription by added (bacterial) RNA polymerase. It is not clear exactly what this assesses, but it could have some limited validity in the sense of distinguishing regions with different contents of endogenous RNA. However, another criterion has been to measure nascent RNA content directly; but more recently it has been reported that redistribution of endogenous RNA may occur during isolation of chromatin fractions (Seidman and Cole, 1977). Attempts to measure transcriptional activity directly must therefore be considered unreliable.

An alternative consideration is to seek differences in protein composition, but since very little is known about the relationship between nonhistone protein content and transcriptional activity, such results are difficult to interpret, even if a clear separation is obtained. Usually the putative active chromatin fractions possess an increased content of nonhistone proteins; sometimes they possess a reduced content of histones. In some cases major changes in composition have been associated with the active fraction, such as the complete absence of particular histones. At face value, this could mean that basic reorganization of structure accompanies gene activation; the alternative is to regard these changes as grotesque and indicating that the fractionation does not reflect in vivo differences, but results from changes induced by the abuses to which chromatin has been subjected dur-

---

*Thermal denaturation has been reported by McConaughy and McCarthy (1972). Centrifugation following mechanical shearing or sonication has been applied by Frenster, Allfrey, and Mirsky (1963), Murphy et al. (1973), Berkowitz and Doty (1975), Doenecke and McCarthy (1975a,b). Two protocols for the use of ectham-cellulose have been developed by Reeck, Simpson, and Sober (1972, 1974), Simpson and Reeck (1973) and by Stratling and O'Malley (1976), Stratling, Van, and O'Malley (1976); it is perhaps an illuminating comment on the consistency of the underlying design that in one protocol the putative active fraction elutes first whereas in the other it elutes last.

ing the fractionation. The removal of proteins might well create a fraction in which DNA was more accessible for transcription by bacterial RNA polymerase, but this would be without biological significance.

What criteria would be acceptable for demonstrating fractionation of chromatin into preparations reflecting biological differences? Nothing less than data to show that the sequences of an active gene are present in one fraction while the sequences of an inactive gene are present in the other can be considered satisfactory. In fact, it is necessary further to show that the partition of genes is altered when chromatin from cells of a different phenotype is used. In other words, a given gene must appear in either the active or inactive fraction depending on its state of expression in the chromatin that has been fractionated. Associated with this distribution should be a reduction of the total complexity of the nonrepetitive DNA contained in the active fraction (since it should largely represent the minority of transcribed sequences); and one might expect satellite DNA to be found exclusively in the inactive fraction.

In the tests that have been made of the distribution of individual genes in fractions obtained by chromatographic techniques, no partition has been detected. In one case, both expressed (C type virus) and nonexpressed (B type virus or globin) genes showed no relative differences in fractionation (Howk et al., 1975). This makes it clear that inactive genes are not found preferentially in the inert fraction; the conclusion about expressed genes is more equivocal, because of the possibility that there might be multiple copies of the C type sequences, not all of which are expressed. The distribution of active (globin) and inactive (keratin) genes has been followed in fractionated chicken erythroid chromatin; again there is no difference to be found (Krieg and Wells, 1976). It is clear, therefore, that fractions considered on the basis of other criteria to differ in transcriptional ability are not in fact distinguished by differing contents of expressed and nonexpressed genes. It is not impossible that the fractionation procedure actually does distinguish regions with biological differences; but whatever these differences might be, they do not appear to be related to gene expression. A solitary caveat in this conclusion is that the genes used for such assays should be shown directly to be controlled at the level of transcription.

More recent methods for chromatin fractionation have relied upon the use of nucleases to try to release preferentially the transcribed regions. A welter of confusion has surrounded the DNAase II technique, originally introduced entirely as an empirical approach, but now considered in the light of the increased susceptibility of transcribed or transcribable sequences to DNAase I (see Chapter 14). The fractionation protocol is to incubate chromatin with DNAase II for 5 minutes and then to centrifuge. The pellet (P1) of undegraded chromatin contains 85% of the original DNA. The supernatant is again fractionated, into the S2 supernatant soluble in 2 mM $MgCl_2$ (which has 11% of the input DNA), while the insoluble pellet (P2) contains the remaining 4% of input material (Marushige and Bonner, 1971;

Gottesfeld et al., 1974). As prepared from rat liver chromatin, fraction S2 has enhanced template activity as detected with E. coli RNA polymerase, is enriched in nonhistone proteins, and apparently lacks H1 and has rather little H4. Taken at face value, this last point would cast doubt on the relevance of the technique.

A further report that raised a question about the protocol is the characterization of the constituent particles of each fraction. The P1 fraction releases 11-13S particles upon further treatment with DNAase II; presumably these are nucleosomes. But the S2 fraction contains only 14.5S and 20S particles (Gottesfeld, Murphy and Bonner, 1975). The possibility that these might be nucleosome dimers or trimers is excluded both by the failure to generate smaller particles on continued incubation with DNAase II and by the measurement of their length of DNA as 170 bp per particle. In any case, this fraction had only 60% of the usual histone content. The conclusion of the authors that these particles represent an alternative structure typical of transcribed regions is strongly at odds with the data discussed in Chapter 14, which suggest the conclusion that transcribed sequences are found in nucleosomes, albeit with altered conformation and increased susceptibility to some nucleases. However, another interpretation of the structure of the S2 fraction has been offered by Gottesfeld and Butler (1977), who found that its content of histones is normal: only the content of nonhistones and RNA is altered. A multimeric series of particles can be found in the S2 fraction; although the particles themselves are larger than the corresponding nucleosome multimers, their content of DNA shows the same 200 bp multiple series. Possibly the size of the particles is increased by the presence of nonhistone proteins, increasing the monomer to 14S. Another distinguishing feature of these particles (prepared from trout testis) is the existence of a much increased proportion of H4 in multiacetylated forms (Davie and Candido, 1978). The RNA content of the S2 fraction is much increased, in line with its candidacy to represent active genes. Which of its features is responsible for its solubility in 2 mM $MgCl_2$ is not known. It should be noted, however, that it must be an open question whether the increased content of nonhistones and RNA was present before fractionation or is generated during preparation.

When this fractionation procedure was applied to rat liver chromatin, the total complexity of the DNA contained in the S2 fraction was reported to be somewhat less than that of the initial DNA (Gottesfeld et al., 1976). The S2 fraction appeared to be enriched in sequences complementary to mRNA, so that 14% of the S2 DNA hybridized in excess RNA (this is 1.5% of input DNA), while 3.5% of the P1 fraction hybridized (corresponding to 3% of input DNA). However, the content of satellite DNA did not appear to be reduced in S2 DNA, which would be surprising in a transcriptionally active fraction. More recently attempts have been made to follow the fractionation of individual genes by this method; it has been reported that about a 5-fold enrichment of the globin gene is obtained in S2 DNA prepared from the chromatin of Friend cells. In some experiments this result was obtained only after the induction of globin synthesis with DMSO; in others both induced and uninduced Friend cells showed the same preferential partition (Gottesfeld and Partington, 1977; Wallace, Dube, and Bonner, 1977). But there was no preferential partition with rat liver. In another report, no preferential distribution could be found in the Friend cell chromatin (Lau et al., 1978). A difficulty in interpreting these results is that the level of control of globin gene expression may differ in different Friend cell lines, and controls to examine the distribution of other gene sequences have not been included. It will be necessary to obtain data on chromatin from other sources, in particular using reciprocal chromatins where the expression of two genes is reversed, before concluding whether the technique is of general application for separating active from inactive genes.

Analogous techniques have been developed using micrococcal nuclease to provide the initial breakage of DNA. After treatment of chick oviduct nuclei with micrococcal nuclease, centrifugation generates a supernatant fraction that is enriched roughly 5-fold for ovalbumin sequences. The pellet is lysed and centrifuged to generate a second pellet, whose content of ovalbumin sequences is the same as the original chromatin, and a supernatant that is roughly 5-fold depleted in ovalbumin sequences (Bloom and Anderson, 1978). A similar approach has been developed by Levy and Dixon (1978) with trout testis nuclei. After treatment with micrococcal nuclease to render soluble about 5–10% of the DNA, nonhistone proteins, especially HMG-T and S, are released into the supernatant. A gentle centrifugation yields a pellet which is resuspended in 1 mM EDTA; upon centrifugation, nucleosomes are found in the soluble supernatant, while undegraded material forms a pellet. The nucleosomal fraction is separated further into a pellet insoluble in 0.1 M NaCl and a soluble supernatant. This last fraction contains about 7% of the input DNA, in the form of nucleosome cores (MN1) containing the four core histones, 140 bp lengths of DNA, and a large amount of the nonhistone protein H6. The DNA of this fraction drives the hybridization of a cDNA probe representing poly(A)$^+$ RNA about 7 times more rapidly than does total DNA. In an RNA-driven reaction, about 35% of the DNA of the fraction reacts (2.5% of input DNA) compared with 4% of total nonrepetitive DNA. The NaCl-insoluble pellet can be dissolved in Tris-HCl and contains both mononucleosomes (MN2) and oligosomes. The MN2 fraction contains all five histones, DNA of length about 170 bp, and lacks H6 (Levy, Connor and Dixon, 1979). Its hybridization properties when annealed with mRNA are very similar to those of the DNA from the MN1 fraction (Levy-Wilson and Dixon, 1979). Cross hybridization shows that MN1 and MN2 DNA sequences overlap but are not identical. Both contain a repetitive component(s) corresponding to about 25% of the total DNA, which complicates the attempt to define their constitution: it will be necessary to use purified nonrepetitive sequence sets to determine their relationship. Since micrococcal nuclease preferentially releases HMG-T, one model is to suppose that transcribed regions are preferentially attacked to release HMG-T and nucleosomes corresponding to the active genes (see also Chapter 14). A similar fractionation has been accomplished by applying the same protocol of micrococcal nuclease treatment to isolated trout testis chromatin (instead of nuclei). The results are generally similar as seen in the protein constitution of the fractions, except that the first supernatant contains histones as well as HMG-T and H6 (Levy-Wilson, Watson, and Dixon, 1979). This implies that there may be a significant difference in the course of the reaction. The data obtained with the MN1 and MN2 fractions raise the possibility that these represent a preferential conversion to nucleosomes or cores of a subset of the sequences in chromatin that may include active genes. To assess the extent to which this procedure may be successful, it is necessary to follow the partition of some individual probes representing active and inactive genes.

To summarize efforts to fractionate chromatin, it is unlikely that early protocols using sheared material have achieved any significant separation of functionally distinct regions. A better rationale can be found for the more recent experiments using DNAase II and micrococcal nuclease. The argument would be that DNAase II resembles DNAase I in preferentially attacking active regions, although presumably it must cleave between rather than within nucleosomes to release fragments. Probably an extensive amount of nicking also occurs, which limits the potential for isolating DNA. Similarly, micrococcal nuclease might be supposed to find targets more readily in transcribed regions whose structure is less condensed; again in order to obtain soluble fractions, it may be necessary to reduce the DNA to rather short lengths. In the sense of isolating intact genes engaged in transcription, these techniques are therefore unlikely to be useful. On the other hand, if it were true that the nucleosomes comprising the entire length of the gene (not just the initiation region) were in a different conformation or were associated with particular nonhistone proteins, then the nature of the difference between active and inactive genes might be seen from such fractionations. In short, the usefulness of this approach may depend on to what extent active genes are distinguished from inactive at the level of the nucleosome rather than at higher levels of organization (see also Chapter 14). In a more general sense, this points to what has always been a difficulty in believing that chromatin fractionation is not a naive expectation. Active genes may be interspersed with inactive regions; and therefore any mass isolation of the active fraction must involve extensive breakage in the DNA.

Models for Gene Regulation

The central issue of this book must remain an unanswered question: how is gene expression controlled? We can consider three aspects of this question, only the first of which can be approached now. What is the level of control?

What are the molecular interactions by which control is exercised? And what is the formal network of control circuits that defines the overall phenotype of any given cell?

The two stages of gene expression on which most attention has been focused as probable points of control are the initiation of transcription and the processing from hnRNA and/or nucleocytoplasmic transport of mRNA. The existence of transcriptional control is evident from the restricted representation of genome sequences in nuclear RNA (see Chapter 25). For some "specialized" genes—such as globin or ovalbumin—this appears qualitatively to be the sole level of control, since nuclear transcripts can be found only in cells in which the messenger RNA is present. (Formally these results demonstrate that control is at the level of production of nuclear RNA precursors; this does not prove directly that the initiation of transcription is involved, but this is the most likely explanation.) But in other instances transcription may be only a first step; gene expression may be controlled at a subsequent stage. The discrepancy between the complexity of hnRNA and mRNA first raised the issue of whether selection of sequences for export to the cytoplasm might be an important level of control; the extent of such control is hard to determine from comparisons of nuclear and cytoplasmic RNA, because an unknown proportion of the reduction in complexity represents the removal of excess (flanking and/or intervening) sequences from the primary transcripts. But comparisons between some populations—for example those of the S. purpuratus gastrula and intestine—show that the mRNA complexity changes extensively while the nuclear RNA is more or less constant. Thus while some genes may be controlled at the level of transcription, others may be regulated by selection of sequences for processing and/or nucleocytoplasmic transport. The generality of these alternatives, the question of whether there is any functional distinction between types of gene controlled at one or the other level, and the relationship between the two types of control system, all remain to be investigated.

Identification of the components of a control system and the definition of their interactions will require the ability to obtain gene expression in vitro. The failure of attempts to obtain meaningful transcription of chromatin in vitro has seen the collapse of one line of approach. The paucity of systems in which transcription can be initiated by eucaryotic RNA polymerase II has been an impediment to the development of in vitro systems that operate faithfully. However, recent progress suggests that it should become possible to establish systems in which the nature of the events involved in initiation can be delineated, which will make it possible to characterize the effects of regulator molecules and of changes in the sequences of the sites that are recognized on DNA. Some systems already have been developed in which accurate splicing occurs in vitro; whether this is used as a control is unknown, but in due course it should become possible to determine whether different cell types have different activities at this level. Little is known about other possible events that might be involved in selection of sequences at the nuclear level.

Some general features of the process of transcription have become clear. Transcription occurs on chromatin which continues to be organized in the form of nucleosomes, although there is a local change in conformation. Whether this occurs in the core itself or at a higher level of organization is not certain (see Chapter 14). To what extent there is a general unfolding of the transcribed region, possibly comparable to the puffing of polytene chromosome bands, is not known. But there is evidence that the changes in nucleosome conformation that are associated with active genes may both precede and follow transcription, so that they are not simple consequences of the act of RNA synthesis. Whether the structure of chromatin is important in the selection of genes for transcription is an interesting question. However, earlier models that proposed changes in the structure of either active genes or regulator sequences have been based on a view of chromatin structure that has been largely invalidated by more recent work (Crick, 1971; Paul, 1972). Transcription probably is initiated in the region immediately preceding the site corresponding to the 5' end of the messenger, which argues against the existence of lengthy 5' sequences that are transcribed but not conserved in mRNA. Thus attention on possible regulator sequences has focused on the region just upstream from the startpoint of the message.

Several formal models for gene regulation in eucaryotes have been proposed, all representing a reformulation of the concept of the model of the bacterial operon proposed by Jacob and Monod (1961). Thus a common feature is the postulate that there exist the two types of sequence identified in bacterial operons: structural genes that are expressed in the form of RNA or protein; and sequences whose function is to be recognized by regulator molecules or by enzymes of nucleic acid synthesis. It has been usual to assume a similar organization to that of bacterial DNA in which the recognition elements lie contiguous with the structural genes that they control.*

Implicit in these models is the view that it is solely sequences of nucleic acid that carry the information necessary for a gene to be recognized for activation. In bacteria, sequences suffering transcription probably are distinguished from the rest of the genome only by the presence of locally unwound sequences of DNA associated with RNA polymerase (see Volume 1 of Gene Expression). Genes are repressed by the interaction of a regulator protein with a short recognition element; genes are activated and transcribed by the interaction of regulator proteins and RNA polymerase with

---

*Sequences whose function is exercised by transcription and translation into a diffusible molecule are described here as *genes*. Sequences which serve only as recognition sites for regulator molecules are described as *elements*. In formal genetic terms, genes are trans active while elements are cis dominant. An alternative nomenclature is to describe all functional units of DNA as genes, whether or not they are transcribed; but this confuses further a situation that is already complicated enough for most tastes.

other short recognition elements. These recognition elements are probably perpetually accessible to regulator proteins and RNA polymerase. Whether a comparable right of access exists in eucaryotic chromosomes is not known; certainly it is fair to argue that the accessibility of a sequence will be determined by analogous reactions between nucleic acid and regulator molecule, but these could be separated temporally from the time of gene activation. Thus in the sense of the immediate concern of how a regulator molecule exerts its effect, the conformation of the chromatin is an important feature unique to eucaryotes whose role is not yet determined. Accepting further that regulation may occur at either the level of transcription or the processing of nuclear RNA, regulator molecules might be found as constituents of chromatin or nuclear ribonucleoprotein particles.

The most detailed of the models for eucaryotic gene regulation was proposed by Britten and Davidson (1969) and has been elaborated by Davidson and Britten (1973, 1979). It postulates the existence of four classes of sequence of eucaryotic DNA whose interactions constitute a system of positive control (in which genes are inactive unless specifically activated; the alternative, seen in the classical system of the E. coli lactose operon, is negative control, in which genes are active unless specifically switched off). *Producer genes* comprise sequences of DNA which code for proteins; these are directly analogous to the *structural genes* of bacterial operons, the term we shall retain here. Adjacent to each structural gene is at least one receptor element; a structural gene can be transcribed only when an activator molecule recognizes an adjacent receptor element.

The regulator molecules which control gene activity are postulated to be *activator RNAs;* however, the model remains formally the same if activator proteins are substituted. The loci which code for the activator RNAs are the *integrator genes;* if the regulator molecules are proteins; it is necessary only to postulate that the transcripts of the integrator genes are translated instead of themselves possessing activator activity. The integrator genes are analogous to the *regulator genes* of bacterial operons.

To provide for the specific control of gene expression, it is necessary for either the synthesis or activity of regulator molecules to respond to the milieu of the cell. The activities of the repressor proteins of bacterial operons are controlled by their interactions with small molecules of the environment. The eucaryotic model proposes that the activities of the integrator genes are controlled by adjacent *sensor elements;* an integrator gene can be transcribed only when its controlling sensor element is activated. Sensor elements are supposed to constitute the targets that are recognized by agents which change the pattern of gene expression; for example, hormone-protein complexes might bind to sensor elements to activate the adjacent integrator genes. The basic control circuit of the model shown in Figure 28.6 therefore comprises a sensor element—integrator gene complex whose synthesis of activator RNA controls the activity of a receptor element—structural gene complex.

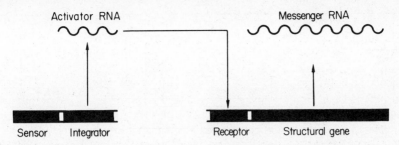

Figure 28.6. Model for control of eucaryotic gene expression. Stimulation of a sensor element causes the adjacent integrator gene to synthesize an activator RNA. The activator RNA recognizes a receptor element and this interaction causes transcription of the adjacent structural gene (as shown) or production of mRNA from a nuclear RNA precursor (not illustrated). After Britten and Davidson (1969).

To account for the concerted control of many genes in each state of differentiation, the model proposes that the receptor elements and integrator genes may be repeated; there may be many copies of any one receptor or integrator sequence in the genome. These two types of repetition allow a small number of control elements to activate a large number of genes.

Because structural genes whose protein products serve related functions may be located at different chromosome sites, their common control requires each to possess a copy of the same receptor sequence. All these genes then may be expressed in response to a single species of activator RNA. (There is no evidence for linkage of related eucaryotic genes into clusters comparable to bacterial operons. The genes for α globin and β globin, for example, lie on different human chromosomes. Although there is a cluster of the β-related genes, presumably reflecting their evolution, these genes are not under coordinate control in the same cell. Possibly the linkage between the genes is part of the control circuit that dictates when each is expressed; but it remains necessary to coordinate this with α-like globin genes located elsewhere.)

Transcription of any gene, or set of genes, may occur in more than one cell phenotype. It must therefore be possible to activate a given structural gene in more than one set of circumstances. This may be achieved by redundancy of either the receptor elements or integrator genes. The model postulates that each structural gene may possess several adjacent receptor elements, any one of which may activate transcription by responding to its appropriate activator RNA. Figure 28.7 shows that redundancy in the receptor elements allows overlapping combinations of structural genes to be controlled by activator RNAs. Production of any particular activator RNA causes transcription of all the structural genes which possess a copy of the receptor element that it recognizes; any structural gene may therefore be included in the *set* recognized by any activator RNA by possession of the appropriate receptor element.

When extensive changes are to be introduced in the pattern of gene ex-

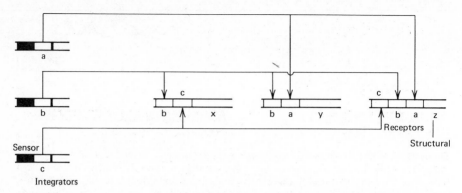

Figure 28.7. Redundancy of receptor sites in the Britten-Davidson control model. Each structural gene may be activated in response to the binding of activator RNA at any one of several adjacent receptor elements. Activator a thus controls structural genes y and z; activator b controls structural genes x, y and z; and activator c controls structural genes x and z. Each structural gene therefore belongs to more than one set.

pression, it may be necessary to activate many sets of genes. Integrator genes might therefore be organized in clusters, each cluster falling under the control of a single sensor element. Activation of a sensor element causes transcription of all the integrator genes under its control; the corresponding activator RNAs then switch on the several sets of structural genes that they recognize. The sets of structural genes controlled by any one sensor element have been termed a *battery*.

Any particular set of genes may be required in response to more than one stimulus. This can be achieved by repetition of the appropriate integrator genes. Figure 28.8 shows that when an integrator sequence is repeated under the control of more than one sensor, its activator RNA may be synthesized in response to more than one set of cellular conditions. Repetition of integrator genes under the control of sensor elements achieves a result similar to the repetition of receptor sites at several structural gene loci; just as it is possible to include one structural gene in more than one set, so it is possible to include one set in more than one battery.

Since all the genes of any one set must always be expressed coordinately, it is reasonable to suppose that each set comprises a comparatively small number of genes whose products serve quite closely related functions. And the number of sets in which any single structural gene may be included might be limited; since it is possible that receptor elements must lie close to the structural genes which they control, there might be a small limit on the number of receptor sites able to activate a structural gene. The number of sets controlled by a sensor gene is potentially high, since many integrator genes might be transcribed into activator RNAs under control of a single sensor element.

The repetition of receptor elements and integrator genes implies that these control functions must form part of the repeated sequences of the ge-

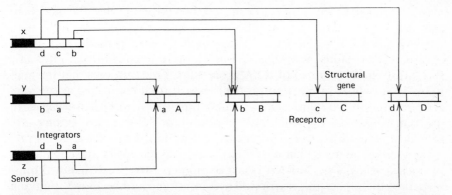

Figure 28.8. Redundancy of integrator elements in the Britten-Davidson control model. Each sensor element causes synthesis of activator RNA at each of the integrator genes adjacent to it. By including the same integrator gene sequence under control of more than one sensor, the set of genes responding to its activator RNA may be activated in response to more than one cellular signal. Thus sensor x controls structural genes B, C and D; sensor y controls structural genes A and B; and sensor z controls structural genes A, B and D.

nome; this model implies that most, perhaps all, of the moderately repetitive component is concerned with coding for, or being recognized by, activator RNA. Although the model makes no prediction about the location in the genome of the groups of integrator genes controlled by each sensor element, it implies that each structural gene—presumably consisting of nonrepetitive DNA—must be adjacent to the repeated sequences of the receptor elements. Whether this is related to the interspersion of nonrepetitive and repetitive sequences observed in eucaryotic genomes (see Chapter 18) seems somewhat doubtful. The average interspersed unit of a 300 bp repetitive sequence and a 1000 bp nonrepetitive sequence is much shorter than would be expected of a mosaic gene.

One advantage to the cell of using RNA rather than protein as the regulator molecule is that the entire control network might be contained in the nucleus; if activator proteins control transcription, the RNA products of the integrator genes would need to be transported to the cytoplasm for translation into the regulator proteins. In fact, the products of the integrator genes would then be messenger RNAs; and the genes themselves would no longer necessarily be expected to show the close sequence relationships predicted for activator RNAs. The genes would therefore not only be larger but also less well related. It follows further that a group of integrator genes under the control of a sensor would become in effect an operon, a cluster of sequences coding for different proteins whose expression is under common control. Returning to the original view that RNA is the regulator, the model predicts that there should be two classes of units of transcription. One comprises the nonrepetitive DNA of a structural gene adjacent to the repetitive DNA of the receptor elements; the other consists of the group of integrator genes under control of a sensor element.

In terms of the control of transcription, one difficulty that the model encounters is to visualize a plausible interaction for the activator RNA with DNA. However, like the Jacob-Monod model before it, the formal model may almost as readily be applied to posttranscriptional control, in this case specifically to the control of RNA processing. This supposes that the activator RNA (whose production probably is still itself controlled at the level of transcription) recognizes a particular hnRNA. A mechanism exists in the form of base pairing; and the formation of the RNA duplex region could provide a signal for processing or for nucleocytoplasmic transport. A prediction of this model is that the repetitive receptor elements must be present in nuclear RNA; since mRNA is largely nonrepetitive, their removal must be part of the processing reaction. Their location could be 5' or 3' flanking or in intervening sequences (although where characterized these show no signs of possessing repetitive components). This model is related to the possibilities that we have discussed in Chapter 26 for using the splicing of intervening sequences as a control mechanism.

To summarize, the model presents a mechanism for controlling gene expression at the level either of transcription or nuclear RNA processing. The lack of regulator mutations in eucaryotes means that there is no evidence for either the location or function of control elements or regulatory genes. The nature of the regulator molecules remains unknown. The identification of repetitive sequence components of the genome with regulatory functions is an intriguing idea, but is not yet proven. The circuitry of the model is plausible, but again there is no information to show to what extent this reflects the formal control network of a eucaryotic genome. Indeed, the reality of such models remains to be seen.

One problem for models of gene regulation is posed by the large amount of eucaryotic DNA. Regulator proteins appear to have not only a high, specific affinity for their recognition sites on DNA, but also display a lower, generalized affinity for all other sequences of DNA. Most of the work to determine the consequences that this has for gene regulation have been performed with the E. coli lactose repressor as a model, but the same general conclusions appear to be true for the mammalian uterine estradiol receptor protein (Lin and Riggs, 1975; Von Hippel et al., 1974; Kao-Hung et al., 1977; Yamamoto and Alberts, 1975). Probably it is true that an inescapable consequence of the ability of a regulator protein to recognize a particular DNA sequence with great affinity will be that it must recognize any other DNA sequence with lower affinity; in other words, there is a limit to how specific its recognition of a DNA sequence can be made. This means that in effect there is competition for the regulator protein between a small number of specific high-affinity sites and a large number of nonspecific low-affinity sites. The ability of the regulator protein to find its specific recognition elements will therefore depend on the ratio of its affinities for the two types of site and on the proportion of total DNA that is represented by the specific sites. In E. coli, for example, the lactose repressor is almost entirely bound to DNA; if the total amount of DNA were increased, recognition of the operators might become impossible. This difficulty might be overcome by increasing the concentration of regulator protein or increasing the number of recognition elements by tandem repetition; also an assumption in posing the problem is that all DNA is equally accessible to the regulator protein and this might not be true in eucaryotic chromatin.

The counterpart of this problem in considering the control of nuclear RNA processing by a putative small RNA regulator molecule is to ask whether a sufficient Rot for reaction would be reached in the nucleus. For example, consider a nucleus of 30 μm³, in which a reaction with a nuclear RNA is driven by an excess of a small activator RNA, say a molecule of 100 bases present in 100 copies. Taking the naive assumption that conditions in the nucleus are comparable to those used in hybridization reactions in vitro, the $Rot_{1/2}$ of such a reaction should be about 5×10^{-5} (see footnote in Chapter 24). This Rot value would be reached in vivo within about 10 seconds of the primary transcript becoming available. Obviously this ignores the reality of conditions in the nucleus, where free diffusion may not apply and where the RNA may be bound to protein; but it suggests very roughly that models of this nature are not a priori implausible.

Species	Tissue Analyzed
Mycoplasma pneumoniae	cultured cells
Escherichia coli	isolated DNA
Bacillus subtilis	spore (not proven that only 1 chromosome present)
Saccharomyces cerevisiae	comparison of cells of different ploidies
Neurospora crassa	conidium
Dictyostelium discoideum	nuclei
Drosophila melanogaster	spermatocyte
Physarum polycephalum	spores
Ascaris lumbicoides	spermatid
Bombyx mori	spermatocyte
Strongylocentrotus purpuratus	spermatocyte
Lytechinus pictus	spermatocyte
Gallus domesticus	spermatozoa and erythrocyte
Bovis domesticus	spermatozoa and various tissues
Homo sapiens	leucocyte
Xenopus laevis	erythrocyte nucleus
Rana pipiens	erythrocyte nucleus
Zea mays	root tip meiosis
Secale cereale	root tip meiosis
Allium cepa	root tip
Lillium longiflorum	root tip meiosis
Protopterus aethiopicus (lungfish)	blood cells

of haploid genomes

Assay	Haploid DNA (pg)	Ref.
sedimentation	0.0000052	A
autoradiography	0.0044	B
phosphorus content and diphenylamine	0.005	C
phosphorus content	0.024	D
DABA and EtBr	0.017	E
optical adsorption	0.017	F
diphenylamine	0.036	G
feulgen (relative to chick erythrocyte taken as 2.5 pg)	0.18	H
	0.3	I
isotope dilution	0.63	J
feulgen (relative to chick erythrocyte taken as 2.5 pg)	0.52	K
chemical but no data given	0.89	L
phosphorus content	0.90	M
phosphorus content	1.25	N
phosphorus content	3.25	O
phosphorus content and diphenylamine	3.4	P
diphenylamine (average of 2 assays)	3.15	Q
diphenylamine and feulgen relative to Xenopus as 6.3 pg	7.6	R
feulgen relative to A. cepa as 16.8 pg	3.9	S
feulgen relative to A. cepa as 16.8 pg	7.9	S
diphenylamine	16.8	T
feulgen relative to A. cepa as 16.8 pg	32.8	S
feulgen relative to mouse as 7 pg	142.0	U

Primary estimates for DNA content are based upon chemical determinations of phosphorus content, derived from the method of Schmidt and Tannhauser (1945), or reactions specific for DNA such as diphenylamine (Dische, 1930; Burton, 1965), diaminobenzoic acid (Kissane and Robins, 1958), ethidium bromide (Klotz and Zimm, 1972). The original series of experiments using the Schmidt-Tannhauser procedure were reported by Mirsky and Ris (1949, 1951, 1952) and by Boivin, Vendrely, and Vendrely (1948) and Vendrely and Vendrely (1949).

Secondary estimates usually are based upon comparing the feulgen reaction of one nucleus with that of another whose DNA content has been chemically determined. Where the absence of primary estimates has made it necessary to quote secondary estimates, the value of DNA content assumed for the standard has been given. Other methods that have been used include sedimentation (for very small genomes, e.g., viruses), autoradiography (for bacterial and organelle genomes), optical adsorption (based on purified DNA), and isotope dilution (in which labeled DNA is added to sample DNA during extraction and content calculated from lack of label). All of these suffer from disadvantages represented by the lack of an absolute standard.

Data have been included in this table only when measurements have been made on cells of known and constant ploidy, e.g., measurements on mammalian liver are not quoted. Data given here in picograms may be converted to base pairs by taking 1 pg = 0.965×10^9 base pairs and to daltons by taking 1 pg = 6.1×10^{11} daltons (subject to some slight inaccuracy depending upon the G + C ratio of the DNA).

The values ascribed to organisms that have been used as standards for comparative measurements are especially important. Bacterial DNA has been used as the standard for comparison in renaturation kinetics (see Chapter 18 where estimates for genome size on this basis are given). The usual estimate for the E. coli genome is based on autoradiography of DNA, which suffers from the disadvantage that there are variations in the relationship of length to mass in DNA spread for electron microscopy. Estimates for B. subtilis are not completely reliable because the chromosome constitution of the spores upon whch chemical measurements were made is not clear. Other attempts to measure bacterial DNA content have been based upon viscoelastic techniques (Klotz and Zimm, 1972), but the scatter in the results is so wide that it is difficult to extract a reliable value. Renaturation analysis relative to a phage DNA control also has been used (Gillis and Ley, 1975); but here unknown variables influence the relative rates of reaction, so that the results are useful only for confirmation. The values derived from these various techniques are consistent with those quoted in the Table, but are not included because of these inherent difficulties.

Standards used for comparisons by the feulgen reaction often are the erythrocytes of X. laevis or chicken. As noted in the table, the value for X. laevis is the average of two determinations, although it is supported by its consistency with some other measurements, e.g., that on R. pipiens. The value for the chicken genome given in the table is that of the original chemical experiments; other analyses since have produced somewhat wide ranging results. The value for Allium cepa depends upon corrections made for cell division in root tip and so cannot be considered accurate within better than 10%. This has been used as standard for an extensive series of measurements on plant genomes by Bennett and Smith (1976); earlier analyses of plants were performed by McLeish and Sunderland (1961). Mammals and other vertebrates have been compared by Feuglen analysis by Atkin et al. (1965); the measurements of Bachmann (1972) and Bachmann et al. (1972) cannot be considered accurate because of the use of liver tissue. An extensive series of relative assays on amphibia has been reported by Hinegardner (1968, 1971, 1973, 1974), using a sensitive fluorescence technique; but these comparisons are converted to absolute DNA values only by reference to a standard for S. purpuratus, which is stated on the basis of unpublished data to have been determined by the DABA technique.

References: A—Neimark and Pene (1965); B—Cairns (1963); C—Fitz-James and Young (1959), Dennis and Wake (1966); D—Ogur et al. (1952), Williamson and Scopes (1961), Shapiro (1968); E—Lauer et al. (1977); F—Minegawa et al. (1959); G—Sussman and Rayner (1971); H—Rasch et al. (1971); I—Mobberg and Rusch (1971); J—Tobler et al. (1972); K—Rasch (1974); L—Hinegardner (1968); M—Mirsky and Ris (1951); N—Mirsky and Ris (1949); O—Mirsky and Ris (1949), Vendrely and Vendrely (1949); P—Mandel et al. (1950); Q—Dawid (1965); S—Thomson et al. (1973); S—Bennett and Smith (1976); T—Van't Hof (1965); U—Pederson (1971).

Species	DNA content of haploid genome, bp	Percent and $Cot_{1/2}$ of slow component
Escherichia coli (bacterium)	4.2×10^6*	100% at 4.0 under defines complexity
Bacillus subtilis (bacterium)	3.0×10^6*	100% at 1.7
Achyla bisexualis (water mold)	4.8×10^7+	82% at 56
Dictyostelium discoideum (slime mold)	5.4×10^7*	69% at 113
Caernorhabditis elegans (nematode worm)	8.0×10^7+	83% at 80
Panagrellus silusiae (nematode)	8.7×10^7+	61% at 83
Ciona intestinalis (sea squirt)	1.4×10^8+	70% at 90
Drosophila melanogaster (fruit fly)	1.4×10^8*	70% at 145
Bombyx mori (silk worm)	5.0×10^8*	55% at 500
Musca domestica (house fly)	8.6×10^8+	34% at 1050
Strongylocentrotus purpuratus (sea urchin	8.6×10^8*	50% at 870
Physar polycephalum (mold)	9.1×10^8+	60% at 750
Antheraea pernyi (silk moth)	9.7×10^8+	35% at 950
Gallus domesticus (chicken)	1.2×10^9*	87% at 900
Gecarcinus lateralis (land crab)	1.3×10^9+	50% at 850
Aplysia californica (mollusc)	1.7×10^9+	50% at 1800
Petroselinum sativum (parsley)	1.9×10^9+	11% at 6250
Cancer borealis (crab)	2.0×10^9+	30% at 3200
Mus musculus (mouse)	2.7×10^9+	60% at 1733 58% at 1500
Rattus norvegicus (rat)	3.0×10^9+	67% at 3100
Homo sapiens (man)	3.3×10^9*	51% at 495
Bovis domesticus (cow)	3.1×10^9*	60% at 4000
Xenopus laevis (toad)	3.1×10^9*	54% at 2850
Crypthecodinium lohnii (dinoflagellate)	3.7×10^9+	40% at 1750
Nicotinia tabacum (tobacco)	4.8×10^9+	33% at 1650
Bufo bufo (amphibian)	6.7×10^9+	20% at 9000
Triticum aestivum (haxaploid wheat)	1.7×10^{10}+	12% at 3700
Triturus cristatus (amphibian)	2.2×10^{10}+	47% at 15,000
Necturus maculosis (amphibian)	5.0×10^{10}+	23% at 60,000

size by reassociation of nonrepetitive DNA

E. coli $Cot_{1/2}$	Kinetic genome size, bp	Chemical/kinetic	Ref.
standard conditions of 4.2×10^6 bp			A
1.8	4.0×10^6	0.8	B
6.0	3.9×10^7	1.2	C
6.5	7.3×10^7	1.4	D
3.2	1.0×10^8	0.8	E
3.8	9.2×10^7	0.9	E
1.8	2.1×10^8	0.7	B
4.2	1.5×10^8	0.9	F
5.0	4.2×10^8	1.2	G
5.4	8.2×10^8	1.0	H
4.0	9.1×10^8	0.9	I
2.7	1.2×10^9	0.8	J
5.4	7.4×10^8	1.3	K
3.6	1.0×10^9	1.2	L
4.2	1.1×10^9	0.9	M
4.2	1.8×10^9	1.0	N
10.5	2.5×10^9	0.8	O
4.5	3.0×10^9	0.7	P
1.8	4.0×10^9	0.7	B
4.0	1.6×10^9	1.7	Q
4.0	3.3×10^9	0.9	R
1.2	1.8×10^9	1.8	S
6.5	2.6×10^9	1.2	A
4.0	3.0×10^9	1.0	T
2.3	3.2×10^9	1.2	U
4.6	1.5×10^9	3.2	V
6.0	6.3×10^9	1.1	W
3.0	5.2×10^9	3.3	X
6.0	1.0×10^{10}	2.2	W
6.0	4.2×10^{10}	1.2	W

The first column shows the DNA content of the haploid genome in base pairs (bp) estimated by chemical assay. Data indicated by an asterisk (*) are taken from the independent assays summarized in Appendix 1; those indicated by a cross (+) are values quoted in the references cited in this table, which must be consulted on the varying degrees of reliability of these estimates. The proportion and kinetics of reassociation for the nonrepetitive component usually have been determined by computer fitted curves following least squares analysis to find the minimum number of independent components in the reassociation of whole DNA. The values given are the observed equivalent $Cot_{1/2}$ points, except for G. lateralis (calculated from the $Cot_{1/2}$ of the slow component of isolated main band DNA, which is 75% of the genome) and for wheat (calculated from the $Cot_{1/2}$ observed for isolated slow component). All reassociations were followed on hydroxyapatite, except D. discoideum (optical density). The $Cot_{1/2}$ for E. coli DNA under the conditions used for reassociation in most cases has been calculated by conversion from the $Cot_{1/2}$ prevailing for standard conditions (4.0) as discussed in the text; in other cases, it is the value of an experimentally determined control. The kinetic genome size is calculated by comparing the $Cot_{1/2}$ of the slow component with that of E. coli DNA; and in almost all cases the ratio of the chemical and kinetic estimates for genome size (which formally is the repetition frequency) is within the experimental error expected for nonrepetitive DNA (ideally the frequency should be 1.0). There is no explanation for the high values of Triturus and man. Tobacco is thought to be tetraploid, which would predict a repetition frequency of 2.0 for the slow component. Wheat is hexaploid; the observed ratio of roughly 3 coincides with the expectation that the corresponding nonrepetitive sequences in each of the three constituent genomes can engage in heterologous as well as homologous reassociation.

References: A—Britten and Kohne (1968); B—Laird (1971); C—Hudspeth et al. (1977); D—Firtel and Bonner (1972); E—Sulston and Brenner (1974); Beauchamp et al. (1978); F—Manning et al. (1975); G—Gage (1974); H—Crain et al. (1976b); I—Graham et al. (1974); J—Foucquet et al. (1974); K—Efstratiadis et al. (1976); L—Eden and Hendrick (1978); M—Holland and Skinner (1977); N—Angerer et al. (1975); O—Kiper and Herzfield (1978); P—Vaughn (1975); Q—Cech and Hearst (1976); R—Pearson et al. (1978); S—Schmid and Deininger (1975); T—Davidson et al. (1973); U—Allen et al. (1975); V—Zimmerman and Goldberg (1977); W—Baldari and Amaldi (1976); X—Smith and Flavell (1975).

Appendix 3   Sequence components of eucaryotic genomes

Species	Proportion of component, %	$Cot_{1/2}$ in whole DNA	Repetition frequency	Complexity of component, bp
A. bisexualis	82	56	1.2	3.2×10^7
	14	0.72	80	7.0×10^4
	2	0.025	2200	3.5×10^2
	5	<0.001		
C. elegans	83	80	0.8	8.0×10^7
	14	0.24	300	3.4×10^5
	3	$<10^{-7}$		
D. melanogaster	70	145	0.9	1.0×10^8
	12	2	70	2.5×10^5
	12	0.006	24000	7.2×10^2
	5	<0.006		
B. mori	55	500	1.2	2.3×10^8
	24	1.5	330	3.8×10^5
	21	0.015	30000	2.6×10^3
C. virginica	38	760*	1.0*	2.5×10^8
	53	20	40	8.7×10^6
	9	<0.01		
A. aurita	36	830*	1.0*	2.5×10^8
	53	4.8	170	2.2×10^6
	4	<0.01		
M. domestica	34	1050	1.0	2.8×10^8
	22	5.2	200	8.8×10^5
	23	0.06	17500	1.1×10^4
S. purpuratus	50	870	0.9	4.6×10^8
	27	83	10	2.3×10^7
	19	5.3	160	1.0×10^6
	10	0.012	6000	1.3×10^3
P. polycephalum	60	750	0.8	7.2×10^8
	30	0.7	1100	3.2×10^5
	10	<0.01		
A. pernyi	35	950	1.3	2.6×10^8
	30	65	15	1.5×10^7
	30	0.6	1600	1.4×10^5
	5	<0.02		
S. solidissima	32	1300*	1.0*	3.7×10^8
	40	40	30	1.6×10^7
	18	3.6	350	6.0×10^5
	4	<0.02		
G. lateralis	40	850	0.9	4.1×10^8
	12	0.18	4700	3.1×10^4
	13	0.006	1.5×10^5	1.0×10^3
	30	<0.0001		

Species	Proportion of component, %	$Cot_{1/2}$ in whole DNA	Repetition frequency	Complexity of component, bp
G. domesticus	87	900	0.9	9.4×10^8
	5	59	17	3.8×10^6
	7	0.6	1700	5.2×10^4
C. lacteus	33	1540*	1.0*	4.6×10^8
	34	42	35	1.3×10^7
	28	1.5	1025	3.8×10^5
	5	<0.01		
A. californica	50	1800	1.0	9.0×10^8
	20	21	85	3.7×10^6
	25	0.4	4600	5.1×10^4
	10	0.00025	7×10^6	
	7	$<10^{-6}$		
P. sativum	11	6250	0.8	2.8×10^8
	16	135	45	8.6×10^6
	48	1.9	3300	3.6×10^5
	13	0.04	1.5×10^5	2.0×10^3
	5	<0.01		
C. borealis	30	3200	0.7	9.0×10^8
	32	1.0	3200	3.0×10^5
	38	<0.01		
L. polyphemus	58	3215*	1.0*	2.7×10^9
	11	67	47	6.3×10^6
	16	>0.01		
	7	<0.01		
M. musculus	58	1500	1.6	9.3×10^8
	11	10	150	1.1×10^6
	14	0.2	7500	2.9×10^4
	8	0.0005	3×10^5	4.0×10^2
	2	$<10^{-5}$		
R. norvegicus	67	3100	0.9	2.2×10^9
	19	262	13	5.2×10^7
	18	0.94	3500	1.8×10^5
	10	<0.01		
H. sapiens	51	495	1.8	9.0×10^8
	22	1.0	500	7.7×10^5
	13	0.01	50000	4.5×10^3
X. laevis	54	2850	1.0	1.6×10^9
	10	27	110	2.8×10^6
	31	1.4	2000	4.5×10^5
	6	0.01	3×10^5	6.3×10^2
	5	<0.01		

Appendix 3 *(Continued)*

Species	Proportion of component, %	$Cot_{1/2}$ in whole DNA	Repetition frequency	Complexity of component, bp
C. cohnii	40	1750	1.2	1.3×10^9
	60	0.5	3500	5.5×10^5
N. tabacum	33	1650	3.2	5.0×10^8
	60	5.4	300	3.0×10^6
	5	0.1	16500	4.5×10^3
	2	<0.01		
B. bufo	20	9000	1.1	1.3×10^9
	10	500	20	3.5×10^7
	50	1.0	9000	3.5×10^5
T. aestivum	12	3700	3.3	6.2×10^8
	83	0.09	4300	1.0×10^6
	4	$<10^{-6}$		
T. cristatus	47	15000	2.2	4.7×10^9
	43	14	11000	4.1×10^5
N. maculosis	23	60000	1.2	9.7×10^9
	10	150	400	1.0×10^7
	47	0.9	65000	3.0×10^5

The properties of the nonrepetitive component in each genome are taken from Appendix 2. The repetitive components represent curves whose sums fit the total data for each genome; but these solutions are only examples of possible fits, and do not describe unique fractions (see text). The repetition frequency of nonrepetitive DNA is based upon the comparison with E. coli; the repetition frequencies of all other components are based on comparison with the nonrepetitive component of the genome (which is therefore assumed for this calculation to be 1.0). Thus in the case of the polyploid genomes of tobacco and wheat, repetition frequencies are expressed relative to each constituent (ancestral) genome. All complexities are those determined by comparison with E. coli DNA (and so in cases where the nonrepetitive component is not shown as 1.0, the repetition frequency implied by the proportion and complexity of each repetitive component will be slightly different from that stated on the basis of comparison with nonrepetitive DNA). When the proportion of zero time binding DNA has been determined, it is given as the amount of DNA renatured at the first Cot point (described as <Cot) (see Chapter 19). This may include some highly repetitive DNA as well as foldback DNA. Where no such component is shown, data were calculated by excluding the zero time binding from the curve before calculation. References are as in Appendix 2, except for Crassostrea virginica (oyster), Aurelia aurita (jellyfish), Spisula solidissima (surf clam), Cerebratulus lacteus (nemertean worm), and Limulus polyphemus (horseshoe crab), whose reassociation properties were described by Goldberg et al. (1975). In these cases, the $Cot_{1/2}$ of the nonrepetitive component was established as that expected from the size of the genome and its repetition frequency is therefore 1.0 by definition; this is indicated by an asterisk.

References

Aaronson, J. F., Rudkin, G. T. and Schultz, J. (1954). A comparison of giant X chromosomes in male and female D. melanogaster by cytophotometry in the ultraviolet. *J. Histochem. Cytochem.* **2,** 458-459.

Abelson, H. T. and Penman, S. (1972). Messenger RNA formation: resistance to inhibition by 3′ deoxycytidine. *Biochim. Biophys. Acta* **277,** 129-133.

Abelson, H. T., Johnson, L. F., Penman, S. and Green, H. (1974). Changes in RNA in relation to growth of the fibroblast. II. The lifetime of mRNA, rRNA, and tRNA in resting and growing cells. *Cell* **1,** 161-166.

Abercrombie, M. (1961). The bases of the locomotory behavior of fibroblasts. *Exp. Cell Res. Suppl.* **8,** 188-198.

Abercrombie, M., Heaysman, J. E. M. and Pegrum, S. (1970a). The locomotion of fibroblasts in cultures. I. Movements of the leading edge. *Exp. Cell Res.* **59,** 393-398.

Abercrombie, M., Heaysman, J. E. M. and Pegrum, S. (1970b). The locomotion of fibroblasts in culture. II. Ruffling. *Exp. Cell Res.* **60,** 437-444.

Abercrombie, M., Heaysman, J. E. M. and Pegrum, S. (1970c). The locomotion of fibroblasts in culture. III. Movements of particles on the dorsal surface of the leading lamella. *Exp. Cell Res.* **62,** 389-398.

Abercrombie, M., Heaysman, J. E. M. and Pegrum, S. M. (1971). The locomotion of fibroblasts in culture. IV. Electron microscopy of the leading lamella. *Exp. Cell Res.* **67,** 359-367.

Abraham, G., Rhodes, D. P. and Bannerjee, A. J. (1975a). Novel initiation of RNA synthesis in vitro by VSV. *Nature* **255,** 37-40.

Abraham, G., Rhodes, D. P. and Bannerjee, A. K. (1975b). Methylated and blocked 5′ termini in VSV in vivo mRNAs. *Cell* **5,** 51-58.

Abraham, K. A. and Pihl, A. (1977). Translation of enzymically decapped mRNA. *Eur. J. Biochem.* **77,** 589-594.

Abrahamson, S., Bender, M. A., Conger, A. D. and Wolff, S. (1973). Uniformity of radiation induced mutation rates among different species. *Nature* **245,** 460-462.

Abuelo, J. G. and Moore, D. E. (1969). The human chromosome. Electron microscopic observations on chromatin fiber organization. *J. Cell Biol.* **41,** 73-90.

Adamson, E. D. and Woodland, H. R. (1974). Histone synthesis in early amphibian development: histone and DNA syntheses are not coordinated. *J. Mol. Biol.* **88,** 263-286.

Adamson, E. D. and Woodland, H. R. (1977). Changes in the rate of histone synthesis during oocyte maturation and very early development of X. laevis. *Devel. Biol.* **57,** 136-149.

Adegoke, J. A. and Taylor, J. H. (1977). Sequence programming of DNA replication over the S phase of Chinese hamster cells. *Exp. Cell Res.* **104,** 47-54.

Adelman, M. R. and Taylor, E. W. (1969a). Isolation of an actomyosin like protein complex from slime mold plasmodium and the separation of the complex into actin and myosin like fractions. *Biochemistry* **8,** 4964-4975.

Adelman, M. R. and Taylor, E. W. (1969b). Further purification and characterization of slime mold myosin and slime mold actin. *Biochemistry* **8,** 4976-4988.

Adelstein, R. S., Pollard, T. D. and Kuehl, W. M. (1971). Isolation and characterization of myosin and two myosin fragments from human blood platelets. *Proc. Nat. Acad. Sci. USA* **68,** 2703-2707.

Adelstein, R. S., Conti, M. A., Johnson, G. S., Pastan, I. and Pollard, T. D. (1972). Isolation and characterization of myosin from cloned mouse fibroblasts. *Proc. Nat. Acad. Sci.* **69,** 3693-3697.

Adesnik, M. and Darnell, J. E. (1972). Biogenesis and characterization of histone mRNA in Hela cells. *J. Mol. Biol.* **67,** 397-406.

Adesnik, M., Salditt, M., Thomas, W. and Darnell, J. E. (1972). Evidence that mRNA molecules (except histone mRNA) contain poly(A) sequences and that the poly(A) has a nuclear function. *J. Mol. Biol.* **71,** 21-30.

Adolph, K. W., Cheng, S. M. and Laemmli, U. K. (1977). Role of nonhistone proteins in metaphase chromosome structure. *Cell* **12,** 805-816.

Adolph, K. W., Cheng, S. M., Paulson, J. R. and Laemmli, U. K. (1977). Isolation of a protein scaffold from mitotic HeLa cell chromosomes. *Proc. Nat. Acad. Sci. USA* **74,** 4937-4941.

Affara, N. A., Jacquet, M., Jakob, H., Jacob, F. and Gros, F. (1977). Comparison of polysomal polyadenylated mRNA from embryonal carcinoma and committed myogenic and erythropoietic cell lines. *Cell* **12,** 509-520.

Akam, M. E., Roberts, D. B., Richards, G. P. and Ashburner, M. (1978). Drosophila: the genetics of two major larval proteins. *Cell* **13,** 215-226.

Akusjarvi, G. and Pettersson, U. (1979a). Sequence analysis of adenovirus DNA. III. The complete nucleotide sequence of the spliced 5′ noncoding region of adenovirus 2 hexon mRNA. *Cell* **16,** 841-850.

Akusjarvi, G. and Pettersson, U. (1979b). Sequence analysis of adenovirus DNA. IV. The genomic sequences encoding the common tripartite leader of late adenovirus mRNA. *J. Virol.* **134,** 143-158.

Albertini, D. F. and Clark, J. I. (1975). Membrane-microtubule interactions: conA capping induced redistribution of cytoplasmic microtubules and colchicine binding proteins. *Proc. Nat. Acad. Sci. USA* **72,** 4976-4980.

Albertini, R. J. and DeMars, R. (1970). Diploid azaguanine resistant mutants of cultured human fibroblasts. *Science* **169,** 482-485.

Albertini, R. J. and DeMars, R. (1973). Somatic cell mutation. Detection and quantification of X ray induced mutation in cultured diploid human fibroblasts. *Mutat. Res.* **18,** 199-224.

Alberty, H., Raba, M. and Gross, H. J. (1978). Isolation from rat liver and sequence of a RNA fragment containing 32 nucleotides from position 5 to 36 from the 3′ end of 18S rRNA. *Nucleic Acids Res.* **5,** 425-434.

Albrecht-Buehler, G. (1977). Phagokinetic tracks of 3T3 cells; parallels between the orientation of track segments and of cellular structures which contain actin or tubulin. *Cell* **12,** 333-339.

Albright, S. C., Nelson, P. P. and Garrard, W. T. (1979). Histone molar ratios among different electrophoretic forms of mono and dinucleosomes. *J. Biol. Chem.* **254,** 1066-1073.

Albring, M., Griffith, J. and Attardi, G. (1977). Association of a protein structure of probable membrane derivation with HeLa cell mitochondrial DNA near its origin of replication. *Proc. Nat. Acad. Sci. USA* **74,** 1348-1352.

Alfageme, C. R., Rudkin, G. T. and Cohen, L. H. (1976). Locations of chromosomal proteins in polytene chromosomes. *Proc. Nat. Acad. Sci. USA* **73,** 2038-2042.

Alfert, M. (1958). Variations in cytochemical properties of cell nuclei. *Exp. Cell Res., Suppl.* **6,** 227-235.

Ali, I. U., Mautner, V., Lanza, R. and Hynes, R. (1977). Restoration of normal morphology, adhesion and cytoskeleton in transformed cells by addition of a transformation sensitive surface protein. *Cell* **11,** 115-126.

Allen, C. and Borisy, G. G. (1974). Structural polarity and directional growth of microtubules of Chlamydomonas flagella. *J. Mol. Biol.* **90,** 381-402.

Allen, J. R., Roberts, T. M., Loeblich, A. R., III and Klotz, L. C. (1975). Characterization of the DNA from the dinoflagellate Crypthecodinium cohnii and implications for nuclear organization. *Cell* **6**, 161-170.

Allet, B. and Rochaix, J. D. (1979). Structure analysis at the ends of the intervening DNA sequences in the chloroplast 23S ribosomal genes of C. reinhardii. *Cell* **18**, 55-60.

Allfrey, V. G. and Mirsky, A. E. (1963). Mechanisms of synthesis and control of protein and RNA synthesis in the cell nucleus. *Cold Spring Harbor Symp. Quant. Biol.* **28**, 247-262.

Allfrey, V. G., Inoue, A., Karn, J., Johnson, E. M. and Vidali, G. (1973). Phosphorylation of DNA-binding nuclear acidic proteins and gene activation in the Hela cell cycle. *Cold Spring Harbor Symp. Quant. Biol.* **38**, 785-801.

Aloni, Y. and Attardi, G. (1971). Expression of the mitochondrial genome in Hela cells. IV. Titration of mitochondrial genes for 16S, 12S and 4S RNA. *J. Mol. Biol.* **55**, 271-276.

Aloni, Y., Dhar, R., Laub, O., Horowitz, M. and Khoury, G. (1977). Novel mechanism for mRNA maturation: the leader sequences of SV40 mRNA are not transcribed adjacent to the coding sequences. *Proc. Nat. Acad. Sci. USA* **74**, 3686-3690.

Alt, F. W., Kellems, R. E. and Schimke, R. T. (1976). Synthesis and degradation of folate reductase in sensitive and methotrexate resistant lines of SI80 cells. *J. Biol. Chem.* **251**, 3063-3074.

Alt, F. W., Kellems, R. E., Bertino, J. R. and Schimke, R. T. (1978). Selective multiplication of dihydrofolate reductase genes in methotrexate resistant variants of cultured murine cells. *J. Biol. Chem.* **253**, 1357-1370.

Altenburger, W., Horz, W. and Zachau, H. G. (1976). Nuclease cleavage of chromatin at 100 nucleotide pair intervals. *Nature* **264**, 517-522.

Altenburger, W., Horz, W. and Zachau, H. G. (1977). Comparative analysis of three guinea pig satellite DNAs by restriction nucleases. *Eur. J. Biochem.* **73**, 393-400.

Alter, B. P., Goff, S. C., Hillman, D. G., Deisseroth, A. B. and Forget, B. G. (1977). Production of mouse globin in heterokaryons of mouse erythroleukemia cells and human fibroblasts. *J. Cell Sci.* **26**, 347-358.

Alwine, J. C., Kemp, D. J. and Stark, G. R. (1977). Method for the detection of specific RNAs in agarose gels by transfer to diazobenzyloxymethyl paper and its hybridization with DNA probes. *Proc. Nat. Acad. Sci. USA* **74**, 5350-5354.

Amalric, F., Merkel, C., Gelfand, R. and Attardi, G. (1978). Fractionation of mitochondrial RNA from HeLa cells by high resolution electrophoresis under strongly denaturing conditions. *J. Mol. Biol.* **118**, 1-26.

Ambler, R. P. and Scott, G. K. (1978). Partial amino acid sequence of penicillinase coded by E. coli plasmid R6K. *Proc. Nat. Acad. Sci. USA* **75**, 3732-3736.

Ambros, V. R., Chen, L. B. and Buchanan, J. M. (1975). Surface ruffles as markers for studies of cell transformation by Rous sarcoma virus. *Proc. Nat. Acad. Sci. USA* **72**, 3144-3148.

Ambrose, E. J. (1961). The movements of fibrocytes. *Exp. Cell Res. Suppl.* **8**, 54-73.

Amos, L. A. and Klug, A. (1974). Arrangement of subunits in flagellar microtubules. *J. Cell Sci.* **14**, 523-537.

Anderson, E. G. (1925). Crossing over in a case of attached X chromosome in D. melanogaster. *Genetics* **10**, 403-417.

Andre, J. and Thiéry, J. P. (1963). Mise en évidence d'une sous-structure fibrillaire dans les filaments axonématique des flagelles. *J. Microscopie* **2**, 71-80.

Angelier, N. and Lacroix, J. C. (1975). Complexes de transcription d'origines nucleolaire et chromosomique d'ovocytes de Pleurodeles waltlii et P. poireti. *Chromosoma* **51**, 323-335.

Angerer, L., Davidson, N., Murphy, W., Lynch, D. and Attardi, G. (1976). An electron microscope study of the relative positions of the 4S and rRNA genes in HeLa mt DNA. *Cell* **9**, 81-90.

Angerer, R. C., Davidson, E. H. and Britten, R. J. (1975). DNA sequence organization in the mollusc Aplysia californica. *Cell* **6**, 29-40.

Angerer, R. C., Davidson, E. H. and Britten, R. J. (1976). Single copy DNA and structural gene sequence relationships among four sea urchin species. *Chromosoma* **56**, 213–226.

Appels, R. and Wells, J. R. E. (1972). Synthesis and turnover of DNA-bound histone during maturation of avian red blood cells. *J. Mol. Biol.* **70**, 425–434.

Appels, R., Bell, P. B. and Ringertz, N. R. (1975). The first division of HeLa × chick erythrocyte heterokaryons. *Exp. Cell Res.* **92**, 78–86.

Appels, R., Bolund, L. and Ringertz, N. R. (1974). Biochemical analysis of reactivated chick erythrocyte nuclei isolated from chick/Hela heterocaryons. *J. Mol. Biol.* **87**, 339–356.

Appels, R., Wells, J. R. E. and Williams, A. F. (1972). Characterization of DNA bound to histones in the cells of the avian erythropoietic series. *J. Cell Sci.* **10**, 47–60.

Appels, R., Bolund, L., Goto, S. and Ringertz, N. R. (1974). The kinetics of protein synthesis by chick erythrocyte nuclei during reactivation in chick-mammalian heterocaryons. *Exp. Cell Res.* **85**, 182–190.

Appels, R., Tallroth, E., Appels, D. M. and Ringertz, N. R. (1975). Differential uptake of protein into the chick nuclei of HeLa × chick erythrocyte heterocaryons. *Exp. Cell Res.* **92**, 70–78.

Arceci, R. J. and Gross, P. R. (1977). Noncoincidence of histone and DNA synthesis in cleavage cycles of early development. *Proc. Nat. Acad. Sci. USA* **74**, 5016–5020.

Arceci, R. J., Senger, D. R. and Gross, P. R. (1976). The programmed switch in lysine rich histone synthesis at gastrulation. *Cell* **9**, 171–178.

Arcos-Teran, L. (1972). DNA replikation und die natur der spat replizierenden orte im X chromosom von D. melanogaster. *Chromosoma* **37**, 233–296.

Arms, K. (1968). Cytonucleoproteins in cleaving eggs of X. laevis. *J. Emb. Exp. Morph.* **20**, 367–374.

Arnberg, A. C., Van Bruggen, E. F. J., Schutgens, R. B. H., Flavell, R. A. and Borst, P. (1972). Multiple D loops in Tetrahymena mitochondrial DNA. *Biochim. Biophys. Acta* **272**, 487–493.

Arnheim, N. and Southern, E. M. (1977). Heterogeneity of the ribosomal genes in mice and men. *Cell* **11**, 363–370.

Arrhigi, F. E. and Hsu, T. C. (1971). Localization of heterochromatin in human chromosomes. *Cytogenetics* **10**, 81–86.

Arrhigi, F. E., Hsu, T. C., Saunders, P. and Saunders, G. F. (1970). Localization of repetitive DNA in the chromosomes of M. agrestis by means of in situ hybridization. *Chromosoma* **32**, 224–236.

Artavanis-Tsakonas, S., Schedl, P., Tschudi, C. Pirrotta, V., Steward, R. and Gehring, W. J. (1977). The 5S genes of D. melanogaster. *Cell* **12**, 1057–1067.

Artavanis-Tsakonas, S., Schedl, P., Mirault, M. E., Moran, L. and Lis, J. (1979). Genes for the 70,000 dalton heat shock protein in two cloned D. melanogaster DNA segments. *Cell* **17**, 8–18.

Artman, M. and Roth, J. S. (1971). Chromosomal RNA: an artefact of preparation? *J. Mol. Biol.* **60**, 291–302.

Ash, J. F. and Singer, S. J. (1976). Concanavalin A induced transmembrane linkage of Con A surface receptors to intracellular myosin containing filaments. *Proc. Nat. Acad. Sci. USA* **73**, 4575–4579.

Ash, J. F., Vogt, P. K. and Singer, S. J. (1976). The reversion from transformed to normal phenotype by inhibition of protein synthesis in rat kidney cells infected with a temperature sensitive Rous sarcoma virus mutant. *Proc. Nat. Acad. Sci. USA* **73**, 3603–3607.

Ashburner, M. (1967). Autosomal puffing patterns in a laboratory stock of D. melanogaster. *Chromosoma* **21**, 398–428.

Ashburner, M. (1969a). The X chromosome puffing pattern of D. melanogaster and D. simulans. *Chromosoma* **27**, 47–63.

Ashburner, M. (1969b). Patterns of puffing activity in the salivary gland chromosomes of Drosophila. IV. Variability of puffing patterns. *Chromosoma* **27**, 156–177.

Ashburner, M. (1970a). Formation and structure of polytene chromosomes during insect development. *Adv. Insect Physiol.* **7**, 1-95.

Ashburner, M. (1970b). Patterns of puffing activity in the salivary gland chromosomes of Drosophila. V. Response to environmental treatments. *Chromosoma* **31**, 356-376.

Ashburner, M. (1972a). Patterns of puffing activity in the salivary gland chromosomes of Drosophila. VI. Induction by ecdysone in salivary glands of D. melanogaster cultured in vitro. *Chromosoma* **38**, 255-282.

Ashburner, M. (1972b). Ecdysone induction of puffing in polytene chromosomes of D. melanogaster. Effects of inhibitors of RNA synthesis. *Exp. Cell Res.* **71**, 433-440.

Ashburner, M. (1973). Temporal control of puffing activity in polytene chromosomes. *Cold Spring Harbor Symp. Quant. Biol.* **38**, 655-662.

Ashburner, M. and Bonner, J. J. (1979). The induction of gene activity in Drosophila by heat shock. *Cell* **17**, 241-254.

Ashihara, T., Chang, S. D. and Baserga, R. (1978). Constancy of the shift up point in two temperature sensitive mammalian lines that arrest in G1. *J. Cell. Physiol.* **96**, 15-22.

Atger, M. and Milgrom, E. (1977). Progesterone induced mRNA. Translation, purification and preliminary characterization of uteroglobin mRNA. *J. Biol. Chem.* **252**, 5412-5418.

Athwal, R. S. and McBride, D. W. (1977). Serial transfer of a human gene to rodent cells by sequential chromosome-mediated gene transfer. *Proc. Nat. Acad. Sci. USA* **74**, 2943-2947.

Atkin, N. B., Mattinson, G., Becak, W. and Ohno, S. (1965). The comparative DNA content of 19 species of placental mammals, reptiles and birds. *Chromosoma* **17**, 1-10.

Atlas, S. J. and Lin, S. (1978). Dihydrocytochalasin B. Biological effects and binding to 3T3 cells. *J. Cell Biol.* **76**, 360-370.

Attardi, G. and Ojala, D. (1971). Mitochondrial miniribosomes in HeLa cells. *Nature New Biol.* **229**, 133-136.

Attardi, G., Aloni, Y., Attardi, B., Ojala, D., Pica-Mattoccia, L., Robberson, L. and Storrie, B. (1970). Transcription of mitochondrial DNA in HeLa cells. *Cold Spring Harbor Symp. Quant. Biol.* **35**, 599-619.

Auerbach, C. (1946). Chemically induced mosaicism in D. melanogaster. *Proc. Roy. Soc. Edin. B* **62**, 211-222.

Auer, G. (1972). Nuclear protein content and DNA-histone interaction. *Exp. Cell Res.* **75**, 231-236.

Auer, G. and Zetterberg, A. (1972). The role of nuclear proteins in RNA synthesis. *Exp. Cell Res.* **75**, 245-253.

Auer, G., Zetterberg, A. and Foley, G. E. (1973). The relationship of DNA synthesis to protein accumulation in the cell nucleus. *J. Cell. Physiol.* **76**, 357-364.

Auer, G., Moore, G. P. M., Ringertz, N. R. and Zetterberg, A. (1973). DNA dependent RNA synthesis in nuclear chromatin of fixed cells. Relationship between dye binding properties, nuclear protein content and RNA polymerase activity. *Exp. Cell Res.* **76**, 229-233.

Augenlicht, L. H. and Baserga, R. (1974). Changes in the GO state of WI38 fibroblasts at different times after confluence. *Exp. Cell Res.* **89**, 255-262.

Augenlicht, L. H. and Lipkin, M. (1976). Appearance of rapidly labeled, high molecular weight RNA in nuclear ribonucleoprotein. Release from chromatin and association with protein. *J. Biol. Chem.* **251**, 2592-2599.

Aula, P. and Saksela, E. (1972). Comparison of areas of quinacrine mustard fluorescence and modified Giemsa staining in human metaphase chromosomes. *Exp. Cell Res.* **71**, 161-167.

Aviv, H., Voloch, Z., Bastos, R. and Levy, S. (1976). Biosynthesis and stability of globin mRNA in cultured erythroleukemic Friend cells. *Cell* **8**, 495-504.

Axel, R. (1975). Cleavage of DNA in nuclei and chromatin with Staphylococcal nuclease. *Biochemistry* **14**, 2921-2925.

Axel, R., Cedar, H. and Felsenfeld, G. (1973). Synthesis of globin RNA from duck reticulocyte chromatin in vitro. *Proc. Nat. Acad. Sci. USA* **70**, 2029-2032.

Axel, R., Feigelson, P. and Schutz, G. (1976). Analysis of the complexity and diversity of mRNA from chicken oviduct and liver. *Cell* **7**, 247-254.

Axel, R., Melchior, W., Jr., Sollner-Webb, B., and Felsenfeld, G. (1974). Specific sites of interaction between histones and DNA in chromatin. *Proc. Nat. Acad. Sci. USA* **71**, 4101-4105.

Axelrod, D. E., Gopalakrishnan, T. V., Willing, M. and Anderson, W. F. (1978). Maintenance of hemoglobin inducibility in somatic cell hybrids of tetraploid (2S) mouse erythroleukemia cells with mouse or human fibroblasts. *Somatic Cell Genet.* **4**, 157-168.

Ayad, S. R., Fox, M. and Winstanley, D. (1969). The use of ficoll gradient centrifugation to produce synchronous mouse lymphoma cells. *Biochem. Biophys. Res. Commun.* **37**, 551-558.

Bachellerie, J. P., Nicoloso, M. and Zalta, J. P. (1977). Nucleolar chromatin in CHO cells. Topographical distribution of rDNA sequences and isolation of ribosomal transcription complexes. *Eur. J. Biochem.* **79**, 23-32.

Bacchetti, S. and Graham, F. L. (1977). Transfer of the gene for thymidine kinase to thymidine kinase deficient human cells by purified herpes simplex viral DNA. *Proc. Nat. Acad. Sci. USA* **74**, 1590-1594.

Bachmann, K. (1972). Genome size in mammals. *Chromosoma* **37**, 85-93.

Bachmann, K., Harrington, B. A. and Craig, J. P. (1972). Genome sizes in birds. *Chromosoma* **37**, 405-416.

Bag, J. and Sarkar, S. (1975). Cytoplasmic nonpolysomal messenger RNP containing actin mRNA in chicken embryonic muscles. *Biochemistry* **14**, 3800-3807.

Baglioni, C. (1962). The fusion of two peptide chains in hemoglobin Lepore and its interpretation as a genetic deletion. *Proc. Nat. Acad. Sci. USA* **48**, 1880-1886.

Bahr, G. F. and Golomb, H. M. (1971). Karyotyping of single human chromosomes from dry mass determined by electron microscopy. *Proc. Nat. Acad. Sci. USA* **68**, 726-730.

Bailey, G. S. and Dixon, G. H. (1973). Histone IIbl from rainbow trout. Comparison in amino acid sequence with calf thymus histone IIbl. *J. Biol. Chem.*, **248**, 5463-5472.

Bajer, A. and Molé-Bajer, J. (1969). Formation of spindle fibers, kinetochore orientation and behavior of the nuclear envelope during mitosis in endosperm. *Chromosoma* **27**, 448-484.

Bajer, A. and Molé-Bajer, J. (1970). Architecture and function of the mitotic spindle. *Adv. Cell Mol. Biol.* **1**, 213-267.

Bajer, A. S. and Molé-Bajer, J. (1972). Spindle dynamics and chromosome movements. *Int. Rev. Cytol., Suppl.* **3**.

Bakay, B., Nyhan, W. L., Croce, C. M. and Koprowski, H. (1975). Reversion in expression of HGPRT following cell hybridization. *J. Cell Sci.* **17**, 567-578.

Bakayev, V. V., Bakayeva, T. G. and Varshavsky, A. J. (1977). Nucleosomes and subnucleosomes: heterogeneity and composition. *Cell* **11**, 619-629.

Bakayev, V. V., Bakayeva, T. G., Schmatchenko, V. V. and Georgiev, G. P. (1978). Non histone proteins in mononucleosomes and subnucleosomes. *Eur. J. Biochem.* **91**, 291-302.

Baker, B. S. (1975). Paternal loss (pal): a meiotic mutant in D. melanogaster causing loss of paternal chromosomes. *Genetics* **80**, 267-296.

Baker, B. S. and Carpenter, A. T. G. (1972). Genetic analysis of sex chromosomal meiotic mutants in D. melanogaster. *Genetics* **71**, 255-286.

Baker, B. S. and Hall, J. C. (1976). Meiotic mutants: genic control of mitotic recombination and chromosome segregation. In M. Ashburner and E. Novitski (Ed.), *The Genetics and Biology of Drosophila*. Academic Press, London, Vol. 1A, pp. 352-435.

Baker, B. S., Carpenter, A. T. C. and Ripoll, P. (1978). The utilization during mitotic cell division of loci controlling meiotic recombination and disjunction in D. melanogaster. *Genetics* **90**, 531-578.

Baker, B. S., Boyd, J. R., Carpenter, A. T. C., Green, M. M., Nguyen, T. D., Ripoll, P. and Smith, P. D. (1976a). Genetic controls of meiotic recombination and somatic DNA metabolism in Drosophila melanogaster. *Proc. Nat. Acad. Sci. USA* **73**, 4140-4144.

Baker, B. S., Carpenter, A. T. C., Esposito, M. S., Esposito, R. E. and Sandler, L. (1976b). The genetic control of meiosis. *Ann. Rev. Genet.* **10**, 53-152.

Baker, C., Isenberg, I., Goodwin, G. H. and Johns, E. W. (1976). Physical studies of the nonhistone chromosomal proteins HMG-1 and HMG-2. *Biochemistry* **15**, 1645-1648.

Baker, C. C., Herisse, J., Coutois, G., Galibert, F. and Ziff, E. (1979). Messenger RNA for the adenovirus 2 DNA binding protein: DNA sequences encoding the first leader and heterogeneity at the mRNA 5′ end. *Cell* **18**, 569-580.

Baker, H. J. and Shapiro, D. J. (1977). Kinetics of estrogen induction of X. laevis vitellogenin mRNA as measured by hybridization to cDNA. *J. Biol. Chem.* **252**, 8428-8434.

Baker, R. M., Brunette, D. M., Mankovitz, R., Thompson, L. Whitmore, G. F., Siminovitch, L. and Till, J. E. (1974). Ouabain resistant mutants of mouse and hamster cells in culture. *Cell* **1**, 9-22.

Balbiani, E. G. (1881). Sur le structure du noyau des cellules salivaires chez les larves de Chironomus. *Zool. Anz.* **4**, 637-641.

Baldacci, P., Royal, A., Cami, B., Perrin, F., Krust, A., Garapin, A. and Kourilsky, P. (1979). Isolation of the lysozyme gene of chicken. *Nucleic Acids Res.* **6**, 2667-2681.

Baldari, C. T. and Amaldi, F. (1976). DNA reassociation kinetics in relation to genome size in four amphibian species. *Chromosoma* **59**, 13-22.

Baldi, M. I., Mattoccia, E. and Tocchini-Valentini, G. P. (1978). DNA supercoiling by X. laevis oocyte extracts: requirement for a nuclear factor. *Proc. Nat. Acad. Sci. USA* **75**, 4873-4876.

Baldwin, J. P., Boseley, P. G., Bradbury, E. M. and Ibel, K. (1975). The subunit structure of the eucaryotic chromosome. *Nature* **253**, 245-248.

Balhorn, R., Oliver, D., Hohmann, P., Chalkley, R. and Granner, D. (1972). Turnover of DNA, histones and lysine rich phosphate in hepatoma tissue culture cells. *Biochemistry* **11**, 3915-3920.

Balhorn, R., Jackson, D., Granner, D. and Chalkley, R. (1975). Phosphorylation of the lysine rich histones throughout the cell cycle. *Biochemistry* **14**, 2504-2511.

Baltimore, D. and Huang, A. S. (1970). Interaction of HeLa cell proteins with RNA. *J. Mol. Biol.* **47**, 263-273.

Bantle, J. A. and Hahn, W. E. (1976). Complexity and characterization of polyadenylated RNA in the mouse brain. *Cell* **8**, 139-150.

Baralle, F. E. (1977a). Complete nucleotide sequence of the 5′ noncoding region of rabbit β globin mRNA. *Cell* **10**, 549-558.

Baralle, F. E. (1977b). Complete nucleotide sequence of the 5′ noncoding region of human α and β globin mRNA. *Cell* **12**, 1085-1095.

Baralle, F. E. and Brownlee, G. G. (1978). AUG is the only recognizable signal sequence in the 5′ noncoding regions of eucaryotic mRNA. *Nature* **274**, 84-87.

Barbosa, E. and Moss, B. (1978). mRNA (nucleoside 2′) methyl transferase from vaccinia virus. Characteristics and substrate specificity. *J. Biol. Chem.* **253**, 7698-7702.

Bard, E., Efron D., Marcus, A. and Perry, R. P. (1974). Translational capacity of deadenylated mRNA. *Cell* **1**, 101-106.

Barnes, W. M. (1978). DNA sequencing by partial ribosubstitution. *J. Mol. Biol.* **119**, 83-100.

Barnett, T. and Rae, P. M. M. (1979). A 9.6 kb intervening sequence in D virilis rDNA and sequence homology in rDNA interruptions of diverse species of Drosophila and other Diptera. *Cell* **16**, 763-775.

Barnett, W. E. and Brown, D. H. (1967). Mitochondrial tRNAs. *Proc. Nat. Acad. Sci. USA* **57**, 452-458.

Barnett, W. E., Brown, D. H. and Epler, J. L. (1967). Mitochondrial specific aminoacyl tRNA synthetases. *Proc. Nat. Acad. Sci. USA* **57**, 1775-1781.

Barnicott, N. A. (1966). A note on the structure of spindle fibers. *J. Cell Sci.* **1**, 217-222.

Barnicott, N. A. (1967). A study of newt mitotic chromosomes by negative staining. *J. Cell Biol.* **32**, 585-602.

Barr, G. C. and Butler, J. A. V. (1963). Histones and gene function. *Nature* **199**, 1170-1172.

Barrell, B. G., Air, G. M. and Hutchison, C. A. (1976). Overlapping genes in ΦX174. *Nature* **264**, 33-40.

Barrett, T., Maryanka, D., Hamlyn, P. H. and Gould, H. J. (1974). Nonhistone proteins control gene expression in reconstituted chromatin. *Proc. Nat. Acad. Sci. USA* **71**, 5057-5061.

Barrieux, A., Ingraham, H. A., David, D. N. and Rosenfeld, M. (1975). Isolation of messenger-like ribonucleoproteins. *Biochemistry* **14**, 1815-1820.

Barsacchi-Pilone, G., Nardi, I., Andronico, F., Batistoni, R. and Durante, M. (1977). Chromosomal location of the rRNA genes in Triturus vulgaris. I. Localization of the DNA sequences complementary to 5S rRNA on mitotic and lampbrush chromosomes. *Chromosoma* **63**, 127-134.

Barski, G., Sorieul, S. and Cornefert, F. (1961). Hybrid type cells in combined cultures of two different mammalian cell strains. *J. Nat. Cancer Inst.* **26**, 1269-1291.

Bartley, J. A. and Chalkley, R. (1972). The binding of DNA and histone in native nucleohistone. *J. Biol. Chem.* **247**, 3647-3655.

Baserga, R. (1968). Biochemistry of the cell cycle: a review. *Cell Tissue Kinet.* **1**, 167-191.

Baserga, R. and Wiebel, F. (1969). The cell cycle of mammalian cells. *Int. Rev. Exp. Pathol.* **7**, 1-31.

Bastia, D., Chiang, K. S., Swift, H. and Siersma, P. (1971). Heterogeneity, complexity and repetition of the chloroplast DNA of C. reinhardii. *Proc. Nat. Acad. Sci. USA* **68**, 1157-1161.

Bastos, R. N. and Aviv, H. (1977a). Theoretical analysis of a model for globin mRNA accumulation during erythropoiesis. *J. Mol. Biol.* **110**, 205-218.

Bastos, R. N. and Aviv, H. (1977b). Globin RNA precursor molecules: biosynthesis and processing in erythroid cells. *Cell* **11**, 641-650.

Bastos, R. N., Volloch, Z. and Aviv, H. (1977). Messenger RNA population analysis during erythroid differentiation: a kinetical approach. *J. Mol. Biol.* **110**, 191-204.

Battey, J. and Clayton, D. A. (1978). The transcription map of mouse mitochondrial DNA. *Cell* **14**, 143-156.

Batts-Young, B. and Lodish, H. F. (1978). Triphosphate residues at the 5' ends of rRNA precursor and 5S RNA from D. discoideum. *Proc. Nat. Acad. Sci. USA* **75**, 74-744.

Batts-Young, B., Maizels, N. and Lodish, H. F. (1977). Precursors of rRNA in the cellular slime mold, D. discoideum. *J. Biol. Chem.* **252**, 3952-3960.

Baudisch, W. and Panitz, R. (1968). Kontrolle eines biochemischen merkmals in den speicheldrusen von Acricotopus lucidus durch einen Balbiani ring. *Exp. Cell Res.* **49**, 470-476.

Bauer, W. and Vinograd, J. (1968). The interaction of closed circular DNA with intercalative dyes. I. The superhelix density of SV40 DNA in the presence and absence of dye. *J. Mol. Biol.* **33**, 141-171.

Baur, E. (1909). Das wesen und die erblichkeitsverhaltnisse der varietates albomarginate hort von P. zonale. *Z. Verebungsl.* **1**, 330-351.

Bazetoux, S., Jouanin, L. and Huguet, T. (1978). Characterization of inverted repeated sequences in wheat nuclear DNA. *Nucleic Acids Res.* **5**, 751-770.

Beard, P. (1978). Mobility of histones on the chromosome of SV40. *Cell* **15**, 955-968.

Beattie, W. G. and Skinner, D. M. (1972). The diversity of satellite DNAs of Crustacea. *Biochim. Biophys. Acta* **281**, 169-178.

Beauchamp, R. S., Pasternak, J. and Straus, N. A. (1979). Characterization of the genome of the free living nematode Panagrellus silusiae: absence of short period interspersion. *Biochemistry* **18**, 245-250.

Beauchamp, R. S., Mitchell, A. R., Buckland, R. A. and Bostock, C. J. (1979). Specific arrangements of human satellite III DNA sequences in human chromosomes. *Chromosoma* **71**, 153-166.

Beaudet, A. L., Roufa, D. J. and Caskey, C. T. (1973). Mutations affecting the structure of HGPRT in cultured Chinese hamster cells. *Proc. Nat. Acad. Sci. USA* **70**, 320-324.

Bech-Hansen, N. T., Till, J. E. and Ling, V. (1976). Pleiotropic phenotype of colchicine resistant CHO cells: cross resistance and collateral sensitivity. *J. Cell. Physiol.* **88**, 23-32.

Becker, H. J. (1959). Due puffs der speicheldrusenchromosomen von D. melanogaster I. Beobachtungen zum verhalten des puffmasters im Normalsta, und bei zwei mutanten, giant und lethal laevae. *Chromosoma* **10**, 654-678.

Becker, H. J. (1962a). Die auslosung der puffbildung, ihre spezifitat und inhre beziehung zur funktion der ringdruse. *Chromosoma* **13**, 341-384.

Becker, H. J. (1962b). Stadienspezifische genaktivierung in speicheldrusen nach transplantation dei D. melanogaster. *Zool. Anz. Suppl.* **25**, 92-101.

Becker, H. J. (1974). Mitotic recombination maps in D. melanogaster. *Naturwissen.* **61**, 441-448.

Becker, H. J. (1975). X ray and TEM induced mitotic recombination in D. melanogaster: unequal and sister strand recombination. *Mol. Gen. Genet.* **138**, 11-24.

Becker, H. J. (1976). Mitotic recombination. In M. Ashburner and E. Novitski (Ed.), *The Genetics and Biology of Drosophila*, Academic Press, London, Vol. 1c, pp. 1020-1089.

Bedbrook, J. R. and Bogorad, L. (1976). Endonuclease recognition sites mapped on Zea mays chloroplast DNA. *Proc. Nat. Acad. Sci. USA* **73**, 4309-4313.

Bedbrook, J. R., Kolodner, R. and Bogorad, L. (1977). Zea mays chloroplast rRNA genes are part of a 22,000 base pair inverted repeat. *Cell* **11**, 739-748.

Bedbrook, J. R., Link, G., Coen, D. M., Bogorad, L. and Rich, A. (1978). Maize plastid gene expressed during photo regulated development. *Proc. Nat. Acad. Sci. USA* **75**, 3060-3064.

Bedi, K. S. and Goldstein, D. J. (1976). Apparent anomalies in nuclear Feulgen-DNA contents. Role of systematic microdensitometric errors. *J. Cell Biol.* **71**, 68-88.

Beebee, T. J. C. and Butterworth, P. H. W. (1975). Transcription of isolated nuclei and nucleoli by exogenous RNA polymerases A and B. *Eur. J. Biochem.* **51**, 537-546.

Beebee, T. J. C. and Butterworth, P. H. W. (1974). Template specificities of X. laevis RNA polymerases. Selective transcription of ribosomal cistrons by RNA polymerase A. *Eur. J. Biochem.* **45**, 395-406.

Beerman, W. (1952). Chromomerrenkonstanz und spezifische modifikationen der chromosomenstruktur in der entwicklung und organ differenzierung von C. tentans. *Chromosoma* **5**, 139-198.

Beerman, W. (1961). Ein Balbiani ring als locus einer speicheldrusen mutation. *Chromosoma* **12**, 1-25.

Beerman, W. (1964). Control of differentiation at the chromosomal level. *J. Exp. Zool.* **157**, 49-62.

Beerman, W. (1965). Differentiation at the level of the chromosomes. In Cell Differentiation and Morphogenesis, ed. W. Beerman. (North Holland: Amsterdam). pp. 24-54.

Beerman, W. (1971). Effect of α-amanitin on puffing and intranuclear RNA synthesis in Chironomus salivary glands. *Chromosoma* **34**, 152-167.

Beerman, W. (1973). Directed changes in the pattern of Balbiani ring puffing in Chironomus: effects of a sugar treatment. *Chromosoma* **41**, 297-326.

Beerman, W. and Bahr, G. F. (1954). The submicroscopic structure of the Balbiani ring. *Exp. Cell Res.* **6**, 195-208.

Beerman, W. and Pelling, C. (1965). H³ thymidin markierung einzelner chromatiden in Riesenchromosomen. *Chromosoma* **16**, 1-21.

Behn, W. and Herrmann, R. G. (1977). Circular molecules in the β satellite DNA of Chlamydomonas reinhardii. *Mol. Gen. Genet.* **157**, 25–30.

Behnke, O. and Forer, A. (1967a). Evidence for four classes of microtubules in individual cells. *J. Cell Sci.* **2**, 169–192.

Behnke, O. and Forer, A. (1967b). Some aspects of microtubules in spermatocyte meiosis in a crane fly (Nephrotoma suturalis Loew). Intranuclear and intrachromosomal microtubules. *C. R. Trav. Lab. Carlsberg* **35**, 437–455.

Behnke, O., Kristinsen, B. I., and Nielsen, L. E. (1971). Electron microscopical observations on actinoid and myosinoid filaments in blood platelets. *J. Ult. Res.* **37**, 351–369.

Bekhor, I., Bonner, J. and Dahmus, G. C. (1969). Hybridization of chromosomal RNA to native DNA. *Proc. Nat. Acad. Sci. USA* **62**, 271–277.

Bekhor, I., Kung, G. M. and Bonner, J. (1969). Sequence specific interaction of DNA and chromosomal protein. *J. Mol. Biol.* **39**, 351–364.

Bell, G. I., DeGennaro, L. J., Gelfand, D. H., Bishop, R. J., Valenzuela, P. and Rutter, W. J. (1977). Ribosomal RNA genes of S. cerevisiae. I. Physical map of the repeating unit and location of the regions coding for 5S, 5.8S, 18S and 25S rRNAs. *J. Biol. Chem.* **252**, 8118–8125.

Bellard, M., Oudet, P., Germond, J. E. and Chambon, P. (1976). Subunit structure of SV40 minichromosome. *Eur. J. Biochem.* **70**, 543–553.

Bellard, M., Gannon, F. and Chambon, P. (1977). Nucleosome structure. III. The structure and transcriptional activity of the chromatin containing the ovalbumin and globin genes in chick oviduct nuclei. *Cold Spring Harbor Symp. Quant. Biol.* **42**, 779–791.

Bello, L. J. (1974). Regulation of thymidine kinase synthesis in human cells. *Exp. Cell Res.* **89**, 263–274.

Benbow, R. M. and Ford, C. C. (1975). Cytoplasmic control of nuclear DNA synthesis during early development of X. laevis: a cell free assay. *Proc. Nat. Acad. Sci. USA* **72**, 2437–2441.

Benbow, R. M., Krauss, M. R. and Reeder, R. H. (1978). DNA synthesis in a multienzyme system from X. laevis eggs. *Cell* **13**, 307–318.

Bendich, A. J. and Anderson, R. S. (1977). Characterization of families of repeated DNA sequences from four vascular plants. *Biochemistry* **16**, 4655–4663.

Bendich, A. J. and McCarthy, B. J. (1970). DNA comparisons among barley, oats, rye and wheat. *Genetics* **65**, 545–565.

Benecke, B. J. and Penman, S. (1977). A new class of small nuclear RNA molecules synthesized by a type I RNA polymerase in Hela cells. *Cell* **12**, 939–946.

Bennett, G. S., Fellini, S. A., Croop, J. M., Otto, J. J., Bryan, J. and Holtzer, H. (1978). Differences among 100 Å filament subunits from different cell types. *Proc. Nat. Acad. Sci. USA* **75**, 4364–4368.

Bennett, M. D. and Smith, J. D. (1976). Nuclear DNA amounts in angiosperms. *Phil. Trans. Roy. Soc. B* **274**, 227–274.

Benoff, S. and Skoultchi, A. I. (1977). X linked control of hemoglobin production in somatic hybrids of mouse erythroleukemic cells and mouse lymphoma or bone marrow cells. *Cell* **12**, 263–274.

Bensch, K. G. and Malawista, S. E. (1969). Microtubular crystals in mammalian cells. *J. Cell Biol.* **40**, 95–107.

Bentley, R. (1861). New American remedies. I. Podophyllum peltatum. *Pharm. J. Trans.* **3**, 456–464.

Benton, W. D. and Davis, R. W. (1977). Screening λ gt recombinant clones by hybridization to single plaques in situ. *Science* **196**, 180–182.

Benyajati, C. and Worcel, A. (1976). Isolation, characterization and structure of the folded interphase genome of D. melanogaster. *Cell* **9**, 393–408.

Benz, E. J., Forget, B. G., Hillman, D. G., Cohen-Solal, M., Pritchard, J., Cavallesco, C., Prensky,

W. and Housman, D. (1978). Variability in the amount of β globin mRNA in β° thalassemia. *Cell* 14, 299-312.

Benzer, S. (1959). On the topography of the genetic fine structure. *Proc. Nat. Acad. Sci. USA* 45, 1607-1620.

Benzer, S. (1961). On the topography of the genetic fine structure. *Proc. Nat. Acad. Sci. USA* 47, 1025-1038.

Berendes, H. D. (1963). The salivary gland chromosomes of D. hydei sturtevant. *Chromosoma* 14, 195-206.

Berendes, H. D. (1965a). Salivary gland functions and chromosomal puffing patterns in D. hydei. *Chromosoma* 17, 35-77.

Berendes, H. D. (1965b). The induction of changes in chromosomal activity in different polytene types of cell in D. hydei. *Devel. Biol.* 11, 371-384.

Berendes, H. D. (1966). Gene activities in the Malpighian tubules of D. hydei at different stages. *J. Exp. Zool.* 162, 209-218.

Berendes, H. D. (1967). The hormone ecdysone as effector of specific changes in the pattern of gene activities of D. hydei. *Chromosoma* 22, 274-293.

Berendes, H.D. (1968). Factors involved in the expression of gene activity in polytene chromosomes. *Chromosoma* 24, 418-437.

Berendes, H. D. (1970). Polytene chromosome structure at the submicroscopic level. I. A map of region X, 1-4E of D. melanogaster. *Chromosoma* 29, 118-130.

Berendes, H.D. and Keyl, H.G. (1967). Distribution of DNA in heterochromatin and euchromatin of polytene nuclei of D. hydei. *Genetics* 57, 1-13.

Berendes, H. D., van Breugel, F. D. and Holt, Th. K. H. (1965). Experimental puffs in salivary gland chromosomes of D. hydei. *Chromosoma* 16, 35-46.

Berezney, R. and Coffey, D. S. (1977). Nuclear matrix. Isolation and characterization of a framework structure from rat liver nuclei. *J. Cell. Biol.* 73, 616-637.

Berget, S. M., Moore, C. M. and Sharp, P. A. (1977). Spliced segments at the 5' terminus of adenovirus 2 late mRNA. *Proc. Nat. Acad. Sci. USA* 74, 3171-3175.

Bergmann, J. E. and Lodish, H. F. (1979). Translation of capped and uncapped VSV and reovirus mRNAs. Sensitivity to m7GpppAm and ionic conditions. *J. Biol. Chem.* 254, 459-468.

Berk, A. J. and Clayton, D. A. (1974). Mechanism of mitochondrial DNA replication in mouse L cells: asynchronous replication of strands, segregation of circular daughter molecules, aspects of topology and turnover of an initiation sequence. *J. Mol. Biol.* 86, 801-824.

Berk, A. J. and Sharp, P.A. (1978a). Spliced early mRNAs of SV40. *Proc. Nat. Acad. Sci. USA* 75, 1274-1278.

Berk, A. J. and Sharp, P. A. (1978b). Structure of the adenovirus 2 early mRNAs. *Cell* 14, 695-672.

Berkowitz, E. M. and Doty, P. (1975). Chemical and physical properties of fractionated chromatin. *Proc. Nat. Acad. Sci. USA* 72, 3328-3332.

Berlowitz, L. (1965). Correlation of genetic activity, heterochromatization and RNA metabolism in mammals. *Proc. Nat. Acad. Sci. USA* 53, 68-73.

Bernard, O., Hozumi, N. and Tonegawa, S. (1978). Sequences of mouse immunoglobulin light chain genes before and after somatic changes. *Cell* 15, 1133-1144.

Bernard, U., Puhler, A., Mayer, F. and Kuntzel, H. (1975a). Denaturation map of the circular mitochondrial genome of N. crassa. *Biochim. Biophys. Acta* 402, 270-278.

Bernard, U., Bade, E. and Kuntzel, H. (1975b). Specific fragmentation of mitochondrial DNA from N. crassa by restriction endonuclease. *Biochem. Biophys. Res. Commun.* 64, 783-789.

Bernardi, G. and Sadron, C. (1964). Studies on DNAase. I. Kinetics of the initial degradation of DNA by DNAase. *Biochemistry* 3, 1411-1418.

Bernardi, G., Faure, S. M., Piperno, G. and Slonimski, P. P. (1970). Mitochondrial DNAs from respiratory sufficient and cytoplasmic respiratory deficient mutant yeast. *J. Mol. Biol.* **48**, 23-42.

Berns, A., Janssen, P. and Bloemendal, H. (1974). The molecular weight of the 14S calf lens mRNA. *Biochem. Biophys. Res. Commun.* **59**, 1157-1164.

Berns, A. J. M., Van Kraaikamp, M., Bloemendal, H. and Lane, C. D. (1972). Calf crystallin synthesis in frog eggs: the translation of lens cell 14S RNA in oocytes. *Proc. Nat. Acad. Sci. USA* **69**, 1606-1609.

Berns, M. W., Rattner, J. B., Brenner, S. and Meredith, S. (1977). The role of the centriolar region in animal cell mitosis. A laser microbeam study. *J. Cell Biol.* **72**, 351-367.

Berridge, M. V. and Lane, C. D. (1976). Translation of Xenopus liver mRNA in Xenopus oocytes: vitellogenin synthesis and conversion to yolk platelet proteins. *Cell* **8**, 283-297.

Berry, R. W. and Shelanski, M. L. (1972). Interactions of tubulin with vinblastine and GTP. *J. Mol. Biol.* **71**, 71-80.

Berstad, P., Farnsworth, V., Weigert, M., Cohn, M. and Hood, L. (1974). Mouse immunoglobulin heavy chains are coded by multiple germ line variable regions. *Proc. Nat. Acad. Sci. USA* **71**, 4096-4100.

Bertazzoni, U., Scovassi, A. I. and Brun, G. M. (1977). Chick embryo DNA polymerase γ. Identity of γ polymerases purified from nuclei and mitochondria. *Eur. J. Biochem.* **81**, 237-248.

Bertazzoni, U., Stefanini, M., Noy, G. P., Giulotto, E., Nuzzo, F. and Spadari, S. (1976). Variations of DNA polymerases α and β during prolonged incubation of human lymphocytes. *Proc. Nat. Acad. Sci. USA* **73**, 785-789.

Bertoletti, R. and Weiss, M. C. (1972a). Expression of differentiated functions in hepatoma cell hybrids. II. Aldolase. *J. Cell. Physiol.* **79**, 211-224.

Bertoletti, R. and Weiss, M. C. (1972b). Expression of differentiated functions in hepatoma cell hybrids. VI. Extinction and reexpression of liver alcohol dehydrogenase. *Biochimie* **54**, 195-201.

Bertrand, A. and Pittenger, G. H. (1972). Complementation among cytoplasmic mutants of N. crassa. *Mol. Gen. Genet.* **117**, 82-90.

Bertrand, H. and Werner, S. (1977). Deficiency of subunits 2 of cytochrome oxidase in the *mi-3* cytoplasmic mutant of N. crassa. *Eur. J. Biochem.* **79**, 599-606.

Betlach, M., Hershfield, V., Chow, L., Brown, W., Goodman, H. M., and Boyer, H. W. (1976). A restriction endonuclease analysis of the bacterial plasmid controlling the EcoRI restriction and modification of DNA. *Fed. Proc.* **35**, 2037-2043.

Bettex-Galland, M. and Luscher, E. F. (1959). Extraction of an actomyosin-like protein from human thrombocytes. *Nature* **184**, 276-277.

Bettex-Galland, M., Portzehl, H. and Luscher, E. F. (1962). Dissociation of thrombosthenin into two components comparable with actin and myosin. *Nature* **193**, 777-778.

Beutler, E., Yeh, M. and Fairbanks, V. F. (1962). The normal human female as a mosaic of X chromosome activity: studies using the gene for G6PD deficiency as a marker. *Proc. Nat. Acad. Sci. USA* **48**, 9-16.

Beyer, A. L., Christensen, M. E., Walker, B. W. and LeStourgeon, W. M. (1977). Identification and characterization of the packaging proteins of core 40S hnRNP particles. *Cell* **11**, 127-138.

Bhattacharyya, B. and Wolff, J. (1974). Promotion of fluorescence upon binding of colchicine to tubulin. *Proc. Nat. Acad. Sci. USA* **71**, 2627-2631.

Bhattacharyya, B. and Wolff, J. (1975). Anion induced increases in the rate of colchicine binding to tubulin. *Biochemistry* **15**, 2283-2288.

Bhattacharyya, B. and Wolff, J. (1976). Tubulin aggregation and disaggregation: mediation by two distinct vinblastine-binding sites. *Proc. Nat. Acad. Sci. USA* **73**, 2375-2378.

Bhisey, A. N. and Freed, J. J. (1971). Cross bridges on the microtubules of cooled interphase HeLa cells. *J. Cell Biol.* **50**, 557-561.

Bhorjee, J. S. and Pederson, T. (1972). Nonhistone chromosomal proteins in synchronized Hela cells. *Proc. Nat. Acad. Sci. USA* **69**, 3345-3349.

Bhorjee, J. S. and Pederson, T. (1973). Chromatin: its isolation from cultured mammalian cells with particular reference to contamination by nuclear ribonucleoprotein particles. *Biochem.* **12**, 2766-2773.

Bibring, T. and Baxandall, J. (1971). Selective extraction of isolated mitotic apparatus. Evidence that typical microtubule protein is extracted by organic mercurial. *J. Cell Biol.* **48**, 324-339.

Bibring, T. and Baxandall, J. (1974). Tubulins 1 and 2. Failure of quantitation in polyacrylamide gel electrophoresis may influence their identification. *Exp. Cell Res.* **86**, 120-126.

Bibring, T., Baxandall, J., Denslow, S. and Walker, B. (1976). Heterogeneity of the alpha subunit of tubulin and the variability of tubulin within a single organism. *J. Cell Biol.* **69**, 301-312.

Bick, M. D. and Davidson, R. L. (1974). Total substitution of BUdR for thymidine in the DNA of a BUdR dependent cell line. *Proc. Nat. Acad. Sci. USA* **71**, 2082-2086.

Bick, M. D. and Davidson, R. L. (1976). Nucleotide analysis of DNA and RNA in cells with thymidine totally replaced by BUdR. *Somatic Cell Genet.* **2**, 63-76.

Biessmann, H., Levy, B. W., and McCarthy, B. J. (1978). In vitro transcription of heat shock specific RNA from chromatin of D. melanogaster cells. *Proc. Nat. Acad. Sci. USA* **75**, 759-763.

Biessmann, H., Gjeset, R. A., Levy, B. W. and McCarthy, B. J. (1976). Fidelity of chromatin transcription in vitro. *Biochemistry* **15**, 4356-4363.

Billeter, M. A. and Hindley, J. (1972). A study of the quantitative variation of histones and their relationship to RNA synthesis during erythropoiesis in the adult chicken. *Eur. J. Biochem.* **28**, 451-462.

Bina-Stein, M. and Simpson, R. T. (1977). Specific folding and contraction of DNA by histones H3 and H4. *Cell* **11**, 609-618.

Bina-Stein, M., Thoren, M., Salzman, N. and Thompson, J. A. (1979). Rapid sequence determination of late SV40 16S mRNA leader by using inhibitors of reverse transcriptase. *Proc. Nat. Acad. Sci. USA* **76**, 731-735.

Binder, L. I. and Rosenbaum, J. L. (1978). The in vitro assembly of flagellar outer doublet tubulin. *J. Cell Biol.* **79**, 500-515.

Binder, L. I., Dentler, W. L. and Rosenbaum, J. L. (1975). Assembly of chick brain tubulin onto flagellar microtubules from Chlamydomonas and sea urchin sperm. *Proc. Nat. Acad. Sci. USA* **72**, 1122-1126.

Bird, A. P. (1978). Use of restriction enzymes to study eucaryotic DNA methylation. II. The symmetry of methylated sites supports semiconservative copying of the methylation pattern. *J. Mol. Biol.* **118**, 49-60.

Bird, A. P. and Southern, E. M. (1978). Use of restriction enzymes to study eucaryotic DNA methylation. I. The methylation pattern in ribosomal DNA from X. laevis. *J. Mol. Biol.* **118**, 27-48.

Bird, A. P., Taggart, M. H. and Smith, B. A. (1979). Methylated and unmethylated DNA compartments in the sea urchin genome. *Cell* **17**, 889-902.

Birkenmeier, E. H., Brown, D. D. and Jordan, E. (1978). A nuclear extract of X. laevis. oocytes that accurately transcribes 5S RNA genes. *Cell* **15**, 1077-1086.

Birky, C., Demko, C. A., Perlman, P. S. and Strausberg, R. (1978). Uniparental inheritance of mitochondrial genes in yeast: dependence on input bias of mitochondrial DNA and preliminary investigations of the mechanism. *Genetics* **89**, 615-651.

Birnboim, H. C. and Coakley, B. V. (1971). Adenylate rich oligonucleotides of ribosomal and ribosomal precursor RNA from Hela cells. *Biochem. Biophys. Res. Commun.* **42**, 1169-1176.

Birnboim, H. C. and Sederoff, R. (1975). Polypyrimidine segments in D. melanogaster DNA. I. Detection of a cryptic satellite containing polypyrimidine/polypurine DNA. *Cell* **5**, 173-182.

Birnie, G. D., MacPhail, E., Young, B. D., Getz, M. J. and Paul, J. (1974). The diversity of the mRNA population in growing Friend cells. *Cell Diff.* **3**, 221-232.

Birnstiel, M. L., Chipchase, M. I. H. and Hyde, B. B. (1966). The nucleolus as a source of ribosomes. *Biochim. Biophys. Acta* **76**, 454-462.

Birnstiel, M. L., Schaffner, W. and Smith, H. D. (1977). DNA sequences coding for the H2B histone of P. miliaris. *Nature* **266**, 603-607.

Birnstiel, M., Telford, J., Weinberg, E. and Stafford, D. (1974). Isolation and some properties of the genes coding for histone proteins. *Proc. Nat. Acad. Sci. USA* **71**, 2900-2904.

Biro, P. A., Carr-Brown, A., Southern, E. M., and Walker, P. M. B. (1975). Partial sequence analysis of mouse satellite DNA: evidence for short range periodicities. *J. Mol. Biol.* **94**, 71-86.

Bishop, J. O. and Freeman, K. B. (1973). DNA sequences neighboring the duck hemoglobin genes. *Cold Spring Harbor Symp. Quant. Biol.* **38**, 707-716.

Bishop, J. O. and Rosbash, M. (1973). Reiteration frequency of duck hemoglobin genes. *Nature New Biol.* **241**, 204-207.

Bishop, J. O., Rosbash, M. and Evans, D. (1974). Polynucleotide sequences in eucaryotic DNA and RNA that form ribonuclease-resistant complexes with poly-U. *J. Mol. Biol.* **85**, 75-86.

Bishop, J. O., Morton, J. C., Rosbash, M. and Richardson, M. (1974). Three abundance classes in Hela cell mRNA. *Nature* **250**, 199-203.

Bitter, G. A. and Roeder, R. G. (1978). Transcription of viral genes by RNA polymerase II in nuclei isolated from adenovirus 2 transformed cells. *Biochemistry* **17**, 2198-2205.

Blackburn, E. H. and Gall, J. G. (1978). A tandemly repeated sequence at the termini of the extrachromosomal rRNA genes in Tetrahymena. *J. Mol. Biol.* **120**, 33-54.

Blanchard, J. M., Brunel, C. and Jeanteur, P. (1978). Phosphorylation in vivo of proteins associated with hnRNA in HeLa cell nuclei. *Eur. J. Biochem.* **86**, 301-310.

Blanchard, J. M., Weber, J., Jelinek, W. and Darnell, J. E. (1978). In vitro RNA-RNA splicing in adenovirus 2 mRNA formation. *Proc. Nat. Acad. Sci. USA* **75**, 5344-5348.

Blatti, S. P., Ingles, C. J., Lindell, T. J., Morris, P. W., Weaver, R. F., Weinberg, F. and Rutter, W. J. (1970). Structure and regulatory properties of eucaryotic RNA polymerase. *Cold Spring Harbor Symp. Quant. Biol.* **35**, 649-658.

Blattner, F. R., Williams, B. G., Blechl, A. E., Faber, H. E. and Smithies, O. (1977). Charon phages: safer derivatives of phage lambda for DNA cloning. *Science* **196**, 161-169.

Blin, N., Sperrazza, J. M., Wilson, F. E., Bieber, D. G., Mickel, F. S. and Stafford, D. W. (1979). Organization of the rRNA gene cluster in Lytechinus variegatus. Restriction analysis and cloning of restriction fragments. *J. Biol. Chem.* **254**, 2716-2721.

Blobel, G. (1972). Protein tightly bound to globin mRNA. *Biochem. Biophys. Res. Commun.* **47**, 88-95.

Blobel, G. (1973). A protein of molecular weight 78,000 bound to a polyadenylate region of eucaryotic mRNA. *Proc. Nat. Acad. Sci. USA* **70**, 924-928.

Blobel, G. and Dobberstein, B. (1975a). Transfer of proteins across membranes. I. Presence of proteolytically processed and unprocessed nascent Ig light chains on membrane bound ribosomes of murine myeloma. *J. Cell Biol.* **67**, 835-851.

Blobel, G. and Dobberstein, B. (1975b). Transfer of proteins across membranes. II. Reconstitution of functional rough microsomes from heterologous components. *J. Cell Biol.* **67**, 852-861.

Bloch, D. P., MacQuigg, R. A., Brack, S. D., and Wu, J. R. (1967). The synthesis of DNA and histone in the onion root meristem. *J. Cell Biol.* **33**, 451-467.

Bloch, P. and Goodman, G. C. (1955). A microphotometric study of the synthesis of DNA and nuclear histone. II. Evidence of differences in the DNA-protein complex of rapidly proliferating and nondividing cells. *J. Biochem. Biophys. Cytol.* **1**, 17-28.

Bloom, K. S. and Anderson, J. N. (1978). Fractionation of hen oviduct chromatin into transcriptionally active and inactive regions after selective micrococcal nuclease digestion. *Cell* **15**, 141-150.

Blose, S. H., Shelanski, M. L. and Chacko, S. (1977). Localization of bovine brain filament antibody on intermediate (10 nm) filaments in guinea pig vascular endothelial cells and chick cardiac muscle cells. *Proc. Nat. Acad. Sci. USA* **74**, 662-665.

Blumenfeld, M. and Forrest, H. S. (1971). Is Drosophila dAT on the Y chromosome? *Proc. Nat. Acad. Sci. USA* **68**, 3145-3149.

Blumenfeld, M. and Forrest, H. (1972). Differential under replication of satellite DNAs during Drosophila development. *Nature New Biol.* **239**, 170-172.

Blumenfeld, M., Fox, A. S. and Forrest, H. S. (1973). A family of three related satellite DNAs in D. virilis. *Proc. Nat. Acad. Sci. USA* **70**, 2772-2775.

Blumenthal, A. B., Kriegstein, H. J. and Hogness, D. S. (1973). The units of DNA replication in D. melanogaster chromosomes. *Cold Spring Harbor Symp. Quant. Biol.* **38**, 205-223.

Boardman, N. K., Francki, R. I. B. and Waldman, S. G. (1966). Protein synthesis by cell free extracts of tobacco leaves. III. Comparison of the physical properties and protein synthesizing activities of 70S chloroplast and 80S cytoplasmic ribosomes. *J. Mol. Biol.* **17**, 470-489.

Bobrow, J. (1973). Acridine orange and the investigation of chromosome banding. *Cold Spring Harbor Symp. Quant. Biol.* **38**, 435-440.

Bobrow, M., Pearson, P. L. and Collacott, H. E. A. C. (1971). Paranucleolar position of the human Y chromosome in interphase nuclei. *Nature* **232**, 556-557.

Boedtker, H., Frischauf, A. M. and Lehrach, H. (1976). Isolation and translation of Calvaria procollagen mRNAs. *Biochemistry* **15**, 4765-4770.

Boersma, D., McGill, S., Mollenkamp, J. and Roufa, D. J. (1979a). Emetine resistance in Chinese hamster cells. Analysis of ribosomal proteins prepared from mutant cells. *J. Biol. Chem.* **254**, 559-567.

Boersma, D., McGill, S. M., Mollenkamp, J. W. and Roufa, D. J. (1979b). Emetine resistance in Chinese hamster cells is linked genetically with an altered 40S ribosomal subunit protein, S20. *Proc. Nat. Acad. Sci. USA* **76**, 415-419.

Boffa, L. C., Vidali, G., Mann, R. S. and Allfrey, V. C. (1978). Suppression of histone deacetylation in vivo and in vitro by sodium butyrate. *J. Biol. Chem.* **253**, 3364-3366.

Bogenhagen, D. and Clayton, D. A. (1974). The number of mitochondrial DNA genomes in mouse L and human HeLa cells. Quantitative isolation of mitochondrial DNA. *J. Biol. Chem.* **249**, 7991-7995.

Bogenhagen, D. and Clayton, D. A. (1977). Mouse L cell mitochondrial DNA molecules are selected randomly for replication throughout the cell cycle. *Cell* **11**, 719-727.

Bogenhagen, D. and Clayton, D. A. (1978a). Mechanism of mitochondrial DNA replication in mouse L cells: kinetics of synthesis and turnover of the initiation sequence. *J. Mol. Biol.* **119**, 49-68.

Bogenhagen, D. and Clayton, D. A. (1978b). Mechanism of mitochondrial DNA replication in mouse L cells: introduction of superhelical turns into newly replicated molecules. *J. Mol. Biol.* **119**, 69-82.

Bogenhagen, D. F., Sakonju, S. and Brown, D. D. (1980). A control region in the center of the 5S RNA gene directs specific initiation of transcription. II. The 3' border of the region. *Cell* **19**, 27-35.

Bogorad, L. (1975). Evolution of organelles and eucaryotic genomes. *Science* **188**, 891-898.

Bohm, L., Hayashi, H., Cary, P. D., Moss, T., Crane-Robinson, C. and Bradbury, E. M. (1977). Sites of histone-histone interaction in the H3-H4 complex. *Eur. J. Biochem.* **77**, 487-494.

Boivin, A., Vendrely, R. and Vendrely, C. (1948). L'acide désoxyribonucléique de noyau cellulaire dépositaire des caractères héréditaires: arguments d'ordre analytique. *C. R. Hebd. Séances. Acad. Sci.* **226**, 1061-1063.

Bolden, A., Noy, G. P. and Weissbach, A. (1977). DNA polymerase of mitochondria is a γ polymerase. *J. Biol. Chem.* **252**, 3351–3356.

Bolivar, F., Rodriguez, R. L., Betlach, M. C. and Boyer, H. W. (1977a). Construction and characterization of new cloning vehicles. I. Amp^r derivatives of the plasmid pMB9. *Gene* **2**, 75–93.

Bolivar, F., Rodriguez, R. L., Greene, P. J., Betlach, M. C., Heyneker, H. L., Boyer, H. W., Crosa, H. J. and Falkow, S. (1977b). Construction and characterization of new cloning vehicles. II. A multipurpose cloning system. *Gene* **2**, 95–113.

Bollum, F. J. (1960). Calf thymus polymerase. *J. Biol. Chem.* **235**, 2399–2403.

Bolotin, M., Coen, D., Deutsch, J., Dujon, B., Netter, P., Petrochilo, E. and Slonimski, P. P. (1971). La recombination des mitochondries chez S. cerevisiae. *Bull. Inst. Pasteur* **69**, 215–239.

Bolotin-Fukuhara, M. and Fukuhara, H. (1976). Modified recombination and transmission of mitochondrial genetic markers in rho minus mutants of S. cerevisiae. *Proc. Nat. Acad. Sci. USA* **73**, 4608–4612.

Bolotin-Fukuhara, M., Faye, G. and Fukuhara, H. (1977). Temperature sensitive respiratory deficient mitochondrial mutants: isolation and genetic mapping. *Mol. Gen. Genet.* **152**, 295–306.

Bols, N. C. and Ringertz, N. R. (1979). A study of metabolic cooperation with established myoblast cell lines. *Exp. Cell Res.* **120**, 15–24.

Bolton, E. T. and McCarthy, B. J. (1962). A general method for the isolation of RNA complementary to DNA. *Proc. Nat. Acad. Sci. USA* **48**, 1390–1397.

Bolund, L., Darzynkiezicz, Z. and Ringertz, N. R. (1969a). Growth of the erythrocyte nuclei undergoing reactivation in heterocaryons. *Exp. Cell Res.* **56**, 406–410.

Bolund, L., Ringertz, N. R. and Harris, H. (1969b). Changes in the cytochemical properties of erythrocyte nuclei reactivated by cell fusion. *J. Cell Sci.* **4**, 71–87.

Bonner, J. and Huang, R. C. C. (1966). Methodology for the study of the template activity of chromosomal nucleohistone. *Biochem. Biophys. Res. Commun.* **22**, 211–217.

Bonner, J., Dahmus, M. E., Fambrough, D., Huang, R. C., Marushige, K. and Tuan, D. Y. H. (1968a). The biology of isolated chromatin. *Science* **159**, 47–56.

Bonner, J., Chalkley, G. R., Dahmus, M., Fambrough, D., Fujimura, F., Huang, R. C., Huberman, J., Jensen, R., Marushige, K., Ohlenbusch, H., Olivera, B. and Widholm, J. (1968b). Isolation and characterization of chromosomal nucleoproteins. *Methods Enzymol.* **12B**, 3–64.

Bonner, J. J. and Pardue, M. L. (1976). The effect of heat shock on RNA synthesis in Drosophila tissues. *Cell* **8**, 43–50.

Bonner, J. J. and Pardue, M. L. (1977). Polytene chromosome puffing and in situ hybridization measure different aspects of RNA metabolism. *Cell* **12**, 227–234.

Bonner, T. I., Brenner, D. J., Neufeld, B. R. and Britten, R. J. (1973). Reduction in the rate of DNA reassociation by sequence divergence. *J. Mol. Biol.* **81**, 123–135.

Bonner, W. M. (1975a). Protein migration into nuclei. I. Frog oocyte nuclei in vivo accumulate microinjected histones, allow entry to small proteins, and exclude large proteins. *J. Cell Biol.* **64**, 421–430.

Bonner, W. M. (1975b). Protein migration into nuclei. II. Frog oocyte nuclei in vivo accumulate a class of microinjected oocyte nuclear proteins and exclude a class of microinjected oocyte cytoplasmic proteins. *J. Cell Biol.* **64**, 431–437.

Bonner, W. M. and Pollard, H. B. (1975). The presence of F3-F2A1 dimers and F1 oligomers in chromatin. *Biochem. Biophys. Res. Commun.* **64**, 282–288.

Borgese, N., Mok, W., Kreibich, G. and Sabatini, D. D. (1974). Ribosomal membrane interaction: in vitro binding of ribosomes to microsomal membranes. *J. Mol. Biol.* **88**, 559–580.

Borisy, G. G., and Olmsted, J. B. (1972). Nucleated assembly of microtubules in porcine brain extracts. *Science* **177**, 1196–1197.

Borisy, G. G. and Taylor, E. W. (1967a). The mechanism of action of colchicine. Binding of ^{3}H-colchicine to cellular protein. *J. Cell Biol.* **34**, 525–533.

Borisy, G. G. and Taylor, E. W. (1967b). The mechanism of action of colchicine. Colchicine binding to sea urchin eggs and the mitotic apparatus. *J. Cell Biol.* **34**, 534-548.

Borst, P. (1972). Mitochondrial nucleic acids. *Ann. Rev. Biochem.* **41**, 333-376.

Borst, P. and Grivell, L. A. (1978). The mitochondrial genome of yeast. *Cell* **15**, 705-724.

Borst, P. and Kroon, A. M. (1969). Mitochondrial DNA: physicochemical properties, replication and genetic function. *Int. Rev. Cytol.* **26**, 108-190.

Borun, T., Scharff, M. D. and Robbins, E. (1967). Rapidly labeled, polyribosome associated RNA having the properties of histone messenger. *Proc. Nat. Acad. Sci. USA* **58**, 1977-1983.

Borun, T. W., Gabrielli, F., Ajiro, K., Zweidler, A. and Baglioni, C. (1975). Further evidence of transcriptional and translational control of histone mRNA during the Hela S3 cycle. *Cell* **4**, 59-68.

Borun, T. W., Ajiro, K., Zweidler, A., Dolby, T. W. and Stephens, R. E. (1977). Studies of human histone mRNA. II. The resolution of fractions containing individual human histone mRNA species. *J. Biol. Chem.* **252**, 173-180.

Bos, J. L., Heyting, C., Borst, P., Arnberg, A. C. and Van Bruggen, E. F. J. (1978). An insert in the single gene for the large rRNA in yeast mitochondrial DNA. *Nature* **275**, 336-3337.

Boseley, P. G., Tuyns, A. and Birnstiel, M. L. (1978). Mapping of the X. laevis 5.8S rDNA by restriction and DNA sequencing. *Nucleic Acids Res.* **5**, 1121-1138.

Boseley, P. G., Bradbury, E. M., Butler-Browne, G. S., Carpenter, B. G. and Stephens, R. M. (1976). Physical studies of chromatin. The recombination of histones with DNA. *Eur. J. Biochem.* **62**, 21-32.

Boseley, P., Moss, T., Machler, M., Portmann, R. and Birnstiel, M. (1979). Sequence organization of the spacer DNA in a ribosomal gene unit of X. laevis. *Cell* **17**, 19-32.

Bostock, C. J. and Christie, S. (1974). Chromosome banding and DNA replication studies on a cell line of Dipodomis merriami. *Chromosoma* **48**, 73-88.

Bostock, C. J. and Christie, S. (1976). Analysis of the frequency of sister chromatid exchange in different regions of chromosomes of the kangaroo rat (Dipodomys ordii). *Chromosoma* **56**, 275-288.

Bostock, C. J., Prescott, D. M. and Hatch, F. T. (1972). Timing of replication of the satellite and main band DNAs in cells of the kangaroo rat. *Exp. Cell Res.* **74**, 487-495.

Bostock, C. J., Christie, S., Lauder, I. J., Hatch, F. T. and Mazrimas, J. A. (1976). S phase patterns of replication of different satellite DNAs in three species of Dipodomys (kangaroo rat). *J. Mol. Biol.* **108**, 417-434.

Bostock, C. J., Gosden, J. R. and Mitchell, A. R. (1978). Localization of a male specific DNA fragment to a sub region of the human Y chromosome. *Nature* **272**, 324-329.

Botchan, M. R. (1974). Bovine satellite I DNA consists of repetitive units 1400 base pairs in length. *Nature* **251**, 288-292.

Botchan, M., Kram, R., Schmid, C. W. and Hearst, J. E. (1971). Isolation and chromosomal localization of highly repeated DNA sequences in D. melanogaster. *Proc. Nat. Acad. Sci. USA* **68**, 1125-1129.

Botchan, P., Reeder, R. H. and Dawid, I. B. (1977). Restriction analysis of the nontranscribed spacers of X. laevis rDNA. *Cell* **11**, 599-607.

Both, G. W., Furuichi, Y., Muthukrishnan, S. and Shatkin, A. J. (1975). Ribosome binding to reovirus mRNA in protein synthesis requires 5' terminal 7-methylguanosine. *Cell* **6**, 185-196.

Both, G. W., Fururichi, Y., Muthukrishnan, S. and Shatkin, A. J. (1976). Effect of 5' terminal structure and base composition on polyribonucleotide binding to ribosomes. *J. Mol. Biol.* **104**, 637-658.

Boublik, M., Bradbury, E. M. and Crane-Robinson, C. (1970a). An investigation of the conformational changes of histones f1 and f2al by proton magnetic resonance spectroscopy. *Eur. J. Biochem.* **14**, 486-497.

Boublik, M., Bradbury, E. M., Crane-Robinson, C. and Johns, E. W. (1970b). An investigation of the conformational changes of histone f2B by high resolution nuclear magnetic resonance. *Eur. J. Biochem.* **17**, 151–159.

Bouchard, R. A. and Swift, H. (1977). Nature of the heterogeneity in mispairing of reannealed middle repetitive fern DNA. *Chromosoma* **61**, 317–334.

Bourgeois, S. and Newby, R. F. (1977). Diploid and haploid states of the glucocorticoid receptor gene of mouse lymphoid cell lines. *Cell* **11**, 423–430.

Bourguignon, L. Y. W., Tokuyasu, K. T. and Singer, S. J. (1978). The capping of lymphocytes and other cells, studied by an improved method for immuno fluorescence staining of frozen sections. *J. Cell. Physiol.* **95**, 239–258.

Bourne, H. R., Coffino, P. and Tomkins, G. M. (1975). Somatic genetic analysis of cyclic AMP action: characterization of unresponsive mutants. *J. Cell. Physiol.* **85**, 611–620.

Boyd, J. B., Golino, M. D. and Setlow, R. B. (1976). The mei^{9a} mutant of D. melanogaster increases mutagen sensitivity and decreases excision repair. *Genetics* **84**, 527–544.

Boyd, J. B., Golino, M. D., Nguyen, T. D. and Green, M. M. (1976). Isolation and characterization of X linked mutants of D. melanogaster which are sensitive to mutagens. *Genetics* **84**, 485–506.

Boyd, Y. L. and Harris, H. (1973). Correction of genetic defects in mammalian cells by the input of small amounts of foreign genetic material. *J. Cell Sci.* **13**, 841–861.

Boynton, J. E., Burton, W. G., Gillham, N. W. and Harris, E. H. (1973). Can a non Mendelian mutation affect both chloroplast and mitochondrial ribosomes. *Proc. Nat. Acad. Sci. USA* **70**, 3463–3467.

Brachet, J. (1940). La localization de l'acide thymonucleic pendant l'oogenesis et la maturation chez les amphibia. *Arch. Biol.* **51**, 151–165.

Brack, C. and Tonegawa, S. (1977). Variable and constant parts of the immunoglobulin light gene of a mouse myeloma cell are 1250 nontranslated bases apart. *Proc. Nat. Acad. Sci. USA* **74**, 5652–5656.

Brack, C., Hirama, M., Lenhard-Schuller, R. and Tonegawa, S. (1978). A complete immuno-globulin gene is created by somatic recombination. *Cell* **15**, 1–14.

Bradbury, E. M. and Crane-Robinson, C. (1971). Physical and conformational studies of histones and nucleohistones. In D. M. P. Phillips (Ed.) *Histones and Nucleohistones*. Plenum Publishing, New York, pp 83–135.

Bradbury, E. M., Inglis, R. J. and Matthews, H. R. (1974a). Control of cell division by very lysine rich histone (F1) phosphorylation. *Nature* **247**, 257–261.

Bradbury, E. M., Crane-Robinson, C., Goldman, H., Rattle, H. W. E. and Stephens, R. N. (1967). Spectroscopic studies of the conformations of histones and protamine. *J. Mol. Biol.* **29**, 507–523.

Bradbury, E., Molgaard, H. V., Stephens, R. M., Bolund, L. and Johns, E. W. (1972a). X ray studies of nucleoproteins depleted of lysine rich histone. *Eur. J. Biochem.* **31**, 474–482.

Bradbury, E. M., Cary, P. D., Crane-Robinson, C., Riches, P. and Johns, E. W. (1972b). Nuclear megnatic resonance and optical spectroscopic studies of conformation and interactions in the cleaved halves of histone f2B. *Eur. J. Biochem.* **26**, 482–489.

Bradbury, E. M., Inglis, R. J., Matthews, H. R. and Sarner, N. (1973). Phosphorylation of very lysine rich histone in Physarum polycephalum. *Eur. J. Biochem.* **33**, 131–139.

Bradbury, E. M., Inglis, J. R., Matthews, H. R. and Langan, T. (1974b). Molecular basis of control of mitotic cell division in eucaryotes. *Nature* **249**, 533–555.

Bradbury, E. M., Cary, P. D., Chapman, G. E., Crane-Robinson, C., Danby, S. E., Rattle, H. W. E., Boublik, M., Palau, J. and Aviles, F. I. (1975a). Studies on the role and mode of operation of the very lysine rich histone H1 (F1) in eucaryotic chromatin. The conformation of histone H1. *Eur. J. Biochem.* **52**, 605–614.

Bradbury, E. M., Chapman, G. E., Danby, S. E., Hartman, P. G. and Riches, P. L. (1975b). Studies on the role and mode of operation of the very lysine rich histone H1 in eucaryote chromatin. The properties of the N terminal and C terminal halves of histone H1. *Eur. J. Biochem.* **57**, 521–528.

Bradbury, E. M., Cary, P. D., Crane-Robinson, C., Rattle, H. W. E., Boublik, M. and Sautiere, P. (1975c). Conformations and interactions of histone H2A. *Biochemistry* **14**, 1876–1886.

Brakel, C. L. and Blumenthal, A. B. (1978). Three forms of DNA polymerase from D. melanogaster embryos. Purification and properties. *Eur. J. Biochem.* **88**, 351–362.

Brakel, C. and Kates, J. R. (1974a). Poly(A) polymerase from vaccinia virus-infected cells. I. Partial purification and characterization. *J. Virol.* **14**, 715–723.

Brakel, C. and Kates, J. R. (1974b). Poly(A) polymerase from vaccinia virus-infected cells. II. Product and primer characterization. *J. Virol.* **14**, 724–732.

Brandhorst, B. P. and Humphries, T. (1972). Stabilities of nuclear and messenger RNA molecules in sea urchin embryos. *J. Cell Biol.* **53**, 474–482.

Brandhorst, B. P. and McConkey, E. H. (1974). Stability of nuclear RNA in mammalian cells. *J. Mol. Biol.* **85**, 451–464.

Brandt, W. F. and Von Holt, C. (1974). The determination of the primary structure of histone f3 from chicken erythrocytes by automatic Edman degradation 2. Sequence analysis of histone f3. *Eur. J. Biochem.* **46**, 419–429.

Brandt, W. F., Strickland, W. N. and Von Holt, C. (1974). The primary structure of H3 from shark erythrocytes. *Febs Lett.* **40**, 349–352.

Brandt, W. F., Strickland, W. N., Strickland, M., Carlisle, L., Woods, D. and Von Holt, C. (1979). A histone programme during the life cycle of the sea urchin. *Eur. J. Biochem.* **94**, 1–10.

Brasch, K., Setterfield, G. and Neelin, J. M. (1972). Effects of sequential extraction of histone proteins on structural organization of avian erythrocyte and liver nuclei. *Exp. Cell Res.* **74**, 27–41.

Braten, T. and Nordby, O. (1973). Ultrastructure of meiosis and centriole behavior in Ulva mutabilis (Foyn). *J. Cell Sci.* **13**, 69–81.

Bratosin, S., Horowitz, M., Laub, O. and Aloni, Y. (1978). Electron microscopic evidence for splicing of SV40 late mRNAs. *Cell* **13**, 783–790.

Braun, J., Fujiwara, K., Pollard, T. D. and Unanue, E. R. (1978a). Two distinct mechanisms for redistribution of lymphocyte surface macromolecules. I. Relationship to cytoplasmic actin. *J. Cell Biol.* **79**, 409–418.

Braun, J., Fujiwara, K., Pollard, T. D. and Unanue, E. R. (1978b). Two distinct mechanisms for redistribution of lymphocyte surface macromolecules. II. Contrasting effects of local anesthetics and a calcium ionophore. *J. Cell Biol.* **79**, 419–426.

Brawerman, G. and Diez, J. (1975). Metabolism of the polyadenylate sequence of nuclear RNA and mRNA in mammalian cells. *Cell* **5**, 271–280.

Brawerman, G. and Eisenstadt, J. M. (1964). DNA from the chloroplast of Euglena gracilis. *Biochim. Biophys. Acta.* **91**, 477–485.

Bray, D. and Thomas C. (1975). The actin content of fibroblasts. *Biochem. J.* **147**, 221–228.

Bray, D. and Thomas, C. (1976). Unpolymerized actin in fibroblasts and brain. *J. Mol. Biol.* **105**, 527–544.

Bray, G. and Brent, T. P. (1972). Deoxynucleoside triphosphate pool fluctuations during the mammalian cell cycle. *Biochim. Biophys. Acta* **269**, 184–191.

Breathnach, R., Mandel, J. L. and Chambon, P. (1977). Ovalbumin gene is split in chicken DNA. *Nature* **270**, 314–319.

Breathnach, R., Benoist, C., O'Hare, K., Gannon, F. and Chambon, P. (1978). Ovalbumin gene: evidence for a leader sequence in mRNA and DNA sequences at the exon-intron boundaries. *Proc. Nat. Acad. Sci. USA* **75**, 4853–4857.

Breindl, M. and Gallwitz, D. (1973). Identification of histone mRNA from Hela cells. Appearance of histone mRNA in the cytoplasm and its translation in a rabbit reticulocyte cell free system. *Eur. J. Biochem.* **32**, 381–391.

Brenner, S. (1974). The genetics of C. elegans. *Genetics* **77**, 71-94.

Brenner, S., Branch, S., Meredith, S. and Berns, M. W. (1977). The absence of centrioles from spindle poles of rat kangaroo (PtK$_2$) cells undergoing a meiotic like reduction division in vitro. *J. Cell Biol.* **72**, 368-379.

Brent, T. P. (1971). Periodicity of DNA synthetic enzymes during the HeLa cell cycle. *Cell Tissue Kinet.* **4**, 297-305.

Brent, T. P., Butler, J. A. V. and Crathorn, A. R. (1965). Variations in phosphokinase activities during the cell cycle in synchronous populations of Hela cells. *Nature* **207**, 176-177.

Brentani, M., Salles, J. M. and Brentani, R. (1977). Determination of the extent of secondary structure in chick embryo procollagen mRNA. *Biochemistry* **16**, 5145-5149.

Bretscher, A. and Weber, K. (1978). Purification of microvilli and an analysis of the protein components of the microfilament core bundle. *Exp. Cell Res.* **116**, 397-408.

Bridges, B. A. and Huckle, J. (1970). Mutagenesis of cultured mammalian cells by X irradiation and ultraviolet light. *Mutat. Res.* **10**, 141-151.

Bridges, B. A., Huckle, J. and Ashwood-Smith, M. J. (1970). X ray mutagenesis of cultured Chinese hamster cells. *Nature* **226**, 184-185.

Bridges, C. B. (1915). A linkage variation in Drosophila. *J. Exp. Zool.* **19**, 1-21.

Bridges, C. B. (1916). Non disjunction as proof of the chromosome theory of heredity. *Genetics* **1**, 1-52.

Bridges, C. B. (1935). Salivary chromosome maps with a key to the banding of the chromosomes of D. melanogaster. *J. Hered.* **26**, 60-64.

Bridges, C. B. (1938). A revised map of the salivary gland X chromosome of D. melanogaster. *J. Hered.* **29**, 11-13.

Bridges, C. B. and Anderson, E. G. (1925). Crossing over in the X chromosomes of triploid females of D. melanogaster. *Genetics* **10**, 418-441.

Briggs, R. and King, T. J. (1960). Nuclear transplantation studies on the early gastrula R. pipiens. *Devel. Biol.* **2**, 252-270.

Brimhall, B., Hollan, S., Jones, R. T., Koler, R. D., Stocklen, Z., and Szelenyi, J. G. (1970). Multiple alpha chain loci for human hemoglobin. *Clin. Res.* **18**, 184-194.

Brinkley, B. R. and Cartwright, J. (1971). Ultrastructural analysis of mitotic spindle elongation in mammalian cells in vitro: direct microtubule counts. *J. Cell Biol.* **50**, 416-431.

Brinkley, B. R. and Stubblefield, E. (1970). Ultrastructure and interaction of the kinetochore and centriole in mitosis and meiosis. In D. M. Prescott, L. Goldstein and E. McConkey (Eds.), *Advances in Cell Biology*, Vol. 1, pp 119-185.

Brinkley, B. R., Stubblefield, E. and Hsu, T. C. (1967). The effects of colcemid inhibition and reversal on the fine structure of the mitotic apparatus of Chinese hamster cells in vitro. *J. Ult. Res.* **19**, 1-18.

Brinkley, B. R., Fuller, G. M. and Highfield, D. P. (1975). Cytoplasmic microtubules in normal and transformed cells in culture: analysis of tubulin antibody immunofluorescence. *Proc. Nat. Acad. Sci. USA* **72**, 4981-4985.

Brinkley, B. R., Stubblefield, E. and Hsu, T. C. (1967). The effects of colcemid inhibition and reversal on the fine structure of the mitotic apparatus of Chinese hamster cells in vitro. *J. Ult. Res.* **19**, 1-18.

Britten, R. J. and Davidson, E. H. (1969). Gene regulation for higher cells: a theory. *Science* **165**, 349-357.

Britten, R. J. and Davidson, E. H. (1971). Repetitive and nonrepetitive DNA sequences and a speculation on the origins of evolutionary novelty. *Quart. Rev. Biol.* **46**, 111-133.

Britten, R. J. and Davidson, E. H. (1976). Studies on nucleic acid reassociation kinetics: empirical equations describing DNA reassociation. *Proc. Nat. Acad. Sci. USA* **73**, 415-419.

Britten, R. J. and Kohne, D. E. (1968). Repeated sequences in DNA. *Science* **161**, 529-540.

Britten, R. J., Cetta, A. and Davidson, E. H. (1978). The single copy DNA sequence polymorphism of the sea urchin S. purpuratus. *Cell* 15, 1175–1186.

Britten, R. J., Graham, D. E. and Neufeld, B. R. (1974). Analysis of repeating DNA sequences by reassociation. *Methods Enzymol.* 29E, 363–406.

Brooks, R. F. (1975). The kinetics of serum induced initiation of DNA synthesis in BHK 21-C13 cells and the influence of exogenous adenosine. *J. Cell. Physiol.* 86, 369–378.

Brooks, R. F. (1977). Continuous protein synthesis is required to maintain the probability of entry into S phase. *Cell* 12, 311–317.

Brosius, J., Palmer, M. L., Kennedy, P. J. and Noller, H. F. (1978). Complete nucleotide sequence of a 16S rRNA gene from E. coli. *Proc. Nat. Acad. Sci. USA* 75, 4801–4805.

Brown, D. D. and Blackler, A. W. (1972). Gene amplification proceeds by a chromosome copy mechanism. *J. Mol. Biol.* 63, 75–84.

Brown, D. D. and Dawid, I. (1968). Specific gene amplification in oocytes. *Science* 160, 272–280.

Brown, D. D. and Gurdon, J. B. (1964). Absence of rRNA synthesis in the anucleolate mutant of X. laevis. *Proc. Nat. Acad. Sci. USA* 51, 139–146.

Brown, D. D. and Gurdon, J. B. (1977). High fidelity transcription of 5S DNA injected into Xenopus oocytes. *Proc. Nat. Acad. Sci. USA* 74, 2064–2068.

Brown, D. D. and Gurdon, J. B. (1978). Cloned single repeating units of 5S DNA direct accurate transcription of 5S RNA when injected into Xenopus oocytes. *Proc. Nat. Acad. Sci. USA* 75, 2849–2853.

Brown, D. D. and Sugimoto, K. (1973). 5S DNAs of X. laevis and X. mulleri. Evolution of a gene family. *J. Mol. Biol.* 78, 397–416.

Brown, D. D. and Weber, C. S. (1968a). Gene linkage by RNA-DNA hybridization. I. Unique DNA sequences homologous to 4S RNA, 5S RNA and rRNA. *J. Mol. Biol.* 34, 661–680.

Brown, D. D. and Weber, C. S. (1968b). Gene linkage by RNA-DNA hybridization. II. Arrangement of the redundant gene sequences for 21S and 18S rRNA. *J. Mol. Biol.* 34, 681–698.

Brown, D. D., Carroll, D. and Brown, R. D. (1977). The isolation and characterization of a second oocyte 5S DNA from X. laevis. *Cell* 12, 1045–1056.

Brown, D. D., Wensink, P. C. and Jordan, E. (1971). Position and some characteristics of 5S DNA from X. laevis. *Proc. Nat. Acad. Sci. USA* 68, 3175–3179.

Brown, D. D., Wensink, P. C. and Jordan, E. (1972). A comparison of the ribosomal DNAs of X. laevis and X. mulleri: the evolution of tandem genes. *J. Mol. Biol.* 63, 57–73.

Brown, I. R. and Church, R. B. (1972). Transcription of nonrepeated DNA during mouse and rabbit development. *Devel. Biol.* 29, 73–84.

Brown, S. W. (1966). Heterochromatin. *Science* 151, 417–425.

Brown, S. W. and Chandra, H. S. (1973). Inactivation system of the mammalian X chromosome. *Proc. Nat. Acad. Sci. USA* 70, 195–199.

Brown, S. W. and Nur, U. (1964). Heterochromatic chromosomes in coccids. *Science* 145, 130–136.

Brown, W. M. and Vinograd, J. (1974). Restriction endonuclease cleavage maps of animal mitochondrial DNAs. *Proc. Nat. Acad. Sci. USA* 71, 4617–4621.

Brown, W. M., Shine, J. and Goodman, H. M. (1978). Human mitochondrial DNA: analysis of 7S DNA from the origin of replication. *Proc. Nat. Acad. Sci. USA* 75, 735–739.

Brown, W. M., Watson, R. M., Vinograd, J., Tait, K. M., Boyer, H. W. and Goodman, H. M. (1976). The structures and fidelity of replication of mouse mitochondrial DNA-pSC101 Eco RI recombinant plasmids grown in E. coli K12. *Cell* 7, 517–530.

Brownlee, G. G. and Cartwright, E. M. (1977). Rapid gel sequencing of RNA by primed synthesis with reverse transcriptase. *J. Mol. Biol.* 114, 93–118.

Brownlee, G. G., Cartwright, E. M. and Brown, D. D. (1974). Sequence studies of the 5S DNA of X. laevis. *J. Mol. Biol.* 89, 703–718.

Brownlee, G. G., Cartwright, E., McShane, T. and Williamson, R. (1972). The nucleotide sequence of somatic 5S RNA from X. laevis. *Febs Lett.* **25**, 8-12.

Brownlee, G. G., Cartwright, E. M., Cowan, N. J., Jarvis, J. M. and Milstein, C. (1973). Purification and sequence of mRNA for immunoglobulin light chains. *Nature New Biol.* **244**, 236-239.

Brun, G. and Weissbach, A. (1978). Initiation of HeLa cell DNA synthesis in a subnuclear system. *Proc. Nat. Acad. Sci. USA* **75**, 5931-5935.

Brutlag, D., Appels, R., Dennis, E. S. and Peacock, W. J. (1977a). Highly repeated DNA in D. melanogaster. *J. Mol. Biol.* **112**, 31-48.

Brutlag, D., Fry, K., Nelson, T. and Hung, P. (1977b). Synthesis of hybrid bacterial plasmids containing highly repeated satellite DNA. *Cell* **10**, 509-519.

Brutlag, D., Carlson, M., Fry, K. and Hsieh, T. S. (1978). DNA sequence organization in Drosophila heterochromatin. *Cold Spring Harbor Symp. Quant. Biol.* **42**, 1137-1146.

Bryan, J. (1972). Vinblastine and microtubules. II. Characterization of two protein subunits from the isolated crystals. *J. Mol. Biol.* **66**, 157-168.

Bryan, J. (1974). Biochemical properties of microtubules. *Fed. Proc.* **33**, 152-157.

Bryan, J. (1976). A quantitative analysis of microtubule elongation. *J. Cell Biol.* **71**, 749-767.

Bryan, J. and Wilson, L. (1971). Are cytoplasmic microtubules heteropolymers? *Proc. Nat. Acad. Sci. USA* **68**, 1762-1766.

Bryan, J., Nagle, B. W. and Doenges, K. H. (1975). Inhibition of tubulin assembly by RNA and other polyanions: evidence for a required protein. *Proc. Nat. Acad. Sci. USA* **72**, 3570-3574.

Bryan, R. N. and Hayashi, M. (1973). Two proteins are bound to most species of polysomal mRNA. *Nature New Biol.* **244**, 271-274.

Buckingham, M. E., Cohen, A. and Gros, F. (1976). Cytoplasmic distribution of pulse labeled poly(A)$^+$ RNA, particularly 26S RNA, during myoblast growth and differentiation. *J. Mol. Biol.* **103**, 611-626.

Buckley, I. K. and Porter, K. R. (1967). Cytoplasmic fibrils in living cultured cells. A light and electron microscope study. *Protoplasma* **64**, 345-349.

Buell, G. N., Wickens, M. P., Payvar, F. and Schimke, R. T. (1978). Synthesis of full length cDNAs from four partially purified oviduct mRNAs. *J. Biol. Chem.* **253**, 2471-2482.

Buhler, J. M., Iborra, F., Sentenac, A. and Fromageot, P. (1976). Structural studies on yeast RNA polymerases. Existence of common subunits in RNA polymerases A(I) and B(II). *J. Biol. Chem.* **251**, 1712-1717.

Bulinski, J. C. and Borisy, G. G. (1979). Self assembly of microtubules in extracts of cultured HeLa cells and the identification of HeLa microtubule associated proteins. *Proc. Nat. Acad. Sci. USA* **76**, 293-297.

Buller, R. E., Schwartz, R. J., Schrader, W. T. and O'Malley, B. (1976). Progesterone-binding components of chick oviduct. In vitro effect of receptor subunits on gene transcription. *J. Biol. Chem.* **251**, 5178-5186.

Bultmann, H. and Laird, C. D. (1973). Mitochondrial DNA from D. melanogaster. *Biochim. Biophys. Acta* **299**, 196-209.

Bunn, C. L., Wallace, D. C. and Eisenstadt, J. M. (1974). Cytoplasmic inheritance of chloramphenicol resistance in mouse tissue culture cells. *Proc. Nat. Acad. Sci. USA* **71**, 1681-1685.

Buongiornò-Nardelli, M., Amaldi, F. and Lava-Sanchez, P. (1972). Amplification as a rectification mechanism for the redundant DNA genes. *Nature New Biol.* **238**, 134-137.

Buongiorno-Nardelli, M., Amaldi, F., Beccari, E. and Junakovic, N. (1977). Size of rDNA repeating units in X. laevis: limited individual heterogeneity and extensive population polymorphism. *J. Mol. Biol.* **110**, 105-118.

Burckhardt, J. and Birnstiel, M. L. (1978). Analysis of histone mRNA of D. melanogaster by two dimensional gel electrophoresis. *J. Mol. Biol.* **118**, 61–80.

Burch, J. W. and McBride, O. W. (1975). Human gene expression in rodent cells after uptake of isolated metaphase chromosomes. *Proc. Nat. Acad. Sci. USA* **72**, 1797–1801.

Burgoyne, L. A., Hewish, D. R. and Mobbs, J. (1974). Mammalian chromatin substructure studies with the calcium magnesium endonuclease and two dimensional polyacrylamide gel electrophoresis. *Biochem. J.* **143**, 67–72.

Burkard, G. and Keller, E. B. (1974). Poly-A polymerase and poly-G polymerase in wheat chloroplasts. *Proc. Nat. Acad. Sci. USA* **71**, 389–393.

Burkholder, G. D. (1975). The ultrastructure of G and C banded chromosomes. *Exp. Cell Res.* **90**, 269–278.

Burkholder, G. D. (1979). An investigation of the mechanism of the reciprocal differential staining of BUdR substituted and unsubstituted chromosome regions. *Exp. Cell Res.* **121**, 209–219.

Burkholder, G. D. and Weaver, M. G. (1977). DNA protein interactions and chromosome binding. *Exp. Cell Res.* **110**, 251–262.

Burkholder, G. D., Okada, T. A. and Comings, D. E. (1972). Whole mount electron microscopy of metaphase I chromosomes and microtubules from mouse oocytes. *Exp. Cell Res.* **75**, 497–511.

Burns, F. J. and Tannock, I. F. (1970). On the existence of a GO phase in the cell cycle. *Cell Tiss. Kinet.* **1**, 167–191.

Burridge, K. and Bray, D. (1975). Purification and structural analysis of myosins from brain and other nonmuscle tissues. *J. Mol. Biol.* **99**, 1–14.

Burstein, Y. and Schechter, I. (1977). Amino acid sequence of the NH_2 terminal extra piece segments of the precursors of mouse immunoglobulin lambda -1 type and kappa type light chains. *Proc. Nat. Acad. Sci. USA* **74**, 716–720.

Burstin, J., Meiss, H. K. and Basilico, C. (1975). A temperature sensitive cell cycle mutant of the BHK cell line. *J. Cell Physiol.* **84**, 397–408.

Burton, D. R., Butler, M. J., Hyde, J. E., Phillips, D., Skidmore, C. J. and Walker, I. O. (1978). The interaction of core histones with DNA: equilibrium binding studies. *Nucleic Acids Res.* **5**, 3643–3654.

Burton, K. (1965). A study of the conditions and mechanism of the diphenylamine reaction for the colorimetric estimation of DNA. *Biochem. J.* **62**, 315–322.

Burton, P. R., and Fernandez, H. L. (1973). Delineation by lanthanum staining of filamentous elements associated with the surfaces of neuronal microtubules. *J. Cell Sci.* **12**, 567–583.

Burton, P. R., Hinkley, R. E. and Pierson, G. B. (1975). Tannic acid stained microtubules with 12, 13 and 15 protofilaments. *J. Cell Biol.* **65**, 227–233.

Burton, W. G., Grabowy, C. T. and Sager, R. (1979). Role of methylation in the modification and restriction of chloroplast DNA in Chlamydomonas. *Proc. Nat. Acad. Sci. USA* **76**, 1390–1394.

Bustin, M. and Cole, R. D. (1969a). A study of the multiplicity of lysine rich histones. *J. Biol. Chem.* **244**, 5286–5290.

Bustin, M. and Cole, R. D. (1969b). Bisection of a lysine rich histone by N-bromosuccinimide. *J. Biol. Chem.* **244**, 5291–5294.

Butler, A. P., Harrington, R. E. and Olins, D. E. (1979). Salt dependent interconversion of inner histone oligomers. *Nucleic Acids Res.* **6**, 1509–1520.

Butler, J. A. V. and Chipperfield, A. R. (1967). Inhibition of RNA polymerase by histones. *Nature* **215**, 1188–1189.

Butterworth, P. H. W., Cox, R. F. and Chesterton, C. J. (1971). Transcription of mammalian chromatin by mammalian DNA dependent RNA polymerase. *Eur. J. Biochem.* **23**, 229–241.

Buzzo, K., Fouts, D. L. and Wolstenholme, D. R. (1978). Eco RI cleavage site variants of mitochondrial DNA molecules from rats. *Proc. Nat. Acad. Sci. USA* **75**, 909–913.

Byers, B. and Goetsch, L. (1975). Electron microscopic observations on the meiotic karyotype of diploid and tetraploid S. cerevisiae. *Proc. Nat. Acad. Sci. USA* **72**, 5056-5060.

Caboche, M. and Bachallerie, J. -P. (1977). RNA methylation and control of eucaryotic RNA biosynthesis. Effects of cycloleucine, a specific inhibitor of methylation, on rRNA maturation. *Eur. J. Biochem.* **74**, 19-30.

Caboche, M. and Mulsant, P. (1978). Selection and preliminary characterization of cycloleucine resistant CHO cells affected in methionine metabolism. *Somatic Cell Genet.* **4**, 407-422.

Cabral, F. and Schatz, G. (1978). Identification of cytochrome c oxidase subunits in nuclear yeast mutants lacking the functional enzyme. *J. Biol. Chem.* **253**, 4396-4401.

Cabral, F., Solios, M., Rudon, Y., Schatz, G., Clavilier, L. and Slonimski, P. P. (1978). Identification of the structural gene for yeast cytochrome c oxidase subunit II on mitochondrial DNA. *J. Biol. Chem.* **253**, 297-304.

Cairns, J. (1963). The chromosome of E. coli. *Cold Spring Harbor Symp. Quant. Biol.* **28**, 43-46.

Cairns, J. (1966). Autoradiography of HeLa cell DNA. *J. Mol. Biol.* **15**, 372-373.

Calderon, D. and Schnedl, W. (1973). A comparison between quinacrine fluorescence banding and ^{3}H thymidine incorporation patterns in human chromosomes. *Humangenetik* **18**, 63-70.

Callan, H. G. (1967). The organization of genetic units in chromosomes. *J. Cell Sci.* **2**, 1-8.

Callan, H. G. (1972). Replication of DNA in the chromosomes of eucaryotes. *Proc. Roy. Soc. B* **181**, 19-41.

Callan, H. G. (1973). DNA replication in the chromosomes of eucaryotes. *Cold Spring Harbor Symp. Quant. Biol.* **38**, 195-203.

Callan, H. G. and Lloyd, L. (1960). Lampbrush chromosomes of crested newts. *Phil. Trans. Roy. Soc. B* **243**, 135-219.

Callan, H. G. and MacGregor, H. C. (1958). Action of DNAase on lampbrush chromosomes. *Nature* **181**, 1479-1480.

Calvet, J. P. and Pederson, T. (1977). Secondary structure of hnRNA: two classes of double stranded RNA in native RNP. *Proc. Nat. Acad. Sci. USA* **74**, 3705-3709.

Calvet, J. P. and Pederson, T. (1978). Nucleoprotein organization of inverted repeat DNA transcripts in hnRNA-ribonucleoprotein particles from HeLa cells. *J. Mol. Biol.* **122**, 361-378.

Calvet, J. P. and Pederson, T. (1979). HnRNA double stranded regions probed in living HeLa cells by crosslinking with the psoralen derivative aminomethyltrixsalen. *Proc. Nat. Acad. Sci. USA* **76**, 755-759.

Camerini-Otero, R. D., Sollner-Webb, B. and Felsenfeld, G. (1976). The organization of histones and DNA in chromatin: evidence for an arginine-rich histone kernal. *Cell* **8**, 333-348.

Cameron, J. R., Loh, E. Y. and Davis, R. W. (1979). Evidence for transposition of dispersed repetitive DNA families in yeast. *Cell* **16**, 739-752.

Campbell, A. M. and Cotter, R. I. (1976). The molecular weight of nucleosome protein by light scattering. *Febs. Lett.* **70**, 209-211.

Campbell, C. E. and Worton, R. G. (1979). Evidence obtained by induced mutation frequency analysis for functional hemizygosity at the emt locus in CHO cells. *Somatic. Cell Genet.* **5**, 51-66.

Campbell, G. R., Littau, V. C., Melera, P. W., Allfrey, V. G. and Johnson, E. M. (1979). Unique sequence arrangement of ribosomal genes in the palindromic rDNA molecule of Physarum polycephalum. *Nucleic Acids Res.* **6**, 1433-1448.

Cameron, J. R., Panasenko, S. M., Lehman, I. R. and Davis, R. W. (1975). In vitro construction of phage λ carrying segments of the E. coli chromosome: selection of hybrids containing the gene for DNA ligase. *Proc. Nat. Acad. Sci. USA* **72**, 3416-3420.

Campo, M. S. and Bishop, J. O. (1974). Two classes of messenger RNA in cultured rat cells: repetitive sequence transcripts and unique sequence transcripts. *J. Mol. Biol.* **90**, 649-664.

Canaani, D., Kahana, C., Mukamel, A. and Groner, Y. (1979). Sequence heterogeneity at the 5′ termini of late SV40 19S and 16S mRNAs. *Proc. Nat. Acad. Sci. USA* **76**, 3078-3082.

Cande, W. Z., Lazarides, E. and McIntosh, J. R. (1977). A comparison of the distribution of actin and tubulin in the mammalian mitotic spindle as seen by indirect immunofluorescence. *J. Cell Biol.* **72**, 552-567.

Cande, W. Z., Snyder, J., Smith, D., Summers, K. and McIntosh, J. R. (1974). A functional mitotic spindle prepared from mammalian cells in culture. *Proc. Nat. Acad. Sci. USA* **71**, 1560-1563.

Candido, E. P. M. and Dixon, G. H. (1971). Sites of an in vivo acetylation in trout testis histone IV. *J. Biol. Chem.* **246**, 3182-3188.

Candido, E. P. M. and Dixon, R. H. (1972a). Acetylation of trout testis histones in vivo. Site of the modification in histone IIbl. *J. Biol. Chem.* **247**, 3868-3873.

Candido, E. P. M. and Dixon, R. H. (1972b). Amino terminal sequences and sites of in vivo acetylation of trout testis histones III and IIb2. *Proc. Nat. Acad. Sci.* **69**, 2015-2019.

Candido, E. P. M., Reeves, R. and Davie, J. R. (1978). Sodium butyrate inhibits histone deacetylation in cultured cells. *Cell* **14**, 105-113.

Capecchi, M. R., Haar, R. A. V., Capecchi, N. E. and Sveda, M. M. (1977). The isolation of a suppressible nonsense mutant in mammalian cells. *Cell* **12**, 371-381.

Carlson, M. and Brutlag, D. (1978a). One of the *copia* genes is adjacent to satellite DNA in D. melanogaster. *Cell* **15**, 733-742.

Carlson, M. and Brutlag, D. (1978b). A gene adjacent to satellite DNA in D. melanogaster. *Proc. Nat. Acad. Sci. USA* **75**, 5898-5902.

Carlson, R. D. and Olins, D. E. (1976). Chromatin model calculations. Arrays of spherical nu bodies. *Nucleic Acids Res.* **3**, 89-100.

Carlsson, S. A., Moore, G. P. M. and Ringertz, N. R. (1973). Nucleocytoplasmic protein migration during the activation of chick erythrocyte nuclei in heterocaryons. *Exp. Cell Res.* **76**, 234-241.

Carlsson, S. A., Luger, O., Ringertz, N. R. and Savage, R. E. (1974). Phenotypic expression in chick erythrocyte x rat myoblast hybrids and in chick myoblast x rat myoblast hybrids. *Exp. Cell Res.* **84**, 47-55.

Carpenter, A. T. C. (1973). A mutant defective in distributive disjunction in D. melanogaster. *Genetics* **73**, 393-428.

Carpenter, A. T. C. (1975a). Electron microscopy of meiosis in D. melanogaster females. II. The recombination nodule—a recombination associated structure at pachytene. *Proc. Nat. Acad. Sci. USA* **72**, 3186-3189.

Carpenter, A. T. C. (1975b). Electron microscopy of meiosis in D. melanogaster females I. Structure, arrangement and temporal change of the synaptonemal complex in wild type. *Chromosoma* **51**, 157-182.

Carpenter, A. T. C. and Sandler, L. (1974). On recombination-defective meiotic mutants in D. melanogaster. *Genetics* **76**, 453-476.

Carpenter, B. G., Baldwin, J. P., Bradbury, E. M. and Ibel, K. (1976). Organization of subunits in chromatin. *Nucleic Acids Res.* **3**, 1739-1746.

Carroll, D. and Brown, D. D. (1976a). Repeating units of X. laevis oocyte-type 5S DNA are heterogeneous in length. *Cell* **7**, 467-476.

Carroll, D. and Brown, D. D. (1976b). Adjacent repeating units of X. laevis 5S DNA can be heterogeneous in length. *Cell* **7**, 477-486.

Carter, S. B. (1967). Effects of cytochalasin on mammalian cells. *Nature* **213**, 261-264.

Carter, T. C. (1955). The estimation of total genetical map lengths from linkage test data. *J. Genet.* **53**, 21-28.

Cary, P. D., Crane-Robinson, C., Bradbury, E. M., Javaherian, K., Goodwin, G. H. and Johns, E. W. (1976). Conformational studies of two nonhistone chromosomal proteins and their interactions with DNA. *Eur. J. Biochem.* **62**, 583-590.

Cashmore, A. R. (1979). Reiteration frequency of the gene coding for the small subunit of RuBP. *Cell* **17**, 383-388.

Caskey, C. T. and Kruth, G. D. (1979). The HGPRT locus. *Cell* **16**, 1-10.

Caspari, E. (1948). Cytoplasmic inheritance. *Adv. Genet.* **2**, 2-66.

Caspersson, T. (1936). Ueber den chemischen aufbau der strukturen des zellkernes. *Acta Med. Skand.* **73**, Suppl. **8**, 1-151.

Caspersson, T., Zech, L. and Johansson, C. (1970). Analysis of human metaphase chromosome set by aid of DNA-binding fluorescent agents. *Exp. Cell Res.* **62**, 490-492.

Caspersson, T., De La Chapelle, A., Schroder, J. and Zech, L. (1972). Quinacrine fluorescence of metaphase chromosomes. Identical patterns in different tissues. *Exp. Cell Res.* **72**, 56-59.

Cattanach, B. M. (1974). Position effect variegation in the mouse. *Genet. Res.* **23**, 291-306.

Cattanach, B. M. (1975). Control of chromosome inactivation. *Ann. Rev. Genet.* **9**, 1-18.

Cattanach, B. M. and Isaacson, J. H. (1965). Genetic control over the inactivation of autosomal genes attached to the X chromosome. *Z. Verebungsl.* **96**, 313-323.

Cattanach, B. M. and Isaacson, J. H. (1967). Controlling elements in the mouse X chromosome. *Genetics* **57**, 331-346.

Cattanach, B. M. and Williams, C. E. (1972). Evidence off nonrandom X chromosome activity in the mouse. *Genet. Res.* **19**, 229-240.

Cattanach, B. M., Perez, J. N. and Pollard, C. E. (1970). Controlling elements in the mouse X chromosome. II. Location in the linkage map. *Genet. Res.* **15**, 183-195.

Cattanach, B. M., Pollard, C. E. and Perez, J. N. (1969). Controlling elements in the mouse X chromosome. I. Interaction with the X linked genes. *Genet. Res.* **14**, 223-235.

Cattanach, B. M., Wolfe, H. G. and Lyon, M. F. (1972). A comparative study of the coats of chimeric mice and those of heterozygotes for X linked genes. *Genet. Res.* **19**, 213-228.

Catterrall, J. F., O'Malley, B. W., Robertson, M. A., Staden, R., Tanaka, Y. and Brownlee, G. G. (1978). Nucleotide sequence homology at 12 exon-intron junctions in the chick ovalbumin gene. *Nature* **275**, 510-514.

Catterall, J. F., Stein, J. P., Lai, E. C., Woo, S. L. C., Mace, M. L., Means, A. R. and O'Malley, B. W. (1979). The ovomucoid gene contains at least six intervening sequences. *Nature* **278**, 323-327.

Cavalier-Smith, T. (1970). Electron microscopic evidence for chloroplast fusion in zygotes of Chlamydomonas reinhardii. *Nature* **228**, 333-335.

Cech, T. R. and Hearst, J. E. (1975). An electron microscopic study of mouse foldback DNA. *Cell* **5**, 429-446.

Cech, T. R. and Hearst, J. E. (1976). Organization of highly repeated sequences in mouse main band DNA. *J. Mol. Biol.* **100**, 227-256.

Cech, T. and Pardue, M. L. (1977). Cross linking of DNA with trimethylpsoralen is a probe for chromatin structure. *Cell* **11**, 631-640.

Cech, T., Potter, D. and Pardue, M. L. (1977). Electron microscopy of DNA cross linked with trimethylpsoralen: a probe for chromatin structure. *Biochemistry* **16**, 5313-5320.

Cech, T. R., Rosenfeld, A. and Hearst, J. E. (1973). Characterization of the most rapidly renaturing sequences in mouse main band DNA. *J. Mol. Biol.* **81**, 299-326.

Cedar, H. (1975). Transcription of DNA and chromatin with calf thymus RNA polymerase B in vitro. *J. Mol. Biol.* **95**, 257-270.

Chae, C. B. and Carter, D. B. (1974). Degradation of chromosomal proteins during dissociation and reconstitution of chromatin. *Biochem. Biophys. Res. Commun.* **57**, 740-746.

Chaganti, R. S. K., Schonberg, S. and German, J. (1977). A manyfold increase in sister chromatid exchanges in Bloom's syndrome lymphocytes. *Proc. Nat. Acad. Sci. USA* **71**, 4508-4512.

Chalkley, R. and Hunter, C. (1975). Histone-histone propinquity by aldehyde fixation of chromatin. *Proc. Nat. Acad. Sci. USA* **72**, 1304-1308.

Chamberlin, M. E., Britten, R. J. and Davidson, E. H. (1975). Sequence organization in Xenopus DNA studied by the electron microscope. *J. Mol. Biol.* **96**, 317-334.

Chamberlin, M. E., Galau, G. A., Britten, R. J. and Davidson, E. H. (1978). Studies on nucleic acid reassociation kinetics. V. Effects of disparity in tracer and driver fragment lengths. *Nucleic Acids Res.* **5**, 2073-2094.

Chambers, C. A., Schell, M. P. and Skinner, D. M. (1978). The primary sequence of a crustacean satellite DNA containing a family of repeats. *Cell* **13**, 97-110.

Chambers, R. (1938). Structural and kinetic aspects of cell division. *J. Cell. Comp. Physiol.* **12**, 149-165.

Chambon, P. (1975). Eucaryotic nuclear RNA polymerases. *Ann. Rev. Biochem.* **44**, 613-638.

Chan, P. H. and Wildman, S. C. (1972). Chloroplast DNA codes for the primary structure of the large subunit of fraction 1 protein. *Biochim. Biophys. Acta* **277**, 677-680.

Chan, V. L., Whitmore, G. F. and Siminovitch, L. (1972). Mammalian cells with altered forms of RNA polymerase II. *Proc. Nat. Acad. Sci. USA* **69**, 3119-3123.

Chandra, H. S. and Brown, S. W. (1975). Chromosome imprinting and the mammalian X chromosome. *Nature* **253**, 165-168.

Chang, A. C. Y., Lansman, R. A., Clayton, D. A. and Cohen, S. N. (1975). Studies of mouse mitochondrial DNA in E. coli: structure and function of the eucaryotic-procaryotic chimeric plasmids. *Cell* **6**, 231-244.

Chang, A. C. Y., Nunberg, J. H., Kaufman, R. J., Erlich, H. A., Schimke, R. T. and Cohen, S. N. (1978). Phenotypic expression in E. coli of a DNA sequence coding for mouse DHFR. *Nature* **275**, 617-623.

Chang, C. -M. and Goldman, R. D. (1973). The localization of actin like fibers in cultured neuroblastoma cells as revealed by meromyosin binding. *J. Cell Biol.* **57**, 867-874.

Chang, J. C. and Kan, Y. W. (1979). β^0 thalassemia, a nonsense mutation in man. *Proc. Nat. Acad. Sci. USA* **76**, 2886-2889.

Chang, J. C., Temple, G. F., Poon, R., Neumann, K. H. and Kan, Y. W. (1977). The nucleotide sequences of the untranslated 5' regions of human α and β globin mRNAs. *Proc. Nat. Acad. Sci. USA* **74**, 5145-5149.

Chang, J. C., Poon, R., Neumann, K. H. and Kan, Y. W. (1978). Effect of histone acetylation on structure and in vitro transcription of chromatin. *Nucleic Acids Res.* **5**, 3523-3548.

Chang, L. M. S. (1973). Low molecular weight DNA polymerase from calf thymus chromatin. II. Initiation and fidelity of homopolymer replication. *J. Biol. Chem.* **248**, 6983-6992.

Chang, L. M. S. (1977). DNA polymerases from bakers' yeast. *J. Biol. Chem.* **252**, 1873-1880.

Chang, L. M. S. and Bollum, F. J. (1972). Low molecular weight DNA polymerase from rabbit bone marrow. *Biochemistry* **11**, 1264-1272.

Chang, L. M. S. and Bollum, F. J. (1973). A comparison of associated enzyme activities in various DNA polymerases. *J. Biol. Chem.* **248**, 3398-3404.

Chang, L. M. S., Brown, M., and Bollum, F. H. (1973). Induction of DNA polymerase in mouse L cells. *J. Mol. Biol.* **74**, 1-8.

Chang, S. E. and Littlefield, J. W. (1976). Elevated dihydrofolate reductase mRNA levels in methotrexate-resistant BHK cells. *Cell* **7**, 391-396.

Chantrenne, H., Burny, A. and Marbaix, G. (1967). The search for mRNA of hemoglobin. *Prog. Nucleic Acid Res.* **7**, 173-194.

Chapman, G. E., Hartman, P. G. and Bradbury, E. M. (1976). Studies on the role and mode of operation of the very lysine rich histone H1 in eucaryote chromatin. The isolation of the globular and nonglobular regions of the histone H1 molecule. *Eur. J. Biochem.* **61**, 69-76.

Chasin, L. A. (1972). Non linkage of induced mutations in Chinese hamster cells. *Nature New Biol.* **240**, 50-52.

Chasin, L. A. (1973). The effect of ploidy on chemical mutagenesis in cultured Chinese hamster cells. *J. Cell. Physiol.* **82**, 299-308.

Chasin, L. A. (1974). Mutations affecting adenine phosphoribosyl transferase activity in Chinese hamster cells. *Cell* 2, 37-42.

Chasin, L. A. and Urlaub, G. (1975). Chromosome wide event accompanies the expression of recessive mutations in tetraploid cells. *Science* 187, 1091-1093.

Chasin, L. A. and Urlaub, G. (1976). Mutant alleles for HGPRT: codominant expression, complementation, and segregation in hybrid Chinese hamster cells. *Somatic Cell Genet.* 2, 453-468.

Chasin, L. A., Feldman, A., Konstam, M. and Urlaub, G. (1974). Reversion of a Chinese hamster cell auxotrophic mutant. *Proc. Nat. Acad. Sci. USA* 71, 718-722.

Chelm, B. K. and Hallick, R. B. (1976). Changes in the expression of the chloroplast genome of Euglena gracilis during chloroplast development. *Biochemistry* 15, 593-599.

Chen, J. H. and Spector, A. (1977). The bicistronic nature of lens α crystallin 14S mRNA. *Proc. Nat. Acad. Sci. USA* 74, 5448-5452.

Chen, S., McDougall, J. K., Creagan, R. P., Lewis, V. and Ruddle, F. (1976). Genetic homology between man and chimpanzee: syntenic relationships of genes for galactokinase and thymidine kinase and adenovirus 12 induced gaps using chimpanzee-mouse somatic cell hybrids. *Somatic Cell Genet.* 2, 205-214.

Chen, S. Y., Ephrussi, B. and Hottinguer, H. (1950). Nature genetiques des mutants a deficience respiratoire de la souche BII de la levure de boulangerie. *Heredity* 4, 337-351.

Cheng, T. -C. and Kazazian, H. H., Jr. (1977). The 5′ terminal structures of murine globin mRNA. *J. Biol. Chem* 252, 1758-1763.

Chen-Kiang, S., Nevins, J. R. and Darnell, J. E. (1979). N⁶ methyl adenosine in Ad2 nuclear RNA is conserved in the formation of mRNA. *J. Mol. Biol.* 135, 733-752.

Chern, C. J. (1977). Detection of active heteropolymeric β-glucoronidase in hybrids between mouse cells and human fibroblasts with β-glucuronidase deficiency. *Proc. Nat. Acad. Sci. USA* 74, 2948-2952.

Chern, C. J., Beutler, E., Kuhl, W., Gilbert, F., Mellman, W. J. and Croce, C. M. (1976). Characterization of heteropolymeric hexosaminidase A in human x mouse hybrid cells. *Proc. Nat. Acad. Sci. USA* 73, 3637-3640.

Chern, C. J., Kennett, R., Engel, E., Mellman, W. J. and Croce, C. M. (1977). Assignment of the structural genes for the α subunit of hexosaminidase A, mannosephosphate isomerase, and pyruvate kinse to the region q22-qter of human chromosome 15. *Somatic Cell Genet.* 3, 553-560.

Chi, J. C., Fellini, S. A. and Holtzer, H. (1975). Differences among myosins synthesized in nonmyogenic cells, presumptive myoblasts, and myoblasts. *Proc. Nat. Acad. Sci. USA* 72, 4999-5003.

Chiang, K. S. and Sueoka, N. (1967). Replication of chloroplast DNA in C. reinhardii during vegetative cell cycle: its mode and regulation. *Proc. Nat. Acad. Sci. USA* 57, 1506-1513.

Chikaraishi, D. M., Deeb, S. S. and Sueoka, N. (1978). Sequence complexity of nuclear RNAs in adult rat tissues. *Cell* 13, 111-120.

Chiu, R. W. and Baril, E. F. (1975). Nuclear DNA polymerases and the Hela cell cycle. *J. Biol. Chem.* 250, 7951-7957.

Choi, K. W. and Bloom, A. D. (1970). Biochemically marked lymphocytoid lines: establishment of Lesch-Nyhan cells. *Science* 170, 89-90.

Chow, L. T. and Broker, T. R. (1978). The spliced structure of adenovirus 2 fiber message and other late mRNAs. *Cell* 15, 497-510.

Chow, L. T., Gelinas, R. E., Broker, T. R. and Roberts, R. J. (1977). An amazing sequence arrangement at the 5′ ends of adenovirus 2 mRNA. *Cell* 12, 1-9.

Christiansen, G. and Griffith, J. (1977). Salt and divalent cations affect the flexible nature of the natural beaded chromatin structure. *Nucleic Acids Res.* 4, 1837-1851.

Christiansen, G., Landers, T., Griffith, J. and Berg, P. (1977). Characterization of components released by alkali disruption of SV40. *J. Virol.* **21**, 1079-1084.

Chu, E. H. Y. (1971). Mammalian cell genetics. III. Characterization of X ray induced forward mutations in Chinese hamster cell cultures. *Mutat. Res.* **11**, 23-34.

Chu, E. H. Y. (1974). Induction and analysis of gene mutations in cultured mammalian somatic cells. *Genetics* **78**, *Suppl.* 115-132.

Chu, E. H. Y. and Malling, H. V. (1968). Mammalian cell genetics. II. Chemical induction of specific locus mutations in Chinese hamster cells in vitro. *Proc. Nat. Acad. Sci. USA* **61**, 1306-1312.

Chu, E. H. Y., Sun, N. C. and Chang, C. C. (1972). Induction of auxotrophic mutations by treatment of Chinese hamster cells with 5 bromodeoxyuridine and black light. *Proc. Nat. Acad. Sci. USA* **69**, 3459-3463.

Chu, E. H. Y., Brimer, P., Jacobson, K. B. and Merriam, E. V. (1969). Mammalian cell genetics. I. Selection and characterization of mutations auxotrophic for L glutamine or resistant to 8 azaguanine in Chinese hamster cells in vitro. *Genetics* **62**, 359-377.

Chua, N. H. and Schmidt, G. W. (1978). Post translational transport into intact chloroplasts of a precursor to the small subunit of ribulose 1,5 biphosphate carboxylase. *Proc. Nat. Acad. Sci. USA* **75**, 6110-6114.

Chua, N. H. and Schmidt, G. W. (1979). Transport of proteins into mitochondria and chloroplasts. *J. Cell Biol.* **81**, 461-483.

Chun, E. L., Vaughan, M. H. and Rich, A. (1963). The isolation and characterization of DNA associated with chloroplast preparations. *J. Mol. Biol.* **7**, 130-141.

Chung, S. Y., Hill, W. E. and Doty, P. (1978). Characterization of the histone core complex. *Proc. Nat. Acad. Sci. USA* **75**, 1680-1684.

Church, K. (1976). Arrangement of chromosome ends and axial core formation during early meiotic prophase in the male grasshopper Brachystola magna by 3D electron microscopic reconstruction. *Chromosoma* **58**, 365-376.

Church, K. and Moens, P. B. (1976). Centromere behavior during interphase and meiotic prophase in Allium fistulosum from 3D electron microscope reconstruction. *Chromosoma* **56**, 249-264.

Church, K. and Wimber, D. E. (1969). Meiosis in Ornithogalum virens (Liliaceae). Meiotic timing and segregation of ^{3}H labeled chromosomes. *Can. J. Genet. Cytol.* **11**, 573-581.

Church, R. B. and McCarthy, B. J. (1967a). Changes in nuclear and cytoplasmic RNA in regenerating liver. *Proc. Nat. Acad. Sci. USA* **58**, 1548-1555.

Church, R. B. and McCarthy, B. J. (1967b). RNA synthesis in regenerating and embryonic liver. II. The synthesis of RNA during embryonic liver development and its relationship to regenerating liver. *J. Mol. Biol.* **23**, 477-486.

Claisse, M. L., Spyridakis, A., Wambier-Kluppel, M. L., Pajot, P. and Slonimski, P. P. (1978). Mosaic organization and expression of the mitochondrial DNA region controlling cytochrome c reductase and oxidase. II. Analysis of proteins translated from the *box* region. In M. Bacila, B. L. Horecker and A. O. M. Stopanni (Eds.), *Biochemistry and Genetics of Yeast*, Academic Press, London, pp 369-390.

Clark, M. F., Matthews, R. E. F. and Ralph, R. K. (1964). Ribosomes and polysomes in Brassica pekinenesis. *Biochim. Biophys. Acta* **91**, 289-304.

Clark, R. J. and Felsenfeld, G. (1971). Structure of chromatin. *Nature New Biol.* **229**, 101-105.

Clark, R. J. and Felsenfeld, G. (1974). Chemical probes of chromatin structure. *Biochemistry* **13**, 3622-3628.

Clark, T. G. and Merriam, R. W. (1977). Diffusible and bound actin in nuclei of X. laevis oocytes. *Cell* **12**, 883-891.

Clark, T. G. and Merriam, R. W. (1978). Actin in Xenopus oocytes. I. Polymerization and gelation in vitro. *J. Cell Biol.* **77**, 427-438.

Clarke, L. and Carbon, J. (1975). Biochemical construction and selection of hybrid plasmids containing specific segments of the E. coli genome. *Proc. Nat. Acad. Sci. USA* **72**, 4361-4365.

Clarke, L. and Carbon, J. (1976). A colony bank containing synthetic Col El hybrid plasmids representative of the entire E. coli genome. *Cell* **9**, 91-99.

Clarke, L. and Carbon, J. (1978). Functional expression of cloned yeast DNA in E. coli: specific complementation of argino succinate lyase (argH) mutations. *J. Mol. Biol.* **120**, 517-532.

Clayton, D. A., Davis, R. W. and Vinograd, J. (1970). Homology and structural relationships between the dimeric and monomeric circular forms of mitochondrial DNA from human leukemic leucocytes. *J. Mol. Biol.* **47**, 137-153.

Clegg, J. B., Weatherall, D. J. and Milner, P. F. (1971). Hemoglobin constant spring—a chain termination mutant? *Nature* **234**, 337-340.

Clements, G. B. and Subak-Sharpe, J. H. (1975a). Reactivation of chick erythrocyte nuclei in BHK derived cells with multiple biochemical lesions. *Exp. Cell Res.* **95**, 15-24.

Clements, G. B. and Subak-Sharpe, J. H. (1975b). Metabolic cooperation between biochemically variant hamster cells and heterocaryons formed between these cells and chick erythrocytes. *Exp. Cell Res.* **95**, 25-30.

Cleveland, D. W., Hwo, S. Y. and Kirschner, M. W. (1977a). Purification of tau, a microtubule associated protein that induces assembly of microtubules from purified tubulin. *J. Mol. Biol.* **116**, 207-226.

Cleveland, D. W., Hwo, S. Y. and Kirschner, M. W. (1977b). Physical and chemical properties of purified tau factor and the role of tau in microtubule assembly. *J. Mol. Biol.* **116**, 227-248.

Clever, U. (1966). Puffing in giant chromosomes of Diptera and the mechanism of its control. In J. Bonner and P. T'so (Eds.), *The Nucleohistones*, Holden-Day, San Francisco, pp 317-331.

Clever, U. (1969). The formation of secretion in the salivary glands of Chironomus. *Exp. Cell Res.* **55**, 317-322.

Clever, U. and Karlson, P. (1960). Induktion von puff-veranderungen in den speicheldrusen-chromosomen von C. tentans durch ecdysone. *Exp. Cell Res.* **20**, 623-626.

Clever, U. and Storbeck, I. (1970). Chromosome activity and cell function in polytenic cells. IV. Polyribosomes and their sensitivity to actinomycin. *Biochim. Biophys. Acta* **217**, 108-119.

Clever, U., Storbeck, I. and Romball, C. G. (1969). Chromosome activity and cell function in polytene cells. I. Protein synthesis at various stages of larval development. *Exp. Cell Res.* **55**, 306-316.

Clive, D., Flamm, W. G., Machesko, M. R. and Bernheim, N. J. (1972). A mutational assay system using the thymidine kinase locus in mouse lymphoma cells. *Mutat. Res.* **16**, 77-87.

Cockburn, A. F., Newkirk, M. J. and Firtel, R. A. (1976). Organization of the ribosomal RNA genes of Dictyostelium discoideum: mapping of the nontranscribed spacer regions. *Cell* **9**, 605-614.

Cockburn, A. F., Taylor, W. C. and Firtel, R. A. (1978). Dictyostelium rDNA consists of nonchromosomal palindromic dimers containing 5S and 36S coding regions. *Chromosoma* **70**, 19-30.

Coen, D., Deutsch, J., Netter, P., Petrochilo, E. and Slonimski, P. P. (1970). Mitochondrial genetics. I. Methodology and phenomenology. *Symp. Soc. Exp. Biol.* **24**, 449-496.

Coen, D. M., Bedbrook, J. R., Bogorad, L. and Rich, A. (1977). Maize chloroplast DNA fragment encoding the large subunit of ribulose biphosphate carboxylase. *Proc. Nat. Acad. Sci. USA* **74**, 5487-5491.

Coffino, P., Bourne, H. R. and Tomkins, G. M. (1975). Somatic genetic analysis of cyclic AMP action: selection of unresponsive mutants. *J. Cell. Physiol.* **85**, 603-610.

Cohen, C., Harrison, S. C. and Stephens, R. E. (1971). X ray diffraction from microtubules. *J. Mol. Biol.* **59**, 375–380.

Cohen, C., DeRosier, D., Harrison, S. C., Stephens, R. E. and Thomas, J. (1975). X ray patterns from microtubules. *Ann. NY Acad. Sci.* **253**, 53–59.

Cohen, J. E., Jasny, B. R. and Tamm, I. (1979). Spatial distribution of initiation sites for mammalian DNA replication: a statistical analysis. *J. Mol. Biol.* **128**, 219–246.

Cohen, I. and Cohen, C. (1972). A tropomyosin-like protein from human platelets. *J. Mol. Biol.* **68**, 383–387.

Cohen, L. H., Newrock, K. M. and Zweidler, A. (1975). Stage specific switches in histone synthesis during embryogenesis of the sea urchin. *Science* **190**, 994–997.

Cohen, L. H., Smits, P. E. M. and Bloemendal, H. (1976). Isolation and characterization of rat lens mRNAs. *Eur. J. Biochem.* **67**, 563–572.

Cohen, L. H., Westerhuis, L. W., Smits, D. P. and Bloemendal, H. (1978). Two structurally closely related polypeptides coded by 14S mRNA isolated from rat lens. *Eur. J. Biochem.* **89**, 251–258.

Cohen, M. M. and Rattazzi, M. C. (1971). Cytological and biochemical correlation of late X chromosomal replication and gene inactivation in the mule. *Proc. Nat. Acad. Sci. USA* **68**, 544–548.

Cohen, M., Jr. (1976a). Ectopic pairing and evolution of 5S rRNA genes in the chromosomes of D. funebris. *Chromosoma* **55**, 349–358.

Cohen, M., Jr. (1976b). Evolution of 5S rRNA genes in the chromosomes of the virilis group of Drosophila. *Chromosoma* **55**, 359–371.

Cohen, S. N. and Chang, A. C. Y. (1977). Revised interpretation of the origin of the pSC101 plasmid. *J. Bacteriol.* **132**, 734–739.

Cohen, S. N., Chang, A. C. Y., Boyer, H. W. and Helling, R. B. (1973). Construction of biologically functional bacterial plasmids in vitro. *Proc. Nat. Acad. Sci. USA* **70**, 3240–3244.

Cohen, W. D. and Rebhun, L. I. (1970). An estimate of the amount of microtubule protein in the isolated mitotic apparatus. *J. Cell Sci.* **6**, 159–176.

Cohn, R. H., Lowry, J. C. and Kedes, L. H. (1976). Histone genes of the sea urchin (S. purpuratus) cloned in E. coli: order, polarity, and strandedness of the five histone-coding and spacer regions. *Cell* **9**, 147–162.

Cohn, R. H. and Kedes, L. H. (1979a). Non allelic histone gene clusters of individual sea urchins (L. pictus): polarity and gene organization. *Cell* **18**, 843–853.

Cohn, R. H. and Kedes, L. H. (1979b). Non allelic histone gene clusters of individual sea urchins (L. pictus): mapping of homologies in coding and spacer DNA. *Cell* **18**, 855–864.

Collins, J. and Hohn, B. (1976). Cosmids: a new type of plasmid cloning vector that is packageable in vitro in phage lambda heads. *Proc. Nat. Acad. Sci. USA* **75**, 4242–4246.

Colson, A. M. and Slonimski, P. P. (1979). Genetic localization of diuren and mucidin resistant mutants relative to a group of loci of the mitochondrial DNA controlling coenzyme QH^2-cytochrome c reductase in S. cerevisiae. *Mol. Gen. Genet.* **167**, 287–298.

Colten, H. R. and Parkman, R. (1972). Biosynthesis of C4 (fourth component) of complement by hybrids of C4-deficient guinea pig cells and HeLa cells. *Science* **176**, 1029–1031.

Comi, P., Giglioni, B., Barbarano, L., Ottolenghi, S., Williamson, R., Novakova, M. and Masera, G. (1977). Transcriptional and post transcriptional defects in β° thalassemia. *Eur. J. Biochem.* **79**, 617–622.

Comings, D. E. (1967). The duration of replication of the inactive X chromosome in humans based on the persistence of the heterochromatic sex body during DNA synthesis. *Cytologia* **6**, 20–37.

Comings, D. E. and Okada, T. A. (1971a). Triple chromosome pairing in triploid chickens. *Nature* **231**, 119–121.

Comings, D. E. and Okada, T. A. (1971b). Whole mount electron microscopy of human meiotic chromosomes. *Exp. Cell Res.* **65**, 99-103.

Comings, D. E. and Okada, T. A. (1971c). Fine structure of the synaptonemal complex. *Exp. Cell Res.* **65**, 104-119.

Comings, D. E. and Okada, T. A. (1972). Architecture of meiotic cells and mechanisms of chromosome pairing. *Adv. Cell Mol. Biol.* **2**, 310-384.

Comings, D. E. and Okada, T. A. (1975). Mechanisms of chromosome banding. VI. Whole mount electron microscopy of banded metaphase chromosomes and comparison with pachytene chromosomes. *Exp. Cell Res.* **93**, 267-274.

Comings, D. E. and Okada, T. E. (1976). Nuclear proteins. III. The fibrillar nature of the nuclear matrix. *Exp. Cell Res.* **103**, 341-360.

Comings, D. E., Avelino, E., Okada, T. A. and Wyandt, H. E. (1973). The mechanism of C and G banding of chromosomes. *Exp. Cell Res.* **77**, 469-493.

Comings, D. E., Kovacs, B. W., Avelino, E., and Harris, D. C. (1975). Mechanisms of chromosome banding. B. Quinacrine banding. *Chromosoma* **50**, 111-146.

Compton, J. L. and McCarthy, B. J. (1978). Induction of the Drosophila heat shock response in isolated polytene nuclei. *Cell* **14**, 191-202.

Compton, J. L., Bellard, M. and Chambon, P. (1976). Biochemical evidence of variability in the DNA repeat length in the chromatin of higher eucaryotes. *Proc. Nat. Acad. Sci. USA* **73**, 4382-4386.

Compton, J. L., Hancock, R., Oudet, P. and Chambon, P. (1976). Biochemical and electron microscopic evidence that the subunit structure of CHO interphase chromatin is conserved in mitotic chromosomes. *Eur. J. Biochem.* **70**, 555-568.

Conconi, F., Rowley, P. T., del Senno, L., Pontremoli, S., and Volpato, S. (1972). Induction of β globin synthesis in the β thalassemia of Ferrara. *Nature New Biol.* **238**, 83-87.

Conde, M. F., Boynton, J. E., Gillham, N. W., Harris, E. H., Tingle, C. L. and Way, W. L. (1975). Chloroplast genes in Chlamydomonas affecting organelles ribosomes. *Mol. Gen. Genet.* **140**, 183-220.

Condeelis, J. S. and Taylor, D. L. (1977). The contractile basis of amoeboid movement. V. The control of gelation, solation and contraction in extracts of D. discoideum. *J. Cell Biol.* **74**, 901-927.

Connolly, J. A., Kalnins, V. I., Cleveland, D. W. and Kirschner, M. W. (1977). Immunofluorescent staining of cytoplasmic and spindle microtubules in mouse fibroblasts with antibody to tau protein. *Proc. Nat. Acad. Sci. USA* **74**, 2437-2440.

Connolly, J. A., Kalnins, V. I., Cleveland, D., and Kirschner, M. W. (1978). Intracellular localization of the high molecular weight microtubule accessory protein by indirect immunofluorescence. *J. Cell. Biol.* **76**, 781-786.

Constantini, F. D., Scheller, R. H., Britten, R. J. and Davidson, E. H. (1978) Repetitive sequence transcripts in the mature sea urchin oocyte. *Cell* **15**, 173-187.

Contreras, R., Rogiers, R., Van de Voorde, A. and Fiers, W. (1977). Overlapping of the VP2-VP3 gene and the VP1 gene in the SV40 genome. *Cell* **12**, 529-538.

Cook, P. R. (1970). Species specificity of an enzyme determined by an erythrocyte nucleus in an interspecific hybrid cell. *J. Cell Sci.* **7**, 1-4.

Cooke, H. J. (1975). Evolution of the long range structure of satellite DNAs in the genus Apodemus. *J. Mol. Biol.* **94**, 87-99.

Cooke, H. (1976). Repeated sequence specific to human males. *Nature* **262**, 182-186.

Cooke, H. J. and McKay, R. D. G. (1978). Evolution of a human Y chromosome specific repeated sequence. *Cell* **13**, 453-460.

Cooke, P. R. (1976). A filamentous cytoskeleton in vertebrate smooth muscle fibers. *J. Cell Biol.* **68**, 539-556.

Cook, P. R. and Brazell, I. A. (1976a). Supercoils in human DNA. *J. Cell Sci.* **19**, 261-280.

Cook, P. R. and Brazell, I. A. (1976b). Conformational constraints in nuclear DNA. *J. Cell Sci.* **22**, 287-302.

Cook, P. R. and Brazell, I. A. (1977). The superhelical density of nuclear DNA from human cells. *Eur. J. Biochem.* **74**, 527-532.

Cook, P. R., Brazell, I. A. and Jost, E. (1976). Characterization of nuclear structures containing superhelical DNA. *J. Cell Sci.* **22**, 303-324.

Cooper, D. W. (1971). A directed genetic change model for X chromosome inactivation in eutherian mammals. *Nature* **230**, 292-294.

Cooper, K. W. (1950). Normal spermatogenesis in Drosophila. In M. Demerec (Ed.), *Biology of Drosophila*, Hafner, New York, pp 1-61.

Corneo, G., Ginelli, E. and Polli, E. (1970). Different satellite DNA of guinea pig and ox. *Biochemistry* **9**, 1565-1570.

Cordell, B., Bell, G., Tischer, E., DeNoto, F., Ullrich, A., Pictet, R., Rutter, W. J. and Goodman, H. M. (1979). Isolation and characterization of a cloned rat insulin gene. *Cell* **18**, 533-543.

Corneo, G., Ginelli, E. and Polli, E. (1971). Renaturation properties and localization in heterochromatin of human satellite DNAs. *Biochim. Biophys. Acta* **247**, 528-534.

Corneo, G., Zardi, L. and Polli, E. (1972). Elution of human satellite DNA on a MAK column: isolation of satellite DNA IV. *Biochim. Biophys. Acta* **269**, 201-204.

Corneo, G., Ginelli, E., Soave, C. and Bernadi, C. (1968). Isolation and characterization of mouse and guinea pig satellite DNA. *Biochemistry* **7**, 4373-4379.

Cornman, I. (1942). Susceptibility of Colchicum and Chlamydomonas to colchicine. *Bot. Gaz.* **104**, 50-61.

Cornman, I. and Cornman, M. E. (1951). The action of podophyllotoxin and its fractions on marine eggs. *Ann. N. Y. Acad. Sci.* **51**, 1443-1487.

Correns, C. (1909). Zur kenntnis der rolle von kern und plasma bei der verebung. *Z. Verebungsl.* **2**, 331-340.

Cory, S. and Adams, J. M. (1975). The modified 5′ terminal sequences in mRNA of mouse myeloma cells. *J. Mol. Biol.* **99**, 519-548.

Cory, S. and Adams, J. M. (1977). A very large repeating unit of mouse DNA containing the 18S, 28S and 5.8S rRNA genes. *Cell* **11**, 795-805.

Cottrell, S., Rabinowitz, M. and Gets, G. S. (1973). Mitochondrial DNA synthesis in a temperature sensitive mutant of DNA replication of S. cerevisiae. *Biochemistry* **12**, 4374-4378.

Counce, S. J. and Meyer, G. F. (1973). Differentiation of the synaptonemal complex and the kinetochore in Locusta spermatocytes studied by whole mount electron microscopy. *Chromosoma* **44**, 231-253.

Cousens, L. S., Gallwitz, D. and Alberts, B. M. (1979). Different accessibilities in chromatin to histone acetylase. *J. Biol. Chem.* **254**, 1716-1723.

Cox, R. F. (1977). Estrogen withdrawal in chick oviduct. Selective loss of high abundance classes of polyadenylated mRNA. *Biochemistry* **16**, 3433-3442.

Cox, R. P., Krauss, M. R., Balis, M. E. and Dancis, J. (1970). Evidence for transfer of enzyme product as the basis of metabolic cooperation between tissue culture fibroblasts of Lesch-Nyhan disease and normal cells. *Proc. Nat. Acad. Sci. USA* **67**, 1573-1579.

Craig, E. A., McCarthy, B. J. and Wadsworth, S. C. (1979). Sequence organization of two recombinant plasmids containing genes for the major heat shock induced protein of D. melanogaster. *Cell* **16**, 575-588.

Craig-Cameron, T. A. and Jones, T. H. (1970). The analysis of exchanges in ³H labeled meiotic chromosomes. I. Schistocerca gregaria. *Heredity* **25**, 223-232.

Crain, W. R., Davidson, E. H. and Britten, R. J. (1976a). Contrasting patterns of DNA sequence arrangement in Apis mellifera (honeybee) and Musca domestica (housefly). *Chromosoma* **59**, 1–12.

Crain, W. R., Eden, F. C., Pearsob, W. R., Davidson, E. H. and Britten, R. J. (1976b). Absence of short period interspersion of repetitive and nonrepetitive sequences in the DNA of D. melanogaster. *Chromosoma* **56**, 309–326.

Crane-Robinson, C., Danby, S. E., Bradbury, E. M, Garel, A., Kovacs, A. -M., Champagne, M., and Daune, M. (1976). Structural studies of chicken erythrocyte histone H5. *Eur. J. Biochem.* **67**, 379–388.

Crane-Robinson, C., Hayashi, H., Cary, P. D., Briand, G., Sautiere, P., Krieger, D., Vidali, G., Lewis, P. N. and Tom-Kun, J. (1977). The location of secondary structure in histone H4. *Eur. J. Biochem.* **79**, 535–548.

Crawford, L. V., Cole, C. N., Smith, A. E., Paucha, E., Tegtmeyer, P., Rundell, K. and Berg, P. (1975). Organization and expression of early genes of SV40. *Proc. Nat. Acad. Sci. USA* **75**, 117–121.

Crawford, R. J., Krieg, P., Harvey, R. P., Hewish, D. A. and Wells, J. R. E. (1979). Histone genes are clustered with a 15 kb repeat in the chicken genome. *Nature* **279**, 132–136.

Creighton, H. B. and McClintock, B. (1931). A correlation of cytological and genetic crossing over in Zea mays. *Proc. Nat. Acad. Sci. USA* **17**, 492–497.

Cremisi, C., Chestier, A. and Yaniv, M. (1977). Preferential association of newly synthesized histones with replicating SV40 DNA. *Cell* **12**, 947–951.

Cremisi, C., Pignatti, P. F., Croissant, O. and Yaniv, M. (1976). Chromatin like structures in polyoma virus and SV40 lytic cycle. *J. Virol.* **17**, 204–211.

Crepau, R. H., McEwen, B. and Edelstein, S. J. (1978). Differences in α and β polypeptide chains of tubulin resolved by electron microscopy with image reconstruction. *Proc. Nat. Acad. Sci. USA* **75**, 5006–5010.

Crews, S., Ojala, D., Posakony, J., Nishiguchi, J. and Attardi, G. (1979). Nucleotide sequence of a region of human mitochondrial DNA containing the precisely identified origin of replication. *Nature* **277**, 192–198.

Crick, F. H. C. (1976). Linking numbers and nucleosomes. *Proc. Nat. Acad. Sci. USA* **73**, 2639–2643.

Crick, F. H. C. and Klug, A. (1975). Kinky helix. *Nature* **255**, 530–533.

Crippa, M., Meza, I. and Dina, D. (1973). Sequence arrangement in mRNA: presence of poly-A and identification of a repetitive fragment at the 5′ end. *Cold Spring Harbor Symp. Quant. Biol.* **38**, 933–942.

Croce, C. (1976). Loss of mouse chromosomes in somatic cell hybrids between HT1080 human fibrosarcoma cells and mouse peritoneal macrophages. *Proc. Nat. Acad. Sci. USA* **73**, 3248–3252.

Croce, C. M. and Koprowski, H. (1974a). Concordant segregation of the expression of SV40 T antigen and human chromosome 7 in mouse-human hybrid subclones. *J. Exp. Med.* **139**, 1350–1353.

Croce, C. M. and Koprowski, H. (1974b). Somatic cell hybrids between peritoneal mouse macrophages and SV40 transformed human cells. I. Positive control of the transformed phenotype by the human chromosome 7 carrying the SV40 genome. *J. Exp. Med.* **140**, 1221–1229.

Croce, C. M. and Koprowski, H. (1975). Assignment of genes for transformation to human chromosome 7 carrying the SV40 genome. *Proc. Nat. Acad. Sci. USA* **72**, 1658–1660.

Croce, C., Girardi, A. J. and Koprowski, H. (1973). Assignment of the T antigen gene of SV40 to human chromosome C7. *Proc. Nat. Acad. Sci. USA* **70**, 3617–3620.

Croce, C. M., Huebner, K. and Koprowski, H. (1974). Chromosome assignment of the T antigen of SV40 in African green monkey cells transformed by adeno 7-SV40 hybrid. *Proc. Nat. Acad. Sci. USA* **71**, 4116–4119.

Croce, C. M., Litwack, G. and Koprowski, H. (1973). Human regulatory gene for inducible TAT in rat-human hybrids. *Proc. Nat. Acad. Sci. USA* **70**, 1268–1272.

Croce, C. M., Bakay, B., Nyhan, W. L. and Koprowski, H. (1973). Reexpression of the rat HGPRT gene in rat-human hybrids. *Proc. Nat. Acad. Sci.* **70**, 2590-2594.

Croce, C. M., Talavera, A., Basilico, C. and Miller, O. J. (1977). Suppression of production of mouse 28S rRNA in mouse-human hybrids segregating mouse chromosomes. *Proc. Nat. Acad. Sci. USA* **74**, 694-697.

Crossen, P. E., Pathak, D. and Arrhigi, F. E. (1975). A high resolution study of the DNA replication patterns of Chinese hamster chromosomes using sister chromatid differential staining technique. *Chromosoma* **52**, 339-347.

Crouse, G. F., Fodor, E. J. B. and Doty, P. (1976). In vitro transcription of chromatin in the presence of a mercurated nucleotide. *Proc. Nat. Acad. Sci. USA* **73**, 1564-1567.

Crouse, H. V. (1960). The controlling element in sex chromosome behavior in Sciara. *Genetics* **45**, 1429-1443.

Cryer, D. R., Goldthwaite, C. D., Zinker, S., Lam, K. B., Storm, E., Hirschberg, R., Blamire, J., Finkelstein, D. B. and Marmur, J. (1973). Studies on nuclear and mitochondrial DNA of S. cerevisiae. *Cold Spring Harbor Symp. Quant. Biol.* **38**, 17-29.

Cunningham, R. S., Bonen, L., Doolittle, W. F. and Gray, M. W. (1976). Unique species of 5S, 18S and 26S RNA in wheat mitochondria. *Febs Lett.* **69**, 116-122.

Curtis, P. J. and Weissmann, C. (1976). Purification of globin mRNA from DMSO-induced Friend cells and detection of a putative globin mRNA precursor. *J. Mol. Biol.* **106**, 1061-1076.

Curtis, P. J., Mantei, N., Van den Berg, J. and Weissmann, C. (1977). Presence of a putative 15S precursor to β globin mRNA but not to α globin mRNA in Friend cells. *Proc. Nat. Acad. Sci. USA* **74**, 3184-3188.

Dahl, D. and Bignami, A. (1973). Immunochemical and immunofluorescence studies of the glial fibrillary acidic protein in vertebrates. *Brain. Res.* **61**, 279-293.

Dahl, D. and Bignami, A. (1976). Immunogenic properties of the glial fibrillary acidic protein. *Brain Res.* **116**, 150-157.

Dahmus, M. E. and McConnell, D. J. (1969). Chromosomal RNA of nonhistone protein of rat liver. *Biochemistry* **8**, 1524-1534.

Dale, R. M. K. and Ward, D. C. (1975). Mercurated polynucleotides: new probes for hybridization and selective polymer fractionation. *Biochemistry* **14**, 2458-2469.

Dales, S. (1972). Concerning the universality of a microtubule antigen in animal cells. *J. Cell Biol.* **52**, 748-754.

Dan, K. and Kojima, N. K. (1963). A study on the mechanism of cleavage in the amphibian egg. *J. Exo. Biol.* **40**, 7-15.

Daneholt, B. (1975). Transcription in polytene chromosomes. *Cell* **4**, 1-9.

Daneholt, B. and Edstrom, J. E. (1967). The content of DNA in individual polytene chromosomes of C. tentans. *Cytogenetics* **6**, 350-356.

D'Anna, J. A. and Isenberg, I. (1972). Fluorescence anisotropy and circular dichroism study of conformational changes in histone IIb2. *Biochemistry* **11**, 4017-4025.

D'Anna, J. A. and Isenberg, I. (1973). A complex of histones IIb2 and IV. *Biochemistry* **12**, 1035-1043.

D'Anna, J. A. and Isenberg, I. (1974a). Conformational changes of histone LAK (f2a2). *Biochemistry* **13**, 2093-2098.

D'Anna, J. A. and Isenberg, I. (1974b). Interactions of histone LAK (f2a2) and histones KAS (f2b) and GRK (f2al). *Biochemistry* **13**, 2098-2104.

D'Anna, J. A. and Isenberg, I. (1974c). Conformational changes of histone ARE (f3). *Biochemistry* **13**, 4987-4992.

D'Anna, J. A. and Isenberg, I. (1974d). A histone cross complexing pattern. *Biochemistry* **13**, 4992-4997.

Danna, K. J. and Nathans, D. (1971). Specific cleavage of SV40 DNA by restriction endonuclease of H. influenzae. *Proc. Nat. Acad. Sci. USA* **68,** 2913-2917.

Danna, K. J., Sack, G. H. and Nathans, D. (1973). Studies of SV40 DNA. VII. A cleavage map of the SV40 genome. *J. Mol. Biol.* **78,** 363-376.

Darlington, C. D. (1932). *Recent Advances in Cytology,* Churchill, London.

Darlington, C. D. (1934). The origin and behavior of chiasmata. VII. Zea mays. *Z. VerebLahre* **67,** 96-114.

Darlington, G. J., Bernhard, H. P. and Ruddle, F. H. (1974a). The expression of hepatic functions in mouse hepatoma x human leucocyte cell hybrids. *Cytogenetics* **13,** 86-88.

Darlington, G. J., Bernhard, H. P. and Ruddle, F. H. (1974b). Human serum albumin phenotype activation in mouse hepatoma-human leucocyte hybrids. *Science* **185,** 859-862.

Darnell, J. E., Wall, R., and Tushinski, R. J. (1971). An adenylic acid rich sequence in mRNA of HeLa cells and its possible relationship to reiterated sites in DNA. *Proc. Nat. Acad. Sci. USA* **68,** 1321-1325.

Darnell, J. E., Pagoulatos, G. N., Lindberg, U. and Balint, R. (1970). Studies on the relationship of mRNA to hnRNA in mammalian cells. *Cold Spring Harbor Symp. Quant. Biol.* **35,** 555-560.

Darnell, J. E., Philipson, L., Wall, R. and Adesnik, M. (1971). Poly(A) sequences: role in conversion of nuclear RNA into mRNA. *Science* **174,** 507-510.

Darnell, J. E., Jelinek, W. R. and Molloy, G. R. (1973). Biogenesis of mRNA: genetic regulation in mammalian cells. *Science* **181,** 1215-1221.

Das, N. K., Micou-Eastwood, J., Ramamurthy, G. and Alfert, M. (1970). Sites of synthesis and processing of rRNA precursors within the nucleolus of Erechis caupo eggs. *Proc. Nat. Acad. Sci. USA* **67,** 968-975.

Dasgupta, R., Shih, D. S., Saris, C. and Kaesberg, P. (1975). Nucleotide sequence of a viral RNA fragment that binds to eucaryotic ribosomes. *Nature* **256,** 624-628.

Davidson, D. (1964). RNA synthesis in roots of Vicia faba. *Exp. Cell Res.* **35,** 317-325.

Davidson, E. H. (1976). *Gene Activity in Early Development.* Academic Press, New York.

Davidson, E. H. and Britten, R. J. (1973). Organization, transcription and regulation in the animal genome. *Quart. Rev. Biol.* **48,** 565-613.

Davidson, E. H. and Britten, R. J. (1979). Regulation of gene expression: possible role of repetitive sequences. *Science* **204,** 1052-1059.

Davidson, E. and Hough, B. R. (1971). Genetic information in oocyte RNA. *J. Mol. Biol.* **56,** 491-506.

Davidson, E. H., Hough, B. R., Amenson, C. S. and Britten, R. J. (1973). General interspersion of repetitive with nonrepetitive sequence elements in the DNA of Xenopus. *J. Mol. Biol.* **77,** 1-24.

Davidson, E. H., Hough, B. R., Klein, W. H. and Britten, R. J. (1975). Structural genes adjacent to interspersed repetitive DNA sequences. *Cell* **4,** 217-238.

Davidson, J. N. and Patterson, D. (1979). Alteration in structure of multifunctional protein from CHO cells defective in pyrimidine biosynthesis. *Proc. Nat. Acad. Sci. USA* **76,** 1731-1735.

Davidson, J. N., Hanson, M. R. and Bogorad, L. (1974). An altered chloroplast ribosome protein in ery-M1 mutants of C. reinhardii. *Mol. Gen. Genet.* **132,** 119-129.

Davidson, R. L. (1972). Regulation of melanin synthesis in mammalian cells: effect of gene dosage on the expression of differentiation. *Proc. Nat. Acad. Sci. USA* **69,** 951-955.

Davidson, R. L. and Bick, M. D. (1973). BUdR dependence—a new mutation in mammalian cells. *Proc. Nat. Acad. Sci. USA* **70,** 138-142.

Davidson, R. L. and Ephrussi, B. (1965). A selective system for the isolation of hybrids between L cells and normal cells. *Nature* **205,** 1170-1171.

Davidson, R. L. and Horn, D. (1974). Reversible "transformation" of bromodeoxyuridine-dependent cells by bromodeoxyuridine. *Proc. Nat. Acad. Sci. USA* **71,** 3338-3342.

Davidson, R. L., Ephrussi, B. and Yamamoto, K. (1966). Regulation of pigment synthesis in mammalian cells as studied by somatic hybridization. *Proc. Nat. Acad. Sci. USA* **56**, 1437-1440.

Davidson, R., Ephrussi, B. and Yamamoto, K. (1968). Regulation of melanin synthesis in mammalian cells as studied by somatic hybridization. I. Evidence for negative control. *J. Cell. Physiol.* **72**, 115-127.

Davidson, R. L., O'Malley, K. A. and Weeler, T. B. (1976). Polyethylene glycol induced mammalian cell hybridization: effect of PEG molecular weight and concentration. *Somat. Cell Genet.* **2**, 271-280.

Davie, J. R. and Candido, P. M. (1978). Acetylated histone H4 is preferentially associated with template active chromatin. *Proc. Nat. Acad. Sci. USA* **75**, 3574-3577.

Davis, B. K. (1971). Genetic analysis of a meiotic mutant resulting in precocious sister centromere separation in D. melanogaster. *Mol. Gen. Genet.* **113**, 251-272.

Davis, D. G. (1969). Chromosome behavior under the influence of claret nondisjunctional in D. melanogaster. *Genetics* **61**, 577-594.

Dawid, I. (1965). DNA in amphibian eggs. *J. Mol. Biol.* **12**, 581-599.

Dawid, I. B. (1972). Mitochondrial RNA in X. laevis. I. The expression of the mitochondrial genome. *J. Mol. Biol.* **63**, 201-216.

Dawid, I. B. and Blackler, A. W. (1972). Maternal and cytoplasmic inheritance of mitochondrial DNA in Xenopus. *Devel. Biol.* **29**, 151-161.

Dawid, I. B. and Botchan, P. (1977). Sequences homologous to ribosomal insertions occur in the Drosophila genome outside the nucleolar organizer. *Proc. Nat. Acad. Sci. USA* **74**, 4233-4237.

Dawid, I. B. and Chase, J. W. (1972). Mitochondrial RNA in X. laevis. II. Molecular weights and other physical properties of mitochondrial rRNA and 4S RNA. *J. Mol. Biol.* **63**, 217-231.

Dawid, I. B. and Wellauer, P. K. (1976). A reinvestigation of 5'-3' polarity in 40S ribosomal RNA precursor of X. laevis. *Cell* **8**, 443-448.

Dawid, I. B., and Wolstenholme, D. R. (1967). Ultracentrifuge and electron microscope studies on the structure of mitochondrial DNA. *J. Mol. Biol.* **28**, 233-245.

Day, J. W. and Grell, R. F. (1976). Synaptonemal complexes during premeiotic DNA synthesis oocytes of D. melanogaster. *Genetics* **83**, 67-79.

Dayhoff, M. O. (1972). *Atlas of Protein Sequence and Structure,* Vol. 5, National Biomedical Research Foundation, Bethesda, Maryland.

Deak, I., Sidebottom, E. and Harris, H. (1972). Further experiments on the role of the nucleolus in the expression of structural genes. *J. Cell Sci.* **11**, 379-392.

Deaven, L. L. and Petersen, D. F. (1973). The chromosomes of CHO, an aneuploid Chinese hamster cell line: G band, C band and autoradiographic analysis. *Chromosoma* **41**, 129-144.

Dedman, J. R., Potter, J. D. and Means, A. R. (1977a). Biological cross reactivity of rat testis phosphodiesterase activator protein and rabbit skeletal muscle troponin C. *J. Biol. Chem.* **252**, 2437-2440.

Dedman, J. R., Potter, J. D., Jackson, R. L., Johnson, D. and Means, A. R. (1977b). Physicochemical properties of rat testis Ca^{2+} dependent regulator protein of cyclic nucleotide phosphodiesterase. *J. Biol, Chem.* **252**, 8415-8422.

Deeley, R. G., Gordon, J. I., Burns, A. T. H., Mullinix, K. P., Bina-Stein, M. and Goldberger, R. F. (1977). Primary activation of the vitellogenin gene in the rooster. *J. Biol. Chem.* **252**, 8310-8319.

Defendi, V. and Manson, L. A. (1963). Analysis of the life cycle in mammalian cells. *Nature* **198**, 359-361.

Defer, N., Crepin, M., Terrious, C., Kruh, J. and Gros, F. (1979). Comparison of nonhistone proteins selectively associated with nucleosomes with proteins released during limited DNAase digestions. *Nucleic Acids Res.* **6**, 953-966.

Degelmann, A., Royer, H. D., and Hollenberg, C. P. (1979). The organization of the rRNA genes of C. tentans and some closely related species. *Chromosoma* **71**, 263-282.

Degnen, G. E., Miller, I. L., Eisenstadt, J. M. and Adelberg, E. A. (1976). Chromosome mediated gene transfer between closely related strains of cultured mouse cells. *Proc. Nat. Acad. Sci. USA* **73**, 2838-2842.

Deininger, P. L. and Schmid, C. W. (1976). An electron microscope study of the DNA sequence organization of the human genome. *J. Mol. Biol.* **106**, 773-790.

Deisseroth, A. and Hendrick, D. (1979). Activation of phenotypic expression of human globin genes from nonerythroid cells by chromosome dependent transfer to tetraploid mouse erythroleukemia cells. *Proc. Nat. Acad. Sci. USA* **76**, 2185-2189.

Deisseroth, A.,. Burk, R., Picciano, D., Minna, J,, Anderson, W. F. and Nienhuis, A. (1975a). Hemoglobin synthesis in somatic cell hybrids: globin gene expression in hybrids between mouse erythroleukemia and human marrow cells or fibroblasts. *Proc. Nat. Acad. Sci. USA* **72**, 1102-1106.

Deisseroth, A., Barker, J., Anderson, W. F. and Nienhuis, A. (1975b). Hemoglobin synthesis in somatic cell hybrids: coexpression of mouse with human or Chinese hamster globin genes in interspecific somatic cell hybrids of mouse erythroleukemic cells. *Proc. Nat. Acad. Sci. USA* **72**, 2682-2686.

Deisseroth, A., Velez, R., Burk, R. D., Minna, J., Anderson, W. F. and Nienhuis, A. (1976). Extinction of globin gene expression in human fibroblast x mouse erythroleukemia cell hybrids. *Somatic Cell Genet.* **2**, 373-384.

Deisseroth, A., Nienhuis, A., Turner, P., Velez, R., Anderson, W. F., Ruddle, F. H., Lawrence, J., Creagan, R. and Kucherlapati, R. (1977). Localization of the human α globin structural gene to chromosome 16 in somatic cell hybrids by molecular hybridization assay. *Cell* **12**, 205-218.

Deisseroth, A., Nienhuis, A., Lawrence, J., Giles, R., Turner, P. and Ruddle, F. H. (1978). Chromosomal localization of human β globin gene on human chromosome 11 in somatic cell hybrids. *Proc. Nat. Acad. Sci. USA* **75**, 1456-1460.

De Jonge, P., Kastelain, R. A. and Planta, R. J. (1978). Nonribosomal nucleotide sequences in 7S RNA, the immediate precursor of 5.8S rRNA in yeast. *Eur. J. Biochem.* **83**, 537-546.

DeLange, R. J. and Smith, E. L. (1971). Histones: structure and function. *Ann. Rev. Biochem.* **40**, 279-314.

DeLange, R. J., Hooper, J. A. and Smith, E. L. (1972). Complete amino acid sequence of calf thymus histone III. *Proc. Nat. Acad. Sci. USA* **69**, 882-884.

DeLange, R. J., Hooper, J. A. and Smith, E. L. (1973). Sequence studies on the cyanogen bromide peptides: complete amino acid sequence of calf thymus histone III. *J. Biol. Chem.* **248**, 3261-3274.

DeLange, R. J., Williams, L. C. and Martinson, H. G. (1979). Identification of interacting amino acids at the histone H2A-H2B binding site. *Biochemistry* **18**, 1942-1946.

DeLange, R. J., Fambrough, D. M., Smith, E. L. and Bonner, J. (1969a). Calf and thymus pea histone IV. II. The complete amino acid sequence of calf thymus histone IV: presence of ϵ-N-acetyllysine. *J. Biol. Chem.* **244**, 319-334.

DeLange, R. J., Fambrough, D. M., Smith, I. L. and Bonner, J. (1969b). Calf and pea histone IV. III. Complete amino acid sequence of pea seedling histone IV: comparison with the homologous calf thymus histone. *J. Biol. Chem.* **244**, 5669-5679.

Delovitch, T. and Baglioni, C. (1973). Immunoglobulin genes: a test of somatic versus germ line hypothesis by RNA/DNA hybridization. *Cold Spring Harbor Symp. Quant. Biol.* **38**, 739-751.

DeMars, R. (1974). Resistance of cultured human fibroblasts and other cells to purine and pyrimidine analogues in relation to mutagenesis detection. *Mutat. Res.* **24**, 335-364.

DeMars, R. and Hooper, J. L. (1960). A method of selecting for auxotrophic mutants of Hela cells. *J. Exp. Med.* **111**, 559-571.

Demerec, M. (1929). Frequency of spontaneous mutations in certain stocks of D. melanogaster. *Genetics* **22**, 469-478.

Demerec, M. (1933). The effect of X ray dosage on sterility and number of lethals in D. melanogaster. *Proc. Nat. Acad. Sci. USA* **19**, 1015-1020.

Demerec, M. (1934). The gene and its role in ontogeny. *Cold Spring Harbor Symp. Quant. Biol.* **2**, 110-117.

De Mey, J., Joniau, M., De Brabander, M., Moens, W. and Geuens, G. (1978). Evidence for unaltered structure and in vivo assembly of microtubules in transformed cells. *Proc. Nat. Acad. Sci. USA* **75**, 1339-1343.

Denhardt, D. T. (1966). A membrane filter technique for the detection of complementary DNA. *Biochem. Biophys. Res. Commun.* **23**, 641-646.

Denis, H. and Wegnez, M. (1973). Synthesis and maturation of 5S RNA in X. laevis oocytes. *Biochimie* **55**, 1137-1151.

Dennis, E. S. and Wake, R. G. (1966). Autoradiography of the B. subtilis chromosome. *J. Mol. Biol.* **15**, 435-439.

Dentler, W. L. and Rosenbaum, J. L. (1977). Flagellar elongation and shortening in Chalmydomonas. III. Structures attached to the tips of flagellar microtubules and their relationship to the directionality of flagellar microtubule assembly. *J. Cell Biol.* **74**, 747-759.

Dentler, W. L., Granett, S., Witman, G. B., and Rosenbaum, J. L. (1974). Directionality of brain microtubule assembly in vitro. *Proc. Nat. Acad. Sci. USA* **71**, 1710-1714.

Derman, E. and Darnell, J. E. (1974). Relationship of chain transcription to poly(A) addition and processing of hnRNA in Hela cells. *Cell* **3**, 255-264.

Derman, E., Goldberg, S. and Darnell, J. E. (1976). hnRNA in Hela cells: distribution of transcript sizes estimated from nascent molecule profile. *Cell* **9**, 465-472.

De Petris, S. (1975). Concanavalin A receptors, immunoglobulins, and θ antigen of the lymphocyte surface. Interactions with concanavalin A and with cytoplasmic structures. *J. Cell Biol.* **6**, 123-146.

Derksen, J., Berendes, H. D. and Willart, E. (1973). Production and release of a locus-specific ribonucleoprotein product in polytene nuclei of D. hydei. *J. Cell Biol.* **59**, 661-668.

DeRobertis, E. M. and Gurdon, J. B. (1977). Gene activation in somatic nuclei after injection into amphibian oocytes. *Proc. Nat. Acad. Sci. USA* **74**, 2470-2474.

DeRobertis, E. M. and Mertz, J. E. (1977). Coupled transcription-translation in DNA-injected Xenopus oocytes. *Cell* **12**, 175-182.

DeRobertis, E. M. and Olson, M. V. (1979). Transcription and processing of cloned yeast tyrosine tRNA genes microinjected into frog oocytes. *Nature* **278**, 137-142.

Deschatrette, J. and Weiss, M. C. (1975). Extinction of liver specific functions in hybrids between differentiated and dedifferentiated rat hepatoma cells. *Somatic Cell Genet.* **1**, 279-292.

Deutsch, J. B., Dujon, P., Netter, E., Petrochilo, E., Slonimski, P. P., Bolotin-Fukuhara, M. and Coen, D. (1974). Mitochondrial genetics VI. The petite mutation in S. cerevisiae. Interrelations between the loss of the rho⁺ factor and the loss of the drug resistance mitochondrial genetic markers. *Genetics* **76**, 195-219.

Dev, V. G., Warburton, D., Miller, O. J., Miller, D. A., Erlanger, B. F. and Beiser, S. M. (1972). Consistent pattern of binding of anti adenosine antibodies to human metaphase chromosomes. *Exp. Cell Res.* **74**, 288-293.

Dev, V. G. R., Tantrahavi, D. A., Miller, D. A. and Miller, O. J. (1977). Nucleolus organizers in M. musculus subspecies and in the RAG mouse cell line. *Genetics* **86**, 389-398.

De Vries, H. (1903). Befruchtung und Bastardierung. In C. S. Gager (Ed.), *Intracellular pangenesis*, English Translation Open Court, Chicago, 1910, pp 217-263.

De Vries, H., De Jonge, J. C., Bakker, H., Meurs, H. and Kroon, A. (1979). The anatomy of the tRNA-rRNA region of the N. crassa mitochondrial DNA. *Nucl. Acids Res.* **6**, 1791-1804.

De Wit, J., Hoeksema, H. L., Halley, D., Hagemeijer, A., Bootsma, D. and Westervald, A. (1977). Regional localization of a β galactosidase locus on human chromosome 22. *Somat. Cell Genet.* **3**, 351-366.

Dick, C. and Johns, E. W. (1969). The biosynthesis of the five main histone fractions of rat thymus. *Biochim. Biophys. Acta* **174**, 380–386.

Dickson, E., Boyd, J. B. and Laird, C. D. (1971). Sequence diversity of polytene chromosome DNA from D. hydei. *J. Mol. Biol.* **61**, 615–628.

Diez, J. and Brawerman, G. (1974). Elongation of the poly(A) segment of mRNA in the cytoplasm of mammalian cells. *Proc. Nat. Acad. Sci. USA* **71**, 4091–4095.

Diggle, J. H., McVittie, J. D. and Peacocke, A. R. (1975). The self association of chicken erythrocyte histones. *Eur. J. Biochem.* **56**, 173–182.

Dippell, R. V. (1976). Effects of nuclease and protease on the ultrastructure of Paramecium basal bodies. *J. Cell Biol.* **69**, 622–637.

Dische, Z. (1930). Ubder einige neue charakeristische farbreaktionen der thymonukleinsaure und ein mikromethode zur bestimmung der selben in tierischen organen mit hilfe dieser reaktionen. *Mikrochemie* **8**, 4–32.

Dixon, G. H. (1972). The basic proteins of trout testis chromatin: aspects of their synthesis, post synthetic modifications and binding to DNA. *Karolinska Symp. Res. Methods Reprod. Biol.* **5**, 128–154.

Dixon, G. H. and Smith, M. (1968). Nucleic acids and protamine in salmon testis. *Prog. Nucleic Acid Res.* **8**, 9–34.

Dobberstein, B., Blobel, G. and Chua, N. H. (1977). In vitro synthesis and processing of a putative precursor for the small subunit of RuBC of C. reinhardii. *Proc. Nat. Acad. Sci. USA* **74**, 1082–1085.

Dobzhansky, T. and Wright, S. (1941). Genetics of natural populations. V. Relations between mutation rate and accumulation of lethals in populations of D. pseudoobscura. *Genetics* **26**, 23–51.

Dodgson, J. B., Strommer, J. and Engel, J. D. (1979). Isolation of the chicken β globin gene and a linked embryonic β-like globin gene from a chicken DNA recombinant library. *Cell* **17**, 879–888.

Doel, M. T. and Carey, N. H. (1976). The translational capacity of deadenylated ovalbumin mRNA. *Cell* **8**, 51–58.

Doenecke, D. and McCarthy, B. J. (1975a). Protein content of chromatin fractions separated by sucrose gradient centrifugation. *Biochemistry* **14**, 1366–1372.

Doenecke, D. and McCarthy, B. J. (1975b). The nature of protein association with chromatin. *Biochemistry* **14**, 1373–1379.

Doenecke, D. and McCarthy, B. J. (1976). Movement of histones in chromatin induced by shearing. *Eur. J. Biochem.* **64**, 405–410.

Doenges, K. H., Biedert, S. and Paweletz, N. (1976). Characterization of a 20S component in tubulin from mammalian brain. *Biochemistry* **15**, 2995–2999.

Doenges, K. H., Nagle, B. W., Uhlmann, A. and Bryan, J. (1977). In vitro assembly of tubulin from nonneural cells (Ehrlich ascites tumor cells). *Biochemistry* **16**, 3455–3459.

Doenges, K. H., Weissinger, M., Fritzsche, R. and Schroeter, D. (1979). Assembly of non neural microtubules in the absence of glycerol and microtubule associated proteins. *Biochemistry* **18**, 1698–1701.

Dott, P. J., Chuang, C. R. and Saunders, G. F. (1976). Inverted repetitive sequences in the human genome. *Biochemistry* **15**, 4120–4125.

Dottin, R. P., Weiner, A. M. and Lodish, H. F. (1976). 5′ terminal nucleotide sequences of the mRNAs of Dictyostelium discoideum. *Cell* **8**, 233–244.

Doty, P., Marmur, J., Eigner, J. and Schildkraut, C. (1960). Strand separation and specific recombination in DNAs: physical studies. *Proc. Nat. Acad. Sci. USA* **46**, 461–476.

Douglas, G. R., McAlpine, P. J. and Hamerton, J. L. (1973). Regional localization of loci for human PGM1 and 6PGD on human chromosome one by use of hybrids of Chinese hamster-human somatic cells. *Proc. Nat. Acad. Sci. USA* **70**, 2737–2740.

Douglas, M. G. and Butow, R. A. (1976). Variant forms of mitochondrial translation products in yeast: evidence for location of determinants on mitochondrial DNA. *Proc. Nat. Acad. Sci. USA* **73**, 1083-1086.

Douglas, M. G., Koh, Y., Ebner, E., Agsteribbe, E. and Schatz, G. (1979). A nuclear mutation conferring aurovertin resistance to yeast mitochondrial ATPase. *J. Biol. Chem.* **254**, 1335-1339.

Douvas, A. S., Harrington, C. A. and Bonner, J. (1975). Major nonhistone proteins of rat liver chromatin: preliminary identification of myosin, actin, tubulin and tropomyosin. *Proc. Nat. Acad. Sci. USA* **72**, 3902-3906.

Doyle, D. and Lauffer, H. (1969). Requirements of RNA synthesis for formation of salivary gland specific proteins in larval C. tentans. *Exp. Cell Res.* **57**, 205-210.

Drake, J. W. (1969). Comparative rates of spontaneous mutation. *Nature* **221**, 1132.

Drews, U., Blecher, S. R., Owen, D. A. and Ohno, S. (1974). Genetically directed preferential X activation seen in mice. *Cell* **1**, 3-8.

Dreyer, W. J. and Bennett, J. C. (1965). The molecular basis of anitbody formation. A paradox. *Proc. Nat. Acad. Sci. USA* **54**, 864-869.

Dubroff, L. M. and Nemer, M. (1975). Molecular classes of heterogeneous nuclear RNA in sea urchin embryos. *J. Mol. Biol.* **95**, 455-476.

Dubroff, L. M. and Nemer, M. (1976). Developmental shifts in the synthesis of hnRNA classes in the sea urchin embryo. *Nature* **260**, 120-124.

Dugaiczyk, A., Woo, S. L. C., Lai, E. C., Mace, M. L., McReynolds, L. and O'Malley, B. W. (1978). The natural ovalbumin gene contains seven intervening sequences. *Nature* **274**, 328-333.

Dugaiczyk, A., Woo, S. L. C., Colbert, D. A., Lai, E. C., Mace, M. L., Jr. and O'Malley, B. W. (1979). The ovalbumin gene: cloning and molecular organization of the entire natural gene. *Proc. Nat. Acad. Sci. USA* **76**, 2253-2257.

Dujon, B., Slonimski, P. P. and Weill, L. (1974). Mitochondrial genetics. IX. A model for recombination and segregation of mitochondrial genomes in S. cerevisiae. *Genetics* **78**, 415-437.

Dujon, B., Bolotin-Fukuhara, M., Coen, D., Deutsch, J., Netter, P., Slonimski, P. P. and Weill, L. (1976). Mitochondrial genetics. XI. Mutations at the mitochondrial locus *w* affecting the recombination of mitochondrial genes in S. cerevisiae. *Mol. Gen. Genet.* **143**, 131-165.

Dunn, A. R. and Hassell, J. A. (1977). A novel method to map transcripts: evidence for homology between an adenovirus mRNA and discrete multiple regions of the viral genome. *Cell* **12**, 23-36.

Dunn, A. R., Matthews, M. B., Chow, L. T., Sambrook, J. and Keller, W. (1978). A supplementary adenoviral leader sequence and its role in messenger translation. *Cell* **15**, 511-526.

DuPraw, E. J. (1965a). Macromolecular organization of nuclei and chromosomes: a folded fiber model based on whole mount electron microscopy. *Nature* **206**, 338-343.

DuPraw, E. J. (1965b). The organization of nuclei and chromosomes in honeybee embryonic cells. *Proc. Nat. Acad. Sci. USA* **53**, 161-165.

DuPraw, E. J. (1966). Evidence for a folded fiber organization in human chromosomes. *Nature* **209**, 577-581.

DuPraw, E. J. (1968). *Cell and Molecular Biology.* Academic Press, New York.

DuPraw, E. J. (1970). *DNA and Chromosomes.* Holt, Rinehart and Winston, New York.

DuPraw, E. J. and Bahr, G. F. (1969). The arrangement of DNA in human chromosomes as investigated by quantitative electron microscopy. *Acta Cytol.* **13**, 188-205.

DuPraw, E. J. and Rae, P. M. M. (1966). Polytene chromosome structure in relation to the folded fiber concept. *Nature* **212**, 598-600.

Dustin, A., Havas, L. and Lits, F. (1937). Action de la colchicine sur les divisions cellulaires chez les vegetaux. *C. R. Assoc. Anat.* **32**, 170-176.

Dutrillaux, B. (1975). Traitements discontinus par le DrdU et coloration par l'acridine orange: obtention de marquages R, Q et intermediaires. *Chromosoma* **52**, 261-273.

Dutrillaux, B. and Lejuene, J. (1971). Sur une nouvelle technique d'analyse du caryotype humain. *C. R. Acad. Sci. Paris* **272**, 2638-2640.

Dworkin, M. B., Rudensey, L. M. and Infante, A. A. (1977). Cytoplasmic nonpolysomal RNP particles in sea urchin embryos and their relationship to protein synthesis. *Proc. Nat. Acad. Sci. USA* **74**, 2231-2235.

Early, P. W., Davis, M. M., Kaback, D. B., Davidson, N. and Hood, L. (1979). Immunoglobulin heavy chain gene organization in mice: analysis of a myeloma genomic clone containing variable and α constant regions, *Proc. Nat. Acad. Sci. USA* **76**, 857-861.

Eccleshall, T. R., Needleman, R. B., Storm, E. M., Buchferer, B. and Marmur, J. (1978). A temperature sensitive yeast mitochondrial mutant with altered cytochrome c oxidase subunit. *Nature* **273**, 67-70.

Edelman, G. M. and Yahara, I. (1976). Temperature sensitive changes in surface modulating assemblies of fibroblasts transformed by mutants of Rous sarcoma virus. *Proc. Nat. Acad. Sci. USA* **73**, 2047-2051.

Edelman, G. M., Yahara, I. and Wang, J. L. (1973). Receptor mobility and receptor cytoplasmic interactions in lymphocytes. *Proc. Nat. Acad. Sci. USA* **70**, 1442-1446.

Eden, F. C. and Hendrick, J. P. (1978). Unusual organization of DNA sequences in the chicken. *Biochemistry* **17**, 5838-5844.

Edenberg, H. J. and Huberman, J. A. (2975). Eucaryotic chromosome replication. *Ann. Rev. Genet.* **9**, 245-284.

Edenberg, H. J., Anderson, S. and DePamphlis, M. L. (1978). Involvement of DNA polymerase α in SV40 DNA replication. *J. Biol. Chem.* **253**, 3273-3280.

Edmonds, M. and Abrams, R. (1960). Polynucleotide biosynthesis: formation of a sequence of adenylate units from ATP by an enzyme from thymus nuclei. *J. Biol. Chem.* **235**, 1142-1149.

Edmonds, M. and Caramela, M. G. (1969). The isolation and characterization of AMP rich polynucleotide synthesized by Ehrlich ascites cells. *J. Biol. Chem.* **244**, 1314-1324.

Edmonds, M., Vaughan, M. H. and Nakazoto, H. (1971). Poly(A) sequences in the hnRNA and rapidly labeled polysomal RNA of HeLa cells: possible evidence for a precursor relationship. *Proc. Nat. Acad. Sci. USA* **68**, 1336-1340.

Edwards, L. J. and Hnilica, L. S. (1968). The specificity of histones in nucleated erythrocytes. *Experientia* **24**, 228-229.

Efstratiadis, A., Kafatos, F. C. and Maniatis, T. (1977). The primary structure of rabbit β globin mRNA as determined from cloned DNA. *Cell* **10**, 571-585.

Efstratiadis, A., Maniatis, T., Kafatos, F. C., Jeffrey, A. and Vournakis, J. N. (1975). Full length and discrete partial reverse transcripts of globin and chorion mRNAs. *Cell* **4**, 367-378.

Efstradiadis, A., Kafatos, F. C., Maxam, A. M. and Maniatis, T. (1976). Enzymatic in vitro synthesis of globin genes. *Cell* **7**, 279-288.

Efstratiadis, A., Crain, E. T., Britten, R. J., Davidson, E. H. and Kafatos, F. C. (1976). DNA sequence organization in the lepidopteran Antheraea permyi. *Proc. Nat. Acad. Sci. USA* **73**, 2289-2293.

Ege, T. and Ringertz, N. R. (1974). Preparation of microcells by enucleation of micronucleate cells. *Exp. Cell Res.* **87**, 378-382.

Ege, T. and Ringertz, N. R. (1975). Viability of cells reconstituted by virus induced fusion of minicells with enuclate cells. *Exp. Cell Res.* **94**, 469-473.

Ege, T., Krondahl, U. and Ringertz, N. R. (1974). Introduction of nuclei and micronuclei into cells and enucleated cytoplasms by Sendai virus induced fusion. *Exp. Cell Res.* **88**, 428-432.

Ege, T., Hamburg, H., Krondahl, U., Ericsson, J. and Ringertz, N. R. (1974). Characterization of minicells (nuclei) obtained by cytochalasin enucleation. *Exp. Cell Res.* **87**, 365-377.

Eicher, E. M. (1970). X-autosome translocation in the mouse: total inactivation versus partial inactivation of the X chromosome. *Adv. Genet.* **15**, 175-259.

Eickbush, T. H. and Moudrianakis, E. N. (1978). The histone core complex: an octamer assembled by two sets of protein-protein interactions. *Biochemistry* **17**, 4955-4964.

Eigsti, O. (1938). A cytological study of colchicine effects in the induction of polyploidy in plants. *Proc. Nat. Acad. Sci. USA* **24**, 56-63.

Eigsti, O. J. and Dustin, P., Jr. (1955). *Colchicine—In Agriculture, Medicine, Biology and Chemistry.* Iowa State College Press, Ames, Iowa.

Elgin, S. C. R. and Bonner, J. (1970). Limited heterogeneity at the major nonhistone chromosome proteins. *Biochemistry* **9**, 4440-4448.

Elgin, S. C. R. and Bonner, J. (1972). Partial fractionation and chemical characterization of the major nonhistone chromosomal proteins. *Biochemistry* **11**, 772-781.

Elgin, S. C. R. and Hood, L. E. (1973). Chromosomal proteins of Drosophila embryos. *Biochemistry* **12**, 4984-4992.

Elgin, S. C. R., Boyd, J. B., Hood, L. E., Wray, W. and Wu, F. C. (1973). A prologue to the study of the nonhistone chromosomal proteins. *Cold Spring Harbor Symp. Quant. Biol.* **38**, 821-833.

Eliceiri, G. L. (1974). Short lived, small RNAs in the cytoplasm of HeLa cells. *Cell* **3**, 11-14.

Eliceiri, G. L. and Green, H. (1969). Ribosomal RNA synthesis in human-mouse hybrid cells. *J. Mol. Biol.* **41**, 253-260.

Ellgaard, E. G. and Clever, U. (1971). RNA metabolism during puff induction in D. melanogaster. *Chromosoma* **36**, 60-78.

Ellison, J. R. and Barr, H. J. (1972). Quinacrine fluorescence of specific chromosome regions. Late replication and high A:T content in Samoaia leonensis. *Chromosoma* **36**, 375-390.

Elsevier, S. M., Kucherlapati, R. S., Nichols, E. A., Creagan, R. P., Giles, R. E., Ruddle, F. H., Willecke, K. and McDougall, J. K. (1974). Assignment of the gene for galactokinase to human chromosome 17 and its regional localization to band q21-22. *Nature* **251**, 633-636.

Elzinga, M., Maron, B. J. and Adelstein, R. S. (1976). Human heart and platelet actins are products of different genes. *Science* **191**, 94-95.

Emerson, S. and Beadle, G. W. (1933). Crossing over near the spindle fiber in attached X chromosomes of D. melanogaster. *Z. VerebLehre* **45**, 129-140.

Endow, S. A. and Gall, J. G. (1975). Differential replication of satellite DNA in polyploid tissues of D. virilis. *Chromosoma* **50**, 175-192.

Endow, S. A., Polan, M. L. and Gall, J. G. (1975). Satellite DNA sequences of D. melanogaster. *J. Mol. Biol.* **96**, 665-692.

Engberg, J. and Pearlman, R. (1972). The amount of rRNA genes in Tetrahymena pyriformis in different physiological states. *Eur. J. Biochem.* **26**, 393-400.

Engberg, J., Andersson, P., Leick, V. and Collins, J. (1976). Free ribosomal DNA molecules from Tetrahymena pyriformis GL are giant palindromes. *J. Mol. Biol.* **104**, 455-470.

Englesberg, E., Bass, R. and Heiser, W. (1976). Inhibition of the growth of mammalian cells in culture by amino acids and the isolation and characterization of L phenylalanine resistant mutants modifying L phenylalanine transport. *Somatic Cell Genet.* **2**, 411-428.

Ensinger, M. J. and Moss, B. (1976). Modification of the 5′ terminus of mRNA by an RNA (guanine-7) methyltransferase from HeLa cells. *J. Biol. Chem.* **251**, 5283-5291.

Ephrussi, B. and Grandchamp, S. (1965). Etudes sur la suppressivite des mutants a deficience respiratoire de la levure. I. Existence au niveau cellulaire de divers "digres de suppressivite." *Heredity* **20**, 1-7.

Ephrussi, B., Hottinguer, H. and Chimenes, A. M. (1949a). Action de l'acriflavine sur les levures. I. Le mutation "petite colonie." *Ann. Inst. Past.* **76**, 351-367.

Ephrussi, B., Hottinguer, H. and Roman, H. (1955). Suppressiveness: a new factor in the genetic determination of the synthesis of respiratory enzymes in yeast. *Proc. Nat. Acad. Sci. USA* **41**, 1065-1071.

Ephrussi, B., Hottinguer, H., and Tavlitzki, J. (1949b). Action de l'acriflavine sur les levures. II. Etudes genetique du mutant "petite colonie." *Ann. Inst. Past.* **76**, 419-450.

Ephrussi, B., Jakob, H. and Grandchamp, S. (1966). Etudes sur la suppressivite des mutants a deficience respiratoire de la levure. II. Etapes de la mutation grande en petite provoquee par le facteur suppressif. *Genetics* **54**, 1-2.

Epplen, J. T., Siebers, J. W. and Vogel, W. (1975). DNA replication patterns of human chromosomes from fibroblasts and amniotic fluid cells revealed by a Giemsa staining technique. *Cytogenet. Cell Genet.* **15**, 177-178.

Epstein, C. J. (1969). Mammalian oocytes: X chromosome activity. *Science* **163**, 1078-1079.

Epstein, C. J. (1972). Expression of the mammalian X chromosome before and after fertilization. *Science* **175**, 1467-1468.

Erickson, H. P. (1974). Microtubule surface lattice and subunit structure and observations on reassembly. *J. Cell Biol.* **60**, 153-167.

Erickson, H. P. (1975). The structure and assembly of microtubules. *Ann. N. Y. Acad. Sci.* **253**, 60-77.

Erickson, H. P. and Voter, W. A. (1976). Polycation induced assembly of purified tubulin. *Proc. Nat. Acad. Sci. USA* **73**, 2813-2817.

Erickson, J. (1965). Meiotic drive in Drosophila involving chromosome breakage. *Genetics* **61**, 557-571.

Ernst, S. G., Britten, R. J. and Davidson, E. H. (1979). Distinct single copy sequence sets in sea urchin nuclear RNAs. *Proc. Nat. Acad. Sci. USA* **76**, 2209-2212.

Estensen, R. D. and Plagemann, P. G. W. (1972). Cytochalasin B: inhibition of glucose and glucosamine transports. *Proc. Nat. Acad. Sci. USA* **69**, 1430-1434.

Etcheverry, T., Colby, D. and Guthrie, C. (1979). A precursor to a minor species of yeast tRNA[Ser] contains an intervening sequence. *Cell* **18**, 11-26.

Evans, D. and Birnstiel, M. L. (1968). Localization of amplified rDNA in the oocyte of X. laevis. *Biochim. Biophys. Acta* **166**, 274-276.

Evans, H. J. (1964). Uptake of ^{3}H thymidine and patterns of replication in nuclei and chromosomes of Vicia faba. *Exp. Cell Res.* **35**, 381-393.

Evans, H. J., Buckland, R. A. and Pardue, M. L. (1974). Location of the genes coding for 18S and 28S rRNA in the human genome. *Chromosoma* **48**, 405-426.

Evans, H. J., Ford, C., Lyon, M. and Gray, J. (1965). DNA replication and genetic expression in female mice with morphologically distinguishable X chromosomes. *Nature* **206**, 900-903.

Evans, M. J. and Lingrel, J. B. (1969a). Hemoglobin mRNA. Distribution of the 9S RNA in polysomes of different sizes. *Biochemistry* **8**, 829-831.

Evans, M. J. and Lingrel, J. B. (1969b). Hemoglobin mRNA. Synthesis of 9S and rRNA during erythroid development. *Biochemistry* **8**, 3000-3005.

Evans, R. M., Fraser, N., Ziff, E., Weber, J., Wilson, M. and Darnell, J. E. (1977). The initiation site for RNA transcription in Ad2 DNA. *Cell* **12**, 733-739.

Evans, R., Weber, J., Ziff, E. and Darnell, J. E. (1979). Premature termination during adenovirus transcription. *Nature* **278**, 367-370.

Faiferman, I. and Pogo, A. O. (1975). Isolation of a nuclear RNP network that contains hnRNA and is bound to the nuclear envelope. *Biochemistry* **14**, 3808-3816.

Fambrough, D. M. and Bonner, J. (1966). On the similarity of plant and animal histones. *Biochemistry* **5**, 2563-2570.

Fan, H. and Penman, S. (1970a). Mitochondrial RNA synthesis during mitosis. *Science* **168**, 135-138.

Fan, H. and Penman, S. (1970b). Regulation of protein synthesis in mammalian cells. II. Inhibition of protein synthesis at the level of initiation during mitosis. *J. Mol. Biol.* **50**, 655-670.

Fan, H. and Penman, S. (1971). Regulation of synthesis and processing of nucleolar components in metaphase arrested cells. *J. Mol. Biol.* **59**, 27–42.

Fansler, B. and Loeb, L. A. (1972). Sea urchin nuclear DNA polymerase. IV. Reversible association of DNA polymerase with nuclei during the cell cycle. *Exp. Cell Res.* **75**, 433–441.

Fansler, B. S., Travaglini, E. C., Loeb, L. A. and Schultz, J. (1970). Structure of D. melanogaster dAT replicated in an in vitro system. *Biochem. Biophys. Res. Commun.* **40**, 1266–1272.

Farace, M. G., Aellen, M. F., Briand, P. A., Faust, C. H., Vassalli, P. and Mach, B. (1976). No detectable reiteration of genes coding for mouse MOPC41 immunoglobulin light chain mRNA. *Proc. Nat. Acad. Sci. USA* **73**, 727–731.

Farber, R. A. and Davidson, R. L. (1978). Altered pattern of replication of human chromosomes in a human fibroblast mouse cell hybrids. *Proc. Nat. Acad. Sci. USA* **75**, 1470–1474.

Farrell, K. W. and Wilson, L. (1978). Microtubule reassembly in vitro of S. purpuratus sperm tail outer doublet tubulin. *J. Mol. Biol.* **121**, 393–410.

Farrell, K. W., Morse, A. and Wilson, L. (1979). Characterization of the in vitro reassembly of tubulin derived from stable S purpuratus outer doublet microtubules. *Biochemistry* **18**, 905–910.

Farrell, S. A. and Worton, R. G. (1977). Chromosome loss is responsible for segregation at the HGPRT locus in Chinese hamster cell hybrids. *Somat. Cell Genet.* **3**, 539–550.

Fauron, C. M. R. and Wolstenholme, D. R. (1976). Structural heterogeneity of mitochondrial DNA molecules within the genus Drosophila. *Proc. Nat. Acad. Sci. USA* **73**, 3623–3627.

Faust, C. H., Diggelman, H. and Mach, B. (1974). Estimation of the number of genes coding for the constant part of the mouse immunoglobulin kappa light chain. *Proc. Nat. Acad. Sci. USA* **71**, 2491–2495.

Faust, C. H., Heim, I. and Moore, J. (1979). Murine myeloma Ig heavy chain mRNA. Isolation, partial purification and characterization of γ1, γ2a, γ2b, γ3, μ and α heavy chain mRNAs. *Biochemistry* **18**, 1106–1119.

Faust, M., Millward, S., Duchastel, A. and Fromson, D. (1976). Methylated constituents of poly(A)$^-$ and poly(A)$^+$ polyribosomal RNA of sea urchin embryos. *Cell* **9**, 597–604.

Fawcett, D. W. (1956). The fine structure of chromosomes in the meiotic prophase of vertebrate spermatocytes. *J. Biochem. Biophys. Cytol.* **2**, 403–406.

Faye, G. and Sor, F. (1977). Analysis of mitochondrial ribosome proteins of S. cerevisiae by two dimensional gel electrophoresis. *Mol. Gen. Genet.* **155**, 27–34.

Faye, G., Kujiwa, C. and Fukuhara, H. (1974). Physical and genetic organization of petite and grande yeast mitochondrial DNA. IV. In vivo transcription products of mitochondrial DNA and localization of 23S rRNA in petite mutants of S. cerevisiae. *J. Mol. Biol.* **88**, 185–203.

Faye, G., Fukuhara, H., Grandchamp, C., Lazowska, J., Michel, F., Casey, J., Getz, G. S., Locker, J., Rabinowitz, M., Bolotin-Fukuhara, M., Coen, D., Deutsch, J., Dujon, B., Netter, P. and Slonimski, P. P. (1973). Mitochondrial nucleic acids in the petite colonie mutants: deletions and repetitions of genes. *Biochimie* **55**, 779–792.

Faye, G., Kujawa, C., Dujon, B., Bolotin-Fukuhara, M., Wolfe, K., Fukuhara, H. and Slonimski, P. P. (1975). Localization of the gene coding for the mitochondrial 16S rRNA using rho$^-$ mutants of S. cerevisiae. *J. Mol. Biol.* **99**, 203–217.

Faye, G., Dennebouy, N., Kujiwara, C. and Jacq, C. (1979). Inserted sequence in the mitochondrial 23S rRNA gene of the yeast S. cerevisiae. *Mol. Gen. Genet.* **168**, 101–110.

Federoff, N. V. (1979a). Deletion mutants of X. laevis 5S rDNA. *Cell* **16**, 551–564.

Federoff, N. V. (1979b). On spacers. *Cell* **16**, 697–710.

Federoff, N. V. and Brown, D. D. (1978). The nucleotide sequence of oocyte 5S DNA in X. laevis. I. the AT rich spacer. *Cell* **13**, 701–716.

Federoff, N., Wellauer, P. K. and Wall, R. (1977). Intermolecular duplexes in hnRNA from HeLa cells. *Cell* **10**, 597–610.

Feit, H., Slusarek, L. and Shelanski, M. L. (1971). Heterogeneity of tubulin subunits. *Proc. Nat. Acad. Sci. USA* **68**, 2028-2031.

Fenwick, R. G. and Caskey, C. T. (1975). Mutant Chinese hamster cells with a thermosensitive HGPRT. *Cell* **5**, 115-122.

Fenwick, R. G., Sawyer, T. H., Kruh, G. D., Astrin, K. H. and Caskey, C. T. (1977). Forward and reverse mutations affecting the kinetics and apparent molecular weight of mammalian HGPRT. *Cell* **12**, 383-391.

Ferrus, A. (1975). Parameters of mitotic recombination in minute mutants of D. melanogaster. *Genetics* **79**, 589-599.

Feulgen, R. and Rossenbeck, H. (1924). Mikroskopisch chemischer nachweis einer nucleinsaure vom typus der thymonucleinsaure und die darauf beruhende elektive farburg von zellkernen in mikroskopischen praparaten. *Hoppe Seyler's Z. Physiol. Chem.* **135**, 203-248.

Fey, G. and Hirt, B. (1974). Fingerprints of polyoma virus proteins and mouse histones. *Cold Spring Harbor Symp. Quant. Biol.* **39**, 235-241.

Fiers, W., Contrearas, R., Haegeman, G., Rogiers, R., Van de Voorde, A., Van Heuverswyn, H., Van Heereweghe, J., Volckaertand, G. and Ysebaer, M. (1978). Complete nucleotide sequence of SV40 DNA. *Nature* **273**, 113-118.

Filipowicz, W., Furuichi, Y., Sierra, J. M., Muthukrishnan, S., Shatkin, A. J. and Ochoa, S. (1976). A protein binding the methylated 5′ terminal sequence, m⁷GpppN, of eucaryotic mRNA. *Proc. Nat. Acad. Sci. USA* **73**, 1559-1563.

Filipski, J., Thiery, J. -P. and Bernardi, G. (1973). An analysis of the bovine genome by Cs_2SO_4-Ag^+ density gradient centrifugation. *J. Mol. Biol.* **80**, 177-197.

Finch, J. T. and Klug, A. (1976). Solenoidal model for superstructure in chromatin. *Proc. Nat. Acad. Sci. USA* **73**, 1897-1901.

Finch, J. T., Noll, M. and Kornberg, R. D. (1975). Electron microscopy of defined lengths of chromatin. *Proc. Nat. Acad. Sci. USA* **72**, 3320-3322.

Finch, J. T., Lutter, L. C., Rhodes, D., Brown, R. S., Rushton, B., Levitt, M. and Klug, A. (1977). Structure of nucleosome core particles of chromatin. *Nature* **269**, 29-35.

Findly, R. C. and Gall, J. G. (1978). Free ribosomal RNA genes in Paramecium are tandemly repeated. *Proc. Nat. Acad. Sci. USA* **75**, 3312-3316.

Fine, R. and Blitz, A. (1975). A chemical comparison of tropomyosins from muscle and nonmuscle tissues. *J. Mol. Biol.* **95**, 447-454.

Fine, R. E. and Bray, D. (1971). Actin in growing nerve cells. *Nature New Biol.* **234**, 115-118.

Fine, R., Blitz, A., Hitchcock, S. and Kaminer, B. (1973). Tropomyosin in brain and growing neurons. *Nature New Biol.* **245**, 182-185.

Firtel, R. A. and Lodish, H. F. (1973). A small nuclear precursor of messenger RNA in the cellular slime mold Dictyostelium discoideum. *J. Mol. Biol.* **79**, 295-314.

Firtel, R. A. and Pederson, T. (1975). Ribonucleoprotein particles containing hnRNA in the cellular slime mold D. discoideum. *Proc. Nat. Acad. Sci. USA* **72**, 301-305.

Firtel, R. A., Cockburn, A., Frankel, G. and Hershfield, V. (1976). Structural organization of the genome of D. discoideum: analysis by EcoRI restriction endonuclease. *J. Mol. Biol.* **102**, 831-852.

Fisher, P. A. and Korn, D. (1977). DNA polymerase α. Purification and structural characterization of the near homogeneous enzyme from human KB cells. *J. Biol. Chem.* **252**, 6528-6535.

Fisher, P. A., Wang, S. F. and Korn, D. (1979). Enzymological characterization of DNA polymerase α. Basic catalytic properties, processivity, and gap utilization of the homogeneous enzyme from human KB cells. *J. Biol. Chem.* **254**, 6128-6135.

Firtel, R. A. and Bonner, J. (1972). Characterization of the genome of the cellular slime mold D. discoideum. *J. Mol. Biol.* **66**, 339-361.

Firtel, R. A. and Lodish, H. F. (1973). A small nuclear precursor of mRNA in the cellular slime mold D. discoideum. *J. Mol. Biol.* **79**, 295-314.

Firtel, R. A. and Kindle, K. (1975). Structural organization of the genome of the cellular slime mold D. discoideum: interspersion of repetitive and single copy DNA sequences. *Cell* 5, 401-412.

Fitch, W. M. (1973). Aspects of molecular evolution. *Ann. Rev. Genet.* 7, 343-380.

Fittler, F. (1977). Analysis of the alpha satellite DNA from African green monkey cells by restriction nucleases. *Eur. J. Biochem.* 74, 343-352.

Fitz-James, P. C. and Young, I. E. (1959). Comparison of species and varieties of the genus Bacillus. *J. Bacteriol.* 78, 743-754.

Flamm, W. G., Bernheim, N. J. and Brubacker, P. E. (1971). Density gradient analysis of newly replicated DNA from synchronized mouse lymphoma cells. *Exp. Cell Res.* 64, 97-104.

Flamm, W. G., Walker, P. M. B. and McCallum, M. (1969). Some properties of the single strands isolated from the DNA of the nuclear satellite of the mouse (Mus musculus). *J. Mol. Biol.* 40, 423-443.

Flavell, A. J., Cowie, A., Legon, S. and Kamen, R. (1979). Multiple 5' terminal cap structures in late polyoma virus RNA. *Cell* 16, 357-373.

Flavell, R. A. and Jones, I. G. (1971). DNA from isolated pellicles of Tetrahymena. *J. Cell Sci.* 9, 719-726.

Flavell, R. A., Kooter, J. M., De Boer, E., Little, P. F. and Williamson, R. (1978). Analysis of the β-δ globin gene loci in normal and HbLepore: direct determination of gene linkage and intergene distance. *Cell* 15, 25-42.

Flavell, R. A., Bernards, R., Kooter, J. M., De Boer, E., Little, P. F. R., Annison, G. and Williamson, R. (1979). The structure of the human β globin gene in β thalassemia. *Nucl. Acids Res.* 6, 2749-2760.

Flemming, W. (1882). Zellsubstanz, Zern and Zelltheilung. Vogel, Leipzig.

Fletcher, J. M. (1979). Light microscope analysis of meiotic prophase chromosomes by silver staining. *Chromosoma* 72, 241-248.

Flint, S. J. and Weintraub, H. M. (1977). An altered subunit configuration associated with the actively transcribed DNA of integrated adenovirus genes. *Cell* 12, 783-794.

Flint, S. J., De Pomerai, D. I., Chesterton, J. and Butterworth, P. H. W. (1974). Template specificity of eucaryotic DNA dependent RNA polymerases. *Eur. J. Biochem.* 42, 567-579.

Flintoff, W. F., Davidson, S. V. and Siminovitch, L. (1976). Isolation and partial characterization of three methotrexate resistant phenotypes from CHO cells. *Somatic Cell Genet.* 2, 245-262.

Foe, V. E., Wilkinson, L. E. and Laird, C. D. (1976). Comparative organization of active transcription units in Oncopeltus fasciatus. *Cell* 9, 131-146.

Ford, D. K. and Yerganian, G. (1958). Observations on the chromosomes of Chinese hamster cells in tissue culture. *J. Nat. Cancer Inst.* 21, 393-424.

Ford, P. J. (1971). Non coordinated accumulation and synthesis of 5S RNA by ovaries of X. laevis. *Nature* 233, 561-564.

Ford, P. J. and Brown, R. D. (1976). Sequences of 5S rRNA from X. mulleri and the evolution of 5S coding sequences. *Cell* 8, 485-493.

Ford, P. J. and Mathieson, T. (1978). The nucleotide sequences of 5.8S rRNA from X. laevis and X. borealis. *Eur. J. Biochem.* 87, 199-213.

Ford, P. J. and Southern, E. M. (1973). Different sequences for 5S RNA in kidney cells and ovaries of X. laevis. *Nature New Biol.* 241, 7-12.

Forer, A. (1966). Characterization of the mitotic traction system and evidence that birefringent spindle fibers neither produce nor transmit force for chromosome movement. *Chromosoma* 19, 44-98.

Forer, A. (1969). Chromosome movements during cell division. In A. Lima-de-Faria (Ed.) *Handbook of Molecular Cytology*. North Holland, Amsterdam, pp 553-604.

Forer, A. and Behnke, O. (1972a). An actin-like component in spermatocytes of a crane fly. The spindle. *Chromosoma* 39, 145-173.

Forer, A. and Behnke, O. (1972b). An actin-like component in spermatocytes of a crane fly. The cell cortex. *Chromosoma* **39**, 175–190.

Forer, A. and Zimmerman, A. M. (1976a). Spindle birefringence of isolated mitotic apparatus analyzed by pressure treatment. *J. Cell Sci.* **20**, 309–328.

Forer, A. and Zimmerman, A. M. (1976b). Spindle birefringence of isolated mitotic apparatus analyzed by treatments with cold, pressure, and diluted isolation medium. *J. Cell Sci.* **20**, 329–340.

Forer, A., Kalnins, V. I. and Zimmerman, A. M. (1976). Spindle birefringence of isolated mitotic apparatus: further evidence for two birefringent spindle components. *J. Cell Sci.* **22**, 115–132.

Forget, B. G., Benz, E. J., Skoultchi, A., Baglioni, C. and Houseman, D. (1974). Absence of mRNA for β globin chains in β° thalassemia. *Nature* **247**, 379–381.

Forget, B. G., Hillman, D. G., Lazarus, H., Barell, E. F., Benz, E. J., Jr., Caskey, C. T., Huisman, T. H. J., Schroeder, W. A. and Housman, D. (1976). Absence of mRNA and gene DNA for β globin chains in hereditary persistence of fetal hemoglobin. *Cell* **7**, 323–330.

Forsheit, A. B., Davidson, N. and Brown, D. (1974). An electron microscope heteroduplex study of the rDNAs of X. laevis and X. mulleri. *J. Mol. Biol.* **90**, 301–314.

Foster, D. N. and Gurney, T., Jr. (1976). Nuclear location of mammalian DNA polymerase activities. *J. Biol. Chem.* **251**, 7893–7898.

Fougére, C. and Weiss, M. C. (1978). Phenotypic exclusion in mouse melanoma-rat hepatoma hybrid cells: pigment and albumin production are not re expressed simultaneously. *Cell* **15**, 843–854.

Fougére, C., Ruiz, F. and Ephrussi, B. (1972). Gene dosage dependence of pigment synthesis in melanoma × fibroblast hybrids. *Proc. Nat. Acad. Sci. USA* **69**, 330–334.

Fouquet, H., Bierweiler, B. and Sauer, H. W. (1974). Reassociation kinetics of nuclear DNA from Physarum polycephalum. *Eur. J. Biochem.* **44**, 407–410.

Fournier, R. E. K. and Ruddle, F. H. (1977a). Microcell mediated transfer of murine chromosomes into mouse, Chinese hamster, and human somatic cells. *Proc. Nat. Acad. Sci. USA* **74**, 319–323.

Fournier, R. E. K. and Ruddle, F. H. (1977b). Stable association of the human transgenome and host murine chromosomes demonstrated with trispecific microcell hybrids. *Proc. Nat. Acad. Sci. USA* **74**, 3937–3941.

Foury, F. and Tzagoloff, A. (1978). Assembly of the mitochondrial membrane system. Genetic complementation of *mit⁻* mutations in mitochondrial DNA of S. cerevisiae. *J. Biol. Chem.* **253**, 3792–3797.

Fox, D. P. (1973). The control of chiasma distribution in the locust Schistocerca gregaria. *Chromosoma* **43**, 289–328.

Francke, U., Denney, R. M. and Ruddle, F. H. (1977). Intrachromosomal gene mapping in man: the gene for tryptophanyl-tRNA synthetase maps in region q21-ter of chromosome 14. *Somat. Cell Genet.* **3**, 381–390.

Frankel, F. R. (1976). Organization and energy dependent growth of microtubules in cells. *Proc. Nat. Acad. Sci. USA* **73**, 2798–2802.

Frankel, G., Cockburn, A. F., Kindle, K. L. and Firtel, R. A. (1977). Organization of the rRNA genes of D. discoideum. Mapping of the transcribed region. *J. Mol. Biol.* **109**, 539–558.

Fraser, J. M. K. and Huberman, J. A. (1977). In vitro HeLa cell DNA synthesis similarity to in vivo replication. *J. Mol. Biol.* **117**, 249–272.

Fraser, N. W., Sehgal, P. B. and Darnell, J. E. (1979). Multiple discrete sites for premature RNA chain termination late in adenovirus 2 infection: enhancement by 5,6 dichloro-1-β-ribofuranosyl-benzimidazole. *Proc. Nat. Acad. Sci. USA* **76**, 2571–2575.

Fraser, N. W., Nevins, J. R., Ziff, E. and Darnell, J. E. (1979). The major late Ad2 transcription unit: termination is downstream from the last poly(A) site. *J. Mol. Biol.* **129**, 643–656.

Fraser, T. H. and Bruce, B. J. (1978). Chicken ovalbumin is synthesized and secreted by E. coli. *Proc. Nat. Acad. Sci. USA* **75**, 5936–5940.

Free, S. J., Rice, P. W. and Metzenberg, R. L. (1979). Arrangement of the genes coding for rRNAs in N. crassa. *J. Bacteriol.* **137**, 1219-1226.

Freese, E. (1958). The arrangement of DNA in the chromosome. *Cold Spring Harbor Symp. Quant. Biol.* **23**, 13-18.

Frenster, J. H., Allfrey, V. G. and Mirsky, A. E. (1963). Repressed and active chromatin isolated from interphase lymphocytes. *Proc. Nat. Acad. Sci. USA* **50**, 1026-1032.

Friderici, K., Kaehler, M. and Rotman, F. (1976). Kinetics of Novikoff cytoplasmic mRNA methylation. *Biochemistry* **15**, 5234-5240.

Fridlender, B., Fry, M., Bolden, A. and Weissbach, A. (1972). A new synthetic RNA dependent DNA polymerase from human tissue culture cells. *Proc. Nat. Acad. Sci. USA* **69**, 452-455.

Friedman, D. L. (1970). DNA polymerase from HeLa cell nuclei: levels of activity during a synchronized cell cycle. *Biochem. Biophys. Res. Commun.* **39**, 100-109.

Friedmann, I., Colwin, A. L. and Colwin, L. H. (1968). Fine structural aspects of fertilization in C. reinhardii. *J. Cell Sci.* **13**, 115-128.

Friedrich, U. and Coffino, P. (1977). Mutagenesis in S49 mouse lymphoma cells: induction of resistance to ouabain, 6-thioguanine and dbcAMP. *Proc. Nat. Acad. Sci. USA* **74**, 679-683.

Friend, C., Patuleia, M. C. and DeHarven, E. (1966). Erythrocytic maturation in vitro of murine (Friend) virus induced leukemic cells. *Nat. Cancer Inst. Monogr.* **22**, 505-520.

Friesen, H. (1936). Spermatogoniales crossing over bei Drosophila. *Z. Indukt. Abstamm. VerebLehre.* **71**, 501-526.

Frigon, R. P. and Timasheff, S. N. (1975a). Magnesium induced self association of calf brain tubulin. I. Stoichiometry. *Biochemistry* **14**, 4559-4566.

Frigon, R. P. and Timasheff, S. N. (1975b). Magnesium induced self association of calf brain tubulin. II. Thermodynamics. *Biochemistry* **14**, 4567-4573.

Fritsch, E. F., Lawn, R. M. and Maniatis, T. (1979). Characterization of deletions which affect the expression of fetal globin genes in man. *Nature* **279**, 598-603.

Frost, J. N. (1961). Autosomal nondisjunction in males of D. melanogaster. *Genetics* **46**, 39-54.

Fujiwara, K. and Pollard, T. D. (1976). Fluorescent antibody localization of myosin in the cytoplasm, cleavage furrow and mitotic spindle of human cells. *J. Cell Biol.* **71**, 848-875.

Fry, K. and Salser, W. (1977). Nucleotide sequences of HSα satellite DNA from kangaroo rat Dipdomys ordii and characterization of similar sequences in other rodents. *Cell* **12**, 1069-1084.

Fry, K., Poon, R., Whitcome, P., Idriss, J., Salser, W., Mazrimas, J. and Hatch, F. (1973). Nucleotide sequence of HS-β satellite DNA from kangaroo rat Dipodomya ordii. *Proc. Nat. Acad. Sci. USA* **70**, 2642-2646.

Fuchs, E. and Green, H. (1979). Multiple keratins of cultured human epidermal cells are translated from different mRNA molecules. *Cell* **17**, 573-582.

Fujiwara, K. and Tilney, L. G. (1975). Substructural analysis of the microtubule and its polymorphic forms. *Ann. N. Y. Acad. Sci.* **253**, 27-50.

Fujiwara, Y. (1967). Role of RNA synthesis in DNA replication of synchronized populations of cultured mammalian cells. *J. Cell. Physiol.* **70**, 291-300.

Fukuhara, H., Moustacchi, E. and Wesolowski, M. (1978). Preferential deletion of a specific region of mitochondrial DNA in S. cerevisiae by ethidium bromide and 3 carbethoxy psoralen. Directional retention of DNA sequences. *Molec. Gen. Genet.* **162**, 191-202.

Fuller, G. M., Brinkley, B. R. and Boughter, J. M. (1975). Immunofluorescence of mitotic spindles by using monospecific antibody against bovine brain tubulin. *Science* **187**, 948-950.

Fulton, C., Kane, R. E. and Stephens, R. E. (1971). Serological similarity of flagellar and mitotic microtubules. *J. Cell Biol.* **50**, 762-773.

Funderud, S., Andreassen, R. and Haugli, F. (1979). DNA replication in Physarum polycephalum: electron microscopic and autoradiographic analysis of replicating DNA from defined stages of the S period. *Nucleic Acids Res.* **6**, 1417-1432.

Furuichi, Y. (1978). Posttranscriptional capping in the biosynthesis of cytoplasmic polyhedrosis virus mRNA. *Proc. Nat. Acad. Sci. USA* **75**, 1086-1090.

Furuichi, Y., LaFiandra, A. and Shatkin, A. J. (1977). 5′ terminal structure and mRNA stability. *Nature* **266**, 235-238.

Furuichi, Y., Morgan, M., Muthukrishnan, S. and Shatkin, A. J. (1975a). Reovirus mRNA contains a methylated, blocked 5′ terminal structure: m⁷G(5′)ppp(5′)GᵐpCp-. *Proc. Nat. Acad. Sci. USA* **72**, 362-366.

Furuichi, Y., Muthukrishnan, S. and Shatkin, A. J. (1975b). 5′ terminal m⁷G(5′)ppp(5′)Gᵐp in vivo: identification in reovirus genome RNA. *Proc. Nat. Acad. Sci. USA* **72**, 742-745.

Furuichi, Y., Morgan, M., Shatkin, A. J., Jelinek, W., Salditt-Georgieff, M. and Darnell, J. E. (1975c). Methylated blocked 5′ termini in HeLa cell mRNA. *Proc. Nat. Acad. Sci. USA* **72**, 1904-1908.

Furuichi, Y., Muthukrishnan, S., Tomasz, J. and Shatkin, A. J. (1976). Mechanism of formation of reovirus mRNA 5′ terminal blocked and methylated sequence, m⁷GpppGᵐpC. *J. Biol. Chem.* **251**, 5043-5053.

Gaddipati, J. P. and Sen, S. K. (1978). DNA replication studies in genus Vicia through fiber autoradiography. *J. Cell Sci.* **29**, 85-92.

Gadski, R. A. and Chae, C. -B. (1977). Effect of proteolysis on transcriptional fidelity of reconstituted chromatin. *Biochemistry* **16**, 3465-3469.

Gage, L. P. (1974). The Bombyx mori genome: analysis of DNA reassociation kinetics. *Chromosoma* **45**, 27-42.

Gage, L. P. and Manning, R. F. (1976). Determination of the multiplicity of the silk fibroin gene and detection of fibroin gene-related DNA in the genome of Bombyx mori. *J. Mol. Biol.* **101**, 327-348.

Gagnon, J., Palmiter, R. D. and Walsh, K. A. (1978). Comparison of the NH₂ terminal sequence of ovalbumin as synthesized in vitro and in vivo. *J. Biol. Chem.* **253**, 7464-7468.

Galau, G. A., Britten, R. J. and Davidson, E. H. (1974). A measurement of the sequence complexity of polysomal mRNA in sea urchin embryos. *Cell* **2**, 9-22.

Galau, G. A., Klein, W. H., Davis, M. M., Wold, B. J., Britten, R. J. and Davidson, E. H. (1976). Structural gene sets active in embryos and adult tissues of the sea urchin. *Cell* **7**, 487-506.

Galau, G. A., Britten, R. J. and Davidson, E. H. (1977a). Studies on nucleic acid reassociation kinetics: rate of hybridization of excess RNA with DNA, compared to the rate of DNA renaturation. *Proc. Nat. Acad. Sci. USA* **74**, 1020-1023.

Galau, G. A., Smith, M. J., Britten, R. J. and Davidson, E. H. (1977b). Studies on nucleic acid reassociation kinetics: retarded rate of hybridization of RNA with excess DNA. *Proc. Nat. Acad. Sci. USA* **74**, 2306-2310.

Galau, G. A., Lipson, E. D., Britten, R. J. and Davidson, E. H. (1977c). Synthesis and turnover of polysomal mRNAs in sea urchin embryos. *Cell* **10**, 415-432.

Gall, J. G. (1954). Lampbrush chromosomes from oocyte nuclei of the newt. *J. Morphol.* **94**, 283-329.

Gall, J. G. (1959). Macromolecular duplication in the ciliated protozoan Euplotes. *J. Biochem. Biophys. Cytol.* **5**, 295-308.

Gall, J. G. (1963). Kinetics of DNAase action on chromosomes. *Nature* **198**, 36-38.

Gall, J. G. (1966). Microtubule fine structure. *J. Cell Biol.* **31**, 639-643.

Gall, J. G. (1968). Differential synthesis of the genes for rRNA during amphibian oogenesis. *Proc. Nat. Acad. Sci. USA* **60**, 553-560.

Gall, J. G. (1978). Early studies on gene amplification. *Harvey Lectures* **71**, 55-70. (Academic Press, New York.)

Gall, J. G. and Atherton, D. D. (1974). Satellite DNA sequences in Drosophila virilis. *J. Mol. Biol.* **85**, 633-664.

Gall, J. G. and Callan, H. G. (1962). ³H uridine incorporation in lampbrush chromosomes. *Proc. Nat. Acad. Sci. USA* **48**, 562–570.

Gall, J. G. and Pardue, M. L. (1969). Formation and detection of RNA-DNA hybrid molecules in cytological preparations. *Proc. Nat. Acad. Sci. USA* **63**, 378–383.

Gall, J. G., Cohen, E. H. and Polan, M. L. (1971). Repetitive DNA sequences in Drosophila. *Chromosoma* **33**, 319–344.

Gallwitz, D. and Mueller, G. C. (1969). Histone synthesis in vitro on HeLa cell microsomes. The nature of the coupling to DNA synthesis. *J. Biol. Chem.* **244**, 5947–5952.

Ganetzky, B. (1977). On the components of segregation distortion in D. melanogaster. *Genetics* **86**, 321–355.

Ganner, E. and Evans, H. J. (1971). The relationship between patterns of DNA replication and of quinacrine fluorescence in the human chromosome complement. *Chromosoma* **35**, 326–341.

Gannon, F., O'Hare, K., Perrin, F., LePennec, J. P., Benoist, C., Cochet, M., Breathnach, R., Royal, A., Garapin, A., Cami, B. and Chambon, P. (1979). Organization and sequences at the 5′ end of a cloned complete ovalbumin gene. *Nature* **278**, 428–434.

Garapin, A. C., LePennec, J. P., Roskam, W., Perrin, F., Cami, B., Krust, A., Breathnach, R., Chambon, P. and Kourilsky, P. (1978a). Isolation by molecular cloning of a fragment of the split ovalbumin gene. *Nature* **273**, 349–353.

Garapin, A. C., Cami, B., Roskam, W., Kourilsky, P., LePennec, J. P., Perrin, F., Gerlinger, P., Cochet, M. and Chambon, M. (1978b). Electron microscopy and restriction enzyme mapping reveal additional intervening sequences in the chicken ovalbumin split gene. *Cell* **14**, 629–639.

Garcia-Bellido, A. (1972). Some parameters of mitotic recombination in D. melanogaster. *Mol. Gen. Genet.* **115**, 54–72.

Gardner, R. L. and Lyon, M. F. (1971). X chromosome inactivation studied by injection of a single cell into the mouse blastocyst. *Nature* **231**, 383–386.

Garel, A. and Axel, R. (1976). Selective digestion of transcriptionally active ovalbumin genes from oviduct nuclei. *Proc. Nat. Acad. Sci. USA* **73**, 3966–3970.

Garel, A., Zolan, M. and Axel, R. (1977). Genes transcribed at diverse rates have a similar conformation in chromatin. *Proc. Nat. Acad. Sci. USA* **74**, 4867–4871.

Gariglio, P., Llopis, R., Oudet, P. and Chambon, P. (1979). The template of the isolated native SV40 transcriptional complexes is a minichromosome. *J. Mol. Biol.* **131**, 75–106.

Garrels, J. I. and Gibson, W. (1976). Identification and characterization of multiple forms of actin. *Cell* **9**, 793–806.

Gartler, S. M., Andina, R. and Gant, N. (1975). Ontogeny of X chromosome inactivation in the female germ line. *Exp. Cell Res.* **91**, 454–457.

Gartler, S. M., Liskay, R. M. and Gant, N. (1973). Two functional X chromosomes in human fetal oocytes. *Exp. Cell Res.* **82**, 464–466.

Gartler, S. M., Liskay, R. M., Campbell, B. K., Sparks, R. and Gant, N. (1972). Evidence for two functional X chromosomes in human oocytes. *Cell Differ.* **1**, 215–218.

Gaskill, P. and Kabat, D. (1971). Unexpectedly large size of globin mRNA. *Proc. Nat. Acad. Sci. USA* **68**, 72–75.

Gaskin, F., Cantor, C. R. and Shelanski, M. L. (1974). Turbidimetric studies of the in vitro assembly and disassembly of porcine neurotubules. *J. Mol. Biol.* **89**, 737–758.

Gatti, M. (1979). Genetic control of chromosome breakage and rejoining in D. melanogaster: spontaneous chromosome aberrations in X linked mutants defective in DNA metabolism. *Proc. Nat. Acad. Sci. USA* **76**, 1377–1381.

Gatti, M., Santini, G., Pimpinelli, S. and Olivieri, G. (1979). Lack of spontaneous sister chromatid exchanges in somatic cells of D. melanogaster. *Genetics* **91**, 255–274.

Gaubatz, J. W. and Chalkley, R. (1977). Distribution of H1 histone in chromatin digested by micrococcal nuclease. *Nucleic Acids Res.* **4**, 3281–3301.

Gautier, F., Bunemann, H. and Grotjahn, L. (1977). Analysis of calf thymus satellite DNA: evidence for specific methylation of cytosine in C-G sequences. *Eur. J. Biochem.* **80**, 175-184.

Gavaudan, P. and Pomriaskinsky-Kobozieff, N. (1937). Sur l'influence de la colchicine sur la carycinese dans les meristemes radiculares de l'Allium cepa. *CR Soc. Biol.* **125**, 705-707.

Gavosto, F., Pegoraro, L., Masera, P. and Rovera, G. (1968). Late DNA replication pattern in human hemapoietic cells. A comparative investigation using a high resolution quantitative autoradiography. *Exp. Cell Res.* **49**, 340-358.

Geahlen, R. L. and Haley, B. E. (1977). Interactions of a photoaffinity analog of GTP with the proteins of microtubules. *Proc. Nat. Acad. Sci. USA* **74**, 4375-4377.

Gedamu, L., Dixon, G. H. and Davies, P. L. (1977). Identification and isolation of protamine mRNP particles from rainbow trout testis. *Biochemistry* **16**, 1383-1390.

Gedamu, L., Iatrou, K. and Dixon, G. H. (1977). Isolation and characterization of trout testis protamine mRNAs lacking poly(A). *Cell* **10**, 443-452.

Gehring, U. and Tomkins, G. M. (1974). A new mechanism for steroid unresponsiveness: loss of nuclear binding activity of a steroid hormone receptor. *Cell* **3**, 301-306.

Geiger, B. (1979). A 130K protein from chicken gizzard: its localization at the termini of microfilament bundles in cultured chicken cells. *Cell* **18**, 193-205.

Geiger, B., Tokuyasu, K. T. and Singer, S. J. (1979). Immunocytochemical localization of α actinin in intestinal epithelial cells. *Proc. Nat. Acad. Sci. USA* **76**, 2833-2837.

Gelbard, A. S., Kim, J. H. and Perez, A. G. (1969). Fluctuations in deoxycytidine monophosphate deaminase activity during the cell cycle in synchronous populations of HeLa cells. *Biochim. Biophys. Acta* **182**, 564-566.

Gelbart, W. M. (1974). A new mutant controlling mitotic chromosome nondisjunction in D. melanogaster. *Genetics* **76**, 51-63.

Gelderman, A., Rake, A. and Britten, R. J. (1971). Transcription of nonrepeated DNA in neonatal and fetal mice. *Proc. Nat. Acad. Sci. USA* **68**, 172-176.

Gelinas, R. E. and Roberts, R. J. (1977). One predominant 5' unadecanucleotide in adenovirus 2 late mRNAs. *Cell* **11**, 533-544.

Gelvin, S., Heizmann, P. and Howell, S. H. (1977). Identification and cloning of the chloroplast gene coding for the large subunit of ribulose biphosphate carboxylase from C. reinhardii. *Proc. Nat. Acad. Sci. USA* **74**, 3193-3197.

Generoso, W. M., Cain, K. T., Krishna, M. and Huff, S. W. (1979). Genetic lesions induced by chemicals in spermatozoa and spermatids of mice are repaired in the egg. *Proc. Nat. Acad. Sci. USA* **76**, 435-437.

Georgiev, G. P. and Samarina, O. P. (1971). D-RNA containing RNP particles. In D. M. Prescott, L. Goldstein and E. McConkey (Ed.) *Adv. Cell Biol.* Appleton Century Crofts, New York, pp 47-110.

German, J. (1969). Bloom's syndrome. I. Genetical and clinical observations of the first 27 patients. *Am. J. Hum. Genet.* **21**, 196-227.

German, J., Crippa, L. P. and Bloom, D. (1974). Bloom's syndrome. III. Analysis of the chromosome aberrations characteristic of this disorder. *Chromosoma* **48**, 361-366.

Germond, J. E., Hirt, B., Oudet, P., Gross-Bellard, M. and Chambon, P. (1975). Folding of the DNA double helix in chromatin-like structures from SV40. *Proc. Nat. Acad. Sci. USA* **72**, 1843-1847.

Germond, J. E., Brutlag, D., Yaniv, M. and Rouviere-Yaniv, J. (1979). The nicking closing enzyme assembles nucleosome-like structures in vitro. *Proc. Nat. Acad. Sci. USA* **76**, 3779-3783.

Gershon, D. and Sachs, L. (1963). Properties of a somatic hybrid between mouse cells with different genotypes. *Nature* **198**, 912-913.

Getz, M. J., Birnie, G. D., Young, B. D., MacPhail, E. and Paul, J. (1975). A kinetic estimation of base sequence complexity of nuclear poly(A)-containing RNA in mouse Friend cells. *Cell* 4, 121–130.

Getz, M. J., Elder, P. K., Benz, E. W., Jr., Stephens, R. E. and Moses, H. L. (1976). Effect of cell proliferation on levels and diversity of poly(A)-containing mRNA. *Cell* 7, 255–266.

Getz, M. J., Reiman, H. M., Siegal, G. P., Quinlan, T. J., Proper, J., Elder, P. K. and Moses, H. L. (1977). Gene expression in chemically transformed mouse embryo cells: selective enhancement of the expression of C type RNA tumor virus genes. *Cell* 11, 909–922.

Gey, G. O., Coffman, W. O. and Kubicek, M. T. (1952). Tissue culture studies of the proliferative capacity of cervical carcinoma and normal epithelium. *Cancer Res.* 12, 264.

Ghangas, G. S. and Milman, G. (1975). Radioimmune determination of HGPRT crossreacting material in erythrocytes of Lesch-Nyhan patients. *Proc. Nat. Acad. Sci. USA* 72, 4147–4150.

Ghosh, P. K., Reddy, V. B., Swinscoe, J., Choudary, P. V., Lebowitz, P. and Weissman, M. (1978a). The 5′ terminal leader sequence of late 16S mRNA from cells infected with SV40. *J. Biol. Chem.* 253, 3643–3647.

Ghosh, P. K., Reddy, V. B., Swinscoe, J., Leibowitz, P. and Weissman, S. M. (1978b). Heterogeneity and 5′ terminal structures of the late RNAs of SV40. *J. Mol. Biol.* 126, 813–846.

Ghysdael, J., Hubert, E., Travnicek, M., Bolognesi, D. P., Burny, A., Clueter, Y., Huez, G., Kettmann, R., Marbaix, G., Portetelle, D. and Chantrenne, H. (1977). Frog oocytes synthesise and completely process the precursor polypeptide to virion structural proteins after microinjection of AMV RNA. *Proc. Nat. Acad. Sci. USA* 74, 3230–3234.

Gianelli, F. and Hamerton, J. L. (1971). Non random late replication of X chromosomes in mules and hinnies. *Nature* 232, 315–319.

Gibbons, I. R. (1965). Chemical dissection of the cilia. *Arch. Biol. (Liege)* 76, 317–352.

Gibbons, I. R. and Grimstone, A. V. (1960). On flagellar structure in certain flagellates. *J. Biophys. Biochem. Cytol.* 7, 697–715.

Gilbert, C. W., Muldal, S. and Lathja, L. G. (1965). Rate of chromosome duplication at the end of the DNA S period in human blood cells. *Nature* 208, 159–161.

Gilbert, W., Maizels, B. and Maxam, A. (1973). Sequences of controlling regions of the lactose operon. *Cold Spring Harbor Symp. Quant. Biol.* 38, 845–855.

Gillespie, D. and Spiegelman, S. (1965). A quantitative assay for DNA-RNA hybrids with DNA immobilized on a membrane. *J. Mol. Biol.* 12, 829–842.

Gillham, N. W. (1963). Transmission and segregation of a nonchromosomal factor controlling streptomycin resistance in diploid Chlamydomonas. *Nature* 200, 294.

Gillham, N. W. (1974). Genetic analysis of the chloroplast and mitochondrial genomes. *Ann. Rev. Genet.* 8, 347–392.

Gillham, N. W. (1978). *Organelle Genetics*. Raven Press, New York.

Gillham, N. W., Boynton, J. E. and Lee, R. W. (1974). Segregation and recombination of non Mendelian genetics in Chlamydomonas. *Genetics* 78, 439–451.

Gillies, C. B. (1972). Reconstruction of the N. crassa pachytene karyotype from serial sections of synaptonemal complexes. *Chromosoma* 36, 119–130.

Gillies, C. B. (1973). Ultrastructural analysis of maize pachytene karyotypes by three dimensional reconstruction of synaptonemal complexes. *Chromosoma* 43, 145–176.

Gillies, C. B. (1974). The nature and extent of synaptonemal complex formation in haploid barley. *Chromosoma* 48, 441–453.

Gillies, C. B. (1975). Synaptonemal complex and chromosome structure. *Ann. Rev. Genet.* 9, 91–110.

Gillies, C. B. (1979). The relationship between synaptonemal complex recombination nodules and crossing over in N. crassa bivalents and translocation quadrivalents. *Genetics* 91, 1–17.

Gillies, C. B., Rasmussen, S. W. and Von Wettstein, D. (1973). The synaptonemal complex in homologous and nonhomologous pairing of chromosomes. *Cold Spring Harbor Symp. Quant. Biol.* **38**, 117-122.

Gillin, F. D., Roufa, D. J., Beaudet, A. L. and Caskey, C. T. (1972). 8 azaguanine resistance in mammalian cells. I. HGPRT. *Genetics* **72**, 239-252.

Gillis, M. and De Ley, J. (1975). Determination of the molecular complexity of double stranded phage genome DNA from initial renaturation rates. The effect of DNA base composition. *J. Mol. Biol.* **98**, 447-464.

Gillum, A. M. and Clayton, D. A. (1978). Displacement loop replication initiation sequence in animal mitochondrial DNA exists as a family of discrete lengths. *Proc. Nat. Acad. Sci. USA* **75**, 677-681.

Gilmore-Hebert, M. and Wall, R. (1978). Immunoglobulin light chain mRNA is processed from large nuclear RNA. *Proc. Nat. Acad. Sci. USA* **75**, 342-345.

Gilmour, R. S. and Paul, J. (1969). RNA transcribed from reconstituted nucleoprotein is similar to natural RNA. *J. Mol. Biol.* **40**, 137-140.

Gilmour, R. S. and Paul, J. (1973). Tissue specific transcription of the globin gene in isolated chromatin. *Proc. Nat. Acad. Sci. USA* **70**, 3440-3442.

Gingold, E. B., Saunders, G. W., Lunkins, H. B. and Linnane, A. W. (1969). Biogenesis of mitochondria. X. Reassortment of the cytoplasmic genetic determinants for respiratory competence and erythromycin resistance in S. cerevisiae. *Genetics* **62**, 735-744.

Giorno, R. and Sauerbier, W. (1976). A radiological analysis of the transcription units for hnRNA in cultured murine cells. *Cell* **9**, 775-784.

Girard, M., Latham, H., Penman, S. and Darnell, J. E. (1965). Entrance of newly formed mRNA and ribosome subunits into HeLa cell cytoplasm. *J. Mol. Biol.* **11**, 187-201.

Gissinger, F. and Chambon, P. (1972). Animal DNA dependent RNA polymerase. 2. Purification of calf thymus A1 enzyme. *Eur. J. Biochem.* **28**, 277-282.

Gjerset, R. A. and McCarthy, B. J. (1977). Limited accessibility of chromatin satellite DNA to RNA polymerase from E. coli. *Proc. Nat. Acad. Sci. USA* **74**, 4337-4340.

Glover, C. V. C. and Gorovsky, M. A. (1978). Histone-histone interactions in a lower eucaryote, Tetrahymena thermophila. *Biochemistry* **17**, 5705-5712.

Glover, C. V. C. and Gorovsky, M. A. (1979). Amino acid sequence of Tetrahymena histone H4 differs from that of higher eucaryotes. *Proc. Nat. Acad. Sci. USA* **76**, 585-589.

Glover, D. M. (1977). Cloned segment of D. melanogaster rDNA containing new types of insertion sequence. *Proc. Nat. Acad. Sci. USA* **74**, 4932-4936.

Glover, D. M. and Hogness, D. S. (1977). A novel arrangement of the 18S and 28S sequences in a repeating unit of D. melanogaster rDNA. *Cell* **10**, 167-176.

Goeddel, D. V., Kleid, D. G., Bolivar, F., Heyneker, H. L., Yansura, D. G., Crea, R., Hirose, T., Kraszewski, A., Itakura, K. and Riggs, A. D. (1979). Expression in E. coli of chemically synthesized genes for human insulin. *Proc. Nat. Acad. Sci. USA* **76**, 106-110.

Goff, S. P. and Berg, P. (1978). Excision of DNA segments introduced into cloning vectors by the poly(dAT) method. *Proc. Nat. Acad. Sci. USA* **75**, 1763-1767.

Goldberg, M. I., Perriard, J. C. and Rutter, W. J. (1977). Purification of rat liver and mouse ascites DNA dependent RNA polymerase I. *Biochemistry* **16**, 1655-1664.

Goldberg, R. B., Galau, G. A., Britten, R. J. and Davidson, E. H. (1973). Nonrepetitive DNA sequence representation in sea urchin embryo mRNA. *Proc. Nat. Acad. Sci. USA* **70**, 3516-3520.

Goldberg, R. B., Crain, W. R., Ruderman, J. V., Moore, G. P., Barnett, T. R., Higgins, R. C., Gelfand, R. A., Galau, G. A., Britten, R. J. and Davidson, E. H. (1975). DNA sequence organization in the genomes of five marine invertebrates. *Chromosoma* **51**, 225-251.

Goldberg, R. B., Hoschek, G., Kamalay, J. C. and Timberlake, W. E. (1978). Sequence complexity of nuclear and polysomal RNA in leaves of the tobacco plant. *Cell* **14**, 123-132.

Goldberg, S., Nevins, J. and Darnell, J. E. (1978). Evidence from UV transcription mapping that late adenovirus type 2 mRNA is derived from a large precursor molecule. *J. Virol.* **25**, 806-810.

Goldberg, S., Schwartz, H. and Darnell, J. E., Jr. (1977). Evidence from UV transcription mapping of HeLa cells that hnRNA is the mRNA precursor. *Proc. Nat. Acad. Sci. USA* **74**, 4520-4523.

Goldblatt, D., Bustin, M. and Sperling, R. (1978). Heterogeneity in the interaction of chromatin subunits with anti histone sera visualized by immune electron microscopy. *Exp. Cell Res.* **112**, 1-14.

Goldenberg, C. J. and Raskas, H. J. (1979). Splicing patterns of nuclear precursors to the mRNA for adenovirus 2 binding protein. *Cell* **16**, 131-138.

Goldknopf, I. L. and Busch, H. (1977). Isopeptide linkage between nonhistone and histone 2A polypeptides of chromosomal conjugate protein A24. *Proc. Nat. Acad. Sci. USA* **74**, 864-868.

Goldknopf, I. L., Taylor, C. W., Baum, R. M., Yeoman, L. C., Olson, M. O. J., Prestayko, A. W. and Busch, H. (1975). Isolation and characterization of protein A24, a histone-like nonhistone chromosomal protein. *J. Biol. Chem.* **250**, 7182-7187.

Goldknopf, I. L., French, M. F., Musso, R. and Busch, H. (1977). Presence of protein A24 in rat liver nucleosomes. *Proc. Nat. Acad. Sci. USA* **74**, 5492-5495.

Goldman, R. (1975). The use of heavy meromyosin binding as an ultrastructural cytochemical method for localizing and determining the possible function of actin like microfilaments in nonmuscle cells. *J. Histochem. Cytochem.* **23**, 529-542.

Goldman, R. D. and Follett, E. A. C. (1969). The structure of the major cell processes of isolated BHK21 fibroblasts. *Exp. Cell Res.* **57**, 263-276.

Goldman, R. and Knipe, D. (1972). Functions of cytoplasmic fibers in nonmuscle cell motility. *Cold Spring Harbor Symp. Quant. Biol.* **37**, 523-534.

Goldman, R. D. and Rebhun, L. I. (1969). The structure and some properties of the isolated mitotic apparatus. *J. Cell Sci.* **4**, 179-209.

Goldman, R. D., Pollack, R. and Hopkins, N. (1973). Preservation of normal behavior by enucleated cells in culture. *Proc. Nat. Acad. Sci. USA* **70**, 750-754.

Goldman, R. D., Schloss, J. A. and Starger, J. M. (1976). Organizational changes of actinlike microfilaments during animal cell movement. In R. Goldman, T. Pollard and J. Rosenbaum (Ed.) *Cell Motility.* Cold Spring Harbor Laboratory, New York, pp 217-245.

Goldman, R. D., Lazarides, E., Pollack, R. and Weber, K. (1975). The distribution of actin in nonmuscle cells. The use of actin antibody in the localization of actin within the microfilament bundles of mouse 3T3 cells. *Exp. Cell Res.* **90**, 333-344.

Goldman, R. D., Pollack, R., Chang, C. M. and Bushell, A. (1975). Properties of enucleated cells. III. Changes in cytoplasmic architecture of enucleated BHK21 cells following trypsinization and replating. *Exp. Cell Res.* **93**, 175-183.

Goldring, E. S. and Peacock, W. J. (1977). Intramolecular heterogeneity of mitochondrial DNA of D. melanogaster. *J. Cell Biol.* **73**, 279-286.

Goldring, E. S., Grossman, L. I., Krupnick, D., Cryer, C. R. and Marmur, J. (1970). The petite mutation in yeast: loss of mitochondrial DNA during induction of petites with ethidium bromide. *J. Mol. Biol.* **52**, 323-335.

Goldstein, P. and Moens, P. B. (1976). Karyotype analysis of Ascaris lumbricoides var suum. Male and female pachytene nuclei by 3D reconstruction from electron microscopy of serial sections. *Chromosoma* **58**, 101-112.

Golomb, H. M. and Bahr, G. F. (1974a). Electron microscopy of human interphase nuclei: determination of total dry mass and DNA packing ratio. *Chromosoma* **46**, 233-246.

Golomb, H. M. and Bahr, G. F. (1974b). Correlation of the fluorescent banding pattern and ultrastructure of a human chromosome. *Exp. Cell Res.* **84**, 121-126.

Goodman, H. M., Olson, M. V. and Hall, B. D. (1977). Nucleotide sequence of a mutant eucaryotic gene: the yeast tyrosine inserting ochre suppressor SUP4-0. *Proc. Nat. Acad. Sci. USA* **74**, 5453-5457.

Goodpasture, C. and Bloom, S. E. (1975). Visualization of nucleolar organizer regions in mammalian chromosomes using silver stain. *Chromosoma* **53**, 37-50.

Goodwin, G. H. and Johns, E. W. (1973). Isolation and characterization of two calf thymus chromatin nonhistone proteins with high contents of acidic and basic amino acids. *Eur. J. Biochem.* **40**, 215-219.

Goodwin, G. H., Sanders, C. and Johns, E. W. (1973). A new group of chromatin associated proteins with a high content of acidic and basic amino acids. *Eur. J. Biochem.* **38**, 14-19.

Goodwin, G. H., Shooter, K. V. and Johns, E. W. (1975). Interaction of a nonhistone chromatin protein (HMG2) with DNA. *Eur. J. Biochem.* **54**, 427-433.

Gopalakrishnan, T. V., Thompson, E. B. and Anderson, W. F. (1977). Extinction of hemoglobin inducibility in Friend erythroleukemia cells by fusion with cytoplasm of enucleated mouse neuroblastoma or fibroblast cells. *Proc. Nat. Acad. Sci. USA* **74**, 1642-1646.

Gordon, W. E. (1978). Immunofluorescent and ultrastructural studies of sarcomeric units in stress fibers of cultured non muscle cells. *Exp. Cell Res.* **117**, 253-260.

Gorovsky, M. A. and Keevert, J. B. (1975). Subunit structure of a naturally occuring chromatin lacking histones F1 and F3. *Proc. Nat. Acad. Sci. USA* **72**, 3536-3540.

Gorovsky, M. A. and Woodard, J. (1967). Histone content of chromosomal loci active and inactive in RNA synthesis. *J. Cell Biol.* **33**, 723-728.

Gorski, J., Morrison, M. R., Merkel, C. G. and Lingrel, J. B. (1974). Size heterogeneity of polyadenylate sequences in mouse globin mRNA. *J. Mol. Biol.* **86**, 363-372.

Gosden, J. R., Mitchell, A. R., Buckland, R. A., Clayton, R. P. and Evans, H. J. (1975). The location of four human satellite DNAs in human chromosomes. *Exp. Cell Res.* **92**, 148-158.

Gosden, J. R., Mitchell, A. R., Seuanez, H. N. and Gosden, C. M. (1977). The distribution of sequences complementary to human satellite DNAs I, II and IV in the chromosomes of chimpanzee, gorilla and orangutan. *Chromosoma* **63**, 253-272.

Goss, S. J. and Harris, H. (1975). New method for mapping genes in human chromosomes. *Nature* **255**, 680-684.

Goss, S. J. and Harris, H. (1977a). Gene transfer by means of cell fusion. I. Statistical mapping of the human X chromosome by analysis of radiation induced gene segregation. *J. Cell Sci.* **25**, 17-38.

Goss, S. J. and Harris, H. (1977b). Gene transfer by means of cell fusion. II. The mapping of 8 loci on human chromosome 1 by statistical analysis of gene assortment in somatic cell hybrids. *J. Cell Sci.* **25**, 39-58.

Gottesfeld, J. M. and Butler, P. J. G. (1977). Structure of transcriptionally active nucleosome subunits. *Nucleic Acids Res.* **4**, 3155-3173.

Gottesfeld, J. M. and Partington, G. A. (1977). Distribution of mRNA: coding sequences in fractionated chromatin. *Cell* **12**, 953-962.

Gottesfeld, J. M., Murphy, R. F. and Bonner, J. (1975). Structure of transcriptionally active chromatin. *Proc. Nat. Acad. Sci. USA* **72**, 4404-4408.

Gottesfeld, J. M., Garrard, W. T., Bagi, G., Wilson, R. F. and Bonner, J. (1974). Partial purification of the template-active fraction of chromatin: a preliminary report. *Proc. Nat. Acad. Sci. USA* **71**, 2193-2197.

Gottesfeld, J. M., Bagi, G., Berg, B. and Bonner, J. (1976). Sequence composition of the template active fraction of rat liver chromatin. *Biochemistry* **15**, 2472-2483.

Gould, R. R. (1975). The basal bodies of Chlamydomonas reinhardii. *J. Cell Biol.* **65**, 65-74.

Gould, R. R. and Borisy, G. G. (1977). The percentriolar material in Chinese hamster ovary cells nucleates microtubule formation. *J. Cell. Biol.* **73**, 601-615.

Gowdridge, B. M. (1956). Heterocaryons between strains of N. crassa with different cytoplasms. *Genetics* **41**, 780-789.

Gowen, J. W. (1933). Meiosis as a genetic character in D. melanogaster. *J. Exp. Zoo.* **65**, 83-106.

Graham, C. F. and Morgan, R. W. (1966). Changes in the cell cycle during early amphibian development. *Devel. Biol.* **14**, 439-460.

Graham, C. F., Arms, K. and Gurdon, J. B. (1966). The induction of DNA synthesis by frog egg cytoplasm. *Devel. Biol.* **14**, 349-381.

Graham, D. E., Neufeld, B. R., Davidson, E. H. and Britten, R. J. (1974). Interspersion of repetitive and nonrepetitive DNA sequences in the sea urchin genome. *Cell* **1**, 127-138.

Grainger, R. M. and Ogle, R. C. (1978). Chromatin structures of the rRNA genes in P. polycephalum. *Chromosoma* **65**, 115-126.

Grainger, R. M. and Wilt, F. H. (1976). Incorporation of ^{13}C ^{15}N labeled nucleosides and measurement of RNA synthesis and turnover in sea urchin embryos. *J. Mol. Biol.* **104**, 589-602.

Gray, P. W. and Hallick, R. B. (1977). Restriction endonuclease map of Euglena gracilis chloroplast DNA. *Biochemistry* **16**, 1665-1671.

Gray, P. W. and Hallick, R. B. (1978). Physical mapping of the Euglena gracilis chloroplast DNA and rRNA gene region. *Biochemistry* **17**, 284-289.

Graziani, F., Caizzi, R. and Gargano, S. (1977). Circular rDNA during ribosomal magnification in D. melanogaster. *J. Mol. Biol.* **112**, 49-64.

Graziano, S. L. and Huang, R. C. C. (1971). Chromatographic separation of chick brain proteins using a SP-Sephadex column. *Biochemistry* **10**, 4770-4777.

Green, H., Wang, R., Kehinde, O. and Meuth, M. (1971). Multiple human TK chromosomes in human-mouse somatic cell hybrids. *Nature New Biol.* **234**, 138-140.

Greenaway, P. J. and Murray, K. (1971). Heterogeneity and polymorphism in chicken erythrocyte histone fraction V. *Nature New Biol.* **229**, 233-238.

Greenberg, H. and Penman, S. (1966). Methylation and processing of ribosomal RNA in HeLa cells. *J. Mol. Biol.* **21**, 527-535.

Greenberg, J. R. (1972). High stability of mRNA in growing cultured cells. *Nature* **242**, 102-104.

Greenberg, J. R. (1976). Isolation of L cell mRNA which lacks poly(A). *Biochemistry* **15**, 3516-3522.

Greenberg, J. R. (1977). Isolation of messenger ribonucleoproteins in Cs_2SO_4 density gradients: evidence that polyadenylated and nonpolyadenylated mRNAs are associated with protein. *J. Mol. Biol.* **108**, 403-416.

Greenberg, J. R. and Perry, R. P. (1971). Hybridization properties of DNA sequences directing the synthesis of mRNA and hnRNA. *J. Cell Biol.* **50**, 774-787.

Greenberg, J. R. and Perry, R. P. (1972). Relative occurrence of poly(A) sequences in mRNA and hnRNA of L cells as determined by poly(U) hydroxyapatite chromatography. *J. Mol. Biol.* **72**, 91-98.

Greenleaf, A. L. and Bautz, E. K. F. (1975). RNA polymerase B from D. melanogaster larvae. Purification and partial characterization. *Eur. J. Biochem.* **60**, 169-180.

Greenleaf, A. L., Plagens, U., Jamrich, M. and Bautz, E. K. F. (1978). RNA polymerase B in heat induced puffs of Drosophila polytene chromosomes. *Chromosoma* **65**, 127-136.

Greenleaf, A. L., Borsett, L. M., Jiamachello, P. F. and Coulter, D. E. (1979). α-Amanitin resistant D. melanogaster with an altered RNA polymerase II. *Cell* **18**, 613-622.

Grell, R. F. (1962a). A new hypothesis on the nature and sequence of meiotic events in the female of D. melanogaster. *Proc. Nat. Acad. Sci. USA* **48**, 165-172.

Grell, R. F. (1962b). A new model for secondary nondisjunction: the role of distributive pairing. *Genetics* **47**, 1737-1754.

Grell, R. F. (1964a). Chromosome size at distributive pairing in D. melanogaster females. *Genetics* 50, 150-166.

Grell, R. F. (1964b). Distributive pairing: the size dependent mechanism for regular segregation of the fourth chromosomes in D. melanogaster. *Proc. Nat. Acad. Sci. USA* 52, 226-232.

Grell, R. F. (1976). Distributive pairing. In M. Ashburner and E. Novitski (Eds.), The Genetics and Biology of Drosophila, Academic Press, London, pp 435-486.

Grell, R. F. (1978). Time of recombination in the D. melanogaster oocyte: evidence from a temperature sensitive recombination-deficient mutant. *Proc. Nat. Acad. Sci. USA* 75, 3351-3354.

Griffith, J. D. (1975). Chromatin structure: deduced from a minochromosome. *Science* 187, 1202-1203.

Griffith, J. D. (1978). DNA structure: evidence from electron microscopy. *Science* 201, 525-527.

Grimes, G. W., Mahler, H. R. and Perlman, P. G. (1974). Nuclear dosage effects on mitochondrial mass and DNA. *J. Cell Biol.* 61, 565-574.

Grimstone, A. V. and Klug, A. (1966). Observations on the substructure of flagellar fibres. *J. Cell Sci.* 1, 351-362.

Grivell, L. A., Reijnders, L. and Borst, P. (1971). Isolation of yeast mitochondrial ribosomes highly active in protein synthesis. *Biochim. Biophys. Acta* 247, 91-103.

Grivell, L. A., Netter, P., Borst, P. and Slonimski, P. P. (1973). Mitochondrial antibiotic resistance in yeast: ribosomal mutants resistant to chloramphenicol, erythromycin and spiramycin. *Biochim. Biophys. Acta* 312, 358-367.

Groner, B., Hynes, N. E., Sippel, A. E., Jeep, S., Huu, M. C. N. and Schutz, G. (1977). Immunoadsorption of specific chicken oviduct polysomes. Isolation of ovalbumin, ovomucoid and lysozyme mRNA. *J. Biol. Chem.* 252, 6666-6674.

Groner, Y., Gilboa, E. and Aviv, H. (1978). Methylation and capping of RNA polymerase II primary transcripts by HeLa nuclear homogenates. *Biochemistry* 17, 977-982.

Groner, Y., Grosfeld, H. and Littauer, U. Z. (1976). 5' capping studies of Artemia salina mRNA and the translational inhibition by cap analogs. *Eur. J. Biochem.* 71, 281-294.

Gross, K., Probst, E., Schaffner, W. and Birnstiel, M. (1976a). Molecular analysis of the histone gene cluster of P. miliaris. I. Fractionation and identification of five individual histone mRNAs. *Cell* 8, 455-470.

Gross, K., Schaffner, W., Telford, J. and Birnstiel, M. (1976b). Molecular analysis of the histone gene cluster of P. miliaris. III. Polarity and asymmetry of the histone coding sequences. *Cell* 8, 479-484.

Gross, K. W., Jacobs-Lorena, M., Baglioni, C. and Gross, P. R. (1973). Cell free translation of maternal messenger RNA from sea urchin eggs. *Proc. Nat. Acad. Sci. USA* 70, 2614-2618.

Grossbach, U. (1969). Chromosomen activat und biochemische Zelldifferenzierung in den speicheldrusen von Comptochironomus. *Chromosoma* 28, 136-187.

Grossbach, U. (1973). Chromosome puffs and gene expression in polytene cells. *Cold Spring Harbor Symp. Quant. Biol.* 38, 619-627.

Grossman, L. I., Watson, R. and Vinograd, J. (1973). The presence of ribonucleotides in mature closed circular mitochondrial DNA. *Proc. Nat. Acad. Sci. USA* 70, 3339-3343.

Groudine, M., Holtzer, H., Scherrer, K., and Therwath, A. (1974). Lineage dependent transcription of globin genes. *Cell* 3, 243-248.

Grouse, L., Chilton, M. D. and McCarthy, B. J. (1972). Hybridization of RNA with unique sequences of mouse DNA. *Biochemistry* 11, 798-805.

Gruenstein, E., Rich, A. and Weihing, R. R. (1975). Actin associated with membranes from 3T3 mouse fibroblasts and HeLa cells. *J. Cell Biol.* 64, 223-234.

Grummt, I., Hall, S. H. and Crouch, R. J. (1979). Localization of an endonuclease specific for double stranded RNA within the nucleolus and its implication in processing ribosomal transcripts. *Eur. J. Biochem.* 94, 437-444.

Grunstein, M. and Hogness, D. S. (1975). Colony hybridization: a method for the isolation of cloned DNAs that contain a specific gene. *Proc. Nat. Acad. Sci. USA* **72**, 3961-3965.

Gruss, P., Lai, C. J., Dhar, R. and Khoury, G. (1979). Splicing as a requirement for biogenesis of functional SV40 mRNA. *Proc. Nat. Acad. Sci. USA* **76**, 4317-4321.

Grzeschik, K. H., Allerdice, P. W., Grzeschick, A., Opitz, J. M., Miller, O. J. and Siniscalco, M. (1972). Cytological mapping of human X linked genes by use of somatic cell hybrids involving an X autosome translocation. *Proc. Nat. Acad. Sci. USA* **69**, 69-73.

Gupta, R. S. and Siminovitch, L. (1976). The isolation and preliminary characterization of somatic cell mutants resistant to the protein synthesis inhibitor emetine. *Cell* **9**, 213-220.

Gupta, R. S. and Siminovitch, L. (1977). The molecular basis of emetine resistance in CHO cells: alteration in the 40S ribosomal subunit. *Cell* **10**, 61-66.

Gupta, R. S. and Siminovitch, L. (1978a). Mutants of CHO cells resistant to the protein synthesis inhibitor emetine: genetic and biochemical characterization of second step mutants. *Somatic Cell Genet.* **4**, 77-94.

Gupta, R. S. and Siminovitch, L. (1978b). An in vitro analysis of the dominance of emetine sensitivity in CHO cells. *J. Biol. Chem.* **253**, 3978-3982.

Gupta, R. S. and Siminovitch, L. (1978c). Genetic and biochemical characterization of mutants of CHO cells resistant to the protein synthesis inhibitor trichodermin. *Somatic Cell Genet.* **4**, 355-374.

Gupta, R. S. and Siminovitch, L. (1978d). Isolation and characterization of mutants of human diploid fibroblasts resistant to diphtheria toxin. *Proc. Nat. Acad. Sci. USA* **75**, 3337-3340.

Gupta, R. S. and Siminovitch, L. (1978e). Diphtheria toxin resistant mutants of CHO cells affected in protein synthesis: a novel phenotype. *Somatic Cell Genet.* **4**, 553-572.

Gupta, R. S. and Siminovitch, L. (1978f). Genetic and biochemical studies with the adenosine analogs toyocamycin and tubercidin: mutation at the adenosine kinase locus in Chinese hamster cells. *Somatic Cell Genet.* **4**, 715-736.

Gupta, R. S., Chan, D. Y. H. and Siminovitch, L. (1978a). Evidence for functional hemizygosity at the emt[r] locus in CHO cells through segregation analysis. *Cell* **14**, 1004-1014.

Gupta, R. S., Chan, D. H. Y. and Siminovitch, L. (1978b). Evidence for variation in the number of functional gene copies at the ama[R] locus in CHO lines. *J. Cell. Physiol.* **97**, 461-468.

Gurdon, J. B. (1960). Factors responsible for the abnormal development of embryos obtained by nuclear transplantation in X. laevis. *J. Emb. Exp. Morph.* **8**, 327-340.

Gurdon, J. B. (1962a). Adult frogs derived from the nuclei of single somatic cells. *Devel. Biol.* **4**, 256-273.

Gurdon, J. B. (1962b). The transplantation of nuclei between two species of Xenopus. *Devel. Biol.* **5**, 68-73.

Gurdon, J. B. (1962c). The developmental capacity of nuclei taken from intestinal epithelium cells of feeding tadpoles. *J. Emb. Exp. Morph.* **10**, 622-640.

Gurdon, J. B. (1963). Nuclear transplantation in amphibia and the importance of stable nuclear changes in promoting cellular differentiation. *Quart. Rev. Biol.* **38**, 54-78.

Gurdon, J. B. (1964). The transplantation of living cell nuclei. *Adv. Morph.* **4**, 1-43.

Gurdon, J. B. (1968). Changes in somatic cell nuclei inserted into growing and maturing amphibian oocytes. *J. Emb. Exp. Morph.* **20**, 401-414.

Gurdon, J. B. (1970). Nuclear transplantation and the control of gene activity in animal development. *Proc. Roy. Soc. B* **176**, 303-314.

Gurdon, J. B. and Brown, D. D. (1965). Cytoplasmic regulation of RNA synthesis and nucleolus formation in developing embryos of X. laevis. *J. Mol. Biol.* **12**, 27-35.

Gurdon, J. B. and Laskey, R. A. (1970). The transplantation of nuclei from single cultured cells into enucleated frogs' eggs. *J. Emb. Exp. Morph.* **24**, 227-248.

Gurdon, J. B. and Speight, V. A. (1969). The appearance of cytoplasmic DNA polymerase activity during the maturation of amphibian oocytes into eggs. *Exp. Cell Res.* **55**, 253–256.

Gurdon, J. B. and Woodland, H. R. (1968). The cytoplasmic control of nuclear activity in animal development. *Biol. Rev.* **43**, 233–267.

Gurdon, J. B., Birnsteil, M. L. and Speight, V. A. (1969). The replication of purified DNA introduced into living egg cytoplasm. *Biochim. Biophys. Acta* **174**, 614–628.

Gurdon, J. B., De Robertis, E. M. and Partington, C. (1976). Injected nuclei in frog oocytes provide a living cell system for the study of transcriptional control. *Nature* **260**, 116–120.

Gurley, L. R., Walters, R. A. and Tobey, R. A. (1973). Histone phosphorylation in late interphase and mitosis. *Biochem. Biophys. Res. Commun.* **50**, 744–750.

Gurley, L. R., Walters, R. A. and Tobey, R. A. (1974). Cell cycle specific changes in histone phosphorylation associated with cell proliferation and chromosome condensation. *J. Cell Biol.* **60**, 356–364.

Gwynn, I., Kemp, R. B., Jones, B. M. and Groschel-Stewart, U. (1974). Ultrastructural evidence for myosin of the smooth muscle type at the surface of trypsin dissociated embryonic chick cells. *J. Cell Sci.* **15**, 279–289.

Hackett, P. B. and Sauerbier, W. (1975). The transcriptional organization of the rRNA genes in mouse L cells. *J. Mol. Biol.* **91**, 235–257.

Haendle, J. (1971a). Rontgeninduzierte mitotische reckombination bei Drosophila. I. Die abhangigkeit von der dosis, der disirate und vom spektrum. *Mol. Gen. Genet.* **113**, 114–131.

Haendle, J. (1971b). Rontgeninduzierte mitotische rekombination bei Drosophila. II. Beweis der existenz und charakterisierung zweier von der art des spektrums adhangiger reaktionen. *Mol. Gen. Genet.* **113**, 132–149.

Haeysman, J. E. M. and Pegrum, S. M. (1973). Early contacts between fibroblasts. *Exp. Cell Res.* **78**, 71–78.

Haff, L. A. and Bogorad, L. (1976). Hybridization of maize chloroplast DNA with tRNAs. *Biochemistry* **15**, 4105–4109.

Haff, L. A. and Keller, E. B. (1975). The polyadenylate polymerases from yeast. *J. Biol. Chem.* **250**, 1838–1847.

Hagele, K. and Kalisch, W. E. (1974). Initial phase of DNA synthesis in D. melanogaster. I. Differential participation in replication of the X chromosomes in males and females. *Chromosoma* **47**, 403–413.

Hagenbuchle, O., Santer, M., Steitz, J. A. and Mans, R. J. (1978). Conservation of the primary structure at the 3′ end of the 18S rRNA from eucaryotic cells. *Cell* **13**, 551–563.

Hager, G. L., Holland, M. J. and Rutter, W. J. (1977). Isolation of RNA polymerases I, II and III from S. cerevisiae. *Biochemistry* **16**, 1–18.

Hagopian, H. K., Riggs, M. G., Swartz, L. A. and Ingram, V. M. (1977). Effect of n butyrate on DNA synthesis in chick fibroblasts and HeLa cells. *Cell* **12**, 855–860.

Hahn, U., Lazarus, C. M., Lunsdorf, H. and Kuntzel, H. (1979). Split gene for mitochondrial 24S rRNA of N. crassa. *Cell* **17**, 201–210.

Hahn, W. E. and Laird, C. D. (1971). Transcription of nonrepeated DNA in mouse brain. *Science* **173**, 158–161.

Haid, A., Schweyer, R. J., Bechmann, H., Kaudewitz, F., Solioz, M. and Schatz, G. (1979). The mitochondrial *cob* region in yeast codes for apocytochrome b and is mosaic. *Eur. J. Biochem.* **94**, 451–464.

Haines, M. E., Carey, N. H. and Palmiter, R. D. (1974). Purification and properties of an ovalbumin mRNA. *Eur. J. Biochem.* **43**, 549–560.

Haldane, J. B. S. (1931). The cytological basis of genetical interference. *Cytologia* **3**, 54–65.

Hall, J. C. (1972). Chromosome segregation influenced by two allele of the meiotic mutant *c(3)G* in D. melanogaster. *Genetics* **71**, 367–400.

Hall, J. C., Gelbart, W. M. and Kankel, D. R. (1976). Mosaic systems. In M. Ashburner and E. Novitski (Ed.), The Genetics and Biology of Drosophila, Academic Press, New York, pp 265-314.

Hall, L., Turnock, G. and Cox, B. J. (1975). Ribosomal RNA genes in the amoebal and plasmodial forms of the slime mold Physarum polycephalum. *Eur. J. Biochem.* **51**, 459-465.

Hallberg, R. L. (1974). Mitochondrial DNA in X. laevis oocytes. I. Displacement loop occurrence. *Devel. Biol.* **38**, 346-355.

Hallick, L. M., Yokota, H. A., Bartholomew, J. C. and Hearst, J. E. (1978). Photochemical addition of the cross linking reagent trimethyl psoralen to intracellular and viral SV40 histone DNA complexes. *J. Virol.* **27**, 127-135.

Hamer, D. H. and Leder, P. (1979). SV40 recombinants carrying a functional RNA splice junction and polyadenylation site from the chromosomal mouse β^{maj} globin gene. *Cell* **17**, 737-748.

Hamer, D. H. and Thomas, C. A., Jr. (1976). Molecular cloning of DNA fragments produced by restriction endonucleases Sal I and Bam I. *Proc. Nat. Acad. Sci. USA* **73**, 1537-1541.

Hamer, D. H., Smith, K. D., Boyer, S. H. and Leder, P. (1979). SV40 recombinants carrying rabbit β globin gene coding sequence. *Cell* **17**, 725-736.

Hamerton, J. L., Richardson, B. J., Gee, P. A., Allen, W. R. and Short, R. V. (1971). Nonrandom X chromosome expression in female mules and hinnies. *Nature* **232**, 312-315.

Hames, B. D. and Perry, R. P. (1977). Homology relationship between the mRNA and hnRNA of mouse L cells. A DNA excess hybridization study. *J. Mol. Biol.* **109**, 437-454.

Hamlin, J. L. and Pardee, A. B. (1976). S phase synchrony in monolayer CHO cultures. *Exp. Cell Res.* **100**, 265-275.

Hamlyn, P. H. and Gould, H. J. (1975). Isolation and identification of separated mRNAs for rabbit α and β globin. *J. Mol. Biol.* **94**, 101-110.

Hammerling, J. (1953). Nucleocytoplasmic relationships in the development of Acetabularia. *Int. Rev. Cytol.* **2**, 475-498.

Hammerling, J. (1963). Nucleocytoplasmic interactions in Acetabularia and other cells. *Ann. Rev. Plant. Physiol.* **14**, 65-92.

Hancock, R. (1978). Assembly of new nucleosomal histones and new DNA into chromatin. *Proc. Nat. Acad. Sci. USA* **75**, 2130-2134.

Hand, R. (1975). Regulation of DNA replication on subchromosomal units of mammalian cells. *J. Cell Biol.* **64**, 89-97.

Hand, R. and Tamm, I. (1973). DNA replication: direction and rate of chain growth in mammalian cells. *J. Cell Biol.* **58**, 410-418.

Hand, R. and Tamm, I. (1974). Initiation of DNA replication in mammalian cells and its inhibition by reovirus infection. *J. Mol. Biol.* **82**, 175-183.

Hanson, C. V., Shen, C. -K. J. and Hearst, J. E. (1976). Cross linking of DNA in situ as a probe for chromatin structure. *Science* **193**, 62-64.

Hanson, D. K., Miller, D. H., Mahler, H. R., Alexander, N. J. and Perlman, P. S. (1979). Regulatory interactions between mitochondrial genes. II. Detailed characterization of novel mutants mapping within one cluster in the *cob2* region. *J. Biol. Chem.* **254**, 2480-2490.

Hanson, M. R., Davidson, J. N., Mets, L. L. and Bogorad, L. (1974). Characterization of chloroplast and cytoplasmic ribosomal proteins of C. reinhardii by two dimensional gel electrophoresis. *Mol. Gen. Genet.* **132**, 105-118.

Harbers, K., Harbers, B. and Spencer, J. H. (1974). Nucleotide clusters in DNAs. X. Sequences of the pyrimidine oligonucleotides of mouse L cell satellite DNA. *Biochem. Biophys. Res. Commun.* **58**, 814-821.

Hardison, R. C., Zeitler, D. P., Murphy, J. M. and Chalkley, R. (1977). Histone neighbours in nuclei and extended chromatin. *Cell* **12**, 417-427.

Hardman, N. and Lack, P. L. (1977). Characterization of the foldback sequences in Physarum polycephalum nuclear DNA using the electron microscope. *Eur. J. Biochem.* **74**, 275-284.

Hardman, N. and Lack, P. L. (1978). Periodic organization of foldback sequences in Physarum polycephalum nuclear DNA. *Nucl. Acids Res.* **5**, 2405-2424.

Harris, E. H., Boynton, J. E., Gillham, N. W., Tingle, C. L. and Fox, S. B. (1977). Mapping of chloroplast genes involved in chloroplast ribosome biogenesis in C. reinhardii. *Mol. Gen. Genet.* **155**, 249-266.

Harris, H. (1963). Nuclear RNA. *Prog. Nuc. Acid Res.* **2**, 20-60.

Harris, H. (1967). The reactivation of the red cell nucleus. *J. Cell Sci.* **2**, 23-32.

Harris, H. (1974). *Nucleus and Cytoplasm.* Oxford University Press, Oxford.

Harris, H. and Cook, P. R. (1969). Synthesis of an enzyme determined by an erythrocyte nucleus in a hybrid cell. *J. Cell Sci.* **5**, 121-134.

Harris, H. and Watkins, J. F. (1965). Hybrid cells derived from mouse and man: artificial heterokaryons of mammalian cells from different species. *Nature* **205**, 640-646.

Harris, H., Sidebottom, E., Grace, D. M. and Bramwell, M. E. (1969). The expression of genetic information: a study with hybrid animal cells. *J. Cell Sci.* **4**, 499-526.

Harris, H., Watkins, J. F., Ford, C. E. and Schoefl, G. I. (1966). Artificial heterokaryons of animal cells from different species. *J. Cell Sci.* **1**, 1-30.

Harris, J. F. and Whitmore, G. F. (1974). Chinese hamster cells exhibiting a temperature dependent alteration in purine transport. *J. Cell. Physiol.* **83**, 43-52.

Harris, J. F. and Whitmore, G. F. (1977). Segregation studies on CHO hybrid cells. I. Spontaneous and mutagen induced segregation events of two recessive drug resistant loci. *Somatic Cell Genet.* **3**, 173-194.

Harris, M. (1971). Mutation rates in cells at different ploidy levels. *J. Cell. Physiol.* **78**, 177-184.

Harris, M. (1973). Anomalous patterns of mutation in cultured mammalian cells. *Genetics Suppl.* **73**, 181-185.

Harris, S. E., Means, A. R., Mitchell, W. M. and O'Malley, B. W. (1973). Synthesis of ^{3}H-DNA complementary to ovalbumin mRNA; evidence for limited copies of the ovalbumin gene in chick oviduct. *Proc. Nat. Acad. Sci. USA* **70**, 3776-3780.

Harris, S. E., Rosen, J. M., Means, A. R. and O'Malley, B. W. (1975). Use of a specific probe for ovalbumin mRNA to quantitate estrogen induced gene transcripts. *Biochemistry* **14**, 2072-2080.

Harrison, P. R., Birnie, G. D., Hell, A., Humphries, S., Young, B. D. and Paul, J. (1974). Kinetic studies of gene frequency. I. Use of a DNA copy of reticulocyte 9S RNA to estimate globin gene dosage in mouse tissues. *J. Mol. Biol.* **84**, 539-554.

Hartl, D. L. (1974). Genetic dissection of segregation distortion. I. Suicide combinations of SD genes. *Genetics* **76**, 477-486.

Hartl, D. L. (1975). Genetic dissection of segregation distortion. II. Mechanism of suppression of distortion by certain inversions. *Genetics* **80**, 539-547.

Hartl, D. L. and Hiraizumi, Y. (1976). Segregation distortion. In M. Ashburner and E. Novitski (Eds.), *The Genetics and Biology of Drosophila,* Academic Press, London, Vol. 1b, pp 616-666.

Hartl, D. L., Hiraizumi, Y. and Crow, J. F. (1967). Evidence for sperm disfunction as the mechanism of segregation distortion in D. melanogaster. *Proc. Nat. Acad. Sci. USA* **58**, 2240-2245.

Hartley, M. R. and Ellis, R. J. (1973). RNA synthesis in chloroplasts. *Biochem. J.* **134**, 249-262.

Hartman, H., Puma, J. D. and Gurney, T., Jr. (1974). Evidence for the association of RNA with the ciliary basal bodies of Tetrahymena. *J. Cell Sci.* **16**, 241-259.

Hartman, P. G., Chapman, G. E., Moss, T. and Bradbury, E. M. (1977). Studies on the role and mode of operation of the very lysine rich histone H1 in eucaryote chromatin. The three structural regions of the histone H1 molecule. *Eur. J. Biochem.* **77**, 45-52.

Harvey, E. B. (1935). The mitotic figure and cleavage planes in the egg of Parechinum microtuberculates as induced by centrifugal force. *Biol. Bull.* **69**, 287-297.

Hastie, N. D. and Bishop, J. O. (1976). The expression of three abundance classes of mRNA in mouse tissues. *Cell* **9**, 761–774.

Hatano, S. and Oosawa, F. (1966). Isolation and characterization of plasmodium actin. *Biochim. Biophys. Acta* **127**, 488–498.

Hatano, S. and Tazawa, M. (1968). Isolation, purification and characterization of myosin B from myxomycete plasmodium. *Biochim. Biophys. Acta* **154**, 507–519.

Hatch, F. T. and Mazrimas, J. A. (1974). Fractionation and characterization of satellite DNAs of the kangaroo rat (D. ordii). *Nucl. Acids Res.* **1**, 559–575.

Hatch, F. T., Bodner, A. J., Mazrimas, J. A. and Moore, D. H. II. (1976). Satellite DNA and cytogenetic evolution. DNA quantity, satellite DNA and karyotypic variations in kangaroo rats (Genus Dipdomys). *Chromosoma* **58**, 155–168.

Hatlen, L. E. and Attardi, G. (1971). Proportion of the HeLa cell genome complementary to tRNA and 5S RNA. *J. Mol. Biol.* **56**, 535–554.

Hay, E. D. and Revel, J. P. (1963). The fine structure of the DNP component of the nucleus. *J. Cell Biol.* **16**, 29–51.

Hayflick, L. (1965). The limited in vitro lifetime of human diploid cell strains. *Exp. Cell Res.* **37**, 614–636.

Haynes, J. R., Kalb, V. F., Rosteck, P. and Lingrel, J. B. (1978). The absence of a precursor larger than 16S to globin mRNA. *Febs Lett.* **91**, 173–177.

Heckle, W. L., Jr., Fenton, R. G., Wood, T. G., Merkel, C. G. and Lingrel, J. B. (1977). Methylated nucleosides in globin mRMA from mouse nucleated erythroid cells. *J. Biol. Chem.* **252**, 1764–1770.

Heckman, J. E. and RajBhandary, U. L. (1979). Organization of tRNA and rRNA genes in N. crassa mitochondria: presence of an intervening sequence in the large rRNA gene. *Cell* **17**, 583–595.

Heidemann, S. R. and Kirschner, M. W. (1975). Aster formation in eggs of X. laevis: induction by isolated basal bodies. *J. Cell Biol.* **67**, 105–117.

Heidemann, S. R., Sander, G. and Kirschner, M. W. (1977). Evidence for a functional role of RNA in centrioles. *Cell* **10**, 337–350.

Heidenhain, M. (1899). Uber die struktur der darmepithelzen. *Arch. Mikrosk, Anat. Entw. Mech.* **54**, 184–224.

Heitz, E. (1928). Das heterochromatin der moose. *Jb. Wiss. Bot.* **69**, 762–818.

Heitz, E. (1931). Nucleolen und chromosomen in der gattung Vicia. *Planta* **15**, 495–505.

Heitz, E. and Bauer, H. (1933). Beweise fur die chromosomennatur der kernschleifen in den knauelkernen von Bibio hortulans. *L. Z. Zellforsch* **17**, 67–82.

Henderson, A. S., Eicher, E. M., Yu, M. T. and Atwood, K. C. (1974). The chromosomal location of rDNA in the mouse. *Chromosoma* **49**, 155–160.

Henderson, A. S., Atwood, K. C., Yu, M. T. and Warburton, D. (1976). The site of 5S RNA genes in primates. I. The great apes. *Chromosoma* **56**, 29–32.

Henderson, S. A. (1963). Chiasma distribution at diplotene in a locust. *Heredity* **18**, 173–190.

Henderson, S. A. and Edwards, R. G. (1968). Chiasma frequency and maternal age in mammals. *Nature* **218**, 22–28.

Henikoff, S., Heywood, J. and Meselson, M. (1974). Orientation of repeating units in Xenopus chromosomal ribosomal DNA: a test of a stochastic model for maintaining intraspecies homogeneity. *J. Mol. Biol.* **85**, 445–450.

Hennig, W. (1972a). Highly repetitive DNA sequences in the genome of D. hydei. I. Preferential localization in the X chromosomal heterochromatin. *J. Mol. Biol.* **71**, 407–417.

Hennig, W. (1972b). Highly repetitive DNA sequences in the genome of D. hydei. II. Occurrence in polytene tissues. *J. Mol. Biol.* **71**, 419–431.

Hennig, W. and Meer, B. (1971). Reduced polyteny of rRNA cistrons in giant chromosomes of D. hydei. *Nature New Biol.* **233**, 70–72.

Hennig, W. and Walker, P. M. B. (1970). Variations in the DNA from two rodent families (Cricetidae and Muridae). *Nature* **225**, 915–919.

Hennig, W., Hennig, I. and Stein, H. (1970). Repeated sequences in the DNA of Drosophila and their localization in giant chromosomes. *Chromosoma* **32**, 31–63.

Henshaw, E. C. (1968). Messenger RNA in rat liver polyribosomes. Evidence that it exists as RNP particles. *J. Mol. Biol.* **36**, 401–412.

Henshaw, E. C. and Loebenstein, J. (1970). Rapidly labeled, polydispersed RNA in rat liver cytoplasm: evidence that it is contained in RNP particles of heterogeneous size. *Biochim. Biophys. Acta* **199**, 405–420.

Hepler, P. K., McIntosh, J. R. and Cleland, S. (1970). Intermicrotubule bridges in mitotic spindles. *J. Cell Biol.* **45**, 438–444.

Hereford, L. M. and Rosbash, M. (1977). Number and distribution of polyadenylated RNA sequences in yeast. *Cell* **10**, 453–462.

Herman, R. and Penman, S. (1977). Multiple decay rates of hnRNA in HeLa cells. *Biochemistry* **16**, 3460–3464.

Herman, R., Weymouth, L. and Penman, S. (1978). Heterogeneous nuclear RNA protein fibers in chromatin depleted nuclei. *J. Cell Biol.* **78**, 663–674.

Herman, R. C., Williams, J. G. and Penman, S. (1976). Message and non-message sequences adjacent to poly(A) in steady-state hnRNA of HeLa cells. *Cell* **7**, 429–438.

Herrick, G. and Wesley, R. D. (1978). Isolation and characterization of a highly repetitious inverted terminal repeat sequence from Oxytricha macronuclear DNA. *Proc. Nat. Acad. Sci. USA* **75**, 2626–2630.

Herrick, G., Spear, B. B. and Veomett, G. (1976). Intracellular localization of mouse DNA polymerase α. *Proc. Nat. Acad. Sci. USA* **73**, 1136–1139.

Herrmann, R. G. (1970). Multiple amounts of DNA related to the size of chloroplasts. I. An autoradiographic study. *Planta* **90**, 80–96.

Herrmann, R. G., Bohnert, H. J., Kowallik, K. V. and Schmidt, J. M. (1975). Size conformation and purity of chloroplast DNA of some higher plants. *Biochim. Biophys. Acta* **378**, 305–317.

Hershfield, V., Boyer, H. W., Yanofsky, C., Lovett, M. A. and Helinski, D. R. (1974). Plasmid ColEl as a molecular vehicle for cloning and amplification of DNA. *Proc. Nat. Acad. Sci. USA* **71**, 3455–3459.

Hershey, N. D., Conrad, S. E., Sodja, A., Cohen, M., Jr., Davidson, N., Ilgen, C. and Carbon, J. (1977). The sequence arrangement of D. melanogaster 5S DNA cloned in recombinant plasmids. *Cell* **11**, 586–598.

Herskowitz, I. H. (1950). An estimate of the number of loci in the X chromosome of D. melanogaster. *Amer. Nat.* **84**, 255–260.

Herzog, W. and Weber, K. (1977). In vitro assembly of tubulin into microtubules in the absence of microtubule associated protein and glycerol. *Proc. Nat. Acad. Sci. USA* **74**, 1860–1864.

Herzog, W. and Weber, K. (1978a). Microtubule formation by pure brain tubulin in vitro. The influence of dextran and poly ethylene glycol. *Eur. J. Biochem.* **91**, 249–262.

Herzog, W. and Weber, K. (1978b). Fractionation of brain microtubule associated proteins: isolation of two different proteins which stimulate tubulin polymerization in vitro. *Eur. J. Biochem.* **92**, 1–14.

Hesslewood, I. P., Holmes, A. M., Wakeling, W. F. and Johnston, I. R. (1978). Studies on the purification and properties of a 6.8S DNA polymerase activity found in calf thymus DNA polymerase α fraction. *Eur. J. Biochem.* **84**, 123–132.

Hewish, D. R. and Burgoyne, L. A. (1973). Chromatin substructure. The digestion of chromatin

DNA at regularly spaced sites by a nuclear DNAase. *Biochem. Biophys. Res. Commun.* **52**, 504–510.

Hewitt, G. M. (1964). Population cytology of British grasshoppers. I. Chiasma variation in Chorthippus brunneus, Chorthippus parallelus and Omocestus viridulus. *Chromosoma* **15**, 212–230.

Hewitt, G. M. (1967). An interchange which raises chiasma frequency. *Chromosoma* **21**, 285–299.

Hewitt, G. M. and John, B. (1965). The influence of numerical and structural chromosome mutations on chiasma conditions. *Heredity* **20**, 123–135.

Heyden, H. W. von and Zachau, H. G. (1971). Characterization of RNA in fractions of calf thymus chromatin. *Biochim. Biophys. Acta* **232**, 651–660.

Heyneker, H. L., Shine, J., Goodman, H. M., Boyer, H. W., Rosenberg, J., Dickerson, R. E., Narang, S. A., Itakura, K., Lin, S. Y. and Riggs, A. D. (1976). Synthetic lac operator DNA is functional in vivo. *Nature* **263**, 748–751.

Heyting, C. and Menke, H. H. (1979). Fine structure of the 21S rRNA region on yeast mitochondrial DNA. III. Physical localization of mitochondrial genetic markers and the molecular nature of ω. *Mol. Gen. Genet.* **168**, 279–292.

Heyting, C., Talen, J. L., Weijers, P. J. and Borst, P. (1979). Fine structure of the 21S rRNA region on yeast mitochondrial DNA. II. The organization of sequences in petite mitochondrial DNAs carrying genetic markers from the 21S region. *Mol. Gen. Genet.* **168**, 251–278.

Hickey, E. D., Weber, L. A. and Baglioni, C. (1976). Inhibition of initiation of protein synthesis by 7-methylguanosine 5' monophosphate. *Proc. Nat. Acad. Sci. USA* **73**, 19–23.

Hickey, E. D., Weber, L. E., Baglioni, C., Kim, C. H. and Sarma, R. H. (1977). A relation between inhibition of protein synthesis and conformation of 5 phosphorylated 7 methylguanine derivatives. *J. Mol. Biol.* **109**, 173–184.

Highfield, D. P. and Dewey, W. C. (1972). Inhibition of DNA synthesis in synchronized Chinese hamster cells treated in G1 or early S phase with cycloheximide or puromycin. *Exp. Cell Res.* **75**, 314–320.

Highfield, P. E. and Ellis, R. J. (1978). Synthesis and transport of the small subunit of chloroplast RuBC. *Nature* **271**, 420–424.

Higuchi, R., Paddock, G. V., Wall, R. and Salser, W. (1976). A general method for cloning eucaryotic structural gene sequences. *Proc. Nat. Acad. Sci. USA* **73**, 3146–3150.

Hill, R. J., Maundrell, K. and Callan, H. G. (1974). Nonhistone proteins of the oocyte nucleus of the newt. *J. Cell Sci.* **15**, 145–161.

Hiller, G. and Weber, K. (1978). Radioimmunoassay for tubulin: a quantitative comparison of the tubulin content of different established tissue culture cells and tissues. *Cell* **14**, 795–804.

Himes, R. H., Burton, P. R., Kersey, R. N. and Pierson, G. B. (1976). Brain tubulin polymerization in the absence of "microtubule associated proteins." *Proc. Nat. Acad. Sci. USA* **73**, 4397–4399.

Hinegardner, R. (1968). Evolution of cellular DNA content in teleost fishes. *Amer. Nat.* **102**, 517–521.

Hinegardner, R. (1971). An improved fluorimetric assay for DNA. *Anal. Biochem.* **39**, 197–201.

Hinegardner, R. (1973). Cellular DNA content of the mollusca. *Comp. Biochem. Physiol.* **47A**, 447–460.

Hinegardner, R. (1974). Cellular DNA content of the Echinodermata. *Comp. Biochem. Physiol.* **49B**, 219–226.

Hinkley, R. and Telser, A. (1974). HMM binding filaments in the mitotic apparatus of mammalian cells. *Exp. Cell Res.* **86**, 161–164.

Hinnebusch, A. G., Clark, V. E. and Klotz, L. C. (1978). Length dependence in reassociation kinetics of radioactive tracer DNA. *Biochemistry* **17**, 1521–1529.

Hiraizumi, Y., Slatko, B., Langley, C. and Nill, A. (1973). Recombination in D. melanogaster male. *Genetics* **73**, 439-444.

Hiramoto, Y. (1956). Cell division without mitotic apparatus in sea urchin eggs. *Exp. Cell Res.* **11**, 630-636.

Hiramoto, Y. (1965). Further studies on cell division without mitotic apparatus in the sea urchin egg. *J. Cell Biol.* **25**, 161-167.

Hiramoto, Y. (1968). The mechanics and mechanism of cleavage in the sea urchin egg. *Symp. Soc. Exp. Biol.* **22**, 311-327.

Hirsch, M. and Penman, S. (1973). Mitochondrial poly(A)-containing RNA: localization and characterization. *J. Mol. Biol.* **80**, 379-392.

Hirsch, M. and Penman, S. (1974). The messenger like properties of the poly(A)$^+$ RNA in mammalian mitochondria. *Cell* **3**, 335-340.

Hirsch, M., Spradling, A. and Penman, S. (1974). The messenger like poly(A)-containing RNA species from the mitochondria of mammals and insects. *Cell* **1**, 31-36.

Hjelm, R. P., Kneale, G. G., Suau, P., Baldwin, J. P., Bradbury, E. M. and Ibel, K. (1977). Small angle neutron scattering studies of chromatin subunits in solution. *Cell* **10**, 139-151.

Ho, N. W. Y. and Gilham, P. T. (1974). Action of micrococcal nuclease on chemically modified DNA. *Biochemistry* **13**, 1082-1087.

Hochman, B. (1971). Analysis of chromosome 4 in D. melanogaster. II. EMS induced lethals. *Genetics* **67**, 235-252.

Hochman, B. (1973). Analysis of a whole chromosome in Drosophila. *Cold Spring Harbor Symp. Quant. Biol.* **38**, 581-589.

Hochman, J., Insel, P. A., Bourne, H. R., Coffino, P. and Tomkins, G. M. (1975). A structural gene mutation affecting the regulatory subunit of cAMP dependent protein kinase in mouse lymphoma cells. *Proc. Nat. Acad. Sci. USA* **72**, 5051-5055.

Hodnett, J. L. and Busch, H. (1968). Isolation and characterization of U rich 7S RNA of rat liver nuclei. *J. Biol. Chem.* **243**, 6334-6344.

Hoffman, H. P. and Avers, C. S. (1973). Mitochondrion of yeast: ultrastructural evidence for the giant, branched organelle per cell. *Science* **181**, 749-751.

Hofstetter, H., Schambock, A., Van den Berg, J. and Weissman, C. (1976). Specific excision of the inserted DNA segment from hybrid plasmids constructed by the poly(dA) poly(dT) method. *Biochim. Biophys. Acta* **454**, 587-591.

Hohmann, P., Tobey, R. A. and Gurley, L. R. (1976). Phosphorylation of distinct regions of f1 histone. Relationship to the cell cycle. *J. Biol. Chem.* **251**, 3685-3692.

Holden, J. A. and Kelley, W. N. (1978). Human HGPRT. Evidence for a tetrameric structure. *J. Biol. Chem.* **253**, 4459-4463.

Holder, J. W. and Lingrel, J. B. (1975). Determination of secondary structure in rabbit globin mRNA by thermal denaturation. *Biochemistry* **14**, 4209-4214.

Holland, C. A. and Skinner, D. M. (1977). The organization of the main component DNA of a crustacean genome with a paucity of middle repetitive sequences. *Chromosoma* **63**, 223-240.

Holland, M. J., Hager, G. L. and Rutter, W. J. (1977). Transcription of yeast DNA by homologous RNA polymerases I and II: selective transcription of ribosomal genes by RNA polymerase I. *Biochemistry* **16**, 16-24.

Hollenberg, C. P. (1976). Proportionate representation of rDNA and Balbiani ring DNA in polytene chromosomes of C. tentans. *Chromosoma* **57**, 185-197.

Hollenberg, C. P., Borst, P. and Van Bruggen, E. F. J. (1970). Mitochondrial DNA. V. A 25 μm closed circular duplex molecule in wild type yeast mitochondria. Structure and genetic complexity. *Biochim. Biophys. Acta* **209**, 1-15.

Hollenberg, C. P., Borst, P., Flavell, R. A., Van Kreijl, C. F., Van Bruggen, E. F. J. and Arnberg, A. C. (1972). The unusual properties of mitochondrial DNA from a low density petite mutant of yeast. *Biochim. Biophys. Acta* **277**, 44-58.

Holmes, A. M., Hesslewood, I. P. and Johnston, I. R. (1974). The occurrence of multiple activities in the high molecular weight DNA polymerase fraction of mammalian tissues. A preliminary study of some of their properties. *Eur. J. Biochem.* **43**, 487-499.

Holmes, A. M., Hesslewood, I. P. and Johnston, I. R. (1976). Evidence that DNA polymerase α of calf thymus contains a subunit of molecular weight 155,000. *Eur. J. Biochem.* **62**, 229-236.

Holmes, D. S. and Bonner, J. (1974). Interspersion of repetitive and single copy sequences in nuclear RNA of high molecular weight. *Proc. Nat. Acad. Sci. USA* **71**, 1108-1112.

Holmes, D. W., Mayfield, J. E., Sander, G. and Bonner, J. (1972). Chromosome RNA: its properties. *Science* **177**, 72-74.

Holt, T. K. H. (1970). Local protein accumulation during gene activation. I. Quantitative measurements on dye binding capacity at subsequent stages of pff formation in D. hydei. *Chromosoma* **32**, 64-78.

Holt, T. K. H. (1971). Local protein accumulation during gene activation. II. Interferometric measurements of the amount of solid material in temperature induced puffs of D. hydei. *Chromosoma* **32**, 428-435.

Honda, B. M., Baillie, D. L. and Candido, E. P. M. (1975). Properties of chromatin subunits from developing trout testis. *J. Biol. Chem.* **250**, 4643-4647.

Honda, B. M., Candido, E. P. M. and Dixon, G. H. (1975a). Histone methylation. Its occurrence in different cell types and relation to histone H4 metabolism in developing trout testis. *J. Biol. Chem.* **250**, 8686-9689.

Honda, B. M., Dixon, G. H. and Candido, E. P. M. (1975b). Sites of in vivo histone methylation in developing trout testis. *J. Biol. Chem.* **250**, 8681-8685.

Honjo, T., Packman, S., Swan, D. and Leder, P. (1974). Organization of immunoglobulin genes: reiteration frequency of the mouse chain constant region gene. *Proc. Nat. Acad. Sci. USA* **71**, 3659-3663.

Honjo, T., Packman, S., Swan, D. and Leder, P. (1976). Quantitation of constant and variable region genes for mouse immunoglobulin lambda chains. *Biochemistry* **15**, 2780-2784.

Honjo, T., Swan, D., Nau, M., Norman, B., Packman, S., Polsky, F. and Leder, P. (1976). Purification and translation of an immunoglobulin lambda chain mRNA from mouse myeloma. *Biochemistry* **15**, 2775-2779.

Honjo, T., Obata, M., Yamakawi-Kataoka, Y., Kataoka, T., Kawakami, T., Takahashi, N. and Mano, Y. (1979). Cloning and complete nucleotide sequence of mouse immunoglobulin γ 1 chain gene. *Cell* **18**, 559-568.

Hoober, J. K. and Blobel, G. (1969). Characterization of the chloroplastic and cytoplasmic ribosomes of C. reinhardii. *J. Mol. Biol.* **41**, 121-138.

Hood, L., Loh, E., Hubert, J., Barstad, P., Eaton, B., Early, P., Fuhrman, J., Johnson, N., Kronenberg, M. and Schilling, J. (1977). The structure and genetics of mouse immunoglobulins: an analysis of NZB myeloma proteins and sets of Balb-c myeloma proteins binding particular haptens. *Cold Spring Harbor Symp. Quant. Biol.* **41**, 817-836.

Hook, E. B. and Brustman, L. D. (1971). Evidence for selective differences between cells with an active horse X chromosome and cells with an active donkey X chromosome in the female mule. *Nature* **232**, 349-350.

Hooper, J. A., Smith, E. L., Sommer, K. R. and Chalkley, R. (1973). Amino acid sequence of histone III of the testes of the carp, Letiobus bubalus. *J. Biol. Chem.* **248**, 3275-3279.

Hopkins, J. M. (1970). Subsidiary components of the flagella of Chlamydomonas reinhardii. *J. Cell Sci.* **7**, 823-839.

Hopper, A. K., Banks, F. and Evangelidis, V. (1978). A yeast mutant which accumulates precursor tRNAs. *Cell* 14, 203–210.

Hopps, H. E., Bernheim, B. C., Nisalak, A., Tjio, J. H. and Smadel, J. E. (1963). Biologic characteristics of a continuous kidney cell line. *J. Immunol.* 91, 416–424.

Hori, T. A. and Lark, K. G. (1974). Autoradiographic studies of the replication of satellite DNA in the kangaroo rat. Autoradiographs of satellite DNA. *J. Mol. Biol.* 88, 221–232.

Horibata, K. and Harris, A. W. (1960). Mouse myelomas and lymphomas in culture. *Exp. Cell Res.* 60, 61–77.

Horn, D. and Davidson, R. L. (1975). Requirement of BUdR for the maintenance of "transformed" characteristics in BUdR dependent cells. *J. Cell. Physiol.* 85, 251–260.

Horowitz, M., Bratosin, S. and Aloni, Y. (1978). Polyoma infected cells contain at least three spliced late RNAs. *Nucl. Acids Res.* 5, 4663–4676.

Horz, W. and Zachau, H. G. (1977). Characterization of distinct segments in mouse satellite DNA by restriction nucleases. *Eur. J. Biochem.* 73, 383–392.

Horz, W., Hess, I. and Zachau, H. G. (1974). Highly regular arrangement of a restriction-nuclease-sensitive site in rodent satellite DNA. *Eur. J. Biochem.* 45, 501–512.

Hotta, Y. and Stern, H. (1971). Analysis of DNA synthesis during meiotic prophase in Lilium. *J. Mol. Biol.* 55, 337–356.

Hotta, Y. and Stern, H. (1976). Persistent discontinuities in late replicating DNA during meiosis in Lilium. *Chromosoma* 55, 171–182.

Hotta, Y., Chandley, A. C. and Stern, H. (1977). Biochemical analysis of meiosis in the male mouse. II. DNA metabolism at pachytene. *Chromosoma* 62, 255–268.

Hough, B. R., Smith, M. J., Britten, R. J. and Davidson, E. H. (1975). Sequence complexity of hnRNA in sea urchin embryos. *Cell* 5, 291–300.

Hough-Evans, B. R., Wold, B. J., Ernst, S. G., Britten, R. J. and Davidson, E. H. (1977). Appearance and persistence of maternal RNA sequences in sea urchin development. *Develop. Biol.* 60, 258–277.

Hourcade, D., Dressler, D. and Wolfson, J. (1973). The amplification of ribosomal RNA genes involves a rolling circle intermediate. *Proc. Nat. Acad. Sci. USA* 70, 2926–2930.

Housman, D. and Huberman, J. A. (1975). Changes in the rate of DNA replication fork movement during S phase in mammalian cells. *J. Mol. Biol.* 94, 173–182.

Howard, A. and Pelc, S. (1953). Synthesis of DNA in normal and irradiated cells and its relation to chromosome breakage. *Heredity Suppl.* 6, 261–273.

Howe, C. and Morgan, C. (1969). Interaction between Sendai virus and human erythrocytes. *J. Virol.* 3, 70–81.

Howell, S. H. and Walker, L. L. (1976). Informational complexity of the nuclear and chloroplast genomes of C. reinhardii. *Biochim. Biophys. Acta* 418, 249–256.

Howk, R. S., Anisowicz, A., Silverman, A. Y., Parks, W. P. and Scolnick, E. M. (1975). Distribution of murine type B and type C viral nucleic acid sequences in template active and template inactive chromatin. *Cell* 4, 321–328.

Howze, G. B., Hsie, A. W. and Olins, A. L. (1976). Nu bodies in mitotic chromatin. *Exp. Cell Res.* 100, 424–428.

Hozier, J. C. and Kraus, R. (1976). Subunit structure of chromosomes in mitotic nuclei of Physarum polycephalum. *Chromosoma* 57, 95–102.

Hozier, J., Renz, M. and Nehls, P. (1977). The chromosome fiber: evidence for an ordered superstructure of nucleosomes. *Chromosoma* 62, 301–318.

Hozumi, N. and Tonegawa, S. (1976). Evidence for somatic rearrangement of immunoglobulin genes coding for variable and constant regions. *Proc. Nat. Acad. Sci. USA* 73, 3628–3632.

Hsie, A. W. and Puck, T. T. (1971). Morphological transformation of Chinese hamster cells by dbcAMP and testosterone. *Proc. Nat. Acad. Sci. USA* **68**, 358-361.

Hsie, A. W., Jones, C. and Puck, T. T. (1971). Further changes in differentiation state accompanying the conversion of Chinese hamster cells to fibroblastic form by dbcAMP and hormones. *Proc. Nat. Acad. Sci. USA* **68**, 1648-1652.

Hsie, A. W., Brimer, P. A., Mitchell, T. J. and Gosslee, D. G. (1975). The dose respose relationship for EMS induced mutations at the HGPRT locus in CHO cells. *Somatic Cell Genet.* **1**, 247-262.

Hsu, T. C. (1964). Mammalian chromosomes in vitro. XVIII. DNA replication sequence in the Chinese hamster. *J. Cell Biol.* **23**, 53-62.

Hsu, T. C. and Arrhigi, F. E. (1971). Distribution of constitutive heterochromatin in mammalian chromosomes. *Chromosoma* **34**, 243-253.

Hsu, T. C. and Pathak, S. (1976). Differential rates of sister chromatid exchange between euchromatin and heterochromatin. *Chromosoma* **58**, 269-274.

Hsu, T. C. and Zenes, M. T. (1964). Mammalian chromosomes in vitro. XVII. Idiogram of the Chinese hamster. *J. Nat. Cancer Inst.* **32**, 857-869.

Hsu, T. C., Spirito, S. E. and Pardue, M. L. (1975). Distribution of 18S and 28S rRNA genes in mammalian genomes. *Chromosoma* **53**, 25-37.

Hsu, T. C., Cooper, J. E. K., Mace, M. L., Jr. and Brinkley, B. R. (1971). Arrangement of centromeres in mouse cells. *Chromosoma* **34**, 73-87.

Huang, R. C. C. and Bonner, J. (1965). Histone bound RNA, a component of native nucleohistone. *Proc. Nat. Acad. Sci. USA* **54**, 960-967.

Huang, R. C. C. and Huang, P. C. (1969). Effect of protein bound RNA associated with chick embryo chromatin on template specificity of the chromatin. *J. Mol. Biol.* **39**, 365-378.

Huang, R. C. C., Bonner, J. and Murray, K. (1967). Physical and biological properties of soluble nucleohistone. *J. Mol. Biol.* **8**, 54-64.

Huberman, J. A. (1973). Structure of chromosome fibers and chromosomes. *Ann. Rev. Biochem.* **42**, 355-378.

Huberman, J. A. and Attardi, G. (1966). Isolation of metaphase chromosomes from HeLa cells. *J. Cell Biol.* **31**, 95-105.

Huberman, J. A. and Attardi, G. (1967). Studies of fractionated HeLa cell metaphase chromosomes. I. The chromosomal distribution of DNA complementary to 28S and 18S rRNA and the cytoplasmic mRNA. *J. Mol. Biol.* **29**, 487-505.

Huberman, J. A. and Riggs, D. A. (1968). On the mechanism of DNA replication in mammalian chromosomes. *J. Mol. Biol.* **32**, 327-341.

Huberman, J. A. and Tsai, A. (1973). Direction of DNA replication in mammalian cells. *J. Mol. Biol.* **75**, 5-12.

Hubscher, U., Kuenzle, C. C. and Spadari, S. (1977). Identity of DNA polymerase γ from synaptosomal mitochondria and rat brain nuclei. *Eur. J. Biochem.* **81**, 249-258.

Hudspeth, M. E. S., Timberlake, W. E. and Goldberg, R. B. (1977). DNA sequence organization in the water mold Achlya. *Proc. Nat. Acad. Sci. USA* **74**, 4332-4336.

Huez, G., Marbaix, G., Hubert, E., Leclercq, M., Nudel, U., Soreq, H., Salomon, R., Lebleu, B., Revel, M. and Littauer, U. (1974). Role of the poly(A) segment in the translation of globin messenger RNA in Xenopus oocytes. *Proc. Nat. Acad. Sci. USA* **71**, 3143-3146.

Huez, G., Marbaix, G., Hubert, E., Cleuter, Y., Leclercq, M., Chantrenne, H., Devos, R., Soreq, H., Nudel, U. and Littauer, U. Z. (1975). Readenylation of poly(A)-free globin mRNA restores its stability in vivo. *Eur. J. Biochem.* **59**, 589-592.

Hughes, S. H., Wahl, G. M. and Capecchi, M. R. (1975). Purification and characterization of mouse HGPRT. *J. Biol. Chem.* **250**, 120-126.

Huisman, T. H. J., Schroeder, W. A., Bannister, W. H. and Grech, J. L. (1972a). Evidence for four nonallelic structural genes for the γ chain of human fetal hemoglobin. *Biochem. Genet.* **7**, 131–139.

Huisman, T. H. J., Wrightstone, R. N., Wilson, J. B., Schroeder, W. A. and Kendall, A. G. (1972b). Hemoglobin Kenya, the product of fusion of γ and β polypeptide chains. *Arch. Biochem. Biophys.* **153**, 850–853.

Hulten, M. (1974). Chiasma distribution at diakinesis in the normal human male. *Hereditas* **76**, 55–78.

Humphries, S., Windass, J. and Williamson, R. (1976). Mouse globin gene expression in erythroid and non erythroid tissues. *Cell* **7**, 267–278.

Humphries, S. E., Young, D. and Carroll, D. (1979). Chromatin structure of the 5S RNA genes of X. laevis. *Biochemistry* **18**, 3223–3231.

Hunt, L. T. and Dayhoff, M. O. (1977). Amino terminal sequence identity of ubiquitin and the nonhistone component of nuclear protein A24. *Biochem. Biophys. Res. Commun.* **74**, 650–655.

Hunter, T. and Garrels, J. I. (1977). Characterization of the mRNAs for α, β and γ actin. *Cell* **12**, 767–781.

Huntley, G. H. and Dixon, G. H. (1972). The primary structure of the N terminal region of histone T. *J. Biol. Chem.* **247**, 4916–4919.

Hutchison, C. A., Newbold, J. E., Potter, S. S. and Edgell, M. H. (1974). Maternal inheritance of mammalian mitochondrial DNA. *Nature* **251**, 536–538.

Hutchinson, M. A., Hunter, T. and Eckhart, W. (1978). Characterization of T antigens in polyoma infected and transformed cells. *Cell* **15**, 65–77.

Hutchison, N. and Pardue, M. L. (1975). The mitotic chromosomes of T. viridescens: localization of C banding regions and DNA sequences complementary to 18S, 28S and 5S rRNAs. *Chromosoma* **53**, 51–69.

Hutton, J. R. and Thomas, C. A., Jr. (1975). The origin of folded DNA rings from D. melanogaster. *J. Mol. Biol.* **98**, 425–438.

Hutton, J. R. and Wetmur, J. G. (1973a). Renaturation of phage φX 174 DNA-RNA hybrid: RNA length effect and nucleation rate constant. *J. Mol. Biol.* **77**, 495–500.

Hutton, J. R. and Wetmur, J. G. (1973b). Length dependence of the kinetic complexity of mouse satellite DNA. *Biochem. Biophys. Res. Commun.* **52**, 1148–1161.

Huxley, H. E. (1969). The mechanism of muscular contraction. *Science* **164**, 1356–1366.

Huxley, H. E. (1973). Muscular contraction and cell motility. *Nature* **243**, 445–449.

Hyams, J. S. and Borisy, G. G. (1978). Nucleation of microtubules in vitro by isolated spindle pole bodies of the yeast S. cerevisiae. *J. Cell Biol.* **78**, 401–414.

Hynes, N. E., Groner, B., Sippel, A. E., Nguyen-Huu, M. C. and Schutz, G. (1977). mRNA complexity and egg white protein mRNA content in mature and hormone withdrawn oviduct. *Cell* **11**, 923–932.

Hynes, R. O. (1974). Role of surface alterations in cell transformation: the importance of proteases and surface proteins. *Cell* **1**, 147–156.

Hynes, R. O. and Destree, A. T. (1978). Relationships between fibronectin and actin. *Cell* **15**, 876–886.

Iatrou, K. and Dixon, G. H. (1977). The distribution of poly(A)$^+$ and poly(A)$^-$ protamine mRNA sequences in the developing trout testis. *Cell* **10**, 433–441.

Imaizumi, T., Diggelmann, H. and Scherrer, K. (1973). Demonstration of globin messenger sequences in giant nuclear precursors or mRNA of avian erythroblasts. *Proc. Nat. Acad. Sci. USA* **70**, 1122–1126.

Infante, A. A. and Nemer, M. (1968). Heterogeneous RNP in the cytoplasm of sea urchin embryos. *J. Mol. Biol.* **32**, 543–565.

Ingles, C. J. (1978). Temperature sensitive RNA polymerase II mutations in CHO cells. *Proc. Nat. Acad. Sci. USA* **75**, 405–409.

Ingles, C. J., Guialis, A., Lam, J. and Siminovitch, L. (1976). Amanitin resistance of RNA polymerase II in mutant Chinese hamster ovary cell lines. *J. Biol. Chem.* **251**, 2729–2734.

Innis, M. A. and Miller, D. L. (1977). Quantitation of rat α fetoprotein mRNA with a cDNA probe. *J. Biol. Chem.* **252**, 8469–8475.

Inoué, S. (1953). Polarization optical studies of the mitotic spindle. I. The demonstration of spindle fibers in living cells. *Chromosoma* **5**, 487–500.

Inoué, S. (1960). On the physical properties of the mitotic spindle. *Ann. N. Y. Acad. Sci.* **90**, 529–530.

Inoué, S. (1964). Organization and function of the mitotic spindle. In R. D. Allen and N. Kamiya (Eds.) *Primitive Motile Systems in Cell Biology*, Academic Press, New York, pp. 549–594.

Inoué, S. and Ritter, H. (1978). Mitosis in Barulanympha. II. Dynamics of a two stage anaphase, nuclear morphogenesis and cytokinesis. *J. Cell. Biol.* **77**, 655–684.

Inoué, S. and Sato, H. (1967). Cell motility by labile association of molecules. The nature of mitotic spindle fibers and their role in chromosome movement. *J. Gen. Physiol.* **50**, *Suppl.* 259–288.

Inoué, S., Borisy, G. G. and Kiehart, D. P. (1974). Growth and lability of Chaetopterus oocyte mitotic spindles isolated in the presence of porcine brain tubulin. *J. Cell Biol.* **62**, 175–184.

Irwin, D., Kumar, A. and Malt, R. A. (1975). Messenger ribonucleoprotein complexes isolated with oligo(dT)-cellulose chromatography from kidney polysomes. *Cell* **4**, 157–166.

Ish-Horowicz, D., Holden, J. J. and Gehring, W. J. (1977). Deletions of two heat activated loci in D. melanogaster and their effects on heat induced protein synthesis. *Cell* **12**, 643–652.

Ish-Horowicz, D., Pinchin, S. M., Gausz, J., Gyurkovics, H., Bencze, G., Goldschmidt-Clermont, M. and Holden, J. J. (1979). Deletion mapping of two D. melanogaster locu that code for the 70,000 dalton heat induced protein. *Cell* **17**, 565–571.

Ishikawa, H., Bischoff, R. and Holtzer, H. (1969). Formation of arrowhead complexes with heavy meromyosin in a variety of cell types. *J. Cell Biol.* **43**, 312–328.

Itakura, K., Hirose, T., Crea, R., Riggs, A. D., Heyneker, H. L., Bolivar, F. and Boyer, H. W. (1977). Expression in E. coli of a chemically synthesized gene for the hormone somatostatin. *Science* **198**, 1056–1062.

Iwai, K., Ishikawa, K. and Hayashi, H. (1970). Amino acid sequence of slightly lysine rich histone. *Nature* **226**, 1056–1058.

Izant, J. G. and Lazarides, E. (1977). Invariance and heterogeneity in the major structural and regulatory proteins of chick muscle cells revealed by two dimensional gel electrophoresis. *Proc. Nat. Acad. Sci. USA* **74**, 1450–1454.

Jackl, G. and Sebald, W. (1975). Identification of two products of mitochondrial protein synthesis associated with mitochondrial ATPase from N. crassa. *Eur. J. Biochem.* **54**, 97–106.

Jackson, D. A., Symons, R. H. and Berg, P. (1972). Biochemical method for inserting new genetic information into DNA of SV40: circular SV40 DNA molecules containing lambda phage genomes and the galactose operon of E. coli. *Proc. Nat. Acad. Sci. USA* **69**, 2904–2909.

Jackson, V. (1978). Studies on histone organization in the nucleosome using formaldehyde as a reversible cross linking agent. *Cell* **15**, 945–953.

Jackson, V., Granner, D. K. and Chalkley, R. (1975). Deposition of histones onto replicating chromosomes. *Proc. Nat. Acad. Sci. USA* **72**, 4440–4444.

Jackson, V., Granner, D. and Chalkley, R. (1976). Deposition of histone onto the replicating chromosome: newly synthesized histone is not found near the replication fork. *Proc. Nat. Acad. Sci. USA* **73**, 2266–2269.

Jackson, V., Shires, A., Chalkley, R. and Granner, D. K. (1975). Studies on highly metabolically active acetylation and phosphorylation of histones. *J. Biol. Chem.* **250**, 4856–4863.

Jackson, V., Shires, A., Tanphaichitr, N. and Chalkley, R. (1976). Modifications to histones immediately after synthesis. *J. Mol. Biol.* **104**, 471-484.

Jacob, E. (1976). Histone gene reiteration in the genome of the mouse. *Eur. J. Biochem.* **65**, 275-284.

Jacob, E., Malacinski, G. and Birnstiel, M. L. (1976). Reiteration frequency of the histone genes in the genome of the amphibian, Xenopus laevis. *Eur. J. Biochem.* **69**, 45-54.

Jacob, F. and Monod, J. (1961). Genetic regulatory mechanisms in the synthesis of proteins. *J. Mol. Biol.* **3**, 318-356.

Jacob, S. T., Sajdel, E. M. and Munro, H. N. (1970). Different responses of soluble whole nuclear RNA polymerase and soluble nucleolar RNA polymerase to divalent cations and to inhibition by α amanitin. *Biochem. Biophys. Res. Commun.* **38**, 765-770.

Jacobs, M., Smith, H. and Taylor, E. W. (1974). Tubulin: nucleotide binding and enzymatic activity. *J. Mol. Biol.* **89**, 455-468.

Jacobs-Lorena, M., Baglioni, C. and Borun, T. W. (1972). Translation of mRNA for histones from HeLa cells by a cell free extract from mouse ascites tumor. *Proc. Nat. Acad. Sci. USA* **69**, 2095-2099.

Jacobson, D. N. and Holt, C. E. (1973). Isolation of rRNA precursors from Physarum polycephalum. *Arch. Biochem. Biophys.* **159**, 342-352.

Jacobson, R. A. and Bonner, J. (1971). Studies of the chromosomal RNA and of the chromosomal RNA binding protein of higher organisms. *Arch. Biochem. Biophys.* **146**, 557-563.

Jacq, B., Jourdan, R. and Jordan, B. R. (1977). Structure and processing of precursor 5S RNA in D. melanogaster. *J. Mol. Biol.* **117**, 785-796.

Jacq, C., Miller, J. R. and Brownlee, G. G. (1977). A pseudogene structure in 5S DNA of X. laevis. *Cell* **12**, 109-120.

Jacquet, M. and Gros, F. (1979). Expression of single copy DNA sequences in nuclear RNA from undifferentiated mouse embryonal carcinoma and differentiated muscle line. *Nucl. Acids Res.* **6**, 1639-1656.

Jacquet, M., Affara, N. A., Robert, B., Jakob, H., Jacob, F. and Gros, F. (1978). Complexity of nuclear and polysomal polyadenylated RNA in a pluripotent embryonal carcinoma cell line. *Biochemistry* **17**, 69-79.

Jaehning, J. A. and Roeder, R. G. (1977). Transcription of specific adenovirus genes in isolated nuclei by exogenous RNA polymerase. *J. Biol. Chem.* **252**, 8753-8761.

Jain, S. K. and Sarkar, S. (1979). Poly(A)-containing mRNP particles of chick embryonic muscles. *Biochemistry* **18**, 745-752.

Jakob, K. M. (1972). RNA synthesis during the DNA synthetic period on the first cell cycle in the root meristem of germinating Vicia faba. *Exp. Cell Res.* **72**, 37-376.

Jamrich, M., Greenleaf, A. L. and Bautz, E. K. F. (1977). Localization of RNA polymerase in polytene chromosomes of D. melanogaster. *Proc. Nat. Acad. Sci. USA* **74**, 2079-2083.

Jamrich, M., Haars, R., Wulf, E. and Bautz, F. A. (1977). Correlation of RNA polymerase B and transcription activation in the chromosomes of D. melanogaster. *Chromosoma* **64**, 319-326.

Janssens, F. A. (1909). Spermatogénèse dans les Batraciens. V. La théori de la chiasmatype. Nouvelles intérpretation des cinès de maturation. *Cellule* **25**, 387-411.

Jeanteur, P., Amaldi, F. and Attardi, G. (1968). Partial sequence analysis of rRNA from HeLa cells. II. Evidence for sequences of non ribosomal type in 45S and 32S rRNA precursors. *J. Mol. Biol.* **33**, 757-776.

Jeffery, W. R. and Brawerman, G. (1975). Association of the poly(A) segment of mRNA with other polynucleotide sequences in mouse sarcoma 180 polysomes. *Biochemistry* **14**, 3445-3450.

Jeffreys, A. J. and Flavell, R. A. (1977a). A physical map of the DNA regions flanking the rabbit β globin gene. *Cell* **12**, 429-439.

Jeffreys, A. J. and Flavell, R. A. (1977b). The rabbit β globin gene contains a large insert in the coding sequence. *Cell* 12, 1097-1108.

Jelinek, W. R. (1977). Specific nucleotide sequences in HeLa cells inverted repeated DNA: enrichment for sequences found in double stranded regions of hnRNA. *J. Mol. Biol.* 115, 591-602.

Jelinek, W. R. (1978). Inverted repeated DNA from CHO cells studied with cloned DNA fragments. *Proc. Nat. Acad. Sci. USA* 75, 2679-2683.

Jelinek, W. and Darnell, J. E. (1972). Double stranded regions on hnRNA from HeLa cells. *Proc. Nat. Acad. Sci. USA* 69, 2537-2541.

Jelinek, W. and Leinwand, L. (1978). Low molecular weight RNAs hydrogen bonded to nuclear and cytoplasmic poly(A)-terminated RNA from cultured CHO cells. *Cell* 15, 205-214.

Jelinek, W., Adesnik, M., Salditt, M., Sheiness, D., Wall, R., Molloy, G., Philipson, L. and Darnell, J. E. (1973a). Further evidence on the nuclear origin and transfer to the cytoplasm of poly(A) sequences in mammalian cell RNA. *J. Mol. Biol.* 75, 515-532.

Jelinek, W., Molloy, G., Salditt, M., Wall, R., Sheiness, D. and Darnell, J. E. (1973b). Origin of mRNA in HeLa cells and the implications for chromosome structure. *Cold Spring Harbor Symp. Quant. Biol.* 38, 891-898.

Jelinek, W. R., Molloy, G., Fernandez-Munoz, R., Salditt, M. and Darnell, J. E. (1974). Secondary structure in hnRNA: involvement of regions from repeated DNA sites. *J. Mol. Biol.* 82, 361-370.

Jendrisak, J. and Guilfoyle, T. J. (1978). Eucaryotic RNA polymerases: comparative subunit structures, immunological properties, and α amanitin sensitivities of the class II enzymes from higher plants. *Biochemistry* 17, 1322-1327.

Jenni, B. and Stutz, E. (1978). Physical mapping of the ribosomal DNA region of Euglena gracilis chloroplast DNA. *Eur. J. Biochem.* 88, 127-134.

Joffe, J., Keene, M. and Weintraub, H. (1977). Histones H2A, H2B, H3 and H4 are present in equimolar amounts in chick erythroblasts. *Biochemistry* 16, 1236-1238.

John, H. A., Birnsteil, M. L. and Jones, K. W. (1969). RNA-DNA hybrids at the cytological level. *Nature* 223, 582-587.

John, H. A., Patrinou-Gourgoulas, M. and Jones, K. W. (1977). Detection of myosin heavy chain mRNA during myogenesis in tissue culture by in vitro and in situ hybridization. *Cell* 12, 501-508.

Johns, E. W. (1964). Preparative methods for histone fractions from calf thymus. *Biochem. J.* 92, 55-59.

Johns, E. W. (1971). The preparation and characterization of histones. In D. M. Phillips (Ed.), *Histones and Nucleohistones*, Plenum Publishing, New York, pp 2-46.

Johns, E. W. and Butler, J. A. V. (1962). Further fractionation of histones from calf thymus. *Biochem. J.* 82, 15-18.

Johns, E. W. and Diggle, J. H. (1969). A method for the large scale preparation of the avian erythrocyte specific histone f2c. *Eur. J. Biochem.* 11, 495-498.

Johns, E. W. and Forrester, S. (1969). Interactions between the lysine rich histone f1 and DNA. *Biochem. J.* 111, 371-374.

Johns, E. W. and Hoare, T. A. (1970). Histones and gene control. *Nature* 226, 650-651.

Johnson, G. S., Friedman, R. M. and Pastan, I. (1971). Restoration of several morphological characteristics of normal fibroblasts in sarcoma cells treated with cyclic AMP and its derivatives. *Proc. Nat. Acad. Sci. USA* 68, 425-429.

Johnson, K. A. and Borisy, G. G. (1977). Kinetic analysis of microtubule self assembly in vitro. *J. Mol. Biol.* 117, 1-32.

Johnson, L. F., Penman, S. and Green, H. (1976). Increasing content of poly(A)$^+$ mRNA of serum stimulated cells in the absence of ribosome synthesis. *J. Cell. Physiol.* 87, 141-146.

Johnson, L. F., Abelson, H. T., Green, H. and Penman, S. (1974). Changes in RNA in relation to growth of the fibroblast. I. Amounts of mRNA, rRNA and tRNA in resting and growing cells. *Cell* 1, 95-100.

Johnson, L. F., Williams, J. G., Abelson, H. T., Green, H. and Penman, S. (1975). Changes in RNA in relation to growth of the fibroblast. III. Posttranscriptional regulation of mRNA formation in resting and growing cells. *Cell* 4, 69-75.

Johnson, L. F., Levis, R., Abelson, H. T., Green, H. and Penman, S. (1976). Changes in RNA in relation to growth of the fibroblast. IV. Alterations in the production and processing of mRNA and rRNA in resting and growing cells. *J. Cell Biol.* 71, 933-937.

Johnson, R. T. and Harris, H. (1969a). DNA synthesis and mitosis in fused cells. I. HeLa homokaryons. *J. Cell Sci.* 5, 603-624.

Johnson, R. T. and Harris, H. (1969b). DNA synthesis and mitosis in fused cells. II. HeLa-chick erythrocyte heterokaryons. *J. Cell Sci.* 5, 625-644.

Johnson, R. T. and Mullinger, A. M. (1975). The induction of DNA synthesis in the chick red cell nucleus in heterokaryons during the first cell cycle after fusion with HeLa cells. *J. Cell Sci.* 18, 455-490.

Johnson, R. T. and Rao, P. N. (1970). Mammalian cell fusion: induction of premature chromosome condensation in interphase nuclei. *Nature* 226, 717-722.

Johnson, R. T. and Rao, P. N. (1971). Nucleocytoplasmic interactions in the achievement of nuclear synchrony in DNA synthesis and mitosis in multinucleate cells. *Biol. Rev.* 46, 97-155.

Johnson, R. T., Mullinger, A. M. and Skaer, R. J. (1975). Perturbation of mammalian cell division: human minisegregants derived from mitotic cells. *Proc. Roy. Soc. B.* 189, 591-602.

Johnson, R. T., Rao, P. N. and Hughes, H. D. (1970). Mammalian cell fusion. III. A HeLa cell inducer of premature chromosome condensation active in cells from a variety of animal species. *J. Cell. Physiol.* 76, 151-158.

Jokelainen, P. T. (1967). The ultrastructure and spatial organization of the metaphase kinetochore in mitotic rat cells. *J. Ult. Res.* 19, 19-44.

Joklik, W. K. and Becker, Y. (1965). Studies on the genesis of polysomes. II. The association of nascent mRNA with the 40S subribosomal particle. *J. Mol. Biol.* 13, 496-510.

Jones, G. E. and Sargent, P. A. (1974). Mutants of cultured Chinese hamster cells deficient in adenine phosphoribosyl transferase. *Cell* 2, 43-54.

Jones, G. H. (1971). The analysis of exchanges in tritium labeled meiotic chromosomes. II. Stethophyma grossum. *Chromosoma* 34, 367-382.

Jones, G. M. T., Rall, S. C. and Cole, R. D. (1974). Extension of the amino acid sequence of a lysine rich histone. *J. Biol. Chem.* 249, 2548-2553.

Jones, K. W. (1970). Chromosomal and nuclear location of mouse satellite DNA in individual cells. *Nature* 225, 912-915.

Jones, K. W. and Corneo, G. (1971). Location of satellite and homogeneous DNA sequences on human chromosomes. *Nature New Biol.* 233, 268-271.

Jones, K. W. and Robertson, F. W. (1970). Localization of reiterated nucleotide sequences in Drosophila or mouse by in situ hybridization of complementary RNA. *Chromosoma* 31, 331-345.

Jorcano, J. L. and Ruiz-Carrillo, A. (1979). H3-H4 tetramer directs DNA and core histone octamer assembly in the nucleosome core particle. *Biochemistry* 18, 768-773.

Jordan, B. R., Jourdan, R. and Jacq, B. (1976). Late steps in the maturation of Drosophila 26S ribosomal RNA: generation of 5.8S and 2S RNAs by cleavages occuring in the cytoplasm. *J. Mol. Biol.* 101, 85-106.

Judd, B. H. and Young, M. W. (1973). An examination of the one cistron: one chromomere concept. *Cold Spring Harbor Symp. Quant. Biol.* 38, 573-579.

Judd, B. H., Shen, M. W. and Kaufman, T. C. (1972). The anatomy and function of a segment of the X chromosome of D. melanogaster. *Genetics* 71, 139-156.

Kacian, D. L., Spiegelman, S., Bank, A., Terada, M., Metafora, S., Dow, L. and Marks, P. (1972). In vitro synthesis of DNA components of human genes for globins. *Nature New Biol.* 235, 167-169.

Kahan, B. and DeMars, R. (1975). Localized derepression on the human inactive X chromosome in mouse-human cell hybrids. *Proc. Nat. Acad. Sci. USA* **72**, 1510-1514.

Kalisch, W. E. and Hagele, K. (1976). Correspondence of banding patterns to ³H thymidine labeling patterns in polytene chromosomes. *Chromosoma* **57**, 19-23.

Kalt, M. R. and Gall, J. G. (1974). Observations on early germ cell development and premeiotic rDNA amplification in X. laevis. *J. Cell Biol.* **62**, 460-470.

Kan, Y. W., Dozy, A. M., Varmus, H. E., Taylor, J. M., Holland, J. P., Lie-Injo, L. E., Ganesan, J. and Todd, D. (1975a). Deletion of α globin genes in hemoglobin H disease demonstrates multiple α globin structural loci. *Nature* **255**, 255-256.

Kan, Y. W., Holland, J. P., Dozy, A. M. and Varmus, H. E. (1975b). Demonstration of nonfunctional β globin mRNA in homozygous β°-thalassemia. *Proc. Nat. Acad. Sci. USA* **72**, 5140-5144.

Kane, R. E. (1962). The mitotic apparatus. Isolation by controlled pH. *J. Cell Biol.* **12**, 47-55.

Kane, R. E. (1967). The mitotic apparatus. Identification of the major soluble component of the glycol isolated mitotic apparatus. *J. Cell Biol.* **32**, 243-253.

Kane, R. E. (1975). Preparation and purification of polymerized actin from sea urchin egg extracts. *J. Cell Biol.* **66**, 305-315.

Kane, R. E. (1976). Actin polymerization and interaction with other proteins in temperature induced gelation of sea urchin egg extracts. *J. Cell Biol.* **71**, 704-714.

Kao, F.-T. (1973). Identification of chick chromosomes in cell hybrids formed between chick erythrocytes and adenine requiring mutants of Chinese hamster cells. *Proc. Nat. Acad. Sci. USA* **70**, 2893-2898.

Kao, F. -T. and Puck, T. T. (1967). Genetics of somatic mammalian cells. IV. Properties of Chinese hamster cells with respect to the requirement for proline. *Genetics* **55**, 513-524.

Kao, F. -T. and Puck, T. T. (1968). Genetics of somatic mammalian cells. VIII. Induction and isolation of nutritional mutants in Chinese hamster cells. *Proc. Nat. Acad. Sci.* **60**, 1275-1281.

Kao, F. -T. and Puck, T. T. (1969). Genetics of somatic mammalian cells. IX. Quantitation of mutagenesis by physical and chemical agents. *J. Cell. Physiol.* **74**, 245-258.

Kao, F. -T. and Puck, T. T. (1970). Genetics of somatic mammalian cells: linkage studies with human-Chinese hamster cell hybrids. *Nature* **228**, 329-333.

Kao, F. T. and Puck, T. T. (1972). Genetics of somatic mammalian cells. XIV. Genetic analysis in vitro of auxotrophic mutants. *J. Cell. Physiol.* **80**, 41-50.

Kao, F. T., Chasin, L. and Puck, T. T. (1969). Genetics of somatic mammalian cells. X. Complementation analysis of glycine requiring mutants. *Proc. Nat. Acad. Sci.* **64**, 1284-1291.

Kao-Huang, Y., Revzin, A., Butler, A. P., O'Connor, P., Noble, D. W. and Von Hippel, P. H. (1977). Nonspecific DNA binding of genome regulating proteins as a biological control mechanism: measurement of DNA bound E. coli lac repressor in vivo. *Proc. Nat. Acad. Sci. USA* **74**, 4228-4232.

Kaplan, D. W. (1953). The influence of Minute upon somatic crossing over in D. melanogaster. *Genetics* **38**, 630-651.

Karkas, J. D., Margulies, L. and Chargaff, E. (1975). A DNA polymerase from embryos of D. melanogaster. Purification and properties. *J. Biol. Chem.* **250**, 8657-8663.

Karn, J., Johnson, E. M., Vidali, G. and Allfrey, V. G. (1974). Differential phosphorylation and turnover of nuclear acidic proteins during the cell cycle of synchronized HeLa cells. *J. Biol. Chem.* **249**, 667-677.

Karn, J., Vidali, G., Boffa, L. C. and Allfrey, V. G. (1977). Characterization of the nonhistone nuclear proteins associated with rapidly labeled hnRNA. *J. Biol. Chem.* **252**, 7307-7322.

Karrer, K. M. and Gall, J. G. (1976). The macronuclear ribosomal DNA of Tetrahymena pyriformis is a palindrome. *J. Mol. Biol.* **104**, 421-454.

Kasamatsu, H., Robberson, D. L. and Vinograd, J. (1971). A novel closed circular mitochondrial DNA with properties of a replicating intermediate. *Proc. Nat. Acad. Sci. USA* **68**, 2252-2257.

Katan, M. B., Harten-Loosbroek, N. V. and Groot, G. S. P. (1976). The cytochrome *bcl* complex of yeast mitochondria. *Eur. J. Biochem.* **70**, 409-417.

Kato, H. (1977). Mechanisms for sister chromatid exchange and their relation to the production of chromosomal aberrations. *Chromosoma* **59**, 179-192.

Kato, H. and Moriwaki, K. (1972). Factors involved in the production of banded structures in mammalian chromosomes. *Chromosoma* **38**, 105-120.

Kato, H. and Yosida, T. H. (1972). Banding patterns of Chinese hamster chromosomes revealed by new techniques. *Chromosoma* **36**, 272-280.

Kaufman, T. C., Shannon, M. P., Shen, M. W. and Judd, B. H. (1975). A revision of the cytology and ontogeny of several deficiencies in the 3A1-3C6 region of the X chromosome of D. melanogaster. *Genetics* **79**, 265-282.

Kaufmann, Y., Milcarek, C., Berissi, H. and Penman, S. (1977). HeLa cell poly(A)⁻ mRNA codes for a subset of poly(A)⁺ mRNA-directed proteins with an actin as a major product. *Proc. Nat. Acad. Sci. USA* **74**, 4801-4805.

Kavenoff, R. and Zimm, B. H. (1973). Chromosome sized DNA molecules from Drosophila. *Chromosoma* **41**, 1-28.

Kedes, L. H. and Birnsteil, M. L. (1971). Reiteration and clustering of DNA sequences complementary to histone mRNA. *Nature New Biol.* **230**, 165-169.

Kedes, L. H., Cohn, R. H., Lowry, J. D., Chang, A. C. Y. and Cohen, S. N. (1975). The organization of sea urchin histone genes. *Cell* **6**, 359-370.

Kedinger, C. and Chambon, P. (1972). Animal DNA dependent RNA polymerases. 3. Purification of calf thymus B1 and B2 enzymes. *Eur. J. Biochem.* **28**, 283-290.

Kedinger, C., Gissinger, F. and Chambon, P. (1974). Animal DNA-dependent RNA polymerases. Molecular structures and immunological properties of calf thymus enzyme A1 and of calf thymus and rat liver enzymes B. *Eur. J. Biochem.* **44**, 421-436.

Kedinger, C., Gissinger, F., Gniazdowski, M., Mandel, J. L. and Chambon, P. (1972). Animal DNA dependent RNA polymerases. I. Large scale solubilization and separation of A and B calf thymus RNA polymerase activities. *Eur. J. Biochem.* **28**, 269-276.

Keith, J. M., Ensinger, M. J. and Moss, B. (1978). HeLa cell RNA methyltransferase specific for the capped 5′ end of mRNA. *J. Biol. Chem.* **253**, 5033-5039.

Kellems, R. E., Alt, F. W. and Schimke, R. T. (1976). Regulation of folate reductase synthesis in sensitive and methotrexate resistant sarcoma cells. *J. Biol. Chem.* **251**, 6987-6993.

Keller, W. and Wendel, I. (1975). Stepwise relaxation of supercoiled SV40 DNA. *Cold Spring Harbor Symp. Quant. Biol.* **39**, 199-208.

Kelley, M. G. and Hartwell, J. L. (1954). The biological effects and the chemical composition of podophyllotoxin. A review. *J. Nat. Cancer Inst.* **14**, 967-1010.

Kempe, T. D., Swyryd, E. A., Bruist, M. and Stark, G. R. (1976). Stable mutants of mammalian cells that overproduce the first three enzymes of pyrimidine nucleotide biosynthesis. *Cell* **9**, 541-550.

Keshgegian, A. and Furth, J. J. (1972). Comparison of transcription of chromatin by calf thymus and E. coli RNA polymerase. *Biochem. Biophys. Res. Commun.* **48**, 757-763.

Keyl, H. G. (1965). A demonstrable local and geometric increase in the chromosomal DNA of Chironomus. *Experientia* **21**, 191-199.

Keyl, H. G. and Pelling, C. (1963). Differentielle DNS replikation in den speicheldrusen chromosomen von C. thummii. *Chromosoma* **14**, 347-359.

Khan, M. S. N. and Maden, B. E. H. (1976). Nucleotide sequences within the rRNAs of HeLa cells, X. laevis and chick embryo fibroblasts. *J. Mol. Biol.* **101**, 235-254.

Khan, M. S. N. and Maden, B. E. H. (1978). Conformation of methylated sequences in HeLa cell 18S rRNA: nuclease S1 as a probe. *Eur. J. Biochem.* **84**, 241–250.

Khoury, G., Gruss, P., Dhar, R. and Lai, C. -J. (1979). Processing and expression of early SV40 mRNA: a role for RNA conformation in splicing. *Cell* **18**, 85–92.

Kiefer, B., Sakai, H., Solari, A. J. and Mazia, D. (1966). The molecular unit of the microtubules of the mitotic apparatus. *J. Mol. Biol.* **29**, 75–79.

Kihlman, B. A. (1971). Molecular mechanisms of chromosome breakage and rejoining. *Adv. Cell. Mol. Biol.* **1**, 59–108.

Kihlman, B. A. and Hartley, B. (1967). Subchromatid exchanges and the folded fiber model of chromosome structure. *Hereditas* **57**, 289–294.

Killander, D. and Zetterberg, A. (1965a). Quantitative cytochemical studies on interphase growth. I. Determination of DNA, RNA and mass content of age determined mouse fibroblasts in vitro and of intercellular variation in generation time. *Exp. Cell Res.* **38**, 272–284.

Killander, D. and Zetterberg, A. (1965b). A quantitative cytochemical investigation of the relationship between cell mass and initiation of DNA synthesis in mouse fibroblasts in vitro. *Exp. Cell Res.* **40**, 12–20.

Kim, H., Binder, L. I. and Rosenbaum, J. L. (1979). The periodic association of MAP2 with brain microtubules in vitro. *J. Cell Biol.* **80**, 266–276.

Kim, M. A., Johanssmann, R. and Grzeschik, K. H. (1975). Giemsa staining of the sites replicating DNA early in human lymphocyte chromosomes. *Cytogenet. Cell Genet.* **15**, 363–371.

Kimmel, A. R. and Firtel, R. A. (1979). A family of short interspersed repeat sequences at the 5′ end of a set of Dictyostelium single copy mRNAs. *Cell* **16**, 787–796.

Kimmel, A. R. and Gorovsky, M. A. (1976). Number of 5S and tRNA genes in macro and micronuclei of T. pyriformis. *Chromosoma* **54**, 327–337.

Kimura, M. and Ohta, T. (1971). Protein polymorphism as a phase of molecular evolution. *Nature* **229**, 467–469.

Kindas-Mugge, I., Lane, C. D. and Kreil, G. (1974). Insect protein synthesis in frog cells: the translation of honey bee promelittin mRNA in Xenopus oocytes. *J. Mol. Biol.* **87**, 451–462.

Kindle, K. L. and Firtel, R. A. (1978). Identification and analysis of Dictyostelium actin genes, a family of moderately repeated genes. *Cell* **15**, 763–778.

Kindle, K. L. and Firtel, R. A. (1979). Evidence that populations of Dictyostelium single copy mRNA transcripts carry common repeat sequences. *Nucl. Acids Res.* **6**, 2403–2422.

King, R. C. (1970). The meiotic behavior of the Drosophila oocyte. *Int. Rev. Cytol.* **28**, 125–168.

King, T. J. and Briggs, R. (1956). Serial transplantation of embryonic nuclei. *Cold Spring Harbor Symp. Quant. Biol.* **21**, 271–290.

Kinkade, J. M., Jr. and Cole, R. D. (1966). A structural comparison of different lysine rich histones of calf thymus. *J. Biol. Chem.* **241**, 5798–5805.

Kinniburgh, A. J. and Ross, J. (1979). Processing of the mouse β globin mRNA precursor: at least two cleavage ligation reactions are necessary to excise the larger intervening sequence. *Cell* **17**, 915–922.

Kinniburgh, A. J., Mertz, J. E. and Ross, J. (1978). The precursor of mouse β globin mRNA contains two intervening RNA sequences. *Cell* **14**, 681–693.

Kiper, M. and Herzfield, F. (1978). DNA sequence organization in the genome of P. sativum. *Chromosoma* **65**, 335–352.

Kirkpatrick, J. B., Hyams, L., Thomas, V. L. and Howley, P. M. (1970). Purification of intact microtubules from brain. *J. Cell Biol.* **47**, 384–394.

Kirk, T. J. O. and Tilney-Bassett, R. A. E. (1967). *The Plastids.* Freeman, San Francisco.

Kirschner, M. W., Honig, L. S. and Williams, R. C. (1975). Quantitative electron microscopy of microtubule assembly in vitro. *J. Mol. Biol.* **99**, 263–276.

Kirschner, M. W., Williams, R. C., Weingarten, M. and Gerhart, J. C. (1974). Microtubules from mammalian brain: some properties of their depolymerization products and a proposed mechanism of assembly and disassembly. *Proc. Nat. Acad. Sci. USA* **71**, 1159-1163.

Kish, V. M. and Pederson, T. (1975). Ribonucleoprotein organization of polyadenylate sequences in HeLa cell heterogeneous nuclear RNA. *J. Mol. Biol.* **95**, 227-238.

Kish, V. M. and Pederson, T. (1976). Poly(A)-rich ribonucleoprotein complexes from HeLa cell mRNA. *J. Biol. Chem.* **251**, 5888-5894.

Kish, V. M. and Pederson, T. (1977). HnRNA secondary structure: oligo(U) sequences base paired with poly(A) and their possible role as binding sites for hnRNA specific proteins. *Proc. Nat. Acad. Sci. USA* **74**, 1426-1430.

Kissane, J. M. and Robins, E. (1958). The fluorometric measurement of DNA in animal tissues with special reference to the central nervous system. *J. Biol. Chem.* **233**, 184-188.

Kit, S. (1961). Equilibrium centrifugation in density gradients of DNA preparations from animal tissues. *J. Mol. Biol.* **3**, 711-716.

Kit, S., Dubbs, D. R., Piekarski, L. J. and Hsu, T. C. (1963). Deletion of thymidine kinase activity from L cells resistant to BUdR. *Exp. Cell Res.* **31**, 297-312.

Kitchingman, G. R., Lai, S. -P. and Westphal, H. (1977). Loop structures in hybrids of early RNA and the separated strands of adenovirus DNA. *Proc. Nat. Acad. Sci. USA* **74**, 4392-4395.

Klebe, R. J., Chen, T. -R. and Ruddle, F. H. (1970). Controlled production of proliferating somatic cell hybrids. *J. Cell Biol.* **45**, 74-82.

Kleene, K. C. and Humphreys, T. (1977). Similarity of hnRNA sequences in blastula and pluteus stage sea urchin embryos. *Cell* **12**, 143-155.

Kleiman, L. and Huang, R. C. C. (1972). Reconstitution of chromatin. The sequential binding of histones to DNA in the presence of salt and urea. *J. Mol. Biol.* **64**, 1-8.

Kleiman, L., Birnie, G. D., Young, B. D. and Paul, J. (1977). Comparison of the base sequence complexities of polysomal and nuclear RNAs in growing Friend erythroleukemia cells. *Biochemistry* **16**, 1218-1222.

Klein, S., Schiff, J. A. and Holowinsky, A. W. (1972). Events surrounding the early development of Euglena chloroplasts. II. Normal development of fine structure and the consequences of preillumination. *Devel. Biol.* **28**, 253-273.

Klein, W. H., Murphy, W., Attardi, G., Britten, R. J. and Davidson, E. H. (1974). Distribution of repetitive and nonrepetitive sequence transcripts in HeLa mRNA. *Proc. Nat. Acad. Sci. USA* **71**, 1785-1789.

Klein, W. H., Thomas, T. L., Lai, C., Scheller, R. H., Britten, R. J. and Davidson, E. H. (1978). Characteristics of individual repetitive sequence families in the sea urchin genome studied with cloned repeats. *Cell* **14**, 889-900.

Kleinsmith, L. J. (1973). Specific binding of phosphorylated non histone chromatin proteins to DNA. *J. Biol. Chem.* **248**, 5648-5653.

Kleinsmith, L. J. and Allfrey, V. G. (1969). Isolation and characterization of a phospho-protein fraction from calf thymus nuclei. *Biochim. Biophys. Acta* **175**, 123-135.

Kleinsmith, L. J., Heidema, J. and. Carroll, A. (1970). Specific binding of rat liver nuclear proteins to DNA. *Nature* **226**, 1025-1027.

Kleinsmith, L. J., Stein, J. and Stein, G. (1976). Dephosphorylation of nonhistone proteins specifically alters the pattern of gene transcription in reconstituted chromatin. *Proc. Nat. Acad. Sci. USA* **73**, 1174-1178.

Klessig, D. F. (1977). Two adenovirus mRNAs have a common 5′ terminal leader sequence encoded at least 10 kb upstream from their main coding regions. *Cell* **12**, 9-21.

Kletzien, R. F. and Perdue, J. F. (1973). The inhibition of sugar transport in chick embryo fibroblasts by cytochalasin B. *J. Biol. Chem.* **248**, 711-719.

Klinger, H. P. and Shin, S. I. (1974). Modulation of the activity of an avian gene transferred into a mammalian cell by cell fusion. *Proc. Nat. Acad. Sci. USA* **71**, 1398-1402.

Klootwijk, J. and Planta, R. J. (1973). Analysis of the methylation sites in yeast rRNA. *Eur. J. Biochem.* **39**, 325-333.

Klootwijk, J., De Jonge, P. and Planta, R. J. (1979). The primary transcript of the ribosomal repeating unit in yeast. *Nucl. Acids Res.* **6**, 27-40.

Klotz, L. C. and Zimm, B. H. (1972). Size of DNA determined by viscoelastic measurements: results on phages, B. subtilis and E. coli. *J. Mol. Biol.* **72**, 779-800.

Klukas, C. K. and Dawid, I. B. (1976). Characterization and mapping of mitochondrial rRNA and mt DNA in D. melanogaster. *Cell* **9**, 615-625.

Knapp, G., Beckmann, J. S., Johnson, P. F., Fuhrman, S. A. and Abelson, J. (1978). Transcription and processing of intervening sequences in yeast tRNA genes. *Cell* **14**, 221-236.

Knapp, G., Ogden, R. C., Peebles, C. L. and Abelson, J. (1979). Splicing of yeast tRNA precursors: structure of the reaction intermediates. *Cell* **18**, 37-45.

Knight, L. A. and Luzzatti, L. (1973). Replication pattern of the X and Y chromosomes in partially synchronized human lymphocyte cultures. *Chromosoma* **40**, 153-166.

Knopf, W. -W., Yamada, M. and Weissbach, A. (1976). HeLa cell DNA polymerase γ: further purification and properties of the enzyme. *Biochemistry* **15**, 4540-4548.

Knowland, J. and Miller, L. (1970). Reduction of rRNA synthesis and rRNA genes in a mutant of X. laevis which organizes only a partial nucleolus. I. Ribosomal RNA synthesis in embryos of different nucleolar types. *J. Mol. Biol.* **53**, 321-328.

Knutton, S. (1979a). Studies of membrane fusion. III. Fusion of erythrocytes with polyethylene glycol. *J. Cell Sci.* **36**, 61-72.

Knutton, S. (1979b). Studies of membrane fusion. IV. Fusion of HeLa cells with Sendai virus. *J. Cell Sci.* **36**, 73-84.

Kobayashi, Y. and Mohri, H. (1977). Microheterogeneity of α and β subunit of tubulin from microtubules of starfish sperm flagella. *J. Mol. Biol.* **116**, 613-618.

Koerner, J. F. and Sinsheimer, R. L. (1957). A DNAase from calf spleen. II. Mode of action. *J. Biol. Chem.* **228**, 1049-1062.

Kohli, J., Hottinger, H., Munz, P., Strauss, A. and Thuriaux, P. (1977). Genetic mapping in S. pombe by mitotic and meiotic analysis and induced haploidization. *Genetics* **87**, 471-489.

Kohne, D. E. (1970). Evolution of higher organism DNA. *Quart. Rev. Biophys.* **3**, 327-375.

Kolodner, R. and Tewari, K. K. (1972). Physicochemical characterization of mitochondrial DNA from pea leaves. *Proc. Nat. Acad. Sci. USA* **69**, 1830-1834.

Kolodner, R. and Tewari, K. K. (1975a). The molecular size and conformation of the chloroplast DNA from higher plants. *Biochim. Biophys. Acta* **402**, 372-390.

Kolodner, R. D. and Tewari, K. K. (1975b). Denaturation mapping studies on the circular chloroplast DNA from pea leaves. *J. Biol. Chem.* **250**, 4888-4895.

Kolodner, R. and Tewari, K. K. (1975c). Presence of displacement loops in the covalently closed circular chloroplast DNA from higher plants. *J. Biol. Chem.* **250**, 8840-8847.

Kolodner, R. and Tewari, K. K. (1979). Inverted repeats in chloroplast DNA from higher plants. *Proc. Nat. Acad. Sci. USA* **76**, 41-45.

Kolodner, R., Warner, R. C. and Tewari, K. K. (1975). The presence of covalently linked ribonucleotide ribonucleotides in the closed circular DNA from higher plants. *J. Biol. Chem.* **250**, 7020-2026.

Kominami, R. and Muramatsu, M. (1977). Heterogeneity of 5′ termini of nucleolar 45S, 32S and 28S RNA in mouse hepatoma. *Nucl. Acids Res.* **4**, 229-240.

Kominami, R., Hamada, H., Fujii-Kuriyama, Y. and Muramatsu, M. (1978). 5′ terminal processing of ribosomal 28S RNA. *Biochemistry* **17**, 3965-3970.

Konkel, D. A. and Ingram, V. M. (1977). RNA aggregation during sulfhydryl agarose chromatography of mercurated RNA. *Nucl. Acids Res.* **4**, 1979-1988.

Konkel, D. A. and Ingram, V. M. (1978). Is there specific transcription from isolated chromatin? *Nucl. Acids Res.* **5**, 1237-1252.

Konkel, D. A., Tilghman, S. M. and Leder, P. (1978). The sequence of the chromosomal mouse β globin major gene: homologies in capping, splicing and poly(A) sites. *Cell* **15**, 1125-1132.

Kootstra, A. and Bailey, G. S. (1978). Primary structure of histone H2B from trout (Salmo trutta) testes. *Biochemistry* **17**, 2504-2509.

Kopecka, H., Crouse, E. J. and Stutz, E. (1977). The Euglena gracilis chloroplast genome: analysis by restriction enzymes. *Eur. J. Biochem.* **72**, 525-536.

Koper-Zwarthoff, C. E., Lockard, R. E., Alzner-deWeerd, B., RajBhandary, U. L. and Bol, J. F. (1977). Nucleotide sequence of 5′ terminus of alfalfa mosaic virus RNA 4 leading into coat protein cistron. *Proc. Nat. Acad. Sci. USA* **74**, 5504-5508.

Korenberg, J. R. and Engels, W. R. (1978). Base ratio, DNA content, and quinacrine brightness of human chromosomes. *Proc. Nat. Acad. Sci. USA* **75**, 3382-3386.

Korenberg, J. R. and Freedlander, E. F. (1974). Giemsa technique for the detection of sister chromatid exchanges. *Chromosoma* **48**, 355-360.

Korge, G. (1975). Chromosome puff activity and protein synthesis in larval salivary glands of D. melanogaster. *Proc. Nat. Acad. Sci. USA* **72**, 4550-4554.

Korge, G. (1977). Direct correlation between a chromosome puff and the synthesis of a larval saliva protein in D. melanogaster. *Chromosoma* **62**, 155-174.

Korn, E. D. (1978). Biochemistry of actomyosin dependent cell motility. *Proc. Nat. Acad. Sci. USA* **75**, 588-599.

Korn, L. J. and Brown, D. D. (1978). Nucleotide sequence of X. borealis oocyte 5S DNA: comparison of sequences that flank several related eucaryotic genes. *Cell* **15**, 1145-1156.

Kornberg, R. D. (1974). Chromatin structure: a repeating unit of histones and DNA. *Science* **184**, 868-871.

Kornberg, R. D. and Thomas, J. O. (1974). Chromatin structure: oligomers of the histones. *Science* **184**, 865-868.

Korwek, E. L., Nakazoto, H., Venkatesan, S. and Edmonds, M. (1976). Poly(U) sequences in mRNA of HeLa cells. *Biochemistry* **15**, 4643-4648.

Kowallik, K. V. and Herrmann, R. G. (1972). Variable amounts of DNA related to the size of chloroplasts. IV. Three dimensional arrangement of DNA in fully differentiated chloroplasts of Beta vulgaris. *J. Cell Sci.* **11**, 357-377.

Kozak, C. A. and Ruddle, F. H. (1977). Assignment of the genes for thymidine kinase and galactokinase to M. musculus chromosome 11 and the preferential segregation of this chromosomes in Chinese hamster-mouse somatic cell hybrids. *Somatic Cell Genet.* **3**, 121-134.

Kozak, L. P., McLean, G. K. and Eicher, E. M. (1974). X linkage of PGK in the mouse. *Biochem. Genet.* **11**, 41-48.

Kozak, M. (1977). Nucleotide sequences of 5′ terminal ribosome-protected initiation regions from two reovirus messages. *Nature* **269**, 390-393.

Kozak, M. (1978). How do eucaryotic ribosomes select initiation regions in mRNA? *Cell* **15**, 1109-1123.

Kozak, M. and Shatkin, A. J. (1977a). Sequences of two 5′ terminal ribosome protected fragments from reovirus mRNAs. *J. Mol. Biol.* **112**, 75-96.

Kozak, M. and Shatkin, A. J. (1977b). Sequences and properties of two ribosome binding sites from the small size class of reovirus mRNA. *J. Biol. Chem.* **252**, 6895-6908.

Kozak, M. and Shatkin, A. J. (1978). Identification of features in 5′ terminal fragments from reovirus mRNA which are important for ribosome binding. *Cell* **13**, 201-212.

Kraemer, P. M., Petersen, D. F. and Van Dilla, M. A. (1971). DNA constancy in heteroploidy and the stem line theory of tumors. *Science* **174**, 714-717.

Kraemer, P. M., Deaven, L. L., Crissman, H. A., Steinkamp, J. A. and Petersen, D. F. (1973). On the nature of heteroploidy. *Cold Spring Harbor Symp. Quant. Biol.* **33**, 133-144.

Kram, R., Botchan, M. and Hearst, J. E. (1972). Arrangement of the highly reiterated DNA sequences in the centric heterochromatin of D. melanogaster. Evidence for interspersed spacer DNA. *J. Mol. Biol.* **64**, 103-118.

Kram, R., Botchan, M. and Hearst, J. E. (1972). Arrangement of the highly reiterated DNA sequences in the centric heterochromatin of D. melanogaster. Evidence for interspersed spacer DNA. *J. Mol. Biol.* **64**, 103-118.

Kramer, F. R. and Mills, D. R. (1978). RNA sequencing with radioactive chain terminating ribonucleotides. *Proc. Nat. Acad. Sci. USA* **75**, 5334-5338.

Kramer, R. A., Phillipsen, P. and Davis, R. W. (1978). Divergent transcription in the yeast rRNA coding region as shown by hybridization to separated strands and sequence analysis of cloned DNA. *J. Mol. Biol.* **123**, 405-416.

Kramers, M. R. and Stebbings, H. (1977). The insensitivity of Vinca rosea to vinblastine. *Chromosoma* **61**, 277-288.

Krause, M. O., Kleinsmith, L. J. and Stein, G. S. (1975). Properties of the genome in normal and SV40 transformed WI38 human diploid fibroblasts. I. Composition and metabolism of nonhistone chromosomal proteins. *Exp. Cell Res.* **92**, 164-174.

Krebs, G. and Chambon, P. (1976). Animal DNA-dependent RNA polymerases. Purification and molecular structure of hen oviduct and liver class B RNA polymerases. *Eur. J. Biochem.* **61**, 15-26.

Krieg, P. and Wells, J. R. E. (1976). The distribution of active genes (globin) and inactive genes (keratin) in fractionated chicken erythroid chromatin. *Biochemistry* **15**, 4549-4558.

Kriegstein, H. J. and Hogness, D. S. (1974). Mechanism of DNA replication in Drosophila chromosomes: structure of replication forks and evidence for bidirectionality. *Proc. Nat. Acad. Sci. USA* **71**, 135-139.

Krondahl, U., Bols, N., Ege, T., Linder, S. and Ringertz, N. R. (1977). Cells reconstituted from cell fragments of two different species multiply and form colonies. *Proc. Nat. Acad. Sci. USA* **74**, 606-609.

Krystal, M. and Arnheim, N. (1978). Length heterogeneity in a region of the human ribosomal gene spacer is not accompanied by extensive population polymorphism. *J. Mol. Biol.* **126**, 91-104.

Krzyzek, R. A., Collett, M. S., Lau, A. F., Perdue, M. L., Leis, J. P. and Faras, A. J. (1978). Evidence for splicing of avian sarcoma virus 5′ terminal genomic sequences onto viral specific RNA in infected cells. *Proc. Nat. Acad. Sci. USA* **75**, 1284-1288.

Kubai, D. F. (1975). The evolution of the mitotic spindle. *Int. Rev. Cytol.* **43**, 167-227.

Kubai, D. F. (1973). Unorthodox mitosis in Trychonympha agilis: kinetochore differentiation and chromosome movement. *J. Cell Sci.* **13**, 511-522.

Kubai, D. F. and Ris, H. (1969). Division of the dinoflagellate Gyrodinium cohnii. A new type of nuclear reproduction. *J. Cell Biol.* **40**, 508-528.

Kucherlapati, R., Hwang, S. P., Shimuzu, N., McDougall, J. K. and Botchan, M. R. (1978). Another chromosomal assignment for an SV40 integration site in human cells. *Proc. Nat. Acad. Sci. USA* **75**, 4460-4464.

Kumar, A. and Lindberg, U. (1972). Characterization of messenger ribonucleoprotein and mRNA from KB cells. *Proc. Nat. Acad. Sci. USA* **69**, 681-685.

Kumar, A. and Pederson, T. (1975). Comparison of proteins bound to hnRNA and mRNA in HeLa cells. *J. Mol. Biol.* **96**, 353-366.

Kumar, A. and Warner, J. R. (1972). Characterization of ribosomal precursor particles from HeLa cell nucleoli. *J. Mol. Biol.* **63**, 233-246.

Kumar, A. and Wu, R. S. (1973). Role of ribosomal RNA transcription in ribosome processing in HeLa cells. *J. Mol. Biol.* **80**, 265-276.

Kunkel, L. M., Smith, K. D., Boyer, S. H., Borgaonkar, D. S., Wachtel, S. S., Miller, O. J., Breg, W. R., Jones, H. W. and Rary, J. M. (1977). Analysis of human Y chromosome specific reiterated DNA in chromosome variants. *Proc. Nat. Acad. Sci. USA* **74**, 1245-1249.

Kunkel, N. S. and Weinberg, E. S. (1978). Histone gene transcripts in the cleavage and mesenchyme blastula embryo of the sea urchin, S. purpuratus. *Cell* **14**, 313-326.

Kunkel, N. S., Hemminki, K. and Weinberg, E. S. (1978). Size of histone gene transcripts in different embryonic stages of the sea urchin, S. purpuratus. *Biochemistry* **17**, 2591-2604.

Kuntzel, H. and Noll, H. (1967). Mitochondrial and cytoplasmic polysomes from N. crassa. *Nature* **215**, 1340-1345.

Kuntzel, H. and Schafer, K. P. (1971). Mitochondrial RNA polymerases from N. crassa. *Nature New Biol.* **231**, 265-269.

Kuntzel, H., Pieniazek, N. J., Pieniazek, D. and Leister, D. E. (1975). Lipophilic proteins encoded by mitochondrial and nuclear genes in N. crassa. *Eur. J. Biochem.* **54**, 567-575.

Kuo, M. T., Sahasrabuddhe, C. G. and Saunders, G. F. (1976). Presence of messenger specifying sequences in the DNA of chromatin subunits. *Proc. Nat. Acad. Sci. USA* **73**, 1572-1575.

Kuriyama, Y. and Luck, D. J. L. (1973). rRNA synthesis in mitochondria of N. crassa. *J. Mol. Biol.* **73**, 425-437.

Kurnit, D. M., Shafit, B. R. and Maio, J. J. (1973). Multiple satellite DNAs in the calf and their relation to the sex chromosomes. *J. Mol. Biol.* **81**, 273-284.

Kurtz, D. T. and Feigelson, P. (1977). Multihormonal induction of hepatic α_{2u} globulin mRNA as measured by hybridization to complementary DNA. *Proc. Nat. Acad. Sci. USA* **74**, 4791-4795.

Kurtz, D. T., Chan, K. M. and Feigelson, P. (1979). Translational control of hepatic α_{2u} globulin synthesis by growth hormone. *Cell* **15**, 743-750.

Kurtz, D. T., Sippel, A. E., Ansah-Yiadom, R. and Feigelson, P. (1976). Effects of sex hormones on the level of the mRNA for the rat hepatic protein α_{2u} globulin. *J. Biol. Chem.* **251**, 3594-3598.

Kusano, T., Long, C. and Green, H. (1971). A new reduced human-mouse somatic cell hybrid containing the human gene for APRT. *Proc. Nat. Acad. Sci. USA* **68**, 82-86.

Kwan, S. -P., Wood, T. G. and Lingrel, J. B. (1977). Purification of a putative precursor of globin mRNA from mouse nucleated erythroid cells. *Proc. Nat. Acad. Sci. USA* **74**, 178-182.

Kwan, S. W. and Brawerman, G. (1972). A particle associated with the polyadenylate segment in mammalian mRNA. *Proc. Nat. Acad. Sci. USA* **69**, 3247-3250.

Labrie, F. (1969). Isolation of an RNA with the properties of hemoglobin messenger. *Nature* **211**, 1217-1222.

Lacy, E. and Axel, R. (1975). Analysis of DNA of isolated chromatin subunits. *Proc. Nat. Acad. Sci. USA* **72**, 3978-3982.

LaFountain,, J. R. (1974). Birefringence and fine structure of spindles of Mephrotoma suturalis at metaphase of first meiotic division. *J. Ult. Res.* **46**, 268-278.

Lai, C. -J. and Khoury, G. (1979). Deletion mutants of SV40 defective in biosynthesis of late viral mRNA. *Proc. Nat. Acad. Sci. USA* **76**, 71-75.

Lai, C. -J., Dhar, R. and Khoury, G. (1978). Mapping the spliced and unspliced late lytic SV40 mRNAs. *Cell* **14**, 971-982.

Lai, E. C., Woo, S. L. C., Dugaiczyk, A., Caterall, J. F. and O'Malley, B. W. (1978). The ovalbumin gene: structural sequences in native chicken DNA are not contiguous. *Proc. Nat. Acad. Sci. USA* **75**, 2205-2209.

Lai, E. C., Woo, L. C., Dugaiczyk, A. and O'Malley, B. W. (1979a). The ovalbumin gene: alleles created by mutations in the intervening sequences of the natural gene. *Cell* **16**, 201-212.

Lai, E. C., Stein, J. P., Catterall, J. F., Woo, S. L. C., Mace, M. L., Means, A. R. and O'Malley, B.

W. (1979b). Molecular structure and nucleotide sequences flanking the natural chicken ovomucoid gene. *Cell* 18, 829-842.

Laine, B., Sautière, P. and Biserte, G. (1976). Primary structure and microheterogeneities of rat chloroleukemia histone H2A. *Biochemistry* 15, 1640-1644.

Laird, C. D. (1971). Chromatid structure: relationship between DNA content and nucleotide sequence diversity. *Chromosoma* 32, 378-406.

Laird, C. D. and McCarthy, B. J. (1968). Nucleotide sequence homology within the genome of D. melanogaster. *Genetics* 60, 323-334.

Laird, C. D. and McCarthy, B. J. (1969). Molecular characterization of the Drosophila genome. *Genetics* 63, 865-882.

Laird, C. D., Chooi, W. Y., Cohen, E. H., Dickson, E., Hutchinson, N. and Turner, S. H. (1973). Organization and transcription of DNA in chromosomes and mitochondria of Drosophila. *Cold Spring Harbor Symp. Quant. Biol.* 38, 311-327.

Laird, C. D., Wilkinson, L. E., Foe, V. E. and Chooi, W. Y. (1976). Analysis of chromatin associated fiber arrays. *Chromosoma* 58, 169-192.

Lake, R. S. (1973). F1 histone phosphorylation in metaphase chromosomes of cultured Chinese hamster cells. *Nature New Biol.* 242, 145-146.

Lake, R. S. and Salzman, N. P. (1972). Occurrence and properties of chromatin associated F1 histone phosphokinase in mitotic Chinese hamster cells. *Biochemistry* 11, 4817-4825.

Lake, R. S., Barban, S. and Salzman, N. P. (1973). Resolution and identification of the core deoxynucleoproteins of the SV40. *Biochem. Biophys. Res. Commun.* 54, 640-647.

Lake, R. S., Goidl, J. A. and Salzman, N. P. (1972). F1 histone modification at metaphase in Chinese hamster cells. *Exp. Cell Res.* 73, 113-121.

Lakhotia, S. C. (1974). EM autoradiographic studies on polytene nuclei of D. melanogaster, III. Localization of nonreplicating chromatin in the chromocentre heterochromatin. *Chromosoma* 46, 145-160.

Lakhotia, S. C. and Jacob, J. (1974). EM autoradiographic studies on polytene nuclei of D. melanogaster. II. Organization and transriptive activity of the chromocentre. *Exp. Cell Res.* 86, 253-263.

Lalley, P. A., Francke, U. and Minna, J. D. (1978). Homologous genes for enolase, phosphogluconate dehydrogenase, phosphoglucomutase and adenylate kinase are syntenic on mouse chromosome 4 and human chromosome 1p. *Proc. Nat. Acad. Sci. USA* 75, 2382-2386.

Lam, D. M. K. and Bruce, W. R. (1971). The biosynthesis of protamine during spermatogenesis of the mouse: extraction, partial characterization and site of synthesis. *J. Cell. Physiol.* 78, 13-24.

Lambert, B., Egyhazi, E., Daneholt, B. and Ringborg, U. (1973a). Quantitative microassay for RNA/DNA hybrids in the study of nucleolar RNA from C. tentans salivary gland cells. *Exp. Cell Res.* 76, 369-380.

Lambowitz, A. M., LaPolla, R. J. and Collins, R. A. (1979). Mitochondrial ribosome assembly in Neurospora. Two dimensional gel electrophoretic analysis of mitochondrial ribosome proteins. *J. Cell Biol.* 82, 17-31.

Lane, C. D., Marbaix, G. and Gurdon, J. B. (1971). Rabbit hemoglobin synthesis in frog eggs: the translation of reticulocyte 9S RNA in frog oocytes. *J. Mol. Biol.* 61, 73-91.

Lane, N. J. and Treherne, J. E. (1970). Lanthanum staining of neurotubes in axons from cockroach ganglia. *J. Cell Sci.* 7, 217-231.

Lansman, R. A. and Clayton, D. A. (1975). Mitochondrial protein synthesis in mouse L cells effect of selective nicking of mitochondrial DNA. *J. Mol. Biol.* 99, 777-793.

Lapeyre, J. N. and Bekhor, I. (1976). Chromosomal protein interactions in chromatin and with DNA. *J. Mol. Biol.* 104, 25-58.

Laskey, R. A., Gurdon, J. B. and Crawford, L. V. (1972). Translation of EMC RNA in oocytes of X. laevis. *Proc. Nat. Acad. Sci. USA* 69, 3665-3669.

Laskey, R. A., Mills, A. D. and Morris, N. R. (1977). Assembly of SV40 chromatin in a cell free system from Xenopus eggs. *Cell* **10**, 237-244.

Laskey, R. A., Honda, B. M., Mills, A. D., Morris, N. R., Wyllie, A. H., Mertz, J. E., de Robertis, E. M. and Gurdon, J. B. (1978a). Chromatin assembly and transcription in eggs and oocytes of X. laevis. *Cold Spring Harbor Symp. Quant. Biol.* **43**, 171-177.

Laskey, R. A., Honda, B. M., Mills, A. D. and Finch, J. T. (1978b). Nucleosomes are assembled by an acidic protein which binds histones and transfers them to DNA. *Nature* **275**, 416-420.

Latham, H. and Darnell, J. E. (1965). Distribution of mRNA in the cytoplasmic polysomes of the HeLa cell. *J. Mol. Biol.* **14**, 1-12.

Latt, S. A. (1973). Microfluorimetric detection of DNA replication in human metaphase chromosomes. *Proc. Nat. Acad. Sci. USA* **70**, 3395-3399.

Latt, S. A. (1974). Sister chromatid exchanges, indices of human chromosome damage and repair: detection by fluorescence and induction by mitomycin C. *Proc. Nat. Acad. Sci. USA* **71**, 3162-3166.

Latt, S. A. (1975). Fluorescence analysis of late DNA replication in human metaphase chromosomes. *Somatic Cell Genet.* **1**, 293-322.

Latt, S. A., Brodie, S. and Munroe, S. H. (1974). Optical studies of complexes of quinacrine with DNA and chromatin: implications for the fluorescence of cytological chromosome preparations. *Chromosoma* **49**, 17-40.

Lau, A. F., Ruddon, R. W., Collett, M. S. and Faras, A. J. (1978). Distribution of the globin gene in active and inactive fractions from Friend erythroleukemia cells. *Exp. Cell Res.* **111**, 269-276.

Lauer, G. D. and Klotz, L. C. (1975). Determination of the molecular weight of S. cerevisiae nuclear DNA. *J. Mol. Biol.* **95**, 309-326.

Lauer, G. D., Roberts, T. M. and Klotz, L. C. (1977). Determination of the nuclear DNA content of S. cerevisiae and implications for the organization of DNA in yeast chromosomes. *J. Mol. Biol.* **113**, 507-526.

Lavi, S. and Groner, Y. (1977). 5′ terminal sequences and coding region of late SV40 mRNAs are derived from noncontiguous segments of the viral genome. *Proc. Nat. Acad. Sci. USA* **74**, 5323-5327.

Lavi, U., Fernandez-Munoz, R. and Darnell, J. E. (1977). Content of N[6] methyl adenylic acid in hnRNA and mRNA of HeLa cells. *Nucleic Acids Res.* **4**, 63-69.

Lawn, R. M. (1978). Gene sized DNA molecules of the Oxytricha macronucleus have the same terminal sequence. *Proc. Nat. Acad. Sci. USA* **74**, 4325-4328.

Lawn, R. M., Fritsch, E. F., Parker, R. C., Blake, G. and Maniatis, T. (1978). The isolation and characterization of linked δ and β globin genes from a cloned library of human DNA. *Cell* **15**, 1157-1174.

Lawrence, J. J., Chan, D. C. F. and Piette, L. H. (1976). Conformational state of DNA in chromatin subunits. Circular dichroism, melting, and ethidium bromide analysis. *Nucl. Acids Res.* **3**, 2878-2894.

Lawson, G. M. and Cole, R. D. (1979). Selective displacement of histone H1 from whole HeLa nuclei: effect on chromatin structure in situ as probed by micrococcal nuclease. *Biochemistry* **18**, 2160-2166.

Laycock, D. G. and Hunt, J. A. (1969). Synthesis of rabbit globin by a bacterial cell free system. *Nature* **221**, 1118-1122.

Lazarides, E. (1975a). Tropomyosin antibody: the specific localization of tropomyosin in nonmuscle cells. *J. Cell Biol.* **65**, 549-561.

Lazarides, E. (1975b). Immunofluorescence studies on the structure of actin filaments in tissue culture cells. *J. Histochem. Cytochem.* **23**, 507-528.

Lazarides, E. (1976). Actin, α-actinin, and tropomyosin interaction in the structural organization of actin filaments in nonmuscle cells. *J. Cell Biol.* **68**, 202-219.

Lazarides, E. (1978). The distribution of desmin (100 Å) filaments in primary cultures of embryonic chick cardiac cells. *Exp. Cell Res.* **112**, 265-274.

Lazarides, E. and Balzer, D. R. (1978). Specificity of desmin to avian and mammalian muscle cells. *Cell* **14**, 429-438.

Lazarides, E. and Burridge, K. (1975). α-Actinin: immunofluorescent localization of a muscle structural protein in nonmuscle cells. *Cell* **6**, 289-298.

Lazarides, E. and Hubbard, B. D. (1976). Immunological characterization of the subunit of the 100 Å filaments from muscle cells. *Proc. Nat. Acad. Sci. USA* **73**, 4344-4348.

Lazarides, E. and Weber, K. (1974). Actin antibody: the specific visualization of actin filaments in nonmuscle cells. *Proc. Nat. Acad. Sci. USA* **71**, 2268-2272.

Lazowska, J. and Slonimksi, P. P. (1976). Electron microscopy analysis of circular repetitive mitochondrial DNA molecules from genetically characterized rho⁻ mutants of S. cerevisiae. *Mol. Gen. Genet.* **146**, 61-78.

Lazowska, J. and Slonimski, P. P. (1977). Site specific recombination in petite colony mutants of S. cerevisiae. I. Electron miscroscopic analysis of the organization of recombinant DNA resulting from end to end joining of two mitochondrial segments. *Mol. Gen. Genet.* **156**, 163-175.

Lazowska, J., Michel, F., Faye, G., Fukuhara, H. and Slonimski, P. P. (1974). Physical and genetic organization of petite and grande yeast mitochondrial DNA. II. DNA-DNA hybridization studies and buoyant density determinations. *J. Mol. Biol.* **85**, 393-410.

Lea, D. E. and Colson, C. A. (1949). The distribution of the numbers of mutants in bacterial populations. *J. Genet.* **49**, 264-285.

Leaver, C. J. and Harmey, M. A. (1973). Plant mitochondrial nucleic acids. *Biochem. Soc. Symp.* **38**, 175-193.

Ledbetter, M. C. and Porter, K. R. (1963). A microtubule in plant cell fine structure. *J. Cell Biol.* **19**, 239-250.

Ledbetter, M. C. and Porter, K. R. (1964). Morphology of microtubules of plant cells. *Science* **144**, 872-874.

Leder, A., Miller, H. I., Hamer, D. H., Seidman, J. G., Norman, B., Sullivan, M. and Leder, P. (1978). Comparison of cloned mouse α and β globin genes conservation of intervening sequence locations and extragenic homology. *Proc. Nat. Acad. Sci. USA* **75**, 6187-6191.

Leder, P., Tiemeier, D. and Enquist, L. (1977). EK2 derivatives of phage lambda useful in the cloning of DNA from higher organisms: the λ gtWES system. *Science* **196**, 175-177.

Lee, C. S. and Thomas, C. A., Jr. (1973). Formation of rings from Drosophila DNA fragments. *J. Mol. Biol.* **77**, 25-56.

Lee, J. C. and Timasheff, S. (1975). The reconstitution of microtubules from purified calf brain tubulin. *Biochemistry* **14**, 5183-5187.

Lee, J. C. and Timasheff, S. N. (1977). In vitro reconstitution of calf brain microtubules: effects of solution variables. *Biochemistry* **16**, 1754-1764.

Lee, J. C. and Yunis, J. J. (1971a). A developmental study of constitutive heterochromatin in Microtus agrestis. *Chromosoma* **32**, 237-250.

Lee, J. C. and Yunis, J. J. (1971b). Cytological variations in the constitutive heterochromatin of Microtus agrestis. *Chromosoma* **35**, 117-124.

Lee, J., Frigon, R. and Timasheff, S. (1973). The chemical characterization of calf brain microtubule protein subunits. *J. Biol. Chem.* **248**, 7253-7262.

Lee, S. L. and Brawerman, G. (1971). Pulse labeled RNA complexes released by dissocoation of rat liver polysomes. *Biochemistry* **10**, 510-516.

Lee, S. Y., Mendecki, J. and Brawerman, G. (1971). A polynucleotide segment rich in adenylic acid in the rapidly labeled polyribosomal RNA component sarcoma 180 ascites cells. *Proc. Nat. Acad. Sci. USA* **68**, 1331-1335.

Leenders, H. J. and Berendes, H. D. (1972). The effect of changes in the respiratory metabolism upon genome activity in Drosophila: I. The induction of gene activity. *Chromosoma* **37**, 433–444.

Lefevre, G., Jr. (1973). The one band-one gene hypothesis: evidence from a cytogenetic analysis of mutant and nonmutant rearrangement breakpoints in D. melanogaster. *Cold Spring Harbor. Symp. Quant. Biol.* **38**, 591–599.

Lefevre, G., Jr. (1976). A photographic representation and interpretation of the polytene chromosomes of D. melanogaster salivary glands. In M. Ashburner and E. Novitski (Eds.), *The Genetics and Biology of Drosophila*, Academic Press, London, pp 32–60.

Lefevre, G. and Wilkins, M. (1966). Cytogenetics of the white locus. *Genetics* **53**, 175–187.

Leffak, I. M., Grainger, R. and Weintraub, H. (1977). Conservative assembly and segregation of nucleosomal histones. *Cell* **12**, 837–845.

Legon, S. (1976). Characterization of the ribosome-protected regions of ¹²⁵I labeled rabbit globin mRNA. *J. Mol. Biol.* **106**, 37–54.

Legon, S., Flavell, A. J., Cowie, A. and Kamen, R. (1979). Amplification in the leader sequence of late polyoma virus mRNAs. *Cell* **16**, 373–388.

Lehmann, H. and Cerrell, R. W. (1968). Differences between α and β chain mutants of human hemoglobin and between α and β thalassemia. Possible duplication of the a gene. *Br. Med. J.* **4**, 748–750.

Leister, D. E. and Dawid, I. B. (1974). Physical properties and protein constituents of cytoplasmic mitochondrial ribosomes of X. laevis. *J. Biol. Chem.* **249**, 5108–5118.

Lemons, R. S., O'Brien, S. J. and Sherr, C. J. (1977). A new genetic locus, *Bevi*, of human chromosome 6 which controls the replication of baboon type C virus in human cells. *Cell* **12**, 251–262.

Lengyel, J. and Penman, S. (1975). hnRNA size and processing as related to different DNA content in two dipterans: Drosophila and Aedes. *Cell* **5**, 281–290.

Lengyel, J. A. and Penman, S. (1977). Differential stability of cytoplasmic RNA in a Drosophila cell line. *Devel. Biol.* **57**, 243–253.

Lenhard-Schuller, R., Hohn, B., Brack, C., Hirama, M. and Tonegawa, S. (1978). DNA clones containing mouse immunoglobulin K genes isolated by in vitro packaging into phage lambda coats. *Proc. Nat. Acad. Sci. USA* **75**, 4709–4713.

Lenk, R., Herman, R. and Penman, S. (1978). Messenger RNA abundance and lifetime: a correlation in Drosophila cells but not in HeLa. *Nucl. Acids Res.* **5**, 3057–3070.

Lenk, R., Ransom, L., Kaufman, Y. and Penman, S. (1977). A cytoskeletal structure with associated polyribosomes obtained from HeLa cells. *Cell* **10**, 67–78.

Leon, W., Fouts, D. L. and Manning, J. (1978). Sequence arrangement of the 16S and 26S rRNA genes in the pathogenic haemoflagellate Leishmania donovani. *Nucl. Acids Res.* **5**, 491–504.

LePennec, J. P., Baldacci, P., Perrin, F., Cami, B., Gerlinger, P., Krust, A., Kourilsky, P. and Chambon, P. (1978). The ovalbumin split gene: molecular cloning of Eco RI fragments c and d. *Nucl. Acids Res.* **5**, 4547–4562.

Lesch, M. and Nyhan, W. L. (1964). A familian disorder of uric acid metabolism and central nervous system function. *Am. J. Med.* **36**, 561–570.

Leung, D. W., Browning, K. S., Heckman, J. E., RajBhandary, U. L. and Clark, J. M. (1979). Nucleotide sequence of the 5′ terminus of satellite tobacco necrosis virus RNA. *Biochemistry* **18**, 1361–1365.

Levan, A. (1938). Effect of colchicine on root mitosis in Allium. *Hereditas* **24**, 471–486.

Levan, A. (1940). The effect of acenaphthene and colchicine on mitosis of Allium and Colchicum. *Hereditas* **26**, 262–276.

Levan, A. and Steineggar, E. (1947). The resistance of Colchicum and Bulbocodium to the c-mitotic action of colchicine. *Hereditas* **33**, 552–556.

Levenson, R. G. and Marcu, K. B. (1976). On the existence of polyadenylated histone mRNA in X. laevis oocytes. *Cell* **9**, 311-322.

Levinson, B. B., Ullman, B. and Martin, D. W. (1979). Pyrimidine pathway variants of cultured mouse lymphoma cells with altered levels of both orotate P transferase and orotidylate decarboxylase. *J. Biol. Chem.* **254**, 4396-4401.

Levis, R. G. and Penman, S. (1977). The metabolism of poly(A)$^+$ and poly(A)$^-$ hnRNA in cultured Drosophila cells studied with a rapid uridine pulse chase. *Cell* **11**, 105-112.

Levis, R. and Penman, S. (1978a). 5' terminal structures of poly(A$^+$) cytoplasmic mRNA and of poly(A$^+$) and poly(A$^-$) hnRNA of cells of the Dipteran D. melanogaster. *J. Mol. Biol.* **120**, 487-516.

Levis, R. and Penman, S. (1978b). Processing steps and methylation in the formation of the rRNA of cultured Drosophila cells. *J. Mol. Biol.* **121**, 219-238.

Levis, R., McReynolds, L. and Penman, S. (1977). Coordinate regulation of protein synthesis and mRNA content during growth arrest of suspension Chinese hamster ovary cells. *J. Cell. Physiol.* **90**, 485-502.

Levitt, M. (1978). How many base pairs per turn does DNA have in solution and in chromatin? Some theoretical calculations. *Proc. Nat. Acad. Sci. USA* **75**, 640-644.

Levy, A. and Jakob, K. M. (1978). Nascent DNA in nucleosome lile structures from chromatin. *Cell* **14**, 259-267.

Levy, S., Sures, I. and Kedes, L. H. (1979). Sequence of the 5' end of S. purpuratus H2B histone mRNA and its location within histone DNA. *Nature* **279**, 737-739.

Levy, S., Wood, P., Grunstein, M. and Kedes, L. (1975). Individual histone mRNAs: identification by template activity. *Cell* **4**, 239-248.

Levy, W. B. and Dixon, G. H. (1977a). Diversity of sequences of polyadenylated cytoplasmic RNA from rainbow trout testis and liver. *Biochemistry* **16**, 958-963.

Levy, W. B. and Dixon, G. H. (1977b). Renaturation kinetics of cDNA complementary to cytoplasmic polyadenylated RNA from rainbow trout testis. Accessibility of transcribed genes to pancreatic DNAase. *Nucl. Acids Res.* **4**, 883-898.

Levy, W. B. and Dixon, G. H. (1978). Partial purification of transcriptionally active nucleosomes from trout testis cells. *Nucl. Acids Res.* **5**, 4155-4164.

Levy, W. B. and McCarthy, B. J. (1975). Messenger RNA complexity in D. melanogaster. *Biochemistry* **14**, 2440-2446.

Levy, W. B. and McCarthy, B. J. (1976). Relationship between nuclear and cytoplasmic RNA in Drosophila cells. *Biochemistry* **15**, 2415-2419.

Levy, W. B., Connor, W. and Dixon, G. H. (1979). A subset of trout testis nucleosomes enriched in transcribed DNA sequences contains high mobility group proteins as major structural components. *J. Biol. Chem.* **254**, 609-620.

Levy, W. B., Wong, N. C. W. and Dixon, G. H. (1977). Selective association of the trout specific H6 protein with chromatin regions susceptible to DNAase I and DNAase II: possible location of HMG-T in the spacer region between core nucleosomes. *Proc. Nat. Acad. Sci. USA* **74**, 2810-2814.

Levy-Wilson, B. and Dixon, G. H. (1979). Limited action of micrococcal nuclease on trout testis nuclei generates two mono nucleosome subsets enriched in transcribed DNA sequences. *Proc. Nat. Acad. Sci. USA* **76**, 1682-1686.

Levy-Wilson, B., Watson, D. C. and Dixon, G. H. (1979). Multiacetylated forms of H4 are found in a putative transcriptionally competent chromatin fraction from trout testis. *Nucl. Acids Res.* **6**, 259-274.

Lewin, A., Morimoto, R., Rabinowitz, M. and Fukuhara, H. (1978). Restriction enzyme analysis of mitochondrial DNAs of petite mutants of yeast: classification of petites and deletion mapping of mitochondrial genes. *Mol. Gen. Genet.* **163**, 257-275.

Lewin, B. (1970). *The Molecular Basis of Gene Expression*, John Wiley, New York.

Lewin, B. (1975a). Units of transcription and translation: the relationship between hnRNA and mRNA. *Cell* **4**, 11–20.

Lewin, B. (1975b). Units of transcription and translation: sequence components of hnRNA and mRNA. *Cell* **4**, 77–94.

Lewis, E. B. (1950). The phenomenon of position effect. *Adv. Genet.* **3**, 75–115.

Lewis, E. B. and Gencarella, W. (1952). Claret and nondisjunction in D. melanogaster. *Genetics* **37**, 600–601.

Lewis, J. B., Anderson, C. W. and Atkins, J. F. (1977). Further mapping of late adenovirus genes by cell free translation of RNA selected by hybridization to specific DNA fragments. *Cell* **12**, 37–44.

Lewis, M., Helmsing, P. J. and Ashburner, M. (1975). Parallel changes in puffing activity and patterns of protein synthesis in salivary glands of Drosophila. *Proc. Nat. Acad. Sci. USA* **72**, 3604–3608.

Lewis, P. N., Bradbury, E. M. and Crane-Robinson, C. (1975). Ionic strength induced structure in histone H4 and its fragments. *Biochemistry* **14**, 3391–3400.

Lewis, W. H. (1938). On the role of a superficial plasmagel layer in division locomotion and changes in form of cells. *Arch. Exp. Zellforsch.* **22**, 270.

Lewis, W. H. (1951). Cell division with special reference to tissue cultures. *Ann. N. Y. Acad, Sci.* **51**, 1287–1294.

Lewis, W. H. and Lewis, M. R. (1924). Behavior of cells in tissue cultures. In E. V. Cowdry (Ed.), *General Cytology*, University of Chicago Press, Illinois, pp 385–447.

Lewis, W. H. and Wright, J. (1978). Ribonucleotide reductase from wild type and hydroxyurea resistant CHO cells. *J. Cell. Physiol.* **97**, 87–98.

Lewis, W. H. and Wright, J. A. (1979). Isolation of hydroxyurea resistant CHO cells with altered levels of ribonucleotide reductase. *Somatic Cell Genet.* **5**, 83–96.

Ley, K. D. and Murphy, M. M. (1973). Synchronization of mitochondrial DNA synthesis in CHO cells deprived of isoleucine. *J. Cell Biol.* **58**, 340–345.

Lezzi, M. (1965). Die wirkung von DNAase auf isolierte polytan-chromosomen. *Exp. Cell Res.* **39**, 289–292.

Li, H. -J. and Bonner, J. (1971). Interaction of histone half molecules with DNA. *Biochemistry* **10**, 1461–1470.

Li, H. J., Chang, C. and Weiskopf, M. (1972). Polylysine binding to histone bound regions in chromatin. *Biochem. Biophys. Res. Commun.* **47**, 883–888.

Li, H. J., Chang, C. and Weiskopf, M. (1973). Helix coil transition in nucleoprotein chromatin structures. *Biochemistry* **12**, 1763–1772.

Li, M. and Tzagoloff, A. (1979). Assembly of the mitochondrial membrane system. Sequences of yeast mitochondrial valine and an unusual threonine tRNA gene. *Cell* **18**, 47–53.

Liau, M. C. and Hurlbert, R. B. (1975a). Interrelationship between synthesis and methylation of rRNA in isolated Novikoff tumor nucleoli. *Biochemistry* **14**, 127–133.

Liau, M. C. and Hurlbert, R. B. (1975b). The topographical order of 18S and 28S ribosomal RNAs within the 45S precursor molecule. *J. Mol. Biol.* **98**, 321–332.

Lie-Injo, L. E., Ganesan, J., Clegg, J. B. and Weatherall, J. D. (1974). Homozygous state for Hb Constant Spring. *Blood* **43**, 251–259.

Liem, R. K. H., Yen, S. H., Salomon, G. D. and Shelanski, M. L. (1978). Intermediate filaments in nervous tissues. *J. Cell Biol.* **79**, 637–645.

Lifschytz, E. and Falk, R. (1968). Fine structure analysis of a chromosome segment in D. melanogaster. Analysis of X ray induced lethals. *Mutat. Res.* **6**, 235–244.

Lifschytz, E. and Falk, R. (1969). Fine structure analysis of a chromosome segment in D. melanogaster. Analysis of EMS induced lethals. *Mutat. Res.* **8**, 147–155.

Lifton, R. P. and Kedes, L. H. (1976). Size and sequence homology of masked maternal and embryonic histone mRNAs. *Devel. Biol.* **48**, 47–55.

Lifton, R. P., Goldberg, M. L., Karp, R. W. and Hogness, D. S. (1978). The organization of the histone genes in D. melanogaster: functional and evolutionary implications. *Cold Spring Harbor Symp. Quant. Biol.* **43**, 1047–1051.

Lim, J. K. and Snyder, L. A. (1974). Cytogenetic and complementation analyses of recessive lethal mutations induced in the X chromosome of Drosophila by three alkylating agents. *Genet. Res.* **24**, 1–10.

Lima de Faria, A. (1959). Differential uptake of ^{3}H thymidine into hetero and euchromatin in Melanoplus and Secale. *J. Biochem. Biophys. Cytol.* **6**, 457–466.

Lima de Faria, A. (1969). DNA replication and gene amplification in heterochromatin. In A. Lima de Faria(Ed.), *Handbook of Molecular Cytology*, North Holland, Amsterdam, pp 277–325.

Lin, D. C. and Lin, S. (1979). Actin polymerization induced by a motility related high affinity cytochalasin binding complex from human erythrocyte membrane. *Proc. Nat. Acad. Sci. USA* **76**, 2345–2349.

Lin, M. S. and Alfi, O. S. (1976). Detection of sister chromatid exchanges by 4'-6-diamidino 2-phenylindole fluorescence. *Chromosoma* **57**, 219–226.

Lin, S. and Spudich, J. A. (1974). Biochemical studies on the mode of action of cytochalasin B. Cytochalasin B binding to red cell membrane in relation to glucose transport. *J. Biol. Chem.* **249**, 5778–5783.

Lin, S., Lin, D. C. and Flanagan, M. D. (1978). Specificity of the effects of cytochalasin B on transport and motile processes. *Proc. Nat. Acad. Sci. USA* **75**, 329–333.

Lin, S., Santi, D. V. and Spudich, J. A. (1974). Biochemical studies on the mode of action of cytochalasin B. Preparation of ^{3}H cytochalasin B and studies on its binding to cells. *J. Biol. Chem.* **249**, 2268–2274.

Lin, S. Y. and Riggs, A. D. (1975). The general affinity of lac repressor for E. coli DNA: implications for gene regulation in procaryotes and eucaryotes. *Cell* **4**, 107–112.

Lindberg, U. and Sundquist, B. (1974). Isolation of messenger ribonucleoproteins from mammalian cells. *J. Mol. Biol.* **86**, 451–468.

Lindegren, C. C. (1932). The genetics of Neurospora. II. Segregation of the sex factors in asci of N. crassa, N. sitophila and N. tetrasperma. *Bull. Torr. Bot. Club* **59**, 119–138.

Lindegren, C. C. (1933). The genetics of Neurospora. III. Pure bred stocks and crossing over in N. crassa. *Bull. Torr. Bot. Club* **60**, 133–154.

Lindell, T. J., Weinberg, F., Morris, P. W., Roeder, R. G. and Rutter, W. J. (1970). Specific inhibition of nuclear RNA polymerase by amanitin. *Science* **170**, 447–448.

Linder, S., Brzeski, H. and Ringertz, N. R. (1979). Phenotypic expression in cybrids derived from teratocarcinoma cells fused with myoblast cytoplasms. *Exp. Cell Res.* **120**, 1–14.

Ling, V. (1977). A membrane altered mutant cold sensitive for growth. *J. Cell. Physiol.* **91**, 209–224.

Ling, V. and Thompson, L. H. (1974). Reduced permeability in CHO cells as a mechanism of resistance to colchicine. *J. Cell. Physiol.* **83**, 103–116.

Link, G., Coen, D. M. and Bogorad, L. (1978). Differential expression of the gene for the large subunit of ribulose biphosphate carboxylase in maize leaf cell types. *Cell* **15**, 725–731.

Linnane, A. W., Saunders, G. W., Gingold, E. B., Lukins, H. B. (1968). The biogenesis of mitochondria. V. Cytoplasmic inheritance of erythromycin resistance in S. cerevisiae. *Proc. Nat. Acad. Sci. USA* **59**, 903–910.

Linney, E. and Hayashi, M. (1973). Two proteins of gene A of ϕX174. *Nature New Biol.* **245**, 6–8.

Lipchitz, L. and Axel, R. (1976). Restriction endonuclease cleavage of satellite DNA in intact bovine nuclei. *Cell* **9**, 355–364.

Lis, J., Prestidge, L. and Hogness, D. S. (1978). A novel arrangement of tandemly repeated genes at a major heat shock site in D. melanogaster. *Cell* 14, 901–919.

Liskay, R. M. (1977). Absence of a measurable G2 phase in two Chinese hamster cell lines. *Proc. Nat. Acad. Sci. USA* 74, 1622–1625.

Little, P. F. R., Flavell, R. A., Kooter, J. M., Annison, G. and Williamson, R. (1979). The structure of the human fetal globin gene locus. *Nature* 278, 227–230.

Littlefield, J. W. (1963). The inosinic acid pyrophosphorylase activity of mouse fibroblasts partially resistant to 8-azaguanine. *Proc. Nat. Acad. Sci. USA* 50, 568–575.

Littlefield, J. W. (1964a). Three degrees of guanylic acid—inosinic acid pyrophosphorylase deficiency in mouse fibroblasts. *Nature* 203, 1142–1144.

Littlefield, J. W. (1964b). Selection of hybrids from matings of fibroblasts in vitro and their presumed recombinants. *Science* 145, 709–710.

Littlefield, J. W. (1969). Hybridization of hamster cells with high and low folate reductase activity. *Proc. Nat. Acad. Sci. USA* 62, 88–95.

Littlefield, J. W., McGovern, A. P. and Margeson, K. B. (1963). Changes in the distribution of polymerase activity during DNA synthesis in mouse fibroblasts. *Proc. Nat. Acad. Sci. USA* 49, 102–107.

Liu, C. P. and Lim, J. K. (1975). Complementation analysis of MMS induced recessive lethal mutations in the zeste-white region of the X chromosome of D. melanogaster. *Genetics* 79, 601–611.

Liu, L. F. and Wang, J. C. (1978). DNA-DNA gyrase complex: the wrapping of the DNA duplex outside the enzyme. *Cell* 15, 979–984.

Livak, K. J., Freund, R., Schweber, M., Wensink, P. C. and Meselson, M. (1978). Sequence organization and transcription at two heat shock loci of Drosophila. *Proc. Nat. Acad. Sci. USA* 75, 5613–5617.

Lizardi, P. M., Williamson, R. and Brown, D. D. (1975). The size of fibroin mRNA and its poly(A) content. *Cell* 4, 199–206.

Lloyd, L. U. (1910). The eclectic alkaloids. *Bull. Lloyd Lib. Bot., Pharm. Materia Medica. Pharm.* ser 2, 12, 7–10.

Lobban, P. E. and Kaiser, A. D. (1973). Enzymatic end to end joining of DNA molecules. *J. Mol. Biol.* 78, 453–471.

Lobban, P. E. and Siminovitch, L. (1975). α-Amanitin resistance: a dominant mutation in CHO cells. *Cell* 4, 167–172.

Lockard, R. E. and Lane, C. (1978). Requirement for 7 methylguanosine in translation of globin mRNA in vivo. *Nucl. Acids Res.* 5, 3237–3248.

Lockard, R. E. and Lingrel, J. B. (1969). The synthesis of mouse hemoglobin beta chains in a rabbit reticulocyte cell free system programmed with mouse reticulocyte 9S mRNA. *Biochem. Biophys. Res. Commun.* 37, 204–212.

Lockard, R. E. and RajBhandary, U. L. (1976). Nucleotide sequences at the 5′ termini of rabbit α and β globin mRNA. *Cell* 9, 747–760.

Locker, J., Rabinowitz, M. and Getz, G. S. (1974a). Tandem inverted repeats in mitochondrial DNA of petite mutants of S. cerevisiae. *Proc. Nat. Acad. Sci. USA* 71, 1366–1370.

Locker, J., Rabinowitz, M. and Getz, G. S. (1974b). Electron microscopic and renaturation kinetic analysis of mitochondrial DNA of cytoplasmic petite mutants of S. cerevisiae. *J. Mol. Biol.* 88, 489–507.

Lockwood, A. H. (1978). Tubulin assembly protein: immunological and immunofluorescent studies on its function and distribution in microtubules and cultured cells. *Cell* 13, 613–627.

Lodish, H. F. and Jacobsen, M. (1972). Equal rates of translation and termination of alpha and beta globin chains. *J. Biol. Chem.* 247, 3622–3629.

Lodish, H. F. and Rose, J. K. (1977). Relative importance of 7 methylguanosine in ribosome binding and translation of VSV mRNA in wheat germ and reticulocyte cell free systems. *J. Biol. Chem.* **252**, 1181-1188.

Lodish, H. F., Firtel, R. A. and Jacobson, A. (1973). Transcription and structure of the genome of the cellular slime mold Dictyostelium discoideum. *Cold Spring. Harbor Symp. Quant. Biol.* **38**, 899-914.

Loeb, L. A., Ewald, J. L. and Agarwal, S. S. (1970). DNA polymerase and DNA replication during lymphocyte transformation. *Cancer Res.* **30**, 2514-2520.

Loening, U. E., Jones, K. W. and Birnsteil, M. L. (1969). Properties of the rRNA precursors in X. laevis: comparison to the precursor in mammals and in plants. *J. Mol. Biol.* **45**, 353-366.

Loewy, A. G. (1952). An actomyosin-like substance from the plasmodium of a myxomycete. *J. Cell. Comp. Physiol.* **40**, 127-156.

Loh, E., Black, B., Riblet, R., Weigert, M., Hood, J. M. and Hood, L. (1979). Myeloma proteins from NZB abd Balb-c mice: structural and functional differences. *Proc. Nat. Acad. Sci. USA* **76**, 1395-1399.

Lohr, D. and Van Holde, K. E. (1975). Yeast chromatin subunit structure. *Science* **188**, 165-166.

Lohr, D., Kovacic, R. T. and Van Holde, K. E. (1977). Quantitative analysis of the digestion of yeast chromatin by staphylococcal nuclease. *Biochemistry* **16**, 463-471.

Lohr, D., Tatchell, K. and Van Holde, K. E. (1977). On the occurrence of nucleosome phasing in chromatin. *Cell* **12**, 829-834.

Lohr, D., Corden, J., Tatchell, K., Kovacic, R. T. and Van Holde, K. E. (1977). Comparative subunit structure of HeLa, yeast and chicken erythrocyte chromatin. *Proc. Nat. Acad. Sci. USA* **74**, 79-83.

Lomedico, P., Rosenthal, N., Efstratiadis, A., Gilbert, W., Kolodner, R. and Tizard, R. (1979). The structure and evolution of the two non allelic rat preproinsulin genes. *Cell* **18**, 545-558.

Louie, A. J. and Dixon, G. H. (1972a). Synthesis, acetylation and phosphorylation of histone IV and its binding to DNA during spermatogenesis in trout. *Proc. Nat. Acad. Sci. USA* **69**, 1975-1979.

Louie, A. J. and Dixon, G. H. (1972b). Kinetics of enzymatic modification of the protamines and a proposal for their binding to chromatin. *J. Biol. Chem.* **247**, 7962-7968.

Louie, A. J., Candido, E. P. M. and Dixon, G. H. (1973). Enzymatic modifications and their possible roles in regulating the binding of basic proteins to DNA and in controlling chromosomal structure. *Cold Spring Harbor Symp. Quant. Biol.* **38**, 803-819.

Louie, A. J., Sung, M. T. and Dixon, G. H. (1973). Modification of histone during spermatogenesis in trout. III. Levels of phosphohistone species and kinetics of phosphorylation of histone IIbl. *J. Biol. Chem.* **248**, 3335-3341.

Loveday, K. S. and Latt, S. A. (1978). Search for DNA interchange corresponding to sister chromatid exchanges in CHO cells. *Nucl. Acids Res.* **5**, 4087-4104.

Lovlie, A. and Braten, T. (1970). On mitosis in the multicellular alga Ulva mutabolis (Foyn). *J. Cell Sci.* **6**, 109-120.

Lowenhaupt, K. and Lingrel, J. B. (1978). A change in the stability of globin mRNA during the induction of murine erythroleukemia cells. *Cell* **14**, 337-344.

Lucas, J. J. and Kates, J. R. (1976). The construction of viable nuclear-cytoplasmic hybrid cells by nuclear transplantation. *Cell* **7**, 397-406.

Lucas, J. J., Szekely, E. and Kates, J. R. (1976). The regeneration and division of mouse L cell karyoplasts. *Cell* **7**, 115-122.

Lucchesi, J. C. (1976). Interchromosomal effects. In M. Ashburner and E. Novitski (Eds.), *The Genetics and Biology of Drosophila*, Academic Press, London, Vol. 1a, pp 315-330.

Luciani, J. M., Morazzani, M. R. and Stahl, A. (1975). Identification of pachytene bivalents in human male meiosis using G banding technique. *Chromosoma* **52**, 275-282.

Luck, D. J. L. and Reich, E. (1964). DNA in mitochondria of N. crassa. *Proc. Nat. Acad. Sci. USA* **52**, 931-938.

Luduena, R. F. and Woodward, D. O. (1973). Isolation and characterization of α and β tubulin from outer doublets of sea urchin sperm and microtubules of chick embryo brain. *Proc. Nat. Acad. Sci. USA* **70**, 3594-3598.

Luduena, R. F., Shooter, E. M. and Wilson, L. (1977). Structure of the tubulin dimer. *J. Biol. Chem.* **252**, 7006-7014.

Lukanidin, E. M., Zakmanzon, E. S., Komaromi, L., Samarinia, O. P. and Georgiev, G. P. (1972). Structure and function of informomers. *Nature New Biol.* **238**, 193-197.

Lukins, H. B., Tate, J. R., Saunders, G. W. and Linnane, A. W. (1973). The biogenesis of mitochondria. 26. Mitochondrial recombination: the segregation of parental and recombinant mitochondrial genotypes during vegetative division in yeast. *Mol. Gen. Genet.* **120**, 17-25.

Luria, S. E. and Delbruck, M. (1943). Mutations of bacteria from virus sensitivity to virus resistance. *Genetics* **28**, 491-511.

Lutter, L. C. (1977). DNAase I produces staggered cuts in the DNA of chromatin. *J. Mol. Biol.* **117**, 53-70.

Lutter, L. C. (1978). Kinetic analysis of DNAase I cleavages in the nucleosome core: evidence for a DNA superhelix. *J. Mol. Biol.* **124**, 391-417.

Lutter, L. C. (1979). Precise location of DNAase I cutting sites in the nucleosome core determined by high resolution gel electrophoresis. *Nucl. Acids Res.* **6**, 41-56.

Luykx, P. (1970). Cellular mechanisms of chromosome distribution. *Int. Rev. Cytol.* suppl. 2.

Lynch, D. C. and Attardi, G. (1976). Amino acid specificity of the tRNA species coded for by HeLa cell mitochondrial DNA. *J. Mol. Biol.* **102**, 125-142.

Lyon, M. F. (1961). Gene action in the X chromosome of the mouse. *Nature* **190**, 372-373.

Lyon, M. F. (1962). Sex chromatin and gene action in the mammalian X chromosome. *Am. J. Hum. Genet.* **14**, 135-148.

Lyon, M. F. (1964). Lack of evidence that inactivation of the mouse X chromosome is incomplete. *Genet. Res.* **8**, 197-203.

Lyon, M. F. (1968). Chromosomal and subchromosomal inactivation. *Ann. Rev. Genet.* **2**, 31-52.

Lyon, M. F. (1971). Possible mechanisms of X chromosome inactivation. *Nature New Biol.* **232**, 229-232.

Lyon, M. F. (1972). X chromosome inactivation and developmental patterns in mammals. *Biol. Rev.* **47**, 1-35.

Lyon, M. F. (1974). Mechanisms and evolutionary origins of variable X chromosome activity in mammals. *Proc. Roy. Soc. B* **187**, 243-268.

Lyon, M. F. (1976). Distribution of crossing over in mouse chromosomes. *Genet. Res.* **28**, 291-300.

Lyon, M. F., Scarle, A. G., Ford, C. E. and Ohno, S. (1964). A mouse translocation suppressing sex linked variegation. *Cytogenetics* **3**, 306-323.

Lyttleton, J. N. (1962). Isolation of ribosomes from spinach chloroplast. *Exp. Cell Res.* **26**, 312-317.

Luzzati, V. and Nicolaieff, A. (1959). Etude par diffusion des rayons X aux petits angles des gels d'DNA at de nucleoproteins. *J. Mol. Biol.* **1**, 127-133.

Luzzati, V. and Nicolaieff, A. (1963). The structure of nucleohistones and nucleo protamines. *J. Mol. Biol.* **7**, 142-163.

Maat, J. and Smith, A. J. H. (1978). A method for sequencing restriction fragments with dideoxynucleoside triphosphates. *Nucl. Acids Res.* **5**, 4537-4546.

Macaya, G., Thiery, J. -P. and Bernardi, G. (1976). An approach to the organization of eucaryotic genomes at a macromolecular level. *J. Mol. Biol.* **108**, 237-254.

Maccecchini, M. L., Rudin, Y., Blobel, G. and Schatz, G. (1979). Import of proteins into mitochondria: precursor forms of the extramitochondrially made F1-ATPase subunits in yeast. *Proc. Nat. Acad. Sci. USA* **76**, 343–347.

MacDonald, R. J., Przybyla, A. E. and Rutter, W. (1977). Isolation and in vitro translation of the mRNA coding for pancreatic amylase. *J. Biol. Chem.* **252**, 5522–5528.

MacGillivray, A. J., Carroll, D. and Paul, J. (1971). The heterogeneity of the nonhistone chromatin proteins from mouse tissues. *Febs Lett.* **13**, 204–207.

MacGillivray, A. J., Cameron, A., Krauze, R. J., Rockwood, D. and Paul, J. (1972). The nonhistone proteins of chromatin. Their isolation and composition in a number of tissues. *Biochim. Biophys. Acta* **277**, 384–402.

MacGregor, H. C., Vlad, M. and Barnett, L. (1977). An investigation of some problems concerning nucleolar organizers in Salamanders. *Chromosoma* **59**, 283–300.

Macino, G. and Tzagoloff, A. (1979). Assembly of the mitochondrial membrane system. The DNA sequence of a mitochondrial ATPase gene in S. cerevisiae. *J. Biol. Chem.* **254**, 4617–4623.

MacIntyre, R. J. and O'Brien, S. J. (1976). Interacting gene enzyme systems in Drosophila. *Ann. Rev. Genet.* **10**, 281–318.

Macleod, A. R., Wong, N. C. W. and Dixon, G. H. (1977). The amino acid sequence of trout testis histone H1. *Eur. J. Biochem.* **78**, 281–292.

Maden, B. E. H. and Robertson, J. S. (1974). Demonstration of the 5.8S ribosomal sequence in HeLa cell ribosomal precursor RNA. *J. Mol. Biol.* **87**, 227–236.

Maden, B. E. H. and Salim, M. (1974). The methylated nucleotide sequences in HeLa cell ribosomal RNA and its precursors. *J. Mol. Biol.* **88**, 133–164.

Maden, B. E. H. and Tartof, K. (1974). Nature of the rRNA transcribed from the X and Y chromosomes of D. melanogaster. *J. Mol. Biol.* **90**, 51–64.

Maden, B. E. H., Salim, M. and Summers, D. F. (1972). Maturation pathway for rRNA in the HeLa cell nucleus. *Nature New Biol.* **237**, 5–9.

Mahler, H. R. and Dawidowicz, K. (1973). Autonomy of mitochondria in S. cerevisiae and their production of mRNA. *Proc. Nat. Acad. Sci. USA* **70**, 111–114.

Maio, J. J. and Schildkraut, C. L. (1969). Isolated mammalian metaphase chromosomes. II. Fractionated chromosomes of mouse and Chinese hamster cells. *J. Mol. Biol.* **40**, 203–216.

Maio, J. J., Brown, F. L. and Musich, P. R. (1977). Subunit structure of chromatin and the organization of eucaryotic highly repetitive DNA: recurrent periodicities and models for the evolutionary origins of repetitive DNA. *J. Mol. Biol.* **117**, 637–656.

Maitland, N. J. and McDougall, J. K. (1977). Biochemical transformation of mouse cells by fragments of herpes simplex virus DNA. *Cell* **11**, 233–241.

Maizels, N. (1976). Dictyostelium 17S, 25S and 5S rDNAs lie within a 38,000 base pair repeated unit. *Cell* **9**, 431–438.

Malawista, S. E. and Weiss, M. C. (1974). Expression of differentiated functions in hepatoma cell hybrids: high frequency of induction of mouse albumin production in rat hepatoma-mouse lymphoblast hybrids. *Proc. Nat. Acad. Sci. USA* **71**, 927–931.

Malcom, D. B. and Sommerville, J. (1974). The structure of chromosome derived RNP in oocytes of Triturus. *Chromosoma* **48**, 137–158.

Malcom, D. B. and Somerville, J. (1977). The structure of nuclear ribonucleoprotein of amphibian oocytes. *J. Cell Sci.* **24**, 143–166.

Maller, T., Poccia, D., Nishioka, D., Kidd, P., Gerhart, T. and Hartman, H. (1976). Spindle formation and cleavage in Xenopus eggs injected with centriole containing fractions from sperm. *Exp. Cell Res.* **99**, 285–294.

Mandel, J. L., Breathnach, R., Gerlinger, P., Le Meur, M., Gannon, F. and Chambon, P. (1978). Organization of coding and intervening sequences in the chicken ovalbumin split gene. *Cell* **14**, 641–653.

Mandel, P., Metais, P. and Cuny, S. (1950). Les quantites de'DNA par lecucoyte chez diverses especes de mammiferes. *C. R. Acad. Sci. Paris* **231**, 1172-1174.

Mandel, R. and Fasman, G. D. (1976). Chromatin and nucleosome structure. *Nucl. Acids Res.* **3**, 1839-1855.

Mandelkow, E., Thomas, J. and Cohen, C. (1977). Microtubule structure at low resolution by X ray diffraction. *Proc. Nat. Acad. Sci. USA* **74**, 3370-3374.

Mangia, F., Abbo-Halbasch, G. and Epstein, C. J. (1975). X chromosome expression during oogenesis in the mouse. *Devel. Biol.* **45**, 366-368.

Maniatis, T., Kee, S. G., Efstratiadis, A. and Kafatos, F. C. (1976). Amplification and characterization of a β globin gene synthesized in vitro. *Cell* **8**, 163-182.

Maniatis, T., Hardison, R. C., Lacy, E., Lauer, J., O'Connell, C., Quon, D., Sim, G. K. and Efstratiadis, E. (1978). The isolation of structural genes from libraries of eucaryotic DNA. *Cell* **15**, 687-701.

Mankovitz, R., Buchwald, M. and Baker, R. M. (1974). Isolation of oubain resistant human diploid fibroblasts. *Cell* **3**, 221-226.

Mannella, C. A., Collins, R. A., Green, M. and Lambowitz, A. M. (1979). Defective splicing of mitochondrial RNA in cytochrome deficient nuclear mutants of N. crassa. *Proc. Nat. Acad. Sci. USA* **76**, 2635-2639.

Mannherz, H. G. and Goody, R. S. (1976). Proteins of contractile systems. *Ann. Rev. Biochem.* **45**, 427-465.

Manning, J. E. and Richards, O. C. (1972). Isolation and molecular weight of circular chloroplast DNA from Euglena gracilis. *Biochim. Biophys. Acta* **378**, 305-317.

Manning, J. E., Schmid, C. W. and Davidson, N. (1975). Interspersion of repetitive and nonrepetitive DNA sequences in the D. melanogaster genome. *Cell* **4**, 141-156.

Manteuil, S., Hamer, D. H. and Thomas, C. A. (1975). Regular arrangement of restriction sites in Drosophila DNA. *Cell* **5**, 413-422.

Manuelides, L. (1978). Chromosomal localization of complex and simple repeated human DNAs. *Chromosoma* **66**, 23-32.

Marahiel, M. A., Imam, G., Nelson, P., Pieniazek, N. J., Stepien, P. P. and Kuntzel, H. (1977). Identification of an intramitochondrially synthesized proteolipid associated with the mito-chondrial ATPase complex as the product of a mitochondrial gene determining oligomycin resistance in Aspergillus nidulans. *Eur. J. Biochem.* **76**, 345-354.

Marbaix, G. and Lane, C. D. (1972). Rabbit hemoglobin synthesis in frog cells. II. Further characterization of the products of translation of reticulocyte 9S RNA. *J. Mol. Biol.* **67**, 517-524.

Marbaix, G., Huez, G., Burny, A., Cleuter, Y., Hubert, E., Leclercq, M., Chantrenne, H., Soreq, H., Nudel, U. and Littauer, U. Z. (1975). Absence of polyadenylate segment in globin mRNA accelerates its degradation in Xenopus oocytes. *Proc. Nat. Acad. Sci. USA* **72**, 3065-3067.

Marcu, K. B., Valbuena, O. and Perry, R. P. (1978a). Isolation, purification, and properties of mouse heavy chain immunoglobulin mRNAs. *Biochemistry* **17**, 1723-1732.

Marcu, K. B., Schibler, U. and Perry, R. P. (1978). The 5' terminal sequences of Ig mRNAs of a mouse myeloma. *J. Mol. Biol.* **120**, 381-400.

Marcum, J. M. and Borisy, G. G. (1978). Characterization of microtubule protein oligomers by analytical centrifugation. *J. Biol. Chem.* **253**, 2825-2833.

Marcum, J. M., Dedman, J. R., Brinkley, B. R. and Means, A. R. (1978). Control of microtubule assembly-disassembly by calcium dependent regulator protein. *Proc. Nat. Acad. Sci. USA* **75**, 3771-3775.

Mardian, J. K. W. and Isenberg, I. (1978). Yeast inner histones and the evolutionary conservation of histone-histone interactions. *Biochemistry* **17**, 3825-3832.

Margolis, R. L. and Wilson, L. (1977). Addition of colchicine tubulin complex to microtubule ends: the mechanism of substoichiometric colchicine poisoning. *Proc. Nat. Acad. Sci. USA* **74**, 3446-3470.

Margolis, R. L. and Wilson, L. (1978). Opposite end assembly and disassembly of microtubules at steady state in vitro. *Cell* **13**, 1-8.

Margulis, L. (1973). Colchicine sensitive microtubules. *Int. Rev. Cytol.* **34**, 333-361.

Marians, K. J., Wu, R., Stawiski, J., Hozumi, T. and Narang, S. A. (1976). Cloned synthetic lac operator DNA is biologically active. *Nature* **263**, 744-747.

Marin, G. and Prescott, D. M. (1964). The frequency of sister chromatid exchanges following exposure to varying doses of ^{3}H thymidine or X rays. *J. Cell Biol.* **21**, 159-167.

Marmur, J. and Doty, P. (1959). Heterogeneity in DNA. I. Dependence on composition of the configurational stability of DNAs. *Nature* **183**, 1427-1428.

Marmur, J., Rownd, R. and Schildkraut, C. L. (1963). Denaturation and renaturation of DNA. *Prog. Nucl. Acid Res.* **1**, 232-300.

Marrotta, C. A., Forget, B. G., Weissman, S. M., Verma, I. M., McCaffrey, R. P. and Baltimore, D. (1974). Nucleotide sequences of human globin mRNA. *Proc. Nat. Acad. Sci. USA* **71**, 2300-2304.

Marotta, C. A., Wilson, J. T., Forget, B. G. and Weissman, S. M. (1977). Human beta globin mRNA. III. Nucleotide sequences derived from complementary DNA. *J. Biol. Chem.* **252**, 5040-5053.

Marsden, M. and Laemmli, U. K. (1979). Metaphase chromosome structure: evidence for a radial loop model. *Cell* **17**, 849-858.

Marshall, C. J., Handmaker, S. D. and Bramwell, M. E. (1975). Synthesis of rRNA in synkaryons and heterokaryons formed between human and rodent cells. *J. Cell Sci.* **17**, 307-325.

Marsland, D. A. (1938). The effects of high hydrostatic pressure upon cell division in Arbacia eggs. *J. Cell. Comp. Physiol.* **12**, 57-70.

Marsland, D. A. (1950). The mechanisms of cell division: temperature-pressure experiments on the cleaving eggs of Arbacia punctulata. *J. Cell. Comp. Physiol.* **36**, 205-227.

Marsland, D. and Landau, J. V. (1954). The mechanisms of cytokinesis: temperature-pressure studies on the cortical gel system in various marine eggs. *J. Exp. Zool.* **125**, 507-539.

Marsland, D., Zimmerman, A. M. and Auclair, W. (1960). Cell division: experimental induction of cleavage furrows in the eggs of Arbacia punctulata. *Exp. Cell Res.* **21**, 179-196.

Martin, D. Z., Todd, R. D., Lang, D., Pei, P. N. and Garrard, W. T. (1977). Heterogeneity in nucleosome spacing. *J. Biol. Chem.* **252**, 8269-8277.

Martin, G. R., Epstein, C. J., Travis, B., Tucker, G., Yatziv, S., Martin, D. W., Clift, S. and Cohen. S. (1978). X chromosome inactivation during differentiation of female teratocarcinoma stem cells in vitro. *Nature* **271**, 329-334.

Martin, M. A. and Hoyer, B. H. (1966). Thermal stabilities and species specificities of reannealed animal DNAs. *Biochemistry* **5**, 2706-2713.

Martin, N. C. and Rabinowitz, M. (1978). Mitochondrial tRNAs in yeast: identification of isoaccepting tRNAs. *Biochemistry* **17**, 1628-1633.

Martin, N. C., Rabinowitz, M. and Fukuhara, H. (1977). Yeast mitochondrial DNA specifies tRNA for 19 amino acids. Deletion mapping of the tRNA genes. *Biochemistry* **16**, 4672-4677.

Martin, S. A. and Moss, B. (1975). Modification of mRNA by mRNA guanylyltransferase and mRNA (guanine-7) methyltransferase from vaccinia virions. *J. Biol. Chem.* **250**, 9330-9335.

Martinson, H. G. and McCarthy, B. J. (1975). Histone histone associations within chromatin. Crosslinking studies using tetranitromethane. *Biochemistry* **14**, 1073-1078.

Martinson, H. G. and McCarthy, B. J. (1976). Histone-histone interactions within chromatin. Preliminary characterization of presumptive H2B-H2A and H2B-H4 binding sites. *Biochemistry* **15**, 4126-4130.

Martinson, H. G. and True, R. J. (1979a). On the mechanism of nucleosome unfolding. *Biochemistry* **18**, 1089-1093.

Martinson, H. G. and True, R. J. (1979b). Amino acid contacts between histones are the same for plants and mammals. Binding site studies using ultraviolet light and tetra mitromethane. *Biochemistry* **18**, 1947-1951.

Martinson, H. G., Shatlar, M. D. and McCarthy, B. J. (1976). Histone-histone interactions within chromatin. Crosslinking studies using ultraviolet light. *Biochemistry* **15**, 2002-2007.

Martinson, H. G., True, R., Lau, C. K. and Mehrabian, M. (1979a). Histone-histone interactions within chromatin. Preliminary location of multiple contact sites between histones 2A, 2B and 4. *Biochemistry* **18**, 1075-1081.

Martinson, H. G., True, R. J. and Burch, J. B. E. (1979b). Specific histone-histone contacts are ruptured when nucleosomes unfold at low ionic strength. *Biochemistry* **18**, 1082-1088.

Martinson, H. G., True, R., Burch, J. B. E. and Kunkel, G. (1979c). Semihistone protein A24 replaces H2A as an integral component of the nucleosome histone core. *Proc. Nat. Acad. Sci. USA* **76**, 1030-1034.

Marushige, K. and Bonner, J. (1971). Fractionation of liver chromatin. *Proc. Nat. Acad. Sci. USA* **68**, 2941-2944.

Marx, K. A., Allen, J. R. and Hearst, J. E. (1976a). Characterization of the repetitive human DNA families. *Biochim. Biophys. Acta* **425**, 129-147.

Marx, K. A., Allen, J. R. and Hearst, J. E. (1976b). Chromosomal localization by in situ hybridization of the repetitious human DNA families and evidence of their satellite DNA equivalents. *Chromosoma* **59**, 23-42.

Mason, J. M. (1976). Orientation disrupter (*ord*): a recombination-defective and disjunction defective meiotic mutant in D. melanogaster. *Genetics* **84**, 545-572.

Mason, T. L. and Schatz, G. (1973). Cytochrome c oxidase from baker's yeast. II. Site of translation of the protein components. *J. Biol. Chem.* **248**, 1355-1360.

Mather, K. (1933). The relation between chiasmata and crossing over in diploid and triploid D. melanogaster. *J. Genet.* **27**, 243-259.

Mather, K. (1936). The determination of position in crossing over. I. D. melanogaster. *J. Genet.* **33**, 207-235.

Mather, K. (1938). Crossing over. *Biol. Rev.* **13**, 252-292.

Mather, K. (1940). The determination of position in crossing over. III. The evidence of metaphase chiasmata. *J. Genet.* **39**, 205-223.

Mathew, C. G. P., Goodwin, G. H. and Johns, E. W. (1979). Studies on the association of the HMG nonhistone chromatin proteins with isolated nucleosomes. *Nucl. Acids Res.* **6**, 167-180.

Mathis, D. J. and Gorovsky, M. A. (1976). Subunit structure of rDNA-containing chromatin. *Biochemistry* **15**, 750-755.

Matsuya, Y., Green, H. and Basilico, C. (1968). Properties and uses of human mouse hybrid cell lines. *Nature* **220**, 1199-1202.

Matthews, K. A. and Hiraizumi, Y. (1978). An analysis of male recombination elements in a natural population of D. melanogaster in South Texas. *Genetics* **88**, 81-91.

Matthews, M. B. (1972). Further studies on the translation of globin mRNA and EMC virus RNA in a cell free system from Krebs II ascites cells. *Biochim. Biophys. Acta* **272**, 108-118.

Mattick, J. S. and Nagley, P. (1977). Comparative studies of the effects of acridines and other petite inducing drugs on the mitochondrial genome of S. cerevisiae. *Mol. Gen. Genet.* **152**, 267-276.

Mattoccia, E., Baldi, M. I., Carrara, G., Fruscolini, P., Benedetti, P. and Tocchini-Valentini, G. P. (1979). Separation of RNA transcription and processing activities from X. laevis germinal vesicles. *Cell* **18**, 643-648.

Mauck, J. C. (1977). Solubilized DNA dependent RNA polymerase activities in resting and growing fibroblasts. *Biochemistry* **16**, 793-796.

Mauck, J. C. and Green, H. (1973). Regulation of RNA synthesis in fibroblasts during transition from resting to growing state. *Proc. Nat. Acad. Sci. USA* **70**, 2819-2822.

Mauck, J. C. and Green, H. (1974). Regulation of pre tRNA synthesis during transition from resting to growing state. *Cell* **3**, 171-178.

Maul, G. G. and Hamilton, T. H. (1967). The intranuclear localization of two DNA dependent RNA polymerase activities. *Proc. Nat. Acad. Sci. USA* **57**, 1371-1378.

Mautner, V. and Hynes, R. O. (1977). Surface distribution of LETS protein in relation to the cytoskeleton of normal and transformed cells. *J. Cell Biol.* **75**, 743-768.

Max, E. E., Seidman, J. G. and Leder, P. (1979). Sequences of five potential recombination sites encoded close to an immunoglobulin K constant region gene. *Proc. Nat. Acad. Sci. USA* **76**, 3450-3454.

Maxam, A. M. and Gilbert, W. (1977). A new method for sequencing DNA. *Proc. Nat. Acad. Sci. USA* **74**, 560-564.

May, E., Kress, M. and May, P. (1978). Characterization of two SV40 early mRNAs and evidence for a nuclear "prespliced" RNA species. *Nucl. Acids Res.* **5**, 3083-3100.

Mayfield, J. E. and Bonner, J. (1971). Tissue differences in rat chromosomal RNA. *Proc. Nat. Acad. Sci. USA* **68**, 2652-2655.

Mayfield, J. E. and Bonner, J. (1972). A partial sequence of nuclear events in regenerating rat liver. *Proc. Nat. Acad. Sci. USA* **69**, 7-10.

Mayfield, J. E., Serunian, L. A., Silver, L. M. and Elgin, S. C. R. (1978). A protein released by DNAase I digestion of Drosophila nuclei is preferentially associated with puffs. *Cell* **14**, 539-544.

Mazia, D. (1961). Mitosis and the physiology of cell division. In J. Brachet and A. E. Mirsky (Ed.). *The Cell*, Academic Press, New York, Vol. 3, pp 77-412.

Mazia, D. and Dan, K. (1952). The isolation and biochemical characterization of the mitotic apparatus of dividing cells. *Proc. Nat. Acad. Sci. USA* **38**, 826-738.

Mazia, D., Harris, P. J. and Bibring, T. (1960). The multiplicity of the mitotic centers and the time course of their duplication and separation. *J. Biochem. Biophys. Cytol.* **7**, 1-20.

Mazrimas, J. A. and Hatch, F. T. (1977). Similarity of satellite DNA properties in the order Rodentia. *Nucl. Acids Res.* **4**, 3215-3227.

McBride, O. W. and Ozer, H. L. (1973). Transfer of genetic information by purified metaphase chromosomes. *Proc. Nat. Acad. Sci. USA* **70**, 1258-1262.

McBride, O. W., Burch, J. W. and Ruddle, F. H. (1978). Cotransfer of thymidine kinase and galactokinase genes by chromosome mediated gene transfer. *Proc. Nat. Acad. Sci. USA* **75**, 914-918.

McBurney, M. W. (1977). Hemoglobin synthesis in cell hybrids formed between teratocarcinoma and Friend erythroleukemia cells. *Cell* **12**, 653-662.

McBurney, M. W. and Adamson, E. D. (1976). Studies on the activity of the X chromosomes in female teratocarcinoma cells in culture. *Cell* **9**, 57-70.

McBurney, M. W., Featherstone, M. S. and Kaplan, H. (1978). Activation of teratocarcinoma derived hemoglobin genes in teratocarcinoma Friend cell hybrids. *Cell* **15**, 1323-1330.

McCarthy, B. J. (1967). The arrangement of base sequences in DNA. *Bacteriol. Rev.* **31**, 215-229.

McCarthy, B. J. and Church, R. B. (1970). The specificity of molecular hybridization reactions. *Ann. Rev. Biochem.* **39**, 131-150.

McCarthy, B. J. and McConaughy, B. L. (1968). DNA/DNA duplex formation and the incidence of partially related base sequences in DNA. *Biochem. Genet.* **2**, 37-53.

McClintock, B. (1934). The relation of a particular chromosomal element to the development of the nucleoli in Zea mays. *Z. Zellforsch.* **21**, 294-328.

McClintock, B. (1940). The stability of broken ends of chromosomes in Zea mays. *Genetics* 26, 234.

McClintock, B. (1965). The control of gene action in maize. *Brook. Symp. Biol.* 18, 162-184.

McConaughy, B. L. and McCarthy, B. J. (1970). The extent of base sequence divergence among the DNAs of various rodents. *Biochem. Genet.* 4, 425-446.

McConaughy, B. L. and McCarthy, B. J. (1972). Fractionation of chromatin by thermal chromatography. *Biochemistry* 11, 998-1002.

McConaughy, B. L., Laird, C. D. and McCarthy, B. J. (1969). Nucleic acid reassociation in formamide. *Biochemistry* 8, 3289-3294.

McConkey, E. H. and Hopkins, J. W. (1964). The relationship of the nucleolus to the synthesis of rRNA in HeLa cells. *Proc. Nat. Acad. Sci. USA* 51, 1197-1204.

McConkey, E. H. and Hopkins, J. W. (1965). Subribosomal particles and the transport of mRNA in HeLa cells. *J. Mol. Biol.* 14, 257-270.

McDonald, J. A. and Kelley, W. N. (1971).. Lesch-Nyhan syndrome: altered kinetic properties of mutant enzyme. *Science* 171, 689-691.

McDonald, K., Pickett-Heaps, J. D., McIntosh, J. R. and Tippit, D. H. (1977). On the mechanism of anaphase spindle elongation in Diatoma vulgare. *J. Cell Biol.* 74, 377-388.

McFarlane, P. W. and Callan, H. G. (1973). DNA replication in the chromosomes of the chicken, Gallus domesticus. *J. Cell Sci.* 13, 821-839.

McGill, M. and Brinkley, B. R. (1975). Human chromosomes and centrioles as nucleation sites for the in vitro assembly of microtubules from bovine brain tubulin. *J. Cell Biol.* 67, 189-199.

McGrogan, M. and Raskas, H. J. (1978). Two regions of the adenovirus 2 genome specify families of late polysomal RNAs containing common sequences. *Proc. Nat. Acad. Sci. USA* 75, 625-629.

McIntosh, J. R. (1964). Bridges between microtubules. *J. Cell Biol.* 61, 166-187.

McIntosh, J. R. and Landis, S. C. (1971). The distribution of spindle microtubules during mitosis in cultured human cells. *J. Cell Biol.* 49, 468-497.

McIntosh, J. R., Hepler, P. K. and Van Wie, D. G. (1969). Model for mitosis. *Nature*, 224, 659-663.

McIntosh, J. R., Cande, Z., Snyder, J. and Vanderslice, K. (1975). Studies on the mechanism of mitosis. *Ann. N. Y. Acad. Sci.* 253, 407-427.

McKean, D. J., Bell, M. and Potter, M. (1978). Mechanisms of antibody diversity: multiple genes encode structurally related mouse k variable regions. *Proc. Nat. Acad. Sci. USA* 75, 3913-3917.

McKenzie, S. L. and Meselson, M. (1977). Translation in vitro of Drosophila heat shock messages. *J. Mol. Biol.* 117, 279-284.

McKenzie, S. L., Henikoff, S. and Meselson, M. (1975). Localization of RNA from heat induced polysomes at puff sites in D. melanogaster. *Proc. Nat. Acad. Sci. USA* 72, 1117-1121.

McKeown, M., Taylor, W. C., Kindle, K. L., Firtel, R. A., Bender, W. and Davidson, N. (1978). Multiple heterogeneous actin genes in Dictyostelium. *Cell* 15, 789-800.

McKnight, G. S. and Schimke, R. T. (1974). Ovalbumin mRNA: evidence that the initial product of transcription is the same size as polysomal ovalbumin messenger. *Proc. Nat. Acad. Sci. USA* 71, 4327-4331.

McKnight, G., Stanley, P. P. and Schimke, R. T. (1975). Induction of ovalbumin mRNA sequences by estrogen and progesterone in chick oviduct as measured by hybridization to complementary DNA. *J. Biol. Chem.* 250, 8105-8110.

McKnight, S. L. and Miller, O. L. (1976). Ultrastructural patterns of RNA synthesis during early embryogenesis of D. melanogaster. *Cell* 8, 305-319.

McKnight, S. L. and Miller, O. L., Jr. (1977). Electron microscopic analysis of chromatin replication in the cellular blastoderm D. melanogaster embryo. *Cell* 12, 795-804.

McKusick, V. A. and Ruddle, F. H. (1977). The status of the gene map of the human chromosomes. *Science* 196, 390-404.

McLeish, J. and Sunderland, N. (1961). Measurements of DNA in higher plants by Feulgen photometry and chemical means. *Exp. Cell Res.* **24**, 527–540.

McMorris, F. A. and Ruddle, F. H. (1974). Expression of neuronal phenotypes in neuroblastoma cell hybrids. *Devel. Biol.* **39**, 226–246.

McMorris, F. A., Kolber, A. R., Moore, B. W. and Perumal, A. S. (1974). Expression of the neuron specific protein 14-3-2 and steroid sulfatase in neuroblastoma cell hybrids. *J. Cell. Physiol.* **84**, 473–480.

McNutt, M. S., Culp, L. A. and Black, P. H. (1973). Contact inhibited revertant cell lines isolated from SV40 transformed cells. IV. Microfilament distribution and cell shape in untransformed, transformed and revertant Balb/c 3T3 cells. *J. Cell Biol.* **56**, 412–428.

McPherson, I. and Stoker, M. (1962). Polyoma transformation of hamster cell clones - an investigation of genetic factors affecting cell competence. *Virology* **16**, 147–151.

McReynolds, L., O'Malley, B. W., Nisbet, A. D., Fothergill, J. E., Givol, D., Fields, S., Robertson, M. and Brownlee, G. G. (1978). Sequence of chicken ovalbumin mRNA. *Nature* **273**, 723–728.

Meagher, R. B., Tait, R. C., Betlach, M. and Boyer, H. W. (1977). Protein expression in E. coli minicells by recombinant plasmids. *Cell* **10**, 521–536.

Mears, J. G., Ramirez, F., Leibowitz, D. and Bank, A. (1978). Organization of human δ and β globin genes in cellular DNA and the presence of intragenic inserts. *Cell* **15**, 15–24.

Meilhac, M. and Chambon, P. (1973). Animal DNA dependent RNA polymerases. Initiation sites on calf thymus DNA. *Eur. J. Biochem.* **35**, 454–463.

Meinke, W., Hall, M. R. and Goldstein, D. A. (1975). Proteins in intracellular SV40 nucleoprotein complexes: comparison with SV40 core proteins. *J. Virol.* **15**, 439–448.

Meiss, H. K. and Basilico, C. (1972). Temperature sensitive mutants of BHK 21 cells. *Nature New Biol.* **239**, 66–68.

Meistrich, M. L., Meyn, R. E. and Barlogie, B. (1977). Synchronization of mouse L-P59 cells by centrifugal elutriation separation. *Exp. Cell Res.* **105**, 169–177.

Melgar, E. and Goldthwaite, D. A. (1968a). DNA nucleases. I. The use of a new method to observe the kinetics of DNA degradation by DNAase I, DNAase II and E. coli endonuclease I. *J. Biol. Chem.* **243**, 4401–4408.

Melgar, E. and Goldthwaite, D. A. (1968b). DNA nucleases. II. The effects of metals on the mechanism of action of DNAase I. *J. Biol. Chem.* **243**, 4409–4416.

Melli, M., Spinelli, G. and Arnold, E. (1977). Synthesis of histone mRNA of HeLa cells during the cell cycle. *Cell* **12**, 167–173.

Melli, M., Whitfield, C., Rao, K. V., Richardson, M. and Bishop, J. O. (1971). DNA-RNA hybridization in great excess. *Nature New Biol.* **231**, 8–12.

Melli, M., Spinelli, G., Wyssling, H. and Arnold, E. (1977). Presence of histone mRNA sequences in high molecular weight RNA of HeLa cells. *Cell* **11**, 651–661.

Mendecki, J., Lee, S. Y. and Brawerman, G. (1972). Characteristics of the poly(A) segment associated with mRNA in mouse sarcoma 180 ascites cells. *Biochemistry* **11**, 792–798.

Mercer, E. H. and Wolpert, L. (1958). Electron microscopy of cleaving sea urchin eggs. *Exp. Cell Res.* **14**, 629–632.

Mercereau-Puijalon, O., Royal, A., Cami, B., Garapin, A., Krust, A., Gannon, F. and Kourilsky, P. (1978). Synthesis of an ovalbumin like protein by E. coli K12 harboring a recombinant plasmid. *Nature* **275**, 505–510.

Merkel, C. G., Kwan, S. -P. and Lingrel, J. B. (1975). Size of the poly(A) region of newly synthesized globin mRNA. *J. Biol. Chem.* **250**, 3725–3728.

Merriam, J. R. and Frost, J. N. (1964). Exchange and nondisjunction of the X chromosomes in female D. melanogaster. *Genetics* **49**, 109–122.

Merriam, R. W. (1969). Movement of cytoplasmic proteins into nuclei induced to enlarge and initiate DNA or RNA synthesis. *J. Cell Sci.* **5**, 333–350.

Merriam, R. W. and Clark, T. G. (1978). Actin in Xenopus oocytes. II. Intracellular distribution and polymerizability. *J. Cell Biol.* **77**, 439-447.

Meselson, M. and Stahl, F. W. (1958). The replication of DNA in E. coli. *Proc. Nat. Acad. Sci. USA* **44**, 671-682.

Meselson, M., Stahl, F. W. and Vinograd, J. (1957). Equilibrium sedimentation of macromolecules in density gradients. *Proc. Nat. Acad. Sci. USA* **43**, 581-588.

Mets, L. and Bogorad, L. (1972). Altered chloroplast ribosomal proteins associated with erythromycin resistant mutants in two genetic systems of C. reinhardii. *Proc. Nat. Acad. Sci. USA* **69**, 3779-3783.

Metz, C. W. (1938). Chromosome behavior, inheritance and sex determination in Sciara. *Am. Nat.* **72**, 485-520.

Meyerink, J. H., Retel, J., Raue, H. A., Planta, R. J., Van der Ende, A. and Van Bruggen, E. F. J. (1978). Genetic organization of the ribosomal transcription units of the yeast S. carlsbergensis. *Nucl. Acids Res.* **5**, 2801-2808.

Meyn, R. E., Hewitt, R. R. and Humphrey, R. M. (1973). Evaluation of S phase synchronization by analysis of DNA replication in 5-BUdR. *Exp. Cell Res.* **82**, 137-142.

Meyuhas, O. and Perry, R. P. (1979). Relationship between size, stability and abundance of the mRNA of mouse L cells. *Cell* **16**, 139-148.

Mezger-Freed, L. (1972). Effects of ploidy and mutagen on BUdR resistance in haploid and diploid frog cells. *Nature New Biol.* **235**, 245-246.

Mezger-Freed, L. (1974). An analysis of survival in haploid and diploid cell cultures after exposure to ICR acridine half mustard compounds mutagenic for bacteria. *Proc. Nat. Acad. Sci. USA* **71**, 4416-4420.

Mezger-Freed, L. (1975). Mutagenesis of haploid cultured frog eggs. *Genetics Suppl.* **79**, 359-372.

Mezger-Freed, L. (1977). Chromosomal evolution in a haploid frog cell line: implications for the origin of karyotypic variants. *Chromosoma* **62**, 1-15.

Michaelis, G., Petrochilo, E. and Slonimski, P. P. (1973). Mitochondrial genetics. III. Recombined molecules of mitochondrial DNA obtained from crosses between cytoplasmic petite mutants S. cerevisiae: physical and genetic characterization. *Mol. Gen. Genet.* **123**, 51-65.

Michaelis, G., Michel, F., Lazowska, J. and Slonimski, P. P. (1976). Recombined molecules of mitochondrial DNA obtained from crosses between cytoplasmic petite of S. cerevisiae: the stoichiometry of parental repeats within the recombined molecule. *Mol. Gen. Genet.* **149**, 125-130.

Michel, F., Lazowska, J., Faye, G., Fukuhara, H. and Slonimski, P. P. (1974). Physical and genetic organization of petite and grande yeast mitochondrial DNA. III. High resolution melting and reassociation studies. *J. Mol. Biol.* **85**, 411-431.

Migeon, B. R. (1972). Stability of X chromosome inactivation in human somatic cells. *Nature* **239**, 87-89.

Migeon, B. R., Smith, S. W. and Leddy, C. L. (1969). The nature of thymidine kinase in the human mouse hybrid cell. *Biochem. Genet.* **3**, 583-590.

Migeon, B. R., Der Kaloustian, V. M., Nyhan, W. L., Young, W. J. and Childs, B. (1968). X linked HGPRT deficiency: heterozygote has two clonal populations. *Science* **160**, 425-427.

Miki-Noumura, T. (1977). Studies on the de novo formation of centrioles: aster formation in the activated eggs of sea urchin. *J. Cell Sci.* **24**, 203-216.

Milcarek, C. and Penman, S. (1974). Membrane bound polysomes in HeLa cells: association of poly(A) with membranes. *J. Mol. Biol.* **89**, 327-338.

Milcarek, C., Price, R. and Penman, S. (1974). The metabolism of a poly(A) minus mRNA fraction in HeLa cells. *Cell* **3**, 1-10.

Miller, C. L. and Ruddle, F. H. (1978). Cotransfer of human X linked markers into murine somatic cells via isolated metaphase chromosomes. *Proc. Nat. Acad. Sci. USA* **75**, 3346-3350.

Miller, C. L., Fuseler, J. W. and Brinkley, B. R. (1977). Cytoplasmic microtubules in transformed mouse x nontransformed human cell hybrids: correlation with in vitro growth. *Cell* 12, 319-331.

Miller, D. A., Allerdice, P. W., Miller, O. J. and Breg, W. R. (1971a). Quinacrine fluorescence patterns of human D group chromosomes. *Nature* 232, 24-27.

Miller, D. A., Kouri, R. E., Dev, V. G., Grewal, M., Hutton, J. J. and Miller, O. J. (1971b). Assignment of four linkage groups to chromosomes in M. musculus and a cytogenetic method for locating their centromeric ends. *Proc. Nat. Acad. Sci. USA* 68, 2699-2702.

Miller, D. A., Miller, O. J., Dev, V. G., Hashmi, S., Tantravahi, R., Medrano, L. and Green, H. (1974). Human chromosome 19 carries a poliovirus receptor gene. *Cell* 1, 167-174.

Miller, D. A., Dev, V. G., Tantravahi, R. and Miller, O. J. (1976a). Suppression of human nucleolus organizer activity in mouse human somatic hybrid cells. *Exp. Cell Res.* 101, 235-242.

Miller, D. L., Gubbins, E. J., Pegg, E. W. and Donelson, J. E. (1977). Transcription and translation of cloned Drosophila DNA fragments in E. coli. *Biochemistry* 16, 1031-1038.

Miller, D. M., Turner, P., Nienhuis, A. W., Axelrod, D. E. and Gopalakrishnan, T. V. (1978). Active conformation of the globin genes in uninduced and induced mouse erythroleukemia cells. *Cell* 14, 511-521.

Miller, J. R., Cartwright, E. M., Brownlee, G. G., Federoff, N. V. and Brown, D. D. (1978). The nucleotide sequence of oocyte 5S DNA in X. laevis. II. The GC rich region. *Cell* 13, 717-725.

Miller, L. (1973). Control of 5S RNA synthesis during early development of anucleolate and partial nucleolate mutants of X. laevis. *J. Cell Biol.* 59, 624-632.

Miller, L. (1974). Metabolism of 5S RNA in the absence of ribosome production. *Cell* 3, 275-282.

Miller, L. and Brown, D. D. (1969). Variations in the activity of nucleolar organizers and the ribosomal gene content. *Chromosoma* 28, 430-444.

Miller, L. and Knowland, J. (1970). Reduction of rRNA synthesis and rRNA genes in a mutant of X. laevis which organizes only a partial nucleolus. II. The number of rRNA genes in animals of different nucleolar types. *J. Mol. Biol.* 51, 329-338.

Miller, O. J. (1976). Is the centromeric heterochromatin of Mus musculus late replicating? *Chromosoma* 55, 165-170.

Miller, O. J., Allerdice, P. W., Miller, D. A., Breg, W. R. and Migeon, B. R. (1971c). Hyman thymidine kinase gene locus: assignment to chromosome 17 in a hybrid of man and mouse cells. *Science* 173, 244-245.

Miller, O. J., Cook, P. R., Khan, P. M., Shin, S. and Siniscalco, M. (1971d). Mitotic separation of two human X linked genes in man mouse somatic cell hybrids. *Proc. Nat. Acad. Sci. USA* 68, 116-120.

Miller, O. J., Miller, D. A., Kouri, R. E., Allerdice, P. W., Dev, V. G., Grewal, M. S. and Hutton, J. J. (1971e). Identification of the mouse karyotype by quinacrine fluorescence and tentative assignment of seven linkage groups. *Proc. Nat. Acad. Sci. USA* 68, 1530-1533.

Miller, O. J., Miller, D. A., Dev, V. G., Tantravahi, R. and Croce, C. M. (1976b). Expression of human and suppression of mouse nucleolus organizer activity in mouse human somatic cell hybrids. *Proc. Nat. Acad. Sci. USA* 73, 4531-4535.

Miller, O. L. (1965). Fine structure of lampbrush chromosomes. *Nat. Cancer Inst. Monog.* 18, 79-100.

Miller, O. L., Beatty, B. R., Hamkalo, B. A. and Thomas, C. A. (1970). Electron microscopic visualization of transcription. *Cold Spring Harbor Symp. Quant. Biol.* 35, 505-512.

Miller, O. L., Jr. and Beatty, B. R. (1969a). Visualization of nucleolar genes. *Science* 164, 955-957.

Miller, O. L., Jr. and Beatty, B. R. (1969b). Extrachromosomal nucleolar genes in amphibian oocytes. *Genetics* 61, 133-143.

Miller, O. L., Jr. and Beatty, B. R. (1969c). Portrait of a gene. *J. Cell Physiol.* suppl. 1 74, 225-232.

Miller, T. E., Huang, C. Y. and Pogo, A. O. (1978a). Rat liver nuclear skeleton and RNP complexes containing hnRNA. *J. Cell Biol.* **76,** 675-691.

Miller, T. E., Huang, C. Y. and Pogo, A. O. (1978b). Rat liver nuclear skeleton and small molecular weight RNA species. *J. Cell Biol.* **76,** 692-704.

Milman, G., Krauss, S. W. and Olsen, A. S. (1977). Tryptic peptide analysis of normal and mutant forms of HGPRT from HeLa cells. *Proc. Nat. Acad. Sci. USA* **74,** 926-930.

Milman, G., Lee, E., Ghangas, G. S., McLaughlin, J. R. and George, M. (1976). Analysis of HeLa cell HGPRT mutants and revertants by two dimensional polyacrylamide gel electrophoresis: evidence for silent gene activation. *Proc. Nat. Acad. Sci. USA* **73,** 4589-4593.

Minegawa, T., Wagner, B. and Strauss, B. (1959). The nucleic acid content of N. crassa. *Arch. Biochem. Biophys.* **80,** 442-445.

Minna, J., Glazer, D. and Nirenberg, M. (1972). Genetic dissection of neural properties using somatic cell hybrids. *Nature New Biol.* **235,** 225-231.

Minna, J., Nelson, P., Peacock, J., Glazer, D. and Nirenberg, M. (1971). Genes for neuronal properties expressed in neuroblastoma x L cell hybrids. *Proc. Nat. Acad. Sci. USA* **68,** 234-239.

Minor, P. D. and Smith, J. A. (1974). Explanation of degree of correlation of sibling generation times in animal cells. *Nature* **248,** 241-243.

Minson, A. C., Wildy, P., Buchan, A. and Darby, G. (1978). Introduction of the Herpes simplex virus thymidine kinase gene into mouse cells using virus DNA or transformed cell DNA. *Cell* **13,** 581-587.

Mirault, M. E., Goldschmidt-Clermont, M., Moran, L., Arrigo, A. P. and Tissieres, A. (1978). The effect of heat shock on gene expression in D. melanogaster. *Cold Spring Harbor Symp. Quant. Biol.* **42,** 819-827.

Mirsky, A. E. (1971). The structure of chromatin. *Proc. Nat. Acad. Sci. USA* **68,** 2945-2948.

Mirsky, A. E. and Ris, H. (1949). Variable and constant components of chromosomes. *Nature* **163,** 666-667.

Mirsky, A. E. and Ris, H. (1951). The composition and structure of isolated chromosomes. *J. Gen. Physiol.* **34,** 475-492.

Mirsky, A. E. and Ris, H. (1952). The DNA content of animal cells and its evolutionary significance. *J. Gen. Physiol.* **34,** 451-462.

Mirsky, A. E. and Silverman, B. (1972). Blocking by histones of accessibility to DNA in chromatin. *Proc. Nat. Acad. Sci. USA* **69,** 2115-2119.

Mirzabekov, A. D., Shick, V. V., Belyavsky, A. V. and Bavykin, S. (1978). Primary organization of nucleosome core particle of chromatin: sequence of histone arrangement along DNA. *Proc. Nat. Acad. Sci. USA* **75,** 4184-4188.

Mitchell, A. R., Seuanez, H. N., Lawrie, S. S., Martin, D. E. and Gosden, J. R. (1977). The location of DNA homologous to human satellite III DNA in the chromosomes of chimpanzee (Pan troglodytes), gorilla (Gorilla gorilla) and orangutang (Pongo pymaeus). *Chromosoma* **61,** 345-358.

Mitchell, M. B. and Mitchell, H. K. (1952). A case of maternal inheritance in N. crassa. *Proc. Nat. Acad. Sci. USA* **38,** 442-449.

Mitchell, M. B., Mitchell, H. K. and Tissieres, A. (1953). Mendelian and non Mendelian factors affecting the cytochrome system in N. crassa. *Proc. Nat. Acad. Sci. USA* **39,** 606-613.

Mitchison, J. M. (1952). Cell membranes and cell division. *Symp. Soc. Exp. Biol.* **6,** 105-127.

Mitchison, J. M. (1971). *The Biology of the Cell Cycle.* Cambridge University Press, England.

Miyaki, M., Koida, K. and Ono, T. (1973). RNAase and alkali sensitivity of closed circular mitochondrial DNA of rat ascites hepatoma cells. *Biochem. Biophys. Res. Commun.* **50,** 252-258.

Mizel, S. B. and Wilson, L. (1972). Inhibition of the transport of several hexoses in mammalian cells by cytochalasin B. *J. Biol. Chem.* **247,** 4102-4105.

Moav, B. and Nemer, M. (1971). Histone synthesis: assignment to a special class of polysomes in sea urchin embryos. *Biochemistry* **10**, 881–888.

Mobberg, J. and Rusch, H. P. (1971). Isolation and DNA content of nuclei of P. polycephalum. *Exp. Cell Res.* **66**, 305–316.

Moehring, J. M. and Moehring, T. J. (1979). Characterization of the diphtheria toxin resistance system in CHO cells. *Somatic Cell Genet.* **5**, 453–468.

Moehring, T. J. and Moehring, J. M. (1977). Selection and characterization of cells resistant to diphtheria toxin and Pseudomonas exotoxin A: presumptive translational mutants. *Cell* **11**, 447–454.

Moehring, T. J., Danley, D. E. and Moehring, J. M. (1979). Codominant translational mutants of CHO cells selected with diphtheria toxin. *Somatic Cell Genet.* **5**, 469–480.

Moens, P. B. (1968). The structure and function of the synaptonemal complex in Llilium longifiorum sporocytes. *Chromosoma* **23**, 418–451.

Moens, P. B. (1969a). The fine structure of meiotic chromosome pairing in the triploid, Lilium tigrinum. *J. Cell Biol.* **42**, 272–279.

Moens, P. B. (1969b). Multiple core complex in grasshopper spermatocytes and spermatids. *J. Cell Biol.* **42**, 542–551.

Moens, P. B. (1969c). The fine structure of meiotic chromosome polarization and pairing in Locusta migratoria spermatocytes. *Chromosoma* **28**, 1–25.

Moens, P. B. (1973). Quantitative electron microscopy of chromosome organization at meiotic prophase. *Cold Spring Harbor Symp. Quant. Biol.* **38**, 99–107.

Mohan, J. and Ritossa, F. M. (1970). Regulation of rRNA synthesis and its bearing on the bobbed phenotype in D. melanogaster. *Devel. Biol.* **22**, 495–512.

Molgaard, H. V., Matthews, H. R. and Bradbury, E. M. (1976). Organization of genes for rRNA in Physarum polycephalum. *Eur. J. Biochem.* **68**, 541–550.

Moll, R. and Wintersberger, E. (1976). Synthesis of yeast histones in the cell cycle. *Proc. Nat. Acad. Sci. USA* **73**, 1863–1867.

Molloy, G. R. and Darnell, J. E. (1973). Characterization of poly(A) regions and the adjacent nucleotides in hnRNA and mRNA from HeLa cells. *Biochemistry* **12**, 2324–2330.

Molloy, G. R., Thomas, W. L. and Darnell, J. E. (1972). Occurrence of uridylate rich oligonucleotide regions in hnRNA of HeLa cells. *Proc. Nat. Acad. Sci. USA* **69**, 3684–3688.

Molloy, G. R., Sporn, M. B., Kelley, D. E. and Perry, R. P. (1972). Localization of poly(A) sequences in mRNA of mammalian cells. *Biochemistry* **11**, 3256–3260.

Molloy, P. L., Linnane, A. W. and Lukins, H. B. (1975). Biogenesis of mitochondria: analysis of deletion of mitochondrial antibiotic resistance markers in petite mutants of S. cerevisiae. *J. Bacteriol.* **122**, 7–18.

Monroy, G., Spencer, E. and Hurwitz, J. (1978). Characteristics of reactions catalyzed by purified guanylyltransferase from vaccinia virus. *J. Biol. Chem.* **253**, 4490–4498.

Montgomery, D. L., Hall, B. D., Gillam, S. and Smith, M. (1978). Identification and isolation of the yeast cytochrome c gene. *Cell* **14**, 673–680.

Moore, P. D. and Holliday, R. (1976). Evidence for the formation of hybrid DNA during mitotic recombination in Chinese hamster cells. *Cell* **8**, 573–579.

Mooseker, M. S. and Tilney, L. G. (1975). Organization of an actin filament-membrane complex. Filament polarity and membrane attachment in the microvilli of intestinal epithelial cells. *J. Cell Biol.* **67**, 725–743.

Mooseker, M. S., Pollard, T. D. and Fujiwara, K. (1978). Characterization and localization of myosin in the brush border of intestinal epithelial cells. *J. Cell Biol.* **79**, 444–453.

Moran, L., Mirault, M. E., Tissieres, A., Lis, J., Schedl, P., Artavanis-Tsakonas, S. and Gehring,

W. J. (1979). Physical map of two D. melanogaster DNA segments containing sequences coding for the 70,000 dalton heat shock protein. *Cell* **17**, 1-8.

Morel, C., Gander, E. S., Herzberg, J. and Scherrer, K. (1973). The duck mRNP complex. Resistance to high ionic strength, particle gel electrophoresis, composition and visualization by dark field electron microscopy. *Eur. J. Biochem.* **36**, 455-464.

Morgan, E. A., Ikemura, T. and Nomura, M. (1977). Identification of spacer tRNA genes in individual rRNA transcription units of E. coli. *Proc. Nat. Acad. Sci. USA* **74**, 2710-2714.

Morgan, E. A., Ikemura, T., Lindahl, L., Fallon, A. M. and Nomura, M. (1978). Some rRNA operons in E. coli have tRNA genes at their distant ends. *Cell* **13**, 335-344.

Morgan, J. L., Holladay, C. R. and Spooner, B. S. (1978). Species dependent immunological differences between vertebrate brain tubulins. *Proc. Nat. Acad. Sci. USA* **75**, 1414-1417.

Morgan, T. H. (1911a). The application of the conception of pure lines to sex limited inheritance and to sexual dimorphism. *Am. Nat.* **45**, 65-78.

Morgan, T. H. (1911b). An attempt to analyze the constitution of the chromosomes on the basis of sex limited inheritance in Drosophila. *J. Exp. Zool.* **11**, 365-414.

Morgan, T. H. and Cattell, E. (1912). Data for the study of sex linked inheritance in Drosophila. *J. Exp. Zoo.* **13**, 79-101.

Morimoto, R., Lewin, A. and Rabinowitz, M. (1977). Restriction cleavage map of mitochondrial DNA from the yeast S. cerevisiae. *Nucl. Acids Res.* **4**, 2331-2351.

Morimoto, R., Lewin, A., Hsu, H. J., Rabinowitz, M. and Fukuhara, H. (1975). Restriction endonuclease analysis of mitochondrial DNA from grande and genetically characterized cytoplasmic petite clones of S. cerevisiae. *Proc. Nat. Acad. Sci. USA* **72**, 3868-3872.

Morimoto, R., Merten, S., Lewin, A., Martin, N. C. and Rabinowitz, M. (1978). Physical mapping of genes on yeast mitochondrial DNA: localization of antibiotic resistance loci and rRNA and tRNA genes. *Mol. Gen. Genet.* **163**, 241-255.

Morishima, A., Grumbach, M. M. and Taylor, J. H. (1962). Asynchronous duplication of human chromosomes and the origin of sex chromatin. *Proc. Nat. Acad. Sci. USA* **48**, 756-763.

Morris, N. R. (1976a). Nucleosome structure in Aspergillus nidulans. *Cell* **8**, 357-364.

Morris, N. R. (1976b). A comparison of the structure of chicken erythrocyte and chicken liver chromatin. *Cell* **9**, 627-632.

Morrow, J. (1970). Genetic analysis of azaguanine resistance in an established mouse cell line. *Genetics* **65**, 279-287.

Morrow, J. F. and Berg, P. (1972). Cleavage of SV40 DNA at a unique site by a bacterial restriction enzyme. *Proc. Nat. Acad. Sci. USA* **69**, 3365-3369.

Morrow, J. F., Cohen, S. N., Chang, A. C. Y., Boyer, H. W., Goodman, H. M. and Helling, R. B. (1974). Replication and transcription of eucaryotic DNA in E. coli. *Proc. Nat. Acad. Sci. USA* **71**, 1743-1747.

Mortimer, R. K. and Hawthorne, D. C. (1973). Genetic mapping in Saccharomyces. IV. Mapping of temperature sensitive genes and use of disomic strains in localizing genes. *Genetics* **74**, 33-54.

Mortimer, R. H. and Hawthorne, D. C. (1975). Genetic mapping in yeast. *Methods Cell Biol.* **11**, 221-233.

Moses, M. J. (1956). Studies on nuclei using correlated cytochemical light and electron microscope techniques. *J. Biochem. Biophys. Cytol.* suppl. **2**, 397-406.

Moses, M. J. (1958). The relation between the axial complex of meiotic prophase chromosomes and chromosome pairing in a Salamander (Plethodon cinercus). *J. Biochem. Biophys. Cytol.* **4**, 633-638.

Moses, M. J. (1968). Synaptonemal complex. *Ann. Rev. Genet.* **2**, 363-412.

Moses, M. J. (1977a). Synaptonemal complex karyotyping in spermatocytes of the Chinese hamster (Cricetulus griseus). I. Morphology of the autosomal complement in spread preparations. *Chromosoma* **60**, 99-126.

Moses, M. J. (1977b). Synaptonemal complex karyotyping in spermatocytes of the Chinese hamster (Crecetulus griseus). II. Morphology of the XY pair in spread preparations. *Chromosoma* **60**, 127-138.

Moses, M. J. and Coleman, J. R. (1964). Structural patterns and the functional organization of chromosomes. In M. Locke (Ed.), *The Role of Chromosomes in Development*, Academic Press, New York, pp 11-49.

Moses, M. J., Slatton, G. H., Gambling, T. M. and Starner, C. F. (1977). Synaptonemal complex karyotyping in spermatocytes of the Chinese hamster. III. Quantitative evaluation. *Chromosoma* **60**, 345-375.

Moss, B. and Koczot, F. (1976). Sequences of methylated nucleotides at the 5' terminus of adenovirus specific RNA. *J. Virol.* **17**, 385-392.

Moss, B., Rosenblum, N. and Paoletti, E. (1973). Polyadenylate polymerase from vaccinia virions. *Nature New Biol.* **245**, 59-63.

Moss, B., Gershowitz, A., Wei, C. M. and Boone, R. (1976). Formation of the guanylylated and methylated 5' terminus of vaccinia virus mRNA. *Virology* **72**, 341-351.

Moss, B., Gershowitz, A., Weber, L. A. and Baglioni, C. (1977). Histone mRNAs contain blocked and methylated 5' terminal sequences but lack methylated nucleosides at internal positions. *Cell* **10**, 113-120.

Moss, T., Cary, P. D., Crane-Robinson, C. and Bradbury, E. M. (1976). Physical studies on the H3/H4 histone tetramer. *Biochemistry* **15**, 2261-2267.

Mott, M. R. and Callan, H. G. (1975). An electron microscope study of the lampbrush chromosomes of the newt Triturus cristatus. *J. Cell Sci.* **17**, 241-262.

Mounolou, J. C., Jakob, H. and Slonimski, P. P. (1966). Mitochondrial DNA from yeast "petite" mutants: specific changes of buoyant density corresponding to different cytoplasmic mutations. *Biochem. Biophys. Res. Commun.* **24**, 218-224.

Moustacchi, E. and Williamson, D. H. (1966). Physiological variations in satellite components of yeast DNA detected by density gradient centrifugation. *Biochem. Biophys. Res. Commun.* **23**, 56-61.

Moyer, S. A., Abraham, G., Adler, R. and Bannerjee, A. K. (1975). Methylated and blocked 5' termini in VSV virus in vivo mRNAs. *Cell* **5**, 59-68.

Mueller, G. C. and Kajiwara, K. (1966a). Early and late replicating DNA complexes in HeLa nuclei. *Biochim. Biophys. Acta* **114**, 108-115.

Mueller, G. C. and Kajiwara, K. (1966b). Actinomycin and fluorophenylalanine, inhibitors of replication in HeLa cells. *Biochim. Biophys. Acta* **119**, 557-565.

Mueller, R. U., Chow, V. and Gander, E. S. (1977). Characterization of the protein moiety of messenger ribonucleoprotein complexes from duck reticulocytes by two dimensional gel electrophoresis. *Eur. J. Biochem.* **77**, 287-297.

Mukherjee, A. B. and Nitowsky, H. M. (1972). Fluorescence of constitutive heterochromatin of Microtus agrestis. *Exp. Cell Res.* **73**, 248-251.

Mukherjee, A. B., Orloff, S., Butler, J. D., Triche, T., Lalley, P. and Schulman, J. D. (1978). Entrapment of metaphase chromosomes into phospho lipid vesicles (lipochromosomes): carrier potential in gene transfer. *Proc. Nat. Acad. Sci. USA* **75**, 1361-1365.

Mukherjee, B. B. and Sinha, A. K. (1964). Single X hypothesis: cytological evidence for random inactivation of X chromosomes in a female mule complement. *Proc. Nat. Acad. Sci. USA* **51**, 252-259.

Mulder, C. and Delius, H. (1972). Specificity of the break produced by restricting endonuclease RI in SV40 DNA, as revealed by partial denaturation mapping. *Proc. Nat. Acad. Sci. USA* **69**, 3215-3219.

Muller, H. J. (1916). The mechanism of crossing over. *Am. Nat.* **50**, 193-221.

Muller, H. J. (1927). The measurement of gene mutation rate in Drosophila, its high variability, and its dependence on temperature. *Genetics* **13**, 279-357.

Muller, H. J. (1950). Our load of mutations. *Am. J. Hum. Gen.* **2**, 111-176.

Muller, H. J. (1956). Further studies bearing on the load of mutations in man. *Acta Gen. Stat. Med.* **6**, 157-168.

Muller, H. J. (1967). The genetic material as the initiator and the organizing basis of life. In R. A. Brink (Ed.), *Heritage from Mendel*, University of Wisconsin Press, Madison, pp 419-447.

Muller, H. J. and Herskowitz, I. H. (1954). Concerning the healing of chromosome ends produced by breakage in D. melanogaster. *Am. Nat.* **88**, 177-208.

Muller, H. J. and Prokofyeva, A. A. (1935). The individual gene in relation to the chromomere and the chromosome. *Proc. Nat. Acad. Sci. USA* **21**, 16-26.

Muller, H. J., Valencia, J. I. and Valencia, R. M. (1949). The frequencies of spontaneous mutation at independent loci in Drosophila. *Records Gen. Soc. Amer.* **18**, 105-106.

Muller, U., Zentgraf, H., Eicken, I. and Keller, W. (1978). Higher order structure of SV40 chromatin. *Science* **201**, 406-415.

Muller, W. (1972). Elektronmikroskopische untersuchungen zum formwechsel der kinetochoren wahrend der spermatocytenteilungen von Pales feruginea. *Chromosoma* **38**, 139-172.

Mulligan, R. C., Howard, B. H. and Berg, P. (1979). Synthesis of rabbit β globin in cultured monkey kidney cells following infection with a SV40 globin-recombinant genome. *Nature* **277**, 108-113.

Mullins, J. I. and Blumenfeld, M. (1979). Satellite Ic: a possible link between the satellite DNAs of D. virilis and D. melanogaster. *Cell* **17**, 615-621.

Munoz, R. F. and Darnell, J. E. (1974). Poly(A) in mRNA does not contribute to secondary structure necessary for protein synthesis. *Cell* **2**, 247-252.

Murphree, S., Stubblefield, E. and Moore, E. C. (1969). Synchronized mammalian cell cultures. III. Variation of ribonucleotide reductase activity during the replication cycle of Chinese hamster fibroblasts. *Exp. Cell Res.* **58**, 118-124.

Murphy, D. B. and Borisy, G. G. (1975). Association of high molecular weight proteins with microtubules and their role in microtubule assembly in vitro. *Proc. Nat. Acad. Sci. USA* **72**, 2696-2700.

Murphy, D. B., Johnson, K. A. and Borisy, G. G. (1977). Role of tubulin associated proteins in microtubule nucleation and elongation. *J. Mol. Biol.* **117**, 33-52.

Murphy, E. C., Jr., Hall, S. H., Shepherd, J. H. and Weiser, R. S. (1973). Fractionation of mouse myeloma chromatin. *Biochemistry* **12**, 3843-3852.

Murphy, W. I., Attardi, B., Tu, C. and Attardi, G. (1975). Evidence for complete symmetrical transcription in vivo of mitochondrial DNA in HeLa cells. *J. Mol. Biol.* **99**, 809-814.

Murray, K. (1965). The basic proteins of cell nuclei. *Ann. Rev. Biochem.* **34**, 209-246.

Murray,, K. (1969). Stepwise removal of histone from native DNP by titration with acid at low temperature and some properties of the resulting partial nucleoproteins. *J. Mol. Biol.* **39**, 125-144.

Murray, N. E. and Murray, K. (1974). Manipulation of restriction targets in phage lambda to form receptor chromosomes for DNA fragments. *Nature* **251**, 476-481.

Musich, P. R., Brown, F. L. and Maio, J. J. (1977). Subunit structure of chromatin and the organization of eucaryotic highly repetitive DNA nucleosomal proteins associated with a highly repetitive mammalian DNA. *Proc. Nat. Acad. Sci. USA* **74**, 3297-3301.

Musich, P. R., Maio, J. J. and Brown, F. L. (1977). Subunit structure of chromatin and the organization of eucaryotic highly repetitive DNA: indications of a phase relation between restriction sites and chromatin subunits in African green monkey and calf nuclei. *J. Mol. Biol.* **117**, 657–678.

Muthukrishnan, S., Both, G. W., Furuichi, Y. and Shatkin, A. J. (1975a). 5′ terminal 7 methylguanosine in eucaryotic mRNA is required for translation. *Nature* **255**, 33–37.

Muthukrishnan, S., Filipowicz, W., Sierra, J. M., Both, G. W., Shatkin, A. J. and Ochoa, S. (1975b). mRNA methylation and protein synthesis in extracts from embryos of brine shrimp, Artemia salina. *J. Biol. Chem.* **250**, 9336–9341.

Muthukrishnan, S., Morgan, M., Bannerjee, A. K. and Shatkin, A. J. (1976). Influence of 5′ terminal m⁷G and 2′-O-methylated residues on mRNA binding to ribosomes. *Biochemistry* **15**, 5761–5768.

Muthukrishnan, S., Moss, B., Cooper, J. A. and Maxwell, E. S. (1978). Influence of 5′ terminal cap structure on the initiation of translation of vaccinia virus mRNA. *J. Biol. Chem.* **253**, 1710–1715.

Nabholz, M., Miggiano, V. and Bodmer, W. (1969). Genetic analysis with human mouse somatic cell hybrids. *Nature* **223**, 358–363.

Nadeau, P., Oliver, D. R. and Chalkley, R. (1978). Effect of inhibition of DNA synthesis on histone synthesis and deposition. *Biochemistry* **17**, 4885–4893.

Nagano, T. and Suzuki, F. (1975). Microtubules with 15 subunits in cockroach epidermal cells. *J. Cell Biol.* **64**, 242–245.

Nagle, B. W., Doenges, K. H. and Bryan, J. (1977). Assembly of tubulin from cultured cells and comparison with the neurotubulin model. *Cell* **12**, 573–586.

Nagley, P. and Linnane, A. W. (1972). Biogenesis of mitochondria. XXI. Studies on the nature of the mitochondrial genome in yeast: the degenerative effect of ethidium bromide on mitochondrial genetic information in a respiratory competent strain. *J. Mol. Biol.* **66**, 181–193.

Nagley, P., Molloy, P. L., Lukins, H. B. and Linnane, A. B. (1974). Studies on mitochondrial gene purification using petite mutants of yeast: characterization of mutants enriched in rRNA cistrons. *Biochem. Biophys. Res. Commun.* **57**, 232–239.

Nakamura, H. and Littlefield, J. W. (1972). Purification, properties and synthesis of dihydrofolate reductase from wild type and methotrexate-resistant hamster cells. *J. Biol. Chem.* **247**, 179–187.

Nakazoto, H., Edmonds, M. and Kopp, D. W. (1974). Differential metabolism of large and small poly-A sequences in the heterogeneous nuclear RNA of HeLa cells. *Proc. Nat. Acad. Sci. USA* **71**, 200–204.

Nass, M. M. K. (1969a). Mitochondrial DNA: advances, problems and goals. *Science* **165**, 25–35.

Nass, M. M. K. (1969b). Mitochondrial DNA. I. Intramitochondrial distribution and structural relations of single and double strand circular DNA. *J. Mol. Biol.* **42**, 521–528.

Nass, M. M. K. (1969c). Mitochondrial DNA. II. Structural and physicochemical properties of isolated DNA. *J. Mol. Biol.* **42**, 529–545.

Nass, M. M. K. and Nass, S. (1963a). Intramitochondrial fibers with DNA characteristics. I. Fixation and staining reactions. *J. Cell Biol.* **19**, 593–611.

Nass, S. and Nass, M. M. K. (1963b). Intramitochondrial fibers with DNA characteristics. II. Enzymatic and other hydrolytic treatments. *J. Cell Biol.* **19**, 613–629.

Nath, K. and Bollon, A. P. (1977). Organization of the yeast rRNA gene cluster via cloning and restriction analysis. *J. Biol. Chem.* **252**, 6562–6570.

Nazar, R. N. and Roy, K. L. (1976). The nucleotide sequence of turtle 5.8S rRNA. *Febs Lett.* **72**, 111–116.

Nazar, R. N. and Roy, K. L. (1978). Nucleotide sequence of rainbow trout ribosomal 5.8S RNA. *J. Biol. Chem.* **253**, 395–399.

Nazar, R. N., Sitz, T. O. and Busch, H. (1976). Sequence homologies in mammalian 5.8S RNA. *Biochemistry* **15**, 505–508.

Nazar, R. N., Owens, T. W., Sitz, T. O. and Busch, H. (1975). Maturation pathway for Novikoff ascites hepatoma 5.8S rRNA. Evidence for its presence in 32S nuclear RNA. *J. Biol. Chem.* **250**, 2475-2481.

Nebel, B. and Ruttle, M. (1938). The cytological and genetical significance of colchicine. *J. Hered.* **29**, 3-9.

Neimark, H. C. and Pene, J. J. (1965). Characterization of pleuropneumonia-like organisms by DNA composition. *Proc. Soc. Exp. Biol. Med.* **118**, 517-519.

Nelson, D., Perry, M. E. and Chalkley, R. (1979). A correlation between nucleosome spacer susceptibility to DNAase I and histone acetylation. *Nucl. Acids Res.* **6**, 561-574.

Nemer, M., Dubroff, L. M. and Graham, M. (1975). Properties of sea urchin embryo mRNA containing and lacking poly(A). *Cell* **6**, 171-178.

Nemer, M., Graham, M. and Dubroff, L. M. (1974). Coexistence of nonhistone mRNA species lacking and containing poly(A) in sea urchin embryos. *J. Mol. Biol.* **89**, 435-454.

Nesbitt, M. N. and Francke, U. (1973). A system of nomenclature for band patterns of mouse chromosomes. *Chromosoma* **41**, 145-158.

Nette, E. G., Sit, H. L., Clavey, W. and King, D. W. (1979). Isolation of viable reconstituted cells from human karyoplasts fused to mouse cytoplasts. *Exp. Cell Res.* **121**, 143-152.

Neuffer, M. G. and Coe, E. H. (1974). Corn (maize). In R. C. King (Ed.), *Handbook of Genetics*, Vol. 2, Plenum, New York, pp 3-30.

Nevins, J. R. and Darnell, J. E. (1978a). Groups of adenovirus 2 mRNAs derived from a large primary transcript: probable nuclear origin and possible common 3′ ends. *J. Virol.* **25**, 811-823.

Nevins, J. R. and Darnell, J. E. (1978b). Steps in the processing of Ad2 mRNA: poly(A)⁺ nuclear sequences are conserved and poly(A)⁺ addition precedes splicing. *Cell* **15**, 1477-1493.

Nevins, J. R. and Joklik, W. K. (1975). Poly(A) sequences of vaccinia virus mRNA: nature, mode of addition and function during translation in vitro and in vivo. *Virology* **63**, 1-14.

Newcombe, H. B. (1949). Origin of bacterial variants. *Nature* **164**, 160.

Newlon, C. S. and Fangman, W. L. (1975). Mitochondrial DNA synthesis in cell cycle mutants of S. cerevisiae. *Cell* **5**, 423-428.

Newlon, C. S., Sonenshein, G. E. and Holt, C. E. (1973). Time of synthesis of genes for rRNA in Physarum. *Biochemistry* **12**, 2338-2345.

Newrock, K. M., Cohen, L. H., Hendricks, M. B., Donnelly, R. J. and Weinberg, E. S. (1978). Stage specific mRNAs coding for subtypes of H2A and h2B histones in the sea urchin embryo. *Cell* **14**, 327-336.

Ng, S. Y., Parker, C. S. and Roeder, R. G. (1979). Transcription of cloned Xenopus 5S RNA genes by X. laevis RNA polymerase III in reconstituted systems. *Proc. Nat. Acad. Sci. USA* **76**, 136-140.

Nguyen-Huu, M. C., Sippel, A. A., Hynes, N. E., Groner, B. and Schutz, G. (1978). Preferential transcription of the ovalbumin gene in isolated hen oviduct nuclei by RNA polymerase B. *Proc. Nat. Acad. Sci. USA* **75**, 686-690.

Nguyen-Huu, M. C., Stratmann, M., Groner, B., Wurtz, T., Land, H., Giesecke, K., Sippel, A. E. and Schutz, G. (1979). Chicken lysozyme gene contains several intervening sequences. *Proc. Nat. Acad. Sci. USA* **76**, 76-80.

Nicklas, R. B. (1967). Chromosome micromanipulation. II. Induced reorientation and the experimental control of segregation in mitosis. *Chromosoma* **21**, 17-50.

Nicklas, R. B. (1972). Mitosis. *Adv. Cell Biol.* **2**, 225-298.

Nicklas, R. B. and Koch, C. A. (1972). Chromosome micromanipulation. IV. Polarized motions within the spindle and models for mitosis. *Chromosoma* **39**, 1-26.

Nicklas, R. B. and Staehly, C. A. (1967). Chromosome micromanipulation. I. The mechanics of chromosome attachment to the spindle. *Chromosoma* **21**, 1-16.

Nicklas, R. B., Brinkley, B. R., Pepper, D. A., Kubai, D. F. and Rickards, G. K. (1979). Electron microscopy of spermatocytes previously studied in life: methods and some observations on micromanipulated chromosomes. *J. Cell Sci.* **35**, 87-104.

Nickol, J. M., Lee, K. L. and Kenney, F. T. (1978). Changes in hepatic levels of tyrosine aminotransferase mRNA during induction by hydrocortisone. *J. Biol. Chem.* **253**, 4009-4015.

Nicolson, G. L. (1973). Temperature dependent mobility of concanavalin A sites on tumor cell surfaces. *Nature New Biol.* **243**, 218-220.

Nicolson, G. L. (1974). The interactions of lectins with animal cell surfaces. *Int. Rev. Cytol.* **39**, 90-190.

Niessing, J. (1975). Three distinct forms of nuclear poly(A) polymerase. *Eur. J. Biochem.* **59**, 127-136.

Niessing, J. and Sekeris, C. E. (1971). Further studies on nuclear RNP particles containing DNA-like RNA from rat liver. *Biochim. Biophys. Acta* **247**, 391-403.

Niles, E. (1978). Isolation of a high specific activity 35S rRNA precursor from T. pyriformis and identification of its 5′ terminus, pppAp. *Biochemistry* **17**, 4839-4844.

Nokin, P., Burny, A., Huez, G. and Marbaix, G. (1976). Globin mRNA from anaemic rabbit spleen. Size of its polyadenylate segment. *Eur. J. Biochem.* **68**, 431-436.

Noll, M. (1974a). Subunit structure of chromatin. *Nature* **251**, 249-251.

Noll, M. (1974b). Internal structure of the chromatin subunit. *Nucl. Acids Res.* **1**, 1573-1578.

Noll, M. (1976). Differences and similarities in chromatin structure of N. crassa and higher eucaryotes. *Cell* **8**, 349-356.

Noll, M. (1977). DNA folding in the nucleosome. *J. Mol. Biol.* **116**, 49-72.

Noll, M. and Kornberg, R. D. (1977). Action of micrococcal nuclease on chromatin and the location of H1. *J. Mol. Biol.* **109**, 393-404.

Noll, M., Thomas, J. O. and Kornberg, R. D. (1975). Preparation of native chromatin and damage caused by shearing. *Science* **187**, 1203-1206.

Nomoto, A., Lee, Y. F. and Wimmer, E. (1976). The 5′ end of poliovirus mRNA is not capped with m⁷G(5′)ppp(5′)Np. *Proc. Nat. Acad. Sci. USA* **73**, 375-380.

Nomura, M., Morgan, E. A. and Jaskunas, S. R. (1977). Genetics of bacterial ribosomes. *Ann. Rev. Genet.* **11**, 297-347.

Norberg, R., Lidman, K. and Fagraeus, A. (1975). Effects of cytochalasin B on fibroblasts, lymphoid cells, and platelets revealed by human antiactin antibodies. *Cell* **6**, 507-512.

Nordstrom, J. L., Roop, D. R., Tsai, M. J. and O'Malley, B. W. (1979). Identification of potential ovomucoid mRNA precursors in chick oviduct nuclei. *Nature* **278**, 328-330.

Novitski, E. (1964). An alternative to the distributive pairing hypothesis in Drosophila. *Genetics* **50**, 1449-1451.

Novitski, E. (1975). Evidence for the single phase pairing theory of meiosis. *Genetics* **79**, 63-71.

Nunberg, J. H., Kaufman, R. J., Schimke, R. T., Urlaub, G. and Chasin, L. A. (1978). Amplified DHFR genes are localized to a homogeneously staining region of a single chromosome in a methotrexate resistant CHO cell line. *Proc. Nat. Acad. Sci. USA* **75**, 5553-5556.

Nur, U. (1967). Reversal of heterochromatization and the activity of the paternal chromosome set in the male mealy bug. *Genetics* **56**, 375-389.

Nuss, D. L. and Fururichi, Y. (1977). Characterization of the m⁷G(5′)pppN pyrophosphatase activity from HeLa cells. *J. Biol. Chem.* **252**, 2815-2821.

Nygaard, A. P. and Hall, B. D. (1963). A method for the detection of RNA-DNA complexes. *Biochem. Biophys. Res. Commun.* **12**, 98-104.

Nygaard, A. P. and Hall, B. D. (1964). Formation and properties of RNA-DNA complexes. *J. Mol. Biol.* **9**, 125-142.

Oates, D. C. and Patterson, D. (1977). Biochemical genetics of Chinese hamster cell mutants with deviant purine metabolism. *Somatic Cell Genet.* **3**, 561-578.

O'Brien, S. J. (1973). On estimating functional gene number in eucaryotes. *Nature New Biol.* **242,** 52-54.

Ockey, C. H. (1972). Distribution of DNA replicator sites in mammalian nuclei. II. Effects of prolonged inhibition of DNA synthesis. *Exp. Cell Res.* **70,** 203-213.

Ockey, C. H. and Allen, T. D. (1975). Distribution of DNA and DNA synthesis in mammalian cells following inhibition with hydroxyurea and 5FUdR. *Exp. Cell Res.* **93,** 275-282.

Ockey, C. H. and Saffhill, R. (1976). The comparative effects of short term DNA inhibition on replicon synthesis in mammalian cells. *Exp. Cell Res.* **103,** 361-374.

O'Farrell, P. Z., Cordell, B., Valanzuela, P., Rutter, W. J. and Goodman, H. M. (1978). Structure and processing of yeast precursor tRNAs containing intervening sequences. *Nature* **274,** 438-444.

Ogawa, Y., Quagliarotti, G., Jordan, J., Taylor, C. W., Starbuck, W. C. and Busch, H. (1969). Structural analysis of the glycin rich arginine rich histone. III. Sequence of the amino termina half of the molecule containing the modified lysine residues and the total sequence. *J. Biol. Chem.* **244,** 4387-4392.

Ogden, R. C., Beckmann, J. S., Abelson, J., Kang, H. S., Soll, D. and Schmidt, O. (1979). In vitro transcription and processing of a yeast tRNA gene containing an intervening sequence. *Cell* **17,** 399-406.

Ogur, M., Minckler, S., Lindegren, G. and Lindegren, C. C. (1952). The nucleic acids in a polyploid series of Saccharomyces. *Arch. Biochem. Biophys.* **40,** 175-184.

Ohi, S., Ramirez, J. L., Upholt, W. B. and Dawid, I. (1978). Mapping of mitochondrial 4S RNA genes in X. laevis by electron microscopy. *J. Mol. Biol.* **121,** 299-310.

Ohlenbusch, H. H., Olivera, B. M., Tuan, D. and Davidson, N. (1967). Selective dissociation of histones from calf thymus nucleoprotein. *J. Mol. Biol.* **25,** 299-315.

Ohno, S. (1971). Simplicity of mammalian regulatory systems inferred by single gene determination of sex phenotypes. *Nature* **234,** 134-137.

Ohno, S. and Cattanach, B. M. (1962). Cytological study of an X chromosome translocation in Mus musculus. *Cytogenetics* **1,** 129-140.

Ohno, S., Geller, L. and Kan, J. (1974). The analysis of Lyon's hypothesis through preferential X activation. *Cell* **1,** 173-184.

Ohno, S., Kaplan, W. D. and Kinosita, R. (1959). Formation of the sex chromatin by a single X chromosome in liver cells of Rattus norvegicus. *Exp. Cell Res.* **18,** 415-418.

Ohno, S., Kaplan, W. D. and Kinosita, R. (1961). X chromosome behavior in germ and somatic cells of Rattus norvegicus. *Exp. Cell Res.* **22,** 535-544.

Ohno, S., Klinger, H. P. and Atkin, N. B. (1962). Human oogenesis. *Cytogenetics* **1,** 42-51.

Ohno, S., Christian, L., Attardi, B. J. and Kan, J. (1973). Modification of expression of the *Tfm* gene of the mouse by a controlling element gene. *Nature New Biol.* **245,** 92-93.

Ohta, N., Sager, R. and Inouye, M. (1975). Identification of a chloroplast ribosomal protein altered by a chloroplast mutation in C. reinhardii. *J. Biol. Chem.* **250,** 3655-3659.

Ohta, T. and Kimura, M. (1971). Functional organization of genetic material as a product of molecular evolution. *Nature* **233,** 118-119.

Ojala, D. and Attardi, G. (1974). Identification and partial characterization of multiple discrete poly(A)-containing RNA components coded for by HeLa mt DNA. *J. Mol. Biol.* **88,** 205-219.

Okada, T. A. and Comings, D. E. (1974). Mechanisms of chromosome banding. III. Similarity between G-bands of mitotic chromosomes and chromomeres of meiotic chromosomes. *Chromosoma* **48,** 65-72.

Okada, Y. (1962a). Analysis of giant polynuclear cell formation caused by HVJ virus from Ehrlich ascites tumor cells. I. Microscopic observation of giant polynuclear cell formation. *Exp. Cell Res.* **26,** 98-107.

Okada, Y. (1962b). Analysis of giant polynuclear cell formation caused by HVJ virus from

Ehrlich ascites tumor cells. III. Relationship between cell condition and fusion reaction or cell degeneration reaction. *Exp. Cell Res.* **26**, 119-128.

Okada, Y. and Murayama, F. (1965). Multinucleated giant cell formation by fusion between cells of two different strains. *Exp. Cell Res.* **40**, 154-158.

Old, J., Clegg, J. B., Weatherall, D. J., Ottolenghi, S., Comi, P., Giglioni, B., Mitcell, J., Tolstoshev, P. and Williamson, R. (1976). A direct estimate of the number of human γ-globin genes. *Cell* **8**, 13-18.

Old, J. M., Proudfoot, N. J., Wood, W. G., Longley, J. I., Clegg, J. B. and Weatherall, D. J. (1978). Characterization of β globin mRNA in β^0 thalassemia. *Cell* **14**, 289-298.

Old, R. W., Callan, H. G. and Gross, K. W. (1977). Localization of histone gene transcripts in newt lampbrush chromosomes by in situ hybridization. *J. Cell Sci.* **27**, 57-80.

Olden, K., Willingham, M. and Pastan, I. (1976). Cell surface myosin in cultured fibroblasts. *Cell* **8**, 383-390.

Olins, A. L. and Olins, D. E. (1974). Spheroid chromatin units (V bodies). *Science* **183**, 330-332.

Olins, A. L., Carlson, R. D. and Olins, D. E. (1975). Visualization of chromatin substructure: nu bodies. *J. Cell Biol.* **64**, 528-537.

Olins, A. L., Carlson, R. D., Wright, E. B. and Olins, D. E. (1976). Chromatin nu bodies: isolation, subfractionation and physical characterization. *Nucl. Acids Res.* **3**, 3271-3291.

Olins, D. E. and Wright, E. B. (1973). Glutaraldehyde fixation of isolated eucaryotic nuclei. Evidence for histone-histone proximity. *J. Cell Biol.* **59**, 304-317.

Oliver, D., Balhorn, R., Granner, D. and Chalkley, R. (1972). Molecular nature of F1 histone phosphorylation in cultured hepatoma cells. *Biochemistry* **11**, 3921-3925.

Oliver, J. M., Ukena, T. E. and Berlin, R. D. (1974). Effects of phagocytosis and colchicine on the distribution of lectin-binding sites on cell surfaces. *Proc. Nat. Acad. Sci. USA* **71**, 394-398.

Olmsted, J. B. and Borisy, G. G. (1973a). Characterization of microtubule assembly in porcine brain extracts by viscometry. *Biochemistry* **12**, 4282-4289.

Olmsted, J. B. and Borisy, G. G. (1973b). Microtubules. *Ann. Rev. Biochem.* **42**, 507-540.

Olmsted, J. B. and Borisy, G. G. (1975). Ionic and nucleotide requirements for microtubule polymerization in vitro. *Biochemistry* **14**, 2996-3005.

Olmsted, J. B., Carlson, K., Klebe, R., Ruddle, F. and Rosenbaum, J. L. (1970). Isolation of microtubule protein from cultured mouse neuroblastoma cells. *Proc. Nat. Acad. Sci. USA* **65**, 129-136.

Olmsted, J. B., Witman, G. B., Carlson, K. and Rosenbaum, J. L. (1971). Comparison of the microtubule proteins of neuroblastoma cells, brain, and Chlamydomonas flagella. *Proc. Nat. Acad. Sci. USA* **68**, 2273-2277.

Olsen, A. S. and Milman, G. (1974). Chinese hamster HGPRT. *J. Biol. Chem.* **249**, 4030-4037.

Olsnes, S. (1971). Characterization of the complex containing rapidly labeled RNA in EDTA treated polysomes from rat liver. *Eur. J. Biochem.* **18**, 242-251.

Olson, M. O. J., Goldknopf, I. L., Guetzow, K. A., James, T. T., Hawkins, T. C., Mays-Rothberg, C. J. and Musch, H. (1977). The NH2 and COOH terminal amino acid sequence of the nuclear protein A24. *J. Biol. Chem.* **251**, 5901-5903.

Orkin, S. H. (1978a). The duplicated human α globin genes lie close together in cellular DNA. *Proc. Nat. Acad. Sci. USA* **75**, 5950-5954.

Orkin, S. H. (1978b). Fidelity of globin RNA synthesis in vitro by isolated nuclei: assymetric gene expression. *Biochemistry* **17**, 487-492.

Orkin, S. H., Old, J., Lazarus, H., Altay, C., Gurgey, A., Weatherall, D. J. and Nathan, D. G. (1979a). The molecular basis of α thalassemias: frequent occurrence of dysfunctional α loci among non Asians with HbH disease. *Cell* **17**, 33-43.

Orkin, S. H., Old, J. M., Weatherall, D. J. and Nathan, D. G. (1979b). Partial deletion of β globin gene DNA in certain patients with β^0 thalassemia. *Proc. Nat. Acad. Sci. USA* **76**, 2400-2404.

Orkwiszeski, K. G., Tedesco, T. A., Mellman, W. J. and Croce, C. M. (1976). Linkage relationship between the genes for thymidine kinase and galactokinase in different primates. *Somatic Cell Genet.* **2**, 21-26.

Osborn, M. and Weber, K. (1976a). Cytoplasmic microtubules in tissue culture cells appear to grow from an organizing center towards the plasma membrane. *Proc. Nat. Acad. Sci. USA* **73**, 867-871.

Osborn, M. and Weber, K. (1976b). Tubulin specific antibody and the expression of microtubules in 3T3 cells after attachment to a substratum: further evidence for the polar growth of cytoplasmic microtubules in vivo. *Exp. Cell Res.* **103**, 331-340.

Osborn, M. and Weber, K. (1977a). The detergent resistant cytoskeleton of tissue culture cells includes the nucleus and the microfilament bundles. *Exp. Cell Res.* **106**, 339-350.

Osborn, M. and Weber, K. (1977b). The display of microtubules in transformed cells. *Cell* **12**, 561-571.

Osborn, M., Francke, W. W. and Weber, K. (1977). Visualization of a system of filaments 7-10 nm thick in cultured cells of an epithelioid line (Pt K2) by immunofluorescence microscopy. *Proc. Nat. Acad. Sci. USA* **74**, 2490-2494.

Ostergren, G. (1950). Considerations on some elementary features of mitosis. *Hereditas* **36**, 1-18.

Ostlund, R. E., Pastan, I. and Adelstein, R. S. (1974). Myosin in cultured fibroblasts. *J. Biol. Chem.* **249**, 3903-3907.

Ottolenghi, S., Lanyon, W. G., Paul, J., Williamson, R., Weatherall, D. J., Clegg, J. B., Pritchard, J., Pootrakul, S. and Boon, W. H. (1974). The severe form of α thalassemia is caused by a haemoglobin gene deletion. *Nature* **251**, 389-391.

Ottolenghi, S., Lanyon, W. G., Williamson, R., Weatherall, D. J., Clegg, J. B. and Pitcher, C. S. (1975). Human globin gene analysis for a patient with $\beta°/\delta\beta°$ thalassemia. *Proc. Nat. Acad. Sci. USA* **72**, 2294-2299.

Ottolenghi, S., Comi, P., Giglioni, B., Tolstoshev, P., Lanyon, W. G., Mitchell, G. J., Williamson, R., Russo, G., Musumeci, S., Schliro, G., Tsistrakis, G. A., Charache, S., Wood, W. G., Clegg, J. B. and Weatherall, D. J. (1976). $\delta\beta$-thalassemia is due to a gene deletion. *Cell* **9**, 71-80.

Oudet, P., Gross-Bellard, M. and Chambon, P. (1975). Electron microscopic and biochemical evidence that chromatin structure is a repeating unit. *Cell* **4**, 281-300.

Oudet, P., Spadafora, C. and Chambon, P. (1978). Nucleosome structure. II. Structure of the SV40 minichromosome and electron microscopic evidence for reversible transitions of the nucleosome structure. *Cold Spring Harbor Symp. Quant. Biol.* **43**, 301-312.

Oudet, P., Germond, J. E., Sures, M., Gallwitz, D., Bellard, M. and Chambon, P. (1978). Nucleosome Structure. I. All four histones are required to form a nucleosome but an H3-H4 subnucleosomal particle is formed with H3-H4 alone. *Cold Spring Harbor Symp. Quant. Biol.* **43**, 287-300.

Overton, G. C. and Weinberg, E. S. (1978). Length and sequence heterogeneity of the histone gene repeat unit of the sea urchin, S. purpuratus. *Cell* **14**, 247-258.

Pace, N. R., Walker, T. A. and Schroeder, E. (1977). Structure of the 5.8S component of the 5.8S-28S rRNA ribosomal RNA junction complex. *Biochemistry* **16**, 5321-5328.

Pachmann, U. and Rigler, R. (1972). Quantum yield of acridines interacting with DNA of defined base sequence. *Exp. Cell Res.* **72**, 602-608.

Padmanaban, G., Hendler, F., Patzer, J., Ryan, R. and Rabinowitz, M. (1975). Translation of RNA that contains poly(A) from yeast mitochondria in an E. coli ribosomal system. *Proc. Nat. Acad. Sci. USA* **72**, 4293-4297.

Pagoulatos, G. N. and Darnell, J. E. (1970). Fractionation of hnRNA: rates of hybridization and chromosomal distribution of reiterated sequences. *J. Mol. Biol.* **54**, 517-536.

Painter, R. B. and Schaeffer, A. W. (1969). Rate of synthesis along replicons of different kinds of mammalian cells. *J. Mol. Biol.* **45**, 467-480.

Painter, R. B., Jermany, D. A. and Rasmussen, R. E. (1966). A method to determine the number of DNA replicating units in cultured mammalian cells. *J. Mol. Biol.* **17**, 47-55.

Painter, R. G., Sheetz, M. and Singer, S. J. (1975). Detection and ultrastructural localization of human smooth muscle myosin-like molecules in human nonmuscle cells by specific antibodies. *Proc. Nat. Acad. Sci. USA* **72**, 1359-1363.

Painter, T. S. (1934). Salivary chromosomes and the attack on the gene. *J. Hered.* **25**, 465-476.

Painter, T. S. and Taylor, A. N. (1942). Nucleic acid storage in the toad's egg. *Proc. Nat. Acad. Sci. USA* **28**, 311-317.

Palmiter, R. D. and Carey, N. H. (1974). Rapid inactivation of ovalbumin mRNA after acute withdrawal of estrogen. *Proc. Nat. Acad. Sci. USA* **71**, 2357-2361.

Palmiter, R. D., Gagnon, J. and Walsh, K. A. (1978). Ovalbumin: a screted protein without a transient hydrophobic leader sequence. *Proc. Nat. Acad. Sci. USA* **75**, 94-98.

Palmiter, R. D., Moore, P. B., Mulvihill, E. R. and Emtage, S. (1976). A significant lag in the induction of ovalbumin mRNA by steroid hormones: a receptor translocation hypothesis. *Cell* **8**, 557-572.

Palmiter, R. D., Mulvihill, E. R., McKnight, G. S. and Senear, A. W. (1978). Regulation of gene expression in the chick oviduct by steroid hormones. *Cold Spring Harbor Symp. Quant. Biol.* **42**, 639-657.

Palter, K. B., Foe, V. E. and Alberts, B. M. (1979). Evidence for the formation of nucleosome like complexes on single stranded DNA. *Cell* **18**, 451-467.

Pan, J., Celma, M. L. and Weissman, S. M. (1977). Studies of low molecular weight RNA from cells infected with adenovirus 2. III. The sequence of the promoter for VA RNA I. *J. Biol. Chem.* **252**, 9047-9054.

Panet, A. and Cedar, H. (1977). Selective degradation of integrated murine leukemia proviral DNA by DNAases. *Cell* **11**, 933-940.

Panet, A., Gorecki, M., Bratosin, S. and Aloni, Y. (1978). Electron microscopic evidence for splicing of Moloney murine leukemia virus RNAs. *Nucl. Acids Res.* **5**, 3219-3230.

Panyim, S. and Chalkley, R. (1969). The heterogeneity of histones. I. A quantitative analysis of calf histones in very long polyacrylamide gels. *Biochemistry* **8**, 3972-3979.

Panyim, S., Bilek, D., and Chalkley, R. (1971). An electrophoretic comparison of vertebrate histones. *J. Biol. Chem.* **246**, 4206-4215.

Paoletti, E. and Moss, B. (1974). Two nucleic acid dependent nucleoside triphosphate phosphohydrolases from vaccinia virus. *J. Biol. Chem.* **249**, 3281-3286.

Pardee, A. B. (1974). A restriction point for control of normal animal cell proliferation. *Proc. Nat. Acad. Sci. USA* **71**, 1286-1290.

Pardee, A. B., Dubrow, R., Hamlin, J. L. and Kletzien, R. F. (1978). Animal cell cycle. *Ann. Rev. Biochem.* **47**, 715-750.

Pardon, J. F., Worcestor, D. L., Wooley, J. C., Tatchell, K., Van Holde, K. E. and Richards, B. M. (1975). Low angle neutron scattering from chromatin subunit particles. *Nucl. Acids Res.* **2**, 2163-2182.

Pardon, J. F. and Wilkins, M. F. H. (1972). A supercoil model for nucleohistone. *J. Mol. Biol.* **68**, 115-124.

Pardon, J. F., Wilkins, M. F. H. and Richards, B. M. (1967). Superhelical model for nucleohistone. *Nature* **215**, 508-509.

Pardon, J. F., Cotter, R. I., Lilley, D. M. J., Worcestor, D. L., Camobell, A. M., Wooley, J. C. and Richards, B. M. (1977). Scattering studies of chromatin subunits. *Cold Spring Harbor Symp. Quant. Biol.* **43**, 11-22.

Pardue, M. L. and Gall, J. G. (1969). Molecular hybridization of radioactive DNA to the DNA of cytological preparations. *Proc. Nat. Acad. Sci. USA* **64**, 600-604.

Pardue, M. L. and Gall, J. G. (1970). Chromosomal localization of mouse satellite DNA. *Science* **168**, 1356-1358.

Pardue, M. L., Brown, D. D. and Birnstiel, M. L. (1973). Location of the genes for 5S rRNA in X. laevis. *Chromosoma* **42**, 191-204.

Pardue, M. L., Kedes, L. H., Weinberg, E. S. and Birnsteil, M. (1977). Localization of sequences coding for histone mRNA in the chromosomes of D. melanogaster. *Chromosoma* **63**, 135-152.

Parker, C. S. and Roeder, R. G. (1977). Selective and accurate transcription of the X. laevis 5S RNA genes in isolated chromatin by purified RNA polymerase III. *Proc. Nat. Acad. Sci. USA* **74**, 44-48.

Parker, M. G. and Mainwaring, W. I. P. (1977). Effects of androgens on the complexity of poly(A) RNA from rat prostate. *Cell* **12**, 401-407.

Parry, D. M. (1973). A meiotic mutant affecting recombination in female D. melanogaster. *Genetics* **73**, 465-486.

Parsons, J. A. and Rustad, R. C. (1968). The distribution of DNA among dividing mitochondria of Tetrahymena pyriformis. *J. Cell Biol.* **37**, 683-693.

Parsons, J. T. and McCarty, K. S. (1968). Rapidly labeled mRNA-protein complex of rat liver nuclei. *J. Biol. Chem.* **243**, 5377-5384.

Patel, G. L. and Thomas, T. L. (1973). Some binding parameters of chromatin acidic proteins with high affinity for DNA. *Proc. Nat. Acad. Sci. USA* **70**, 2524-2528.

Paterson, B. M. and Bishop, J. O. (1977). Changes in the mRNA population of chick myoblasts during myogenesis in vitro. *Cell* **12**, 751-765.

Paterson, B. M. and Rosenberg, M. (1979). Efficient translation of procaryotic mRNAs in a eucaryotic cell free system requires addition of a cap structure. *Nature* **279**, 692-695.

Pathak, S. and Hsu, T. C. (1976). Chromosomes and DNA of Mus: the behavior of constitutive heterochromatin in spermatogenesis of M. dunni. *Chromosoma* **57**, 227-234.

Pathak, S. and Hsu, T. C. (1979). Silver stained structures in mammalian meiotic prophase. *Chromosoma* **70**, 195-204.

Patterson, D. (1975). Biochemical genetics of Chinese hamster cell mutants with deviant purine metabolism: biochemical analysis of eight mutants. *Somatic Cell Genet.* **1**, 91-110.

Patterson, D. (1976a). Biochemical genetics of Chinese hamster cell mutants with deviant purine metabolism. III. Isolation and characterization of a mutant unable to convert IMP to AMP. *Somatic Cell Genet.* **2**, 41-54.

Patterson, D. (1976b). Biochemical genetics of Chinese hamster cell mutants with deviant purine metabolism. IV. Isolation of a mutant which accumulates adenylo succinic acid and succinyl-aminoimidazole carboxamide ribotide. *Somatic Cell Genet.* **2**, 189-203.

Patterson, D. and Carnright, D. V. (1977). Biochemical genetic analysis of pyrimidine biosynthesis in mammalian cells. I. Isolation of a mutant defective in the early steps of de novo pyrimidine synthesis. *Somatic Cell Genet.* **3**, 483-496.

Patterson, D., Kao, F. -T. and Puck, T. T. (1974). Genetics of somatic mammalian cells: biochemical genetics of Chinese hamster cell mutants with deviant purine metabolism. *Proc. Nat. Acad. Sci. USA* **71**, 2057-2061.

Patterson, D., Waldren, C. and Walker, C. (1976). Isolation and characterization of temperature sensitive mutants of CHO cells after treatment with UV and X irradiation. *Somatic Cell Genet.* **2**, 113-124.

Patthy, L. and Smith, E. L. (1975). Histone III. VI. Two forms of calf thymus histone III. *J. Biol. Chem.* **250**, 1919-1920.

Patthy, L., Smith, E. L. and Johnson, J. (1973). Histone III. V. The amino acid sequence of pea embryo histone III. *J. Biol. Chem.* **248**, 6834-6840.

Paucha, E. and Smith, A. E. (1978). The sequences between 0.59 and 0.54 map units on SV40 DNA code for the unique regions of small t antigen. *Cell* **15**, 1011-1020.

Paucha, E., Mellor, A., Harvey, R., Smith, A. E., Hewick, R. M. and Waterfield, M. D. (1978). Large and small tumor antigens from SV40 have identical amino termini mapping at 0.65 map units. *Proc. Nat. Acad. Sci. USA* **75**, 2165-2169.

Paul, J. and Gilmour, R. S. (1968). Organ specific restriction of transcription in mammalian chromatin. *J. Mol. Biol.* **34**, 305-316.

Paulson, J. R. and Laemmli, U. K. (1977). The structure of histone depleted metaphase chromosomes. *Cell* **12**, 817-828.

Peacock, J. H., McMorris, F. A. and Nelson, P. G. (1973). Electrical excitability and chemosensitivity of mouse neuroblastoma x mouse or human fibroblast hybrids. *Exp. Cell Res.* **79**, 199-212.

Peacock, W. J. (1970). Replication, recombination and chiasmata in Goniae australasiae (Orthoptera: Acrididae). *Genetics* **65**, 593-617.

Peacock, W. J., Lohe, A. R., Gerlach, W. L., Dunsmuir, P., Dennis, E. S. and Appels, R. (1978). Fine structure and evolution of DNA in heterochromatin. *Cold Spring Harbor Symp. Quant. Biol.* **42**, 1121-1135.

Peacock, W. J., Brutlag, D., Goldring, E., Appels, R., Hinton, C. W. and Lindsley, D. L. (1973). The organization of highly repeated DNA sequence in D. melanogaster chromosomes. *Cold Spring Harbor Symp. Quant. Biol.* **38**, 405-421.

Pearson, P. L., Bobrow, M. and Vosa, C. G. (1970). Technique for identifying Y chromosomes in human interphase nuclei. *Nature* **226**, 78-80.

Pearson, W. R., Wu, J. -R. and Bonner, J. (1978). Analysis of rat repetitive DNA sequences. *Biochemistry* **17**, 51-59.

Pease, D. (1963). The ultrastructure of flagellar fibrils. *J. Cell Biol.* **18**, 313-326.

Peattie, D. A. (1979). Direct chemical method for sequencing RNA. *Proc. Nat. Acad. Sci. USA* **76**, 1760-1764.

Peavy, D. E., Taylor, J. M. and Jefferson, L. S. (1978). Correlation of albumin production rates and albumin mRNA levels in livers of normal, diabetic and insulin treated diabetic rats. *Proc. Nat. Acad. Sci. USA* **75**, 5879-5883.

Pedersen, R. A. (1971). DNA content, ribosomal gene multiplicity and cell size in fish. *J. Exp. Zool.* **177**, 65-78.

Pederson, T. (1974a). Proteins associated with hnRNA in eucaryotic cells. *J. Mol. Biol.* **83**, 163-184.

Pederson, T. (1974b). Gene activation in eucaryotes: are nuclear acid proteins the cause or effect? *Proc. Nat. Acad. Sci. USA* **71**, 617-621.

Pederson, T. (1977). Isolation and characterization of chromatin from the cellular slime mold, D. discoideum. *Biochemistry* **16**, 2771-2777.

Pederson, T. and Bhorjee, J. S. (1975). A special class of nonhistone protein tightly complexed with template inactive DNA in chromatin. *Biochemistry* **14**, 3238-3242.

Pederson, T. and Bhorjee, J. S. (1979). Evidence for a role of RNA in eucaryotic chromosome structure. Metabolically stable small nuclear RNA species are covalently linked to chromosomal DNA in HeLa cells. *J. Mol. Biol.* **128**, 451-480.

Pederson, T. and Kumar, A. (1971). Relationship between protein synthesis and ribosome assembly in HeLa cells. *J. Mol. Biol.* **61**, 655-668.

Peebles, C. L., Ogden, R. C., Knapp, G. and Abelson, J. (1979). Splicing of yeast tRNA precursors: a two stage reaction. *Cell* **18**, 27-37.

Pelham, H. R. B. and Jackson, R. J. (1976). An efficient mRNA dependent translation system from reticulocyte lysates. *Eur. J. Biochem.* **67**, 247-256.

Pellegrini, M., Manning, J. and Davidson, N. (1977). Sequence arrangement of the rDNA of D. melanogaster. *Cell* **10**, 213-224.

Pellicer, A., Wigler, M., Axel, R. and Silverstein, S. (1978). The transfer and stable integration of the HSV thymidine kinase gene into mouse cells. *Cell* 14, 133–142.

Pelling, C. (1964). Ribonukein saure synthese der reisenchromosomen. *Chromosoma* 15, 71–122.

Pelling, C. (1966). A replicative and synthetic chromosomal unit—the modern concept of the chromomere. *Proc. Roy. Soc. B* 164, 279–289.

Pene, J. J., Knight, E. and Darnell, J. E. (1969). Characterization of a new low molecular weight RNA in HeLa cell ribosomes. *J. Mol. Biol.* 33, 609–623.

Penman, S. (1966). RNA metabolism in the cell nucleus. *J. Mol. Biol.* 17, 117–130.

Penman, S., Rosbash, M. and Penman, M. (1970). Messenger and hnRNA in HeLa cells: differential inhibition by cordycepin. *Proc. Nat. Acad. Sci. USA* 67, 1878–1885.

Penman, S., Vesco, C. and Penman, M. (1968). Localization and kinetics of formation of nuclear heterodisperse RNA, cytoplasmic heterodisperse RNA, and polyribosome associated mRNA in HeLa cells. *J. Mol. Biol.* 34, 49–69.

Penningroth, S. M. and Kirschner, M. W. (1977). Nucleotide binding and phosphorylation in microtubule assembly in vitro. *J. Mol. Biol.* 115, 643–674.

Pera, F. and Mattias, P. (1976). Labelling of DNA and differential sister chromatid staining of BrdU treatment in vivo. *Chromosoma* 57, 13–18.

Perdue, J. F. (1973). The distribution, ultrastructure, and chemistry of microfilaments in cultured chick embryo fibroblasts. *J. Cell Biol.* 58, 265–283.

Perkins, D. D. and Barry, E. G. (1977). The cytogenetics of Neurospora. *Adv. Genet.* 19, 133–285.

Perlman, P. S., Douglas, M. G., Strausberg, R. L. and Butow, R. A. (1977). Localization of genes for variant forms of mitochondrial proteins on mitochondrial DNA of S. cerevisiae. *J. Mol. Biol.* 115, 675–694.

Perlman, S., Phillips, C. and Bishop, J. O. (1976). A study of foldback DNA. *Cell* 8, 33–42.

Perry, M. M., John, H. A. and Thomas, N. S. T. (1971). Actin like filaments in the cleavage furrow of newt eggs. *Exp. Cell Res.* 65, 249–253.

Perry, R. P. and Kelley, D. E. (1968). Messenger RNA—protein complexes and newly synthesized ribosomal subunits: analysis of free particles and components of polyribosomes. *J. Mol. Biol.* 35, 37–60.

Perry, R. P. and Kelley, D. E. (1970). Inhibition of RNA synthesis by actinomycin D: characteristic dose response of different RNA species. *J. Cell Physiol.* 76, 127–140.

Perry, R. P. and Kelley, D. E. (1972). The production of rRNA from high molecular precursors. III. Hydrolysis of pre ribosomal and rRNA by a 3′-OH specific exoribonuclease. *J. Mol. Biol.* 70, 265–280.

Perry, R. P. and Kelley, D. E. (1973). Messenger RNA turnover in mouse L cells. *J. Mol. Biol.* 79, 681–696.

Perry, R. P. and Kelley, D. E. (1974). Existence of methylated mRNA in mouse L cells. *Cell* 1, 37–42.

Perry, R. P. and Kelley, D. E. (1976). Kinetics of formation of 5′ terminal caps in mRNA. *Cell* 8, 433–442.

Perry, R. P., Kelley, D. E. and Latorre, J. (1974). Synthesis and turnover of nuclear and cytoplasmic poly-A in mouse L cells. *J. Mol. Biol.* 82, 315–332.

Perry, R. P., Cheng, T. T., Freed, J. J., Greenberg, J. R., Kelley, D. E. and Tartof, K. D. (1970). Evolution of the transcription unit of rRNA. *Proc. Nat. Acad. Sci. USA* 65, 609–616.

Perry, R. P., Kelley, D. E., Friderici, K. and Rottman, F. (1975). The methylated constituents of L cell mRNA: evidence for an unusual cluster at the 5′ terminus. *Cell* 4, 387–394.

Perry, R. P., Kelley, D. E., Schibler, U., Huebner, K. and Croce, C. M. (1979). Selective suppression of the transcription of ribosomal genes in mouse-human hybrid cells. *J. Cell. Physiol.* 98, 553–560.

Petersen, D. F. and Anderson, E. C. (1964). Quantity production of synchronized mammalian cells in suspension culture. *Nature* **203**, 642-643.

Peterson, J. A. and Weiss, M. C. (1972). Expression of differentiated functions in hepatoma cell hybrids: induction of mouse albumin production in rat hepatoma mouse fibroblast hybrids. *Proc. Nat. Acad. Sci. USA* **69**, 571-575.

Peterson, J. L. and McConkey, E. H. (1976). Nonhistone chromosomal proteins from HeLa cells. A survey by high resolution, two dimensional electrophoresis. *J. Biol. Chem.* **251**, 548-554.

Peterson, S. P. and Berns, M. W. (1979). Mitosis in flat PTK2 human hybrid cells. *Exp. Cell Res.* **120**, 223-236.

Petes, T. D. (1979). Yeast ribosomal DNA genes are located on chromosome XII. *Proc. Nat. Acad. Sci. USA* **76**, 410-414.

Petes, T. D. and Botstein, D. (1977). Simple mendelian inheritance of the reiterated ribosomal DNA of yeast. *Proc. Nat. Acad. Sci. USA* **74**, 5091-5095.

Petes, T. D. and Fangman, W. L. (1972). Sedimentation properties of yeast chromosomal DNA. *Proc. Nat. Acad. Sci. USA* **69**, 1188-1191.

Petes, T. D., Byers, B. and Fangman, W. L. (1973). Size and structure of yeast chromosomal DNA. *Proc. Nat. Acad. Sci. USA* **70**, 3072-3076.

Petes, T. D., Hereford, L. and Skryabin, K. G. (1978). Characterization of two types of yeast rRNA genes. *J. Bacteriol.* **134**, 295-305.

Pett, D. M., Estes, M. K. and Pagano, J. S. (1975). Structural proteins of SV40. I. Histone characteristics of low molecular weight polypeptides. *J. Virol.* **15**, 379-385.

Pfeffer, T. A., Asnes, C. F. and Wilson, L. (1976). Properties of tubulin in unfertilized sea urchin eggs. *J. Cell Biol.* **69**, 599-607.

Phillips, D. M. (1966). Substructure of flagellar tubules. *J. Cell Biol.* **31**, 635-638.

Phillips, D. M. P. and Johns, E. W. (1965). A fractionation of the histones of group f2a from calf thymus. *Biochem. J.* **94**, 127-130.

Philipps, G. and Gigot, C. (1977). DNA associated with nucleosomes in plants. *Nucl. Acids Res.* **4**, 3617-3626.

Phillips, S. G. and Rattner, J. B. (1976). Dependence of centriole formation on protein synthesis. *J. Cell Biol.* **70**, 9-19.

Philippsen, P., Thomas, M., Kramer, R. A. and Davis, R. W. (1978). Unique arrangement of coding sequences for 5S, 5.8S, 18S, and 25S rRNA in S. cerevisiae as determined by R loop and hybridization analysis. *J. Mol. Biol.* **123**, 387-404.

Philipson, L., Wall, R., Glickman, R. and Darnell, J. E. (1971). Addition of polyadenylate sequences to virus specific RNA during adenovirus replication. *Proc. Nat. Acad. Sci. USA* **68**, 2806-2809.

Philipson, L., Andersson, P., Olshevsky, U., Weinberg, R., Baltimore, D. and Gesteland, R. (1978). Translation of MuLV and MSV RNAs in nuclease treated reticulocyte extracts: enhancement of the *gag-pol* polypeptide with yeast suppressor tRNA. *Cell* **13**, 189-200.

Pica-Mattoccia, L. and Attardi, G. (1972). Expression of the mitochondrial genome in HeLa cells. IX. Replication of mitochondrial DNA in relationship to the cell cycle in HeLa cells. *J. Mol. Biol.* **64**, 465-484.

Pickett-Heaps, J. D. (1969). The evolution of the mitotic apparatus: an attempt at comparative ultrastructural cytology in dividing plant cells. *Cytobios* **1**, 257-280.

Pickett-Heaps, J. D. (1971). The autonomy of the centriole: fact or fallacy. *Cytobios* **3**, 205-214.

Pickett-Heaps, J. D. (1975). Aspects of spindle evolution. *Ann. N. Y. Acad. Sci.* **253**, 352-361.

Pickett-Heaps, J. D. and Tippit, D. H. (1978). The diatom spindle in perspective. *Cell* **14**, 455-468.

Piper, P. W., Celis, J., Kaltoft, K., Leer, J. C., Nielsen, O. F. and Westergaard, O. (1976).

Tetrahymena rRNA gene chromatin is digested by micrococcal nuclease at sites which have the same regular spacing on the DNA as corresponding sites in the bulk nuclear chromatin. *Nucl. Acids Res.* **3**, 493–505.

Piperno, G. and Luck, D. J. (1976). Phosphorylation of axonemal proteins in C. reinhardii. *J. Biol. Chem.* **251**, 2161–2167.

Pittinger, T. H. (1956). Synergism of two cytoplasmically inherited mutants in N. crassa. *Proc. Nat. Acad. Sci. USA* **42**, 747–752.

Plagemann, P. G. (1970). Vinblastine sulfate: metaphase arrest, inhibition of RNA synthesis, and cytotoxicity in Novikoff rat hepatoma cells. *J. Nat. Cancer Inst.* **45**, 589–595.

Plagens, U., Greenleaf, A. L. and Bautz, E. K. F. (1976). Distribution of RNA polymerase on Drosophila polytene chromosomes as studied by indirect immunofluorescence. *Chromosoma* **59**, 157–166.

Planck, S. R. and Mueller, G. C. (1977). DNA chain growth in isolated HeLa nuclei. *Biochemistry* **16**, 2778–2782.

Platz, R. D. and Hnilica, L. S. (1973). Phosphorylation of nonhistone chromatin proteins during sea urchin development. *Biochem. Biophys. Res. Commun.* **54**, 222–227.

Plaut, W. (1969). On ordered DNA replication in polytene chromosomes. *Genetics Suppl.* **61**, 239–244.

Plaut, W., Nash, D. and Fanning, T. (1966). Ordered replication of DNA in polytene chromosomes of D. melanogaster. *J. Mol. Biol.* **16**, 85–93.

Pledger, W. J., Stiles, C. D., Antoniades, H. N. and Scher, C. D. (1977). Induction of DNA synthesis in Balbc/3T3 cells by serum components: reevaluation of the commitment process. *Proc. Nat. Acad. Sci. USA* **74**, 4481–4485.

Pledger, W. J., Stiles, C. D., Antoniades, H. N. and Scher, C. D. (1978). An ordered sequence of events is required before Balc/3T3 cells become committed to DNA synthesis. *Proc. Nat. Acad. Sci. USA* **75**, 2839–2843.

Polan, M. K., Gall, J. G., Friedman, S. and Gehring, W. (1973). Isolation and characterization of mitochondrial DNA from D. melanogaster. *J. Cell Biol.* **56**, 580–589.

Polisky, B. and McCarthy, B. J. (1975). Location of histones on SV40 DNA. *Proc. Nat. Acad. Sci. USA* **72**, 2895–2899.

Pollack, R. and Rifkin, D. (1975). Actin-containing cables within anchorage dependent rat embryo cells are dissociated by plasmin and trypsin. *Cell* **6**, 495–506.

Pollack, R., Osborn, M. and Weber, K. (1975). Patterns of organization of actin and myosin in normal and transformed cultured cells. *Proc. Nat. Acad. Sci. USA* **72**, 994–998.

Pollack, R., Goldman, R. D., Conlon, S. and Chang, C. (1974). Properties of enucleated cells. II. Characteristic overlapping of transformed cells is reestablished by enucleates. *Cell* **3**, 51–54.

Pollard, T. D. (1975). Functional implications of the biochemical and structural properties of cytoplasmic contractile proteins. In S. Inoue and R. E. Stephens (Ed.), *Molecules and Cell Movement*, Raven Press, New York.

Pollard, T. D. (1976). The role of actin in the temperature dependent gelation and contraction of extracts of Acanthamoeba. *J. Cell Biol.* **68**, 579–601.

Pollard, T. D. and Korn, E. D. (1971). Filaments of Amoeba proteus. II. Binding of heavy meromyosin by thin filaments in motile cytoplasmic extracts. *J. Cell Biol.* **48**, 216–219.

Pollard, T. D. and Weihing, R. R. (1974). Actin and myosin and cell movement. *CRC Crit. Rev. Biochem.* **2**, 1–65.

Ponder, B. A. J. and Crawford, L. V. (1977). The arrangement of nucleosomes in nucleoprotein complexes from polyoma virus and SV40. *Cell* **11**, 35–49.

Pontecorvo, G. (1971). Induction of directional chromosome elimination in somatic cell hybrids. *Nature* **230**, 367–369.

Pontecorvo, G. (1975). Production of mammalian somatic cell hybrids by means of polyethylene glycol treatment. *Somatic Cell Genet.* **1**, 397–400.

Pontecorvo, G., Riddle, P. N. and Hales, A. (1977). Time and mode of fusion of human fibroblasts treated with polyethylene glycol. *Nature* **265**, 257–258.

Porter, K. R., Puck, T. T., Hsie, A. W. and Kelley, D. (1974). An electron microscopic study of the effects of Bt₂cAMP on CHO cells. *Cell* **2**, 145–162.

Portmann, R., Schaffner, W. and Birnstiel, M. (1976). Partial denaturation mapping of cloned histone DNA from the sea urchin Psammechinus miliaris. *Nature* **264**, 31–33.

Poste, G. (1972). Enucleation of mammalian cells by cytochalasin B. I. Characterization of anucleate cells. *Exp. Cell Res.* **73**, 273–286.

Poste, G. and Reeve, P. (1972). Enucleation of mammalian cells by cytochalasin B. II. Formation of hybrid cells and heterokaryons by fusion of anucleate and nucleated cells. *Exp. Cell Res.* **73**, 287–294.

Potter, S. S., Newbold, J. E., Hutchison, C. A., III and Edgell, M. H. (1975). Specific cleavage analysis of mammalian mitochondrial DNA. *Proc. Nat. Acad. Sci. USA* **72**, 4496–4500.

Potter, S. S., Brorein, W. J., Dunsmuir, P. and Rubin, G. M. (1979). Transposition of elements of the 412, copia and 297 dispersed repeated gene families in Drosophila. *Cell* **17**, 415–428.

Poyton, R. O. and Groot, G. S. P. (1975). Biosynthesis of polypeptides of cytochrome *c* oxidase by isolated mitochondria. *Proc. Nat. Acad. Sci. USA* **72**, 172–176.

Premkumar, E., Shoyab, M. and Williamson, A. R. (1974). Germ line basis for antibody diversity: immunoglobulin V$_H$- and C$_H$ gene frequencies measured by DNA-DNA hybridization. *Proc. Nat. Acad. Sci. USA* **71**, 99–103.

Prescott, D. M. (1963). Cellular sites of RNA synthesis. *Prog. Nucleic Acid Res.* **3**, 33–57.

Prescott, D. M. (1966). The synthesis of total macronuclear protein, histone and DNA during the cell cycle on Euplotes eurystomus. *J. Cell Biol.* **31**, 1–9.

Prescott, D. M. (1970). The structure and replication of eucaryotic chromosomes. *Adv. Cell Biol.* **1**, 57–117.

Prescott, D. M. (1976). *Reproduction of Eucaryotic Cells.* Academic Press, New York.

Prescott, D. M. and Bender, M. A. (1962). Synthesis of RNA and protein during mitosis in mammalian tissue culture cells. *Exp. Cell Res.* **26**, 260–268.

Prescott, D. M. and Bender, M. A. (1963a). Autoradiographic study of the chromatid distribution of labeled DNA in two types of mammalian cells in vitro. *Exp. Cell Res.* **29**, 430–442.

Prescott, D. M. and Bender, M. A. (1963b). Synthesis and behavior of nuclear proteins during the cell life cycle. *J. Cell. Physiol. Suppl.* **62**, 175–194.

Prescott, D. M., Myerson, D. and Wallace, J. (1972). Enucleation of mammalian cells with cytochalasin B1. *Exp. Cell Res.* **71**, 480–485.

Prescott, D. M., Bostock, C. J., Hatch, F. T. and Mazrimas, J. A. (1973). Location of satellite DNAs in the chromosomes of the kangaroo rat (D. ordii). *Chromosoma* **42**, 205–213.

Prestayko, A. W., Lewis, B. C. and Busch, H. (1972). Endoribonuclease activity associated with nucleolar RNP particles from Novikoff hepatoma. *Biochim. Biophys. Acta* **269**, 90–103.

Prestayko, A. W., Tonato, M. and Busch, H. (1970). Low molecular weight RNA associated with 28S nucleolar RNA. *J. Mol. Biol.* **47**, 506–515.

Prestayko, A. W., Klomp, G. R., Schmoll, D. J. and Busch, H. (1974). Comparison of proteins of ribosomal subunits and nucleolar preribosomal particles from Novikoff hepatoma ascites cells by 2D gel electrophoresis. *Biochemistry* **13**, 1945–1951.

Pribnow, D. (1975). Phage T7 early promoters: nucleotide sequences of two RNA polymerase binding sites. *J. Mol. Biol.* **99**, 419–444.

Price, P. A. (1975). The essential role of Ca²⁺ in the activity of bovine pancreatic DNAase. *J. Biol. Chem.* **250**, 1981–1986.

Price, R. P., Ransom, L. and Penman, S. (1974). Identification of a small subfraction of hnRNA with the characteristics of a precursor to mRNA. *Cell* **2**, 253-258.

Price, R. and Penman, S. (1972). A distinct RNA polymerase activity, synthesizing 5.5S, 5S and 4S RNA in nuclei from adenovirus 2 infected HeLa cells. *J. Mol. Biol.* **70**, 435-450.

Pring, D. R. (1974). Maize mitochondria: purification and characterization of ribosomes and ribosomal RNA. *Plant. Physiol.* **53**, 677-687.

Prives, C., Gilboa, E., Revel, M. and Winicour, E. (1977). Cell free translation of SV40 early mRNA coding for viral T antigen. *Proc. Nat. Acad. Sci. USA* **74**, 457-461.

Procunier, J. D. and Dunn, R. J. (1978). Genetic and molecular organization of the 5S locus and mutants in D. melanogaster. *Cell* **15**, 1087-1094.

Procunier, J. D. and Tartof, K. D. (1975). Genetic analysis of the 5S RNA genes in D. melanogaster. *Genetics* **81**, 515-523.

Procunier, J. D. and Tartof, K. D. (1976). Restriction map of 5S RNA genes of D. melanogaster. *Nature* **263**, 255-256.

Procunier, J. D. and Tartof, K. D. (1978). A genetic locus having trans and contiguous cis functions that controls the disproportionate replication of rRNA genes in D. melanogaster. *Genetics* **88**, 67-79.

Prosser, J., Moar, M., Bobrow, M. and Jones, K. W. (1973). Satellite sequences in chimpanzee (Pan troglodytes). *Biochim. Biophys. Acta* **319**, 122-134.

Proudfoot, N. J. (1977). Complete 3' noncoding region sequences of rabbit and human β globin mRNAs. *Cell* **10**, 559-570.

Proudfoot, N. J. and Brownlee, G. G. (1976). 3' noncoding region sequences in eucaryotic mRNA. *Nature* **263**, 211-214.

Proudfoot, N. J., Gillam, S., Smith, M. and Longley, J. I. (1977). Nucleotide sequence of the 3' terminal third of rabbit α-globin mRNA: comparison with human α-globin mRNA. *Cell* **11**, 807-816.

Prudhommeau, C. (1972). Irradiation UV des cellules polaires de l'oeuf chez D. melanogaster. III. Etude de la recombinasson mitotique induite chez le male. *Mutat. Res.* **14**, 53-64.

Prunell, A. and Bernardi, G. (1974). The mitochondrial genome of wild type yeast cells. IV. Genes and spacers. *J. Mol. Biol.* **86**, 825-841.

Prunell, A. and Bernardi, G. (1977). The mitochondrial genome of wild type yeast cells. VI. Genome organization. *J. Mol. Biol.* **110**, 53-74.

Prunell, A., Kornberg, R. D., Lutter, L., Klug, A., Levitt, M. and Crick, F. H. C. (1979). Periodicity of DNAase I digestion of chromatin. *Science* **204**, 855-858.

Puck, T. T. and Kao, F. -T. (1967). Genetics of somatic mammalian cells. V. Treatment with BUdR and visible light for isolation of nutritionally deficient mutants. *Proc. Nat. Acad. Sci. USA* **58**, 1227-1234.

Puck, T. T., Cieciura, S. J. and Robinson, A. (1958). Genetics of somatic mammalian cells. III. Long term cultivation of euploid cells from human and animal subjects. *J. Exp. Med.* **108**, 945-956.

Puck, T. T., Waldren, C. A. and Hsie, A. W. (1972). Membrane dynamics in the action of dbcAMP and testosterone on mammalian cells. *Proc. Nat. Acad. Sci. USA* **69**, 1943-1947.

Puckett, L. and Darnell, J. E. (1977). Essential factors in the kinetic analysis of RNA synthesis in HeLa cells. *J. Cell. Physiol.* **90**, 521-534.

Puckett, L. D. and Snyder, L. A. (1975). Biochemical evidence for position-effect suppression of ribosomal RNA synthesis in D. melanogaster. *Exp. Cell Res.* **95**, 31-38.

Puro, J. (1964). Temporal distribution of X ray induced recessive lethals and recombinants in post sterile broods of Drosophila males. *Mutat. Res.* **1**, 268-278.

Puszkin, S. and Berl, S. (1972). Actomyosin like protein from brain. Separation and characterization of the actin like component. *Biochem. Biophys. Acta* **256**, 695-709.

Quastler, H. and Sherman, F. G. (1959). Cell population kinetics in the intestinal epithelium of the mouse. *Exp. Cell Res.* **17,** 420–438.

Quinlan, T. J., Kinniburgh, A. J. and Martin, T. E. (1977). Properties of a nuclear polyadenylate protein complex from mouse acites cells. *J. Biol. Chem.* **252,** 1156–1161.

Quinlan, T. J., Beeler, G. W., Cox, R. F., Elder, P. K., Moses, H. L. and Getz, M. J. (1978). The concept of mRNA abundance classes: a critical re evaluation. *Nucl. Acids Res.* **5,** 1611–1626.

Rabbitts, T. H. (1976). Bacterial cloning of plasmids carrying copies of rabbit globin mRNA. *Nature* **260,** 221–225.

Rabbitts, T. H. (1978). Evidence for splicing of interrupted Ig variable and constant region sequences in nuclear RNA. *Nature* **275,** 291–296.

Rabbitts, T. H. and Forster, A. (1978). Evidence for noncontiguous variable and constant region genes in both germ line and myeloma DNA. *Cell* **13,** 319–327.

Rabbits, T. H. and Milstein, C. (1975). Mouse immunoglobulin genes: studies on the reiteration frequency of light chain genes by hybridization procedures. *Eur. J. Biochem.* **52,** 125–133.

Rabbitts, T. H., Jarvis, J. M. and Milstein, C. (1975). Demonstration that a mouse immunoglobulin light chain mRNA hybridizes exclusively with unique DNA. *Cell* **6,** 5–12.

Rabinowitz, M., Sinclair, J., DeSalle, L., Haselkorn, R. and Swift, H. H. (1965). Isolation of DNA from mitochondria of chick embryo heart and liver. *Proc. Nat. Acad. Sci. USA* **53,** 1126–1133.

Radloff, R., Bauer, W. and Vinograd, J. (1967). A dye buoyant density method for the detection and isolation of closed circular duplex DNA: the closed circular DNA in HeLa cells. *Proc. Nat. Acad. Sci. USA* **57,** 1514–1521.

Rae, P. M. M. (1966). Whole mount electron microscopy of Drosophila salivary chromosomes. *Nature* **212,** 139–142.

Raj, N. B. K., Ro-Choi, T. S. and Busch, H. (1975). Nuclear ribonucleoprotein complexes containing U1 and U2 RNA. *Biochemistry* **14,** 4380–4385.

Rall, S. C. and Cole, R. D. (1971). Amino acid sequence and sequence variability of the amino terminal regions of lysine rich histones. *J. Biol. Chem.* **246,** 7175–7190.

Rall, S. C., Okinaka, R. T. and Strniste, G. F. (1977). Histone composition of nucleosomes from cultured Chinese hamster cells. *Biochemistry* **16,** 4940–4943.

Rambach, A. and Hogness, D. S. (1977). Translation of D. melanogaster sequences in E. coli. *Proc. Nat. Acad. Sci. USA* **74,** 5041–5045.

Rambach, A. and Tiollais, P. (1974). Phage lambda having Eco RI endonuclease sites only in the nonessential region of the genome. *Proc. Nat. Acad. Sci. USA* **71,** 3927–3930.

Ramirez, F., Natta, C., O'Donnell, J. V., Canale, V., Bailey, G., Sanguensermsri, T., Maniatis, G. M., Marks, P. A. and Bank, A. (1975). Relative numbers of human globin genes assayed with purified α and β complementary human DNA. *Proc. Nat. Acad. Sci. USA* **72,** 1550–1554.

Ramirez, J. L. and Dawid, I. B. (1978). Mapping of mitochondrial DNA in X. laevis and X. borealis: the positions of ribosomal genes and D loops. *J. Mol. Biol.* **119,** 133–146.

Randall, J. and Disbrey, C. (1965). Evidence for the presence of DNA at basal body sites in Tetrahymena pyriformis. *Proc. Roy. Soc. B* **162,** 473–491.

Randall, L. L. and Hardy, S. J. S. (1977). Synthesis of exported proteins by membrane bound polysomes from E. coli. *Eur. J. Biochem.* **75,** 43–53.

Rao, P. N. and Johnson, R. T. (1970). Mammalian cell fusion: studies on the regulation of DNA synthesis and mitosis. *Nature* **225,** 159–164.

Rao, P. N. and Johnson, R. T. (1972). Premature chromosome condensation: a mechanism for the elimination of chromosomes in virus fused cells. *J. Cell Sci.* **10,** 495–514.

Rao, P. N., Hittelman, W. N. and Wilson, B. A. (1975). Mammalian cell fusion. VI. Regulation of mitosis in binucleate HeLa cells. *Exp. Cell Res.* **90,** 40–46.

Rao, P. N., Sunkara, P. S. and Wilson, B. A. (1977). Regulation of DNA synthesis: age dependent cooperation among G1 cells upon fusion. *Proc. Nat. Acad. Sci. USA* **74,** 2869–2873.

Rapaport, R. (1971). Cytokinesis in animal cells. *Int. Rev. Cytol.* **31**, 169-212.

Rasch, E. M. (1974). The DNA content of sperm and hemocyte nuclei of the silkworm, Bombyx mori. *Chromosoma* **45**, 1-26.

Rasch, E. M., Barr, H. J. and Rasch, R. W. (1971). The DNA content of sperm of D. melanogaster. *Chromosoma* **33**, 1-18.

Rastl, E. and Dawid, I. B. (1979). Expression of the mitochondrial genome in X. laevis: a map of transcripts. *Cell* **18**, 501-510.

Rasmussen, P. S., Murray, K. and Luck, J. M. (1962). On the complexity of calf thymus histone. *Biochemistry* **1**, 79-89.

Rasmussen, S. W. (1975). Synaptonemal polycomplexes in D. melanogaster. *Chromosoma* **49**, 321-331.

Rasmussen, S. W. (1976). The meiotic prophase in Bombyx mori females analyzed by three dimensional reconstructions of synaptonemal complexes. *Chromosoma* **54**, 245-293.

Rathke, P. C., Seib, E., Weber, K., Osborn, M. and Franke, W. W. (1977). Rod like elements from actin containing microfilament bundles observed in cultured cells after treatment with cytochalasin A (CA). *Exp. Cell Res.* **105**, 253-262.

Rattner, J. B. and Berns, M. W. (1976). Centriole behavior in early mitosis of rat kangaroo cells (PTK2). *Chromosoma* **54**, 387-395.

Rattner, J. B. and Hamkalo, B. A. (1978a). Higher order structure in metaphase chromosomes. I. The 250 Å fiber. *Chromosoma* **69**, 363-372.

Rattner, J. B. and Hamkalo, B. A. (1978b). Higher order structure in metaphase chromosomes. II. The relationship between the 250Å fiber, superbeads and beads on a string. *Chromosoma* **69**, 373-380.

Rattner, J. B. and Phillips, S. G. (1973). Independence of centriole formation and DNA synthesis. *J. Cell Biol.* **57**, 359-372.

Rattner, J. B., Branch, A. and Hamkalo, B. A. (1975). Electron microscopy of whole mount metaphase chromosomes. *Chromosoma* **52**, 329-338.

Ratzkin, B. and Carbon, J. (1977). Functional expression of cloned yeast DNA in E. coli. *Proc. Nat. Acad. Sci. USA* **74**, 487-491.

Rawson, J. R. Y. and Boerma, C. (1976a). Influence of growth conditions upon the number of chloroplast DNA molecules in Euglena gracilis. *Proc. Nat. Acad. Sci. USA* **73**, 2401-2404.

Rawson, J. R. Y. and Boerma, C. L. (1976b). A measurement of the fraction of chloroplast DNA transcribed during chloroplast development in Euglena gracilis. *Biochemistry* **15**, 588-592.

Rawson, J. R. and Stutz, E. (1969). Isolation and characterization of Euglena gracilis cytoplasmic and chloroplast ribosomes and their RNA counterparts. *Biochim. Biophys. Acta* **190**, 368-380.

Ray, D. S. and Hanawalt, P. C. (1964). Properties of the satellite DNA associated with the chloroplasts of Euglena gracilis. *J. Mol. Biol.* **9**, 812-824.

Ray, M., Gee, P. A., Richardson, B. J. and Hamerton, J. L. (1972). G6PD expression and X chromosome late replication in fibroblast clones from a female mule. *Nature* **237**, 396-397.

Rebhun, L. I. and Sharpless, T. K. (1964). Isolation of spindles from the surf clam Spisula solidissima. *J. Cell Biol.* **22**, 488-491.

Rebhun, L. I., Rosenbaum, J., Lefebvre, P. and Smith, G. (1974). Reversible restoration of the birefringence of cold treated isolated mitotic apparatus of surf clam eggs with chick brain tubulin. *Nature* **249**, 113-115.

Reddi, O. S., Reddy, G. M. and Rao, M. S. (1965). Induction of crossing over in Drosophila males by means of an ovarian extract. *Nature* **208**, 203.

Reddy, R., Ro-Choi, T. S., Henning, D. and Busch, H. (1974). Primary sequence of U1 nuclear RNA of Novikoff hepatoma ascites cells. *J. Biol. Chem.* **249**, 6486-6495.

Reddy, V. B., Thimmappaya, B., Dhar, R., Subramanian, K. N., Zain, B. S., Pan, J., Ghosh, P. K., Celma, M. L. and Weissman, S. M. (1978). The genome of SV40. *Science* **200**, 494-502.

Reeck, G. R., Simpson, R. T. and Sober, H. A. (1972). Resolution of a spectrum of nucleoprotein species in sonicated chromatin. *Proc. Nat. Acad. Sci. USA* **69**, 2317-2321.

Reeck, G. R., Simpson, R. T. and Sober, H. A. (1974). The distribution of histones and nonhistone proteins in the Ectham-cellulose fractions of chromatin from several tissues. *Eur. J. Biochem.* **49**, 407-414.

Reeder, R. H. (1973). Transcription of chromatin by bacterial RNA polymerase. *J. Mol. Biol.* **80**, 229-241.

Reeder, R. H. and Roeder, R. G. (1972). Ribosomal RNA synthesis in isolated nuclei. *J. Mol. Biol.* **67**, 433-442.

Reeder, R. H., Higashnikagawa, T. and Miller, O., Jr. (1976a). The 5'-3' polarity of the Xenopus ribosomal RNA precursor molecule. *Cell* **8**, 449-454.

Reeder, R. H., Brown, D. D., Wellauer, P. K. and Dawid, I. B. (1976b). Patterns of rDNA spacer lengths are inherited. *J. Mol. Biol.* **105**, 507-516.

Reeder, R. H., Sollner-Webb, B. and Wahn, H. L. (1977). Sites of transcription initiation in vivo on X. laevis rDNA. *Proc. Nat. Acad. Sci. USA* **74**, 5402-5406.

Reeves, R. (1976). Ribosomal genes of X. laevis: evidence of nucleosomes in transcriptionally active chromatin. *Science* **194**, 529-532.

Reeves, R. (1977). Analysis and reconstruction of Xenopus ribosomal chromatin nucleosomes. *Eur. J. Biochem.* **75**, 545-560.

Reeves, R. (1978). Nucleosome structure of Xenopus oocyte amplified ribosomal genes. *Biochemistry* **17**, 4908-4915.

Reeves, R. and Jones, A. (1976). Genomic transcriptional activity and the structure of chromatin. *Nature* **260**, 495-500.

Reff, M. E. and Davidson, R. L. (1979). In vitro DNA dependent synthesis of globin RNA sequences from erythroleukemic cell chromatin. *Nucl. Acids Res.* **6**, 275-288.

Reich, E. and Luck, D. J. L. (1966). Replication and inheritance of mitochondrial DNAs. *Proc. Nat. Acad. Sci. USA* **55**, 1600-1608.

Reider, C. L. (1978). RNP staining of centrioles and kinetochores in newt lung cell spindles. *J. Cell Biol.* **80**, 1-9.

Reijnders, L., Sloof, P. and Borst, P. (1973). The molecular weights of the mitochondrial rRNAs of S. carlsbergensis. *Eur. J. Biochem.* **35**, 266-269.

Rein, A. and Penman, S. (1969). Species specificity of the low molecular weight nuclear RNAs. *Biochim, Biophys. Acta* **190**, 1-9.

Reis, R. J. S. and Biro, P. A. (1978). Sequence and evolution of mouse satellite DNA. *J. Mol. Biol.* **121**, 357-374.

Renaud, F. L., Rowe, A. J. and Gibbons, I. R. (1968). Some properties of the protein forming the outer fiber of cilia. *J. Cell Biol.* **38**, 79-90.

Renkawitz-Pohl, R. and Kunz, W. (1975). Underreplication of satellite DNAs in polyploid ovarian tissue of D. virilis. *Chromosoma* **49**, 375-382.

Renz, M., Nehls, P. and Hozier, J. (1977). Involvement of histone H1 in the organization of the chromosome fiber. *Proc. Nat. Acad. Sci. USA* **74**, 1879-1883.

Rhoades, M. M. (1950). Meiosis in maize. *J. Hered.* **41**, 58-67.

Rhodes, D. P., Moyer, S. A. and Bannerjee, A. K. (1974). In vitro synthesis of methylated mRNA by virion associated RNA polymerase of VSV. *Cell* **3**, 327-334.

Ricciuti, F. C. and Ruddle, F. H. (1973a). Assignment of nucleoside phosphorylase to D14 and localization of X linked loci in man by somatic cell genetics. *Nature New Biol.* **241**, 180-182.

Ricciuti, F. C. and Ruddle, F. H. (1973b). Assignment of three gene loci (PGK, HGPRT, G6PD) to the long arm of the human X chromosome by somatic cell genetics. *Genetics* **74**, 661-678.

Ricco, G., Mazza, U., Turi, R. M., Pich, P. G., Canaschella, C., Gaglio, G. and Bernini, L. F.

(1976). Significance of a new type of human fetal hemoglobin carrying a replacement Ile →Thr at position 75 of the γ chain. *Human Genet.* **32**, 305–313.

Rice, N. R. and Straus, N. A. (1973). Relatedness of mouse satellite DNA to DNA of various Mus species. *Proc. Nat. Acad. Sci. USA* **70**, 3546–3550.

Richards, K., Guilley, H., Jonard, G. and Hirth, L. (1978). Nucleotide sequence at the 5′ extremity of TMV RNA. I. The noncoding region (nucleotides 1-68). *Eur. J. Biochem.* **84**, 513–519.

Richardson, B. J., Czuppon, A. B. and Sharman, G. B. (1971). Inheritance of G6P DH variation in kangaroos. *Nature New Biol.* **230**, 154–155.

Richardson, C. C., Lehman, I. R. and Kornberg, A. (1964). A DNA polymerase exonuclease from E. coli. II. Characterization of the exonuclease activity. *J. Biol. Chem.* **239**, 251–258.

Rickards, G. K. (1975). Prophase chromosome movements in living house cricket spermatocytes and their relationship to premetaphase, anaphase and granule movement. *Chromosoma* **49**, 407–455.

Rickwood, D. and MacGillivray, A. J. (1975). Improved techniques for the fractionation of nonhistone proteins of chromatin on hydroxyapatite. *Eur. J. Biochem.* **51**, 593–601.

Rifkin, M. R., Wood, D. D. and Luck, D. J. L. (1967). rRNA and ribosomes from mitochondria of N. crassa. *Proc. Nat. Acad. Sci. USA* **58**, 1025–1032.

Riggs, M. G., Whittaker, R. G., Neumann, J. R. and Ingram, V. M. (1977). n-Butyrate causes histone modification in HeLa and Friend erythroleukemia cells. *Nature* **268**, 462–464.

Riley, D. and Weintraub, H. (1978). Nucleosomal DNA is digested to repeats of 10 bases by exonuclease III. *Cell* **13**, 281–293.

Riley, D. and Weintraub, H. (1979). Conservative segregation of parental histones during replication in the presence of cycloheximide. *Proc. Nat. Acad. Sci. USA* **76**, 328–332.

Ringertz, N. R. and Savage, R. E. (1976). *Cell Hybrids.* Academic Press, New York.

Ringertz, N. R., Krondahl, U. and Coleman, J. R. (1978). Reconstitution of cells by fusion of cell fragments. I. Myogenic expression after fusion of minicells from rat myoblasts (L6) with mouse fibriblast (A9) cytoplasm. *Exp. Cell Res.* **113**, 233–246.

Ringertz, N. R., Carlsson, S. A., Ege, T. and Bolund, L. (1971). Detection of human and chick nuclear antigens in nuclei of chick erythrocytes during reactivation in heterokaryons with HeLa cells. *Proc. Nat. Acad. Sci. USA* **68**, 3228–3232.

Ringo, D. L. (1967a). The arrangement of subunits in flagellar fibers. *J. Ult. Res.* **17**, 266–277.

Ringo, D. L. (1967b). Flagellar motion and fine structure of the flagellar apparatus in Chlamydomonas. *J. Cell Biol.* **33**, 543–571.

Ris, H. (1956). A study of chromosomes with the electron microscope. *J. Biochem. Biophys. Cytol.* Suppl. **2**, 385–392.

Ris, H. (1969). The molecular organization of chromosomes. In A. Lima de Faria (Ed.), *Handbook of Molecular Cytology*, North Holland: Amsterdam, pp 221–250.

Ris, H. and Chandler, B. L. (1963). The ultrastructure of genetic systems in procaryotes and eucaryotes. *Cold Spring Harbor Symp. Quant. Biol.* **28**, 1–8.

Ris, H. and Kubai, D. F. (1974). An unusual mitotic mechanism in the parasitic protozoan Syndinium. *J. Cell Biol.* **60**, 702–720.

Ris, H. and Plaut, W. (1962). Ultrastructure of DNA containing areas in the chloroplast of Chlamydomonas. *J. Cell Biol.* **13**, 383–391.

Ritossa, F. M. (1968). Unstable redundancy of genes for rRNA. *Proc. Nat. Acad. Sci. USA* **60**, 509–516.

Ritossa, F. M. (1964). Experimental activation of specific loci in polytene chromosomes of Drosophila. *Exp. Cell Res.* **35**, 601–607.

Ritossa, F. M. (1972). Procedure for magnification of lethal deletions of genes for rRNA. *Nature New Biol.* **240**, 109–111.

Ritossa, F. M. (1973). Crossing over between X and Y chromosomes during rDNA magnification in D. melanogaster. *Proc. Nat. Acad. Sci. USA* **70**, 1950-1954.

Ritossa, F. M. and Scala, G. (1969). Equilibrium variations in the redundancy of rDNA in D. melanogaster. *Genetics* Suppl. **61**, 305-317.

Ritossa, F. M. and Spiegelman, S. (1965). Localization of DNA complementary to rRNA in the nucleolus organizer region of D. melanogaster. *Proc. Nat. Acad. Sci. USA* **53**, 737-745.

Ritossa, F. M., Atwood, K. C. and Spiegelman, S. (1966). A molecular explanation of the bobbed mutants of Drosophila as partial deficiencies of rDNA. *Genetics* **54**, 819-834.

Ritossa, F., Malva, C., Boncinelli, E., Graziani, F. and Polito, L. (1971). The first steps of magnification of DNA complementary to rRNA in D. melanogaster. *Proc. Nat. Acad. Sci. USA* **68**, 1580-1584.

Rizzoni, M. and Palitti, F. (1973). Regulatory mechanism of cell division. I. Colchicine induced endoreduplication. *Exp. Cell Res.* **77**, 450-458.

Robberson, D. L. and Clayton, D. A. (1972). Replication of mitochondrial DNA in mouse L cells and their thymidine kinase⁻ derivatives: displacement replication on a covalently closed circular template. *Proc. Nat. Acad. Sci. USA* **69**, 3810-3814.

Robberson, D. L., Clayton, D. A. and Morrow, J. F. (1974). Cleavage of replicating forms of mitochondrial DNA by Eco RI endonuclease. *Proc. Nat. Acad. Sci. USA* **71**, 4447-4451.

Robberson, D. L., Kasamatsu, H. and Vinograd, J. (1972). Replication of mitochondrial DNA. Circular replicative intermediates in mouse L cells. *Proc. Nat. Acad. Sci. USA* **69**, 737-741.

Robberson, D., Aloni, Y., Attardi, G. and Davidson, N. (1971). Expression of the mitochondrial genome in HeLa cells. VI. Size determination of mitochondrial rRNA by electron microscopy. *J. Mol. Biol.* **60**, 473-484.

Robbins, E. and Borun, T. (1967). The cytoplasmic synthesis of histone in HeLa cells and its temporal relationship to DNA replication. *Proc. Nat. Acad. Sci. USA* **57**, 409-416.

Robbins, E. and Gonatas, N. K. (1964). Histochemical and ultrastructural studies on HeLa cell cultures exposed to spindle inhibitors with special reference to the interphase cell. *J. Histochem. Cytochem.* **12**, 704-711.

Robbins, E. and Shelanski, M. (1969). Synthesis of a colchicine binding protein during the HeLa cell life cycle. *J. Cell Biol.* **43**, 371-373.

Robbins, E., Jentzsch, G. and Micali, A. (1968). The centriole cycle in synchronized HeLa cells. *J. Cell Biol.* **36**, 329-339.

Robbins, L. G. (1971). Nonexchange alignment: a meiotic process revealed by a synthetic meiotic mutant of D. melanogaster. *Mol. Gen. Genet.* **110**, 144-166.

Roberts, B. E. and Paterson, B. M. (1973). Efficient translation of TMV RNA and rabbit globin 9S RNA in a cell free system from commercial wheat germ. *Proc. Nat. Acad. Sci. USA* **70**, 2330-2334.

Roberts, P. A. (1976). The genetics of chromosome aberration. In M. Ashburner and E. Novitski (Eds.), *The Genetics and Biology of Drosophila*, Academic Press, New York, pp 68-173.

Robertson, H. D., Dickson, E. and Jelinek, W. (1977). Determination of nucleotide sequences from double stranded regions of HeLa cell nuclear RNA. *J. Mol. Biol.* **115**, 571-590.

Robertson, M. A., Staden, R., Tanaka, Y., Catterall, J. F., O'Malley, B. W. and Brownlee, G. G. (1979). Sequence of three introns in the chick ovalbumin gene. *Nature* **278**, 370-372.

Robinson, J. H., Smith, J. A., Totty, N. F. and Riddle, P. N. (1976). Transition probability and the hormonal and density dependent regulation of cell proliferation. *Nature* **262**, 298-300.

Rochaix, J. D. (1978). Restriction endonuclease map of the chloroplast DNA of C. reinhardii. *J. Mol. Biol.* **126**, 597-618.

Rochaix, J. D. and Malnoe, P. (1978). Anatomy of the chloroplast ribosomal DNA of C. reinhardii. *Cell* **15**, 661-670.

Rochaix, J. D., Bird, A. and Bakken, A. (1974). Ribosomal RNA gene amplification by rolling circles. *J. Mol. Biol.* **87**, 473-488.

Ro Choi, T. S., Moriyama, Y., Choi, Y. C. and Busch, H. and Busch, H. (1970). Nuclear 4.5S RNA of the Novikoff hepatoma. *J. Biol. Chem.* **245,** 1970-1977.

Ro-Choi, T., Reddy, R., Henning, D., Takano, T., Taylor, C. and Busch, H. (1972). Nucleotide sequence of 4.5S RNA of hepatoma cell nuclei. *J. Biol. Chem.* **247,** 3205-3222.

Rodman, T. C. (1967). DNA replication in salivary gland nuclei of D. melanogaster at successive larval and prepupal stages. *Genetics* **55,** 375-386.

Rodriguez, R. L., Bolivar, F., Goodman, H. M., Boyer, H. W. and Betlach, M. (1976). Construction and characterization of cloning vehicles. In D. P. Nierlich, W. J. Rutter and C. F. Fox (Eds.), *Molecular Mechanisms in the Control of Gene Expression,* Academic Press, New York, pp 471-477.

Roeder, R. J. (1974). Multiple forms of DNA dependent RNA polymerase in X. laevis. I. Isolation and partial characterization. *J. Biol. Chem.* **249,** 241-248.

Roeder, R. G. (1976). Eucaryotic nuclear RNA polymerases. In R. Losick and M. Chamberlin (Eds.) *RNA Polymerase,* Cold Spring Harbor Laboratory, New York, pp 285-329.

Roeder, R. G. and Rutter, W. J. (1969). Multiple forms of DNA dependent RNA polymerase in eucaryotic organisms. *Nature* **224,** 234-237.

Roeder, R. G. and Rutter, W. J. (1970). Specific nucleolar and nucleoplasmic RNA polymerases. *Proc. Nat. Acad. Sci. USA* **65,** 675-682.

Roeder, R. G., Reeder, R. H. and Brown, D. D. (1970). Multiple forms of RNA polymerase in X. laevis: their relationship to RNA synthesis in vivo and their fidelity of transcription in vitro. *Cold Spring Harbor Symp. Quant. Biol.* **35,** 727-737.

Roewekamp, W. G., Hofer, E. and Sekeris, C. E. (1976). Translation of mRNA from rat liver polysomes into tyrosine aminotransferase and tryptophan oxygenase in a protein synthesizing system from wheat germ. *Eur. J. Biochem.* **70,** 259-268.

Romeo, G. and Migeon, B. R. (1970). Genetic inactivation of the galactosidase locus in carriers of Fabry's disease. *Science* **170,** 180-181.

Roop, D. R., Nordstrom, J. L., Tsai, S. Y., Tsai, M. J. and O'Malley, B. W. (1978). Transcription of structural and intervening sequences in the ovalbumin gene and identification of potential ovalbumin mRNA precursors. *Cell* **15,** 671-685.

Roos, V. P. (1973). Light and electron microscopy of rat kangaroo cells in mitosis. I. Formation and breakdown of the mitotic apparatus. *Chromosoma* **40,** 43-82.

Roscoe, D. H., Robinson, H. and Carbonell, A. W. (1973). DNA synthesis and mitosis in a temperature sensitive Chinese hamster cell line. *J. Cell. Physiol.* **82,** 333-338.

Rosbash, M., Campo, M. S. and Gummerson, K. S. (1975). Conservation of cytoplasmic poly(A)-containing RNA in mouse and rat. *Nature* **258,** 682-686.

Rose, K. M., Morris, H. P. and Jacob, S. T. (1975). Mitochondrial poly(A) polymerase from a poorly differentiated hepatoma: purification and characteristics. *Biochemistry* **14,** 1025-1032.

Rose, J. K. (1978). Complete sequences of the ribosome recognition sites in VSV mRNAs: recognition by the 40S and 80S complexes. *Cell* **14,** 345-353.

Rose, J. K. and Lodish, H. F. (1976). Translation in vitro of VSV mRNA lacking 5' terminal 7-methylguanosine. *Nature* **262,** 32-37.

Rosenbaum, J. L. and Child, F. M. (1967). Flagellar regeneration in protozoan flagellates. *J. Cell Biol.* **34,** 345-364.

Rosenbaum, J. L., Moulder, J. E. and Ringo, D. L. (1969). Flagellar elongation and shortening in Chlamydomonas. The use of cycloheximide and colchicine to study the synthesis and assembly of flagellar proteins. *J. Cell Biol.* **41,** 600-619.

Rosenberg, M. and Paterson, B. M. (1979). Efficient cap dependent translation of polycistronic procaryotic mRNAs is restricted to the first gene in the operon. *Nature* **279,** 696-700.

Rosenblith, J. Z., Ukena, T. E., Yin, H. H., Berlin, R. D. and Karnovsky, M. J. (1973). A comparative evaluation of the distribution of Con A binding sites on the surfaces of normal, virally transformed and protease treated fibroblasts. *Proc. Nat. Acad. Sci. USA* **70**, 1625-1629.

Rosenstraus, M. and Chasin, L. A. (1975). Isolation of mammalian cell mutants deficient in glucose-6-phosphate dehydrogenase activity: linkage to hypoxanthine phosphoribosyl transferase. *Proc. Nat. Acad. Sci. USA* **72**, 493-497.

Rosenstraus, M. J. and Chasin, L. A. (1977). Mutants of CHO cells with altered G6PDH activity. *Somatic Cell Genet.* **3**, 323-334.

Ross, A. (1968). The substructure of centriole subfibers. *J. Ult. Res.* **23**, 537-539.

Ross, J. (1976). A precursor of globin mRNA. *J. Mol. Biol.* **106**, 403-420.

Ross, J. and Knecht, D. A. (1978). Precursors of α and β globin mRNAs. *J. Mol. Biol.* **119**, 1-20.

Ross, J. and Sautner, D. (1976). Induction of globin mRNA accumulation by hemin in cultured erythroleukemic cells. *Cell* **8**, 513-520.

Ross, J., Aviv, H., Scolnick, E. and Leder, P. (1972). In vitro synthesis of DNA complementary to purified rabbit globin mRNA. *Proc. Nat. Acad. Sci. USA* **69**, 264-268.

Ross, J., Gielen, J., Packman, S., Ikawa, Y. and Leder, P. (1974). Globin gene expression in cultured erythro leukemic cells. *J. Mol. Biol.* **87**, 697-714.

Rothenberg, E., Donoghue, D. J. and Baltimore, D. (1978). Analysis of a 5' leader sequence on murine leukemia virus 21S RNA: heteroduplex mapping with long reverse transcriptase products. *Cell* **13**, 435-451.

Roufa, D. J., Sadow, B. N. and Caskey. C. T. (1973). Derivation of TK⁻ clones from revertant TK⁺ mammalian cells. *Genetics* **75**, 515-530.

Rougeon, F. and Mach, B. (1976). Stepwise biosynthesis in vitro of globin genes from globin mRNA by DNA polymerase of avian myeloblastosis virus. *Proc. Nat. Acad. Sci. USA* **73**, 3418-3422.

Rougeon, F. and Mach, B. (1977). Cloning and amplification of rabbit alpha and beta gene sequences into E. coli plasmids. *J. Biol. Chem.* **252**, 2209-2217.

Rouvière-Yaniv, J., Yaniv, M. and Germond, J. E. (1979). E. coli DNA binding protein HU forms nucleosome like structure with circular double stranded DNA. *Cell* **17**, 265-274.

Rovera, G. and Baserga. R. (1971). Early changes in the synthesis of acidic nuclear proteins in human diploid fibroblasts stimulated to synthesize DNA by changing the medium. *J. Cell. Physiol.* **77**, 201-212.

Rovera, G. and Baserga, R. (1973). Effect of nutritional changes on chromatin template activity and nonhistone chromosomal protein synthesis in WI38 and 3T6 cells. *Exp. Cell Res.* **78**, 118-126.

Rovera, G., Baserga, R. and Defendi, V. (1972). Early increase in nuclear acidic protein synthesis after SV40 infection. *Nature New. Biol.* **237**, 240-241.

Rowley, J. D. and Bodmer, W. F. (1971). Relationship of centromeric heterochromatin to the fluorescent banding patterns of metaphase chromosomes in the mouse. *Nature* **231**, 503-505.

Roy, A. K. and Zubay, G. (1966). RNA synthesis stimulated by sonicated nucleohistone. *Biochim. Biophys. Acta* **129**, 403-405.

Royal, A., Garapin, A., Cami, B., Perrin, F., Mandel, J. L., LeMeur, M., Bregegegre, F., Gannon, F., LePennec, J. P., Chambon, P. and Kourilsky, P. (1979). The ovalbumin gene region: common features in the organization of three genes expressed in chicken oviduct under hormonal control. *Nature* **279**, 125-131.

Rozek, C. E., Orr, W. C. and Timberlake, W. E. (1978). Diversity and abundance of polyadenylated RNA from Achlya ambisexualis. *Biochemistry* **17**, 716-722.

Rubenstein, J. L. R., Brutlag, D. and Clayton, D. A. (1977). The mitochondrial DNA of D. melanogaster exists in two distinct and stable superhelical forms. *Cell* **12**, 471-482.

Rubenstein, P. A. and Spudich, J. A. (1977). Actin microheterogeneity in chick embryo fibroblasts. *Proc. Nat. Acad. Sci. USA* **74**, 120-123.

Rubin, C. S., Dancis, J., Yip, Y. C., Nowinski, R. C. and Balis, M. E. (1971). Purification of IMP: pyrophosphate phosphoribosyl transferases, catalytically incompetent enzymes in Lesch-Nyhan diseases. *Proc. Nat. Acad. Sci. USA* **68**, 1461-1464.

Rubin, G. M. (1973). The nucleotide sequence of S. cerevisiae 5.8S rRNA. *J. Biol. Chem.* **248**, 3860-3875.

Rubin, G. M. (1974). Three forms of the 5.8S rRNA species in S. cerevisiae. *Eur. J. Biochem.* **41**, 197-202.

Rubin, G. M. and Hogness, D. S. (1975). Effect of heat shock on the synthesis of low molecular weight RNAs in Drosophila: accumulation of a novel form of 5S RNA. *Cell* **6**, 207-214.

Rubin, G. M. and Sulston, J. E. (1973). Physical linkage of the 5S cistrons to the 18S and 28S rRNA cistrons in S. cerevisiae. *J. Mol. Biol.* **79**, 521-530.

Rubin, M. S. and Tzagoloff, A. (1973). Assembly of the mitochondrial membrane systems. X. Mitochondrial synthesis of three of the subunit proteins of yeast cytochrome oxidase. *J. Biol. Chem.* **248**, 4275-4279.

Rubin, R. L. and Moudrianakis, E. N. (1975). The f3-f2a1 complex as a unit in the self assembly of nucleoproteins. *Biochemistry* **14**, 1718-1727.

Rubin, R. W., Goldstein, L. and Ko, C. (1978). Differences between nucleus and cytoplasm in the degree of actin polymerization. *J. Cell Biol.* **77**, 698-701.

Rubinstein, L. and Clever, U. (1972). Chromosome activity and cell function in polytenic cells. V. Developmental changes in RNA synthesis and turnover. *Devel. Biol.* **27**, 519-537.

Ruckert, J. (1882). Zur entwickelungsgeschichte des ovarialeies bei selachiern. *Anat. Anz.* **7**, 107-158.

Ruddle, F. H. (1973). Linkage analysis on man by somatic cell genetics. *Nature* **242**, 165-169.

Ruddle, F. and Creagan, R. P. (1975). Parasexual approaches to the genetics of man. *Ann. Rev. Genet.* **9**, 407-486.

Ruddle, F. H., Chapman, V. M., Ricciuti, F., Murnane, M., Klebe, R. and Khan, P. M. (1971). Linkage relationships of seventeen human gene loci as determined by man mouse somatic cell hybrids. *Nature New Biol.* **232**, 69-73.

Ruderman, J. V. and Gross, P. R. (1974). Histones and histone synthesis in sea urchin development. *Devel. Biol.* **36**, 286-298.

Ruderman, J. V. and Pardue, M. L. (1978). A portion of all major classes of histone mRNA in amphibian oocytes is polyadenylated. *J. Biol. Chem.* **253**, 2018-2026.

Ruderman, J. V., Baglioni, C. and Gross, P. R. (1974). Histone mRNA and histone synthesis during embryogenesis. *Nature* **247**, 36-38.

Rudkin, G. T. (1969). Nonreplicating DNA in Drosophila. *Genetics* **61**, *suppl.* 227-238.

Ruiz-Carrillo, A. and Jorcano, J. L. (1979). An octamer of core histones in solution: central role of the H3-H4 tetramer in self assembly. *Biochemistry* **18**, 760-767.

Ruiz-Carrillo, A., Wangh, L. J. and Allfrey, V. G. (1975). Processing of newly synthesized histone molecules. *Science* **190**, 117-127.

Runge, M. S., Detrich, H. W. and Williams, R. C. (1979). Identification of the major 68,000 dalton protein of microtubule preparations as a 10 nm filament protein and its effects on microtubule assembly in vitro. *Biochemistry* **18**, 1689-1697.

Russell, L. B. (1961). Genetics of mammalian sex chromosomes. *Science* **133**, 1795-1803.

Russell, L. B. (1963). Mammalian X chromosome action: inactivation limited in spread and in region of origin. *Science* **140**, 976-978.

Russell, L. B. (1964). Another look at the single active X hypothesis. *Trans. N. Y. Acad. Sci.* **26**, 726-736.

Russell, L. B. and Montgomery, C. S. (1970). Comparative studies in X autosome translocation in the mouse. II. Inactivation of autosomal loci, segregation, and mapping of autosomal breakpoints in five T(X:1)'S'. *Genetics* **64**, 281-312.

Russell, W. L. (1967). Repair mechanisms in radiation mutation in the mouse. *Brookhaven Symp. Biol.* **20**, 179-189.

Russell, W. L., Russell, L. B. and Kelly, E. M. (1958). Radiation dose rate and mutation frequency. *Science* **128**, 1546-1550.

Russev, G. and Tsanev, R. (1979). Nonrandom segregation of histones during chromatin replication. *Eur. J. Biochem.* **93**, 123-146.

Ryan, R., Grant, D., Chiang, K. S. and Swift, H. (1978). Isolation and characterization of mitochondrial DNA from C. reinhardii. *Proc. Nat. Acad. Sci. USA* **75**, 3268-3272.

Ryffel, G. U. and McCarthy, B. J. (1975). Complexity of cytoplasmic RNA in different mouse tissues measured by hybridization of polyadenylated RNA to complementary DNA. *Biochemistry* **14**, 1379-1385.

Sager, R. (1954). Mendelian and non Mendelian inheritance of streptomycin resistance in C. reinhardii. *Proc. Nat. Acad. Sci. USA* **40**, 356-363.

Sager, R. (1972). *Cytoplasmic Genes and Organelles.* Academic Press, New York.

Sager, R. (1977). Genetic analysis of chloroplast DNA in Chlamydomonas. *Adv. Genet.* **19**, 287-340.

Sager, R. and Ishida, M. R. (1963). Chloroplast DNA in Chlamydomonas. *Proc. Nat. Acad. Sci. USA* **50**, 725-730.

Sager, R. and Kitchin, R. (1975). Selective silencing of eucaryotic DNA. *Science* **189**, 426-433.

Sager, R. and Lane, D. (1972). Molecular basis of maternal inheritance. *Proc. Nat. Acad. Sci. USA* **69**, 2410-2413.

Sager, R. and Ramanis, Z. (1963). The particulate nature of nonchromosomes genes in Chlamydomonas. *Proc. Nat. Acad. Sci. USA* **50**, 260-268.

Sager, R. and Ramanis, Z. (1965). Recombination of nonchromosomal genes in Chlamydomonas. *Proc. Nat. Acad. Sci. USA* **53**, 1053-1061.

Sager, R. and Ramanis, Z. (1967). Biparental inheritance of nonchromosomal genes induced by UV irradiation. *Proc. Nat. Acad. Sci. USA* **58**, 931-937.

Sager, R. and Ramanis, Z. (1968). The pattern of segregation of cytoplasmic genes in Chlamydomonas. *Proc. Nat. Acad. Sci. USA* **61**, 324-331.

Sager, R. and Ramanis, Z. (1970). A genetic map of non Mendelian genes in Chlamydomonas. *Proc. Nat. Acad. Sci. USA* **65**, 593-600.

Sager, R. and Ramanis, Z. (1974). Mutations that alter the transmission of chloroplast genes in Chlamydomonas. *Proc. Nat. Acad. Sci. USA* **71**, 4698-4702.

Sager, R. and Ramanis, Z. (1976a). Chloroplast genetics of Chlamydomonas. I. Allelic segregation ratio. *Genetics* **83**, 303-321.

Sager, R. and Ramanis, Z. (1976b). Chloroplast genetics of Chlamydomonas. II. Mapping by cosegregation frequency analysis. *Genetics* **83**, 323-340.

Saglio, G., Ricco, G., Mazza, U., Camaschella, C., Pich, P. G., Gianni, A. M., Gianazza, E., Righetti, P. G., Giglioni, B., Comi, P., Gusmeroli, M. and Ottolenghi, S. (1979). Human γ^T globin chain is a variant of γ^A chain. *Proc. Nat. Acad. Sci. USA* **76**, 3420-3424.

Sahar, E. and Latt, S. A. (1978). Enhancement of banding patterns in human metaphase chromosomes by energy transfer. *Proc. Nat. Acad. Sci. USA* **75**, 5650-5654.

Sakai, H. (1966). Studies on sulphydryl groups during cell division of sea urchin eggs. VIII. Some properties of mitotic apparatus proteins. *Biochim. Biophys. Acta* **112**, 132-145.

Sakonju, S., Bogenhagen, D. and Brown, D. D. (1980). A control region in the center of the 5S RNA gene directs specific initiation of transcription. II. The 5′ border of the region. *Cell* **19**, 13-25.

Sakano, H., Rogers, J. H., Huppi, K., Brack, C., Traunecker, A., Maki, R., Wall, R. and Tonegawa, S. (1979a). Domains and the hinge region of an immunoglobulin heavy chain are encoded in separate DNA segments. *Nature* **277**, 627-632.

Sakano, H., Hupi, K., Heinrich, G. and Tonegawa, S. (1979b). Sequences at the somatic recombination sites of immunoglobulin light chain genes. *Nature* **280**, 288-293.

Salditt-Georgieff, M., Jelinek, W., Darnell, J. E., Furuichi, Y., Morgan, M. and Shatkin, A. (1976). Methyl labeling of HeLa cell hnRNA: a comparison with mRNA. *Cell* **7**, 227-238.

Salmon, E. D. (1975a). Pressure induced depolymerization of spindle microtubules. I. Changes in birefringence and spindle length. *J. Cell Biol.* **65**, 603-614.

Salmon, E. D. (1975b). Pressure induced depolymerization of spindle microtubules. II. Thermodynamics of in vivo spindle assembly. *J. Cell Biol.* **66**, 114-127.

Salmon, E. D. (1975c). Pressure induced depolymerization of brain microtubules in vitro. *Science* **189**, 884-886.

Salser, W., Bowen, S., Browne, D., El Adli, F., Federoff, N., Fry, K., Heindell, H., Paddock, G., Poon, R., Wallace, B. and Whitcome, P. (1976). Investigation of the organization of mammalian chromosomes at the DNA sequence level. *Fed. Proc.* **35**, 23-35.

Sampson, J., Matthews, M. B., Osborn, M. and Borghetti, A. F. (1972). Hemoglobin mRNA translation in cell free system from rat and mouse liver and Landschutz ascites cells. *Biochemistry* **11**, 3636-3640.

Sanders, J. P. M. and Borst, P. (1977). The organization of genes in yeast mitochondrial DNA. IV. Analysis of dAdT clusters in yeast mitochondrial DNA by poly-U sephadex chromatography. *Mol. Gen. Genet.* **157**, 263-280.

Sanders, J. P. M., Flavell, R. A., Borst, P. and Mol, J. N. M. (1973). Nature of the base sequence conserved in the mitochondrial DNA of a low density petite. *Biochim. Biophys. Acta* **312**, 441-457.

Sanders, J. P. M., Borst, P. and Weijers, P. J. (1975a). The organization of genes in yeast mitochondrial DNA. II. The physical map of Eco RI and Hind II + III fragments. *Mol. Gen. Genet.* **143**, 53-64.

Sanders, J. P. M., Heyting, C. and Borst, P. (1975b). The organization of genes in yeast mitochondria DNA. I. The genes for large and small rRNA are far apart. *Biochem. Biophys. Res. Commun.* **65**, 699-707.

Sanders, J. P. M., Heyting, C., Verbeet, M. P., Meijlink, F. C. P. W. and Borst, P. (1977). The organization of genes in yeast mitochondrial DNA. III. Comparison of the physical maps of the mitochondrial DNAs from three wild type Saccharomyces strains. *Mol. Gen. Genet.* **157**, 239-261.

Sanders, L. A. (1974). Isolation and characterization of the nonhistone chromosomal proteins of developing avian erythroid cells. *Biochemistry* **13**, 527-534.

Sanders, L. A. and McCarty, K. S. (1972). Isolation and purification of histones from avian erythrocytes. *Biochemistry* **11**, 4216-4221.

Sandler, L. and Szauter, P. (1978). The effect of recombination defective meiotic mutants on fourth chromosome crossing over in D. melanogaster. *Genetics* **90**, 699-712.

Sandler, L., Hiraizumi, Y. and Sandler, I. (1957). Meiotic drive in natural populations of D. melanogaster. I. The cytogenetic basis of segregation distortion. *Genetics* **44**, 233-250.

Sandler, L., Lindsley, D. L., Nicoletti, B. and Trippa, G. (1968). Mutants affecting meiosis in natural populations of D. melanogaster. *Genetics* **60**, 525-558.

Sanford, K. K., Earle, W. R. and Likely, G. D. (1948). The growth in vitro of single isolated tissue culture cells. *J. Nat. Cancer Inst.* **9**, 229-246.

Sanger, F. and Coulson, A. R. (1975). A rapid method for determining sequences in DNA by primed synthesis with DNA polymerase. *J. Mol. Biol.* **94**, 441-448.

Sanger, F., Nicklen, S. and Coulson, A. R. (1977). DNA sequencing with chain terminating inhibitors. *Proc. Nat. Acad. Sci. USA* **74**, 5463-5467.

Sanger, F., Coulson, A. R., Kossel, H. and Fischer, D. (1973). Use of DNA polymerase I primed by a synthetic oligonucleotide to determine a nucleotide sequence in phage f1 DNA. *Proc. Nat. Acad. Sci. USA* **70**, 1209-1213.

Sanger, J. W. (1975a). Changing patterns of actin localization during cell division. *Proc. Nat. Acad. Sci. USA* **72**, 1913-1916.

Sanger, J. W. (1975b). Intracellular localization of actin with fluorescently labelled heavy meromyosin. *Cell Tissue Res.* **161**, 431-444.

Sanger, J. W. (1975c). Presence of actin during chromosomal movement. *Proc. Nat. Acad. Sci.* **72**, 2451-2455.

Satlin, A., Kucherlapati, R. and Ruddle, F. H. (1975). Assignment of the gene for human UMPK to chromosome 1 using somatic cell hybrid clone panels. *Cytogenet. Cell Genet.* **15**, 146-152.

Sato, H., Ellis, G. W. and Inoué, S. (1975). Microtubular origin of mitotic spindle form birefringence. Demonstration of the applicability of Wiener's equation. *J. Cell Biol.* **67**, 501-517.

Sato, K. (1975). A leukemic cell mutant with a thermolabile ala-tRNA synthetase. *Nature* **257**, 813-814.

Sato, K., Slesinski, R. S. and Littlefield, J. W. (1972). Chemical mutagenesis at the phosphoribosyltransferase locus in cultured human lymphoblasts. *Proc. Nat. Acad. Sci. USA* **69**, 1244-1248.

Sautière, P., Breynaert, M. D., Moschetto, Y. and Biserte, G. (1970). Sequence complete des acides amines de l'histone riche en glycine et en arginine de thymus du porc. *C. R. Acad. Sci. Paris* **271**, 364-365.

Sautière, P., Tyrou, D., Moschetto, Y. and Biserte, G. (1971). Structure primaire de l'histone riche en glycine et en arginine isolee de la tumeur de chloroleucemie du rat. *Biochimie* **53**, 479-483.

Sautière, P., Briand, G., Kmiecik, D., Loy, O., Biserte, G., Garel, A. and Champagne, M. (1976). Chicken erythrocyte histone 5. III. Sequence of the amino terminal half of the molecule (111 residues). *Febs Lett.* **63**, 164-166.

Sawicki, S. G., Jelinek, W. and Darnell, J. E. (1977). 3′ terminal addition to HeLa cell nuclear and cytoplasmic poly(A). *J. Mol. Biol.* **113**, 219-236.

Schachat, F. H. and Hogness, D. S. (1973). Repetitive sequences in isolated Thomas circle from D. melanogaster. *Cold Spring Harbor Symp. Quant. Biol.* **38**, 371-381.

Schachner, M., Hedley-Whyte, E. T., Hsu, D. W., Schoonmaker, G. and Bignami, A. (1977). Ultrastructural localization of glian fibrillary acidic protein in mouse cerebellum by immunoperoxidase labeling. *J. Cell Biol.* **75**, 67-73.

Schafer, K. P. and Kuntzel, H. (1972). Mitochondrial genes in Neurospora: a single cistron for rRNA. *Biochem. Biophys. Res. Commun.* **46**, 1312-1319.

Schafer, K. P., Bugge, G., Grandi, M. and Kuntzel, H. (1971). Transcription of mitochondrial DNA in vitro from N. crassa. *Eur. J. Biochem.* **21**, 478-488.

Schaffhausen, B. S. and Benjamin, T. L. (1976). Deficiency in histone acetylation in nontransforming host range mutants of polyoma virus. *Proc. Nat. Acad. Sci. USA* **73**, 1092-1096.

Schaffner, W., Gross, K., Telford, J. and Birnstiel, M. (1976). Molecular analysis of the histone gene cluster of P. miliaris. II. The arrangement of the five histone coding and spacer sequences. *Cell* **8**, 471-478.

Schaffner, W., Kunz, G., Daetwyler, H., Telford, J., Smith, H. O. and Birnstiel, M. L. (1978). Genes and spacers of cloned sea urchin histone DNA analyzed by sequencing. *Cell* **14**, 655-673.

Schalet, A. and Lefevre, G., jr. (1976). The proximal region of the X chromosome. In M. Ashburner and E. Novitski (Eds.), *The Genetics and Biology of Drosophila*, Academic Press, London, pp 848-902.

Scharff, M. D. and Robbins, E. (1966). Polyribosome disaggregation during metaphase. *Science* **151**, 992-995.

Schatz, G. and Mason, T. L. (1974). The biosynthesis of mitochondrial proteins. *Ann. Rev. Biochem.* **43**, 51-87.

Schechter, I. (1974). Use of antibodies for the isolation of biologically pure mRNA from fully functional eucaryotic cells. *Biochemistry* **13**, 1875-1885.

Schedl, P., Artavanis-Tsakonas, S., Steward, R., Gehring, W. J., Mirault, M. E., Goldschmidt-Clermont, M., Moran, L. and Tissieres, A. (1978). Two hybrid plasmids with D. melanogaster sequences complementary to mRNA coding for the major heat shock protein. *Cell* **14**, 921–929.

Scheer, U. (1978). Changes of nucleosome frequency in nucleolar and non nucleolar chromatin as a function of transcription: an electron microscopic study. *Cell* **13**, 535–549.

Scheller, R. H., Dickerson, R. E., Boyer, H. W., Riggs, A. D. and Itakura, K. (1977). Chemical synthesis of restriction enzyme recognition sites useful for cloning. *Science* **196**, 177–180.

Scheller, R. H., Constantini, F. D., Kozlowski, M. R., Britten, R. J. and Davidson, E. H. (1978). Specific representation of cloned repetitive DNA sequences in sea urchin RNAs. *Cell* **15**, 189–203.

Schendl, W. (1971). Analysis of the human karyotype using a reassociation technique. *Chromosoma* **34**, 448–454.

Scherrer, K., Spohr, G., Granboulan, N., Morel, C., Grosclaude, J. and Chezzi, C. (1970). Nuclear and cytoplasmic messenger like RNA and their relation to the active messenger RNA in polyribosomes of HeLa cells. *Cold Spring Harbor Symp. Quant. Biol.* **35**, 539–554.

Schibler, U. and Perry, R. P. (1976). Characterization of the 5′ termini of hnRNA in mouse L cells: implications for processing and cap formation. *Cell* **9**, 121–130.

Schibler, U. and Perry, R. P. (1977). The 5′ termini of hnRNA: a comparison among molecules of different sizes and ages. *Nucl. Acids Res.* **4**, 4133–4150.

Schibler, U., Kelley, D. E. and Perry, R. P. (1977). Comparison of methylated sequences in mRNA and hnRNA from mouse L cells. *J. Mol. Biol.* **115**, 695–714.

Schibler, U., Marcu, K. B. and Perry, R. P. (1978). The synthesis and processing of the mRNAs specifying heavy and light chain immunoglobulins in MPC11 cells. *Cell* **15**, 1495–1510.

Schibler, U., Hagenbuchle, O., Wyler, T., Weber, R., Boseley, P., Telford, J. and Birnstiel, M. L. (1976). The arrangement of 18S and 28S rRNAs within the 40S precursor molecule of X. laevis. *Eur. J. Biochem.* **68**, 471–480.

Schindler, R., Ramseier, L., Schaer, J. C. and Grieder, A. (1970). Studies on the division cycle of mammalian cells. III. Preparation of synchronously dividing cell populations by isotonic sucrose gradient centrifugation. *Exp. Cell Res.* **59**, 90–96.

Schlaeger, E. J. and Knippers, R. (1979). DNA histone interaction in the vicinity of replication points. *Nucleic Acids Res.* **6**, 645–656.

Schlaepfer, W. W. and Freeman, L. A. (1978). Neurofilament proteins of rat peripheral nerve and spinal cord. *J. Cell Biol.* **78**, 653–662.

Schlegel, R. A. and Rechsteiner, M. C. (1975). Microinjection of thymidine kinase and BSA into mammalian cells by fusion with red blood cells. *Cell* **5**, 371–379.

Schloss, J. A., Milsted, A. and Goldman, R. D. (1977). Myosin subfragment binding for the localization of actin-like microfilaments in cultured cells. A light and electron microscope study. *J. Cell Biol.* **74**, 794–815.

Schmid, C. W. and Deininger, P. L. (1975). Sequence organization of the human genome. *Cell* **6**, 345–358.

Schmid, C. W., Manning, J. E. and Davidson, N. (1975). Inverted repeat sequences in the Drosophila genome. *Cell* **5**, 159–172.

Schmidt, G. and Tannhauser, S. J. (1945). A method for the detection of DNA, RNA and phosphorylated proteins in animal tissues. *J. Biol. Chem.* **161**, 83–89.

Schmitt, H. and Atlas, D. (1976). Specific affinity labelling of tubulin with bromocolchicine. *J. Mol. Biol.* **102**, 743–758.

Schneeberger, E. E. and Harris, H. (1966). An ultrastructural study of interspecific cell fusion induced by inactivated Sendai virus. *J. Cell Sci.* **1**, 401–406.

Schneider, I. (1972). Cell lines derived from late embryonic stages of D. melanogaster. *J. Emb. Exp. Morphol.* **27**, 353–365.

Schneider, J. A. and Weiss, M. C. (1971). Expression of differentiated functions in hepatoma cell hybrids. I. TAT in hepatoma-fibroblast hybrids. *Proc. Nat. Acad. Sci. USA* **68**, 127–131.

Schneller, J. M., Faye, G., Kujawa, C. and Stahl, A. J. C. (1975). Number of genes and base composition of mt DNA from S. cerevisiae. *Nucl. Acids Res.* **2**, 831–838.

Schochetman, G. and Perry, R. P. (1972). Characterization of the mRNA released from L cell polyribosomes as a result of temperature shock. *J. Mol. Biol.* **63**, 577–590.

Schor, S. L., Johnson, R. T. and Mullinger, A. M. (1975). Perturbation of mammalian cell division. II. Studies on the isolation and characterization of human mini segregant cells. *J. Cell Sci.* **19**, 281–303.

Schreck, R. R., Warburton, D., Miller, O. J., Beiser, S. M. and Erlanger, B. F. (1973). Chromosome structure as revealed by a combined chemical and immunochemical procedure. *Proc. Nat. Acad. Sci. USA* **70**, 804–807.

Schreck, R. R., Erlanger, B. F. and Miller, O. J. (1974). The use of antinucleoside antibodies to probe the organization of chromosomes denatured by UV irradiation. *Exp. Cell Res.* **88**, 31–39.

Schreck, R. R., Dev, V. G., Erlanger, B. F. and Miller, O. J. (1977). The structural organization of mouse metaphase chromosomes. *Chromosoma* **62**, 337–350.

Schreier, M. H. and Staehelin, T. (1973). Initiation of mammalian protein synthesis: the importance of ribosomes and initiation factor quality for the efficiency of in vitro systems. *J. Mol. Biol.* **73**, 329–350.

Schroeder, T. E. (1970). The contractile ring. I. Fine structure of dividing mammalian (HeLa) cells and the effects of cytochalasin B. *Z. Zellforsch.* **109**, 431–449.

Schroeder, T. E. (1973). Actin in dividing cells: contractile ring filaments bind heavy meromyosin. *Proc. Nat. Acad. Sci. USA* **70**, 1688–1692.

Schroeder, W. A., Huisman, T. H. J., Shelton, R., Shelton, J. B., Klaihauer, E. F., Dozy, A. M. and Robberson, B. (1968). Evidence for multiple structural genes for the γ chain of human fetal hemoglobin. *Proc. Nat. Acad. Sci. USA* **60**, 537–544.

Schultz, L. D. and Hall, B. D. (1976). Transcription in yeast: α-amanitin sensitivity and other properties which distinguish between RNA polymerases I and III. *Proc. Nat. Acad. Sci. USA* **73**, 1029–1033.

Schutz, G., Killewich, L., Chen, G. and Feigelson, P. (1975). Control of the mRNA for hepatic tryptophan oxygenase during hormonal and substrate induction. *Proc. Nat. Acad. Sci. USA* **72**, 1017–1020.

Schvartzman, J. B. and Cortes, F. (1977). Sister chromatid exchange in Allium cepa. *Chromosoma* **62**, 119–131.

Schwartz, A. G., Cook, P. R. and Harris, H. (1971). Correction of a genetic defect in a mammalian cell. *Nature New Biol.* **230**, 5–8.

Schwartz, H. and Darnell, J. E. (1976). The association of protein with the poly(A) of HeLa cell mRNA: evidence for a transport role of a 75,000 molecular weight polypeptide. *J. Mol. Biol.* **104**, 833–852.

Schwartz, L. B. and Roeder, R. G. (1974). Purification and subunit structure of DNA dependent RNA polymerase I from the mouse myeloma, MOPC 315. *J. Biol. Chem.* **249**, 5898–5906.

Schwartz, L. B. and Roeder, R. G. (1975). Purification and subunit structure of DNA dependent RNA polymerase II from the mouse plasmacytoma, MOPC 315. *J. Biol. Chem.* **250**, 3221–3228.

Schwartz, L. B., Sklar, V. E. F., Jaehning, J. A., Weinmann, R. and Roeder, R. G. (1974). Isolation and partial characterization of the multiple forms of DNA-dependent RNA polymerase in the mouse myeloma, MOPC 315. *J. Biol. Chem.* **249**, 5889–5897.

Schwartz, S. A., Panem, S. and Kirsten, W. H. (1975). Distribution and virogenic effects of BUdR in synchronized rat embryo cells. *Proc. Nat. Acad. Sci. USA* **72**, 1829–1833.

Schwartzbach, S. D., Hecker, L. I. and Barnett, W. E. (1976). Transcriptional origin of Euglena chloroplast tRNAs. *Proc. Nat. Acad. Sci. USA* **73**, 1984–1988.

Schweizer, E., MacKechnie, C. and Halvorson, H. D. (1969). The redundancy of ribosomal and transfer RNA genes in S. cerevisiae. *J. Mol. Biol.* **40**, 261-278.

Scott, A. C. and Wells, J. R. E. (1976). Reiteration frequency of the gene for tissue specific histone H5 in the chicken genome. *Nature* **259**, 635-639.

Scott, N. S. (1976). Precursors of chloroplast rRNA in Euglena gracilis. *Phytochemistry* **15**, 1207-1213.

Scott, S. EM. and Sommerville, J. (1974). Location of nuclear proteins on the chromosomes of newt oocytes. *Nature* **250**, 680-682.

Scott, W. A. and Wigmore, D. J. (1978). Sites in SV40 chromatin which are preferentially cleaved by endonucleases. *Cell* **15**, 1511-1518.

Seabright, M. (1972). The use of proteolytic enzymes for the mapping of structural rearrangements in the chromosomes of man. *Chromosoma* **36**, 204-210.

Seale, R. L. (1976a). Studies on the mode of segregation of histone nu bodies during replication in HeLa cells. *Cell* **9**, 423-430.

Seale, R. L. (1976b). Temporal relationships of chromatin protein synthesis, DNA synthesis, and assembly of deoxyribonucleoprotein. *Proc. Nat. Acad. Sci. USA* **73**, 2270-2274.

Seale, R. L. (1978). Nucleosomes associated with newly replicated DNA have an altered conformation. *Proc. Nat. Acad. Sci. USA* **75**, 2717-2721.

Seale, R. L. and Aronson, A. I. (1973). Chromatin associated proteins of the developing sea urchin embryo. II. Acid soluble proteins. *J. Mol. Biol.* **75**, 647-658.

Seale, R. L. and Simpson, R. T. (1975). Effects of cycloheximide on chromatin biosynthesis. *J. Mol. Biol.* **94**, 479-501.

Sealy, L. and Chalkley, R. (1978a). The effect of sodium butyrate on histone modification. *Cell* **14**, 115-122.

Sealy, L. and Chalkley, R. (1978b). DNA associated with hyperacetylated histone is preferentially digested by DNAase I. *Nucl. Acids Res.* **5**, 1863-1876.

Sears, E. R. (1976). Genetic control of chromosome pairing in wheat. *Ann. Rev. Genet.* **10**, 31-52.

Sebald, W. (1977). Biogenesis of mitochondrial ATPase. *Biochim. Biophys. Acta* **463**, 1-27.

Sederoff, R., Lowenstein, L. and Birnboim, H. C. (1975). Polypyrimidine segments in D. melanogaster DNA. II. Chromosome location and nucleotide sequence. *Cell* **5**, 183-194.

Sedwick, W. D., Wang, T. S. F. and Korn, D. (1975). Cytoplasmic DNA polymerase. Structure and properties of the highly purified enzyme from human KB cells. *J. Biol. Chem.* **250**, 7045-7056.

Seeburg, P. H., Shine, J., Martial, J., Ullrich, A., Baxter, J. D. and Goodman, H. M. (1977a). Nucleotide sequence of part of the gene for human chorionic somatomammotropin: purification of DNA complementary to predominant mRNA species. *Cell* **12**, 157-165.

Seeburg, P. H., Shine, J., Martial, J. A., Baxter, J. D. and Goodman, H. M. (1977b). Nucleotide sequence and amplification in bacteria of structural gene for rat growth hormone. *Nature* **270**, 486-494.

Seeburg, P. H., Shine, J., Martial, J. A., Ivarie, R. D., Morris, J. A., Ullrich, A., Baxter, J. D. and Goodman, H. M. (1978). Synthesis of growth hormone by bacteria. *Nature* **276**, 795-798.

Seegmiller, J. E., Rosenbloom, F. M. and Kelley, W. N. (1967). Enzyme defect associated with a sex linked human neurological disorder and excessive purine biosynthesis. *Science* **155**, 1682-1684.

Segal, S., Levine, A. J. and Khoury, G. (1979). Evidence for non spliced SV40 RNA in undifferentiated murine teratocarcinoma stem cells. *Nature*, **280**, 335-337.

Sehgal, P. B., Darnell, J. E. and Tamm, I. (1976). The inhibition by DRB of hnRNA and mRNA production in HeLa cells. *Cell* **9**, 473-480.

Sehgal, P. B., Fraser, N. W. and Darnell, J. E. (1979). Early ad-2 transcription units: only proximal RNA continues to be made in the presence of DRB. *Virology* **94**, 185-191.

Sehgal, P. B., Soreq, H. and Tamm, I. (1978). Does 3′ terminal poly(A) stabilize human fibroblast interferon mRNA in oocytes of X. laevis. *Proc. Nat. Acad. Sci. USA* **75,** 5030-5033.

Seidman, M. M. and Cole, R. D. (1977). Chromatin fractionation related to cell type and chromosome condensation but perhaps not to transcriptional activity. *J. Biol. Chem.* **252,** 2630-2639.

Seidman, J. G. and Leder, P. (1978). The arrangement and rearrangement of antibody genes. *Nature* **276,** 790-794.

Seidman, J. G., Leder, A., Edgell, M. H., Polsky, F., Tilghman, S. M., Tiemeier, D. C. and Leder, P. (1978a). Multiple related immunoglobulin variable region genes identified by cloning and sequence analysis. *Proc. Nat. Acad. Sci. USA* **75,** 3881-3885.

Seidman, J. G., Leder, A., Nau, M., Norman, B. and Leder, P. (1978b). Antibody diversity. *Science* **202,** 11-16.

Seitz-Mayr, G., Wolf, K. and Kaudewitz, F. (1978). Extrachromosomal inheritance in S. pombe. VII. Studies by zygote clone analysis on transmission, segregation, recombination and uniparental inheritance of mitochondrial markers conferring resistance to antimycin, chloramphenicol and erythromycin. *Molec. Gen. Genet.* **164,** 309-320.

Selker, E. and Yanofsky, C. (1979). Nucleotide sequence and conserved features of the 5.8S rRNA coding region of N. crassa. *Nucl. Acids Res.* **6,** 2561-2568.

Selman, G. G. and Perry, M. M. (1970). Ultrastructural changes in the surface layers of the newt's egg in relation to the mechanism of its cleavage. *J. Cell Sci.* **6,** 207-227.

Selsing, E., Wells, R. D., Alden, C. J. and Arnott, S. (1979). Bent DNA: visualization of a base paired and stacked A-B conformation junction. *J. Biol. Chem.* **254,** 5417-5422.

Senior, M. B., Olins, A. D. and Olins, D. E. (1975). Chromatin fragments resembling nu bodies. *Science* **187,** 173-175.

Setterfield, G., Sheinin, R., Dardick, I., Kiss, G. and Dubsky, M. (1978). Structure of interphase nuclei in relation to the cell cycle. Chromatin organization in mouse L cells temperature sensitive for DNA replication. *J. Cell Biol.* **77,** 246-263.

Sevall, J. S., Cockburn, A., Savage, M. and Bonner, J. (1975). DNA protein interactions of the rat liver nonhistone chromosomal proteins. *Biochemistry* **14,** 782-789.

Seyedin, S. M. and Kistler, W. S. (1979a). H1 subfractions of mammalian testes. 1. Organ specificity in the rat. *Biochemistry* **18,** 1371-1375.

Seyedin, S. M. and Kistler, W. S. (1979b). H1 subfractions of mammalian tests. 2. Organ specificity in mice and rabbits. *Biochemistry* **18,** 1376-1379.

Sgaramella, V., Van de Sande, J. H. and Khorana, H. G. (1970). Studies on polynucleotides. C. A novel joining reaction catalyzed by the T4 polynucleotide ligase. *Proc. Nat. Acad. Sci. USA* **67,** 1468-1475.

Shafritz, D. A., Weinstein, J. A., Safer, B., Merrick, W. C., Weber, L. A., Hickey, E. D. and Baglioni, C. (1976). Evidence for role of $m^7G^{5'}$ phosphate group in recognition of eucaryotic mRNA by initiation factor IF-M$_3$. *Nature* **261,** 291-294.

Shall, S. (1973). Selection synchronization by velocity sedimentation separation of mouse fibroblast cells grown in suspension culture. *Methods Cell Biol.* **7,** 269-285.

Shannon, M. P., Kaufman, T. C., Shen, M. W. and Judd, B. H. (1972). Lethality patterns and morphology of selected lethal and semi lethal mutations in the zest-white region of D. melanogaster. *Genetics* **72,** 615-638.

Shapiro, D. J. and Baker, H. J. (1977). Purification and characterization of X. laevis vitellogenin mRNA. *J. Biol. Chem.* **252,** 5244-5250.

Shapiro, D. J. and Schimke, R. T. (1975). Immunochemical isolation and characterization of ovalbumin mRNA. *J. Biol. Chem.* **250,** 1759-1765.

Shapiro, H. S. (1968). DNA content per cell of various organisms. In H. A. Sober (Ed.), *Handbook of Biochemistry*, Chemical Rubber Co, Cleveland, Ohio, pp H52-H61.

Sharman, G. B. (1971). Late DNA replication in the paternally derived X chromosomes of female kangaroos. *Nature* **230**, 231-232.

Sharp, J. D., Capecchi, N. E. and Capecchi, M. R. (1973). Altered enzymes in drug resistant variants of mammalian tissue culture cells. *Proc. Nat. Acad. Sci. USA* **70**, 3145-3149.

Sharp, P. A., Sugden, B. and Sambrook, J. (1973). Detection of two restriction endonuclease activities in Haemophilus parainfluenzae using analytical agarose-ethidium bromide electrophoresis. *Biochemistry* **12**, 3055-3063.

Shatkin, A. J. (1974). Methylated mRNA synthesis in vitro by purified reovirus. *Proc. Nat. Acad. Sci. USA* **71**, 3204-3207.

Shatkin, A. J. (1976). Capping of eucaryotic mRNAs. *Cell* **9**, 645-653.

Shaw, B. R., Corden, J. L., Sahasrabuddhe, C. G. and Van Holde, K. E. (1974). Chromatographic separation of chromatin subunits. *Biochem. Biophys. Res. Commun.* **61**, 1193-1198.

Shaw, B. R., Herman, T. M., Kovacic, R. T., Beaudreau, G. S. and Van Holde, K. E. (1976). Analysis of subunit organization in chicken erythrocyte chromatin. *Proc. Nat. Acad. Sci. USA* **73**, 505-509.

Shaw, D. D. (1972). Genetic and environmental components of chiasma control. II. The response to selection in Schistocerca. *Chromosoma* **37**, 297-308.

Shaw, D. D. (1974). Genetic and environmental components of chiasma control. III. Genetic analysis of chiasma frequency in two selected lines of Schistocerca gregaria Forsk. *Chromosoma* **46**, 365-374.

Shaw, D. D. and Knowles, G. R. (1976). Comparative chiasma analysis using a computerized optical digitiser. *Chromosoma* **59**, 103-128.

Shaw, L. M. J. and Huang, R. C. C. (1970). A description of two procedures which avoid the use of extreme pH conditions for the resolution of components isolated from chromatins prepared from pig cerebellar and pituitary nuclei. *Biochemistry* **9**, 4530-4541.

Shay, J. W. (1977). Selection of reconstituted cells from karyoplasts fused to chloramphenicol resistant cytoplasts. *Proc. Nat. Acad. Sci. USA* **74**, 2461-2464.

Shay, J. W., Porter, K. R. and Prescott, D. M. (1974). The surface morphology and fine structure of CHO (Chinese hamster ovary) cells following enucleation. *Proc. Nat. Acad. Sci. USA* **71**, 3059-3063.

Shay, J. W., Gershenbaum, M. R. and Porter, K. R. (1975). Enucleation of CHO cells by means of cytochalasin B and centrifugation: the topography of enucleation. *Exp. Cell Res.* **94**, 47-55.

Shearer, R. W. and McCarthy, B. J. (1967). Evidence for RNA molecules restricted to the cell nucleus. *Biochemistry* **6**, 283-289.

Shearer, R. W. and McCarthy, B. J. (1970). Characterization of mRNA molecules restricted to the nucleus in mouse L cells. *J. Cell. Physiol.* **75**, 97-106.

Shearn, A. (1974). Complementation analysis of late lethal mutants of D. melanogaster. *Genetics* **77**, 115-125.

Sheiness, D. and Darnell, J. E. (1973). Poly(A) segment in mRNA becomes shorter with age. *Nature New Biol.* **241**, 265-268.

Sheinin, R. (1976). Preliminary characterization of the temperature sebsitive defect in DNA replication in a mutant mouse L cell. *Cell* **7**, 49-57.

Shelanski, M. L. and Taylor, E. W. (1967). Isolation of a protein subunit from microtubules. *J. Cell Biol.* **34**, 549-554.

Shelanski, M. L. and Taylor, E. W. (1968). Properties of the protein subunit of central pair and outer doublet microtubules of sea urchin flagella. *J. Cell Biol.* **38**, 304-315.

Shelanski, M. L., Gaskin, F. and Cantor, C. R. (1973). Microtubule assembly in the absence of added nucleotides. *Proc. Nat. Acad. Sci. USA* **70**, 765-768.

Sheldon, R. and Kates, J. (1974). Mechanism of poly(A) synthesis by vaccinia virus. *J. Virol.* **14**, 214-224.

Shen, C. K. J. and Hearst, J. E. (1978). Chromatin structures of main bands and satellite DNAs in D. melanogaster nuclei as probed by photochemical cross linking of DNA with trioxsalen. *Cold Spring Harbor Symp. Quant. Biol.* **43**, 179-189.

Shen, C. K., Wiesehahn, G. and Hearst, J. E. (1976). Cleavage patterns of D. melanogaster satellite DNA by restriction enzymes. *Nucleic Acids Res.* **3**, 931-951.

Shepherd, G. R., Hardin, J. M. and Noland, B. J. (1971). Methylation of lysine residues of histone fractions in synchronized mammalian cells. *Arch. Biochem. Biophys.* **143**, 1-5.

Shepherd, G. R., Hardin, J. M. and Noland, B. J. (1972). Dephosphorylation of histone fractions of cultured mammalian cells. *Arch. Biochem. Biophys.* **153**, 599-602.

Shepherd, H. S., Boynton, J. E. and Gillham, N. W. (1979). Mutations in nine chloroplast loci of Chlamydomonas affecting different photosynthetic functions. *Proc. Nat. Acad. Sci. USA* **76**, 1353-1357.

Shepherd, J. and Maden, B. E. H. (1972). Ribosome assembly in HeLa cells. *Nature* **236**, 211-214.

Sherline, P., Leung, J. T. and Kipnis, D. M. (1975). Binding of colchicine to purified microtubule protein. *J. Biol. Chem.* **250**, 5481-5486.

Sherman, F. and Slonimski, P. P. (1963). Respiration deficient mutants of yeast. II. Biochemistry. *Biochim. Biophys. Acta* **90**, 1-15.

Sherman, F. and Stewart, J. W. (1971). Genetics and biosynthesis of cytochrome *c. Ann. Rev. Genet.* **5**, 257-296.

Sherman, F., Stewart, J. W., Margoliash, E., Parker, J. and Campbell, W. (1966). The structural gene for yeast cytochrome c. *Proc. Nat. Acad. Sci. USA* **55**, 1498-1504.

Sherod, D., Johnson, G. and Chalkley, R. (1974). Studies on the heterogeneity of lysine rich histones in dividing cells. *J. Biol. Chem.* **249**, 3923-3931.

Shibata, H., Ro-Choi, T. S., Reddy, R., Choi, Y. C., Henning, D. and Busch, H. (1975). The primary nucleotide sequence of nuclear U2 RNA. The 5' terminal portion of the molecule. *J. Biol. Chem.* **250**, 3909-3920.

Shields, R. and Smith, J. A. (1977). Cells regulate their proliferation through alterations in transition probability. *J. Cell. Physiol.* **91**, 345-356.

Shih, T. Y. and Bonner, J. (1970a). Thermal denaturation and template properties of DNA complexes with purified histone fractions. *J. Mol. Biol.* **48**, 469-487.

Shimotohno, K., Kodama, Y., Hashimoto, J. and Miura, K.-I. (1977). Importance of 5' terminal blocking structure to stabilize mRNA in eucaryotic protein synthesis. *Proc. Nat. Acad. Sci. USA* **74**, 2734-2738.

Shin, S. I. (1974). Nature of mutations conferring resistance to 8-azaguanine in mouse cell lines. *J. Cell Sci.* **14**, 235-251.

Shin, S., Caneva, R., Schildkraut, C. L., Klinger, H. P. and Siniscalco, M. (1973). Cells with phosphoribosyl transferase activity recovered from mouse cells resistant to 8 azaguanine. *Nature New Biol.* **241**, 194-196.

Shine, J. and Dalgarno, L. (1974). The 3' terminal sequence of E. coli 16S rRNA: complementarity to nonsense triplets and ribosome binding sites. *Proc. Nat. Acad. Sci. USA* **71**, 1342-1346.

Shine, J., Seeburg, P. H., Martial, J. A., Baxter, J. D. and Goodman, H. M. (1977). Construction and analysis of recombinant DNA for human chorionic somatomammotropin. *Nature* **270**, 494-500.

Shooter, K. V., Goodwin, G. H. and Johns, E. W. (1974). Interactions of a purified non histone chromosomal protein with DNA and histone. *Eur. J. Biochem.* **47**, 263-270.

Shows, T. B. and Brown, J. A. (1975). Human X-linked genes regionally mapped utilizing X autosome translocations and somatic cell hybrids. *Proc. Nat. Acad. Sci. USA* **72**, 2125-2129.

Shure, M. and Vinograd, J. (1976). The number of superhelical turns in native virion SV40 DNA and minicol DNA determined by the band counting method. *Cell* **8**, 215-226.

Sibley, C. H. and Tomkins, G. M. (1974a). Isolation of lymphoma cell variants resistant to killing by glucocorticoids. *Cell* **2**, 213–220.

Sibley, C. H. and Tomkins, G. M. (1974b). Mechanisms of steroid resistance. *Cell* **2**, 221–227.

Sidebottom, E. and Harris, H. (1969). The role of the nucleolus in the transfer of RNA from nucleus to cytoplasm. *J. Cell Sci.* **5**, 351–364.

Siev, M., Weinberg, R. and Penman, S. (1969). The selective interruption of nucleolar RNA synthesis in HeLa cells by cordycepin. *J. Cell Biol.* **41**, 510–520.

Silagi, S. G. (1967). Hybridization of a malignant melanoma cell line with L cells in vitro. *Cancer Res.* **27**, 1953–1960.

Silver, L. M. and Elgin, S. C. R. (1976). A method for determination of the in situ distribution of chromosomal proteins. *Proc. Nat. Acad. Sci. USA* **73**, 423–427.

Silver, L. M. and Elgin, S. C. R. (1977). Distribution patterns of three subfractions of Drosophila nonhistone chromosomal proteins: possible correlations with gene activity. *Cell* **11**, 971–983.

Silver, L.. M. and Elgin, S. C. R. (1978). Production and characterization of antisera against three individual NHC proteins: a case of a generally distributed NHC protein. *Chromosoma* **68**, 101–114.

Silverman, B. and Mirsky, A. E. (1973). Accessibility of DNA in chromatin to DNA polymerase and RNA polymerase. *Proc. Nat. Acad. Sci. USA* **70**, 1326–1330.

Siminovitch, L. (1976). On the nature of hereditable variation in cultured somatic cells. *Cell* **7**, 1–12.

Simon, R. H., Camerini-Otero, R. D. and Felsenfeld, G. (1978). An octamer of histones H3 and H4 forms a compact complex with DNA of nucleosome size. *Nucl. Acids Res.* **5**, 4805–4818.

Simpson, R. T. (1978a). Structure of the chromatosome, a chromatin particle containing 160 base pairs of DNA and all the histones. *Biochemistry* **17**, 5524–5531.

Simpson, R. T. (1978b). Structure of chromatin containing extensively acetylated H3 and H4. *Cell* **13**, 691–699.

Simpson, R. T. and Reeck, G. R. (1973). A comparison of the proteins of condensed and extended chromatin fractions of rabbit liver and calf thymus. *Biochemistry* **12**, 3853–3858.

Simpson, R. T. and Whitlock, J. P. (1976a). Chemical evidence that chromatin DNA exists as 160 base pair beads interspersed with 40 base pair bridges. *Nucl. Acids Res.* **3**, 117–127.

Simpson, R. T. and Whitlock, J. P., Jr. (1976b). Mapping DNAase I susceptible sites in nucleosomes labeled at the 5′ ends. *Cell* **9**, 347–354.

Sinclair, J. H., Stevens, B. J., Gross, N. and Rabinowitz, M. (1967). The constant size of circular mitochondrial DNA in several organisms and different organs. *Biochim. Biophys. Acta* **145**, 528–531.

Singer, B., Sager, R. and Ramanis, Z. (1976). Chloroplast genetics of Chlamydomonas. III. Closing the circle. *Genetics* **83**, 341–354.

Singer, R. H. and Penman, S. (1972). Stability of HeLa cell mRNA in actinomycin. *Nature* **242**, 100–102.

Singer, R. H. and Penman, S. (1973). Messenger RNA in HeLa cells: kinetics of formation and decay. *J. Mol. Biol.* **78**, 321–334.

Singer, I. I. (1979). The fibronexus: a transmembrane association of fibronectin-containing fibers and bundles of 5nm microfilaments in hamster and human fibroblasts. *Cell* **16**, 675–686.

Singh, L., Purdom, I. F. and Jones, K. W. (1976). Satellite DNA and evolution of sex chromosomes. *Chromosoma* **59**, 43–62.

Singh, L., Purdom, I. F. and Jones, K. W. (1979). Behavior of sex chromosome associated satellite DNAs in somatic and germ cells in snakes. *Chromosoma* **71**, 167–182.

Sippel, A. E., Hynes, N., Groner, B. and Schutz, G. (1977a). Frequency distribution of messenger sequences within polysomal mRNA and nuclear RNA from rat liver. *Eur. J. Biochem.* **77**, 141–152.

Sippel, A. E., Groner, B., Hynes, N. and Schutz, G. (1977b). Size distribution of rat liver nuclear RNA containing mRNA sequences. *Eur. J. Biochem.* **77,** 153-164.

Sitz, T. O., Kuo, S. C. and Nazar, R. N. (1978). Multimer forms of eucaryotic 5.8S rRNA. *Biochemistry* **17,** 5811-5814.

Sivolap, Y. M. and Bonner, J. (1971). Association of chromosomal RNA with repetitive DNA. *Proc. Nat. Acad. Sci. USA* **68,** 387-389.

Skaer, R. J. (1977). Interband transcription in Drosophila. *J. Cell Sci.* **26,** 251-266.

Skinner, D. M. (1967). Satellite DNAs in the creabs Gercarcinus lateralis and Cancer pagurus. *Proc. Nat. Acad. Sci. USA* **58,** 103-110.

Skinner, D. M., Beattie, W. G., Blattner, F. R., Stark, B. P. and Dahlberg, J. E. (1974). The repeat sequence of a hermit crab satellite DNA is $(-T-A-G-G)_N$. $(-A-T-C-C)_n$. *Biochemistry* **13,** 3930-3937.

Sklar, V. E. F. and Roeder, R. G. (1976). Purification and subunit structure of DNA-dependent RNA polymerase III from the mouse plasmacytoma, MOPC 315. *J. Biol. Chem.* **251,** 1064-1073.

Sklar, V. E. F. and Roeder, R. G. (1977). Transcription of specific genes in isolated nuclei by exogenous RNA polymerases. *Cell* **10,** 405-414.

Sklar, V. E. F., Jaehning, J. A., Gage, L. P. and Roeder, R. G. (1976). Purification and subunit structure of DNA dependent RNA polymerase III from the posterior silk gland of B. mori. *J. Biol. Chem.* **251,** 3794-3800.

Skoultchi, A. and Gross, P. R. (1973). Maternal messenger RNA: Detection by molecular hybridization. *Proc. Nat. Acad. Sci. USA* **70,** 2840-2844.

Skryabin, K. G., Maxam, A. M., Petes, T. D. and Hereford, L. (1978). Location of the 5.8S rRNA gene of S. cerevisiae. *J. Bacteriol.* **134,** 306-309.

Slater, I. and Slater, D. W. (1974). Polyadenylation and transcription following fertilization. *Proc. Nat. Acad. Sci.* **71,** 1103-1107.

Slatko, B. E. (1978a). Evidence for newly induced genetic activity responsible for male recombination induction in D. melanogaster. *Genetics* **90,** 105-124.

Slatko, B. E. (1978b). Parameters of male and female recombination influenced by the T-007 second chromosome in D. melanogaster. *Genetics* **90,** 257-276.

Slatko, B. E. and Hiraizumi, Y. (1975). Elements causing male crossing over in D. melanogaster. *Genetics* **81,** 313-324.

Sloboda, R. D., Dentler, W. L. and Rosenbaum, J. L. (1976). Microtubule associated proteins and the stimulation of tubulin assembly in vitro. *Biochemistry* **15,** 4497-4505.

Sloboda, R. D., Rudolph, S. A., Rosenbaum, J. L. and Greengard, P. (1975). Cyclic AMP dependent endogenous phosphorylation of a microtubule associated protein. *Proc. Nat. Acad. Sci. USA* **72,** 177-181.

Slonimski, P. P. and Tzagoloff, A. (1976). Localization in yeast mitochondrial DNA of mutations expressed in a deficiency of cytochrome oxidase and for coenzyme QH_2 cytochrome reductase. *Eur. J. Biochem.* **61,** 27-41.

Slonimski, P. P., Pajot, P., Jacq, C., Foucher, M., Perrodin, G., Kochko, A. and Lamouroux, A. (1978a). Mosaic organization and expression of the mitochondrial DNA region controlling cytochrome c reductase and oxidase. I. Genetic, physical and complementation maps of the *box* region. In M. Bacila, B. L. Horecker and A. O. M. Stoppani (Eds.), *Biochemistry and Genetics of Yeast,* Academic Press, London, pp 339-368.

Slonimski, P. P., Claisse, M. L., Foucher, M., Jacq, C., Kochko, A., Lamouroux, A., Pajot, P., Perrodin, G., Spyridakis, A. and Wambier-Kluppel, M. L. (1978b). Mosaic organization and expression of the mitochondrial DNA region controlling cytochrome c reductase and oxidase. III. A model of structure and function. In M. Bacila, B. L. Horecker, and A. O. M. Stoppani (Eds.), *Genetics and Biochemistry of Yeast,* Academic Press, London, pp 391-402.

Small, J. V. and Sobieszek, A. (1977). Studies on the function and composition of the 10 nm filaments of vertebrate smooth muscle. *J. Cell Sci.* **23**, 243–268.

Smart, J. E. and Bonner, J. (1971a). Selective dissociation of histones from chromatin by sodium deoxycholate. *J. Mol. Biol.* **58**, 651–660.

Smart, J. E. and Bonner, J. (1971b). Studies on the role of histones in the structure of chromatin. *J. Mol. Biol.* **58**, 661–674.

Smart, J. E. and Bonner, J. (1971c). Studies on the role of histones in relation to the template activity and precipitability of chromatin at physiological ionic strengths. *J. Mol. Biol.* **58**, 675–684.

Smart, J. E. and Ito, Y. (1978). Three species of polyoma virus tumor antigens share common peptides probably near the amino termini of the proteins. *Cell* **15**, 1427–1437.

Smerdon, M. J. and Isenberg, I. (1976a). Conformational changes in subfractions of calf thymus histone H1. *Biochemistry* **15**, 4233–4241.

Smerdon, M. J. and Isenberg, I. (1976b). Interactions between the subfractions of calf thymus H1 and nonhistone chromosomal proteins HMG1 and HMG2. *Biochemistry* **15**, 4242–4247.

Smith, B. J. and Wigglesworth, N. M. (1973). A temperature sensitive function in a Chinese hamster cell line affecting DNA synthesis. *J. Cell. Physiol.* **82**, 339–348.

Smith, D., Tauro, P., Schweizer, E. and Halvorson, O. (1968). The replication of mitochondrial DNA during the cell cycle in Saccharomyces lactis. *Proc. Nat. Acad. Sci. USA* **60**, 936–942.

Smith, D. B. and Flavell, R. B. (1975). Characterization of the wheat genome by renaturation kinetics. *Chromosoma* **50**, 223–242.

Smith, D. H., Clegg, J. B., Weatherall, D. J. and Gilles, H. M. (1973). Hereditary persistence of fetal hemoglobin associated with $\gamma\beta$ fusion variant, hemoglobin Kenya. *Nature New Biol.* **246**, 184–186.

Smith, G. P. (1973). Unequal crossover and the evolution of multigene families. *Cold Spring Harbor Symp. Quant. Biol.* **38**, 507–513.

Smith, G. P. (1976). Evolution of repeated DNA sequences by unequal crossovers. *Science* **191**, 528–535.

Smith, G. P. (1977). Significance of hybridization kinetic experiments for theories of antibody diversity. *Cold Spring Harbor Symp. Quant. Biol.* **41**, 863–875.

Smith, H. J. and Bogorad, L. (1974). The polypeptide subunit structure of the DNA dependent RNA polymerase of Z. mays chloroplasts. *Proc. Nat. Acad. Sci. USA* **71**, 4839–4842.

Smith, H. O. and Birnstiel, M. L. (1976). A simple method for DNA restriction site mapping. *Nucl. Acids Res.* **3**, 2387–2398.

Smith, J. A. and Martin, L. (1973). Do cells cycle? *Proc. Nat. Acad. Sci. USA* **70**, 1263–1267.

Smith, K. and Lingrel, J. B. (1978). Sequence organization of the β globin mRNA precursor. *Nucl. Acids Res.* **5**, 3295–3302.

Smith, K. D., Church, R. B. and McCarthy, B. J. (1969). Template specificity of isolated chromatin. *Biochemistry* **8**, 4271–4277.

Smith, M. M. and Huang, R. C. C. (1976). Transcription in vitro of immunoglobulin kappa light chain genes in isolated mouse myeloma nuclei and chromatin. *Proc. Nat. Acad. Sci. USA* **73**, 775–779.

Smith, M., Brown, N. L., Air, G. M., Barrell, B. G., Coulson, A. R., Hutchison, C. A. and Sanger, F. (1977). DNA sequence at the C termini of the overlapping genes A and B in phage ϕX174. *Nature* **265**, 702–705.

Smith, M., Leung, D. W., Gillam, S., Astell, C. R., Montogomery, D. L. and Hall, B. D. (1979). The sequence of the gene for iso-l-cytochrome c in S. cerevisiae. *Cell* **16**, 753–761.

Smith, M. J., Britten, R. J. and Davidson, E. H. (1975). Studies on nucleic acid reassociation kinetics: reactivity of single stranded tails in DNA-DNA renaturation. *Proc. Nat. Acad. Sci. USA* **72**, 4805–4809.

Smith, M. J., Hough, B. R., Chamberlin, M. E. and Davidson, E. H. (1974). Repetitive and nonrepetitive sequences in sea urchin nuclear heterogeneous nuclear RNA. *J. Mol. Biol.* **85**, 103–126.

Smith, P. A. and King, R. C. (1968). Genetic control of synaptonemal complexes in D. melanogaster. *Genetics* **60**, 335–351.

Smith, W. P., Tai, P. C. and Davis, B. D. (1978). Nascent peptide as sole attachment of polysomes to membranes in bacteria. *Proc. Nat. Acad. Sci. USA* **75**, 814–817.

Smith, W. P., Tai, P. C., Thompson, R. C. and Davis, B. D. (1977). Extracellular labeling of nascent polypeptides traversing the membrane of E. coli. *Proc. Nat. Acad. Sci. USA* **74**, 2830–2834.

Smith-Sonneborn, J. and Plaut, W. (1967). Evidence for the presence of DNA in the pellicle of Paramecium. *J. Cell Sci.* **2**, 225–234.

Smith-Sonneborn, J. and Plaut, W. (1969). Studies on the autonomy of pellicular DNA in Paramecium. *J. Cell. Sci.* **5**, 365–372.

Smuckler, E. A. and Tata, J. R. (1972). Nearest neighbour base frequency of the RNA formed by rat liver DNA dependent RNA polymerase A and B with homologous RNA. *Biochem. Biophys. Res. Commun.* **49**, 16–22.

Snell, W. J., Dentler, W. L., Haimo, L. T., Binder, L. I. and Rosenbaum, J. L. (1974). Assembly of chick brain tubulin onto isolated basal bodies of Chlamydomonas reinhardi. *Science* **185**, 357–359.

Snow, M. H. L. and Callan, H. G. (1969). Evidence for a polarized movement of the lateral loops of newt lampbrush chromosomes during oogenesis. *J. Cell Sci.* **5**, 1–25.

Snyder, J. A. and McIntosh, J. R. (1975). Initiation and growth of mitotic tubules from mitotic centers in lysed mammalian cells. *J. Cell Biol.* **67**, 744–760.

Snyder, J. A. and McIntosh, J. R. (1976). Biochemistry and physiology of microtubules. *Ann. Rev. Biochem.* **45**, 699–720.

So, M., Gill, R. and Falkow, S. (1975). The generation of a ColEl amp^r cloning vehicle which allows detection of inserted DNA. *Molec. Gen. Genet.* **142**, 239–249.

Sobell, H. M., Tsai, C.-C., Gilbert, S. G., Jain, S. C. and Sakore, T. D. (1976). Organization of DNA in chromatin. *Proc. Nat. Acad. Sci. USA* **73**, 3068–3072.

Soeiro, R. and Darnell, J. E. (1969). Competitive hybridization by presaturation of HeLa cell DNA. *J. Mol. Biol.* **44**, 551–562.

Soeiro, R. and Darnell, J. E. (1970). A comparison between hnRNA and polysomal mRNA in HeLa cells by RNA-DNA hybridization. *J. Cell Biol.* **44**, 467–475.

Soeiro, R., Vaughan, M. H., Warner, J. R. and Darnell, J. E. (1968). The turnover of nuclear DNA-like RNA in HeLa cells. *J. Cell Biol.* **39**, 112–118.

Solari, A. J. (1970a). The spatial relationship of the X and Y chromosomes during meiotic prophase in mouse spermatocytes. *Chromosoma* **29**, 217–236.

Solari, A. J. (1970b). The behavior of chromosomal axes during diplotene in mouse spermatocytes. *Chromosoma* **31**, 217–230.

Solari, A. J. (1971). Experimental changes in the width of the chromatin fibers from chicken erythrocytes. *Exp. Cell Res.* **67**, 161–170.

Solari, A. J. (1972). Ultrastructure and composition of the synaptonemal complex in spreads and negatively stained spermatocytes of the golden hamster and the albino rat. *Chromosoma* **39**, 237–263.

Solari, A. J. (1974). The behavior of the XY pair in mammals. *Int. Rev. Cytol.* **38**, 273–317.

Solari, A. J. and Moses, M. J. (1973). The structure of the central region in the synaptonemal complexes of hamster and cricket spermatocytes. *J. Cell Biol.* **56**, 145–152.

Sollner-Webb, B. and Felsenfeld, G. (1975). A comparison of the digestion of nuclei and chromatin by staphyloccocal nuclease. *Biochemistry* **14**, 2915–2920.

Sollner-Webb, B. and Felsenfeld, G. (1977). Pancreatic DNAase cleavage sites in nuclei. *Cell* **10**, 537–547.

Sollner-Webb, B. and Reeder, R. H. (1979). The nucleotide sequence of the initiation and termination sites for rRNA transcription in X. laevis. *Cell* **18**, 485–499.

Sollner-Webb, B., Camerini-Otero, R. D. and Felsenfeld, G. (1976). Chromatin structure as probed by nucleases and proteases: evidence for the central role of histones H3 and H4. *Cell* **9**, 179–194.

Sollner-Webb, B., Melchior, W. and Felsenfeld, G. (1978). DNAase I, DNAase II and staphylococcal nuclease cut at different, yet symmetrically located, sites in the nucleosome core. *Cell* **14**, 611–627.

Solomon, W., Bobrow, M., Goodfellow, P. N., Bodmer, W. F., Swallow, D. N., Povey, S. and Noel, B. (1976). Human gene mapping using an X/autosome translocation. *Somatic Cell Genet.* **2**, 125–140.

Somers, D. G., Pearson, M. L. and Ingles, C. J. (1975). Isolation and characterization of an α amanitin resistant rat myoblast mutant cell line possessing α amanitin resistant RNA polymerase II. *J. Biol. Chem.* **250**, 4825–4831.

Sommer, S., Lavi, U. and Darnell, J. E. (1978). The absolute frequency of labeled N^6 methyl A in HeLa cell mRNA decreases with label time. *J. Mol. Biol.* **124**, 487–500.

Sonenberg, N. and Shatkin, A. J. (1977). Reovirus mRNA can be covalently crosslinked via the 5′ cap to proteins in initiation complexes. *Proc. Nat. Acad. Sci. USA* **74**, 4288–4292.

Sonenberg, N., Morgan, M. A., Merrick, W. C. and Shatkin, A. J. (1978). A polypeptide in eukaryotic initiation factors that cross links specifically to the 5′ terminal cap in mRNA. *Proc. Nat. Acad. Sci. USA* **75**, 4843–4847.

Sonenshein, G. E. and Brawerman, G. (1977). Entry of mRNA into polysomes during recovery from starvation in mouse sarcoma 180 cells. *Eur. J. Biochem.* **73**, 307–312.

Sonnenberg, B. P. and Zubay, G. (1965). Nucleohistone as a primer for RNA synthesis. *Proc. Nat. Acad. Sci. USA* **54**, 415–420.

Sorieul, S. and Ephrussi, B. (1961). Karyological demonstration of hybridization of mammalian cells in vitro. *Nature* **190**, 653–654.

Sotirov, N. and Johns, E. W. (1972). Quantitative differences in the content of the histone f2c between chicken erythrocytes and erythroblasts. *Exp. Cell Res.* **73**, 13–16.

Southern, D. I. (1967). Chiasma distribution in Traxaline grasshoppers. *Chromosoma* **22**, 164–191.

Southern, E. M. (1970). Base sequence and evolution of guinea pig alpha satellite DNA. *Nature* **227**, 794–798.

Southern, E. M. (1971). Effects of sequence divergence on the reassociation properties of repetitive DNAs. *Nature New Biol.* **232**, 82–83.

Southern, E. M. (1975a). Long range periodicities in mouse satellite DNA. *J. Mol. Biol.* **94**, 51–70.

Southern, E. M. (1975b). Detection of specific sequences among DNA fragments separated by gel electrophoresis. *J. Mol. Biol.* **98**, 503–518.

Spadafora, C., Oudet, P. and Chambon, P. (1978). The same amount of DNA is organized in in vitro assembled nucleosomes irrespective of the origin of the histones. *Nucl. Acids Res.* **5**, 3479–3490.

Spadari, S. and Weissbach, A. (1974a). HeLa cell R-DNA polymerases. Separation and characterization of two enzymatic activities. *J. Biol. Chem.* **249**, 5809–5815.

Spadari, S. and Weissbach, A. (1974b). The inter relation between DNA synthesis and various DNA polymerase activities in synchronized HeLa cells. *J. Mol. Biol.* **86**, 11–20.

Spadari, S. and Weissbach, A. (1975). RNA primed DNA synthesis: specific catalysis by HeLa cell DNA polymerase α. *Proc. Nat. Acad. Sci. USA* **72**, 503–507.

Spooner, B. S., Yamada, K. M. and Wessells, N. K. (1971). Microfilaments and cell locomotion. *J. Cell Biol.* **49**, 595-613.

Spadafora, C., Bellard, M., Compton, J. L. and Chambon, P. (1976). The DNA repeat lengths in chromatins from sea urchin sperm and gastrula cells are markedly different. *Febs Lett.* **69**, 281-285.

Spandidos, D. A. and Siminovitch, L. (1977a). Transfer of codominant markers by isolated metaphase chromosomes in CHO cells. *Proc. Nat. Acad. Sci. USA* **74**, 3480-3485.

Spandidos, D. A. and Siminovitch, L. (1977b). Linkage of markers controlling consecutive biochemical steps in CHO cells as demonstrated by chromosome transfer. *Cell* **12**, 235-242.

Spandidos, D. A. and Siminovitch, L. (1977c). Transfer of anchorage dependence by isolated metaphase chromosomes in hamster cells. *Cell* **12**, 675-682.

Spear, B. B. (1974). The genes for ribosomal RNA in diploid and polytene chromosomes of D. melanogaster. *Chromosoma* **48**, 159-180.

Spear, B. B. and Gall, J. G. (1973). Independent control of ribosomal gene replication in polytene chromosomes of D. melanogaster. *Proc. Nat. Acad. Sci. USA* **70**, 1359-1363.

Spector, D. H. and Baltimore, D. (1975). Poly(A) on poliovirus. III. In vitro addition of poly(A) to poliovirus RNAs. *J. Virol.* **15**, 1432-1439.

Spelsberg, T. C. and Hnilica, L. S. (1969). The effects of acidic proteins and RNA on the histone inhibition of the DNA dependent RNA synthesis in vitro. *Biochim. Biophys. Acta* **195**, 63-75.

Spelsberg, T. C. and Hnilica, L. S. (1971a). Proteins of chromatin in template restriction. I. RNA synthesis in vitro. *Biochim. Biophys. Acta* **228**, 202-211.

Spelsberg, T. C. and Hnilica, L. S. (1971b). Proteins of chromatin in template restriction. II. Specificity of RNA synthesis. *Biochim. Biophys. Acta* **228**, 212-222.

Spelsberg, T. C., Hnilica, L. S. and Ansevin, A. T. (1971). Proteins of chromatin in template restriction. II. The macromolecules in specific restriction of the chromatin DNA. *Biochim. Biophys. Acta.* **228**, 550-562.

Spencer, E., Loring, D., Hurwitz, J. and Monroy, G. (1978). Enzymatic conversion of 5′ phosphate terminated RNA to 5′ di and triphosphate terminated RNA. *Proc. Nat. Acad. Sci. USA* **75**, 4793-4797.

Spencer, R. A., Hauschka, T. S., Amos, D. B. and Ephrussi, B. (1964). Codominance of isoantigens in somatic hybrids of murine cells grown in vitro. *J. Nat. Cancer Inst.* **33**, 893-903.

Sperling, L. and Klug, A. (1977). X ray studies on "native" chromatin. *J. Mol. Biol.* **112**, 253-264.

Spirin, A. S. (1972). Non ribosomal RNP particles (informosomes) of animal cells. In L. Bosch. (Ed.), *The Mechanism of Protein Synthesis and its Regulation*, North Holland, Amsterdam, pp 515-538.

Spofford, J. B. (1976). Position effect variegation in Drosophila. In M. Ashburner and E. Novitski (Eds.), *The Genetics and Biology of Drosophila*, Academic Press, London, Vol. 1c, pp 955-1019.

Spohr, G., Imaizumi, T. and Scherrer, K. (1974). Synthesis and processing of nuclear precursor mRNA in avian erythroblasts and HeLa cells. *Proc. Nat. Acad. Sci. USA* **71**, 5009-5013.

Spohr, G., Granboulan, N., Morel, C. and Scherrer, K. (1970). Messenger RNA in HeLa cells: an investigation of free and polyribosome bound cytoplasmic mRNP particles by kinetic labeling and electron microscopy. *Eur. J. Biochem.* **17**, 296-318.

Spradling, A., Hui, H. and Penman, S. (1975). Two very different components of mRNA in an insect cell line. *Cell* **4**, 131-140.

Spradling, A. L., Pardue, M. L. and Penman, S. (1977). Messenger RNA in heat shocked Drosophila cells. *J. Mol. Biol.* **109**, 559-588.

Spradling, A., Penman, S. and Pardue, M. L. (1975). Analysis of Drosophila mRNA by in situ hybridization: sequences transcribed in normal and heat shocked cultured cells. *Cell* **4**, 395-404.

Spradling, A., Penman, S., Campo, M. S. and Bishop, J. O. (1974). Repetitious and unique sequences in the heterogeneous nuclear and cytoplasmic mRNA of mammalian and insect cells. *Cell* **3**, 23–30.

Sprague, K. U., Steitz, J. A., Grenley, R. M. and Stocking, C. E. (1977). 3′ terminal sequences of 16S rRNA do not explain translation specificity differences between E. coli and B. stearo-thermophilus ribosomes. *Nature* **267**, 462–465.

Sripati, C. E., Groner, Y. and Warner, J. (1976). Methylated blocked 5′ termini of yeast mRNA. *J. Biol. Chem.* **251**, 2898–2904.

Sriprakash, K. S., Molloy, P. L., Nagley, P., Lukins, H. B. and Linnane, A. W. (1976). Biogenesis of mitochondria. XLI. Physical mapping of mitochondrial genetic markers in yeast. *J. Mol. Biol.* **104**, 485–503.

Stalder, J., Seebeck, T. and Braun, R. (1978). Degradation of the rDNA genes by DNAase I in Physarum polycephalum. *Eur. J. Biochem.* **90**, 391–396.

Stamatoyannopoulos, G., Weitkamp, L. R., Kotsakis, P. and Akrivakis, A. (1978). The linkage relationships of the β and δ globin genes. *Hemoglobin* **1**, 561–570.

Stanley, P., Caillibot, V. and Siminovitch, L. (1975a). Stable alterations at the cell membrane of CHO cells resistant to the cytotoxicity of PHA. *Somatic Cell Genet.* **1**, 3–26.

Stanley, P., Narasimhan, S., Siminovitch, L. and Schachter, H. (1975b). Chinese hamster ovary cells selected for resistance to the cytotoxicity of PHA are deficient in a UDP-*N*-acetylglucos-amine-glycoprotein *N*-acetylglucosaminyl-transferase activity. *Proc. Nat. Acad. Sci. USA* **72**, 3323–3327.

Stanners, C. P. and Till, J. E. (1960). DNA synthesis in individual L strain mouse cell. *Biochim. Biophys. Acta* **37**, 406–419.

Starbuch, W. C., Mauritzen, C. M., Taylor, C. W., Saroja, I. S. and Busch, H. (1968). A large scale procedure for isolation of the glycine rich arginine rich histone and the arginine rich lysine rich histone in a highly purified form. *J. Biol. Chem.* **243**, 2038–2047.

Starger, J. M., Brown, W. E., Goldman, A. E. and Goldman, R. D. (1978). Biochemical and immunological analysis of rapidly purified 10 nm filaments from baby hamster kidney cells. *J. Cell Biol.* **78**, 93–109.

Stavnezer, J., Huang, R. C. C., Stavnezer, R. and Bishop, M. (1974). Isolation of mRNA for an immunoglobulin kappa chain and enumeration of the genes for the constant region of kappa chain in the mouse. *J. Mol. Biol.* **88**, 43–64.

Stavrianopoulos, J. G., Karkas, J. D. and Chargaff, E. (1972). DNA polymerase of chicken embryo: purification and properties. *Proc. Nat. Acad. Sci. USA* **69**, 1781–1785.

Stedman, E. and Stedman, E. (1950). Cell specificity of histones. *Nature* **166**, 780–781.

Stefos, K. and Arrhigi, F. E. (1974). Repetitive DNA of Gallus domesticus and its cytological locations. *Exp. Cell Res.* **83**, 9–14.

Steggles, A. W., Wilson, G. N., Kantor, J. A., Picciano, D. J., Falvey, A. K. and Anderson, W. F. (1974). Cell free transcription of mammalian chromatin: transcription of globin mRNA sequences from bone marrow chromatin with mammalian RNA polymerase. *Proc. Nat. Acad. Sci. USA* **71**, 1219–1223.

Stein, A., Bina-Stein, M. and Simpson, R. T. (1977). Crosslinked histone octamer as a model of the nucleosome core. *Proc. Nat. Acad. Sci. USA* **74**, 2780–2784.

Stein, G., Park, W., Thrall, C., Mans, R. and Thiessen, W. E. (1975a). Regulation of cell cycle stage specific transcription of histone genes from chromatin by non histone chromosomal proteins. *Nature* **257**, 764–767.

Stein, G. S., Mans, R. J., Gabbay, E. J., Stein, J. L., Davis, J. and Adawadkar, P. D. (1975b). Evidence for fidelity of chromatin reconstitution. *Biochemistry* **14**, 1859–1865.

Stein, J. L., Stein, G. S. and McGuire, P. M. (1977). Histone mRNA from HeLa cells: evidence for modified 5′ termini. *Biochemistry* **16**, 2207–2212.

Steinberg, R. A., O'Farrell, P. H., Friedrich, U. and Coffino, P. (1977). Mutations causing charge alterations in regulatory subunits of the cAMP dependent protein kinase of cultured S49 lymphoma cells. *Cell* **10**, 381–391.

Steinberg, R. A., Van Daalen Wetters, T. and Coffino, P. (1978). Kinase negative mutants of S49 mouse lymphoma cells carry a trans dominant mutation affecting expression of cAMP dependent protein kinase. *Cell* **15**, 1351–1362.

Steinmetz, M., Streeck, R. E. and Zachau, H. G. (1975). Nucleosome formation abolishes base specific binding of histones. *Nature* **258**, 447–449.

Steinmetz, M., Streeck, R. E. and Zachau, H. G. (1978). Closely spaced nucleosome cores in reconstituted histone-DNA complexes and histone H1-depleted chromatin. *Eur. J. Biochem.* **83**, 615–628.

Steitz, J. A. and Jakes, K. (1975). How ribosomes select initiator regions in mRNA. Base pair formation between the 3′ terminus of 16S rRNA and the mRNA during initiation of protein synthesis in E. coli. *Proc. Nat. Acad. Sci. USA* **72**, 4734–4738.

Stephens, R. E. (1967). The mitotic apparatus. Physical and chemical characterization of the 22S protein component and its subunits. *J. Cell Biol.* **32**, 255–275.

Stephens, R. E. (1970). Thermal fractionation of outer fiber doublet microtubules into A and B subfiber components. *J. Mol. Biol.* **47**, 353–363.

Stern, C. (1931). Zytologisch genetische untersuchungen als beweise für die Morgansche theorie des faktorenaustauschs. *Biol. Zbl.* **51**, 547–587.

Stern, C. (1936). Somatic crossing over and segregation in D. melanogaster. *Genetics* **21**, 625–730.

Stern, H. and Hotta, Y. (1969). Biochemistry of mitosis. In A. Lima de Faria (Ed.), *Handbook of Molecular Cytology*, North Holland: Amsterdam, pp 520–539.

Stern, H., Westergaard, M. and Von Wettstein, D. (1975). Presynaptic events in meiocytes of Lilium longiflorum and their relation to crossing over: a preselection hypothesis. *Proc. Nat. Acad. Sci. USA* **72**, 961–965.

Stevenin, J., Gallinaro-Matringe, H., Gattoni, R. and Jacob, M. (1977). Complexity of the structure of particles containing hnRNA as demonstrated by RNAase treatment. *Eur. J. Biochem.* **74**, 589–602.

Stevenin, J., Gattoni, R., Gallinaro-Matringe, H. and Jacob, M. (1978). Nuclear RNP particles contain specific proteins and unspecific non histone nuclear proteins. *Eur. J. Biochem.* **84**, 541–550.

Stevenin, J., Gattoni, R., Divilliers, G. and Jacob, M. (1979). Rearrangements in the course of pre messenger ribonucleoproteins: a warning. *Eur. J. Biochem.* **95**, 593–606.

Stevens, B. J. and Swift, H. (1966). RNA transport from nucleus to cytoplasm in Chironomus salivary glands. *J. Cell Biol.* **51**, 55–57.

Stevens, F. C., Walsh, M., Ho, H. C., Teo, T. S. and Wang, J. H. (1976). Comparison of calcium binding proteins. Bovine heart and bovine brain activators of cyclic nucleotide phosphodiesterase and rabbit skeletal muscle troponin C. *J. Biol. Chem.* **251**, 4495–4500.

Stevens, R. E., Renaud, F. L. and Gibbons, I. R. (1967). Guanine nucleotide associated with the protein of the outer fibers of flagella and cilia. *Science* **156**, 1606–1608.

Stevens, W. L. (1936). The analysis of interference. *J. Genet.* **32**, 51–64.

Steward, D. L., Shaeffer, J. R. and Humphrey, R. M. (1968). Breakdown and assembly of polysomes in synchronized Chinese hamster cells. *Science* **161**, 791–793.

Stiles, C. D., Lee, K.-L. and Kenney, F. T. (1976). Differential degradation of mRNA in mammalian cells. *Proc. Nat. Acad. Sci. USA* **73**, 2634–2638.

Stiles, C. D., Isberg, R. R., Pledger, W. J., Antoniades, H. N. and Scher, C. D. (1979). Control of the Balb-3T3 cell cycle by nutrients and serum factors: analysis using platelet derived growth factor and platelet poor plasma. *J. Cell. Physiol.* **99**, 395–406.

Stockert, J. C. and Lisanti, J. A. (1972). Acridine orange differential fluorescence of fast and slow reassociating chromosomal DNA after in situ DNA denaturation and reassociation. *Chromosoma* **37**, 117–130.

Stoker, M. (1975). The effects of topoinhibition and cytochalasin B on metabolic cooperation. *Cell* **6**, 253–258.

Stoker, M. and MacPherson, I. (1964). Syrian hamster fibroblast cell line BHK 21 and its derivatives. *Nature* **203**, 1355–1357.

Stolarsky, L. and Kemper, B. (1978). Characterization and partial purification of parathyroid hormone mRNA. *J. Biol. Chem.* **253**, 7194–7201.

Storms, R. and Hastings, P. J. (1977). A fine structure analysis of meiotic pairing in Chlamydomonas reinhardii. *Exp. Cell Res.* **104**, 39–46.

Storrie, B. and Edelson, P. J. (1977). Distribution of Con A in fibroblasts: direct endycytosis versus surface capping. *Cell* **11**, 707–719.

Storti, R. V. and Rich, A. (1976). Chick cytoplasmic actin and muscle actin have different structural genes. *Proc. Nat. Acad. Sci. USA* **73**, 2346–2350.

Storti, R. V., Coen, D. M. and Rich, A. (1976). Tissue specific forms of actin in the developing chick. *Cell* **8**, 521–528.

Stossel, T. P. and Hartwig, J. H. (1976). Interactions of actin, myosin and a new actin binding protein of rabbit pulmonary macrophages. II. Role in cytoplasmic movement and phagocytosis. *J. Cell Biol.* **68**, 602–619.

Strair, R. K., Skoultchi, A. I. and Shafritz, D. A. (1977). A characterization of globin mRNA sequences in the nucleus of duck immature red blood cells. *Cell* **12**, 133–141.

Strair, R. K., Yap, S. H. and Shafritz, D. A. (1977). Use of molecular hybridization to purify and analyze albumin mRNA from rat liver. *Proc. Nat. Acad. Sci. USA* **74**, 4346–4350.

Stratling, W. H. and O'Malley, B. W. (1976). Studies on the structure and function of chick oviduct chromatin. 2. Biochemical characterization of two chromatin fractions isolated by ECTHAM-cellulose chromatography. *Eur. J. Biochem.* **66**, 435–442.

Stratling, W. H., Van, N. T. and O'Malley, B. W. (1976). Studies on the structure and function of chick oviduct chromatin. I. Fractionation by ECTHAM-cellulose chromatography and physico-chemical characterization. *Eur. J. Biochem.* **66**, 423–435.

Stratling, W. H., Muller, U. and Zentgraf, H. (1978). The higher order repeat structure of chromatin is built of globular particles containing eight nucleosomes. *Exp. Cell Res.* **117**, 301–312.

Straus, N. A. (1971). Comparative renaturation kinetics in amphibia. *Proc. Nat. Acad. Sci.* **68**, 799–802.

Strausberg, R. L. and Butow, R. A. (1977). Expression of petite mitochondrial DNA in vivo: zygotic gene rescue. *Proc. Nat. Acad. Sci. USA* **74**, 2715–2719.

Strausberg, R. L., Vincent, R. D., Perlman, P. S. and Butow, R. A. (1978). Asymmetric gene conversion at inserted segments on yeast mitochondrial DNA. *Nature* **276**, 577–582.

Strauss, A. W., Bennett, C. D., Donohue, A. M., Rodkey, J. A. and Alberts, A. W. (1977). Rat liver pre proalbumin. Complete amino acid sequence of the pre piece. Analysis of the direct translation product of albumin mRNA. *J. Biol. Chem.* **252**, 6846–6856.

Strickland, M., Strickland, W. N., Brandt, W. F. and Von Holt, C. (1977a). The complete amino acid sequence of histone H2B(1) from sperm of the sea urchin Parechinus angulosus. *Eur. J. Biochem.* **77**, 263–276.

Strickland, W. N., Strickland, M., Brandt, W. R. and Von Holt, C. (1977b). The complete amino acid sequence of histone H2B(2) from sperm of the sea urchin Parechinus angulosus. *Eur. J. Biochem.* **77**, 277–286.

Strickland, M., Strickland, W. N., Brandt, W. F., Holt, C., Wittman-Liebold, B. and Lehmann, A. (1978). The complete amino acid sequence of histone H2B(3) from sperm of the sea urchin Parechinus angulosus. *Eur. J. Biochem.* **89**, 443–452.

Strobel, E., Dunsmuir, P. and Rubin, G. M. (1979). Polymorphisms in the chromosomal locations of elements of the 412, copia and 297 dispersed repeated gene families in Drosophila. *Cell* 17, 429-440.

Struhl, K. and Davis, R. W. (1977). Production of a functional eucaryotic enzyme in E. coli: cloning and expression of the yeast structural gene for imidazole glycerol phosphate dehydratase (*his3*). *Proc. Nat. Acad. Sci. USA* 74, 5255-5259.

Struhl, K., Cameron, J. R. and Davis, R. W. (1976). Functional genetic expression of eucaryotic DNA in E. coli. *Proc. Nat. Acad. Sci. USA* 73, 1471-1475.

Stubblefield, E. (1966). Mammalian chromosomes in vitro. XIX. Chromosomes of Don C, a Chinese hamster fibroblast strain with part of autosome 1b translocated to the Y chromosome. *J. Nat. Cancer Inst.* 37, 799-817.

Stubblefield, E. and Brinkley, B. R. (1967). Architecture and function of the mammalian centriole. In K. B. Warren (Ed.), *Formation and Fate of Cell Organelles*, Symposium of the International Society of Cell Biology. Vol. 6, pp 175-218 (Academic Press, New York).

Stubblefield, E. and Murphree, S. (1967). Synchronized mammalian cell cultures. II. Thymidine kinase activity in colcemid synchronized fibroblasts. *Exp. Cell Res.* 48, 652-656.

Sturtevant, A. H. (1913). The linear arrangement of six sex linked factors in Drosophila, as shown by their mode of association. *J. Ezp. Zool.* 14, 43-59.

Sturtevant, A. H. (1929). The claret mutant type of D. simulans: a study of chromosome elimination and cell lineage. *Z. Wiss. Zool.* 135, 323-356.

Stutz, E. (1970). The kinetic complexity of Euglena gracilis chloroplasts. *Febs Lett.* 8, 25-28.

Stutz, E. and Noll, H. (1967). Characterization of cytoplasmic and chloroplast polysomes in plants. Evidence for three classes of rRNA in nature. *Proc. Nat. Acad. Sci. USA* 57, 774-787.

Suau, P., Kneale, G. G., Braddock, G. W., Baldwin, J. P. and Bradbury, E. M. (1977). A low resolution model for the chromatin core particle by neutron scattering. *Nucleic Acids Res.* 4, 3769-3786.

Subak-Sharpe, H., Burk, R. R. and Pitts, J. D. (1969). Metabolic cooperation between biochemically marked mammalian cells in tissue culture. *J. Cell Sci.* 4, 353-367.

Subirana, J. A. (1973). Studies on the thermal denaturation of nucleohistone. *J. Mol. Biol.* 74, 363-386.

Sueoka, N. (1961). Variation and heterogeneity of base composition of DNAs: a compilation of old and new data. *J. Mol. Biol.* 3, 31-40.

Sueoka, N., Chiang, K. S. and Kates, J. R. (1967). DNA replication in meiosis of Chlamydomonas reinhardii. I. Isotopic transfer experiments with a strain producing 8 zoospores. *J. Mol. Biol.* 25, 47-66.

Sugino, A., Goodman, H. M., Heyneker, H. L., Shine, J., Boyer, H. W. and Cozzarelli, N. R. (1977). Interaction of phage T4 RNA and DNA ligases in joining of duplex DNA at base paired ends. *J. Biol. Chem.* 252, 3987-3994.

Sulkowski, E. and Laskowski, M. (1962). Mechanism of action of micrococcal nuclease on DNA. *J. Biol. Chem.* 237, 2620-2626.

Sulston, J. E. and Brenner, S. (1974). The DNA of Caenorhabditis elegans. *Genetics* 77, 95-104.

Summers, K. E. and Gibbons, I. R. (1971). ATP induced sliding of tubules in trypsin treated flagella of sea urchin sperm. *Proc. Nat. Acad. Sci. USA* 68, 3092-3096.

Summers, K. E. and Gibbons, I. R. (1973). Effects of trypsin digestion on flagellar structures and their relationship to motility. *J. Cell Biol.* 58, 618-629.

Sumner, A. T. and Evans, H. J. (1973). Mechanisms involved in the banding of chromosomes with quinacrine and Giemsa. II. The interaction of the dyes with the chromosomal components. *Exp. Cell Res.* 81, 223-236.

Sumner, A. T., Evans, H. J. and Buckland, R. A. (1971). New technique for distinguishing between human chromosomes. *Nature New Biol.* 232, 31-32.

Sumner, A. T., Evans, H. J. and Buckland, R. A. (1973). Mechanisms involved in the banding of chromosomes with quinacrine and Giemsa. I. The effects of fixation in methanol-acetic acid. *Exp. Cell Res.* **81**, 214-222.

Sun, T. T. and Green, H. (1978). Immunofluorescent staining of keratin fibers in cultured cells. *Cell* **14**, 469-476.

Sung, M. T. (1977). Phosphorylation and dephosphorylation of histone V (H5) controlled condensation of avian erythrocyte chromatin. *Biochemistry* **16**, 286-290.

Sung, M. T. and Dixon, G. H. (1970). Modification of histones during spermiogenesis in trout: a molecular mechanism for altering histone binding to DNA. *Proc. Nat. Acad. Sci. USA* **67**, 1616-1623.

Sung, M. T. and Freedlender, E. F. (1978). Sites of in vivo phosphorylation of histone H5. *Biochemistry* **17**, 1884-1889.

Sung, M. T., Dixon, G. H. and Smithies, O. (1971). Phosphorylation and synthesis of histones in regenerating rat liver. *J. Biol. Chem.* **246**, 1358-1364.

Sung, M. T., Harford, J., Bundman, M. and Vidalakas, G. (1977). Metabolism of histones in avian erythroid cells. *Biochemistry* **16**, 279-285.

Sunkara, P. S., Wright, D. A. and Rao, P. N. (1979). Mitotic factors from mammalian cells induce germinal vesicle breakdown and chromosome condensation in amphibian oocytes. *Proc. Nat. Acad. Sci. USA* **76**, 2799-2802.

Sures, I., Lowry, J. and Kedes, L. H. (1978). The DNA sequence of sea urchin (S. purpuratus) H2A, H2B and H3 histone coding and spacer regions. *Cell* **15**, 1033-1044.

Sures, I., Maxam, A., Cohn, R. H. and Kedes, L. H. (1976). Identification and location of the histone H2A and H3 genes by sequence analysis of sea urchin (S. purpuratus) DNA cloned in E. coli. *Cell* **9**, 495-502.

Surrey, S. and Nemer, M. (1976). Methylated blocked 5' terminal sequences of sea urchin embryo mRNA classes containing and lacking poly(A). *Cell* **9**, 589-596.

Surzycki, S. J. and Gillham, N. W. (1971). Organelle mutations and their expression in Chlamydomonas reinhardii. *Proc. Nat. Acad. Sci. USA* **68**, 1301-1306.

Sussman, J. L. and Trifonov, E. N. (1978). Possibility of nonkinked packing of DNA in chromatin. *Proc. Nat. Acad. Sci. USA* **75**, 103-107.

Sussman, R. and Rayner, E. P. (1971). Physical characterization of DNAs in D. discoideum. *Arch. Biochem. Biophys.* **144**, 127-137.

Sutcliffe, J. G. (1978a). Nucleotide sequence of the ampicillin resistance gene of E. coli plasmid pBR322. *Proc. Nat. Acad. Sci. USA* **75**, 3737-3741.

Sutcliffe, J. G. (1978b). pBR322 restriction map derived from the DNA sequence: accurate DNA size markers up to 4361 base pairs long. *Nucleic Acids Res.* **5**, 2721-2728.

Sutton, W. D. and McCallum, M. (1971). Mismatching and the reassociation rate of mouse satellite DNA. *Nature New Biol.* **232**, 83-85.

Sutton, W. D. and McCallum, M. (1972). Related satellite DNAs in the genus Mus. *J. Mol. Biol.* **71**, 633-656.

Sutton, W. S. (1903). The chromosomes in heredity. *Biol. Bull. (Woods Hole)* **4**, 231-248.

Suyama, Y. (1966). Mitochondrial DNA of tetrahymena. Its partial physical characterization. *Biochemistry* **5**, 2214-2221.

Suzuki, Y., Gage, L. P. and Brown, D. D. (1972). The genes for silk fibroin in Bombyx mori. *J. Mol. Biol.* **70**, 637-657.

Swann, M. M. (1952). The nucleus in fertilization, mitosis and cell division. *Symp. Soc. Exp. Biol.* **6**, 89-106.

Swanson, R. F. and Dawid, I. B. (1970). The mitochondrial ribosomes of X. laevis. *Proc. Nat. Acad. Sci. USA* **66**, 117-124.

Swartz, M. N., Trautner, T. A. and Kornberg, A. (1962). Enzymatic synthesis of DNA. XI. Further studies on nearest neighbour base sequences in DNAs. *J. Biol. Chem.* **237**, 1961-1967.

Swift, H. H. (1950). The DNA content of animal nuclei. *Physiol. Zool.* **23**, 169-199.

Swift, H. (1962). Nucleic acids and cell morphology in Dipteran salivary glands. In J. M. Allen (Ed.) *The Molecular Control of Cellular Activity*. McGraw Hill, New York, pp 73-125.

Swift, H. (1964). The histones of polytene chromosomes. In P. T'so and J. Bonner (Eds.), *The Nucleohistones*. Holden Day, San Francisco, pp 169-183.

Swift, H., Adams, B. J. and Larsen, K. (1964). Electron microscope cytochemistry of nucleic acids in Drosophila salivary glands and Tetrahymena. *J. Roy. Microscop. Soc.* **83**, 161-167.

Syverton, J. T. and McLaren, L. C. (1957). Human cells in continuous culture. I. Derivation of cell strains from oesophagus, plate, liver and lung. *Cancer Res.* **17**, 923-925.

Szollosi, D., Calarco, P. and Donahue, R. P. (1972). Absence of centrioles in the first and second meiotic spindles of mouse oocytes. *J. Cell Sci.* **11**, 521-541.

Szpirer, C. (1974). Reactivation of chick erythrocyte nuclei in heterokaryons with rat hepatoma cells. *Exp. Cell Res.* **83**, 47-54.

Szybalski, W., Szybalska, E. H. and Ragni, G. (1962). Genetic studies with human cell lines. *Nat. Cancer Inst. Monogr.* **7**, 75-89.

Talavera, A. and Basilico, C. (1978). Requirements of BHK cells for the exit from different quiescent states. *J. Cell. Physiol.* **97**, 429-440.

Tan, C. H. and Miyagi, M. (1970). Specificity of transcription of chromatin in vitro. *J. Mol. Biol.* **50**, 641-653.

Tanphaichitr, N., Moore, K. C., Granner, D. and Chalkley, R. (1976). Relationship between chromosome condensation and metaphase lysine-rich histone phosphorylation. *J. Cell Biol.* **69**, 43-50.

Tantravahi, R., Miller, D. A., Dev, V. G. and Miller, O. J. (1976). Detection of nucleolus organizer regions in chromosomes of human, chimpanzee, gorilla, orangutan and gibbon. *Chromosome* **56**, 15-28.

Tarnowka, M. A. and Baglioni, C. (1979). Regulation of protein synthesis in mitotic HeLa cells. *J. Cell. Physiol.* **99**, 359-369.

Tartof, K. D. (1971). Increasing the multiplicity of rRNA genes in D. melanogaster. *Science* **171**, 294-297.

Tartof, K. D. (1973). Regulation of rRNA gene multiplicity in D. melanogaster. *Genetics* **73**, 57-70.

Tartof, K. D. (1974). Unequal mitotic sister chromatid exchange as the mechanism of ribosomal RNA gene magnification. *Proc. Nat. Acad. Sci. USA* **71**, 1272-1276.

Tartof, K. D. (1979). Evolution of transcribed and spacer sequences in the rRNA genes of Drosophila. *Cell* **17**, 607-614.

Tartof, K. D. and Dawid, I. B. (1976). Similarities and differences in the structure of X and Y chromosome rRNA genes of Drosophila. *Nature* **263**, 27-30.

Tartof, K. D. and Perry, R. P. (1970). The 5S RNA genes of D. melanogaster. *J. Mol. Biol.* **51**, 171-183.

Tatchell, K. and Van Holde, K. E. (1977). Reconstitution of chromatin core particles. *Biochemistry* **16**, 5295-5302.

Tatchell, K. and Van Holde, K. E. (1978). Compact oligomers and nucleosome phasing. *Proc. Nat. Acad. Sci. USA* **75**, 3583-3587.

Tatchell, K. and Van Holde, K. E. (1979). Nucleosome reconstitution: effect of DNA length on nucleosome structure. *Biochemistry* **18**, 2871-2879.

Tatsumi, K. and Strauss, B. (1978). Production of DNA bifilarly substituted with BUdR in the first round of synthesis: branch migration during isolation of cellular DNA. *Nucleic Acids Res.* **5**, 331-348.

Taub, M. and Englesberg, E. (1976). Isolation and characterization of 5 fluorotryptophan resistant mutants with altered L tryptophan transport. *Somatic Cell Genet.* **2**, 441-452.

Taub, M. and Englesberg, E. (1978). 5-fluorotryptophan resistant mutants affecting the A and L transport systems in the mouse L cell line A9. *J. Cell. Physiol.* **97**, 477-486.

Taylor, E. W. (1965). The mechanism of colchicine inhibition of mitosis. I. Kinetics of inhibition and the binding of ^{3}H-colchicine. *J. Cell Biol.* **25**, 145-160.

Taylor, E. W. (1965). Control of DNA synthesis in mammalian cells in culture. *Exp. Cell Res.* **40**, 316-332.

Taylor, E. W. (1972). Chemistry of muscle contraction. *Ann. Rev. Biochem.* **41**, 577-616.

Taylor, J. H. (1957). The time and mode of duplication of chromosomes. *Am. Nat.* **91**, 209-222.

Taylor, J. H. (1958). Sister chromatid exchanges in tritium labeled chromosomes. *Genetics* **43**, 515-529.

Taylor, J. H. (1963). The replication and organization of DNA in chromosomes. In J. H. Taylor (Ed.), *Molecular Genetics*, Academic Press, New York, pp 65-113.

Taylor, J. H. (1965). Distribution of ^{3}H labeled DNA among chromosomes during meiosis. I. Spermatogenesis in the grasshopper. *J. Cell Biol.* **25**, 57-67.

Taylor, J. H., Woods, P. and Hughes, W. (1957). The organization and duplication of chromosomes as revealed by autoradiographic studies using ^{3}H labeled thymidine. *Proc. Nat. Acad. Sci. USA* **43**, 122-128.

Taylor, J. M. and Schimke, R. T. (1974). Specific binding of albumin antibody to rat liver polysomes. *J. Biol. Chem.* **249**, 3597-3607.

Taylor, J. M. and Tse, T. P. H. (1976). Isolation of rat liver albumin mRNA. *J. Biol. Chem.* **251**, 7461-7467.

Taylor, J. M., Dozy, A., Kan, Y. W., Varmus, H. E., Lie-Injo, L. E., Ganesan, J. and Todd, D. (1974). Genetic lesion in homozygous α thalassemia (hydrops fetalis). *Nature* **251**, 392-393.

Tchurikov, N. A., Ilyin, Y. V., Ananiev, E. V. and Georgiev, G. P. (1978). The properties of gene Dm 225, a representative of dispersed repetitive genes in D. melanogaster. *Nucleic Acids Res.* **5**, 2169-2188.

Tease, C. and Jones, G. H. (1978). Analysis of exchanges in differentially stained meiotic chromosomes of Locusta migratoria after BUdR substitution and FPG staining. I. Crossover exchanges in monochiasmate bivalents. *Chromosoma* **69**, 163-178.

Tekamp, P. A., Valenzuela, P., Maynard, T., Bell, G. I. and Rutter, W. J. (1979). Specific gene transcription in yeast nuclei and chromatin by added homologous RNA polymerases I and III. *J. Biol. Chem.* **254**, 955-963.

Telford, J. L., Kressmann, A., Koski, R. A., Grosschedl, R., Muller, F., Clarkson, S. G. and Birnstiel, M. L. (1979). Delimitation of a promoter for RNA polymerase III by means of a functional test. *Proc. Nat. Acad. Sci. USA* **76**, 2590-2594.

Telzer, B. R. and Rosenbaum, J. L. (1979). Cell cycle dependent, in vitro assembly of microtubules onto the pericentriolar material of HeLa cells. *J. Cell Biol.* **81**, 484-488.

Telzer, B. R., Moses, M. J. and Rosenbaum, J. L. (1975). Assembly of microtubules onto kinetochores of isolated mitotic chromosomes of HeLa cells. *Proc. Nat. Acad. Sci. USA* **72**, 4023-4027.

Temple, G. F. and Housman, D. E. (1972). Separation and translation of the mRNAs coding for alpha and beta chains of rabbit globin. *Proc. Nat. Acad. Sci. USA* **69**, 1574-1577.

Temple, G. F., Chang, J. C. and Kan, Y. W. (1977). Authentic β globin mRNA sequences in homozygous β° thalassemia. *Proc. Nat. Acad. Sci. USA* **74**, 3047-3051.

Teng, C. S., Teng, C. T. and Allfrey, V. G. (1971). Studies of nuclear acidic proteins. Evidence for their phosphorylation, tissue specificity, selective binding to DNA and stimulatory effects on transcription. *J. Biol. Chem.* **246**, 3957-3609.

Teng, C. T., Teng, C. S. and Allfrey, V. G. (1970). Species specific interactions between nuclear phosphoproteins and DNA. *Biochem. Biophys. Res. Commun.* **41**, 690-696.

Teo, T. S. and Wang, J. H. (1973). Mechanism of activation of a cyclic AMP phospho diesterase from bovine heart by calcium ions. Identification of the protein activator as a Ca^{2+} binding protein. *J. Biol. Chem.* **248**, 5950-5955.

Terasima, T. and Tolmach, L. J. (1963). Growth and nucleic acid synthesis in synchronously dividing populations of HeLa cells. *Exp. Cell Res.* **30**, 344-362.

Terasima, T. and Yasukawa, M. (1966). Synthesis of G1 proteins preceding DNA synthesis in cultured mammalian cells. *Exp. Cell Res.* **44**, 669-671.

Terasima, T. and Tolmach, L. J. (1963). Growth and nucleic acid synthesis in synchronously dividing populations of HeLa cells. *Exp. Cell Res.* **30**, 344-362.

Terpstra, P., Holtrop, M. and Kroon, A. M. (1977). A complete cleavage map of N. crassa mt DNA obtained with endonucleases EcoRI and BamHI. *Biochim. Biophys. Acta* **475**, 571-588.

Ter Schegget, J., Flavell, R. A. and Borst, P. (1971). DNA synthesis by isolated mitochondria. III. Characterization of D loop DNA, a novel intermediate in mitochondrial DNA synthesis. *Biochim. Biophys. Acta* **254**, 1-14.

Tewari, K. K., Jayamaran, J. and Mahler, M. R. (1965). Separation and characterization of mitochondrial DNA from yeast. *Biochem. Biophys. Res. Commun.* **21**, 141-150.

Thiery, J.-P., Macaya, G. and Bernardi, G. (1976). An analysis of eucaryotic genomes by density gradient centrifugation. *J. Mol. Biol.* **108**, 219-236.

Thoma, F. and Koller, T. (1977). Influence of histone H1 on chromatin structure. *Cell* **12**, 101-107.

Thomas, C. A., Hamkalo, B. A., Misra, D. N. and Lee, C. S. (1970). Cyclization of eucaryotic DNA fragments. *J. Mol. Biol.* **51**, 621-631.

Thomas, D. Y. and Wilkie, D. (1968a). Inhibition of mitochondrial synthesis in yeast by erythromycin: cytoplasmic and nuclear factors controlling resistance. *Genetic Res.* **11**, 33-41.

Thomas, D. Y. and Wilkie, D. (1968b). Recombination of mitochondrial drug resistance factors in S. cerevisiae. *Biochem. Biophys. Res. Commun.* **30**, 368-372.

Thomas, J. R. and Tewari, K. K. (1974). Conservation of 70S rRNA genes in the chloroplast DNAs of higher plants. *Proc. Nat. Acad. Sci. USA* **71**, 3147-3151.

Thomas, J. O. and Butler, P. J. G. (1977). Characterization of the octamer of histones free in solution. *J. Mol. Biol.* **116**, 769-782.

Thomas, J. O. and Furber, V. (1976). Yeast chromatin structure. *Febs Lett.* **66**, 274-280.

Thomas, J. O. and Kornberg, R. D. (1975). An octamer of histones in chromatin and free in solution. *Proc. Nat. Acad. Sci. USA* **72**, 2626-2630.

Thomas, J. O. and Thompson, R. J. (1977). Variation in chromatin structure in two cell types from the same tissue: a short DNA repeat length in cerebral cortex neurons. *Cell* **10**, 633-640.

Thomas, M., Cameron, J. R. and Davis, R. W. (1974). Viable molecular hybrids of phage lambda and eucaryotic DNA. *Proc. Nat. Acad. Sci. USA* **71**, 4579-4583.

Thomas, T. L. and Patel, G. L. (1976a). Optimal conditions and specificity of interaction of a distinct class of nonhistone chromosomal proteins with DNA. *Biochemistry* **15**, 1481-1489.

Thomas, T. L. and Patel, G. L. (1976b). DNA unwinding component of the nonhistone chromatin proteins. *Proc. Nat. Acad. Sci. USA* **73**, 4364-4368.

Thompson, E. B. and Gelehrter, T. D. (1971). Expression of TAT activity in somatic cell heterokaryons: evidence for negative control of enzyme expression. *Proc. Nat. Acad. Sci. USA* **68**, 2589-2593.

Thompson, J. A., Radonovich, M. F. and Salzman, N. P. (1979). Characterization of the 5′ terminal structure of SV40 early mRNAs. *J. Virol.* **31**, 437-446.

Thompson, L. H. and Baker, R. M. (1973). Isolation of mutants of cultured mammalian cells. In D. M. Prescott (Ed.), *Methods in Cell Biology,* **VI,** Academic Press, New York, pp 209–281.

Thompson, L. H. and Lindl, P. A. (1976). A CHO cell mutant with a defect in cytokinesis. *Somatic Cell Genet.* **2,** 387–400.

Thompson, L. H., Harkins, J. L. and Stanners, C. P. (1973). A mammalian cell mutant with a temperature sensitive leu-tRNA synthetase. *Proc. Nat. Acad. Sci.* **70,** 3094–3098.

Thompson, L. H., Lofgren, D. J. and Adair, G. M. (1977). CHO cell mutants for arg, asp, gln, his and met tRNA synthetase: identification and initial characterization. *Cell* **11,** 157–168.

Thompson, L. H., Lofgren, D. J. and Adair, G. M. (1978). Evidence for structural gene alterations affecting aminoacyl-tRNA synthetases in CHO cell mutants and revertants. *Somatic Cell Genet.* **4,** 426–436.

Thompson, L. H., Stanners, C. P. and Siminovitch, L. (1975). Selection by ^{3}H amino acids of CHO mutants with altered leucyl and asparagyl tRNA synthetases. *Somat. Cell Genet.* **1,** 187–208.

Thompson, L. H., Mankovitz, R., Baker, R. M., Till, J. E., Siminovitch, L. and Whitmore, G. F. (1970). Isolation of temperature sensitive mutants of L cells. *Proc. Nat. Acad. Sci. USA* **66,** 377–384.

Thompson, L. H., Mankovitz, R., Baker, R. M., Wright, J. A., Till, J. E., Siminovitch, L. and Whitmore, G. F. (1971). Selective and nonselective isolation of temperature sensitive mutants of mouse L cells and their characterization. *J. Cell. Physiol.* **78,** 431–440.

Thomson, K. S., Gall, J. G. and Coggins, L. W. (1973). Nuclear DNA contents of Coelacanth erythrocytes. *Nature* **241,** 126.

Tibbetts, C. J. B. and Vinograd, J. (1973a). Properties and mode of action of a partially purified DNA from the mitochondria of HeLa cells. *J. Biol. Chem.* **248,** 3367–3379.

Tibbetts, C. J. B. and Vinograd, J. (1973b). Synthesis of mitochondrial DNA with a partially purified DNA polymerase from the mitochondria of HeLa cells. *J. Biol. Chem.* **248,** 3380–3385.

Tiemeier, D. C., Tilghman, S. M., Polsky, F. I., Seidman, J. G., Leder, A., Edgell, M. H. and Leder, P. (1978). A comparison of two cloned mouse β globin genes and their surrounding and intervening sequences. *Cell* **14,** 237–246.

Tilghman, S. M., Tiemeier, D. C., Polsky, F., Edgell, M. H., Seidman, J. G., Leder, A., Enquist, L. W., Norman, B. and Leder, P. (1977). Cloning specific segments of the mammalian genome: phage lambda containing mouse globin and surrounding gene sequences. *Proc. Nat. Acad. Sci. USA* **74,** 4406–4410.

Tilghman, S. M., Tiemeier, D. C., Seidman, J. G., Peterlin, B. M., Sullivan, M., Maizel, J. V. and Leder, P. (1978a). Intervening sequence of DNA identified in the structural portion of a mouse β globin gene. *Proc. Nat. Acad. Sci. USA* **75,** 725–729.

Tilghman, S. M., Curtis, P. J., Tiemeier, D. C., Leder, P. and Weissman, C. (1978b). The intervening sequence of a mouse globin β gene is transcribed within the 15S β globin mRNA precursor. *Proc. Nat. Acad. Sci. USA* **75,** 1309–1313.

Tilney, L. G. (1976a). The polymerization of actin. II. How nonfilamentous actin becomes nonrandomly distributed in sperm: evidence for the association of this actin with membranes. *J. Cell Biol.* **69,** 51–72.

Tilney, L. G. (1976b). The polymerization of actin. III. Aggregates of nonfilamentous actin and its associated proteins: a storage form of actin. *J. Cell Biol.* **69,** 73–89.

Tilney, L. G. (1978). Polymerization of actin. V. A new organelle, the actomere, that initiates the assembly of actin filaments in Thyone sperm. *J. Cell Biol.* **77,** 551–564.

Tilney, L. G. and Cardell, R. R. (1970). Factors controlling the reassembly of the microvillus border of the small intestine of the salamander. *J. Cell Biol.* **47,** 408–422.

Tilney, L. G. and Detmers, P. (1975). Actin in erythrocyte ghosts and its association with spectrin. Evidence for a nonfilamentous form of these two molecules in situ. *J. Cell Biol.* **66,** 508–520.

Tilney, L. G. and Gibbons, J. R. (1968). Differential effects of antimitotic agents on the stability and behavior of cytoplasmic and ciliary microtubules. *Protoplasma* **65,** 167–179.

Tilney, L. G. and Kallenbach, N. (1979). Polymerization of actin. VI. The polarity of the actin filaments in the acrosomal process and how it might be determined. *J. Cell Biol.* **81**, 608-623.

Tilney, L. G. and Marsland, D. (1969). A fine structural analysis of cleavage induction and furrowing in the eggs of Arbacia punctulata. *J. Cell Biol.* **42**, 170-184.

Tilney, L. G. and Mooseker, M. S. (1976). Actin filament membrane attachment: are membrane particles involved? *J. Cell Biol.* **71**, 402-416.

Tilney, L. G. and Porter, K. R. (1965). Studies on the microtubules of Heliozoa. I. Fine structure of Actinosphaerium with particular reference to axial rod structure. *Protoplasma* **60**, 317-344.

Tilney, L. G. and Porter, K. R. (1967). Studies on the microtubules of Heliozoa. II. The effect of low temperature on these structures in the formation and maintenance of the axopodia. *J. Cell Biol.* **34**, 327-343.

Tilney, L. G., Hiramoto, Y. and Marsland, D. (1966). Studies on the microtubules in Heliozoa. III. A pressure analysis of the role of these structures in the formation and maintenance of the axopodia of Actinosphaerium nucleofilum. *J. Cell Biol.* **29**, 77-95.

Tilney, L. G., Hatano, S., Ishikawa, H. and Mooseker, M. S. (1973a). The polymerization of actin: its role in the generation of the acrosomal process of certain echinoderm sperm. *J. Cell Biol.* **59**, 109-126.

Tilney, L. G., Bryan, J., Bush, D., Fujiwara, K., Mooseker, M. S., Murphy, D. B. and Snyder, D. H. (1973b). Microtubules: evidence for 13 protofilaments. *J. Cell Biol.* **59**, 267-275.

Tilney, L. G., Kiehart, D. P., Sardet, C. and Tilney, M. (1978). Polymerization of actin. IV. Role of Ca^{2+} and H^+ in the assembly of actin and in membrane fusion in the acrosomal reaction of echinoderm sperm. *J. Cell Biol.* **77**, 536-550.

Timberlake, W. E., Shumard, D. S. and Goldberg, R. B. (1977). Relationship between nuclear and polysomal RNA populations of Achlya: a simple eucaryotic system. *Cell* **10**, 623-632.

Tippit, D. H. and Pickett-Heaps, J. D. (1977). Mitosis in the pennate diatom Surirella ovalis. *J. Cell. Biol.* **73**, 705-727.

Tischfield, J. A. and Ruddle, F. H. (1974). Assignment of the gene for APRT to human chromosome 16 by mouse-somatic cell hybridization. *Proc. Nat. Acad. Sci. USA* **71**, 45-49.

Tissières, A., Mitchell, H. K. and Tracy, U. M. (1974). Protein synthesis in salivary glands of D. melanogaster in relation to chromosome puffs. *J. Mol. Biol.* **84**, 389-398.

Tjio, J. H. and Puck, T. T. (1958). Genetics of somatic mammalian cells. II. Chromosomal constitution of cells in tissue culture. *J. Exp. Med.* **108**, 259-268.

Tobey, R. A. and Crissman, H. A. (1972). Preparation of large quantities of synchronized mammalian cells in late G1 in the pre DNA replicative phase of the cell cycle. *Exp. Cell Res.* **75**, 460-464.

Tobey, R. A., Anderson, E. C. and Petersen, D. F. (1967a). The effect of thymidine on the duration of G1 in Chinese hamster cells. *J. Cell Biol.* **35**, 53-59.

Tobey, R. A., Anderson, E. C. and Petersen, D. F. (1967b). Properties of mitotic cells prepared by mechanically shaking monolayer cultures of Chinese hamster cells. *J. Cell. Physiol.* **70**, 63-68.

Tobler, H., Smith, K. D. and Ursprung, H. (1972). Molecular aspects of chromatin elimination in Ascaris lumbicoides. *Devel. Biol.* **27**, 190-203.

Todaro, G. and Green, H. (1963). Quantitative studies on the growth of mouse embryo cells in culture and their development into established lines. *J. Cell Biol.* **17**, 299-313.

Todaro, G. J., Lazar, G. U. and Green, H. (1965). The initiation of cell division in a contact inhibited mammalian cell line. *J. Cell. Physiol.* **66**, 325-333.

Todd, R. D. and Garrard, W. T. (1979). Overall pathway of mononucleosome production. *J. Biol. Chem.* **254**, 3074-3083.

Tokuyasu, K. T., Peacock, W. J. and Hardy, R. W. (1976). Dynamics of speriogenesis in D. melanogaster. VII. Effects of SD chromosome. *J. Ult. Res.* **58**, 96-107.

Tolstoshev, P., Mitchell, J., Lanyon, G., Williamson, R., Ottolenghi, S., Comi, P., Giglioni, B., Masera, G., Modell, B., Weatherall, D. J. and Clegg, J. B. (1976). Presence of gene for β globin in homozygous β thalassaemia. *Nature* **259**, 95-98.

Tolstoshev, P., Williamson, R., Eskdale, J., Verdier, G., Godet, J., Nigon, V., Trabuchet, G. and Benabadji, M. (1977). Demonstration of two α globin genes per human haploid genome for normals and Hb J Mexico. *Eur. J. Biochem.* **78**, 161-166.

Tonegawa, S. (1976). Reiteration frequency of immunoglobulin light chain genes: further evidence for somatic generation of antibody diversity. *Proc. Nat. Acad. Sci. USA* **73**, 203-207.

Tonegawa, S., Steinberg, C., Dube, S. and Bernardini, A. (1974). Evidence for somatic generation of antibody diversity. *Proc. Nat. Acad. Sci. USA* **71**, 4027-4031.

Tonegawa, S., Brack, C., Hozumi, N. and Schuller, R. (1977). Cloning of an immunoglobulin variable region gene from mouse embryo. *Proc. Nat. Acad. Sci. USA* **74**, 3518-3522.

Tonegawa, S., Maxam, A. M., Tizard, R., Bernard, O. and Gilbert, W. (1978a). Sequence of a mouse germ line gene for a variable region of an immunoglobulin light chain. *Proc. Nat. Acad. Sci. USA* **75**, 1485-1489.

Tonegawa, S., Brack, C., Hozumi, N. and Pirrotta, V. (1978b). Organization of immunoglobulin genes. *Cold Spring Harbor Symp. Quant. Biol.* **42**, 921-931.

Toniolo, D., Meiss, H. K. and Basilico, C. (1973). A temperature sensitive mutation affecting 28S rRNA production in mammalian cells. *Proc. Nat. Acad. Sci. USA* **70**, 1273-1277.

Tourian, A., Johnson, R. T., Burg, K., Nicolson, S. W. and Sperling, K. (1978). Transfer of human chromosomes via human minisegregant cells into mouse cells and the quantitation of the expression of HGPRT. *J. Cell Sci.* **30**, 193-210.

Towle, H. C., Tsai, M.-J., Tsai, S. Y. and O'Malley, B. W. (1977). Effect of estrogen on gene expression in the chick oviduct. Preferential initiation and assymetrical transcription of specific chromatin genes. *J. Biol. Chem.* **252**, 2396-2404.

Tres, L. L. (1977). Extensive pairing of the XY bivalent in mouse spermatocytes as visualized by whole mount electron microscopy. *J. Cell Sci.* **25**, 1-16.

Truett, M. A. and Gall, J. G. (1977). The replication of rDNA in the macronucleus of tetrahymena. *Chromosoma* **64**, 295-304.

Tsai, M. J., Michaelis, G. and Criddle, R. S. (1971). DNA dependent RNA polymerase from yeast mitochondria. *Proc. Nat. Acad. Sci. USA* **68**, 473-477.

Tsai, M.-J., Schwartz, R. J., Tsai, S. Y. and O'Malley, B. W. (1975). Effect of estrogen on gene expression in the chick oviduct. V. Changes in the number of RNA polymerase binding and initiation sites in chromatin. *J. Biol. Chem.* **250**, 5175-5182.

Tsai, M.-J., Towle, H. C., Harris, S. E. and O'Malley, B. W. (1976). Effect of estrogen on gene expression in the chick oviduct. Comparative aspects of RNA chain initiation in chromatin using homologous versus E. coli RNA polymerase. *J. Biol. Chem.* **251**, 1960-1968.

Tsuijimoto, Y. and Suzuki, Y. (1979). Continuous DNA sequence of B. mori fibroin gene including the 5′ flanking, mRNA coding, entire intervening and fibroin protein coding regions. *Cell* **18**, 591-600.

Tuan, D., Biro, P. A., DeRiel, J. K., Lazarus, H. and Forget, B. G. (1979). Restriction endonuclease mapping of the human γ gene loci. *Nucleic Acids Res.* **6**, 2519-2544.

Tuan, D. Y. H. and Bonner, J. (1969). Optical absorbance and optical rotatory dispersion studies on calf thymus nucleohistone. *J. Mol. Biol.* **45**, 59-76.

Tucker, R. W., Sanford, K. K. and Frankel, F. R. (1978). Tubulin and actin in paired non neoplastic and spontaneously transformed neoplastic cell line in vitro: fluorescent antibody studies. *Cell* **13**, 629-642.

Turner, F. R. (1968). An ultrastructural study of plant spermatogenesis. *J. Cell Biol.* **37**, 370-392.

Tzagaloff, A. and Meagher, P. (1972). Assembly of the mitochondrial membrane system. VI.

Mitochondrial synthesis of subunit proteins of the rutamycin sensitive ATPase. *J. Biol. Chem.* **247**, 594–603.

Tzagoloff, A., Akai, A. and Foury, F. (1976b). Assembly of the mitochondrial membrane system. XVI. Modified forms of the ATPase proteolipid in oligomycin resistant mutants of S. cerevisiae. *Febs Lett.* **65**, 391–395.

Tzagoloff, A., Akai, A. and Needleman, R. B. (1975a). Properties of cytoplasmic mutants of S. cerevisiae with specific lesions in cytochrome oxidase. *Proc. Nat. Acad. Sci. USA* **72**, 2054–2057.

Tzagoloff, A., Akai, A. and Needleman, R. B. (1975b). Assembly of the mitochondrial membrane system: isolation of nuclear and cytoplasmic mutants of S. cerevisiae with specific defects in mitochondrial functions. *J. Bacteriol.* **122**, 826–831.

Tzagoloff, A., Akai, A. and Needleman, R. B. (1975c). Assembly of the mitochondrial membrane system. Characterization of nuclear mutants of S. cerevisiae with defects in mitochondrial ATPase and respiratory enzymes. *J. Biol. Chem.* **250**, 8228–8235.

Tzagoloff, A., Akai, A. and Sierra, M. F. (1972). Assembly of the mitochondrial membrane system. VII. Synthesis and integration of F1 subunits into the rutamycin sensitive ATPase. *J. Biol. Chem.* **247**, 6511–6516.

Tzagoloff, A., Foury, A. and Akai, A. (1976a). Assembly of the mitochondrial membrane system. XVIII. Genetic loci on mitochondrial DNA involved in cytochrome *b* biosynthesis. *Molec. Gen. Genet.* **149**, 33–42.

Tzagoloff, A., Rubin, M. S. and Sierra, M. F. (1973). Biosynthesis of mitochondrial enzymes. *Biochim. Biophys. Acta* **301**, 71–104.

Tzagoloff, A., Akai, A., Needleman, R. B. and Zulch, G. (1975d). Assembly of the mitochondrial membrane system. Cytoplasmic mutants of S. cerevisiae with lesions in enzymes of the respiratory chain and in the mitochondrial ATPase. *J. Biol. Chem.* **250**, 8236–8242.

Udem, S. A. and Warner, J. R. (1972). Ribosomal RNA synthesis in S. cerevisiae. *J. Mol. Biol.* **65**, 227–242.

Ukena, T. E. and Berlin, R. D. (1972). Effect of colchicine and vinblastine on the topographical separation of membrane functions. *J. Exp. Med.* **136**, 1–7.

Ukena, T. E., Borysenko, J. Z., Karnovsky, M. J. and Berlin, R. D. (1974). Effects of colchicine, cytochalasin B, and 2 deoxyglucose on the topographical organization of surface bound concanavalin A in normal and transformed fibroblasts. *J. Cell Biol.* **61**, 70–82.

Upchurch, K. S., Levya, A., Arnold, W. J., Holmes, E. W. and Kelley, W. N. (1975). HGPRT deficiency: association of reduced catalytic activity with reduced levels of immunologically detectable enzyme protein. *Proc. Nat. Acad. Sci. USA* **72**, 4142–4146.

Usher, D. C. and Reiter, H. (1977). Catabolism of thymidine during the lymphocyte cell cycle. *Cell* **12**, 365–370.

Valbuena, O., Marcu, K. B., Weigert, M. and Perry, R. P. (1978). Multiplicity of germline genes specifying a group of related mouse K chains with implications for the generation of immunoglobulin diversity. *Nature* **276**, 780–784.

Valenzuela, P., Bell, G. I., Venegas, A., Sewell, E. T., Masiarz, F. R., DeGennaro, L. J., Weinberg, F. and Rutter, W. J. (1977). Ribosomal RNA genes of S. cerevisiae. II. Physical map and nucleotide sequence of the 5S rRNA gene and adjacent intergenic regions. *J. Biol. Chem.* **252**, 8126–8135.

Valenzuela, P., Venegas, A., Weinberg, F., Bishop, R. and Rutter, W. J. (1978). Structure of yeast phe tRNA genes: an intervening DNA segment within the region coding for the tRNA. *Proc. Nat. Acad. Sci. USA* **75**, 190–194.

Vallee, R. B. and Borisy, G. G. (1978). The non tubulin component of microtubule protein oligomers. Effect on self association and hydrodynamic properties. *J. Biol. Chem.* **253**, 2834–2845.

Van, N. T., Monahan, J. J., Woo, S. L., Means, A. R. and O'Malley, B. W. (1977). Comparative studies on the secondary structure of ovalbumin mRNA and its complementary DNA transcript. *Biochemistry* **16**, 4090–4100.

Van Bruggen, E. F. J., Runner, C. M., Borst, P., Ruttenberg, G. J. C. M., Kroon, A. M. and Stekhove, F. M. A. H. (1968). Mitochondrial DNA. III. Electron microscopy of DNA released from mitochondria by osmotic shock. *Biochim. Biophys. Acta.* **161**, 402–414.

Vandenberg, J. L., Cooper, D. W. and Sharman, G. B. (1973). PGK A polymorphism in the wallaby M. parryi: activity of both X chromosomes in muscle. *Nature New Biol.* **243**, 47–48.

Van den Berg, J., Van Ooyen, A., Mantei, N., Schambock, A., Grosveld, G., Flavell, R. A. and Weissmann, C. (1978). Comparison of cloned rabbit and mouse β globin genes showing strong evolutionary divergence of two homologous pairs of introns. *Nature* **276**, 37–43.

Van den Broek, H. W. J., Nooden, L. D., Sevall, J. S. and Bonner, J. (1973). Isolation, purification and fractionation of nonhistone chromosomal proteins. *Biochemistry* **12**, 229–236.

Vandekerckhove, J. and Weber, K. (1978a). Mammalian cytoplasmic actins are the products of at least two genes and differ in primary structure in at least 25 identified positions from skeletal muscle actins. *Proc. Nat. Acad. Sci. USA* **75**, 1106–1110.

Vandekerckhove, J. and Weber, K. (1978b). Actin amino acid sequences. Comparison of actins from calf thymus, bovine brain, and SV40 transformed mouse 3T3 cells with rabbit skeletal muscle actin. *Eur. J. Biochem.* **90**, 451–462.

Vandekerckhove, J. and Weber, K. (1978c). At least six different actins are expressed in a higher mammal: an analysis based on the amino acid sequence of the amino terminal tryptic peptide. *J. Mol. Biol.* **126**, 783–802.

Van Diggelen, O. P., Donahue, T. F. and Shin, S. I. (1979). Basis for differential cellular sensitivity to 8 azaguanine and 6 thioguanine. *J. Cell. Physiol.* **98**, 59–72.

Van Holde, K. E., Sahsrabuddhe, C. G., Shaw, B. R., Van Bruggen, E. F. J. and Amberg, A. C. (1974). Electron microscopy of chromatin subunit particles. *Biochem. Biophys. Res. Commun.* **60**, 1365–1370.

Van Lente, F., Jackson, J. F. and Weintraub, H. (1975). Identification of specific cross linked histones after treatment of chromatin with formaldehyde. *Cell* **5**, 45–50.

Van Ommen, G. J. B., Groot, G. S. P. and Grivell, L. A. (1979). Transcription maps of the mitochondrial DNAs of two strains of Saccharomyces: evidence for transcription of strain specific insertions and complex RNA maturation including splicing. *Cell* **18**, 511–523.

Van Ommen, G. J. B., Groot, G. S. P. and Borst, P. (1977). Fine structure physical mapping of 4S RNA genes on mitochondrial DNA of S. cerevisiae. *Molec. Gen. Genet.* **154**, 255–262.

Van Zeeland, A. A., Van Diggelen, M. C. and Simons, J. W. I. (1972). The role of metabolic cooperation in selection of HGPRT deficient mutants from diploid mammalian cell strains. *Mutat. Res.* **14**, 355–363.

Van't Hof, J. (1965). Relationships between mitotic cycle variation, S period duration, and the average rate of DNA synthesis in the root meristem of several plants. *Exp. Cell Res.* **39**, 48–58.

Van't Hof, J. (1975). DNA fiber replication in chromosomes of a higher plant (Pisum sativum). *Exp. Cell Res.* **93**, 95–104.

Van't Hof, J. (1976a). DNA fiber replication of chromosomes of pea root cells terminating S. *Exp. Cell Res.* **99**, 47–56.

Van't Hof, J. (1976b). Replicon size and rate of fork movement in early S of higher plant cells (Pisum sativum). *Exp. Cell Res.* **103**, 395–404.

Van't Hof, J., Bjerknes, C. A. and Clinton, D. H. (1978). Replicon properties of chromosomal DNA fibers and the duration of DNA synthesis of sunflower root tip meristem cells at different temperatures. *Chromosoma* **66**, 161–172.

Van't Hof, J., Kuniyuki, A. and Bjerknes, C. A. (1978). The size and number of replicon families of chromosomal DNA of Arabidopsis thaliana. *Chromosoma* **68**, 269–285.

Vapnek, D., Hautala, J. A., Jacobson, J. W., Giles, N. H. and Kushner, S. R. (1977). Expression in E. coli K12 of the structural gene for catabolic dehydroquinase of N. crassa. *Proc. Nat. Acad. Sci. USA* **74**, 3508–3512.

Varshavsky, A. J., Sundin, O. H. and Bohn, M. J. (1978). SV40 viral minichromosome: preferential exposure of the origin of replication as probed by restriction endonucleases. *Nucleic Acids Res.* **5**, 3469-3479.

Varshavsky, A. J., Bakayev, V. V. and Georgiev, G. P. (1976). Heterogeneity of chromatin subunits in vitro and location of H1. *Nucleic Acids Res.* **3**, 477-492.

Varshavsky, A. J., Sundin, O. and Bohn, M. (1979). A stretch of late SV40 viral DNA about 400 bp long which includes the origin of replication is specifically exposed in SV40 minichromosomes. *Cell* **16**, 453-466.

Varshavsky, A. J., Bakayev, V. V., Chumackov, P. M. and Georgiev, G. P. (1976). Mini-chromosome of SV40: presence of histone H1. *Nucleic Acids Res.* **3**, 2101-2113.

Varshavsky, A. J., Nedospasov, S. A., Schmatchenko, V. V., Bakayev, V. V., Chumackov, P. M. and Georgiev, G. P. (1977). Compact form of SV40 viral minichromosome is resistance to nuclease: possible implications for chromatin structure. *Nucleic Acids Res.* **4**, 3303-3325.

Vaughan, M. H., Warner, J. R. and Darnell, J. E. (1967). Ribosomal precursor particles in the HeLa cell nucleus. *J. Mol. Biol.* **25**, 235-251.

Vaughn, J. C. (1975). DNA reassociation kinetics and chromosome structure in the crabs Cancer borealis and Libinia emarginata. *Chromosoma* **50**, 243-258.

Vendrely, R. and Vendrely, C. (1949). La teneur du noyan cellulaire en DNA a travers les organes, les individus et les especes animales. *Experientia* **5**, 327-329.

Veomett, G., Prescott, D. M., Shay, J. and Porter, K. R. (1974). Reconstruction of mammalian cells from nuclear and cytoplasmic components separated by treatment with cytochalasin B. *Proc. Nat. Acad. Sci. USA* **71**, 1999-2001.

Verma, I. M., Temple, G. F., Fan, H. and Baltimore, D. (1972). In vitro synthesis of DNA complementary to rabbit reticulocyte 10S RNA. *Nature New Biol.* **235**, 163-167.

Vidali, G., Boffa, L. C. and Allfrey, V. G. (1977). Selective release of chromosomal proteins during limited DNAase I digestion of avian erythrocyte chromatin. *Cell* **12**, 409-415.

Vidali, G., Boffa, L. C., Bradbury, E. M. and Allfrey, V. G. (1978). Butyrate suppression of histone deactylation leads to accumulation of multi acetylated forms of histones H3 and H4 and increased DNAase I sensitivity of the associated DNA sequences. *Proc. Nat. Acad. Sci. USA* **75**, 2239-2243.

Villa-Komaroff, L., Efstratiadis, A., Broome, S., Lomedico, P., Tizard, R., Naber, S. P., Chick, W. L. and Gilbert, W. (1978). A bacterial clone synthesizing proinsulin. *Proc. Nat. Acad. Sci. USA* **75**, 3727-3731.

Villarreal, L. P., White, R. T. and Berg, P. (1979). Mutational alterations within the SV40 leader segment generated altered 16S and 19S mRNAs. *J. Virol.* **29**, 209-219.

Vinograd, J., Lebowitz, J. and Watson, R. (1968). Early and late helix coil transitions in closed circular DNA. The number of superhelical turns in polyoma DNA. *J. Mol. Biol.* **33**, 173-197.

Vlad, M. (1977). Quantitative studies of rDNA in amphibians. *J. Cell Sci.* **24**, 109-119.

Vogel, A., Raines, E., Kariya, B., Rivest, M.-J. and Ross, R. (1978). Coordinate control of 3T3 cell proliferation by platelet derived growth factor and plasma components. *Proc. Nat. Acad. Sci. USA* **75**, 2810-2814.

Vogt, V. M. and Braun, R. (1976a). Repeated structure of chromatin in metaphase nuclei of Physarum polycephalum. *Febs Lett.* **64**, 190-194.

Vogt, V. M. and Braun, R. (1976b). Structure of ribosomal DNA in Physarum polycephalum. *J. Mol. Biol.* **106**, 567-588.

Vogt, V. M. and Braun, R. (1977). The replication of ribosomal DNA in Physarum poly-cephalum. *Eur. J. Biochem.* **80**, 557-566.

Volckaert, G., Van de Voorde, A. and Fiers, W. (1978). Nucleotide sequence of the SV40 small t gene. *Proc. Nat. Acad. Sci. USA* **75**, 2160-2164.

Vollenweider, H. J., James, A. and Szybalski, W. (1978). Discrete length classes of DNA depend on mode of dehydration. *Proc. Nat. Acad. Sci. USA* **75**, 710-714.

Von Hippel, P. H., Revzin, A., Gross, C. A. and Wang, A. C. (1974). Non specific DNA binding of genome regulating proteins as a biological control mechanism. I. The lac operon: equilibrium aspects. *Proc. Nat. Acad. Sci. USA* **71**, 4808-4812.

Von Kap-herr, C. and Mukherjee, B. B. (1977). Stability of inactive X chromosome in mouse embryoid body—mule cell and transformed mouse cell—mule heterokarons. *Exp. Cell Res.* **104**, 369-376.

Vosa, C. G. (1970). The discriminating fluorescence patterns of the chromosomes of D. melanogaster. *Chromosoma* **31**, 446-451.

Vournakis, J. N., Gelinas, R. E. and Kafatos, F. C. (1974). Short polyadenylic acid sequences in insect chorion mRNA. *Cell* **3**, 265-274.

Waalwijk, C. and Flavell, R. A. (1978). DNA methylation at a CCGG sequence in the large intron of the rabbit β globin gene: tissue specific variations. *Nucleic Acids Res.* **5**, 4531-4542.

Wahl, G. M., Hughes, S. H. and Capecchi, M. R. (1975). Immunological characterization of HGPRT mutants of mouse L cells: evidence for mutations at different loci in the HGPRT gene. *J. Cell. Physiol.* **85**, 307-320.

Wahli, W., Dawid, I. B., Wyler, T., Jaggi, R. B., Weber, R. and Ryffel, G. U. (1979). Vitellogenin in X. laevis is encoded in a small family of genes. *Cell* **16**, 535-550.

Wake, R. G. (1973). Circularity of the B. subtilis chromosome and further studies on its bidirectional replication. *J. Mol. Biol.* **77**, 569-575.

Walbot, V. (1977). The dimorphic chloroplasts of the C4 plant Panicum maximum contain identical genomes. *Cell* **11**, 729-737.

Walbot, V. and Dure, L. S. (1976). Developmental biochemistry of cotton seed embryogenesis and germination. VII. Characterization of the cotton genome. *J. Mol. Biol.* **101**, 503-536.

Waldeck, W., Fohring, B., Chowdhury, K., Gruss, P. and Sauer, G. (1978). Origin of DNA replication in papovavirus chromatin is recognized by endogenous endonuclease. *Proc. Nat. Acad. Sci. USA* **75**, 5964-5968.

Walker, J. M., Hastings, J. R. B. and Johns, E. W. (1977). The primary structure of a nonhistone chromosomal protein. *Eur. J. Biochem.* **76**, 461-468.

Walker, P. M. B. (1971a). Origin of satellite DNA. *Nature* **229**, 306-308.

Walker, P. M. B. (1971b). Repetitive DNA in higher organisms. *Prog. Biophys. Mol. Biol.* **23**, 145-190.

Wall, R., Philipson, L. and Darnell, J. E. (1972). Processing of adenovirus specific nuclear RNA during virus replication. *Virology* **50**, 27-34.

Wallace, D. C., Bunn, C. L. and Eisenstadt, J. (1975). Cytoplasmic transfer of chloramphenicol resistance in human tissue culture cells. *J. Cell Biol.* **67**, 174-188.

Wallace, H. and Birnsteil, M. L. (1966). Ribosomal cistrons and the nucleolar organizer. *Biochim. Biophys. Acta* **114**, 296-310.

Wallace, R. B., Dube, S. K. and Bonner, J. (1977). Localization of the globin gene in the template active fraction of chromatin of Friend leukemia cells. *Science* **198**, 1166-1168.

Wang, E. and Goldberg, A. R. (1976). Changes in microfilament organization and surface topography upon transformation of chick embryo fibroblasts with RSV. *Proc. Nat. Acad. Sci. USA* **73**, 4065-4069.

Wang, J. C. (1979). Helical repeat of DNA in solution. *Proc. Nat. Acad. Sci. USA* **76**, 200-203.

Wang, K., Ash, F. and Singer, S. (1975). Filamin: a new high molecular weight protein found in smooth muscle and nonmuscle cells. *Proc. Nat. Acad. Sci. USA* **72**, 4483-4486.

Wang, R. J. and Yin, L. (1976). Further studies on a mutant mammalian line defective in mitosis. *Exp. Cell Res.* **101**, 331-336.

Wang, T. S. F., Sedwick, W. D. and Korn, D. (1974). Nuclear DNA polymerase. Purification and properties of the homogeneous enzyme from human KB cells. *J. Biol. Chem.* **249**, 841-850.

Wang, T. S. F., Sedwick, W. D. and Korn, D. (1975). Nuclear DNA polymerase. Further observations on the structure and properties of the enzyme from human KB cells. *J. Biol. Chem.* **250**, 7040-7044.

Waqar, M. A., Evans, M. J. and Huberman, J. A. (1978). Effect of 2'3'dideoxythymidine 5' triphosphate on HeLa cell in vitro DNA synthesis: evidence that DNA polymerase α is the only polymerase required for cellular DNA replication. *Nucleic Acids Res.* **5**, 1933-1946.

Warner, F. D. (1967). New observations on flagellar fine structure. The relationship between matrix structure and the microtubule component of the axoneme. *J. Cell Biol.* **47**, 159-182.

Warner, F. D. and Satir, P. (1973). The substructure of ciliary microtubules. *J. Cell Sci.* **12**, 313-326.

Warner, J. R., Soeiro, R., Birnboim, H. C., Girard, M. and Darnell, J. E. (1966a). Rapidly labeled HeLa cell nuclear RNA. I. Identification by zone sedimentation of a heterogeneous fraction separate from ribosomal precursor RNA. *J. Mol. Biol.* **19**, 349-361.

Warner, J. R., Girard, M., Latham, H. and Darnell, J. E. (1966b). Ribosome formation in HeLa cells in the absence of protein synthesis. *J. Mol. Biol.* **19**, 373-382.

Wasmuth, J. J. and Caskey, C. T. (1976a). Biochemical characterization of azetidine carboxylic acid-resistant Chinese hamster cells. *Cell* **8**, 71-78.

Wasmuth, J. J. and Caskey, C. T. (1976b). Selection of temperature sensitive CHL asp-tRNA synthetase mutants using the toxic lysine analog, S-2-aminoethyl-L-cysteine. *Cell* **9**, 655-662.

Watson, B., Gormley, I. P., Gardner, S. E., Evans, H. J. and Harris, H. (1972). Reappearance of murine HGPRT activity in mouse A9 cells after attempted hybridization with human cell lines. *Exp. Cell Res.* **75**, 401-409.

Watson, D. C., Peters, E. H. and Dixon, G. H. (1977). The purification, characterization and partial sequence determination of a trout testis nonhistone protein, HMG-T. *Eur. J. Biochem.* **74**, 53-60.

Weatherall, D. J. and Clegg, J. B. (1975). The α chain termination mutants and their relationship to the α thalassemias. *Phil. Trans. Roy. Soc.* **271**, 411-455.

Weatherall, D. J. and Clegg, J. B. (1976). Molecular genetics of human hemoglobin. *Ann. Rev. Genet.* **10**, 157-178.

Weatherall, D. J. and Clegg, J. B. (1979). Recent developments in the molecular genetics of human hemoglobin. *Cell* **16**, 467-479.

Weatherall, D. J., Clegg, J. B., Milner, P. F., Marsh, G. W., Bolton, F. G. and Serjeant, S. R. (1976). Linkage relationships between β and δ structural loci and African forms of β thalassemia. *J. Med. Genet.* **13**, 20-26.

Weatherbee, J. A., Luftig, R. B. and Weihing, R. R. (1978). In vitro polymerization of microtubules from HeLa cells. *J. Cell Biol.* **78**, 47-57.

Weaver, S., Haigwood, N. L., Hutchison, C. A. and Edgell, M. H. (1979). DNA fragments of the M. musculus β globin haplotypes Hbb, and Hbbd. *Proc. Nat. Acad. Sci. USA* **76**, 1385-1389.

Weber, K. and Groeschel-Stewart, U. (1974). Antibody to myosin: the specific localization of myosin containing filaments in nonmuscle cells. *Proc. Nat. Acad. Sci. USA* **71**, 4561-4564.

Weber, K., Rathke, P. C., Osborn, M. and Francke, W. W. (1976). Distribution of actin and tubulin in cells and in glycerinated cell models after treatment with cytochalasin B. *Exp. Cell Res.* **102**, 285-297.

Weber, K., Bibring, T. and Osborn, M. (1975). Specific visualization of tubulin containing structures in tissue culture cells by immunofluorescence. Cytoplasmic microtubules, vinblastine-induced paracrystals and mitotic figures. *Exp. Cell Res.* **95**, 111-120.

Weber, K., Pollack, R. and Bibring, T. (1975). Antibody against tubulin: the specific visualization of cytoplasmic microtubules in tissue culture cells. *Proc. Nat. Acad. Sci. USA* **72**, 459-463.

Weber, L. A., Hickey, E. D., Nuss, D. L. and Baglioni, C. (1977). 5' terminal 7 methylguanosine and mRNA function influence of K⁺ concentration on translation in vitro. *Proc. Nat. Acad. Sci. USA* **74**, 3254-3258.

Weber, L. A., Feman, E. R., Hickey, E. D., Williams, M. C. and Baglioni, C. (1976). Inhibition of HeLa mRNA translation by 7-methylguanosine 5′ monophosphate. *J. Biol. Chem.* **251**, 5657-5662.

Wegnez, M., Monier, R. and Denis, H. (1972). Sequence heterogeneity of 5S RNA in X. laevis. *Febs Lett.* **25**, 13-30.

Wei, C. M. and Moss, B. (1974). Methylation of newly synthesized viral mRNA by an enzyme in vaccinia virus. *Proc. Nat. Acad. Sci. USA* **71**, 3014-3018.

Wei, C. M. and Moss, B. (1975). Methylated nucleotides block 5′ terminus of vaccinia virus mRNA. *Proc. Nat. Acad. Sci. USA* **72**, 318-322.

Wei, C. M. and Moss, B. (1977a). Nucleotide sequences at the N⁶ methyladenosine sites of HeLa cell mRNA. *Biochemistry* **16**, 1672-1676.

Wei, C.-M. and Moss, B. (1977b). 5′ terminal capping of RNA by guanylyl transferase from HeLa cell nuclei. *Proc. Nat. Acad. Sci. USA* **74**, 3758-3761.

Wei, C.-M., Gershowitz, A. and Moss, B. (1975). Methylated nucleotides block 5′ terminus of HeLa cell. *Cell* **4**, 379-386.

Wei, C. M., Gershowitz, A. and Moss, B. (1976). 5′ terminal and internal methylated nucleotide sequences in HeLa cell mRNA. *Biochemistry* **15**, 397-401.

Weigert, M. and Riblet, R. (1977). Genetic control of antibody variable regions. *Cold Spring Harbor Symp. Quant. Biol.* **41**, 837-846.

Weigert, M., Gatmaitan, L., Loh, E., Schilling, J. and Hood, L. (1978). Rearrangement of genetic information may produce immunoglobulin diversity. *Nature* **276**, 785-789.

Weihing, R. R. (1976). Cytochalasin B inhibits actin related gelation of HeLa cell extracts. *J. Cell Biol.* **71**, 303-307.

Weihing, R. R. (1977). Effects of myosin and heavy meromyosin on actin related gelation of HeLa cell extracts. *J. Cell Biol.* **75**, 95-103.

Weil, P. A. and Blatti, S. P. (1975). Partial purification and properties of calf thymus DNA dependent RNA polymerase III. *Biochemistry* **14**, 1636-1643.

Weil, P. A., Luse, D. S., Segall, J. and Roeder, R. G. (1979a). Selective and accurate initiation of transcription at the adenovirus 2 major late promoter in a soluble cell free system dependent upon purified RNA polymerase II and purified DNA templates. *Cell* **18**, 469-484.

Weil, P. A., Sagall, J., Harris, B., Ng, S. and Roeder, R. G. (1979b). Faithful transcription of eucaryotic genes by RNA polymerase III in systems reconstituted with purified templates. *J. Biol. Chem.* **254**, 6163-6173.

Weinberg, E. S., Birnsteil, M. L., Purdom, I. F. and Williamson, R. (1972). Genes coding for polysomal 9S RNA of sea urchins: conservation and divergence. *Nature* **240**, 225-228.

Weinberg, R. A. and Penman, S. (1968). Small molecular weight monodisperse nuclear RNA. *J. Mol. Biol.* **38**, 289-304.

Weinberg, R. and Penman, S. (1969). Metabolism of small molecular weight monodisperse nuclear RNA. *Biochim. Biophys. Acta* **190**, 10-29.

Weinberg, R. A. and Penman, S. (1970). Processing of 45S nucleolar RNA. *J. Mol. Biol.* **47**, 169-178.

Weinberg, E. S., Overton, G. C., Shutt, R. H. and Reeder, R. H. (1975). Histone gene arrangement in the sea urchin, S. purpuratus. *Proc. Nat. Acad. Sci. USA* **72**, 4815-4819.

Weiner, A. M. and Weber, K. (1971). Natural readthrough at the UGA termination signal of Qβ coat protein cistron. *Nature New Biol.* **234**, 206-208.

Weiner, A. M. and Weber, K. (1973). A single UGA codon functions as a natural termination signal in the phage Qβ coat protein. *J. Mol. Biol.* **80**, 887-889.

Weingarten, M. D., Suter, M. M., Littman, D. R. and Kirschner, M. W. (1974). Properties of the depolymerization products of microtubules from mammalian brain. *Biochemistry* **13**, 5529-5537.

Weingarten, M. D., Lockwood, A. H., Hwo, S.-Y. and Kirschner, M. W. (1975). A protein factor essential for microtubule assembly. *Proc. Nat. Acad. Sci. USA* **72**, 1858-1862.

Weinmann, R. and Roeder, R. G. (1974). Role of DNA-dependent RNA polymerase III in the transcription of the tRNA and 5S RNA genes. *Proc. Nat. Acad. Sci. USA* **71**, 1790-1794.

Weinmann, R., Brendler, T. G., Raskas, H. J. and Roeder, R. G. (1976). Low molecular weight viral RNAs transcribed by RNA polymerase III during adenovirus 2 infection. *Cell* **7**, 557-566.

Weinstock, R., Sweet, R., Weiss, M., Cedar, H. and Axel, R. (1978). Intragenic spacers interrupt the ovalbumin gene. *Proc. Nat. Acad. Sci. USA* **75**, 1299-1303.

Weintraub, H. (1973). The assembly of newly replicated DNA into chromatin. *Cold Spring Harbor Symp. Quant. Biol.* **38**, 247-256.

Weintraub, H. (1975). Release of discrete subunits after nuclease and trypsin digestion of chromatin. *Proc. Nat. Acad. Sci. USA* **72**, 1212-1216.

Weintraub, H. (1976). Cooperative alignment of nu bodies during chromosome replication in the presence of cycloheximide. *Cell* **9**, 419-422.

Weintraub, H. (1978). The nucleosome repeat length increases during erythropoiesis in the chick. *Nucleic Acids Res.* **5**, 1179-1188.

Weintraub, H. and Groudine, M. (1976). Chromosomal subunits in active genes have an altered conformation. *Science* **193**, 848-856.

Weintraub, H. and Van Lente, F. (1974). Dissection of chromosome structure with trypsin and nucleases. *Proc. Nat. Acad. Sci. USA* **71**, 4249-4253.

Weintraub, H., Palter, K. and Van Lente, F. (1975). Histones H2A, H2B, H3 and H4 form a tetrameric complex in solutions of high salt. *Cell* **6**, 85-110.

Weintraub, H., Worcel, A. and Alberts, B. (1976). A model for chromatin based upon two symmetrically paired half nucleosomes. *Cell* **9**, 409-418.

Weisblum, B. and De Haseth, P. L. (1972). Quinacrine, a chromosome stain specific for dAT rich regions in DNA. *Proc. Nat. Acad. Sci. USA* **69**, 629-632.

Weisbrod, S. and Weintraub, H. (1979). Isolation of a subclass of nuclear proteins responsible for conferring a DNAase I sensitive structure on globin chromatin. *Proc. Nat. Acad. Sci. USA* **76**, 630-634.

Weischet, W. O., Tatchell, K., Van Holde, K. E. and Klump, H. (1978). Thermal denaturation of nucleosomal core particles. *Nucleic Acids Res.* **5**, 139-160.

Weischet, W. O., Allen, J. R., Riedel, G. and Van Holde, K. E. (1979). The effects of salt concentration and H1 depletion on the digestion of calf thymus chromatin by micrococcal nuclease. *Nucleic Acids Res.* **6**, 1843-1862.

Weisenberg, R. C. (1972). Microtubule formation in vitro in solutions containing low calcium concentrations. *Science* **177**, 1104-1105.

Weisenberg, R. C. and Rosenfeld, A. C. (1975). In vitro polymerization of microtubules into asters and spindles in homogenates of surf clam eggs. *J. Cell Biol.* **64**, 146-158.

Weisenberg, R. C., Borisy, G. G. and Taylor, E. W. (1968). The colchicine binding protein of mammalian brain and its relation to microtubules. *Biochemistry* **7**, 4466-4478.

Weisenberg, R. C., Deery, W. J. and Dickinson, P. J. (1976). Tubulin-nucleotide interactions during the polymerization and depolymerization of microtubules. *Biochemistry* **15**, 4248-4254.

Weiss, B. (1976). Endonuclease II of E. coli is exonuclease III. *J. Biol. Chem.* **251**, 1896-1901.

Weiss, B. G. (1969). The dependence of DNA synthesis on protein synthesis in HeLa S3 cells. *J. Cell. Physiol.* **73**, 85-90.

Weiss, M. C. and Chaplain, M. (1971). Expression of differentiation functions in hepatoma cell hybrids: reappearance of TAT inducibility after the loss of chromosomes. *Proc. Nat. Acad. Sci. USA* **68**, 3026–3030.

Weiss, M. C. and Ephrussi, B. (1966). Studies of interspecific (rat x mouse) somatic hybrids. I. Isolation, growth and evolution of the karyotype. *Genetics* **54**, 1095–1109.

Weiss, M. C. and Green, H. (1967). Human mouse hybrid cell lines containing partial complements of human chromosomes and functioning human genes. *Proc. Nat. Acad. Sci. USA* **58**, 1104–1111.

Weiss, S. R., Hackett, P. B., Oppermann, H., Ulrich, A., Levintow, L. and Bishop, J. M. (1978). Cell free translation of avian sarcoma virus RNA: suppression of the *gag* termination codon does not augment synthesis of the joint gag/pol product. *Cell* **15**, 607–614.

Weisbeek, P. J., Borrias, W. E., Langeveld, S. A., Baas, P. D. and Van Arkel, G. A. (1977). Phage ϕX174: gene A overlaps gene B. *Proc. Nat. Acad. Sci. USA* **74**, 2504–2508.

Weissbach, A. (1975). Vertebrate DNA polymerases. *Cell* **5**, 101–108.

Weissbach, A., Schlachbach, A., Fridlender, B. and Bolden, A. (1971). DNA polymerase from human cells. *Nature New Biol.* **231**, 167–170.

Wellauer, P. K. and Dawid, I. B. (1973). Secondary structure maps of RNA: processing of HeLa rRNA. *Proc. Nat. Acad. Sci. USA* **70**, 2827–2831.

Wellauer, P. K. and Dawid, I. B. (1974). Secondary structure maps of ribosomal RNA and DNA. I. Processing of X. laevis ribosomal RNA and structure of single stranded rDNA. *J. Mol. Biol.* **89**, 379–396.

Wellauer, P. K. and Dawid, I. B. (1977). The structural organization of rDNA in D. melanogaster. *Cell* **10**, 193–212.

Wellauer, P, K. and Dawid, I. B. (1978). Ribosomal DNA in D. melanogaster. II. Heteroduplex mapping of cloned and uncloned rDNA. *J. Mol. Biol.* **126**, 769–782.

Wellauer, P. K. and Dawid, I. B. (1979). Isolation and sequence organization of human ribosomal DNA. *J. Mol. Biol.* **128**, 289–304.

Wellauer, P. K. and Reeder, R. H. (1975). A comparison of the structural organization of amplified ribosomal DNA from Xenopus mulleri and Xenopus laevis. *J. Mol. Biol.* **94**, 151–162.

Wellauer, P. K., Dawid, I. B. and Tartof, K. D. (1978). X and Y chromosomal rRNA of Drosophila: comparison of spacers and insertions. *Cell* **14**, 269–278.

Wellauer, P. K., Dawid, I. B., Kelley, D. E. and Perry, R. P. (1974a). Secondary structure maps of ribosomal RNA. II. Processing of mouse L cell rRNA and variations in the processing pathway. *J. Mol. Biol.* **89**, 397–407.

Wellauer, P. K., Reeder, R. H., Carroll, D., Brown, D. D., Deutch, A., Higashinakagawa, T. and Dawid, I. B. (1974b). Amplified ribosomal DNA from X. laevis has heterogeneous spacer lengths. *Proc. Nat. Acad. Sci. USA* **71**, 2823–2827.

Wellauer, P. K., Dawid, I. B., Brown, D. D. and Reeder, R. H. (1976a). The molecular basis for length heterogeneity in ribosomal DNA from X. laevis. *J. Mol. Biol.* **105**, 461–486.

Wellauer, P. K., Reeder, R. H., Dawid, I. B. and Brown, D. D. (1976b). The arrangement of length heterogeneity in repeating units of amplified and chromosomal rDNA from X. laevis. *J. Mol. Biol.* **105**, 487–506.

Wells, R. and Sager, R. (1971). Denaturation and renaturation kinetics of chloroplast DNA from C. reinhardii. *J. Mol. Biol.* **58**, 611–622.

Welsh, M. J., Dedman, J. R., Brinkley, B. R. and Means, A. R. (1978). Calcium dependent regulator protein: localization in mitotic apparatus of eucaryotic cells. *Proc. Nat. Acad. Sci. USA* **75**, 1867–1871.

Welsh, M. J., Dedman, J. R., Brinkley, B. R. and Means, A. R. (1979). Tubulin and calmodulin. Effects on microtubule and microfilament inhibitors on localization in the mitotic apparatus. *J. Cell Biol.* **81**, 624–634.

Wensink, P. C. (1978). Sequence homology within families of D. melanogaster middle repetitive DNA. *Cold Spring Harbor Symp. Quant. Biol.* **42**, 1033-1039.

Wensink, P. C. and Brown, D. D. (1971). Denaturation map of the rDNA of X. laevis. *J. Mol. Biol.* **60**, 235-248.

Wensink, P. C., Finnegan, D. J., Donelson, J. E. and Hogness, D. (1974). A system for mapping DNA sequences in the chromosomes of D. melanogaster. *Cell* **3**, 315-325.

Westergaard, M. and Von Wettstein, D. (1972). The synaptonemal complex. *Ann. Rev. Genet.* **6**, 71-110.

Wetmur, J. G. and Davidson, N. (1968). Kinetics of renaturation of DNA. *J. Mol. Biol.* **31**, 349-370.

Wettstein, R. and Sotelo, J. (1967). Electron microscopic serial reconstruction of the spermatocytes. I. Nuclei at pachytene. *J. Microscopie* **6**, 557-576.

Whalen, R. G., Butler-Browne, G. S. and Gros, F. (1976). Protein synthesis and actin heterogeneity in calf muscle cells in culture. *Proc. Nat. Acad. Sci. USA* **73**, 2018-2022.

Wheeler, L. L. and Altenburg, L. C. Hoechst 33258 banding of D. nasutoides metaphase chromosomes. *Chromosoma* **62**, 351-360.

White, R., Pasztor, L. M. and Hu, F. (1975). Mouse satellite DNA in noncentromeric heterochromatin of cultured cells. *Chromosoma* **50**, 275-282.

White, R. L. and Hogness, D. S. (1977). R loop mapping of the 18S and 28S sequences in the long and short repeating units of D. melanogaster rDNA. *Cell* **10**, 177-192.

Whitehouse, H. L. K. (1967). A cycloid model for the chromosome. *J. Cell Sci.* **2**, 9-22.

Whitehouse, H. L. K. (1973). Towards an Understanding of the Mechanism of Heredity. St Martins, London.

Whiteway, M. S. and Lee, R. W. (1977). Chloroplast DNA content increases with nuclear ploidy in Chlamydomonas. *Molec. Gen. Genet.* **157**, 11-16.

Whitfield, P. R. and Spencer, D. (1968). Buoyant density of tobacco and spinach chloroplast DNA. *Biochim. Biophys. Acta* **157**, 333-343.

Whitfeld, P. R., Herrmann, R. G. and Bottomley, W. (1978). Mapping the rRNA genes on spinach chloroplast DNA. *Nucleic Acids Res.* **5**, 1741-1752.

Whitfield, C. D., Buchsbaum, B., Bostedor, R. and Chu, E. H. Y. (1978). Inverse relationship between galactokinase activity and 2-deoxygalactose resistance in CHO cells. *Somatic Cell Genet.* **4**, 699-714.

Whitlock, J. P., Jr. and Simpson, R. T. (1976). Removal of histone H1 exposes a 50 base pair DNA segment between nucleosomes. *Biochemistry* **15**, 3307-3313.

Whitlock, J. P., Jr. and Simpson, R. T. (1977). Localization of the sites along nucleosome DNA which interact with NH2 terminal histone regions. *J. Biol. Chem.* **252**, 6516-6520.

Whitlock, J. P. and Stein, A. (1978). Folding of DNA by histones which lack their NH2 terminal regions. *J. Biol. Chem.* **253**, 3857-3861.

Whitlock, J. P., Rushizky, G. W. and Simpson, R. T. (1977). DNAase sensitive sites in nucleosomes. Their relative susceptibilities depend on nuclease used. *J. Biol. Chem.* **252**, 3003-3006.

Wiche, G. and Cole, R. D. (1976a). Reversible in vitro polymerization of tubulin from a cultured cell line (rat glial cell clone C6). *Proc. Nat. Acad. Sci. USA* **73**, 1227-1231.

Wiche, G. and Cole, R. D. (1976b). An improved preparation of highly specific tubulin antibodies. *Exp. Cell Res.* **99**, 15-22.

Wiche, G., Honig, L. S., Cole, R. D. (1979). Microtubule protein preparations from C6 glial cells and their spontaneous polymer formation. *J. Cell Biol.* **80**, 553-563.

Wickett, R. R., Li, H. J. and Isenberg, I. (1972). Salt effects on histone IV conformation. *Biochemistry* **11**, 2952-2957.

Widnell, C. C. and Tata, J. R. (1966). Studies on the stimulation by ammonium sulfate of the DNA dependent RNA polymerase of isolated rat liver nuclei. *Biochim. Biophys. Acta* 123, 478-492.

Wienand, U. and Feix, G. (1978). Electrophoretic fractionation and translation in vitro of poly(A) containing RNA from maize endosperm. Evidence for two mRNAs coding for zein protein. *Eur. J. Biochem.* 92, 605-612.

Wiesehahn, G. P., Hyde, J. E. and Hearst, J. E. (1977). The photoaddition of trimethylpsoralen to D. melanogaster nuclei: a probe for chromatin substructure. *Biochemistry* 16, 925-931.

Wigler, M., Silverstein, S., Lee, L.-S., Pellicer, A., Cheng, Y.-C. and Axel, R. (1977). Transfer of purified herpes virus thymidine kinase gene to cultured mouse cells. *Cell* 11, 223-232.

Wigler, M., Pellicer, A., Silverstein, S. and Axel, R. (1978). Biochemical transformation of single copy eucaryotic genes using DNA as donor. *Cell* 14, 725-731.

Wigler, M., Sweet, R., Sim, G. K., Wold, B., Pellicer, A., Lacy, E., Maniatis, T., Silverstein, S. and Axel, R. (1979a). Transformation of mammalian cells with genes from procaryotes and eucaryotes. *Cell* 16, 777-786.

Wigler, M., Pellicer, A., Silverstein, S., Axel, R., Urlaub, G. and Chasin, L. (1979b). DNA mediated transfer of the APRT locus into mammalian cells. *Proc. Nat. Acad. Sci. USA* 76, 1373-1376.

Wild, M. A. and Gall, J. G. (1979). An intervening sequence in the gene coding for 25S rRNA of Tetrahymena pigmentosa. *Cell* 16, 565-573.

Wilhelm, J. A. and McCarty, K. A. (1970). The uptake and turnover of acetate in HeLa cell histone fractions. *Cancer Res.* 30, 418-425.

Wilkie, D. and Thomas, D. Y. (1973). Mitochondrial genetic analysis by zygote cell lineages in S. cerevisiae. *Genetics* 73, 367-377.

Wilkins, M. F. H., Zubay, G. and Wilson, H. R. (1959). X ray diffraction studies of the molecular structure of nucleohistone and chromosomes. *J. Mol. Biol.* 1, 179-185.

Willecke, K. and Ruddle, F. H. (1975). Transfer of human gene for HGPRT via isolated human metaphase chromosomes into mouse L cells. *Proc. Nat. Acad. Sci. USA* 72, 1792-1796.

Willecke, K., Lange, R., Kruger, A. and Reber, T. (1976). Cotransfer of two linked human genes into cultured mouse cells. *Proc. Nat. Acad. Sci. USA* 73, 1274-1278.

Willecke, K., Reber, T., Kucherlapati, R. S. and Ruddle, F. H. (1977). Human mitochondrial thymidine kinase is coded for by a gene on chromosome 16 of the nucleus. *Somatic Cell Genet.* 3, 237-246.

Williams, A. F. (1972a). Deoxythymidine metabolism in avian erythroid cells. *J. Cell Sci.* 11, 777-784.

Williams, A. F. (1972b). DNA polymerase in avian erythroid cells. *J. Cell Sci.* 11, 785-798.

Williams, J. G. and Penman, S. (1975). The messenger RNA sequences in growing and resting mouse fibroblasts. *Cell* 6, 197-206.

Williams, J. G., Hoffman, R. and Penman, S. (1977). The extensive homology between mRNA sequences of normal and SV40 transformed human fibroblasts. *Cell* 11, 901-908.

Williamson, D. H. (1970). The effect of environmental and genetic factors on the replication of mitochondrial DNA in yeast. *Symp. Soc. Exp. Biol.* 24, 247-276.

Williamson, D. H. and Scopes, A. W. (1961). Nucleic acids and proteins in different sized yeast cells. *Exp. Cell Res.* 24, 151-153.

Williamson, R., Crossley, J. and Humphries, S. (1974). Translation of mouse globin mRNA from which the poly-A sequence has been removed. *Biochemistry* 13, 703-707.

Willingham, M. C. and Pastan, I. (1975). Cyclic AMP and cell morphology in cultured fibroblasts. Effects on cell shape, microfilament and microtubule distribution, and orientation to substratum. *J. Cell Biol.* 67, 146-159.

Willingham, M., Ostlund, R. and Pastan, I. (1974). Myosin is a component of the cell surface of cultured cells. *Proc. Nat. Acad. Sci. USA* **71**, 4144-4148.

Willingham, M. C., Yamada, K. M., Yamada, S. S., Pouyssegur, J. and Pastan, I. (1977). Microfilament bundles and cell shape are related to adhesiveness to substratum and are dissociable from growth control in cultured fibroblasts. *Cell* **10**, 375-380.

Willis, M. V., Baseman, J. B. and Amos, H. (1974). Noncoordinate control of RNA synthesis in eucaryotic cells. *Cell* **3**, 185-188.

Wilson, A. C., Carlson, S. S. and White, T. J. (1977). Biochemical evolution. *Ann. Rev. Biochem.* **46**, 573-640.

Wilson, D. A. and Thomas, C. A., Jr. (1974). Palindromes in chromosomes. *J. Mol. Biol.* **84**, 115-138.

Wilson, E. B. (1925). The cell in development and heredity. Macmillan, London.

Wilson, F. E., Blin, N. and Stafford, D. W. (1978). A denaturation map of sea urchin rDNA. *Chromosoma* **58**, 247-254.

Wilson, F. E., Blin, N. and Stafford, D. (1978). Arrangement of the rRNA sequences in the rDNA of L. variegatus. *Chromosoma* **65**, 373-382.

Wilson, G. N., Steggles, A. W. and Nienhuis, A. W. (1975). Strand selective transcription of globin genes in rabbit erythroid cells and chromatin. *Proc. Nat. Acad. Sci. USA* **72**, 4835-4839.

Wilson, J. T., de Riel, J. K., Forget, B. G., Marotta, C. A. and Weissman, S. M. (1977). Nucleotide sequence of 3' untranslated portion of human α globin mRNA. *Nucleic Acids Res.* **4**, 2353-2368.

Wilson, L. (1970). Properties of colchicine binding protein from chick embryo brain. Interactions with vinca alkaloids and podophyllotoxin. *Biochemistry* **9**, 4999-5007.

Wilson, L. (1975). Microtubules as drug receptors: pharmacological properties of microtubule protein. *Ann. NY Acad. Sci.* **253**, 213-231.

Wilson, L. and Meza, I. (1973). The mechanism of action of colchicine. Colchicine binding properties of sea urchin sperm tail outer doublet tubulin. *J. Cell Biol.* **58**, 709-719.

Wilson, L., Anderson, K. and Chin, D. (1976). Nonstoichiometric poisoning of microtubule polymerization: a model for the mechanism of action of the Vinca alkaloids, podophyllotoxin and colchicine. In R. Goldman, R. Pollack and J. Rosenbaum (Eds.), *Cell Motility*, Cold Spring Harbor, New York, pp 1051-1064.

Wilson, L., Creswell, K. and Chin, D. (1975). The mechanism of action of vinblastine. Binding of acetyl-^{3}H vinblastine to embryonic chick brain tubulin and tubulin from sea urchin sperm tail outer doublet microtubules. *Biochemistry* **14**, 5586-5592.

Wilson, L., Morse, A. N. C. and Bryan, J. (1978). Characterization of acetyl-^{3}H-labeled vinblastine binding to vinblastine tubulin crystals. *J. Mol. Biol.* **121**, 255-268.

Wilson, M. C. and Melli, M. (1977). Determination of the number of histone genes in human DNA. *J. Mol. Biol.* **110**, 511-536.

Wilson, R. K., Starbuck, W. D., Taylor, C. W., Jordon, J. and Busch, H. (1970). Structure of the glycine rich arginine rich histone of the Novikoff hepatoma. *Cancer Res.* **30**, 2942-2951.

Wilson, L., Bryan, J., Ruby, A. and Mazia, D. (1970). Precipitation of proteins by vinblastine and calcium ions. *Proc. Nat. Acad. Sci. USA* **66**, 807-814.

Wilson, M. C., Sawicki, S. G., White, P. A. and Darnell, J. E. (1978). A correlation between the rate of poly(A) shortening and half life of mRNA in adenovirus transformed cells. *J. Mol. Biol.* **126**, 23-36.

Wilt, F. H. (1973). Polyadenylation of maternal RNA of sea urchin eggs after fertilization. *Proc. Nat. Acad. Sci. USA* **70**, 2345-2349.

Wilt, F. H. (1977). The dynamics of maternal poly(A)-containing mRNA in fertilized sea urchin eggs. *Cell* **11**, 673-681.

Wimber, D. E. and Steffensen, D. M. (1970). Localization of 5S RNA genes on Drosophila chromosomes by RNA-DNA hybridization. *Science* **170**, 639-640.

Winicov, I. and Perry, R. P. (1974). Characterization of a nucleolar endonuclease possibly involved in ribosomal RNA maturation. *Biochemistry* 13, 2908-2914.

Winicov, I. and Perry, R. P. (1976). Synthesis, methylation and capping of nuclear RNA by a subcellular system. *Biochemistry* 15, 5039-5045.

Winters, M. A. and Edmonds, M. (1973a). A poly(A) polymerase from calf thymus. Purification and properties of the enzyme. *J. Biol. Chem.* 248, 4756-4762.

Winters, M. A. and Edmonds, M. (1973b). A poly(A) polymerase from calf thymus. Characterization of the reaction product and the primer requirement. *J. Biol. Chem.* 248, 4763-4768.

Wintersberger, E. (1978). Yeast DNA polymerases: antigenic relationship, use of RNA primer and associated exonuclease activity. *Eur. J. Biochem.* 84, 167-172.

Wise, G. E. and Prescott, D. M. (1973). Ultrastructure of enucleated mammalian cells in culture. *Exp. Cell Res.* 81, 63-72.

Wissinger, W. and Wang, R. J. (1978). Studies on cell division in mammalian cells. IV. A temperature sensitive cell line defective in post metaphase chromosome movement. *Exp. Cell Res.* 112, 89-94.

Witman, G. B., Carlson, K. and Rosenbaum, J. L. (1972). Chlamydomonas flagella. II. The distribution of tubulins 1 and 2 in the outer doublet microtubules. *J. Cell Biol.* 54, 540-555.

Witman, G. B., Cleveland, D. W., Weingarten, M. D. and Kirschner, M. W. (1976). Tubulin requires tau for growth onto microtubule initiating sites. *Proc. Nat. Acad. Sci. USA* 73, 4070-4074.

Wold, B. J., Klein, W. H., Hough-Evans, B. R., Britten, R. J. and Davidson, E. H. (1978). Sea urchin embryo mRNA sequences expressed in the nuclear RNA of adult tissues. *Cell* 14, 941-950.

Wolf, K. and Seitz-Mayr, G. (1978). Extrachromosomal inheritance in S. pombe. VIII. Extent of cytoplasmic mixing in zygotes estimated by tetrad analysis of crosses involving mitochondrial markers conferring resistant to ant, chm and ery. *Molec. Gen. Genet.* 164, 321-330.

Wolfe, J. (1970). Structural analysis of basal bodies of the isolated oral apparatus of Tetrahymena pyriformis. *J. Cell Sci.* 6, 679-700.

Wolff, S. and Perry, P. (1974). Differential Giemsa staining of sister chromatids and the study of sister chromatid exchanges without autoradiography. *Chromosoma* 48, 341-353.

Wolff, S. and Perry, P. (1975). Insights on chromosome structure from sister chromatid exchange ratios and the lack of both isolabelling and heterolabelling as determined by the FPG technique. *Exp. Cell Res.* 93, 23-30.

Wolff, S., Bodycote, J. and Rodin, B. (1978). Chromosomal isolabeling caused by three rounds of synthesis in late replicating regions. *Chromosoma* 69, 179-184.

Wolstenholme, D. R. (1973). Replicating molecules from eggs of D. melanogaster. *Chromosoma* 43, 1-18.

Wolstenholme, D. R. and Fauron, C. M. R. (1976). A partial map of the circular mitochondrial genome of D. melanogaster. *J. Cell Biol.* 71, 434-448.

Wong-Staal, F., Mendelsohn, J. and Goulian, M. (1973). Ribonucleotides in closed circular mitochondrial DNA from HeLa cells. *Biochem. Biophys. Res. Commun.* 53, 140-148.

Woo, S. L. C., Rosen, J. M., Liarakos, C. D., Robberson, D. L., Choi, Y. C., Busch, H., Means, A. R. and O'Malley, B. W. (1975). Physical and chemical characterization of purified ovalnumin mRNA. *J. Biol. Chem.* 250, 7027-7039.

Woodcock, C. L. F., Safer, J. P. and Stanchfield, J. E. (1976). Structural repeating units in chromatin. I. Evidence for their general occurrence. *Exp. Cell Res.* 97, 101-110.

Woodcock, C. L. F., Sweetman, H. E. and Frado, L.-L. (1976). Structural repeating units in chromatin. II. Their isolation and partial characterization. *Exp. Cell Res.* 97, 111-119.

Woodrum, D. T., Rich, S. A. and Pollard, T. D. (1975). Evidence for biased bidirectional polymerization of actin filaments using HMM prepared by an improved method. *J. Cell Biol.* 67, 231-237.

Worcel, A. and Benyajati, C. (1977). Higher order coiling of DNA in chromatin. *Cell* 12, 83-100.

Worcel, A., Han, S. and Wong, M. L. (1978). Assembly of newly replicated chromatin. *Cell* 15, 969-968.

Worton, R. G., Ho, C. C. and Duff, C. (1977). Chromosome stability in CHO cells. *Somatic Cell Genet.* 3, 27-46.

Wouters, D., Sautiere, P. and Biserte, G. (1978). Primary structure of histone H2A from gonad of the sea urchin Psammechinus miliaris. *Eur. J. Biochem.* 90, 231-240.

Wright, J. A. and Lewis, W. H. (1974). Evidence of a common site of action for the antitumor agents, hydroxyurea and guanazole. *J. Cell Physiol.* 83, 437-440.

Wright, R. E. and Lederberg, J. (1957). Extranuclear transmission in yeast heterokaryons. *Proc. Nat. Acad. Sci. USA* 43, 919-923.

Wright, W. E. and Hayflick, L. (1972). Formation of enucleate and multinucleate cells in normal and SV40 transformed WI-38 by cytochalasin B. *Exp. Cell Res.* 74, 187-194.

Wu, C., Wong, Y. C. and Elgin, S. C. R. (1979). The chromatin structure of specific genes. II. Disruption of chromatin structure during gene activity. *Cell* 16, 807-814.

Wu, C., Bingham, P. M., Livak, K. J., Holmgren, R. and Elgin, S.C. R. (1979). The chromatin structure of specific genes. I. Evidence for higher order domains of defined DNA sequence. *Cell* 16, 797-806.

Wu, F. C., Elgin, S. C. R. and Hood, L. E. (1973). Nonhistone chromosomal proteins of rat tissues. A comparative study by gel electrophoresis. *Biochem.* 12, 2792.

Wu, G.-J. (1978). Adenovirus DNA directed transcription of 5.5S RNA in vitro. *Proc. Nat. Acad. Sci. USA* 75, 2175-2179.

Wu, G. J. and Dawid, I. B. (1972). Purification and properties of mitochondrial DNA dependent RNA polymerase from ovaries of X. laevis. *Biochemistry* 11, 3589-3595.

Wu, M., Holmes, D. S., Davidson, N., Cohn, R. H. and Kedes, L. H. (1976). The relative positions of sea urchin histone genes on the chimeric plasmids pSp2 and pSp17 as studied by electron microscopy. *Cell* 9, 163-170.

Wullems, G. J., Van der Horst, J. and Bootsma, D. (1975). Incorporation of isolated chromosomes and induction of HGPRT in Chinese hamster cells. *Somatic Cell Genet.* 1, 137-152.

Wullems, G. J., Van der Horst, J. and Bootsma, D. (1976a). Expression of human HGPRT in Chinese hamster cells treated with isolated human chromosomes. *Somatic Cell Genet.* 2, 155-164.

Wullems, G. J., Van der Horst, J. and Bootsma, D. (1976b). Transfer of the human X chromosome to human Chinese hamster cell hybrids via isolated HeLa metaphase chromosomes. *Somatic Cell Genet.* 2, 359-372.

Wullems, G. J., Van der Horst, J. and Bootsma, C. (1977). Transfer of the human genes coding for thymidine kinase and galactokinase to Chinese hamster cells and human-Chinese hamster cell hybrids. *Somatic Cell Genet.* 3, 281-294.

Wurtz, E. A., Boynton, J. E. and Gillham, N. W. (1977). Perturbation of chloroplast DNA amounts and chloroplast gene transmission in C. reinhardii by FUdR. *Proc. Nat. Acad. Sci. USA* 74, 4552-4556.

Wurtz, E. A., Sears, B. B., Rabert, D. K., Shepherd, H. S., Gillham, N. W. and Boynton, J. E. (1979). A specific increase in chloroplast gene mutations following growth of Chlamydomonas in 5-fluorodeoxyuridine. *Molec. Gen. Genet.* 170, 235-242.

Wyandt, H. E. and Hecht, F. (1973a). Human Y chromatin. I. Dispersion and condensation. *Exp. Cell Res.* 81, 453-461.

Wyandt, H. E. and Hecht, F. (1973b). Human Y chromatin. II. DNA replication. *Exp. Cell Res.* 81, 462-467.

Yahara, I. and Edelman, G. M. (1975). Modulation of lymphocyte receptor mobility by locally bound Con A. *Proc. Nat. Acad. Sci. USA* 72, 1579-1583.

Yamada, K. M. (1978). Immunological characterization of a major transformation-sensitive fibroblast cell surface glycoprotein. Localization, redistribution and role in cell shape. *J. Cell Biol.* **78**, 520-541.

Yamada, K. M., Spooner, B. S. and Wessells, N. K. (1970). Axon growth: roles of microfilaments and microtubules. *Proc. Nat. Acad. Sci. USA* **66**, 1206-1212.

Yamada, K. M., Spooner, B. S. and Wessells, N. K. (1971). Ultrastructure and function of growth cones and axons of cultured nerve cells. *J. Cell Biol.* **49**, 614-635.

Yamamoto, K. R. and Alberts, B. (1975). The interaction of estradiol-receptor protein with the genome: an argument for the existence of undetected specific sites. *Cell* **4**, 301-310.

Yamamoto, M., Jonas, D. and Seifart, K. (1977). Transcription of ribosomal 5S RNA by RNA polymerase C in isolated chromatin from HeLa cells. *Eur. J. Biochem.* **80**, 243-254.

Yang, N. S., Manning, R. F. and Gage, L. P. (1976). The blocked and methylated 5' terminal sequence of a specific cellular messenger: the mRNA for silk fibroin of Bombyx mori. *Cell* **7**, 339-348.

Yang, Y.-Z. and Perdue, J. F. (1972). Contractile proteins of cultured cells. Isolation and characterization of an actin like protein from cultured cells. *J. Biol. Chem.* **247**, 4503-4509.

Yao, M.-C. and Gall, J. G. (1977). A single integrated gene for rRNA in a eucaryote, Tetrahymena pyriformis. *Cell* **12**, 121-132.

Yao, M. C., Kimmel, A. R. and Gorovsky, M. A. (1974). A small number of cistrons for the rRNA in the germinal nucleus of a eucaryote, Tetrahymena pyriformis. *Proc. Nat. Acad. Sci. USA* **71**, 3082-3086.

Yasmineh, W. G. and Yunis, J. J. (1969). Satellite DNA in mouse autosomal heterochromatin. *Biochem. Biophys. Res. Commun.* **35**, 779-782.

Yasmineh, W. G. and Yunis, J. J. (1971). Satellite DNA in calf heterochromatin. *Exp. Cell Res.* **64**, 41-48.

Yen, A. and Pardee, A. B. (1977). Location of 3T3 cell cycle arrests induced by low serum and isoleucine deprivation. *Exp. Cell Res.* **114**, 389-396.

Yen, A. and Pardee, A. B. (1978). Exponential 3T3 cells escape in mid G1 from their high serum requirement. *Exp. Cell Res.* **116**, 103-114.

Yen, A. and Riddle, V. G. H. (1979). Plasma and platelet associated factors act in G1 to maintain proliferation and to stabilize arrested cells in a viable quiescent state. A temporal map of control points in the G1 phase. *Exp. Cell Res.* **120**, 349-358.

Yen, S.-H., Dahl, D., Schachner, M. and Shelanski, M. L. (1976). Biochemistry of the filaments of brain. *Proc. Nat. Acad. Sci. USA* **73**, 529-533.

Yeoman, L. C., Olson, M. D. J., Sugano, N., Jordan, J. J., Taylor, C. W., Starbuck, W. C. and Busch, H. (1972). Amino acid sequence of the center of the arginine-lysine rich histone from calf thymus. The total sequence. *J. Biol. Chem.* **247**, 6018-6024.

Yeoman, L. C., Taylor, C. W., Jordan, J. J. and Busch, H. (1975). Differences in chromatin proteins of growing and nongrowing tissues. Exp. Cell Res. **91**, 207-215.

Yerganian, G. and Nell, M. B. (1966). Hybridization of dwarf hamster cells by ultraviolet inactivated Sendai virus. *Proc. Nat. Acad. Sci. USA* **55**, 1066-1073.

Yoshikawa-Fukuda, M. (1967). The intermediate state of ribosome formation in animal cells in culture. *Biochim. Biophys. Acta* **145**, 651-663.

Young, B. D., Birnie, G. D. and Paul, J. (1976). Complexity and specificity of polysomal poly(A)⁺ RNA in mouse tissues. *Biochemistry* **15**, 2823-2828.

Young, E. T. and Sinsheimer, R. L. (1965). A comparison of the initial actions of spleen DNAase and pancreatic DNAase. *J. Biol. Chem.* **240**, 1274-1280.

Young, M. W. and Judd, B. H. (1978). Nonessential sequences, genes and the polytene chromosome bands of D. melanogaster. *Genetics* **88**, 723-742.

Young, R. A., Macklis, R. and Steitz, J. A. (1979). Sequence of the 16S-23S spacer region in two rRNA operons of E. coli. *J. Biol. Chem.* **254**, 3264-3261.

Yu, S. S., Li, H. J., Goodwin, G. H. and Johns, E. W. (1977). Interaction of non histone chromosomal proteins HMG1 and HMG2 with DNA. *Eur. J. Biochem.* **78**, 497–502.

Yunis, J. J. and Yasmineh, W. G. (1971). Heterochromatin, satellite DNA and cell function. *Science* **174**, 1200–1209.

Yunis, J. J. and Yasmineh, W. G. (1972). Model for mammalian constitutive heterochromatin. *Adv. Cell Mol. Biol.* **2**, 1–46.

Yunis, J. J., Kuo, M. T. and Saunders, G. F. (1977). Localization of sequences specifying mRNA to light staining G bands of human chromosomes. *Chromosoma* **61**, 335–344.

Yunis, J., Roldan, L., Yasmineh, W. G. and Lee, J. C. (1971). Staining of satellite DNA in metaphase chromosomes. *Nature* **231**, 532–533.

Zain, S., Sambrook, J., Roberts, R. J., Keller, W., Fried, M. and Dunn, A. R. (1979). Nucleotide sequence analysis of the leader segments in a cloned copy of adenovirus 2 fiber mRNA. *Cell* **16**, 851–861.

Zasloff, M. and Felsenfeld, G. (1977). Analysis of in vitro transcription of duck reticulocyte chromatin using mercury substituted ribonucleoside triphosphates. *Biochemistry* **16**, 5135–5144.

Zehavi-Willner, T. and Lane, C. D. (1977). Subcellular compartmentation of albumin and globin made in oocytes under the direction of injected mRNA. *Cell* **11**, 683–693.

Zelenka, P. and Piatigorsky, J. (1974). Isolation and in vitro translation of δ-crystallin mRNA from embryonic chick lens fiber. *Proc. Nat. Acad. Sci. USA* **71**, 1896–1900.

Zelenka, P. and Piatigorsky, J. (1976). Reiteration frequency of δ-crystallin DNA in lens and non lens tissues of chick embryos. δ-crystallin gene is not amplified during lens cell differentiation. *J. Biol. Chem.* **251**, 4294–4298.

Zetterberg, A. (1966a). Synthesis and accumulation of nuclear and cytoplasmic proteins during interphase in mouse fibroblasts in vitro. *Exp. Cell Res.* **42**, 500–511.

Zetterberg, A. (1966b). Nuclear and cytoplasmic nucleic acid content cytoplasmic protein synthesis during interphase in mouse fibroblasts in vitro. *Exp. Cell Res.* **43**, 517–525.

Zetterberg, A. (1966c). Protein migration between cytoplasm and cell nucleus during interphase in mouse fibroblast in vitro. *Exp. Cell Res.* **43**, 526–536.

Zetterberg, A. (1970). Nuclear and cytoplasmic growth during interphase in mammalian cells. *Adv. Cell Biol.* **1**, 211–232.

Zetterberg, A. and Killander, D. (1965). Quantitative cytophotometric and autoradiographic studies on the rate of protein synthesis in mouse fibroblasts in vitro. *Exp. Cell Res.* **40**, 1–11.

Zhimulev, I. F. and Belyaeva, F. A. (1975). [3]H uridine labeling patterns in the D. melanogaster salivary gland chromosomes X, 2R and 3L. *Chromosoma* **49**, 219–232.

Zickler, D. (1973). Fine structure of chromosome pairing in ten Ascomycetes: meiotic and premeiotic (mitotic) synaptonemal complexes. *Chromosoma* **40**, 401–416.

Zickler, D. (1977). Development of the synaptonemal complex and the "recombination nodules" during meiotic prophase in the 7 bivalents of the fungi Sordaria macrospora auersw. *Chromosoma* **61**, 289–316.

Zieve, G. and Penman, S. (1976). Small RNA species of the HeLa cell: metabolism and subcellular localization. *Cell* **8**, 19–32.

Zieve, G., Benecke, B.-J. and Penman, S. (1977). Synthesis of two classes of small RNA species in vivo and in vitro. *Biochemistry* **16**, 4520–4525.

Ziff, E. B. and Evans, R. M. (1978). Coincidence of the promoter and capped 5′ terminus of RNA from the adenovirus 2 major late transcription unit. *Cell* **15**, 1463–1475.

Ziff, E. and Fraser, N. (1978). Adenovirus type 2 late mRNAs: structural evidence for 3′ coterminal species. *J. Virol.* **25**, 887–906.

Zigmond, S. H., Otto, J. J. and Bryan, J. (1979). Organization of myosin in a submembraneous sheath in well spread human fibroblasts. *Exp. Cell Res.* **119**, 205–220.

Zimmering, S. (1976). Genetic and cytogenetic aspects of altered segregation phenomena in

Drosophila. In M. Ashburner and E. Novitski (Eds.), *The Genetics and Biology of Drosophila*, Academic Press, London and New York, **1b,** 569–615.

Zimmering, S., Sandler, L. and Nicoletti, B. (1970). Mechanisms of meiotic drive. *Ann. Rev. Genet.* **4,** 409–436.

Zimmerman, E. F. (1968). Secondary methylation of rRNA in HeLa cells. *Biochemistry* **7,** 3156–3163.

Zimmerman, E. F. and Holler, B. W. (1977). Methylation of 45S rRNA precursor in HeLa cells. *J. Mol. Biol.* **23,** 149–161.

Zimmerman, J. L. and Goldberg, R. B. (1977). DNA sequence organization in the genome of Nictiana tabacum. *Chromosoma* **59,** 227–252.

Zimmerman, S. B. and Pheiffer, B. H. (1979). Helical parameters of DNA do not change when DNA fibers are wetted: X ray diffraction study. *Proc. Nat. Acad. Sci. USA* **76,** 2703–2707.

Zubay, G. and Doty, P. (1959). The isolation and properties of DNP particles containing single nucleic acid molecules. *J. Mol. Biol.* **1,** 1–20.

Zylber, E. A. and Penman, S. (1971a). Product of RNA polymerase in HeLa cell nuclei. *Proc. Nat. Acad. Sci. USA* **68,** 2861–2865.

Zylber, E. A. and Penman, S. (1971b). Synthesis of 5S and 4S RNA in metaphase arrested HeLa cells. *Science* **172,** 947–949.

Zylber, E., Vesco, C. and Penman, S. (1969). Selective inhibition of the synthesis of mitochondrial associated RNA by ethidium bromide. *J. Mol. Biol.* **44,** 195–204.

Index

An asterisk indicates that a topic is indexed separately